STUDENT'S SOLUTIONS MANUAL

DAVID ATWOOD

Rochester Community and Technical College

A GRAPHICAL APPROACH TO ALGEBRA & TRIGONOMETRY

SIXTH EDITION

John Hornsby

University of New Orleans

Margaret L. Lial

American River College

Gary Rockswold

Minnesota State University, Mankato

PEARSON

Boston Columbus Indianapolis New York San Francisco Upper Saddle River
Amsterdam Cape Town Dubai London Madrid Milan Munich Paris Montreal Toronto
Delhi Mexico City São Paulo Sydney Hong Kong Seoul Singapore Taipei Tokyo

The author and publisher of this book have used their best efforts in preparing this book. These efforts include the development, research, and testing of the theories and programs to determine their effectiveness. The author and publisher make no warranty of any kind, expressed or implied, with regard to these programs or the documentation contained in this book. The author and publisher shall not be liable in any event for incidental or consequential damages in connection with, or arising out of, the furnishing, performance, or use of these programs.

Reproduced by Pearson from electronic files supplied by the author.

Publishing as Pearson, 75 Arlington Street, Boston, MA 02116.

 Printed in the United States of America.

ISBN-13: 978-0-321-90065-4
ISBN-10: 0-321-90065-0

2 3 4 5 6 V069 17 16 15 14

www.pearsonhighered.com

PEARSON

Contents

Chapter 1: Linear Functions, Equations, and Inequalities

1.1: Real Numbers and the Rectangular Coordinate System

1. (a) The only natural number is 10.

   (b) The whole numbers are 0 and 10.

   (c) The integers are $-6, -\frac{12}{4}$ (or -3), 0, 10.

   (d) The rational numbers are $-6, -\frac{12}{4}(\text{or } -3), -\frac{5}{8}, 0, .31, .\overline{3}$, and 10.

   (e) The irrational numbers are $-\sqrt{3}$, 2π and $\sqrt{17}$.

   (f) All of the numbers listed are real numbers.

   (f) All of the numbers listed are real numbers.

3. (a) There are no natural numbers listed.

   (b) There are no whole numbers listed.

   (c) The integers are $-\sqrt{100}(\text{or } -10)$ and -1.

   (d) The rational numbers are $-\sqrt{100}$ (or -10), $-\frac{13}{6}, -1, 5.23, 9.\overline{14}, 3.14$, and $\frac{22}{7}$.

   (e) There are no irrational numbers listed.

   (f) All of the numbers listed are real numbers..

5. The number 16,351,000,000,000 is a natural number, integer, rational number, and real number.

7. The number -25 is an integer, rational, and real number.

9. The number $\frac{7}{3}$ is a rational and real number.

11. The number $5\sqrt{2}$ is a real number.

13. Natural numbers would be appropriate because population is only measured in positive whole numbers.

15. Rational numbers would be appropriate because shoes come in fraction sizes.

17. Integers would be appropriate because temperature is given in positive and negative whole numbers.

19. −4 −3 −2 −1 0 1

21. 0 $\frac{5}{3}$
   −.5 .75 3.5

23. A rational number can be written as a fraction, $\frac{p}{q}$, $q \neq 0$, where p and q are integers. An irrational number cannot be written in this way.

25. The point $\left(2, \frac{5}{7}\right)$ is in Quadrant I. See Figure 25-33.

27. The point $(-3,-2)$ is in Quadrant III. See Figure 25-33.

29. The point $(0,5)$ is located on the y-axis, therefore is not in a quadrant. See Figure 25-33.

31. The point $(-2,4)$ is in Quadrant II. See Figure 25-33.

33. The point $(-2,0)$ is located on the x-axis, therefore is not in a quadrant. See Figure 25-33.

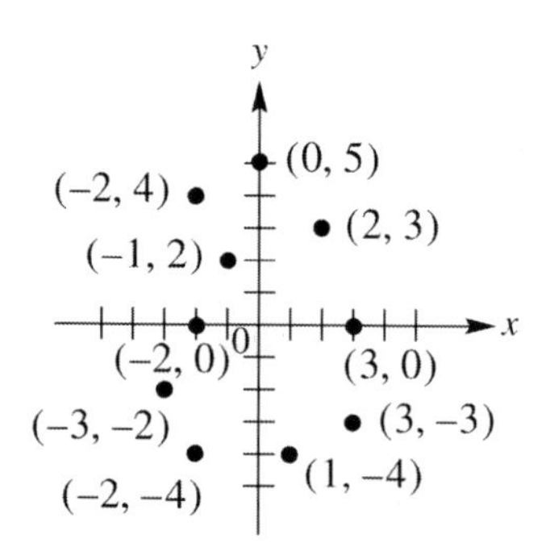

Figure 25-33

35. If $xy>0$, then either $x>0$ and $y>0 \Rightarrow$ Quadrant I, or $x<0$ and $y<0 \Rightarrow$ Quadrant III.

37. If $\frac{x}{y}<0$, then either $x>0$ and $y<0 \Rightarrow$ Quadrant IV, or $x<0$ and $y>0 \Rightarrow$ Quadrant II.

39. Any point of the form $(0,b)$ is located on the y-axis.

41. $[-5,5]$ by $[-25,25]$

43. $[-60,60]$ by $[-100,100]$

45. $[-500,300]$ by $[-300,500]$

47. See Figure 47.

49. See Figure 49.

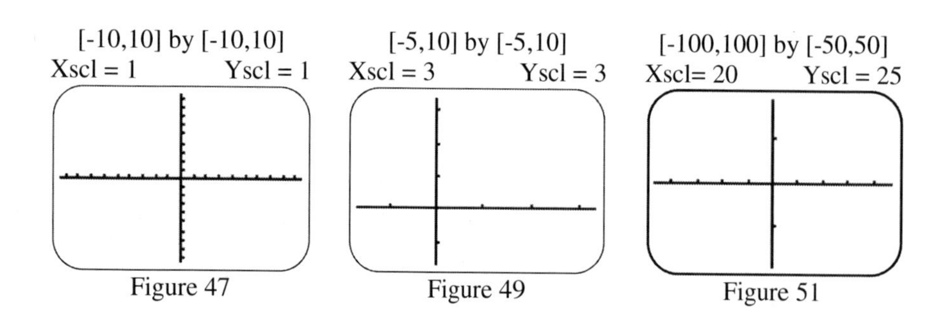

Figure 47 Figure 49 Figure 51

51. See Figure 51.

53. There are no tick marks, which is a result of setting Xscl and Yscl to 0.

55. $\sqrt{58} \approx 7.615773106 \approx 7.616$

57. $\sqrt[3]{33} \approx 3.20753433 \approx 3.208$

59. $\sqrt[4]{86} \approx 3.045261646 \approx 3.045$

61. $19^{1/2} \approx 4.35889844 \approx 4.359$

63. $46^{1.5} \approx 311.9871792 \approx 311.987$

65. $(5.6-3.1)/(8.9+1.3) \approx .25$

67. $\sqrt{(\pi \wedge 3+1)} \approx 5.66$

69. $3(5.9)^2 - 2(5.9)+6 = 98.63$

71. $\sqrt{(4-6)^2+(7+1)^2)} \approx 8.25$

73. $\sqrt{(\pi-1)}/\sqrt{(1+\pi)} \approx .72$

75. $2/(1-\sqrt[3]{5}) \approx -2.82$

77. $a^2+b^2=c^2 \Rightarrow 8^2+15^2=c^2 \Rightarrow 64+225=c^2 \Rightarrow 289=c^2 \Rightarrow c=17$

79. $a^2+b^2=c^2 \Rightarrow 13^2+b^2=85^2 \Rightarrow 169+b^2=7225 \Rightarrow b^2=7056 \Rightarrow b=84$

81. $a^2+b^2=c^2 \Rightarrow 5^2+8^2=c^2 \Rightarrow 25+64=c^2 \Rightarrow 89=c^2 \Rightarrow c=\sqrt{89}$

83. $a^2+b^2=c^2 \Rightarrow a^2+(\sqrt{13})^2=(\sqrt{29})^2 \Rightarrow a^2+13=29 \Rightarrow a^2=16 \Rightarrow a=4$

85. (a) $d=\sqrt{(2-(-4))^2+(5-3)^2}=\sqrt{(6)^2+(2)^2}=\sqrt{36+4}=\sqrt{40}=2\sqrt{10}$

(b) $M=\left(\frac{-4+2}{2},\frac{3+5}{2}\right)=\left(\frac{-2}{2},\frac{8}{2}\right)=(-1,4)$

87. (a) $d=\sqrt{(6-(-7))^2+(-2-4)^2}=\sqrt{(13)^2+(-6)^2}=\sqrt{169+36}=\sqrt{205}$

(b) $M=\left(\frac{-7+6}{2},\frac{4+(-2)}{2}\right)=\left(\frac{-1}{2},\frac{2}{2}\right)=\left(-\frac{1}{2},1\right)$

89. (a) $d=\sqrt{(2-5)^2+(11-7)^2}=\sqrt{(-3)^2+(4)^2}=\sqrt{9+16}=\sqrt{25}=5$

(b) $M=\left(\frac{5+2}{2},\frac{7+11}{2}\right)=\left(\frac{7}{2},\frac{18}{2}\right)=\left(\frac{7}{2},9\right)$

91. (a) $d=\sqrt{(-3-(-8))^2+((-5)-(-2))^2}=\sqrt{(5)^2+(-3)^2}=\sqrt{25+9}=\sqrt{34}$

(b) $M=\left(\frac{-8+(-3)}{2},\frac{-2+(-5)}{2}\right)=\left(\frac{-11}{2},\frac{-7}{2}\right)=\left(-\frac{11}{2},-\frac{7}{2}\right)$

93. (a) $d=\sqrt{(6.2-9.2)^2+(7.4-3.4)^2}=\sqrt{(-3)^2+(4)^2}=\sqrt{9+16}=\sqrt{25}=5$

(b) $M=\left(\frac{9.2+6.2}{2},\frac{3.4+7.4}{2}\right)=\left(\frac{15.4}{2},\frac{10.8}{2}\right)=(7.7,5.4)$

95. (a) $d=\sqrt{(6x-13x)^2+(x-(-23x))^2}=\sqrt{(-7x)^2+(24x)^2}=\sqrt{49x^2+576x^2}=\sqrt{625x^2}=25x$

(b) $M=\left(\frac{13x+6x}{2},\frac{-23x+x}{2}\right)=\left(\frac{19x}{2},\frac{-22x}{2}\right)=\left(\frac{19}{2}x,-11x\right)$

97. Using the midpoint formula we get: $\left(\frac{7+x_2}{2},\frac{-4+y_2}{2}\right)=(8,5)\Rightarrow\left(\frac{7+x_2}{2}\right)=8\Rightarrow 7+x_2=16\Rightarrow x_2=9$ and $\frac{-4+y_2}{2}=5\Rightarrow -4+y_2=10\Rightarrow y_2=14$. Therefore the coordinates are: $Q(19,14)$.

99. Using the midpoint formula we get: $\left(\frac{5.64+x_2}{2},\frac{8.21+y_2}{2}\right)=(-4.04,1.60)\Rightarrow\frac{5.64+x_2}{2}=-4.04\Rightarrow$ $5.64+x_2=-8.08\Rightarrow x_2=-13.72$ and $\frac{8.21+y_2}{2}=1.60\Rightarrow 8.21+y_2=3.20\Rightarrow y_2=-5.01$. Therefore the coordinates are: $Q(-13.72,-5.01)$.

101. $M=\left(\frac{2007+2011}{2},\frac{17+36}{2}\right)=\left(\frac{4018}{2},\frac{53}{2}\right)=(2009,26.5)$; the revenue was about \$26.5 billion.

103. In 2005, $M=\left(\frac{2003+2007}{2},\frac{18,810+21,203}{2}\right)=\left(\frac{4010}{2},\frac{40,013}{2}\right)=(2005,20,006.5)$; poverty level was approximately \$20,007. In 2009, $M=\left(\frac{2007+2011}{2},\frac{21,203+22,350}{2}\right)=\left(\frac{4018}{2},\frac{43,553}{2}\right)=$ $(2009,21,776.5)$; poverty level was approximately \$21,777.

105. (a) See Figure 105.

(b) $d=\sqrt{(50-0)^2+(0-40)^2}=\sqrt{(50)^2+(-40)^2}=\sqrt{2500+1600}=\sqrt{4100}\approx 64.0$ miles.

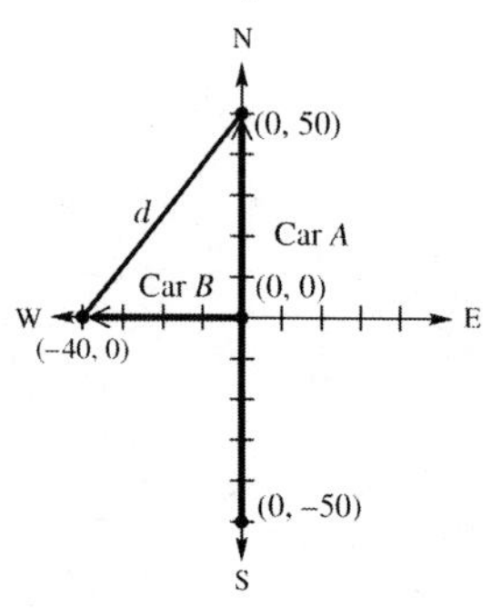

Figure 105

107. Using the area of a square produces: $(a+b)^2=a^2+2ab+b^2$. Now, using the sum of the small square and the four right triangles produces $c^2+4\left(\frac{1}{2}ab\right)=c^2+2ab$. Therefore $a^2+2ab+b^2=c^2+2ab$, and subtracting $2ab$ from both sides produces $a^2+b^2=c^2$.

1.2: Introduction to Relations and Functions

1. The interval is $(-1,4)$.

−1 0 4

3. The interval is $(-\infty,0)$.

−1 0

5. The interval is $[1,2)$.

7. $(-4,3) \Rightarrow \{x \mid -4 < x < 3\}$

9. $(-\infty,-1] \Rightarrow \{x \mid x \le -1\}$

11. $\{x \mid -2 \le x < 6\}$

13. $\{x \mid x \le -4\}$

15. A parenthesis is used if the symbol is $<$, $>$, $-\infty$, or ∞ or . A square bracket is used if the symbol is $\le$ or $\ge$.

17. See Figure 17

19. See Figure 19

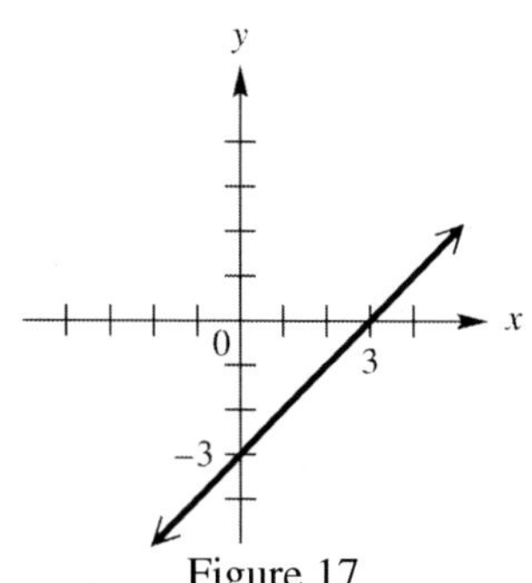

Figure 17

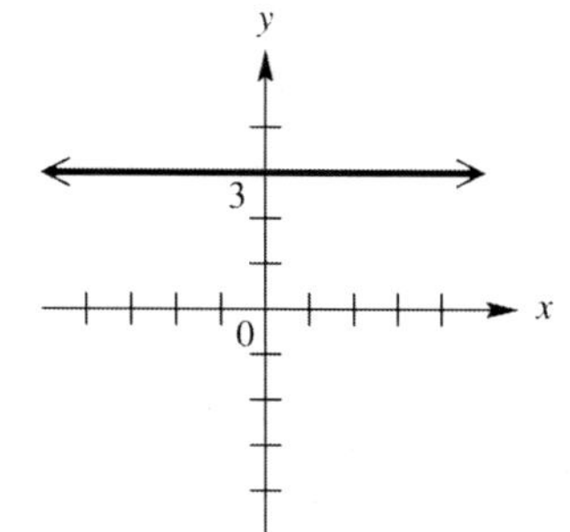

Figure 19

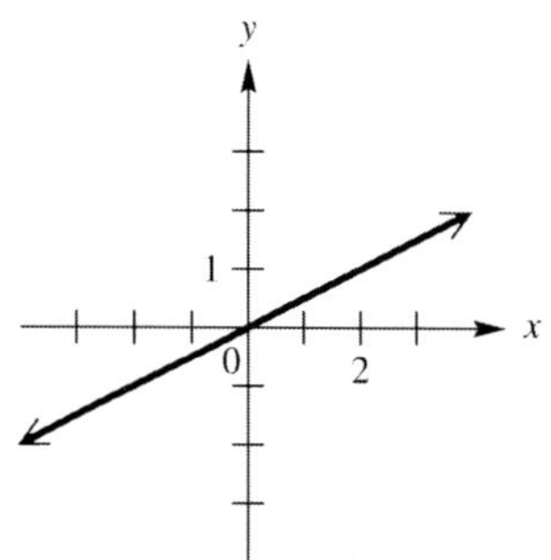

Figure 21

21. See Figure 21

23. See Figure 23

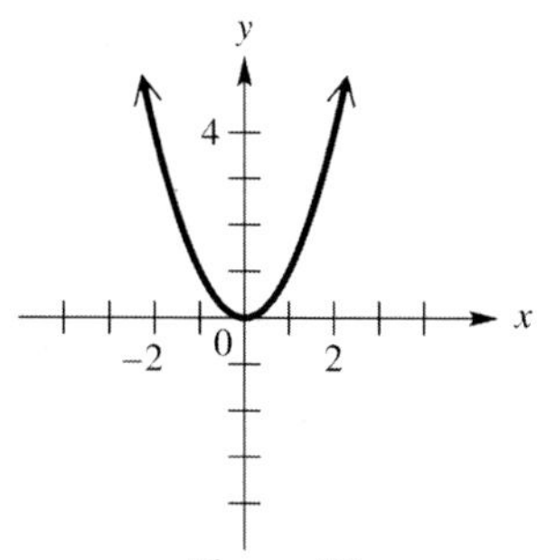

Figure 23

25. The relation is a function. Domain: $\{5,3,4,7\}$ Range: $\{1,2,9,6\}$.

27. The relation is a function. Domain: $\{1,2,3\}$, Range: $\{6\}$.

29. The relation is not a function. Domain: $\{4,3,-2\}$, Range: $\{1,-5,3,7\}$.

31. The relation is a function. Domain: $\{11,12,13,14\}$, Range: $\{-6,-7\}$.

33. The relation is a function. Domain: $\{0,1,2,3,4\}$, Range: $\{\sqrt{2},\sqrt{3},\sqrt{5},\sqrt{6},\sqrt{7}\}$.

35. The relation is a function. Domain: $(-\infty,\infty)$, Range: $(-\infty,\infty)$.

37. The relation is not a function. Domain: $[-4,4]$, Range: $[-3,3]$.

39. The relation is a function. Domain: $[2,\infty)$, Range: $[0,\infty)$.

41. The relation is not a function. Domain: $[-9,\infty)$, Range: $(-\infty,\infty)$.

43. The relation is a function. Domain: $\{-5,-2,-1,-.5,0,1.75,3.5\}$, Range: $\{-1,2,3,3.5,4,5.75,7.5\}$.

45. The relation is a function. Domain: $\{2,3,5,11,17\}$ Range: $\{1,7,20\}$.

47. From the diagram, $f(-2)=2$.

49. From the diagram, $f(11)=7$.

51. f(1) is undefined since 1 is not in the domain of the function.

53. $f(-2)=3(-2)-4=-6-4=-10$

55. $f(1)=2(1)^2-(1)+3=2-1+3=4$

57. $f(4)=-(4)^2+(4)+2=-16+4+2=-10$

59. $f(9)=5$

61. $f(-2)=\sqrt{(-2)^3+12}=\sqrt{-8+12}=\sqrt{4}=2$

63. $f(8)=|5-2(8)|=|-11|=11$

65. Given that $f(x)=5x$, then $f(a)=5a$, $f(b+1)=5(b+1)=5b+5$, and $f(3x)=5(3x)=15x$

67. Given that $f(x)=2x-5$, then $f(a)=2a-5$, $f(b+1)=2(b+1)-5=2b+2-5=2b-3$, and

$f(3x)=2(3x)-5=6x-5$

69. Given that $f(x)=1-x^2$, then $f(a)=1-a^2$, $f(b+1)=1-(b+1)^2=1-(b^2+2b+1)=-b^2-2b$, and

$f(3x)=1-(3x)^2=1-9x^2$

71. Since $f(-2)=3$, the point $(-2,3)$ lies on the graph of f.

73. Since the point $(7,8)$ lies on the graph of f, $f(7)=8$.

75. From the graph: (a) $f(-2)=0$, (b) $f(0)=4$, (c) $f(1)=2$, and (d) $f(4)=4$.

77. From the graph: (a) $f(-2)$ is undefined, (b) $f(0)=-2$, (c) $f(1)=0$, and (d) $f(4)=2$.

79. (a) – (f) Answers will vary. Refer to the definitions in the text.

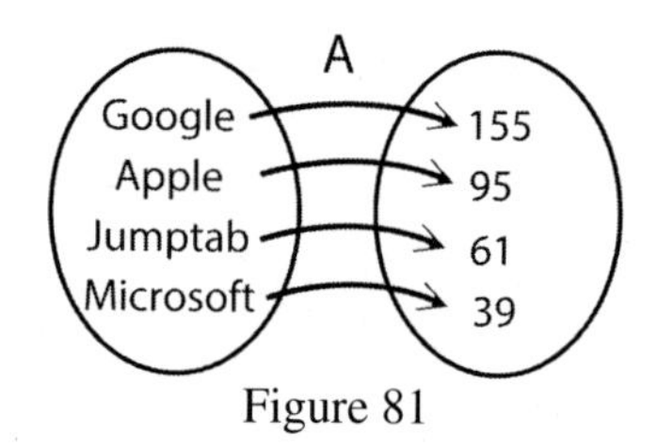

Figure 81

81. (a) $A=\{(\text{Google},155),(\text{Apple},95),(\text{Jumptab},61),(\text{Microsoft},39)\}$, The U.S. mobile advertising revenue in 2011 for Google was $155,000,000 dollars.

(b) See Figure 81.

(c) $D = \{\text{Google, Apple, Jumptab, Microsoft}\}$, $R = \{155, 95, 61, 39\}$

Reviewing Basic Concepts (Sections 1.1 and 1.2)

1. See Figure 1.

2. The distance is $d = \sqrt{(6-(-4))^2 + (-2-5)^2} = \sqrt{100+49} = \sqrt{149}$.

   The midpoint is $M = \left(\frac{-4+6}{2}, \frac{5-2}{2}\right) = \left(1, \frac{3}{2}\right)$.

3. $\frac{\sqrt{5+\pi}}{(\sqrt[3]{3}+1)} \approx 1.168$

4. $d = \sqrt{(12-(-4))^2 + (-3-27)^2} = \sqrt{256+900} = \sqrt{1156} = 34$

5. Using Pythagorean Theorem, $11^2 + b^2 = 61^2 \Rightarrow b^2 = 61^2 - 11^2 \Rightarrow b^2 = 3600 \Rightarrow b = 60$ inches.

6. The set $\{x \mid -2 < x \le 5\}$ is the interval $(-2,5]$. The set $\{x \mid x \ge 4\}$ is the interval $[4,\infty)$.

7. The relation is not a function because it does not pass the vertical line test. Domain: $[-2,2]$,

   Range: $[-3,3]$.

8. See Figure 8.

9. Given $f(x) = 3-4x$ then $f(-5) = 3-4(-5) = 23$ and $f(a+4) = 3-4(a+4) = 3-4a-16 = -4a-13$

10. From the graph, $f(2) = 3$ and $f(-1) = -3$.

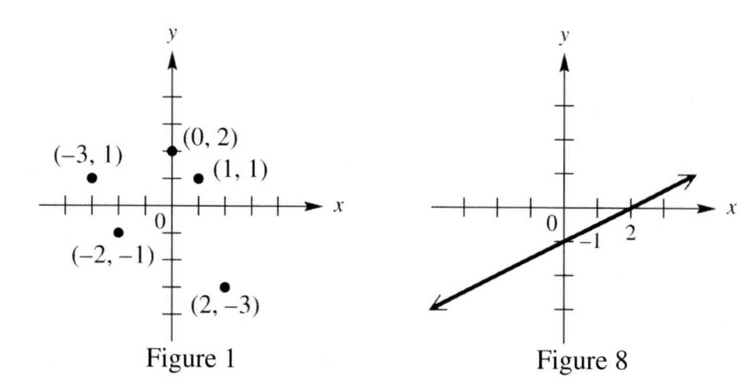

Figure 1 Figure 8

1.3: Linear Functions

1. The graph is shown in Figure 1.

   (a) x-intercept: 4 (b) y-intercept: -4 (c) Domain: $(-\infty,\infty)$ (d) Range: $(-\infty,\infty)$

   (e) The equation is in slope-intercept form, therefore $m = 1$.

3. The graph is shown in Figure 3.

   (a) x-intercept: 2 (b) y-intercept: -6 (c) Domain: $(-\infty,\infty)$ (d) Range: $(-\infty,\infty)$

(e) The equation is in slope-intercept form, therefore $m = 3$.

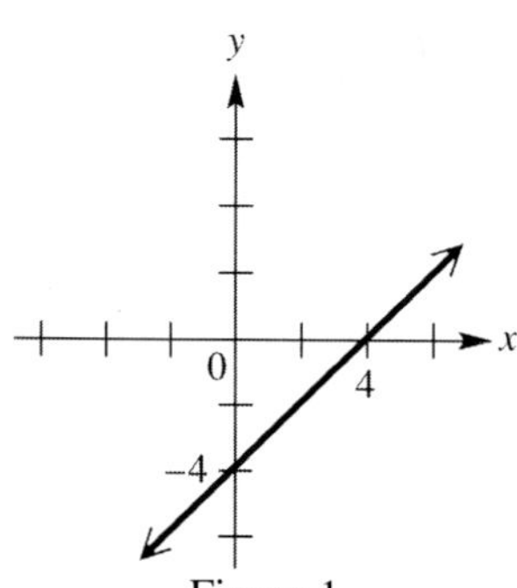

Figure 1

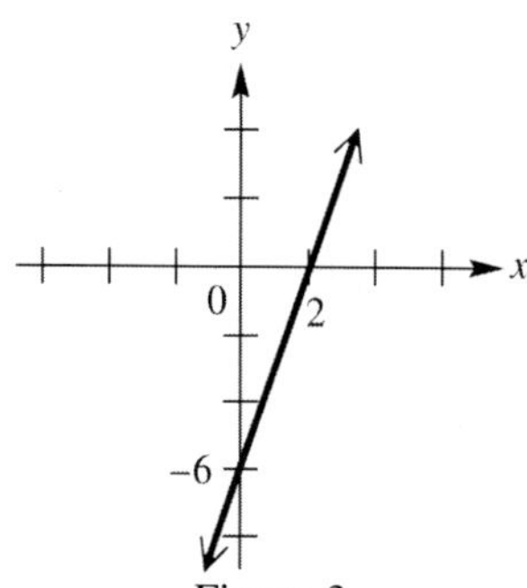

Figure 3

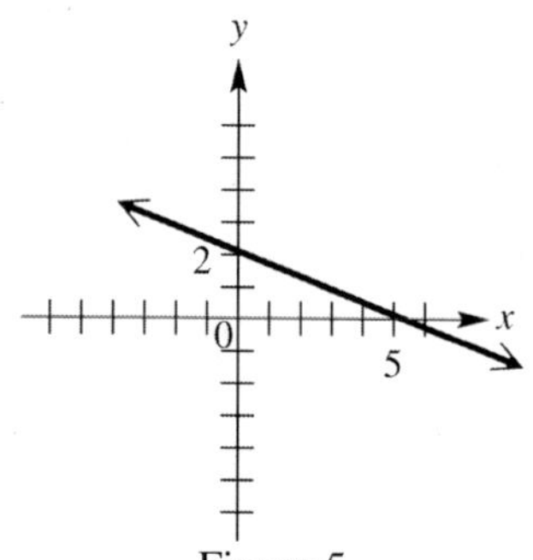

Figure 5

5. The graph is shown in Figure 5.

(a) x-intercept: 5 (b) y-intercept: 2 (c) Domain: $(-\infty, \infty)$ (d) Range: $(-\infty, \infty)$

(e) The equation is in slope-intercept form, therefore $m = -\dfrac{2}{5}$.

7. The graph is shown in Figure 7.

(a) x-intercept: 0 (b) y-intercept: 0 (c) Domain: $(-\infty, \infty)$ (d) Range: $(-\infty, \infty)$

(e) The equation is in slope-intercept form, therefore $m = 3$.

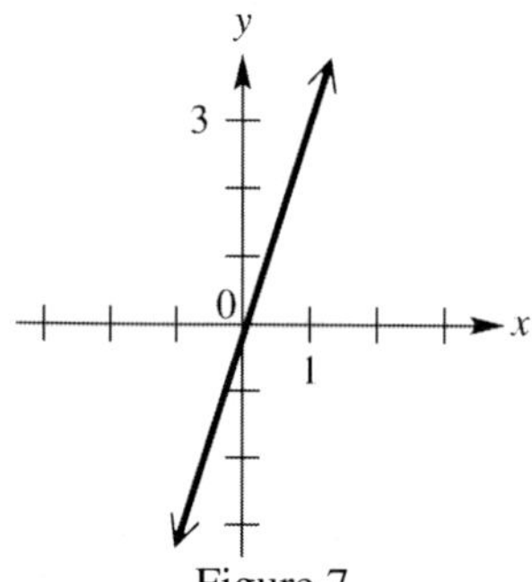

Figure 7

9. (a) $f(-2) = (-2) + 2 = 0$ and $f(4) = (4) + 2 = 6$

(b) The x-intercept is -2 and corresponds to the zero of f. See Figure 9.

(c) $x + 2 = 0 \Rightarrow x = -2$

11. (a) $f(-2) = 2 - \dfrac{1}{2}(-2) = 3$ and $f(4) = 2 - \dfrac{1}{2}(4) = 0$

(b) The x-intercept is 4 and corresponds to the zero of f. See Figure 11.

(c) $2 - \dfrac{1}{2}x = 0 \Rightarrow \dfrac{1}{2}x = 2 \Rightarrow x = 4$

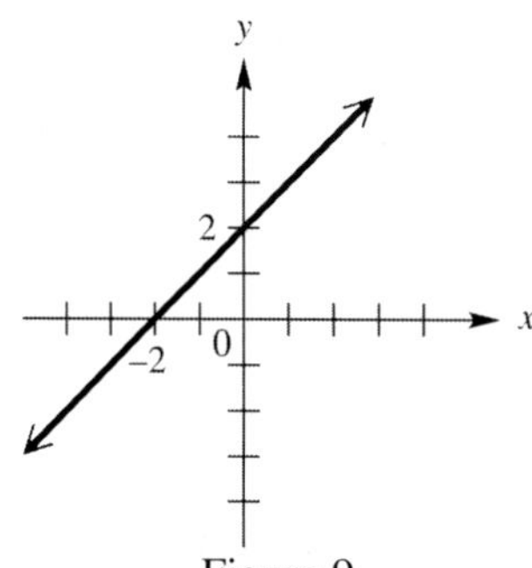

Figure 9

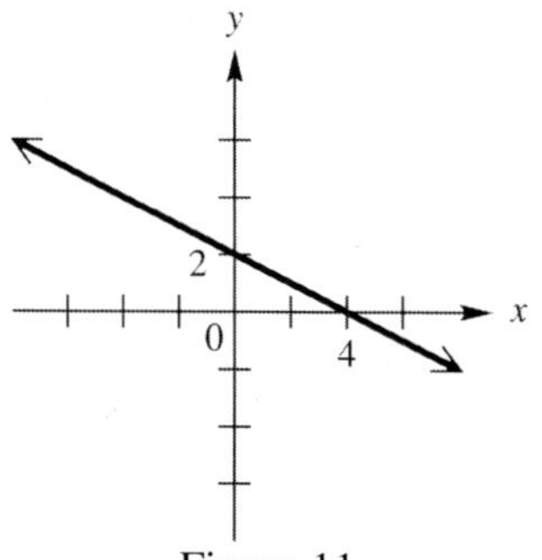

Figure 11

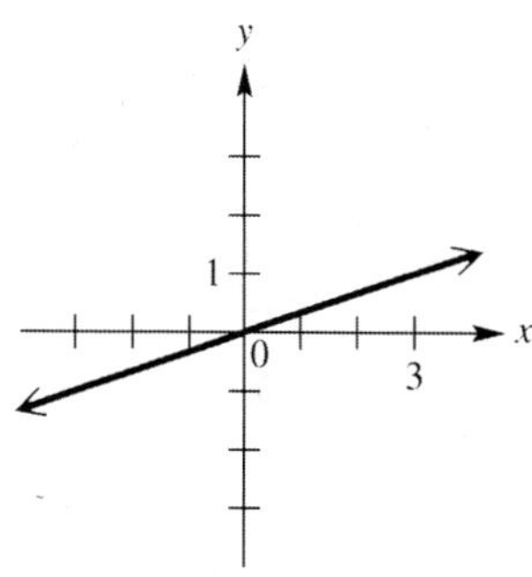

Figure 13

13. (a) $f(-2) = \frac{1}{3}(-2) = -\frac{2}{3}$ and $f(4) = \frac{1}{3}(4) = \frac{4}{3}$

 (b) The x-intercept is 0 and corresponds to the zero of f. See Figure 13.

 (c) $\frac{1}{3}x = 0 \Rightarrow x = 0$

15. (a) $f(-2) = .4(-2) + .15 = -.65$ and $f(4) = .4(4) + .15 = 1.75$

 (b) The x-intercept is $-.375$ and corresponds to the zero of f. See Figure 15.

 (c) $.4x + .15 = 0 \Rightarrow .4x = -.15 \Rightarrow x = -.375$

17. (a) $f(-2) = \frac{2-(-2)}{4} = 1$ and $f(4) = \frac{2-(4)}{4} = -\frac{1}{2}$

 (b) The x-intercept is 2 and corresponds to the zero of f. See Figure 17.

 (c) $\frac{2-x}{4} = 0 \Rightarrow 2 - x = 0 \Rightarrow x = 2$

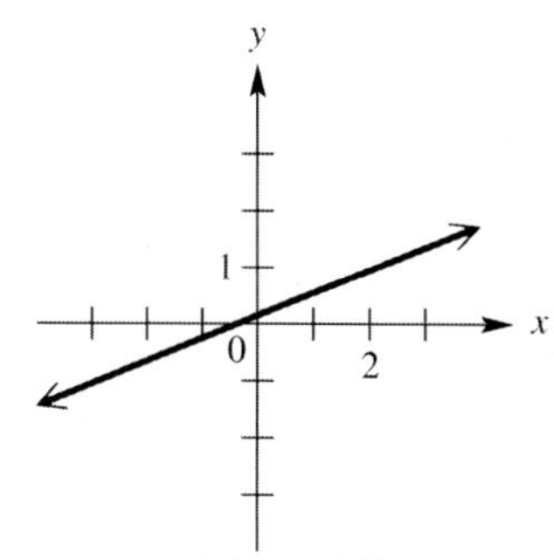

Figure 15

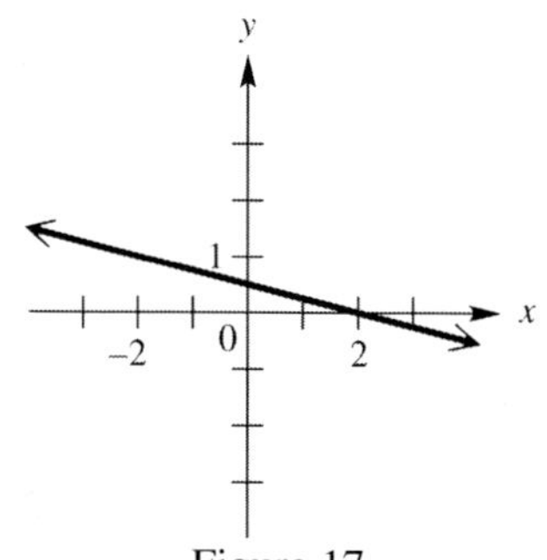

Figure 17

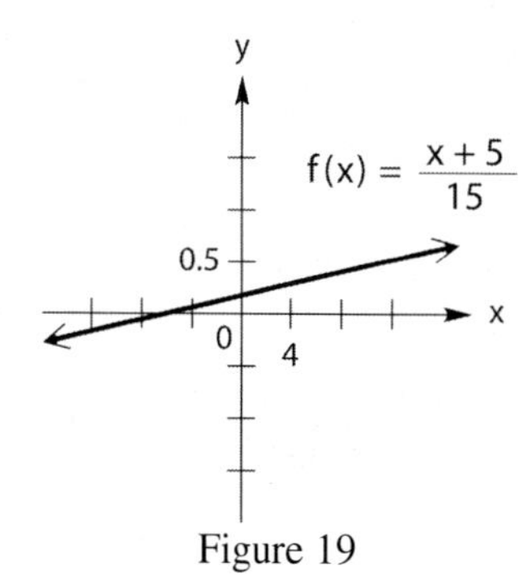

Figure 19

19. (a) $f(-2) = \frac{-2+5}{15} = \frac{3}{15} = \frac{1}{5}$ and $f(4) = \frac{4+5}{15} = \frac{9}{15} = \frac{3}{5}$

 (b) The x-intercept is –5 and corresponds to the zero of f. See Figure 19

 (c) $\frac{x+5}{15} = 0 \Rightarrow x + 5 = 0 \Rightarrow x = -5$

21. The graph of $y = ax$ always passes through (0, 0).

23. The graph is shown in Figure 23.

 (a) x-intercept: none (b) y-intercept: –3 (c) Domain: $(-\infty, \infty)$ (d) Range: $\{-3\}$

 (e) The slope of all horizontal line graphs or constant functions is $m = 0$.

25. The graph is shown in Figure 25.

(a) x-intercept: -1.5 (b) y-intercept: none (c) Domain: $\{-1.5\}$ (d) Range: $(-\infty,\infty)$

(e) All vertical line graphs are not functions, therefore the slope is undefined.

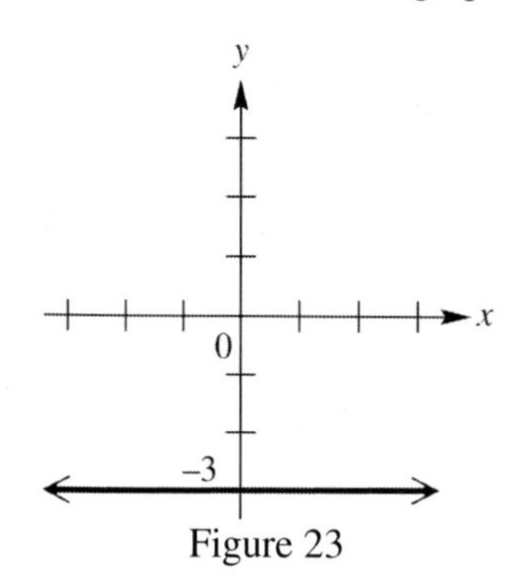

Figure 23

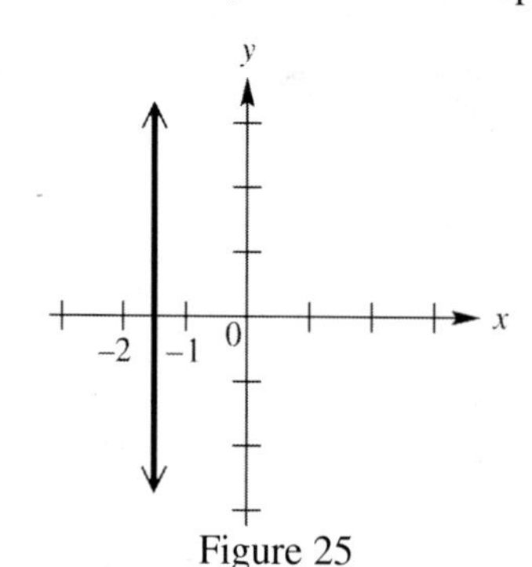

Figure 25

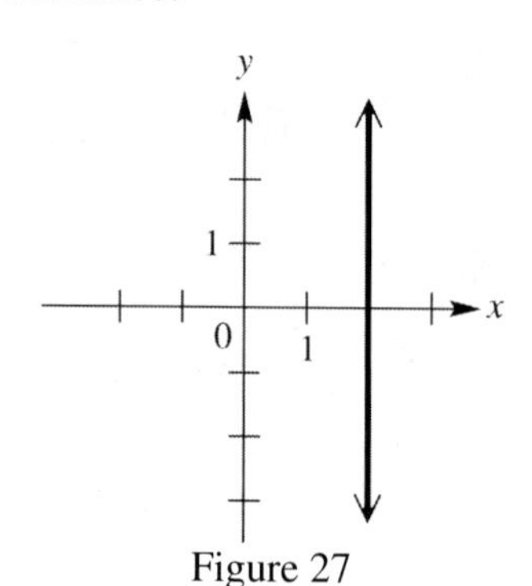

Figure 27

27. The graph is shown in Figure 27.

(a) x-intercept: 2 (b) y-intercept: none (c) Domain: $\{2\}$

(d) Range: $(-\infty,\infty)$ (e) All vertical line graphs are not functions, therefore the slope is undefined.

29. All functions in the form $f(x) = a$ are constant functions.

31. This is a horizontal line graph, therefore $y = 3$.

33. This is a vertical line graph on the y-axis, therefore $x = 0$.

35. Window B gives the more comprehensive graph. See Figures 35a and 35b.

[-10,10] by [-10,10]
Xscl = 1 Yscl = 1

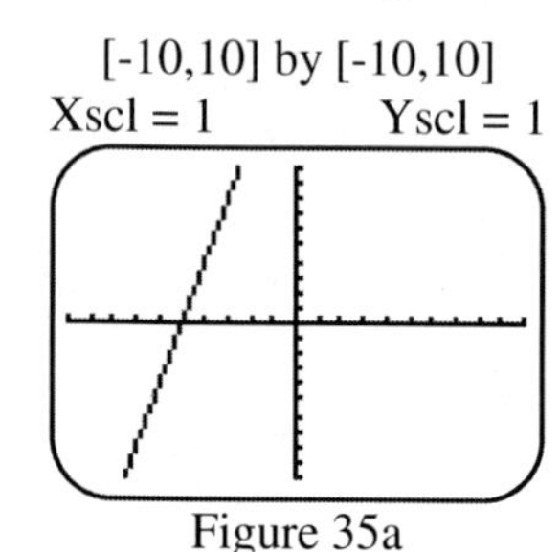
Figure 35a

[-10,10] by [-5,25]
Xscl = 1 Yscl = 5

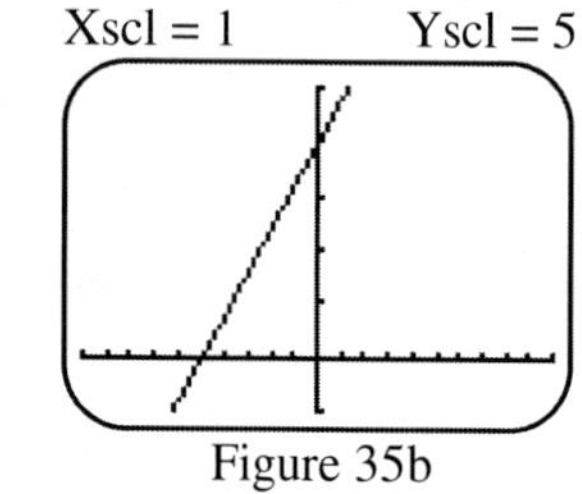
Figure 35b

[-3,3] by [-5,5]
Xscl= 1 Yscl = 1

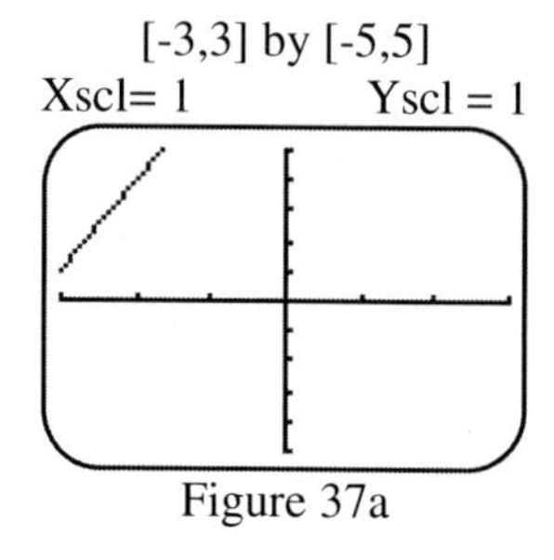
Figure 37a

[-5,5] by [-10,14]
Xscl = 1 Yscl= 2

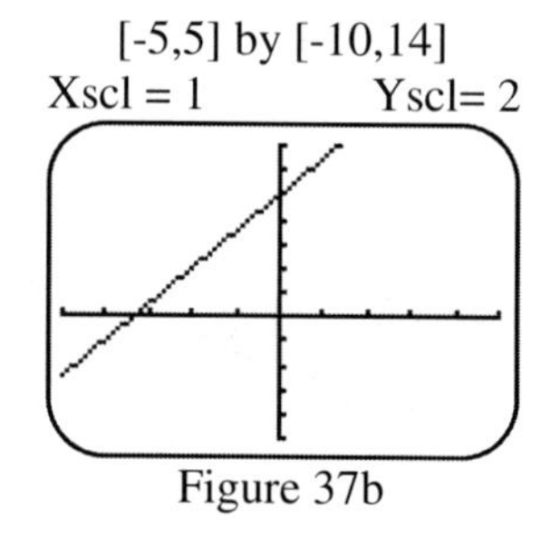
Figure 37b

37. Window B gives the more comprehensive graph. See Figures 37a and 37b.

39. $m = \dfrac{6-1}{3-(-2)} = \dfrac{5}{5} = 1$

41. $m = \dfrac{4-(-3)}{8-(-1)} = \dfrac{7}{9}$

43. $m = \dfrac{5-3}{-11-(-11)} = \dfrac{2}{0} \Rightarrow$ undefined slope

45. $m = \dfrac{9-9}{\frac{1}{2}-\frac{2}{3}} = \dfrac{0}{-\frac{1}{6}} \Rightarrow 0$

47. $m = \dfrac{-\frac{2}{3}-\frac{1}{6}}{\frac{1}{2}-\left(-\frac{3}{4}\right)} = \dfrac{-\frac{5}{6}}{\frac{5}{4}} \Rightarrow -\dfrac{2}{3}$

49. The average rate of change is evaluated as $m = \dfrac{y_2 - y_1}{x_2 - x_1} = \dfrac{20-4}{0-4} = -\dfrac{16}{4} = -4$. The value of the machine is decreasing \$4000 each year during these years.

51. The average rate of change is evaluated as $m = \dfrac{y_2 - y_1}{x_2 - x_1} = \dfrac{3-3}{4-0} = \dfrac{0}{4} = 0$. The percent of pay raise is not changing but will remain constant at 3% per year.

53. Since $m = 3$ and $b = 6$, graph A most closely resembles the equation.

55. Since $m = -3$ and $b = -6$, graph C most closely resembles the equation.

57. Since $m = 3$ and $b = 0$, graph H most closely resembles the equation.

59. Since $m = 0$ and $b = 3$, graph B most closely resembles the equation.

61. (a) The graph passes through (0,1) and (1,-1) $\Rightarrow m = \dfrac{-1-1}{1-0} = \dfrac{-2}{1} = -2$. The y-intercept is $(0,1)$ and the x-intercept is $\left(\dfrac{1}{2},0\right)$.

(b) Using the slope and y-intercept, the formula is $f(x) = -2x+1$.

(c) The x-intercept is the zero of $f \Rightarrow \dfrac{1}{2}$.

63. (a) The graph passes through (0, 2) and (3,1) $\Rightarrow m = \dfrac{1-2}{3-0} = \dfrac{-1}{3} = -\dfrac{1}{3}$.. The y-intercept is $(0,2)$ and the x-intercept is $(6,0)$.

(b) Using the slope and y-intercept, the formula is $f(x) = -\dfrac{1}{3}x+2$.

(c) The x-intercept is the zero of $f \Rightarrow 6$.

65. (a) The graph passes through (0, 300) and (2, −100) $\Rightarrow m = \dfrac{-100-300}{2-0} = \dfrac{-400}{2} = -200$.

The y-intercept is $(0,300)$ and the x-intercept is $\left(\dfrac{3}{2},0\right)$.

(b) Using the slope and y-intercept, the formula is $f(x) = -200x+300$.

(c) The x-intercept is the zero of $f \Rightarrow \dfrac{3}{2}$.

67. Using (0, 2) and (1, 6), $m = \dfrac{6-2}{1-0} = \dfrac{4}{1} = 4$. From the table, the y-intercept is $(0,2)$. Using these two answers and slope-intercept form, the equation is $f(x) = 4x+2$.

69. Using (0, −3.1) and (.2, −3.38), $m = \dfrac{-3.38-(-3.1)}{.2-0} = \dfrac{-.28}{.2} = -1.4$. From the table, the y-intercept is $(0,-3.1)$. Using these two answers and slope-intercept form, the equation is $f(x) = -1.4x - 3.1$.

71. The graph of a constant function with positive k is a horizontal graph above the x-axis. Graph A

73. The graph of an equation of the form $x = k$ with $k > 0$ is a vertical line right of the y-axis. Graph D

75. Using (−1, 3) with a rise of 3 and a run of 2, the graph also passes through (1, 6). See Figure 75.

77. Using (3, −4) with a rise of −1 and a run of 3, the graph also passes through (6, −5). See Figure 77.

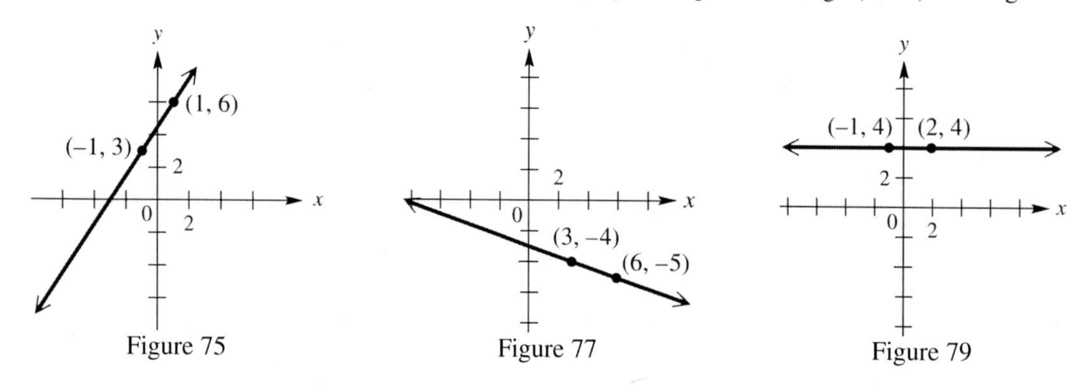

Figure 75 Figure 77 Figure 79

79. Using (−1, 4) with slope of 0, the graph is a horizontal line which also passes through (2, 4). See Figure 79.

81. Using (0, −4) with a rise of 3 and a run of 4, the graph also passes through (4, −1). See Figure 81.

83. Using (−3, 0) with undefined slope, the graph is a vertical line which also passes through (−3, 2). See Figure 83.

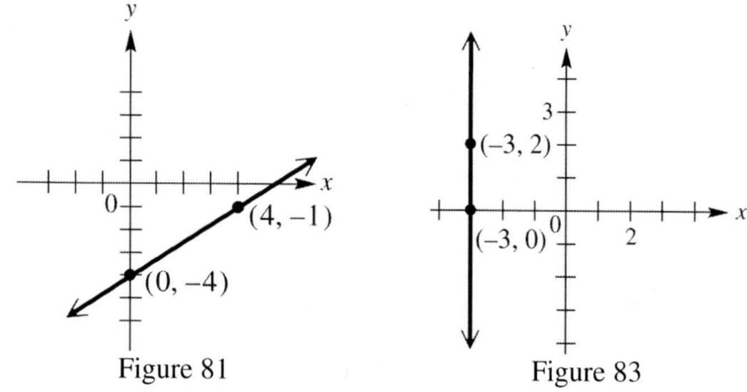

Figure 81 Figure 83

85. (a) Using the points (0, 2000) and (4, 4000), $m = \dfrac{4000-2000}{4-0} = \dfrac{2000}{4} = 500$. The y-intercept is $(0, 2000)$. The formula is $f(x) = 500x + 2000$.

(b) Water is entering the pool at a rate of 500 gallons per hour. The pool contains 2000 gallons initially.

(c) From the graph $f(7) = 5500$ gallons. By evaluating, $f(7) = 500(7) + 2000 = 5500$ gallons.

87. (a) The rain fell at a rate of $\frac{1}{4}$ inches per hour, so $m = \frac{1}{4}$. The initial amount of rain at noon was 3 inches, so $b = 3$. The equation $f(x) = \frac{1}{4}x + 3$.

(b) By 2:30 P.M. ($x = 2.5$), the total rainfall was $f(2.5) = \frac{1}{4}(2.5) + 3 = 3.625$ in.

89. (a) $f(15) = \frac{15}{5} = 3$, The delay of a bolt of lightning 3 miles away is 15 seconds.

(b) See Figure 89.

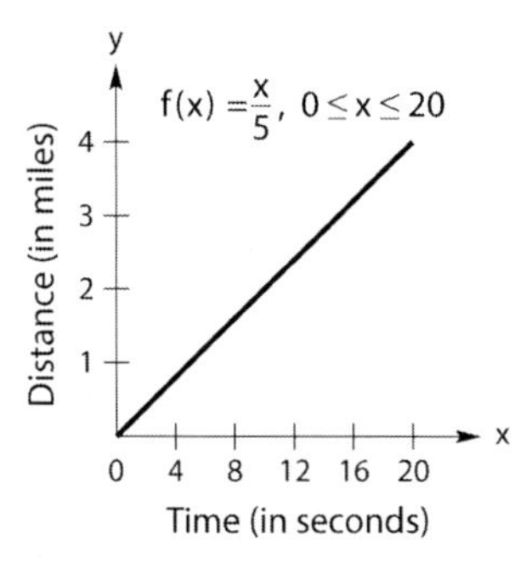

Figure 89

91. $f(x) = 0.075x$, $f(86) = 0.075(86) = 6.45$, The tax on \$86 is \$6.45.

93. The increase of \$192 per credit can be shown as the slope and the fixed fees of \$275 can be shown as the y-intercept. The function is $f(x) = 192x + 275$. $f(11) = 192(11) + 275 = \$2387$

95. (a) Since the average rate of change has been 0.9 degrees per decade we will write the slope as 0.09 degrees per year. The function is $W(x) = 0.09x$.

(b) $W(15) = 0.09(15) = 1.35$, In 15 years the Antarctic has warmed 1.35 degrees farenheit, on average.

1.4: Equations of Lines and Linear Models

1. Using Point-Slope Form yields $y - 3 = -2(x - 1) \Rightarrow y - 3 = -2x + 2 \Rightarrow y = -2x + 5$.

3. Using Point-Slope Form yields $y - 4 = 1.5(x - (-5)) \Rightarrow y - 4 = 1.5x + 7.5 \Rightarrow y = 1.5x + 11.5$.

5. Using Point-Slope Form yields $y - 1 = -.5(x - (-8)) \Rightarrow y - 1 = -.5x - 4 \Rightarrow y = -.5x - 3$.

7. Using Point-Slope Form yields $y - (-4) = 2\left(x - \frac{1}{2}\right) \Rightarrow y + 4 = 2x - 1 \Rightarrow y = 2x - 5$.

9. Using Point-Slope Form yields $y - \frac{2}{3} = \frac{1}{2}\left(x - \frac{1}{4}\right) \Rightarrow y - \frac{2}{3} = \frac{1}{2}x - \frac{1}{8} \Rightarrow y = \frac{1}{2}x + \frac{13}{24}$.

11. Use the points to (–4, –6) and (6, 2) find the slope: $m=\frac{2-(-6)}{6-(-4)} \Rightarrow m=\frac{4}{5}$. Now using Point-Slope Form yields $y-2=\frac{4}{5}(x-6) \Rightarrow y-2=\frac{4}{5}x-\frac{24}{5} \Rightarrow y=\frac{4}{5}x-\frac{14}{5}$.

13 Use the points (–12, 8) and (8, –12) to find the slope: $m=\frac{-12-8}{8-(-12)} \Rightarrow m=\frac{-20}{20} \Rightarrow m=-1$. Now using Point-Slope Form yields $y-8=-1(x+12) \Rightarrow y-8=-x-12 \Rightarrow y=-x-4$.

15. Use the points (4, 8) and (0, 4) to find the slope: $m=\frac{4-8}{0-4} \Rightarrow m=\frac{-4}{-4} \Rightarrow m=1$. Now using Slope-Intercept Form yields $b=4 \Rightarrow y=x+4$.

17. Use the points (3, –8) and (5, –3) to find the slope: $m=\frac{-3-(-8)}{5-3} \Rightarrow m=\frac{5}{2}$. Now using Point-Slope Form yields $y-(-8)=\frac{5}{2}(x-3) \Rightarrow y+8=\frac{5}{2}x-\frac{15}{2} \Rightarrow y=\frac{5}{2}x-\frac{31}{2}$.

19. Use the points (2, 3.5) and (6, –2.5) to find the slope: $m=\frac{-2.5-3.5}{6-2} \Rightarrow m=\frac{-6}{4} \Rightarrow m=-1.5$. Now using Point-Slope Form yields $y-3.5=-1.5(x-2) \Rightarrow y-3.5=-1.5x+3 \Rightarrow y=-1.5x+6.5$.

21. Use the points (0, 5) and (10, 0) to find the slope: $m=\frac{0-5}{10-0} \Rightarrow m=\frac{-5}{10} \Rightarrow m=-\frac{1}{2}$. Now using Point-Slope Form yields $y-5=-\frac{1}{2}(x-0) \Rightarrow y-5=-\frac{1}{2}x \Rightarrow y-\frac{1}{2}x+5$.

23. Use the points (–5, –28) and (–4, –20) to find the slope: $m=\frac{-20-(-28)}{-4-(-5)} \Rightarrow m=\frac{8}{1} \Rightarrow m=8$. Now using Point-Slope Form yields $y-(-20)=8(x-(-4)) \Rightarrow y+20=8x+32 \Rightarrow y=8x+12$.

25. Use the points (2, - 5) and (4, – 11) to find the slope: $m=\frac{-11-(-5)}{4-2} \Rightarrow m=\frac{-6}{2} \Rightarrow m=-3$. Now using Point-Slope Form yields $y-(-5)=-3(x-2) \Rightarrow y+5=-3x+6 \Rightarrow y=-3x+1$.

27. To find the x-intercept set $y=0$, then $x-0=4 \Rightarrow x=4$. Therefore (4, 0) is the x-intercept. To find the y-intercept set $x=0$, then $0-y=4 \Rightarrow y=-4$. Therefore (0, –4) is the y-intercept. See Figure 27.

29. To find the x-intercept set $y=0$, then $3x-0=6 \Rightarrow 3x=6 \Rightarrow x=2$. Therefore (2, 0) is the x-intercept. To find the y-intercept set $x=0$, then $3(0)-y=6 \Rightarrow y=-6$. Therefore (0, –6) is the y-intercept. See Figure 29.

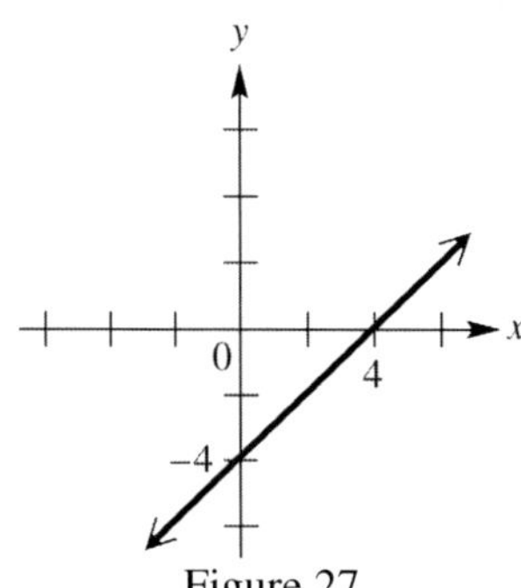

Figure 27

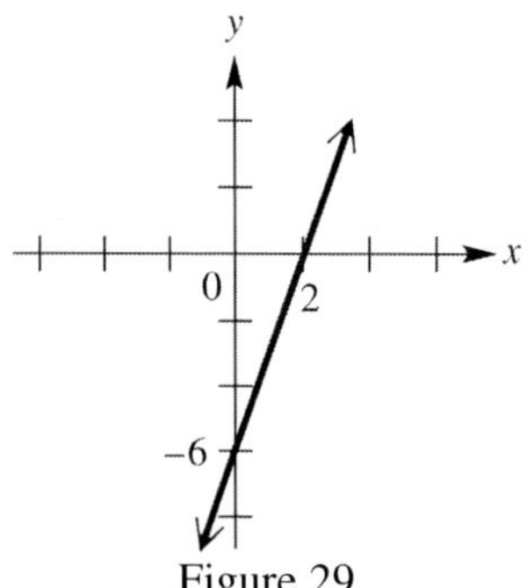

Figure 29

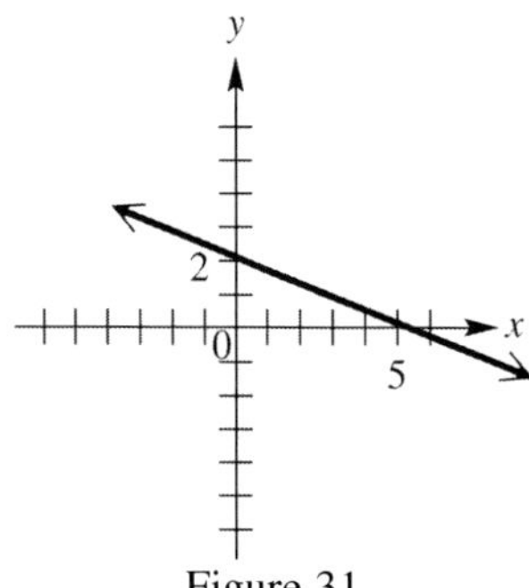

Figure 31

31. To find the x-intercept: set $y = 0$, then $2x + 5(0) = 10 \Rightarrow 2x = 10 \Rightarrow x = 5$. Therefore $(5, 0)$ is the x-intercept. To- find the y-intercept: set $x = 0$, then $2(0) + 5y = 10 \Rightarrow 5y = 10 \Rightarrow y = 2$. Therefore $(0, 2)$ is the y-intercept. See Figure 31.

33. To find a second point set $x = 1$, then $y = 3(1) \Rightarrow y = 3$. A second point is $(1, 3)$. See Figure 33.

35. To find a second point set $x = 4$, then $y = -.75(4) \Rightarrow y = -3$. A second point is $(4, -3)$. See Figure 35.

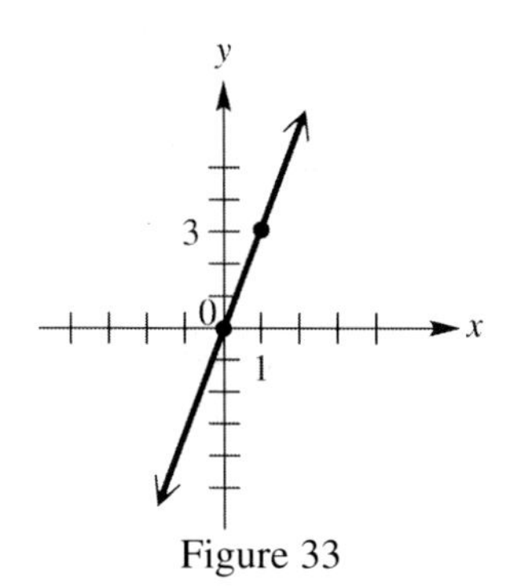

Figure 33

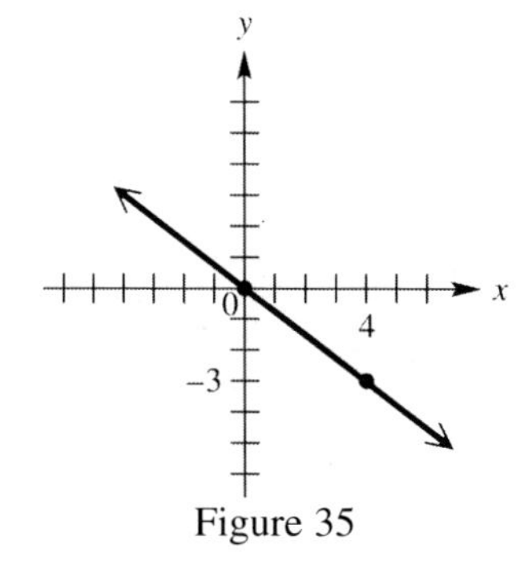

Figure 35

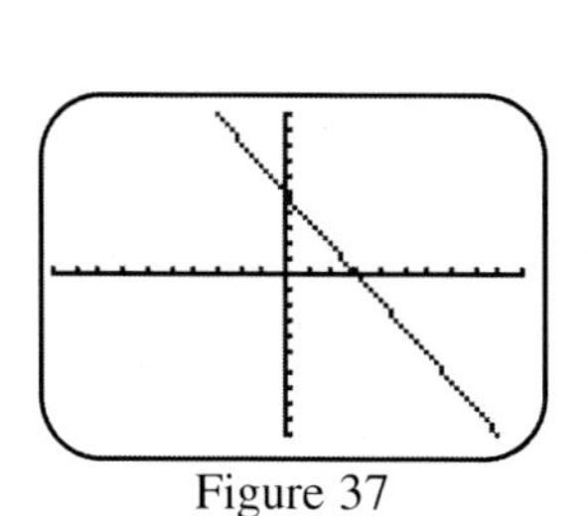
Figure 37

37. $5x + 3y = 15 \Rightarrow 3y = -5x + 15 \Rightarrow y = -\frac{5}{3}x + 5$. See Figure 37.

39. $-2x + 7y = 4 \Rightarrow 7y = 2x + 4 \Rightarrow y = \frac{2}{7}x + \frac{4}{7}$. See Figure 39.

41. $1.2x + 1.6y = 5.0 \Rightarrow 12x + 16y = 50 \Rightarrow 16y = -12x + 50 \Rightarrow y = -\frac{12}{16}x + \frac{50}{16} \Rightarrow y = -\frac{3}{4}x + \frac{25}{8}$. See Figure 41.

[-5,5] by [-5,5]
Xscl = 1 Yscl = 1

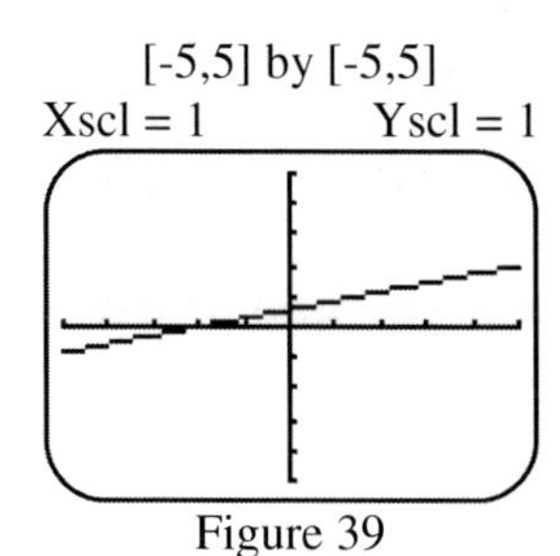
Figure 39

[-6,6] by [-4,4]
Xscl = 1 Yscl = 1

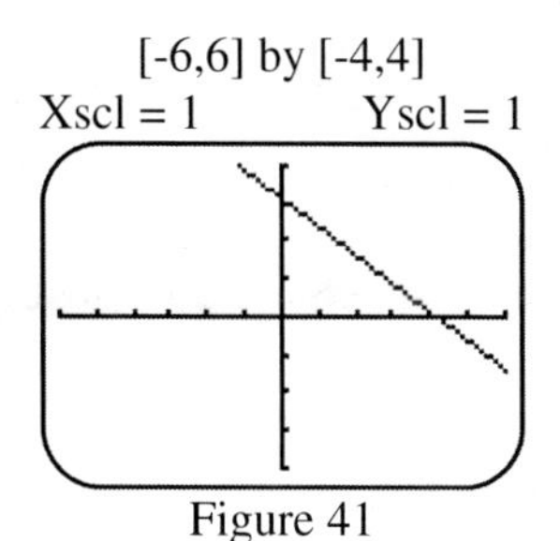
Figure 41

43. Put into slope-intercept form to find slope: $x+3y=5 \Rightarrow 3y=-x+5 \Rightarrow y=-\frac{1}{3}x+\frac{5}{3} \Rightarrow m=-\frac{1}{3}$.

Since parallel lines have equal slopes, use $m=-\frac{1}{3}$ and $(-1, 4)$ in point-slope form to find the equation:

$y-4=-\frac{1}{3}(x-(-1)) \Rightarrow y-4=-\frac{1}{3}x-\frac{1}{3} \Rightarrow y=-\frac{1}{3}x+\frac{11}{3}$.

45. Put into slope-intercept form to find slope: $3x+5y=1 \Rightarrow 5y=-3x+1 \Rightarrow y=-\frac{3}{5}x+\frac{1}{5} \Rightarrow m=-\frac{3}{5}$. Since perpendicular lines have negative reciprocal slopes, use $m=\frac{5}{3}$ and $(1, 6)$ in point-slope form to find the equation: $y-6=\frac{5}{3}(x-1) \Rightarrow y-6=\frac{5}{3}x-\frac{5}{3} \Rightarrow y=\frac{5}{3}x+\frac{13}{3}$.

47. The equation $y=-2$ has a slope $m=0$. A line perpendicular to this would have an undefined slope which would have an equation in the form $x=a$. An equation in the form $x=a$ through $(-5, 7)$ is $x=-5$.

49. The equation $y=-.2x+6$ has a slope $m=-0.2$. Since parallel lines have equal slopes, use $m=-2$ and $(-5, 8)$ in point-slope form to find the equation $y-8=-0.2(x-(-5)) \Rightarrow y-8=-0.2x-1 \Rightarrow y=-0.2x+7$.

51. Put into slope-intercept form to find slope: $2x+y=6 \Rightarrow y=-2x+6 \Rightarrow m=-2$. Since perpendicular lines have negative reciprocal slopes, use $m=\frac{1}{2}$ and the origin $(0, 0)$ in point-slope form to find the equation

$y-0=\frac{1}{2}(x-0) \Rightarrow y=\frac{1}{2}x$.

53. The equation $x=3$ has an undefined slope. A line perpendicular to this would have a slope $m=0$, which would have an equation in the form $y=b$. An equation in the form $y=b$ through $(1, 2)$ is $y=2$.

55. We will first find the slope of the line through the given points: $m=\frac{\frac{2}{3}-\frac{1}{2}}{-3-(-5)}=\frac{\frac{1}{6}}{2} \Rightarrow m=\frac{1}{12}$. Since perpendicular lines have negative reciprocal slopes, use $m=-12$ and the point (-2,4) in point-slope form to find the equation $y-4=-12(x-(-2)) \Rightarrow y=-12x-20$.

57. The slope of the perpendicular bisector will have a negative reciprocal slope and will pass through the midpoint of the line segment joined by the two points. We will first find the slope of the line through the given points: $m=\frac{10-2}{2-(-4)}=\frac{8}{6} \Rightarrow m=\frac{4}{3}$. The midpoint of the line segment is $\left(\frac{-4+2}{2}, \frac{2+10}{2}\right)=(-1,6)$. Use $m=-\frac{3}{4}$ and the point (-1,6) in point-slope form to find the equation $y-6=-\frac{3}{4}(x-(-1)) \Rightarrow y=-\frac{3}{4}x+\frac{21}{4}$.

59. (a) The Pythagorean Theorem and its converse.

(b) Using the distance formula from (0, 0) to (x_1, m_1x_1) yields: $d(0,P)=\sqrt{(x_1)^2+(m_1x_1)^2}$.

(c) Using the distance formula from (0, 0) to (x_2, m_2x_2) yields: $d(0,Q)=\sqrt{(x_2)^2+(m_2x_2)^2}$.

(d) Using the distance formula from (x_1, m_1x_1) to (x_2, m_2x_2) yields:

$d(P,Q)=\sqrt{(x_2-x_1)^2+(m_2x_2-m_1x_1)^2}$.

(e) Using Pythagorean Theorem yields: $[d(0,P)]^2+[d(0,Q)]^2=[d(P,Q)]^2 \Rightarrow$

$(x_1)^2+(m_1x_1)^2+(x_2)^2+(m_2x_2)^2=(x_1-x_2)^2+(m_1x_1-m_2x_2)^2 \Rightarrow (x_1)^2+(m_1x_1)^2+(x_2)^2+(m_2x_2)^2=$

$(x_1)^2-2x_1x_2+(x_2)^2+(m_1x_1)^2-2m_1m_2x_1x_2+(m_2x_2)^2 \Rightarrow 0=-2m_1m_2x_1x_2-2x_1x_2$.

(f) $0=-2x_1x_2-2m_1m_2x_1x_2 \Rightarrow 0=-2x_1x_2(1+m_1m_2)$

(g) By the zero-product property, for $-2x_1x_2(1+m_1m_2)=0$ either $-2x_1x_2=0$ or $1+m_1m_2=0$.

Since $x_1 \neq 0$ and $x_2 \neq 0$, $-2x_1x_2 \neq 0$, and it follows that $1+m_1m_2=0 \Rightarrow m_1m_2=-1$.

(h) The product of the slopes of two perpendicular lines, neither of which is parallel to an axis, is -1.

61. (a) Use the given points to find slope, then $m=\frac{161-128}{4-1}=\frac{33}{3} \Rightarrow m=11$. Now use point-slope form to find the equation: $y-128=11(x-1) \Rightarrow y-128=11x-11 \Rightarrow y=11x+117$.

(b) From the slope the biker is traveling 11 mph.

(c) At $x=0$, $y=11(0)+117 \Rightarrow y=117$, therefore 117 miles from the highway.

(d) Since at 1 hour and 15 minutes $x=1.25$, then $y=11(1.25)+117 \Rightarrow y=130.75$, so 130.75 miles away.

63. (a) Use the points (2007, 18), (2010, 24) to find slope, then $m=\frac{24-18}{2010-2007}=\frac{6}{3} \Rightarrow m=2$. Now use point-slope form to find the equation: $y-18=2(x-2007) \Rightarrow y-18=2x-4014 \Rightarrow y=2x-3996$.

(b) y = 2(2013) – 3996 =30. There was approximately $30 billion in betting revenue in 2013.

65. (a) Since the plotted points form a line, it is a linear relation. See Figure 65.

(b) Using the first two points find the slope: $m=\frac{0-(-40)}{32-(-40)}=\frac{40}{72}=\frac{5}{9}$, now use slope-intercept form to find the function: $C(x)-0=\frac{5}{9}(x-32) \Rightarrow C(x)=\frac{5}{9}(x-32)$. The slope of $\frac{5}{9}$ means that the Celsius temperature changes 5° for every 9° change in Fahrenheit temperature.

(c) $C(83)=\frac{5}{9}(83-32)=28\frac{1}{3}°C$

67. (a) The slope is $\frac{37-6}{2011-2005}=\frac{31}{6}$:. Using point-slope form produces the equation:

$y-6=\frac{31}{6}(x-2005)$.

(b) Every year from 2005 to 2011, Google advertising revenue increased by about $5.2 billion on average.

(c) 2007 Revenue: $y-6=\frac{31}{6}(2007-2005)\Rightarrow y=\frac{31}{6}(2)+6\Rightarrow y\approx 16.3$ billion

2009 Revenue: $y-6=\frac{31}{6}(2009-2005)\Rightarrow y=\frac{31}{6}(4)+6\Rightarrow y\approx 26.67$ billion

The 2007 value compares favorably and the 2009 value is too high.

69. (a) Enter the years in L_1 and enter tuition and fees in L_2. The regression equation is: $y\approx 586.89x-1{,}147{,}738$.

(b) See Figure 69.

(c) At $x=2005$, $y\approx 586.89(2005)-1{,}147{,}738\Rightarrow \$28{,}976$ this is close to the actual value of \$29,307.

[-50,250] by [-50,110]
Xscl = 50 Yscl = 50

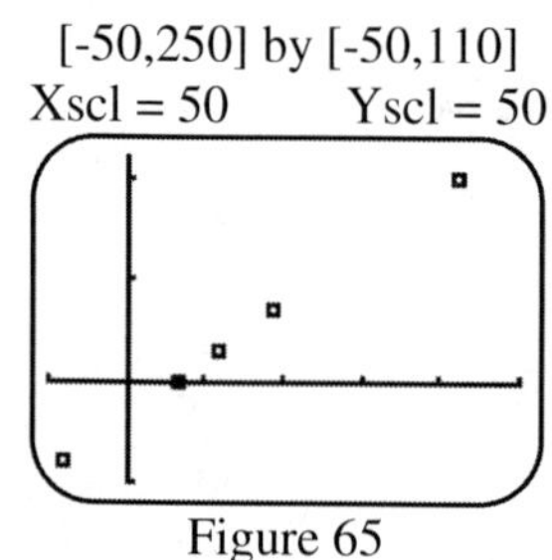

Figure 65

[1975,2015] by [13000,32000]
Xscl = 10 Yscl = 10

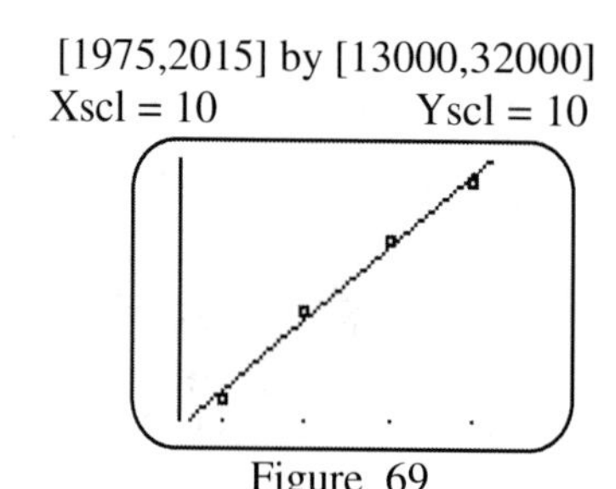

Figure 69

71. (a) Enter the distance in L_1 and enter velocity in L_2. The regression equation is: $y\approx 0.06791x-16.32$.

(b) At $y=37{,}000$, $y\approx 0.06791(37{,}000)-16.32\approx 2500$ or approximately 2500 light-years.

73. Enter the Gestation Period in L_1 and enter Life Span in L_2. The regression equation is: $y\approx .101x+11.6$ and the correlation coefficient is: $r\approx .909$. There is a strong positive correlation, because .909 is close to 1.

Reviewing Basic Concepts (Sections 1.3 and 1.4)

1. Since $m=1.4$ and $b=-3.1$, slope-intercept form gives the function: $f(x)=1.4x-3.1$.

   $f(1.3)=1.4(1.3)-3.1\Rightarrow f(1.3)=-1.28$

2. See Figure 2. x-intercept: $\frac{1}{2}$, y-intercept: 1, slope: -2, domain: $(-\infty,\infty)$, range: $(-\infty,\infty)$

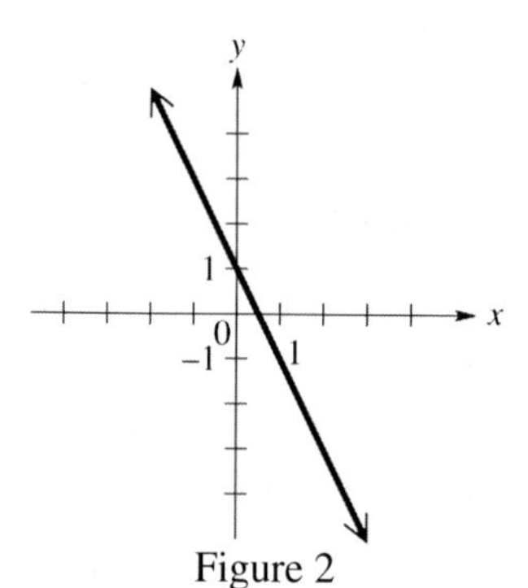

Figure 2

3. $m = \dfrac{6-4}{5-(-2)} = \dfrac{2}{7}$

4. Vertical line graphs are in the form $x = a$; through point $(-2, 10)$ would be $x = -2$.

   Horizontal line graphs are in the form $y = b$; through point $(-2, 10)$ would be $y = 10$.

5. See Figures 5a and 5b.

[-10,10] by [-10,10]
Xscl = 1 Yscl = 1

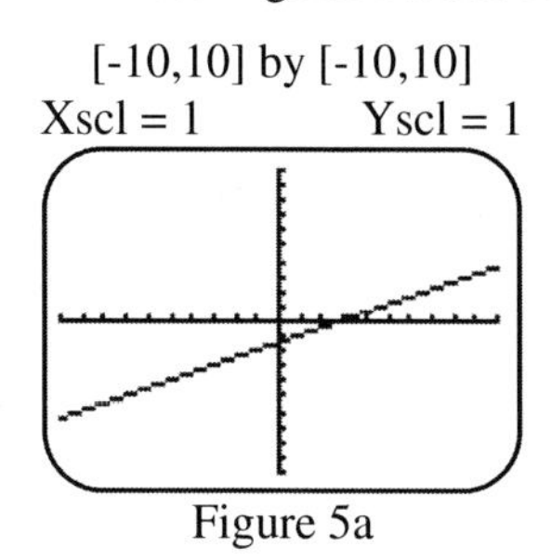

Figure 5a

[-10,10] by [-10,10]
Xscl = 1 Yscl = 1

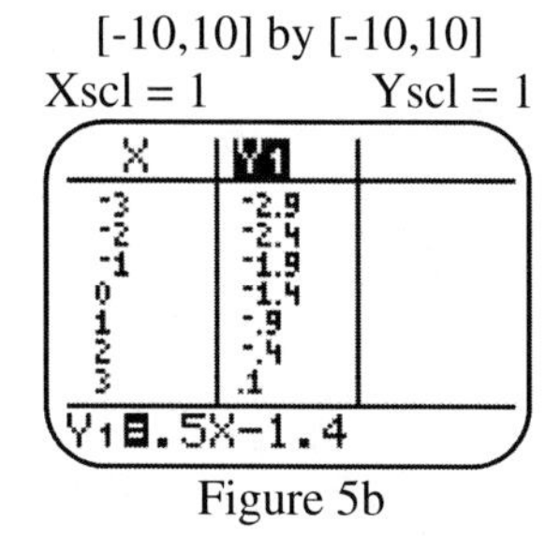

Figure 5b

[1940,2020] by [2,4]
Xscl= 10 Yscl = .5

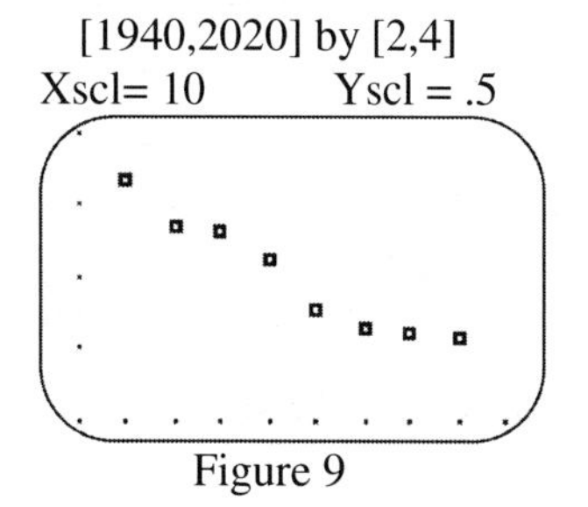

Figure 9

6. The line of the graph rises 2 units for each 1 unit to the right, therefore the slope is: $m = \dfrac{2}{1} = 2.$

   The y-intercept is: $b = -3$. The slope-intercept form of the equation is: $y = 2x - 3.$

7. The slope is: $m = \dfrac{4-2}{(-2)-5} = \dfrac{2}{-7} = -\dfrac{2}{7}$; now using point-slope form the equation is:

   $$y - 4 = -\frac{2}{7}(x+2) \Rightarrow y - 4 = -\frac{2}{7}x - \frac{4}{7} \Rightarrow y = -\frac{2}{7}x + \frac{24}{7}.$$

8. Find the given equation in slope-intercept form: $3x - 2y = 5 \Rightarrow -2y = -3x + 5 \Rightarrow y = \dfrac{3}{2}x - \dfrac{5}{2}.$

   The slope of this equation is $m = \dfrac{3}{2}$, therefore the slope of a perpendicular line will be the negative

   reciprocal: $m = -\dfrac{2}{3}$. Using point-slope form yields the equation:

   $$y - 3 = -\frac{2}{3}(x+1) \Rightarrow y - 3 = -\frac{2}{3}x - \frac{2}{3} \Rightarrow y = -\frac{2}{3}x + \frac{7}{3}.$$

9. (a) See Figure 9.

   (b) As x increases, y decreases, therefore a negative correlation coefficient.

   (c) Enter the years in L_1 and enter people per household in L_2. The regression equation is: $y \approx -0.0165x + 35.6$ and the correlation coefficient is: $r = -0.9648$

   (d) The regression equation is: $y \approx -0.0165(1975) + 35.6 \Rightarrow y \approx 3.01$, which is close to the actual value 2.94.

1.5: Linear Equations and Inequalities

1. $-3x - 12 = 0 \Rightarrow -3x = 12 \Rightarrow x = -4$

3. $5x = 0 \Rightarrow x = 0$

5. $2(3x-5)+8(4x+7)=0 \Rightarrow 6x-10+32x+56=0 \Rightarrow 38x=-46 \Rightarrow x=-\frac{46}{38} \Rightarrow x=-\frac{23}{19}$

7. $3x+6(x-4)=0 \Rightarrow 3x+6x-24=0 \Rightarrow 9x=24 \Rightarrow x=\frac{24}{9} \Rightarrow x=\frac{8}{3}$

9. $1.5x+2(x-3)+5.5(x+9)=0 \Rightarrow 1.5x+2x-6+5.5x+49.5=0 \Rightarrow 9x=-43.5 \Rightarrow$

$x=\frac{-43.5}{9} \Rightarrow x=-\frac{29}{6}$

11. The solution to $y_1 = y_2$ is the intersection of the lines or $x=\{10\}$.

13. The solution to $y_1 = y_2$ is the intersection of the lines or $x=\{1\}$.

15. When $y_1 = y_2$, $y=0$. $y=0$ when the graph crosses the x-axis or at the zero $x=\{3\}$.

17. When $x=10$ is substituted into each function the result is 20.

19. There is no real solution if $y_1 - y_2$ yields a contradiction, $y=b$, where $b \neq 0$. This equation is called a contradiction and the solution set is: $\varnothing$.

21. $2x-5=x+7 \Rightarrow x-5=7 \Rightarrow x=12$ **Check:** $2(12)-5=12+7 \Rightarrow 19=19$ The graphs of the left and right sides of the equation intersect when $x=12$. The solution set is $\{12\}$.

23. $0.01x+3.1=2.03x-2.96 \Rightarrow 3.1=2.02x-2.96 \Rightarrow 6.06=2.02x \Rightarrow x=3$

Check: $0.01(3)+3.1=2.03(3)-2.96 \Rightarrow .03+3.1=6.09-2.96 \Rightarrow 3.13=3.13$

The graphs of the left and right sides of the equation intersect when $x=3$. The solution set is $\{3\}$.

25. $-(x+5)-(2+5x)+8x=3x-5 \Rightarrow -x-5-2-5x+8x=3x-5 \Rightarrow 2x-7=3x-5 \Rightarrow -2=x$

Check: $-(-2+5)-(2+5(-2))+8(-2)=3(-2)-5 \Rightarrow -2-5-2+10-16=-6-5 \Rightarrow -11=-11$. The graphs of the left and right sides of the equation intersect when $x=-2$. The solution set is $\{-2\}$.

27. $\frac{2x+1}{3}+\frac{x-1}{4}=\frac{13}{2} \Rightarrow 12\left(\frac{2x+1}{3}+\frac{x-1}{4}\right)=12\left(\frac{13}{2}\right) \Rightarrow 8x+4+3x-3=78 \Rightarrow 11x+1$

$=78 \Rightarrow 11x=77 \Rightarrow x=7$ **Check:** $\frac{2(7)+1}{3}+\frac{7-1}{4}=\frac{13}{2} \Rightarrow 5+\frac{6}{4}=\frac{13}{2} \Rightarrow \frac{13}{2}=\frac{13}{2}$

The graphs of the left and right sides of the equation intersect when $x=7$. The solution set is $\{7\}$.

29. $\frac{1}{2}(x-3)=\frac{5}{12}+\frac{2}{3}(2x-5) \Rightarrow 12\left[\frac{1}{2}(x-3)=\frac{5}{12}+\frac{2}{3}(2x-5)\right] \Rightarrow 6x-18=5+16x-40 \Rightarrow$

$-10x=-17 \Rightarrow x=\frac{17}{10}$ **Check:** $\frac{1}{2}\left(\frac{17}{10}-3\right)=\frac{5}{12}+\frac{2}{3}\left(2\left(\frac{17}{10}\right)-5\right) \Rightarrow \frac{1}{2}\left(-\frac{13}{10}\right)=\frac{5}{12}+\frac{2}{3}\left(-\frac{16}{10}\right)$

$\Rightarrow \frac{13}{20}=\frac{5}{12}+\left(\frac{32}{30}\right) \Rightarrow \frac{78}{120}=\frac{50}{120}+\left(-\frac{128}{120}\right) \Rightarrow \frac{78}{120}=-\frac{78}{120}$. The graphs of the left and right sides of the equation intersect when $x=\frac{17}{10}$. The solution set is $\left\{\frac{17}{10}\right\}$.

31. $0.1x - 0.05 = -0.07x \Rightarrow 0.17x = 0.05 \Rightarrow 17x = 5 \Rightarrow x = \frac{5}{17}$

Check: $1\left(\frac{5}{17}\right) - 0.05 = -0.07\left(\frac{5}{17}\right) \Rightarrow 10\left(\frac{5}{17}\right) - 5 = -7\left(\frac{5}{17}\right) \Rightarrow \frac{50}{17} - 5 = -\frac{35}{17} = -\frac{35}{17}$

The graphs of the left and right sides of the equation intersect when $x = \frac{5}{17}$. The solution set is $\left\{\frac{5}{17}\right\}$.

33. $0.40x + 0.60(100 - x) = 0.45(100) \Rightarrow 0.40x + 60 - 0.60x = 45 \Rightarrow -0.20x = -15 \Rightarrow 20x = -1500 \Rightarrow x = 75$

Check: $0.40(75) + 0.60(100 - 75) = 0.45(100) \Rightarrow 30 + 15 = 45 \Rightarrow 45 = 45$. The graphs of the left and right sides of the equation intersect when $x = 75$. The solution set is $\{75\}$.

35. $2[x - (4 + 2x) + 3] = 2x + 2 \Rightarrow 2[x - 4 - 2x + 3] = 2x + 2 \Rightarrow 2[-x - 1] = 2x + 2 \Rightarrow$

$-2x - 2 = 2x + 2 \Rightarrow -4x = 4 \Rightarrow x = -1$

Check: $2[-1 - (4 + 2(-1)) + 3] = 2(-1) + 2 \Rightarrow 2[-1 - 2 + 3] = 0 \Rightarrow 2[0] = 0 \Rightarrow 0 = 0$

The graphs of the left and right sides of the equation intersect when $x = -1$. The solution set is $\{-1\}$.

37. $\frac{5}{6}x - 2x + \frac{1}{3} = \frac{1}{3} \Rightarrow 6\left(\frac{5}{6}x - 2x + \frac{1}{3} = \frac{1}{3}\right) \Rightarrow 5x - 12x + 2 = 2 \Rightarrow -7x = 0 \Rightarrow x = 0$

Check: $\frac{5}{6}(0) - 2(0) + \frac{1}{3} = \frac{1}{3} \Rightarrow \frac{1}{3} = \frac{1}{3}$

The graphs of the left and right sides of the equation intersect when $x = 0$. The solution set is $\{0\}$.

39. $5x - (8 - x) = 2[-4 - (3 + 5x - 13)] \Rightarrow 6x - 8 = 2[-5x + 6] \Rightarrow 6x - 8 = -10x + 12 \Rightarrow$

$16x = 20 \Rightarrow x = \frac{20}{16} = \frac{5}{4}$

Check: $5\left(\frac{5}{4}\right) - \left(8 - \frac{5}{4}\right) = 2\left[-4 - \left(3 + 5\left(\frac{5}{4}\right) - 13\right)\right] \Rightarrow \frac{25}{4} - \frac{27}{4} = 2\left[-4 - \left(\frac{25}{4}\right) - 10\right] \Rightarrow$

$-\frac{2}{4} = 2\left[6 - \frac{25}{4}\right] \Rightarrow -\frac{1}{2} = 2\left[-\frac{1}{4}\right] \Rightarrow -\frac{1}{2} = -\frac{1}{2}$. The graphs of the left and right sides of the equation intersect when $x = \frac{5}{4}$.The solution set is $\left\{\frac{5}{4}\right\}$.

41. When $x = 4$, both Y_1 and Y_2 have a value of 8. Therefore the solution set is $\{4\}$.

43. Graph $Y_1 = 4(0.23 + \sqrt{5})$ and $Y_2 = \sqrt{2}x + 1$ as shown in Figure 43. The graphs intersect when $x \approx 16.07$. Therefore the solution set is $\{16.07\}$.

45. Graph $Y_1 = 2\pi x + \sqrt[3]{4}$ and $Y_2 = 0.5\pi x - \sqrt{28}$ as shown in Figure 45. The graphs intersect when $x \approx -1.46$. Therefore the solution set is $\{-1.46\}$.

[-10,30] by [-10,30] [-15,5] by [-15,5] [-10,10] by [-10,10]

Xscl = 2 Yscl = 2

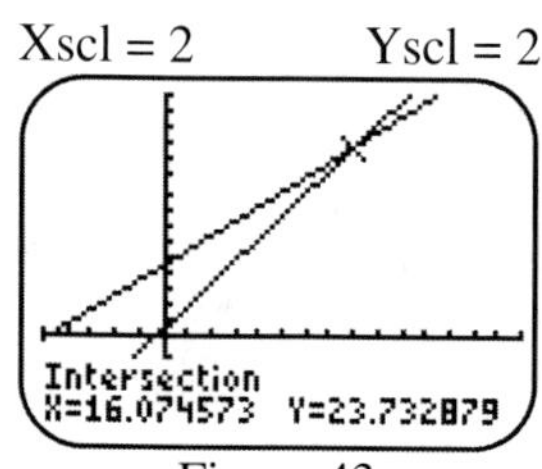

Figure 43

Xscl= 1 Yscl = 1

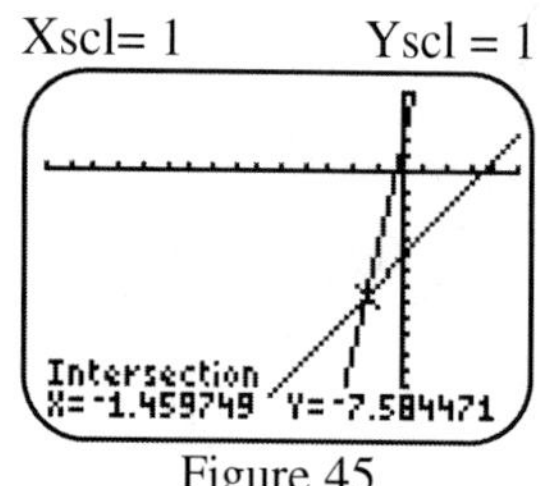

Figure 45

Xscl = 1 Yscl = 1

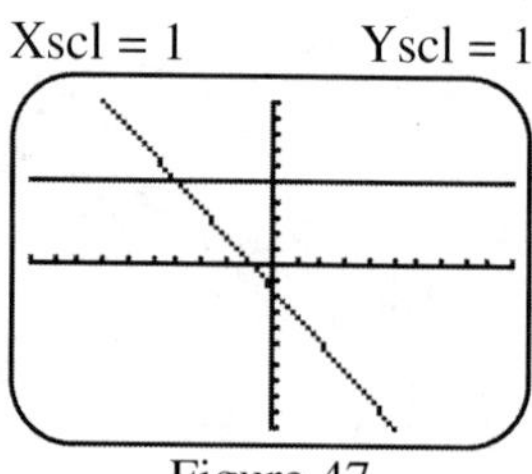

Figure 47

47. Graph $Y_1 = 0.23\left(\sqrt{3} + 4x\right) - 0.82(\pi x + 2.3)$ and $Y_2 = 5$ as shown in Figure 47. The graphs intersect when $x \approx -3.92$. Therefore the solution set is $\{-3.92\}$.

49. $5x + 5 = 5(x+3) - 3 \Rightarrow 5x + 5 = 5x + 15 - 3 \Rightarrow 5x + 5 = 5x + 12 \Rightarrow 5 = 12 \Rightarrow$ Contradiction. The solution set is $\varnothing$ The table of $Y_1 = 5x + 5$ and $Y_2 = 5(x+3) - 3$ never produces the same answers, therefore supports the Contradiction.

51. $6(2x+1) = 4x + 8\left(x + \frac{3}{4}\right) \Rightarrow 12x + 6 = 4x + 8x + 6 \Rightarrow 12x + 6 = 12x + 6 \Rightarrow 6 = 6 \Rightarrow$ Identity. The solution set is $(-\infty, \infty)$ The table of $Y_1 = 6(2x+1)$ and $Y_2 = 4x + 8\left(x + \frac{3}{4}\right)$ produces all the same answers, therefore supports the Identity.

53. $7x - 3[5x - (5 + x)] = 1 - 4x \Rightarrow 7x - 3[4x - 5] = 1 - 4x \Rightarrow 7x - 12x + 15 = 1 - 4x \Rightarrow$
$-5x + 15 = 1 - 4x \Rightarrow -x = -14 \Rightarrow x = 14 \Rightarrow$ Conditional.

The solution set is 14. The table of $Y_1 = 7x - 3[5x - (5 + x)]$ and $Y_2 = 1 - 4x$ shows that the answers are the same when x = 14.

55. $0.2(5x - 4) - 0.1(6 - 3x) = 0.4 \Rightarrow x - 0.8 - 0.6 + 0.3x = 0.4 \Rightarrow 1.3x - 1.4 = 0.4 \Rightarrow 1.3x = 1.8 \Rightarrow$

$x = \frac{18}{13} \Rightarrow$ Conditional. The solution set is $\frac{18}{13}$ The table of $Y_1 = 0.2(5x - 4) - 0.1(6 - 3x)$ and $Y_2 =$ 0.4 shows that the answers are the same when $x = \frac{18}{13}$.

57. $-4[6 - (-2 + 3x)] = 21 + 12x \Rightarrow -4[8 - 3x] = 21 + 2x \Rightarrow -32 + 12x = 21 + 2x \Rightarrow -32 = 21 \Rightarrow$

Contradiction. The solution set is $\varnothing$. The table of $Y_1 = -4[6 - (-2 + 3x)]$ and $Y_2 = 21 + 12x$ never produces the same answers, therefore supports the Contradiction.

59. $\frac{1}{2}x - 2(x-1) = 2 - \frac{3}{2}x \Rightarrow \frac{1}{2}x - 2x + 2 = 2 - \frac{3}{2}x \Rightarrow -\frac{3}{2} + 2 = -\frac{3}{2} + 2 \Rightarrow 2 = 2 \Rightarrow$ Identity.

The solution set is $(-\infty, \infty)$. The table of $Y_1 = \frac{1}{2}x - 2(x-1)$ and $Y_2 = 2 - \frac{3}{2}x$ produces all the same answers, therefore supports the Identity.

61. $\frac{x-1}{2}=\frac{3x-2}{6}\Rightarrow 6\left[\frac{x-1}{2}=\frac{3x-2}{6}\right]\Rightarrow 3(x-2)=3x-2\Rightarrow 3x-6=3x-3\Rightarrow -6=-3\Rightarrow$

Contradiction. The solution set is ∅. The table of $Y_1=\frac{x-1}{2}$ and $Y_2=\frac{3x-2}{6}$ never produces the same answers, therefore supports the Contradiction.

63. For the given functions, $f(x)=g(x)$ when the graphs intersect or when $x=3$. The solution is $\{3\}$.

65. For the given functions, $f(x)<g(x)$ when the graph of $f(x)$ is below the graph of $g(x)$ or when $x>3$. The solution is $(3,\infty)$.

67. For the given inequality, $y_1-y_2\geq 0\Rightarrow f(x)-g(x)\geq 0\Rightarrow f(x)\geq g(x)$ when the graph of $f(x)$ is above or intersects the graph of $g(x)$ or when $x\leq 3$. The solution is $(-\infty,3]$.

69. For the given functions, $f(x)\leq f(x)$ when the graph of is $f(x)$ below or intersects the graph $g(x)$ of or when $x\geq 3$. The solution is $[3,\infty)$.

71. For the given functions, $f(x)\leq 2$ when the graph of $f(x)$ is below or equal to 2 or when $x\geq 3$. The solution is $[3,\infty)$.

73. (a) The function $f(x)>0$ when the graph is above the x-axis for the interval $(20,\ \infty)$.

(b) The function $f(x)<0$ when the graph is below the x-axis for the interval $(-\infty,\ 20)$.

(c) The function $f(x)\geq 0$ when the graph intersects or is above the x-axis for the interval $[20,\ \infty)$.

(d) The function $f(x)\leq 0$ when the graph intersects or is below the x-axis for the interval $(-\infty,\ 20]$.

75. (a) If the solution set of $f(x)\geq g(x)$ is $[4,\infty)$, then $f(x)=g(x)$ at the intersection of the graphs, $x=4$ or $\{4\}$.

(b) If the solution set of $f(x)\geq g(x)$ is $[4,\infty)$, then $f(x)>g(x)$ is the same, but does not include the intersection of the graphs for the interval $(4,\ \infty)$.

(c) If the solution set of $f(x)\geq g(x)$ is $[4,\infty)$, then $f(x)<g(x)$ is left of the intersection of the graphs for the interval: $(-\infty,\ 4)$.

77. (a) $3x-6=0\Rightarrow 3x=6\Rightarrow x=2,$ Interval Notation: $\{2\}$

(b) $3x-6>0\Rightarrow 3x>6\Rightarrow x>2,$ Interval Notation: $(2,\infty)$

(c) $3x-6<0\Rightarrow 3x<6\Rightarrow x<2,$ Interval Notation: $(-\infty,2)$

79. (a) $1-2x=0\Rightarrow -2x=-1\Rightarrow x=\frac{1}{2},$ Interval Notation: $\left\{\frac{1}{2}\right\}$

(b) $1-2x\leq 0\Rightarrow -2x\leq -1\Rightarrow x\geq\frac{1}{2},$ Interval Notation: $\left[\frac{1}{2},\infty\right)$

(c) $1-2x \geq 0 \Rightarrow -2x \geq -1 \Rightarrow x \leq \frac{1}{2}$, Interval Notation : $\left(-\infty, \frac{1}{2}\right]$

81. (a) $x+12=4x \Rightarrow -3x=-12 \Rightarrow x=4$, Interval Notation : $\{4\}$

(b) $x+12>4x \Rightarrow -3x>-12 \Rightarrow x<4$, Interval Notation : $(-\infty, 4)$

(c) $x+12<4x \Rightarrow -3x<-12 \Rightarrow x>4$, Interval Notation : $(4, \infty)$

83. (a) $9-(x+1)<0 \Rightarrow -x+8<0 \Rightarrow -x<-8 \Rightarrow x>8 \Rightarrow$ the interval is $(8, \infty)$. The graph of $y_1 = 9-(x+1)$ is below the x-axis for the interval $(8, \infty)$.

(b) If $9-(x+1)<0$ for $(8, \infty)$, then $9-(x+1) \geq 0$ for the interval $(-\infty, 8]$. The graph of $y_1 = 9-(x+1)$ intersects or is above the x-axis for the interval $(-\infty, 8]$.

85. (a) $2x-3>x+2 \Rightarrow x-3>2 \Rightarrow x>5 \Rightarrow$ the interval is $(5, \infty)$. The graph of $y_1 = 2x-3$ is above the graph of $y_2 = x+2$ for the interval $(5, \infty)$.

(b) If $2x-3>x+2$ for $(5, \infty)$, then $2x-3 \leq x+2$ for the interval $(-\infty, 5]$. The graph $y_1 = 2x-3$ intersects or is below the graph $y_2 = x+2$ for the interval $(-\infty, 5]$.

87. (a) $10x+5-7x \geq 8(x+2)+4 \Rightarrow 3x+5 \geq 8x+20 \Rightarrow -5x \geq 15 \Rightarrow x \leq -3 \Rightarrow$ the interval is $(-\infty, -3]$. The graph of $y_1 = 10x+5-7x$ intersects or is above the graph of $y_2 = 8(x+2)+4$ for the interval $(-\infty, -3]$.

(b) If $10x+5-7x \geq 8(x+2)+4$ for $(-\infty, -3)$, then $10x+5-7x < 8(x+2)+4$ for the interval $(-3, \infty)$. The graph of $y_1 = 10x+5-7x$ is below the graph of $y_2 = 8(x+2)+4$ for the interval $(-3, \infty)$.

89. (a) $x+2(-x+4)-3(x+5)<-4 \Rightarrow x-2x+8-3x-15<-4 \Rightarrow -4x<3 \Rightarrow x>-\frac{3}{4} \Rightarrow$ the interval is $\left(-\frac{3}{4}, \infty\right)$. The graph of $y_1 = x+2(-x+4)-3(x+5)$ is below the graph of $y_2 = -4$ for the interval $\left(-\frac{3}{4}, \infty\right)$.

(b) If $x+2(-x+4)-3(x+5)<-4$ for $\left(-\frac{3}{4}, \infty\right)$, then $x+2(-x+4)-3(x+5) \geq -4$ for the interval $\left(-\infty, -\frac{3}{4}\right]$. The graph of $y_1 = x+2(-x+4)-3(x+5)$ intersects or is above the graph $y_2 = -4$ for the interval $\left(-\infty, -\frac{3}{4}\right]$.

91. $\frac{1}{3}x-\frac{1}{5}x\le 2\Rightarrow\left[\frac{1}{3}x-\frac{1}{5}x\le 2\right]\Rightarrow 5x-3x\le 30\Rightarrow 2x\le 30\Rightarrow x\le 15\Rightarrow(-\infty,15]$. The graph of

$y_1=\frac{1}{3}x-\frac{1}{5}x$ intersects or is below the graph of $y_2=2$ for the interval $(-\infty,15]$.

93. $\frac{x-2}{2}-\frac{x+6}{3}>-4\Rightarrow 6\left[\frac{x-2}{2}-\frac{x+6}{3}>-4\right]\Rightarrow 3x-6-(2x+12)>-24\Rightarrow$

$x-18>-24\Rightarrow x>-6\Rightarrow(-6,\infty)$. The graph of $y_1=\frac{x-2}{2}-\frac{x+6}{3}$ is above the graph of

$y_2=5$ for the interval: $(-6,\infty)$.

95. $0.6x-2(0.5x+.2)\le 0.4-0.3x\Rightarrow .6x-1x-0.4\le 0.4-0.3x\Rightarrow 10[0.6x-1x-0.4\le 0.4-0.3x]\Rightarrow$

$6x-10x-4\le 4-3x\Rightarrow -4x-4\le 4-3x\Rightarrow -x\le 8\Rightarrow x\ge -8\Rightarrow[-8,\infty)$. The graph of

$y_1=0.6x-2(0.5x+.2)$ intersects or is below the graph of $y_2=0.4-0.3x$ for the interval $[-8,\infty)$.

97. $-\frac{1}{2}x+0.7x-5>0\Rightarrow 10\left[-\frac{1}{2}x+0.7x-5>0\right]\Rightarrow -5x+7x-50>0\Rightarrow 2x>50\Rightarrow$

$x>25\Rightarrow(25,\infty)$. The graph of $y_1=-\frac{1}{2}x+.7x-5$ is above the graph of $y_2=0$ for the interval $(25,\infty)$.

99. $-4(3x+2)\ge -2(6x+1)\Rightarrow -12x-8\ge -12x-2\Rightarrow -8\ge -2;$ since this is false the solution is $\varnothing$.

The graph of $y_1=-4(3x+2)$ never intersects or is above the graph of $y_2=-2(6x+1)$, therefore the solution is $\varnothing$.

101. (a) As time increases, distance increases, therefore the car is moving away from Omaha.

(b) The distance function $f(x)$ intersects the 100 mile line at 1 hour and the 200 mile line at 3 hours.

(c) Using the answers from (b) the interval is [1, 3].

(d) Because x hours is $0\le x\le 6$, the interval is (1, 6].

103. $4\le 2x+2\le 10\Rightarrow 2\le 2x\le 8\Rightarrow 1\le x\le 4\Rightarrow[1,4]$ The graph of $y_2=2x+2$ is between the graphs of $y_1=4$ and $y_3=10$. for the interval [1, 4].

105. $-10>3x+2>-16\Rightarrow -12>3x>-18\Rightarrow -4>x>-6\Rightarrow -6<x<-4\Rightarrow(-6,-4)$

The graph of $y_2=3x+2$ is between the graphs of $y_1=-10$ and $y_3=-16$ for the interval $(-6,-4)$.

107. $-3\le\frac{x-4}{-5}<4\Rightarrow 15\ge x-4>-20\Rightarrow 19\ge x>-16\Rightarrow -16<x19\Rightarrow(-16,19]$ The graph of

$y_2=\frac{x-4}{-5}$ is between the graphs of $y_1=-3$ and $y_3=4$ for the interval $(-16,19]$.

109. $-\frac{1}{2}<x-4<\frac{1}{2}\Rightarrow\frac{7}{2}<x<\frac{9}{2}\Rightarrow\left(\frac{7}{2},\frac{9}{2}\right)$. The graph of $y_2=x-4$ is between $y_1=-\frac{1}{2}$ and $y_3=\frac{1}{2}$ for the interval $\left(\frac{7}{2},\frac{9}{2}\right)$.

111. $-4\le\frac{1}{2}x-5\le4\Rightarrow1\le\frac{1}{2}x\le9\Rightarrow2\le x\le18\Rightarrow[2,18]$. The graph of $y_2=\frac{1}{2}x-5$ is on or between $y_1=-4$ and $y_3=4$ for the interval $(2,18)$.

113. $\sqrt{2}\le\frac{2x+1}{3}\le\sqrt{5}\Rightarrow3\sqrt{2}\le2x+1\le3\sqrt{5}\Rightarrow3\sqrt{2}-1\le2x\le3\sqrt{5}-1\Rightarrow$

$\frac{3\sqrt{2}-1}{2}\le x\le\frac{3\sqrt{5}-1}{2}\Rightarrow\left[\frac{3\sqrt{2}-1}{2},\frac{3\sqrt{5}-1}{2}\right]$; The graph of $y_2=\frac{2x+1}{3}$ is on or between the graphs of $y_1=\sqrt{2}$ and $y=\sqrt{5}.$ for the interval $\left[\frac{3\sqrt{2}-1}{2},\frac{3\sqrt{5}-1}{2}\right]$.

115. (a) The graph of $T(x)=65-19x$ intersects the graph of $D(x)=50-5.8x$ at $\approx(1.136,43.41)$. Since the x-coordinate is altitude, the clouds will not form below 1.14 miles or for the interval: [0, 1.14).

(b) Clouds will not form when air temperature is above dew point temperature or $T(x)>D(x)$.

Then $65-19x>50-5.8x\Rightarrow-13.2x>-15\Rightarrow x<\frac{15}{13.2}$ for the interval $\left[0,\frac{15}{13.2}\right)$.

117. Since $C=2\pi r$ and radius is in the range $1.99\le r\le2.01$, circumference is in the range $2\pi(1.99)\le2\pi r\le2\pi(2.01)\Rightarrow3.98\pi\le C\le4.02\pi.$

119. The graph of $y_1=3.7x-11.1$ crosses the x-axis at $x=3$. There is one solution to this equation. Because a linear equation can only cross the x-axis in one location, there is only one solution to any linear equation.

120. $3.7x-11.1<0\Rightarrow3.7x<11.1\Rightarrow x<3\Rightarrow(-\infty,3)$ $3.7x-11.1>0\Rightarrow3.7x>11.1\Rightarrow x>3\Rightarrow(3,\infty)$

The value of $x=3$ given by the equation represents the boundary between the sets of real numbers given by the inequality solutions $(-\infty,3)$ and $(3,\infty)$.

121. The graph of $y_1=-4x+6$ crosses the x-axis at $x=1.5$.

$-4x+6<0\Rightarrow-4x<-6\Rightarrow x>\frac{-6}{-4}\Rightarrow x>1.5\Rightarrow(1.5,\infty)$

$-4x+6>0\Rightarrow-4x>-6\Rightarrow x<\frac{-6}{-4}\Rightarrow x>1.5\Rightarrow(-\infty,1.5)$

122. (a) If $a\ne0,$ then $ax+b=0\Rightarrow ax=-b\Rightarrow x=-\frac{b}{a}$

(b) If $a>0,$ a positive slope, then $ax+b<0\Rightarrow ax<-b\Rightarrow x<\frac{-b}{a}\Rightarrow\left(-\infty,\frac{-b}{a}\right)$

If $a > 0$, a positive slope, then $ax + b < 0 \Rightarrow ax > -b \Rightarrow x > \frac{-b}{a} \Rightarrow \left(\frac{-b}{a}, \infty\right)$

(c) If $a < 0$, a negative slope, then $ax + b < 0 \Rightarrow ax < -b \Rightarrow x > \frac{-b}{a} \Rightarrow \left(\frac{-b}{a}, \infty\right)$

If $a < 0$, a positive slope, then $ax + b > 0 \Rightarrow ax > -b \Rightarrow x < \frac{-b}{a} \Rightarrow \left(-\infty, \frac{-b}{a}\right)$

1.6: Applications of Linear Functions

1. $.75(40) = 30L$

3. When combining a 26% acid solution to a 32% acid solution the result will be a solution with between 26% and 32% acid. (A) 36% is not in between these percent values and not a possible concentration.

5. If x is the second number, then $6x - 3$ is the first number. The equation with the sum of these two numbers equal to 32 is: (D) $(6x - 3) + x = 32$.

7. If $P = 2L + 2W$ then $P = 2L + 2(19) \Rightarrow 98 = 2L + 38 \Rightarrow 60 = 2L \Rightarrow L = 30$ cm.

9. Let x = width and $2x - 2.5$ = length. If $P = 2W + 2L$, then

   $40.6 = 2x + 2(2x - 2.5) \Rightarrow 40.6 = 6x - 5 \Rightarrow 45.6 = 6x \Rightarrow x = 7.6$. The width is 7.6 cm.

11. Let x = the original square side length and $2x - 3$ = the new square side length. If $P = 4s$, then

    $4(x + 3) = 2x + 40 \Rightarrow 4x + 12 = 2x + 40 \Rightarrow 2x = 28 \Rightarrow x = 14$. The original side length is 14 cm.

13. With an aspect ratio of 4:3, let x = width and $\frac{4}{3}x$ = length. If $P = 2W + 2L$, then

    $98 = 2x + 2\left(\frac{4}{3}x\right) \Rightarrow 98 = 2x + \frac{8}{3}x \Rightarrow 98 = \frac{14}{3}x \Rightarrow 294 = 14x \Rightarrow x = 21$. The width is 21 inches and

    the length is $\frac{4}{3}(21) = 28$ inches. Use the Pythagorean theorem to find the diagonal

    $c^2 = (21)^2 + (28)^2 \Rightarrow c^2 = 441 + 784 \Rightarrow c^2 = 1225 \Rightarrow c = 35$. The television is advertised as a 35 inch screen.

15. Let x = the short side length and $2x$ = the longer two side lengths. If $P = s + s + s$, then

    $30 = x + 2x + 2x \Rightarrow 30 = 5x \Rightarrow x = 6$. The shortest side is 6 cm long.

17. Let x = the number of hours traveled at 70 mph, and (6-x) = the number of hours traveled at 55 mph. Since D = RT and the total distance traveled by the car was 372 miles, then 55(6-x) + 70x = $372 \Rightarrow 330 - 55x + 70x = 372 \Rightarrow 15x = 42 \Rightarrow x = 2.8$. The car traveled for 2.8 hours at 70 mph and (6-2.8)=3.2 hours at 55 mph.

19. Let x = gallons of 5% acid solution. Then

    $5(.10) + x(.05) = (x + 5)(.07) \Rightarrow .50 + .05x = .07x + .35 \Rightarrow .15 = .02x \Rightarrow 15 = 2x \Rightarrow x = 7.5$.

Mix in 7.5 gallons of 5% acid solution.

21. Let $x =$ gallons of pure alcohol. Then

$20(.15) + x(1.00) = (x + 20)(.25) \Rightarrow 3 + x = .25x + 5 \Rightarrow .75x = 2 \Rightarrow x = 2.67$ or $2\frac{2}{3}$.

Mix in $2\frac{2}{3}$ gallons of pure alcohol.

23. Let $x =$ milliliters of water. Then $8(.06) + x(0) = (x + 8)(.04) \Rightarrow .48 = .04x + .32 \Rightarrow$

$0.16 = 0.04 \Rightarrow 4x = 16 \Rightarrow x = 4$. Mix in 4 milliliters of water.

25. Let $x =$ liters of fluid to be drained and pure antifreeze added. The $16(0.80) - x(0.80) + x(1.00) = 16(0.90) \Rightarrow$

$12.8 - 0.80x + x = 14.4 \Rightarrow 0.20x = 1.6 \Rightarrow 20x = 160 \Rightarrow x = 8$. Drain and add in 8 liters of antifreeze.

27. Let $x =$ gallons of 94-octane gasoline. Then

$400(0.99) + x(0.94) = (x + 400)(0.97) \Rightarrow 396 + 0.94x = 0.97x + 388$

$8 = 0.03x \Rightarrow 3x = 800 \Rightarrow x = 266.67$ or $266\frac{2}{3}$. Mix in $266\frac{2}{3}$ gallons of 94-octane gasoline.

29. (a) $F(2008) = -\frac{5}{44}(2008) + 276.18 \approx 48$, IN 2008 the winning men's 100-meter freestyle time was about 48 seconds.

(b) $53 = -\frac{5}{44}x + 276.18 \Rightarrow x \approx 1963.98$. The years would have been from 1948 to about 1964.

31. (a) Use the points (2011, 192) and (2014, 249) to find the slope, $m = \frac{249 - 192}{2014 - 2011} = \frac{57}{3} = 19$ and the point (2014, 249) to find the equation of the line. $S(x) - 249 = 19(x - 2014) \Rightarrow S(x) = 19x - 38017$

(b) Sales increased , on average by $19 billion per year.

(c) $325 = 19x - 38017 \Rightarrow x = 2018$ The sales will reach $325 billion in 2018.

33. (a) If the fixed cost $= \$200$ and the variable cost $= \$.02$ the cost function is: $C(x) = 0.02x + 200$.

(b) If she gets paid $.04 per envelope stuffed and $x =$ number of envelopes, the revenue function is $R(x) = 0.04x$.

(c) $R(x) = C(x)$ when $0.02x + 200 = 0.04x \Rightarrow 200 = 0.02x \Rightarrow x = 10{,}000$.

(d) Graph $C(x)$ and $R(x)$, see Figure 33. Rebecca takes a loss when stuffing less than 10,000 envelopes and makes a profit when stuffing over 10,000 envelopes.

35. (a) If the fixed cost $= \$2300$ and the variable cost $= \$3.00$ the cost function is: $C(x) = 3.00x + 2300$.

(b) If he gets paid $5.50 per delivery and $x =$ number of deliveries, the revenue function is: $R(x) = 5.50x$.

(c) $R(x) = C(x)$ when $3.00x + 2300 = 5.50x \Rightarrow 2300 = 2.50x \Rightarrow x = 920$.

(d) Graph $C(x)$ and $R(x)$. See Figure 35. Tom takes a loss when making fewer than 920 deliveries and makes a profit when making over 920 deliveries.

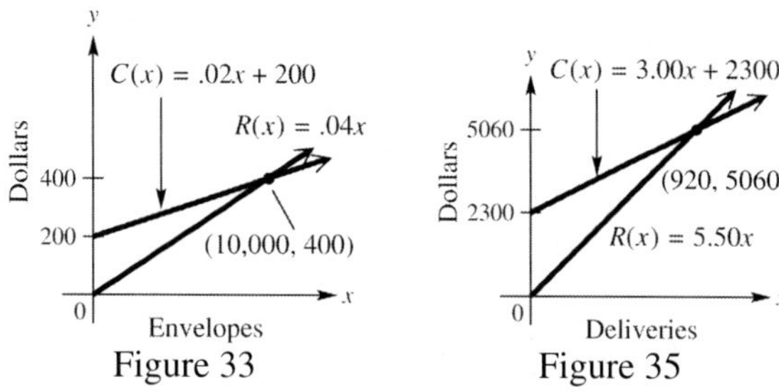

Figure 33 Figure 35

37. If $y = kx$, $x = 3$, and $y = 7.5$, then $7.5 = k(3) \Rightarrow k = 2.5$. Now, with

$k = 2.5$ and $x = 8$, $y = 2.5(8) \Rightarrow y = 20$ when $x = 8$.

39. If $y = kx$, $x = 25$, and $y = 1.5$, then $1.50 = k(25) \Rightarrow k = 0.06$. Now, with $k = 0.06$

and $y = 5.10$, $5.10 = 0.06x \Rightarrow x = \$85$ when $y = \$5.10$

41. Let $y =$ pressure and $x =$ depth for the direct proportion: $y = kx$. Then $13 = k(30) \Rightarrow k = \frac{13}{30}$.

Now use $k = \frac{13}{30}$ and a depth of 70 feet to find the pressure: $y = \frac{13}{30}(70) \Rightarrow y = \frac{91}{3} \Rightarrow y = 30\frac{1}{3}\text{ lb/in}^2$.

43. Let $t =$ tuition and $c =$ credits taken for the direct proportion $t = kc$. Then

$720.50 = k(11) \Rightarrow k = 65.5$. Now use the constant of variation $k = 65.5$ and 16 credits to find the

tuition: $y = 65.5(16) \Rightarrow y = \$1048$.

45. First use proportion to find the radius of the water at a depth of 44 feet $\frac{5}{11} = \frac{x}{6} \Rightarrow 11x = 30 \Rightarrow x \approx 2.727$.

Now use the cone volume formula to find the water's volume:

$V = \frac{1}{3}\pi r^2 h \Rightarrow V = \frac{1}{3}\pi(2.727)^2(6) \approx 46.7\text{ ft}^3$.

47. Since the triangles are similar, use a proportion to solve: $\frac{1.75}{2} = \frac{45}{x} \Rightarrow 1.75x = 90 \Rightarrow x \approx 51.43$ or

$x = 51\frac{3}{7}$ feet tall.

49. Let $w =$ weight, $d =$ distance, and use the direct proportion $w = kd$ to find k:

$3 = k(2.5) \Rightarrow k = 1.2$. Now use $k = 1.2$ and a weight of 17 pounds to find the stretch

length: $17 = 1.2(d) \Rightarrow d = 14.17$ or $14\frac{1}{6}$ in.

51. With direct proportion $y_1 = kx_1$ and $y_2 = kx_2$, then $k = \frac{y_1}{x_1} = \frac{y_2}{x_2}$. Now let $y_1 = 250$ tagged

trout, $y_2 = 7$ tagged trout, $x_2 = 350$ sample trout, and $x_1 =$ total trout. Therefore

$\frac{250}{x_1} = \frac{7}{350} \Rightarrow 7x_1 = 87{,}500 \Rightarrow x_1 = 12{,}500$ is the estimate for total population of trout.

53. (a) Let $x =$ number of heaters produced and $y =$ cost. Then (10, 7500) and (20, 13900) are two points on the graph of the linear function. Find the slope: $m = \dfrac{13900-7500}{20-10} = \dfrac{6400}{10} = 640$.

Now use point-slope form to find the linear function:

$y-7500 = 640(x-10) \Rightarrow y-7500 = 640x-640 \Rightarrow y = 640x+1100.$

(b) $y = 640(25)+1100 \Rightarrow y = 16,000+1100 \Rightarrow y = \$17,100.$

(c) Graph $y = 640x+1100$ and locate the point (25, 17,100) on the graph.

55. (a) Let $x =$ number of years after 2002 and $y =$ value, then (0, 120,000) and (10, 146,000) are two points on the graph of the linear function. Find the slope: $m = \dfrac{146,000-120,000}{10-0} = \dfrac{26,000}{10} = 2,600.$

The y-intercept is: 120,000. Therefore the linear function is: $y = 2,600x+120,000.$

(b) $y = 2,600(7)+120,000 \Rightarrow y = 18200+120,000 \Rightarrow y = \$138,200$ value of the house.

(c) The value of the house increased, on average, \$2,600 per year.

57. (a) Surface area $= 4\pi r^2 \Rightarrow 4\pi(3960)^2 \approx 197,000,000 \text{ mi}^2$.

(b) $(197,000,000)(0.71) \approx 140,000,000 \text{ mi}^2$

(c) $\dfrac{680,000}{140,000,000} \approx 0.00486$ miles. Converted to feet is $(0.00486)(5280) \approx 25.7$ feet.

(d) Since this height is greater than the heights of both Boston and San Diego these cities would be flooded.

(e) We know from above that oceans cover approximately 140,000,000 square miles of the earth. The Antarctic ice cap contains 6,300,000 cubic miles of water.

$\dfrac{6,300,000}{140,000,000} = 0.045$ miles. Converted to feet is $(0.045)(5280) \approx 238$ feet.

59. (a) $y = \dfrac{5}{3}(27)+455 \Rightarrow y = 45+455 \Rightarrow y = 500\text{cm}^3.$

(b) $605 = \dfrac{5}{3}x+455 \Rightarrow 150 = \dfrac{5}{3}x \Rightarrow x = 90°\text{ C}.$

(c) $0 = \dfrac{5}{3}x+455 \Rightarrow -455 = \dfrac{5}{3}x \Rightarrow x = -273°\text{C}.$

61. $I = PRT \Rightarrow \dfrac{I}{RT} = P$ or $P = \dfrac{I}{RT}$

63. $P = 2L+2W \Rightarrow P-2L = 2W \Rightarrow W = \dfrac{P-2L}{2}$ or $W = \dfrac{P}{2}-L$

65. $A = \dfrac{1}{2}h(b_1+b_2) \Rightarrow 2A = h(b_1+b_2) \Rightarrow h = \dfrac{2A}{b_1+b_2}$

67. $S = 2LW + 2WH + 2HL \Rightarrow S - 2LW = H(2W + 2L) \Rightarrow \dfrac{S - 2LW}{2W + 2L}$

69. $V = \dfrac{1}{3}\pi r^2 h \Rightarrow 3V = \pi r^2 h \Rightarrow h = \dfrac{3V}{\pi r^2}$

71. $S = \dfrac{n}{2}(a_1 + a_n) \Rightarrow 2S = n(a_1 + a_n) \Rightarrow n = \dfrac{2S}{a_1 + a_n}$

73. $s = \dfrac{1}{2}gt^2 \Rightarrow 2s = gt^2 \Rightarrow g = \dfrac{2s}{t^2}$

75. Let P = the amount put into the short-term note, then $240,000 - P$ = the amount put into the long-term note. With \$13,000 one year interest income, solve:

$P(.06)(1) + (240,000 - P)(0.05)(1) = 13,000 \Rightarrow 0.06P + 120,000 - 0.05P = 13,000 \Rightarrow$

$0.01P = 1,000 \Rightarrow P \Rightarrow 100,000.$ The short-term note was \$100,000 and the long-term was \$140,000.

77. Let P = the amount deposited at 2.5% interest rate, then $2P$ = the amount deposited at 3% interest rate. With a one year interest income of \$850, solve:

$0.025P(1) + 0.03(2P)(1) = 850 \Rightarrow 0.025P + 0.06P = 850 \Rightarrow 0.085P = 850 \Rightarrow P = 10,000.$

Therefore, \$10,000 was deposited at 2.5% and \$20,000 was deposited at 3%.

79. After taxes, Marietta was able to invest 70% of the original winnings. This is $0.70(200,000) = \$140,000$. Now let P = amount invested at 1.5%, then $140,000 - P$ = the amount invested at 4%. With a one year interest income of \$4350, solve: $0.015P + 0.04(140,000 - P) = 4350 \Rightarrow 0.015P + 5600 - 0.04P = 4350 \Rightarrow$ $-0.025P = -1250 \Rightarrow P = 50,000.$ Therefore, \$50,000 was invested at 1.5% and \$90,000 was invested at 4%.

Reviewing Basic Concepts (Sections 1.5 and 1.6)

1. $3(x-5)+2 = 1-(4+2x) \Rightarrow 3x-15+2 = 1-4-2x \Rightarrow 3x-13 = -2x-3 \Rightarrow 5x = 10 \Rightarrow x = 2.$ The graphs of the left and right sides of the equation intersect at $x = 2$; this supports the solution set: $\{2\}$.

2. Graph $y_1 = \pi(1-x)$ and $y_2 = .6(3x-1)$. See Figure 2. The graphs intersect at $x = .757$.

3. $0 = \dfrac{1}{3}(4x-2)+1 \Rightarrow 0 = \dfrac{4}{3}x - \dfrac{2}{3} + 1 \Rightarrow -\dfrac{1}{3} = \dfrac{4}{3}x \Rightarrow -\dfrac{1}{4}.$ Graph $y_1 = \dfrac{1}{3}(4x-2)+1.$ See Figure 3.

The graph intersects the x-axis at $x = -\dfrac{1}{4}$.

[-10,10] by [-10,10]

Xscl = 1 Yscl = 1

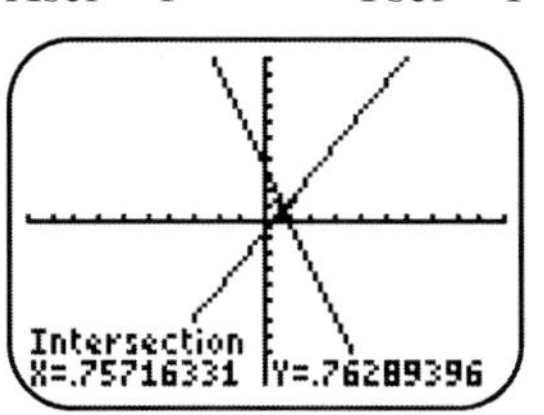

Figure 2

[-10,10] by [-10,10]

Xscl = 1 Yscl = 1

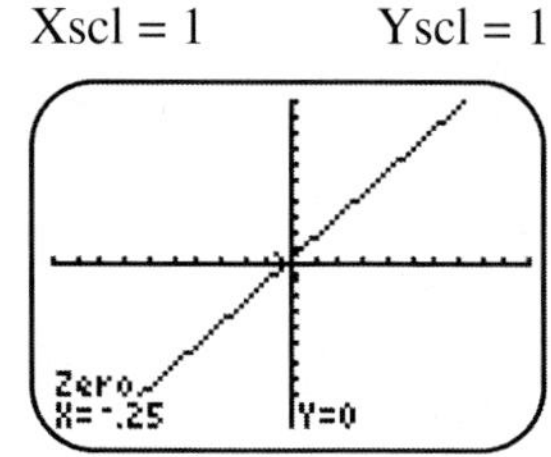

Figure 3

4. (a) $4x-5=-2(3-2x)+3 \Rightarrow 4x-5=-6+4x+3 \Rightarrow -5=-3$. Since this is false, the equation is a contradiction and the solution set is: $\varnothing$.

   (b) $5x-9=5(-2+x)+1 \Rightarrow 4x-5=-6+4x+3 \Rightarrow -5=-3$. Since this is true the equation is an identity and the solution set is: $(-\infty,\infty)$.

   (c) $5x-4=3(6-x) \Rightarrow 5x-4=18-3x \Rightarrow 8x=22 \Rightarrow x=\frac{11}{4}$. The equation is a conditional equation and the solution set is: $\left\{\frac{11}{4}\right\}$.

5. $2x+3(x+2)<1-2x \Rightarrow 2x+3x+6<1-2x \Rightarrow 7x<-5 \Rightarrow x<-\frac{5}{7}$. The solution set is: $\left(-\infty,\frac{5}{7}\right)$.

   Graph $y_1=2x+3(x+2)$ and $y_2=1-2x$; the graph of y_1 is below the graph of y_2 when $x<-\frac{5}{7}$, which supports the original solution.

6. $-5\le 1-2x<6 \Rightarrow -6 \Rightarrow -2x<5 \Rightarrow 3\ge x>-\frac{5}{2} \Rightarrow -\frac{5}{2}<x\le 3 \Rightarrow \left(-\frac{5}{2},3\right]$

7. (a) The graphs intersect at $x=2 \Rightarrow \{2\}$

   (b) The graph of $f(x)$ intersects or is below the graph of $g(x)$ for $x\ge 2$, on $[2,\infty)$.

8. Since the triangles formed by the shadows are similar, use proportion to solve.

   $\frac{x}{27}=\frac{6}{4} \Rightarrow 4x=162 \Rightarrow x=40.5$ ft.

9. (a) Since the income from each disc is \$5.50, a function R for revenue from selling x discs is: $R(x)=5.50x$.

   (b) Since the cost of producing each disc is \$1.50 and there is a one-time equipment cost of \$800, a function C for cost of recording x discs is: $C(x)=1.50x+800$.

   (c) Solve $R(x)=C(x)$: $5.5x=1.5x+800 \Rightarrow 4x=800 \Rightarrow x=200$ discs.

10. $V=\pi r^2 h \Rightarrow h=\frac{V}{\pi r^2}$

Chapter 1 Review Exercises

1. Use the distance formula: $d=\sqrt{(-1-5)^2+(16-(-8))^2} \Rightarrow d \Rightarrow \sqrt{(-6)^2+24^2} \Rightarrow d=\sqrt{36+576} \Rightarrow$ $d=\sqrt{612}=6\sqrt{17}$.

2. Use the midpoint formula: Midpoint $=\left(\frac{-1+5}{2},\frac{16-8}{2}\right)=(2,4)$.

3. Use the slope formula: $m = \frac{16-(-8)}{-1-5} \Rightarrow m = \frac{24}{-6} = -4.$ 4. Use point-slope form and slope from ex. 3: $y-16 = -4(x-(-1)) \Rightarrow y-16 = -4x-4 \Rightarrow y = -4x+12.$

5. Change to slope-intercept form: $3x+4y=144 \Rightarrow 4y=-3x+144 \Rightarrow y = -\frac{3}{4}x+36 \Rightarrow m = -\frac{3}{4}.$

6. For the x-intercept, $y=0$. Therefore, $3x+4(0)=144 \Rightarrow 3x=144 \Rightarrow x=48.$

7. For the y-intercept, $x=0$. Therefore, $3(0)+4y=144 \Rightarrow 4y=144 \Rightarrow y=36.$

8. One possible window is: $[-10, 50]$ by $[-40, 40]$.

9. Since $f(3)=6$ and $f(-2)=1$, $(3,6)$ and $(-2,1)$ are points on the graph of the line. Using these points, find the slope: $m = \frac{6-1}{3-(-2)} = \frac{5}{5} = 1.$ Use point-slope form to find the function:

$f(x)-6 = 1(x-3) \Rightarrow f(x)-6 = x-3 \Rightarrow f(x) = x+3.$ Now solve for $f(8)$: $f(8) = 8+3 = 11.$

10. The slope of the given equation is -4. A line perpendicular to this will have a slope of $\frac{1}{4}$. Using this and point-slope form produces: $y-4 = \frac{1}{4}(x-(-2)) \Rightarrow y-4 = \frac{1}{4}x+\frac{1}{2} \Rightarrow y = \frac{1}{4}x+\frac{9}{2}.$

11. (a) $m = \frac{-4-5}{2-(-1)} \Rightarrow m = \frac{-9}{3} = -3$

(b) Use point-slope form: $y-5 = -3(x-(-1)) \Rightarrow y-5 = -3x-3 \Rightarrow y = -3x+2$

(c) Midpoint $= \left(\frac{-1+2}{2}, \frac{5+(-4)}{2}\right) = \left(\frac{1}{2}, \frac{1}{2}\right)$ or $(0.5, 0.5)$

(d) $d = \sqrt{(2-(-1))^2+(-4-5)^2} \Rightarrow d\sqrt{3^2+(-9)^2} \Rightarrow d = \sqrt{9+81} = \sqrt{90} = 3\sqrt{10}$

12. (a) $m = \frac{1.5-(-3.5)}{-1-(-3)} \Rightarrow m = \frac{5}{2} = 2.5$

(b) Use point-slope form: $y-1.5 = 2.5(x-(-1) \Rightarrow y-1.5 = 2.5x+2.5 \Rightarrow y = 2.5x+4$

(c) Midpoint $= \left(\frac{-1+(-3)}{2}, \frac{1.5+(-3.5)}{2}\right) = \left(-\frac{4}{2}, -\frac{2}{2}\right) = (-2,-1)$

(d) $d = \sqrt{(-1-(-3))^2+(1.5-(-3.5))^2} \Rightarrow d = \sqrt{2^2+(5)^2} \Rightarrow d = \sqrt{4+25} = \sqrt{29}$

13. C most closely represents: $m<0, b<0.$

14. F most closely represents: $m>0, b<0.$

15. A most closely represents: $m<0, b>0.$

16. B most closely represents: $m>0, b>0.$

17. E most closely represents: $m=0.$

18. D most closely represents: $b=0.$

19. The rate of change is evaluated as $m = \frac{62.9 - 66.7}{2009 - 2001} = -\frac{3.8}{8} = -0.475$. The graph confirms that the line through the ordered pairs falls from left to right and therefore has a negative slope. Thus, the number of basic cable subscribers decreased by an average of 0.475 million (or 475,000) each year from 2001 to 2009. See Figure 19.

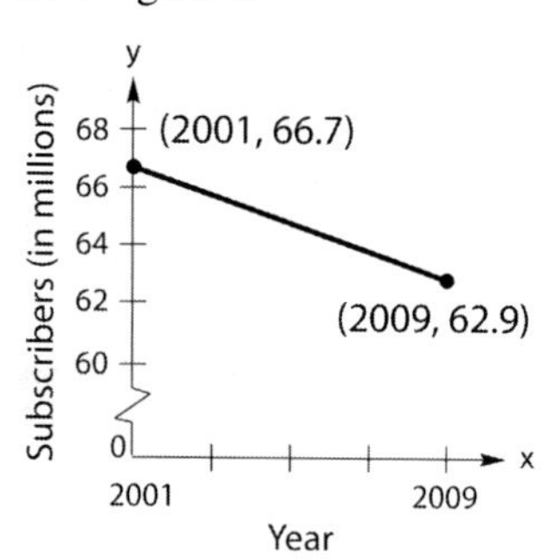

Figure 19

20. False, the slopes are different. Although the difference is small, the lines are not parallel and will intersect.

21. $f(x) = g(x)$ when the graphs intersect or when $x = -3$. I $\{-3\}$ is the best match.

22. $f(x) > g(x)$ when the graph of $f(x)$ is above the graph of $g(x)$ or when $x > -3$. K $(-3, \infty)$ is the best match.

23. $f(x) < g(x)$ when the graph of $f(x)$ is below the graph of $g(x)$ or when $x < -3$. B $(-\infty, 3)$ is the best match.

24. $g(x) \geq f(x)$ when the graph of $g(x)$ intersects or is above the graph of $f(x)$ or when $x \leq -3$. A $(-\infty, -3]$ is the best match.

25. $y_2 - y_1 = 0 \Rightarrow y_2 = y_1 \Rightarrow g(x) = f(x)$ when the graphs intersect or when $x = -3.1$. I $\{-3\}$ is the best match.

26. $f(x) < 0$ when the graph of $f(x)$ is below the x-axis or when $x < -5$. M $(-\infty, -5)$ is the best match.

27. $g(x) > 0$ when the graph of $g(x)$ is above the x-axis or when $x < -2$. O $(-\infty, -2)$ is the best match.

28. $y_2 - y_1 < 0 \Rightarrow y_2 < y_1 \Rightarrow g(x) < f(x)$ when the graph of $g(x)$ is below the graph of $f(x)$ or when $x > -3$. K $(-3, \infty)$ is the best match.

29. $5[3 + 2(x-6)] = 3x + 1 \Rightarrow 5[2x - 9] = 3x + 1 \Rightarrow 10x - 45 = 3x + 1 \Rightarrow 7x = 46 \Rightarrow x = \frac{46}{7}$.

The graphs of $y_1 = 5[3 + 2(x-6)]$ and $y_2 = 3x + 1$ intersect at: $\left\{\frac{46}{7}\right\}$, which supports the result.

30. $\frac{x}{4} - \frac{x+4}{3} = -2 \Rightarrow 12\left[\frac{x}{4} - \frac{x+4}{3} = -2\right] \Rightarrow 3x - 4(x+4) = -24 \Rightarrow -x - 16 = -24 \Rightarrow -x = -8 \Rightarrow x = 8$.

The graphs of $y_1 = \frac{x}{4} - \frac{x+4}{3}$ and $y_2 = -2$ intersect at: $\{8\}$, which supports the result.

31. $-3x-(4x+2)=3 \Rightarrow -7x-2=3 \Rightarrow -7x=5 \Rightarrow x=-\frac{5}{7}$. The graphs

of $y_1=-3x-(4x+2)$ and $y_2=3$ intersect at: $\left\{-\frac{5}{7}\right\}$, which supports the result.

32. $-2x+9+4x=2(x-5)-3 \Rightarrow 2x+9=2x-10 \Rightarrow 9=-10$. This is false, therefore this is a contradiction and the solution is: $\varnothing$. The graphs of $y_1=-2x+9+4x$ and $y_2=2(x-5)-3$ are parallel and do not intersect, therefore no solution, $\varnothing$, which supports the result.

33. $0.5x+0.7(4-3x)=0.4x \Rightarrow 0.5x+2.8-2.1x=0.4x \Rightarrow 5x+28-21x=4x \Rightarrow -20x=-28 \Rightarrow$

$x=\frac{7}{5}$. The graphs of $y_1=.5x+.7(4-3x)$ and $y_2=.4x$ intersect at: $\left\{\frac{7}{5}\right\}$, which supports the result.

34. $\frac{x}{4}-\frac{5x-3}{6}=2-\frac{7x+18}{12} \Rightarrow 12\left[\frac{x}{4}-\frac{5x-3}{6}=2-\frac{7x+18}{12}\right] \Rightarrow 3x-2(5x-3)=$

$24-(7x+18) \Rightarrow -7x+6=-7x+6 \Rightarrow 6=6$. This is true, therefore an identity and the solution is:

$(-\infty,\infty)$. The graphs of $y_1=\frac{x}{4}-\frac{5x-3}{6}$ and $y_2=2-\frac{7x+18}{12}$ are the same line; the solution is

$(-\infty,\infty)$, which supports the result.

35. $x-8<1-2x \Rightarrow 3x<9 \Rightarrow x<3$. The graph of $y_1=x-8$ is below the graph of $y_2=1-2x$ for the interval: $(-\infty,3)$, which supports the result.

36. $\frac{4x-1}{3} \ge \frac{x}{5}-1 \Rightarrow 15\left[\frac{4x-1}{3} \ge \frac{x}{5}-1\right] \Rightarrow 5(4x-1) \ge 3x-15 \Rightarrow 20x-5 \ge 3x-15 \Rightarrow$

$17x \ge -10 \Rightarrow x \ge -\frac{10}{17}$. The graph of $y_1=\frac{4x-1}{3}$ intersects or is above the graph of $y_2=\frac{x}{5}-1$

for the interval: $\left[-\frac{10}{17},\infty\right)$, which supports the result.

37. $-6 \le \frac{4-3x}{7} < 2 \Rightarrow -42 \le 4-3x < 14 \Rightarrow -46 \le -3x < 10 \Rightarrow \frac{46}{3} \ge x > -\frac{10}{3} \Rightarrow -\frac{10}{3} < x \le \frac{46}{3}$ or

for the interval: $\left(-\frac{10}{3},\frac{46}{3}\right]$.

38. (a) Graph $y_1=5\pi+(\sqrt{3})x-6.24(x-8.1)+(\sqrt[3]{9})x$. See Figure 38. Find the x- intercept: $x \approx \{-3.81\}$.

(b) $f(x)<0$ when the graph of $f(x)$ is below the x-axis. This happens for the interval: $(-\infty,-3.81)$.

(c) $f(x) \ge 0$ when the graph of $f(x)$ intersects or is above the x-axis. This happens for the interval: $[-3.81,\infty)$.

[-10,10] by [-10,10]
Xscl = 1 Yscl = 1

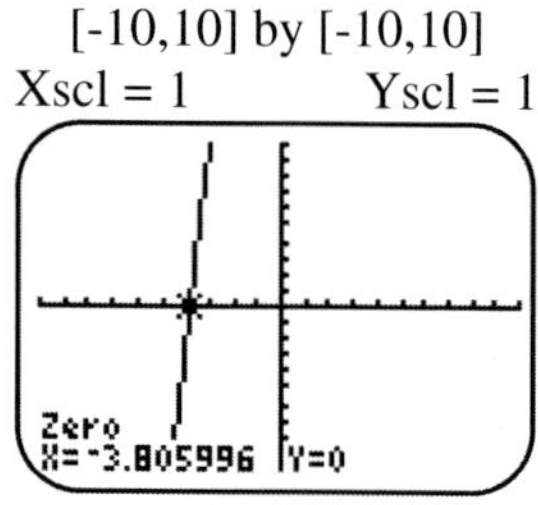

Figure 38

39. It costs \$30 to produce each CD and there is a one-time advertisement cost, therefore: $C(x) = 30x + 150$.

40. Each tape is sold for \$37.50, therefore: $R(x) = 37.50x$.

41. $C(x) = R(x)$ when $30x + 150 = 37.50x \Rightarrow 150 = 7.50x \Rightarrow x = 20$.

42. When the graph of $R(x)$ is below $C(x)$ the company is losing money, when $R(x)$ intersects $C(x)$ the company breaks even, and when $R(x)$ is above $C(x)$ the company makes money. This happens as follows: losing money when $x < 20$, breaking even when $x = 20$, and making money when $x > 20$.

43. $A = \dfrac{24f}{B(p+1)} \Rightarrow AB(p+1) = 24f \Rightarrow f = \dfrac{AB(p+1)}{24}$.

44. $A = \dfrac{24f}{B(p+1)} \Rightarrow AB(p+1) = 24f \Rightarrow B = \dfrac{24f}{A(p+1)}$.

45. (a) $f(x) = -3.52(5) + 58.6 \Rightarrow f(x) = -17.6 + 58.6 = 41^\circ$ F.

(b) $-15 = -3.52x + 58.6 \Rightarrow -73.6 = -3.52x \Rightarrow x \approx 20.9$ or about 21,000 feet.

(c) Graph $y_1 = -3.52x + 58.6$. Find the coordinates of the point where $x = 5$ to support the answer in (a). Find the coordinates of the point where $y = -15$ to support the answer in (b).

46. (a) Linear regression gives the model: $f(x) \approx 0.12331x - 244.75$. Answers may vary.

(b) $f(1990) \approx 0.12331x - 244.75 = \$0.63$ million The cost of a 30 second Super Bowl ad in 1990 was approximately \$0.63 million. This value is within about 0.17 million of the actual cost.

(c) $4 \approx 0.12331x - 244.75 \Rightarrow x \approx 2017$ Thus, the cost for a 30 second Super Bowl ad could reach \$4 million 2017.

47. Let x = bat speed and y = ball travel distance; then (50, 320) and (80, 440) are two points on the graph of the function. Use the points to find slope: $m = \dfrac{440 - 320}{80 - 50} \Rightarrow m = \dfrac{120}{30} = 4$. Now use point-slope form to find the equation for the model: $y - 320 = 4(x - 50) \Rightarrow y - 320 = 4x - 200 \Rightarrow y = 4x + 120$. Because the slope is 4, the ball will travel 4 feet further for each additional 1 mph in bat speed.

48. Since surface area is: $A = 2(lw) + 2(lh) + 2(wh)$ we can solve $496 = 2(18)(8) + 2(18)h + 2(8)h \Rightarrow 496 = 288 + 36h + 16h \Rightarrow 208 = 52h \Rightarrow h = 4$. The height of the box is 4 feet.

49. Since there are 5280 feet in a mile, there are $5280 \times 26.2 = 138,336$ feet in a marathon. Since there are 3.281 feet in a meter, there are $100 \times 3.281 = 328.1$ feet in a 100 meter dash. Now use a proportion to solve:

$\frac{9.58}{328.1} = \frac{x}{138,336} \Rightarrow 328.1x = 1325258.88 \Rightarrow x \approx 4039.19$ seconds to run a marathon. Divide by 60 to get minutes run: $4039.19 \div 60 = 67.32$ minutes run or 1 hour 7 minutes and 19 seconds.

50. $C = \frac{5}{9}(864 - 32) \Rightarrow C = \frac{5}{9}(832) \Rightarrow C = \frac{4160}{9} \Rightarrow 462\frac{2}{9}\,^{\circ}$C.

51. Find the constant of variation: $4k = 3000 \Rightarrow k = 750$. Use this to find the pressure: $10(750) = 7500$ kg / m^2.

52. (a) Use any two points to find slope: $m = \frac{1.8 - 3}{1 - 0} \Rightarrow m = \frac{-1.2}{1} = -1.2$ Now use point-slope form to find the equation: $y - 3 = -1.2(x - 0) \Rightarrow y = -1.2x + 3$.

(b) $y = -1.2(-1.5) + 3 \Rightarrow y = 1.8 + 3 \Rightarrow y = 4.8$ when $x = -1.5$.

$y = -1.2(3.5) + 3 \Rightarrow y = -4.2 + 3 \Rightarrow y = -1.2$ when $x = 3.5$.

53. (a) Enter the years in L_1 and enter test scores in L_2. The regression equation is: $f(x) = 1.2x - 1886.4$

(b) $y = 1.2(2012) - 1886.4 \Rightarrow y = 2414.4 - 1886.4 \Rightarrow y = 528$.

(c) Over time data can change its pattern or character. Answers may vary.

54. Let $x =$ the amount of 5% solution to be added, then solve:

$120(.20) + x(.05) = (120 + x)(.10) \Rightarrow 24 + .05x = 12 + .10x \Rightarrow 12 = .05x \Rightarrow x = 240$ mL of 5% solution needs to be added.

55. The company will at least break even when $R(x) \geq C(x)$, therefore solve:

$8x \geq 3x + 1500 \Rightarrow 5x \geq 1500 \Rightarrow x \geq 300$ or for the interval: $[300, \infty)$. 300 or more DVD's need to be sold to at least break even.

56. Let m = mental age and c = chronological age, then $IQ = \frac{100m}{c}$.

(a) $130 = \frac{100m}{7} \Rightarrow 910 = 100m \Rightarrow m = 9.1$ years.

(b) $IQ = \frac{100(20)}{16} \Rightarrow IQ = \frac{2000}{16} \Rightarrow IQ = 125$ years.

57. (a) Enter the heights in L_1 and enter weights in L_2. The regression equation is: $y \approx 4.512x - 154.4$

(b) $y = 4.51(75) - 154.4 \Rightarrow y = 338.25 - 154.4 \Rightarrow y \approx 184$.

58. (a) Enter the heights in L_1 and enter weights in L_2. The regression equation is: $y \approx 4.465x - 133.3$

(b) $y = 4.465(80) - 133.3 \Rightarrow y = 357.2 - 133.3 \Rightarrow y \approx 224$.

Chapter 1 Test

1. (a) The number $\frac{4}{2}=2$ is a natural number, integer, rational number, and real number.

   (b) The number π is a real number.

   (c) The number $\sqrt{2}$ is a real number.

   (d) The number $0.25=\frac{1}{4}$ is a rational number and real number.

2. Use technology to approximate the following.

   (a) $\sqrt{5}\approx 2.236$

   (b) $\sqrt[3]{7}\approx 1.913$

   (c) $3^{1/4}\approx 1.316$

   (d) $\frac{1-1.1^2}{2+\pi^2}\approx -0.018$

3. $d=\sqrt{(-2-4)^2+(4-(-3))^2}=\sqrt{6^2+7^2}=\sqrt{36+49}=\sqrt{85}$, $M=\left(\frac{-2+4}{2},\frac{4+(-3)}{2}\right)=\left(\frac{2}{2},\frac{1}{2}\right)=\left(1,\frac{1}{2}\right)$

4. (a) $f(-2)=4-7(-2)=4+14=18$

   (b) $f(b)=4-7b$

   (c) $f(a+h)=4-7(a+h)=4-7a-7h$

5. The set of ordered pairs $\{(3,4),(2,-5),(1,0),(4,-5)\}$ represents a function since each x-coordinate corresponds to only one y-coordinate.

6. (a) $m=\frac{4-1}{1-(-5)}=\frac{3}{6}=\frac{1}{2}$

   (b) $m=\frac{9-3}{4-4}=\frac{6}{0}\Rightarrow$ undefined

   (c) $m=\frac{5-5}{1.7-1.2}=\frac{0}{0.5}=0$

7. (a) Domain: $(-\infty,\infty)$, Range: $[2,\infty)$, x-intercept: none, y-intercept: (0, 3)

   (b) Domain: $(-\infty,\infty)$, Range: $(-\infty,0]$, x-intercept: (3,0), y-intercept: $(0,-3)$

   (c) Domain: $[-4,\infty)$, Range: $[0,\infty)$, x-intercept: $(-4,0)$, y-intercept: $(0,2)$

8. (a) $f(x)=g(x)$ when the graph of $f(x)$ intersects the graph of $g(x)$ therefore $\{-4\}$.

   (b) $f(x)<g(x)$ when the graph of $f(x)$is below the graph of $g(x)$, for the interval: $(-\infty,-4)$.

   (c) $f(x)\geq g(x)$ when the graph of $f(x)$intersects or is above the graph of $g(x)$, for the interval: $[-4,\infty)$.

   (d) $f(\, y_2-y_1)=0\Rightarrow y_2=y_1\Rightarrow g(x)=f(x)$ when the graph of $g(x)$ intersects the graph of $f(x)$, therefore $\{-4\}$.

9. (a) $y_1 = 0$ when the graph of $f(x)$ intersects the x-axis, therefore $\{5.5\}$.

(b) $y_1 < 0$ when the graph of $f(x)$ is below the x-axis, for the interval $(-\infty, 5.5]$.

(c) $y_1 > 0$ when the graph of $f(x)$ is above the x-axis, for the interval $(5.5, \infty)$.

(d) $y_1 \leq 0$ when the graph of $f(x)$ intersects or is below the x-axis, for the interval $(-\infty, 5.5]$.

10. (a) $3(x-4)-2(x-5) = -2(x+1)-3 \Rightarrow 3x-12-2x+10 = -2x-2-3 \Rightarrow x-2 = -2x-5$

$\Rightarrow 3x = -3 \Rightarrow x = -1$. Check: $3(-1-4)-2(-1-5) = -2(-1+1)-3 \Rightarrow$

$3(-5)-2(-6) = -2(0)-3 \Rightarrow -15+12 = 0-3 \Rightarrow -3 = -3$

(b) Graph $y_1 = 3(x-4)-2(x-5)$ and $y_2 = -2(x+1)-3$. See Figure 10. $f(x) > g(x)$ for the interval: $(-1, \infty)$ because the graph of $y_1 = f(x)$ is above the graph of $y_2 = g(x)$ for domain values greater than -1.

(c) See Figure 10. $f(x) < g(x)$ for the interval $(-\infty, -1)$ because the graph of $y_1 = f(x)$ is below the graph of $y_2 = g(x)$ for domain values less than -1.

11. (a) $-\frac{1}{2}(8x+4)+3(x-2) = 0 \Rightarrow -4x-2+3x-6 = 0 \Rightarrow -x-8 = 0 \Rightarrow x = -8$ or $\{-8\}$.

(b) $-\frac{1}{2}(8x+4)+3(x-2) \leq 0 \Rightarrow -4x-2+3x-6 \leq 0 \Rightarrow -x-8 \leq 0 \Rightarrow x \geq -8$ or for the interval: $[-8, \infty)$.

(c) Graph $y_1 -\frac{1}{2}(8x+4)+3(x-2)$. See Figure 11. The x-intercept is -8 supporting the result in part (a). The graph of the linear function lies below or on the x-axis for domain values greater than or equal to -8, supporting the results in part (b).

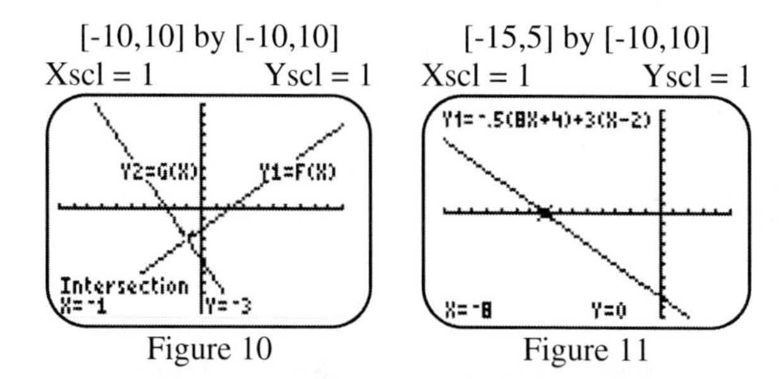

Figure 10 Figure 11

12. (a) Since x represents the number of years since 2007, we will use the points $(0, 13837)$ and $(5, 15082)$ find the slope: $a = \frac{15{,}082-13{,}837}{5-0} = \frac{1245}{5} = 249$. The y-intercept is the point (0, 13837) thus the value of b is 13837. The linear function is $f(x) = 249x + 13{,}837$.

(b) The number of stations increased, on average , by 249 per year.

(c) Since x represents the number of years since 2007 we will let $x = 7$.

$f(7) = 249(7) + 13{,}837 = 15{,}580$. The number of radio stations in 2014 is about 15,580.

13. (a) Since the given line has a slope of -2 and parallel lines have equal slopes, our new line has a slope of -2. Now use point-slope form: $y-5=-2\left(x-(-3)\right) \Rightarrow y-5=-2x-6 \Rightarrow y=-2x-1$.

(b) The equation: $-2x+y=0 \Rightarrow y=2x$ has a slope of 2. Since perpendicular lines have slopes whose product equals -1, our new line has a slope of $-\frac{1}{2}$. Now use point-slope form:

$$y-5=-\frac{1}{2}\left(x-(-3)\right) \Rightarrow y-5=-\frac{1}{2}x-\frac{3}{2} \Rightarrow y=-\frac{1}{2}x+\frac{7}{2}.$$

14. For the x-intercept $y=0$, therefore: $3x-4(0)=6 \Rightarrow 3x=6 \Rightarrow x=2$. The x-intercept is $(2,0)$. For the y-intercept $x=0$, therefore: $3(0)-4y=6 \Rightarrow -4y=6 \Rightarrow y=-\frac{6}{4}=-\frac{3}{2}$. The y-intercept is: $\left(0,-\frac{3}{2}\right)$. Using the intercepts: $\left(0,-\frac{3}{2}\right)$ and $(2,0)$, the slope is $m=\dfrac{0-\left(-\frac{3}{2}\right)}{2-0}=\dfrac{\frac{3}{2}}{2} \Rightarrow m=\frac{3}{4}$.

15. The equation of the horizontal line passing through $(-3,7)$ is $y=7$. The equation of the vertical line passing through $(-3,7)$ is $x=-3$.

16. (a) Enter the wind speed in L_1 and enter degrees in L_2. The regression equation is: $Y \approx -.246x+35.7$ and the correlation coefficient is: $r \approx -.96$.

(b) $y \approx -.246(40)+35.7 \Rightarrow y \approx -9.84+35.7 \Rightarrow y \approx 25.9\ ^{\circ}F$.

17. Let x be the number of hours the car traveled at 60 mph and then $4-x$ will be the number of hours the car traveled at 74 mph. Using the formula D = RT, we know the distance traveled at 60 mph is $60x$, and the distance traveled at 74 mph is $74(4-x)$. Since the total distance traveled is 275 miles we have the equation $60x+74(4-x)=275 \Rightarrow 60x+296-74x=275 \Rightarrow -14x=-21 \Rightarrow x=1.5$ and $4-x=2.5$. Therefore, the car traveled for 1.5 hours at 60 mph and 2.5 hours at 74 mph.

18. Since the load is directly proportional to the width we have $y=kx$, where y is the number of pounds that can be supported and x is the width in inches. Then, $510=k(2.25) \Rightarrow k=\frac{510}{2.25}=226\frac{2}{3}$ and

$$y=\left(226\frac{2}{3}\right)(3.1)=702\frac{2}{3} \text{ pounds.}$$

Chapter 2: Analysis of Graphs and Functions

2.1: Graphs of Basic Functions and Relations; Symmetry

1. $(-\infty,\infty)$.

3. $(0,0)$

5. increases

7. x-axis

9. odd

11. The domain can be all real numbers; therefore, the function is continuous for the interval $(-\infty,\infty)$.

13. The domain can only be values where $x\geq 0$; therefore, the function is continuous for the interval $[0,\infty)$.

15. The domain can be all real numbers except -3; therefore, the function is continuous for the interval $(-\infty,-3)\cup(-3,\infty)$.

17. (a) The function is increasing for the interval $(3,\infty)$

    (b) The function is decreasing for the interval $(-\infty,3)$

    (c) The function is never constant; therefore, none.

    (d) The domain can be all real numbers; therefore, the interval $(-\infty,\infty)$.

    (e) The range can only be values where $y\geq 0$; therefore, the interval $[0,\infty)$.

19. (a) The function is increasing for the interval $(-\infty,1)$

    (b) The function is decreasing for the interval $(4,\infty)$

    (c) The function is constant for the interval $(1,4)$

    (d) The domain can be all real numbers; therefore, the interval $(-\infty,\infty)$.

    (e) The range can only be values where $y\leq 3$; therefore, the interval $(-\infty,3]$.

21. (a) The function is never increasing; therefore, none

    (b) The function is decreasing for the intervals $(-\infty,-2)$ and $(3,\infty)$

    (c) The function is constant for the interval $(-2,3)$.

    (d) The domain can be all real numbers; therefore, the interval $(-\infty,\infty)$.

    (e) The range can only be values where $y\leq 1.5$ or $y\geq 2$; therefore, the interval $(-\infty,1.5]\cup[2,\infty)$.

23. Graph $f(x)=x^5$. See Figure 23. As x increases for the interval $(-\infty,\infty)$, y increases; therefore, the function is increasing.

25. Graph $f(x)=x^4$. See Figure 25. As x increases for the interval $(-\infty,0)$ y decreases; therefore, the function is decreasing on $(-\infty,0)$

[-10,10] by [-10,10]
Xscl = 1 Yscl = 1

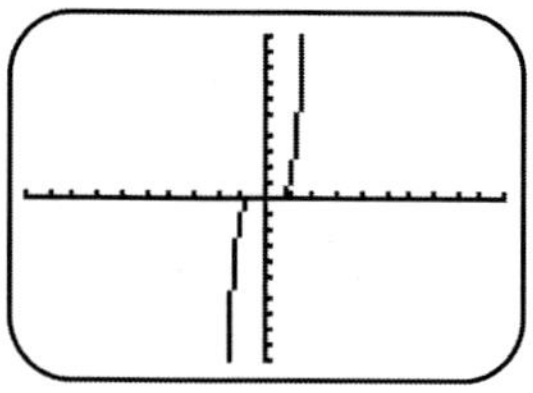

Figure 23

[-10,10] by [-10,10]
Xscl = 1 Yscl = 1

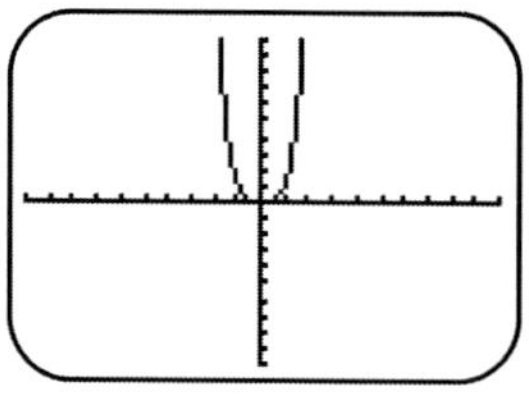

Figure 25

[-10,10] by [-10,10]
Xscl= 1 Yscl = 1

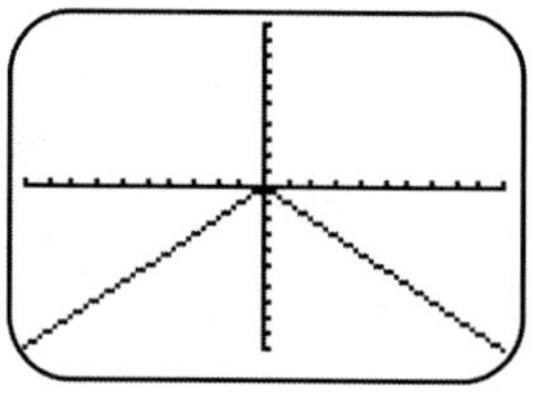

Figure 27

[-10,10] by [-10,10]
Xscl = 1 Yscl= 1

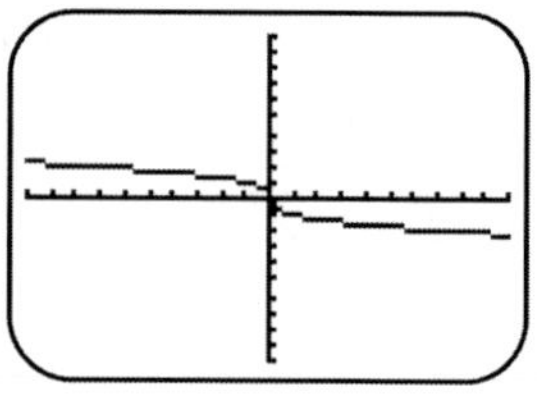

Figure 29

27. Graph $f(x) = -|x|$. See Figure 27. As x increases for the interval $(-\infty, 0)$, y increases; therefore, the function is increasing on $(-\infty, 0)$.

29. Graph $f(x) = -\sqrt[3]{x}$. See Figure 29. As x increases for the interval $(-\infty, \infty)$, y decreases; therefore, the function is decreasing.

31. Graph $f(x) = 1 - x^3$. See Figure 31. As x increases for the interval $(-\infty, \infty)$, y decreases; therefore, the function is decreasing.

33. Graph $f(x) = 2 - x^2$. See Figure 33. As x increases for the interval $(-\infty, 0)$ y increases; therefore, the function is increasing on $(-\infty, 0)$.

[-10,10] by [-10,10]
Xscl = 1 Yscl = 1

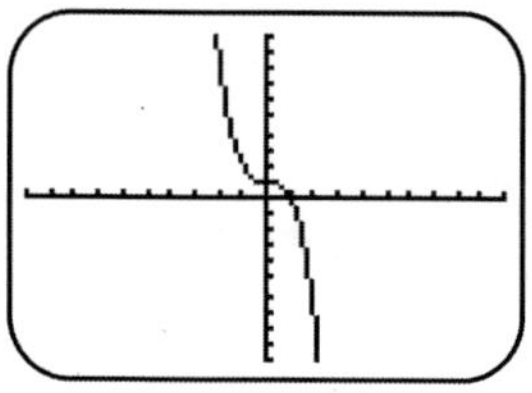

Figure 31

[-10,10] by [-10,10]
Xscl = 1 Yscl = 1

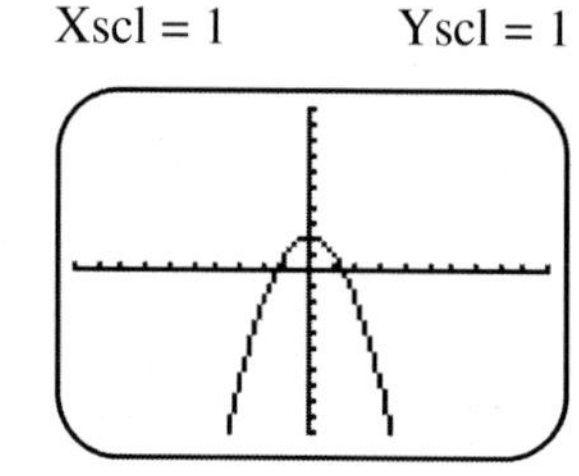

Figure 33

35. (a) No (b) Yes (c) No
37. (a) Yes (b) No (c) No
39. (a) Yes (b) Yes (c) Yes
41. (a) No (b) No (c) Yes
43. (a) Since $f(-x) = f(x)$, this is an even function and is symmetric with respect to the y-axis. See Figure 43a.
    (b) Since $f(-x) = -f(x)$, this is an odd function and is symmetric with respect to the origin. See Figure 43b.

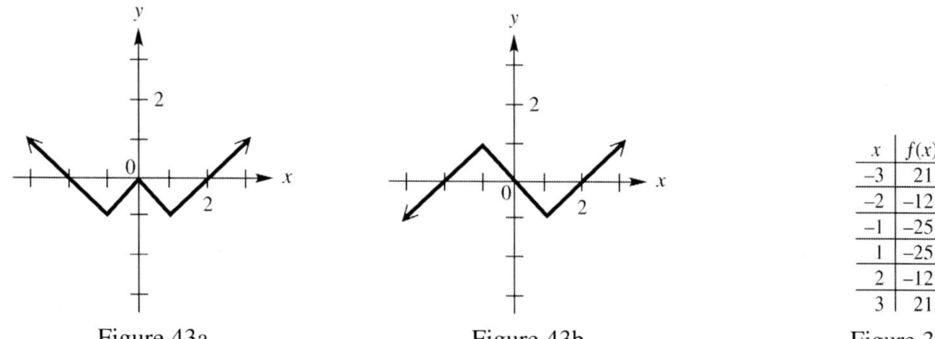

Figure 43a Figure 43b

x	$f(x)$
–3	21
–2	–12
–1	–25
1	–25
2	–12
3	21

Figure 34

45. If f is an even function then $f(-x) = f(x)$ or opposite domains have the same range. See Figure 45
47. This is an even function since opposite domains have the same range.
49. This is an odd function since opposite domains have the opposite range.
51. This is neither even nor odd since the opposite domains are neither the opposite or same range.
53. If $f(x) = x^4 - 7x^2 + 6,$ then $f(-x) = (-x)^4 - 7(-x)^2 + 6 \Rightarrow f(-x) = x^4 - 7x^2 + 6.$ Since $f(-x) = f(x),$ the function is even.
55. If $f(x) = 3x^3 - x,$ then $f(-x) = 3(-x)^3 - (-x) \Rightarrow f(-x) = -3x^3 + x$ and $-f(x) = -(3x^3 - x) \Rightarrow -f(x) = -3x^3 + x.$ Since $f(-x) = -f(x),$ the function is odd.
57. If $f(x) = x^6 - 4x^4 + 5$ then $f(-x) = (-x)^6 - 4(-x)^4 + 5 \Rightarrow f(-x) = x^6 - 4x^4 + 5.$ Since $f(-x) = f(x),$ the function is even.
59. If $f(x) = 3x^5 - x^3 + 7x,$ then $f(-x) = 3(-x)^5 - (-x)^3 + 7(-x) \Rightarrow f(-x) = -3x^5 + x^3 - 7x$ and $-f(x) = -(3x^5 - x^3 + 7x) \Rightarrow -f(x) = -3x^5 + x^3 - 7x.$ Since $f(-x) = -f(x),$ the function is odd.
61. If $f(x) = |5x|,$ then $f(-x) = |5(-x)| \Rightarrow f(-x) = |5x|.$ Since $f(-x) = f(x),$ the function is even.
63. If $(-3,11)$ and $(2,9)$ then $f(-x) = \dfrac{1}{2(-x)} \Rightarrow f(-x) = -\dfrac{1}{2x}$ and $-f(x) = -\left(\dfrac{1}{2x}\right) \Rightarrow -f(x) = -\dfrac{1}{2x}$ Since $f(-x) = -f(x),$ the function is odd.
65. If $f(x) = -x^3 + 2x,$ then $f(-x) = -(-x)^3 + 2(-x) \Rightarrow f(-x) = x^3 - 2x$ and $-f(x) = -(-x^3 + 2x) \Rightarrow -f(x) = x^3 - 2x.$ Since $f(-x) = -f(x),$ the function is symmetric with respect to the origin. Graph $f(x) = -x^3 + 2x;$ the graph supports symmetry with respect to the origin.
67. If $f(x) = 0.5x^4 - 2x^2 + 1,$ then $f(-x) = 0.5(-x)^4 - 2(-x)^2 + 1 \Rightarrow f(-x) = 0.5x^4 - 2x^2 + 1.$ Since $f(-x) = f(x),$ the function is symmetric with respect to the y-axis. Graph $f(x) = 0.5x^4 - 2x^2 + 1;$ the graph supports symmetry with respect to the y-axis.

69. If $f(x)=x^3-x+3$, then $f(-x)=(-x)^3-(-x)+3 \Rightarrow f(-x)=-x^3+x+3$ and $-f(x)=-(x^3-x+3) \Rightarrow -f(x)=-x^3+x-3$. Since $f(x) \neq f(-x) \neq -f(x)$, the function is not symmetric with respect to the y-axis or the origin.

71. If $f(x)=x^6-4x^3$, then $f(-x)=(-x)^6-4(-x)^3 \Rightarrow f(-x)=x^6+4x^3$ and $-f(x)=-(x^6-4x^3) \Rightarrow -f(x)=-x^6+4x^3$. Since $f(x) \neq f(-x) \neq -f(x)$, the function is not symmetric with respect to the y-axis or the origin. Graph $f(x)=x^6-4x^3$; the graph supports no symmetry with respect to the y-axis or the origin.

73. If $f(x)=-6$, then $f(-x)=-6$, Since $f(-x)=f(x)$, the function is symmetric with respect to the y-axis. Graph $f(x)=-6$; the graph supports symmetry with respect to the y-axis.

75. If $f(x)=\frac{1}{4x^3}$, then $f(-x)=\frac{1}{4(-x)^3} \Rightarrow f(-x)=-\frac{1}{4x^3}$ and $-f(x)=-\left(\frac{1}{4x^3}\right) \Rightarrow -f(x)=-\frac{1}{4x^3}$. Since $f(-x)=-f(x)$, the function is symmetric with respect to the origin. Graph $f(x)=\frac{1}{4x^3}$; the graph supports symmetry with respect to the origin.

2.2: Vertical and Horizontal Shifts of Graphs

1. The equation $y=x^2$ shifted 3 units upward is $y=x^2+3$.

3. The equation $y=\sqrt{x}$ shifted 4 units downward is $y=\sqrt{x}-4$.

5. The equation $y=|x|$ shifted 4 units to the right is $y=|x-4|$.

7. The equation $y=x^3$ shifted 7 units to the left is $y=(x+7)^3$.

9. The equation $y=x^2$ shifted 2 units downward and 3 units right is $y=(x-3)^2-2$.

11. The equation $y=\sqrt{x}$ shifted 3 units upward and 6 units to the left is $y=\sqrt{x+6}+3$.

13. The equation $y=x^2$ shifted 500 units upward and 2000 units right is $y=(x-2000)^2+500$.

15. Shift the graph of f 4 units upward to obtain the graph of g.

17. The equation $y=x^2-3$ is $y=x^2$ shifted 3 units downward; therefore, graph B.

19. The equation $y=(x+3)^2$ is $y=x^2$ shifted 3 units to the left; therefore, graph A.

21. The equation $y=|x+4|-3$ is $y=|x|$ shifted 4 units to the left and 3 units downward; therefore, graph B.

23. The equation $y=(x-3)^3$ is $y=x^3$ shifted 3 units to the right; therefore, graph C.

25. The equation $y=(x+2)^3-4$ is $[-a,-b]$. shifted 2 units to the left and 4 units downward; therefore, graph B.

27. Using $Y_2=Y_1+k$ and $x=0$, we get $-5=-3+k \Rightarrow k=-2$.

29. From the graphs, $(6,2)$ is a point on Y_1 and $(6,-1)$ a point on Y_2. Using $Y_2 = Y_1 + k$ and $x = 6$, we get $-1 = 2 + k \Rightarrow k = -3$.

31. For the equation $y = x^2$, the Domain is $(-\infty,\infty)$ and the Range is $[0,\infty)$. Shifting this 3 units downward gives us: (a) Domain: $(-\infty,\infty)$ (b) Range: $[-3,\infty)$.

33. For the equation $y = |x|$, the Domain is $(-\infty,\infty)$ and the Range is $[0,\infty)$. Shifting this 4 units to the left and 3 units downward gives us: (a) Domain: $(-\infty,\infty)$ (b) Range: $[-3,\infty)$.

35. For the equation $y = x^3$, the Domain is $(-\infty,\infty)$ and the Range is $(-\infty,\infty)$. Shifting this 3 units to the right gives us: (a) Domain: $(-\infty,\infty)$ (b) Range: $(-\infty,\infty)$

37. For the equation $y = x^2$, the Domain is $(-\infty,\infty)$ and the Range is $[0,\infty)$. Shifting this 1 unit to the right and 5 units downward gives us: (a) Domain: $(-\infty,\infty)$ (b) Range: $[-5,\infty)$.

39. For the equation $y = \sqrt{x}$, the Domain is $[0,\infty)$. and the Range is $[0,\infty)$. Shifting this 4 units to the right gives us: (a) Domain: $[4,\infty)$. (b) Range: $[0,\infty)$.

41. For the equation $y = x^3$, the Domain is $(-\infty,\infty)$ and the Range is $(-\infty,\infty)$. Shifting this 1 unit to the right and 4 units upward gives us: (a) Domain: $(-\infty,\infty)$ (b) Range: $(-\infty,\infty)$

43. The graph of $y = f(x)$ is the graph of the equation $y = x^2$ shifted 1 unit to the right. See Figure 43.

45. The graph of $y = x^3 + 1$ is the graph of the equation $y = x^3$ shifted 1 unit upward. See Figure 45.

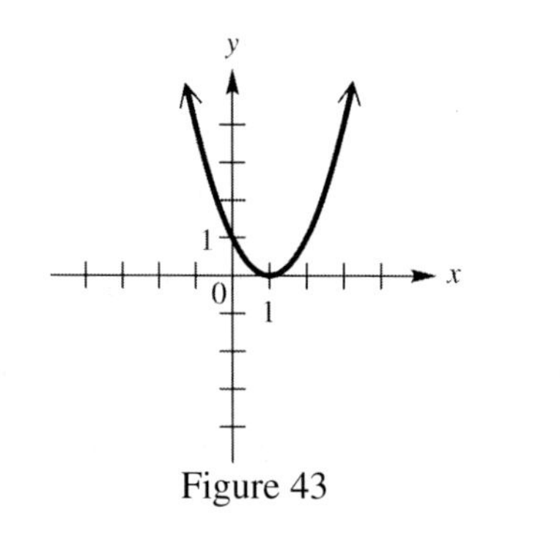

Figure 43

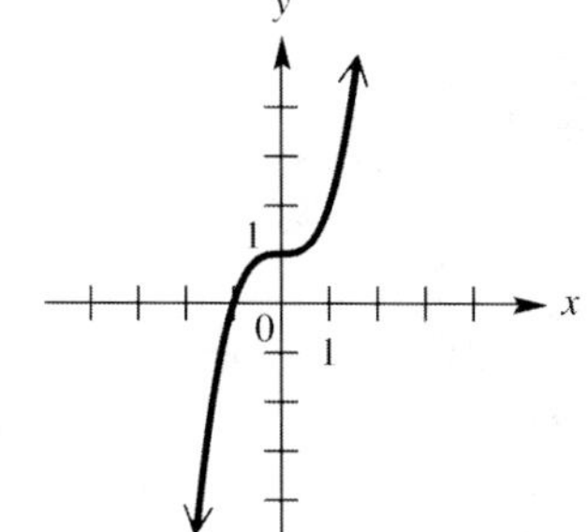

Figure 45

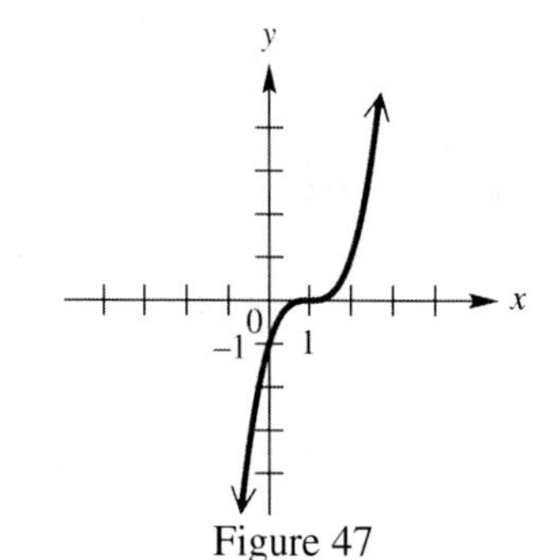

Figure 47

47. The graph of $y = (x-1)^3$ is the graph of the equation $y = x^3$ shifted 1 unit to the right. See Figure 47.

49. The graph of $y = \sqrt{x-2} - 1$ is the graph of the equation $y = \sqrt{x}$ shifted 2 units to the right and 1 unit downward. See Figure 49.

51. The graph of $f(x)$ is the graph of the equation $y = x^2$ shifted 2 units to the left and 3 units upward. See Figure 51.

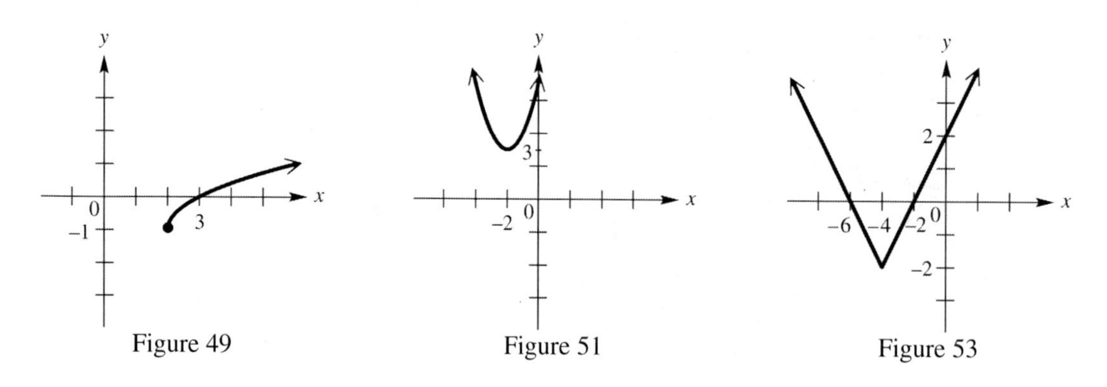

Figure 49 Figure 51 Figure 53

53. The graph of $y=|x+4|-2$ is the graph of the equation $y=|x|$ shifted 4 units to the left and 2 units downward. See Figure 53.

55. Since h and k are positive, the equation is $y=x^2$ shifted to the right and down; therefore, B.

57. Since h and k are positive, the equation is $y=x^2$ shifted to the left and up; therefore, A.

59. The equation $y=f(x)+2$ is $y=f(x)$ shifted up 2 units or add 2 to the y-coordinate of each point as follows: $(-3,2)\Rightarrow(-3,0);(-1,4)\Rightarrow(-1,6);(5,0)\Rightarrow(5,2)$. See Figure 59.

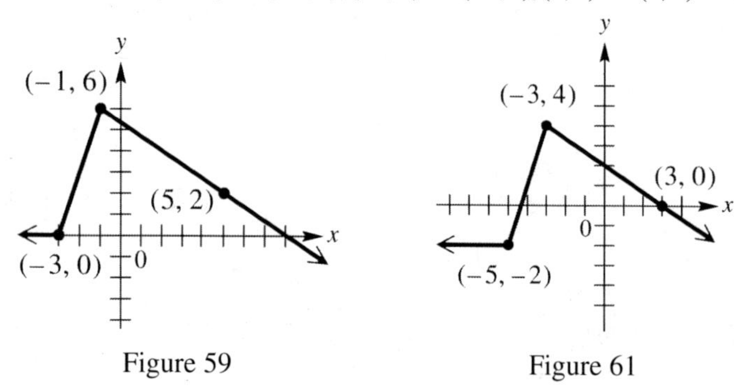

Figure 59 Figure 61

61. The equation $y=f(x+2)$ is $y=f(x)$ shifted left 2 units or subtract 2 from the x-coordinate of each point as follows: $(-3,-2)\Rightarrow(-5,-2);(-1,4)\Rightarrow(-3,4);(5,0)\Rightarrow(3,0)$. See Figure 61.

63. The graph is the basic function $y=x^2$ translated 4 units to the left and 3 units up; therefore, the new equation is $y=(x+4)^2+3$. The equation is now increasing for the interval: (a) $(-4,\infty)$ and decreasing for the interval: (b) $(-\infty,-4)$.

65. The graph is the basic function $y=x^3$ translated 5 units down; therefore, the new equation is $y=x^3-5$. The equation is now increasing for the interval: (a) $(-\infty,\infty)$ and does not decrease; therefore: (b) none.

67. The graph is the basic function $y=\sqrt{x}$ translated 2 units to the right and 1 unit up; therefore, the new equation is $y=\sqrt{x-2}\ +\ 1$. The equation is now increasing for the interval: (a) $(2,\infty)$ and does not decrease; therefore: (b) none.

69. (a) $f(x)=0$: $\{3,4\}$

(b) $f(x) > 0$: for the intervals $(-\infty, 3) \cup (4, \infty)$.

(c) $f(x) < 0$: for the interval $(3, 4)$.

71. (a) $f(x) = 0$: $\{-4, 5\}$

(b) $f(x) \geq 0$: for the intervals $(-\infty, -4] \cup [5, \infty)$

(c) $f(x) \leq 0$: for the interval $[-4, 5]$.

72. (a) $f(x) = 0$: never; therefore: $\varnothing$.

(b) $f(x) \geq 0$: for the interval $[1, \infty)$.

(c) $f(x) \leq 0$: never; therefore: $\varnothing$.

73. The translation is 3 units to the left and 1 unit up; therefore, the new equation is $y = |x+3| + 1$. The form $y = |x-h| + k$ will equal $y = |x+3| + 1$ when: $h = -3$ and $k = 1$.

74. The equation $y = x^2$ has a Domain: $(-\infty, \infty)$ and a Range: $[0, \infty)$. After the translation the Domain is still: $(-\infty, \infty)$ but now the Range is $(38, \infty)$, a positive or upward shift of 38 units. Therefore, the horizontal shift can be any number of units, but the vertical shift is up 38. This makes h any real number and $k = 38$.

75. (a) $B(4) = 66.25(4) + 160 = 425$; In 2010, 425,000 bankruptcies were filed.

(b) We will use the point (2006, 160) and the slope of 66.25 in the point slope form for the equation of a line. $y - y_1 = m(x - x_1) \Rightarrow y - 160 = 66.25(x - 2006) \Rightarrow y = 66.25(x - 2006) + 160$

(c) $y = 66.25(2010 - 2006) + 160 = 66.25(4) + 160 = 425$, In 2010, 425,000 bankruptcies were filed.

(d) $293 = 66.25(x - 2006) + 160 \Rightarrow 133 = 66.25(x - 2006) \Rightarrow \dfrac{133}{66.25} = x - 2006 \Rightarrow x = 2006 + \dfrac{133}{66.25}$.

There will be 293 thousand bankruptcies in 2008.

77. $U(2011) = 13(2011 - 2006)^2 + 115 = 13(25) + 115 = 440$; The average U.S. household spent \$440 on Apple products in 2011.

79. (a) Enter the year in L_1 and enter tuition and fees in L_2. The year 2000 corresponds to $x = 0$ and so on. The regression equation is $y \approx 402.5x + 3460$.

(b) Since $x = 0$ corresponds to 2000, the equation when the exact year is entered is $y = 402.5(x - 2000) + 3460$

(c) $y \approx 402.5(2009 - 2000) + 3460 \Rightarrow y \approx \$7100$

81. See Figure 81.

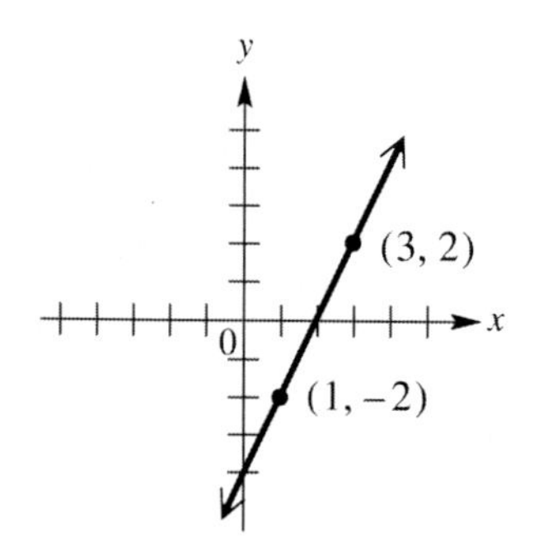

Figure 81

82. $m = \frac{2-(-2)}{3-1} \Rightarrow m = \frac{4}{2} = 2$

83. Using slope-intercept form yields: $y_1 - 2 = 2(x-3) \Rightarrow y_1 - 2 = 2x - 6 \Rightarrow y_1 = 2x - 4$

84. $(1, -2+6)$ and $(3, 2+6) \Rightarrow (1, 4)$ and $(3, 8)$

85. $m = \frac{8-4}{3-1} \Rightarrow m = \frac{4}{2} = 2$

86. Using slope-intercept form yields: $y_2 - 4 = 2(x-1) \Rightarrow y_2 - 4 = 2x - 2 \Rightarrow y_2 = 2x + 2$.

87. Graph $y_1 = 2x - 4$ and $y_2 = 2x + 2$ See Figure 87. The graph y_2 can be obtained by shifting the graph of y_1 upward 6 units. The constant 6, comes from the 6 we added to each y-value in Exercise 84.

[-10,10] by [-10,10]
Xscl = 1 Yscl = 1

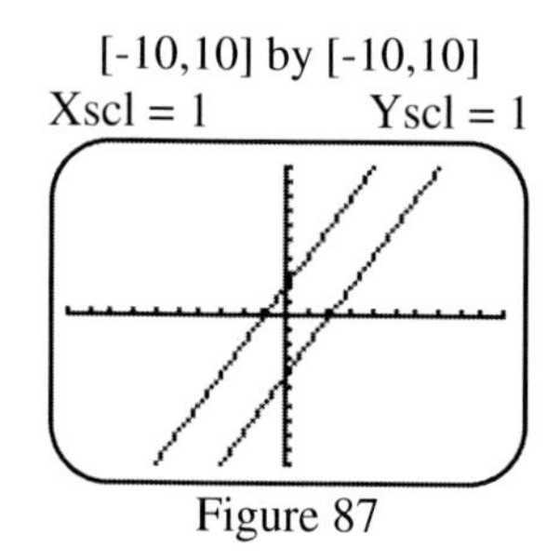

Figure 87

88. c; c; the same as; c; upward (or positive vertical)

2.3: Stretching, Shrinking, and Reflecting Graphs

1. The function $y = x^2$ vertically stretched by a factor of 2 is $y = 2x^2$.
3. The function $y = \sqrt{x}$ reflected across the y-axis is $y = \sqrt{-x}$.
5. The function $y = |x|$ vertically stretched by a factor of 3 and reflected across the x-axis is $y = -3|x|$.
7. The function $y = x^3$ vertically shrunk by a factor of 0.25 and reflected across the y-axis is $y = 0.25(-x^3)$ or $y = -0.25x^3$.
9. Graph $y_1 = x$, $y_2 = x + 3$ (y_1 shifted up 3 units), and $y_3 = x - 3$ (y_1 shifted down 3 units). See Figure 9.
11. Graph $y_1 = |x|$, $y_2 = |x - 3|$ (y_1 shifted right 3 units), and $y_3 = |x + 3|$ (y_1 shifted left 3 units). See Figure 11

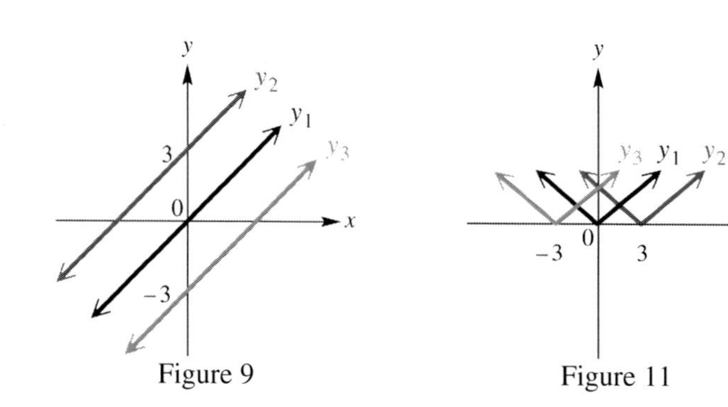

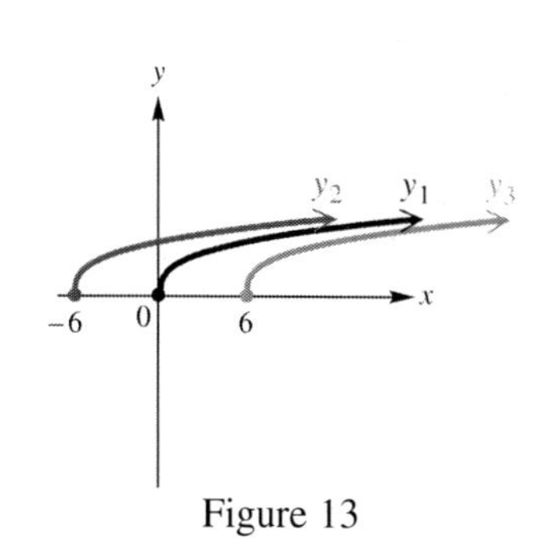

13. Graph $y_1 = \sqrt{x}$, $y_2 = \sqrt{x+6}$ (y_1 shifted left 6 units), and $y_3 = \sqrt{x-6}$ (y_1 shifted right 6 units). See Figure 13.

15. Graph $y_1 = \sqrt[3]{x}$, $y_2 = -\sqrt[3]{x}$ (y_1 reflected across the x-axis), and $y_3 = -2\sqrt[3]{x}$ (y_1 reflected across the x-axis and stretched vertically by a factor of 2). See Figure 15.

17. Graph $y_1 = |x|$, $y_2 = -2|x-1|+1$ (y_1 reflected across the x-axis, stretched vertically by a factor of 2, shifted right 1 unit, and shifted up 1 unit), and $y_3 = -\frac{1}{2}|x| - 4$ (y_1 reflected across the x-axis, shrunk by factor of $\frac{1}{2}$, and shifted down 4 units). See Figure 17.

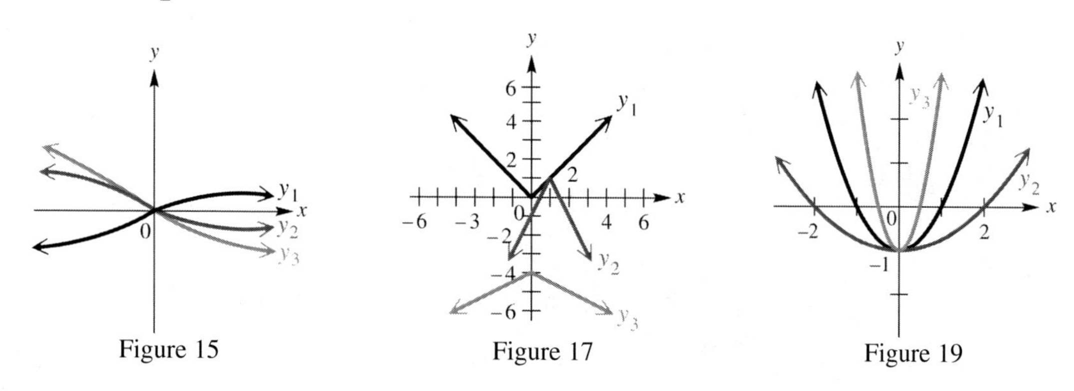

19. Graph $y_1 = x^2 - 1$ (which is $y = x^2$ shifted down 1 unit), $y_2 = \left(\frac{1}{2}x\right)^2 - 1$ (y_1 shrunk vertically by A factor of $\frac{1}{2}$), and $y_3 = (2x)^2 - 1$ (y_1 stretched vertically by a factor of 2^2 or 4). See Figure 19.

21. Graph $y_1 = \sqrt[3]{x}$, $y_2 = \sqrt[3]{-x}$ (y_1 reflected across the y-axis), and $y_3 = \sqrt[3]{-(x-1)}$ (y_1 reflected across the y-axis and shifted right 1 unit). See Figure 21.

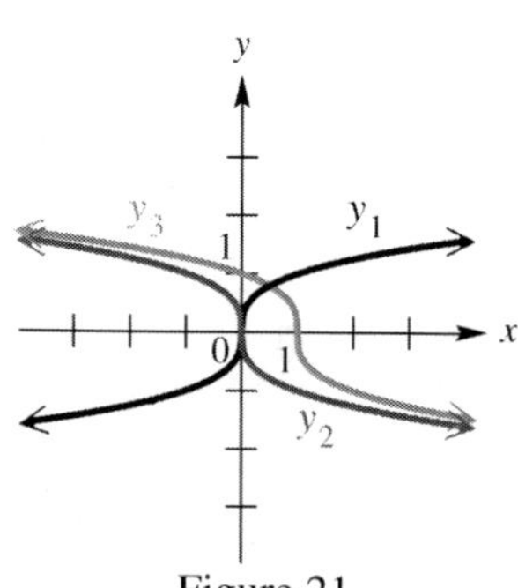

Figure 21

23. The graph $y = f(x) = x^2$ has been reflected across the x-axis, shifted 5 units to the right, and shifted 2 units downward; therefore, the equation of $g(x)$ is $g(x) = -(x-5)^2 - 2$.

25. The graph $y = f(x) = \sqrt{x}$ has been reflected across the y-axis and shifted 1 unit upward; therefore, the equation of $g(x)$ is $g(x) = \sqrt{-x} + 1$.

27. 4; x

29. 2; left; $\frac{1}{4}$; x; 3; downward (or negative)

31. 3; right; 6

33. The function $y = x^2$ is vertically shrunk by a factor of $\frac{1}{2}$ and shifted 7 units down; therefore, $y = \frac{1}{2}x^2 - 7$.

35. The function $y = \sqrt{x}$ is shifted 3 units right, vertically stretched by a factor of 4.5, and shifted 6 units down; therefore, $y = 4.5\sqrt{x-3} - 6$.

37. The function $f(x) = \sqrt{x-3} + 2$ is $f(x) = \sqrt{x}$ shifted 3 units right and 2 units upward. See Figure 37.

39. The function $f(x) = \sqrt{2x} = \sqrt{2}\sqrt{x}$ is $f(x) = \sqrt{x}$ stretched vertically by a factor of $\sqrt{2}$. See Figure 39.

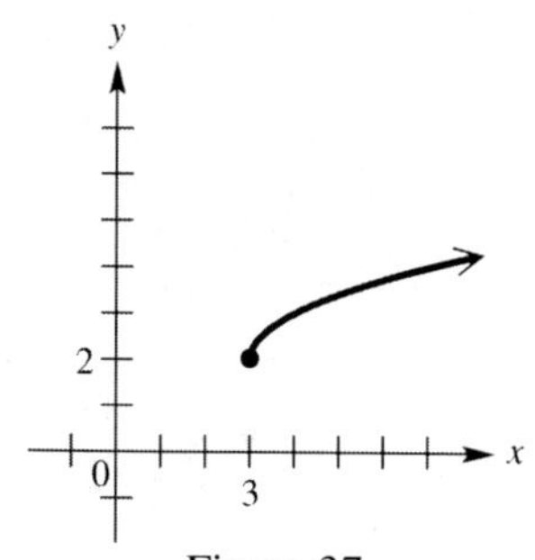

Figure 37

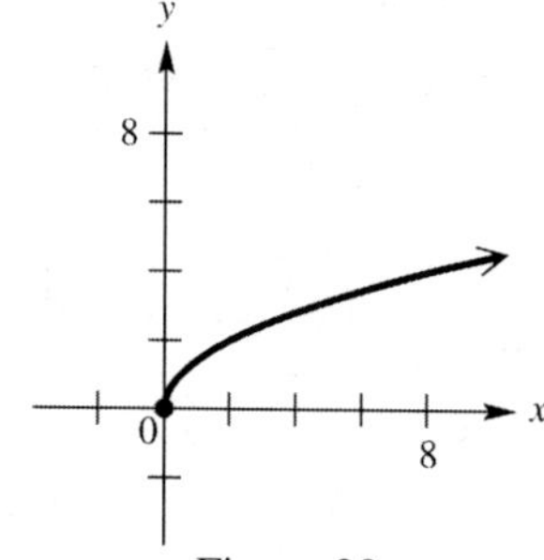

Figure 39

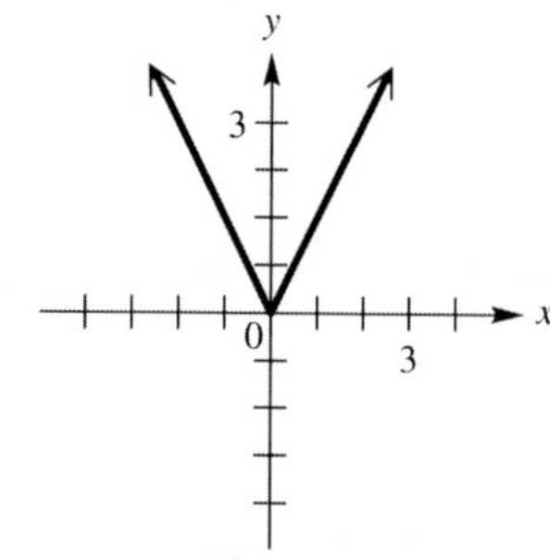

Figure 41

41. The function $f(x) = |2x| = 2|x|$ is $f(x) = |x|$ stretched vertically by a factor of 2. See Figure 41.

43. The function $f(x) = 1 - \sqrt{x}$ is $f(x) = \sqrt{x}$ reflected across the x-axis and shifted 1 unit upward. See Figure 43.

45. The function $f(x) = -\sqrt{1-x} = -\sqrt{-(x-1)}$ is $f(x) = \sqrt{x}$ reflected across both the x-axis, the y-axis and shifted 1 unit right. See Figure 45.

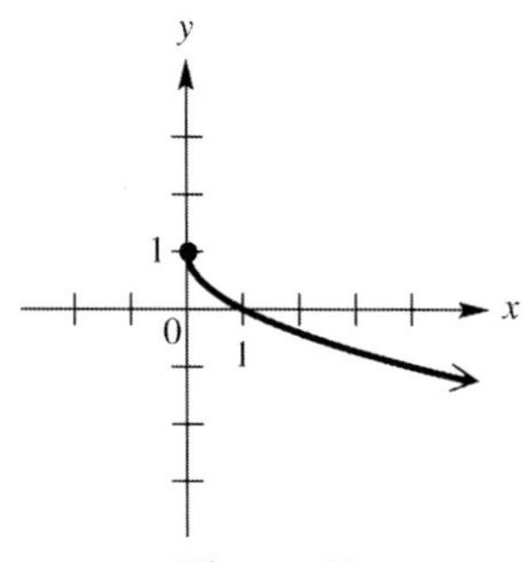

Figure 43

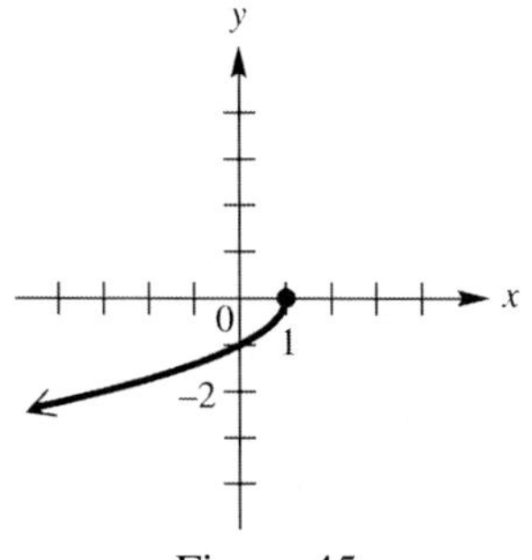

Figure 45

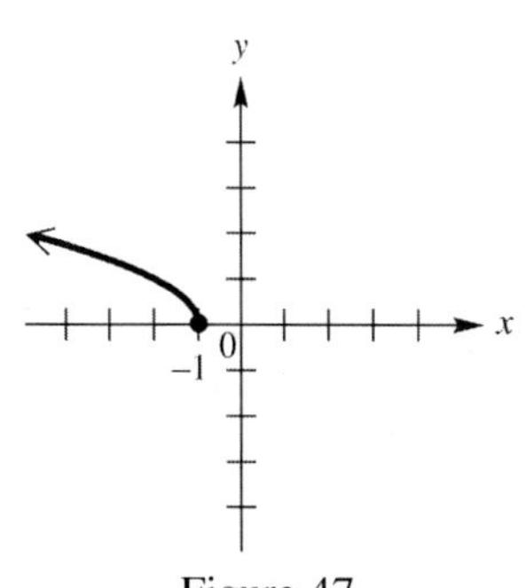

Figure 47

47. The function $f(x)=\sqrt{-(x+1)}$ is $f(x)=\sqrt{x}$ reflected across the y-axis and shifted 1 unit left. See Figure 47.

49. The function $f(x)=(x-1)^3$ is $f(x)=x^3$ shifted 1 unit right. See Figure 49.

51. The function $f(x)=-x^3$ is $f(x)=x^3$ reflected across the x-axis. See Figure 51.

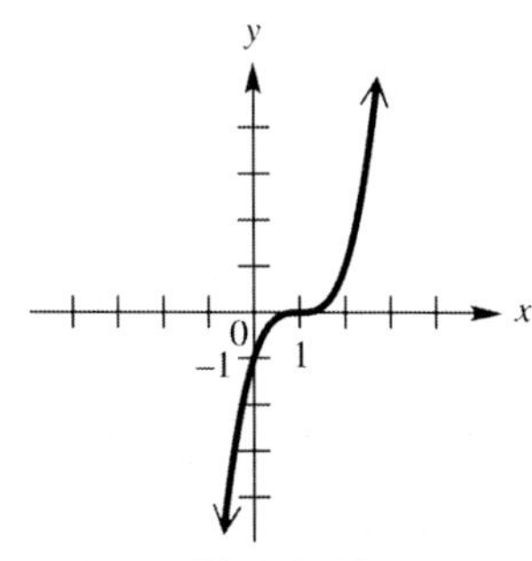

Figure 49

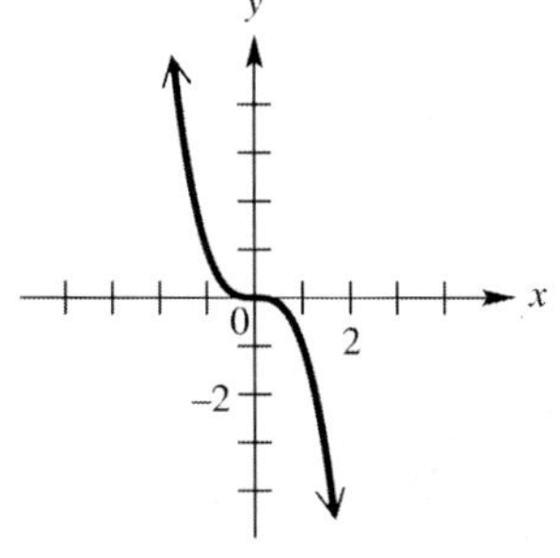

Figure 51

53. (a) The equation $y=-f(x)$ is $y=f(x)$ reflected across the x-axis. See Figure 53a.

(b) The equation $y=f(-x)$ is $y=f(x)$ reflected across the y-axis. See Figure 53b.

(c) The equation $y=2f(x)$ is $y=f(x)$ stretched vertically by a factor of 2. See Figure 53c.

(d) From the graph $f(0)=1$.

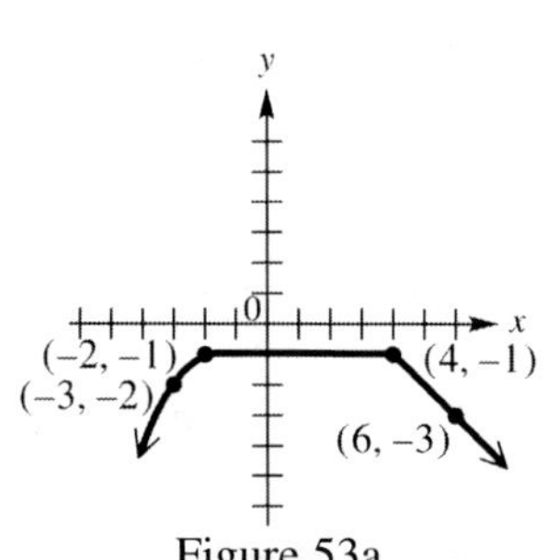

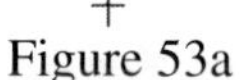
Figure 53a

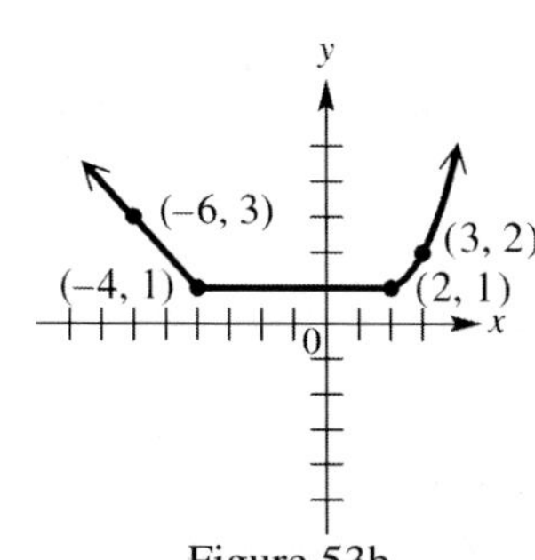

Figure 53b

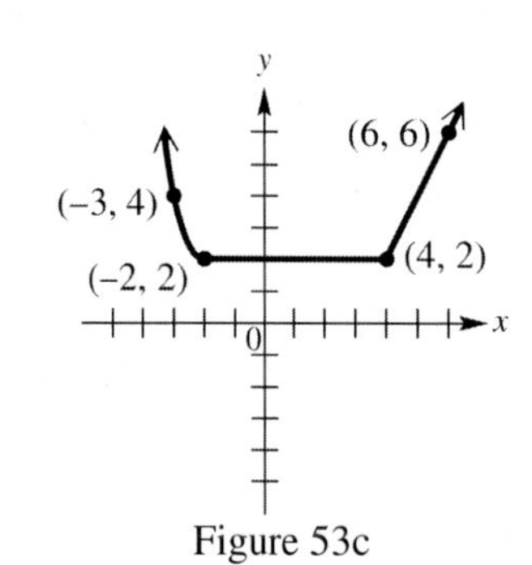

Figure 53c

55. (a) The equation $f(x)$ is $y=f(x)$ reflected across the x-axis. See Figure 55a.

(b) The equation $y=f(-x)$ is $y=f(x)$ reflected across the y-axis. See Figure 55b.

(c) The equation $y=f(x+1)$ is $y=f(x)$ shifted 1 unit to the left. See Figure 55c.

(d) From the graph, there are two x-intercepts, $(-1,0)$ and $(4,0)$.

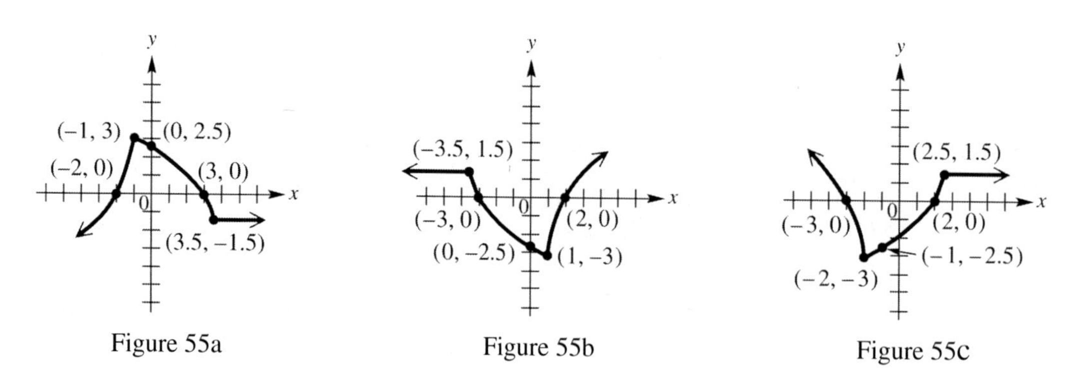

Figure 55a Figure 55b Figure 55c

57. (a) The equation $y=-f(x)$ is $y=f(x)$ reflected across the x-axis. See Figure 57a.

    (b) The equation $y=f\left(\frac{1}{3}x\right)$ is $y=f(x)$ stretched horizontally by a factor of 3. See Figure 57b.

    (c) The equation $y=0.5f(x)$ is $y=f(x)$ shrunk vertically by a factor of 0.5. See Figure 57c.

    (d) From the graph, symmetry with respect to the origin.

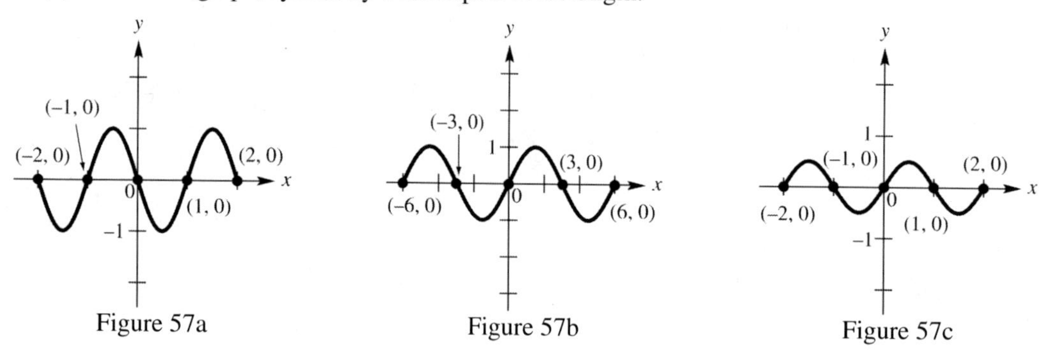

Figure 57a Figure 57b Figure 57c

59. (a) The equation $y=f(x)+1$ is $y=f(x)$ shifted 1 unit upward. See Figure 59a.

    (b) The equation -30° is $y=f(x)$ reflected across the x-axis and shifted 1 unit down. See Figure 59b.

    (c) The equation $y=2f\left(\frac{1}{2}x\right)$ is $y=f(x)$ stretched vertically by a factor of 2 and horizontally by a factor of 2. See Figure 59c.

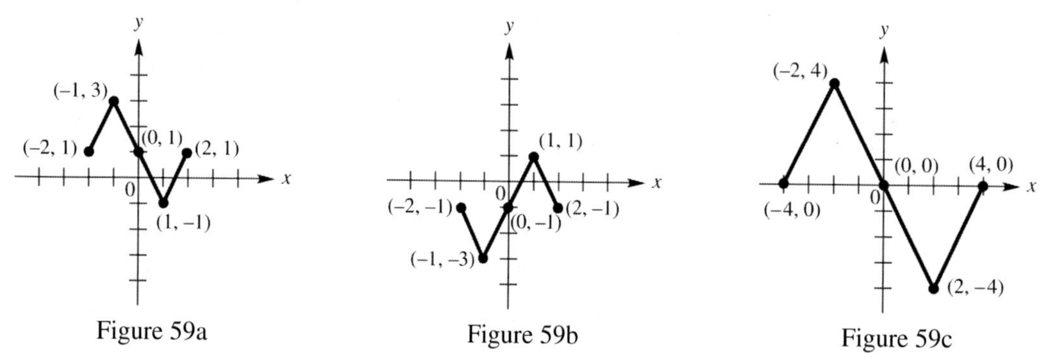

Figure 59a Figure 59b Figure 59c

61. (a) The equation $y=f(2x)+1$ is x shrunk horizontally by a factor of $\{2\}$ and shifted 1 unit upward. See Figure 61a.

(b) The equation $y = 2f\left(\frac{1}{2}x\right)+1$ is $y = f(x)$ stretched vertically by a factor of 2, stretched horizontally by a factor of 2, and shifted 1 unit upward. See Figure 61b.

(c) The equation $y = \frac{1}{2}f(x-2)$ is $y = f(x)$ shrunk vertically by a factor of $\frac{1}{2}$ and shifted 2 units to the right. See Figure 61c.

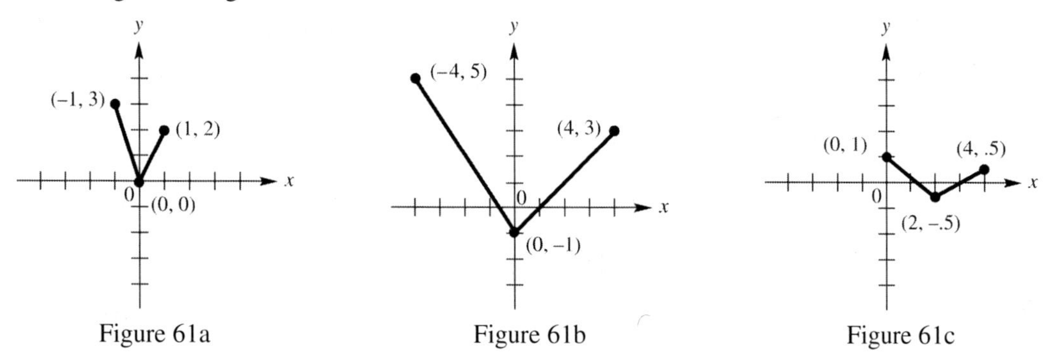

63. (a) If $(r, 0)$ is the x-intercept of $y = f(x)$ and $y = -f(x)$ is $y = f(x)$ reflected across the x-axis, then $(r, 0)$ is also the x-intercept of $y = -f(x)$.

(b) If $(r, 0)$ is the x-intercept of $y = f(x)$ and $y = f(-x)$ is $f(x)$ reflected across the y-axis, then $(-r, 0)$ is the x-intercept of $y = -f(x)$

(c) If $(r, 0)$ is the x-intercept of $y = f(x)$ and $y = -f(-x)$ is $y = f(x)$ reflected across both the x-axis and y-axis, then $(-r, 0)$, is the x-intercept of $y = -f(-x)$.

65. Since $y = f(x-2)$ is $y = f(x)$ shifted 2 units to the right, the domain of $f(x-2)$ is $[-1+2, 2+2]$ or $[1, 4]$, and the range is the same: $[0, 3]$.

67. Since $-f(x)$ is $f(x)$ reflected across the x-axis, the domain of $-f(x)$ is the same: $[-1, 2]$, and the range is $[-3, 0]$. .

69. Since $f(2x)$ is $f(x)$ shrunk horizontally by a factor of $\frac{1}{2}$, the domain of $f(2x)$ is $\left[\frac{1}{2}(-1), \frac{1}{2}(2)\right]$ or $\left[-\frac{1}{2}, 1\right]$, and the range is the same: $[0, 3]$.

71. Since $3f\left(\frac{1}{4}x\right)$ is $f(x)$ stretched horizontally by a factor of 4, the domain of $3f\left(\frac{1}{4}x\right)$ is $[4(-1), 4(2)]$ or $[-4, 8]$, and stretched vertically by a factor of 3, the range is $[3(0), 3(3)]$ or $[0, 9]$.

73. Since $f(-x)$ is $f(x)$ reflected across the y-axis, the domain of $f(-x)$ is $[-(-1), -(2)] = [1, -2]$ or $[-2, 1]$; and the range is the same: $[0, 3]$.

75. Since $f(-3x)$ is $f(x)$ reflected across the y-axis and shrunk horizontally by a factor of $\frac{1}{3}$, the domain of $f(-3x)$ is $\left[-\frac{1}{3}(-1),-\frac{1}{3}(2)\right]=\left[\frac{1}{3},-\frac{2}{3}\right]$ or $\left[-\frac{2}{3},\frac{1}{3}\right]$, and the range is the same: $[0,3]$.

77. Since $y=\sqrt{x}$ has an endpoint of (0, 0), and the graph of $y=10\sqrt{x-20}+5$ is the graph of $y=\sqrt{x}$ shifted 20 units right, stretched vertically by a factor of 10, and shifted 5 units upward, the endpoint of $y=10\sqrt{x-20}+5$ is $(0+20,10(0)+5)$ or $(20,5)$. Therefore, the domain is $[20,\infty)$, and the range is $[5,\infty)$.

79. Since $y=\sqrt{x}$ has an endpoint of (0, 0), and the graph of $y=-.5\sqrt{x+10}+5$ is the graph of $y=\sqrt{x}$ shifted 10 units left, reflected across the x-axis, shrunk vertically by a factor of .5, and shifted 5 units upward, the endpoint of $y=-.5\sqrt{x+10}+5$ is $(0-10,-.5(0)+5)$ or $(-10,5)$. Therefore, the domain is $[-10,\infty)$, and the range, because of the reflection across the x-axis, is $(-\infty,5]$.

81. The graph of $y=-f(x)$ is $y=f(x)$ reflected across the x-axis; therefore, $y=-f(x)$ is decreasing for the interval (a,b).

83. The graph of $y=-f(-x)$ is $y=f(x)$ reflected across both the x-axis and y-axis; therefore, $y=-f(-x)$ is increasing for the interval $(-b,-a)$

85. (a) the function is increasing for the interval: $(-1,2)$.

 (b) the function is decreasing for the interval: $(-\infty,-1)$.

 (c) the function is constant for the interval: $(2,\infty)$.

87. (a) the function is increasing for the interval: $(1,\infty)$.

 (b) the function is decreasing for the interval: $(-2,1)$.

 (c) the function is constant for the interval: $(-\infty,-2)$.

89. From the graph, the point on y_2 is approximately $(8,10)$.

91. Use two points on the graph to find the slope. Two points are $(-2,-1)$ and $(-1,1)$; therefore, the slope is $m=\frac{1-(-1)}{-1-(-2)}=\frac{2}{1}\Rightarrow m=2$. The stretch factor is 2 and the graph has been shifted 2 units to the left and 1 unit down; therefore, the equation is $y=2|x+2|-1$.

93. Use two points on the graph to find the slope. Two points are $(0,2)$ and $(1,-1)$; therefore, the slope is $m=\frac{-1-2}{1-0}=\frac{-3}{1}\Rightarrow m=-3$. The stretch factor is 3, the graph has been reflected across the x-axis, and shifted 2 units upward; therefore, the equation is $y=-3|x|+2$.

95. Use two points on the graph to find the slope. Two points are (0,-4) and (3,0); therefore, the slope is

$m=\frac{-4-0}{0-3}=\frac{4}{3}\Rightarrow m=\frac{4}{3}$. The stretch factor is $\frac{4}{3}$ and the graph has been shifted 4 units down.; therefore,

the equation is $y=\frac{4}{3}|x|-4$.

97. Since $y=f(x)$ is symmetric with respect to the y-axis, for every (x,y) on the graph, $(-x,y)$ is also on the graph. Reflection across the y-axis reflect onto itself and will not change the graph. It will be the same.

Reviewing Basic Concepts (Sections 2.1—2.3)

1. (a) The function $f(x)=|x|$ shifted up one unit yields the function $f(x)=|x|+1$. Therefore, this function has a domain of $(-\infty,\infty)$ and a range of of $[1,\infty)$. The function is increasing from $(0,\infty)$ and decreasing from $(-\infty,0)$.

   (b) The function $f(x)=x^2$ shifted to the right 2 units yields the function $f(x)=(x-2)^2$. Therefore, this function has a domain of $(-\infty,\infty)$ and a range of of $[0,\infty)$. The function is increasing from $(2,\infty)$ and decreasing from $(-\infty,2)$.

   (c) The function $f(x)=\sqrt{x}$ reflected over the x-axis yields the function $f(x)=-\sqrt{x}$. Therefore, this function has a domain of $[0,\infty)$ and a range of of $(-\infty,0]$. The function is never increasing and decreasing from $(0,\infty)$.

2. (a) If $y=f(x)$ is symmetric with respect to the origin, then another function value is $f(-3)=-6$.

   (b) If $y=f(x)$ is symmetric with respect to the y-axis, then another function value is $f(-3)=6$.

   (c) If $f(-x)=-f(x), y=f(x)$ is symmetric with respect to both the x-axis and y-axis, then another function value is $f(-3)=-6$.

   (d) If $y=f(-x),\ y=f(x)$ is symmetric with respect to the y-axis, then another function value is $f(-3)=6$.

3. (a) The equation $y=(x-7)^2$ is $y=x^2$ shifted 7 units to the right: B.

   (b) The equation $y=x^2-7$ is $y=x^2$ shifted 7 units downward: D.

   (c) The equation $y=7x^2$ is $y=x^2$ stretches vertically by a factor of 7: E.

   (d) The equation $y=(x+7)^2$ is $y=x^2$ shifted 7 units to the left: A.

   (e) The equation $y=\left(\frac{1}{3}x\right)^2$ is $y=x^2$ stretches horizontally by a factor of 3: C.

4. (a) The equation $y=x^2+2$ is $y=x^2$ shifted 2 units upward: B.

   (b) The equation $y=x^2-2$ is $y=x^2$ shifted 2 units downward: A.

   (c) The equation $y=(x+2)^2$ is $y=x^2$ shifted 2 units to the left: G.

   (d) The equation $y=(x-2)^2$ is $y=x^2$ shifted 2 units to the right: C.

(e) The equation $y = 2x^2$ is $y = x^2$ stretched vertically by a factor of 2: F.

(f) The equation $y = -x^2$ is $y = x^2$ reflected across the x-axis D.

(g) The equation $y = (x-2)^2 + 1$ is $y = x^2$ shifted 2 units to the right and 1 unit upward: H.

(h) The equation $y = (x+2)^2 + 1$ is $y = x^2$ shifted 2 units to the left and 1 unit upward: E.

5. (a) The equation $y = |x| + 4$ is $y = |x|$ shifted 4 units upward. See Figure 5a.

(b) The equation $y = |x+4|$ is $y = |x|$ shifted 4 units to the left. See Figure 5b.

(c) The equation $y = |x-4|$ is $y = |x|$ shifted 4 units to the right. See Figure 5c.

(d) The equation $y = |x+2| - 4$ is $y = |x|$ shifted 2 units to the left and 4 units down. See Figure 5d.

(e) The equation $y = -|x-2| + 4$ is $y = |x|$ reflected across the x-axis, shifted 2 units to the right, and 4 units upward. See Figure 5e.

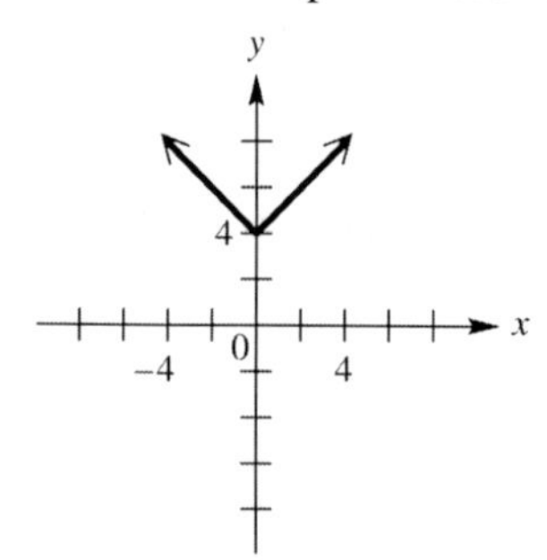

Figure 5a

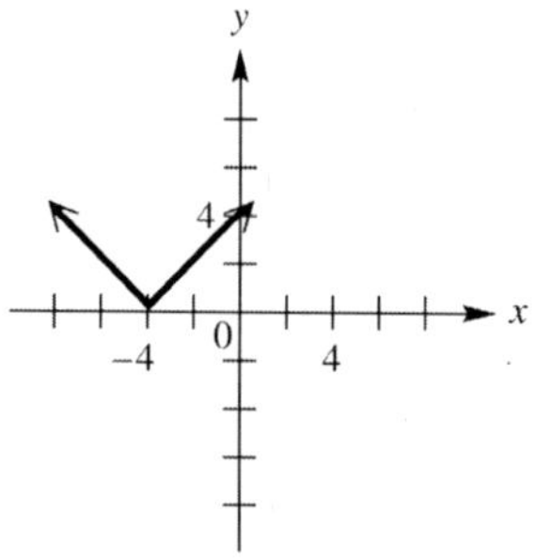

Figure 5b

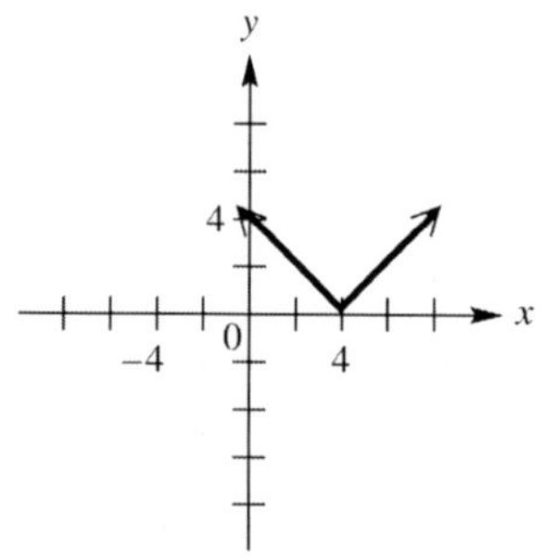

Figure 5c

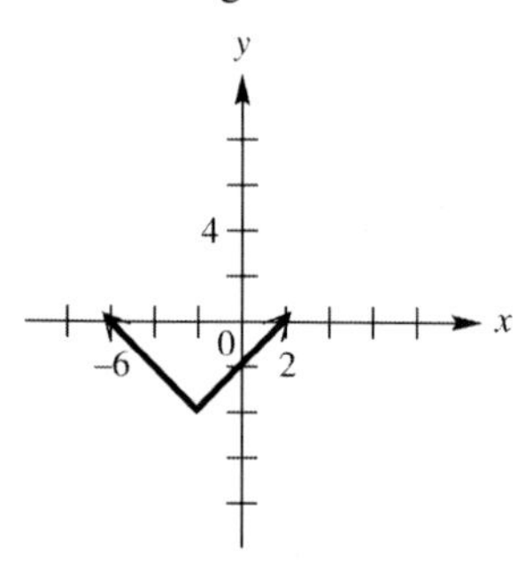

Figure 5d

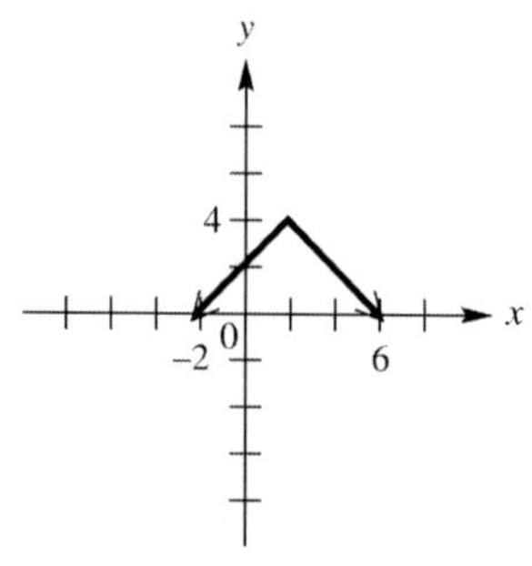

Figure 5e

6. (a) The graph is the function $f(x) = |x|$ reflected across the x-axis, shifted 1 unit left and 3 units upward Therefore, the equation is $y = -|x+1| + 3$.

(b) The graph is the function $g(x) = \sqrt{x}$ reflected across the x-axis, shifted 4 units left and 2 units upward. Therefore, the equation is $y = -\sqrt{x+4} + 2$.

(c) The graph is the function $g(x) = \sqrt{x}$ stretches vertically by a factor of 2, shifted 4 units left and 4 units downward. Therefore, the equation is $y = 2\sqrt{x+4} - 4$.

(d) The graph is the function $f(x) = |x|$ shrunk vertically by a factor of $\frac{1}{2}$, shifted 2 units right and 1 unit downward. Therefore, the equation is $y = \frac{1}{2}|x-2| - 1$.

7. (a) The graph of $g(x)$ is the graph $f(x)$ shifted 2 units upward. Therefore, $c = 2$.

   (b) The graph of $g(x)$ is the graph $f(x)$ shifted 4 units to the left. Therefore, $c = 4$.

8. The graph of $y = F(x+h)$ is a horizontal translation of the graph of $y = F(x)$. The graph of $y = F(x)+h$ is not the same as the graph of $y = F(x+h)$, because the graph of $y = F(x)+h$ is a vertical translation of the graph of $y = F(x)$.

9. (a) If f is even, then $f(x) = f(-x)$. See Figure 9a.

   (b) If f is odd, then $f(-x) = -f(-x)$. See Figure 9b.

x	$f(x)$
-3	4
-2	-6
-1	5
1	5
2	-6
3	4

Figure 9a

x	$f(x)$
-3	4
-2	-6
-1	5
1	-5
2	6
3	-4

Figure 9b

10. (a) $R(x) = 5(7) + 2 = 37$, In 2011, Google's ad revenues were \$37 billion.

    (b) Using the point (2004, 2) and the slope of 5 with the point slope formula we will have $y - 2 = 5(x - 2004) \Rightarrow y = 5(x - 2004) + 2$.

    (c) $y = 5(2011 - 2004) + 2 = 5(7) + 2 = 37$, In 2011, Google's ad revenues were \$37 billion.

    (d) $27 = 5(x - 2004) + 2 \Rightarrow 25 = 5(x - 2004) \Rightarrow 5 = x - 2004 \Rightarrow x = 2009$

2.4: Absolute Value Functions

1. We reflect the graph of $y = f(x)$ across the x-axis for all points for which $y < 0$. Where $y \geq 0$, the graph remains unchanged. See Figure 1.

3. We reflect the graph of $y = f(x)$ across the x-axis for all points for which $y < 0$. Where $y \geq 0$, the graph remains unchanged. See Figure 3.

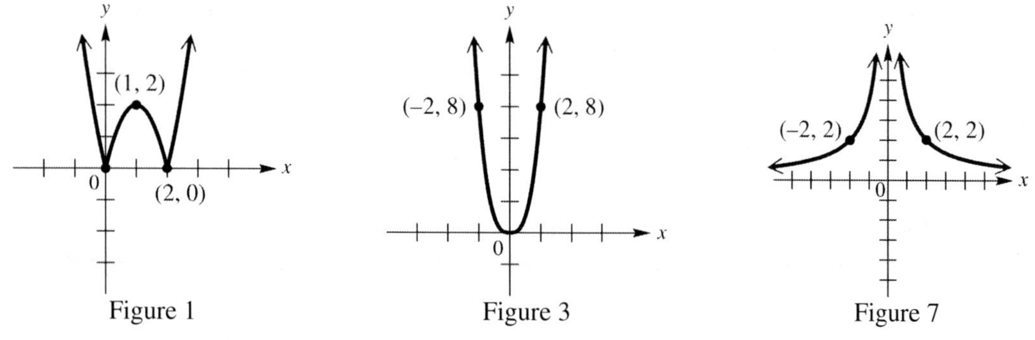

Figure 1 Figure 3 Figure 7

5. Since for all y, $y \geq 0$, the graph remains unchanged. That is, $y = |f(x)|$ has the same graph as $y = f(x)$.

7. We reflect the graph of $y = f(x)$ across the x-axis for all points for which $y < 0$. Where $y \geq 0$, the graph remains unchanged. See Figure 7.

9. We reflect the graph of $y = f(x)$ across the x-axis for all points for which $y < 0$. Where $y \geq 0$, the graph remains unchanged. See Figure 9.

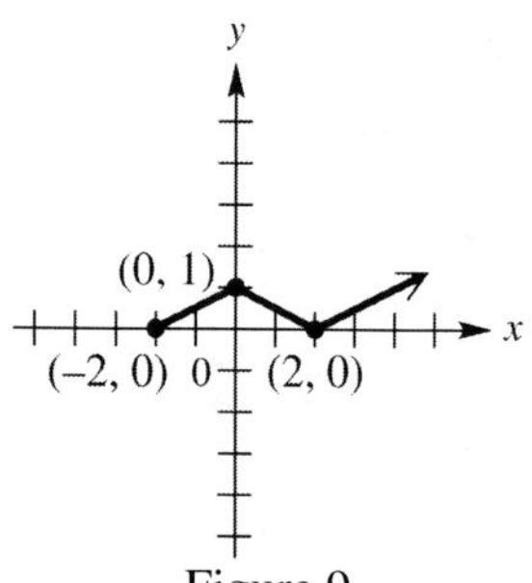

Figure 9

11. If $f(a) = -5$, then $|f(a)| = |-5| = 5$.

13. If $f(x) = -x^2$, then $y = |f(x)| \Rightarrow y = |-x^2| \Rightarrow y = x^2$. Therefore, the range of $y = |f(x)|$ is $[0, \infty)$.

15. If the range of $y = f(x)$ is $(-\infty, -2]$, the range of $y = |f(x)|$ is $[2, \infty)$ since all negative values of y are reflected across the x-axis.

17. From the graph of $y = (x+1)^2 - 2$ the domain of $f(x)$ is $(-\infty, \infty)$, and the range is $[-2, \infty)$.

    From the graph of $y = |(x+1)^2 - 2|$ the domain of $|f(x)|$ is $(-\infty, \infty)$, and the range is $[0, \infty)$.

19. From the graph of $y = -1 - (x-2)^2$ the domain of $f(x)$ is $(-\infty, \infty)$, and the range is $(-\infty, -1]$. From the graph of $y = |-1 - (x-2)^2|$ the domain of $|f(x)|$ is $(-\infty, \infty)$, and the range is $[1, \infty)$.

21. From the graph, the domain of $f(x)$ is $[-2, 3]$, and the range is $[-2, 3]$. For the function $y = |f(x)|$, we reflect the graph of $y = f(x)$ across the x-axis for all points for which $y < 0$, and, where $y \geq 0$, the graph remains unchanged. Therefore, the domain of $y = |f(x)|$ is $[-2, 3]$, and the range is $[0, 3]$.

23. From the graph, the domain of $f(x)$ is $[-2, 3]$, and the range is $[-3, 1]$. For the function $y = |f(x)|$, we reflect the graph of $y = f(x)$ across the x-axis for all points for which $y < 0$, and, where $y \geq 0$, the graph remains unchanged. Therefore, the domain of $y = |f(x)|$ is $[-2, 3]$, and the range is $[0, 3]$.

25. (a) The function $y = f(-x)$ is the function $y = f(x)$ reflected across the y-axis. See Figure 25a.

    (b) The function $y = -f(-x)$ is the function $y = f(x)$ reflected across both the x-axis and y-axis. See Figure 25b.

    (c) For the function $y = |-f(-x)|$ we reflect the graph of $y = -f(-x)$ (ex. b) across the x-axis for all points for which $y < 0$, and where $y \geq 0$, the graph remains unchanged. See Figure 25c.

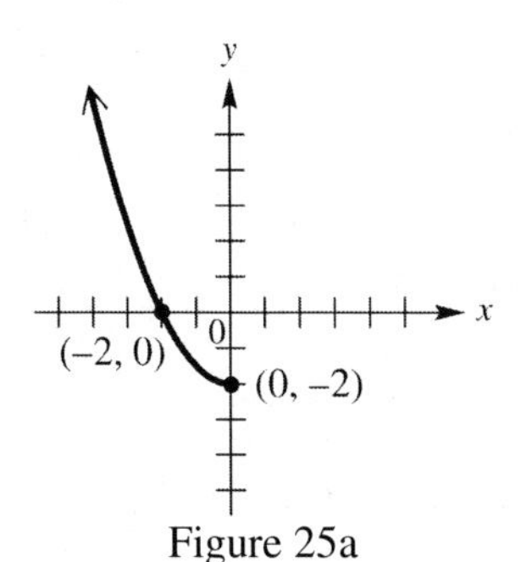

Figure 25a

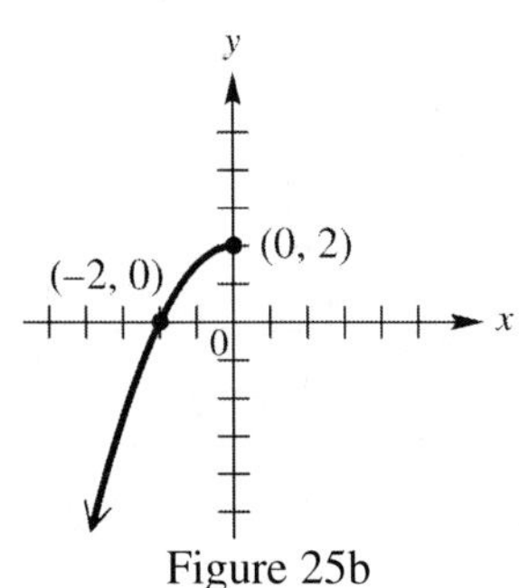

Figure 25b

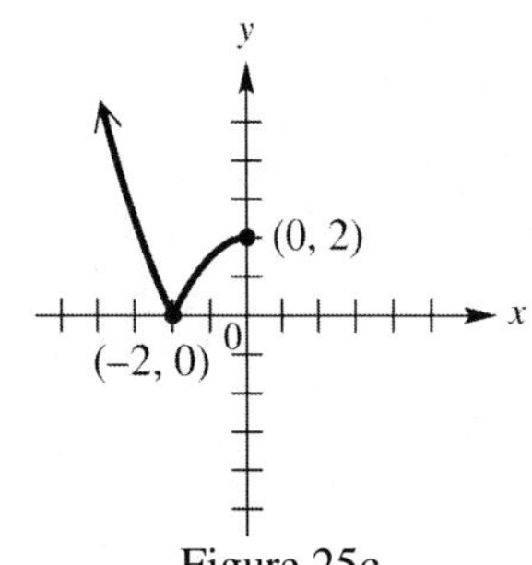

Figure 25c

27. The graph of $y = |f(x)|$ can not be below the x-axis; therefore, Figure A shows the graph of $y = f(x)$, while Figure B shows the graph of $y = |f(x)|$.

29. (a) From the graph, $y_1 = y_2$ at the coordinates $(-1,5)$ and $(6,5)$; therefore, the solution set is $\{-1,6\}$.

    (b) From the graph, $y_1 < y_2$ for the interval $(-1,6)$.

    (c) From the graph, $y_1 > y_2$ for the intervals $(-\infty,-1)\cup(6,\infty)$.

31 (a) From the graph, $y_1 = y_2$ at the coordinate $(4,1)$; therefore, the solution set is $\{4\}$.

    (b) From the graph, $y_1 < y_2$ never occurs; therefore, the solution set is $\varnothing$.

    (c) From the graph, $y_1 > y_2$ for all values for x except 4; therefore, the solution set is the intervals $(-\infty,4)\cup(4,\infty)$.

33. The V-shaped graph is that of $f(x) = |.5x+6|$, since this is typical of the graphs of absolute value functions of the form $f(x) = |ax+b|$

34. The straight line graph is that of $g(x) = 3x-14$ which is a linear function.

35. The graphs intersect at $(8,10)$, so the solution set is $\{8\}$.

36. From the graph, $f(x) > g(x)$ for the interval $(-\infty,8)$.

37. From the graph, $f(x) < g(x)$ for the interval $(8,\infty)$.

38. If $|.5x+6|-(3x-14)=0$ then $|.5x+6| = 3x-14$. Therefore, the solution is the intersection of the graphs, or $\{8\}$.

39. (a) $|x+4| = 9 \Rightarrow x+4=9$ or $x+4=-9 \Rightarrow x=5$ or $x=-13$. The solution set is $\{-13,5\}$, which is supported by the graphs of $y_1 = |x+4|$ and $y_2 = 9$.

    (b) $|x+4| > 9 \Rightarrow x+4>9$ or $x+4<-9 \Rightarrow x>5$ or $x<-13$. The solution is $(-\infty,-13) \cup (5,\infty)$, which is supported by the graphs of $y_1 = |x+4|$ and $y_2 = 9$.

    (c) $|x+4| < 9 \Rightarrow -9 < x+4 < 9 \Rightarrow -13 < x < 5$. The solution is $(-13,5)$, which is supported by the graphs of $|x+4|$ and $y_2 = 9$.

41. (a) $|7-2x|=3 \Rightarrow 7-2x=3$ or $7-2x=-3 \Rightarrow -2x=-4$ or $-2x=-10 \Rightarrow x=2$ or $x=5$. The solution set is $\{2,5\}$, which is supported by the graphs of $y_1=|7-2x|$ and $y_2=3$.

(b) $|7-2x|\ge 3 \Rightarrow 7-2x\ge 3$ or $7-2x\le -3 \Rightarrow -2x\ge -4$ or $-2x\le -10 \Rightarrow x\le 2$ or $x\ge 5$. The solution set is $(-\infty,2]\cup[5,\infty)$, which is supported by the graphs of $y_1=|7-2x|$ and $y_2=3$.

(c) $|7-2x|\le 3 \Rightarrow -3\le 7-2x\le 3 \Rightarrow -10\le -2x\le -4 \Rightarrow 5\ge x\ge 2$ or $2\le x\le 5$. The solution is $[2,5]$, which is supported by the graphs of $y_1=|7-2x|$ and $y_2=3$.

43. (a) $|2x+1|+3=5 \Rightarrow 2x+1=2$ or $2x+1=-2 \Rightarrow 2x=1$ or $2x=-3 \Rightarrow x=\frac{1}{2}$ or $x=-\frac{3}{2}$. The solution set is $\left\{-\frac{3}{2},\frac{1}{2}\right\}$, which is supported by the graphs of $y_1=|2x+1|+3$ and $y_2=5$.

(b) $|2x+1|+3\le 5 \Rightarrow -2\le 2x+1\le 2 \Rightarrow -3\le 2x\le 1 \Rightarrow -\frac{3}{2}\le x\le \frac{1}{2}$. The solution is $\left[-\frac{3}{2},\frac{1}{2}\right]$, which is supported by the graphs of $y_1=|2x+1|+3$ and $y_2=5$.

(c) $|2x+1|+3\ge 5 \Rightarrow 2x+1\ge 2$ or $2x+1\le -2 \Rightarrow 2x\ge 1$ or $2x\le -3 \Rightarrow x\ge \frac{1}{2}$ or $x\le -\frac{3}{2}$. The solution is $\left(-\infty,-\frac{3}{2}\right]\cup\left[\frac{1}{2},\infty\right)$, which is supported by the graphs of $y_1=|2x+1|+3$ and $y_2=5$.

45. (a) $|5-7x|=0 \Rightarrow 5-7x=0 \Rightarrow 7x=5 \Rightarrow x=\frac{5}{7}$. The solution set is $\left\{\frac{5}{7}\right\}$, which is supported by the graphs of $y_1=|5-7x|$ and $y_2=0$.

(b) $|5-7x|\ge 0 \Rightarrow 5-7x\ge 0$ or $5-7x\le 0 \Rightarrow 7x\ge 5$ or $7x\le 5 \Rightarrow x\ge \frac{5}{7}$ or $x\le \frac{5}{7}$. The solution is $(-\infty,\infty)$, which is supported by the graphs of $y_1=|5-7x|$ and $y_2=0$.

(c) $|5-7x|\le 0 \Rightarrow 0\le 5-7x\le 0 \Rightarrow 5\ge 7x\ge 5 \Rightarrow \frac{5}{7}\ge x\ge \frac{5}{7}$. The solution set is $\left\{\frac{5}{7}\right\}$, which is supported by the graphs of $y_1=|5-7x|$ and $y_2=0$.

47. (a) Absolute value is always positive; therefore, the solution set is $\varnothing$, which is supported by the graphs of $y_1=\left|\sqrt{2x}-3.6\right|$ and $y_2=-1$.

(b) Absolute value is always positive, and so cannot be less than or equal to -1; therefore, the solution set is $\varnothing$, which is supported by the graphs of $y_1=\left|\sqrt{2x}-3.6\right|$ and $y_2=-1$.

(c) Absolute value is always positive, and so is always greater than -1; therefore, the solution is $(-\infty,\infty)$, which is supported by the graphs of $y_1=\left|\sqrt{2x}-3.6\right|$ and $y_2=-1$.

49. $3|4-3x|-4=8 \Rightarrow 3|4-3x|=12 \Rightarrow |4-3x|=4 \Rightarrow 4-3x=4$ or $4-3x=-4 \Rightarrow -3x=0$ or

$-3x=-8 \Rightarrow x=0$ or $x=\frac{8}{3}$. Therefore, the solution set is $\left\{0, \frac{8}{3}\right\}$.

51. $\frac{1}{2}\left|-2x+\frac{1}{2}\right|=\frac{3}{4} \Rightarrow \left|-2x+\frac{1}{2}\right|=\frac{3}{2} \Rightarrow -2x+\frac{1}{2}=\frac{3}{2}$ or $-2x+\frac{1}{2}=-\frac{3}{2} \Rightarrow -2x=1$ or

$-2x=-2 \Rightarrow x=-\frac{1}{2}$ or $x=1$. Therefore, the solution set is $\left\{-\frac{1}{2}, 1\right\}$.

53. $4.2|.5-x|+1=3.1 \Rightarrow 4.2|.5-x|=2.1 \Rightarrow |.5-x|=.5 \Rightarrow .5-x=.5$ or $.5-x=-.5 \Rightarrow -x=0$ or

$-x=-1 \Rightarrow x=0$ or $x=1$. Therefore, the solution set is $\{0,1\}$.

55. $|15-x|<7 \Rightarrow -7<15-x<7 \Rightarrow 22>x>8$ or $8<x<22$. Therefore, the solution is $(8,22)$.

57. $|2x-3|>1 \Rightarrow 2x-3>1$ or $2x-3<-1 \Rightarrow 2x>4$ or $\left(\frac{f}{g}\right)(-3)=\frac{-3(-3)-4}{(-3)^2}=\frac{5}{9}$. or $x<1$. Therefore,

the solution is $(-\infty,1)\cup(2,\infty)$.

59. $|-3x+8|\geq 3 \Rightarrow -3x+8\geq 3$ or $-3x+8\leq -3 \Rightarrow -3x\geq -5$ or $-3x\leq -11 \Rightarrow x\leq \frac{5}{3}$ or $x\geq \frac{11}{3}$.

Therefore, the solution is $\left(-\infty, \frac{5}{3}\right]\cup\left[\frac{11}{3}, \infty\right)$.

61. $\left|6-\frac{1}{3}x\right|>0 \Rightarrow 6-\frac{1}{3}x>0$ or $6-\frac{1}{3}x<0 \Rightarrow -\frac{1}{3}x>-6$ or $-\frac{1}{3}x<-6 \Rightarrow x<18$ or $x>18$. Therefore,

the solution is every real number except 18: $(-\infty,18)\cup(18,\infty)$.

63. Absolute value is always positive, and so cannot be less than or equal to -6; therefore, the solution set is $\varnothing$.

65. Absolute value is always positive, and so is always greater than -5; therefore, the solution is $(-\infty,\infty)$.

67. (a) $3x+1=2x-7 \Rightarrow x+1=-7 \Rightarrow x=-8$ $3x+1=-(2x-7) \Rightarrow 3x+1=-2x+7 \Rightarrow 5x=6 \Rightarrow x=\frac{6}{5}$.

Therefore, the solution set is $\left\{-8, \frac{6}{5}\right\}$.

(b) Graph $y_1=|3x+1|$ and $y_2=|2x-7|$. See Figure 67. From the graph, $|f(x)|>|g(x)|$ when $y_1>y_2$

which is for the interval $(-\infty,-8)\cup\left(\frac{6}{5},\infty\right)$.

(c) Graph $y_1=|3x+1|$ and $y_2=|2x-7|$ See Figure 67. From the graph, $|f(x)|<|g(x)|$ when $y_1<y_2$

which is for the interval $\left(-8, \frac{6}{5}\right)$.

[-20,20] by [-10,50]
Xscl = 2 Yscl = 5

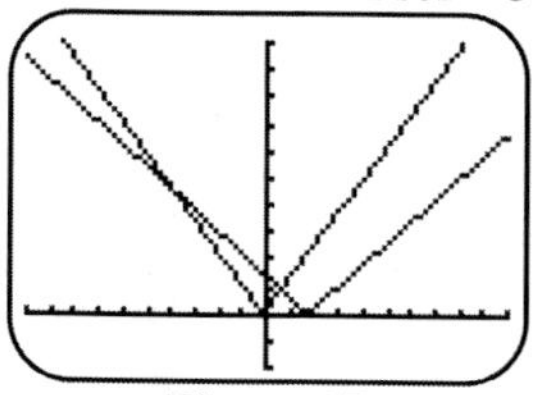

Figure 67

[-10,10] by [-4,16]
Xscl = 1 Yscl = 1

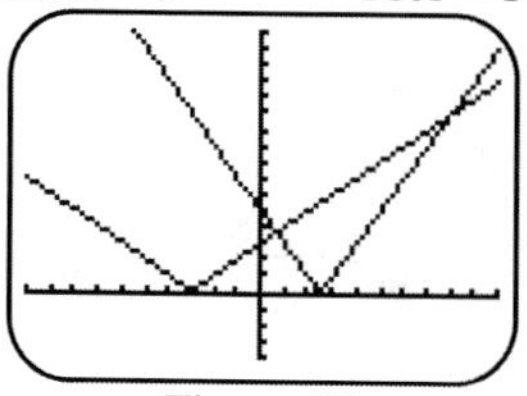

Figure 69

69. (a) $-2x+5=x+3 \Rightarrow -3x=-2 \Rightarrow x=\frac{2}{3}$ or $-2x+5=-(x+3) \Rightarrow -2x+5=-x-3 \Rightarrow$

$-x=-8 \Rightarrow x=8$. Therefore, the solution set is $\left\{\frac{2}{3}, 8\right\}$.

(b) Graph $y_1=|-2x+5|$ and $y_2=|x+3|$. See Figure 69. From the graph, $|f(x)|>|g(x)|$ when $y_1>y_2$, which it is for the interval $\left(-\infty, \frac{2}{3}\right) \cup (8, \infty)$.

(c) Graph $y_1=|-2x+5|$ and $y_2=|x+3|$. See Figure 69. From the graph, $|f(x)|<|g(x)|$ when $y_1<y_2$, which it is for the interval $\left(\frac{2}{3}, 8\right)$.

[-6,6] by [-2,10]
Xscl = 1 Yscl = 1

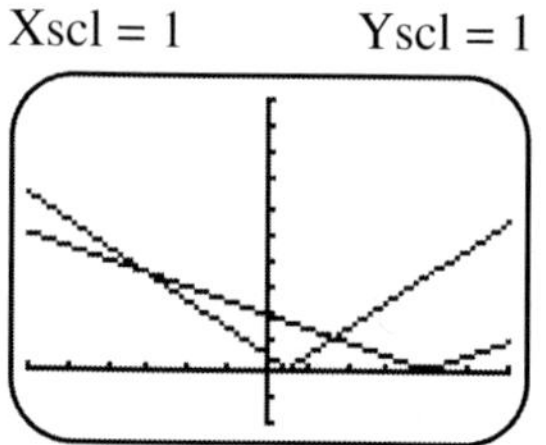

Figure 71

[-10,10] by [-4,16]
Xscl = 1 Yscl = 2

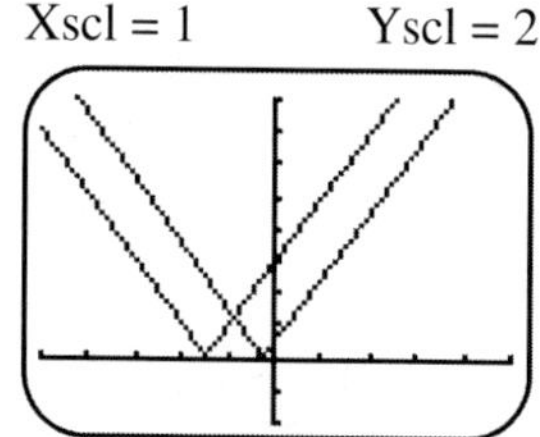

Figure 73

71. (a) $x-\frac{1}{2}=\frac{1}{2}x-2 \Rightarrow \frac{1}{2}x=-\frac{3}{2} \Rightarrow x=-3$ or $x-\frac{1}{2}=-\left(\frac{1}{2}x-2\right) \Rightarrow x-\frac{1}{2}=-\frac{1}{2}x+2 \Rightarrow$

$\frac{3}{2}x=\frac{5}{2} \Rightarrow x=\frac{5}{3}$. Therefore, the solution set is $\left\{-3, \frac{5}{3}\right\}$.

(b) Graph $y_1=\left|x-\frac{1}{2}\right|$ and $y_2=\left|\frac{1}{2}x-2\right|$. From the graph, $|f(x)|>|g(x)|$ when $y_1>y_2$, which it is for the interval $(-\infty,-3) \cup \left(\frac{5}{3}, \infty\right)$.

(c) Graph $y_1=\left|x-\frac{1}{2}\right|$ and $y_2=\left|\frac{1}{2}x-2\right|$. See Figure 71. From the graph $|f(x)|<|g(x)|$ when $y_1<y_2$, which it is for the interval $\left(-3, \frac{5}{3}\right)$.

73. (a) $4x+1=4x+6 \Rightarrow 1=6 \Rightarrow \varnothing$ or $4x+1=-(4x+6) \Rightarrow 4x=1=-4x-6 \Rightarrow 8x=-7 \Rightarrow -\frac{7}{8}$.

Therefore, the solution set is $\left\{-\frac{7}{8}\right\}$.

(b). Graph $y_1=|4x+1|$ or $y_2=|4x+6|$. See Figure 73. From the graph $|f(x)|>|g(x)|$ when $y_1>y_2$, which it is for the interval $\left(-\infty,\frac{7}{8}\right)$.

(c) Graph $y_1=|4x+1|$ or $y_2=|4x+6|$. See Figure 73. From the graph, $|f(x)|<|g(x)|$ when $y_1>y_2$, which is for the interval $\left(-\frac{7}{8},\infty\right)$.

75. (a) $0.25x+1=0.75x-3 \Rightarrow -0.50x=-4 \Rightarrow x=8$ or $0.25x+1=-(0.75x-3) \Rightarrow 0.25x+1=-0.75x+3 \Rightarrow$ $x=2$. Therefore, the solution set is $\{2,8\}$.

(b) Graph $y_1=|.25x+1|$ and $y_2=|.75x-3|$. See Figure 75. From the graph, $|f(x)|>|g(x)|$ when $y_1<y_2$, which it is for the interval $(2,8)$.

(c) Graph $y_1=|.25x+1|$ and $y_2=|.75x-3|$. See Figure 75. From the graph, $|f(x)|<|g(x)|$ when $y_1<y_2$, which it is for the interval $(-\infty,2)\cup(8,\infty)$.

[-20,20] by [-4,16]
Xscl = 2 Yscl = 1

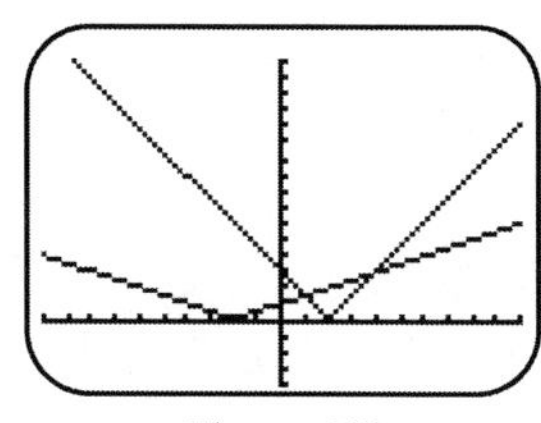

Figure 75

[-10,10] by [-10,10]
Xscl= 1 Yscl = 1

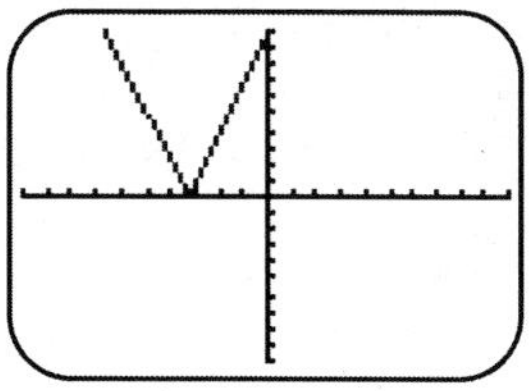

Figure 77

77. (a) $3x+10=-(-3x-10) \Rightarrow 3x+10=3x+10 \Rightarrow$ there are an infinite number of solutions.

Therefore, the solution set is $(-\infty,\infty)$.

(b) Graph $y_1=|3x+10|$ and $y_2=|-3x-10|$. See Figure 77. From the graph, $|f(x)|>|g(x)|$ when $y_1>y_2$, for which there is no solution.

(c) Graph $y_1=|3x+10|$ and $y_2=|-3x-10|$. See Figure 77. From the graph, $|f(x)|<|g(x)|$ when $y_1<y_2$, for which there is no solution.

79. Graph $y_1=|x+1|+|x-6|$ and $y_2=11$. See Figure 79. From the graph, the lines intersect at $(-3,11)$ and $(2,9)$. Therefore, the solution set is $\{-3,8\}$.

81. Graph $y_1=|x|+|x-4|$ and $y_2=8$. See Figure 81. From the graph, the lines intersect at $(-2,8)$ and $(6,8)$. Therefore, the solution set is $\{-2,6\}$.

[-10,10] by [-4,16]

Xscl = 1 Yscl = 1

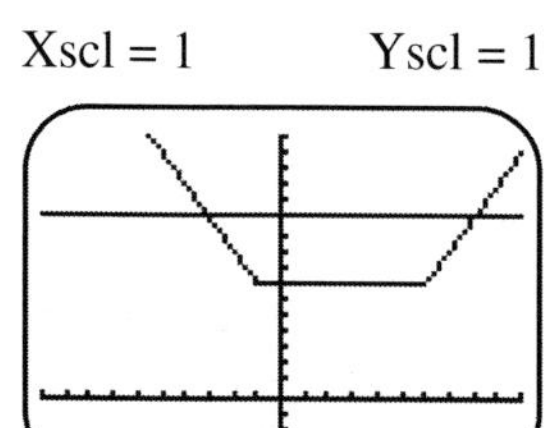

Figure 79

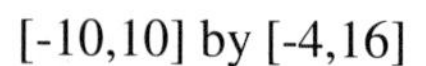

Xscl= 1 Yscl = 1

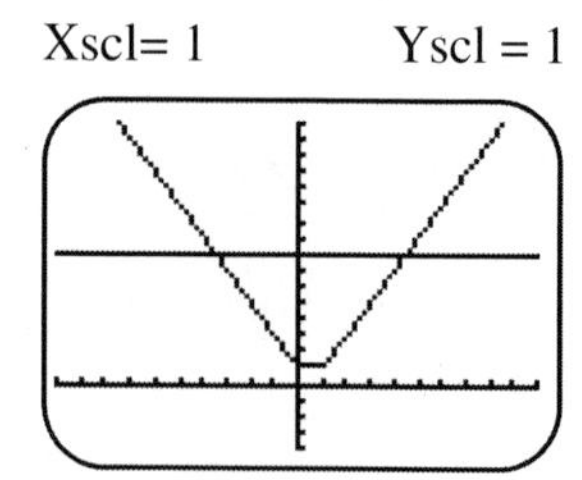

Figure 81

83. (a) $|T-50|\le 22 \Rightarrow -22\le T-50\le 22 \Rightarrow 28\le T\le 72.$

    (b) The average monthly temperatures in Boston vary between a low of 28° F and a high of 72° F. The monthly averages are always within 22° of 50° F.

85. (a) $|T-61.5|\le 12.5 \Rightarrow -12.5\le T-61.5\le 12.5 \Rightarrow 49\le T\le 74..$

    (b) The average monthly temperatures in Buenos Aires vary between a low of 49° F (possibly in July) and a high of 74° F (possibly in January). The monthly averages are always within 12.5° of 61.5° F.

87. $|x-8.0|\le 1.5 \Rightarrow -1.5\le x-8.0\le 1.5 \Rightarrow 6.5\le x\le 9.5$; therefore, the range is the interval $[6.5,9.5]$.

89. (a) $P_d=|116-125| \Rightarrow P_d=|-9|=9$.

    (b) $17=|P-130| \Rightarrow P-130=17$ or $P-130=-17 \Rightarrow P=147$ or $\text{P}=113$.

91. If the difference between y and 1 is less than .1, then $|y-1|<.1 \Rightarrow |2x+1-1|<.1 \Rightarrow$ $|2x|<.1 \Rightarrow -.1<2x<.1 \Rightarrow -.05<x<.05$. The open interval of x is $(-.05,.05)$.

93. If the difference between y and 3 is less than .001, then $|y-3|<.001 \Rightarrow |4x-8-3|<.001 \Rightarrow$ $|4x-11|<.001 \Rightarrow -.001<4x-11<.001 \Rightarrow 10.999<4x<11.001 \Rightarrow 2.74975<x<2.75025$. The open interval of x is $(2.74975,2.75025)$.

95. If $|2x+7|=6x-1$ then $|2x+7|-(6x-1)=0$. Graph $y_1=|2x+7|-(6x-1)$, See Figure 95. The x-intercept is 2; therefore, the solution set is $\{2\}$.

97. If $|x-4|>.5x-6$ then $|x-4|-(.5x-6)>0$. Graph $y_1=|x-4|-(.5x-6)$, See Figure 97. The equation is >0, or the graph is above the x-axis, for the interval: $(-\infty,\infty)$.

[-10,10] by [-10,10]

Xscl = 1 Yscl = 1

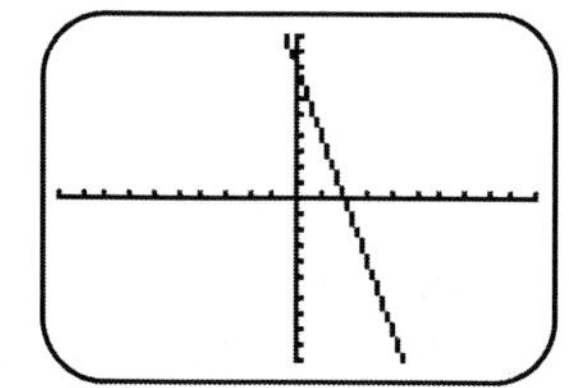

Figure 95

[-10,10] by [-10,10]

Xscl = 1 Yscl = 1

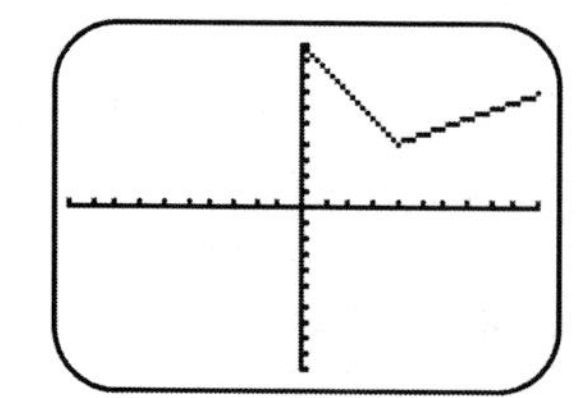

Figure 97

[-10,10] by [-1,6]

Xscl= 1 Yscl = 1

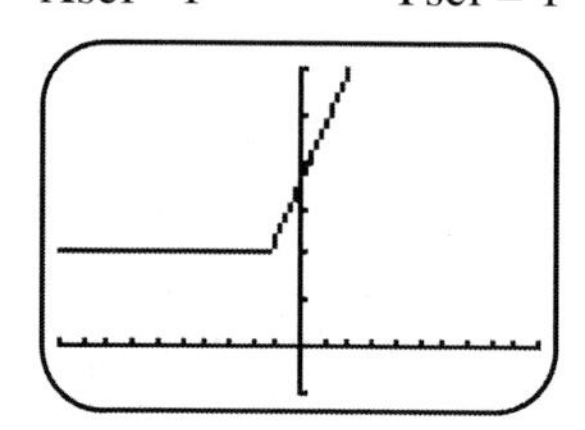

Figure 99

99. If $|3x+4| < -3x-14$ then $|3x+4|-(-3x-14) < 0$. Graph $y_1 = |3x+4|-(-3x-14)$, See Figure 99. The equation is < 0, or the graph is below the x-axis, never or for the solution set: $\varnothing$.

2.5: Piecewise-Defined Functions

1. (a) From the graph, the speed limit is 40 mph.
   (b) From the graph, the speed limit is 30 mph for 6 miles.
   (c) From the graph, $f(5) = 40$ mph; $f(13) = 30$ mph; and $f(19) = 55$ mph.
   (d) From the graph, the graph is discontinuous at $x = 4, 6, 8, 12, \text{and } 16$. The speed limit changes at each discontinuity.
3. (a) From the graph, the Initial amount was: 50,000 gal.; and the final amount was: 30,000 gal.
   (b) From the graph, during the first and fourth days.
   (c) From the graph, $f(2) = 45{,}000$ gal; $f(4) = 40{,}000$ gal.
   (d) From the graph, between days 1 and 3 the water dropped: $\frac{50{,}000-40{,}000}{2} = \frac{10{,}000}{2} = 5{,}000$ gal./day.
5. (a) $f(-5) = 2(-5) = -10$ (b) $f(-1) = 2(-1) = -2$
   (c) $f(0) = 0-1 = -1$ (d) $f(3) = 3-1 = 2$
7. (a) $f(-5) = 2+(-5) = -3$ (b) $f(-1) = -(-1) = 1$
   (c) $f(0) = -(0) = 0$ (d) $f(3) = 3(3) = 9$
9. Yes, continuous. See Figure 9.
11. Not continuous. See Figure 11

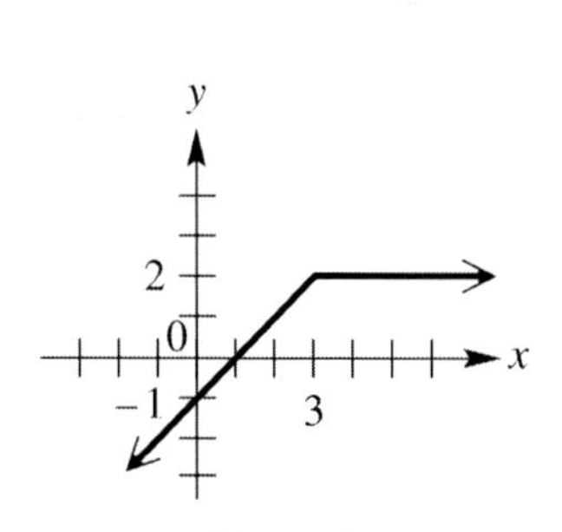

Figure 9

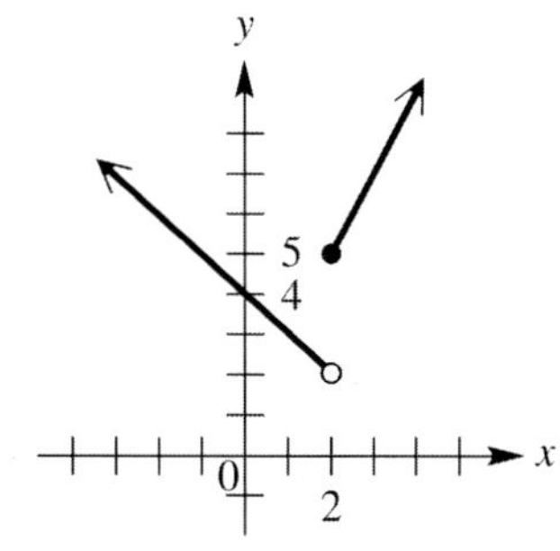

Figure 11

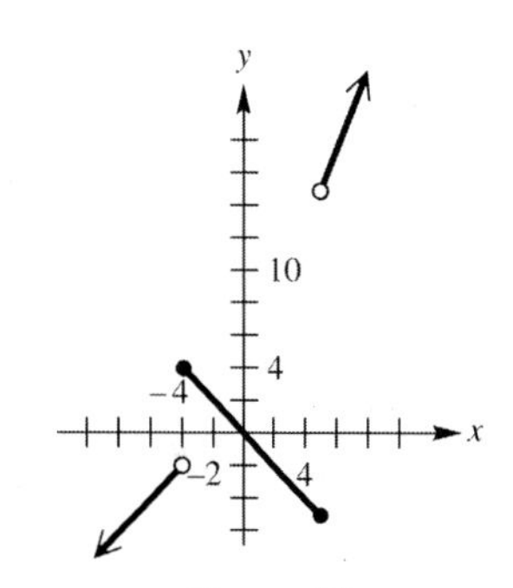

Figure 13

13. Not continuous. See Figure 13.
15. Not continuous. See Figure 15.
17. Yes, continuous. See Figure 17.

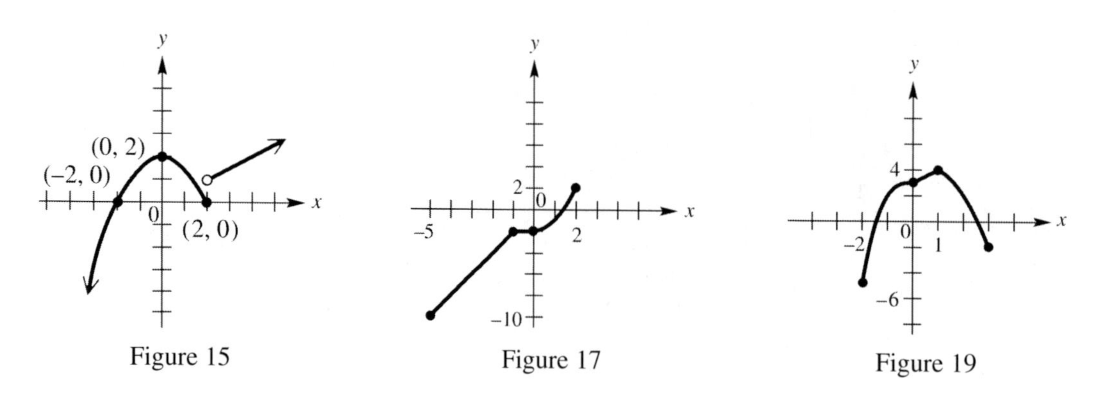

Figure 15 Figure 17 Figure 19

19. Yes, continuous. See Figure 19.

21. Look for a $y = x^2$ graph if $x \geq 0$; and a linear graph if $x < 0$. Therefore: B.

23. Look for a horizontal graph above the x-axis if $x \geq 0$; and a horizontal graph below the x-axis if $x < 0$. Therefore: D.

25. Graph $y_1 = (x-1)(x \leq 3) + (2)(x > 3)$, See Figure 25.

27. Graph $y_1 = (4-x)(x < 2) + (1+2x)(x \geq 2)$, See Figure 27.

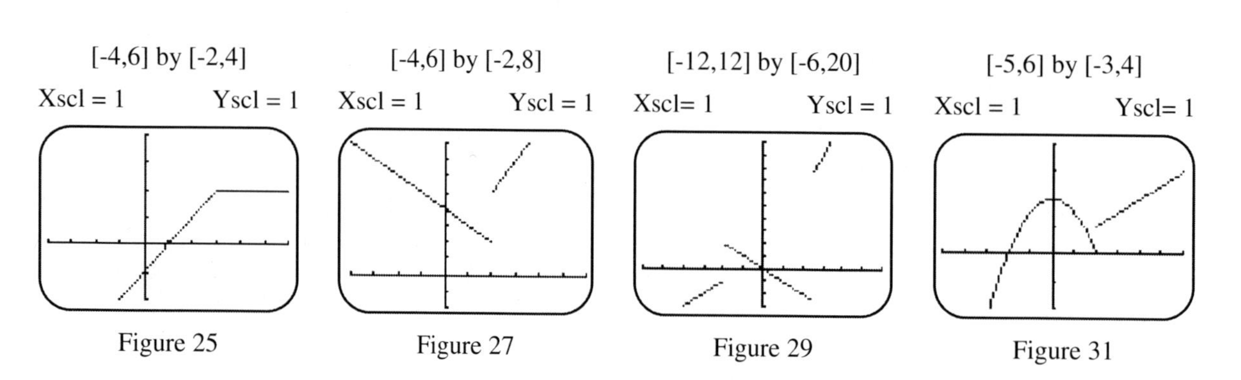

Figure 25 Figure 27 Figure 29 Figure 31

29. Graph $y_1 = (2+x)(x < -4) + (-x)(-4 \leq x \text{ and } x \leq 5) + (3x)(x > 5)$, See Figure 29.

31. Graph $y_1 = \left(-\dfrac{1}{2}x + 2\right)(x \leq 2) + \left(\dfrac{1}{2}x\right)(x > 2)$, See Figure 31.

33. From the graph, the function is $f(x) = \begin{cases} 2 & \text{if } x \leq 0 \\ -1 & \text{if } x > 1 \end{cases}$; domain : $(-\infty, 0] \cup (1, \infty)$; range: $\{-1, 2\}$.

35. From the graph, the function is $f(x) = \begin{cases} x & \text{if } x \leq 0 \\ 2 & \text{if } x > 0 \end{cases}$; domain : $(-\infty, \infty)$; range: $(-\infty, 0] \cup \{2\}$.

37. From the graph, the function is $f(x) = \begin{cases} \sqrt[3]{x} & \text{if } x < 1 \\ x+1 & \text{if } x \geq 1 \end{cases}$; domain : $(-\infty, \infty)$; range : $(-\infty, 1) \cup [2, \infty)$.

39. There is an overlap of intervals since the number 4 satisfies both conditions. To be a function, every x-value is used only once.

41. The graph of $y = [\![x]\!]$ is shifted 1.5 units downward.

43. The graph of $y = [\![x]\!]$ is reflected across the x-axis.

45. Graph $y = [\![x]\!] - 1.5$, See Figure 45.

47. Graph $y = -[\![x]\!]$, See Figure 47.

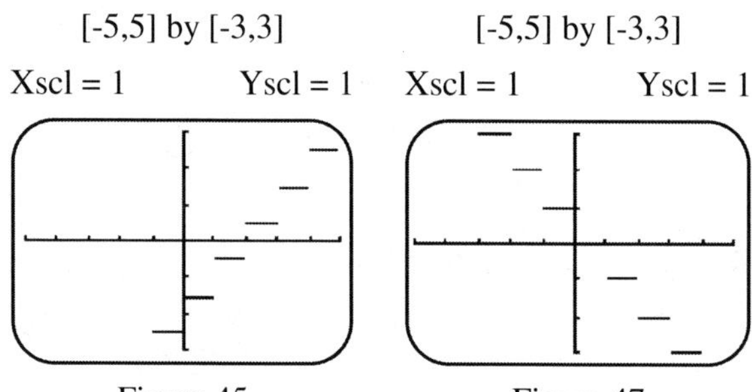

Figure 45 Figure 47

49. When $0 \le x \le 3$ the slope is 5, which means the inlet pipe is open and the outlet pipe is closed; when $3 < x \le 5$ the slope is 2, which means both pipes are open; when $5 < x \le 8$ the slope is 0, which means both pipes are closed; when $8 < x \le 10$ the slope is -3, which means the inlet pipe is closed and the outlet pipe is open.

51. (a) $f(1.5) = 1.12, f(3) = 1.32$, It costs \$1.12 to mail 1.5 oz and \$1.32 to mail 3 oz.

(b) Domain: $(0,5]$; Range: $\{0.92, 1.12, 1.32, 1.52, 1.72\}$. See figure 51.

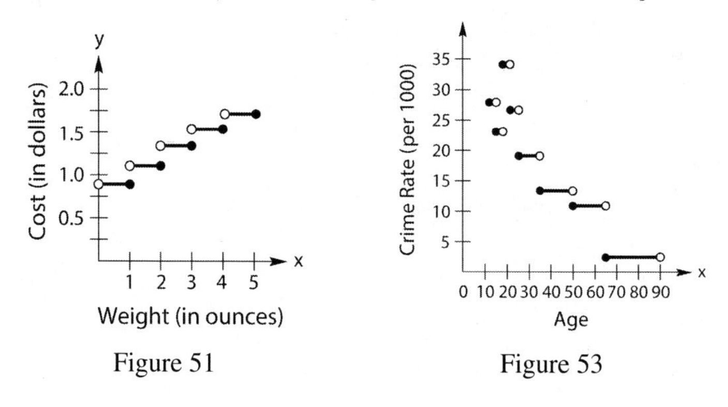

Figure 51 Figure 53

53. (a) From the table, graph the piecewise function. See Figure 53.

(b) The likelihood of being a victim peaks from age 16 up to age 20, then decreases.

55. (a) From the graph, the highest speed is 55 mph and the lowest speed is 30 mph.

(b) There are approximately 12 miles of highway with a speed of 55 mph.

(c) Fromt the graph, $f(4) = 40; f(12) = 30; f(18) = 55$.

57. (a) A 3.5 minute call would round up to 4 minutes. A 4 minute call would cost: $.50 + 3(.25) = \$1.25$.

(b) We use a piecewise defined function where the cost increases after each whole number as follows:

$$f(x)=\begin{cases} .50 & \text{if } 0<x\le 1 \\ .75 & \text{if } 1<x\le 2 \\ 1.00 & \text{if } 2<x\le 3 \\ 1.25 & \text{if } 3<x\le 4 \\ 1.50 & \text{if } 4<x\le 5 \end{cases}.$$ Another possibility is $f(x)=\begin{cases} .50 & \text{if } 0<x\le 1 \\ .50-.25[\![1-x]\!] & \text{if } 1<x\le 5 \end{cases}$.

59. For x in the interval $(0,2]$, $y=25$. For x in the interval $(2,3]$, $y=25+3=28$. For x in the interval $(3,4]$, $y=28+3=31$ and so on. The graph is a step function. In this case, the first step has a different width. See Figure 59.

61. Sketch a piecewise function that fills a tank at a rate of 5 gallons a minute for the first 20 minutes (the time it takes to fill the 100 gallon tank) and then drains the tank at a rate of 2 gallons per minute for 50 minutes (the time it takes to drain the 100 gallon tank). See Figure 61.

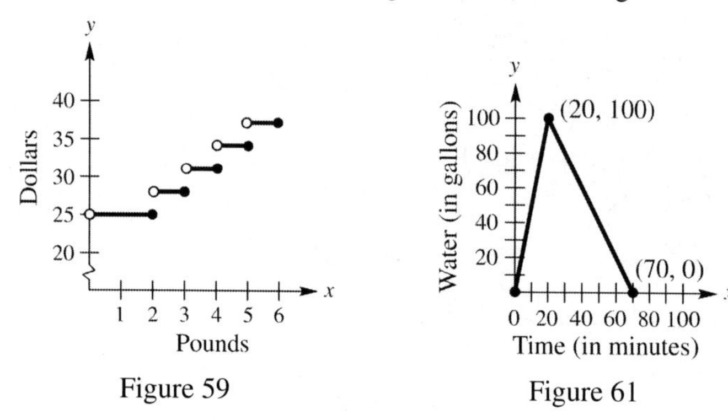

Figure 59 Figure 61

2.6: Operations and Composition

1. $x^2+(2x-5)=x^2+2x-5 \Rightarrow E.$

3. $x^2(2x-5)=2x^3-5x^2 \Rightarrow F.$

5. $(2x-5)^2=4x^2-20x+25 \Rightarrow A.$

7. $(f\circ g)(3)=f(g(3))=f(2(3)-1)=f(5)=5^2+3(5)=40$

9. $(f\circ g)(x)=f(g(x))=f(2x-1)=(2x-1)^2+3(2x-1)=4x^2-4x+1+6x-3=4x^2+2x-2$

11. $(f+g)(3)=f(3)+g(3)=((3)^2+3(3))+(2(3)-1)=23$

13. $(f\cdot g)(4)=f(4)\cdot g(4)=\left((4)^2+3(4)\right)\cdot\left(2(4)-1\right)=196$

15. $\left(\dfrac{f}{g}\right)(-1)=\dfrac{f(-1)}{g(-1)}=\dfrac{(-1)^2+3(-1)}{2(-1)-1}=\dfrac{2}{3}$

17. $(f-g)(2)=f(2)-g(2)=\left((2)^2+3(2)\right)-\left(2(2)-1\right)=7$

19. $(g-f)(-2)=g(-2)-f(-2)=\left(2(-2)-1\right)-\left((-2)^2+3(-2)\right)=-3$

21. $\left(\dfrac{g}{f}\right)(0)=\dfrac{f(0)}{g(0)}=\dfrac{2(0)-1}{(0)^2+3(0)}=\dfrac{-1}{0}\Rightarrow$ undefined

23. (a) $(f+g)(x)=(4x-1)+(6x+3)=10x+2$, $(f-g)(x)=(4x-1)-(6x+3)=-2x-4$

$(fg)(x)=(4x-1)(6x+3)=24x^2+12x-6x-3=24x^2+6x-3$

(b) All values can replace x in all three equations; therefore, the domain is $(-\infty,\infty)$ in all cases.

(c) $\left(\frac{f}{g}\right)(x)=\frac{4x-1}{6x+3}$; all values can replace x, except $-\frac{1}{2}$; therefore, the domain is $\left(-\infty,-\frac{1}{2}\right)\cup\left(-\frac{1}{2},\infty\right)$.

(d) $(f\circ g)(x)=f[g(x)]=4(6x+3)-1=24x+12-1=24x+11$; all values can be input for x; therefore, the domain is $(-\infty,\infty)$

(e) $(g\circ f)(x)=g[f(x)]=6(4x-1)+3=24x-6+3=24x-3$; all values can replace x; therefore, the domain is $(-\infty,\infty)$.

25. (a) $(f+g)(x)=|x+3|+2x$, $(f-g)(x)=|x+3|-2x$, $(fg)(x)=|x+3|(2x)$

(b) All values can replace x in all three equations; therefore, the domain is $(-\infty,\infty)$ in all cases.

(c) $\left(\frac{f}{g}\right)(x)=\frac{|x+3|}{2x}$; all values can replace x, except 0; therefore, the domain is $(-\infty,0)\cup(0,\infty)$

(d) $(f\circ g)(x)=f[g(x)]=|(2x)+3|=|2x+3|$; all values can replace x; therefore, the domain is $(-\infty,\infty)$.

(e) $(g\circ f)(x)=g[f(x)]=2(|x+3|)=2|x+3|$; all values can replace x; therefore, the domain is $(-\infty,\infty)$

27. (a) $(f+g)(x)=\sqrt[3]{x+4}+(x^3+5)=\sqrt[3]{x+4}+x^3+5$, $(f-g)(x)=\sqrt[3]{x+4}-(x^3+5)=\sqrt[3]{x+4}-x^3-5$

$(fg)(x)=\left(\sqrt[3]{x+4}\right)(x^3+5)$

(b) All values can replace x in all three equations; therefore, the domain is $(-\infty,\infty)$ in all cases.

(c) $\left(\frac{f}{g}\right)(x)=\frac{\sqrt[3]{x+4}}{(x^3+5)}$ all values can replace x, except $\sqrt[3]{-5}$, so the domain is $(-\infty,\sqrt[3]{-5})\cup(\sqrt[3]{-5},\infty)$.

(d) $(f\circ g)(x)=f[g(x)]=\sqrt[3]{(x^3+5)+4}=\sqrt[3]{x^3+9}$; all values can replace x, so the domain is $(-\infty,\infty)$.

(e) $(g\circ f)(x)=g[f(x)]=\left(\sqrt[3]{x+4}\right)^3+5=x+4+5=x+9$; all values can replace x, so the domain is $(-\infty,\infty)$.

29. (a) $(f+g)(x)=\sqrt{x^2+3}+(x+1)=\sqrt{x^2+3}+x+1$, $(f-g)(x)=\sqrt{x^2+3}-(x+1)=\sqrt{x^2+3}-x-1$

$(fg)(x)=\left(\sqrt{x^2+3}\right)(x+1)$

(b) All values can replace x in all three equations; therefore, the domain is $(-\infty,\infty)$ in all cases.

(c) $\left(\frac{f}{g}\right)(x)=\frac{\sqrt{x^2+3}}{(x+1)}$; all values can replace x, except -1; therefore, the domain is $(-\infty,-1)\cup(-1,\infty)$.

(d) $(f\circ g)(x)=f[g(x)]=\sqrt{(x+1)^2+3}=\sqrt{x^2+2x+1+3}=\sqrt{x^2+2x+4}$; all values can replace x; therefore, the domain is $(-\infty,\infty)$.

(e) $(g \circ f)(x) = g[f(x)] = \left(\sqrt{x^2+3}\right)+1 = \sqrt{x^2+3}+1$; all values can replace x; therefore, the domain is, $(-\infty, \infty)$.

31. (a) From the graph, $4+(-2)=2$.

(b) From the graph, $1-(-3)=4$.

(c) From the graph, $(0)(-4)=0$.

(d) From the graph, $\frac{1}{-3}=-\frac{1}{3}$.

33. (a) From the graph, $0+3=3$.

(b) From the graph, $-1-4=-5$.

(c) From the graph, $(2)(1)=2$.

(d) From the graph, $\frac{3}{0} \Rightarrow$ undefined.

35. (a) From the table, $7+(-2)=5$.

(b) From the table, $10-5=5$.

(c) From the table, $(0)(6)=0$.

(d) From the table, $\frac{5}{0}=$ undefined.

37. See Table 37.

x	$(f+g)(x)$	$(f-g)(x)$	$(fg)(x)$	$\left(\frac{f}{g}\right)(x)$
-2	6	-6	0	0
0	5	5	0	undefined
2	5	9	-14	-3.5
4	15	5	50	2

Figure 37

39. $M(2004) \approx 260$ and $F(2004) \approx 400 \Rightarrow T(2004) = M(2004) + F(2004) = 260 + 400 = 660$

41. The slopes of the line segments for the period 2000-2004 are much steeper than the slopes of the corresponding line segments for the period 2004-2008. Thus, the number of associate's degrees increased more rapidly during the period 2000-2004.

43. $(T-S)(2000) = T(2000) - S(2000) = 19 - 13 = 6$, This represents the billions of dollars spent for general science in 2000.

45. In space and other technologies spending was almost static in the years 1995-2000.

47. (a) $(f \circ g)(4) = f[g(4)]$, so from the graph find $g(4)=0$. Now find $f(0)=-4$; therefore, $(f \circ g)(4)=-4$.

(b) $(g \circ f)(3) = g[f(3)]$, so from the graph find $f(3)=2$. Now find $g(2)=2$; therefore, $(g \circ f)(3)=2$.

(c) $(f \circ f)(2) = f[f(2)]$, so from the graph find $f(2)=0$. Now find $f(0)=-4$; therefore, $(f \circ f)(2)=-4$.

49. (a) $(f \circ g)(1) = f[g(1)]$, so from the graph find $g(1)=2$. Now find $f(2)=-3$; therefore, $(f \circ g)(1)=-3$.

(b) $(g \circ f)(-2) = g[f(-2)]$, so from the graph find $f(-2) = -3$. Now find $g(-3) = -2$; therefore, $(g \circ f)(-2) = -2$.

(c) $(g \circ g)(-2) = g[g(-2)]$, so from the graph find $g(-2) = -1$. Now find $g(-1) = 0$; therefore, $(g \circ g)(-1) = 0$.

51. (a) $(g \circ f)(1) = g[f(1)]$, so from the table find $f(1) = 4$. Now find $g(4) = 5$; therefore, $(g \circ f)(1) = 5$.

(b) $(f \circ g)(4) = f[g(4)]$, so from the table find $g(4) = 5$. $f(5)$ is undefined; therefore, $(f \circ g)(4)$ is undefined.

(c) $(f \circ f)(3) = f[f(3)]$, so from the table find $f(3) = 1$. Now find $f(1) = 4$; therefore, $(f \circ f)(3) = 4$.

53. From the table, $g(3) = 4$ and $f(4) = 2$.

55. Since $Y_3 = Y_1 \circ Y_2$ and $X = -1$, we solve $Y_1[Y_2(-1)]$. First solve $Y_2 = (-1)^2 = 1$, now solve $Y_1 = 2(1) - 5 = -3$; therefore, $Y_3 = -3$.

57. Since $Y_3 = Y_1 \circ Y_2$ and $X = 7$, we solve $Y_1[Y_2(7)]$. First solve $Y_2 = (7)^2 = 49$, now solve $Y_1 = 2(49) - 5 = 93$; therefore, $Y_3 = 93$.

59. (a) $(f \circ g)(x) = f[g(x)] = (x^2 + 3x - 1)^3$; all values can be input for x; therefore, the domain is $(-\infty, \infty)$.

(b) $(g \circ f)(x) = g[f(x)] = (x^3)^2 + 3(x^3) - 1 = x^6 + 3x^3 - 1$; all values can be input for x; therefore, the domain is $(-\infty, \infty)$.

(c) $(f \circ f)(x) = f[f(x)] = (x^3)^3 = x^9$; all values can be input for x; therefore, the domain is $(-\infty, \infty)$.

61. (a) $(f \circ g)(x) = f[g(x)] = \left(\sqrt{1-x}\right)^2 = 1 - x$; all values can be input for x; therefore, the domain is $(-\infty, \infty)$.

(b) $(g \circ f)(x) = g[f(x)] = \left(\sqrt{1-(x^2)}\right) = \sqrt{1-x^2}$; only values where $x^2 \le 1$ can be input for x; therefore, the domain is $[-1, 1]$.

(c) $(f \circ f)(x) = f[f(x)] = (x^2)^2 = x^4$; all values can be input for x; therefore, the domain is $(-\infty, \infty)$.

63. (a) $(f \circ g)(x) = f[g(x)] = \dfrac{1}{(5x)+1} = \dfrac{1}{5x+1}$; all values can be input for x, except $-\dfrac{1}{5}$; therefore, the domain is $\left(-\infty, -\dfrac{1}{5}\right) \cup \left(-\dfrac{1}{5}, \infty\right)$.

(b) $(g \circ f)(x) = g[f(x)] = 5\left(\dfrac{1}{x+1}\right) = \dfrac{5}{x+1}$; all values can be input for x, except, -1; therefore, the domain is $(-\infty. -1) \cup (-1, \infty)$.

(c) $(f \circ f)(x) = f[f(x)] = \dfrac{1}{\left(\dfrac{1}{x+1}\right)+1} = \dfrac{1}{\dfrac{1}{x+1}+\dfrac{x+1}{x+1}} = \dfrac{1}{\dfrac{x+2}{x+1}} = \dfrac{x+1}{x+2}$; all values can be input for x, except those that make $\dfrac{x+1}{x+2} = 0$ or undefined. That would be -1 and -2; therefore, the domain is $(-\infty,-2)\cup(-2,-1)\cup(-1,\infty)$.

65. (a) $(f \circ g)(x) = f[g(x)] = 2(4x^3 - 5x^2) + 1 = 8x^3 - 10x^2 + 1$; all values can be input for x; therefore, the domain is $(-\infty,\infty)$.

(b) $(g \circ f)(x) = g[f(x)] = 4(2x+1)^3 - 5(2x+1)^2 = 4(8x^3 + 12x^2 + 6x + 1) - 5(4x^2 + 4x + 1) =$ $32x^3 + 48x^2 + 24x + 4 - (20x^2 + 20x + 5) = 32x^3 + 28x^2 + 4x - 1$; all values can be input for x; therefore, the domain is $(-\infty,\infty)$.

(c) $(f \circ f)(x) = f[f(x)] = 2(2x+1) + 1 = 4x + 3$; all values can be input for x; therefore, the domain is $(-\infty,\infty)$.

67. (a) $(f \circ g)(x) = f(g(x)) = f(5) = 5$, all values can be input for x, thefore the domain is $(-\infty,\infty)$.

(b) $(g \circ f)(x) = g(f(x)) = g(5) = 5$, all values can be input for x, therefore the domain is $(-\infty,\infty)$.

(c) $(f \circ f)(x) = f(f(x)) = f(5) = 5$, all values can be input for x, therefore the domain is $(-\infty,\infty)$.

69. $(f \circ g)(x) = f[g(x)] = 4\left(\dfrac{1}{4}(x-2)\right) + 2 = x - 2 + 2 = x$, $(g \circ f)(x) = g[f(x)] = \dfrac{1}{4}((4x+2)-2) = \dfrac{1}{4}(4x) = x$

71. $(f \circ g)(x) = f\left[g(x)\right] = \sqrt[3]{5\left(\dfrac{1}{5}x^3 - \dfrac{4}{5}\right) + 4} = \sqrt[3]{\left(x^3 - 4\right) + 4} = \sqrt[3]{x^3} = x$

$(g \circ f)(x) = g\left[f(x)\right] = \dfrac{1}{5}\left(\sqrt[3]{5x+4}\right)^3 - \dfrac{4}{5} = \dfrac{1}{5}(5x+4) - \dfrac{4}{5} = x + \dfrac{4}{5} - \dfrac{4}{5} = x$

73. Graph $y_1 = \sqrt[3]{x-6}$, $y_2 = x^3 + 6$, and $y_3 = x$ in the same viewing window. See Figures 73. The graph of y_2 can be obtained by *reflecting* the graph of y_1 across the line $y_3 = x$.

[-10,10] by [-10,10]
Xscl = 1 Yscl = 1

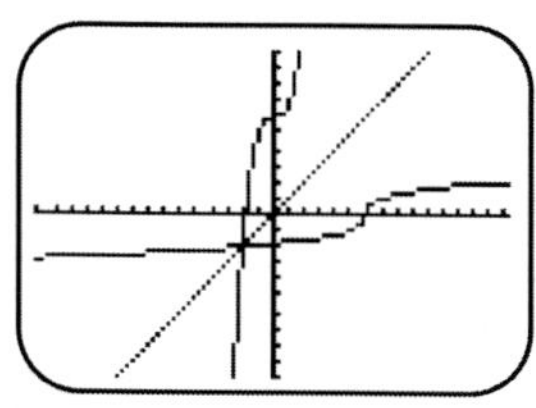

Figure 73

75. $f(x) = x^2 - 4 \Rightarrow f(x+h) = (x+h)^2 - 4 = x^2 + 2xh + h^2 - 4$;

$f(x) + f(h) = \left(x^2 - 4\right) + \left(h^2 - 4\right) = x^2 + h^2 - 8$

77. $f(x)=3x-x^2 \Rightarrow f(x+h)=3(x+h)-(x+h)^2=3x+3h-\left(x^2+2xh+h^2\right)=-x^2-2xh+3x-h^2+3h$;

$f(x)+f(h)=\left(3x-x^2\right)+\left(3h-h^2\right)=-x^2+3x-h^2+3h$

79. Using $\dfrac{f(x+h)-f(x)}{h}$ gives: $\dfrac{4(x+h)+3-(4x+3)}{h}=\dfrac{4x+4h+3-4x-3}{h}=\dfrac{4h}{h}=4.$

81. Using $\dfrac{f(x+h)-f(x)}{h}$ gives: $\dfrac{-6(x+h)^2-(x+h)+4-\left(-6x^2-x+4\right)}{h}$

$=\dfrac{-6\left(x^2+2xh+h^2\right)-x-h+4+6x^2+x-4}{h}=\dfrac{-6x^2-12xh-6h^2-x-h+4+6x^2+x-4}{h}$

$=\dfrac{-12xh-6h^2-h}{h}=-12x-6h-1.$

83. Using $\dfrac{f(x+h)-f(x)}{h}$ gives: $\dfrac{(x+h)^3-x^3}{h}=\dfrac{x^3+3x^2h+3xh^2+h^3-x^3}{h}=\dfrac{3x^2h+3xh^2+h^3}{h}=3x^2+3xh+h^2.$

85. Using $\dfrac{f(x+h)-f(x)}{h}$ gives: $\dfrac{1-(x+h)^2-(1-x^2)}{h}=\dfrac{1-\left(x^2+2xh+h^2\right)-1+x^2}{h}=$

$\dfrac{1-x^2-2xh-h^2-1+x^2}{h}=\dfrac{-2xh-h^2}{h}=-2x-h$

87. Using $\dfrac{f(x+h)-f(x)}{h}$ gives: $\dfrac{3(x+h)^2-(3x^2)}{h}=\dfrac{3\left(x^2+2xh+h^2\right)-3x^2}{h}=$

$\dfrac{3x^2+6xh+3h^2-3x^2}{h}=\dfrac{6xh+3h^2}{h}=6x+3h$

89. Using $\dfrac{f(x+h)-f(x)}{h}$ gives: $\dfrac{\frac{1}{2(x+h)}-\frac{1}{2x}}{h}=\dfrac{\frac{2x-(2x+2h)}{2x(2x+2h)}}{h}=\dfrac{\frac{-2h}{2x(2x+2h)}}{h}=\dfrac{-1}{2x(x+h)}$

91. One possible solution is $f(x)=x^2$ and $g(x)=6x-2$. Then $(f\circ g)(x)=f[g(x)]=(6x-2)^2$.

93. One possible solution is $f(x)=\sqrt{x}$ and $g(x)=x^2-1$. Then $(f\circ g)(x)=f[g(x)]=\sqrt{x^2-1}$.

95. One possible solution is $f(x)=\sqrt{x}+12$ and $g(x)=6x$. Then $(f\circ g)(x)=f[g(x)]=\sqrt{6x}+12$.

97. (a) With a cost of \$10 to produce each item and a fixed cost of \$500, the cost function is $C(x)=10x+500$.

(b) With a selling price of \$35 for each item, the revenue function is $R(x)=35x$.

(c) The profit function is $P(x)=R(x)-C(x)\Rightarrow P(x)=35x-(10x+500)\Rightarrow P(x)=25x-500$.

(d) A profit is shown when $P(x)>0\Rightarrow 25x-500>0\Rightarrow 25x>500\Rightarrow x>20$. Therefore, 21 items must be produced and sold to realize a profit.

(e) Graph $y_1=25x-500$. The smallest whole number for which $P(x)>0$ is 21. Use a window of $[0,30]$ by $[-1000,500]$, for example.

99. (a) With a cost of \$100 to produce each item and a fixed cost of \$2700, the cost function is $C(x)=100x+2700$.

(b) With a selling price of \$280 for each item, the revenue function is $R(x)=280x$.

(c) The profit function is $P(x)=R(x)-C(x)\Rightarrow P(x)=280x-(100x+2700)\Rightarrow P(x)=180x-2700$.

(d) A profit is shown when $P(x) > 0 \Rightarrow 180x - 2700 > 0 \Rightarrow 180x > 2700 \Rightarrow x > 15$. Therefore, 16 items must be produced and sold to realize a profit.

(e) Graph $y_1 = 180x - 2700$, the smallest whole number for which $P(x) > 0$ is 16. Use a window of $[0,30]$ by $[-3000,500]$, for example.

101. (a) If $V(r) = \frac{4}{3}\pi r^3$, then a 3 inch increase would be $V(r) = \frac{4}{3}\pi(r+3)^3$, and the volume gained would be $V(r) = \frac{4}{3}\pi(r+3)^3 - \frac{4}{3}\pi r^3$.

(b) Graph $y_1 = \frac{4}{3}\pi(x+3)^3 - \frac{4}{3}\pi x^3$ in the window $[0,10]$ by $[0,1500]$. See Figure 101. Although this appears to be a portion of a parabola, it is actually a cubic function.

(c) From the graph in exercise 91b, an input value of $x = 4$ results in a gain of $y \approx 1168.67$.

(d) $V(4) = \frac{4}{3}\pi(4+3)^3 - \frac{4}{3}\pi(4)^3 = \frac{4}{3}\pi(343) - \frac{4}{3}\pi(64) = \frac{1372}{3}\pi - \frac{256}{3}\pi = \frac{1116}{3}\pi = 372\pi \approx 1168.67.$

103. (a) If $x =$ width, then $2x =$ length. Since the perimeter formula is $P = 2W + 2L$ our perimeter function is $P(x) = 2(x) + 2(2x) = 2x + 4x \Rightarrow P(x) = 6x$. This is a linear function.

(b) Graph $P(x) = 6x$ in the window $[0,10]$ by $[1,100]$. See Figure 103b. From the graph when $x = 4$, $y = 24$. The 4 represents the width of a rectangle and 24 represents the perimeter.

(c) If $x = 4$ is the width of a rectangle then $2x = 8$ is the length. See Figure 103c. Using the standard perimeter formula yields $P = 2(4) + 2(8) = 24$. This compares favorably with the graph result in part b.

(d) (Answers may vary.) If the perimeter y of a rectangle satisfying the given conditions is 36, then the width x is 6. See Figure 103d.

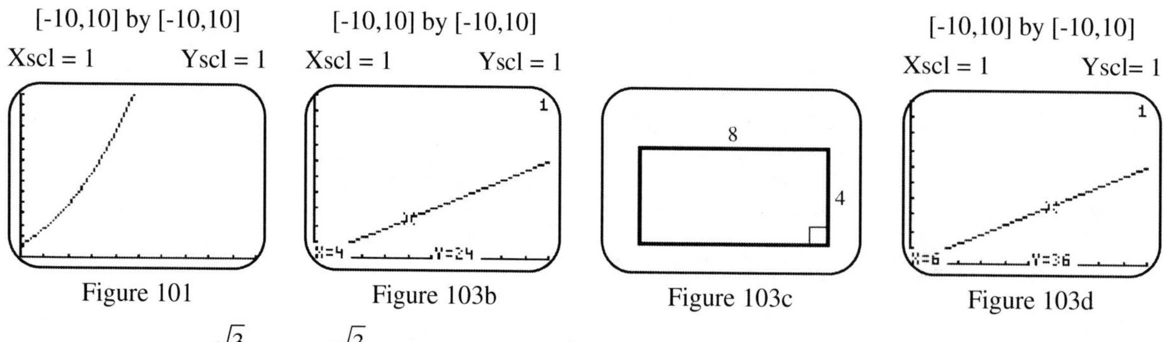

Figure 101 Figure 103b Figure 103c Figure 103d

105. (a) $A(2x) = \frac{\sqrt{3}}{4}(2x)^2 = \frac{\sqrt{3}}{4}(4x^2) \Rightarrow A(2x) = \sqrt{3}x^2$

(b) $A(x) = \frac{\sqrt{3}}{4}(16)^2 = \frac{\sqrt{3}}{4}(256) \Rightarrow A(x) = 64\sqrt{3}$ square units.

(c) On the graph of $y = \frac{\sqrt{3}}{4}x^2$, locate the point where $x = 16$ to find $y \approx 110.85$, an approximation for $64\sqrt{3}$.

107. (a) $A(2100) = 42$, ; The average age of a person in 2100 is projected to be 42 years. $T(2100) = 430$, In 2100, the living world's population will have a combined life experience of 430 billion years.

(b) $\frac{T(2100)}{A(2100)} = \frac{430}{42} \approx 10.2$, The world population will be about 10.2 billion in 2100.

(c) $P(x)$ gives the world's population during year x.

109. (a) $(f+g)(2010) = 13.0 + 74.3 = 87.3$

(b) The function $(f+g)(x)$ computes the total SO_2 and Carbon Monoxide during year x.

(c) Add functions f and g.

x	1970	1980	1990	2000	2010
$(f+g)(x)$	235.2	211.3	177.3	130.8	87.3

111. (a) The function h is the subtraction of function f from g. Therefore, $h(x) = g(x) - f(x)$.

(b) $h(1996) = g(1996) - f(1996) = 841 - 694 = 147$, $h(2006) = g(2006) = f(2006) = 1165 - 1012 = 153$

(c) Using the points (1996, 147) and (2006, 153) from part b, the slope is $m = \frac{153-147}{2006-1996} = \frac{6}{10} = .6$. Now using point slope form: $y - 147 = .6(x - 1996) \Rightarrow y = .6(x - 1996) + 147$.

Reviewing Basic Concepts (Sections 2.4—2.6)

1. (a) $\left|\frac{1}{2}x + 2\right| = 4 \Rightarrow \frac{1}{2}x + 2 = 4 \Rightarrow \frac{1}{2}x = 2 \Rightarrow x = 4$ or $\frac{1}{2}x + 2 = -4 \Rightarrow \frac{1}{2}x = -6 \Rightarrow x = -12$.

Therefore, the solution set is $\{-12, 4\}$.

(b) $\left|\frac{1}{2}x + 2\right| > 4 \Rightarrow \frac{1}{2}x + 2 > 4 \Rightarrow \frac{1}{2}x > 2 \Rightarrow x > 4$ or $\frac{1}{2}x + 2 < -4 \Rightarrow \frac{1}{2}x < -6 \Rightarrow x < -12$. Therefore, the solution interval is $(-\infty, -12) \cup (4, \infty)$.

(c) $\left|\frac{1}{2}x + 2\right| \le 4 \Rightarrow -4 \le \frac{1}{2}x + 2 \le 4 \Rightarrow -6 \le \frac{1}{2}x \le 2 \Rightarrow -12 \le x \le 4$.

Therefore, the solution interval is [-12,4].

2. For the graph of $y = |f(x)|$, we reflect the graph of $y = f(x)$ across the x-axis for all points for which $y < 0$. Where $y \ge 0$, the graph remains unchanged. See Figure 2.

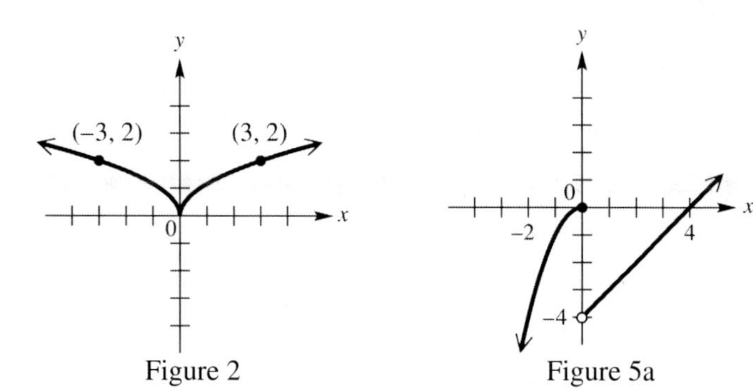

Figure 2

Figure 5a

3. $|2x + 4| = |1 - 3x| \Rightarrow 2x + 4 = 1 - 3x \Rightarrow 5x = -3 \Rightarrow x = -\frac{3}{5}$ or $2x + 4 = -(1 - 3x) \Rightarrow 2x + 4 = 3x - 1 \Rightarrow 5 = x$

The solution set is $\left\{-\frac{3}{5}, 5\right\}$.

4. (a) $f(-3) = 2(-3) + 3 = -3$ (b) $f(0) = (0)^2 + 4 = 4$ (c) $f(2) = (2)^2 + 4 = 8$

5. (a) See Figure 5a.

   (b) Graph $y_1 = (-x^2)*(x \le 0) + (x-4)*(x > 0)$ in the window [-10,10] by [-10,10]. See Figure 5b.

[-10,10] by [-10,10]
Xscl = 1 Yscl = 1

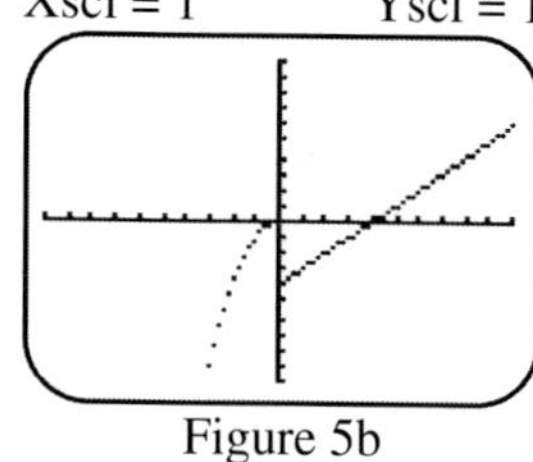

Figure 5b

6. (a) $(f+g)(x) = (-3x-4) + (x^2) = x^2 - 3x - 4$. Therefore, $(f+g)(1) = (1)^2 - 3(1) - 4 = -6$.

   (b) $(f-g)(x) = (-3x-4) - (x^2) = -x^2 - 3x - 4$. Therefore, $(f-g)(3) = -(3)^2 - 3(3) - 4 = -22$.

   (c) $(fg)(x) = (-3x-4)(x^2) = -3x^3 - 4x^2$. Therefore, $(fg)(1) = -3(-2)^3 - 4(-2)^2 = 24 - 16 = 8$.

   (d) $\left(\frac{f}{g}\right)(x) = \frac{-3x-4}{x^2}$. Therefore, $\left(\frac{f}{g}\right)(-3) = \frac{-3(-3)-4}{(-3)^2} = \frac{5}{9}$.

   (e) $(f \circ g)(x) = f[g(x)] = -3(x)^2 - 4 \Rightarrow (f \circ g)(x) = -3x^2 - 4$

   (f) $(g \circ f)(x) = g[f(x)] = (-3x-4)^2 \Rightarrow (g \circ f)(x) = 9x^2 + 24x + 16$

7. One of many possible solutions for $(f \circ g)(x) = h(x)$ is $f(x) = x^4$ and $g(x) = x + 2$. Then $(f \circ g)(x) = f[g(x)] = (x+2)^4$.

8. $$\frac{-2(x+h)^2 + 3(x+h) - 5 - (-2x^2 + 3x - 5)}{h} = \frac{-2(x^2 + 2xh + h^2) + 3x + 3h - 5 + 2x^2 - 3x + 5}{h} =$$

$$\frac{-2x^2 - 4xh - 2h^2 + 3x + 3h - 5 + 2x^2 - 3x + 5}{h} = \frac{-4xh - 2h^2 + 3h}{h} = -4x - 2h + 3.$$

9. (a) At 4% simple interest the equation for interest earned is $y_1 = .04x$.

   (b) If he invested x dollars in the first account, then he invested x+500 in the second account. The equation for the amount of interest earned on this account is $y_2 = .025(x + 500) \Rightarrow y_2 = .025x + 12.5$.

   (c) It represents the total interest earned in both accounts for 1 year.

   (d) Graph $y_1 + y_2 = .04x + (.025x + 12.5) \Rightarrow y_1 + y_2 = .04x + .025x + 12.5$ in the window [0,1000] by [0,100]. See Figure 9. An input value of $x = 250$, results in \$28.75 earned interest.

   (e) At $x = 250$, $y_1 + y_2 = .04(250) + .025(250) + 12.5 = 10 + 6.25 + 12.5 = \$28.75$.

[0,1000] by [-20,100]
Xscl = 100 Yscl = 10

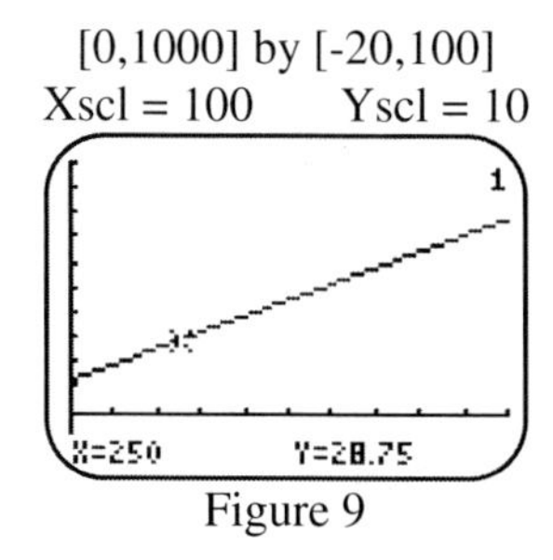

Figure 9

10. If the radius is r, then the height is $2r$ and the equation is
$S = \pi r\sqrt{r^2 + (2r)^2} = \pi r\sqrt{r^2 + 4r^2} = \pi r\sqrt{5r^2} \Rightarrow S = \pi r^2\sqrt{5}.$

Chapter 2 Review Exercises

The graphs for exercises 1–10 can be found in the "Function Capsule" boxes located in section 2.1 in the text.

1. True Both $f(x) = x^2$ and $f(x) = |x|$ have the interval: $[0, \infty)$ as the range.
2. True Both $f(x) = x^2$ and $f(x) = |x|$ increase on the interval: $[0, \infty)$.
3. False The function $f(x) = \sqrt{x}$ has the domain: $[0, \infty)$ and $f(x) = \sqrt[3]{x}$ the domain: $(-\infty, \infty)$.
4. False The function $f(x) = \sqrt[3]{x}$ increases on its entire domain.
5. True The function $f(x) = x$ has a domain and range of: $(-\infty, \infty)$
6. False The function $f(x) = \sqrt{x}$ is not defined on $(-\infty, 0)$, so certainly cannot be continuous.
7. True All of the functions show increases on the interval: $[0, \infty)$
8. True Both $f(x) = x$ and $f(x) = x^3$ have graphs that are symmetric with respect to the origin.
9. True Both $f(x) = x^2$ and $f(x) = |x|$ have graphs that are symmetric with respect to the y-axis.
10. True No graphs are symmetric with respect to the x-axis.
11. Only values where $x \ge 0$ can be input for x, therefore the domain of $f(x) = \sqrt{x}$ is: $[0, \infty)$
12. Only positive solution are possible in absolute value functions, therefore the range of $f(x) = \sqrt{x}$ is: $[0, \infty)$
13. All solution are possible in cube root functions, therefore the range of $f(x) = \sqrt[3]{x}$ is: $(-\infty, \infty)$.
14. All values can be input for x, therefore the domain of $f(x) = x^2$ is: $(-\infty, \infty)$.
15. The function $f(x) = \sqrt[3]{x}$ increases for all inputs for x, therefore the interval is: $(-\infty, \infty)$.
16. The function $f(x) = |x|$ increases for all inputs where $x \ge 0$, therefore the interval is: $[0, \infty)$
17. The equation is the equation $y = \sqrt{x}$. Only values where $x \ge 0$ can be input for x, therefore the domain of $y = \sqrt{x}$ is: $[0, \infty)$
18. The equation $y^2 = x$ is the equation $y = \sqrt{x}$ Square root functions have both positive and negative solutions and all solution are possible, therefore the range of $y = \sqrt{x}$ is: $(-\infty, \infty)$.
19. The graph of $f(x) = (x + 3) - 1$ is the graph $y = x$ shifted 3 units to the left and 1 unit downward.

See Figure 19.

20. The graph of $f(x) = -\frac{1}{2}x + 1$ is the graph reflected $y = x$ across the x-axis, vertically shrunk by a factor of $\frac{1}{2}$, and shifted 1 unit upward. See Figure 20.

21. The graph of $f(x) = (x+1)^2 - 2$ is the graph $y = x^2$ shifted 1 unit to the left and 2 units downward. See Figure 21.

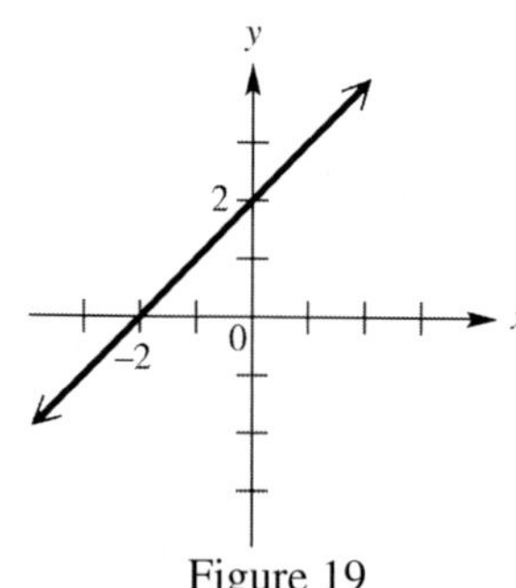

Figure 19

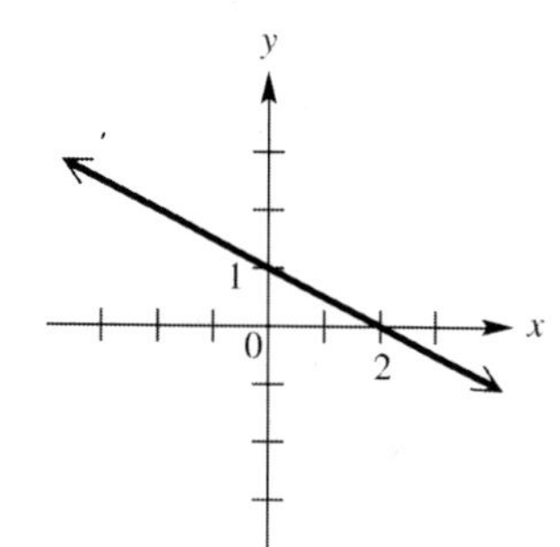

Figure 20

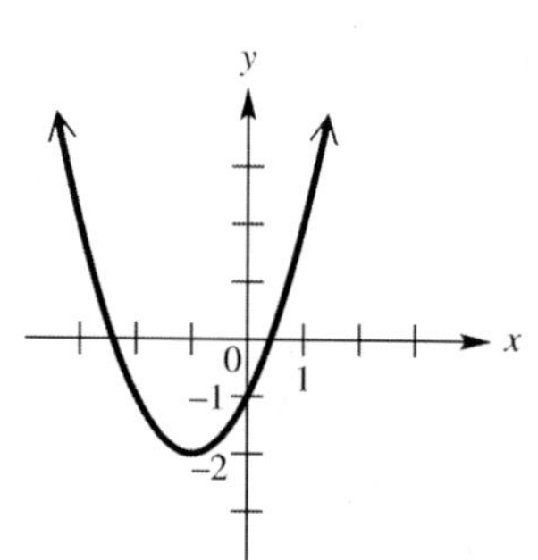

Figure 21

22. The graph of $f(x) = -2x^2 + 3$ is the graph $y = x^2$ reflected across the x-axis, vertically stretched by a factor of 2, and shifted 3 units upward. See Figure 22.

23. The graph of $f(x) = -x^3 + 2$ is the graph $y = x^3$ reflected across the x-axis and shifted 2 units upward. See Figure 23.

24. The graph of $f(x) = (x-3)^3$ is the graph $y = x^3$ shifted 3 units to the right. See Figure 24.

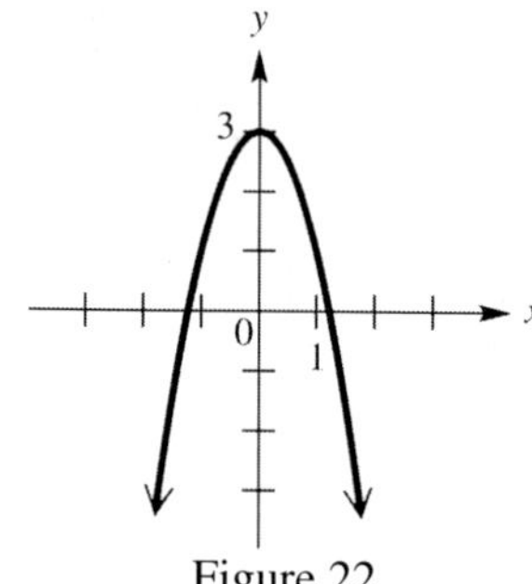

Figure 22

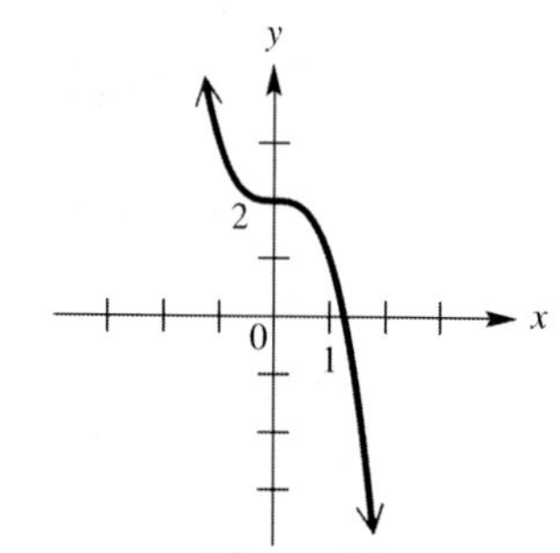

Figure 23

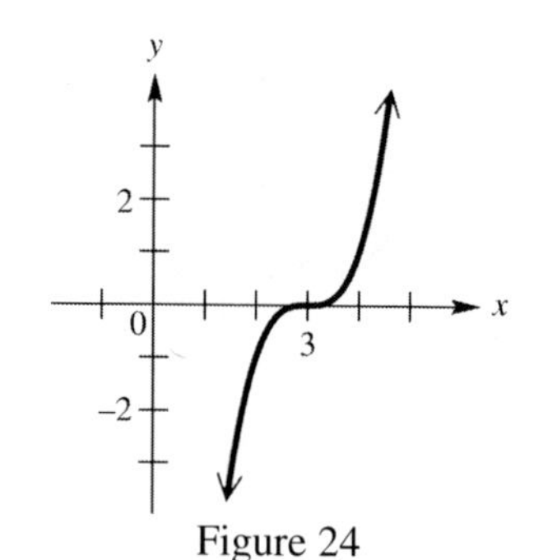

Figure 24

25. The graph of $f(x) = \sqrt{\frac{1}{2}x}$ is the graph $y = \sqrt{x}$ horizontally stretched by a factor of 2. See Figure 25.

26. The graph of $f(x) = \sqrt{x-2} + 1$ is the graph $y = \sqrt{x}$ shifted 2 units to the right and 1 unit upward. See Figure 26.

27. The graph of $f(x) = 2\sqrt[3]{x}$ is the graph $y = \sqrt[3]{x}$ vertically stretched by a factor of 2. See Figure 27.

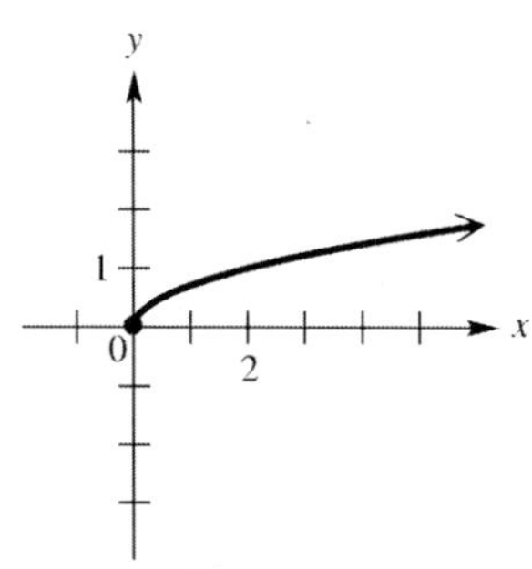

Figure 25

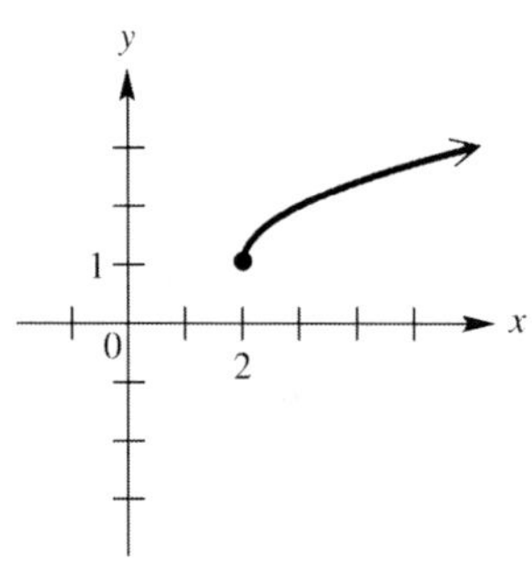

Figure 26

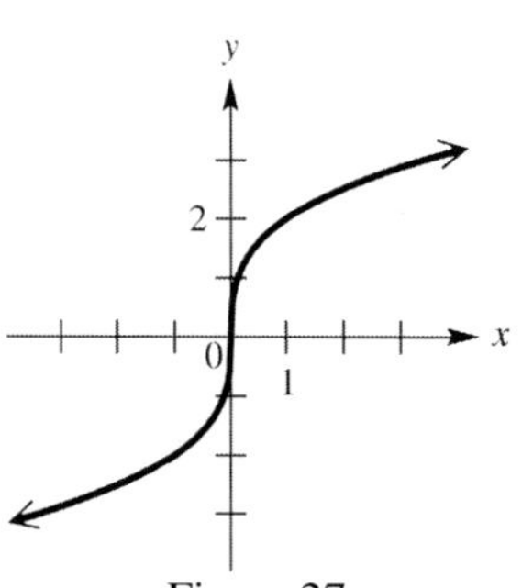

Figure 27

28. The graph of $f(x)=\sqrt[3]{x}-2$ is the graph $y=\sqrt[3]{x}$ shifted 2 units downward. See Figure 28.

29. The graph of $f(x)=|x-2|+1$ is the graph $y=|x|$ shifted 2 units right and 1 unit upward. See Figure 29.

30. The graph of $f(x)=|-2x+3|$ is the graph $y=|x|$ horizontally shrunk by a factor of $\frac{1}{2}$, shifted $\left(\frac{1}{2}\right)(3)$ or $\frac{3}{2}$ units to the left, and reflected across the y-axis. See Figure 30.

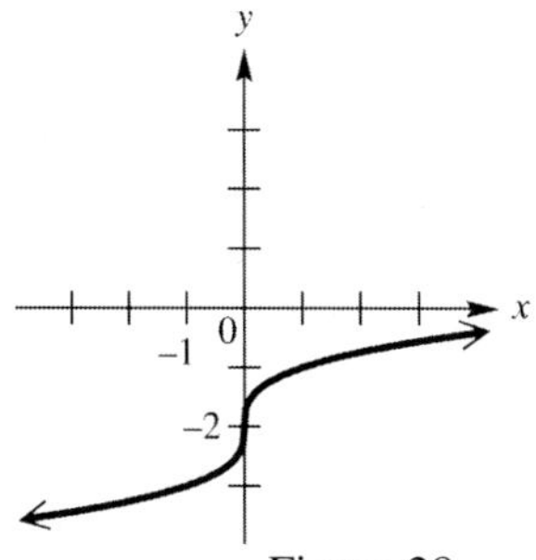

Figure 28

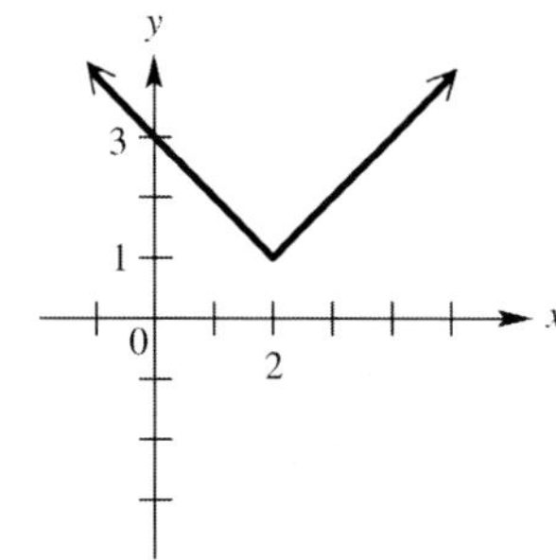

Figure 29

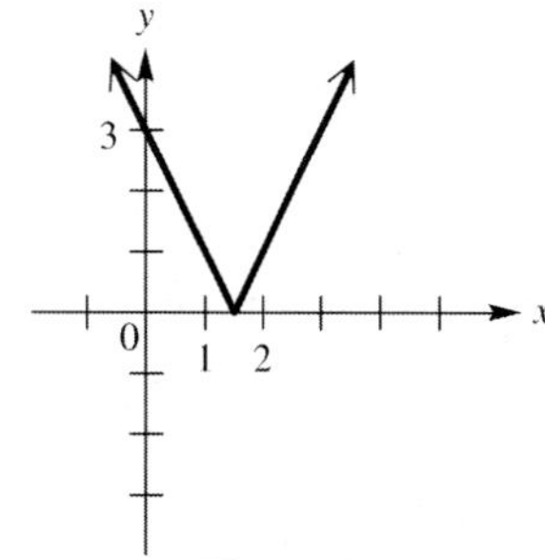

Figure 30

31. (a) From the graph, the function is continuous for the intervals: $(-\infty,-2),[-2,1]$ and $(1,\infty)$

(b) From the graph, the function is increasing for the interval: $[-2,1]$

(c) From the graph, the function is decreasing for the interval: $(-\infty,-2)$

(d) From the graph, the function is constant for the interval: $(1,\infty)$

(e) From the graph, all values can be input for x, therefore the domain is: $(-\infty,\infty)$

(f) From the graph, the possible values of y or the range is: $\{-2\}\cup[-1,1]\cup(2,\infty)$

32. $x=y^2-4\Rightarrow y^2=x+4\Rightarrow y=\sqrt{x+4}$ and $y=-\sqrt{x+4}$.

33. From the graph, the relation is symmetric with respect to the x-axis, y-axis, and origin. The relation is not a function since some inputs x have two outputs y.

34. If $F(x)=x^3-6$,then $F(-x)=(-x)^3-6\Rightarrow F(-x)=-x^{-3}-6$ and $-F(x)=-(x^3-6)\Rightarrow -F(x)=-x^3+6$.

Since $F(x)\neq F(-x)\neq -F(x)$ the function has no symmetry and is neither an ever nor an odd function.

35. If $f(x)=|x|+4$,then $f(-x)=|(-x)|+4 \Rightarrow f(-x)=|x|+4$ and $-f(x)=-|x|-4$ Since $f(-x)=f(x)$ the function is symmetric with respect to the *y*-axis and is an even function.

36. If $f(x)=\sqrt{x-5}$ then $f(-x)=\sqrt{(-x)-5}$ and $-f(x)=-\sqrt{x-5}$ .Since $f(x)\neq f(-x)\neq -f(x)$, the function has no symmetry and is neither an even nor an odd function.

37. If $y^2=x-5$ then $y=\pm\sqrt{x-5}$. Since $f(x)=-\sqrt{x-5}$ is the reflection of $f(x)=-\sqrt{x-5}$ across the *x*-axis, the relation has symmetry with respect to the *x*-axis. Also, one *x* inputs can produce two *y* outputs the relation is not a function.

38. If $f(x)=3x^4+2x^2+1$ then $f(-x)=3(-x)^4+2(-x)^2+1 \Rightarrow f(-x)=3x^4+2x^2+1$ and $-f(x)=-3x^4-2x^2-1$. Since $f(-x)=f(x)$ the function is symmetric with respect to the *y*-axis and is an even function.

39. True, a graph that is symmetrical with respect to the *x*-axis means that for every (x, y) there is also $(x, -y)$.

40. True, since an even function and one that is symmetric with respect to the *y*-axis both contain the points (x, y) and $(-x, y)$.

41. True, since an odd function and one that is symmetric with respect to the origin both contain the points (x, y) and $(-x, -y)$.

42. False, for an even function, if (a,b) is on the graph, then $(-a,b)$ is on the graph and not $(a,-b)$. For example, $f(x)=x^2$ is even, and (2, 4) is on the graph, but (2, −4) is not.

43. False, for an odd function, if (a,b) is on the graph, then $(-a,-b)$ is on the graph and not $(-a,b)$ For example, $f(x)=x^3$ is odd, and $(2,8)$ is on the graph, but $(-2,8)$ is not.

44. True, if $(x,0)$ is on the graph of $f(x)=0$ then $(-x,0)$ is on the graph.

45. The graph of $y=-3(x+4)^2-8$ is the graph of $y=x^2$ shifted 4 units to the left, vertically stretched by a factor of 3, reflected across the *x*-axis, and shifted 8 units downward.

46. The equation $y=\sqrt{x}$ reflected across the *y*-axis is: $f(x)=\sqrt{-x}$ then reflected across the *x*-axis is: $y=-\sqrt{-x}$ now vertically shrunk by a factor of $\frac{2}{3}$ is: $y=-\frac{2}{3}\sqrt{-x}$, and finally shifted 4 units upward is: $y=-\frac{2}{3}\sqrt{-x}+4$.

47. Shift the function *f* upward 3 units. See Figure 47.

48. Shift the function *f* to the right 2 units. See Figure 48.

49. Shift the function *f* to the left 3 units and downward 2 units. See Figure 49.

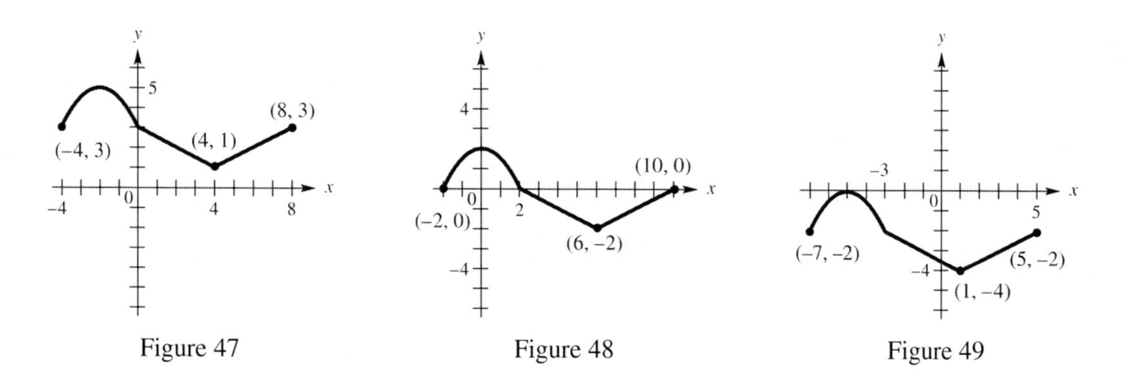

50. For values where $f(x) > 0$ the graph remains the same. For values where $f(x) < 0$ reflect the graph across the x-axis. See Figure 50.

51. Horizontally shrink the function f by a factor of $\frac{1}{4}$. See Figure 51.

52. Horizontally stretch the function f by a factor of 2. See Figure 52.

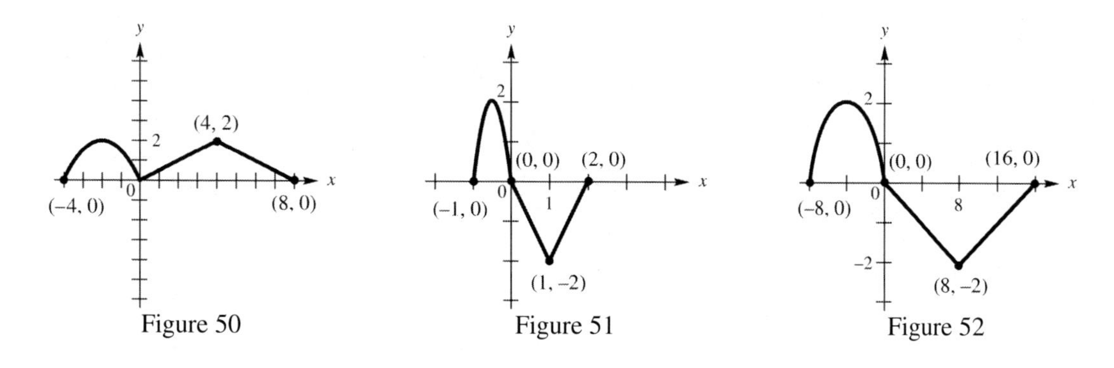

53. The function is shifted upward 4 units, therefore the domain remains the same: $[-3,4]$ and the range is increased by 4 and is: $[2,9]$.

54. The function is shifted left 10 units, therefore the domain is decreased by 10 and is $[-13,-6]$; and the function is stretched vertically by a factor of 5, therefore the range is multiplied by 5 and is: $[-10,25]$.

55. The function is horizontally shrunk by a factor of $\frac{1}{2}$, therefore the domain is divided by 2 and is: $\left[-\frac{3}{2},2\right]$; and the function is reflected across the x-axis, therefore the range is opposites of the original and is: $[-5,2]$.

56. The function is shifted right 1 unit, therefore the domain is increased by 1 and is: $[-2,5]$; and the function is also shifted upward 3 units, therefore the range is increased by 3 and is: $[1,8]$.

57. We reflect the graph of $y = f(x)$ across the x-axis for all points for which $y < 0$. Where $y \ge 0$, the graph remains unchanged. See Figure 57.

58. We reflect the graph of $y = f(x)$ across the x-axis for all points for which < 0. Where $y \ge 0$, the graph remains unchanged. See Figure 58.

59. Since the range is $\{2\}, y \geq 0,$ so the graph remains unchanged.

60. Since the range is $\{-2\}, y < 0,$ so we reflect the graph across the x-axis. See Figure 60.

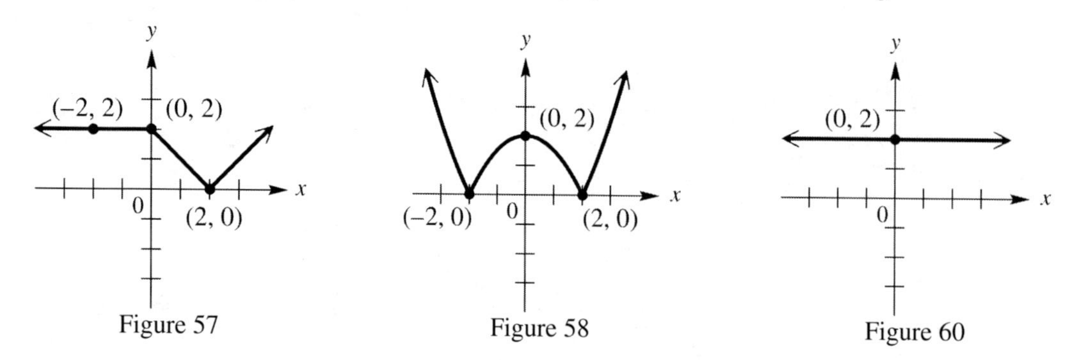

Figure 57 Figure 58 Figure 60

61. $|4x+3| = 12 \Rightarrow 4x+3 = 12 \Rightarrow 4x = 9 \Rightarrow x = \frac{9}{4}$ or $4x+3 = -12 \Rightarrow 4x = -15 \Rightarrow x = -\frac{15}{4}$, therefore the solution set is: $\left\{-\frac{15}{4}, \frac{9}{4}\right\}$.

62. $|-2x-6| + 4 = 1 \Rightarrow |-2x-6| = -3.$ Since an absolute value equation can not have a solution less than zero, the solution set is: $\varnothing$

63. $|5x+3| = |x+11| \Rightarrow 5x+3 = x+11 \Rightarrow 4x = 8 \Rightarrow x = 2$ or $5x+3 = -(x+11) \Rightarrow 6x = -14 \Rightarrow x = -\frac{14}{6} = -\frac{7}{3},$

therefore the solution set is: $\left\{-\frac{7}{3}, 2\right\}$.

64. $|2x+5| = 7 \Rightarrow 2x+5 = 7 \Rightarrow 2x = 2 \Rightarrow x = 1$ or $2x+5 = -7 \Rightarrow 2x = -12 \Rightarrow x = -6,$ therefore the solution set is: $\{-6, 1\}$.

65. $|2x+5| \leq 7 \Rightarrow -7 \leq 2x+5 \leq 7 \Rightarrow -12 \leq 2x \leq 2 \Rightarrow -6 \leq x \leq 1,$ therefore the interval is: $\{-6, 1\}$.

66. $|2x+5| \geq 7 \Rightarrow 2x+5 \geq 7 \Rightarrow 2x \geq 2 \Rightarrow x \geq 1$ or $2x+5 \leq -7 \Rightarrow 2x \leq -12 \Rightarrow x \leq -6,$ therefore the solution is the interval: $(-\infty, -6] \cup [1, \infty)$.

67. $|5x-12| 0 \Rightarrow 5x-12 > 0 \Rightarrow 5x > 12 \Rightarrow x > \frac{12}{5}$ or $5x-12 < 0 \Rightarrow 5x = 12 \Rightarrow x < \frac{12}{5},$ therefore the solution is the interval: $\left(-\infty, \frac{12}{5}\right) \cup \left(\frac{12}{5}, \infty\right)$ or $\left\{x \mid x \neq \frac{12}{5}\right\}$.

68. Since an absolute value equation can not have a solution less than zero, the solution set is: $\varnothing$

69. $2|3x-1| + 1 = 21 \Rightarrow 2|3x-1| = 20 \Rightarrow |3x-1| = 10 \Rightarrow 3x-1 = 10 \Rightarrow 3x = 11 \Rightarrow x = \frac{11}{3}$ or

$3x-1 = -10 \Rightarrow 3x = -9 \Rightarrow x = -3,$ therefore the solution set is: $\left\{-3, \frac{11}{3}\right\}$.

70. $|2x+1|=|-3x+1| \Rightarrow 2x+1=-3x+1 \Rightarrow 5x=0 \Rightarrow x=0$ or $2x+1=-(-3x+1) \Rightarrow -x=-2 \Rightarrow x=2,$ therefore the solution set is: $\{0,2\}$.

71. The x-coordinates of the points of intersection of the graphs are -6 and 1.Thus, $\{-6,1\}$ is the solution set of $y_1 = y_2$ The graph of y_1 lies on or below the graph of y_2 between -6 and 1, so the solution set of $y_1 \le y_2$ is $[-6,1]$ The graph of y_1 lies above the graph of y_2 everywhere else, so the solution set of $y_1 \ge y_2$ is $(-\infty,-6]\cup[1,\infty)$.

72. Graph $y_1 = |x+1|+|x-3|$ and $\{-3,-1\}$ See Figure 72. The intersections are $x=-3$ and $x=5,$ therefore the solution set is: $\{-3,5\}$. **Check:** $|(-3)+1|+|(-3)-3|=8 \Rightarrow |-2|+|-6|=8 \Rightarrow 2+6=8 \Rightarrow 8=8$ and $|(5)+1|+|(5)-3|=8 \Rightarrow |6|+|2|=8 \Rightarrow 6+2=8 \Rightarrow 8=8$

[-10,10] by [-4,16]

Xscl = 1 Yscl = 1

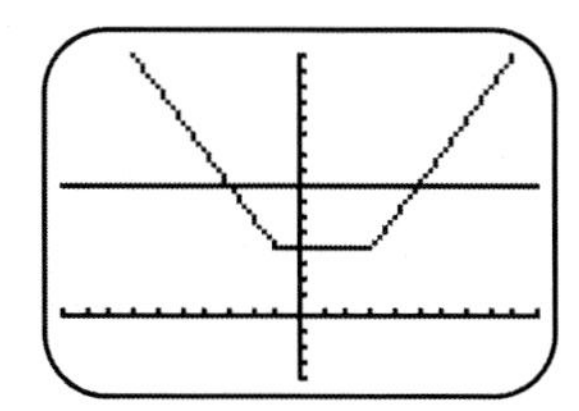

Figure 72

73. Initially, the car is at home. After traveling 30 mph for 1 hr, the car is 30 mi away from home. During the second hour the car travels 20 mph until it is 50 mi away. During the third hour the car travels toward home at 30 mph until it is 20 mi away. During the fourth hour the car travels away from home at 40 mph until it is 60 mi away from home. During the last hour, the car travels 60 mi at 60 mph until it arrives home.

74. See Figure 74

75. See Figure 75

76. See Figure 76

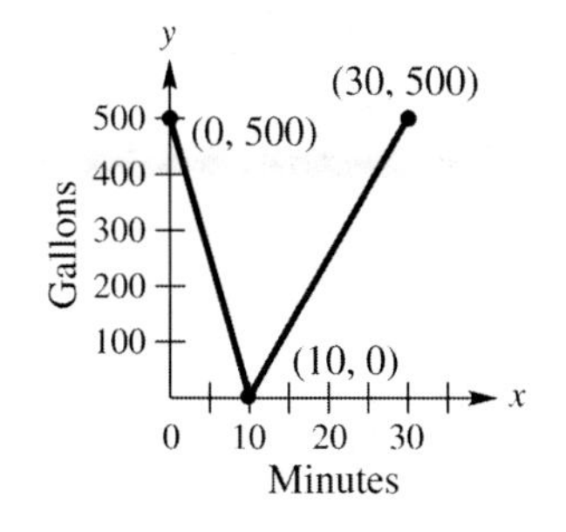

Figure 74

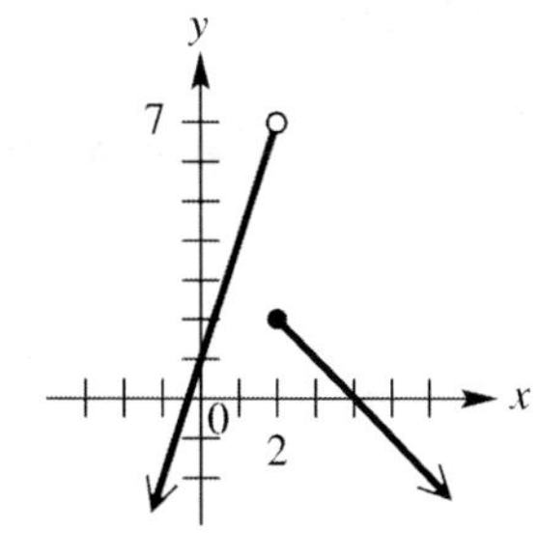

Figure 75

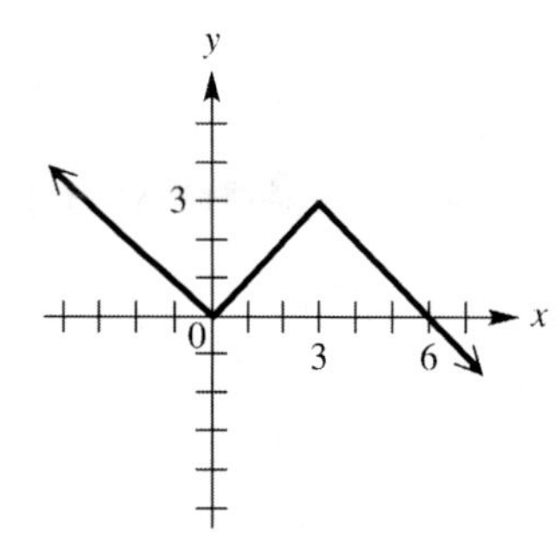

Figure 76

77. Graph $y_1 = (3x+1)*(x<2)+(-x+4)*(x \geq 2)$ in the window $[-10,10]$ by

$|4x+8| > 4 \Rightarrow 4x+8 > 4 \Rightarrow 4x > -4 \Rightarrow -1$ See Figure 77.

78. See Figure 78.

[-10,10] by [-10,10]
Xscl = 1 Yscl = 1

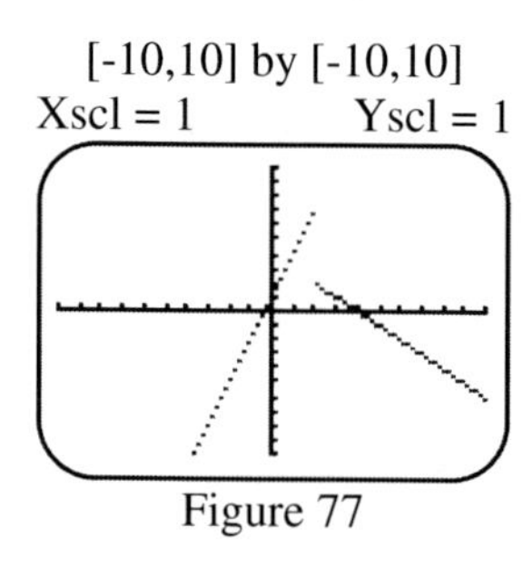

Figure 77

[-5,5] by [-5,5]
Xscl = 1 Yscl = 1

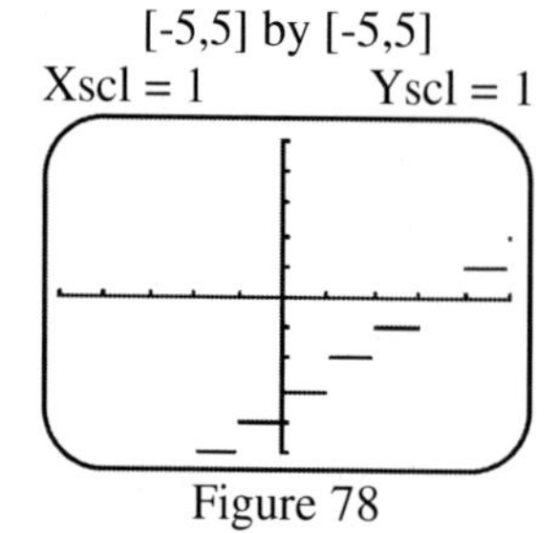

Figure 78

79. From the graphs $(f+g)(1) = 2+3 = 5$

80. From the graphs $(f-g)(0) = 1-4 = -3$

81. From the graphs $(fg)(-1) = (0)(3) = 0$

82. From the graphs $\left(\frac{f}{g}\right)(2) = \frac{3}{2}$

83. From the graphs $(f \circ g)(2) = f[g(2)] = f(2) = 3$

84. From the graphs $(g \circ f)(2) = g[f(2)] = g(3) = 2$

85. From the graphs $(g \circ f)(-4) = g[f(-4)] = g(2) = 2$

86. From the graphs $(f \circ g)(-2) = f[g(-2)] = f(2) = 3$

87. From the table $(f+g)(1) = 7+1 = 8$

88. From the table $(f-g)(3) = 9-1 = 0$

89. From the table $(fg)(-1) = (3)(-2) = -6$

90. From the table $\left(\frac{f}{g}\right)(0) = \frac{5}{0}$, which is undefined.

91. From the tables $(g \circ f)(-2) = g[f(-2)] = g(1) = 2$

92. From the graphs $(f \circ g)(3) = f[g(3)] = f(-2) = 1$

93. $\frac{2(x+h)+9-(2x+9)}{h} = \frac{2x+2h+9-2x-9}{h} = \frac{2h}{h} = 2$

94. $\frac{(x+h)^2-5(x+h)+3-(x^2-5x+3)}{h} = \frac{x^2+2xh+h^2-5x-5h+3-x^2+5x-3}{h} = \frac{2xh+h^2-5h}{h} = 2x+h-5$

95. One of many possible solutions for $(f \circ g)(x) = h(x)$ is: $f(x) = x^2$ and $g(x) = x^3 - 3x$ Then $(f \circ g)(x) = (x^3 - 3x)^2$.

96. One of many possible solutions for $(f \circ g)(x) = h(x)$ is $f(x) = \frac{1}{x}$ and $g(x) = x - 5$ Then

$(f \circ g)(x) = f[g(x)] = \frac{1}{x-5}$

97. If $V(r) = \frac{4}{3}\pi r^3$, then a 4 inch increase would be: $V(r) = \frac{4}{3}\pi(r+4)^3$, and the volume gained would be:

$V(r) = \frac{4}{3}\pi(r+4)^3 - \frac{4}{3}\pi r^3$.

98. (a) Since $h = d, r = \frac{d}{2}$ and the formula for the volume of a can is: $V = \pi r^2 h$, the function is:

$V(d) = \pi\left(\frac{d}{2}\right)^2 d \Rightarrow V(d) = \frac{\pi d^3}{4}$

(b) Since $h = d, r = \frac{d}{2}, c = 2\pi r$ and the formula for the surface area of a can is: $A = 2\pi rh + 2\pi r^2$, the

function is: $S(d) = 2\pi\left(\frac{d}{2}\right)d + 2\pi\left(\frac{d}{2}\right)^2 \Rightarrow S(d) = \pi d^2 + \frac{\pi d^2}{2} \Rightarrow S(d) = \frac{3\pi d^2}{2}$

99. The function for changing yards to inches is: $f(x) = 36x$ and the function for changing miles to yards is: $g(x) = 1760x$ The composition of this which would change miles into inches is:

$f[g(x)] = 36[1760(x)] \Rightarrow (f \circ g)(x) = 63,360x$.

100. If $x =$ width, then length $= 2x$ A formula for Perimeter can now be written as: $P = x + 2x + x + 2x$ and the function is: $P(x) = 6x$ This is a linear function.

Chapter 2 Test

1. (a) D, only values where $x \geq 0$ can be input into a square root function.

(b) D, only values where $y \geq 0$ can be the solution to a square root function.

(c) C, all values can be input for x in a squaring function.

(d) B, only values where $y \geq 3$ can be the solution to $f(x) = x^2 + 3$

(e) C, all values can be input for x in a cube root function.

(f) C, all values can be a solution in a cube root function.

(g) C, all values can be input for x in an absolute value function.

(h) D, only values where $y \geq 0$ can be the solution to an absolute value function.

(i) D, if $x = y^2$ then $y = \sqrt{x}$ and only values where $x \geq 0$ can be input into a square root function.

(j) C, all values can be a solution in this function.

2. (a) This is $f(x)$ shifted 2 units upward. See Figure 2a.

(b) This is $f(x)$ shifted 2 units to the left. See Figure 2b.

(c) This is $f(x)$ reflected across the x-axis. See Figure 2c.

(d) This is $f(x)$ reflected across the y-axis. See Figure 2d.

(e) This is $f(x)$ vertically stretched by a factor of 2. See Figure 2e.

(f) We reflect the graph of $y = f(x)$ across the x-axis for all points for which $\$2.75$ Where $y \geq 0$ the graph remains unchanged. See Figure 2f.

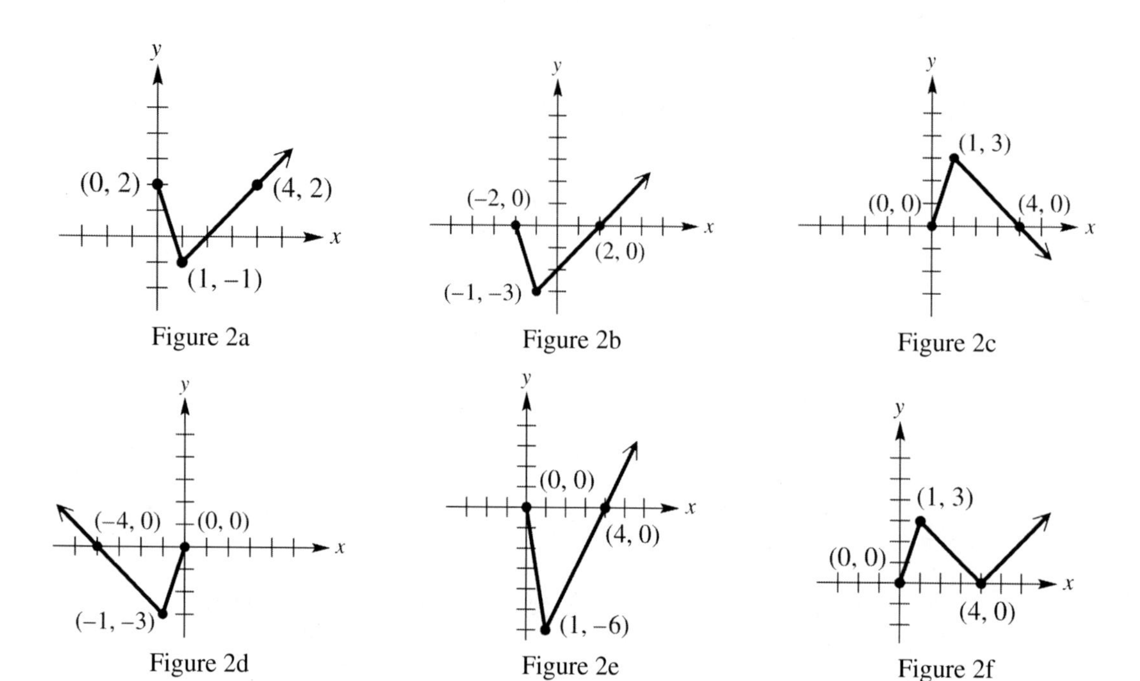

Figure 2a Figure 2b Figure 2c
Figure 2d Figure 2e Figure 2f

3. (a) Since $y = f(2x)$ is $y = f(x)$ horizontally shrunk by a factor of $\frac{1}{2}$, the point $(-2,4)$ on $y = f(x)$ becomes the point $(-1,4)$ on the graph of $y = f(2x)$.

   (b) Since $y = f\left(\frac{1}{2}x\right)$ is $y = f(x)$ horizontally stretched by a factor of 2, the point $(-2,4)$ on $y = f(x)$ becomes the point $(-4,4)$ on the graph of $y = f\left(\frac{1}{2}x\right)$.

4. (a) The graph of $f(x) = -x(x-2)^2 + 4$ is the basic graph $f(x) = x^2$ reflected across the x-axis, shifted 2 units to the right, and shifted 4 units upward. See Figure 4a.

   (b) The graph of $f(x) = -2\sqrt{-x}$ is the basic graph $f(x) = \sqrt{x}$ reflected across the y-axis and vertically stretched by a factor of 2. See Figure 4b.

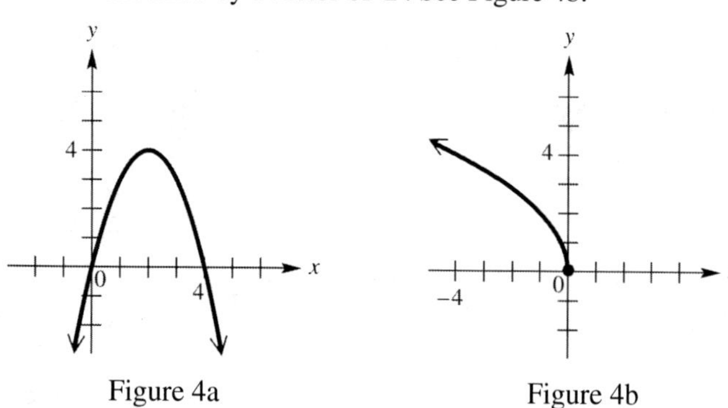

Figure 4a Figure 4b

5. (a) If the graph is symmetric with respect to the y-axis, then $(x, y) \Rightarrow (-x, y)$ therefore $(3,6) \Rightarrow (-3,6)$.

   (b) If the graph is symmetric with respect to the x-axis, then $(x, y) \Rightarrow (-x, -y)$ therefore $(3,6) \Rightarrow (-3,-6)$.

   (c) See Figure 5. We give an actual screen here. The drawing should resemble it.

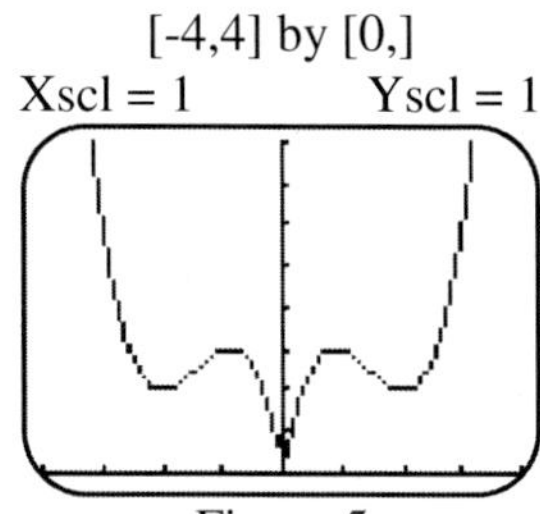

Figure 5

6. (a) Shift the graph of $y = \sqrt[3]{x}$ to the left 2 units, vertically stretch by a factor of 4, and shift 5 units downward.

   (b) Graph $y = |x|$ reflected across the x-axis, vertically shrunk by a factor of $\frac{1}{2}$ shifted 3 units to the right, and shifted up 2 units. See Figure 6. From the graph the domain is: $(-\infty, \infty)$; and the range is: $(-\infty, 2]$.

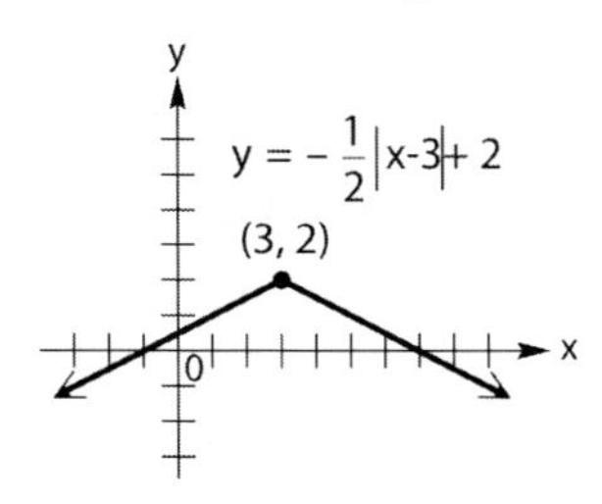

Figure 6

7. (a) From the graph, the function is increasing for the interval: $(-\infty, -3)$

   (b) From the graph, the function is decreasing for the interval: $(4, \infty)$

   (c) From the graph, the function is constant for the interval: $(-3, 4)$

   (d) From the graph, the function is continuous for the intervals: $(-\infty, -3), (-3, 4), (4, \infty)$.

   (e) From the graph, the domain is: $(-\infty, \infty)$

   (f) From the graph, the range is: $(-\infty, 2)$

8. (a) $|4x+8| = 4 \Rightarrow 4x+8 = 4 \Rightarrow 4x = -4 \Rightarrow x = -1$ or $4x+8 = -4 \Rightarrow 4x = -12 \Rightarrow x = -3$, therefore the solution set is: $\{-3, -1\}$

   (b) $|4x+8| < 4 \Rightarrow -4 < 4x+8 < 4 \Rightarrow -12 < 4x < -4 \Rightarrow -3 < x < -1$, therefore the solution is: $(-3, -1)$.

   (c) $|4x+8| > 4 \Rightarrow 4x+8 > 4 \Rightarrow 4x > -4 \Rightarrow x > -1$ or $4x+8 < -4 \Rightarrow 4x < -12 \Rightarrow x < -3$ therefore the solution is: $(-\infty, -3) \cup (-1, \infty)$.

9. (a) $(f-g)(x) = 2x^2 - 3x + 2 - (-2x+1) \Rightarrow (f-g)(x) = 2x^2 - x + 1$

(b) $\left(\dfrac{f}{g}\right)(x)=\dfrac{2x^2-3x+2}{-2x+1}$

(c) The domain can be all values for x, except any that make $g(x)=0$. Therefore

$-2x+1\neq 0 \Rightarrow -2x\neq -1 \Rightarrow x\neq \dfrac{1}{2}$ or the interval: $\left(-\infty,\dfrac{1}{2}\right)\cup\left(\dfrac{1}{2},\infty\right)$

(d) $(f\circ g)(x)=f\left[g(x)\right]=2(-2x+1)^2-3(-2x+1)+2=2(4x^2-4x+1)+6x-3+2=$

$8x^2-8x+2+6x-3+2=8x^2-2x+1$

(e) $(g\circ f)(x)=g\left[f(x)\right]=-2(2x^2-3x+2)+1=-4x^2+6x-4+1=-4x^2+6x-3$

(f) $\dfrac{2(x+h)^2-3(x+h)+2-(2x^2-3x+2)}{h}=\dfrac{2(x^2+2xh+h^2)-3x-3h+2-2x^2+3x-2}{h}=$

$\dfrac{2x^2+4xh+2h^2-3x-3h+2-2x^2+3x-2}{h}=\dfrac{4xh+2h^2-3h}{h}=4x+2h-3$

10. (a) See Figure 10a.

(b) Graph $y_1=(-x^2+3)*(x\leq 1)+\left(\sqrt[3]{x}+2\right)+\left(\sqrt[3]{x}+2\right)*(x>1)$ in the window $[-4.7,4.7]$ by $[-5.1,5.1]$

See Figure 10b.

(c) The graph is not connected at $x=1$ and thus f is not continuous when $x=1$.

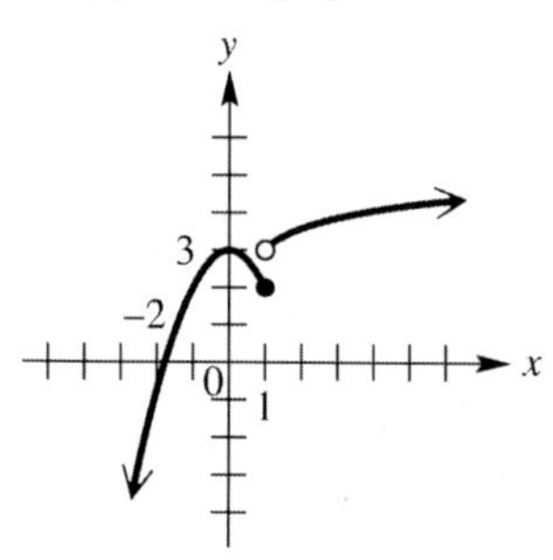

Figure 10a

[-4.7,4.7] by [-5.1,5.1]
Xscl = 1 Yscl = 1

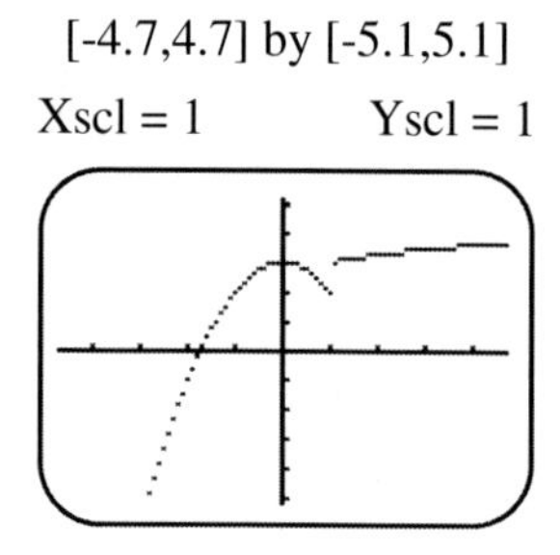

Figure 10b

[0,10] by [0,6]
Xscl = 1 Yscl = 1

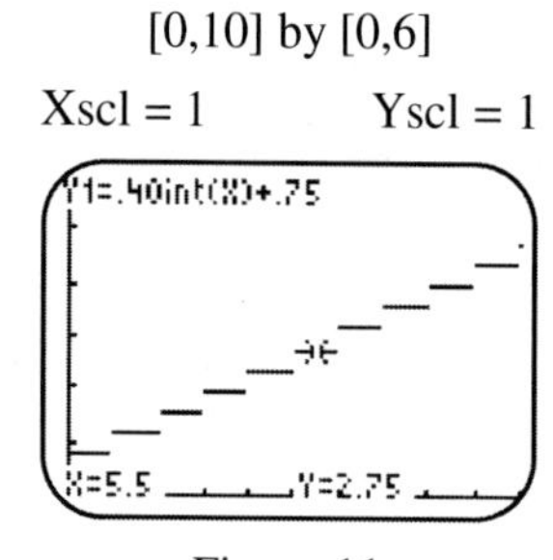

Figure 11

[0,1,000] by [-4,000,4000]
Xscl = 50 Yscl = 500

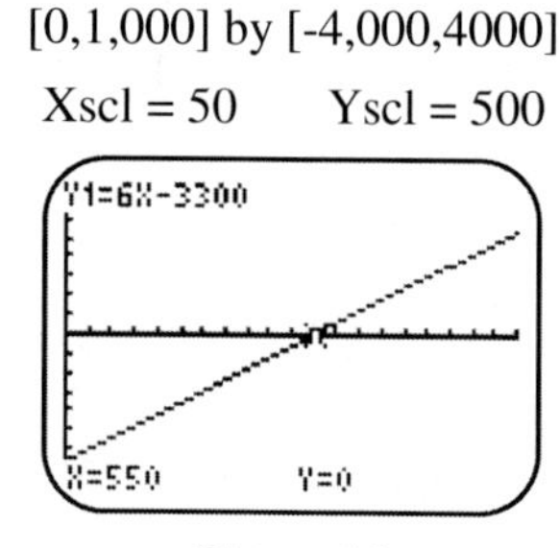

Figure 12

11. (a) See Figure 11.

(b) Set $x=5.5$ then $\$2.75$ is the cost of a 5.5 minute call. See the display at the bottom of the screen.

12. (a) With an initial set-up cost of \$3300 and a production cost of $\$4.50$ the function is: $C(x)=3300+4.50x$

(b) With a selling price of $\$10.50$ the revenue function is: $R(x)=10.50x$

(c) $P(x) = R(x) - C(x) \Rightarrow P(x) = 10.50x - (3300 + 4.50x) \Rightarrow P(x) = 6x - 3300$

(d) To make a profit $P(x) > 0$, therefore $6x - 3300 > 0 \Rightarrow 6x > 3300 \Rightarrow x > 550$

Tyler needs to sell 551 before he earns a profit.

(e) Graph $y_1 = 6x - 3300$, See Figure 12. The first integer x-value for which $P(x) > 0$ is 551

Chapter 3: Polynomial Functions

3.1: Complex Numbers

1. The complex number $-9i$ can be written $0-9i$.

   (a) The real part is 0. (b) The imaginary part is -9. (c) The number is pure imaginary.

3. The complex number π can be written $\pi+0i$.

   (a) The real part is π. (b) The imaginary part is 0. (c) The number is real.

5. The complex number $3+7i$ is written in standard form.

   (a) The real part is 3. (b) The imaginary part is 7. (c) The number is nonreal complex.

7. The complex number $i\sqrt{7}$ can be written $0+\sqrt{7}i$.

   (a) The real part is 0. (b) The imaginary part is $\sqrt{7}$. (c) The number is pure imaginary.

9. The complex number $\sqrt{-7}$ can be written $0+\sqrt{7}i$.

   (a) The real part is 0. (b) The imaginary part is $\sqrt{7}$. (c) The number is pure imaginary.

11. $3i+5i=(3+5)i=8i$

13. $(-7i)(1+i)=-7i-7i^2=-7i-7(-1)=7-7i$

15. True

17. True

19. False. *Every* real number is a complex number.

21. $\sqrt{-100}=i\sqrt{100}=10i$

23. $-\sqrt{-400}=-i\sqrt{400}=-20i$

25. $-\sqrt{-39}=-i\sqrt{39}$

27. $5+\sqrt{-4}=5+i\sqrt{4}=5+2i$

29. $9-\sqrt{-50}=9-i\sqrt{50}=9-i\sqrt{25.2}=9-5i\sqrt{2}$

31. $i\sqrt{-9}=i^2\sqrt{9}=-3$

33. $\sqrt{-13}\cdot\sqrt{-13}=i\sqrt{13}\cdot i\sqrt{13}=13i^2=-13$

35. $\sqrt{-3}\cdot\sqrt{-8}=i\sqrt{3}\cdot i\sqrt{8}=\sqrt{24}i^2=-2\sqrt{6}$

37. $\dfrac{\sqrt{-30}}{\sqrt{-10}}=\dfrac{i\sqrt{30}}{i\sqrt{10}}=\dfrac{i\sqrt{10\cdot 3}}{i\sqrt{10}}=\dfrac{i\sqrt{10}\cdot\sqrt{3}}{i\sqrt{10}}=\sqrt{3}$

39. $\dfrac{\sqrt{-24}}{\sqrt{8}}=\dfrac{i\sqrt{24}}{\sqrt{8}}=\dfrac{i\sqrt{8\cdot 3}}{\sqrt{8}}=\dfrac{i\sqrt{8}\cdot\sqrt{3}}{\sqrt{8}}=i\sqrt{3}$

41. $\dfrac{\sqrt{-10}}{\sqrt{-40}}=\dfrac{i\sqrt{10}}{i\sqrt{40}}=\dfrac{i\sqrt{10}}{i\sqrt{10\cdot 4}}=\dfrac{i\sqrt{10}}{i\sqrt{10}\cdot\sqrt{4}}=\dfrac{1}{\sqrt{4}}=\dfrac{1}{2}$

43. $\dfrac{\sqrt{-6}\cdot\sqrt{-2}}{\sqrt{3}}=\dfrac{i\sqrt{6}\cdot i\sqrt{2}}{\sqrt{3}}=\dfrac{i^2\sqrt{3}\cdot\sqrt{2}\cdot\sqrt{2}}{\sqrt{3}}=\dfrac{i^2\sqrt{2}\cdot\sqrt{2}}{1}=-\sqrt{4}=-2$

45. $(3+2i)+(4-3i)=(3+4)+(2-3)i=7-i$

47. $(-2+3i)-(-4+3i)=(-2-(-4))+(3-3)i=2+0i=2$

49. $(3-8i)+(2i+4)=(3+4)+(-8+2)i=7-6i$

51. $(2-5i)-(3+4i)-(-2+i)=2-5i-3-4i+2-i=(2-3+2)+(-5i-4i-i)=1-10i$

53. $(-6+5i)+(4-4i)+(2-i)=(-6+4+2)+(5+(-4)+(-1))i=0+0i=0$

55. $(2+i)(3-2i)=6-4i+3i-2i^2=6-i-2(-1)=8-i$

57. $(2+4i)(-1+3i)=-2+6i-4i+12i^2=-2+2i+12(-1)=-14+2i$

59. $(-3+2i)^2=(-3+2i)(-3+2i)=9-6i-6i+4i^2=9-12i+4(-1)=5-12i$

61. $(3+i)(-3-i)=-9-3i-3i-i^2=-9-6i-(-1)=-8-6i$

63. $(2+3i)(2-3i)=4-9i^2=4-9(-1)=13$

65. $\left(\sqrt{6}+i\right)\left(\sqrt{6}-i\right)=6-i^2=6-(-1)=7$

67. $i(3-4i)(3+4i)=i(9-16i^2)=i(9-16(-1))=25i$

69. $3i(2-i)^2=3i(4-4i+i^2)=3i(4-4i+(-1))=3i(3-4i)=9i-12i^2=9i-12(-1)=12+9i$

71. $(2+i)(2-i)(4+3i)=(4-i^2)(4+3i)=(4-(-1))(4+3i)=5(4+3i)=20+15i$

73. $i^5=i^4\cdot i^1=1^1\cdot i=i$

75. $i^{15}=(i^4)^3\cdot i^3=1^3\cdot(-i)=-i$

77. $i^{64}=(i^4)^{16}=1^{16}=1$

79. $i^{-6}=(i^6)^{-1}=[(i^4)^1\cdot i^2]^{-1}=[1^1\cdot(-1)]^{-1}=(-1)^{-1}=-1$

81. $\dfrac{1}{i^9}=i^{-9}=(i^9)^{-1}=[(i^4)^2\cdot i]^{-1}=[1^2\cdot i]^{-1}=(i)^{-1}=\dfrac{1}{i}=\dfrac{1}{i}\cdot\dfrac{i}{i}=\dfrac{i}{i^2}=\dfrac{i}{-1}=-i$

83. $\dfrac{1}{i^{-51}}=i^{51}=(i^4)^{12}\cdot i^3=1^{12}\cdot(-i)=-i$

85. $\dfrac{-1}{-i^{12}}=\dfrac{1}{i^{12}}=\dfrac{1}{(i^4)^3}=\dfrac{1}{1^3}=\dfrac{1}{1}=1$

87. $\left(\dfrac{\sqrt{2}}{2}+\dfrac{\sqrt{2}}{2}i\right)^2=\left(\dfrac{\sqrt{2}}{2}\right)^2+2\left(\dfrac{\sqrt{2}}{2}\right)\left(\dfrac{\sqrt{2}}{2}i\right)+\left(\dfrac{\sqrt{2}}{2}i\right)^2=\dfrac{2}{4}+\dfrac{2}{2}i+\dfrac{2}{4}i^2=\dfrac{1}{2}+i-\dfrac{1}{2}=i$

89. The conjugate of $5-3i$ is $5+3i$.

91. The conjugate of $-18i=0-18i$ is $0+18i=18i$.

93. The conjugate of $-\sqrt{8}=-\sqrt{8}+0i$ is $-\sqrt{8}-0i=-\sqrt{8}$.

95. $\dfrac{3}{-i}=\dfrac{3}{-i}\cdot\dfrac{i}{i}=\dfrac{3i}{-i^2}=\dfrac{3i}{-(-1)}=\dfrac{3i}{1}=3i$

97. $\dfrac{-10}{i}=\dfrac{-10}{i}\cdot\dfrac{-i}{-i}=\dfrac{10i}{-(-1)}=\dfrac{10i}{1}=10i$

99. $\dfrac{1-3i}{1+i}=\dfrac{1-3i}{1+i}\cdot\dfrac{1-i}{1-i}=\dfrac{1-i-3i+3i^2}{1^2-i^2}=\dfrac{-2-4i}{1-(-1)}=\dfrac{-2-4i}{2}=-1-2i$

101. $\dfrac{-3+4i}{2-i}=\dfrac{-3+4i}{2-i}\cdot\dfrac{2+i}{2+i}=\dfrac{-6-3i+8i+4i^2}{2^2-i^2}=\dfrac{-10+5i}{4-(-1)}=\dfrac{-10+5i}{5}=-2+i$

103. $\dfrac{4-3i}{4+3i}=\dfrac{4-3i}{4+3i}\cdot\dfrac{4-3i}{4-3i}=\dfrac{16-12i-12i+9i^2}{4^2-(3i)^2}=\dfrac{7-24i}{16-(-9)}=\dfrac{7-24i}{25}=\dfrac{7}{25}-\dfrac{24}{25}i$

105. The method involves multiplying by 1, the multiplicative identity.

3.2: Quadratic Functions and Graphs

1. Since $a>0$, the parabola opens upward. The vertex is $(4,-3)$. The graph is shown in B.

3. Since $a>0$, the parabola opens upward. The vertex is $(-4,-3)$. The graph is shown in D.

5. (a) $P(x)=x^2-2x-15 \Rightarrow P(x)+15=x^2-2x \Rightarrow P(x)+15+1=x^2-2x+1 \Rightarrow$

   $P(x)+16=(x-1)^2 \Rightarrow P(x)=(x-1)^2-16$

   (b) The vertex is $(1,-16)$.

   (c) See Figure 5a and 5b.

7. (a) $P(x)=-x^2-3x+10 \Rightarrow -P(x)=x^2+3x-10 \Rightarrow -P(x)+10=x^2+3x \Rightarrow$

   $-P(x)+10+\dfrac{9}{4}=x^2+3x+\dfrac{9}{4} \Rightarrow -P(x)+\dfrac{49}{4}=\left(x+\dfrac{3}{2}\right)^2 \Rightarrow -P(x)=\left(x+\dfrac{3}{2}\right)^2-\dfrac{49}{4} \Rightarrow$

   $P(x)=-\left(x+\dfrac{3}{2}\right)^2+\dfrac{49}{4}$

   (b) The vertex is $\left(-\dfrac{3}{2},\dfrac{49}{4}\right)$ or $(-1.5,12.25)$.

   (c) See Figure 7a and 7b.

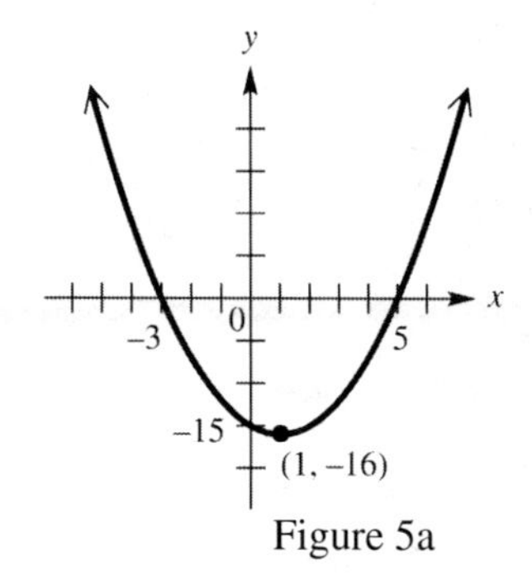

Figure 5a

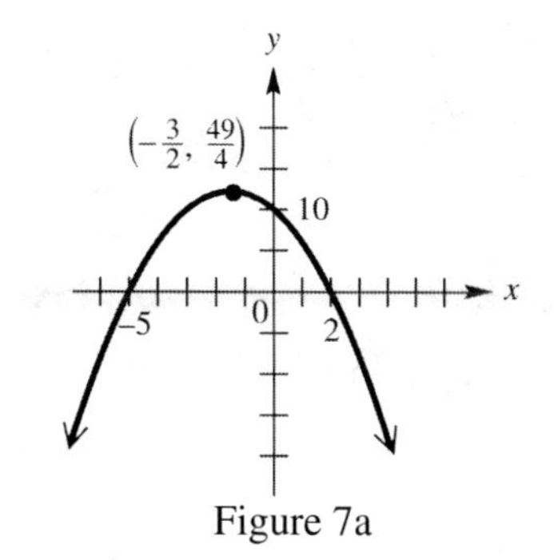

Figure 7a

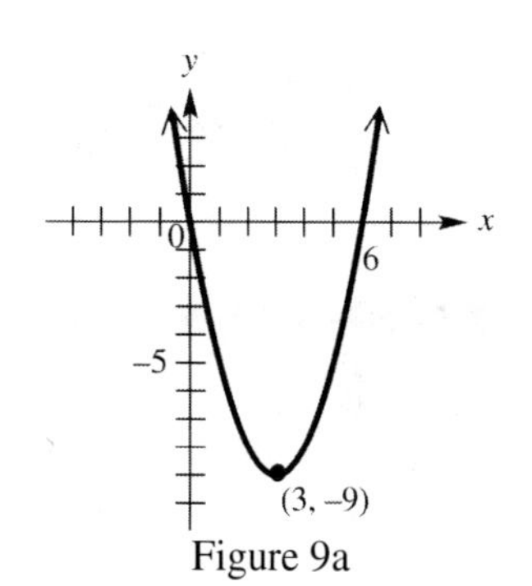

Figure 9a

[-10,10] by [-20,10]
Xscl = 1 Yscl = 5

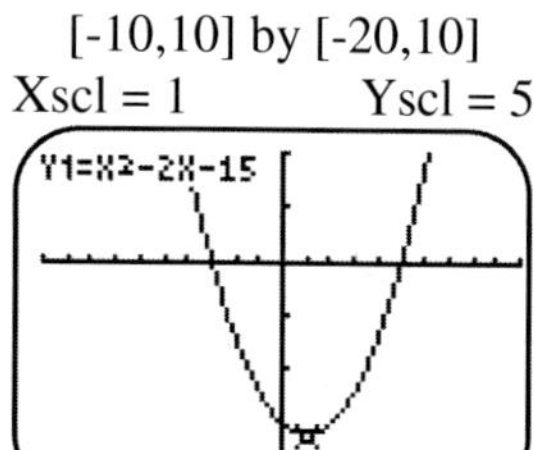

Figure 5b

[-10,10] by [-5,15]
Xscl = 1 Yscl = 1

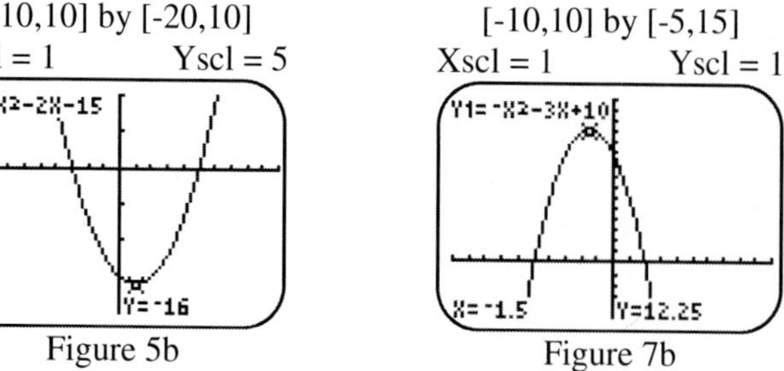

Figure 7b

[-10,10] by [-15,5]
Xscl = 1 Yscl = 1

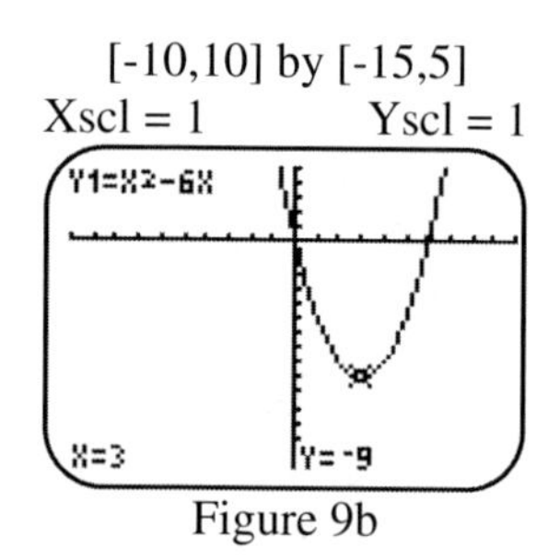

Figure 9b

9. (a) $P(x)=x^2-6x \Rightarrow P(x)+9=x^2-6x+9 \Rightarrow P(x)+9=(x-3)^2 \Rightarrow P(x)=(x-3)^2-9$

(b) The vertex is $(3,-9)$.

(c) See Figure 9a and 9b.

11. (a). $P(x)=2x^2-2x-24 \Rightarrow \dfrac{P(x)}{2}=x^2-x-12 \Rightarrow \dfrac{P(x)}{2}+12=x^2-x \Rightarrow$

$$\frac{P(x)}{2}+12+\frac{1}{4}=x^2-x+\frac{1}{4} \Rightarrow \frac{P(x)}{2}+\frac{49}{4}=\left(x-\frac{1}{2}\right)^2 \Rightarrow \frac{P(x)}{2}=\left(x-\frac{1}{2}\right)^2-\frac{49}{4} \Rightarrow P(x)=2\left(x-\frac{1}{2}\right)^2-\frac{49}{2}$$

(b) The vertex is $\left(\dfrac{1}{2},-\dfrac{49}{2}\right)$ or $(0.5,-24.5)$.

(c) See Figure 11a and 11b.

13. (a) $P(x)=-2x^2+6x \Rightarrow \dfrac{P(x)}{-2}=x^2-3x \Rightarrow \dfrac{P(x)}{-2}+\dfrac{9}{4}=x^2-3x+\dfrac{9}{4} \Rightarrow$

$$\frac{P(x)}{-2}+\frac{9}{4}=\left(x-\frac{3}{2}\right)^2 \Rightarrow \frac{P(x)}{-2}=\left(x-\frac{3}{2}\right)^2-\frac{9}{4} \Rightarrow P(x)=-2\left(x-\frac{3}{2}\right)^2+\frac{9}{2}$$

(b) The vertex is $\left(\dfrac{3}{2},\dfrac{9}{2}\right)$ or $(1.5,4.5)$.

(c) See Figure 13a and 13b.

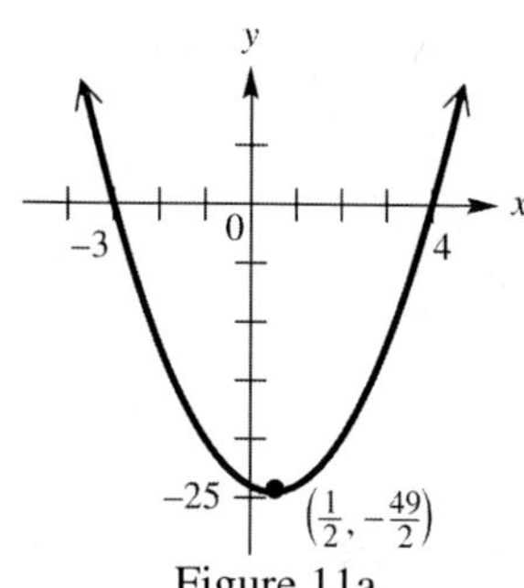

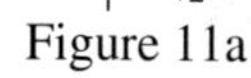

Figure 11a

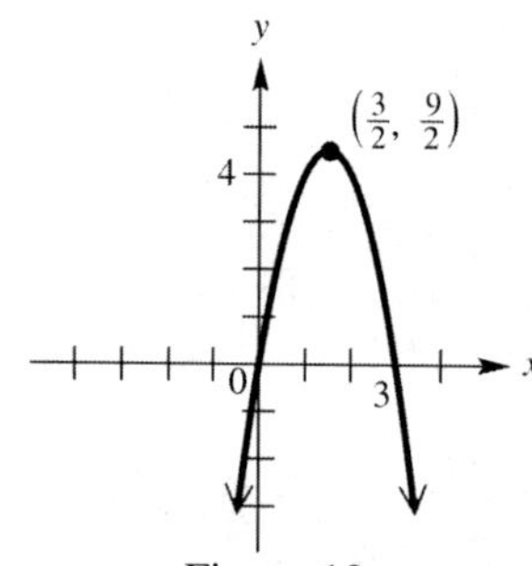

Figure 13a

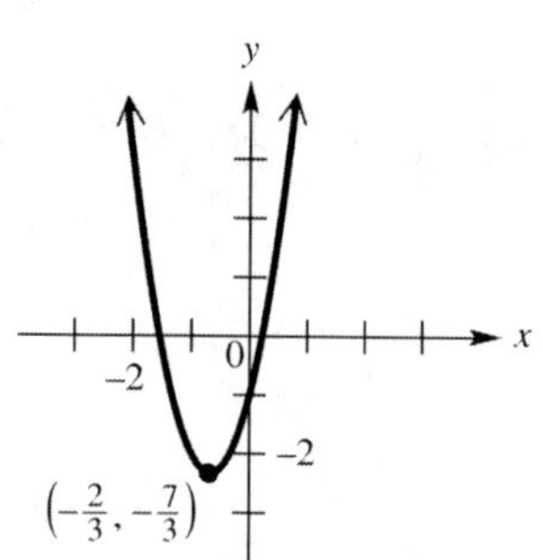

Figure 15a

[-10,10] by [-30,10]
Xscl = 1 Yscl = 5

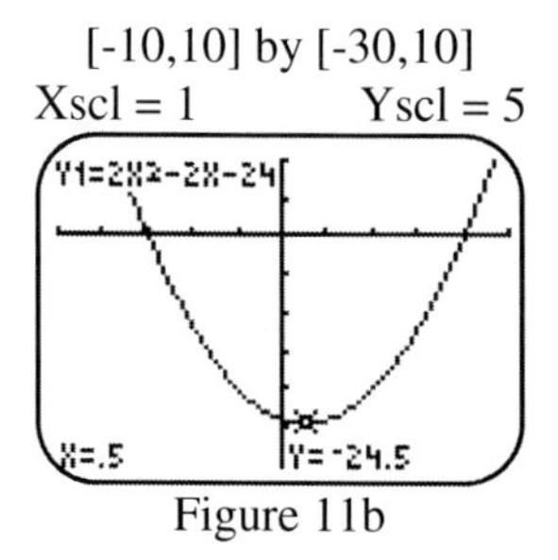

Figure 11b

[-5,5] by [-10,10]
Xscl = 1 Yscl = 1

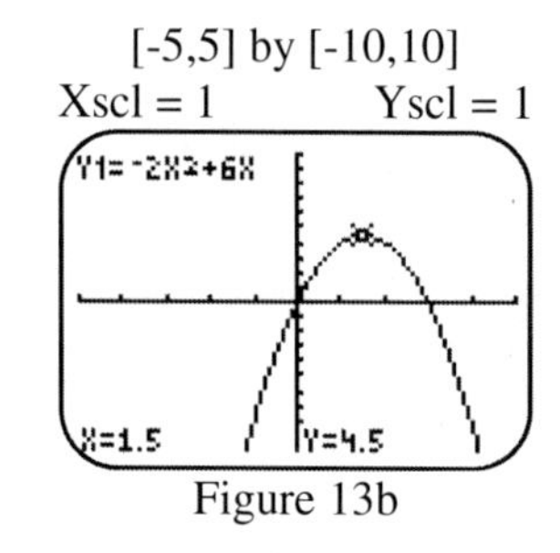

Figure 13b

[-5,5] by [-5,15]
Xscl = 1 Yscl = 1

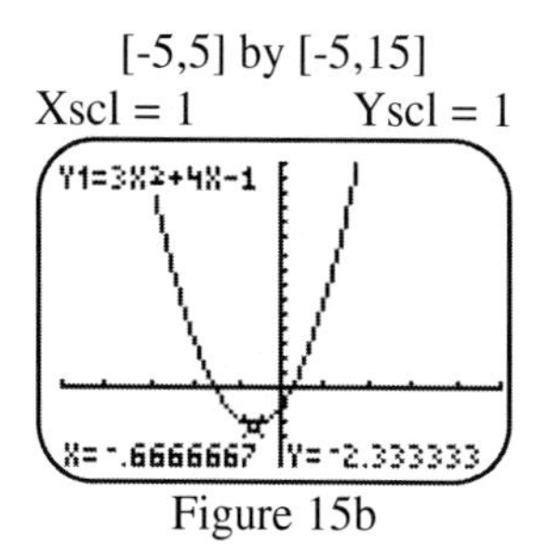

Figure 15b

15. (a) $P(x) = 3x^2 + 4x - 1 \Rightarrow \frac{P(x)}{3} = x^2 + \frac{4}{3}x - \frac{1}{3} \Rightarrow \frac{P(x)}{3} + \frac{1}{3} = x^2 + \frac{4}{3}x \Rightarrow$

$\frac{P(x)}{3} + \frac{1}{3} + \frac{4}{9} = x^2 + \frac{4}{3}x + \frac{4}{9} \Rightarrow \frac{P(x)}{3} + \frac{7}{9} = \left(x + \frac{2}{3}\right)^2 \Rightarrow \frac{P(x)}{3} = \left(x + \frac{2}{3}\right)^2 - \frac{7}{9} \Rightarrow P(x) = 3\left(x + \frac{2}{3}\right)^2 - \frac{7}{3}$

(b). The vertex is $\left(-\frac{2}{3}, -\frac{7}{3}\right)$ or $(-0.67, -2.33)$.

(c) See Figure 15a and 15 b.

17. (a) Since $a > 0$, the parabola opens upward. The vertex is (4, −2). The graph is given in D.

(b) Since $a > 0$, the parabola opens upward. The vertex is (2, −4). The graph is given in B.

(c) Since $a < 0$, the parabola opens downward. The vertex is (4, −2). The graph is given in C.

(d) Since $a < 0$, the parabola opens downward. The vertex is (2, −4). The graph is given in A.

19. (a) $(2,0)$ (b) $D:(-\infty,\infty),\ R:[0,\infty)$ (c) $x = 2$ (d) $[2,\infty)$ (e) $(-\infty,2]$ (f) Min.: $P(2) = 0$

21. (a) $(-3,-4)$ (b) $D:(-\infty,\infty),\ R:[-4,\infty)$ (c) $x = -3$ (d) $[-3,\infty)$ (e) $(-\infty,-3]$ (f) Min.: $f(-3) = -4$

23. (a) $(-3,2)$ (b) $D:(-\infty,\infty),\ R:(-\infty,2]$ (c) $x = -3$ (d) $(-\infty,-3]$ (e) $[-3,\infty)$ (f) Max.: $P(-3) = 2$

25. (a) $x = -\frac{b}{2a} = -\frac{-10}{2(1)} = 5;\ y = P(5) = (5)^2 - 10(5) + 21 = -4 \Rightarrow$ Vertex: (5, -4)

(b) See Figure 25.

27. (a) $x = -\frac{b}{2a} = -\frac{4}{2(-1)} = 2;\ y = -(2)^2 + 4(2) - 2 = 2 \Rightarrow$ Vertex: (2, 2)

(b) See Figure 27.

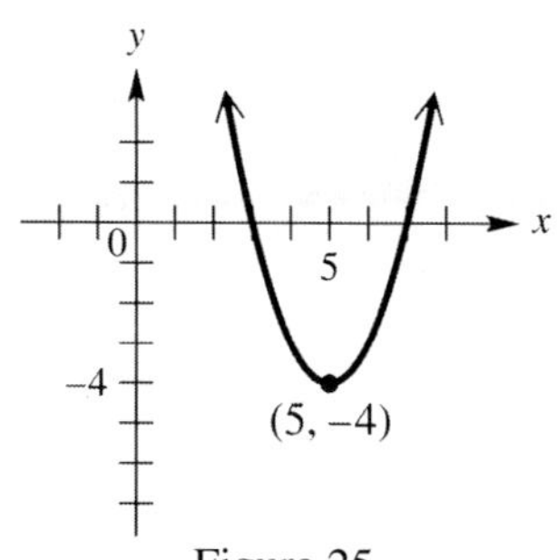

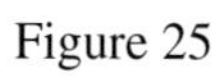
Figure 25

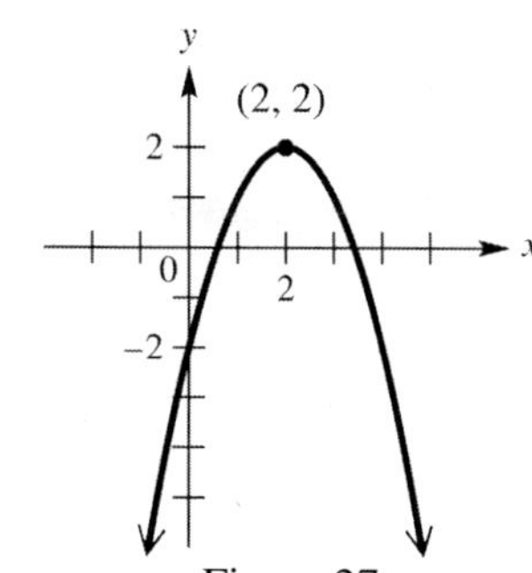

Figure 27

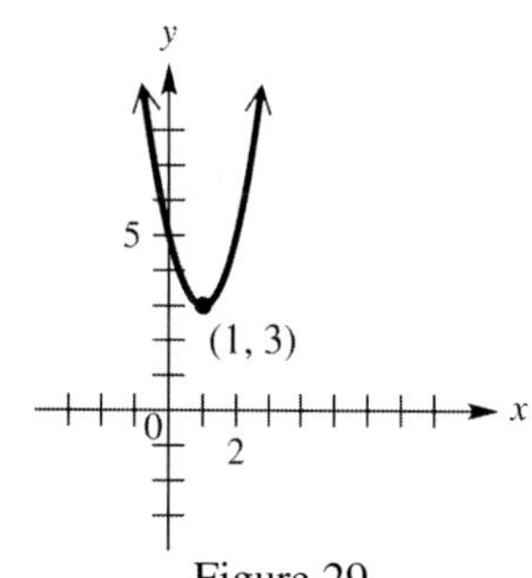

Figure 29

29. (a) $x = -\frac{b}{2a} = -\frac{(-4)}{2(2)} = 1;\ y = f(1) = 2(1)^2 - 4(1) + 5 = 3 \Rightarrow$ Vertex: (1, 3)

(b) See Figure 29.

31. (a) $x = -\frac{b}{2a} = -\frac{24}{2(-3)} = 4;\ y = f(4) = -3(4)^2 + 24(4) - 46 = 2 \Rightarrow$ Vertex: (4, 2)

(b) See Figure 31.

33. (a) $x = -\frac{b}{2a} = -\frac{-2}{2(2)} = \frac{1}{2};\ y = f\left(\frac{1}{2}\right) = 2\left(\frac{1}{2}\right)^2 - 2\left(\frac{1}{2}\right) + 1 = \frac{1}{2} \Rightarrow$ Vertex: $\left(\frac{1}{2}, \frac{1}{2}\right)$.

(b) See Figure 33.

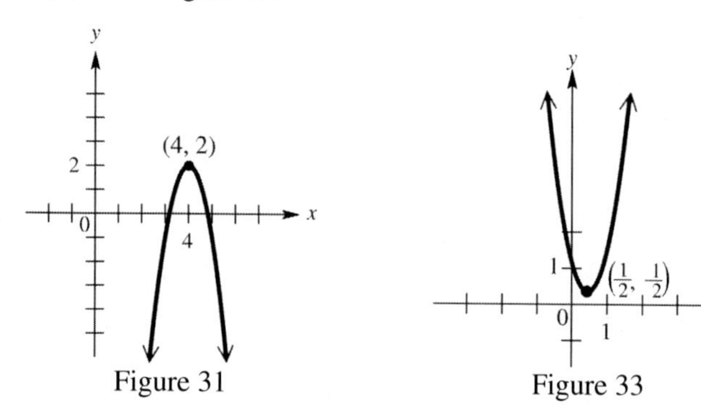

Figure 31 Figure 33

35. The graph is shown in Figure 35.

(a) The vertex is approximately (2.71, 5.20)

(b) The x-intercepts are approximately $(-1.33, 0)$ and $(6.74, 0)$.

37. The graph is shown in Figure 37.

(a) The vertex is approximately (1.12, 0.56).

(b) There are no x-intercepts.

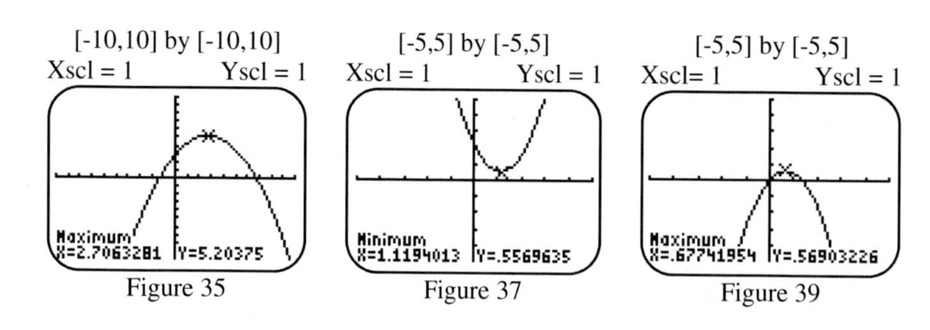

Figure 35 Figure 37 Figure 39

39. The graph is shown in Figure 39.

(a) The vertex is approximately (0.68, 0.57).

(b) The x-intercepts are approximately $(0, 0)$ and $(1.35, 0)$.

41. The minimum value of 3 is found at the vertex.

43. The graph would not intersect the line $y = 1$. There are no solutions to $f(x) = 1$.

45. (a) From the symmetry of the y-values in the table, the vertex is $(4,-12)$.

(b) Since all other y-values are larger than –12, the vertex is a minimum point.

(c) The minimum value of the function is –12 and is located at the vertex.

(d) Since the function is quadratic with a minimum value of –12, the range is $[-12,\infty)$.

47. (a) From the symmetry of the y-values in the table, the vertex is $(1.5,2)$.

(b) Since all other y-values are smaller than 2, the vertex is a maximum point.

(c) The maximum value of the function is 2 and is located at the vertex.

(d) Since the function is quadratic with a maximum value of 2, the range is $(-\infty,2]$.

49. Since the data are in the shape of a parabola that opens downward, a quadratic model with $a<0$ is appropriate.

51. Since the data are in the shape of a parabola that opens upward, a quadratic model with $a>0$ is appropriate.

53. Since the data are in the shape of a line that rises from left to right, a linear model with $m>0$ is appropriate.

55. Let $h=-1,\ k=-4,\ x=5$ and $P(x)=104$ in $P(x)=a(x-h)^2+k$ and determine the value of a.

$104=a(5+1)^2-4\Rightarrow 104=a(6)^2-4\Rightarrow 108=36a\Rightarrow a=3$

$P(x)=3(x+1)^2-4\Rightarrow P(x)=3(x^2+2x+1)-4\Rightarrow P(x)=3x^2+6x-1$

57. Let $h=8,\ k=3,\ x=10$ and $P(x)=5$ in $P(x)=a(x-h)^2+k$ and determine the value of a.

$5=a(10-8)^2+3\Rightarrow 5=a(2)^2+3\Rightarrow 2=4a\Rightarrow a=0.5$. $P(x)=0.5(x-8)^2+3\Rightarrow$

$P(x)=0.5(x^2-16x+64)+3\Rightarrow P(x)=0.5x^2-8x+35$ or $P(x)=\frac{1}{2}x^2-8x+35$

59. Let $h=-4,\ k=-2,\ x=2$ and $P(x)=-26$ in $P(x)=a(x-h)^2+k$ and determine the value of a.

$-26=a(2+4)^2-2\Rightarrow -26=a(6)^2-2\Rightarrow -24=36a\Rightarrow a=-\frac{2}{3}$

$P(x)=-\frac{2}{3}(x+4)^2-2\Rightarrow P(x)=-\frac{2}{3}(x^2+8x+16)-2\Rightarrow P(x)=-\frac{2}{3}x^2-\frac{16}{3}x-\frac{38}{3}$

61. (a) Since the parabola opens upward and the vertex is (4,90), the function decreases from 0 to 4 and increases from 4 to 8. Therefore, the heart rate decreases during the first 4 minutes and increases during the next 4 minutes.

(b) Since parabola opens upward the minimum occurs at the vertex (4, 90). Therefore, the minimum heart rate is 90 bpm after 4 minutes.

63. (a) The data are not linear since the data increase and then decrease.

(b) Using coordinate (2,120), $f(x)=a(x-2)^2+120$. To find , use $(0,84)\Rightarrow$

$84=a(0-2)^2+120\Rightarrow -36=a(4)\Rightarrow a=-9$; $f(x)=-9(x-2)^2+120$.

(c) $D=\{x\,|\,0\le x\le 4\}$

65. (a) The value of t cannot be negative because t represents elapsed time after the rock is launched.

(b) The original height of the rock is $s_0=0$ which represents ground level.

(c) Since $v_0 = 90$ and $s_0 = 0$, $s(t) = -16t^2 + v_0 t + s_0 \Rightarrow s(t) = -16t^2 + 90t$.

(d) $s(1.5) = -16(1.5)^2 + 90(1.5) = 99$ feet

(e) $x = -\frac{b}{2a} = -\frac{90}{2(-16)} = 2.8125$; $y \approx s(2.8125) = -16(2.8125)^2 + 90(2.8125) = 126.5625$. The vertex is (2.8125, 126.5625). The rock reaches a maximum height of 126.5625 feet after 2.8125 seconds. A graph of $y = -16x^2 + 90x$ (not shown) also gives a vertex of (2.8125, 126.5625).

(f) A graph of $y = -16x^2 + 90x$ (not shown) has x-intercepts at (0, 0) and (5.625, 0). The rock will hit the ground after 5.625 seconds.

67. (a) The graphs of $-16x^2 + 150x$ and $y_2 = 355$ are shown in Figure 67a. The graphs do not intersect, which indicates that the ball does not reach a height of 355 feet.

(b) The graphs of $y_1 = -16x^2 + 250x + 30$ and $y_2 = 355$ are shown in Figure 67b. The graphs intersect when $x \approx 1.43$ and $x \approx 14.19$ which indicates that the ball is 355 feet high at about 1.4 and 14.2 seconds.

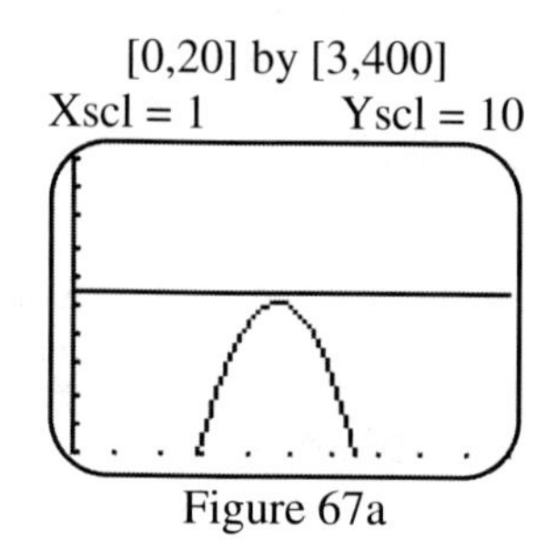

Figure 67a

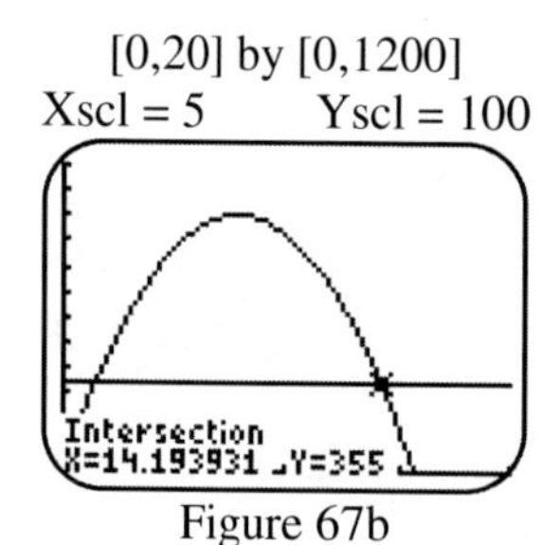

Figure 67b

69. See Figure 69.

71. See Figure 71.

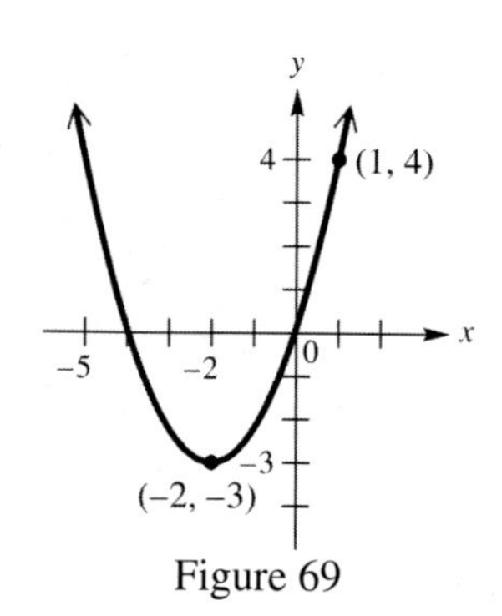

Figure 69

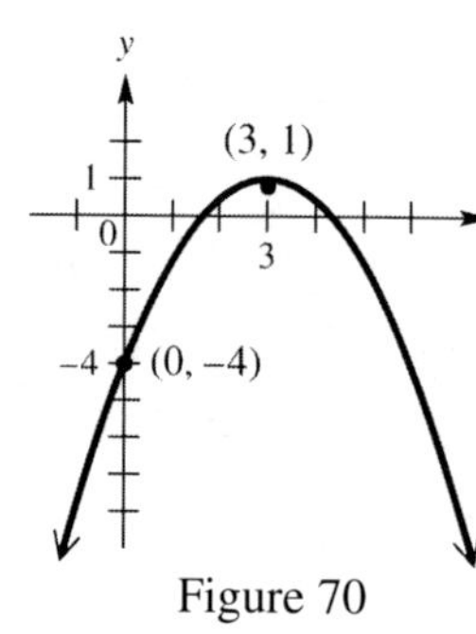

Figure 70

3.3: Quadratic Equations and Inequalities

1. $x^2 = 4 \Rightarrow \sqrt{x^2} = \sqrt{4} \Rightarrow x = \pm 2$; therefore, G.

3. $x^2 + 2 = 0 \Rightarrow x^2 = -2 \Rightarrow \sqrt{x^2} = \sqrt{-2} \Rightarrow x = \pm i\sqrt{2}$; therefore, C.

5. $x^2=-8 \Rightarrow \sqrt{x^2}=\sqrt{-8} \Rightarrow x=\pm 2i\sqrt{2}$; therefore, H.

7. $x-2=0 \Rightarrow x=2$; therefore, D.

9. Equation D, which is in the form $ab=0$, is set up for the zero product property. Either $3x+1=0 \Rightarrow x=-\frac{1}{3}$ or $x-7=0 \Rightarrow x=7$. The solution set is $\left\{-\frac{1}{3},7\right\}$.

11. Equation C, which has an x^2 coefficient of 1, does not require step 1 of the method of completing the square. $x^2+x=12 \Rightarrow x^2+x+\frac{1}{4}=12+\frac{1}{4} \Rightarrow \left(x+\frac{1}{2}\right)^2=\frac{49}{4} \Rightarrow x+\frac{1}{2}=\pm\frac{7}{2} \Rightarrow x=-\frac{1}{2}\pm\frac{7}{2}$. The solution set is $\{-4,3\}$.

13. $x^2=16 \Rightarrow x=\pm\sqrt{16} \Rightarrow x=\{\pm 4\}$; For graphical support, show that the graphs of $y_1=x^2$ and $y_2=16$ intersect when $x=\pm 4$.

15. $2x^2=90 \Rightarrow x^2=45 \Rightarrow x=\pm\sqrt{45} \Rightarrow x=\pm 3\sqrt{5}$; For graphical support, show that the graphs of $y_1=2x^2$ and $y_2=90$ intersect when $x=\pm 3\sqrt{5}\approx\pm 6.7$.

17. $x^2=-16 \Rightarrow x=\pm i\sqrt{16} \Rightarrow x=\pm 4i$; For graphical support, show that the graphs of $y_1=x^2$ and $y_2=-16$ do not intersect; therefore, no real solutions.

19. $x^2=-18 \Rightarrow x=\pm i\sqrt{18} \Rightarrow x=\pm 3i\sqrt{2}$; For graphical support, show that the graphs of $y_1=x^2$ and $y_2=-18$ do not intersect; therefore, no real solutions.

21. $(3x-1)^2=12 \Rightarrow 3x-1=\pm\sqrt{12} \Rightarrow 3x-1=\pm 2\sqrt{3} \Rightarrow 3x=\pm 2\sqrt{3}+1 \Rightarrow x=\frac{1\pm 2\sqrt{3}}{3}$. For graphical support, show that the graphs of $y_1=(3x-1)^2$ and $y_2=12$ intersect when $x=\frac{1\pm 2\sqrt{3}}{3}$; $x\approx -0.8$ or 1.5.

23. $(5x-3)^2=-3 \Rightarrow 5x-3=\pm\sqrt{-3} \Rightarrow 5x-3=\pm i\sqrt{3} \Rightarrow 5x=\pm i\sqrt{3}+3 \Rightarrow x=\frac{3}{5}\pm\frac{\sqrt{3}}{5}i$. For graphical support, show that the graphs of $y_1=(5x-3)^2$ and $y_2=-3$ do not intersect; therefore, there are no real solutions.

25. $x^2=2x+24 \Rightarrow x^2-2x-24=0 \Rightarrow (x-6)(x+4)=0 \Rightarrow x-6=0$ or $x+4=0 \Rightarrow x=-4$ or 6. For graphical support, show that the graph of $y_1=x^2-2x-24=0$ has x-intercepts $x=-4$ and 6.

27. $3x^2-2x=0 \Rightarrow x(3x-2)=0 \Rightarrow x=0$ or $3x-2=0 \Rightarrow x=0$ or $\frac{2}{3}$. For graphical support, show that the graph of $y_1=3x^2-2x$ has x-intercepts $x=0$ and $\frac{2}{3}$.

29. $x(14x+1)=3 \Rightarrow 14x^2+x-3=0 \Rightarrow (2x+1)(7x-3)=0 \Rightarrow 2x+1=0$ or $7x-3=0 \Rightarrow x=-\frac{1}{2}$ or $\frac{3}{7}$. For graphical support, show that the graph of $y_1=14x^2+x-3$ has x-intercepts $x=-\frac{1}{2}$ and $\frac{3}{7}$.

31. $-4+9x-2x^2=0 \Rightarrow (-4+x)(1-2x)=0 \Rightarrow -4+x=0$ or $1-2x=0 \Rightarrow x=\frac{1}{2}$ or 4. For graphical support, show that the graph of $y_1=-4+9x-2x^2$ has x-intercepts $x=\frac{1}{2}$ and 4.

33. $\frac{1}{3}x^2-\frac{1}{6}x=24 \Rightarrow \frac{1}{3}x^2+\frac{1}{3}x-24=0 \Rightarrow x^2-x-72=0 \Rightarrow (x-9)(x+8)=0 \Rightarrow x-9=0$ or $x+8=0 \Rightarrow x=-8$ or 9. For graphical support, show that the graph of $y_1=\frac{1}{3}x^2-\frac{1}{3}x-24$ has x-intercepts $x=-8$ and 9.

35. $(x+2)(x-1)=7x+5 \Rightarrow x^2+x-2=7x+5 \Rightarrow x^2-6x-7=0 \Rightarrow (x-7)(x+1)=0 \Rightarrow x-7=0$ or $x+1=0 \Rightarrow x=-1$ or 7. For graphical support, show that the graph of $y_1=x^2-6x-7$ has x-intercepts $x=-1$ and 7.

37. Use the quadratic formula to solve $x^2-2x-4=0$; therefore, $a=1$, $b=-2$ and $c=-4$

$x=\frac{-(-2)\pm\sqrt{(-2)^2-4(1)(-4)}}{2(1)}=\frac{2\pm\sqrt{4+16}}{2}=\frac{2\pm\sqrt{20}}{2}=\frac{2\pm2\sqrt{5}}{2}=\frac{2\left(1\pm\sqrt{5}\right)}{2} \Rightarrow x=1\pm\sqrt{5}$. For graphical support, show that the graph of $y_1=x^2-2x-4$ has x-intercepts $x=1\pm\sqrt{5}$; $x\approx-1.24$ and 3.24.

39. Use the quadratic formula to solve $y_1=2x^2+2x=-1 \Rightarrow 2x^2+2x+1=0$; therefore, $a=2$, $b=2$ and $c=1$

$x=\frac{-2\pm\sqrt{(2)^2-4(2)(1)}}{2(2)}=\frac{-8\pm\sqrt{4-8}}{4}=\frac{-2\pm\sqrt{-4}}{4}=\frac{-2\pm2i}{4}=\frac{2(-1\pm i)}{4} \Rightarrow x=-\frac{1}{2}\pm\frac{1}{2}i$. For graphical support, show that the graph of $y_1=2x^2+2x+1$ has no x-intercepts; therefore, there are no real solutions.

41. Use the quadratic formula to solve $x(x-1)=1 \Rightarrow x^2-x-1=0$; therefore, $a=1$, $b=-1$ and $c=-1$

$x=\frac{-(-1)\pm\sqrt{(-1)^2-4(1)(-1)}}{2(1)}=\frac{1\pm\sqrt{1+4}}{2} \Rightarrow x=\frac{1\pm\sqrt{5}}{2}$. For graphical support, show that the graph of $y_1=x^2-x-1$ has x-intercepts $x=\frac{1\pm\sqrt{5}}{2}$; $x\approx-0.62$ and 1.62.

43. Use the quadratic formula to solve $x^2-5x=x-7 \Rightarrow x^2-6x+7=0$; therefore, $a=1$, $b=-6$, and $c=7$.

$x=\frac{-(-6)\pm\sqrt{(-6)^2-4(1)(7)}}{2(1)}=\frac{6\pm\sqrt{36-28}}{2}=\frac{6\pm\sqrt{8}}{2}=\frac{6\pm2\sqrt{2}}{2}=\frac{2\left(3\pm\sqrt{2}\right)}{2} \Rightarrow x=3\pm\sqrt{2}$. For graphical support, show that the graph of $y_1=x^2-6x+7$ has x-intercepts $x=3\pm\sqrt{2}$; $x\approx1.59$ and 4.41.

45. Use the quadratic formula to solve $4x^2-12x=-11 \Rightarrow 4x^2-12x+11=0$; therefore, $a=4$, $b=-12$ and $c=11$. $x=\frac{-(-12)\pm\sqrt{(-12)^2-4(4)(11)}}{2(4)}=\frac{12\pm\sqrt{144-176}}{8}=\frac{12\pm\sqrt{-32}}{8}=\frac{12\pm4i\sqrt{2}}{8}=\frac{4\left(3\pm i\sqrt{2}\right)}{8} \Rightarrow$

$x=\frac{3}{2}\pm\frac{\sqrt{2}}{2}i.$ For graphical support, show that the graph of $y_1=4x^2-12x+11$ has no x-intercepts; therefore, there are no real solutions.

47. Use the quadratic formula to solve $\frac{1}{3}x^2+\frac{1}{4}x-3=0\Rightarrow 12\left(\frac{1}{3}x^2+\frac{1}{4}x-3=0\right)\Rightarrow 4x^2+3x-36=0;$ therefore,

$a=4, b=3$ and $c=-36.$ $x=\frac{-3\pm\sqrt{(3)^2-4(4)(-36)}}{2(4)}=\frac{-3\pm\sqrt{9+576}}{8}=\frac{-3\pm\sqrt{585}}{8}\Rightarrow x=\frac{-3\pm 3\sqrt{65}}{8}.$ For graphical support, show that the graph of $y_1=4x^2+3x-36$ has x-intercepts

$x=\frac{-3\pm 3\sqrt{65}}{8};$ $x\approx -3.4$ and 2.6.

49. Use the quadratic formula to solve $(3-x)^2=25\Rightarrow 9-6x+x^2=25\Rightarrow x^2-6x-16=0;$ therefore,

$a=1, b=-6$ and $c=-16.$ $x=\frac{-(-6)\pm\sqrt{(-6)^2-4(1)(-16)}}{2(1)}=\frac{6\pm\sqrt{36+64}}{2}=\frac{6\pm\sqrt{100}}{2}\Rightarrow\frac{6\pm 10}{2}\Rightarrow$

$x=-2$ or 8. For graphical support, show that the graph of $y_1=x^2-6x-16$ has x-intercepts $x=-2$ and 8.

51. Use the quadratic formula to solve $2x^2-4x=1\Rightarrow 2x^2-4x-1=0;$ therefore $a=2,$ $b=-4$ and $c=-1.$

$x=\frac{-(-4)\pm\sqrt{(-4)^2-4(2)(-1)}}{2(2)}=\frac{4\pm\sqrt{16+8}}{4}=\frac{4\pm\sqrt{24}}{4}\Rightarrow\frac{4\pm 2\sqrt{6}}{4}=\frac{2\left(2\pm\sqrt{6}\right)}{4}\Rightarrow x=\frac{2\pm\sqrt{6}}{2}.$ For graphical support, show that the graph of $y_1=2x^2-4x-1$ has x-intercepts $x=\frac{2\pm\sqrt{6}}{2};$ $x\approx -0.2$ and 2.2.

53. Use the quadratic formula to solve $x^2=-1-x\Rightarrow x^2+x+1=0;$ therefore, $a=1,$ $b=1$ and $c=1.$

$x=\frac{-1\pm\sqrt{(1)^2-4(1)(1)}}{2(1)}=\frac{-1\pm\sqrt{1-4}}{2}=\frac{-1\pm\sqrt{-3}}{2}\Rightarrow x=-\frac{1}{2}\pm\frac{\sqrt{3}}{2}i.$ For graphical support, show that the graph of $y_1=x^2+x+1$ has no x-intercepts; therefore, there are no real solutions.

55. Use the quadratic formula to solve $4x^2-20x+25=0;$ therefore, $a=4,$ $b=-20$ and $c=25.$

$x=\frac{-(-20)\pm\sqrt{(-20)^2-4(4)(25)}}{2(4)}=\frac{20\pm\sqrt{400-400}}{8}=\frac{20\pm 0}{8}\Rightarrow x=\frac{20}{8}=\frac{5}{2}.$ For graphical support, show that the graph of $y_1=4x^2-20x+25$ has x-intercept $x=\frac{5}{2}=2.5.$

57. Use the quadratic formula to solve $-3x^2+4x+4=0;$ therefore, $a=-3,$ $b=4$ and $c=4.$

$x=\frac{-4\pm\sqrt{(4)^2-4(-3)(4)}}{2(-3)}=\frac{-4\pm\sqrt{16+48}}{-6}=\frac{-4\pm\sqrt{64}}{-6}=\frac{-4\pm 8}{-6}=\frac{-2(2\pm 4)}{-6}=\frac{2\pm 4}{3}\Rightarrow x=-\frac{2}{3}$ or 2. For graphical support, show that the graph of $y_1=-3x^2+4x+4$ has x-intercepts $x=-\frac{2}{3}$ or 2; $x\approx -0.7$ and 2.

59. Use the quadratic formula to solve $(x+5)(x-6)=(2x-1)(x-4) \Rightarrow x^2-x-30=2x^2-9x+4 \Rightarrow$

$x^2-8x+34=0$; therefore, $a=1$, $b=-8$, and $c=34$. $x=\dfrac{-(-8)\pm\sqrt{(-8)^2-4(1)(34)}}{2(1)}=\dfrac{8\pm\sqrt{64-136}}{2}=$

$\dfrac{8\pm\sqrt{-72}}{2}=\dfrac{8\pm 6i\sqrt{2}}{2}=\dfrac{2\left(4\pm 3i\sqrt{2}\right)}{2}\Rightarrow x=4\pm 3i\sqrt{2}$. For graphical support, show that the graph of

$y_1=2x^2-9x+4$ has no x-intercepts; therefore, there are no real solutions.

61. $x^2-2x=2\Rightarrow x^2-2x+1=2+1\Rightarrow(x-1)^2=3\Rightarrow x-1=\pm\sqrt{3}\Rightarrow x=1\pm\sqrt{3}$

63. $2x^2+6x-3=0\Rightarrow x^2+3x-\dfrac{3}{2}=0\Rightarrow x^2+3x=\dfrac{3}{2}\Rightarrow x^2+3x+\dfrac{9}{4}=\dfrac{3}{2}+\dfrac{9}{4}\Rightarrow\left(x+\dfrac{3}{2}\right)^2=\dfrac{15}{4}\Rightarrow$

$\left(x+\dfrac{3}{2}\right)=\pm\sqrt{\dfrac{15}{4}}\Rightarrow x=-\dfrac{3}{2}\pm\dfrac{\sqrt{15}}{2}\Rightarrow x=\dfrac{-3\pm\sqrt{15}}{2}$

65. $x(x-1)=3\Rightarrow x^2-x=3\Rightarrow x^2-x+\dfrac{1}{4}=3+\dfrac{1}{4}\Rightarrow\left(x-\dfrac{1}{2}\right)^2=\dfrac{13}{4}\Rightarrow x-\dfrac{1}{2}=\pm\sqrt{\dfrac{13}{4}}\Rightarrow x=\dfrac{1\pm\sqrt{13}}{2}$

67. $2x^2-x+3=0\Rightarrow x^2-\dfrac{1}{2}x+\dfrac{3}{2}=0\Rightarrow x^2-\dfrac{1}{2}x=-\dfrac{3}{2}\Rightarrow x^2-\dfrac{1}{2}x+\dfrac{1}{16}=-\dfrac{3}{2}+\dfrac{1}{16}\Rightarrow\left(x-\dfrac{1}{4}\right)^2=\dfrac{-23}{16}\Rightarrow$

$\left(x-\dfrac{1}{4}\right)=\pm\sqrt{\dfrac{-23}{16}}\Rightarrow x=\dfrac{1}{4}\pm i\dfrac{\sqrt{23}}{4}$

69. For the equation $x^2+8x+16=0$; $a=1$, $b=8$, and $c=16$; therefore, the discriminant is

$(8)^2-4(1)(16)=64-64=0$. Because the discriminant is 0, there is 1 real solution. Since a, b, and c are nonzero integers and the discriminant is a square of an integer the solution is rational.

71. For the equation $4x^2=6x+3\Rightarrow 4x^2-6x-3=0$; $a=4$, $b=-6$, and $c=-3$; therefore, the discriminant is

$(-6)^2-4(4)(-3)=36+48=84$. Because the discriminant is positive, there are 2 real solutions. Since a, b, and c are nonzero integers and the discriminant is not the square of an integer, the solutions are irrational.

73. For the equation $9x^2+11x+4=0; a=9$, $b=11$, and $c=4$; therefore, the discriminant is

$(11)^2-4(9)(4)=121-144=-23$. Because the discriminant is negative, there are no real solutions.

75. If $x=4$ or 5, then $(x-4)(x-5)=0\Rightarrow x^2-9x+20=0$; and $a=1$, $b=-9$, and $c=20$.

77. If $x=1-\sqrt{2}$ or $1+\sqrt{2}$, then $\left(x-\left(1-\sqrt{2}\right)\right)\left(x-\left(1+\sqrt{2}\right)\right)=0\Rightarrow\left(x-1+\sqrt{2}\right)\left(x-1-\sqrt{2}\right)=0\Rightarrow$

$x^2-x-x\sqrt{2}-x+1+\sqrt{2}+x\sqrt{2}-\sqrt{2}-2=0\Rightarrow x^2-2x-1=0$; and $a=1$, $b=-2$, and $c=-1$.

79. If $x=2i$ or $-2i$, then $(x-2i)(x+2i)=0\Rightarrow x^2-4i^2=0\Rightarrow x^2+4=0$; and $a=1, b=0, c=4$.

81. If $x=2-\sqrt{5}$ or $2+\sqrt{5}$, then $\left(x-\left(2-\sqrt{5}\right)\right)\left(x-\left(2+\sqrt{5}\right)\right)=0\Rightarrow\left(x-2+\sqrt{5}\right)\left(x-2-\sqrt{5}\right)=0\Rightarrow$

$x^2-2x-x\sqrt{5}-2x+4+2\sqrt{5}+x\sqrt{5}-2\sqrt{5}-5=0\Rightarrow x^2-4x-1=0$; and $a=1, b=-4, c=-1$.

83. The graph of the function $f(x) = ax^2 + bx + c$ is a parabola, $a < 0$ will make the parabola open downward, and $b^2 - 4ac = 0$ gives us 1 real solution or 1 x-intercept. See Figure 83.

85. The graph of the function $f(x) = ax^2 + bx + c$ is a parabola, $a < 0$ will make the parabola open downward, and $b^2 - 4ac < 0$ gives us no real solutions or no x-intercepts. See Figure 85.

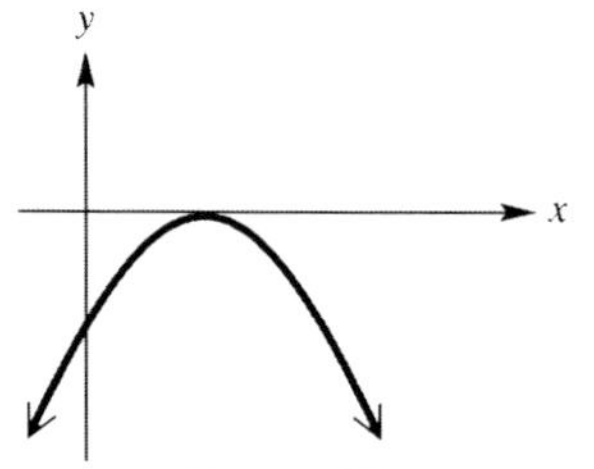

Figure 83

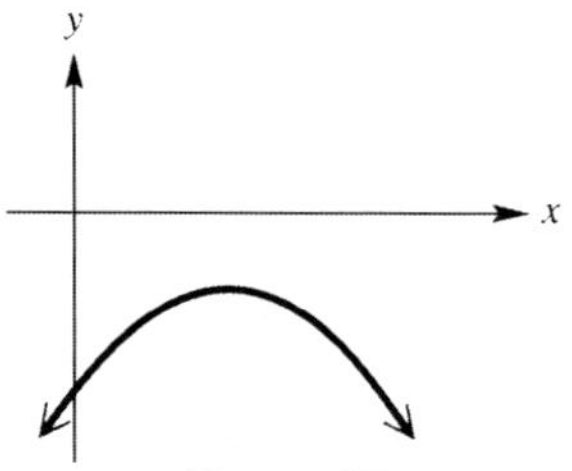

Figure 85

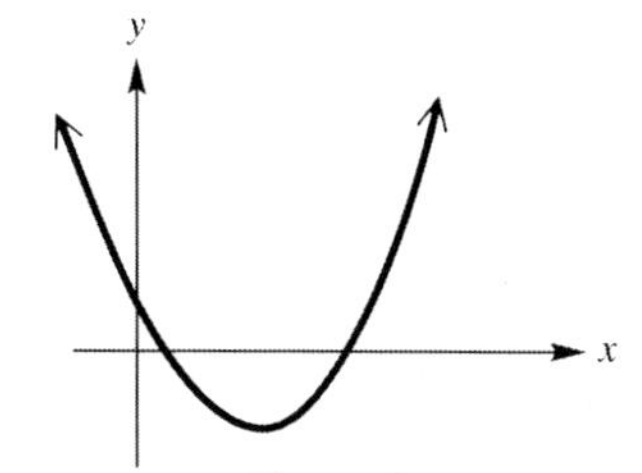

Figure 87

87. The graph of the function $f(x) = ax^2 + bx + c$ is a parabola, $a > 0$ will make the parabola open upward, and $b^2 - 4ac > 0$ gives us 2 real solutions or 2 x-intercepts. See Figure 87.

89. From the graph, $f(x) = 0$ when 2 or 4.

91. From the graph, $f(x) > 0$ for the interval $(-\infty, 2) \cup (4, \infty)$.

93. From the graph, $g(x) < 0$ for the interval $(-\infty, 3) \cup (3, \infty)$.

95. From the graph, $h(x) > 0$ always, but no x-intercepts; therefore, the solution is $(-\infty, \infty)$.

97. From the graph, $h(x) = 0$ has no real solutions, because $h(x)$ has no x-intercepts. The graph is completely above the x-axis; therefore, there are two nonreal complex solutions.

99. The x-coordinate of the vertex of the graph of $y = f(x)$ is the midpoint of the x-intercepts or $\frac{2+4}{2} = 3$.

101. From the graph, $y = g(x)$ will have a y-intercept and it will be negative.

103. (a) First, set $x^2 + 4x + 3 = 0$, then $(x+1)(x+3) = 0$ and the endpoints will be $x = \{-3, -1\}$. The numbers -3 and -1 divide the number line into 3 intervals. The interval $(-\infty, -3)$ has a positive product, the interval $(-3, -1)$ has a negative product, and the interval $(-1, \infty)$ has a positive product. Therefore, $x^2 + 4x + 3 \geq 0$ for the interval $(-\infty, -3] \cup [-1, \infty)$. The graph of $y_1 = x^2 + 4x + 3$ intersects or is above the x-axis for the interval $(-\infty, -3] \cup [-1, \infty)$.

(b) First, set $x^2 + 4x + 3 = 0$, then $(x+1)(x+3) = 0$ and the endpoints will be $x = \{-3, -1\}$. The numbers -3 and -1 divide the number line into 3 intervals. The interval $(-\infty, -3)$ has a positive product, the interval $(-3, -1)$ has a negative product, and the interval $(-1, \infty)$ has a positive product. Therefore,

$x^2+4x+3<0$ for the interval $(-3,-1)$. The graph of $y_1=x^2+4x+3$ is below the x-axis for the interval $(-3,-1)$.

105. (a) First, $2x^2-9x>-4 \Rightarrow 2x^2-9x+4>0$, now set $2x^2-9x+4=0$ then $(2x-1)(x-4)=0$ and the endpoints will be $x=\frac{1}{2}$ and $x=4$. The numbers $\frac{1}{2}$ and 4 divide the number line into 3 intervals. The interval $\left(-\infty,\frac{1}{2}\right)$ has a positive product, the interval $\left(\frac{1}{2},4\right)$ has a negative product, and the interval $(4,\infty)$ has a positive product. Therefore, $2x^2-9x>-4$ for the interval $\left(-\infty,\frac{1}{2}\right)\cup(4,\infty)$. The graph of $y_1=2x^2-9x+4$ is above the x-axis for the interval $\left(-\infty,\frac{1}{2}\right)\cup(4,\infty)$.

(b) First, $2x^2-9x\le-4 \Rightarrow 2x^2-9x+4\le0$, now set $2x^2-9x+4=0$ then $(2x-1)(x-4)=0$ and the endpoints will be $x=\frac{1}{2}$ and $x=4$. The numbers $\frac{1}{2}$ and 4 divide the number line into 3 intervals. The interval $\left(-\infty,\frac{1}{2}\right)$ has a positive product, the interval $\left(\frac{1}{2},4\right)$ has a negative product, and the interval $(4,\infty)$ has a positive product. Therefore, $2x^2-9x\le-4$ for the interval $\left(\frac{1}{2},4\right)$. The graph of $y_1=2x^2-9x+4$ intersects or is below the x-axis for the interval $\left[\frac{1}{2},4\right]$.

107. (a) First, set $-x^2-x=0$, then $-x(x+1)=0$ and the endpoints will be $x=-1$ and $x=0$. The numbers -1 and 0 divide the number line into 3 intervals. The interval $(-\infty,-1)$ has a negative product, the interval $(-1,0)$ has a positive product, and the interval $(0,\infty)$ has a negative product. Therefore, $-x^2-x\le0$ for the interval $(-\infty,-1]\cup[0,\infty)$. The graph of $y_1=-x^2-x$ intersects or is below the x-axis for the interval $(-\infty,-1]\cup[0,\infty)$.

(b) First, set $-x^2-x=0$, then $-x(x+1)=0$ and the endpoints will be $x=-1$ and $x=0$. The numbers -1 and 0 divide the number line into 3 intervals. The interval $(-\infty,-1)$ has a negative product, the interval $(-1,0)$ has a positive product, and the interval $(0,\infty)$ has a negative product. Therefore, $-x^2-x>0$ for the interval $(-1,0)$. The graph of $y_1=-x^2-x$ is above the x-axis for the interval $(-1,0)$.

109. (a) First, set $x^2-x+1=0$, then by the quadratic formula $x=\frac{-(-1)\pm\sqrt{(-1)^2-4(1)(1)}}{2(1)}=$ $\frac{1\pm\sqrt{-3}}{2}\Rightarrow x=\frac{1}{2}\pm\frac{\sqrt{3}}{2}i$. The inequality has no real solutions, the graph is a parabola opening upward

which does not intersect, but is completely above the x-axis. Therefore, $x^2 - x + 1 < 0$ never happens and the solution is $\varnothing$.

(b) First, set $x^2 - x + 1 = 0$, then by the quadratic formula $x = \frac{-(-1) \pm \sqrt{(-1)^2 - 4(1)(1)}}{2(1)} =$

$\frac{1 \pm \sqrt{-3}}{2} \Rightarrow x = \frac{1}{2} \pm \frac{\sqrt{3}}{2} i.$ The graph of this equation is a parabola opening upward, has no real solutions, and does not intersect, but is completely above the x-axis. Therefore, $x^2 - x + 1 \geq 0$ for all values of x and the solution is the interval $(-\infty, \infty)$.

111. (a) First, $2x + 1 \geq x^2 \Rightarrow x^2 - 2x - 1 \leq 0$, now set $x^2 - 2x - 1 = 0$, then by the quadratic formula

$x = \frac{-(-2) \pm \sqrt{(-2)^2 - 4(1)(-1)}}{2(1)} = \frac{2 \pm \sqrt{8}}{2} = \frac{2(1 \pm \sqrt{2})}{2} \Rightarrow x = 1 \pm \sqrt{2}$, and the endpoints will be $x = 1 - \sqrt{2}$ and $x = 1 + \sqrt{2}$. The numbers $1 - \sqrt{2}$ and $1 + \sqrt{2}$ divide the number line into 3 intervals. The interval $(-\infty, 1 - \sqrt{2})$ has a positive product, the interval $(1 - \sqrt{2}, 1 + \sqrt{2})$ has a negative product, and the interval $(1 + \sqrt{2}, \infty)$ has a positive product. Therefore, $x^2 - 2x - 1 \leq 0$ for the interval $[1 - \sqrt{2}, 1 + \sqrt{2}]$. The graph of $y_1 = x^2 - 2x - 1$ intersects or is below the x-axis for the interval $[1 - \sqrt{2}, 1 + \sqrt{2}]$.

(b) First, $2x + 1 < x^2 \Rightarrow x^2 - 2x - 1 > 0$, now set $x^2 - 2x - 1 = 0$, then by the quadratic formula

$x = \frac{-(-2) \pm \sqrt{(-2)^2 - 4(1)(-1)}}{2(1)} = \frac{2 \pm \sqrt{8}}{2} = \frac{2(1 \pm \sqrt{2})}{2} \Rightarrow x = 1 \pm \sqrt{2}$, and the endpoints will be $x = 1 - \sqrt{2}$ and $x = 1 + \sqrt{2}$. The numbers $1 - \sqrt{2}$ and $1 + \sqrt{2}$ divide the number line into 3 intervals. The interval $(-\infty, 1 - \sqrt{2})$ has a positive product, the interval $(1 - \sqrt{2}, 1 + \sqrt{2})$ has a negative product, and the interval $(1 + \sqrt{2}, \infty)$ has a positive product. Therefore, $x^2 - 2x - 1 > 0$ for the interval $(-\infty, 1 - \sqrt{2}) \cup (1 + \sqrt{2}, \infty)$. The graph of $y_1 = x^2 - 2x - 1$ is above the x-axis for the interval $(-\infty, 1 - \sqrt{2}) \cup (1 + \sqrt{2}, \infty)$.

113. (a) First, $x - 3x^2 > -1 \Rightarrow 3x^2 - x - 1 < 0$, now set $3x^2 - x - 1 = 0$, then by the quadratic formula

$x = \frac{1 \pm \sqrt{(-1)^2 - 4(3)(-1)}}{2(3)} \Rightarrow x = \frac{1 \pm \sqrt{13}}{6}$, and the endpoints will be $x = \frac{1 - \sqrt{13}}{6}$ and $x = \frac{1 + \sqrt{13}}{6}$. The numbers $\frac{1 - \sqrt{13}}{6}$ and $\frac{1 + \sqrt{13}}{6}$ divide the number line into 3 intervals. The interval $\left(-\infty, \frac{1 - \sqrt{13}}{6}\right)$ has a positive product, the interval $\left(\frac{1 - \sqrt{13}}{6}, \frac{1 + \sqrt{13}}{6}\right)$ has a negative product, and the interval $\left(\frac{1+\sqrt{13}}{6},\infty\right)$ has positive product. Therefore, $3x^2-x-1<0$ for the interval $\left(\frac{1-\sqrt{13}}{6},\frac{1+\sqrt{13}}{6}\right)$. The graph of $y_1=3x^2-x-1$ is below the x-axis for the interval $\left(\frac{1-\sqrt{13}}{6},\frac{1+\sqrt{13}}{6}\right)$.

(b) First, $x-3x^2\le -1\Rightarrow 3x^2-x-1\ge 0,$ now set $3x^2-x-1=0,$ then by the quadratic formula $x=\frac{1\pm\sqrt{(-1)^2-4(3)(-1)}}{2(3)}\Rightarrow x=\frac{1\pm\sqrt{13}}{6},$ and the endpoints will be $x=\frac{1-\sqrt{13}}{6}$ and $x=\frac{1+\sqrt{13}}{6}$. The numbers $\frac{1-\sqrt{13}}{6}$ and $\frac{1+\sqrt{13}}{6}$ divide the number line into 3 intervals. The interval $\left(-\infty,\frac{1-\sqrt{13}}{6}\right)$ has a positive product, the interval $\left(\frac{1-\sqrt{13}}{6},\frac{1+\sqrt{13}}{6}\right)$ has a negative product, and the interval $\left(\frac{1+\sqrt{13}}{6},\infty\right)$ has positive product. Therefore, $3x^2-x-1\ge 0$ for the intervals $\left(-\infty,\frac{1-\sqrt{13}}{6}\right)$ or $\left(\frac{1+\sqrt{13}}{6},\infty\right)$. The graph of $y_1=3x^2-x-1$ is above the x-axis for the interval $\left(-\infty,\frac{1-\sqrt{13}}{6}\right)\cup\left(\frac{1+\sqrt{13}}{6},\infty\right)$.

115. $s=\frac{1}{2}gt^2\Rightarrow\frac{2s}{g}=t^2\Rightarrow t=\pm\sqrt{\frac{2s}{g}}\Rightarrow t=\pm\frac{\sqrt{2sg}}{g}$

117. $a^2+b^2=c^2\Rightarrow a^2=c^2-b^2\Rightarrow a=\pm\sqrt{c^2-b^2}$

119. $S=4\pi r^2\Rightarrow\frac{S}{4\pi}=r^2\Rightarrow r=\pm\sqrt{\frac{S}{4\pi}}\Rightarrow r=\pm\frac{\sqrt{S\pi}}{2\pi}$

121. $V=e^3\Rightarrow e=\sqrt[3]{V}$

123. $F=\frac{kMv^4}{r}\Rightarrow\frac{Fr}{kM}=v^4\Rightarrow v=\pm\sqrt[4]{\frac{Fr}{kM}}\Rightarrow v=\frac{\pm\sqrt[4]{FrkMkMkM}}{kM}\Rightarrow v=\pm\frac{\sqrt[4]{Frk^3M^3}}{kM}$

125. $P=\frac{E^2R}{(r+R)^2}\Rightarrow P(r+R)^2=E^2R\Rightarrow P(r^2+2rR+R^2)-E^2R=0\Rightarrow \mathrm{Pr}^2+2\,\mathrm{Pr}\,R+\mathrm{P}R^2-E^2R=0$

$\Rightarrow PR^2+(2\,\mathrm{Pr}-E^2)R+\mathrm{Pr}^2=0$. Now use the quadratic formula:

$$\text{R}=\frac{-\left(2\text{Pr}-E^2\right)\pm\sqrt{4P^2r^2-4E^2\text{ Pr}+E^4-4(P)\left(\text{Pr}^2\right)}}{2(P)}=\frac{E^2-2\text{Pr}\pm\sqrt{E^4-4E^2\text{ Pr}}}{2P}\Rightarrow$$

$$R=\frac{E^2-2\text{Pr}\pm E\sqrt{E^2-4\text{Pr}}}{2P}$$

127. Use the quadratic formula: $x=\frac{-y\pm\sqrt{y^2-4(1)\left(y^2\right)}}{2(1)}=\frac{-y\pm\sqrt{-3y^2}}{2}\Rightarrow x=-\frac{y}{2}\pm\frac{\sqrt{3}}{2}yi$. For $y^2+xy+x^2=0$

use the quadratic formula: $y=\frac{-x\pm\sqrt{x^2-4(1)\left(x^2\right)}}{2(1)}=\frac{-x\pm\sqrt{-3x^2}}{2}\Rightarrow\ y=-\frac{x}{2}\pm\frac{\sqrt{3}}{2}xi.$

129. $3y^2+4xy-9x^2=-1\Rightarrow-9x^2+4yx+\left(3y^2+1\right)=0.$ Now use the quadratic formula:

$$x=\frac{-4y\pm\sqrt{(4y)^2-4(-9)\left(3y^2+1\right)}}{2(-9)}=\frac{-4y\pm\sqrt{16y^2+108y^2+36}}{-18}=$$

$$\frac{-4y\pm\sqrt{124y^2+36}}{-18}=\frac{-4y\pm\sqrt{4\left(31y^2+9\right)}}{-18}=\frac{-4y\pm2\sqrt{31y^2+9}}{-18}\Rightarrow x=\frac{2y\pm\sqrt{31y^2+9}}{9}$$

$3y^2+4xy-9x^2=-1\Rightarrow3y^2+4xy+\left(-9x^2+1\right)=0.$ Now use the quadratic formula:

$$x=\frac{-4x\pm\sqrt{(4x)^2-4(3)\left(-9x^2+1\right)}}{2(3)}=\frac{-4x\pm\sqrt{16x^2+108x^2-12}}{6}=$$

$$\frac{-4x\pm\sqrt{124x^2-12}}{6}=\frac{-4x\pm\sqrt{4\left(31x^2-3\right)}}{6}=\frac{-4x\pm2\sqrt{31x^2-3}}{6}\Rightarrow y=\frac{-2x\pm\sqrt{31x^2-3}}{3}$$

131. (a) $f(0)=\frac{4}{5}(0-10)^2+80=160$ and $f(2)=\frac{4}{5}(2-10)^2+80=131.2$. Initially when the person stops exercising the heart rate is 160 beats per minute, and after 2 minutes the heart rate has dropped to about 131 beats per minute.

(b) Graph $Y_1=100, Y_2=0.8(x-10)^2+80$ and $Y_3=120$ (Not shown). The person's heart rate is between 100 and 120 when the graph of Y_2 is between Y_1 and Y_3. This occurs between approximately 2.9 minutes and 5 minutes after the person stops exercising.

Reviewing Basic Concepts (Sections 3.1—3.3)

1. $(5+6i)-(2-4i)-3i=5+6i-2+4i-3i=3+7i$

2. $i(5+i)(5-i)=i\left(25-i^2\right)=i\left(25-(-1)\right)=26i$

3. $\frac{-10-10i}{2+3i}\cdot\frac{2-3i}{2-3i}=\frac{-20+30i-20i+30i^2}{4-9i^2}=\frac{-20+10i-30}{4+9}=-\frac{50}{13}+\frac{10}{13}i$

4. See Figure 4.

[-10,10] by [-10,10]
Xscl = 1 Yscl = 1

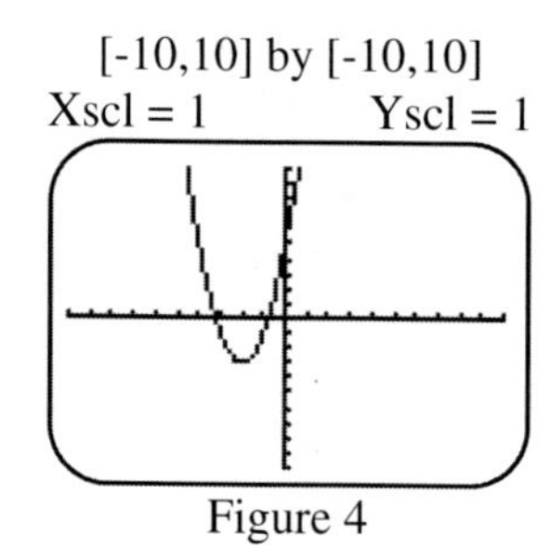

Figure 4

5. Use the vertex formula: $\frac{-b}{2a}=\frac{-8}{2(2)}=-2,$ now input $x=-2$ into the equation: $2(-2)^2+8(-2)+5=-3.$ The vertex is $(-2,-3)$ and it is a minimum because $a=2,$ which is positive; therefore, the parabola opens up.

6. Since the vertex is $(-2,-3)$ and the graph is a parabola opening up, the axis of symmetry is $x=-2.$

7. From the graph, the domain is $(-\infty,\infty)$ and the range is $[-3,\infty).$

8. $9x^2=25 \Rightarrow x^2=\frac{25}{9} \Rightarrow x=\pm\frac{5}{3}$

9. $3x^2-5x=2 \Rightarrow 3x^2-5x-2=0 \Rightarrow (3x+1)(x-2)=0 \Rightarrow 3x+1=0$ or $x-2=0 \Rightarrow x=-\frac{1}{3}$ and $x=2.$

10. For $-x^2+x+3=0$ use the quadratic formula: $x=\frac{-1\pm\sqrt{1^2-4(-1)(3)}}{2(-1)}=\frac{-1\pm\sqrt{13}}{-2} \Rightarrow x=\frac{1\pm\sqrt{13}}{2}.$

11. First, set $3x^2-5x-2=0,$ then $g(x)$ and the endpoints will be $x=-\frac{1}{3}$ and $x=2.$ The numbers $-\frac{1}{3}$ and 2 divide the number line into 3 intervals. The interval $\left(-\infty,-\frac{1}{3}\right)$ has a positive product, the interval $\left(-\frac{1}{3},2\right)$ has a negative product, and the interval $(2,\infty)$ has a positive product. Therefore, $3x^2-5x-2\le 0$ for the interval $\left[-\frac{1}{3},2\right].$ The graph of $y_1=3x^2-5x-2$ intersects or is below the x-axis for the interval $\left[-\frac{1}{3},2\right].$

12. First, set $x^2-x-3=0$ then by the quadratic formula $x=\frac{-(-1)\pm\sqrt{(-1)^2-4(1)(-3)}}{2(1)} \Rightarrow x=\frac{1\pm\sqrt{13}}{2}$ and the endpoints will be $x=\frac{1-\sqrt{13}}{2}$ and $x=\frac{1+\sqrt{13}}{2}.$ The numbers $\frac{1-\sqrt{13}}{2}$ and $\frac{1+\sqrt{13}}{2}$ divide the number line into 3 intervals. The interval $\left(-\infty,\frac{1-\sqrt{13}}{2}\right)$ has a positive product, the interval $\left(\frac{1-\sqrt{13}}{2},\frac{1+\sqrt{13}}{2}\right)$ has a negative product, and the interval $\left(\frac{1+\sqrt{13}}{2},\infty\right)$ has a positive product. Therefore, $x^2-x-3>0$ for the

interval $\left(-\infty, \frac{1-\sqrt{13}}{2}\right) \cup \left(\frac{1+\sqrt{13}}{2}, \infty\right)$. The graph of $y_1 = x^2 - x - 3$ is above the x-axis for the interval

$\left(-\infty, \frac{1-\sqrt{13}}{2}\right) \cup \left(\frac{1+\sqrt{13}}{2}, \infty\right)$.

13. First, $x(3x-1) \le -4 \Rightarrow 3x^2 - x + 4 \le 0$, now set $3x^2 - x + 4 = 0$, then by the quadratic formula $x = \frac{-(-1) \pm \sqrt{(-1)^2 - 4(3)(4)}}{2(3)} = \frac{1 \pm \sqrt{-47}}{6} \Rightarrow x = \frac{1}{6} \pm \frac{\sqrt{47}}{6} i$. The inequality has no real solutions, the graph is a parabola opening upward which does not intersect, but is completely above the x-axis. Therefore, $3x^2 - x + 4 \le 0$ never happens and the solution is $\varnothing$.

Section 3.4 Applications of Quadratic Functions and Models

1. The y-coordinate of the vertex is maximum y-value. Using the vertex formula yields:

   $x = -\frac{b}{2a} = -\frac{32}{2(-16)} = 1$. If $x = 1$, , then $y = -16(1)^2 + 32(1) + 100 = 116 \Rightarrow$ the maximum y-value is 116.

3. The y-coordinate of the vertex is minimum y-value. Using the vertex formula yields:

   $x = -\frac{b}{2a} = -\frac{-24}{2(3)} = 4$. If $x = 4$, , then $y = 3(4)^2 - 24(4) + 50 = 2 \Rightarrow$ the minimum y-value is 2.

5. $-4x^2 + 5x = 1 \Rightarrow 4x^2 - 5x + 1 = 0 \Rightarrow (4x-1)(x-1) = 0 \Rightarrow (4x-1) = 0 \Rightarrow x = \frac{1}{4}$ and $(x-1) = 0 \Rightarrow x = 1$.

7. $\frac{1}{2}x^2 + 3 = 6x \Rightarrow \frac{1}{2}x^2 - 6x + 3 = 0$. Using the quadratic equation we have the following:

   $$\frac{6 \pm \sqrt{(-6)^2 - 4\left(\frac{1}{2}\right)(3)}}{2\left(\frac{1}{2}\right)} = 6 \pm \sqrt{30}$$

9. A, The area of a rectangle is $L \cdot W = A \Rightarrow x(2x+2) = 40{,}000$.

11. (a) $30 - x$ would be the other number.

    (b) Since both numbers are positive and their sum is 30, the restrictions are $0 < x < 30$.

    (c) Multiplying $x(30-x)$ would give the function $P(x) = 30x - x^2 \Rightarrow P(x) = -x^2 + 30x$.

    (d) Using the Vertex formula yields $x = \frac{-b}{2a} = \frac{-30}{2(-1)} \Rightarrow x = 15$. If $x = 15$, then $y = 30 - (15) \Rightarrow y = 15$.

    Also, if $x = 15$ the maximum product is $P(15) = -(15)^2 + 30(15) = -225 + 450 = 225$. The graph of $y = -x^2 + 30x$ has a vertex point of $(15, 15)$, which supports the result.

13. Perimeter of fence $= 2l + 2w = 1000 \Rightarrow l = 500 - w$. If $A = l \cdot w$ then $A = (500 - w)w = 500w - w^2$. This is a parabola opening downward and by the vertex formula, the maximum area occurs when $w = -\frac{b}{2a} = -\frac{500}{2(-1)} = 250$.

    The dimensions that maximize area are 250 ft. by 250 ft.

15. (a) $640-2x$ would be the other side.

(b) Since all three numbers are positive, their sum is 640, and two of them are x, the restrictions are $0<2x<640 \Rightarrow 0<x<320$.

(c) Multiplying $x(640-2x)$ would give the function $A(x)=640x-2x^2 \Rightarrow A(x)=-2x^2+640x$.

(d) See Figure 15. From the graph, between 57.04 ft and 85.17 ft or between 234.17 ft and 262.96 ft, will give an area between 30,000 and 40,000 feet.

(e) Using the Vertex formula yields $x=\frac{-b}{2a}=\frac{-640}{2(-2)} \Rightarrow x=160$. If $x=160$, then $y=640-2(160) \Rightarrow$ $y=320$. Also, if $x=160$, the maximum product is $A(160)=-2(160)^2+640(160)=-51,200+102,400$ $\Rightarrow A(x)=51,200$. The graph of $y=-2x^2+640x$ has a vertex point of $(160,320)$, which supports the result.

[0,320] by [0,55,000]
Xscl = 20 Yscl = 10,000

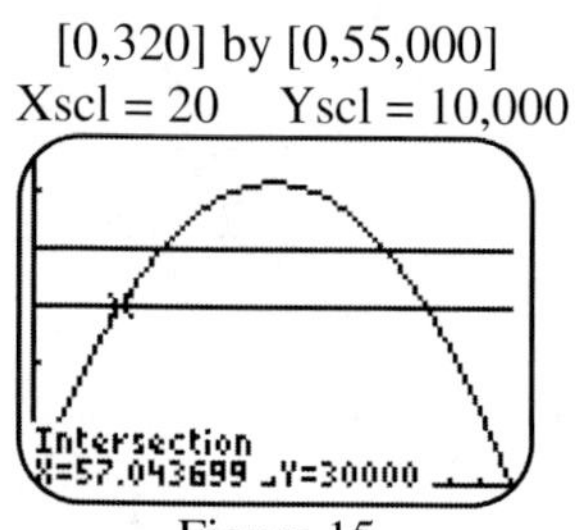

Figure 15

17. (a) $s(1)=-16(1)^2+44(1)+4=32$; the baseball is 32 feet high after 1 second.

(b) For $s(t)=-16t^2+44t+4$, the vertex formula gives $t=-\frac{b}{2a}=-\frac{44}{2(-16)}=1.375$ and $f(1.375)=-16(1.375)^2+44(1.375)+4=34.25$; the maximum height is 34.25 feet.

19. The height when it hits the ground will be $0 \Rightarrow 75-16t^2=0 \Rightarrow 16t^2=75 \Rightarrow t^2=\frac{75}{16} \Rightarrow$

$t=\sqrt{\frac{75}{16}} \Rightarrow t=\frac{\sqrt{75}}{4} \Rightarrow t \approx 2.2$ seconds.

21. (a) If the width is x units long, then the length is twice the width or $2x$ units long.

(b) The width will be $x-2(2)=x-4$ units long, and the length will be $2x-2(2)=2x-4$ units long. Since both measurements are positive and 4 inches are removed from each measurement, the restrictions are $x>4$.

(c) Since volume for the box is $V=L\cdot W\cdot H, V=(2x-4)(x-4)(2)$ and the function is $V(x)=4x^2-24x+32$.

(d) $320 = 4x^2 - 24x + 32 \Rightarrow 4x^2 - 24x - 288 = 0 \Rightarrow 4(x^2 - 6x - 72) = 0 \Rightarrow 4(x-12)(x+6) = 0$

$\Rightarrow x = -6$ or 12. Since length cannot be negative, $x = 12$. If $x = 12$, the dimensions are 8 in by 20 in. From the graph of $y = 4x^2 - 24x + 32$, when $x = 12$, $y = 320$. This support our analytical result.

(e) From the graph (not shown) of $y = 4x^2 - 24x + 32$, when $400 < y < 500, 13.0 < x < 14.2$ in.

23. The surface area of the can is $V = \pi r^2 + \pi r^2 + 2\pi r(h) \Rightarrow 54.19 = 2\pi r^2 + 2(4.25)\pi r \Rightarrow$

$6.28x^2 + 26.69x - 54.19 = 0$ now use the quadratic formula: $\dfrac{-26.69 \pm \sqrt{(26.69)^2 - 4(6.28)(-54.19)}}{2(6.28)} =$

$\dfrac{-26.69 \pm \sqrt{2073.6}}{12.56} = \dfrac{-26.69 \pm 45.537}{12.56} \Rightarrow r = -5.751$ or 1.5. Since length cannot be negative, r = 1.5 in.

25. If $A = s^2$, then $800 = s^2 \Rightarrow s = \sqrt{800} = 20\sqrt{2}$. Since the lawn is square, the diagonal of the lawn would equal $d^2 = (20\sqrt{2})^2 + (20\sqrt{2})^2 \Rightarrow d^2 = 1600 \Rightarrow d = 40$. The radius of the circular pattern is half the diagonal; therefore, the radius is r = 20 ft.

27. If we use Pythagorean Theorem, we get $h^2 + 12^2 = (2h+3)^2 \Rightarrow$

$h^2 + 144 = 4h^2 + 12h + 9 \Rightarrow 3h^2 + 12h - 135 = 0 \Rightarrow (3h+27)(h-5) = 0 \Rightarrow x = -9.5$. Since length cannot be negative, the height of the dock is 5 ft.

29. If we use Pythagorean Theorem, we get $(8+2)^2 + (9+4)^2 = x^2 \Rightarrow 100 + 169 = x^2 \Rightarrow$

$x^2 = 269 \Rightarrow x = \sqrt{269} \Rightarrow x \approx 16.4$. Since a 16 ft ladder will be too short, the ladder must be at least 17 ft.

31. Let $x =$ length of the picture and let $(x-4) =$ the width of the picture. Then $A = l \cdot w \Rightarrow A = x(x-4) \Rightarrow$

$320 = x^2 - 4x \Rightarrow x^2 - 4x - 320 = 0$. $\Rightarrow (x-20)(x+16) = 0 \Rightarrow x = 20$. Thus, the length and width of the picture is 20 and16 inches respectively. The dimensions of the frame will include an additional 4 inches to both the length and width to yield the final dimensions of 20 inches by 24 inches.

33. (a) Since x equals the loss of 1 apartment for each \$20 increase, the number of apartments rented is $80 - x$.

(b) Since each increase is \$20 and x equals the number of increases, the rent per apartment is $400 + 20x$.

(c) Since revenue is the number of apartments rented times rent per apartment, we multiply a and b.

$(80 - x)(400 + 20x) = 32{,}000 + 1200x - 20x^2 \Rightarrow \text{R}(x) = -20x^2 + 1200x + 32{,}000.$

(d) $37{,}500 = -20x^2 + 1200x + 32{,}000 \Rightarrow -20x^2 + 1200x - 5{,}500 = 0 \Rightarrow$

$-20(x^2 - 60x + 275) = 0 \Rightarrow -20(x-55)(x-5) = 0 \Rightarrow x = 5$ or 55.

(e) Use the vertex formula: $x = \dfrac{-b}{2a} = \dfrac{-1200}{-40} = 30$. If $x = 30$, then the rent per apartment is

$r(30) = 400 + 20(30) = 400 + 600 = \$1000.$

35. (a) If $f(x) = 10$ and $x = 15$, then $10 = \dfrac{-16(15)^2}{0.434v^2} + 1.15(15) + 8 \Rightarrow 10 = \dfrac{-3600}{0.434v^2} + 25.25 \Rightarrow$

$-15.25 = \dfrac{-3600}{0.434v^2} \Rightarrow -6.6185v^2 = -3600 \Rightarrow v^2 = \dfrac{3600}{6.6185} \Rightarrow v \approx 23.32\text{ft/sec.}$

(b) If $v = 23.32$, then $y = \dfrac{-16x^2}{0.434(23.32)^2} + 1.15x + 8 \Rightarrow y = -.06779x^2 + 1.15x + 8$. Graph this equation, the graph does pass through points $(0,8)$ and $(15,10)$.

(c) Use the vertex formula: $x = \dfrac{-b}{2a} = \dfrac{-1.15}{2(-0.06779)} \approx 8.48$. If $x \approx 8.48$, then the maximum height is $f(8.48) = -0.06779(8.48)^2 + 1.15(8.48) + 8 \Rightarrow f(8.48) = 12.88\text{ ft.}$

37. (a) If $x = 2$, then $h(2) = -0.5(2)^2 + 1.25(2) + 3 = -2 + 2.5 + 3 \Rightarrow h(2) = 3.5\text{ft}$.

(b) If $h(x) = 3.25$.then $3.25 = -0.5x^2 + 1.25x + 3 \Rightarrow -0.5x^2 + 1.25x - 0.25 = 0 \Rightarrow$

$4(-0.5x^2 + 1.25x + 0.25 = 0) \Rightarrow -2x^2 + 5x - 1 = 0 \Rightarrow 2x^2 - 5x + 1 = 0$ Now use the quadratic formula: $x = \dfrac{5 \pm \sqrt{(-5)^2 - 4(2)(1)}}{2(2)} = \dfrac{5 \pm \sqrt{17}}{4} \Rightarrow x = 0.219$ or 2.281. Therefore, the frog is 3.25 feet above the ground at approximately 0.2 ft and 2.3 ft.

(c) Using the vertex formula yields $x = \dfrac{-b}{2a} = \dfrac{-1.25}{2(-0.5)} \Rightarrow x = 1.25$ ft as the horizontal distance.

(d) From part (c) the maximum height was reached when the horizontal distance was $x = 1.25$; therefore, $h(1.25) = -0.5(1.25)^2 + 1.25(1.25) + 3 = -0.78125 + 1.5625 + 3 \Rightarrow h(1.25) \approx 3.78$ ft high.

39. If $f(x) = 800$, then $800 = \dfrac{1}{10}x^2 - 3x + 22 \Rightarrow .1x^2 - 3x - 778 - 0$. Now use the quadratic formula:

$x = \dfrac{3 \pm \sqrt{9 - 4(0.1)(-778)}}{2(0.1)} = \dfrac{3 \pm \sqrt{320.2}}{0.2} \Rightarrow x \approx 104.5$ ft/sec. Converted is $x = \dfrac{104.5(60)(60)}{5280} = 71.25$ mph.

41. (a) See Figure 41a.

(b) Using the defined function and the vertex (7, 12) yields $f(x) = a(x-7) + 12$. Now use (1, 108) to find the function; $108 = a(1-7)^2 + 12 \Rightarrow 96 = 36a \Rightarrow a = \dfrac{8}{3}$; therefore, the function is $f(x) = \dfrac{8}{3}(x-7)^2 + 12$.

(c) See Figure 41c. From the graph, we see that there is a good fit.

(d) Using the regression feature yields $g(x) \approx 2.72x^2 - 38.93x + 149.46$.

(e) For February, $x = 2$. Therefore, $f(2) = \dfrac{8}{3}(2-7)^2 + 12 \approx 79$, and

$g(2) = 2.72(2)^2 - 38.93(2) + 149.46 \approx 82$.

For June, $x = 6$. Therefore, $f(6) = \dfrac{8}{3}(6-7)^2 + 12 \approx 15$, and $g(6) = 2.72(6)^2 - 38.93(6) + 149.46 \approx 14$.

[0,11] by [0,125]
Xscl = 1 Yscl = 10

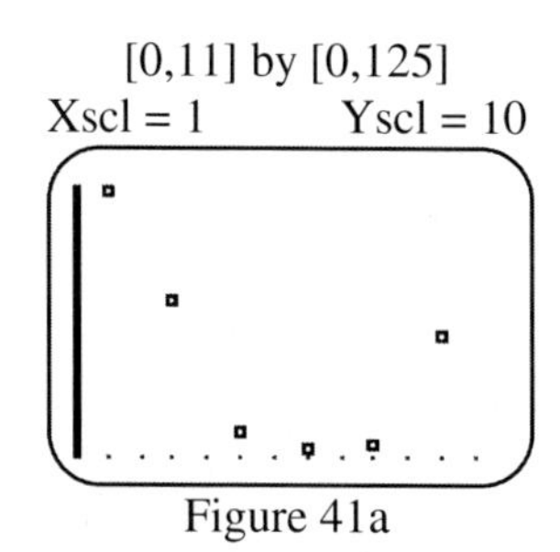

Figure 41a

[0,11] by [0,125]
Xscl = 1 Yscl = 10

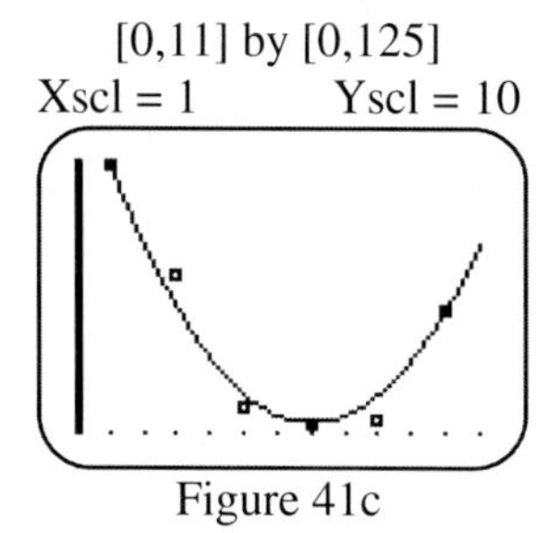

Figure 41c

43. (a) See Figure 43a.

(b) $f(45) = 0.056057(45)^2 + 1.06657(45) = 113.515425 + 47.99565 \Rightarrow x \approx 161.5$ feet. When the speed is 45 mph, the stopping distance is 161.5 feet.

(c) See Figure 43c. The model is quite good, although the stopping distances are a little low for the higher speeds.

[15,75] by [-10,450]
Xscl = 10 Yscl = 50

[15,75] by [-10,450]
Xscl = 10 Yscl = 50

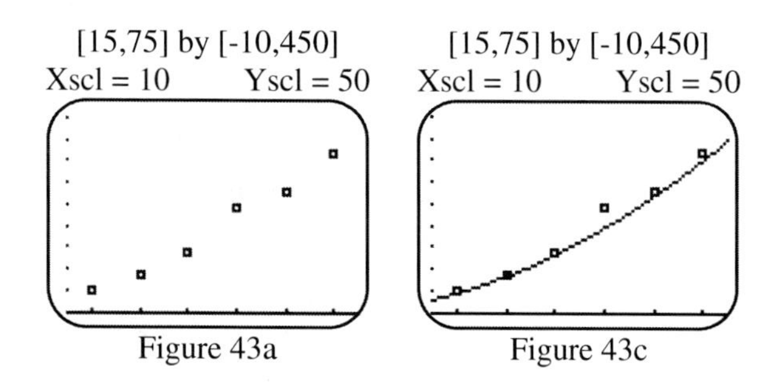

Figure 43a Figure 43c

3.5: Higher-Degree Polynomial Functions and Graphs

1. With three extrema, the minimum degree of *f* is 4.
3. The points (a,b) and (c,d) are local maxima, and (e,t) is a local minimum.
5. The highest point of the graph, *(a, b)* is an absolute maximum.
7. The function *f*, has local maximum values of *b* and *d*; a local minimum value of *t*; and an absolute maximum value of *b*.
9. $P(x)$ is a positive odd-degree polynomial graph; therefore,
11. $P(x)$ is a negative odd-degree polynomial graph; therefore,
13. $P(x)$ is a positive even-degree polynomial graph; therefore,
15. $P(x)$ is a negative even-degree polynomial graph; therefore,
17. $P(x)$ is a positive even-degree polynomial graph; therefore,
19. $P(x)$ is a negative odd-degree polynomial graph; therefore,
21. The graph of $f(x) = x^n$ for $n \in$ {positive odd integers} will take the shape of the graph of $f(x) = x^3$, but gets steeper as *n* and *x* increase.

23. See Figure 23. As the odd exponent n gets larger, the graph *flattens out* in the window $[-1,1]$ by $[-1,1]$.

    The graph of $y = x^7$ will be between $y = x^5$ and the x-axis in this window.

[-1,1] by [-1,1]
Xscl = 1 Yscl = 1

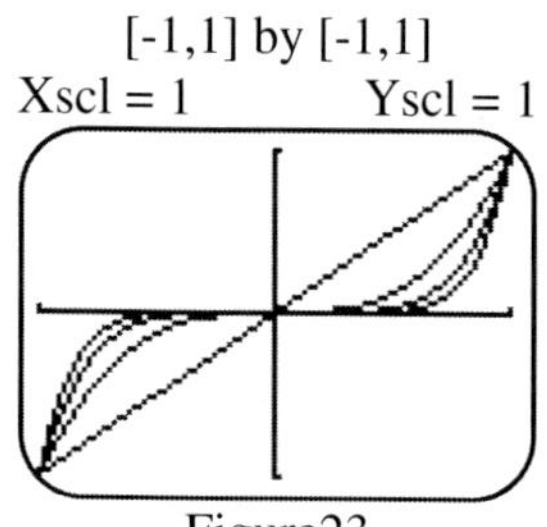

Figure23

25. Graphing in various windows produces a local maximum of (2, 3.67), and a local minimum of (3, 3.5).
27. Graphing in various windows produces a local maximum of (−3.33, −1.85), and a local minimum of (−4, −2).
29. Graphing in various windows produces two x-intercepts, $(2.10,0)$ and $(2.15,0)$.
31. Graphing in various windows produces no x-intercept.
33. $P(x)$ is a positive odd-degree polynomial graph; therefore, D
35. $P(x)$ is a negative even-degree polynomial graph; therefore, B
37. A. The third-degree function can have at most 2 local extrema and a positive lead coefficient will yield a right side end behavior opening up.
39. From the graph, the function graphed in C has 1 real zero.
41. B and D. A third-degree function can have at most 2 local extrema.
43. From the graph, the function graphed in A has 1 *negative* real zero.
45. From the graph, the function graphed in B has a range of approximately $[-100,\infty)$.
47. False. A polynomial function of degree 3 will have at most 3 x-intercepts.
49. True. With a positive y-intercept and negative even-degree end behavior, it must have 2 x-intercepts.
51. True. With a negative even-degree end behavior, which is shifted up 5 units, it will have 2 x-intercepts.
53. False. With odd-degree end behavior, it will have at least 1 and at most 5 x-intercepts. Thus it may have only 1.
55. Shift the graph of $y = x^4$ three units to the left, stretch vertically by a factor of 2, and shift downward 7 units. See Figures 55a and 55b.
56. Shift the graph of $y = x^4$ one unit to the left, stretch vertically by a factor of 3, reflect across the x-axis, and shift upward 12 units. See Figures 56a and 56b.

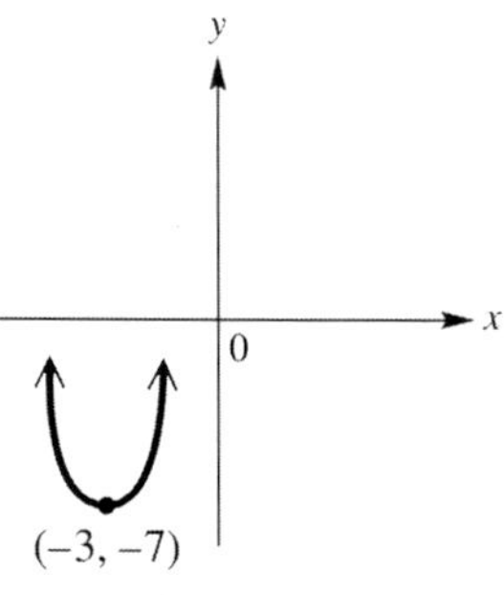

Figure 55a

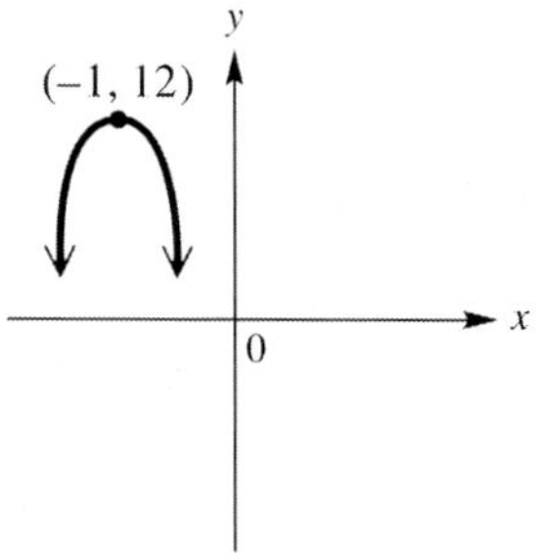

(−1, 12)

Figure 56a

[-10,10] by [-10,10]
Xscl = 1 Yscl = 1

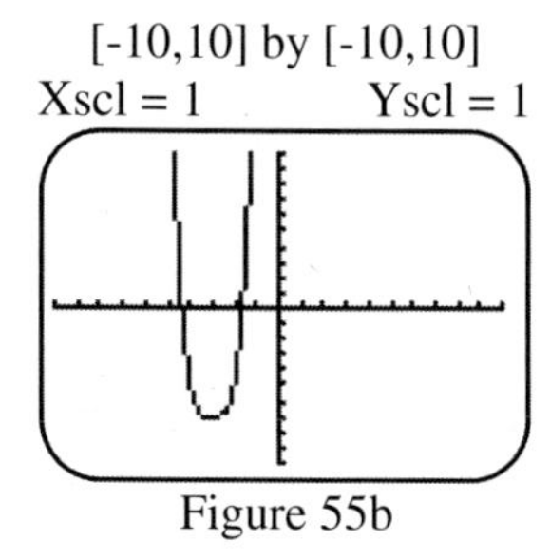

Figure 55b

[-6,6] by [-10,13]
Xscl = 1 Yscl = 1

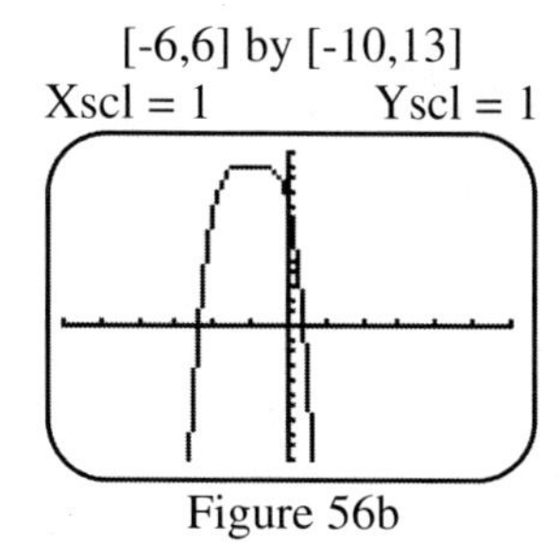

Figure 56b

57. Shift the graph of $y = x^3$ one unit to the right, stretch vertically by a factor of 3, reflect across the x-axis, and shift upward 12 units. See Figures 57a and 57b.

58. Shift the graph of $y = x^5$ one unit to the right, stretch vertically by a factor of .5, and shift upward 13 units. See Figures 58a and 58b.

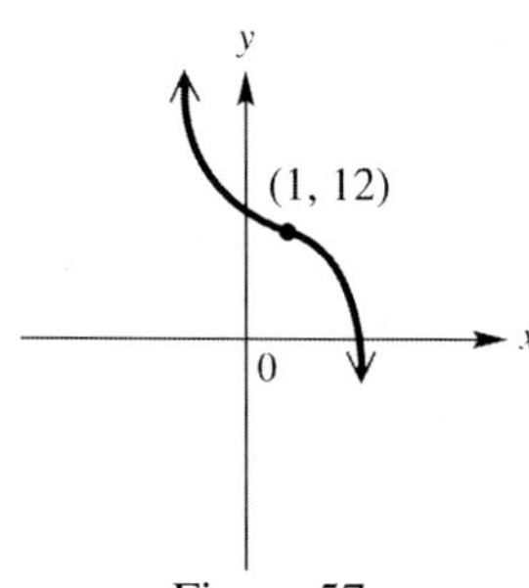

Figure 57a

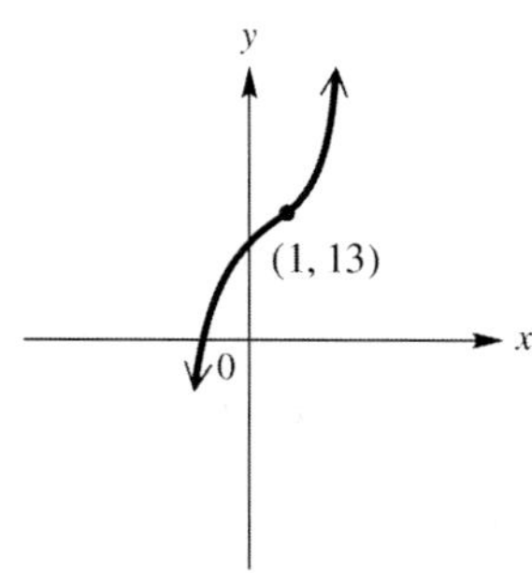

Figure 58a

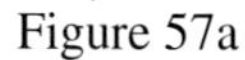

[-6,6] by [-100,100]
Xscl = 1 Yscl = 10

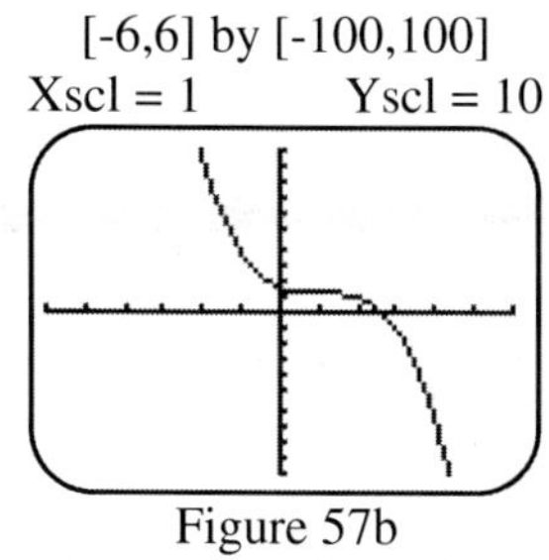

Figure 57b

[-6,6] by [-100,100]
Xscl = 1 Yscl = 10

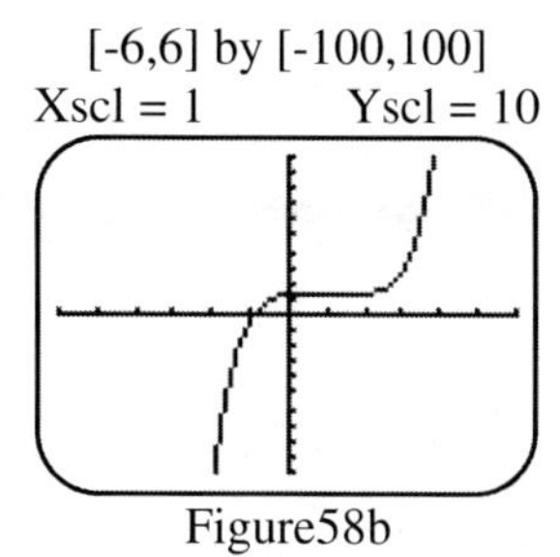

Figure58b

59. See Figure 59 for the graph of this function.
    (a) Because it is a polynomial function, its domain is $(-\infty, \infty)$.

(b) From the graph of P, there is one local minimum, which by the calculator is $(-4.74, -27.03)$. Because the function is of odd-degree, it is not an absolute minimum point.

(c) From the graph of P, there is one local maximum, which by the calculator is $(0.07, 84.07)$. Because the function is of odd-degree, it is not an absolute minimum point.

(d) Because the polynomial function is of odd-degree, its range is $(-\infty,\infty)$.

(e) Using the calculator we find the x-intercepts are $(-6,0)$, $(-3.19,0)$ and $(2.19,0)$; the y-intercept is $(0,84)$.

(f) From the graph and our results, the function is increasing for the interval $(-4.74,\ 0.07)$.

(g) From the graph and our results, the function is decreasing for the intervals $(-\infty,-4.74)$ and $(0.07,\infty)$.

$(-\infty,-1.52)$ and $(2.85,\infty)$.

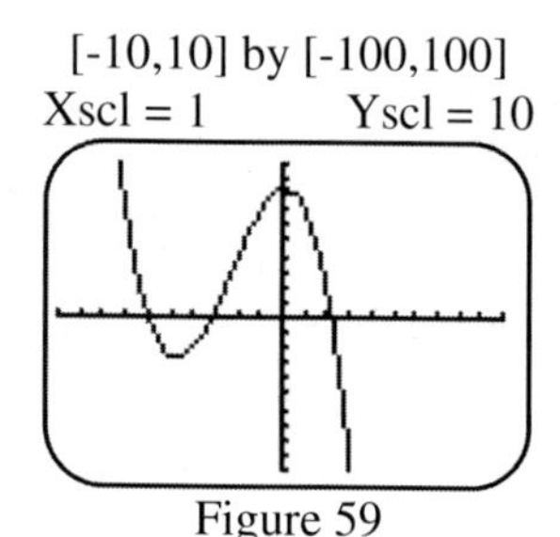

Figure 59

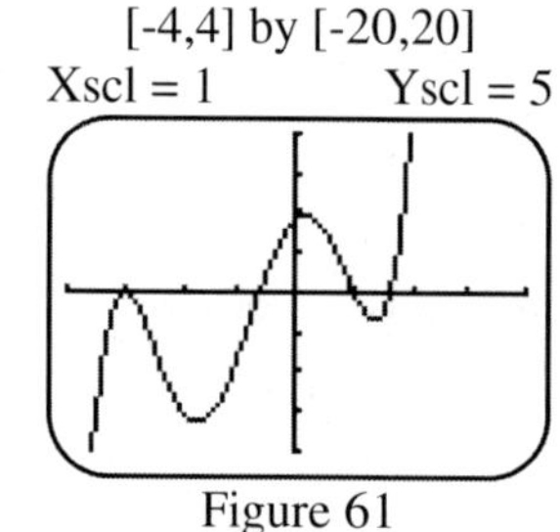

Figure 61

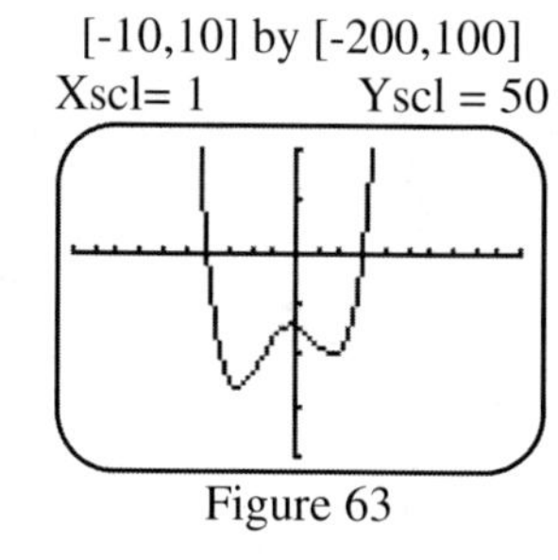

Figure 63

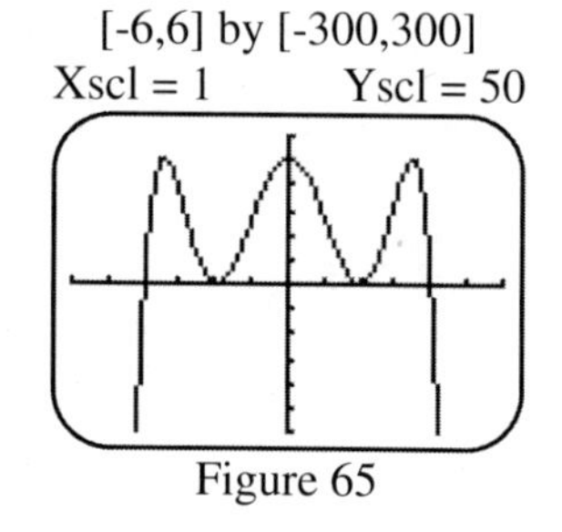

Figure 65

61. See Figure 61 for the graph of this function.

(a) Because it is a polynomial function, its domain is $(-\infty,\infty)$.

(b) From the graph of P, there are 2 local minimum points, which by the calculator are $(-1.73, -16.39)$ and $(1.35, -3.49)$. Because the function is of odd-degree, neither is an absolute minimum point.

(c) From the graph of P, there are 2 local maximum points, which by the calculator are $(-3, 0)$ and $(0.17, 9.52)$. Because the function is of odd-degree, neither is an absolute maximum point.

(d) Because the polynomial function is of odd-degree, its range is $(-\infty,\infty)$.

(e) Using the calculator, we find the x-intercepts are $(-3,0)$, $(-0.62,0)$, $(1,0)$, and $(1.62,0)$; the y-intercept is $(0,9)$.

(f) From the graph and our results, the function is increasing on $(-\infty,-3),(-1.73,0.17)$ and $(1.35,\infty)$.

(g) From the graph and our results, the function is decreasing for the intervals $(-3,-1.73)$ and $(0.17,1.35)$.

63. See Figure 63 for the graph of this function.

(a) Because it is a polynomial function, its domain is $(-\infty,\infty)$.

(b) From the graph of P, there is one absolute minimum point, which by the calculator is $(-2.63,\ -132.69)$, and one local minimum point, which by the calculator is $(1.68, -99.90)$.

(c) From the graph of P, there is one local maximum point, which by the calculator is $(-0.17, -71.48)$. Because the function is of positive even-degree, there is not an absolute maximum point.

(d) Because the positive even-degree function has an absolute minimum, given in (b), its range is $(-132.69, \infty)$.

(e) Using the calculator, we find the x-intercepts are $(-4,0)$, and $(3,0)$; the y-intercept is $(0,-72)$.

(f) From the graph and our results, the function is increasing for the intervals, $(-2.63, \ -0.17)$ and $(1.68, \infty)$.

(g) From the graph and our results, the function is decreasing for the intervals, $(-\infty, -2.63)$ and $(-0.17, 1.68)$.

65. See Figure 65 for the graph of this function.

(a) Because it is a polynomial function, its domain is $(-\infty, \infty)$.

(b) From the graph of P, there is two local minimum points, which by the calculator are $(-2, 0)$ and $(2, 0)$. Because the function is of negative even-degree, there is not an absolute minimum point.

(c) From the graph of P, there are 3 absolute maximum points, which by the calculator are $(-3.46, 256)$, $(0, 256)$, and $(3.46, 256)$.

(d) Because the negative even-degree function has an absolute maximum, its range is $(-\infty, 256]$.

(e) Using the calculator we find the x-intercepts are $(-4,0)$, $(-2,0)$, $(2,0)$, and $(4,0)$; the y-intercept is $(0, 256)$.

(f) From the graph and our results, the function is increasing on $(-\infty, -3.46)$, $(-2,0)$ and $(2, 3.46)$.

(g) From the graph and our results, the function is decreasing on $(-3.46, \ -2)$, $(0,2)$, and $(3.46, \infty)$.

67. There are many possible valid windows, through experimentation one window is $[-10, 10]$ by $[-40, 10]$.

69. There are many possible valid windows, through experimentation one window is $[-10, 20]$ by $[-1500, 500]$.

71. There are many possible valid windows, through experimentation one window is $[-10, 10]$ by $[-20, 500]$.

73. (a) Let x = 3 correspond to March. $f(3) = 0.0145(3)^4 - 0.426(3)^3 + 3.53(3)^2 - 6.23(3) + 72 \approx 74.8$. The average high temperature in March is 74.8^o.

Let x = 7 correspond to July. $f(7) = 0.0145(7)^4 - 0.426(7)^3 + 3.53(7)^2 - 6.23(7) + 72 \approx 90.1$ The average high temperature in July is 90.1^o.

(b) From the graph of $f(x)$ we can see that the average temperature is 80 in April and October.

Reviewing Basic Concepts (Sections 3.4 and 3.5)

1. (a) The total length of the fence must equal $2L + 2x = 300 \Rightarrow 2L = 300 - 2x \Rightarrow L = 150 - x$.

(b) Since width equals x and length equals $150 - x$, the function is $A(x) = x(150 - x)$.

(c) Since length and width must both be positive, the restrictions are $0 < x < 150$.

(d) $5000 = x(150 - x) \Rightarrow x^2 - 150x + 5000 = 0 \Rightarrow (x - 50)(x - 100) = 0 \Rightarrow x = 50 \ or \ 100$. Using either value will yield a garden, which is 50 m by 100 m.

2. (a) See Figure 2a.

(b) Using the defined function and the point (51, .1) yields $f(x) = a(x-51)+0.1$. Now use (101, 4.7) to find the function: $4.7 = a(101-51)^2+0.1 \Rightarrow 4.6 = 2500a \Rightarrow a = 0.0018$; therefore, the function is $f(x) = 0.0018(x-51)^2+0.1$.

(c) Using the regression feature yields $g(x) \approx 0.0026x^2 - 0.3139x + 9.426$.

(d) Graph the data and two equations from (b) and (c), See Figure 2d. The regression function fits better because it passes through more data points. Neither function would fit the data for $x < 51$.

3. $P(x)$ is a positive odd-degree polynomial graph; therefore,

4. A polynomial of degree 3 can have at most 2 extrema, and at most 3 x-intercepts.

5. $P(x)$ is a positive even-degree polynomial graph; therefore, .

6. See Figure 6.

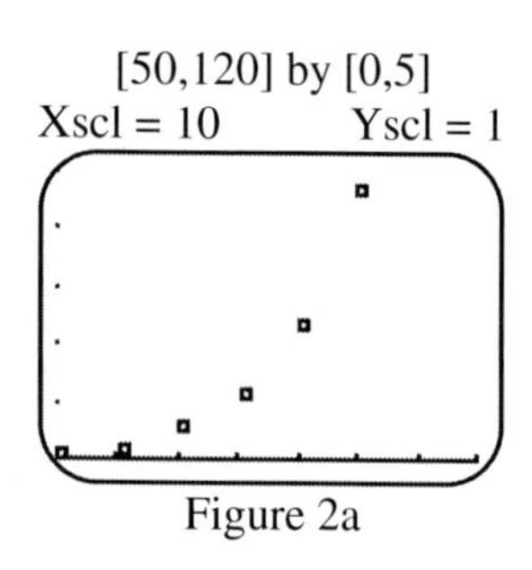

Figure 2a

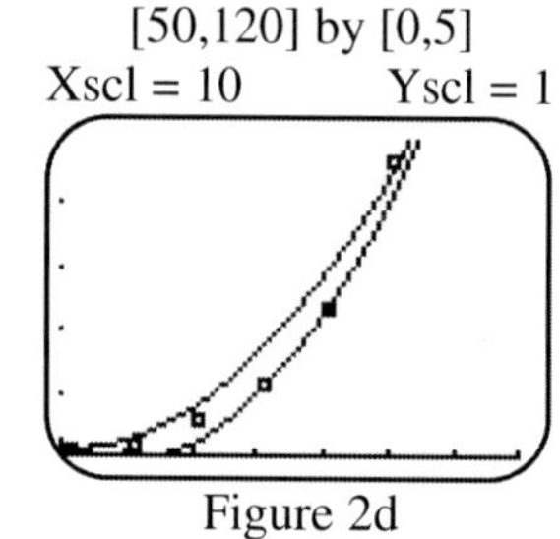

Figure 2d

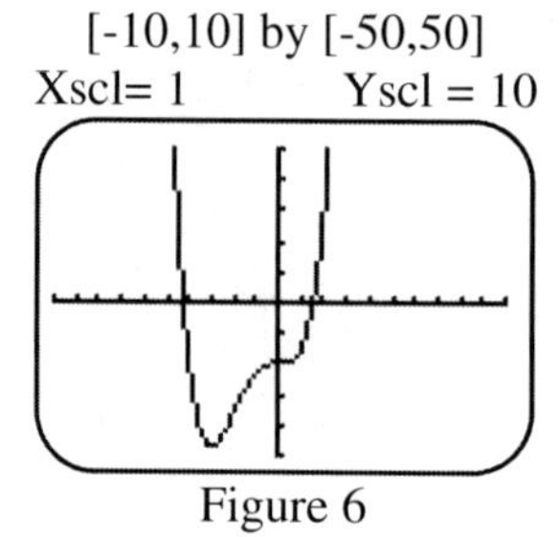

Figure 6

7. From the graph of P, there is one extrema, an absolute minimum point, which by the calculator is $(-3, -47)$.

8. Using the calculator, we find the x-intercepts are $(-4.26, 0)$ and $(1.53, 0)$; the y-intercept is $(0, -20)$.

3.6: Topics in the Theory of Polynomial Functions (I)

1. $\dfrac{10x^6}{5x^3} = 2x^3$

3. $\dfrac{8x^9}{3x^7} = \dfrac{8}{3}x^2$

5. $\dfrac{2x^6+3x^3}{2x} = \dfrac{2x^6}{2x} + \dfrac{3x^3}{2x} = x^5 + \dfrac{3}{2}x^2$

7. $\dfrac{8x^3-5x}{2x} = \dfrac{8x^3}{2x} - \dfrac{5x}{2x} = 4x^2 - \dfrac{5}{2}$

9. $P(1) = 3(1)^2 - 2(1) - 6 \Rightarrow P(1) = -5$ and $P(2) = 3(2)^2 - 2(2) - 6 \Rightarrow P(2) = 2$. These answers differ in sign; therefore, there is a real zero between them. Graphed on a calculator, the real zero is approximately 1.79.

11. $P(2) = 2(2)^3 - 8(2)^2 + (2) + 16 \Rightarrow P(2) = 2$ and $P(2.5) = 2(2.5)^3 - 8(2.5)^2 + (2.5) + 16 \Rightarrow P(2.5) = -0.25$.

These answers differ in sign; therefore, there is a real zero between them. By graphing on the calculator, the real zero is approximately 2.39.

13. $P(1.5) = 2(1.5)^4 - 4(1.5)^2 + 3(1.5) - 6 \Rightarrow P(1.5) = -0.375$ and $P(2) = 2(2)^4 - 4(2)^2 + 3(2) - 6 \Rightarrow P(2) = 16$.

These answers differ in sign; therefore, there is a real zero between them. By graphing on the calculator, the real zero is approximately 1.52.

15. $P(2.7) = -(2.7)^4 + 2(2.7)^3 + (2.7) + 12 \Rightarrow P(2.7) = 0.9219$ and $P(2.8) = -(2.8)^4 + 2.(2.8)^3 + (2.8) + 12 \Rightarrow$

$P(2.8) = -2.7616$. These answers differ in sign; therefore, there is a real zero between them. By graphing on the calculator, the real zero is approximately 2.73.

17. $P(-1.6) = (-1.6)^5 - 2(-1.6)^3 + 1 \Rightarrow P(-1.6) = -1.29376$ and $P(-1.5) = (-1.5)^5 - 2(-1.5)^3 + 1 \Rightarrow$

$P(-1.5) = 0.15625$. These answers differ in sign; therefore, there is a real zero between them. By graphing on the calculator, the real zero is approximately -1.51.

19. There is at least one zero between 2 and 2.5.

21. $$\begin{array}{r}
x^2 - 3x - 2 \\
x+5 \overline{)\, x^3 + 2x^2 - 17x - 10} \\
\underline{x^3 + 5x^2 \qquad\qquad\quad} \\
-3x^2 - 17x - 10 \\
\underline{-3x^2 - 15x \qquad} \\
-2x - 10 \\
\underline{-2x - 10} \\
0
\end{array} \Rightarrow \frac{x^3 + 2x^2 - 17x - 10}{x+5} = x^2 - 3x - 2$$

23. $$\begin{array}{r}
3x^2 + 4x \qquad \\
x-5 \overline{)\, 3x^3 - 11x^2 - 20x + 3} \\
\underline{3x^3 - 15x^2 \qquad\qquad\quad} \\
4x^2 - 20x + 3 \\
\underline{4x^2 - 20x \qquad} \\
3
\end{array} \Rightarrow \frac{3x^3 - 11x^2 - 20x + 3}{x-5} = 3x^2 + 4x + \frac{3}{x-5}$$

25.
$$
\begin{array}{r}
x^3 - x^2 - 6x \\
x-2\overline{)x^4 - 3x^3 - 4x^2 + 12x} \\
\underline{x^4 - 2x^3} \\
-x^3 - 4x^2 + 12x \\
\underline{-x^3 + 2x^2} \\
-6x^2 + 12x \\
\underline{-6x^2 + 12x} \\
0
\end{array}
\Rightarrow \frac{x^4 - 3x^3 - 4x^2 + 12x}{x-2} = x^3 - x^2 - 6x
$$

27.
$$
\begin{array}{r}
x^2 + 3x + 3 \\
x-1\overline{)x^3 + 2x^2 + 0x - 3} \\
\underline{x^3 - x^2} \\
3x^2 + 0x - 3 \\
\underline{3x^2 - 3x} \\
3x - 3 \\
\underline{3x - 3} \\
0
\end{array}
\Rightarrow \frac{x^3 + 2x^2 - 3}{x-1} = x^2 + 3x + 3
$$

29.
$$
\begin{array}{r}
-2x^2 + 2x - 3 \\
x+1\overline{)-2x^3 + 0x^2 - x - 2} \\
\underline{-2x^3 - 2x^2} \\
2x^2 - x - 2 \\
\underline{2x^2 + 2x} \\
-3x - 2 \\
\underline{-3x - 3} \\
1
\end{array}
\Rightarrow \frac{-2x^3 - x - 2}{x+1} = -2x^2 + 2x - 3 + \frac{1}{x+1}
$$

31.
$$
\begin{array}{r}
x^4 + x^3 + x^2 + x + 1 \\
x-1\overline{)x^5 + 0x^4 + 0x^3 + 0x^2 + 0x - 1} \\
\underline{x^5 - x^4} \\
x^4 + 0x^3 + 0x^2 + 0x - 1 \\
\underline{x^4 - x^3} \\
x^3 + 0x^2 + 0x - 1 \\
\underline{x^3 - x^2} \\
x^2 + 0x - 1 \\
\underline{x^2 - x} \\
x - 1 \\
\underline{x - 1} \\
0
\end{array}
\Rightarrow \frac{x^5 - 1}{x-1} = x^4 + x^3 + x^2 + x + 1
$$

33. $$\begin{array}{r|rrr} 3 & 1 & -4 & 3 \\ & & 3 & -3 \\ \hline & 1 & -1 & 0 \end{array}$$

Therefore, $P(3) = 0$.

35. $$\begin{array}{r|rrrr} -2 & 5 & 2 & -1 & 5 \\ & & -10 & 16 & -30 \\ \hline & 5 & -8 & 15 & -25 \end{array}$$

Therefore, $P(-2) = -25$.

37. $$\begin{array}{r|rrr} 2 & 1 & -5 & 1 \\ & & 2 & -6 \\ \hline & 1 & -3 & -5 \end{array}$$

Therefore, $P(2) = -5$.

39. $$\begin{array}{r|rrrr} 0.5 & 1 & 0 & -1 & 4 \\ & & 0.5 & 0.25 & -.375 \\ \hline & 1 & 0.5 & -0.75 & 3.625 \end{array}$$

Therefore, $P(0.5) = 3.625$.

41. $$\begin{array}{r|rrrrr} \sqrt{2} & 1 & 0 & -1 & 0 & -3 \\ & & \sqrt{2} & 2 & \sqrt{2} & 2 \\ \hline & 1 & \sqrt{2} & 1 & \sqrt{2} & -1 \end{array}$$

Therefore, $P\left(\sqrt{2}\right) = -1$.

43. $$\begin{array}{r|rrrr} \sqrt[3]{4} & -1 & 0 & 1 & 4 \\ & & -\sqrt[3]{4} & -\sqrt[3]{16} & \sqrt[3]{4}-4 \\ \hline & -1 & -\sqrt[3]{4} & 1-\sqrt[3]{16} & \sqrt[3]{4} \end{array}$$

Therefore, $P(\sqrt[3]{4}) = \sqrt[3]{4}$.

45. $$\begin{array}{r|rrr} 2 & 1 & 2 & -8 \\ & & 2 & 8 \\ \hline & 1 & 4 & 0 \end{array}$$

Yes; since $P(2) = 0$, 2 is a zero.

47. $$\begin{array}{r|rrrr} 4 & 2 & -6 & -9 & 6 \\ & & 8 & 8 & -4 \\ \hline & 2 & 2 & -1 & 2 \end{array}$$

No; since $P(4) \neq 0$, 4 is not a zero.

49. $$\begin{array}{r|rrrr} -0.5 & 4 & 12 & 7 & 1 \\ & & -2 & -5 & -1 \\ \hline & 4 & 10 & 2 & 0 \end{array}$$

Yes; since $P(-0.5) = 0$, -0.5 is a zero.

51. $$\begin{array}{r|rrrr} -5 & 8 & 50 & 47 & 15 \\ & & -40 & -50 & 15 \\ \hline & 8 & 10 & -3 & 30 \end{array}$$

No; since $P(-5) \neq 0$, -5 is not a zero.

53. $$\begin{array}{r|rrrrrrr} \sqrt{6} & -2 & 0 & 5 & 0 & -3 & 0 & 270 \\ & & -2\sqrt{6} & -12 & -7\sqrt{6} & -42 & -45\sqrt{6} & -270 \\ \hline & -2 & -2\sqrt{6} & -7 & -7\sqrt{6} & -45 & -45\sqrt{6} & 0 \end{array}$$

Yes; since $P(\sqrt{6}) = 0, \sqrt{6}$ is a zero.

55. Since the x-intercepts are -3, 1, and 4, the linear factors are $(x-(-3)),(x-1),(x-4)$ or $(x+3),(x-1),(x-4)$.

56. Since the x-intercepts are -3, 1, and 4, the solutions to $P(x)=0$ are -3, 1, and 4.

57. Since the x-intercepts are -3, 1 and 4; the zeros are -3, 1, and 4

58. Use synthetic division and divide by 2:

$$\begin{array}{r|rrrr} 2 & 1 & -2 & -11 & 12 \\ & & 2 & 0 & -22 \\ \hline & 1 & 0 & -11 & -10 \end{array}$$

The remainder is -10; therefore, $P(2)=-10$.

59. Using the x-intercepts and graph, $P(x)>0$ for the $(-3,1)\cup(4,\infty)$.

60. Using the x-intercepts and graph, $P(x)<0$ for the $(-\infty,-3)\cup(1,4)$.

61. First, use synthetic division to factor out the given zero.

$$\begin{array}{r|rrrr} 3 & 1 & -2 & -5 & 6 \\ & & 3 & 3 & -6 \\ \hline & 1 & 1 & -2 & 0 \end{array}$$

Now completely factor the resulting quadratic expression: $x^2+x-2 \Rightarrow (x-1)(x+2)$.

The other zeros are -2 and 1.

63. First, use synthetic division to factor out the given zero.

$$\begin{array}{r|rrrr} 1 & 1 & 0 & -2 & 1 \\ & & 1 & 1 & -1 \\ \hline & 1 & 1 & -1 & 0 \end{array}$$

Now solve the quadratic equation, $x^2+x-1=0$, using the quadratic formula:

$x=\dfrac{-1\pm\sqrt{1^2-4(1)(-1)}}{2(1)}=\dfrac{-1\pm\sqrt{5}}{2}$ The other zeros are $\dfrac{-1-\sqrt{5}}{2}$ and $\dfrac{-1+\sqrt{5}}{2}$.

65. First, use synthetic division to factor out the given zero.

$$\begin{array}{r|rrrr} -2 & 3 & 5 & -3 & -2 \\ & & -6 & 2 & 2 \\ \hline & 3 & -1 & -1 & 0 \end{array}$$

Now solve the quadratic equation, $3x^2-x-1=0$, using the quadratic formula:

$x=\dfrac{1\pm\sqrt{(-1)^2-4(3)(-1)}}{2(3)}=\dfrac{1\pm\sqrt{13}}{6}$ The other zeros are $\dfrac{1-\sqrt{13}}{6}$ and $\dfrac{1+\sqrt{13}}{6}$.

67. First, use synthetic division to factor out the first given zero. Then use synthetic division to factor out the second given zero from the resulting expression: $x^3-6x^2-5x+30$.

$$\begin{array}{r|rrrrr} -6 & 1 & 0 & -41 & 0 & 180 \\ & & -6 & 36 & 30 & -180 \\ \hline & 1 & -6 & -5 & 30 & 0 \end{array} \Rightarrow \begin{array}{r|rrrr} 6 & 1 & -6 & -5 & 30 \\ & & 6 & 0 & -30 \\ \hline & 1 & 0 & -5 & 0 \end{array}$$

Finally, solve the resulting equation $x^2 - 5 = 0 \Rightarrow x^2 = 5 \Rightarrow x = \pm\sqrt{5}$.

Therefore, the other zeros are $-\sqrt{5}$ and $\sqrt{5}$.

69. First, use synthetic division to factor out the given zero.

$$\begin{array}{r|rrrr} 8 & -1 & 8 & 3 & -24 \\ & & -8 & 0 & 24 \\ \hline & -1 & 0 & 3 & 0 \end{array}$$

Now solve the resulting equation: $-x^2 + 3 = 0 \Rightarrow -x^2 = -3 \Rightarrow x^2 = 3 \Rightarrow x = \pm\sqrt{3}$

Therefore, the other zeros are $-\sqrt{3}$ and $\sqrt{3}$.

71. First, use synthetic division to factor out the given zero.

$$\begin{array}{r|rrrr} 2 & 2 & -3 & -17 & 30 \\ & & 4 & 2 & -30 \\ \hline & 2 & 1 & -15 & 0 \end{array}$$

Now completely factor the resulting quadratic expression: $2x^2 + x - 15 \Rightarrow (2x-5)(x+3)$.

The linear factors are $P(x) = (x-2)(2x-5)(x+3)$.

73. First, use synthetic division to factor out the given zero.

$$\begin{array}{r|rrrr} -4 & 6 & 25 & 3 & -4 \\ & & -24 & -4 & 4 \\ \hline & 6 & 1 & -1 & 0 \end{array}$$

Now completely factor the resulting quadratic expression: $6x^2 + x - 1 \Rightarrow (3x-1)(2x+1)$. Therefore, the linear factors are $P(x) = (x+4)(3x-1)(2x+1)$.

75. First, use synthetic division to factor out the given zero.

$$\begin{array}{r|rrrr} -3 & -6 & -13 & 14 & -3 \\ & & 18 & -15 & 3 \\ \hline & -6 & 5 & -1 & 0 \end{array}$$

Now completely factor the resulting quadratic expression: $-6x^2 + 5x - 1 \Rightarrow (-3x+1)(2x-1)$.

The linear factors are $P(x) = (x+3)(-3x+1)(2x-1)$.

77. First, use synthetic division to factor out the given zero.

$$\begin{array}{r|rrrr} -5 & 1 & 5 & -3 & -15 \\ & & -5 & 0 & 15 \\ \hline & 1 & 0 & -3 & 0 \end{array}$$

Now completely factor the resulting quadratic expression: $x^2 - 3 = (x + \sqrt{3})(x - \sqrt{3})$.

The linear factors are $P(x) = (x+5)(x+\sqrt{3})(x-\sqrt{3})$.

79. First, use synthetic division to factor out the given zero.

$$\begin{array}{r|rrrr} -1 & 1 & -2 & -7 & -4 \\ & & -1 & 3 & 4 \\ \hline & 1 & -3 & -4 & 0 \end{array}$$

Now completely factor the resulting quadratic expression: $x^2 - 3x - 4 \Rightarrow (x-4)(x+1)$.

The linear factors are $P(x) = (x+1)(x-4)(x+1)$ or $(x+1)^2(x-4)$.

81.
$$\begin{array}{r} x^3 + 2 \\ 3x-7\overline{\smash{)}3x^4 - 7x^3 + 6x - 16} \\ \underline{3x^4 - 7x^3} \\ 6x - 16 \\ \underline{6x - 14} \\ -2 \end{array} \Rightarrow \frac{3x^4 - 7x^3 + 6x - 16}{3x-7} = x^3 + 2 + \frac{-2}{3x-7}$$

83.
$$\begin{array}{r} 5x^2 - 12 \\ x^2+2\overline{\smash{)}5x^4 - 2x^2 + 6} \\ \underline{5x^4 + 10x^2} \\ -12x^2 + 6 \\ \underline{-12x^2 - 24} \\ 30 \end{array} \Rightarrow \frac{5x^4 - 2x^2 + 6}{x^2+2} = 5x^2 - 12 + \frac{30}{x^2+2}$$

85.
$$\begin{array}{r} 4x + 5 \\ 2x^2-3\overline{\smash{)}8x^3 + 10x^2 - 12x - 15} \\ \underline{8x^3 - 12x} \\ 10x^2 - 15 \\ \underline{10x^2 - 15} \\ 0 \end{array} \Rightarrow \frac{8x^3 + 10x^2 - 12x - 15}{2x^2+3} = 4x + 5$$

87. $$\begin{array}{r} x^2-2x+4 \\ 2x^2+3x+2\overline{)2x^4-x^3+4x^2+8x+7} \\ \underline{2x^4+3x^3+2x^2} \\ -4x^3+2x^2+8x \\ \underline{-4x^3-6x^2-4x} \\ 8x^2+12x+7 \\ \underline{8x^2+12x+8} \\ -1 \end{array} \Rightarrow \frac{2x^4-x^3+4x^2+8x+7}{2x^2+3x+2}=x^2-2x+4+\frac{-1}{2x^2+3x+2}$$

89. $$\begin{array}{r} \frac{1}{2}x \\ 2x+1\overline{)x^2+\frac{1}{2}x-1} \\ \underline{x^2+\frac{1}{2}x} \\ -1 \end{array} \Rightarrow \frac{x^2+\frac{1}{2}x-1}{2x+1}=\frac{1}{2}x+\frac{-1}{2x+1}$$

91. $$\begin{array}{r} \frac{1}{2}x-\frac{1}{2} \\ 2x^2-1\overline{)x^3-x^2+0x+1} \\ \underline{x^3\quad -\frac{1}{2}x} \\ -x^2+\frac{1}{2}x+1 \\ \underline{-x^2\quad +\frac{1}{2}} \\ \frac{1}{2}x+\frac{1}{2} \end{array} \Rightarrow \frac{x^3-x^2+1}{2x^2-1}=\frac{1}{2}x-\frac{1}{2}+\frac{\frac{1}{2}x+\frac{1}{2}}{2x^2-1}$$

3.7: Topics in the Theory of Polynomial Functions (II)

1. With the given zeros 4 and $2+i$, the conjugate of $2+i$ or $2-i$ is also a zero. Therefore,

   $P(x)=(x-4)(x-(2+i))(x-(2-i))=(x-4)(x-2-i)(x-2+i)=(x-4)(x^2-4x+5)$

   $=x^3-4x^2+5x-4x^2+16x-20 \Rightarrow P(x)=x^3-8x^2+21x-20.$

3. With the given zeros 5 and i, the conjugate of i or $-i$ is also a zero. Therefore,

   $P(x)=(x-5)(x-i)(x-(-i))=(x-5)(x^2-i^2)=(x-5)(x^2+1) \Rightarrow P(x)=x^3-5x^2+x-5.$

5. With the given zeros 0 and $3+i$, the conjugate of $3+i$ or $3-i$ is also a zero. Therefore,

   $P(x)=(x-0)(x-(3+i))(x-(3-i))=(x)(x-3-i)(x-3+i)=(x)(x^2-6x+10)$

   $\Rightarrow P(x)=x^3-6x^2+10x.$

7. With the given zeros $-3, -1,$ and 4, the function $P(x)$ is $P(x)=a(x-(-3))(x-(-1))(x-4)=$

   $a(x+3)(x+1)(x-4)=a(x+3)(x^2-3x-4) \Rightarrow P(x)=a(x^3-13x-12)$

Since $P(2) = 5$ we can solve for a:

$$5 = a((2)^3 - 13(2) - 12) \Rightarrow 5 = a(8 - 26 - 12) \Rightarrow -30a = 5 \Rightarrow a = -\frac{5}{30} = -\frac{1}{6}$$

Therefore, $P(x) = -\frac{1}{6}(x^3 - 13x - 12) \Rightarrow P(x) = -\frac{1}{6}x^3 + \frac{16}{3}x + 2.$

9. With the given zeros $-2, 1$ and 0, the function $P(x)$ is

$$P(x) = a(x - (-2))(x - 1)(x - 0) = a(x + 2)(x - 1)(x) \Longrightarrow P(x) = ax(x^2 + x - 2)$$

Since $P(-1) = -1$, we can solve for a: $-1 = a(-1)((-1)^3 + (-1) - 2) \Rightarrow -1 = -a(1 - 1 - 2)$

$\Rightarrow -a(-2) = -1 \Rightarrow 2a = -1 \Rightarrow a = -\frac{1}{2}$. Therefore, $P(x) = -\frac{1}{2}x(x^2 + x - 2) \Rightarrow P(x) = -\frac{1}{2}x^3 - \frac{1}{2}x^2 + x.$

11. With the given zeros 4, $1 + i$, and the conjugate of $1 + i$ or $1 - i$ also a zero, the function $P(x)$ is

$$P(x) = a(x - 4)(x - (1 + i))(x - (1 - i)) = a(x - 4)(x - 1 - i)(x - 1 + i) =$$

$a(x - 4)(x^2 - 2x + 2) = a(x^3 - 2x^2 + 2x - 4x^2 + 8x - 8) \Rightarrow P(x) = a(x^3 - 6x^2 - 10x - 8)$. Since $P(2) = 4$,

we can solve for a: $4 = a((2)^3 - 6(2)^2 + 10(2) - 8) \Rightarrow 4 = a(8 - 24 + 20 - 8) \Rightarrow -4a = 4 \Rightarrow a = -\frac{1}{1} = -1.$

Therefore, $P(x) = -1(x^3 - 6x^2 + 10x - 8) \Rightarrow P(x) = -x^3 + 6x^2 - 10x + 8.$

13. First, use synthetic division to factor out the given zero.

$$\begin{array}{r|rrrr} 3 & 1 & -1 & -4 & -6 \\ & & 3 & 6 & 6 \\ \hline & 1 & 2 & 2 & 0 \end{array}$$

Now solve the resulting quadratic equation, $x^2 + 2x + 2 = 0$, using the quadratic formula:

$$x = \frac{-2 \pm \sqrt{(2)^2 - 4(1)(2)}}{2(1)} = \frac{-2 \pm \sqrt{-4}}{2} = \frac{2 \pm 2i}{2} = -1 \pm i.$$

Therefore, the other zeros are $-1 - i$ and $-1 + i$.

15. First, use synthetic division to factor out the first given zero. Then use synthetic division to factor out the second given zero from the resulting expression $x^3 - x^2 - 7x + 3$

$$\begin{array}{r|rrrrr} -3 & 1 & 2 & -10 & -18 & 9 \\ & & -3 & 33 & 21 & -9 \\ \hline & 1 & -1 & -7 & 3 & 0 \end{array} \Rightarrow \quad \begin{array}{r|rrrr} 3 & 1 & -1 & -7 & 3 \\ & & 3 & 6 & -3 \\ \hline & 1 & 2 & -1 & 0 \end{array}$$

Finally, solve the resulting quadratic equation, $x^2 + 2x - 1 = 0$, using the quadratic formula:

$$x = \frac{-2 \pm \sqrt{(2)^2 - 4(1)(-1)}}{2(1)} = \frac{-2 \pm \sqrt{8}}{2} = \frac{-2 \pm 2\sqrt{2}}{2} = -1 \pm \sqrt{2}.$$

Therefore, the other zeros are $-1 + \sqrt{2}$ and $-1 - \sqrt{2}$.

17. With the given zero $3i$, the conjugate $-3i$ is also a zero. Use synthetic division to factor out these zeros.

$$\begin{array}{r|rrrrr} 3i & 1 & -1 & 10 & -9 & 9 \\ & & 3i & -9-3i & 9+3i & -9 \\ \hline & 1 & -1+3i & 1-3i & 3i & 0 \end{array} \Rightarrow \quad \begin{array}{r|rrrr} -3i & 1 & -1+3i & 1-3i & 3i \\ & & -3i & 3i & -3i \\ \hline & 1 & -1 & 1 & 0 \end{array}$$

Finally, solve the resulting quadratic equation, $x^2 - x + 1 = 0$, using the quadratic formula:

$$x = \frac{1 \pm \sqrt{(-1)^2 - 4(1)(1)}}{2(1)} = \frac{1 \pm \sqrt{-3}}{2} = \frac{1}{2} \pm \frac{\sqrt{3}}{2}i.$$

Therefore, the other zeros are $-3i$, $\frac{1}{2} + \frac{\sqrt{3}}{2}i$, and $\frac{1}{2} - \frac{\sqrt{3}}{2}i$.

19. With the given zeros 5 and –4, $P(x) = a(x-5)(x-(-4)) = a(x-5)(x+4) \Rightarrow P(x) = a(x^2 - x - 20)$.

There are many possible solutions. For example, when $a = 1$, $P(x) = x^2 - x - 20$.

21. With the given zeros –3, 2, i, the conjugate of i or $-i$ is also zero. The function $P(x)$ is

$P(x) = a(x-(-3))(x-2)(x-i)(x-(-i)) = a(x+3)(x-2)(x-i)(x+i) =$

$\Rightarrow P(x) = a(x^2 + x - 6)(x^2 + 1) = a(x^4 + x^3 - 5x^2 + x - 6)$. There are many possible solutions. For example, when $a = 1$, $P(x) = x^4 + x^3 - 5x^2 + x - 6$.

23. With the given zeros $1+\sqrt{3}$, $1-\sqrt{3}$, and 1,

$P(x) = a\left(x - \left(1+\sqrt{3}\right)\right)\left(x - \left(1-\sqrt{3}\right)\right)(x-1) = a\left(x-1-\sqrt{3}\right)\left(x-1+\sqrt{3}\right)(x-1) =$

$a\left(x^2 - x + \sqrt{3}x - x + 1 - \sqrt{3} - \sqrt{3}x + \sqrt{3} - 3\right)(x-1) = a(x^2 - 2x - 2)(x-1) =$

$a(x^3 - x^2 - 2x^2 + 2x - 2x + 2) \Rightarrow P(x) = a(x^3 - 3x^2 + 2)$. There are many possible solutions. For example, when $a = 1$, $P(x) = x^3 - 3x^2 + 2$.

25. With the given zeros −1, 2, and 3 + 2i, the conjugate of 3 + 2i or 3 − 2i is also a zero.

$P(x) = a(x-(3+2i))(x-(3-2i))(x-(-1))(x-2) = a(x-3-2i)(x-3+2i)(x+1)(x-2) =$

$a(x^2 - 3x + 2ix - 3x + 9 - 6i - 2ix + 6i - 4i^2)(x^2 - x - 2) = a(x^2 - 6x + 13)(x^2 - x - 2) =$

$a(x^4 - x^3 - 2x^2 - 6x^3 + 6x^2 + 12x + 13x^2 - 13x - 26) \Rightarrow P(x) = a(x^4 - 7x^3 + 17x^2 - x - 26)$. There are many possible solutions. For example, when $a = 1$, $P(x) = x^4 - 7x^3 + 17x^2 - x - 26$.

27. With the given zeros –1, and $6-3i$, the conjugate of $6-3i$ or $6+3i$ also a zero. The function is

$P(x) = a(x-(6-3i))(x-(6+3i))(x-(-1)) = a(x-6+3i)(x-6-3i)(x+1) =$

$a(x^2 - 6x - 3ix - 6x + 36 + 18i + 3ix - 18i - 9i^2)(x+1) = a(x^2 - 12x + 45)(x+1) =$

$a(x^3-12x^2+45x+x^2-12x+45) \Rightarrow P(x)=a(x^3-11x^2+33x+45)$. There are many possible solutions. For example, when $a=1$, $P(x)=x^3-11x^2+33x+45$.

29. With the given zeros -3 (multiplicity 2) and $2+i$, the conjugate of $2+i$ or $2-i$ also a zero.

$P(x)=a(x-(2+i))(x-(2-i))(x-(-3))^2=a(x-2-i)(x-2+i)(x+3)^2=$

$a(x^2-2x+ix-2x+4-2i-ix+2i-i^2)(x^2+6x+9)=a(x^2-4x+5)(x^2+6x+9)=$

$a(x^4+6x^3+9x^2-4x^3-24x^2-36x+5x^2+30x+45) \Rightarrow P(x)=a(x^4+2x^3-10x^2-6x+45)$. There are many possible solutions. For example, when $a=1$, $P(x)=x^4+2x^3-10x^2-6x+45$.

31 - 39. See Figures 31-39

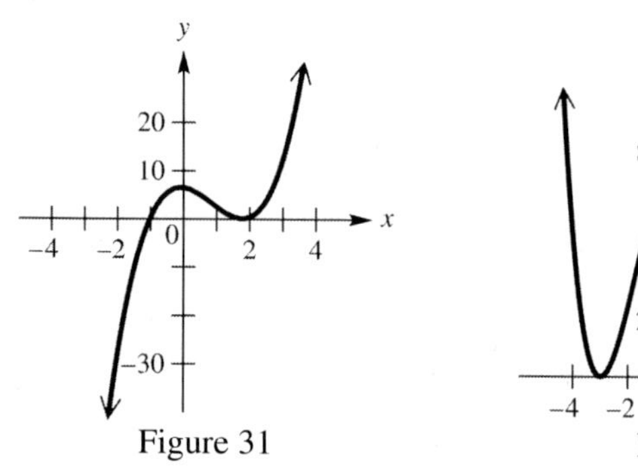

Figure 31

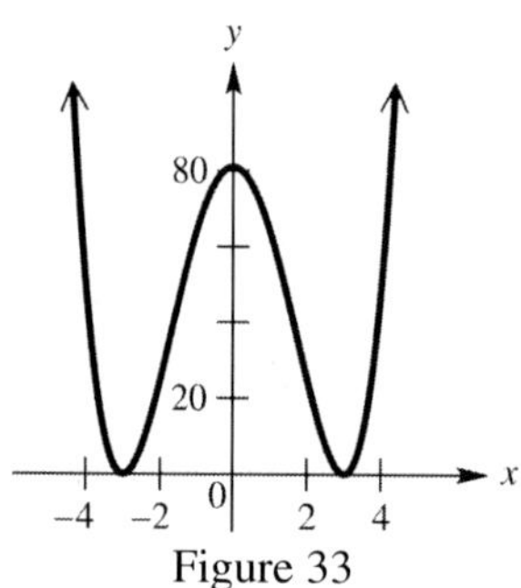

Figure 33

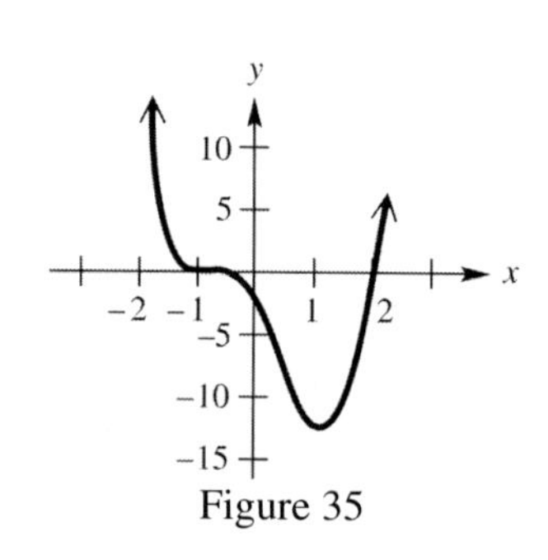

Figure 35

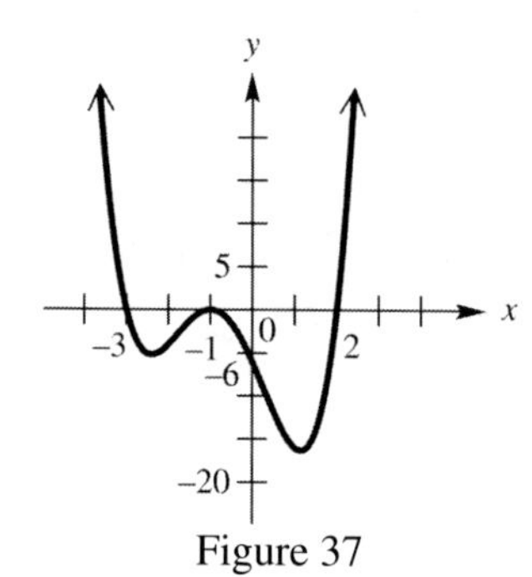

Figure 37

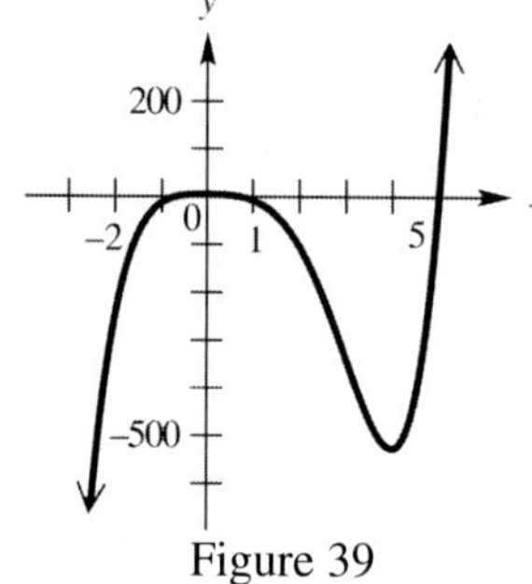

Figure 39

41. Since the graph crosses the x-axis at $(-1,0)$, $(0,0)$, and $(2,0)$ and by the factor theorem the three factors are $(x-(-1))$, $(x-0)$, and $(x-2)$. The general form of the function is $f(x)=a(x)(x+1)(x-2)$. We will find the value of a using the point $(1, 2)$. $2=a(1)(1+1)(1-2) \Rightarrow 2=-2a \Rightarrow a=-1$. The function is $f(x)=-(x)(x+1)(x-2)$.

43. Since the graph crosses the x-axis at $(-2,0)$, $(-1,0)$, and $(1,0)$ and by the factor theorem the three factors are $(x-(-2))$, $(x-(-1))$, and $(x-1)$. The general form of the function is $f(x)=a(x+2)(x+1)(x-1)$. We will find the value of a using the point $(0, -1)$. $-1=a(0+2)(0+1)(0-1) \Rightarrow -1=-2a \Rightarrow a=\frac{1}{2}$. The function is

$f(x)=\frac{1}{2}(x+2)(x+1)(x-1)$.

45. Since the graph crosses the x-axis at $(-4,0)$, $(-1,0)$, $(1,0)$, and $(2,0)$ and by the factor theorem the four factors are $(x-(-4))$, $(x-(-1))$, $(x-1)$, $(x-2)$. The general form of the function is $f(x)=a(x+4)(x+1)(x-1)(x-2)$. We will find the value of a using the point (0, 2).

$2=a(0+4)(0+1)(0-1)(0-2)\Rightarrow 2=8a\Rightarrow a=\frac{1}{4}$. The function is

$f(x)=\frac{1}{4}(x+4)(x+1)(x-1)(x-2)$.

47. First, use synthetic division to factor out the first –2. Then use synthetic division to factor out the second –2 from the resulting expression: x^3+0x^2-7x-6.

$$\begin{array}{r|rrrrr} -2 & 1 & 2 & -7 & -20 & -12 \\ & & -2 & 0 & 14 & 12 \\ \hline & 1 & 0 & -7 & -6 & 0 \end{array} \Rightarrow \qquad \begin{array}{r|rrrr} -2 & 1 & 0 & -7 & -6 \\ & & -2 & 4 & 6 \\ \hline & 1 & -2 & -3 & 0 \end{array}$$

Finally, completely factor the resulting quadratic expression: $x^2-2x-3=(x-3)(x+1)$. Therefore, the zeros are –2, 3, and –1, and the factored form is $P(x)=(x+2)^2(x-3)(x+1)$.

49. Since non real complex zeros occur in pairs, there will be 0, 2, or 4 non real complex zeros, leading to 5, 3, or 1 real zeros.

51. (a) Not Possible. With a zero of $1+i$, its conjugate $1-i$ must also be a zero, and this would give 4 zeros to a degree 3 function, which is not possible.

(b) Possible. Since non real complex zeros occur in pairs, there can be 4 non real complex zeros.

(c) Not Possible. We cannot have a multiplicity of 6 for a degree 5 function.

(d) Possible. With $1+2i$ having a multiplicity of 2, its conjugate $1-2i$ must also have a multiplicity of 2. This means the function has a minimum of 4 zeros which is possible.

53. (a) Using the Rational Zeros Theorem yields $\frac{p}{q}=\pm\frac{1,2,5,10}{1}=\pm1\pm2\pm5\pm10$.

(b) A graph (not shown) would indicate that there are no zeros less than -2 or greater than 5.

(c) First, use synthetic division to factor using values discovered in steps (a) and (b)

$$\begin{array}{r|rrrr} -2 & 1 & -2 & -13 & -10 \\ & & -2 & 8 & 10 \\ \hline & 1 & -4 & -5 & 0 \end{array}$$

Now completely factor the resulting quadratic expression: $x^2-4x-5\Rightarrow(x-5)(x+1)$. Therefore, the rational zeros are $-2, -1$ and 5.

(d) The factored form of $P(x)$ is $P(x)=(x+2)(x+1)(x-5)$.

55. (a) Using the Rational Zeros Theorem yields $\frac{p}{q} = \pm\frac{1,2,3,5,6,10,15,30}{1}$

(b) A graph (not shown) would indicate that there are no zeros less than -5 or greater than 2.

(c) First, use synthetic division to factor using values discovered in steps (a) and (b).

$$\begin{array}{r|rrrr} -5 & 1 & 6 & -1 & -30 \\ & & -5 & -5 & 30 \\ \hline & 1 & 1 & -6 & 0 \end{array}$$

Now completely factor the resulting quadratic expression: $x^2 + x - 6 \Rightarrow (x+3)(x-2)$. Therefore, the rational zeros are $-5, -3$ and 2.

(d) The factored form of $P(x)$ is $P(x) = (x+5)(x+3)(x-2)$.

57. (a) Using the Rational Zeros Theorem yields

$$\frac{p}{q} = \pm\frac{1,2,3,4,6,12}{1,2,3,6} = \pm1, \pm2, \pm3, \pm4, \pm6, \pm12, \pm\frac{1}{2}, \pm\frac{3}{2}, \pm\frac{1}{3}, \pm\frac{2}{3}, \pm\frac{4}{3}, \pm\frac{1}{6}.$$

(b) A graph (not shown) would indicate that there are no zeros less than -4 or greater than $\frac{3}{2}$.

(c) First, use synthetic division to factor using values discovered in steps (a) and (b).

$$\begin{array}{r|rrrr} -4 & 6 & 17 & -31 & -12 \\ & & -24 & 28 & 12 \\ \hline & 6 & -7 & -3 & 0 \end{array}$$

Now completely factor the resulting quadratic expression: $6x^2 - 7x - 3 \Rightarrow (3x+1)(2x-3)$. Therefore, the rational zeros are $-4, -\frac{1}{3}$, and $\frac{3}{2}$.

(d) The factored form of $P(x)$ is $P(x) = (x+4)(3x+1)(2x-3)$.

59. Using the Rational Zeros Theorem yields

$\frac{p}{q} = \pm\frac{1,2,3,6}{1,2,3,4,6,12} = \pm1, \pm2, \pm3, \pm6, \pm\frac{1}{2}, \pm\frac{3}{2}, \pm\frac{1}{3}, \pm\frac{2}{3}, \pm\frac{1}{6}, \pm\frac{1}{12}, \pm\frac{1}{4}, \pm\frac{3}{4}$. Use synthetic division to factor using values from above.

$$\begin{array}{r|rrrr} -\frac{3}{2} & 12 & 20 & -1 & -6 \\ & & -18 & -3 & 6 \\ \hline & 12 & 2 & -4 & 0 \end{array}$$

Now completely factor the resulting quadratic expression: $12x^2 + 2x - 4 \Rightarrow 2(2x-1)(3x+2)$. The factored form of $P(x)$ is $P(x) = (3x+2)(2x+3)(2x-1)$.

61. Using the Rational Zeros Theorem yields $\frac{p}{q}=\pm\frac{1,2,3,4,6,12}{1,2,3,4,6,8,12,24}=$

$\pm1,\pm2,\pm3,\pm4,\pm6,\pm12,\pm\frac{1}{2},\pm\frac{3}{2},\pm\frac{1}{3},\pm\frac{2}{3},\pm\frac{4}{3},\pm\frac{1}{4},\pm\frac{3}{4},\pm\frac{1}{6},\pm\frac{1}{8},\pm\frac{3}{8},\pm\frac{1}{12},\pm\frac{1}{24}.$

Factor out a 2, $P(x)=24x^3+40x^2-2x-12 \Rightarrow P(x)=2(12x^3+20x^2-x-6)$. Now use synthetic division to factor using values from above.

$$\begin{array}{r|rrrr} -\frac{3}{2} & 12 & 20 & -1 & -6 \\ & & -18 & -3 & 6 \\ \hline & 12 & 2 & -4 & 0 \end{array}$$

Now completely factor the resulting quadratic expression: $12x^2+2x-4 \Rightarrow 2(3x+2)(2x-1)$. The factored form of $P(x)$ is $P(x)=2(2x+3)(3x+2)(2x-1)$.

63. To eliminate the fractions, multiply the function by 2; therefore, $P(x)=2\left(x^3+\frac{1}{2}x^2-\frac{11}{2}x-5\right)=$

$2x^3+x^2-11x-10$. Now, use the Rational Zeros Theorem to identify possible zeros:

$\frac{p}{q}=\pm10,\pm5,\pm1,\pm\frac{1}{2},\pm\frac{5}{2}$. Next, from the calculator graph of the equation, we choose to check -1 by synthetic division.

$$\begin{array}{r|rrrr} -1 & 2 & 1 & -11 & -10 \\ & & -2 & 1 & 10 \\ \hline & 2 & -1 & -10 & 0 \end{array}$$

Now, completely factor the resulting quadratic expression: $2x^2-x-10 \Rightarrow (2x-5)(x+2)$. Therefore, the rational zeros are $-2,-1,$ and $\frac{5}{2}$.

65. To eliminate the fractions, multiply the function by 12; therefore,

$P(x)=12\left(\frac{1}{6}x^4-\frac{11}{12}x^3+\frac{7}{6}x^2-\frac{11}{12}x+1\right)=2x^4-11x^3+14x^2-11x+12$. Now, use the Rational Zeros Theorem to identify possible zeros: $\frac{p}{q}=\pm1,\ \pm2,\pm3,\pm4,\pm6,\pm12,\pm\frac{1}{2},\pm\frac{3}{2}$. Next, from the calculator, graph of the equation, we choose to check $\frac{3}{2}$ and 4 by synthetic division. (The second from the result of the first)

$$\begin{array}{r|rrrrr} \frac{3}{2} & 2 & -11 & 14 & -11 & 12 \\ & & 3 & -12 & 3 & -12 \\ \hline & 2 & -8 & 2 & -8 & 0 \end{array} \Rightarrow \begin{array}{r|rrrr} 4 & 2 & -8 & 2 & -8 \\ & & 8 & 0 & 8 \\ \hline & 2 & 0 & 2 & 0 \end{array}$$

Now, completely factor the resulting quadratic expression: $2x^2+2 \Rightarrow 2\left(x^2+1\right) \Rightarrow x^2=-1 \Rightarrow x=\pm i.$

Since $\pm i$ are not real numbers, the rational zeros are $\frac{3}{2}$ and 4.

67. $P(x)=6x^4-5x^3-11x^2+10x-2$. Now use the Rational Zeros Theorem to

identify possible zeros: $\frac{p}{q}=\pm 1, \pm 2, \pm\frac{1}{2}, \pm\frac{1}{3}, \pm\frac{2}{3}, \pm\frac{1}{6}$. Next, from the calculator, graph of the equation, we

choose to check $\frac{1}{2}$ and $\frac{1}{3}$ by synthetic division. (The second from the result of the first)

$$\begin{array}{r|rrrrr} \frac{1}{2} & 6 & -5 & -11 & 10 & -2 \\ & & 3 & -1 & -6 & 2 \\ \hline & 6 & -2 & -12 & 4 & 0 \end{array} \Rightarrow \begin{array}{r|rrrr} \frac{1}{3} & 6 & -2 & -12 & 4 \\ & & 2 & 0 & -4 \\ \hline & 6 & 0 & -12 & 0 \end{array}$$

Now completely factor the resulting quadratic expression: $6x^2-12 \Rightarrow 6\left(x^2-2\right) \Rightarrow x^2=2 \Rightarrow x=\pm\sqrt{2}$

The factored form of $P(x)$ is $P(x)=(2x-1)(3x-1)\left(x-\sqrt{2}\right)\left(x+\sqrt{2}\right)$.

69. $P(x)=21x^4+13x^3-103x^2-65x-10$. Now use the Rational Zeros Theorem to

identify possible zeros:

$\frac{p}{q}=\pm 1, \pm 2, \pm 5, \pm 10, \pm\frac{5}{21}, \pm\frac{5}{7}, \pm\frac{5}{3}, \pm\frac{2}{21}, \pm\frac{2}{7}, \pm\frac{2}{3}, \pm\frac{10}{21}, \pm\frac{10}{7}, \pm\frac{10}{3}, \pm\frac{1}{21}, \pm\frac{1}{7}, \pm\frac{1}{3}$. Next, from the

calculator, graph of the equation, we choose to check $-\frac{2}{7}$ and $-\frac{1}{3}$ by synthetic division. (The second from

the result of the first)

$$\begin{array}{r|rrrrr} -\frac{2}{7} & 21 & 13 & -103 & -65 & -10 \\ & & -6 & -2 & 30 & 10 \\ \hline & 21 & 7 & -105 & -35 & 0 \end{array} \Rightarrow \begin{array}{r|rrrr} -\frac{1}{3} & 21 & 7 & -105 & -35 \\ & & -7 & 0 & 35 \\ \hline & 21 & 0 & -105 & 0 \end{array}$$

Now completely factor the resulting quadratic expression: $21x^2-105 \Rightarrow 21\left(x^2-5\right) \Rightarrow x^2=5 \Rightarrow x=\pm\sqrt{5}$

The factored form of $P(x)$ is $P(x)=(7x+2)(3x+1)\left(x-\sqrt{5}\right)\left(x+\sqrt{5}\right)$.

71. With the given zero i, the conjugate $-i$ is also a zero. Now use synthetic division to factor out these zeros.

$$\begin{array}{r|rrrrrr} i & 1 & -1 & 5 & -5 & 4 & -4 \\ & & i & -1-i & 1+4i & -4+4i & 4 \\ \hline & 1 & -1+i & 4-i & -4+4i & -4i & 0 \end{array} \Rightarrow \begin{array}{r|rrrrr} -i & 1 & -1+i & 4-i & -4+4i & -4i \\ & & -i & i & -4i & 4i \\ \hline & 1 & -1 & 4 & -4 & 0 \end{array}$$

From the graph of the resulting polynomial we can see that 1 is a zero.

$$\begin{array}{r|rrrr} 1 & 1 & -1 & 4 & -4 \\ & & 1 & 0 & 4 \\ \hline & 1 & 0 & 4 & 0 \end{array}$$

The resulting polynomial $x^2+4 \Rightarrow x = \pm 2i$. The factors of P(x) are $(x-1)(x-i)(x+i)(x-2i)(x+2i)$.

73. With the given zero $-2i$,the conjugate $2i$ is also a zero. Now use synthetic division to factor out these zeros.

$$\begin{array}{r|rrrrr} -2i & 1 & 1 & 2 & 4 & -8 \\ & & -2i & -4-2i & -4+4i & 8 \\ \hline & 1 & 1-2i & -2-2i & 4i & 0 \end{array} \Rightarrow \begin{array}{r|rrrr} 2i & 1 & 1-2i & -2-2i & 4i \\ & & 2i & 2i & -4i \\ \hline & 1 & 1 & -2 & 0 \end{array}$$

Factoring the resulting polynomial we have $x^2+x-2=(x+2)(x-1)$. The factors of P(x) are $(x+2)(x-1)(x-2i)(x+2i)$.

75. With the given zero $1+i$, the conjugate $1-i$ is also a zero. Now use synthetic division to factor out these zeros.

$$\begin{array}{r|rrrrr} 1+i & 1 & -2 & 3 & -2 & 2 \\ & & 1+i & -2 & 1+i & -2 \\ \hline & 1 & -1+i & 1 & -1+i & 0 \end{array} \Rightarrow \begin{array}{r|rrrr} 1-i & 1 & -1+i & 1 & -1+i \\ & & 1-i & 0 & 1-i \\ \hline & 1 & 0 & 1 & 0 \end{array}$$

Factoring the resulting polynomial we have $x^2+1 \Rightarrow x = \pm i$. The factors of P(x) are $(x-(1+i))(x-(1-i))(x-i)(x+i)$.

77. Because the function $P(x)$ has coefficient signs: + − + +, which is 2 sign changes, the function has 2 or 0 possible positive real zeros. Because $P(-x)=2(-x)^3-4(-x)^2+2(-x)+7=-2x^3-4x^2-2x+7$ has coefficient signs: − − − +, which is 1 sign change, the function has 1 negative real zero. From graphing the function, $P(x)$ actually has 0 positive and 1 negative real zeros.

79. Because the function $P(x)$ has coefficient signs: + + + −, which is 1 sign change, the function has 1 positive real zero. Because $P(-x)=5(-x)^4+3(-x)^2+2(-x)-9=5x^4+3x^2-2x-9$ has coefficient signs: + + − −, which is 1 sign change, the function has 1 negative real zero.

81. Because the function $P(x)$ has coefficient signs: + + − + +, which is 2 sign changes, the function has 2 or 0 possible positive real zeros. Because $P(-x)=(-x)^5+3(-x)^4-(-x)^3+2(-x)+3=$

$-x^5+3x^4+x^3-2x+3$ has coefficient signs: $-\ +\ +\ -\ +$, which is 3 sign changes, the function has 3 or 1 possible negative real zeros. From graphing the function, $P(x)$ actually has 0 positive and 1 negative real zero.

83. Using the Boundedness Theorem, we divide synthetically by $x-2$.

$$\begin{array}{r|rrrr} 2 & 1 & -1 & 3 & -8 & 8 \\ & & 2 & 2 & 10 & 4 \\ \hline & 1 & 1 & 5 & 2 & 12 \end{array}$$

The result is all nonnegative; therefore, no real zero greater than 2.

85. Using the Boundedness Theorem, we divide synthetically by $x-(-2)$.

$$\begin{array}{r|rrrr} -2 & 1 & 1 & -1 & 0 & 3 \\ & & -2 & 2 & -2 & 4 \\ \hline & 1 & -1 & 1 & -2 & 7 \end{array}$$

The result alternates in sign; therefore, no real zero less than -2.

87. Using the Boundedness Theorem, we divide synthetically by $x-1$.

$$\begin{array}{r|rrrr} 1 & 3 & 2 & -4 & 1 & -1 \\ & & 3 & 5 & 1 & 2 \\ \hline & 3 & 5 & 1 & 2 & 1 \end{array}$$

The result is all nonnegative; therefore, no real zero greater than 1.

89. Using the Boundedness Theorem, we divide synthetically by $x-2$.

$$\begin{array}{r|rrrrr} 2 & 1 & 0 & -3 & 0 & 1 & 2 \\ & & 2 & 4 & 2 & 4 & 10 \\ \hline & 1 & 2 & 1 & 2 & 5 & 12 \end{array}$$

The result is all nonnegative; therefore, no real zero greater than 2.

91. From the graph, the function has zeros $-6, 2$ and 5, the function $P(x)$ is

$P(x)=a(x-(6))(x-2)(x-5)=a(x+6)(x-2)(x-5)=a(x+6)(x^2-7x+10)\Rightarrow$

$P(x)=a(x^3-x^2-32x+60)$. Since $\mathrm{P}(0)=30$, we can solve for a:

$$30=a((0)^3-(0)^2-42(0)+60)\Rightarrow 30=a(0-0-0+60)\Rightarrow 60a=30\Rightarrow a=\frac{30}{60}=\frac{1}{2}.$$

Therefore, $P(x)=\frac{1}{2}(x^3-x^2-32x+60)\Rightarrow P(x)=\frac{1}{2}x^3-\frac{1}{2}x^2-16x+30.$

93. (a) Because the function $P(x)$ has coefficient signs: $-\ -\ +\ +$, which is 1 sign change, the function has 1 positive real zero. Because $P(-x)=-2(-x)^4-(-x)^3+(-x)+2=-2x^4+x^2-x+2$ has coefficient signs: $-\ +\ -\ +$, which is 3 sign changes, the function has 3 or 1 negative real zero.

(b) By the Rational Zero Theorem, the possible rational zeros are $\frac{p}{q}=\pm\frac{1,2}{1,2}=\pm1,\pm2,\pm\frac{1}{2}$.

(c) From part (b) and a calculator graph of the equation, we choose 1 and -1 to check by synthetic division. (The second from the result of the first)

$$\begin{array}{r|rrrrr} 1 & -2 & -1 & 0 & 1 & 2 \\ & & -2 & -3 & -3 & -2 \\ \hline & -2 & -3 & -3 & -2 & 0 \end{array} \Rightarrow \begin{array}{r|rrrr} -1 & -2 & -3 & -3 & -2 \\ & & 2 & 1 & 2 \\ \hline & -2 & -1 & -2 & 0 \end{array}$$

The resulting polynomial $-2x^2 - x - 2$ has no real zeros, so the rational zeros are 1 and -1.

(d) No other real zeros, the remaining zeros are imaginary.

(e) Using the quadratic equation yields

$$x = \frac{-(-1) \pm \sqrt{(-1)^2 - 4(-2)(-2)}}{2(-2)} = \frac{1 \pm \sqrt{-15}}{-4} \Rightarrow x = -\frac{1}{4} + i\frac{\sqrt{15}}{4} \text{ and } -\frac{1}{4} - i\frac{\sqrt{15}}{4}.$$

(f) The rational zeros are the x-intercepts; therefore, (1, 0) and $(-1, 0)$.

(g) Find $P(0)$: $P(0) = -2(0)^4 - (0)^3 + (0) + 2 \Rightarrow P(0) = 2$. The y-intercept is (0, 2).

(h)

$$\begin{array}{r|rrrrr} 4 & -2 & -1 & 0 & 1 & 2 \\ & & -8 & -36 & -144 & -572 \\ \hline & -2 & -9 & -36 & -143 & -570 \end{array}$$

Therefore, $f(4) = -570$; $(4, -570)$.

(i) $P(x)$ is an negative even-degree polynomial graph; therefore,

(j) See Figure 93.

94. (a) Because the function $P(x)$ has coefficient signs: + + + + +, which is 0 sign changes, the function has 0 positive real zeros. Because $P(-x) = 4(-x)^5 + 8(-x)^4 + 9(-x)^3 + 27(-x)^2 + 27(-x) =$ $-4x^5 + 8x^4 - 9x^3 + 27x^2 - 27x$ has coefficient signs: − + − + −, which is 4 sign changes, the function has 4, 2, or 0 negative real zeros.

(b) By using the Rational Zero Theorem, after factoring out an x, the possible rational zeros are

$$\frac{p}{q} = \pm\frac{1,3,9,27}{1,2,4} = 0, \pm1, \pm3, \pm9, \pm27, \pm\frac{1}{2}, \pm\frac{3}{2}, \pm\frac{9}{2}, \pm\frac{27}{2}, \pm\frac{1}{4}, \pm\frac{3}{4}, \pm\frac{9}{4}, \pm\frac{27}{4}.$$

(c) Factoring out the x gives 0 as one of the zeros, now from part (b) and a calculator graph of the equation, we choose $-\frac{3}{2}$ (multiplicity 2, because it appears tangent) to check by synthetic division. (The second from the result of the first)

$$\begin{array}{r|rrrrr} -\frac{3}{2} & 4 & 8 & 9 & 27 & 27 \\ & & -6 & -3 & -9 & -27 \\ \hline & 4 & 2 & 6 & 18 & 0 \end{array} \Rightarrow \begin{array}{r|rrrr} -\frac{3}{2} & 4 & 2 & 6 & 18 \\ & & -6 & 6 & -18 \\ \hline & 4 & -4 & 12 & 0 \end{array}$$

The resulting polynomial $4x^2 - 4x + 12$ has no real zeros, so the rational zeros are 0 and $-\frac{3}{2}$ (multiplicity 2).

(d) No other real zeros, the remaining zeros are imaginary.

(e) Using the quadratic formula after factoring $4(x^2-x+3)$ yields

$$x=\frac{-(-1)\pm\sqrt{(-1)^2-4(1)(3)}}{2(1)}=\frac{1\pm\sqrt{-11}}{2}\Rightarrow x=\frac{1}{2}+i\frac{\sqrt{11}}{2} \text{ and } \frac{1}{2}-i\frac{\sqrt{11}}{2}.$$

(f) The rational zeros are the x-intercepts; therefore, (0, 0) and $\left(-\frac{3}{2},0\right)$.

(g) Find $P(0)$: $P(0)=4(0)^5+8(0)^4+9(0)^3+27(0)^2+27(0)\Rightarrow P(0)=0$. The y-intercept is (0, 0).

(h)

4)	4	8	9	27	27	0
		16	96	420	1788	7260
	4	24	105	447	1815	7260

Therefore, $f(4)=7260$; $(4,7260)$.

(i) $P(x)$ is an positive odd-degree polynomial graph; therefore, ↗

(j) See Figure 94.

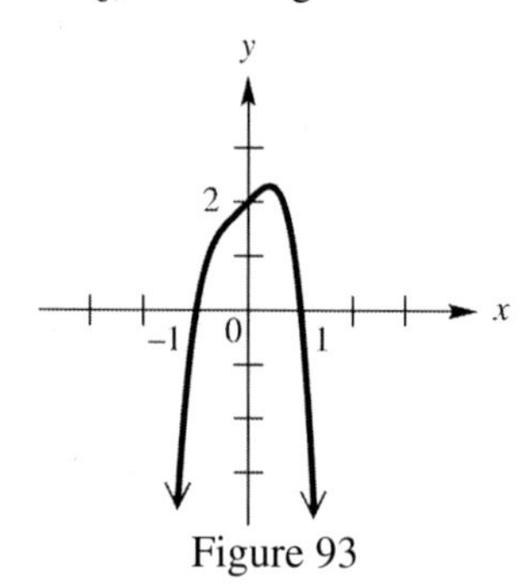

Figure 93

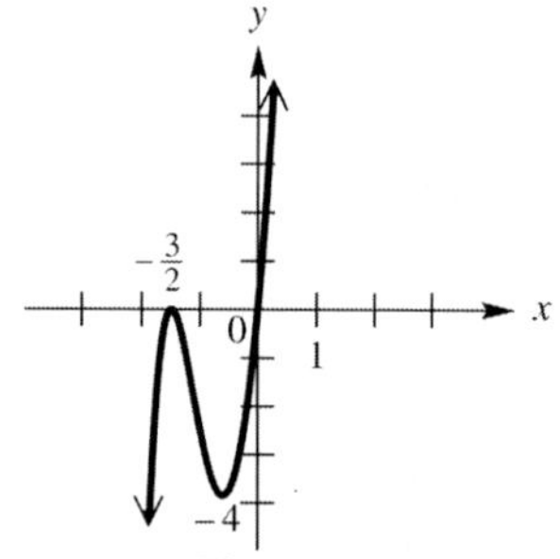

Figure 94

95. (a) Because the function $P(x)$ has coefficient signs: $+--$, which is 1 sign change, the function has 1 positive real zero. Because $P(-x)=3(-x)^4-14(-x)^2-5=3x^4-14x^2-5$ has coefficient signs: $+--$, which is 1 sign change, the function has 1 negative real zero.

(b) By the Rational Zeros Theorem, the possible rational zeros are $\frac{p}{q}=\pm\frac{1,5}{1,3}=\pm1,\pm5,\pm\frac{1}{3},\pm\frac{5}{3}$.

(c) Set equal to 0 and factor the function: $0=3x^4-14x^2-5\Rightarrow 0=(3x^2+1)(x^2-5)$. The factor $3x^2+1$ will yield imaginary zeros and x^2-5 yields irrational zeros; therefore, there are no rational zeros.

(d) From part (c), the factor $\left\{\pm\frac{5}{2},\pm1\right\}=\{-2.5,-1,1,2.5\}$; therefore, $-\sqrt{5}$ and $\sqrt{5}$ are real zeros.

(e) Using the quadratic formula on the factor $3x^2+1=0$ found in part (c), yields

$$x=\frac{-(0)\pm\sqrt{(0)^2-4(3)(1)}}{2(3)}=\frac{0\pm\sqrt{-12}}{6}=\frac{\pm2i\sqrt{3}}{6}\Rightarrow x=-i\frac{\sqrt{3}}{3} \text{ and } i\frac{\sqrt{3}}{3}.$$

(f) The rational zeros are the x-intercepts; therefore, 1 and $\left(-\sqrt{5},0\right)$ and $\left(\sqrt{5},0\right)$.

(g) Find $P(0)$: $P(0)=3(0)^4-14(0)^2-5\Rightarrow P(0)=-5$. The y-intercept is $(0,-5)$.

(h)

$$\begin{array}{r|rrrrr} 4 & 3 & 0 & -14 & 0 & -5 \\ & & 12 & 48 & 136 & 144 \\ \hline & 3 & 12 & 34 & 136 & 536 \end{array}$$

Therefore, $f(4) = 539$; $(4,\ 539)$.

(i) $P(x)$ is a positive even-degree polynomial graph; therefore,

(j) See Figure 95.

96. (a) Because the function $P(x)$ has coefficient signs: $- - + + - -$, which is 2 sign changes, the function has 2 or 0 positive real zeros. Because $P(-x) = -(-x)^5 - (-x)^4 + 10(-x)^3 + 10(-x)^2 - 9(-x) - 9 =$ $x^5 - x^4 - 10x^3 + 10x^2 + 9x - 9$ has coefficient signs: $+ - - + + -$, which is 3 sign changes, the function has 3 or 1 negative real zeros.

(b) By using the Rational Zeros Theorem, the possible rational zeros are $\frac{p}{q} = \pm\frac{1,3,9}{1} = \pm 1,\ \pm 3,\ \pm 9$.

(c) From part (b) and a calculator graph of the equation, we choose $-3, 1$ and 3 to check by synthetic division.

$$\begin{array}{r|rrrrrr} -3 & -1 & -1 & 10 & 10 & -9 & -9 \\ & & 3 & -6 & -12 & 6 & 9 \\ \hline & -1 & 2 & 4 & -2 & -3 & 0 \end{array} \Rightarrow \begin{array}{r|rrrrr} 1 & -1 & 2 & 4 & -2 & -3 \\ & & -1 & 1 & 5 & 3 \\ \hline & -1 & 1 & 5 & 3 & 0 \end{array} \Rightarrow$$

$$\begin{array}{r|rrrr} 3 & -1 & 1 & 5 & 3 \\ & 0 & -3 & -6 & -3 \\ \hline & -1 & -2 & -1 & 0 \end{array}$$

Factor the resulting expression to obtain the last 2 zeros. $-x^2 - 2x - 1 = -(x+1)(x+1) = 0 \Rightarrow x = -1$ (multiplicity 2). Therefore, the rational zeros are $-3, -1$ (multiplicity 2), and 3.

(d) No other real zeros, all are rational and given in part (c).

(e) No complex zeros, all are rational.

(f) The rational zeros are the x-intercepts; therefore, $(-3,0), (-1,0)$ (multiplicity 2), and $(3,0)$.

(g) Find $P(0): P(0) = -(0)^5 - (0)^4 + 10(0)^3 + 10(0)^2 - 9(0) - 9 \Rightarrow P(0) = -9$. The y-intercept is $(0,-9)$.

(h)

$$\begin{array}{r|rrrrrr} 4 & -1 & -1 & 10 & 10 & -9 & -9 \\ & & -4 & 20 & -40 & -120 & -516 \\ \hline & -1 & -5 & -10 & -30 & -129 & -525 \end{array}$$

Therefore, $f(4) = -525$; $(4,\ \text{-}525)$.

(i) $P(x)$ is a negative odd-degree polynomial graph; therefore,

(j) See Figure 96.

97. (a) Because the function $P(x)$ has coefficient signs, $- + - + -$, which is 4 sign changes, the function has 4, 2 or 0 positive real zeros. Because $P(-x) = -3x^4 - 22x^3 - 55x^2 - 52x - 12$ has coefficient signs, $-----$, which is 0 sign changes, the function has 0 negative real zeros.

(b) By the Rational Zeros Theorem, the possible rational zeros are

$$\frac{p}{q} = \pm\frac{1,\ 2,\ 3,\ 4,\ 6,\ 12}{1,\ 3} = \pm 1, \pm 2,\ \pm 3, \pm 4, \pm 6, \pm 12, \pm\frac{1}{3}, \pm\frac{2}{3}, \pm\frac{4}{3}$$

(c) From part (b) and a calculator graph of the equation we choose $\frac{1}{3}$ and 3 to check by synthetic division.

$$\frac{1}{3}\overline{)\begin{array}{rrrrr}-3 & 22 & -55 & 52 & -12 \\ & -1 & 7 & -16 & 12 \\ \hline -3 & 21 & -48 & 36 & 0\end{array}} \Rightarrow 3\overline{)\begin{array}{rrrr}-3 & 21 & -48 & 36 \\ & -9 & 36 & -36 \\ \hline -3 & 12 & -12 & 0\end{array}}$$

Factor the resulting expression to obtain the last 2 zeros. $-3x^2+12x-12=-3(x-2)(x-2)=0 \Rightarrow x=2$ (with a multiplicity of 2). Therefore, the rational zeros are $\left(\frac{1}{3},0\right)$, $(2,0)$ (multiplicity 2), and $(3,0)$.

(d) No other real zeros, all are rational.

(e) No complex zeros, all are rational.

(f) The rational zeros are the x-intercepts; therefore, the answers are $\frac{1}{3}$, 2 (multiplicity 2), and 3.

(g) Find $P(0): P(0)=-3(0)^4+22(0)^3-55(0)^2+52(0)-12 \Rightarrow P(0)=-12$. The y-intercept is -12.

(h) $$4\overline{)\begin{array}{rrrrr}-3 & 22 & -55 & 52 & -12 \\ & -12 & 40 & -60 & -32 \\ \hline -3 & 10 & -15 & -8 & -44\end{array}}$$ Therefore, $f(4)=-44$; $(4, -44)$.

(i) $P(x)$ is a negative even-degree polynomial graph; therefore, it is $\cap$

(j) See Figure 97.

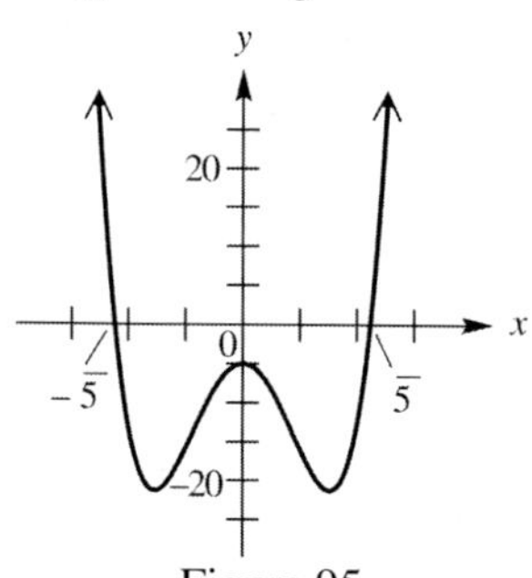

Figure 95

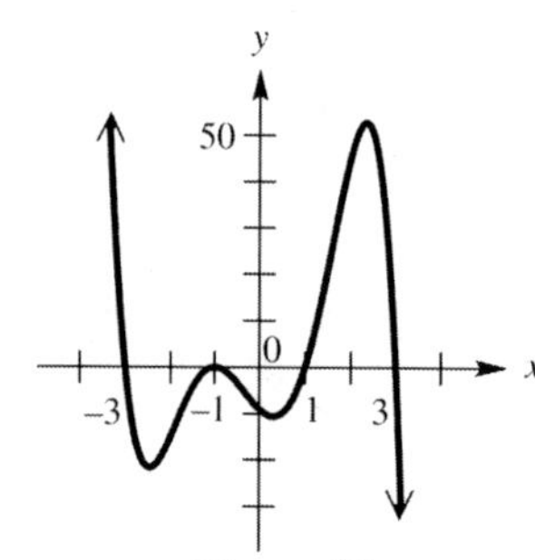

Figure 96

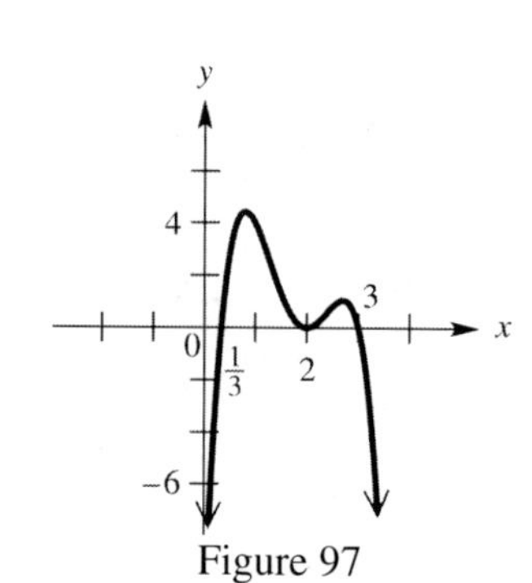

Figure 97

98. The function has 2 irrational zeros of $\pm\sqrt{5}$, which can be approximated to ± 2.236.

3.8: Polynomial Equations and Inequalities; Further Applications and Models

1. $x^3-25x=0 \Rightarrow x(x^2-25)=0 \Rightarrow x(x+5)(x-5)=0 \Rightarrow x=0,(x+5)=0,(x-5)=0 \Rightarrow x=0, x=-5, x=5.$

3. $x^4-x^2=2x^2+4 \Rightarrow x^4-3x^2-4=0 \Rightarrow (x^2-4)(x^2+1)=0 \Rightarrow (x^2-4)=0,(x^2+1)=0 \Rightarrow$

   $x=\pm 2$. The solution to x^2+1 is not a real number.

5. $x^3-3x^2-18x=0 \Rightarrow x(x^2-3x-18)=0 \Rightarrow x(x-6)(x+3)=0 \Rightarrow x=0,(x-6)=0,(x+3)=0 \Rightarrow$

   $x=0, x=6, x=-3$

7. $2x^3=4x^2-2x \Rightarrow 2x^3-4x^2+2x=0 \Rightarrow 2x(x^2-2x+1)=0 \Rightarrow 2x(x-1)^2=0 \Rightarrow 2x=0,(x-1)^2=0 \Rightarrow x=0, x=1.$

9. $12x^3 = 17x^2 + 5x \Rightarrow 12x^3 - 17x^2 - 5x = 0 \Rightarrow x\left(12x^2 - 17x - 5\right) = 0 \Rightarrow x\left(4x+1\right)\left(3x-5\right) = 0 \Rightarrow$

$x = 0, x = -\dfrac{1}{4}, x = \dfrac{5}{3}.$

11. $2x^3 + 4 = x\left(x+8\right) \Rightarrow 2x^3 + 4 = x^2 + 8x \Rightarrow 2x^3 - x^2 - 8x + 4 = 0 \Rightarrow x^2\left(2x-1\right) - 4\left(2x-1\right) = 0 \Rightarrow \left(x^2 - 4\right)\left(2x-1\right) = 0$

$\Rightarrow \left(x^2 - 4\right) = 0, \left(2x-1\right) = 0 \Rightarrow x = \pm 2, x = \dfrac{1}{2}.$

13. First, factor the equation: $7x^3 + x = 0 \Rightarrow x(7x^2 + 1) = 0 \Rightarrow x = 0$ and $7x^2 + x = 0$. Now solve:

$7x^2 + 1 = 0 \Rightarrow 7x^2 = -1 \Rightarrow x^2 = -\dfrac{1}{7} \Rightarrow x = \pm\dfrac{\sqrt{-1}}{\sqrt{7}} \Rightarrow x = \pm\dfrac{i}{\sqrt{7}} \Rightarrow x = \pm\dfrac{i}{\sqrt{7}} \cdot \dfrac{\sqrt{7}}{\sqrt{7}} \Rightarrow x = \pm\dfrac{i\sqrt{7}}{7}$. The solution set is $\left\{0, \pm\dfrac{i\sqrt{7}}{7}\right\}$.

15. First, factor the equation.

$3x^3 + 2x^2 - 3x - 2 = 0 \Rightarrow x^2(3x+2) - 1(3x+2) = 0 \Rightarrow (x^2 - 1)(3x+2) = 0 \Rightarrow (x+1)(x-1)(3x+2) = 0$.

Now solve. $x + 1 = 0 \Rightarrow x = -1;\ x - 1 = 0 \Rightarrow x = 1;\ 3x + 2 = 0 \Rightarrow x = -\dfrac{2}{3}$. The solution set is $\left\{-1, -\dfrac{2}{3}, 1\right\}$.

17. First, factor the equation: $x^4 - 11x^2 + 10 = 0 \Rightarrow (x^2 - 1)(x^2 - 10) = 0 \Rightarrow (x+1)(x-1)(x^2 - 10) = 0$. Now solve: $x + 1 = 0 \Rightarrow x = -1;\ x - 1 = 0 \Rightarrow x = 1;\ x^2 - 10 = 0 \Rightarrow x^2 = 10 \Rightarrow x = \pm\sqrt{10}$. The solution set is $\left\{\pm 1, \pm\sqrt{10}\right\}$.

19. First, factor the equation: $4x^4 - 25x^2 + 36 = 0 \Rightarrow (4x^2 - 9)(x^2 - 4) = 0$. Now solve:

$4x^2 - 9 = 0 \Rightarrow 4x^2 = 9 \Rightarrow x^2 = \dfrac{9}{4} \Rightarrow x = \pm\sqrt{\dfrac{9}{4}} \Rightarrow x = \pm\dfrac{3}{2};\ x^2 - 4 = 0 \Rightarrow x^2 = 4 \Rightarrow\ x = \pm\sqrt{4} \Rightarrow x = \pm 2.$ The solution set is $\left\{\pm\dfrac{3}{2},\ \pm 2\right\} = \{-2, -1.5,\ 1.5, 2\}$. When graphed on the calculator, these values are supported as the x-intercepts of $y = 4x^4 - 29x^2 + 25$.

21. First, factor the equation: $x^4 - 15x^2 - 16 = 0 \Rightarrow (x^2 - 16)(x^2 + 1) = 0$. Now solve:

$x^2 - 16 = 0 \Rightarrow x^2 = 16 \Rightarrow x = \pm\sqrt{16} \Rightarrow x = \pm 4;\ x^2 + 1 = 0 \Rightarrow x^2 = -1 \Rightarrow x = \pm\sqrt{-1} \Rightarrow x = \pm i.$ The solution set is $\{-4, 4, -i, i\}$. When graphed on the calculator, the real values -4 and 4 are supported as the only x-intercepts of $y = x^4 - 15x^2 - 16$.

23. First, factor the equation: $x^3 - x^2 - 64x + 64 = 0 \Rightarrow x^2(x-1) - 64(x-1) = 0 \Rightarrow$

$(x^2 - 64)(x-1) = 0 \Rightarrow (x+8)(x-8)(x-1) = 0.$ Now solve: $x + 8 = 0 \Rightarrow x = -8;\ x - 8 = 0 \Rightarrow x = 8;$

$x - 1 = 0 \Rightarrow x = 1.$ The solution set is $\{-8, 1, 8\}$. When graphed on the calculator, these values are supported as the x-intercepts of $y = x^3 - x^2 - 64x + 64$.

25. First, factor the equation: $-2x^3 - x^2 + 3x = 0 \Rightarrow -x(2x^2 + x - 3) = 0 \Rightarrow -x(2x+3)(x-1) = 0$. Now solve:

$-x = 0 \Rightarrow x = 0$; $2x + 3 = 0 \Rightarrow x = -\frac{3}{2}$; $x - 1 = 0 \Rightarrow x = 1$. The solution set is $\{-1.5, 0, 1\}$. When graphed on the calculator, these values are supported as the x-intercepts of $-2x^3 - x^2 + 3x = 0$.

27. First, factor the equation: $x^3 + x^2 - 7x - 7 = 0 \Rightarrow x^2(x+1) - 7(x+1) = 0 \Rightarrow (x^2 - 7)(x+1) = 0$. Now solve:

$x + 1 = 0 \Rightarrow x = -1$; $x^2 - 7 = 0 \Rightarrow x^2 = 7 \Rightarrow x = \pm\sqrt{7}$. The solution set is $\{-1, \pm\sqrt{7}\}$, When graphed on the calculator, these values are supported as the x-intercepts of $y = x^3 + x^2 - 7x - 7$.

29. First, factor the equation by $-x: -3x^3 - x^2 + 6x = 0 \Rightarrow -x(3x^2 + x - 6) = 0$. Now one solution is $-x = 0 \Rightarrow x = 0$, and we use the quadratic formula on the remaining factor to find the other solutions:

$x = \frac{-1 \pm \sqrt{1^2 - 4(3)(-6)}}{2(3)} \Rightarrow x = \frac{-1 \pm \sqrt{73}}{6}$. The solution set is $\left\{0, \frac{-1-\sqrt{73}}{6}, \frac{-1+\sqrt{73}}{6}\right\}$. When graphed on the calculator, these values are supported as the x-intercepts of $y = -3x^3 - x^2 + 6x$.

31. First, factor the equation by $3x: 3x^3 + 3x^2 + 3x = 0 \Rightarrow 3x(x^2 + x + 1) = 0$. Now one solution is $3x = 0 \Rightarrow x = 0$ and we use the quadratic formula on the remaining factor to find the other solutions:

$x = \frac{-1 \pm \sqrt{1^2 - 4(1)(1)}}{2(1)} \Rightarrow x = \frac{-1 \pm \sqrt{-3}}{2} \Rightarrow x = \frac{-1 \pm i\sqrt{3}}{2}$. The solution set is $\left\{0, -\frac{1}{2} - \frac{\sqrt{3}}{2}i, -\frac{1}{2} + \frac{\sqrt{3}}{2}i\right\}$.

When graphed on the calculator, the real value 0 is the only x-intercept of $y = 3x^3 + 3x^2 + 3x$.

33. First, factor the equation: $x^4 + 17x^2 + 16 = 0 \Rightarrow (x^2 + 16)(x^2 + 1) = 0$. Now solve: $x^2 + 16 = 0 \Rightarrow$

$x^2 = -16 \Rightarrow x = \pm\sqrt{-16} \Rightarrow x = \pm 4i$; $x^2 + 1 = 0 \Rightarrow x^2 = -1 \Rightarrow$ $x = \pm\sqrt{-1} \Rightarrow x = \pm i$. The solution set is $\{-4i, -i, i, 4i\}$. There are no real solutions, a conclusion which is supported when graphed on the calculator, with no x-intercepts for $y = x^4 + 17x^2 + 16$.

35. First, factor the equation: $x^6 + 19x^3 - 216 = 0 \Rightarrow (x^3 - 8)(x^3 + 27) = 0 \Rightarrow$

$(x-2)(x^2 + 2x + 4)(x+3)(x^2 - 3x + 9) = 0$. Now solve: $x - 2 = 0 \Rightarrow x = 2$; $x + 3 = 0 \Rightarrow x = -3$. We use the quadratic formula on the remaining factors to find the other solutions:

$$x = \frac{-2 \pm \sqrt{2^2 - 4(1)(4)}}{2(1)} = \frac{-2 \pm \sqrt{-12}}{2} = \frac{-2 \pm 2i\sqrt{3}}{2} \Rightarrow -1 \pm i\sqrt{3};$$

$x = \frac{-(-3) \pm \sqrt{(-3)^2 - 4(1)(9)}}{2(1)} = \frac{3 \pm \sqrt{-27}}{2} \Rightarrow x = \frac{3 \pm 3i\sqrt{3}}{2} = \frac{3}{2} \pm \frac{3\sqrt{3}}{2}i$. The solution set is

$\left\{-3, 2, -1 - i\sqrt{3}, -1 + i\sqrt{3}, \frac{3}{2} - \frac{3\sqrt{3}}{2}i, \frac{3}{2} + \frac{3\sqrt{3}}{2}i\right\}$. When graphed on the calculator, the real values -3 and 2 are supported as the only x-intercepts of $y = x^6 + 19x^3 - 216$.

37. See Figure 37. From the graph the following is found:

(a) $P(x)=0:\{-2,\ 1,\ 4\}$.

(b). $P(x)<0:(-\infty,-2)\cup(1,4)$.

(c). $P(x)>0:(-2,1)\cup(4,\infty)$.

39. See Figure 39. From the graph the following is found:

(a). $P(x)=0:\{-2.5,\ 1,\ 3\text{ (multiplicity 2)}\}$.

(b). $P(x)<0:(-2.5,1)$.

(c). $P(x)>0:(-\infty,-2.5)\cup(1,3)\cup(3,\infty)$.

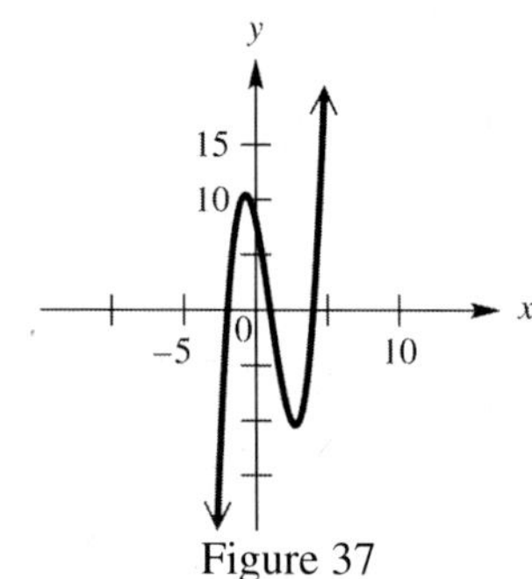

Figure 37

Figure 39

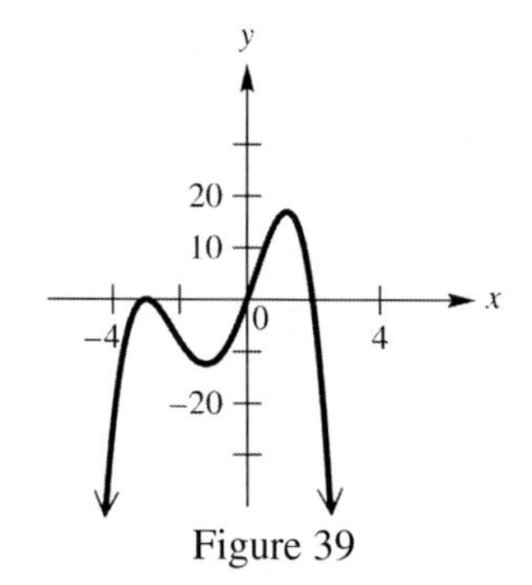

Figure 39

41. See Figure 41. From the graph the following is found:

(a) $P(x)=0:\{-3\text{ (multiplicity 2)},\ 0,\ 2\}$.

(b) $P(x)\geq 0:[-3]\cup[0,2]$.

(c) $P(x)\leq 0:(-\infty,0]\cup[2,\infty)$.

43. (a) $3\left(x^2+4\right)+2x(3x-12)=0\Rightarrow 3x^2+12+6x^2-24x=0\Rightarrow 9x^2-24x+12=0\Rightarrow$

$$3\left(3x^2-8x+4\right)=0\Rightarrow(3x-2)(x-2)=0\Rightarrow 3x-2=0\text{ or }x-2=0\Rightarrow x=\frac{2}{3},2$$

(b) The graph of $y=3\left(x^2+4\right)+2x(3x-12)$ or equivalently $y=9x^2-24x+12$ is a parabola that opens upward with zeros $\frac{2}{3}$ and 2. The parabola is below the x-axis between its zeros, so the solution set to the inequality is the interval $\left(\frac{2}{3},2\right)$. Note that it is not necessary to actually graph the parabola to solve the inequality.

45. (a) $3(x+1)^2(2x-1)^4+8(x+1)^3(2x-1)^3=0\Rightarrow\left[(x+1)^2(2x-1)^3\right][3(2x-1)+8(x+1)]=0\Rightarrow$

$$\left[(x+1)^2(2x-1)^3\right][(6x-3)+(8x+8)]=0\Rightarrow(x+1)^2(2x-1)^3(14x+5)=0\Rightarrow$$

$$(x+1)^2=0,\ (2x-1)^3=0,\ (14x+5)=0\Rightarrow x=-1,-\frac{5}{14},\frac{1}{2}$$

(b) The graph of $y=(x+1)^2(2x-1)^3(14x+5)$ is a degree six polynomial that is above the x-axis from $\left(-\infty,-\frac{5}{14}\right)$ and $\left(\frac{1}{2},\infty\right)$. The solution set to the inequality is the interval $\left(-\infty,-\frac{5}{14}\right]\cup\left[\frac{1}{2},\infty\right)$. Note that it is helpful to actually graph the cubic to solve the inequality.

47. (a) $3kx^2-7x=0\Rightarrow x(3kx-7)=0\Rightarrow x=0$ or $3kx-7=0\Rightarrow 3kx=7\Rightarrow x=\frac{7}{kx}$,

The solution set is $\left\{0,\frac{7}{3k}\right\}$.

(b) The graph of $y=3kx^2-7x$ is a parabola that opens upward (since k is positive) with zeros 0 and $\frac{7}{3k}$.

The parabola is below the x-axis between the zeros, so the solution set to the inequality is the interval $\left(0,\frac{7}{3k}\right)$.

49. From the calculator graph of $y_1=0.86x^3-5.24x^2+3.55x+7.84$, the solution set is $\{-0.88, 2.12, 4.86\}$.

51. From the calculator graph of $y_1=-\sqrt{7}x^3+\sqrt{5}x^2+\sqrt{17}$, the solution set is $\{1.52\}$.

53. Given the equation $2.45x^4-3.22x^3=-0.47x^2+6.54x+3$, rearrange to get the calculator graph of $y_1=2.54x^4-3.22x^3+0.47x^2-6.54x-3$. The solution set is $\{-0.40, 2.02\}$.

55. $x^2=-1\Rightarrow x=\pm\sqrt{-1}\Rightarrow x=\pm i\Rightarrow\{-i, i\}$.

57. First, factor the equation $x^3=-1\Rightarrow x^3+1=0\Rightarrow(x+1)(x^2-x+1)=0$. Now one solution is $x+1=0\Rightarrow x=-1$, and we use the quadratic formula on the remaining factor to find the other solutions.

$x=\frac{-(-1)\pm\sqrt{(-1)^2-4(1)(1)}}{2(1)}\Rightarrow x=\frac{1\pm i\sqrt{3}}{2}$. The solution set is $\left\{-1,\ \frac{1}{2}-\frac{\sqrt{3}}{2}i,\ \frac{1}{2}+\frac{\sqrt{3}}{2}i\right\}$.

59. First, factor the equation $x^3=27\Rightarrow x^3-27=0\Rightarrow(x-3)(x^2+3x+9)=0$. Now one solution is $x-3=0\Rightarrow x=3$. We use the quadratic formula on the remaining factor to find the other solutions.

$x=\frac{-3\pm\sqrt{3^2-4(1)(9)}}{2(1)}\Rightarrow x=\frac{-3\pm\sqrt{-27}}{2}=\frac{-3\pm 3i\sqrt{3}}{2}$. The solution set is $\left\{3,\ -\frac{3}{2}-\frac{3\sqrt{3}}{2}i,\ -\frac{3}{2}+\frac{3\sqrt{3}}{2}i\right\}$.

61. First, factor the equation $x^4=16\Rightarrow x^4-16=0\Rightarrow(x^2-4)(x^2+4)=0\Rightarrow(x-2)(x+2)(x^2+4)=0$. Now two solutions are $x-2=0\Rightarrow x=2$, and $x+2=0\Rightarrow x=-2$. We solve the remaining factor for the last two solutions. $x^2+4=0\Rightarrow x^2=-4\Rightarrow x=\pm\sqrt{-4}\Rightarrow x=\pm 2i$. The solution set is $\{-2, 2, -2i, 2i\}$.

63. First, factor the equation $x^3=-64\Rightarrow x^3+64=0\Rightarrow(x+4)(x^2-4x+16)=0$. Now one solution is $x+4=0\Rightarrow x=-4$. We use the quadratic formula on the remaining factor to find the other solutions.

$x=\frac{-(-4)\pm\sqrt{(-4)^2-4(1)(16)}}{2(1)}=\frac{4\pm\sqrt{-48}}{2}\Rightarrow x=\frac{4\pm 4i\sqrt{3}}{2}=2\pm 2i\sqrt{3}$. The solution set is $\{-4,\ 2-2i\sqrt{3},\ 2+2i\sqrt{3}\}$.

65. $x^2 = -18 \Rightarrow x = \sqrt{-18} \Rightarrow x = \pm 3i\sqrt{2} \Rightarrow \{-3i\sqrt{2},\ 3i\sqrt{2}\}.$

67. (a) First, find f(x) for $d = 0.8$. $f(x) = \frac{\pi}{3}x^3 - 5\pi x^2 + \frac{500\pi(0.8)}{3} \Rightarrow f(x) = \frac{\pi}{3}x^3 - 5\pi x^2 + \frac{400\pi}{3}$. Now graph this equation on a graphing calculator. We find the smallest positive zero is at $x \approx 7.1286$ cm. or $x \approx 7.13$ cm. The ball floats partly above the surface.

(b) First, find $f(x)$ for $d = 2.7$. $f(x) = \frac{\pi}{3}x^3 - 5\pi x^2 + \frac{500\pi(2.7)}{3} \Rightarrow f(x) = \frac{\pi}{3}x^3 - 5\pi x^2 + \frac{1350\pi}{3}$. Now graph this equation on a graphing calculator. We find the graph has no zero; the sphere sinks below the surface because it is more dense than water.

(c) First, find $f(x)$ for $d = 1$. $f(x) = \frac{\pi}{3}x^3 - 5\pi x^2 + \frac{500\pi(1)}{3} \Rightarrow f(x) = \frac{\pi}{3}x^3 - 5\pi x^2 + \frac{500\pi}{3}$. Now graph this equation on a graphing calculator. We find the smallest positive zero is at $x = 10$ cm; the balloon floats even with the surface.

69. (a) Since the box has sides $x > 0$ and since the box must have some width, we have $12 - 2x > 0 \Rightarrow -2x > -12 \Rightarrow x < 6$. Therefore, the restrictions are $0 < x < 6$.

(b) Since $V = lwh$, length $= 18 - 2x$, and width $= 12 - 2x$, the function is $V(x) = (18 - 2x)(12 - 2x)(x) \Rightarrow V(x) = x(4x^2 - 60x + 216) \Rightarrow V(x) = 4x^3 - 60x^2 + 216x$.

(c) By graphing the function from (b) on a graphing calculator, we can find that the maximum volume is at $x \approx 2.35$ in., which produces a volume of $V(2.35) \approx 4(2.35)^3 - 60(2.35)^2 + 216(2.35) \approx 228.16\text{ in}^3$.

(d) By graphing the function from (b) on a graphing calculator, we can find the volume to be greater than 80 in^3 for the interval $0.42 < x < 5$.

71. By graphing the equation on a graphing calculator, we find the smallest positive zero which is less than 10 is $x \approx 2.61$ in.

73. (a) A length 1 inch less than the hypotenuse (x) is $x - 1$.

(b) Using the Pythagorean Theorem yields $l^2 + (x-1)^2 = x^2 \Rightarrow l^2 = x^2 - (x-1)^2 \Rightarrow$ $l^2 = x^2 - (x^2 - 2x + 1) \Rightarrow l^2 = 2x - 1 \Rightarrow l = \sqrt{2x-1}$.

(c) Since $A = \frac{1}{2}bh = 84$, the equation is $\frac{1}{2}\left(\sqrt{2x-1}\right)(x-1) = 84 \Rightarrow$

$(\sqrt{2x-1})(x-1) = 168 \Rightarrow [(\sqrt{2x-1})(x-1)]^2 = 168^2 \Rightarrow (2x-1)(x-1)^2 = 28,224 \Rightarrow$

$(2x-1)(x^2 - 2x + 1) = 28,224 \Rightarrow 2x^3 - 5x^2 + 4x - 28,225 = 0$.

(d) By graphing the equation from (c) on a graphing calculator, we get an x-intercept of 25. Therefore, the hypotenuse is $x = 25$ in., one leg is $x - 1 = 24$ in., and the other leg is $\sqrt{2x-1} = \sqrt{49} = 7$ in.

75. (a) Since $l = 11 - 2x$, $w = 8.5 - 2x$, $h = x$, and $V = lwh$, we get an equation of $V = (11 - 2x)(8.5 - 2x)(x) \Rightarrow V = (93.5 - 17x - 22x + 4x^2)(x) \Rightarrow V = 4x^3 - 39x^2 + 93.5x$. Because all sides of the box must be positive, the restrictions on this equation will be $8.5 - 2x > 0 \Rightarrow -2x > -8.5 \Rightarrow$

$x < 4.25 \Rightarrow 0 < x < 4.25$. Now using the table feature on the graphing calculator yields a maximum value, within the restrictions, of $x \approx 1.59$. The volume is $4(1.59)^3 - 39(1.59)^2 + 93.5(1.59) \approx 66.15 \text{ in}^3$.

(b) Using the table feature on the graphing calculator for the equation $V = 4x^3 - 39x^2 + 93x$ yields a volume greater than 40 in^3 when x is in the following range: $0.54 \text{ in} < x < 2.92$ in.

77. (a) See Figure 77a.

(b) From the regression feature, the best-fitting quadratic function is $C(x) \approx 0.0035x^2 - 49x + 22$. See Figure 77b.

(c) From the regression feature, the best-fitting cubic function is $C(x) \approx 0.000068x^3 + 0.00987x^2 - 0.653x + 23$. See Figure 77c.

(d) Using the graphs of each, the cubic function is a slightly better fit.

(e) Using the cubic function and graph, $C(x) > 10$ when $0 \le x < 31.92$.

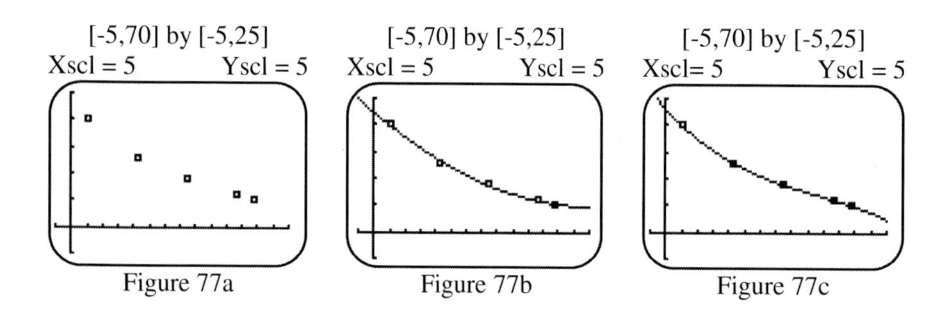

Figure 77a Figure 77b Figure 77c

Reviewing Basic Concepts (Sections 3.6—3.8)

1. $P(3) = 2(3)^4 - 7(3)^3 + 29(3) - 30 \Rightarrow P(3) = 162 - 189 + 87 - 30 \Rightarrow P(3) = 30$.

2. Since $P(2) = 2(2)^4 - 7(2)^3 + 29(2) - 30 \Rightarrow P(2) = 32 - 56 + 58 - 30 \Rightarrow P(2) = 4$, $P(2)$ is not a zero.

3. Since $2+i$ is a zero, its conjugate $2-i$ is also a zero. First, use synthetic division to factor out these two zeros from the equation and the resulting equation.

$$\begin{array}{r|rrrrr} 2+i & 2 & -7 & 0 & 29 & -30 \\ & & 4+2i & -8+i & -17-6i & 30 \\ \hline & 2 & -3+2i & -8+i & 12-6i & 0 \end{array} \Rightarrow \begin{array}{r|rrrr} 2-i & 2 & -3+2i & -8+i & 12-6i \\ & & 4+2i & 2-i & -12+6i \\ \hline & 2 & 1 & -6 & 0 \end{array}$$

Now completely factor the resulting expression, $2x^2 + x - 6 = (2x-3)(x+2)$. The linear factors are $P(x) = (2x-3)(x+2)(x-2-i)(x-2+i)$.

4. Since the zeros are $\frac{3}{2}$, and i, and the conjugate of i is $-i$, the function is $P(x) = a\left(x - \frac{3}{2}\right)(x-i)(x+i) \Rightarrow$

$P(x) = a\left(x - \frac{3}{2}\right)(x^2+1) \Rightarrow P(x) = a\left(x^3 - \frac{3}{2}x^2 + x - \frac{3}{2}\right)$. If $P(x) = 15$, then $15 = a\left(3^3 - \frac{3}{2}(3)^2 + 3 - \frac{3}{2}\right) \Rightarrow$

$15 = a\left(27 - \frac{27}{2} + 3 - \frac{3}{2}\right) \Rightarrow 15 = a(30 - 15) \Rightarrow 15 = 15a \Rightarrow a = 1$. Therefore, the function is

$P(x) = x^3 - \frac{3}{2}x^2 + x - \frac{3}{2}$.

5. Since the zeros are $-4, -4,\ 1+2i$, and its conjugate $1-2i$, the function is

$P(x) = a(x+4)^2(x-1-2i)(x-1+2i) \Rightarrow\ P(x) = a(x^2+8x+16)(x^2-2x+5) \Rightarrow$

$P(x) = a(x^4+6x^3+5x^2+8x+80)$. If $a=1$, then $P(x) = x^4+6x^3+5x^2+8x+80$. (Answers may vary).

6. For the function the possible zeros are $x = \pm\dfrac{1, 2, 5, 10}{1, 2} \Rightarrow x = \pm1,\ \pm2,\ \pm5,\ \pm10,\ \pm\dfrac{1}{2},\ \pm\dfrac{5}{2}$. From a calculator graph, we can choose $x=-2$ to check by synthetic division.

$$\begin{array}{r|rrrr} -2 & 2 & 1 & -11 & -10 \\ & & -4 & 6 & 10 \\ \hline & 2 & -3 & -5 & 0 \end{array}$$

Now solve the resulting quadratic equation: $2x^2-3x-5=0 \Rightarrow (2x-5)(x+1)=0 \Rightarrow\ 2x-5=0$ and $x+1=0 \Rightarrow$

$x = \dfrac{5}{2},\ -1$, and -2.

7. Use the quadratic formula on the equation $x^2 = \dfrac{12 \pm \sqrt{(-12)^2-4(3)(1)}}{2(3)} = \dfrac{12\pm\sqrt{132}}{6} = \dfrac{12\pm2\sqrt{33}}{6} = \dfrac{6\pm\sqrt{33}}{3}$.

Therefore, $x = \pm\sqrt{\dfrac{6\pm\sqrt{33}}{3}}$.

8. (a) From the regression feature, the best-fitting quadratic function is

$P(x) \approx 0.0004815x^3 + 0.0303x^2 - 0.1989x + 3.1$.

(b) $P(34) \approx 0.0004815(34)^3 + 0.0303(34)^2 - 0.1989(34) + 3.1 \approx 50.3$ This value is within 300 of the actual value of 50 thousand.

(c) Using calculator graphs of $y_1 = 0.0004815x^3 + 0.0303x^2 - 0.1989x + 3.1.$ and $y_2 = 100$, the intersection is $x \approx 45$ or approximately the year 2005.

Chapter 3 Review Exercises

1. $(17-i) + (1-3i) = 18-4i.$

2. $(17-i) - (1-3i) = 17-i-1+3i = 16+2i.$

3. $(17-i)(1-3i) = 17-51i-i+3i^2 = 17-52i-3 = 14-52i.$

4. $(17-i)^2 = (17-i)(17-i) = 289-17i+i^2 = 289-34i-1 = 288-34i.$

5. $\dfrac{1}{1-3i}\bullet\dfrac{1+3i}{1+3i} = \dfrac{1+3i}{1-9i^2} = \dfrac{1+3i}{1-9i^2} = \dfrac{1+3i}{10} = \dfrac{1}{10}+\dfrac{3}{10}i.$

6. $\dfrac{17-i}{1-3i}\bullet\dfrac{1+3i}{1+3i} = \dfrac{17+51i-i-3i^2}{1-9i^2} = \dfrac{20+50i}{10} = 2+5i.$

7. Since all real numbers can be input for x, the domain is $(-\infty, \infty)$.

8. By using the vertex formula, we first find the x-coordinate of the vertex. $x = \frac{-b}{2a} = \frac{-(-6)}{2(2)} = \frac{6}{4} = \frac{3}{2}$. Now solve for y. $P\left(\frac{3}{2}\right) = 2\left(\frac{3}{2}\right)^2 - 6\left(\frac{3}{2}\right) - 8 \Rightarrow P\left(\frac{3}{2}\right) = \frac{9}{2} - \frac{18}{2} - \frac{16}{2} = -\frac{25}{2}$. The vertex is $\left(\frac{3}{2}, -\frac{25}{2}\right)$.

9. Since the function is a positive even-degree equation, the end behavior is $\nwarrow\nearrow$

10. At the x-intercept, $P(x) = 0$. Therefore, $2x^2 - 6x - 8 = 0 \Rightarrow x^2 - 3x - 4 = 0 \Rightarrow$ $(x-4)(x+1) = 0 \Rightarrow x = -1,\ 4$. The x-intercepts are $(-1,0)$ and $(4,0)$.

11. At the y-intercept, $x = 0$. Therefore, $P(0) = 2(0)^2 - 6(0) - 8 \Rightarrow P(0) = -8$. The y-intercept is $(-8,0)$.

12. From a calculator graph of $P(x) = 2x^2 - 6x - 8$, the range is $\left[-\frac{25}{2}, \infty\right)$.

13. From a calculator graph of $P(x) = 2x^2 - 6x - 8$, the graph is increasing for the interval $\left[\frac{3}{2}, \infty\right)$, and decreasing for the interval $\left(-\infty, \frac{3}{2}\right]$.

14. (a) $2x^2 - 6x - 8 = 0 \Rightarrow x^2 - 3x - 4 = 0 \Rightarrow (x-4)(x+1) = 0 \Rightarrow x = -1$ or 4.

    (b) First, set $2x^2 - 6x - 8 = 0$, then $x^2 - 3x - 4 = 0 \Rightarrow (x-4)(x+1) = 0$, and the solution set will be $\{-1, 4\}$. The numbers -1 and 4 divide the number line into 3 intervals. The interval $(-\infty, -1)$ has a positive product, the interval $(-1, 4)$ has a negative product, and the interval $(4, \infty)$ has a positive product. Therefore, $2x^2 - 6x - 8 > 0$ for the interval $(-\infty, 1) \cup (4, \infty)$.

    (c) First, set $2x^2 - 6x - 6 - 8 = 0$, then $x^2 - 3x - 4 = 0 \Rightarrow (x-4)(x+1) = 0$ and the solution set will be $\{-1, 4\}$. The numbers -1 and 4 divide the number line into 3 intervals. The interval $(-\infty, -1)$ has a positive product, the interval $(-1, 4)$ has a negative product, and the interval $(4, \infty)$ has a positive product. Therefore, $2x^2 - 6x - 8 \le 0$ for the interval $[-1, 4]$.

15. The graph intersects the x-axis at -1 and 4, supporting the answer in (a). It lies above the x-axis when $x < -1$ or $x > 4$, supporting the answer in (b). It lies below the x-axis when x is between -1 and 4 inclusive, supporting the answer in (c).

16. From problem 8, we know the vertex of $P(x) = 2x^2 - 6x - 8$ is $\left(\frac{3}{2}, -\frac{25}{2}\right)$. The vertical line of symmetry through this point has the equation $x = \frac{3}{2}$.

17. The discriminate is $b^2 - 4ac = (5.47)^2 - 4(-2.64)(3.54) = 29.9209 + 37.3824 = 67.3033$. Since the discriminant is greater than 0, there are two x-intercepts.

18. From the calculator graph of $P(x) = -2.64x^2 + 5.47x + 3.54$, the x-intercepts are $x = 0.52$ and 2.59.

19. From the calculator graph of $P(x) = -2.64x^2 + 5.47x + 3.54$, the vertex is $(1.04, 6.37)$.

20. Using the vertex formula yields $x = \frac{-b}{2a} = \frac{-5.47}{2(-2.64)} \approx 1,04$. Now solve for

$P(1.04) = -2.64(1.04)^2 + 5.47(1.04) + 3.54 \approx 6.37$. The vertex is $(1.04, 6.37)$.

21. (a) From the graph, the maximum value of $f(x)$ is 4.

(b) From the graph, the maximum value of 4 is reached when $x = 1$.

(c) From the graph, there would be two intersections with $y = 2$; therefore, there are two real solutions.

(d) From the graph, there would be no intersections with $y = 6$; therefore, there are no real solutions.

22. At zero seconds, the height is $s(0) = -16(0)^2 + 800(0) + 600 = 600$. The projectile was fired from 600 feet.

23. From the calculator graph of the equation, the vertex is $(25, 10600)$. The maximum height is reached after 25 sec.

24. From the calculator graph of the equation, the vertex is $(25, 10600)$. The maximum height is 10,600 feet.

25. From the calculator graph of the equation, and $y = 5000$, the graph of the equation is above $y = 5000$ for the interval (6.3, 43.7). Therefore, the projectile is above 5,000 feet between 6.3 seconds and 43.7 seconds.

26. From the calculator graph of the equation, the x-intercept (when the projectile hits the ground) is $x = 50.739$. The projectile will be in the air for approximately 50.7 seconds.

27. (a) Let x = the width and $3x$ = the length of the cardboard. Then the base of the box will be $w = (x-8)$, and $l = (3x-8)$, and its height is $h = 4$. Since $v = lwh$, we get a function

$V(x) = (x-8)(3x-8)(4) \Rightarrow V(x) = (3x^2 - 32x + 64)(4) \Rightarrow V(x) = 12x^2 - 128x + 256$.

(b) If the volume is 2496, then $2496 = 12x^2 - 128x + 256 \Rightarrow 0 = 12x^2 - 128x - 2240 \Rightarrow$

$0 = 3x^2 - 32x - 560$. Using the quadratic formula to solve this yields

$x = \frac{-(-32) \pm \sqrt{(-32)^2 - 4(3)(-560)}}{2(3)} = \frac{32 \pm 88}{6} \Rightarrow x = 20, -\frac{28}{3}$. Since we can not have a negative distance we throw out $x = -\frac{28}{3}$. Therefore, the original dimensions are 20 in. by 60 in.

(c) Graphing $y_1 = V(x)$ and $y_2 = 2496$ on a graphing calculator shows that the graphs intersect at $x = 20$.

28. (a) See Figure 28.

(b) Using the form $P(x) = a(x-h)^2 + k$ and a vertex of $(0, 353)$ yields $P(x) = a(x-0)^2 + 353$. Now solve for a using a second point of (285, 2000). $2000 = a(285-0)^2 + 353 \Rightarrow$

$1647 = 81225a \Rightarrow a \approx 0.0203$. Therefore, the function is $P(x) = 0.0203x^2 + 353$.

(c) Using the regression feature of the graphing calculator for the data yields

[-28.5,313.5] by [73,2280]

Xscl = 50 Yscl = 100

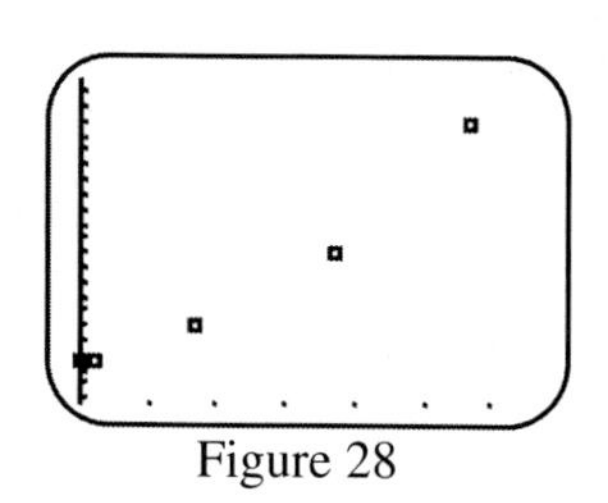

Figure 28

29. First, find $P(-2) = -3(-2)^3 - (-2)^2 + 2(-2) - 4 = 24 - 4 - 4 - 4 \Rightarrow P(-2) = 12$ and $P(-1) = -3(-1)^3 - (-1)^2 + 2(-1) - 4 = 3 - 1 - 2 - 4 \Rightarrow P(-1) = -4$. Since $P(-2) = 12$ and $P(-1) = -4$ differ in sign, the intermediate value theorem assures us that there is a real zero between –2 and –1.

30. (a) $$\begin{array}{r|rrrr} 3 & 1 & 1 & -11 & -10 \\ & & 3 & 12 & 3 \\ \hline & 1 & 4 & 1 & -7 \end{array}$$ Therefore, $Q(x) = x^2 + 4x + 1;\ R = -7.$

(b) $$\begin{array}{r|rrrr} -2 & 3 & 8 & 5 & 10 \\ & & -6 & -4 & -2 \\ \hline & 3 & 2 & 1 & 8 \end{array}$$ Therefore, $Q(x) = 3x^2 + 2x + 1;\ R = 8.$

31. $$\begin{array}{r} 2x^2 - 2x + 2 \\ 3x+1\overline{)\,6x^3 - 4x^2 + 4x + 3} \\ \underline{6x^3 + 2x^2} \qquad\qquad \\ -6x^2 + 4x \qquad \\ \underline{-6x^2 - 2x} \qquad \\ 6x + 3 \\ \underline{6x + 2} \\ 1 \end{array} \Rightarrow \frac{6x^3 - 4x^2 + 4x + 3}{3x+1} = 2x^2 - 2x + 2 + \frac{1}{3x+1}$$

32. $$\begin{array}{r} 2x + 1 \\ x^2 - 3x + 1\overline{)\,2x^3 - 5x^2 + 0x + 1} \\ \underline{2x^3 - 6x^2 + 2x} \qquad \\ x^2 - 2x + 1 \\ \underline{x^2 + 3x + 1} \\ x \end{array} \Rightarrow \frac{2x^3 - 5x^2 + 1}{x^2 - 3x + 1} = 2x + 1 + \frac{x}{x^2 - 3x + 1}$$

33. $$\begin{array}{r|rrrr} 2 & -1 & 5 & -7 & 1 \\ & & -2 & 6 & -2 \\ \hline & -1 & 3 & -1 & -1 \end{array}$$ Therefore, $P(2) = -1.$

34. $$\begin{array}{r|rrrr} 2 & 2 & -3 & 7 & -12 \\ & & 4 & 2 & 18 \\ \hline & 2 & 1 & 9 & 6 \end{array}$$ Therefore, $P(2) = 6.$

35. $$\begin{array}{r|rrrrr} 2 & 5 & 0 & -12 & 2 & -8 \\ & & 10 & 20 & 16 & 36 \\ \hline & 5 & 10 & 8 & 18 & 28 \end{array}$$ Therefore, $P(2) = 28$.

36. $$\begin{array}{r|rrrrrr} 2 & 1 & 0 & 0 & 4 & -2 & -4 \\ & & 2 & 4 & 8 & 24 & 44 \\ \hline & 1 & 2 & 4 & 12 & 22 & 40 \end{array}$$ Therefore, $P(2) = 40$.

37. The conjugate of $7+2i$ must also be a zero; therefore, $7-2i$ is also a zero.

38. With the given zeros of –1, 4, and 7, the function $P(x)$ is

$P(x) = a(x-(-1))(x-4)(x-7) = a(x+1)(x-4)(x-7) = a(x+1)(x^2-11x+28) \Rightarrow$

$P(x) = a(x^3-10x^2+17x+28)$. One of many possible functions, if $a=1$, is $P(x) = x^3-10x^2+17x+28$.

39. With the given zeros of 8, 2, and 3, the function $P(x)$ is

$P(x) = a(x-8)(x-2)(x-3) = a(x-8)(x-2)\ (x-3) = a(x-8)(x^2-5x+6) \Rightarrow$

$P(x) = a(x^3-13x^2+46x-48)$. One of many possible functions, if $a=1$, is $P(x) = x^3-13x^2+46x-48$.

40. With the given zeros of $\sqrt{3}, -\sqrt{3}, 2,$ and 3, the function $P(x)$ is

$P(x) = a(x-\sqrt{3})(x-(-\sqrt{3})(x-2)(x-3) = a(x-\sqrt{3})(x+\sqrt{3})(x-2)(x-3) =$

$a(x^2-3)(x^2-5x+6) \Rightarrow a(x^4-5x^3+3x^2+15x-18)$. One of many possible functions, if $a=1$, is

$P(x) = x^4-5x^3+3x^2+15x-18$.

41. With the given zeros of $-2+\sqrt{5}, -2-\sqrt{5},$ –2 and 1, the function $P(x)$ is

$P(x) = a(x-(-2+\sqrt{5}))(x-(-2-\sqrt{5}))(x-(-2))(x-1) =$
$a(x+2-\sqrt{5})(x+2+\sqrt{5})(x+2)(x-1) = a(x^2+4x-1)(x^2+x-2) = a(x^4+5x^3+x^2-9x+2)$. One of many possible functions, if $a=1$, is $P(x) = x^4+5x^3+x^2-9x+2$.

42. Use synthetic division to check if –1 is a zero.

$$\begin{array}{r|rrrrr} -1 & 2 & 1 & -4 & 3 & 1 \\ & & -2 & 1 & 3 & -6 \\ \hline & 2 & -1 & -3 & 6 & -5 \end{array}$$ Since $P(-1) = -5$ and not 0, –1 is not a zero.

43. Use synthetic division to check if $x+1$ or $x=-1$ is a zero.

$$\begin{array}{r|rrrr} -1 & 1 & 2 & 3 & 2 \\ & & -1 & -1 & -2 \\ \hline & 1 & 1 & 2 & 0 \end{array}$$ Since $P(-1) = 0$, -1 or $x+1$ is a factor.

44. With the given zeros 3, 1, $-1-3i$, the conjugate of $-1-3i$ is also a zero, so the function $P(x)$ is

$P(x) = a(x-3)(x-1)(x-(-1-3i))(x-(-1+3i)) =$

$a(x-3)(x-1)(x+1+3i)(x+1-3i) = a(x^2-4x+3)(x^2+2x+10) = a(x^4-2x^3+5x^2-34x+30)$.

Since $P(2) = -36,$ we can solve for a.

$-36 = a((2)^4 - 2(2)^3 + 5(2)^2 - 34(2) + 30) \Rightarrow -36 = a(16 - 16 + 20 - 68 + 30) \Rightarrow -18a = -36 \Rightarrow a = 2.$

Therefore, $P(x) = 2(x^4 - 2x^3 + 5x^2 - 34x + 30) \Rightarrow P(x) = 2x^4 - 4x^3 + 10x^2 - 68x + 60.$

45. With the given zero $1-i,$ the conjugate $1+i$ is also a zero. Use synthetic division to factor out these zeros.

$$\begin{array}{r|rrrrr} 1-i & 1 & -3 & -8 & 22 & -24 \\ & & 1+i & -3+i & -10+12i & 24 \\ \hline & 1 & -2-i & -11+i & 12+12i & 0 \end{array} \Rightarrow \begin{array}{r|rrrr} 1+i & 1 & -2-i & -11+i & 12+12i \\ & & 1+i & -1-i & -12-12i \\ \hline & 1 & -1 & -12 & 0 \end{array}$$

Finally, solve the resulting quadratic equation, $x^2 - x - 12 = 0 \Rightarrow x^2 - x - 12 = 0 \Rightarrow$ $(x-4)(x+3) = 0 \Rightarrow x = -3,\ 4.$ The zeros are $x = \{-3, 4, 1-i, 1+i\}.$

46. With the given zeros 1 and $2i,$ the conjugate of $2i,$ or $-2i,$ is also a zero. Use synthetic division to factor out these zeros.

$$\begin{array}{r|rrrrr} 1 & 2 & -1 & 7 & -4 & -4 \\ & & 2 & 1 & 8 & 4 \\ \hline & 2 & 1 & 8 & 4 & 0 \end{array} \Rightarrow \begin{array}{r|rrrr} 2i & 2 & 1 & 8 & 4 \\ & & 4i & -8+2i & -4 \\ \hline & 2 & 1+4i & 2i & 0 \end{array} \Rightarrow \begin{array}{r|rrr} -2i & 2 & 1+4i & 2i \\ & & -4i & -2i \\ \hline & 2 & 1 & 0 \end{array}$$

Finally, solve the resulting linear equation: $2x + 1 = 0 \Rightarrow 2x = -1 \Rightarrow x = -\frac{1}{2}.$ The zeros are $x = -\frac{1}{2}, 1, 2i, -2i.$

47. Using the Rational Zeros Theorem, the possible zeros are $x = \pm\frac{1,2,4,8}{1,3} = \pm 1, \pm 2, \pm 4, \pm 8, \pm\frac{1}{3}, \pm\frac{2}{3}, \pm\frac{4}{3}, \pm\frac{8}{3}.$ From graphing the equation, we choose to check possible zeros -2, 4, and $\frac{1}{3}$ by synthetic division.

$$\begin{array}{r|rrrrrr} 2 & 3 & -4 & -26 & -21 & -14 & 8 \\ & & -6 & 20 & 12 & 18 & 8 \\ \hline & 3 & -10 & -6 & -9 & 4 & 0 \end{array} \Rightarrow \begin{array}{r|rrrrr} 4 & 3 & -1 & -6 & -9 & 4 \\ & & 12 & 8 & 8 & -4 \\ \hline & 3 & 2 & 2 & -1 & 0 \end{array} \Rightarrow \begin{array}{r|rrrr} \frac{1}{3} & 3 & 2 & 2 & -1 \\ & & 1 & 1 & 1 \\ \hline & 3 & 3 & 3 & 0 \end{array}$$

Finally, solve the resulting quadratic equation, $x^2 + x + 1 = 0,$ using the quadratic formula:

$x = \frac{-1 \pm \sqrt{1^2 - 4(1)(1)}}{2(1)} = \frac{-1 \pm \sqrt{-3}}{2}.$ This results in non-real solutions; therefore, the rational solutions are $x = -2, \frac{1}{3},$ and 4.

48. Because the function $P(x)$ has coefficient signs, $++---,$ which is 1 sign change, the function has 1 positive real zero. Because $P(-x) = 3(-x)^4 + (-x)^3 - (-x)^2 - 2(-x) - 1 = 3x^4 - x^3 - x^2 + 2x - 1$ has coefficient signs, $+--+-,$ which is 3 sign changes, the function has 3 or 1 possible negative real zeros.

49. Using the Boundedness Theorem, we divide synthetically by $x-2$.

$$\begin{array}{r|rrrrr} 2 & 2 & 3 & -5 & 8 & -10 \\ & & 4 & 14 & 18 & 52 \\ \hline & 2 & 7 & 9 & 26 & 42 \end{array}$$

The result is entirely non-negative; therefore, there can be no real zero greater than 2. Again using the Boundedness Theorem, we divide synthetically by $x-(-4)$.

$$\begin{array}{r|rrrrr} -4 & 2 & 3 & -5 & 8 & -10 \\ & & -8 & 20 & -60 & 208 \\ \hline & 2 & -5 & 15 & -52 & 198 \end{array}$$

The result alternates in sign; therefore, there is no real zero less than –4.

50. Graphing the equation on a calculator shows x-intercepts of –1.62, 0.62, and 3. Therefore, there are 3 real solutions and the integer root is 3.

51. Using synthetic division we can factor out the root 3.

$$\begin{array}{r|rrrr} 3 & 1 & -2 & -4 & 3 \\ & & 3 & 3 & -3 \\ \hline & 1 & 1 & -1 & 0 \end{array}$$

Using the result of the synthetic division the factors are $(x-3)(x^2+x-1)$.

52. The remaining zeros can be found by solving, $x^2+x-1=0$, using the quadratic formula.

$x=\dfrac{-1\pm\sqrt{1^2-4(1)(-1)}}{2(1)}=\dfrac{-1\pm\sqrt{5}}{2}$. The solution set is $\left\{\dfrac{-1-\sqrt{5}}{2},\dfrac{-1+\sqrt{5}}{2}\right\}$.

53. The x-intercepts from the graph are approximately equal to the solutions found in #50. $\dfrac{-1-\sqrt{5}}{2}\approx-1.62$, and $\dfrac{-1+\sqrt{5}}{2}\approx 0.62$.

54. (a) From the graph and the answers of #50, $P(x)>0$ for the interval $\left(\dfrac{-1-\sqrt{5}}{2},\dfrac{-1+\sqrt{5}}{2}\right)\cup(3,\infty)$.

(b) From the graph and the answers of #50, $P(x)\le 0$ for the interval $\left(-\infty,\dfrac{-1-\sqrt{5}}{2}\right]\cup\left[\dfrac{-1+\sqrt{5}}{2},3\right)$.

55. First, factor out an x. $x^3-2x^2+5x=0\Rightarrow x(x^2-2x+5)\Rightarrow x=0$ is a zero. Now solve the quadratic equation using the quadratic formula: $x=\dfrac{-2\pm\sqrt{2^2-4(1)(5)}}{2(1)}=\dfrac{-2\pm\sqrt{-16}}{2}=\dfrac{-2\pm 4i}{2}=-1\pm 2i$. Since two of the zeros are not real, the only x-intercept is $(0,0)$.

56. The zeros are –2, 1, and 3. The 3 has multiplicity 2, because it appears tangent to the x-axis. The factored form is now: $P(x)=(x+2)(x-1)(x-3)^2$.

57. Since its end behavior has both ends going in the same direction, either up or down, the degree is even.

58. Since its end behavior has one end going up and the other going down, the degree is odd.

59. Since its end behavior has the right end approaching positive infinity, the lead coefficient is positive.

60. Since $g(x)$ has no x-intercepts, it has no real solutions.

61. From the graph, $f(x)<0$ is true for the interval $(-\infty,a)\cup(b,c)$.

62. From the graph, $f(x)>g(x)$ is true for the interval (d,h).

63. If $f(x)-g(x)=0$ then $f(x)=g(x)$. This happens at the intersection of the graphs, which is at $\{d,h\}$.

64. If $r+pi$ is a solution, then its conjugate $r-pi$ is also a solution.

65. Since $f(x)$ has three real zeros, and a polynomial of degree 3 can have at most three zeros, there can be no other zeros, real or non-real complex.

66. False. A 7th degree function can have at most 7 x-intercepts.

67. True. A 7th degree function can have at most 6 local extrema.

68. True. $f(0)=3(0)^7-8(0)^6+9(0)^5+12(0)^4-18(0)^3+26(0)^2-(0)+500 \Rightarrow f(0)=500$.

69. True. The end behavior of an odd-degree polynomial with a positive lead coefficient is down to the left. Because the function has a positive y-intercept, the graph of the equation must cross the negative x-axis at least one time. Therefore, it must have at least one negative x-intercept.

70. True. An even-degree polynomial with a positive lead coefficient will have the end behavior of both ends going upward; therefore, a graph with a negative y-intercept must cross the x-axis at least 2 times and will have at least 2 real zeros.

71. False. The conjugate must also be a zero, but the conjugate of $-\frac{1}{2}+i\frac{\sqrt{3}}{2}$ is $-\frac{1}{2}-i\frac{\sqrt{3}}{2}$, and not $\frac{1}{2}+i\frac{\sqrt{3}}{2}$.

72. From the graph, there are two local maxima.

73. From the graph, there is one local minimum that lies on the x-axis. It has coordinates $(2,0)$.

74. Use synthetic division to factor out $x-5$.
$$\begin{array}{r|rrrrrr} 5 & -2 & 15 & -21 & -32 & 60 & 0 \\ & & -10 & 25 & 20 & -60 & 0 \\ \hline & -2 & 5 & 4 & -12 & 0 & 0 \end{array}$$
From this result, the quotient is the equation $Q(x)=-2x^4+5x^3+4x^2-12x$.

75. Because all real numbers can be input for x, the range is $(-0.97,-54.15)$.

76. Using the calculator functions, this local minimum point has coordinates $(-0.97,-54.15)$.

77. We first factor by grouping and then solve. $3x^3+2x^2-21x-14=0 \Rightarrow x^2(3x+2)-7(3x+2)=0 \Rightarrow$ $(3x+2)(x^2-7)=0 \Rightarrow 3x+2=0 \Rightarrow 3x=-2 \Rightarrow x=-\frac{2}{3}$ or $x^2-7=0 \Rightarrow x^2=7 \Rightarrow x=\pm\sqrt{7}$. Therefore, the solution set is $\left\{-\sqrt{7},-\frac{2}{3},\sqrt{7}\right\}$. Graphing this equation shows x-values of ±2.65 and –0.67, which are equal to the solution set, and therefore support our analytic solution.

78. An even-degree polynomial with a negative lead coefficient has an end behavior of $\frown$
The function has zeros $x=-1$ (multiplicity 2,) 2, and 3; therefore, there are x-intercepts at those values of x.

The graph will be tangent to the x-axis at $x=-1$ (multiplicity 2) and the y-intercept is $y=-6$. See Figure 78.

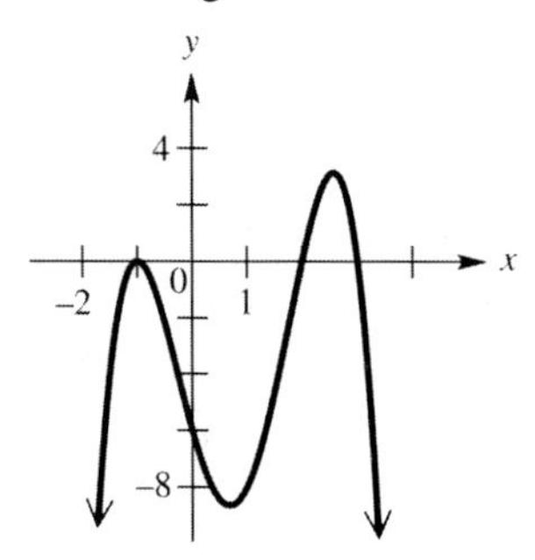

Figure 78

(a) $P(x)=0$ at the x-intercepts; therefore, $x=\{1, 2, 3\}$.

(b) From the graph, $P(x)>0$ for the interval $(2, 3)$.

(c) From the graph, $P(x)<0$ for the interval $(-\infty,-1)\cup(-1,2)\cup(3,\infty)$.

79. If $x=$ cube side length and $V=s^3$, the volume of the cube after the top is cut off is $V=(x)(x)(x-2)=32 \Rightarrow x^3-2x^2-32=0$. After graphing for possible roots, we use synthetic division to check 4 as a root.

$$\begin{array}{r|rrrr} 4 & 1 & -2 & 0 & -32 \\ & & 4 & 8 & 32 \\ \hline & 1 & 2 & 8 & 0 \end{array}$$

Now solve the resulting equation, $x^2-2x+8=0$. $x=\dfrac{-2\pm\sqrt{2^2-4(1)(8)}}{2(1)}=\dfrac{-2\pm\sqrt{-28}}{2}$. Both of these roots are non-real; therefore, the only real root is 4, and the dimensions of the original cube are 4 in. by 4 in. by 4 in.

80. (a) Since, from a calculator graph of the equation, the zero is at approximately 9.26. Therefore, the restrictions on t are $0 \le t \le 9.26$.

(b) See Figure 80.

(c) From the graph, the object reaches its highest point at approximately $t=4.08$ seconds.

(d) From the graph, the object reaches its highest point at approximately $s(t)=131.63$ meters.

(e) The object reaches the ground when $s(t)=0$; therefore, the equation is $-4.9t^2+40t+50=0$. We solve this using the quadratic formula.

$t=\dfrac{-40\pm\sqrt{40^2-4(-4.9)(50)}}{2(-4.9)}=\dfrac{-40\pm\sqrt{2580}}{-9.8}$; $t\approx-1.10$ or 9.26. Since we cannot have negative time, the object reaches the ground in approximately 9.26 sec.

81. For the year 2010 $x=45$; therefore, $y=-0.00002(45)^3+0.0022(45)^2-0.096(45)+3.05\approx1.36$ million.

82. (a) See Figure 82a.

(b) Using the capabilities of the graphing calculator, the quadratic function is $f(x)\approx-0.011x^2+0.869x+11.9$.

(c) Using the capabilities of the calculator, the cubic function is
$f(x) \approx -0.00087x^3 + 0.0456x^2 - 0.219x + 17.8.$

(d) See Figure 82b, for the quadratic function, and Figure 82c, for the cubic function.

(e) Both functions approximate the data well. Although the value of R^2 is closer to 1 for the cubic function, the quadratic function is probably better for prediction because it is unlikely that the percent of out-of-pocket spending would decrease after 2025 (as the cubic function shows) unless changes were made in Medicare law.

(f) A linear model would not really be appropriate because the data points lie in a curved nonlinear path.

[0,10] by [0,200]
Xscl = 1 Yscl = 10

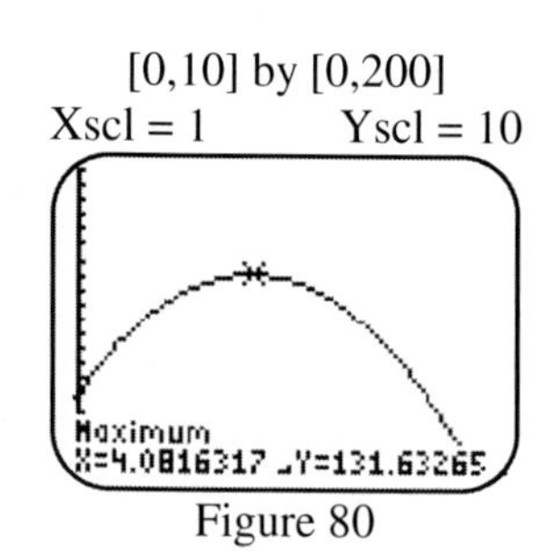

Figure 80

[0,40] by [0,40]
Xscl = 5 Yscl = 5

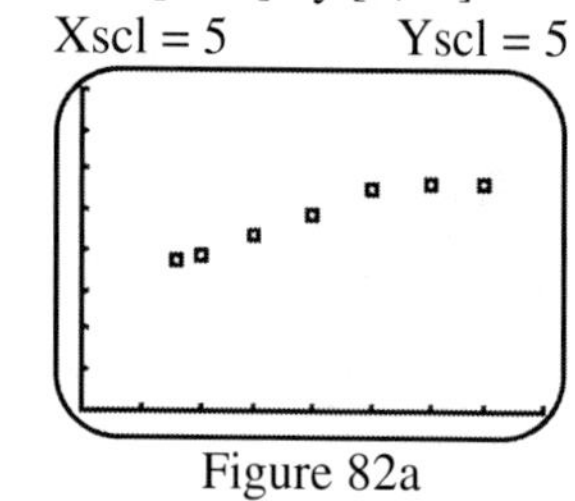
Figure 82a

[0,40] by [0,40]
Xscl= 5 Yscl = 5

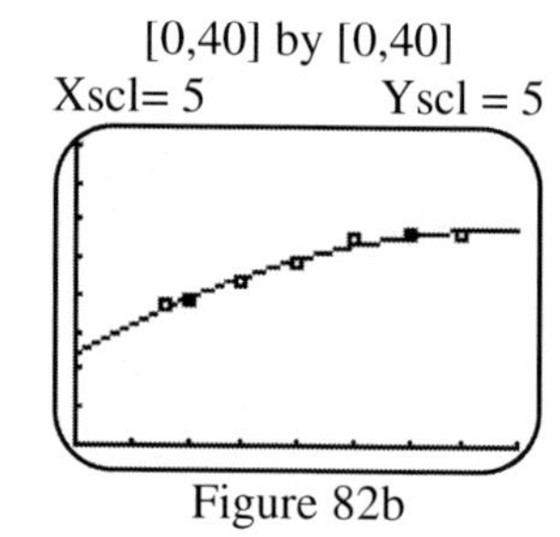
Figure 82b

[0,40] by [0,40]
Xscl = 5 Yscl= 5

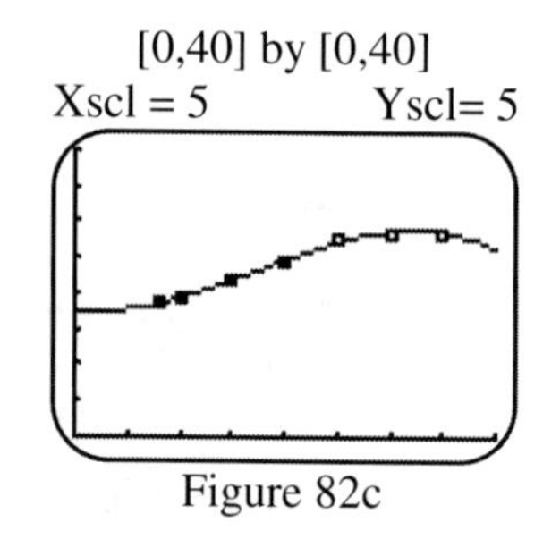
Figure 82c

83. Answers will vary. Example, given 7; x^7

84. Answers will vary.

85. Answers will vary.

86. Answers will vary. Domain: $(-\infty,\infty)$; Range: $(-\infty,\infty)$.

Chapter 3 Test

1. (a) $(8-7i)-(-12+2i) = 20-9i.$

(b) $\dfrac{11+10i}{2+3i}\cdot\dfrac{2-3i}{2-3i} = \dfrac{22-33i+20i-30i^2}{4-9i^2} = \dfrac{22-13i+30}{4+9} = \dfrac{52-13i}{13} = 4-i.$

(c) $i^{65} = \left(i^4\right)^{16}(i) = 1^{16}(i) = i.$

(d) $2i(3\ \ i)^2 = 2i\left(9-6i+i^2\right) = 2i(9-6i-1) - 2i(8-6i) = 16i-12i^2 = 12+16i.$

(e) $\sqrt{-36} = \sqrt{36}\sqrt{-1} = 6i$

(f) $\sqrt{-5}\cdot\sqrt{-20} = \sqrt{5}\cdot\sqrt{-1}\cdot\sqrt{4}\cdot\sqrt{5}\cdot\sqrt{-1} = \sqrt{5}\cdot i\cdot 2\cdot\sqrt{5}\cdot i = 5\cdot 2\cdot(-1) = -10$

2. (a) Use the vertex formula to find the x-coordinate. $x = -\dfrac{b}{2a} = -\dfrac{-4}{2(-2)} = -1.$ Now solve for $f(-1)$ to find the y-coordinate of the vertex. $f(-1) = -2(-1)^2 - 4(-1) + 6 = -2+4+6 = 8.$ The vertex is $(-1,8)$.

(b) See Figure 2b. Finding the maximum on the graph gives the vertex: $(-1,8)$.

(c) First, we find the zeros by factoring.
$-2x^2 - 4x + 6 = 0 \Rightarrow -2\left(x^2+2x-3\right) = 0 \Rightarrow -2(x+3)(x-1) = 0 \Rightarrow x = -3$ or 1. Now support this on the graphing calculator using a graph and a table, See Figure 2c.

(d) For the y-intercept we let $x=0$; therefore, $f(0)=-2(0)-4(0)+6 \Rightarrow f(0)=6$.

(e) All values can be input for x. The domain is $(-\infty,\infty)$; since $(-1,8)$ is a maximum, the range is $(-\infty,8]$.

(f) From the graph, the function is increasing for the interval $(-\infty,-1]$, and decreasing for the interval $[-1,\infty)$.

3. $6x^2-15x+6=0 \Rightarrow 3(2x^2-5x+2)=0 \Rightarrow (2x-1)(x-2)=0 \Rightarrow x=\frac{1}{2},2$

4. $x^2-4x=2 \Rightarrow x^2-4x+4=2+4 \Rightarrow (x-2)^2=6 \Rightarrow x-2=\pm\sqrt{6} \Rightarrow x=2\pm\sqrt{6}$

5. (a) We solve the equation using the quadratic formula. $x=\frac{-3\pm\sqrt{3^2-4(3)(-2)}}{2(3)}=\frac{-3\pm\sqrt{9+24}}{6} \Rightarrow$

$$x=\left\{\frac{-3-\sqrt{33}}{6},\frac{-3+\sqrt{33}}{6}\right\}.$$

(b) See Figure 5. From this graph, (i) $f(x)<0$ for the interval $\left(\frac{-3-\sqrt{33}}{6},\frac{-3+\sqrt{33}}{6}\right)$, and (ii) $f(x)\geq 0$ for the interval $\left(-\infty,\frac{-3-\sqrt{33}}{6}\right]\cup\left[\frac{-3+\sqrt{33}}{6},\infty\right)$.

[-9.4,9.4] by [-4.1,8.1]
Xscl = 1 Yscl = 1

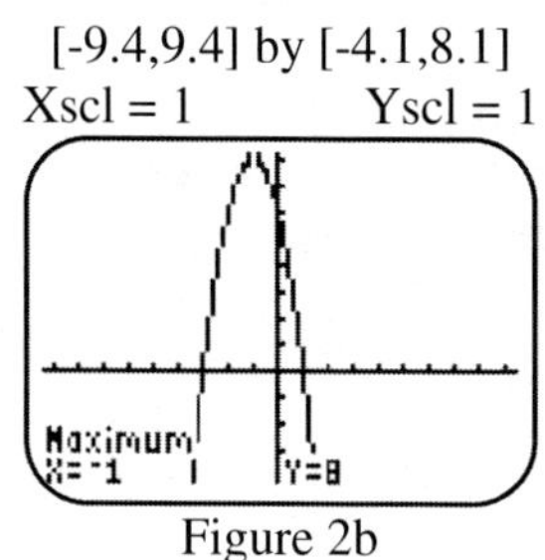

Figure 2b

[-9.4,9.4] by [-4.1,8.1]
Xscl = 1 Yscl = 1

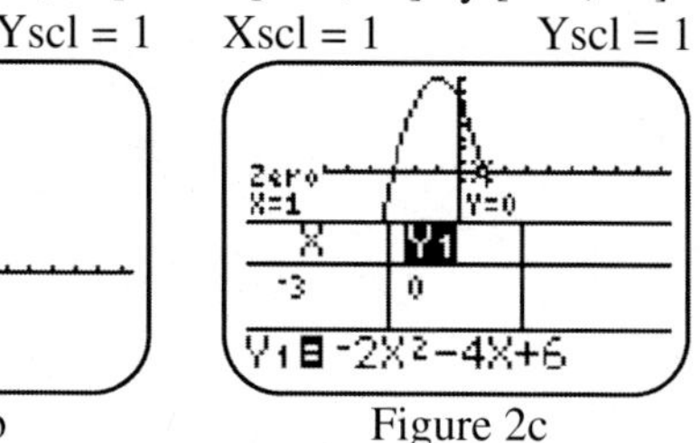

Figure 2c

[-4.7,4.7] by [-3.1,3.1]
Xscl= 1 Yscl = 1

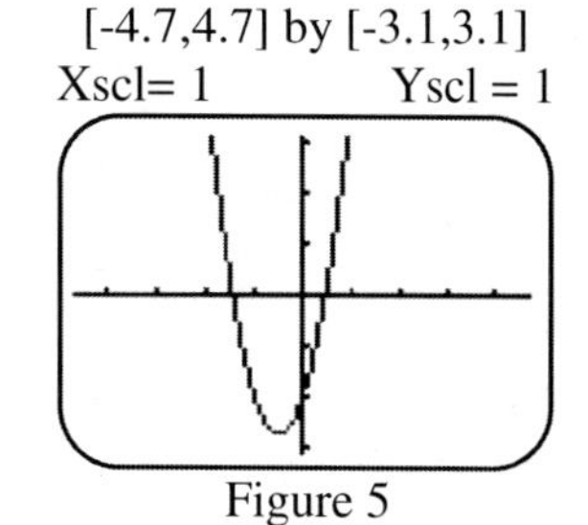

Figure 5

6. Since $V=lwh$, $720=(11+x)(3x)(x) \Rightarrow 720=3x^3+33x^2 \Rightarrow 0=3x^3+33x^2-720 \Rightarrow 0=x^3+11x^2-240$.

After graphing on a calculator, we use synthetic division to check 4 as a factor.

4)	1	11	0	−240
		4	60	240
	1	15	60	0

Now solve the resulting equation, $x^2+15x+60=0$, using the quadratic formula:

$x=\frac{-15\pm\sqrt{15^2-4(1)(60)}}{2(1)}=\frac{-15\pm\sqrt{-15}}{2}$. These solutions will both be nonreal; therefore, the only real solution is 4, and the dimensions are $11+4$, $3(4)$, and 4 or 15 in. by 12 in. by 4 in.

7. (a) Using the capabilities of the graphing calculator, the quadratic function is $f(x)=0.00019838x^2-0.79153x+791.46$

(b). $f(1975) \approx 1.99$, This value is very close to the actual value of 2.01.

8. (a) Use synthetic division to factor out the given zeros.

$$3\overline{)\begin{array}{rrrrrrr} 1 & -5 & 3 & 1 & 40 & -24 & -72 \\ & 3 & -6 & -9 & -24 & 48 & 72 \\ \hline 1 & -2 & -3 & -8 & 16 & 24 & 0 \end{array}} \Rightarrow 3\overline{)\begin{array}{rrrrrr} 1 & -2 & -3 & -8 & 16 & 24 \\ & 3 & 3 & 0 & -24 & -24 \\ \hline 1 & 1 & 0 & -8 & -8 & 0 \end{array}} \Rightarrow$$

$$-1\overline{)\begin{array}{rrrrr} 1 & 1 & 0 & -8 & -8 \\ & -1 & 0 & 0 & 8 \\ \hline 1 & 0 & 0 & -8 & 0 \end{array}} \Rightarrow 2\overline{)\begin{array}{rrrr} 1 & 0 & 0 & -8 \\ & 2 & 4 & 8 \\ \hline 1 & 2 & 4 & 0 \end{array}}$$

Now solve the resulting equation, $x^2+2x+4=0$, using the quadratic formula.

$x=\dfrac{-2\pm\sqrt{2^2-4(1)(4)}}{2(1)}=\dfrac{-2\pm\sqrt{-12}}{2}=\dfrac{-2\pm 2i\sqrt{3}}{2}=-1\pm i\sqrt{3}$. The other zeros of $f(x)$ are $x=-1-i\sqrt{3},-1+i\sqrt{3}$.

(b) An even-degree polynomial with a positive lead coefficient has end-behavior $\cup$. See Figure 8.

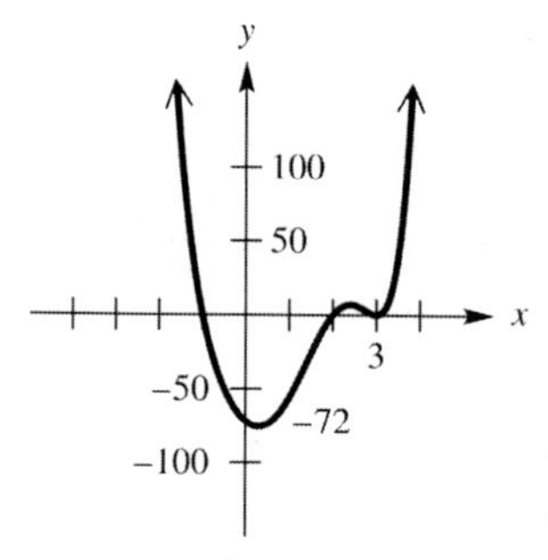

Figure 8

9. (a) We can find the zeros by factoring. $4x^4-21x^2-25=0\Rightarrow(4x^2-25)(x^2+1)=0\Rightarrow$

$4x^2-25=0\Rightarrow 4x^2=25\Rightarrow x^2=\dfrac{25}{4}\Rightarrow x=\pm\sqrt{\dfrac{25}{4}}=\pm\dfrac{5}{2}$ or

$x^2+1=0\Rightarrow x^2=-1\Rightarrow x=\pm\sqrt{-1}\Rightarrow x=\pm i$. The zeros are $x=-\dfrac{5}{2},\dfrac{5}{2},-i,i$.

(b) Use a graph and table to support the results. See Figure 9.

(c) It is symmetric with respect to the y-axis.

(d) From the graph and the results, (i) $f(x)\geq 0$ for the interval $\left(-\infty,-\dfrac{5}{2}\right]\cup\left[\dfrac{5}{2},\infty\right)$, and (ii) $f(x)<0$ for the interval $\left(-\dfrac{5}{2},\dfrac{5}{2}\right)$.

[-4,4] by [-60,60]

Xscl = 1 Yscl = 10

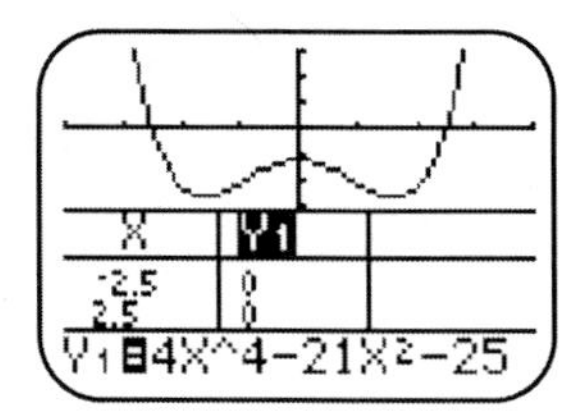

Figure 9

10. (a) Using the Rational Zeros Theorem, the possible zeros are $x = \pm\dfrac{1, 2, 11, 22}{1, 3} =$

$\pm 1, \pm 2, \pm 11, \pm 22, \pm\dfrac{1}{3}, \pm\dfrac{2}{3}, \pm\dfrac{11}{3}$, and $\pm\dfrac{22}{3}$.

(b) From a calculator graph of the equation, we choose to check -2 and $\dfrac{1}{3}$ by synthetic division.

$$\begin{array}{r|rrrrr} -2 & 3 & 5 & -35 & -55 & 22 \\ & & -6 & 2 & 66 & -22 \\ \hline & 3 & -1 & -33 & 11 & 0 \end{array} \quad \Rightarrow \quad \begin{array}{r|rrrr} \frac{1}{3} & 3 & -1 & -33 & 11 \\ & & 1 & 0 & -11 \\ \hline & 3 & 0 & -33 & 0 \end{array}$$

Now solve the resulting equation. $3x^2 - 33 = 0 \Rightarrow x^2 - 11 = 0 \Rightarrow x^2 = 11 \Rightarrow x = \pm\sqrt{11}.$ Since these are not rational, the rational zeros are $-2, \dfrac{1}{3}$.

(c) Since $f(3) = 3(3)^4 + 5(3)^3 + 35(3)^2 - 55(3) + 22 = -80$ and $f(4) = 3(4)^4 + 5(4)^3 + 35(4)^2 - 55(4) + 22 = 330$ differ in sign, the intermediate value theorem assures us that there is a real zero between 3 and 4.

(d) Because the function has coefficient signs $++--+,$ which is 2 sign changes, the function has 2 or 0 positive real zeros. Because $f(-x) = 3(-x)^4 + 5(-x)^3 - 35(-x)^2 - 55(-x) + 22 =$ $3x^4 - 5x^3 - 35x^2 + 55x + 22$ has coefficient signs, $+--++,$ which is 2 sign changes, the function has 2 or 0 possible negative real zeros.

(e) Using the Boundedness Theorem we divide synthetically by $x - (-5)$.

$$\begin{array}{r|rrrrr} -5 & 3 & 5 & -35 & -55 & 22 \\ & & -15 & 50 & -75 & 650 \\ \hline & 3 & -10 & 15 & -130 & 672 \end{array}$$

The result alternates signs; therefore, there is no real zero less than -5. Also using the Boundedness Theorem, we divide synthetically by $x - 4$.

$$\begin{array}{r|rrrrr} 4 & 3 & 5 & -35 & -55 & 22 \\ & & 12 & 68 & 132 & 308 \\ \hline & 3 & 17 & 33 & 77 & 330 \end{array}$$

The result is all non-negative; therefore, there is no real zero greater than 4.

11. (a) Using the capabilities of the calculator and graphing the equation yields real solutions of $\{0.189, 1, 3.633\}$.

(b) A 5th degree equation has 5 solutions. Since there are 3 real solutions, there must be 2 non-real complex solutions.

12. (a) $\dfrac{8x^3-4x^2}{2x}=\dfrac{8x^3}{2x}-\dfrac{4x^2}{2x}=4x^2-2x$

(b)
$$\begin{array}{r} 3x^2-2x-2 \\ x-1\overline{\smash{)}3x^3-5x^2+0x+6} \\ \underline{3x^3-3x^2} \\ -2x^2+0x \\ \underline{-2x^2+2x} \\ -2x+6 \\ \underline{-2x+2} \\ 4 \end{array} \Rightarrow \frac{3x^3-5x^2+6}{x-1}=3x^2-2x-2+\frac{4}{x-1}$$

(c)
$$\begin{array}{r} x^3+2x-1 \\ 2x-1\overline{\smash{)}2x^4-x^3+4x^2-4x+3} \\ \underline{2x^4-x^3} \\ 4x^2-4x \\ \underline{-4x^2-2x} \\ -2x+3 \\ \underline{-2x+1} \\ 2 \end{array} \Rightarrow \frac{2x^4-x^3+4x^2-4x+3}{2x-1}=x^3+2x-1+\frac{2}{2x-1}$$

(d)
$$\begin{array}{r} x^2-2 \\ x^2+2\overline{\smash{)}x^4+0x^2-2x+6} \\ \underline{x^4+2x^2} \\ -2x^2-2x \\ \underline{-2x^2\quad\quad -4} \\ -2x+4+6 \end{array} \Rightarrow \frac{x^4-2x+6}{x^2+2}=x^2-2+\frac{-2x+10}{x^2+2}$$

13. The cubic polynomial has zeros of 4 and $2i$ will also have $-2i$ (conjugate pairs theorem). Therefore, we will have $f(x)=a(x-4)(x-(2i))(x-(-2i))-a(x-4)(x-2i)(x+2i)$. Now use the given point to find the value of a, $-15=a(1-4)(1-2i)(1+2i)\Rightarrow -15=a(-3)(5)\Rightarrow a=1$. The polynomial is $f(x)=(x-4)(x-2i)(x+2i)=(x-4)(x^2+4)=x^3-4x^2+4x-16$

14. $x^3+3x=0\Rightarrow x(x^2+3)=0\Rightarrow x=0\ or\ (x^2+3)=0\Rightarrow x^2=-3\Rightarrow x=\pm i\sqrt{3}$ The solution set is $\{0,\pm i\sqrt{3}\}$.

Chapter 4: Rational, Power, and Root Functions

4.1: Rational Functions and Graphs I

1. The only value for x that cannot be used as input is 0. The domain is $(-\infty,0)\cup(0,\infty)$.

   It is not possible for this function to output the value 0. The range is $(-\infty,0)\cup(0,\infty)$.

3. The function decreases everywhere it is defined, $(-\infty,0)\cup(0,\infty)$. It never increases and is never constant.

5. Because the function is undefined when $x=3$, the vertical asymptote has the equation $x=3$. As $|x|$ increases without bound, the graph of the function will move closer and closer to the graph of $y=2$.

7. Because $f(-x)=f(x)$, the function is even. The graph has symmetry with respect to the y-axis.

9. Graphs A, B, and C have domain $(-\infty,3)\cup(3,\infty)$ because each has a vertical asymptote at $x=3$.

11. Graph A has range $(-\infty,3)\cup(3,\infty)$ because it exists above and below the horizontal asymptote at $y=0$.

13. The only graph that would intersect the line $y=3$ exactly one time is graph A.

15. Graphs A, C, and D have the x-axis as a horizontal asymptote.

17. Let $g(x)=y=\frac{1}{x}$, then $f(x)=2g(x)$. To obtain the graph of f, stretch the graph of $y=\frac{1}{x}$ vertically by a factor of 2. See Figures 17a and 17b. The domain is $(-\infty,0)\cup(0,\infty)$. The range is $(-\infty,0)\cup(0,\infty)$.

19. Let $g(x)=y=\frac{1}{x}$, then $f(x)=g(x+2)$. To obtain the graph of f, shift the graph of $y=\frac{1}{x}$ to the left 2 units. See Figures 19a and 19b. The domain is $(-\infty,-2)\cup(-2,\infty)$. The range is $(-\infty,0)\cup(0,\infty)$.

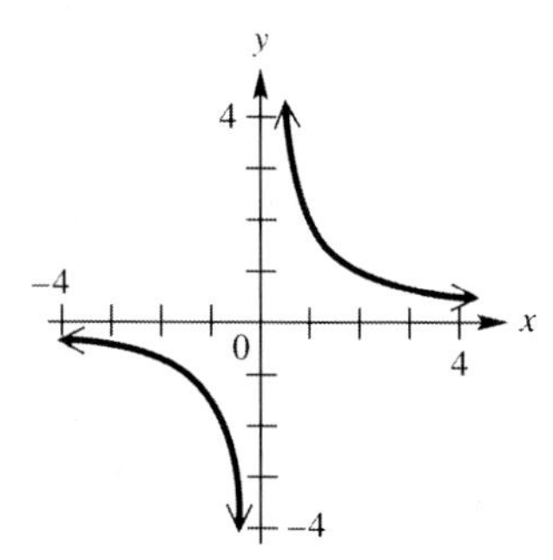

Figure 17a

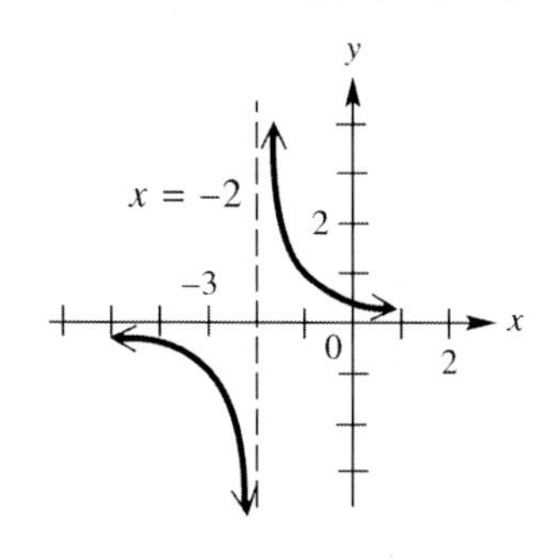

Figure 19a

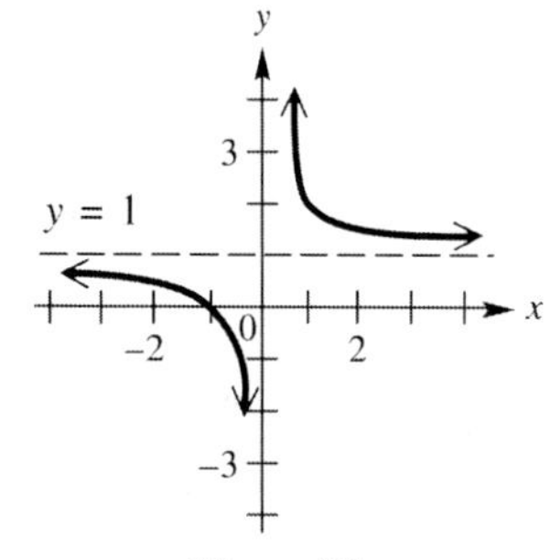

Figure 21a

[-4.7,4.7] by [-3.1,31.]
Xscl = 1 Yscl = 1

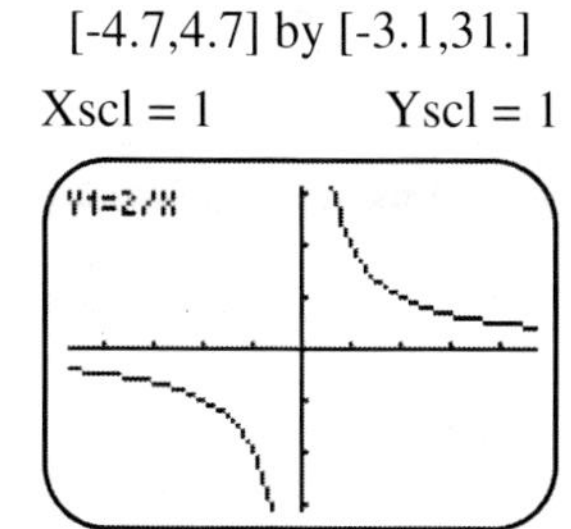

Figure 17b

[-6.7,2.7] by [-3.1,3.1]
Xscl = 1 Yscl = 1

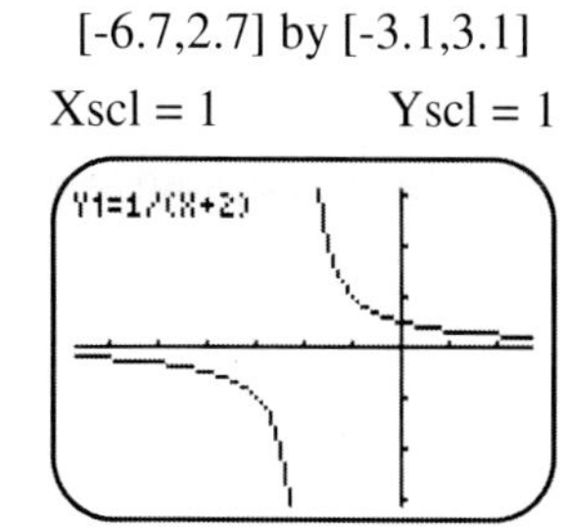

Figure 19b

[-4.7,4.7] by [-2.1,4.1]
Xscl = 1 Yscl = 1

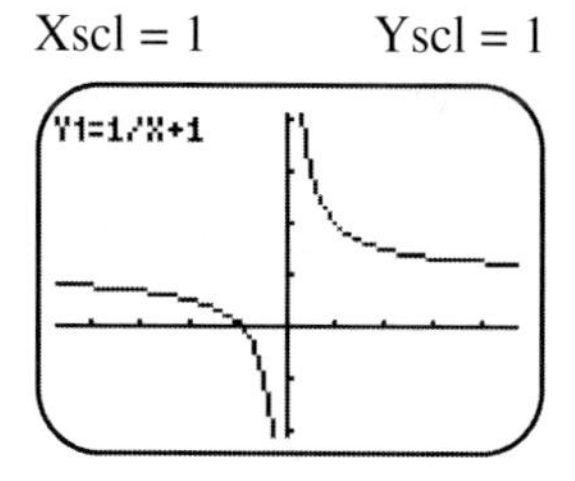

Figure 21b

21. Let $g(x) = y = \frac{1}{x}$, then $f(x) = g(x)+1$. To obtain the graph of *f*, shift the graph of $y = \frac{1}{x}$ upward 1 unit. See Figures 21a and 21b. The domain is $(-\infty,0)\cup(0,\infty)$. The range is $(-\infty,1)\cup(1,\infty)$.

23. Let $g(x) = y = \frac{1}{x}$, then $f(x) = g(x-1)+1$. To obtain the graph of *f*, shift the graph of $y = \frac{1}{x}$ to the right 1 unit and upward 1 unit. See Figures 23a and 23b. The domain is $(-\infty,1)\cup(1,\infty)$. The range is $(-\infty,1)\cup(1,\infty)$.

25. Let $g(x) = y = \frac{1}{x^2}$, then $f(x) = g(x)-2$. To obtain the graph of *f*, shift the graph of $y = \frac{1}{x^2}$ downward 2 units. See Figures 25a and 25b. The domain is $(-\infty,0)\cup(0,\infty)$. The range is $(-2,\infty)$.

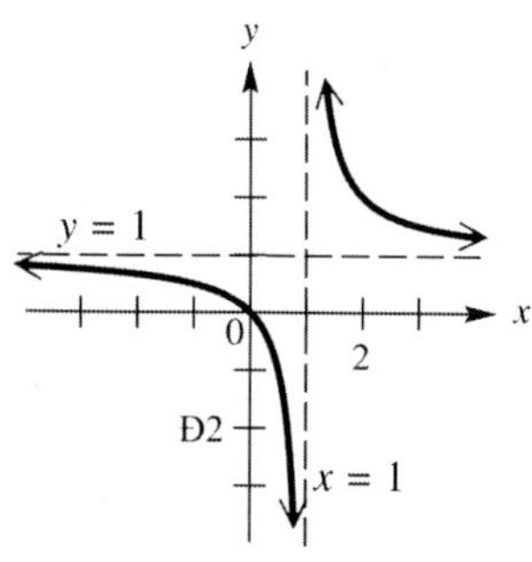

Figure 23a

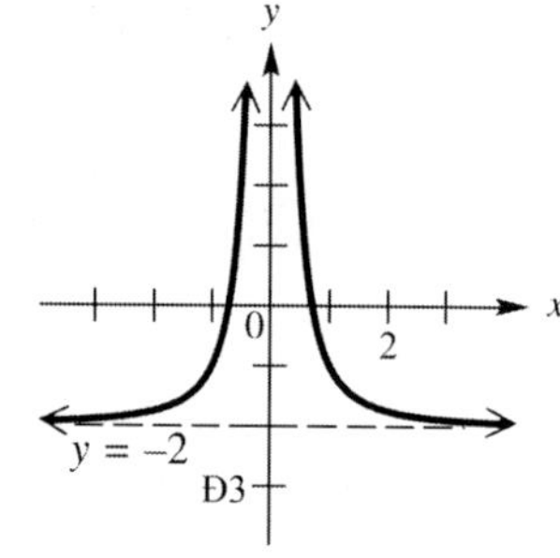

Figure 25a

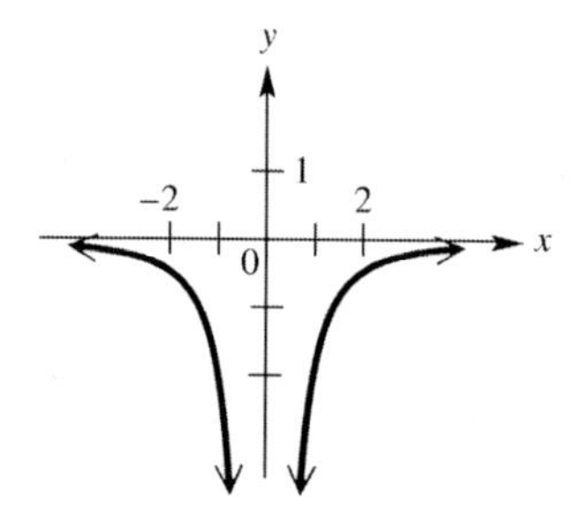

Figure 27a

[-4.7,4.7] by [-3.1,3.1]
Xscl = 1 Yscl = 1

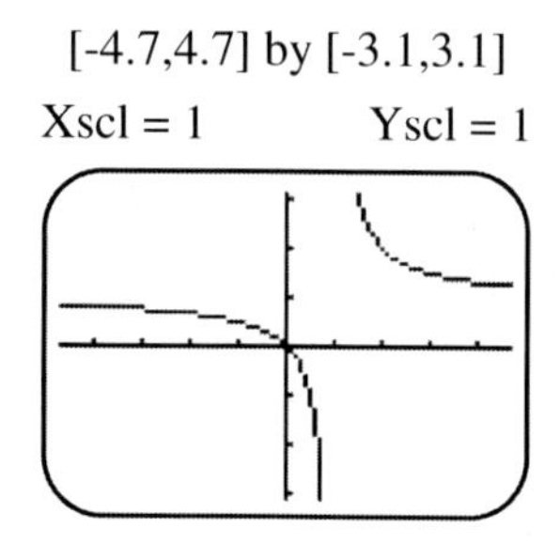
Figure 23b

[-4.7,4.7] by [-3.1,3.1]
Xscl = 1 Yscl = 1

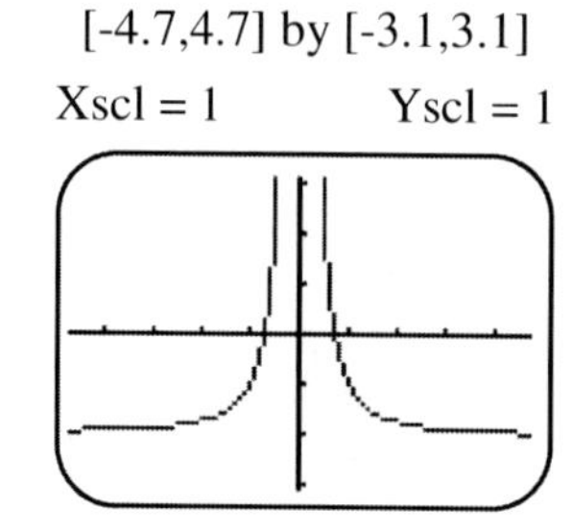
Figure 25b

[-4.7,4.7] by [-3.1,3.1]
Xscl = 1 Yscl = 1

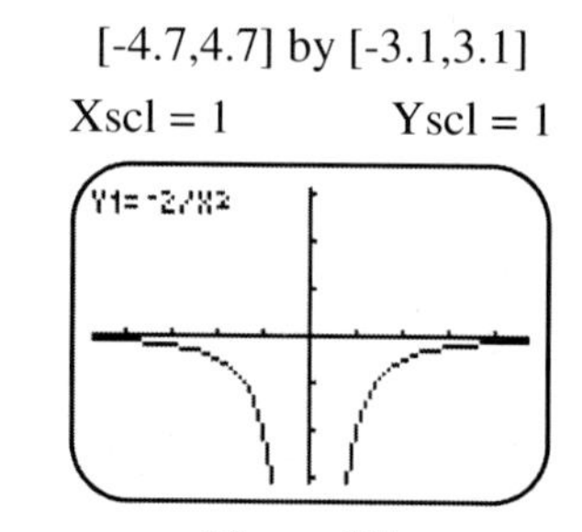

Figure 27b

27. Let $g(x) = y = \frac{1}{x^2}$, then $f(x) = -2g(x)$. To obtain the graph of *f*, stretch the graph of $y = \frac{1}{x^2}$ vertically by a factor of 2 and reflect it across the *x*-axis. See Figures 27a and 27b. The domain is $(-\infty,0)\cup(0,\infty)$. The range is $(-\infty,0)$

29. Let $g(x) = y = \frac{1}{x^2}$, then $f(x) = g(x-3)$. To obtain the graph of *f*, shift the graph of $y = \frac{1}{x^2}$ to the right 3 units. See Figures 29a and 29b. The domain is $(-\infty,3)\bigcup(3,\infty)$. The range is $(0,\infty)$.

31. Let $g(x) = y = \frac{1}{x^2}$, then $f(x) = -g(x+2) - 3$. To obtain the graph of *f*, shift the graph of $y = \frac{1}{x^2}$ to the left 2 units, reflect it across the *x*-axis and shift it downward 3 units. See Figures 31a and 31b. The domain is $(-\infty, -2) \cup (-2, \infty)$. The range is $(-\infty, -3)$.

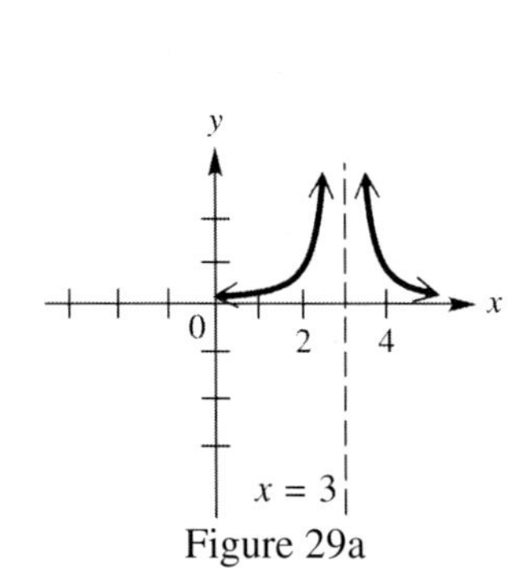

Figure 29a

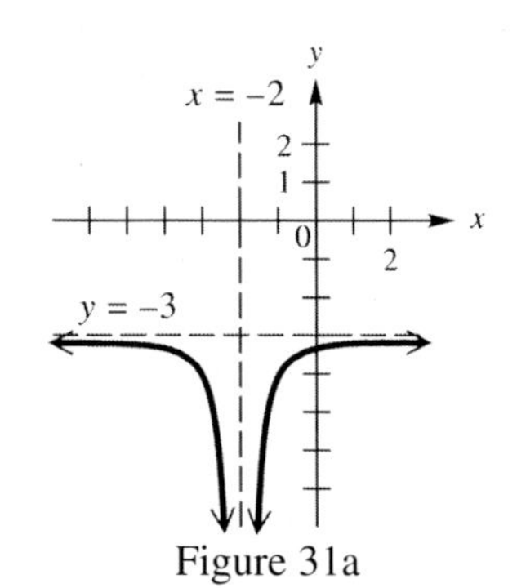

Figure 31a

[-1.7,7.7] by [-1.1,5.1]
Xscl = 1 Yscl = 1

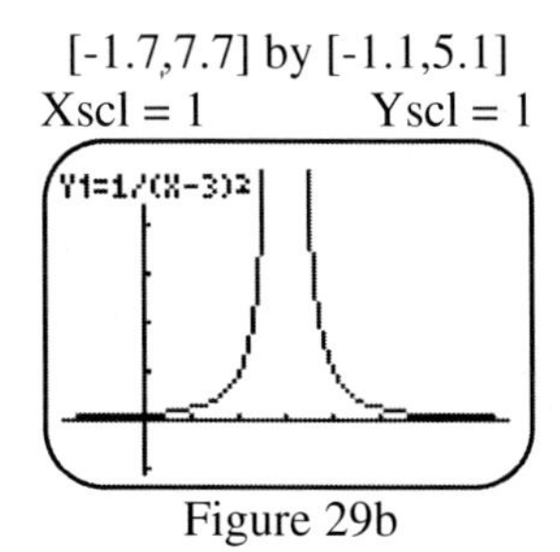

Figure 29b

[-6.7,2.7] by [-5.1,1.1]
Xscl = 1 Yscl = 1

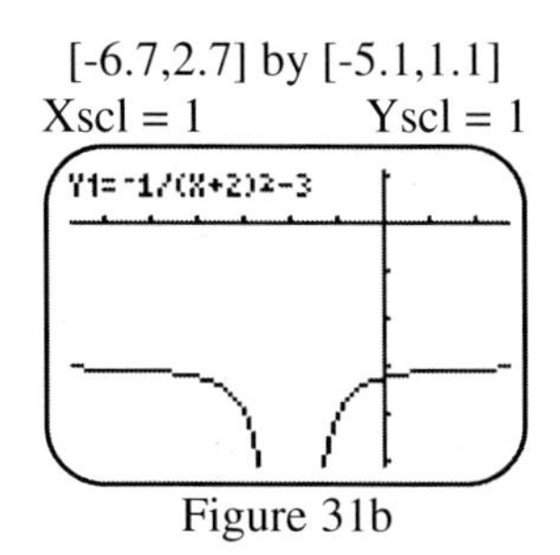

Figure 31b

33. The graph of *f* is obtained by shifting the graph of $y = \frac{1}{x^2}$ to the right 2 units. This is shown in graph C.

35. The graph of *f* is obtained by shifting the graph of $y = \frac{1}{x}$ to the right 2 units and reflecting it across the *x*-axis. This is shown in graph B.

37. The vertical asymptote shifts 2 units left to $x = -1$. The horizontal asymptote shifts 1 unit down to $y = 1$. The domain shifts 2 units left to $(-\infty, -1) \cup (-1, \infty)$. The range shifts 1 unit down to $(-\infty, 1) \cup (1, \infty)$.

39. Perform long division: $x-2\overline{)x-1}$ with quotient 1; subtract $\underline{x-2}$; remainder 1 $\Rightarrow \frac{x-1}{x-2} = 1 + \frac{1}{x-2}$. The graph of $y = 1 + \frac{1}{x-2}$ is obtained by shifting the graph of $y = \frac{1}{x}$ to the right 2 units and 1 unit upward. A sketch is shown in Figure 39a. A calculator graph of $y_1 = (x-1)/(x-2)$ is shown in Figure 39b.

41. Perform long division: $x+3\overline{\smash{)}-2x-5}$ with quotient -2; $\underline{-2x-6}$; remainder 1 $\Rightarrow \dfrac{-2x-5}{x+3}=-2+\dfrac{1}{x+3}$. The graph of $y=-2+\dfrac{1}{x+3}$ is obtained by shifting the graph of $y=\dfrac{1}{x}$ to the left 3 units and 2 units downward. A sketch is shown in Figure 41a. A calculator graph of $y_1=(-2x-5)/(x+3)$ is shown in Figure 41b.

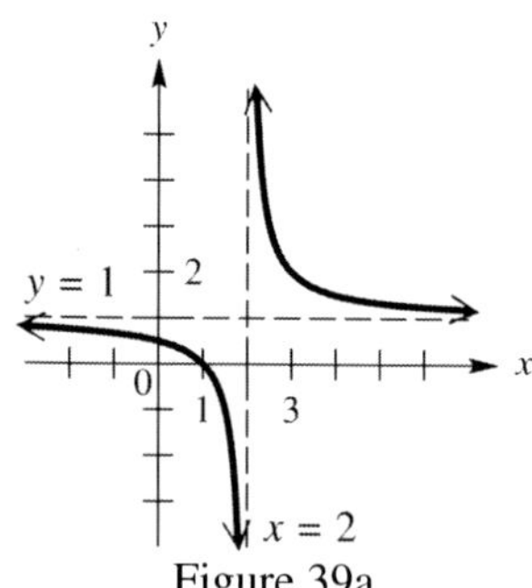

Figure 39a

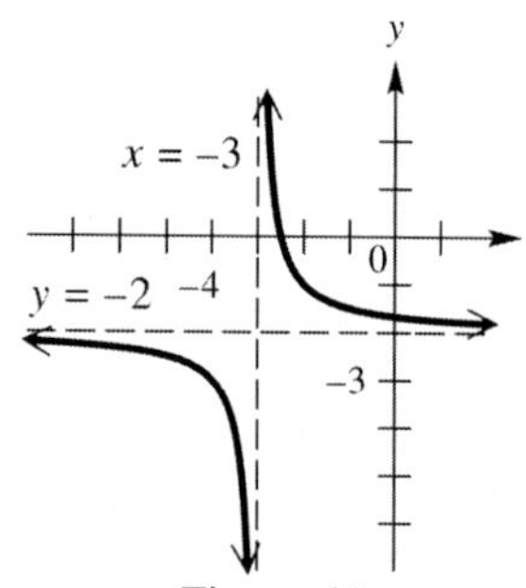

Figure 41a

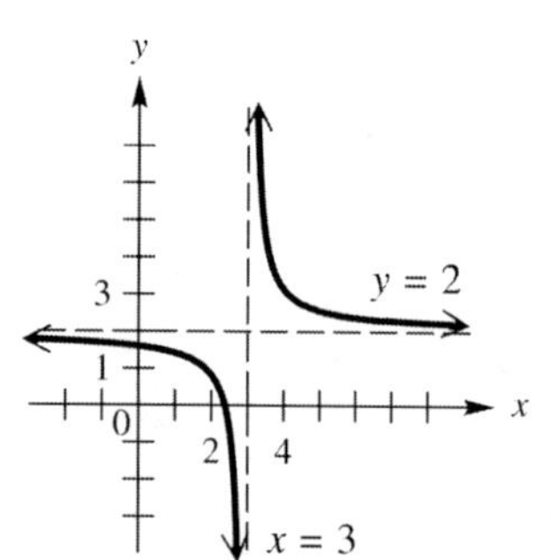

Figure 43a

[-2.7,6.7] by [-2.1,4.1]
Xscl = 1 Yscl = 1

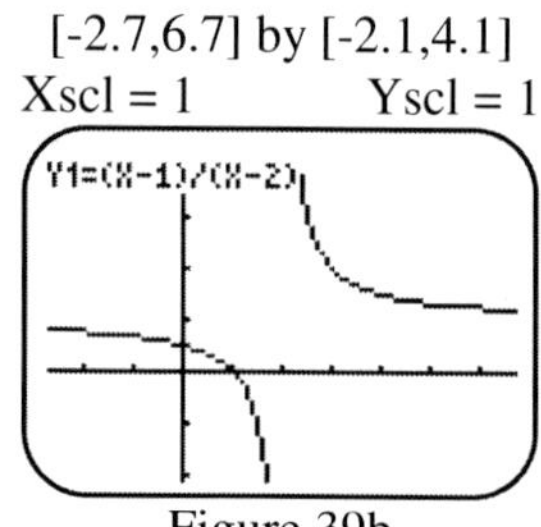

Figure 39b

[-7.7,1.7] by [-5.1,1.1]
Xscl = 1 Yscl = 1

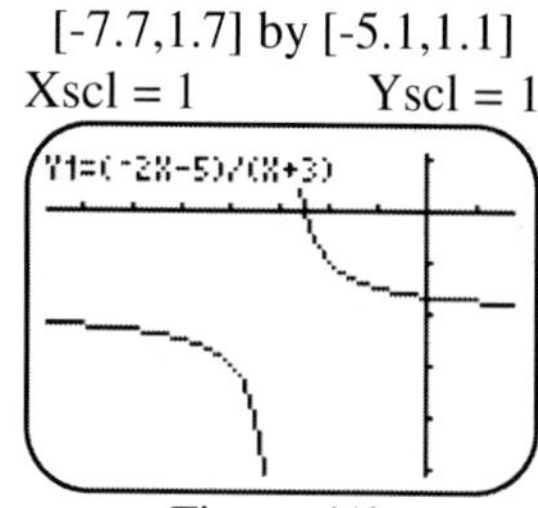

Figure 41b

[-1.7,7.7] by [-1.1,5.1]
Xscl = 1 Yscl = 1

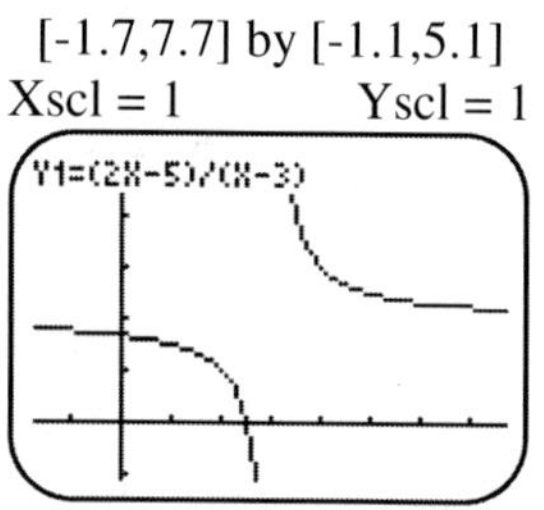

Figure 43b

43. Perform long division: $x-3\overline{\smash{)}2x-5}$ with quotient 2; $\underline{2x-6}$; remainder 1 $\Rightarrow \dfrac{2x-5}{x-3}=2+\dfrac{1}{x-3}$. The graph of $y=2+\dfrac{1}{x-3}$ is obtained by shifting the graph of $y=\dfrac{1}{x}$ to the right 3 units and 2 units upward. A sketch is shown in Figure 43a. A calculator graph of $y_1=(2x-5)/(x-3)$ is shown in Figure 43b.

4.2: Rational Functions and Graphs II

1. Because -1 is a zero of the denominator but not of the numerator, the line $x=-1$ is a vertical asymptote: D.
3. The degree of the numerator is less than the degree of the denominator, so the x-axis is a horizontal asymptote: G.
5. Because $f(x)=\dfrac{x^2-16}{x+4}=\dfrac{(x-4)(x+4)}{x+4}=x-4$ for all $x\neq-4$, there is a hole at $x=-4$: E.

7. Because $f(x)=\dfrac{x^2+3x+4}{x-5}=x+8+\dfrac{44}{x-5}$, the line $y=x+8$ is an oblique asymptote: F.

9. Since 5 is a zero of the denominator but not of the numerator, the line $x=5$ is a vertical asymptote and the domain is $(-\infty,5)\cup(5,\infty)$. Since $f(x)\to 0$ as $|x|\to\infty$, the line $y=0$ is a horizontal asymptote. Since the degree of the numerator is less than the degree of the denominator, there is no oblique asymptote.

11. Since $-\dfrac{1}{2}$ is a zero of the denominator but not of the numerator, the line $x=-\dfrac{1}{2}$ is a vertical asymptote and the domain is $\left(-\infty,-\dfrac{1}{2}\right)\cup\left(-\dfrac{1}{2},\infty\right)$. Since the degrees of the numerator and denominator are equal, the horizontal asymptote is $y=\dfrac{-3}{2}=-\dfrac{3}{2}$. Since the degree of the numerator is equal to the degree of the denominator, there is no oblique asymptote.

13. Since -3 is a zero of the denominator but not of the numerator, the line $x=-3$ is a vertical asymptote and the domain is $(-\infty,-3)\cup(-3,\infty)$. Since the degree of the numerator is greater than the degree of the denominator, there is no horizontal asymptote. Since $\dfrac{x^2-1}{x+3}=x-3+\dfrac{8}{x+3}$, the oblique asymptote is $y=x-3$.

15. Since -2 and $\dfrac{5}{2}$ are zeros of the denominator but not of the numerator, the lines $x=-2$ and $x=\dfrac{5}{2}$ are vertical asymptotes and the domain is $(-\infty,-2)\cup\left(-2,\dfrac{5}{2}\right)\cup\left(\dfrac{5}{2},\infty\right)$. Since the degrees of the numerator and denominator are equal, the horizontal asymptote is $y=\dfrac{1}{2}$. Since the degree of the numerator is equal to the degree of the denominator, there is no oblique asymptote.

17. Function A, because the denominator can never be equal to 0.

19. From the graph, the vertical asymptote is $x=2$, the horizontal asymptote is $y=4$, and there is no oblique asymptote. The function is defined for all $x\neq 2$, therefore the domain is $(-\infty,2)\cup(2,\infty)$.

21. From the graph, the vertical asymptotes are $x=-2$ and $x=2$, the horizontal asymptote is $y=-4$, and there is no oblique asymptote. The function is defined for all $x\neq\pm 2$, therefore the domain is $(-\infty,-2)\cup(-2,2)\cup(2,\infty)$.

23. From the graph, the vertical asymptote is $x=1$, the horizontal asymptote is $y=0$ and there is no oblique asymptote. The function is defined for all $x\neq 1$, therefore the domain is $(-\infty,1)\cup(1,\infty)$.

25. From the graph, the vertical asymptote is $x=-1$, there is no horizontal asymptote, and the oblique asymptote passes through the points $(0,-1)$ and $(1,0)$. Thus, the equation of the oblique asymptote is $y=x-1$. The function is defined for all $x\neq -1$, therefore the domain is $(-\infty,-1)\cup(-1,\infty)$.

27. From the graph, there is no vertical asymptote, the horizontal asymptote is $y = 0$, and there is no oblique asymptote. The function is defined for all x, therefore the domain is $(-\infty, \infty)$.

29. See Figure 29.

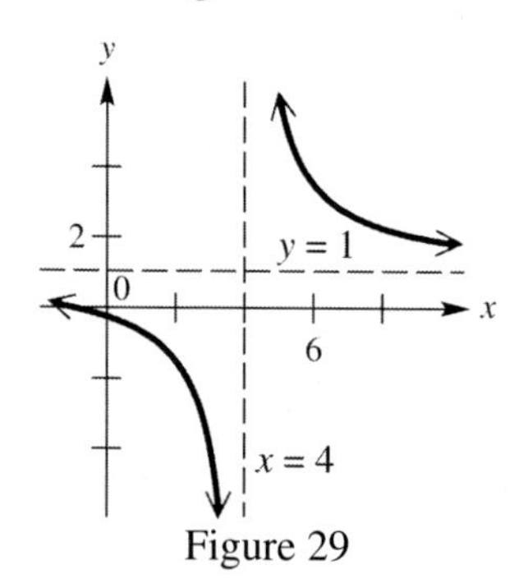

Figure 29

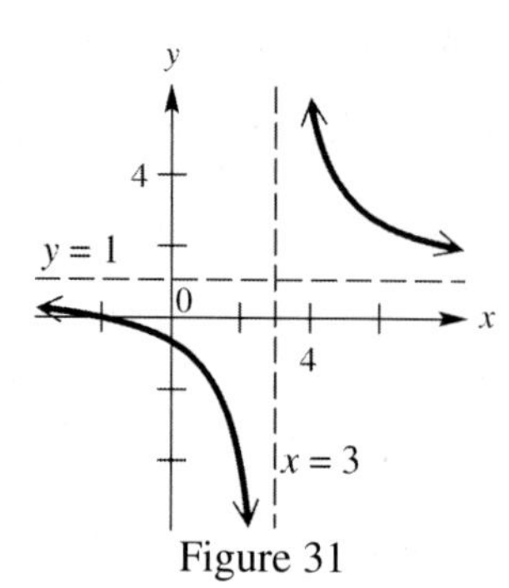

Figure 31

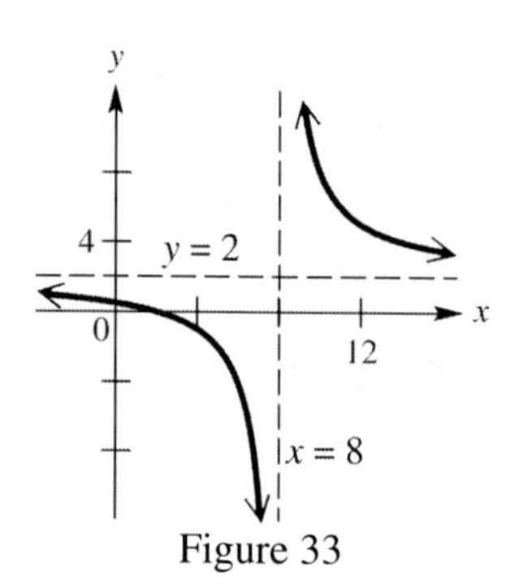

Figure 33

31. See Figure 31.
33. See Figure 33.
35. See Figure 35.
37. See Figure 37.

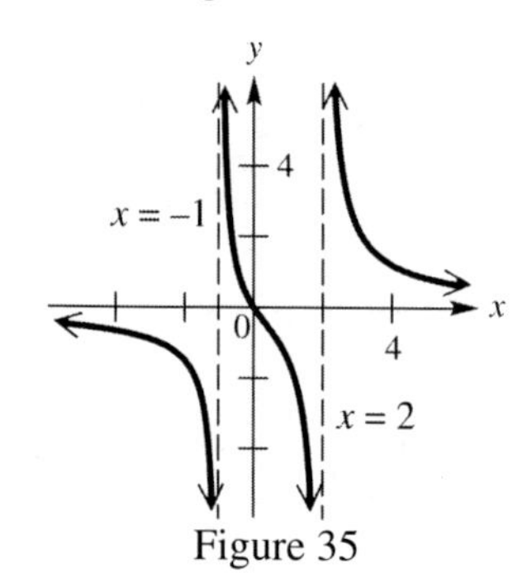

Figure 35

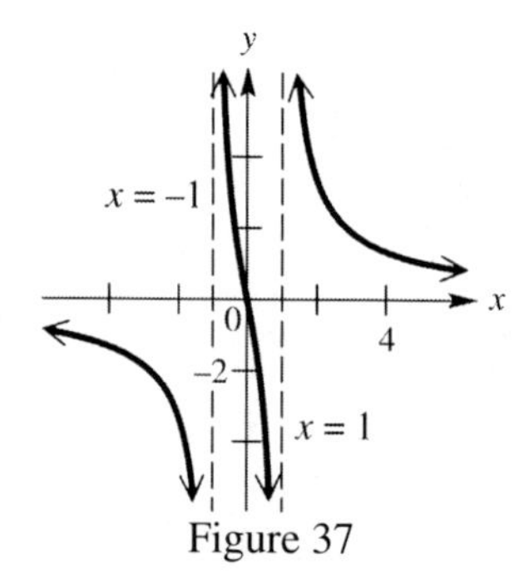

Figure 37

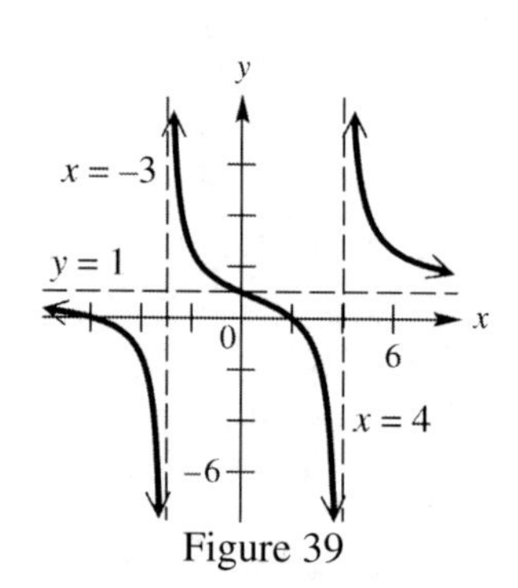

Figure 39

39. See Figure 39.
41. See Figure 41.
43. See Figure 43.

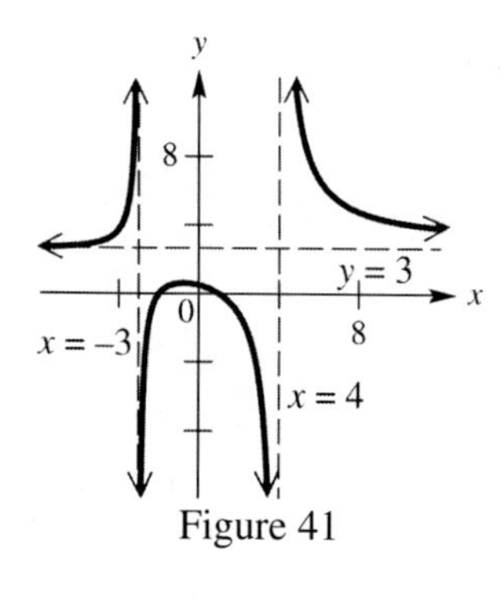

Figure 41

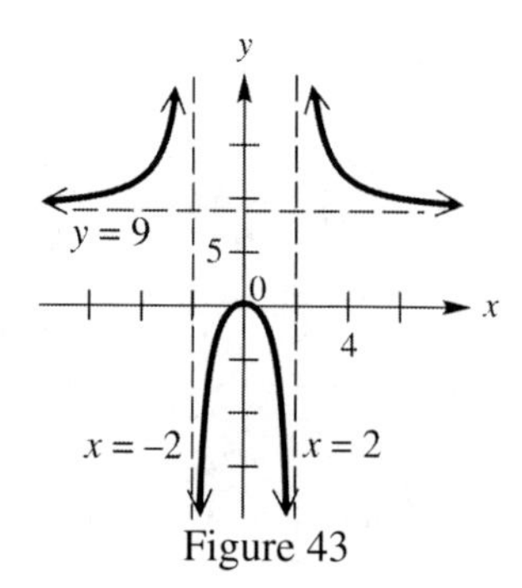

Figure 43

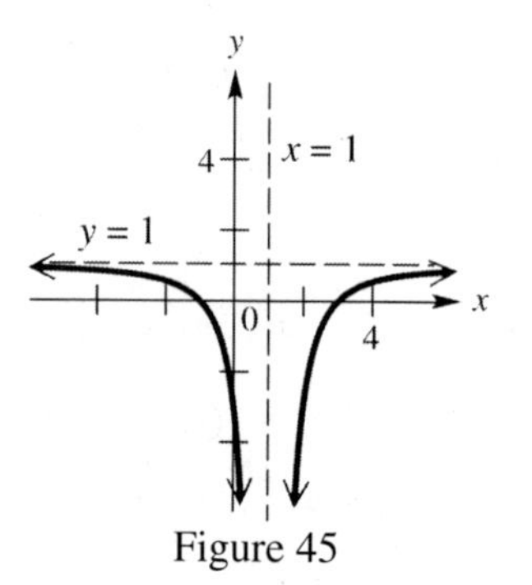

Figure 45

45. See Figure 45.

47. See Figure 47.

49. See Figure 49.

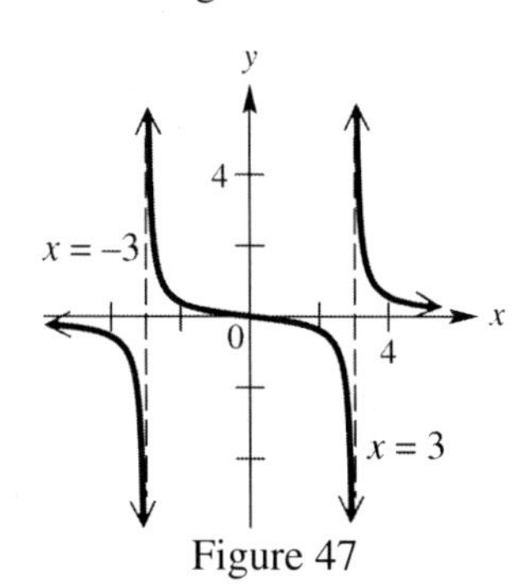

Figure 47

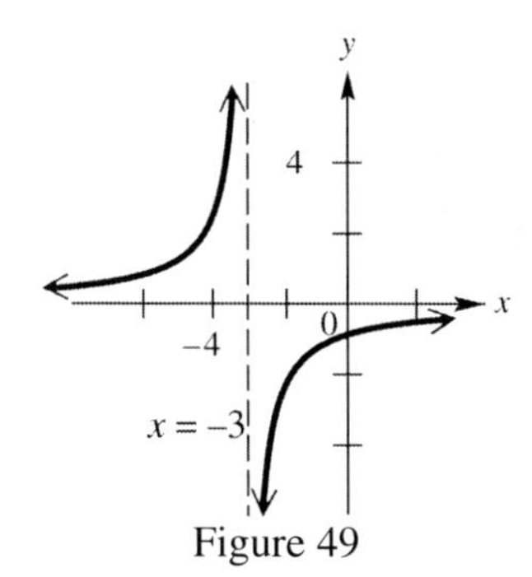

Figure 49

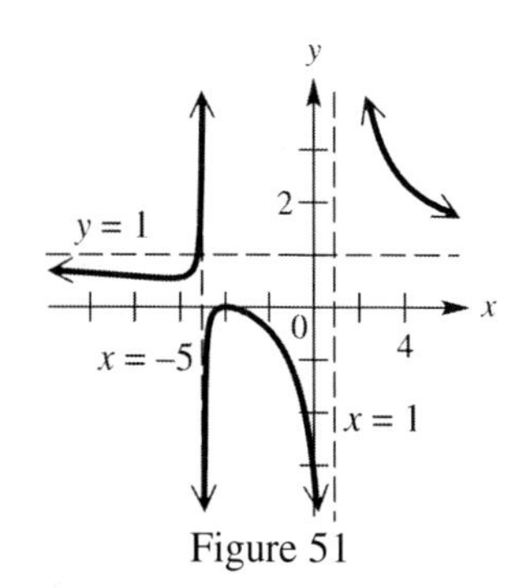

Figure 51

51. See Figure 51.

53. See Figure 53.

55. See Figure 55.

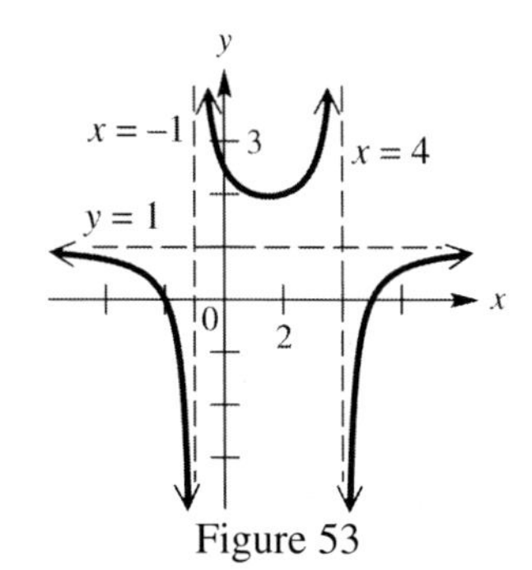

Figure 53

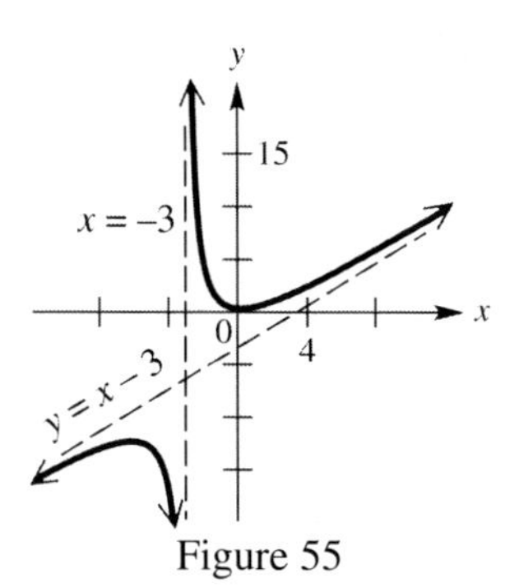

Figure 55

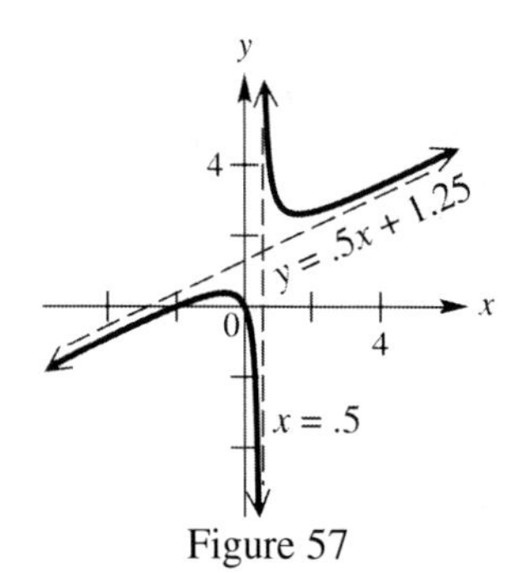

Figure 57

57. See Figure 57.

59. See Figure 59.

61. See Figure 61

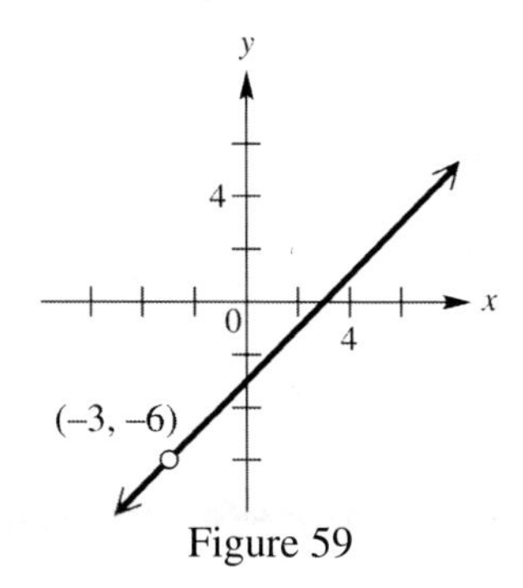

Figure 59

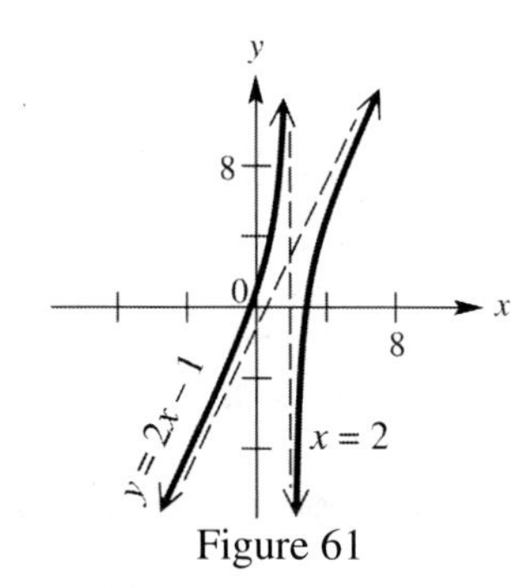

Figure 61

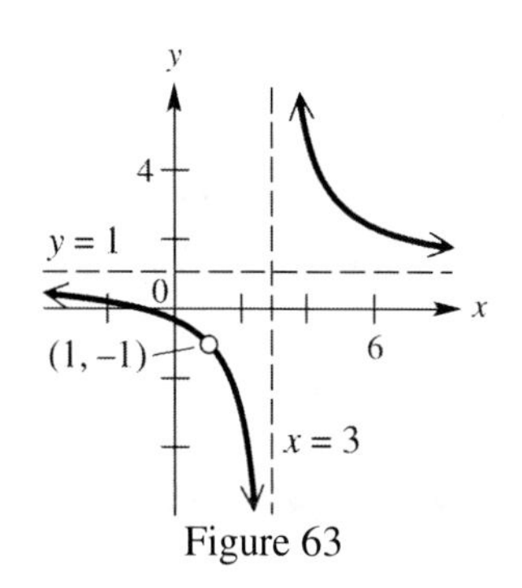

Figure 63

63. See Figure 63.

65. See Figure 65.

67. See Figure 67.

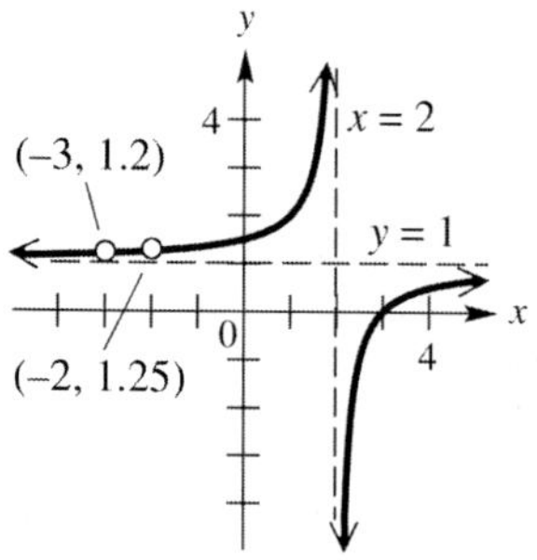

Figure 65

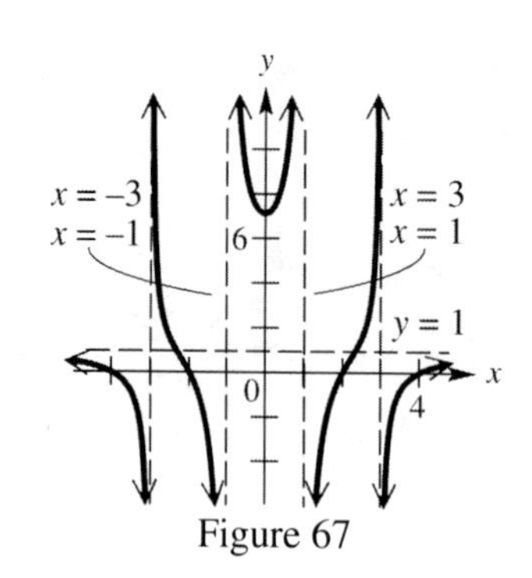

Figure 67

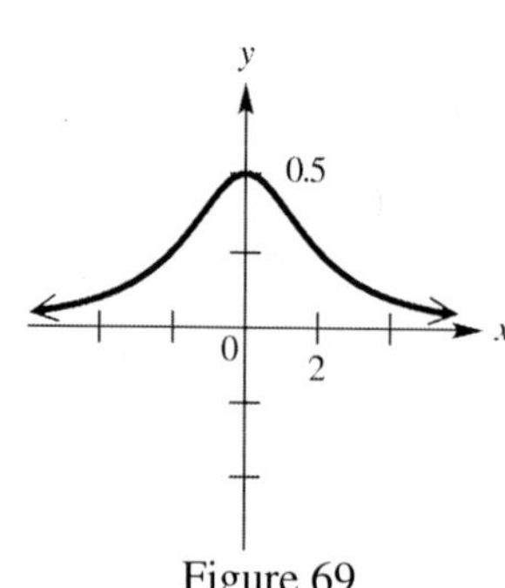

Figure 69

69. The domain is $(-\infty,\infty)$ because the denominator as no real zeros, the range is the interval $\left(0,\frac{1}{2}\right]$, and the graph is symmetric with respect to the y-axis. There are no vertical asymptotes and the x-axis is the horizontal asymptote. See Figure 69.

71. $D:(-\infty,\infty)$, $R:(-1,0]$ The function is symmetric to the y-axis and the HA is y = - 1. See Figure 71.

73. $D:(-\infty,\infty)$, $R:[0,1]$ The function is symmetric to the y-axis and the HA is y = 0. See Figure 73.

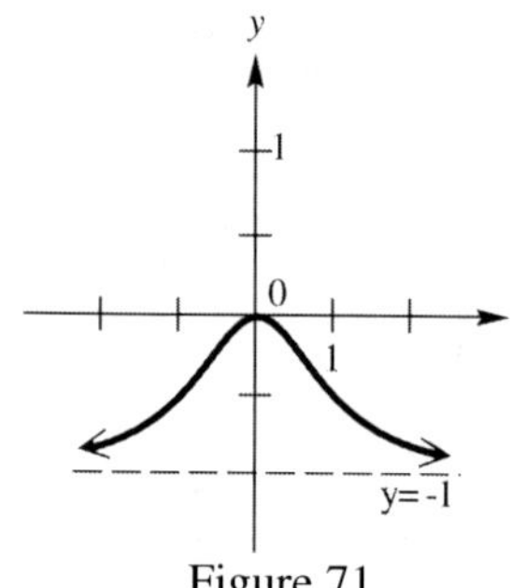

Figure 71

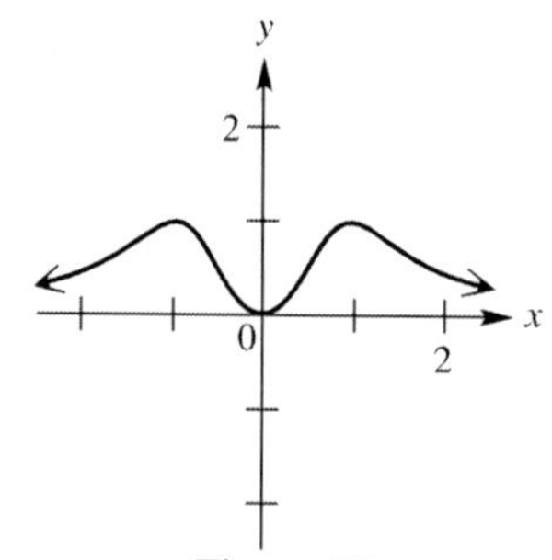

Figure 73

75. (a) The x-intercepts are the solutions of $3x^3+2x^2-12x-8=0 \Rightarrow (3x+2)(x^2-4)=0 \Rightarrow x=-\frac{2}{3},\pm 2.$ The y-intercept is $f(0)=-2.$

(b) The denominator has no real zeros because the discriminant is equal to -15. Therefore, there are no vertical asymptotes for $f(x)$.

(c) When the numerator is divided by the denominator using long division the result is $3x-1+\frac{-23x-4}{x^2+x+4}$, so the equation of the oblique asymptote is y=3x-1.

(d) The graph in Figure 75 correctly suggests that both the domain and range are both $(-\infty,\infty)$.

77. (a) The x-intercepts are the solutions of $x^3+4x^2-x-4=0 \Rightarrow (x-4)(x^2-1)=0 \Rightarrow x=4,\pm 1.$ The y-intercept is $f(0)=1.$

(b) The denominator has no real zeros because the discriminant is equal to -28. Therefore, there are no vertical asymptotes for $f(x)$.

(c) When the numerator is divided by the denominator using long division the result is

$$-\frac{1}{2}x-\frac{3}{2}+\frac{-2x-10}{-2x^2-2x-4},$$ so the equation of the oblique asymptote is $-\frac{1}{2}x-\frac{3}{2}.$

(d) The graph in Figure 77 correctly suggests that both the domain and range are both $(-\infty,\infty)$.

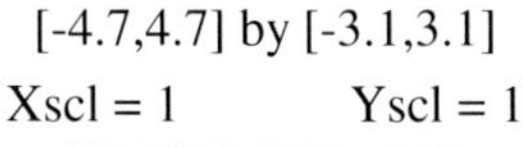

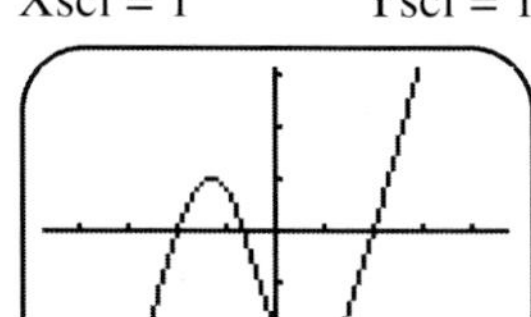

Figure 75

[-4.7,4.7] by [-3.1,3.1]

Xscl = 1 Yscl = 1

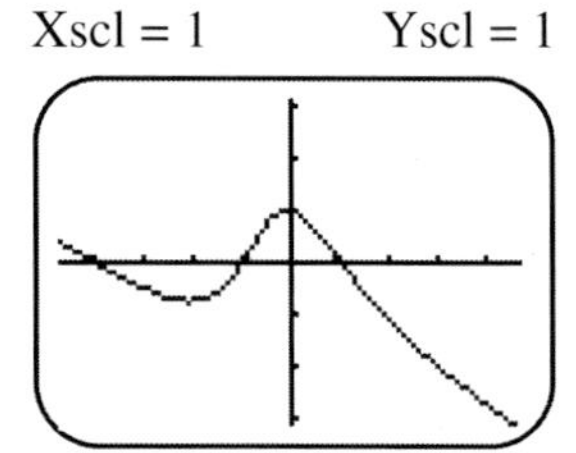

Figure 77

79. The expression is equal to 0 when $x = 4$, thus $p = 4$. The expression is undefined when $x = 2$, thus $q = 2$.

81. The expression is equal to 0 when $x = -2$, thus $p = -2$. The expression is undefined when $x = -1$, thus $q = -1$.

83. A vertical asymptote of $x = 2$ indicates that the denominator should contain the factor $(x-2)$. A hole at $x = -2$ indicates that both the numerator and denominator should contain the factor $(x+2)$. An x-intercept of 3 indicates that the numerator should contain the factor $(x-3)$. One possible equation for the function is $f(x)=\frac{(x-3)(x+2)}{(x-2)(x+2)}=\frac{x^2-x-6}{x^2-4}$. Other functions are possible.

85. Vertical asymptotes of $x = 0$ and $x = 4$ indicate that the denominator should contain the factors x and $(x-4)$. An x-intercept of 2 indicates that the numerator should contain the factor $(x-2)$. Since there is a horizontal asymptote at $y = 0$, the degree of the numerator should be less than the degree of the denominator. One possible equation for the function is $f(x)=\frac{x-2}{x(x-4)}=\frac{x-2}{x^2-4x}$. Other functions are possible.

87. (a) Reflect the graph across the x-axis. See Figure 87a.
    (b) Reflect the graph across the y-axis. See Figure 87b.

88. (a) Reflect the graph across the x-axis. See Figure 88a.
    (b) Reflect the graph across the y-axis. See Figure 88b.

89. (a) Reflect the graph across the x-axis. See Figure 89a.
    (b) Reflect the graph across the y-axis. See Figure 89b.

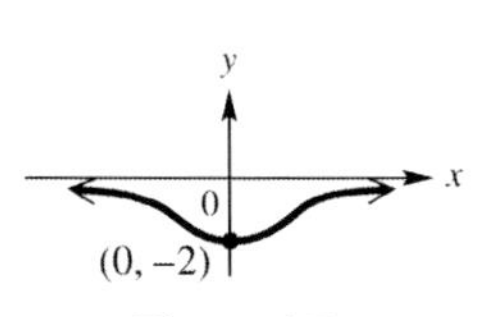

Figure 87a

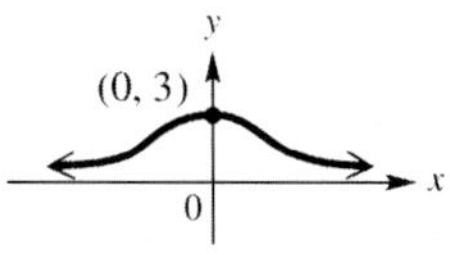

(0, 3)

Figure 88a

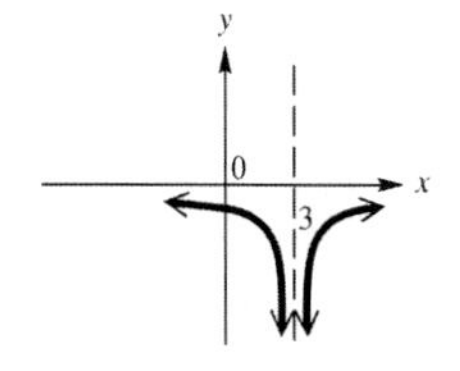

Figure 89a

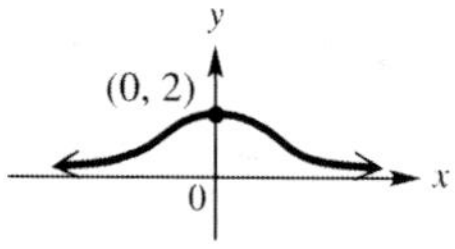

(Same as $y = f(x)$)

Figure 87b

(Same as $y = f(x)$)

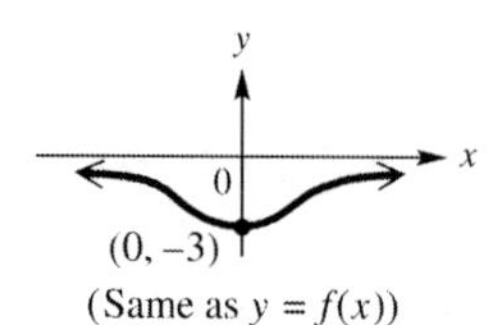

Figure 88b

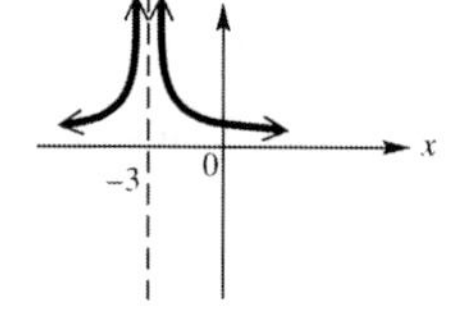

Figure 89b

90. (a) Reflect the graph across the x-axis. See Figure 90a.

(b) Reflect the graph across the y-axis. See Figure 90b.

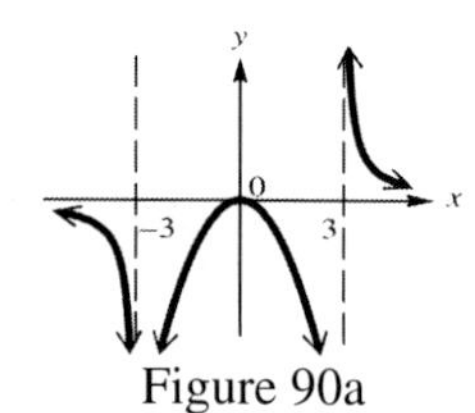

Figure 90a

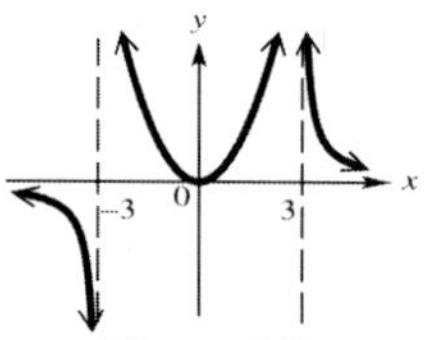

Figure 90b

91. Perform long division:

$$
\begin{array}{r}
-2x-8 \\
-x+4\,\overline{)\,2x^2+0x+3} \\
\underline{2x^2-8x} \\
8x+\;3 \\
\underline{8x-32} \\
35
\end{array}
\Rightarrow \frac{2x^2+3}{4-x} = -2x-8+\frac{35}{4-x}
$$

The oblique asymptote is $y = -2x - 8$.

The graphs of $y_1 = (2x^2+3)/(4-x)$ and $y_2 = -2x-8$ are shown in Figure 91.

93. Perform long division:

$$
\begin{array}{r}
-x+3 \\
x+2\,\overline{)\,-x^2+x+0} \\
\underline{-x^2-2x} \\
3x+0 \\
\underline{3x+6} \\
-6
\end{array}
\Rightarrow \frac{x-x^2}{x+2} = -x+3+\frac{-6}{x+2}
$$

The oblique asymptote is $y = -x + 3$.

The graphs of $y_1 = (x-x^2)/(x+2)$ and $y_2 = -x+3$ are shown in Figure 93.

[-18.8,18.8] by [-50,25]
Xscl = 1 Yscl = 1

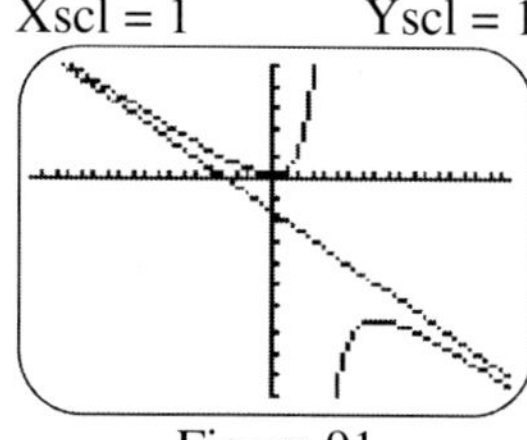

Figure 91

[-3,8] by [-10,10]
Xscl = 1 Yscl = 1

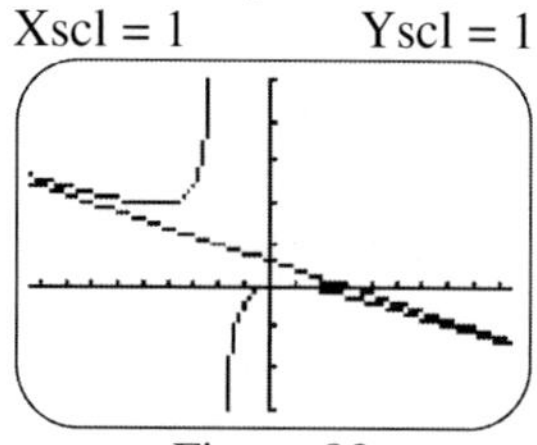

Figure 93

[-50,50] by [0,1000]
Xscl = 10 Yscl = 100

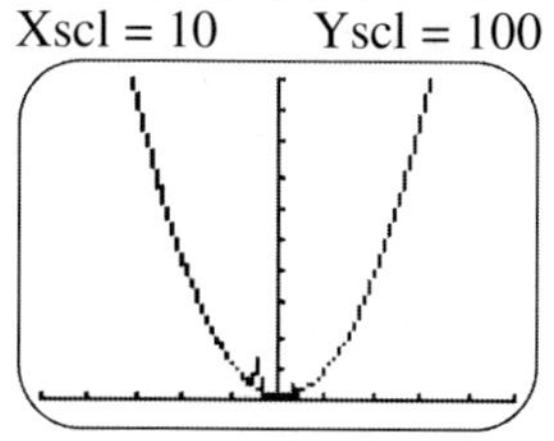

Figure 97

95. (a) Disregard the remainder to get the equation of the oblique asymptote, $y = x + 1$.

(b) Set the original function equal to the expression for the oblique asymptote and solve:

$$\frac{x^5+x^4+x^2+1}{x^4+1}=x+1 \Rightarrow x^5+x^4+x^2+1=(x^4+1)(x+1) \Rightarrow$$

$$x^5+x^4+x^2+1=x^5+x^4+x+1 \Rightarrow x^2+1=x+1 \Rightarrow x^2-x=0 \Rightarrow x(x-1)=0 \Rightarrow x=0,1.$$

The graph crosses the oblique asymptote at $x=0$ and $x=1$

(c) For large values of $x, x+1<\dfrac{x^5+x^4+x^2+1}{x^4+1}$. Thus, the function approaches its asymptote from above.

97.

$$\begin{array}{r} x^2-x+8 \\ x^2+x-12\overline{\big)x^4+0x^3-5x^2+0x+4} \\ \underline{x^4+x^3-12x^2} \\ -x^3+7x^2+0x \\ \underline{-x^3-x^2+12x} \\ 8x^2-12x+4 \\ \underline{8x^2+8x-96} \\ -20x+100 \end{array} \Rightarrow \frac{x^4-5x^2+4}{x^2+x-12}=x^2-x+8+\frac{-20x+100}{x^2+x-12}$$

The graphs of $y_1=(x^4-5x+4)/(x^2+x-12)$ and $y_2=x^2-x+8$ are shown in figure 97. In this window, the two graphs seem to overlap (coincide), suggesting that as $|x|\to\infty$, the graph of f approaches the graph of g, giving an asymptotic effect.

4.3: Rational Equations, Inequalities, Models and Applications

1. (a) The graph never intersects the x-axis. The solution set of $f(x)=0$ is $\varnothing$.

   (b) The graph is below the x-axis on the interval $(-\infty,-2)$. Thus $f(x)<0$ on $(-\infty,-2)$.

   (c) The graph is above the x-axis on the interval $(-2,\infty)$. Thus $f(x)>0$ on $(-2,\infty)$.

3. (a) The graph intersects the x-axis at $x=-1$. The solution set of $f(x)=0$ is $\{-1\}$.

   (b) The graph is below the x-axis on the interval $(-1,0)$. Thus $f(x)<0$ on $(-1,0)$.

   (c) The graph is above the x-axis on the interval $(-\infty,-1)\cup(0,\infty)$. Thus $f(x)>0$ on $(-\infty,-1)\cup(0,\infty)$.

5. (a) The graph intersects the x-axis at $x=0$. The solution set of $f(x)=0$ is $\{0\}$.

   (b) The graph is below the x-axis on the interval $(-2,0)\cup(2,\infty)$. Thus $f(x)<0$ on $(-2,0)\cup(2,\infty)$.

   (c) The graph is above the x-axis on the interval $(-\infty,-2)\cup(0,2)$. Thus $f(x)>0$ on $(-\infty,-2)\cup(0,2)$.

7. (a) The graph intersects the x-axis at $x=0$. The solution set of $f(x)=0$ is $\{0\}$.

   (b) The graph is below the x-axis on the interval $(-1,0)\cup(0,1)$. Thus $f(x)<0$ on $(-1,0)\cup(0,1)$.

   (c) The graph is above the x-axis on the interval $(-\infty,-1)\cup(1,\infty)$. Thus $f(x)>0$ on $(-\infty,-1)\cup(1,\infty)$.

9. (a) The graph intersects the x-axis at $x=0$. The solution set of $f(x)=0$ is $\{0\}$.

(b) The graph is below the x-axis on the interval $(1,2)\cup(2,3)$.. Thus $f(x)<0$ on $(1,2)\cup(2,3)$.

(c) The graph is above the x-axis on the interval $(-\infty,-2)\cup(-2,0)\cup(0,1)\cup(3,\infty)$. Thus $f(x)>0$ on the interval $(-\infty,-2)\cup(-2,0)\cup(0,1)\cup(3,\infty)$.

11. (a) The graph intersects the x-axis at $x=2.5$. The solution set of $f(x)=0$ is $\{2.5\}$.

(b) The graph is below the x-axis on the interval $(2.5,3)$. Thus $f(x)<0$ on $(2.5,3)$.

(c) The graph is above the x-axis on the interval $(-\infty,2.5)\cup(3,\infty)$. Thus $f(x)>0$ on $(-\infty,2.5)\cup(3,\infty)$.

13. Multiply through by the LCD, $(x-1)(x+1)$. $\frac{2x}{x^2-1}\cdot(x-1)(x+1)=\frac{2}{x+1}\cdot(x-1)(x+1)-\frac{1}{x-1}\cdot(x-1)(x+1)\Rightarrow$

$2x=2(x-1)-(x+1)\Rightarrow 2x=2x-2-x-1\Rightarrow 2x=x-3\Rightarrow x=-3$. The solution set is $\{-3\}$.

15. Multiply through by the LCD, $x(x-3)(x+3)$.

$\frac{4}{x^2-3x}\cdot x(x-3)(x+3)-\frac{1}{x^2-9}\cdot x(x-3)(x+3)=0\cdot x(x-3)(x+3)\Rightarrow 4(x+3)-x=0\Rightarrow$

$4x+12-x=0\Rightarrow 3x=-12\Rightarrow x=-4$. The solution set is $\{-4\}$.

17. Multiply through by the LCD, x^2.

$1\cdot x^2-\frac{13}{x}\cdot x^2+\frac{36}{x^2}\cdot x^2=0\cdot x^2\Rightarrow x^2-13x+36=0\Rightarrow(x-4)(x-9)=0\Rightarrow x=4$ or 9.

The solution set is $\{4,9\}$.

19. Multiply through by the LCD, x^2.

$1\cdot x^2+\frac{3}{x}\cdot x^2=\frac{5}{x^2}\cdot x^2\Rightarrow x^2+3x=5\Rightarrow x^2+3x-5=0\Rightarrow$

$x=\frac{-3\pm\sqrt{3^2-4(1)(-5)}}{2(1)}=\frac{-3\pm\sqrt{29}}{2}$. The solution set is $\left\{\frac{-3\pm\sqrt{29}}{2}\right\}$.

21. Multiply through by the LCD, $x(2-x)$.

$\frac{x}{2-x}\cdot x(2-x)+\frac{2}{x}\cdot x(2-x)-5\cdot x(2-x)=0\cdot x(2-x)\Rightarrow x^2+2(2-x)-5x(2-x)=0\Rightarrow$

$6x^2-12x+4=0\Rightarrow 3x^2-6x+2=0$.

$x=\frac{-(-6)\pm\sqrt{(-6)^2-4(3)(2)}}{2(3)}=\frac{6\pm\sqrt{12}}{6}=\frac{2(3\pm\sqrt{3})}{2(3)}=\frac{3\pm\sqrt{3}}{3}$. The solution set is $\left\{\frac{3\pm\sqrt{3}}{3}\right\}$.

23. Rewrite the equation as $\frac{1}{x^4}-\frac{3}{x^2}-4=0$ and multiply through by the LCD, x^4.

$\frac{1}{x^4}\cdot x^4-\frac{3}{x^2}\cdot x^4-4\cdot x^4=0\cdot x^4\Rightarrow 1-3x^2-4x^4=0\Rightarrow 4x^4+3x^2-1=0\Rightarrow$

$(4x^2-1)(x^2+1)=0\Rightarrow(2x+1)(2x-1)(x^2+1)=0\Rightarrow 2x+1=0$ or $2x-1=0$ or $x^2+1=0\Rightarrow$

$2x=-1$ or $2x=1$ or $x^2=-1 \Rightarrow x=-\frac{1}{2}$ or $x=\frac{1}{2}$ or $x=\pm i$. The solution set is $\left\{\pm\frac{1}{2}, \pm i\right\}$.

25. Multiply through by the LCD, $(x+2)(x+7)$.

$$\frac{1}{x+2}\cdot(x+2)(x+7)+\frac{3}{x+7}\cdot(x+2)(x+7)=\frac{5}{x^2+9x+14}\cdot(x+2)(x+7) \Rightarrow$$

$1(x+7)+3(x+2)=5 \Rightarrow x+7+3x+6=5 \Rightarrow 4x=-8 \Rightarrow x=-2.$

This solution is extraneous. The solution set is $\varnothing$.

27. Multiply through by the LCD, $(x-3)(x+3)$.

$$\frac{x}{x-3}\cdot(x-3)(x+3)+\frac{4}{x+3}\cdot(x-3)(x+3)=\frac{18}{x^2-9}\cdot(x-3)(x+3) \Rightarrow$$

$x(x+3)+4(x-3)=18 \Rightarrow x^2+3x+4x-12=18 \Rightarrow x^2+7x-30=0 \Rightarrow (x+10)(x-3)=0 \Rightarrow x=-10$ or $x=3$.

The solution $x=3$ is extraneous. The solution set is $\{-10\}$.

29. Rewrite the equation as $\frac{9}{x}+\frac{4}{6x-3}=\frac{2}{6x-3}$ and multiply through by the LCD, $x(6x-3)$.

$$\frac{9}{x}\cdot x(6x-3)+\frac{4}{6x-3}\cdot x(6x-3)=\frac{2}{6x-3}\cdot x(6x-3) \Rightarrow 9(6x-3)+4x=2x \Rightarrow$$

$54x-27+4x=2x \Rightarrow 56x=27 \Rightarrow x=\frac{27}{56}$. The solution set is $\left\{\frac{27}{56}\right\}$.

31. (a) $\frac{x-3}{x+5}=0 \Rightarrow x-3=0(x+5) \Rightarrow x-3=0 \Rightarrow x=3$. The solution set is $\{3\}$.

(b) Graph $y_1=(x-3)/(x+5)$ as shown in Figure 31. With vertical asymptote $x=-5$, the graph is below or intersects the x-axis on the interval $(-5,3]$. The solution set is $(-5,3]$.

(c) Graph $y_1=(x-3)/(x+5)$ as shown in Figure 31. With vertical asymptote $x=-5$, the graph is above or intersects the x-axis on the interval $(-\infty,-5)\cup[3,\infty)$. The solution set is $(-\infty,-5)\cup[3,\infty)$.

33. (a) $\frac{x-1}{x+2}=1 \Rightarrow x-1=1(x+2) \Rightarrow x-1=x+2 \Rightarrow -1=2$ (false). The solution set is $\varnothing$.

(b) Graph $y_1=(x-1)/(x+2)-1$ as shown in Figure 33. With vertical asymptote $x=-2$, the graph is above the x-axis on the interval $(-\infty,-2)$. The solution set is $(-\infty,-2)$.

(c) Graph $y_1=(x-1)/(x+2)-1$ as shown in Figure 33. With vertical asymptote $x=-2$, the graph is below the x-axis on the interval $(-2,\infty)$. The solution set is $(-2,\infty)$.

[-10,10] by [-5,8]
Xscl = 1 Yscl = 1

[-10,10] by [-5,10]
Xscl = 1 Yscl = 1

[-5,5] by [-5,5]
Xscl= 1 Yscl = 1

[-3,3] by [-20,20]
Xscl = 1 Yscl= 5

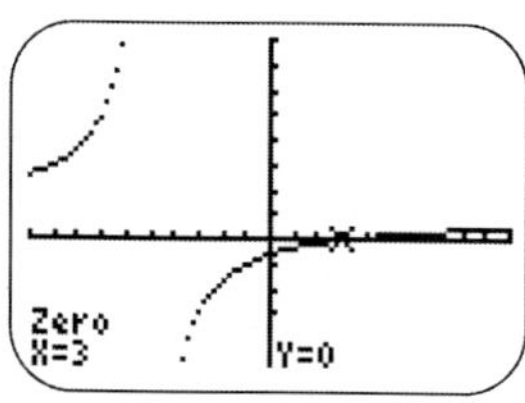

Figure 31

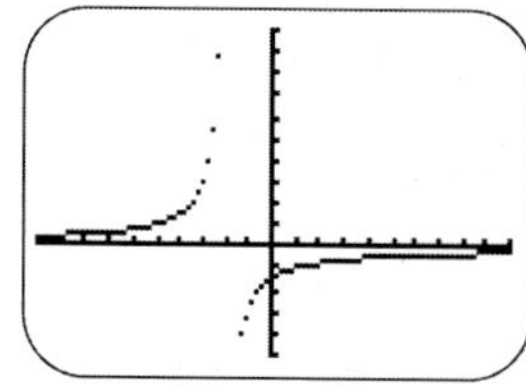
Figure 33

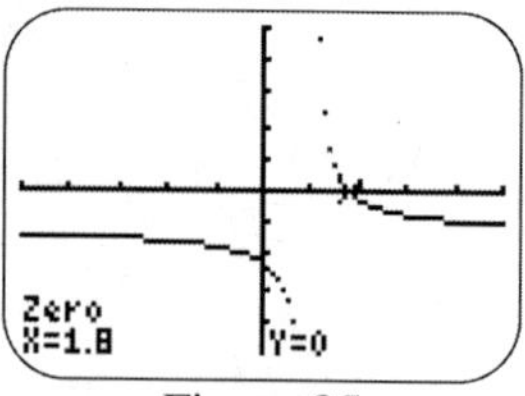

Figure 35

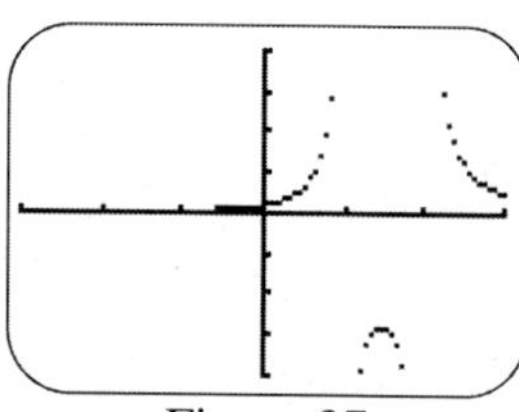
Figure 37

35. (a) $\frac{1}{x-1}=\frac{5}{4}\Rightarrow 5(x-1)=4(1)\Rightarrow 5x-5=4\Rightarrow 5x=9\Rightarrow x=\frac{9}{5}$. The solution set is $\left\{\frac{9}{5}\right\}$.

(b) Graph $y_1=1/(x-1)-5/4$ as shown in Figure 35. With vertical asymptote $x=1$, the graph is below the x-axis on the interval $(-\infty,1)\cup\left(\frac{9}{5},\infty\right)$. The solution set is $(-\infty,1)\cup\left(\frac{9}{5},\infty\right)$.

(c) Graph $y_1=1/(x-1)-5/4$ as shown in Figure 35. With vertical asymptote $x=1$, the graph is above the x-axis on the interval $\left(1,\frac{9}{5}\right)$. The solution set is $\left(1,\frac{9}{5}\right)$.

37. (a) $\frac{4}{x-2}=\frac{3}{x-1}\Rightarrow 4(x-1)=3(x-2)\Rightarrow 4x-4=3x-6\Rightarrow x=-2$. The solution set is $\{-2\}$.

(b) Graph $y_1=4/(x-2)-3/(x-1)$ as shown in Figure 37. With vertical asymptotes $x=1$ and $x=2$, the graph is below or intersects the x-axis on $(-\infty,-2]\cup(1,2)$.. The solution set is $(-\infty,-2]\cup(1,2)$.

(c) Graph $y_1=4/(x-2)-3/(x-1)$ as shown in Figure 37. With vertical asymptotes $x=1$ and $x=2$, the graph is above or intersects the x-axis on $[-2,1)\cup(2,\infty)$. The solution set is $[-2,1)\cup(2,\infty)$.

39. (a) $\frac{1}{(x-2)^2}=0\Rightarrow 1=0(x-2)^2\Rightarrow 1=0$ (false). The solution set is $\varnothing$.

(b) Graph $y_1=1/(x-2)^2$ as shown in Figure 39. The graph is never below the x-axis. The solution set is $\varnothing$.

(c) Graph $y_1=1/(x-2)^2$ as shown in Figure 39. With vertical asymptote $x=2$, the graph is above the x-axis on the interval $(-\infty,2)\cup(2,-\infty)$. The solution set is $(-\infty,2)\cup(2,\infty)$.

41. (a) $\frac{5}{x+1}=\frac{12}{x+1}\Rightarrow 5(x+1)=12(x+1)\Rightarrow 5x+5=12x+12\Rightarrow -7x=7\Rightarrow x=-1$. This solution is extraneous. The solution set is $\varnothing$.

(b) Graph $y_1=5/(x+1)-12/(x+1)$ as shown in Figure 41. With vertical asymptote $x=-1$, the graph is above the x-axis on the interval $(-\infty,-1)$. The solution set is $(-\infty,-1)$.

(c) Graph $y_1=5/(x+1)-12/(x+1)$ as shown in Figure 41. With vertical asymptote $x=-1$, the graph is below the x-axis on the interval $(-1,\infty)$. The solution set is $(-1,\infty)$.

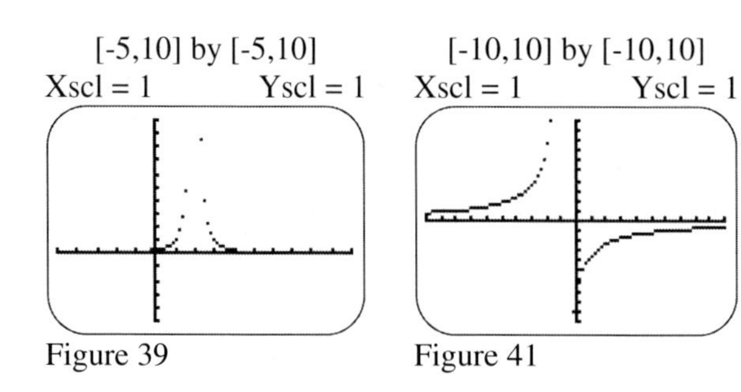

Figure 39 Figure 41

43. (a) The equation $\frac{(x-2)(2)-(2x+1)(1)}{(x-2)^2}=0$ is true only when $(x-2)(2)-(2x+1)(1)=0$. Solving we have

$(x-2)(2)-(2x+1)(1)=0 \Rightarrow 2x-4-2x-1=0 \Rightarrow -5 \neq 0$, The solution is $\varnothing$.

(b) The quotient changes sign only when x-values make the numerator or denominator 0. From part (a) we know that there are no x-values that make the numerator equal to 0. Solve $(x-2)^2=0$ to find the value(s) that make the denominator 0. $(x-2)^2=0 \Rightarrow x-2=0 \Rightarrow x=2$

Interval	Test Value	Positive or Negative Result
$(-\infty,2)$	0	Negative
$(2,\infty)$	3	Negative

The solution is $(-\infty,2)\cup(2,\infty)$.

45. (a) The equation $\frac{(x^2+1)(2x)-(x^2-1)(2x)}{(x^2+1)^2}=0$ is true only when $(x^2+1)(2x)-(x^2-1)(2x)=0$.

Solving we have $(x^2+1)(2x)-(x^2-1)(2x)=0 \Rightarrow (2x)(x^2+1-x^2+1)=0 \Rightarrow (2x)(2)=0 \Rightarrow x=0$

The solution is 0.

(b) The quotient changes sign only when x-values make the numerator or denominator 0. From part (a) we know that $x=0$. Solve $(x^2+1)^2=0$ to find the value(s) that make the denominator 0.

$(x^2+1)^2=0 \Rightarrow x^2+1=0 \Rightarrow x^2=-1 \Rightarrow x=\pm i$ Thus, there are no real numbers that make the denominator equal to zero.

Interval	Test Value	Positive or Negative Result
$(-\infty,0]$	−2	Negative
$[0,\infty)$	2	Positive

The solution is $[0,\infty)$.

47. (a) The equation $\frac{(2x+1)(2x)-(x^2+1)(2)}{(2x+1)^2}=0$ is true only when $(2x+1)(2x)-(x^2+1)(2)=0$.

Solving we have $(2x+1)(2x)-(x^2+1)(2)=0 \Rightarrow 4x^2+2x-2x^2-2=0 \Rightarrow 2x^2+2x-2=0 \Rightarrow$

$$x=\frac{-(2)\pm\sqrt{(2)^2-4(2)(-2)}}{2(2)}=\frac{-2\pm\sqrt{4+16}}{4}=\frac{-2\pm\sqrt{20}}{4}=\frac{-2\pm2\sqrt{5}}{4}=\frac{-1\pm\sqrt{5}}{2}$$

The solution is $\frac{-1\pm\sqrt{5}}{2}$.

(b) The quotient changes sign only when x-values make the numerator or denominator 0. From part (a) we know that $x=\frac{-1\pm\sqrt{5}}{2}$. Solve $(2x+1)^2=0$ to find the value(s) that make the denominator 0.

$(2x+1)^2=0 \Rightarrow 2x+1=0 \Rightarrow 2x=-1 \Rightarrow x=-\frac{1}{2}$.

Interval	Test Value	Positive or Negative Result
$\left(-\infty,\frac{-1-\sqrt{5}}{2}\right)$	-2	Positive
$\left(\frac{-1-\sqrt{5}}{2},-\frac{1}{2}\right)$	-1	Negative
$\left(-\frac{1}{2},\frac{-1+\sqrt{5}}{2}\right)$	0	Negative
$\left(\frac{-1+\sqrt{5}}{2},\infty\right)$	2	Positive

The solution is $\left(\frac{-1-\sqrt{5}}{2},-\frac{1}{2}\right)\cup\left(-\frac{1}{2},\frac{-1+\sqrt{5}}{2}\right)$.

49. The numerator is negative and the denominator is always positive; therefore, the quotient is always negative. Because the inequality requires that the rational expression be less than 0, the solution set is $(-\infty,\infty)$.

51. The numerator is negative and the denominator is always positive; therefore, the quotient is always negative. Because the inequality requires that the rational expression be greater than 0, the solution set is $\varnothing$.

53. The numerator is always positive and the denominator is negative; therefore, the quotient is always negative. Because the inequality requires that the rational expression be greater than or equal to 0, the solution set is $\varnothing$.

55. The numerator is always positive and the denominator is always positive; therefore, the quotient is always positive. Because the inequality requires that the rational expression be greater than 0, the solution set is $(-\infty,\infty)$.

57. The numerator is always positive or zero and the denominator is always positive; therefore, the quotient is always positive or zero. Because the inequality requires that the rational expression be less than or equal to 0, the solution set is $\{1\}$. Note that when $x=1$, the rational expression equals 0 and the inequality is satisfied.

59. The graph of $y_1 = (3-2x)/(1+x)$ has a vertical asymptote at $x=-1$ and an x-intercept at $x=\frac{3}{2}$. The graph of y_1 is below the x-axis on the interval $(-\infty, -1)\cup\left(\frac{3}{2}, \infty\right)$.

61. The graph of $y_1 = \frac{(x+1)(x-2)}{x+3}$ has a vertical asymptote at $x=-3$ and $x-$intercepts at $x=-1$ and $x=2$. The graph of $P(3)=0$ is below the $x-$axis on the interval $(-\infty,-3)\cup(-1,2)$.

63. The graph of $y_1 = (x+1)^2/(x-2)$ has a vertical asymptote at $x=2$ and an x-intercept at $x=-1$. The graph of y_1 intersects or is below the x-axis on the interval $(-\infty, 2)$. Note that 2 can not be included.

65. The graph of $y_1 = \frac{2x-5}{(x+1)(x-1)}$ has vertical asymptotes at $x=-1$ and $x=1$ and an x-intercept at $x=\frac{5}{2}$.

The graph of y_1 is on or above the x-axis on the interval $(-1,1)\cup\left[\frac{5}{2},\infty\right)$.

67. First, rewrite the inequality: $\frac{1}{x-3}\le\frac{5}{x-3}\Rightarrow\frac{1}{x-3}-\frac{5}{x-3}\le 0\Rightarrow\frac{-4}{x-3}\le 0$

The graph of $y_1=-4/(x-3)$ has a vertical asymptote at $x=3$ and no x-intercept.

The graph of y_1 intersects or is below the x-axis on the interval $(3,\infty)$. Note that 3 can not be included.

69. First, rewrite the inequality: $2-\frac{5}{x}+\frac{2}{x^2}\ge 0\Rightarrow\frac{2x^2-5x+2}{x^2}\ge 0\Rightarrow\frac{(2x-1)(x-2)}{x^2}\ge 0$

The graph of $y_1=(2x-1)(x-2)/x^2$ has a vertical asymptote at $x=0$ and x-intercepts of $x=\frac{1}{2}$ and $x=2$.

The graph of y_1 intersects or is above the x-axis on the interval $(-\infty,0)\cup\left(0,\frac{1}{2}\right]\cup[2,\infty)$.

71. (a) The graph of $y_1=(\sqrt{2}x+5)/(x^3-\sqrt{3})$ is shown in Figure 71. The x-intercept is approximately -3.54. The solution set of the equation is $\{-3.54\}$.

(b) There is a vertical asymptote when $x^3-\sqrt{3}=0\Rightarrow x^3=\sqrt{3}\Rightarrow x=\sqrt[3]{\sqrt{3}}\approx 1.20$. Thus, the graph is above the x-axis on the interval $(-\infty,-3.54)\cup(1.20,\infty)$. See Figure 71.

(c) The graph is below the x-axis on the interval $(-3.54, 1.20)$. See Figure 71.

73. (a) The graph is shown in Figure 73. The equation of the horizontal asymptote is $y=\frac{10}{1}\Rightarrow y=10$.

(b) The initial insect population occurs when $x=0$. Here $f(0)=\frac{10(0)+1}{(0)+1}=1$ million insects.

(c) After several months, the insect population levels off at 10 million insects.

(d) The horizontal asymptote $y=10$ represents the limiting population after a long time.

[-4.7,4.7] by [-3.1,3.1]
Xscl = 1 Yscl = 1

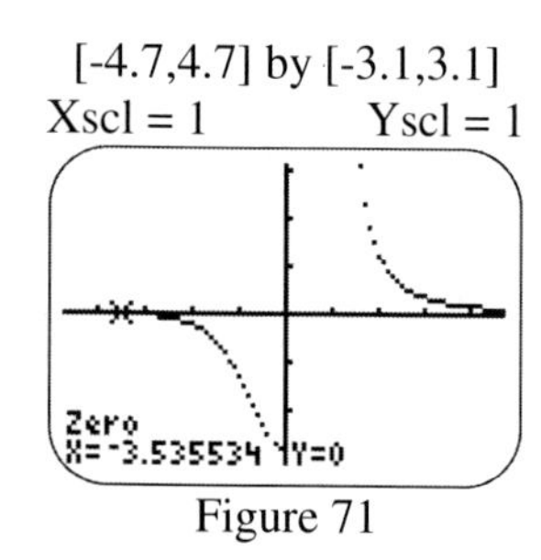

Figure 71

[0,14] by [0,14]
Xscl = 1 Yscl = 1

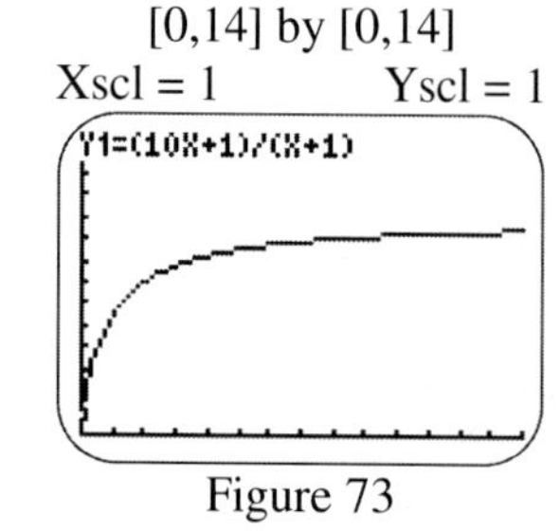

Figure 73

[0,600] by [0,50]
Xscl = 100 Yscl = 5

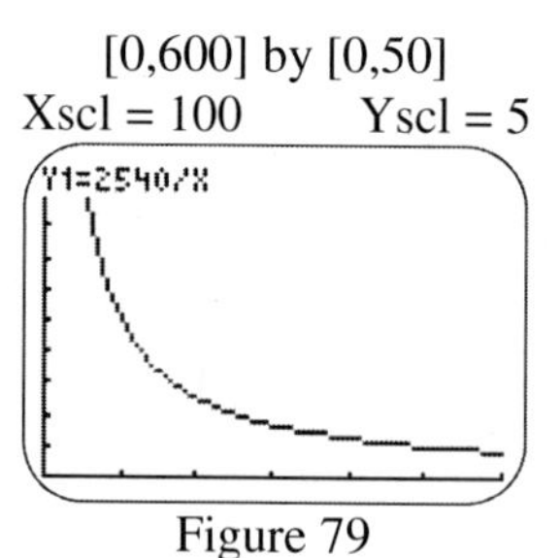

Figure 79

75. (a) $0.25 = \dfrac{x-5}{x^2-10x} \Rightarrow 0.25(x^2-10x) = x-5 \Rightarrow 0.25x^2 - 2.5x = x-5 \Rightarrow 0.25x^2 - 3.5x + 5 = 0 \Rightarrow$

$$x = \frac{-(-3.5) \pm \sqrt{(-3.5)^2 - 4(0.25)(5)}}{2(0.25)} = \frac{3.5 \pm \sqrt{7.25}}{0.5} \approx 1.6 \text{ or } 12.4;$$ Since $x > 10$, $x \approx 12.4$ cars/min.

(b) Since 2 attendants can serve only 10 cars/min, 3 attendants are needed.

77. A surface area of 280 square inches indicates that $2lw + 2lh + 2hw = 280$ or $lw + lh + hw = 140$. Solving this equation for h gives $h = \dfrac{140 - lw}{l + w}$. Since the length is twice the width, $l = 2w$. Then, by substitution, $h = \dfrac{140 - (2w)w}{2w + w} \Rightarrow h = \dfrac{140 - 2w^2}{3w}$. Since the volume is 196 cubic inches, $lwh = 196$. Substituting for l and h yields $(2w)w\left(\dfrac{140 - 2w^2}{3w}\right) = 196 \Rightarrow \dfrac{280w - 4w^3}{3} = 196 \Rightarrow 280w - 4w^3 = 588 \Rightarrow$

$4w^3 - 280w + 588 = 0 \Rightarrow w^3 - 70w + 147 = 0$. Graphing this equation as $y_1 = x^\wedge 3 - 70x + 147$ in $[0,10]$ by $[-150, 450]$ shows the possible values for the width as the x-intercepts (figure not shown). The possible values for the width are 7 inches and approximately 2.266 inches.

If $w \approx 2.266$, then $l \approx 2(2.266) \approx 4.532$ and $h = \dfrac{140 - 2(2.266)^2}{3(2.266)} \approx 19.084$. In this case, the dimensions are approximately 2.266 by 4.532 by 19.084 inches. If $w = 7$, then $l = 14$ and $h = \dfrac{140 - 2(7)^2}{3(7)} = 2$ inches. In this case the dimensions are 7 by 14 by 2 inches.

79. (a) $f(400) = \dfrac{2540}{400} = 6.35$ in. A curve designed for 60 miles per hour with a radius of 400 feet should have the outer rail elevated 6.35 inches.

(b) See Figure 79. As the radius x of the curve increases, the elevation of the outer rail decreases.

(c) The horizontal asymptote is $y = 0$. As the radius of the curve increases without bound $(x \to \infty)$, the tracks become straight and no elevation or banking $(y \to 0)$ is necessary.

(d) $12.7 = \dfrac{2540}{x} \Rightarrow 12.7x = 2540 \Rightarrow x = \dfrac{2540}{12.7} = 200$ ft.

81. (a) $D(0.05)=\dfrac{2500}{30(0.3+0.05)}\approx 238$. The braking distance for a car going 50 mph on a 5% uphill grade is about 238 ft.

(b) As the uphill grade x increases, the braking distance decreases. This agrees with driving experience.

(c) $220=\dfrac{2500}{30(0.3+x)}\Rightarrow 6600(0.3+x)=2500\Rightarrow 0.3+x=\dfrac{2500}{6000}\Rightarrow x=\dfrac{2500}{6600}-0.3\Rightarrow x\approx 0.079$. The grade associated with a braking distance of 220 feet is about 7.9% uphill.

83. Here $r=\dfrac{km^2}{s}$. By substitution, $12=\dfrac{k(6)^2}{4}\Rightarrow 48=36k\Rightarrow k=\dfrac{48}{36}\Rightarrow k=\dfrac{4}{3}$. That gives $r=\dfrac{\frac{4}{3}m^2}{s}$. When $m=4$ and $s=10$, $r=\dfrac{\frac{4}{3}(4)^2}{10}=\dfrac{\frac{64}{3}}{10}=\dfrac{64}{30}=\dfrac{32}{15}$.

85. Here $a=\dfrac{kmn^2}{y^3}$. By substitution, $9=\dfrac{k(4)(9)^2}{(3)^3}\Rightarrow 243=324k\Rightarrow k=\dfrac{243}{324}\Rightarrow k=\dfrac{3}{4}$. That gives $a=\dfrac{\frac{3}{4}mn^2}{y^3}$.

When $m=6$, $n=2$ and $y=5$, $a=\dfrac{\frac{3}{4}(6)(2)^2}{(5)^3}=\dfrac{18}{125}$.

87. For $k>0$, if y varies directly as x, when x increases, y increases, and when x decreases, y decreases.

89. If y is inversely proportional to x, $y=\dfrac{k}{x}$. If x doubles, $y=\dfrac{k}{2x}\Rightarrow y=\dfrac{1}{2}\cdot\dfrac{k}{x}$. That is, y becomes half as much.

91. If y is directly proportional to the third power of x, $y=kx^3$. If x triples, $y=k(3x)^2\Rightarrow y=27\cdot kx^3$. That is, y becomes 27 times as much.

93. Here $BMI=\dfrac{kw}{h^2}$. By substitution, $24=\dfrac{k(177)}{(72)^2}\Rightarrow 177k\Rightarrow k=\dfrac{124,416}{177}\Rightarrow k\approx 703$. That gives $BMI\approx\dfrac{703w}{h^2}$. When $w=130$ and $h=66$, $BMI\approx\dfrac{703(130)}{(66)^2}\approx 21$.

95. Here $R=\dfrac{k}{d^2}$. By substitution, $0.5=\dfrac{k}{(2)^2}\Rightarrow 2=k$. That gives $R=\dfrac{2}{d^2}$. When $d=3$, $R=\dfrac{2}{(3)^2}\approx\dfrac{2}{9}$ ohm.

97. Here $W=\dfrac{k}{d^2}$. By substitution, $160=\dfrac{k}{(4000)^2}\Rightarrow 2,560,000,000=k$. That gives $W=\dfrac{2,560,000,000}{d^2}$. Note that 8000 miles above Earth's surface, $d=12,000$. When $d=12,000$, $W=\dfrac{2,560,000,000}{(12,000)^2}=17.8$ lb.

99. Here $V=kr^2h$. By substitution, $300=k(3)^2(10.62)\Rightarrow\dfrac{300}{95.58}=k\Rightarrow k\approx 3.1387$. That gives $V=3.1387r^2h$. When $r=4$ and $h=15.92$, $V\approx 3.1387(4)^2(15.92)\approx 799.5$. (Note that the actual value of k is π.)

101. Tooney can clean the entire mess in 15 minutes and can complete $\frac{t}{15}$ of the mess in t minutes. Mudcat can clean the entire mess in12 minutes and can complete $\frac{t}{12}$ of the mess in t minutes. Together they can complete $\frac{t}{15}+\frac{t}{12}$ of the mess in t minutes. The job is complete when the fraction of the mess reaches 1.

$\frac{t}{15}+\frac{t}{12}=1 \Rightarrow 12t+15t=180 \Rightarrow 27t=180 \Rightarrow t=6$ minutes and 40 seconds.

103. Mrs. Schmulen can grade tests in 5 hours and can complete $\frac{t}{5}$ of the tests in t hours. Mr. Elwyn and Mrs. Schmulen can grade the tests together in 3 hours and can complete $\frac{t}{3}$ of the tests in t hours.

The job is complete when the fraction of the completed tests reaches 1.

$\frac{t}{3}-\frac{t}{5}=1 \Rightarrow 5t-3t=15 \Rightarrow 2t=15 \Rightarrow t=7$ hours and 30 minutes.

105. The inlet pipe can fill the vat in 5 hours and can fill $\frac{t}{5}$ of the vat in t hours. The outlet pipe can empty the vat in 10 hours and can empty $\frac{t}{10}$ of the vat in t hours. The job is complete when the fraction of the full tank reaches 1. $\frac{t}{5}-\frac{t}{10}=1 \Rightarrow 10t-5t=50 \Rightarrow 5t=50 \Rightarrow t=10$ hours.

107. Let x = the amount of time required to empty the pool then (x-20) will be the amount of time required to fill the pool. Then the rate of emptying the pool is given by $\frac{1}{x}$ and the rate of filling of the pool is $\frac{1}{x-20}$.

When both pipes are open it takes 4 hours (or 240 minutes) to fill the pool. The rate of filling the pool when both pipes are open is given by $\frac{1}{240}$.

$\frac{1}{x-20}-\frac{1}{x}=\frac{1}{240} \Rightarrow 240x-240(x-20)=x(x-20) \Rightarrow x^2-20x-4800=0 \Rightarrow (x-80)(x+60)=0 \Rightarrow x=80.$

Therefore, it will take 80 minutes to empty with the outlet pipe and 60 minutes to fill with the inlet pipe.

Reviewing Basic Concepts (Sections 4.1—4.3)

1. A sketch of the graph is shown in Figure 1a. A graph of $y_1=1/(x+2)-3$ is shown in Figure 1b.
2. For the function to be defined, the denominator cannot equal 0. $x^2-1 \neq 0 \Rightarrow x^2 \neq 1 \Rightarrow x \neq \pm 1$. The domain is $(-\infty,-1)\cup(-1,1)\cup(1,\infty)$.
3. A vertical asymptote occurs when the denominator equals 0. $x-6=0 \Rightarrow x=6$.
4. Since the degrees of the numerator and the denominator are the same, the ratio of the leading coefficients of the numerator and denominator is used to find the equation of the horizontal asymptote. $y=\frac{1}{1} \Rightarrow y=1$.

5. Perform long division:

$$\begin{array}{r} x-2 \\ x+3\overline{)x^2+\ \ x+5} \\ \underline{x^2+3x} \\ -2x+5 \\ \underline{-2x-6} \\ 11 \end{array} \quad \Rightarrow \quad \frac{x^2+x+5}{x+3}=x-2+\frac{11}{x+3}$$

The oblique asymptote is $y=x-2$.

6. A sketch of the graph is shown in Figure 6a. A graph of $y_1=(3x+6)/(x-4)$ $y_1=1/(x+2)-3$ is shown in Figure 6b.

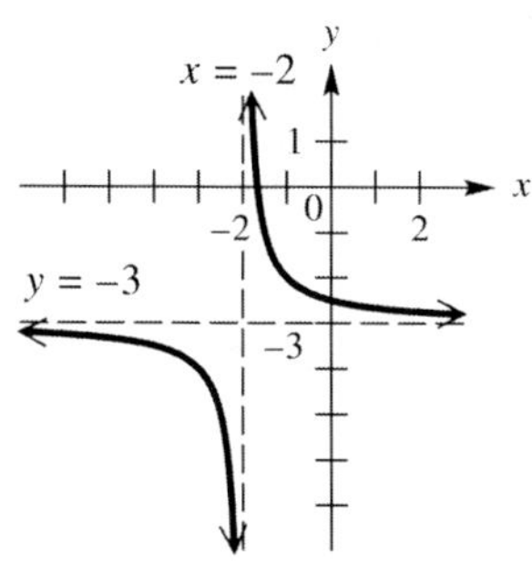

Figure 1a

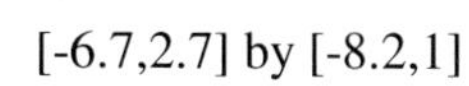
[-6.7,2.7] by [-8.2,1]

Xscl = 1 Yscl = 1

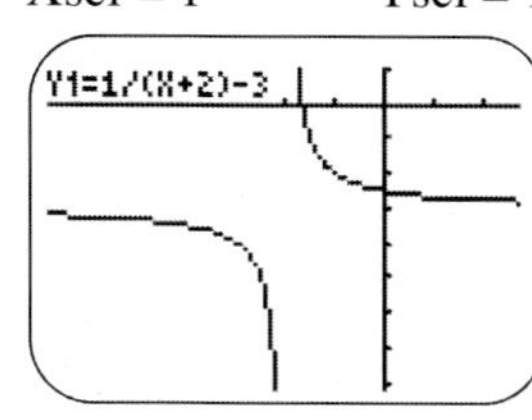

Figure 1b

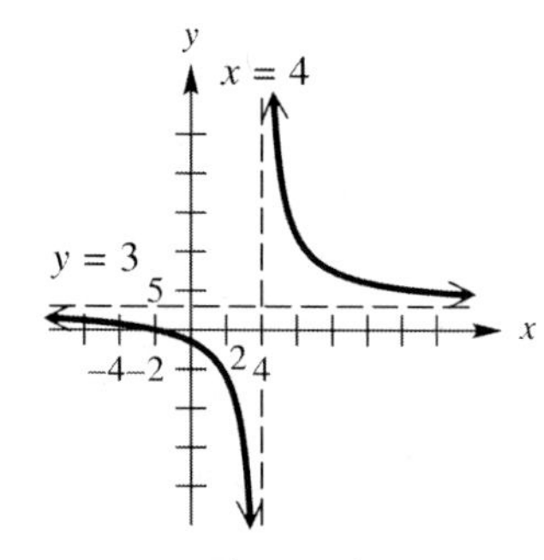

Figure 6a

[-6.2,12.4] by [-18.6,18.6]

Xscl = 1 Yscl=1

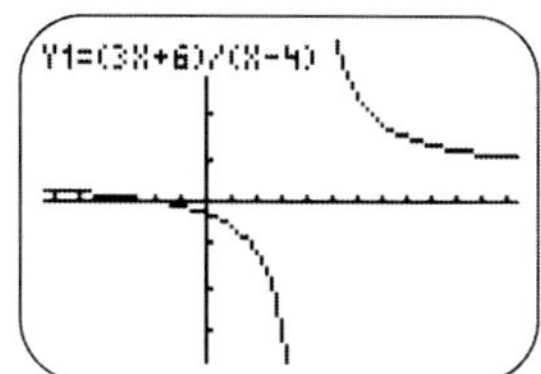

Figure 6b

7. (a) The graph intersects the x-axis at $x=-4$. The solution set is $\{-4\}$.

   (b) The graph is above the x-axis on the interval $(-\infty,-4)\cup(2,\infty)$. The solution set is $(-\infty,-4)\cup(2,\infty)$.

   (c) The graph is below the x-axis on the interval $(-4,2)$. The solution set is $(-4,2)$.

8. First, rewrite the inequality. $\dfrac{x+4}{3x+1}>1\Rightarrow\dfrac{x+4}{3x+1}-1>0\Rightarrow\dfrac{x+4-3x-1}{3x+1}>0\Rightarrow\dfrac{-2x+3}{3x+1}>0$. The graph of $y_1=(-2x+3)/(3x+1)$ has a vertical asymptote at $x=-\dfrac{1}{3}$ and an x-intercept at $x=\dfrac{3}{2}$. The graph of y_1 is above the x-axis on the interval $\left(-\dfrac{1}{3},\dfrac{3}{2}\right)$.

9. The base of the parallelogram varies inversely as its height. The constant of variation is 24.

10. Here $S = \frac{kWT^2}{L}$ By substitution, $600 = \frac{k(4)(4)^2}{50} = \frac{64k}{50} \Rightarrow k = 468.75$. When $W = 2$, $T = 2$, $L = 50$

$S = \frac{468.75(2)(2)^2}{(50)^2} = 75$ pounds.

4.4: Functions Defined by Powers and Roots

1. Since $13^2 = 169, \sqrt{169} = 13$.

3. Since $(-2)^5 = -32, \sqrt[5]{-32} = -2$.

5. $81^{3/2} = (\sqrt{81})^3 = 9^3 = 729$.

7. $125^{-2/3} = \frac{1}{125^{2/3}} = \frac{1}{\left(\sqrt[3]{125}\right)^2} = \frac{1}{5^2} = \frac{1}{25}$.

9. $(-1000)^{2/3} = (\sqrt[3]{-1000})^2 = (-10)^2 = 100$.

11. $8^{2/3} = (8^{1/3})^2 = 2^2 = 4$.

13. $16^{3/4} = (16^{1/4})^{-3} = (2)^{-3} = \frac{1}{2^3} = \frac{1}{8}$.

15. $-81^{0.5} = -81^{1/2} = -\sqrt{81} = -9$.

17. $64^{1/6} = \sqrt[6]{64} = 2$.

19. $(-9^{3/4})^2 = (9^{3/4})^2 = 9^{3/2} = (\sqrt{9})^3 = 3^3 = 27$.

21. $\sqrt[3]{2x} = (2x)^{1/3}$.

23. $\sqrt[3]{z^5} = z^{5/3}$.

25. $(\sqrt[4]{y})^{-3} = (y^{1/4})^{-3} = y^{-3/4} = \frac{1}{y^{3/4}}$.

27. $\sqrt{x} \cdot \sqrt[3]{x} = x^{1/2} \cdot x^{1/3} = x^{1/2+1/3} = x^{5/6}$.

29. $\sqrt{y \cdot \sqrt{y}} = (y \cdot y^{1/2})^{1/2} = (y^{3/2})^{1/2} = y^{3/4}$.

31. $\sqrt[3]{(-4)} \approx -1.587401052$.

33. $\sqrt[3]{(-125)} = -5$.

35. $\sqrt[3]{(-17)} \approx -2.571281591$.

37. $\sqrt[6]{(\pi^2)} \approx 1.464591888$.

39. $13^{-1/3} \approx 0.4252903703$.

41. $32^{0.2} = 2$.

43. $(5/6)^{-1.3} \approx 1.267463962.$

45. $\pi^{-3} \approx 0.0322515344.$

47. $f(x) = x^{1.62} \Rightarrow f(1.2) = 1.2^{1.62} \approx 1.34.$

49. $f(x) = x^{3/2} - x^{1/2} \Rightarrow f(50) = 50^{3/2} - 50^{1/2} \approx 346.48.$

51. See Figure 51.

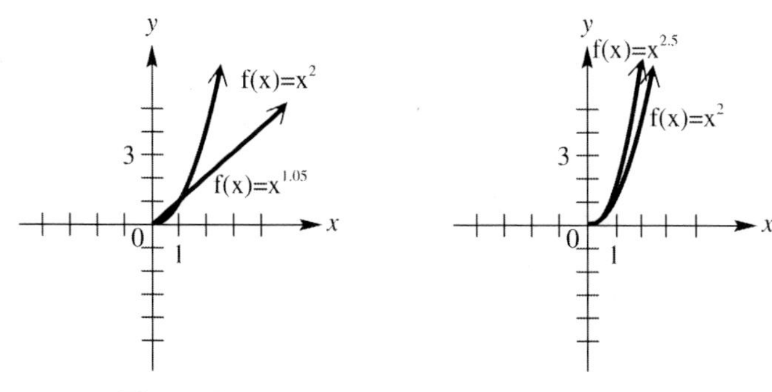

Figure 51 Figure 53

53. See Figure 53.

55. (a) $16^{-3/4} = \dfrac{1}{16^{3/4}} = \dfrac{1}{(\sqrt[4]{16})^3} = \dfrac{1}{(2)^3} = \dfrac{1}{8} = 0.125.$

(b) $16^{-3/4} = (\sqrt[4]{16})^{-3} = 0.125$ and $16^{-3/4} = \sqrt[4]{16^{-3}} = 0.125.$ Other expressions are possible.

(c) A calculator will show that $0.125 = \dfrac{1}{8}.$

57. See Figure 57.

58. See Figure 58.

59. See Figure 59.

60. See Figure 60.

[0,100] by [0,3]
Xscl = 10 Yscl= 1

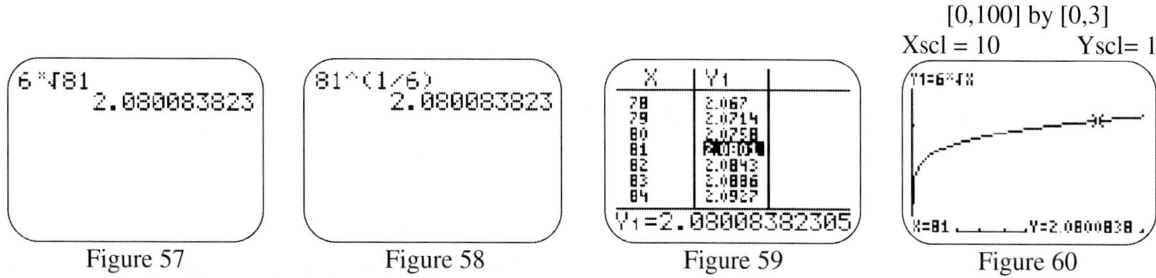

Figure 57 Figure 58 Figure 59 Figure 60

61. $S(4) = 1.27(4)^{2/3} \approx 3.2\ ft^2.$

63. $f(15) = 15^{1.5} \approx 58.1\ yr.$

65. (a) From the hint, $a(1)^b = 1960 \Rightarrow a = 1960.$

(b) Since $f(3) = 525, 1960(3)^b = 525 \Rightarrow 3^b = \dfrac{525}{1960} \Rightarrow 3^b \approx 0.278.$ By trial and error, $b \approx -1.2.$

(c) $f(4) = 1960(4)^{-1.2} \approx 371$. If the zinc ion concentration reaches 371 mg per L, a rainbow trout will survive, on average, 4 min.

67. $f(2) = 0.445(2)^{1.25} \approx 1.06$ g.

69. Using the Power Regression feature on a graphing calculator yields $a \approx 874.54$ and $b \approx -0.49789$.

71. $S(0.5) = 0.2(0.5)^{\frac{2}{3}} \approx 0.126$. The surface area of the wings of a 0.5 kg bird is approximately 0.126 sq. meters.

73. $5 + 4x \geq 0 \Rightarrow 4x \geq -5 \Rightarrow x \geq -\frac{5}{4}$. The domain is $\left[-\frac{5}{4}, \infty\right)$.

75. $6 - x \geq 0 \Rightarrow 6 \geq x \Rightarrow x \leq 6$. The domain is $(-\infty, 6)$.

77. The cube root function is defined for all values of x. The domain is $(-\infty, \infty)$.

79. The graph of $y = 49 - x^2 = (7 + x)(7 - x)$ is a parabola that opens downward with x-intercepts -7 and

$8 = \frac{k\left(1^4\right)}{9^2} \Rightarrow 648 = k \Rightarrow k = 648$. This graph intersects or is above the x-axis for values of x in the interval $[-7, 7]$. That is, $49 - x^2 \geq 0$ on the interval $[-7, 7]$. The domain is $[-7, 7]$.

81. The graph of $y = x^3 - x = x(x+1)(x-1)$ is a cubic graph with x-intercepts $-1, 0$ and 1. This graph intersects or is above the x-axis for values of x in the interval $[-1, 0] \cup [1, \infty)$. That is, $x^3 - x \geq 0$ on the interval $[-1, 0] \cup [1, \infty)$. The domain is $[-1, 0] \cup [1, \infty)$.

83. The graph is shown in Figure 83.
   (a) The range is $[0, \infty)$.
   (b) The function is increasing on the interval $\left(-\frac{5}{4}, \infty\right)$.
   (c) The function is never decreasing.
   (d) The graph intersects the x-axis when $x = -1.25$. The solution set of $f(x) = 0$ is $\{-1.25\}$.

85. The graph is shown in Figure 85.
   (a) The range is $(-\infty, 0]$.
   (b) The function is increasing on the interval $(-\infty, 6)$.
   (c) The function is never decreasing.
   (d) The graph intersects the x-axis when $x = 6$. The solution set of $f(x) = 0$ is $\{6\}$.

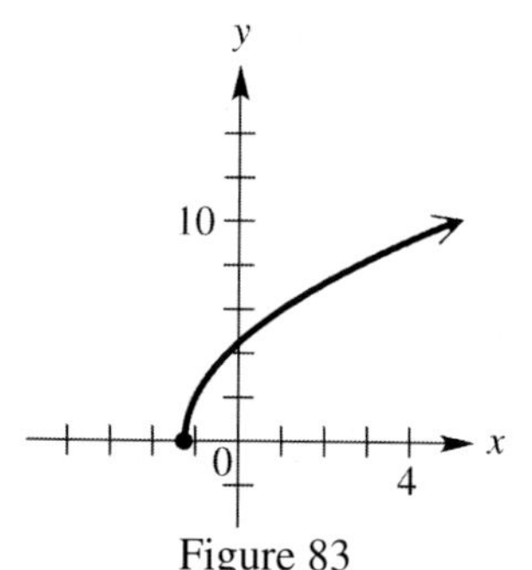

Figure 83

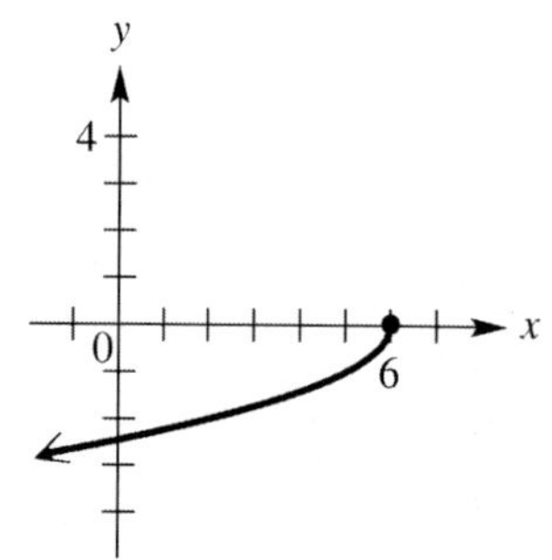

Figure 85

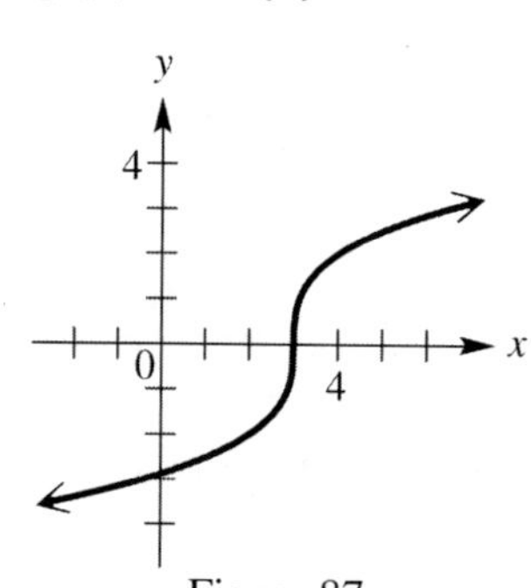

Figure 87

87. The graph is shown in Figure 87.
    (a) The range is $(-\infty,\infty)$.
    (b) The function is increasing on the interval $(-\infty,\infty)$.
    (c) The function is never decreasing.
    (d) The graph intersects the x-axis when $x = 3$. The solution set of $f(x) = 0$ is $\{3\}$.

89. The graph is shown in Figure 89.
    (a) The range is $[0, 7]$.
    (b) The function is increasing on the interval $(-7, 0)$.
    (c) T he function is decreasing on the interval $(0, 7)$.
    (d) The graph intersects the x-axis when $x = -7$ or $x = 7$. The solution set of $f(x) = 0$ is $\{-7, 7\}$.

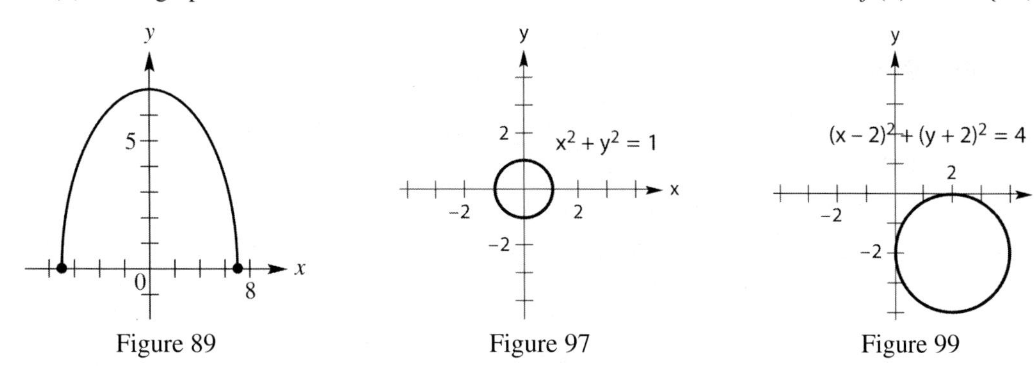

Figure 89 Figure 97 Figure 99

91. Since $y = \sqrt{9x+27} = \sqrt{9(x+3)} = 3\sqrt{x+3}$, the graph can be obtained by shifting the graph of $y = \sqrt{x}$ to the left 3 units and vertically stretching by a factor of 3.

93. Since $y = \sqrt{4x+16} + 4 = \sqrt{4(x+4)} + 4 = 2\sqrt{x+4} + 4$, the graph can be obtained by shifting the graph of $y = \sqrt{x}$ to the left 4 units, vertically stretching by a factor of 2, and shifting upward 4 units.

95. Since $y = \sqrt{18-9x} + 1 = \sqrt{9(2-x)} + 1 = 3\sqrt{2-x} + 1$ the graph can be obtained by reflecting the graph of $y = \sqrt{x}$ across the y-axis, shifting 2 units to the right, vertically stretching by a factor of 3, and shifting upward 1 unit.

97. See Figure 97.

99. See Figure 99.

101. See Figure 101.

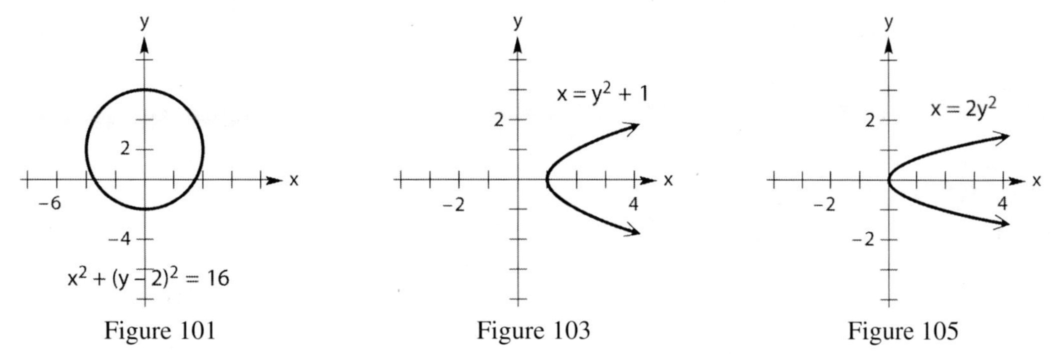

Figure 101 Figure 103 Figure 105

103. See Figure 103.

105. See Figure 105.

107. See Figure 107.

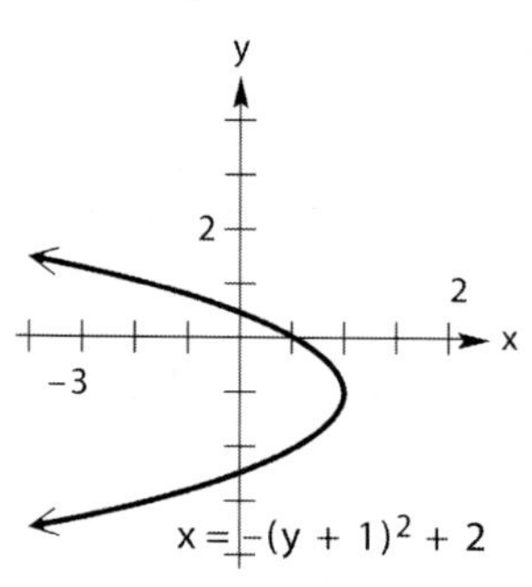

Figure 107

109. The graph is a circle. See Figure 109.

$x^2 + y^2 = 100 \Rightarrow y^2 = 100 - x^2 \Rightarrow y = \pm\sqrt{100 - x^2}$; Thus $y_1 = \sqrt{100 - x^2}$ and $y_2 = -\sqrt{100 - x^2}$.

111. The graph is a (shifted) circle. See Figure 111.

$(x-2)^2 + y^2 = 9 \Rightarrow y^2 = 9 - (x-2)^2 \Rightarrow y = \pm\sqrt{9 - (x-2)^2}$; Thus $y_1 = \sqrt{9 - (x-2)^2}$ and $y_1 = -\sqrt{9 - (x-2)^2}$.

113. The graph is a horizontal parabola. See Figure 113.

$x = y^2 + 6y + 9 \Rightarrow x = (y+3)^2 \Rightarrow \pm\sqrt{x} = y + 3 \Rightarrow y = -3 \pm \sqrt{x}$; Thus $y_1 = -3 + \sqrt{x}$ and $y_2 = -3 - \sqrt{x}$.

[-15,15] by [-10,10]

Xscl = 1 Yscl = 1

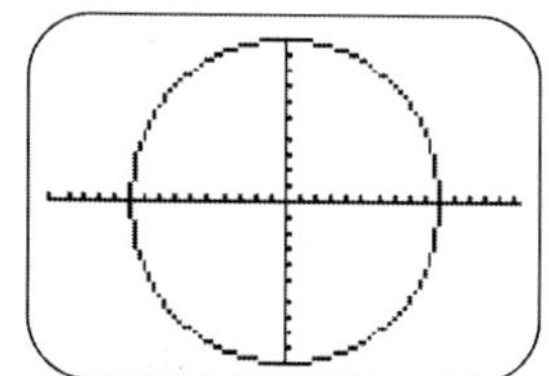

Figure 109

[-9.4,9.4] by [-6.2,6.2]

Xscl = 1 Yscl = 1

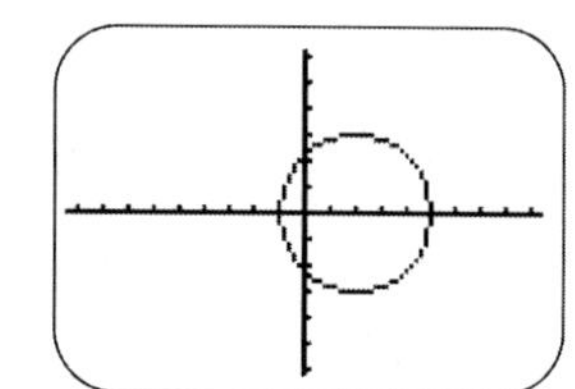

Figure 111

[-10,10] by [-10,10]

Xscl= 1 Yscl = 1

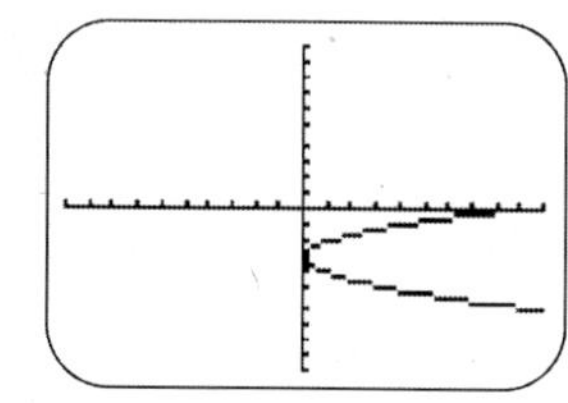

Figure 113

[-10,10] by [-10,10]

Xscl = 1 Yscl= 1

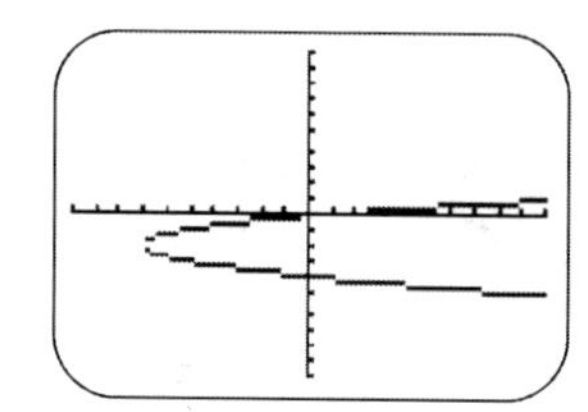

Figure 115

115. The graph is a horizontal parabola. See Figure 115.

$x = 2y^2 + 8y + 1 \Rightarrow 0.5x = y^2 + 4y + 0.5 \Rightarrow 0.5x = (y^2 + 4y + 4) + 0.5 - 4 \Rightarrow 0.5x = (y+2)^2 - 3.5 \Rightarrow$

$0.5x + 3.5 = (y+2)^2 \Rightarrow \pm\sqrt{0.5x + 3.5} = y + 2 \Rightarrow y = -2 \pm \sqrt{0.5x + 3.5}$

Thus $y_1 = -2 + \sqrt{0.5x + 3.5}$ and $y_2 = -2 - \sqrt{0.5x + 3.5}$.

4.5: Equations, Inequalities, and Applications Involving Root Functions

1. A sketch is shown in Figure 1. There is one real solution.

$$\sqrt{x} = 2x - 1 \Rightarrow (\sqrt{x})^2 = (2x-1)^2 \Rightarrow x = 4x^2 - 4x + 1 \Rightarrow 4x^2 - 5x + 1 = 0 \Rightarrow (4x-1)(x-1) = 0 \Rightarrow x = \frac{1}{4} \text{ or } 1.$$

The solution set is $\{1\}$. The value $\frac{1}{4}$ is extraneous.

3. A sketch is shown in Figure 3. There is one real solution.

$\sqrt{x} = -x+3 \Rightarrow (\sqrt{x})^2 = (-x+3)^2 \Rightarrow x = x^2 - 6x + 9 \Rightarrow x^2 - 7x + 9 = 0 \Rightarrow$

$$x = \frac{-(-7) \pm \sqrt{(-7)^2 - 4(1)(9)}}{2(1)} = \frac{7 \pm \sqrt{13}}{2}.$$

The solution set is $\{11\}$. The value $\left\{\frac{7-\sqrt{13}}{2}\right\}$ is extraneous.

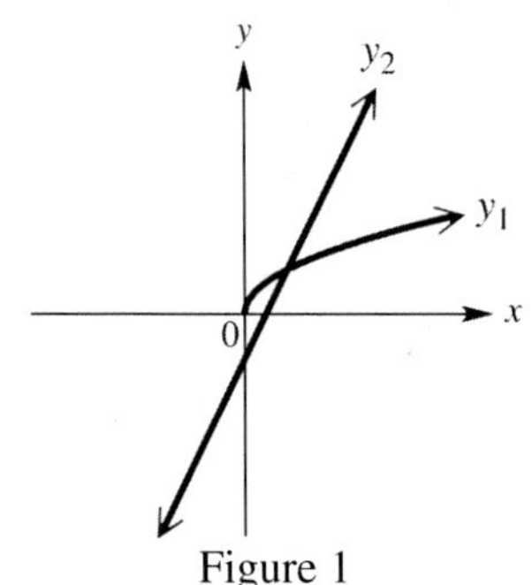

Figure 1

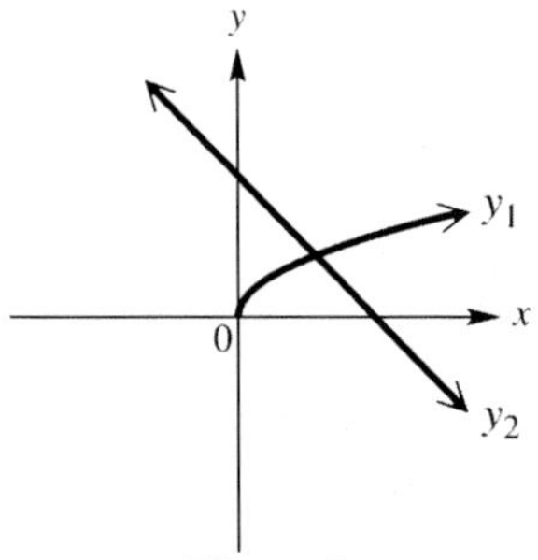

Figure 3

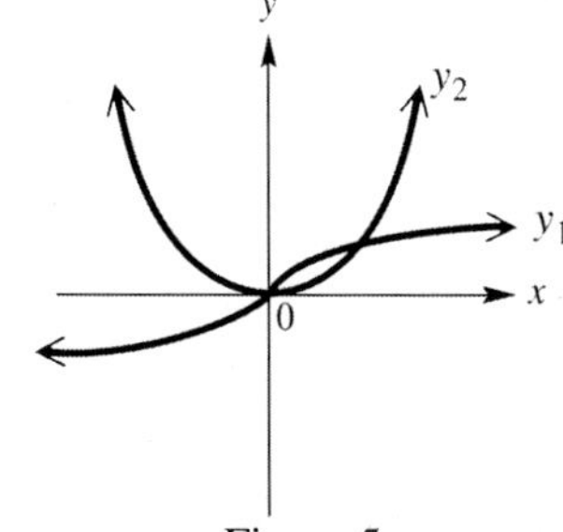

Figure 5

5. A sketch is shown in Figure 5. There are two real solutions. There are no extraneous values.

$\sqrt[3]{x} = x^2 \Rightarrow (\sqrt[3]{x})^3 = (x^2)^3 \Rightarrow x = x^6 \Rightarrow x^6 - x = 0 \Rightarrow x(x^5 - 1) = 0 \Rightarrow x = 0$ or 1.

The solution set is $\{0,1\}$.

7. Substitute $x = \frac{3}{2}$: $15\left(\frac{3}{2}\right)^{-2} - 19\left(\frac{3}{2}\right)^{-1} + 6 = 0 \Rightarrow 15\left(\frac{2}{3}\right)^2 - 19\left(\frac{2}{3}\right) + 6 = 0 \Rightarrow 15\left(\frac{4}{9}\right) - 19\left(\frac{2}{3}\right) + 6 = 0 \Rightarrow$

$\left(\frac{20}{3}\right) - \left(\frac{38}{3}\right) + \left(\frac{18}{3}\right) = 0 \Rightarrow \frac{0}{3} = 0$ The value of $x = \frac{3}{2}$ is a correct solution.

Substitute $x = \frac{5}{3}$: $15\left(\frac{5}{3}\right)^{-2} - 19\left(\frac{5}{3}\right)^{-1} + 6 = 0 \Rightarrow 15\left(\frac{3}{5}\right)^2 - 19\left(\frac{3}{5}\right) + 6 = 0 \Rightarrow 15\left(\frac{9}{25}\right) - 19\left(\frac{3}{5}\right) + 6 = 0 \Rightarrow$

$\left(\frac{27}{5}\right) - \left(\frac{57}{5}\right) + \frac{30}{5} = 0 \Rightarrow \frac{0}{5} = 0$ The value of $x = \frac{5}{3}$ is a correct solution.

9. $\sqrt{3x-8} = x-4 \Rightarrow 3x-8 = (x-4)^2 \Rightarrow 3x-8 = x^2 - 8x + 16 \Rightarrow x^2 - 11x + 24 = 0 \Rightarrow$

$(x-8)(x-3) = 0 \Rightarrow x = 8$ or 3

Check: $\sqrt{3(8)-8} = 8-4 \Rightarrow \sqrt{16} = 4 \Rightarrow 4 = 4$.

Check: $\sqrt{3(3)-8} = 3-4 \Rightarrow \sqrt{1} = -1 \Rightarrow 1 \neq -1$. The solution set is $\{8\}$.

11. $\sqrt{x+5} + 1 = x \Rightarrow \sqrt{x+5} = x-1 \Rightarrow x+5 = (x-1)^2 \Rightarrow$

$x+5 = x^2 - 2x + 1 \Rightarrow x^2 - 3x - 4 = 0 \Rightarrow (x-4)(x+1) = 0 \Rightarrow x = -1$ or $x = 4$.

Check: $\sqrt{-1+5} + 1 = -1 \Rightarrow \sqrt{4} + 1 = -1 \Rightarrow x = -1 \Rightarrow 3 \neq -1$ (not a solution).

Check: $\sqrt{4+5} + 1 = 4 \Rightarrow \sqrt{9} + 1 = 4 \Rightarrow 4 = 4$. The solution set is $\{4\}$.

13. $\sqrt{2x+3}-\sqrt{x+1}=1 \Rightarrow \sqrt{2x+3}=\sqrt{x+1}+1 \Rightarrow 2x+3=\left(\sqrt{x+1}+1\right)^2 \Rightarrow$

$2x+3=x+12\sqrt{x+1}+1 \Rightarrow 2x+3=x+2+2\sqrt{x+1} \Rightarrow x+1=2\sqrt{x+1} \Rightarrow$

$(x+1)^2=4(x+1) \Rightarrow x^2+2x+1=4x+4 \Rightarrow x^2-2x-3=0 \Rightarrow (x-3)(x+1)=0 \Rightarrow \; x=-1$ or $x=3$

Check: $\sqrt{2(-1)+3}-\sqrt{(-1)+1}=1 \Rightarrow \sqrt{1}-\sqrt{0}=1 \Rightarrow 1=1.$

Check: $\sqrt{2(3)+3}-\sqrt{3+1}=1 \Rightarrow \sqrt{9}-\sqrt{4}=1 \Rightarrow 3-2=1 \Rightarrow 1=1$; The solution set is $\{-1,3\}$.

15. $\sqrt[3]{x+1}=-3 \Rightarrow x+1=(-3)^3 \Rightarrow x+1=-27 \Rightarrow x=-28.$

Check: $\sqrt[3]{-28+1}=\sqrt[3]{-27}=-3.$ The solution set is $\{-28\}$.

17. $\sqrt[3]{3x^2+7}=\sqrt[3]{7-4x} \Rightarrow 3x^2+7=7-4x \Rightarrow 3x^2+4x=0 \Rightarrow x(3x+4)=0 \Rightarrow x=0$ or $x=-\dfrac{4}{3}$

Check: $\sqrt[3]{3(0)^2+7}=\sqrt[3]{7-4(0)} \Rightarrow \sqrt[3]{7}=\sqrt[3]{7}$

Check: $\sqrt[3]{3\left(-\dfrac{4}{3}\right)^2+7}=\sqrt[3]{7-4\left(-\dfrac{4}{3}\right)} \Rightarrow \sqrt[3]{\dfrac{37}{3}}=\sqrt[3]{\dfrac{37}{3}}$ The solution set is $\left\{0,-\dfrac{4}{3}\right\}$.

19. $\sqrt[4]{x-2}+4=6 \Rightarrow \sqrt[4]{x-2}=2 \Rightarrow x-2=(2)^4 \Rightarrow x-2=16 \Rightarrow x=18.$

Check: $\sqrt[4]{18-2}+4=6 \Rightarrow \sqrt[4]{16}+4=6 \Rightarrow 6=6.$ The solution set is $\{18\}$.

21. $x^{2/5}=4 \Rightarrow (x^{2/5})^{5/2}=4^{5/2} \Rightarrow x=(\sqrt{4})^5 \Rightarrow x=(\pm 2)^5 \Rightarrow x=\pm 32.$

Check: $(\pm 32)^{2/5}=(\sqrt[5]{\pm 32})^2=(\pm 2)^2=4.$ The solution set is $\{-32,32\}$.

23. $2x^{1/3}-5=1 \Rightarrow 2x^{1/3}=6 \Rightarrow x^{1/3}=3 \Rightarrow (x^{1/3})^3=3^3 \Rightarrow x=27.$

Check: $2(27)^{1/3}-5=2\sqrt[3]{27}-5=2(3)-5=6-5=1.$ The solution set is $\{27\}$.

25. $x^{-2}+3x^{-1}+2=0 \Rightarrow (x^{-1})^2+3(x^{-1})+2=0$, let $u=x^{-1}$, then $u^2+3u+2=0 \Rightarrow (u+2)(u+1)=0 \Rightarrow u=-2$

or $u=-1$. Because $u=x^{-1}$, it follows that $x=u^{-1}$. Thus $x=(-2)^{-1} \Rightarrow x=\dfrac{1}{(-2)} \Rightarrow x=-\dfrac{1}{2}$ or

$x=(-1)^{-1} \Rightarrow x=\dfrac{1}{(-1)}=-1$. Therefore, $x=-1$ or $-\dfrac{1}{2}$.

Check: $(-1)^{-2}+3(-1)^{-1}+2=1-3+2=0$; $\left(-\dfrac{1}{2}\right)^{-2}+3\left(-\dfrac{1}{2}\right)^{-1}+2=4-6+2=0.$

The solution set is $\left\{-1,-\dfrac{1}{2}\right\}$.

27. $5x^{-2}+13x^{-1}=28 \Rightarrow 5(x^{-1})^2+13(x^{-1})-28=0$, let $u=x^{-1}$, then

$5u^2+13u-28=0 \Rightarrow (5u-7)(u+4)=0 \Rightarrow u=\frac{7}{5}$ or $u=-4$. Because $u=x^{-1}$, it follows that $x=u^{-1}$. Thus

$x=\left(\frac{7}{5}\right)^{-1} \Rightarrow x=\frac{5}{7}$ or $x=(-4)^{-1} \Rightarrow x=-\frac{1}{4}$. Therefore, $x=-\frac{1}{4}$ or $\frac{5}{7}$.

Check: $5\left(-\frac{1}{4}\right)^{-2}+13\left(-\frac{1}{4}\right)^{-1}=80-52=28$; $5\left(\frac{5}{7}\right)^{-2}+13\left(\frac{5}{7}\right)^{-1}=\frac{49}{5}+\frac{91}{5}=\frac{140}{5}=28$.

The solution set is $\left\{-\frac{1}{4},\frac{5}{7}\right\}$.

29. $x^{2/3}-x^{1/3}-6=0 \Rightarrow (x^{1/3})^2-x^{1/3}-6=0;$, let $u=x^{1/3}$, then $u^2-u-6=0 \Rightarrow (u-3)(u+2)=0 \Rightarrow u=-2$ or $u=3$. Because $u=x^{1/3}$, it follows that $x=u^3$. Thus $x=(-2)^3 \Rightarrow x=-8$ or $x=3^3 \Rightarrow x=27$. Therefore, $x=-8$ or 27.

Check: $(-8)^{2/3}-(-8)^{1/3}-6=4+2-6=0$; $(27)^{2/3}-(27)^{1/3}-6=9-3-6=0$. The solution set is $\{-8,27\}$.

31. $x^{3/4}-x^{1/2}-x^{1/4}+1=0 \Rightarrow (x^{1/4})^3-(x^{1/4})^2-(x^{1/4})+1=0;$ let $u=x^{1/4}$, then

$u^3-u^2-u+1=0 \Rightarrow (u^3-u^2)-(u-1)=0 \Rightarrow u^2(u-1)-1(u-1)=0 \Rightarrow$

$(u^2-1)(u-1)=0 \Rightarrow (u+1)(u-1)(u-1)=0 \Rightarrow u=-1$ or $u=1$. Because $u=x^{1/4}$ it follows that $x=u^4$. Thus $x=(-1)^4 \Rightarrow x=1$ or $x=(1)^4 \Rightarrow x=1$. Therefore, $x=1$.

Check: $(1)^{3/4}-(1)^{1/2}-(1)^{1/4}+1=1-1-1+1=0$; The solution set is $\{1\}$.

33. (a) $\sqrt{3x+7}=2 \Rightarrow (\sqrt{3x+7})^2=2^2 \Rightarrow 3x+7=4 \Rightarrow 3x=-3 \Rightarrow x=\frac{-3}{3} \Rightarrow x=-1$. The solution set is $\{-1\}$.

A graph of $y_1=\sqrt{3x+7}$ and $y_2=2$ is shown in Figure 33.

(b) From the graph, $y_1>y_2$ on the interval $(-1,\infty)$.

(c) Since the graph of y_1 begins when $3x+7=0 \Rightarrow x=-\frac{7}{3}$, $y_1<y_2$ on the interval $\left[-\frac{7}{3},-1\right)$.

35. (a) $\sqrt{4x+13}=2x-1 \Rightarrow (\sqrt{4x+13})^2=(2x-1)^2 \Rightarrow 4x+13=4x^2-4x+1 \Rightarrow$

$4x^2-8x-12=0 \Rightarrow x^2-2x-3=0 \Rightarrow (x+1)(x-3)=0 \Rightarrow x=-1$ or 3 (-1 is extraneous). The solution set is $\{3\}$. A graph of $y_1=\sqrt{4x+13}$ and $y_2=2x-1$ is shown in Figure 35.

(b) Since the graph of y_1 begins when $4x+13=0 \Rightarrow x=-\frac{13}{4}$, $y_1>y_2$ on the interval $\left[-\frac{13}{4},3\right)$.

(c) From the graph, $y_1<y_2$ on the interval. $(3,\infty)$.

[-4.7,4.7] by [-2.1,4.1]
Xscl = 1 Yscl = 1

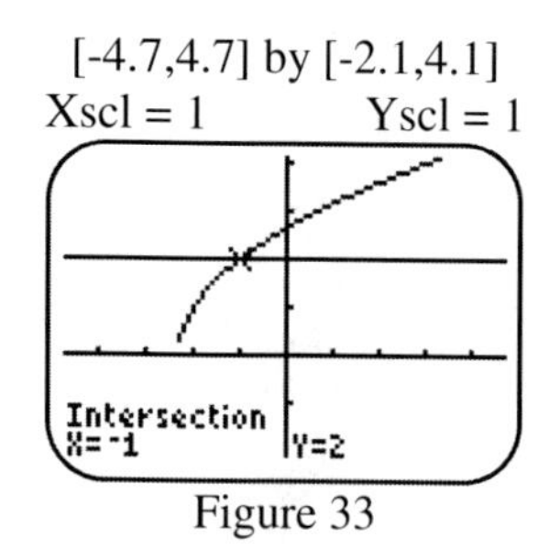

Figure 33

[-4.7,4.7] by [-4.2,8.2]
Xscl = 1 Yscl = 1

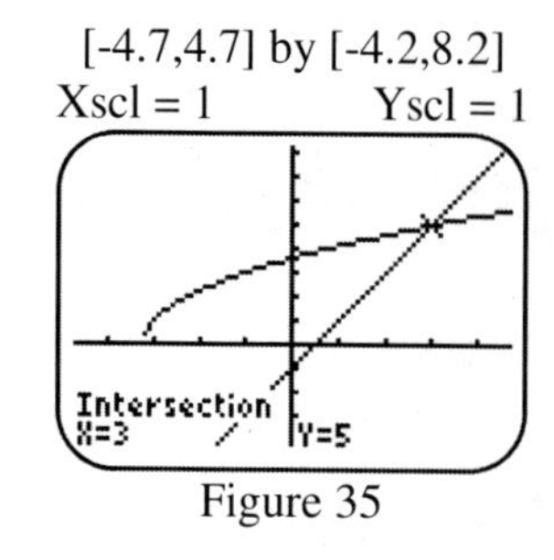

Figure 35

[-2.7,6.7] by [-3.2,9.2]
Xscl= 1 Yscl = 1

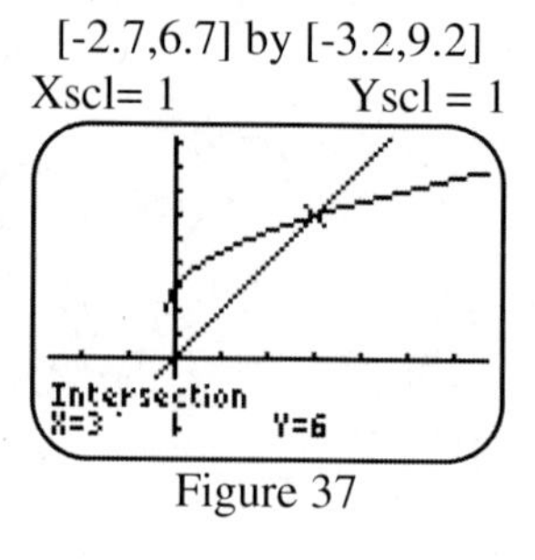

Figure 37

[-3.4,15.4] by [-2.2,10.2]
Xscl = 1 Yscl= 1

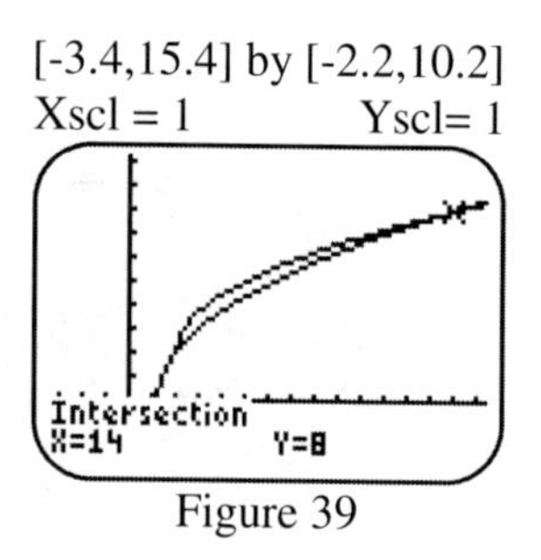

Figure 39

37. (a) $\sqrt{5x+1}+2=2x \Rightarrow \sqrt{5x+1}=2x-2 \Rightarrow (\sqrt{5x+1})^2=(2x-2)^2 \Rightarrow$

$5x+1=4x^2-8x+4 \Rightarrow 4x^2-13x+3=0 \Rightarrow (4x-1)(x-3)=0 \Rightarrow x=\frac{1}{4}$ or 3 ($\frac{1}{4}$ is extraneous).

The solution set is $\{3\}$. A graph of $y_1=\sqrt{5x+1}+2$ and $y_2=2x$ is shown in Figure 37.

(b) Since the graph of y_1 begins when $5x+1=0 \Rightarrow x=-\frac{1}{5}$, $y_1>y_2$ on the interval $\left[-\frac{1}{5},3\right)$.

(c) From the graph, $y_1<y_2$ on the interval $(3,\infty)$.

39. (a) $\sqrt{3x-6}+2=\sqrt{5x-6} \Rightarrow (\sqrt{3x-6}+2)^2=(\sqrt{5x-6})^2 \Rightarrow$

$3x-6+4\sqrt{3x-6}+4=5x-6 \Rightarrow 4\sqrt{3x-6}=2x-4 \Rightarrow (4\sqrt{3x-6})^2=(2x-4)^2 \Rightarrow$

$16(3x-6)=4x^2-16x+16 \Rightarrow 48x-96=4x^2-16x+16 \Rightarrow 4x^2-64x+112=0 \Rightarrow$

$x^2-16x+28=0 \Rightarrow (x-2)(x-14)=0 \Rightarrow x=2$ or 14. The solution set is $\{2,14\}$. A graph of $y_1=\sqrt{3x-6}+2$ and $y_2=\sqrt{5x-6}$ is shown in Figure 39.

(b) From the graph, $y_1>y_2$ on the interval $(2,14)$.

(c) From the graph, $y_1<y_2$ on the interval $(14,\infty)$.

41. (a) $\sqrt[3]{x^2-2x}=\sqrt[3]{x} \Rightarrow (\sqrt[3]{x^2-2x})^3=(\sqrt[3]{x})^3 \Rightarrow x^2-2x=x \Rightarrow x^2-3x=0 \Rightarrow x(x-3)=0 \Rightarrow x=0$ or $x=3$.

The solution set is $\{0,3\}$. A graph of $y_1=\sqrt[3]{x^2-2x}$ and $y_2=\sqrt[3]{x}$ is shown in Figure 41.

(b) From the graph, $y_1>y_2$ on the interval $(-\infty,0)\cup(3,\infty)$.

(c) From the graph, $y_1<y_2$ on the interval $(0,3)$.

43. (a) $\sqrt[4]{3x+1}=1 \Rightarrow (\sqrt[4]{3x+1})^4=1^4 \Rightarrow 3x+1=1 \Rightarrow 3x=0 \Rightarrow x=0$. The solution set is $\{0\}$. A graph of $y_1=\sqrt[4]{3x+1}$ and $y_2=1$ is shown in Figure 43.

(b) From the graph, $y_1>y_2$ on the interval $(0,\infty)$.

(c) Since the graph of y_1 begins when $3x+1=0 \Rightarrow x=-\frac{1}{3}$, $y_1<y_2$ on the interval $\left[-\frac{1}{3},0\right)$.

[-4.7,4.7] by [-3.1,3.1]
Xscl = 1 Yscl = 1

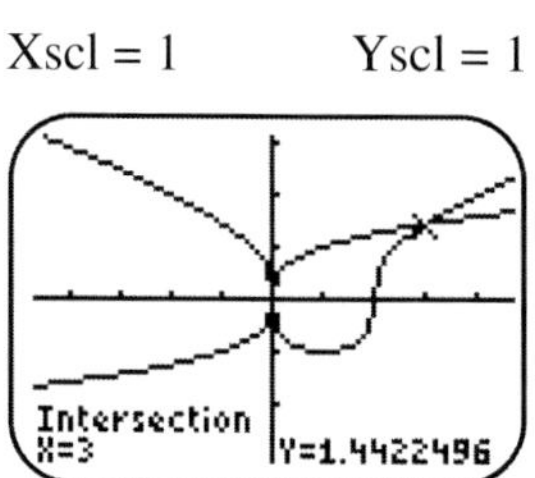

Figure 41

[-4.7,4.7] by [-3.1,3.1]
Xscl = 1 Yscl = 1

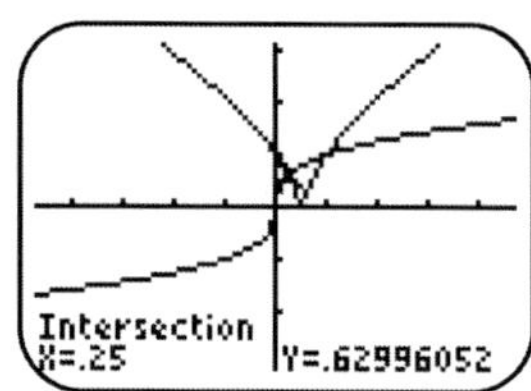

Figure 42

[-4.7,4.7 by [-3.1,3.1]
Xscl= 1 Yscl = 1

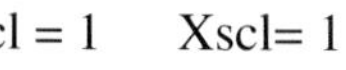
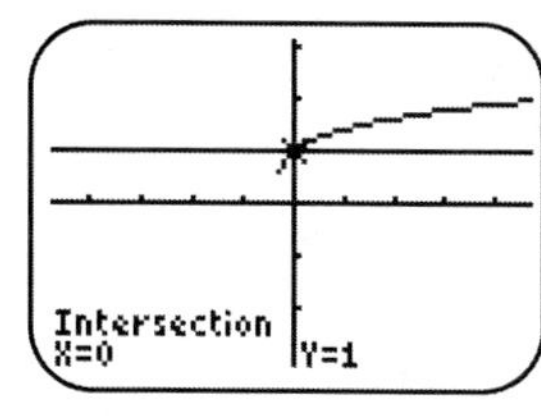

Figure 43

[-5,45] by [-1,3]
Xscl = 5 Yscl= 1

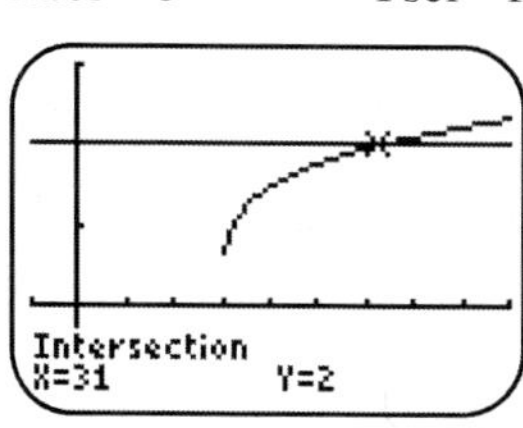

Figure 44

45. (a) $(2x-5)^{1/2}-2=(x-2)^{1/2} \Rightarrow [(2x-5)^{1/2}-2]^2=[(x-2)^{1/2}]^2 \Rightarrow$
$2x-5-4(2x-5)^{1/2}+4=x-2 \Rightarrow x+1=4(2x-5)^{1/2} \Rightarrow (x+1)^2=[4(2x-5)^{1/2}]^2 \Rightarrow$
$x^2+2x+1=16(2x-5) \Rightarrow x^2+2x+1=32x-80 \Rightarrow x^2-30x+81=0 \Rightarrow (x-3)(x-27)=0 \Rightarrow x=3$ or 27 (3 is extraneous). The solution set is $\{27\}$. A graph of $y_1=(2x-5)^{1/2}-2$ and $y_2=(x-2)^{1/2}$ is shown in Figure 45.

(b) From the graph, $y_1 \geq y_2$ on the interval $[27,\infty)$.

(c) Since the graph of y_1 begins when $2x-5=0 \Rightarrow x=\frac{5}{2}$, $y_1 \leq y_2$ on the interval $\left[\frac{5}{2},27\right]$.

47. (a) $(x+6x)^{1/4}=2 \Rightarrow [(x^2+6x)^{1/4}]^4=2^4 \Rightarrow x^2+6x=16 \Rightarrow x^2+6x-16=0 \Rightarrow (x+8)(x-2)=0 \Rightarrow x=-8$ or 2. The solution set is $\{-8,2\}$. A graph of $y_1=(x^2+6x)^{1/4}$ and $y_2=2$ is shown in Figure 47.

(b) From the graph, $y_1 > y_2$ on the interval $(-\infty,-8)\cup(2,\infty)$.

(c) Since $x^2+6x=0 \Rightarrow x(x+6)=0 \Rightarrow x=0$ or -6, the graph of y_1 has endpoints when $x=-6$ and 0. From the graph, $y_1 < y_2$ on the interval $(-8,-6]\cup[0,2)$.

[-5,45] by [-2,8]
Xscl = 5 Yscl = 2

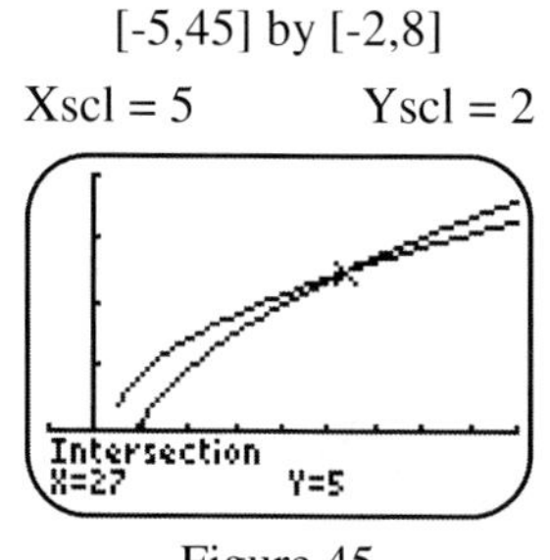

Figure 45

[-11,5] by [-1,4]
Xscl = 1 Yscl = 1

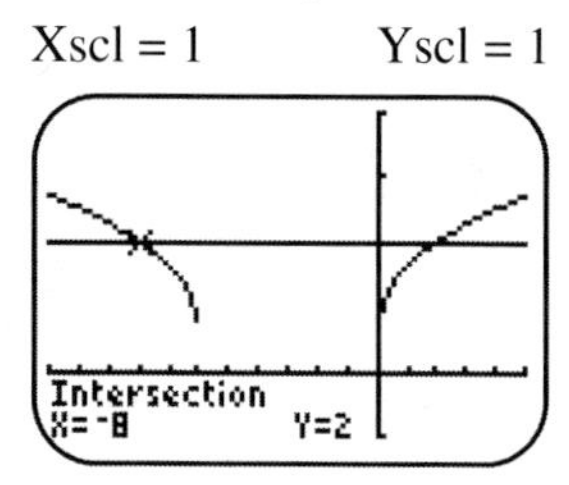

Figure 47

[-2,2] by [-2,2]
Xscl= 1 Yscl = 1

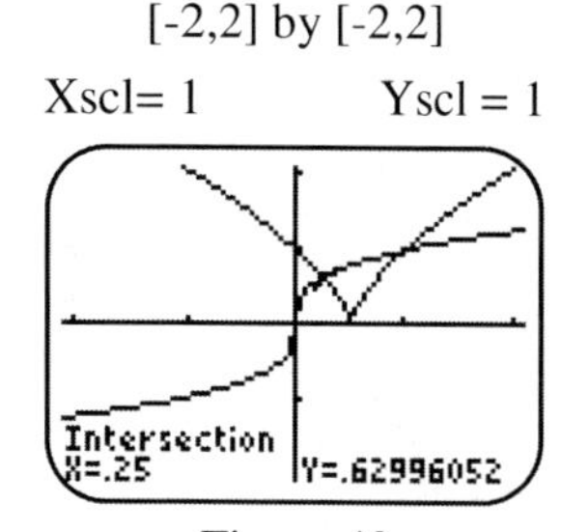

Figure 49

[5,10] by [-100,100]
Xscl= 1 Yscl = 10

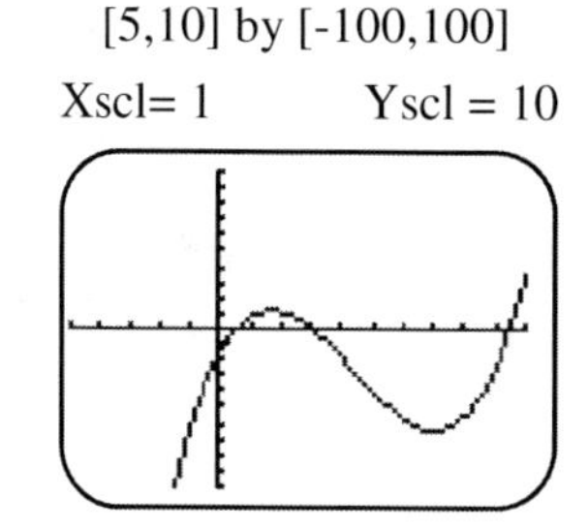

Figure 55

49. (a) $(2x-1)^{2/3}=x^{1/3} \Rightarrow [(2x-1)^{2/3}]^3=(x^{1/3})^3 \Rightarrow (2x-1)^2=x$
$\Rightarrow 4x^2-4x+1=x \Rightarrow 4x^2-5x+1=0 \Rightarrow (4x-1)(x-1)=0 \Rightarrow x=\frac{1}{4}$ or 1. The solution set is $\left\{\frac{1}{4},1\right\}$.
A graph of $y_1=(2x-1)^{2/3}$ and $y_2=x^{1/3}$ is shown in Figure 49.

(b) From the graph, $y_1 > y_2$ on the interval $\left(-\infty,\frac{1}{4}\right)\cup(1,\infty)$.

(c) From the graph, $\sqrt{18-2}=\sqrt{16}=4\neq 14-18;$ on the interval $\left(\frac{1}{4},1\right)$.

51. $\sqrt[3]{4x-4}=\sqrt{x+1}\Rightarrow(4x-4)^{1/3}=(x+1)^{1/2}$.

52. The LCD for $\frac{1}{3}$ and $\frac{1}{2}$ is 6.

53. $\left[(4x-4)^{1/3}\right]^6=\left[(x+1)^{1/2}\right]^6\Rightarrow(4x-4)^2=(x+1)^3$.

54. $(4x-4)^2=(x+1)^3\Rightarrow 16x^2-32x+16=x^3+3x^2+3x+1\Rightarrow x^3-13x^2+35x-15=0$.

55. The graph crosses the x-axis 3 times so the equation has 3 real roots. See Figure 55.

56.
$$\begin{array}{r|rrrr} 3 & 1 & -13 & 35 & -15 \\ & & 3 & -30 & 15 \\ \hline & 1 & -10 & 5 & 0 \end{array}$$
The value in the remainder position is 0; therefore, $P(3)=0$.

57. From the synthetic division shown above, $P(x)=(x-3)(x^2-10x+5)$.

58. If $x^2-10x+5=0,$ then $x=\frac{-(-10)\pm\sqrt{(-10)^2-4(1)(5)}}{2(1)}=\frac{10\pm\sqrt{80}}{2}=5\pm 2\sqrt{5}\Rightarrow\left\{5\pm 2\sqrt{5}\right\}$.

59. The three solutions are $3, 5+2\sqrt{5},$ and $5-2\sqrt{5}$.

60. The graph of $y_3=y_1-y_2$, where $y_1=\sqrt[3]{4x-4}$, and $y_2=\sqrt{x+1}$ is shown in Figure 60.

There are 2 real solutions.

[-5,10] by [-100,100]
Xscl = 1 Yscl = 10

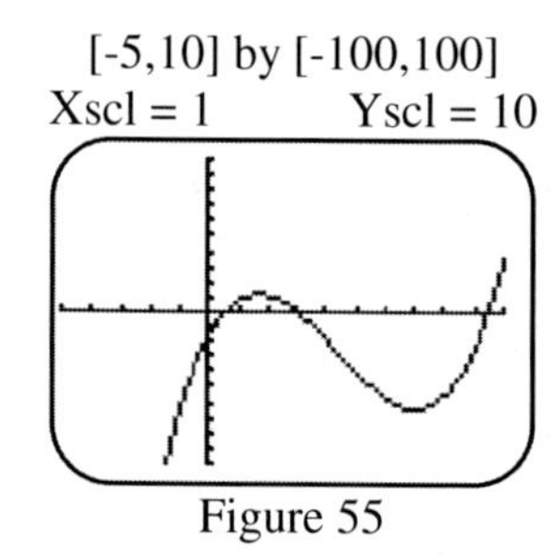

Figure 55

[-2,20] by [-0.5,0.5]
Xscl = 1 Yscl = 0.1

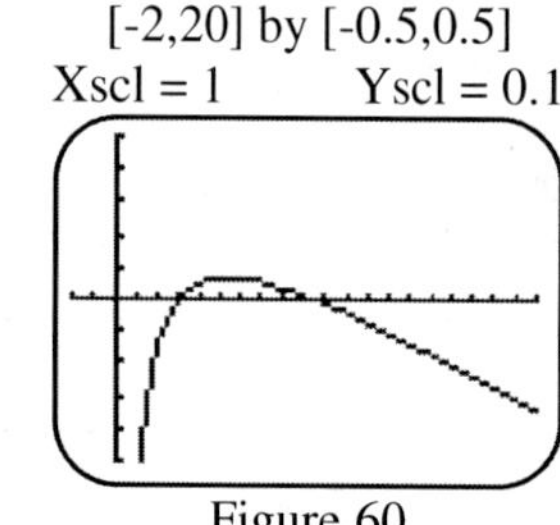

Figure 60

61 If the value $5-2\sqrt{5}\approx 0.528$ is substituted for x in the expression $\sqrt[3]{4x-4}$, the result is negative. Since the right side of the original equation will never yield a negative value, the root $5-2\sqrt{5}\approx 0.528$ is extraneous. The solution set is $\left\{3, 5+2\sqrt{5}\right\}$. The calculator figure shown above supports this result.

62. The solution set of the original equation is a subset of the solution set found in the previous exercise. The extraneous solution was obtained when each side of the original equation was raised to the sixth power.

63. (a) $$\frac{\frac{2}{3}(x-2)x^{-1/3}-x^{2/3}}{(x-2)^2}=\frac{\frac{3}{1}}{\frac{3}{1}}\left[\frac{\frac{2}{3}(x-2)x^{-1/3}-x^{2/3}}{(x-2)^2}\right]=\frac{2(x-2)x^{-1/3}-3x^{2/3}}{3(x-2)^2}=\frac{x^{-1/3}\left[2(x-2)-3x\right]}{3(x-2)^2}=$$
$$\frac{2x-4-3x}{3x^{1/3}(x-2)^2}=\frac{-x-4}{3x^{1/3}(x-2)^2}$$

(b) Starting with the simplified expression from part (a) we will solve $\dfrac{-x-4}{3x^{1/3}(x-2)^2}=0$. The expression

$\dfrac{-x-4}{3x^{1/3}(x-2)^2}$ will equal 0 only when the numerator is equal to 0. $-x-4=0 \Rightarrow x=-4$

65. (a) $$\frac{\frac{1}{3}(x^2+1)x^{-2/3}-2x^{4/3}}{(x^2+1)^2}=\frac{\frac{3}{1}}{\frac{3}{1}}\left[\frac{\frac{1}{3}(x^2+1)x^{-2/3}-2x^{4/3}}{(x^2+1)^2}\right]=\frac{(x^2+1)x^{-2/3}-6x^{4/3}}{3(x^2+1)^2}=$$

$$\frac{(x^2+1)x^{-2/3}-6x^{4/3}}{3(x^2+1)^2}=\frac{x^{-2/3}\left[(x^2+1)-6x^2\right]}{3(x^2+1)^2}=\frac{\left[(x^2+1)-6x^2\right]}{3x^{2/3}(x^2+1)^2}=\frac{1-5x^2}{3x^{2/3}(x^2+1)^2}$$

(b) Starting with the simplified expression from part (a) we will solve $\dfrac{1-5x^2}{3x^{2/3}(x^2+1)^2}=0$. The expression

$\dfrac{1-5x^2}{3x^{2/3}(x^2+1)^2}$ will equal 0 only when the numerator is equal to 0. Solve as follows:

$$1-5x^2=0 \Rightarrow 5x^2=1 \Rightarrow x^2=\frac{1}{5} \Rightarrow x=\pm\sqrt{\frac{1}{5}}$$

67. (a) $$\frac{x^{1/4}-x^{-3/4}}{x}=\frac{x^{-3/4}[x-1]}{x}=\frac{x-1}{x^{3/4}(x)}=\frac{x-1}{x^{7/4}}$$

(b) Starting with the simplified expression from part (a) we will solve $\dfrac{x-1}{x^{7/4}}=0$. The expression

$\dfrac{x-1}{x^{7/4}}$ will equal 0 only when the numerator is equal to 0. $x-1=0 \Rightarrow x=1$

69. (a) $$\frac{(x^2+1)^{1/2}-\frac{1}{2}x(x^2+1)^{-1/2}(2x)}{x^2+1}=\frac{\frac{2}{1}}{\frac{2}{1}}\left[\frac{(x^2+1)^{1/2}-\frac{1}{2}x(x^2+1)^{-1/2}(2x)}{x^2+1}\right]=$$

$$\frac{2(x^2+1)^{1/2}-x(x^2+1)^{-1/2}(2x)}{2(x^2+1)}=\frac{(x^2+1)^{-1/2}\left[2(x^2+1)-x(2x)\right]}{2(x^2+1)}=\frac{\left[2(x^2+1)-x(2x)\right]}{2(x^2+1)^{1/2}(x^2+1)}=$$

$$\frac{\left[(2x^2+2)-(2x^2)\right]}{2(x^2+1)^{1/2}(x^2+1)}=\frac{2}{2(x^2+1)^{1/2}(x^2+1)}=\frac{1}{(x^2+1)^{3/2}}$$

(b) Starting with the simplified expression from part (a) we will solve $\dfrac{1}{(x^2+1)^{3/2}}=0$. The expression

$\frac{1}{\left(x^2+1\right)^{3/2}}$ will equal 0 only when the numerator is equal to 0. Since the numerator is a constant there is no solution to this equation.

71. $\sqrt{\sqrt{x}} = x \Rightarrow x^{1/4} = x \Rightarrow \left[x^{1/4}\right]^4 = x^4 \Rightarrow x = x^4 \Rightarrow x - x^4 = 0 \Rightarrow x(1-x^3) = 0 \Rightarrow x = 0$ or 1.

The solution set is $\{0,1\}$.

73. $\sqrt{\sqrt{28x+8}} = \sqrt{3x+2} \Rightarrow (28x+8)^{1/4} = (3x+2)^{1/2} \Rightarrow \left[(28x+8)^{1/4}\right]^4 = \left[(3x+2)^{1/2}\right]^4 \Rightarrow$

$28x+8 = (3x+2)^2 \Rightarrow 28x+8 = 9x^2+12x+4 \Rightarrow 9x^2-16x-4 = 0 \Rightarrow (9x+2)(x-2) = 0 \Rightarrow x = -\frac{2}{9}$ or 2. The solution set is $\left\{-\frac{2}{9}, 2\right\}$.

75. $\sqrt[3]{\sqrt{32x}} = \sqrt[3]{x+6} \Rightarrow (32x)^{1/6} = (x+6)^{1/3} \Rightarrow \left[(32)^{1/6}\right]^6 = \left[(x+6)^{1/3}\right]^6 \Rightarrow 32x = (x+6)^2 \Rightarrow$

$32x = x^2+12x+36 \Rightarrow x^2-20x+36 = 0 \Rightarrow (x-2)(x-18) = 0 \Rightarrow x = 2$ or 18. The solution set is $\{2,18\}$.

77. $v = \frac{350}{\sqrt{6000}} \approx 4.5$ km per sec

79. $p = 2\pi\sqrt{\frac{5}{32}} \approx 2.5$ sec

81. $s = 30\sqrt{\frac{900}{97}} \approx 91$ mph

83. (a) Since the distance from C to D is 20 feet, the distance from P to C is $20-x$.

(b) The value must be between 0 and 20. That is, $0 < x < 20$.

(c) $(AP)^2 = (DP)^2 + (AD)^2 \Rightarrow (AP)^2 = x^2 + 12^2 \Rightarrow AP = \sqrt{x^2+12^2}$;

$(BP)^2 = (CP)^2 + (BC)^2 \Rightarrow (BP)^2 = (20-x)^2 + 16^2 \Rightarrow BP = \sqrt{(20-x)^2+16^2}$

(d) $f(x) = \sqrt{x^2+12^2} + \sqrt{(20-x)^2+16^2},\ 0 < x < 20.$

(e) The graph of $y_1 = \sqrt{x^2+12^2} + \sqrt{(20-x)^2+16^2}$ is shown in Figure 73. Here $f(4) \approx 35.28$. When the stake is 4 feet from the 12-foot pole, approximately 35.28 feet of wire will be required.

(f) Using the calculator, $f(x)$ is a minimum (about 34.41 feet) when $x \approx 8.57$ feet.

(g) This problem examined how the total amount of wire used can be expressed in terms of the distance from the stake at P to the base of the 12-foot pole. We find that the amount of wire used can be minimized when the stake is approximately 8.57 feet from the 12-foot pole.

[0,20] by [0,50]
Xscl = 1 Yscl = 10

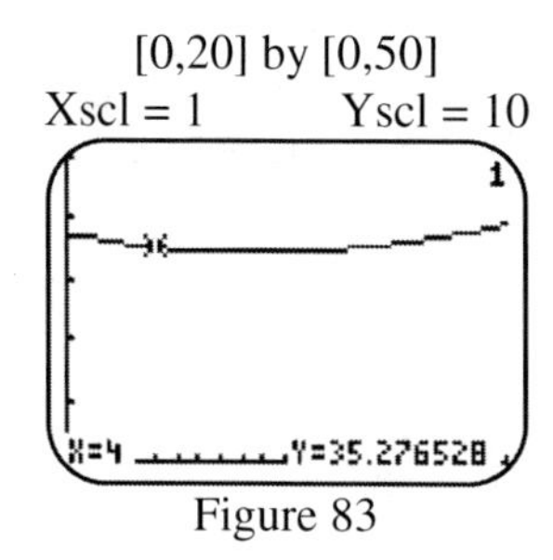

Figure 83

85. Using the Pythagorean theorem, the diagonal distance traveled on the river is $d=\sqrt{x^2+3^2}$. Noting that the rate of travel on land is 5 mph, the rate of travel on water is 2 mph, and that $t=\frac{d}{r}$, the time needed to travel on land is $\frac{8-x}{5}$ and the time needed to travel on the river is $\frac{\sqrt{x^2+9}}{2}$. The total travel time is given by the function $y_1=\frac{8-x}{5}+\frac{\sqrt{x^2+9}}{2}$. By graphing this function (not shown), we find that the minimum value for time occurs when $x\approx 1.31$. The hunter should travel $8-1.31=6.69$ miles along the river.

87. We will refer to the original position of the *Inspiration* as the origin. At time x hours past noon, the *Celebration* is $60-30x$ miles south of the origin and the *Inspiration* is $20x$ miles west of the origin. Using the Pythagorean theorem, the distance between the ships is given by the function $y_1=\sqrt{(60-30x)^2+(20x)^2}$. By graphing this function (not shown), we find that the distance between the ships is a minimum when $x\approx 1.38$ hours. At the time 1.38 hours past noon (about 1:23 P.M.) the ships are about 33.28 miles apart.

Reviewing Basic Concepts (Sections 4.4 and 4.5)

1. As the exponent increases in value, the curve rises more rapidly for $x\geq 1$. See Figure 1.
2. $S(0.75)=0.3(0.75)^{3/4}\approx 0.24\,m^2$
3. See Figure 3a and Figure 3b.

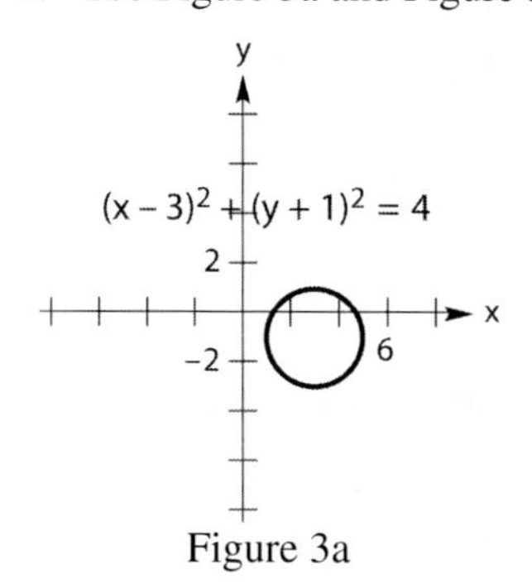

Figure 3a

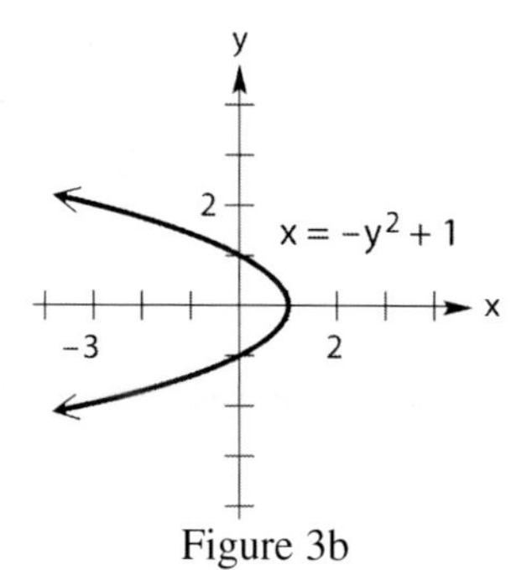

Figure 3b

4. Solving the equation for y yields $x^2+y^2=16\Rightarrow y^2=16-x^2\Rightarrow y=\pm\sqrt{16-x^2}$. The expressions are $y_1=\sqrt{16-x^2}$ and $y_2=-\sqrt{16-x^2}$ which are graphed in a square window in Figure 4.

5. Solving the equation for y yields $x = y^2 + 4y + 4 + 2 \Rightarrow x - 2 = (y+2)^2 \Rightarrow$

$\pm\sqrt{x-2} = y + 2 \Rightarrow -2 \pm \sqrt{x-2} = y \Rightarrow y = -2 \pm \sqrt{x-2}$. The expressions are

$y_1 = -2 + \sqrt{x-2}$ and $y_2 = -2 - \sqrt{x-2}$ which are graphed in a square window in Figure 5.

6. $\sqrt{3x+4} = 8 - x \Rightarrow \left(\sqrt{3x+4}\right)^2 = (8-x)^2 \Rightarrow 3x + 4 = 64 - 16x + x^2 \Rightarrow$

$x^2 - 19x + 60 = 0 \Rightarrow (x-4)(x-15) = 0 \Rightarrow x = 4,\ 15$. The solution 15 is extraneous. The solution set is $\{4\}$.

7. The graph of $y_1 = \sqrt{3x+4}$ is above the graph of $y_2 = 8 - x$ on the interval $(4, \infty)$. See Figure 7-8.

[0,10] by [0,10]
Xscl = 1 Yscl = 1

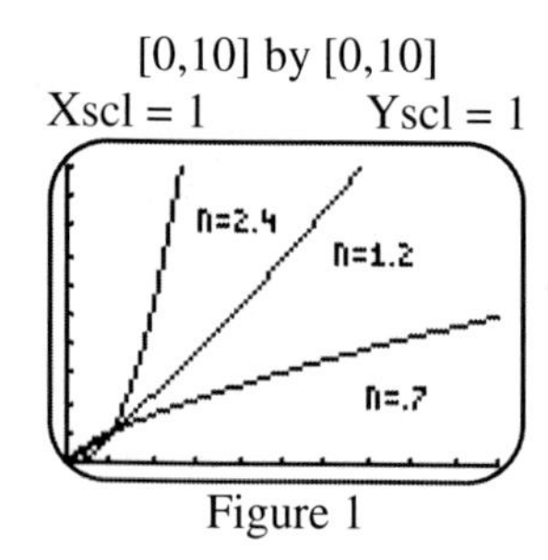

Figure 1

[-9.4,9.4] by [-6.2,6.2]
Xscl = 1 Yscl = 1

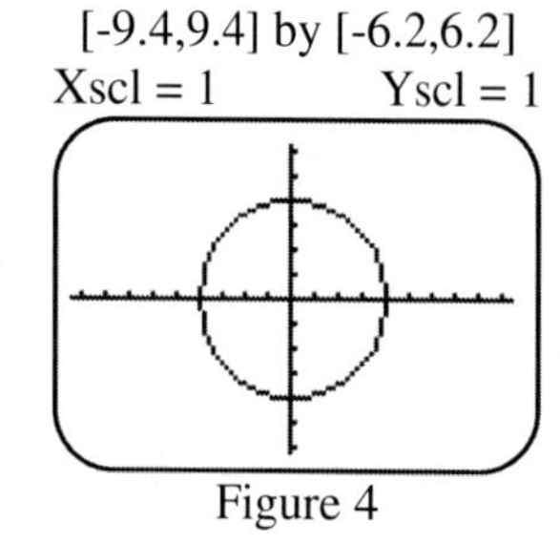

Figure 4

[-9.4,9.4] by [-6.2,6.2]
Xscl= 1 Yscl = 1

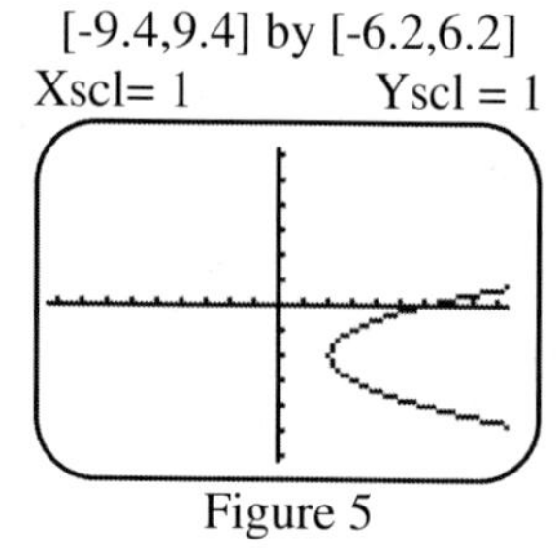

Figure 5

[-3,12] by [-3,12]
Xscl = 1 Yscl= 1

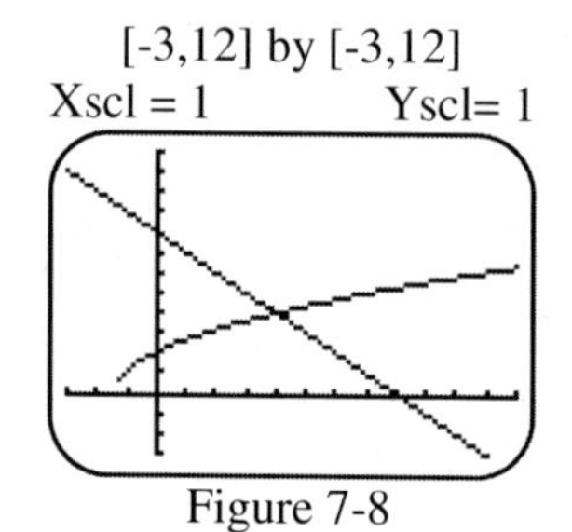

Figure 7-8

8. Noting that the graph of $y_1 = \sqrt{3x+4}$ does not start until $3x + 4 = 0 \Rightarrow 3x = -4 \Rightarrow x = -\frac{4}{3}$, the graph of

$y_1 = \sqrt{3x+4}$ is below the graph of $y_2 = 8 - x$ on the interval $\left[-\frac{4}{3}, 4\right)$. See Figure 7-8.

9. $\sqrt{3x+4} + \sqrt{5x+6} = 2 \Rightarrow \sqrt{3x+4} = 2 - \sqrt{5x+6} \Rightarrow \left(\sqrt{3x+4}\right)^2 = \left(2 - \sqrt{5x+6}\right)^2 \Rightarrow$

$3x + 4 = 4 - 4\sqrt{5x+6} + 5x + 6 \Rightarrow 4\sqrt{5x+6} = 2x + 6 \Rightarrow \left(4\sqrt{5x+6}\right)^2 = (2x+6)^2 \Rightarrow$

$16(5x+6) = 4x^2 + 24x + 36 \Rightarrow 80x + 96 = 4x^2 + 24x + 36 \Rightarrow 4x^2 - 56x - 60 = 0 \Rightarrow$

$x^2 - 14x - 15 = 0 \Rightarrow (x+1)(x-15) = 0 \Rightarrow x = -1,\ 15$. The solution 15 is an extraneous solution.

The solution set is $\{-1\}$.

10. (a) cat: $f(24) = \dfrac{1607}{\sqrt[4]{24^3}} \approx 148$ beats per minute; person: $f(66) = \dfrac{1607}{\sqrt[4]{66^3}} \approx 69$ beats per minute

(b) $400 = \dfrac{1607}{\sqrt[4]{66^3}} \Rightarrow 400\sqrt[4]{x^3} = 1607 \Rightarrow \sqrt[4]{x^3} = \dfrac{1607}{400} \Rightarrow x^{3/4} = \dfrac{1607}{400} \Rightarrow x = \left(\dfrac{1607}{400}\right)^{4/3} \approx 6.4$ inches.

Chapter 4 Review Exercises

1. (a) The graph of $y = -\frac{1}{x} + 6$ can be obtained by reflecting the graph of $y = \frac{1}{x}$ across the x-axis and shifting upward 6 units.

   (b) A sketch of the graph is shown in Figure 1b.

   (c) A calculator graph is shown in Figure 1c.

2. (a) The graph of $y = \frac{4}{x} - 3$ can be obtained by stretching the graph of $y = \frac{1}{x}$ by a factor of 4 and shifting downward 3 units.

(b) A sketch of the graph is shown in Figure 2b.

(c) A calculator graph is shown in Figure 2c.

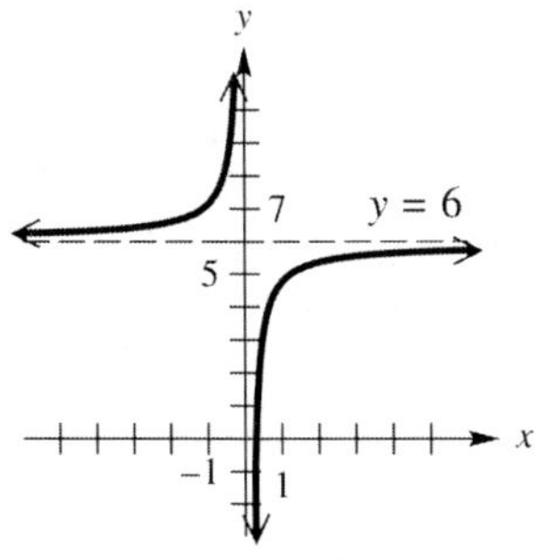

Figure 1b

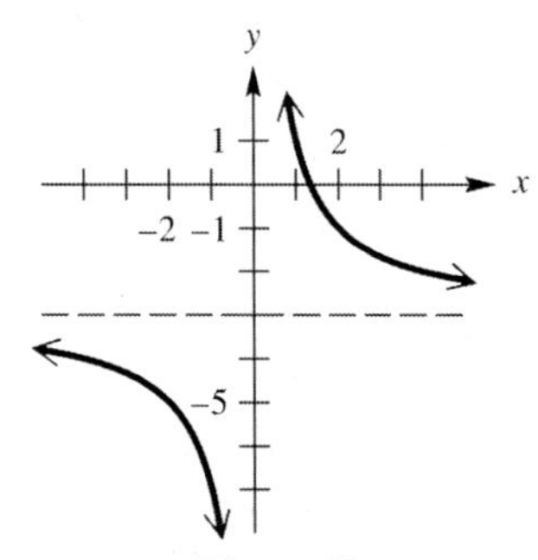

Figure 2b

[-4.7,4.7] by [-3.1,12.4]

Xscl = 1 Yscl = 1

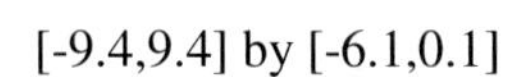

Xscl = 1 Yscl = 1

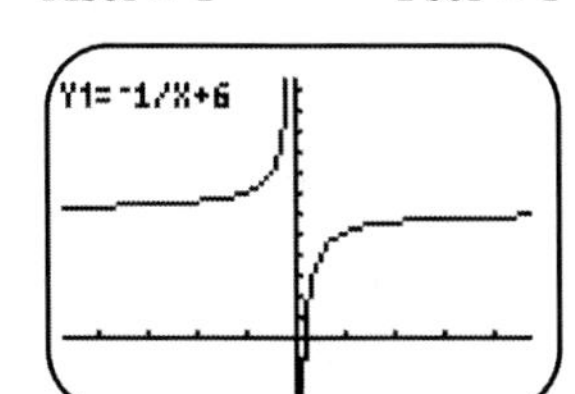

Figure 1c

Y1=4/X-3

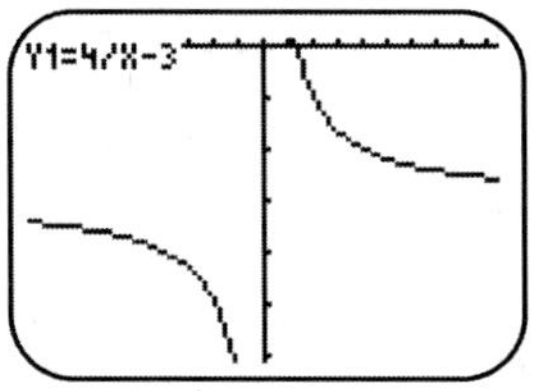

Figure 2c

3. (a) The graph of $y=-\dfrac{1}{(x-2)^2}$ can be obtained by reflecting the graph of $y=\dfrac{1}{x^2}$ across the x-axis and shifting to the right 2 units.

(b) A sketch of the graph is shown in Figure 3b.

(c) A calculator graph is shown in Figure 3c.

4. (a) The graph of $y=\dfrac{2}{x^2}+1$ can be obtained by stretching the graph of $y=\dfrac{1}{x^2}$ by a factor of 2 and shifting upward 1 unit.

(b) A sketch of the graph is shown in Figure 4b.

(c) A calculator graph is shown in Figure 4c.

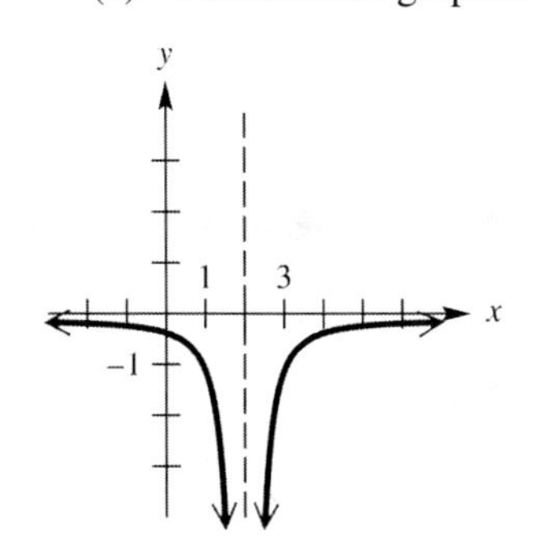

Figure 3b

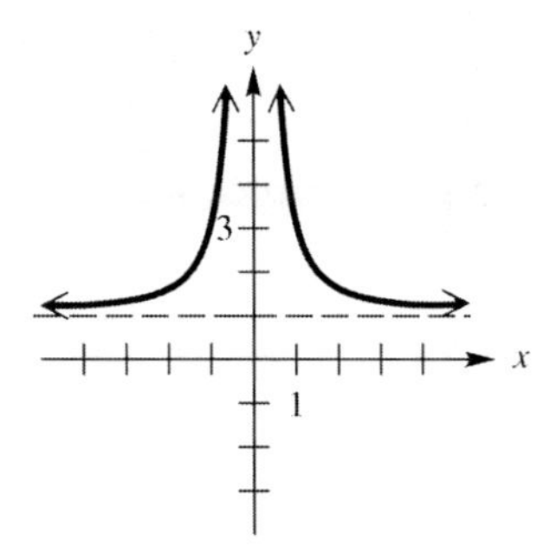

Figure 4b

[-2.7,6.7] by [-3.1,3.1]

Xscl = 1 Yscl = 1

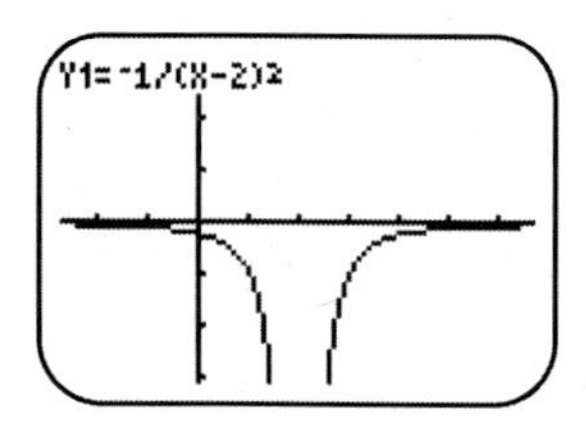

Figure 3c

[-9.4,9.4] by [-6.1,0.1]

Xscl = 1 Yscl = 1

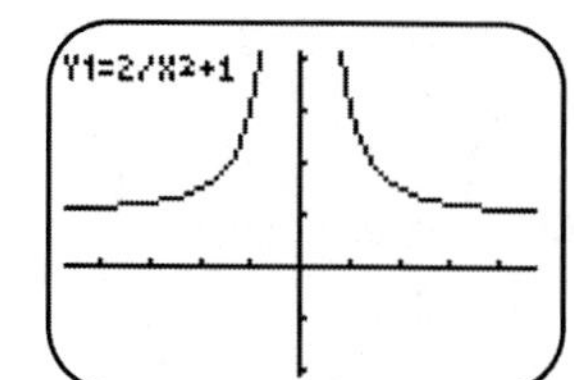

Figure 4c

5. When the degree of the numerator is exactly 1 greater than the degree of the denominator, the graph of a rational function defined by an expression written in lowest terms will have an oblique asymptote.
6. See Figure 6.
7. See Figure 7.
8. See Figure 8.

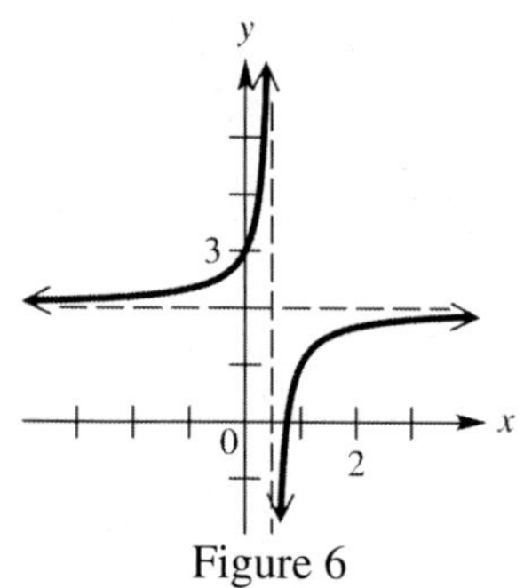

Figure 6

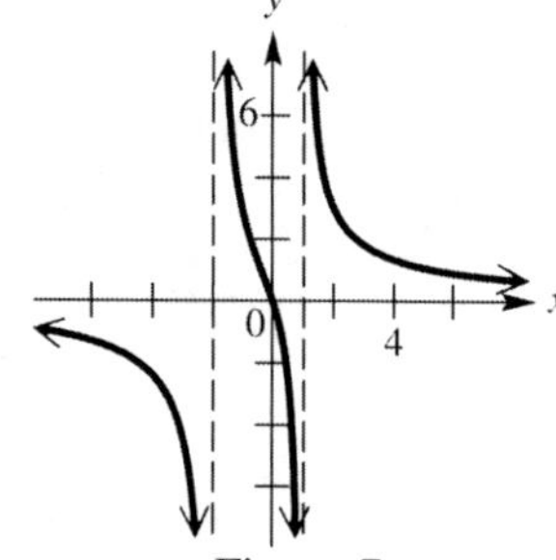

Figure 7

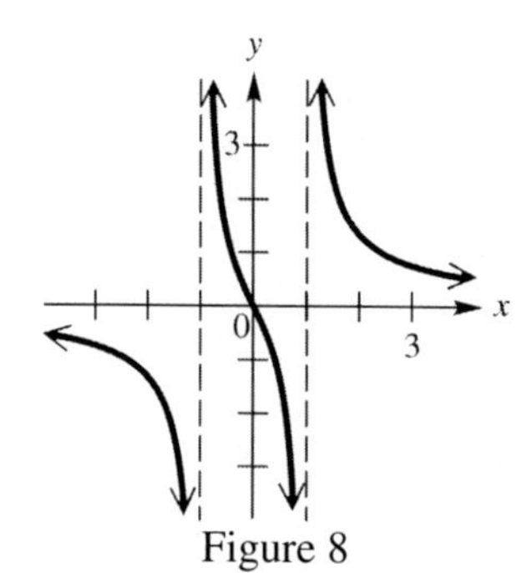

Figure 8

9. See Figure 9.
10. See Figure 10.
11. See Figure 11.

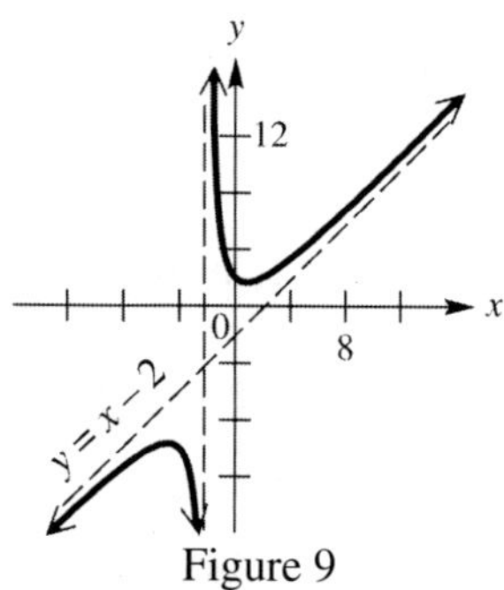

Figure 9

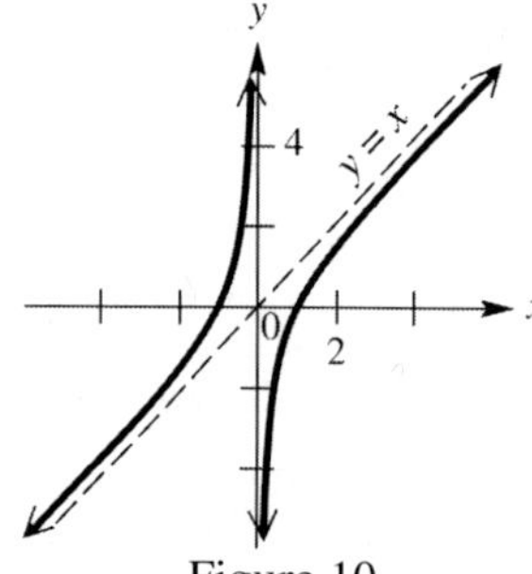

Figure 10

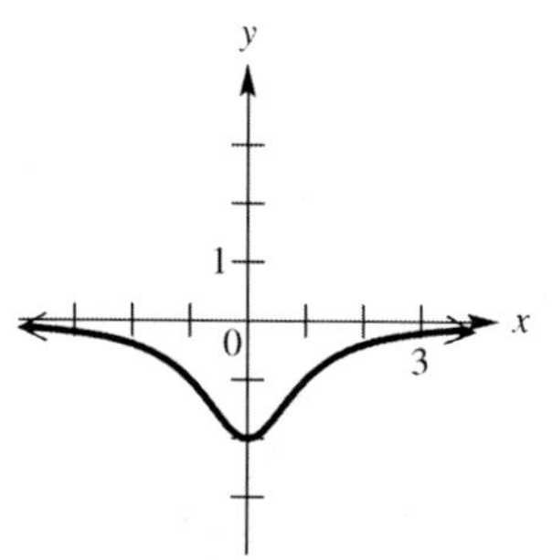

Figure 11

12. See Figure 12.
13. See Figure 13.

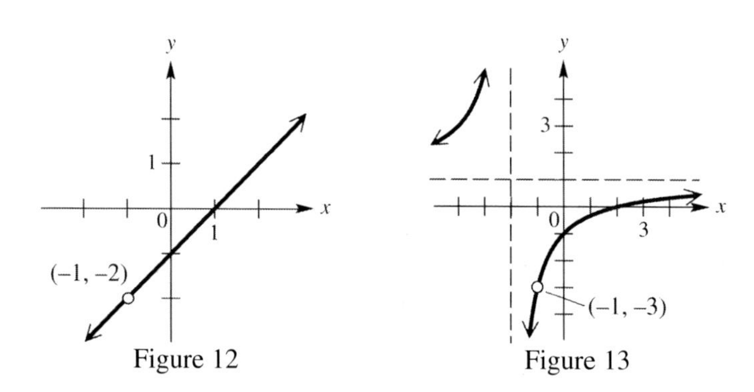

Figure 12

Figure 13

14. A vertical asymptote of $x = 0$ indicates that the denominator should contain the factor x. Since there are no x-intercepts, the numerator can never equal zero. Since there is a horizontal asymptote at $y = 0$, the degree of the numerator is less than the degree of the denominator. The end behavior of the graph suggests that the degree of the denominator is even. Since the graph passes through the point $\left(2, \frac{1}{4}\right)$, one possible equation for the function is $f(x) = \frac{1}{x^2}$. Other functions are possible.

15. A vertical asymptote of $x = 1$ indicates that the denominator should contain the factor $(x-1)$. A horizontal asymptote of $y = -3$ indicates that the degrees of the numerator and denominator should be equal and the leading coefficient of the numerator should be -3 times that of the denominator. An x-intercept of 2, indicates that the numerator should contain the factor $(x-2)$. One possible equation for the function is $f(x) = \frac{-3(x-2)}{x-1} = \frac{-3x+6}{x-1}$. Other functions are possible.

16. An x-intercept of -2 indicates that the numerator should contain the factor $(x+2)$. A "hole" when $x = 2$ indicates that both the numerator and denominator should contain the factor $(x-2)$. One possible equation for the function is $f(x) = \frac{(x+2)(x-2)}{x-2} = \frac{x^2-4}{x-2}$. Other functions are possible.

17. Since there are no x-intercepts the numerator can never equal zero. Since there is a horizontal asymptote at y = 0, the degree of the numerator is less than the degree of the denominator. The end behavior of the graph suggest that the degree of the denominator is even. Since the graph goes through (0,-1) a possible graph is $y = -\frac{1}{x^2+1}$.

18. (a) $\frac{5}{2x+5} = \frac{3}{x+2} \Rightarrow 5(x+2) = 3(2x+5) \Rightarrow 5x+10 = 6x+15 \Rightarrow x = -5$; the solution set is $\{-5\}$.

(b) Graph $y_1 = 5/(2x+5) - 3/(x+2)$ as shown in Figure 18. With vertical asymptotes $x = -\frac{5}{2}$ and $x = -2$, the graph is below the x-axis on $\left(-5, -\frac{5}{2}\right) \cup (-2, \infty)$. The solution set is $\left(-5, -\frac{5}{2}\right) \cup (-2, \infty)$.

(c) Graph $y_1 = 5/(2x+5) - 3/(x+2)$ as shown in Figure 18. With vertical asymptotes $x = -\frac{5}{2}$ and $x = -2$, the graph is above the x-axis on $(-\infty, -5) \cup \left(-\frac{5}{2}, -2\right)$. The solution set is $(-\infty, -5) \cup \left(-\frac{5}{2}, -2\right)$.

19. (a) $\frac{3x-2}{x+1} = 0 \Rightarrow 3x - 2 = 0 \Rightarrow 3x = 2 \Rightarrow x = \frac{2}{3}$. The solution set is $\left\{\frac{2}{3}\right\}$.

(b) Graph $y_1 = (3x-2)/(x+1)$ as shown in Figure 17. With vertical asymptote $x = -1$, the graph is below the x-axis on the interval $\left(-1, \frac{2}{3}\right)$. The solution set is $\left(-1, \frac{2}{3}\right)$.

(c) Graph $y_1 = (3x-2)/(x+1)$ as shown in Figure 17. With vertical asymptote $x = -1$, the graph is above the x-axis on the interval $(-\infty, -1) \cup \left(\frac{2}{3}, \infty\right)$. The solution set is $(-\infty, -1) \cup \left(\frac{2}{3}, \infty\right)$.

20. (a) Multiply through by the LCD, x^2.

$$1 \cdot x^2 - \frac{5}{x} \cdot x^2 + \frac{6}{x^2} \cdot x^2 = 0 \cdot x^2 \Rightarrow x^2 - 5x + 6 = 0 \Rightarrow (x-2)(x-3) = 0 \Rightarrow x = 2, 3$$

This solution set is $\{2, 3\}$.

(b) Graph $y_1 = 1 - 5/x + 6/x^2$ as shown in Figure 20. With vertical asymptote $x = 0$, the graph is below or intersects the x-axis on $[2, 3]$. The solution set is $[2, 3]$.

(c) Graph $y_1 = 1 - 5/x + 6/x^2$ as shown in Figure 20. With vertical asymptote $x = 0$, the graph is above or intersects the x-axis on $(-\infty, 0) \cup (0, 2] \cup [3, \infty)$. The solution set is $(-\infty, 0) \cup (0, 2] \cup [3, \infty)$.

21. (a) Multiply through by the LCD, $(x-2)(x+1)$.

$$\frac{3}{x-2} \cdot (x-2)(x+1) + \frac{1}{x+1} \cdot (x-2)(x+1) = \frac{1}{x^2 - x - 2} \cdot (x-2)(x+1) \Rightarrow$$

$3(x+1) + 1(x-2) = 1 \Rightarrow 3x + 3 + x - 2 = 1 \Rightarrow 4x = 0 \Rightarrow x = 0$; This solution set is $\{0\}$.

(b) Graph $y_1. = \frac{3}{x-2} + \frac{1}{x+1} - \frac{1}{x^2 - x - 2}$ as shown in Figure 19. With vertical asymptotes $x = -1$ and $x = 2$, the graph is below or intersects the x-axis on $(-\infty, -1) \cup [0, 2)$. The solution set is $(-\infty, -1) \cup [0, 2)$.

(c) Graph $y_1 = \frac{3}{x-2} + \frac{1}{x+1} - \frac{1}{x^2 - x - 2}$ as shown in Figure 19. With vertical asymptotes $x = -1$ and $x = 2$, the graph is above or intersects the x-axis on $(-1, 0] \cup (2, \infty)$. The solution set is $(-1, 0] \cup (2, \infty)$.

[-6.9,2.5] by [-10,30]
Xscl = 1 Yscl = 5

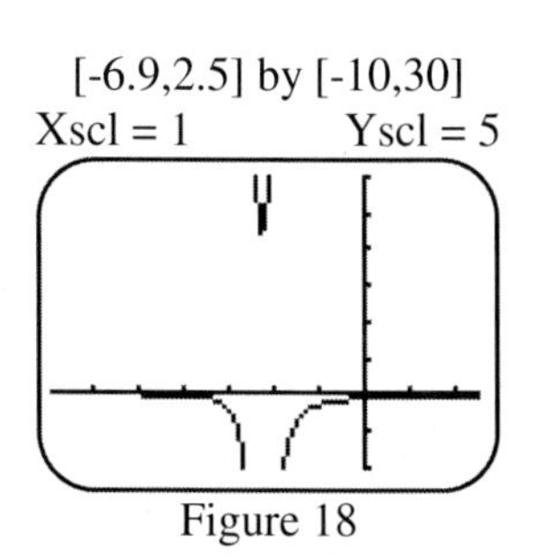

Figure 18

[-9.4,9.4 by [-3.1,6.2]
Xscl = 1 Yscl = 1

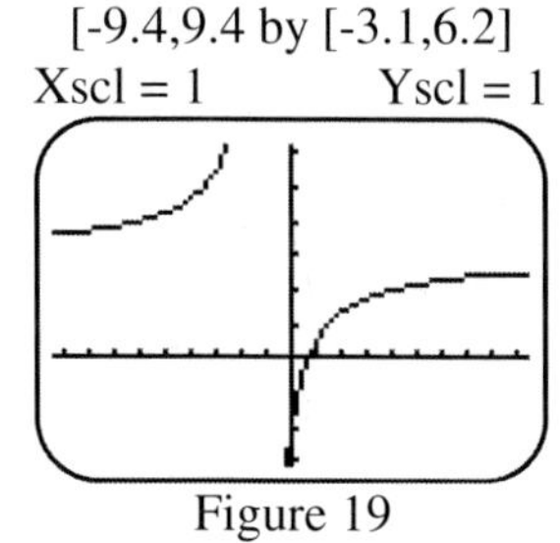

Figure 19

[-9.4,9.4] by [-6.2,6.2]
Xscl= 1 Yscl = 1

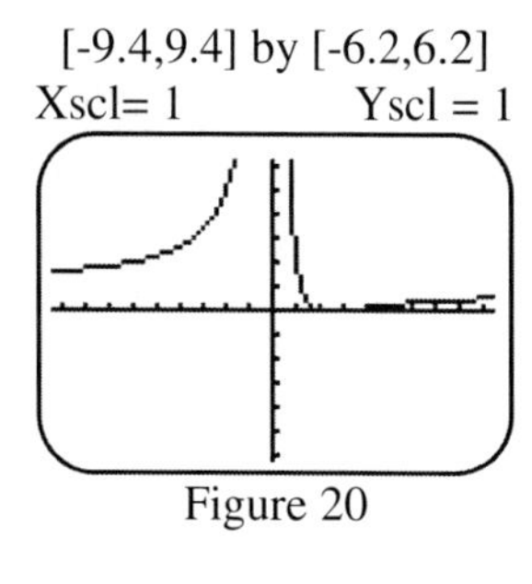

Figure 20

[-9.4,9.4] by [-6.2,6.2]
Xscl = 1 Yscl= 1

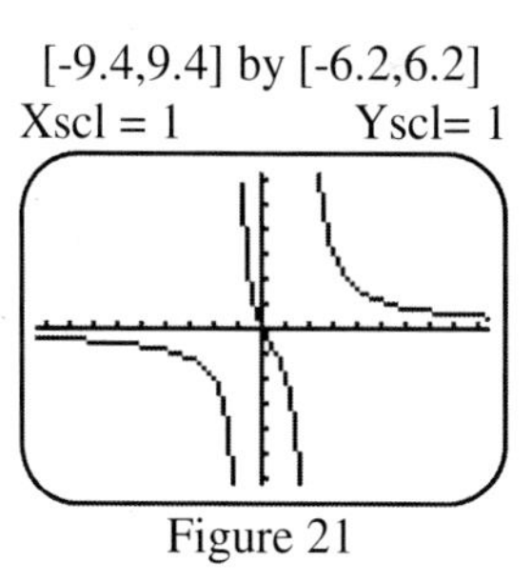

Figure 21

22. The graph intersects the x-axis when $x = -2$. The solution set is $\{-2\}$.

23. The graph is below the x-axis on the interval $(-2,-1)$. The solution set is $(-2,-1)$.

24. The graph is above the x-axis on the interval $(-\infty,-2)\cup(-1,\infty)$. The solution set is $(-\infty,-2)\cup(-1,\infty)$.

25. $|f(x)|>0$ for all $x \neq -2$ and $x \neq -1$. The solution set is $(-\infty,-2)\cup(-2,-1)\cup(-1,\infty)$.

26. (a) See Figure 26.
    (b) 127.3 thousand dollars or \$127,300

[0,100] by [0,150]
Xscl = 10 Yscl = 10

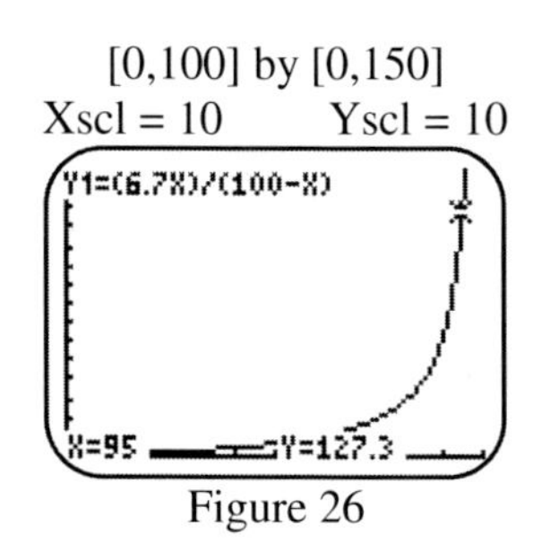

Figure 26

27. (a) Since $0 \le x < 40$, the denominator is positive. Multiplying by the LCD will not change the inequality.
$\dfrac{x^2}{1600-40x} \le 8 \Rightarrow x^2 \le 8(1600-40x) \Rightarrow x^2 \le 12{,}800 - 320x \Rightarrow x^2 + 302x - 12{,}800 \le 0$ The left side of this inequality is a quadratic polynomial whose positive zero can be obtained using the quadratic formula. The zero is approximately 35.96. The graph of this parabola is below the x-axis for values of x in the interval [0, 36]. Note that rounding to a whole number is appropriate since x represents a number of cars. The solution set is [0, 36].
    (b) The average line length is less than or equal to 8 cars when the average arrival rate is 36 cars per hour or less.

28. (a) When the coefficient of friction becomes smaller, the braking distance increases.
    (b) Since $0 < x \le 1$, the denominator is positive. Multiplying by the LCD will not change the inequality.
$\dfrac{120}{x} \ge 400 \Rightarrow 120 \ge 400x \Rightarrow \dfrac{120}{400} \ge x \Rightarrow x \le 0.3$; The solution set is $(0,0.3]$.

29. Here $y = \dfrac{k}{x}$. By substitution, $5 = \dfrac{k}{6} \Rightarrow k = 30$. That is, $y = \dfrac{30}{x}$. When $x = 15$, $y = \dfrac{30}{15} = 2$.

30. Here $z = \frac{k}{t^3}$. By substitution, $0.08 = \frac{k}{5^3} \Rightarrow 0.08\left(5^3\right) = k \Rightarrow k = 10$. That is, $z = \frac{10}{t^3}$.

When $t = 2$, $z = \frac{10}{2^3} = 1.25$.

31. Here $m = \frac{knp^2}{q}$. By substitution, $20 = \frac{k(5)\left(6^2\right)}{18} \Rightarrow 360 = 180k \Rightarrow k = \frac{360}{180} \Rightarrow k = 2$. That is, $m = \frac{2np^2}{q}$.

When $n = 7$, $p = 11$, and $q = 2$, $m = \frac{2(7)\left(11^2\right)}{2} = 847$.

32. The height of this cone varies *inversely* as the *square* of the *radius* of its base. The constant of variation is $\frac{300}{\pi}$.

33. Here $I = \frac{k}{d^2}$. By substitution, $70 = \frac{k}{5^2} \Rightarrow 70\left(5^2\right) = k \Rightarrow k = 1750$. That is, $I = \frac{1750}{d^2}$.

When $d = 12$, $I = \frac{1750}{12^2} \approx 12.15$ candela.

34. Here $R = \frac{k}{d^2}$. By substitution, $0.4 = \frac{k}{0.01^2} \Rightarrow 0.4\left(0.01^2\right) = k \Rightarrow k = 0.00004$. That is, $R = \frac{0.00004}{d^2}$. When $d = 0.03$, $I = \frac{0.00004}{0.03^2} = \frac{0.00004}{0.0009} = \frac{4}{90} = \frac{2}{45}$ ohm.

35. Here $I = kPt$. By substitution, $110 = k(1000)(2) \Rightarrow \frac{110}{2000} = k \Rightarrow k = 0.055$. That is, $I = 0.055Pt$. When $P = 5000$ and $t = 5$, $I = 0.055(5000)(5) = \$1375$.

36. Here $F = \frac{kws^2}{r}$. By substitution, $3000 = \frac{k(2000)\left(30^2\right)}{500} \Rightarrow 1{,}500{,}000 = 1{,}800{,}000k \Rightarrow k = \frac{1.5}{1.8} \Rightarrow k = \frac{5}{6}$.

That is, $F = \frac{\frac{5}{6}ws^2}{r}$. When $w = 2000$, $s = 60$, and $r = 800$, $F = \frac{\frac{5}{6}(2000)\left(60^2\right)}{(800)} = 7500$ lb.

37. Here $L = \frac{kd^4}{h^2}$. By substitution, $8 = \frac{k\left(1^4\right)}{9^2} \Rightarrow k = 648$. That is, $L = \frac{648d^4}{h^2}$. When $h = 12$, and $d = \frac{2}{3}$, $L = \frac{648\left(\frac{2}{3}\right)^4}{12^2} = \frac{128}{144} = \frac{8}{9}$ metric tons.

38. Here $L = \frac{kwh^2}{l}$. By substitution, $400 = \frac{k(12)(15)^2}{8} \Rightarrow 3200 = 2700k \Rightarrow k = \frac{3200}{2700} = \frac{32}{27}$. That is,

$L = \frac{\frac{32}{27}wh^2}{l}$. When $l = 16, w = 24$ and $h = 8$, $L = \frac{\frac{32}{27}(24)\left(8^2\right)}{16} = \frac{\frac{49{,}152}{27}}{16} = \frac{49{,}152}{432} = \frac{1024}{9} = 113\frac{1}{9}kg$.

39. Here $w = \frac{k}{d^2}$. By substitution, $90 = \frac{k}{6400^2} \Rightarrow 90\left(6400^2\right) = k \Rightarrow k = 3{,}686{,}400{,}000$.

That is, $w = \frac{3,686,400,000}{d^2}$. When $d = 7200$, $w = \frac{3,686,400,000}{7200^2} = \frac{3,686,400,000}{51,840,000} = \frac{640}{9} = 71\frac{1}{9}$ kg.

40. Louise can clean the site in 5 hours and can complete $\frac{t}{5}$ of the grooming in t hours. Keith can clean the site in 7 hours and can complete $\frac{t}{7}$ of the grooming in t hours. Together they can complete $\frac{t}{7} + \frac{t}{7}$ of the cleaning in t hours. The job is complete when the fraction of the cleaning reaches 1. $\frac{t}{5} + \frac{t}{7} = 1 \Rightarrow 7t + 5t = 35 \Rightarrow 12t = 35 \Rightarrow t = 2\frac{11}{12}$ or 2 hours and 55 minutes.

41. Terry and Carrie can clean the house in 7 hours and can clean $\frac{t}{7}$ of the hours in t hours. Daniel can make a mess of the house in 2 hours and can make a mess of $\frac{t}{2}$ of the house in t hours. The job is complete when the fraction of the messy house reaches 1. $\frac{t}{2} - \frac{t}{7} = 1 \Rightarrow 7t - 2t = 14 \Rightarrow 5t = 14 \Rightarrow t = 2\frac{4}{5}$ or 2 hours and 48 minutes.

42. Jack and Kevin can clean the dishes in 15 minutes and can clean $\frac{t}{15}$ of the dishes in t minutes. Jack can clean the dishes in 35 minutes and can clean $\frac{t}{35}$ of the dishes in t minutes. The job is complete when the fraction of the clean dishes reaches 1. $\frac{t}{15} - \frac{t}{35} = 1 \Rightarrow 35t - 15t = 525 \Rightarrow 20t = 525 \Rightarrow t = 26\frac{1}{4}$ or 26 minutes and 15 seconds.

43. This function yields the cube root graph shifted a units to the left. See Figure 43.

44. This function yields the square root graph stretched and reflected across the x-axis. See Figure 44.

45. This function yields the cube root graph shifted up b units. See Figure 45.

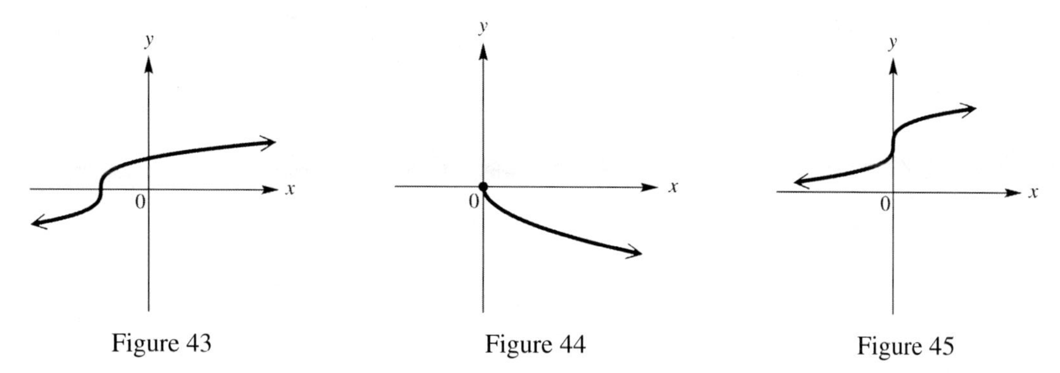

Figure 43 Figure 44 Figure 45

46. This function yields the cube root graph shifted a units to the right. See Figure 46.

47. This function yields the cube root graph stretched, reflected across the x-axis, and shifted down b units. See Figure 47.

48. This function yields the square root graph shifted a units to the left and b units up. See Figure 48.

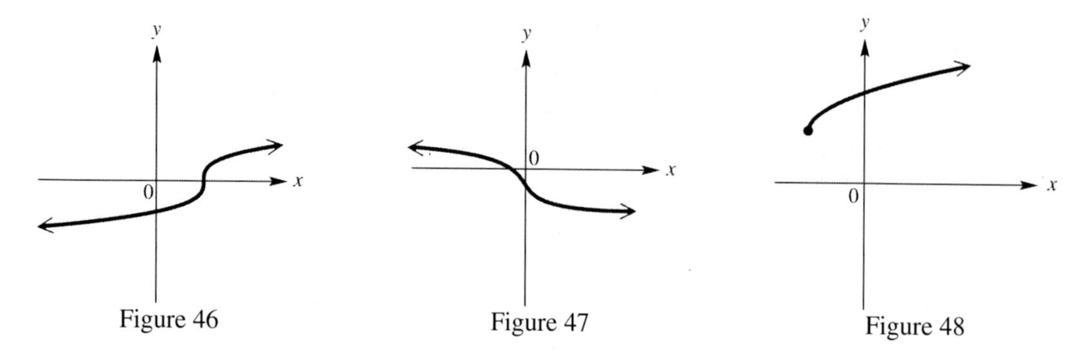

Figure 46 Figure 47 Figure 48

49. Since $12^2 = 144$, $\sqrt{144} = 12$.

50. Since $(-4)^3 = -64$, $\sqrt[3]{-64} = -4$.

51. Since $\left(\frac{1}{3}\right)^3 = \frac{1}{27}$, $\sqrt[3]{\frac{1}{27}} = \frac{1}{3}$.

52. Since $3^4 = 81$, $\sqrt[4]{81} = 3$.

53. Since $\left(-\frac{2}{3}\right)^5 = -\frac{32}{243}$, $\sqrt[5]{\frac{32}{243}} = -\frac{2}{3}$.

54. Since $(-2)^5 = -32$, $(-32)^{1/5} = -2$. Therefore, $-(-32)^{1/5} = -(-2) = 2$.

55. $36^{-3/2} = \frac{1}{36^{3/2}} = \frac{1}{\left(\sqrt{36}\right)^3} = \frac{1}{6^3} = \frac{1}{216}$

56. $-1000^{2/3} = -\left(\sqrt[3]{1000}\right)^2 = -(10)^2 = -100$

57. $(-27)^{-4/3} = \frac{1}{(-27)^{4/3}} = \frac{1}{\left(\sqrt[3]{-27}\right)^4} = \frac{1}{(-3)^4} = \frac{1}{81}$

58. $16^{3/4} = \left(\sqrt[4]{16}\right)^3 = 2^3 = 8$

59. $\sqrt[4]{84.6} \approx 2.429260411$

60. $\sqrt[4]{\frac{1}{16}} = 0.5$

61. $\left(\frac{1}{8}\right)^{4/3} = 0.0625$

62. $12^{1/3} \approx 2.289428485$

63. $2x - 4 \geq 0 \Rightarrow 2x \geq 4 \Rightarrow x \geq \frac{4}{2} \Rightarrow x \geq 2$; The domain is $[2, \infty)$.

64. The graph of f is a square root graph reflected across the x-axis. The range is $(-\infty, 0]$.

65. (a) None. The function is never increasing.

(b) The function is decreasing over its entire domain, $(2, \infty)$.

66. Solving the equation for y yields $x^2 + y^2 = 25 \Rightarrow y^2 = 25 - x^2 \Rightarrow y = \pm\sqrt{25 - x^2}$. The expressions are $y_1 = \sqrt{25 - x^2}$ and $y_2 = -\sqrt{25 - x^2}$ which are graphed in a square window in Figure 66.

67. (a) $\sqrt{5+2x} = x+1 \Rightarrow \left(\sqrt{5+2x}\right)^2 = (x+1)^2 \Rightarrow 5+2x = x^2+2x+1 \Rightarrow$

$x^2 - 4 = 0 \Rightarrow (x+2)(x-2) = 0 \Rightarrow x = -2$ or 2 (-2 is extraneous)

The solution set is $\{2\}$. A graph of $y_1 = \sqrt{5+2x}$ and $y_2 = x+1$ is shown in Figure 67.

(b) Since the graph of y_1 begins when $5+2x = 0 \Rightarrow x = -2.5$, $y_1 > y_2$ on the interval $[-2.5, 2)$.

(c) From the graph, $y_1 < y_2$ on the interval $(2, \infty)$.

68. (a) $\sqrt{2x+1} - \sqrt{x} = 1 \Rightarrow \sqrt{2x+1} = \sqrt{x}+1 \Rightarrow \left(\sqrt{2x+1}\right)^2 = \left(\sqrt{x}+1\right)^2 \Rightarrow$

$2x+1 = x+2\sqrt{x}+1 \Rightarrow x = 2\sqrt{x} \Rightarrow x^2 = \left(2\sqrt{x}\right)^2 \Rightarrow x^2 = 4x \Rightarrow x^2 - 4x = 0 \Rightarrow x(x-4) = 0 \Rightarrow x = 0$ or 4

The solution set is $\{0, 4\}$. A graph of $y_1 = \sqrt{2x+1} - \sqrt{x}$ and $y_2 = 1$ is shown in Figure 68.

(b) From the graph, $y_1 > y_2$ on the interval $(4, \infty)$.

(c) From the graph, $y_1 < y_2$ on the interval $(0, 4)$.

69. (a) $\sqrt[3]{6x+2} = \sqrt[3]{4x} \Rightarrow \left(\sqrt[3]{6x+2}\right)^3 = \left(\sqrt[3]{4x}\right)^3 \Rightarrow 6x+2 = 4x \Rightarrow 2x = -2 \Rightarrow x = -1$

The solution set is $\{-1\}$. A graph of $y_1 = \sqrt[3]{6x+2}$ and $y_2 = \sqrt[3]{4x}$ is shown in Figure 69.

(b) From the graph, $y_1 \geq y_2$ on the interval $[-1, \infty)$.

(c) From the graph, $y_1 \leq y_2$ on the interval $(-\infty, -1]$.

[-9.4,9.4] by [-6.1,6.1]
Xscl = 1 Yscl = 1

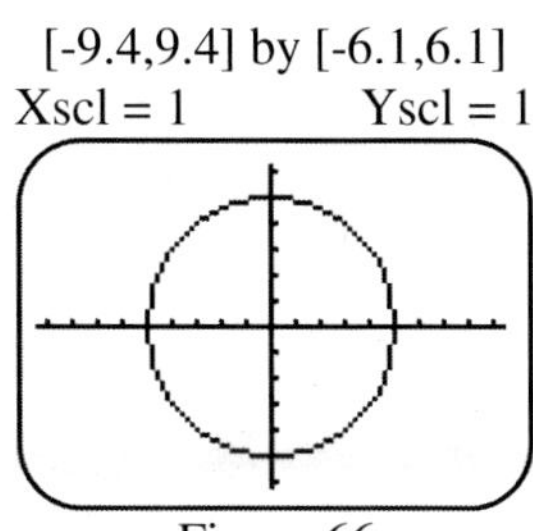

Figure 66

[-9.4,9.4] by [-6.1,6.1]
Xscl = 1 Yscl = 1

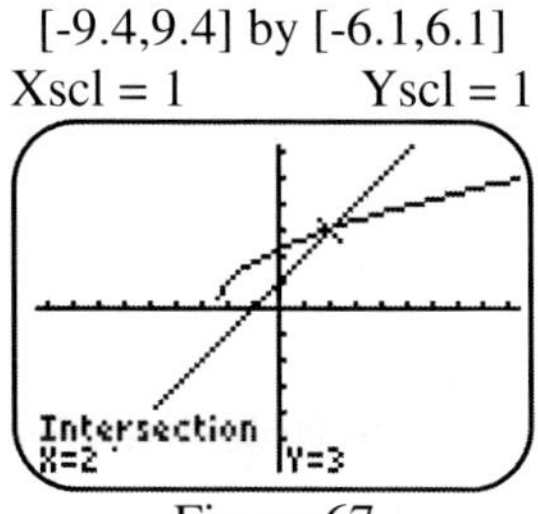

Figure 67

[-2.7,6.7] by [-2,2]
Xscl= 1 Yscl = 1

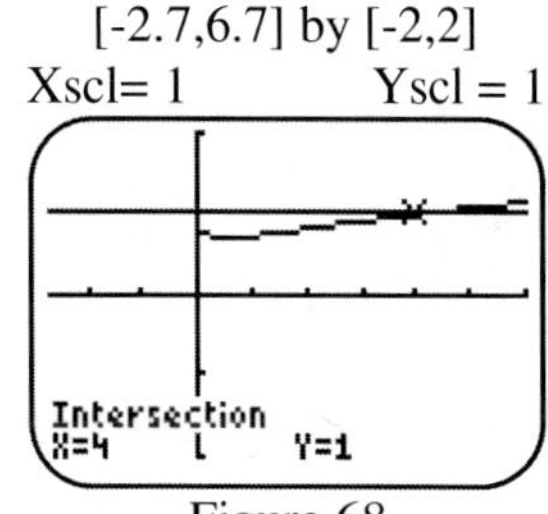

Figure 68

[-4.7,4.7] by [-3.1,3.1]
Xscl = 1 Yscl= 1

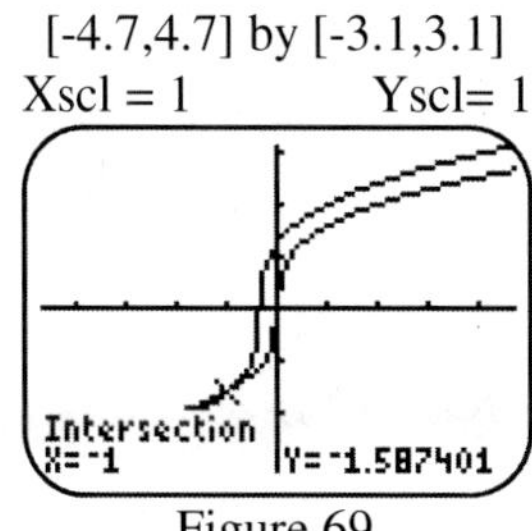

Figure 69

70. (a) $(x-2)^{2/3} - x^{1/3} = 0 \Rightarrow (x-2)^{2/3} = x^{1/3} \Rightarrow \left[(x-2)^{2/3}\right]^3 = \left[x^{1/3}\right]^3 \Rightarrow$

$(x-2)^2 = x \Rightarrow x^2 - 4x + 4 = x \Rightarrow x^2 - 5x + 4 = 0 \Rightarrow (x-1)(x-4) = 0 \Rightarrow x = 1, 4$

The solution set is $\{1, 4\}$. A graph of $y_1 = (x-2)^{2/3}$ and $y_2 = 0$ is shown in Figure 70.

(b) From the graph, $y_1 \geq y_2$ on the interval $(-\infty,1]\cup[4,\infty)$.

(c) From the graph, $y_1 \leq y_2$ on the interval $[1,4]$.

[-2.7,6.7] by [-6.2,6.2]
Xscl = 1 Yscl = 1

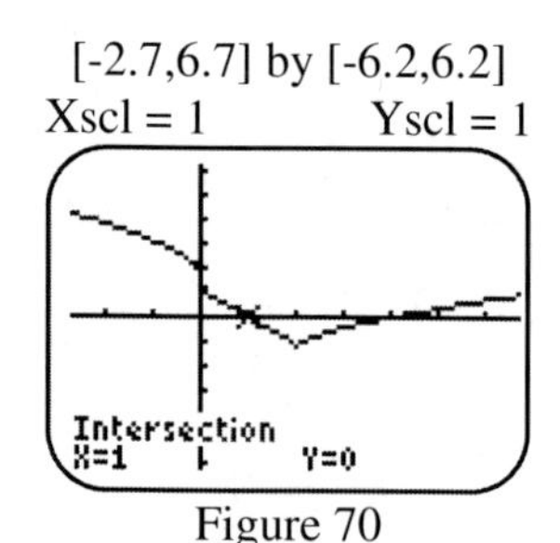

Figure 70

71. $x^5 = 1024 \Rightarrow \left(x^5\right)^{1/5} = 1024^{1/5} \Rightarrow x = 4$ The solution set is $\{4\}$. **Check:** $4^5 = 1024$

72. $x^{1/3} = 4 \Rightarrow \left(x^{1/3}\right)^3 = 4^3 \Rightarrow x = 64$ The solution set is $\{64\}$. **Check:** $64^{1/3} = 4$

73. $\sqrt{x-2} = x-4 \Rightarrow \left(\sqrt{x-2}\right)^2 = (x-4)^2 \Rightarrow x-2 = x^2-8x+16 \Rightarrow x^2-9x+18=0 \Rightarrow$

$(x-3)(x-6)=0 \Rightarrow x=3$ or 6

The solution $x=3$ is extraneous. **Check:** $\sqrt{3-2}=\sqrt{1}=1\neq 3-4$

The solution set is $\{6\}$. **Check:** $\sqrt{6-2}=\sqrt{4}=2=6-4$

74. $x^{3/2} = 27 \Rightarrow \left(x^{3/2}\right)^{2/3} = 24^{2/3} \Rightarrow x = 9$ The solution set is $\{9\}$. **Check:** $9^{3/2} = 27$

75. $2x^{1/4}+3=6 \Rightarrow 2x^{1/4}=3 \Rightarrow x^{1/4}=\frac{3}{2} \Rightarrow \left(x^{1/4}\right)^4 \Rightarrow x=\frac{81}{16}$ The solution set is $\left\{\frac{81}{16}\right\}$.

Check: $2\left(\frac{81}{16}\right)^{1/4}+3=2\left(\frac{3}{2}\right)+3=3+3=6$

76. $\sqrt{x-2}=14-x \Rightarrow \left(\sqrt{x-2}\right)^2=(14-x)^2 \Rightarrow x-2=196-28x+x^2 \Rightarrow$

$x^2-29x+198=0 \Rightarrow (x-11)(x-18)=0 \Rightarrow x=11$ or 18

The solution $x=18$ is extraneous. **Check:** $\sqrt{18-2}=\sqrt{16}=4\neq 14-18$

The solution set is $\{11\}$. **Check:** $\sqrt{11-2}=\sqrt{9}=3=14-11$

77. $\sqrt[3]{2x-3}+1=4 \Rightarrow \sqrt[3]{2x-3}=3 \Rightarrow \left(\sqrt[3]{2x-3}\right)^3=3^3 \Rightarrow 2x-3=27 \Rightarrow 2x=30 \Rightarrow x=15$

The solution set is $\{15\}$. **Check:** $\sqrt[3]{2(15)-3}+1=\sqrt[3]{30-3}+1=\sqrt[3]{27}+1=3+1=4$

78. $x^{1/3}+3x^{1/3}=-2 \Rightarrow 4x^{1/3}=-2 \Rightarrow x^{1/3}=-\frac{1}{2} \Rightarrow \left(x^{1/3}\right)^3=\left(-\frac{1}{2}\right)^3 \Rightarrow x=-\frac{1}{8}$

The solution set is $\left\{-\frac{1}{8}\right\}$. **Check:** $\left(-\frac{1}{8}\right)^{1/3}+3\left(-\frac{1}{8}\right)^{1/3}=-\frac{1}{2}+3\left(-\frac{1}{2}\right)=-\frac{1}{2}-\frac{3}{2}=-\frac{4}{2}=-2$

79. $2x^{-2}-5x^{-1}=3\Rightarrow 2\left(x^{-1}\right)^2-5\left(x^{-1}\right)=3$

Let $u=x^{-1}$, then $2u^2-5u-3=0\Rightarrow(2u+1)(u-3)=0\Rightarrow u=-\frac{1}{2}$ or 3.

If $u=x^{-1}$, then $x=u^{-1}$, thus $x=-2$ or $\frac{1}{3}$

The solution set is $\left\{-2,\frac{1}{3}\right\}$. **Check:** $2(-2)^{-2}-5(-2)^{-1}=3\Rightarrow 2\left(\frac{1}{4}\right)-5\left(-\frac{1}{2}\right)=3\Rightarrow\frac{1}{2}+\frac{5}{2}=3;$

$2\left(\frac{1}{3}\right)^{-2}-5\left(\frac{1}{3}\right)^{-1}=3\Rightarrow 2(9)-5(3)=3\Rightarrow 18-15=3$

80. $x^{-3}+2x^{-2}+x^{-1}=0\Rightarrow(x^{-1})^3+2(x^{-1})^2+(x^{-1})=0$

Let $u=x^{-1}$, then $u^3+2u^2+u=0\Rightarrow u(u^2+2u+1)=0\Rightarrow u(u+1)(u+1)=0\Rightarrow u=-1$ or 0.

If $u=x^{-1}$, then $x=u^{-1}$, thus $x=(-1)^{-1}$ or $(0)^{-1}\Rightarrow x=\frac{1}{-1}$ or $\frac{1}{0}$ Since $\frac{1}{0}$ is undefined, $x=-1$.

The solution set is $\{-1\}$. **Check:** $(-1)^{-3}+2(-1)^{-2}+(-1)^{-1}=0\Rightarrow -1+2+(-1)=0\Rightarrow 0=0$

81. $x^{2/3}-4x^{1/3}-5=0\Rightarrow(x^{1/3})^2-4(x^{1/3})-5=0$

Let $u=x^{1/3}$, then $u^2-4u-5=0\Rightarrow (u-5)(u+1)=0\Rightarrow u=-1$ or 5.

If $u=x^{1/3}$, then $x=u^3$, thus $x=-1$ or 125.

The solution set is $\{-1, 125\}$. **Check:** $(-1)^{2/3}-4(-1)^{1/3}-5=0\Rightarrow 1-(-4)-5=0\Rightarrow 0=0;$

$(125)^{2/3}-4(125)^{1/3}-5=0\Rightarrow 25-4(5)-5=0\Rightarrow 0=0$

82. $x^{3/4}-16x^{1/4}=0\Rightarrow(x^{1/4})^3-16(x^{1/4})=0$

Let $u=x^{1/4}$, then $u^3-16u=0\Rightarrow u\left(u^2-16\right)=0\Rightarrow u(u+4)(u-4)=0\Rightarrow u=-4,\ 0,\ 4.$

If $u=x^{1/4}$, then $x=u^4$; thus $x=0$ or 256.

The solution set in $\{0,256\}$. **Check:** $(0)^{3/4}-16(0)^{1/4}=0\Rightarrow 0-0=0;$

$(256)^{3/4}-16(256)^{1/4}=0\Rightarrow 64-16(4)=0\Rightarrow 0=0$

83. $\sqrt{x+1}+1=\sqrt{2x}\Rightarrow(\sqrt{x+1}+1)^2=2x\Rightarrow x+1+2\sqrt{x+1}+1=2x\Rightarrow 2\sqrt{x+1}=x-2\Rightarrow$

$4(x+1)=x^2-4x+4\Rightarrow 4x+4=x^2-4x+4\Rightarrow x^2-8x=0\Rightarrow x(x-8)=0\Rightarrow x=0$ or 8

The solution $x=0$ is extraneous. **Check:** $\sqrt{0+1}+1=\sqrt{2(0)}\Rightarrow 1+1=0\Rightarrow 2\neq 0$

The solution set is $\{8\}$. **Check:** $\sqrt{8+1}+1=\sqrt{2(8)}\Rightarrow\sqrt{9}+1=\sqrt{16}\Rightarrow 4=4$

84. $\sqrt{x-2}=5-\sqrt{x+3}\Rightarrow(\sqrt{x-2})^2=(5-\sqrt{x+3})^2\Rightarrow x-2=25-10\sqrt{x+3}+x+3\Rightarrow$

$-30=-10\sqrt{x+3}\Rightarrow(-30)^2=\left(-10\sqrt{x+3}\right)^2\Rightarrow 900=100(x+3)\Rightarrow 9=x+3\Rightarrow x=6$

The solution set is $\{6\}$. **Check:** $\sqrt{6-2}=5-\sqrt{6+3}\Rightarrow\sqrt{4}=5-\sqrt{9}\Rightarrow 2=5-3\Rightarrow 2=2$

85. (a) If the length L of the pendulum increases, so does the period of oscillation T.

(b) There are a number of ways to find n and k. One way is to realize that $k = \dfrac{L}{T^n}$ for some integer n. The ratio should be the constant k for each data point when the correct n is found. Another way is to use regression.

(c) By trial and error, $k \approx 0.81;\ n = 2.$

(d) $5 = 0.81(T^2) \Rightarrow \dfrac{5}{0.81} = T^2 \Rightarrow \sqrt{\dfrac{5}{0.81}} = T \Rightarrow T \approx 2.48$ sec

(e) Since $T = \sqrt{\dfrac{L}{k}}$, when length doubles T increases by a factor of $\sqrt{2} \approx 1.414.$

86. Since the volume is $\pi r^2 h = 27\pi$, solving for h yields $h = \dfrac{27}{r^2}$. Substituting this value in the formula for surface area yields $S = 2\pi r\left(\dfrac{27}{r^2}\right) + \pi r^2 \Rightarrow S = \dfrac{54\pi}{r} + \pi r^2$. By graphing $y_1 = \dfrac{54\pi}{r} + \pi r^2$ (not shown), the minimum value for surface area occurs when $r = 3$ inches.

Chapter 4 Test

1. (a) See Figure 1a.

   (b) The graph is obtained by reflecting the graph of $y = \dfrac{1}{x}$ across the x-axis or the y-axis.

   (c) See Figure 1c.

2. (a) See Figure 2a.

   (b) The graph is obtained by reflecting the graph of $y = \dfrac{1}{x^2}$ across the x-axis and shifting it downward 3 units.

   (c) See Figure 2c

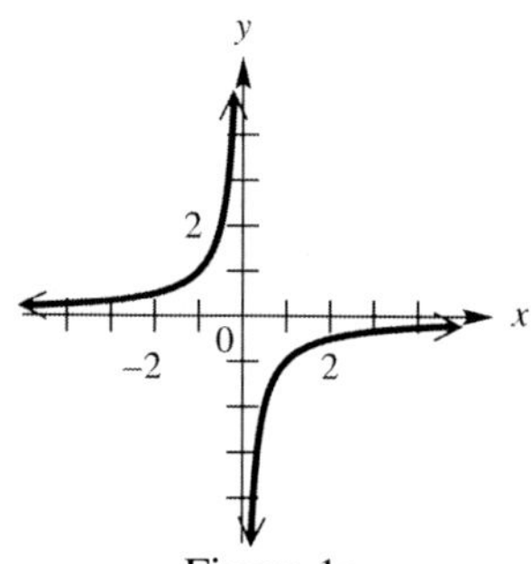

Figure 1a

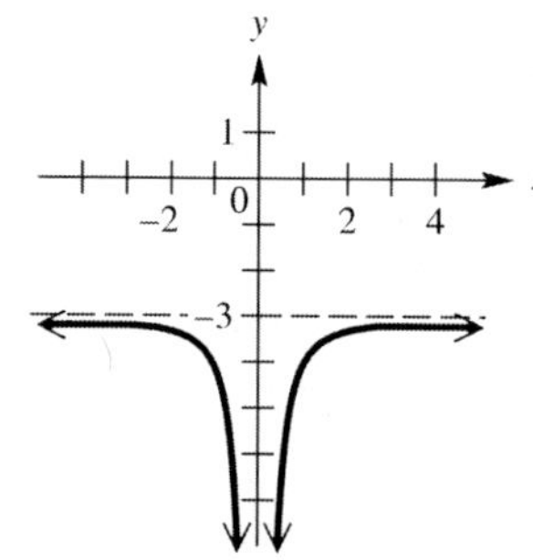

Figure 2a

[-10,10] by [-10,10]
Xscl = 1 Yscl = 1

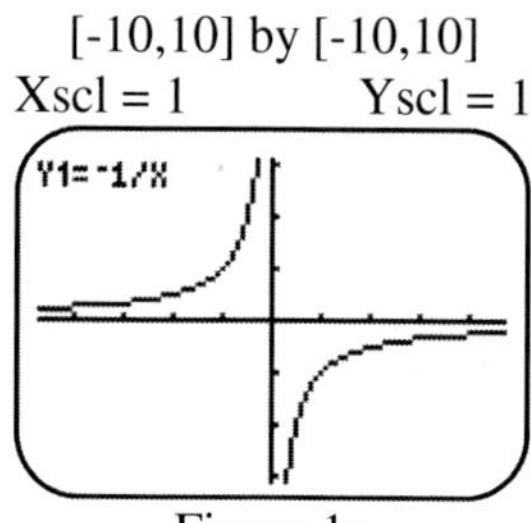

Figure 1c

[-10,10] by [-10,10]
Xscl = 1 Yscl = 1

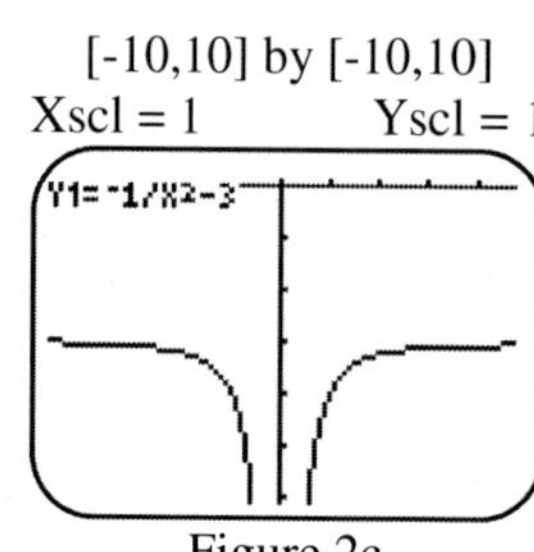

Figure 2c

3. (a) The domain is the set of values that do not make the denominator equal to zero.
$x^2-3x-4=0 \Rightarrow (x-4)(x+1)=0 \Rightarrow x=4,-1$ The domain is $(-\infty,-1)\cup(-1,4)\cup(4,\infty)$.

(b) Vertical asymptotes occur for values of x that make the denominator equal to zero, $x=-1$ and $x=4$.

(c) Since the degrees of the numerator and denominator are equal, the horizontal asymptote can be found using the ratio of the leading coefficients, $y=\frac{1}{1}=1$.

(d) The y-value of the y-intercept occurs at $f(0)=\frac{0^2+0-6}{0^2-3(0)-4}=1.5$. The intercept is $(0,1.5)$.

(e) The x-intercepts occur for values of x that make the numerator equal to zero, -3 and 2. The intercepts are $(-3,0)$ and $(2,0)$

(f) $\frac{x^2+x-6}{x^2-3x-4}=1 \Rightarrow x^2+x-6=x^2-3x-4 \Rightarrow x-6=-3x-4 \Rightarrow 4x=2 \Rightarrow x=0.5$

The coordinates of the intersection point are (0.5, 1).

(g) See Figure 3.

4. (a) Since $\frac{2x^2+x-3}{x-2}=2x+5+\frac{7}{x-2}$, the oblique asymptote is $y=2x+5$.

(b) See Figure 4.

5. (a) Since $\frac{x^2-16}{x+4}=\frac{(x+4)(x-4)}{x+4}=x-4$ for $x\neq -4$, The "hole" occurs when $x=-4$.

(b) See Figure 5.

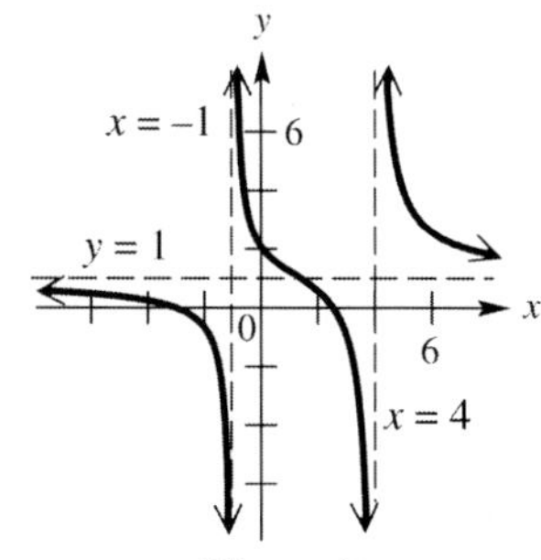

Figure 3

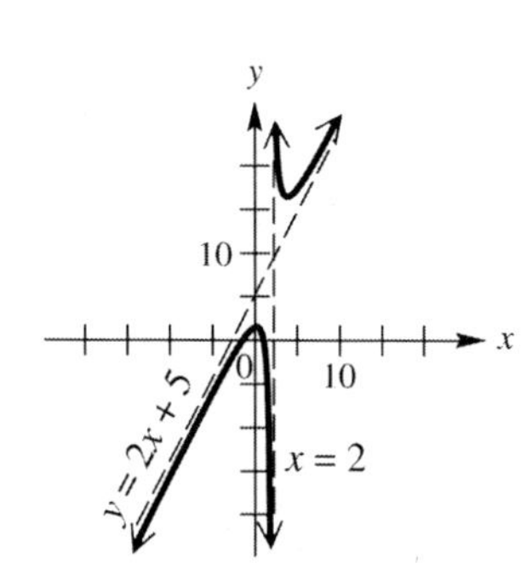

Figure 4

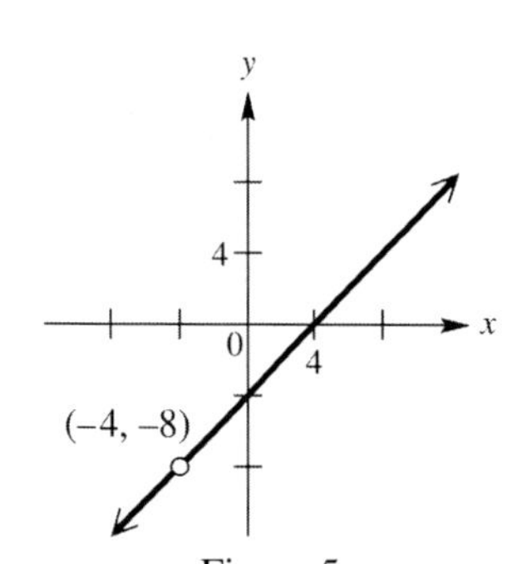

Figure 5

6. (a) Multiply each side of the equation by the LCD, $(x-2)(x+2)$.

$$(x-2)(x+2)\cdot\frac{3}{x-2}+(x-2)(x+2)\cdot\frac{21}{x^2-4}=\frac{14}{x+2}\cdot(x-2)(x+2) \Rightarrow$$

$3(x+2)+21=14(x-2) \Rightarrow 3x+6+21=14x-28 \Rightarrow 11x=55 \Rightarrow x=5$. The solution set is $\{5\}$.

(b) Graph $y_1=3/(x-2)+21/(x^2-4)-14/(x+2)$ as shown in Figure 6. With vertical asymptotes $x=-2$ and $x=2$, and x-intercept 5, the graph is above or intersects the x-axis on the interval $(-\infty,-2)\cup(2,5]$. The solution set is $(-\infty,-2)\cup(2,5]$.

[-9.4,9.4] by [-20,10]
Xscl = 1 Yscl = 2

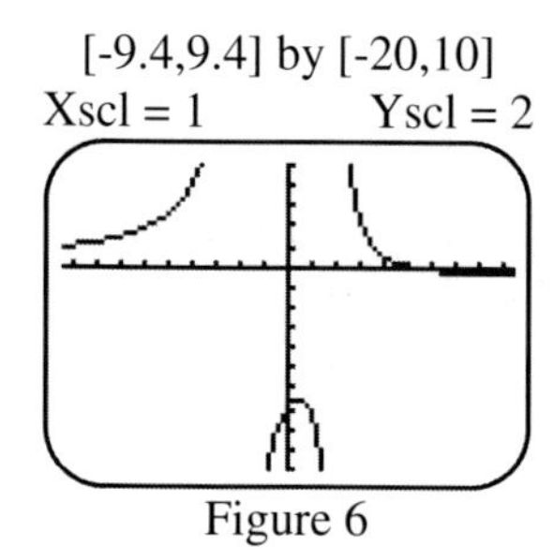

Figure 6

7. (a) $W(30) = \frac{1}{40-30} = \frac{1}{10}$; $W(39) = \frac{1}{40-39} = \frac{1}{1} = 1$; $W(39.9) = \frac{1}{40-39.9} = \frac{1}{0.1} = 10$.

When the rate is 30 vehicles per minute, the average wait time is $\frac{1}{10}$ minute (6 seconds). The other results are interpreted similarly.

(b) The vertical asymptote occurs when the denominator equals zero, at $x = 40$. As x approaches 40, W gets larger and larger without bound. See Figure 7.

(c) $5 = \frac{1}{40-x} \Rightarrow 5(40-x) = 1 \Rightarrow 200 - 5x = 1 \Rightarrow 5x = 199 \Rightarrow x = \frac{199}{5} = 39.8$

8. Here $p = \frac{k\sqrt[3]{w}}{h}$. By substitution, $100 = \frac{k\sqrt[3]{48,820}}{78.7} \Rightarrow 7870 = \sqrt[3]{48,820}k \Rightarrow k = \frac{7870}{\sqrt[3]{48,820}} \Rightarrow k \approx 215.3$. That is, $p = \frac{215.3\sqrt[3]{w}}{h}$. When $w = 5,430$ and $h = 88.9$, $p = \frac{215.3\sqrt[3]{54,430}}{88.9} \approx 92$. The individual is undernourished.

[0,40] by [-0.5,1]
Xscl = 5 Yscl = 1

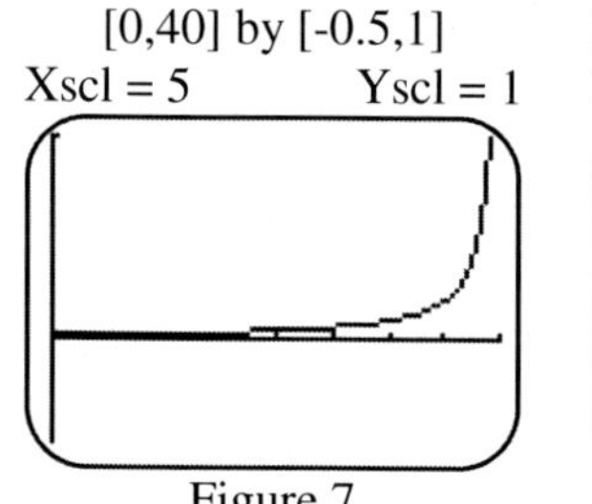

Figure 7

[-5,42] by [-1000,10,000]
Xscl = 5 Yscl = 1000

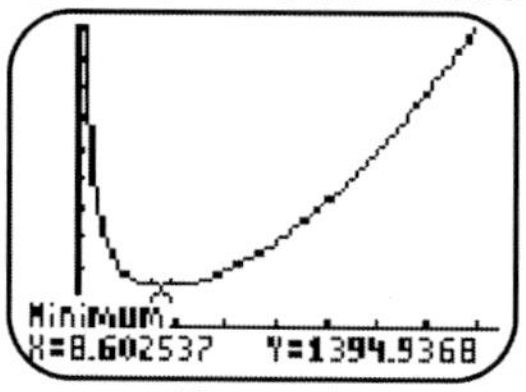

Minimum
X=8.602537 Y=1394.9368

Figure 9

9. Graph $y_1 = (8000 + 2\pi x^3)/x$ as shown in Figure 9. When the radius is approximately 8.6 cm, the amount of aluminum needed will be a minimum of approximately 1394.9 cm^2.

10. (a) $\sqrt[3]{-27} = \sqrt[3]{-1^3 \cdot 3^3} = -3$

(b) $25^{-\frac{3}{2}} = \left(25^{\frac{1}{2}}\right)^{-3} = 5^{-3} = \frac{1}{5^3} = \frac{1}{125}$

11. $\left(\sqrt[3]{x}\right)^{-4} = x^{-\frac{4}{3}} = \frac{1}{x^{4/3}}$

12. See Figure 12a and Figure 12b.

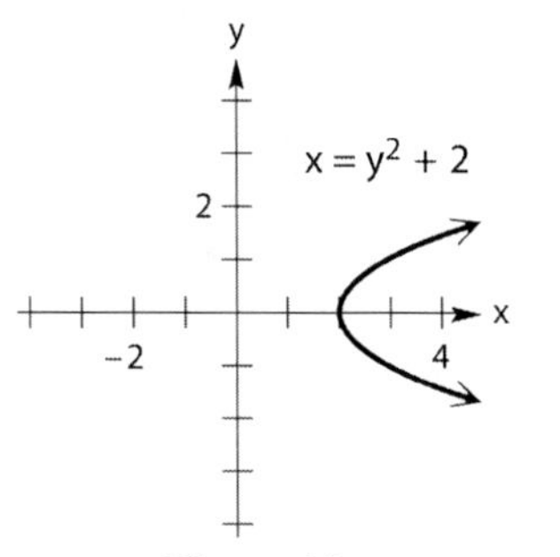

Figure 12a

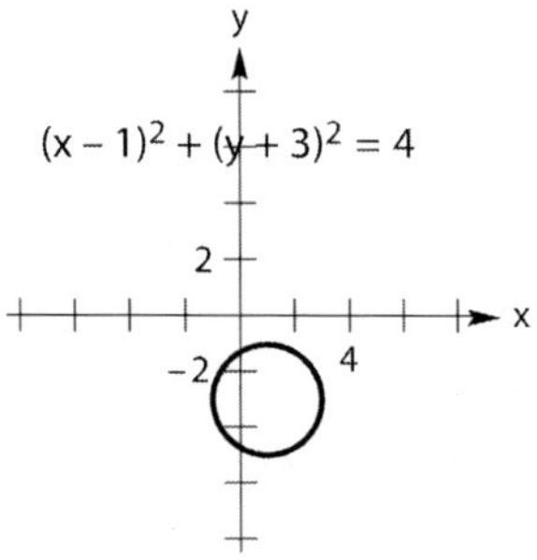

y
(x − 1)² + (y + 3)² = 4
2
−2
4
x

Figure 12b

13. The graph of $y_1 = -\sqrt{5-x}$ is shown in Figure 13.

(a) $5-x \geq 0 \Rightarrow 5 \geq x \Rightarrow x \leq 5$; The domain is $(-\infty, 5]$.

(b) From the graph, the range is $(-\infty, 0]$.

(c) This function *increases* over its entire domain.

(d) The graph intersects the x-axis when $x = 5$. The solution set is $\{5\}$.

(e) The graph below the x-axis for all values in the domain except 5. The solution set is $(-\infty, 5)$.

14. (a) $\sqrt{4-x} = x+2 \Rightarrow \left(\sqrt{4-x}\right)^2 = (x+2)^2 \Rightarrow 4-x = x^2+4x+4 \Rightarrow x^2+5x = 0 \Rightarrow x(x+5) = 0 \Rightarrow x = 0,\ -5.$

The solution $x = -5$ is extraneous. The solution set is $\{0\}$. The graph of $y_1 = \sqrt{4-x}$ and $y_2 = x+2$ shown in Figure 14 supports this result.

(b) The graph of y_1 is above the graph of y_2 on the interval $(-\infty, 0)$. The solution set is $(-\infty, 0)$.

(c) The graph of y_1 is below or intersects the graph of y_2 on the interval $[0, 4]$. The solution set is $[0, 4]$.

[-10,10] by [-10,10]
Xscl = 1 Yscl = 1

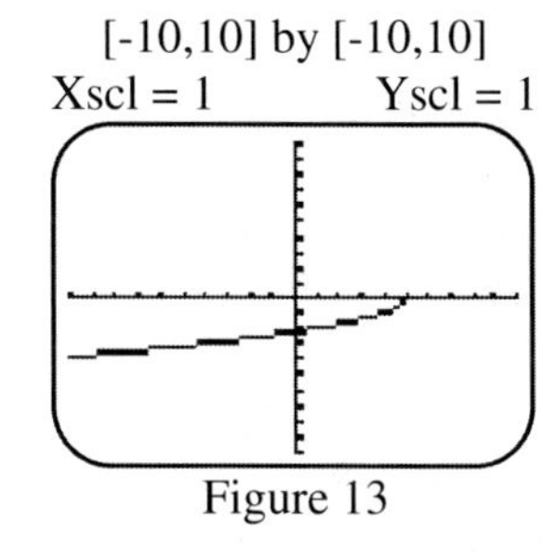

Figure 13

[-10,10] by [-10,10]
Xscl = 1 Yscl = 1

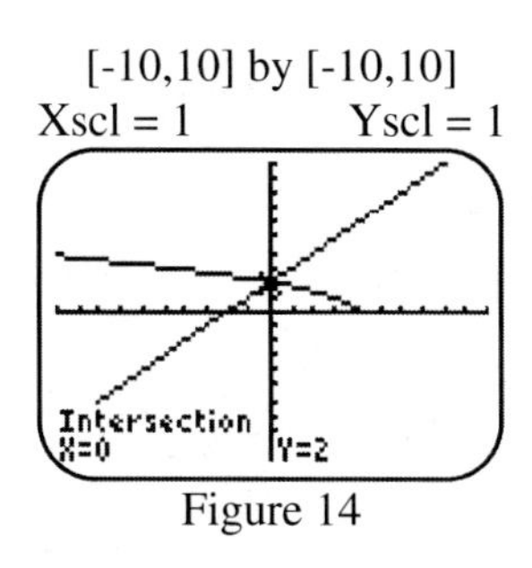

Figure 14

[0,600] by [70,0000,100,000]
Xscl = 100 Yscl = 10,000

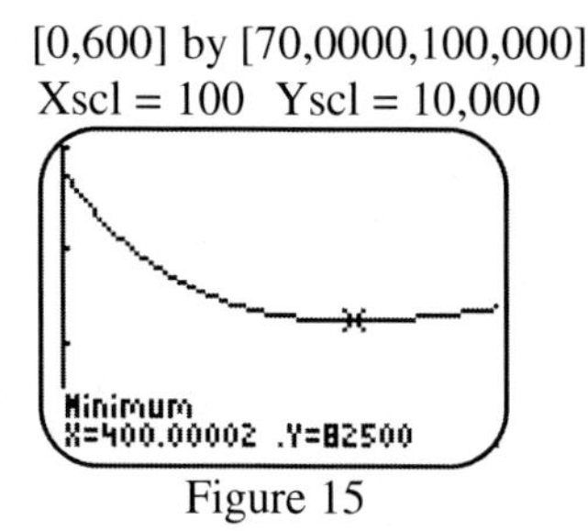

Figure 15

15. Let S be a point on land between Q and R such that the distance from Q to S is x yards. Then the distance from S to R is given by $600 - x$. By the Pythagorean theorem, the distance from P to S is $\sqrt{x^2+300^2}$. The total cost of the cable is given by $C = 125\sqrt{x^2+300^2} + 100(600-x)$. The graph of this function is shown in Figure 15. The minimum cost is achieved when the cable is laid underwater from P to S, which is located 400 yards away from Q in the direction of R.

16. $z=\dfrac{kx^2}{y}\Rightarrow 10=\dfrac{k\left(4^2\right)}{2}\Rightarrow k=\dfrac{5}{4},\ z=\dfrac{5/4(6)^2}{8}=5.625$

Chapter 5: Inverse, Exponential, and Logarithmic Functions

5.1: Inverse Functions

1. Different x-values always produce different y-values; therefore, yes, it is one-to-one.
3. Choosing 2 and -2 as values for x yields $f(2)=2^2=4$ and $f(-2)=(-2)^2=4$. Since different values of x produce the same value for $f(x)$, the function is not one-to-one.
5. Choosing 6 and -6 as values for x yields $f(6)=\sqrt{36-6^2}=0$ and $f(-6)=\sqrt{36-(6)^2}=0$ .Since different values of x produce the same value for $f(x)$, the function is not one-to-one.
7. Every horizontal line will intersect the graph at exactly one point; therefore, yes, it is one-to-one.
9. There are horizontal lines that will intersect the graph at more than one point; therefore, it is not one-to-one.
11. Every horizontal line will intersect the graph at exactly one point; therefore, yes, it is one-to-one.
13. A certain horizontal line intersects the whole horizontal graph (more than one point); therefore, it is not one-to-one.
15. Every horizontal line will intersect the graph at exactly one point; therefore, yes, it is one-to-one.
17. Choosing 4 and 0 as values for x yields $f(4)=(4-2)^2=4$ and $f(0)=(0-2)^2=4$. Since different values of x produce the same value for $f(x)$, the function is not one-to-one.
19. Different x-values always produce different y-values; therefore, yes, it is one-to-one.
21. Different x-values always produce different y-values; therefore, yes, it is one-to-one.
23. Different x-values always produce different y-values; therefore, yes, it is one-to-one.
25. Since different x-values greater than zero all produce the same $f(x)=3$, it is not one-to-one.
27. For a function to have an inverse, it must be one-to-one.
29. If f and g are inverses, then $(f\circ g)(x)=x$, and $(g\circ f)(x)=x$.
31. If the point (a,b) lies on the graph of f, and f has an inverse, then the point (b,a) lies on the graph f^{-1}.
33. If the function f has an inverse, then the graph of f^{-1} may be obtained by reflecting the graph of f across the line with equation $y=x$.
35. If $f(-4)=16$ and $f(4)=16$, then f does not have an inverse because it is not one-to-one.
37. $(f\circ g)(x)=3\left(\frac{x+7}{3}\right)-7=x+7-7=x$ and $(g\circ f)(x)=\frac{(3x-7)+7}{3}=\frac{3x}{3}=x$. Since $(f\circ g)(x)=x$ and $(g\circ f)(x)=x$, the function f and g are inverses.
39. $(f\circ g)(x)=\left(\sqrt[3]{x-4}\right)^3+4=x-4+4=x$ and $(g\circ f)(x)=\sqrt[3]{(x^3+4)-4}=\sqrt[3]{x^3}=x$. Since $(f\circ g)(x)=x$ and $(g\circ f)(x)=x$, the function f and g are inverses.
41. $(f\circ g)(x)=-(-\sqrt[5]{x})^5=-(-x)=x$ and $(g\circ f)(x)=-\sqrt[5]{(-x^5)}=-(-x)=x$. Since $(f\circ g)(x)=x$ and $(g\circ f)(x)=x$, the function f and g are inverses.

43. Every x-value in f corresponds to only one y-value, and every y-value corresponds to only one x-value, so f is a one-to-one function. The inverse function is found by interchanging the x- and y-values in each ordered pair; therefore, $f^{-1} = \{(4,10),(5,20),(6,30),(7,40)\}$.

45. Every x-value in f corresponds to only one y-value. However, the y-value 5 corresponds to two x-values, 1 and 3. Because some y-values corresponds to more than one x-value, f is not one-to-one.

47. Every x-value in f corresponds to only one y-value, and every y-value corresponds to only one x-value, so f is a one-to-one function. The inverse function is found by interchanging the x- and y-values in each ordered pair. Therefore, $f^{-1} = \{(0^2,0),(1^2,1),(2^2,2),(3^2,3),(4^2,4)\}$.

49. Untying your shoelaces

51. Leaving a room

53. Unwrapping a package

55. For the function $y = 3x - 4$, the inverse is $x = 3y - 4 \Rightarrow x + 4 = 3y \Rightarrow y = \frac{x+4}{3} \Rightarrow f^{-1}(x) = \frac{x+4}{3}$. For the graphs, see Figure 55. Since all real numbers can be input for x in f and f^{-1}, and since all real numbers can be solutions for $f(x)$ and $f^{-1}(x)$, the domain and range of both f and f^{-1} are $(-\infty, \infty)$.

57. For the function $y = x^3 + 1$, the inverse is $x = y^3 + 1 \Rightarrow x - 1 = y^3 \Rightarrow y = \sqrt[3]{x-1} \Rightarrow f^{-1}(x) = \sqrt[3]{x-1}$. For the graphs, see Figure 57. Since all real numbers can be input for x in f and f^{-1}, and all real numbers can be solutions for $f(x)$ and $f^{-1}(x)$, the domain and range of both f and f^{-1} are $(-\infty, \infty)$.

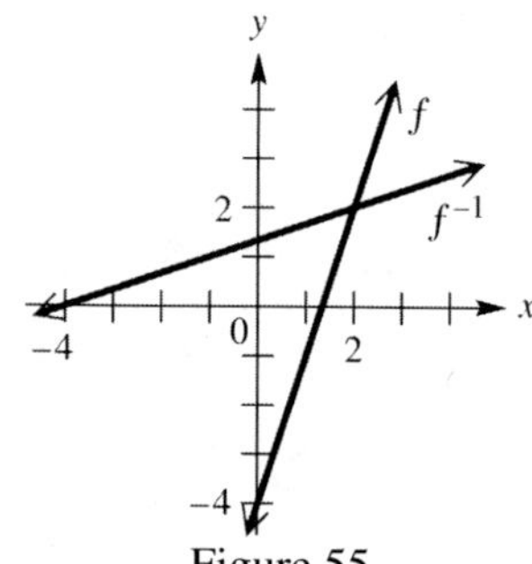

Figure 55

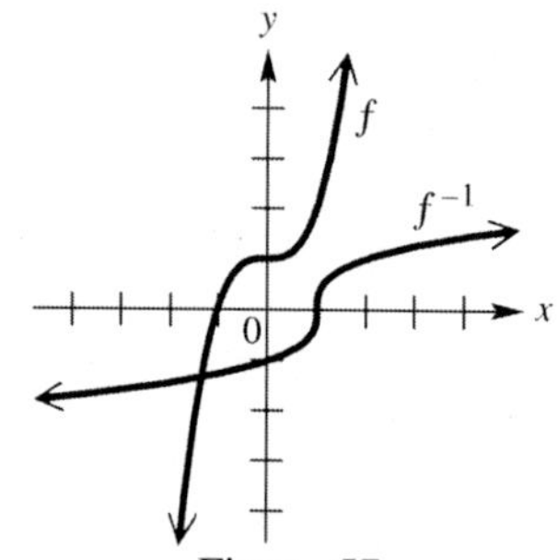

Figure 57

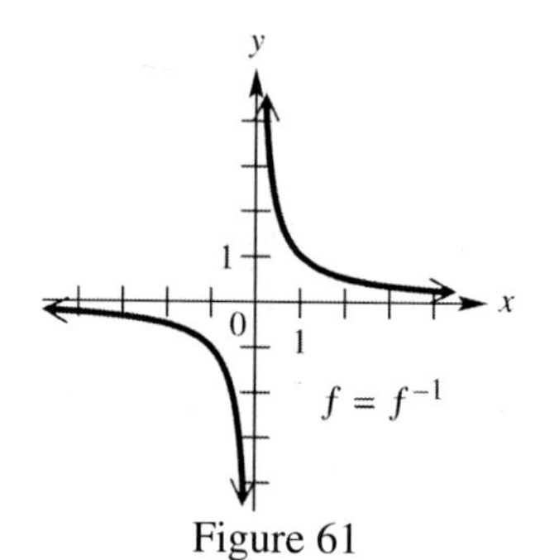

Figure 61

59. Since when $y = 4$, $x = -2$ or 2, the function is not one-to-one.

61. For the function $y = \frac{1}{x}$, the inverse is $x = \frac{1}{y} \Rightarrow y = \frac{1}{x} \Rightarrow f^{-1}(x) = \frac{1}{x}$ For the graphs, see Figure 61. Since all real numbers except 0 can be input for x in f and f^{-1}, and all real numbers except 0 can be solutions for $f(x)$, the domain and range of both f and f^{-1} are $(-\infty, 0) \cup (0, \infty)$.

63. For the function $y = \dfrac{2}{x+3}$, the inverses is $x = \dfrac{2}{y+3} \Rightarrow (y+3)x = 2 \Rightarrow y+3 = \dfrac{2}{x} \Rightarrow y = \dfrac{2}{x} - 3 \Rightarrow$

$y = \dfrac{2-3x}{x} \Rightarrow f^{-1}(x) = \dfrac{2-3x}{x}$. For the graphs, see Figure 63. All real numbers except -3 can be input for x in f and all real numbers except 0 can be input for x in f^{-1}. Therefore, the domain of f is equal to the range of $f^{-1} = (-\infty,-3) \cup (-3,\infty)$, and the domain of f^{-1} is equal to the range of $f = (-\infty,0) \cup (0,\infty)$.

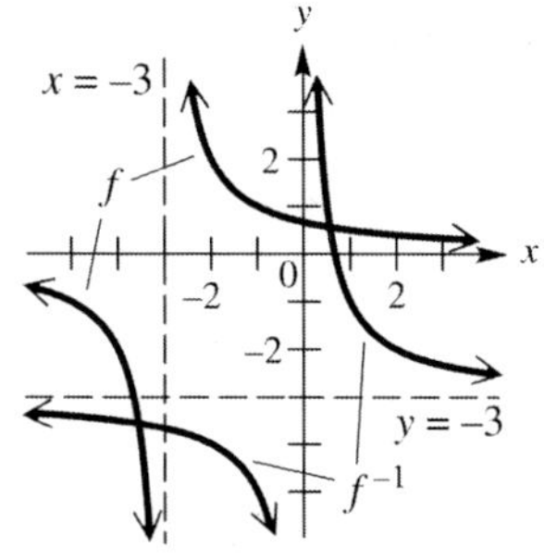

Figure 63

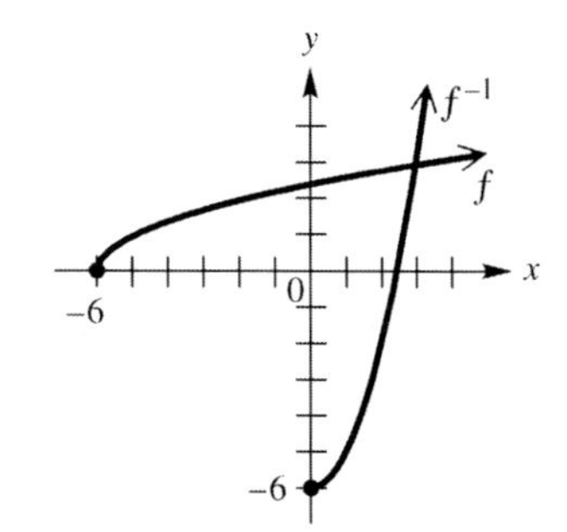

Figure 65

65. For the function $f(x) = \sqrt{6+x},\ x \ge -6$, the inverse is $x = \sqrt{6+y},\ x \ge 0$ (A square root can not equal a negative number.) $\Rightarrow x^2 = 6 + y \Rightarrow y = x^2 - 6 \Rightarrow f^{-1}(x) = x^2 - 6,\ x \ge 0$. For the graph, see Figure 65. All real numbers $x \ge -6$ can be input for x in x and all real numbers $x \ge 0$ can be input for x in f^{-1}. Also, all real numbers $x \ge 0$ can be a solution for f and all real numbers $x \ge -6$ can be a solution for f^{-1}. Therefore, the domain of f is equal to the range of $f^{-1} = [-6,\infty)$, and the domain of f^{-1} is equal to the range of $f = [0,\infty)$.

67. $y = \dfrac{4x}{x+1} \Rightarrow y(x+1) = 4x \Rightarrow xy + y = 4x \Rightarrow xy - 4x = -y \Rightarrow x(y-4) = -y \Rightarrow x = \dfrac{-y}{y-4} \Rightarrow$

$f^{-1}(x) = \dfrac{-x}{x-4} \Rightarrow f^{-1}(x) = \dfrac{x}{4-x}$

69. $y = \dfrac{1-2x}{3x} \Rightarrow 3xy = 1 - 2x \Rightarrow 3xy + 2x = 1 \Rightarrow x(3y+2) = 1 \Rightarrow x = \dfrac{1}{3y+2} \Rightarrow f^{-1}(x) = \dfrac{1}{3x+2}$

71. $y = \sqrt{x^2 - 4}, x \ge 2 \Rightarrow y^2 = x^2 - 4 \Rightarrow x^2 = y^2 + 4 \Rightarrow x = \sqrt{y^2 + 4} \Rightarrow f^{-1}(x) = \sqrt{x^2 + 4}$

73. $y = 5x^3 - 7 \Rightarrow 5x^3 = y + 7 \Rightarrow x^3 = \dfrac{y+7}{5} \Rightarrow x = \sqrt[3]{\dfrac{y+7}{5}} \Rightarrow f^{-1}(x) = \sqrt[3]{\dfrac{x+7}{5}}$

75. $y = \dfrac{x}{4+3x} \Rightarrow y(4+3x) = x \Rightarrow 4y + 3xy = x \Rightarrow 3xy - x = -4y \Rightarrow x(3y-1) = -4y \Rightarrow$

$x = \dfrac{-4y}{3y-1} \Rightarrow f^{-1}(x) = \dfrac{-4x}{3x-1}$

77. $y=\dfrac{3-x}{2x+1} \Rightarrow y(2x+1)=3-x \Rightarrow 2xy+y=3-x \Rightarrow 2xy+x=3-y \Rightarrow x(2y+1)=3-y \Rightarrow x=\dfrac{3-y}{2y+1} \Rightarrow$

$f^{-1}(x)=\dfrac{3-x}{2x+1}$

79. $f(2)=(2)^3=8.$

81. $f(-2)=(-2)^3=-8.$

83. If $f(x)=x^3$, then the inverse is $x=y^3 \Rightarrow y=\sqrt[3]{x} \Rightarrow f^{-1}(x)=\sqrt[3]{x}$, and $f^{-1}(0)=\sqrt[3]{0}=0.$

85. For inverses, when $f^{-1}(4)$, then $f(x)=4$ Therefore, from the graph, $x=4.$

87. For inverses, when $f^{-1}(0)$, then $f(x)=0.$ Therefore, from the graph $x=2.$

89. For inverses, when $f^{-1}(-3)$, then $f(x)=-3.$ Therefore, from the graph $x=-2.$

91. The graphs are reflections of each other through $x=y$; therefore, the functions are inverses.

93. The graphs are not reflections of each other through $x=y$; therefore, the functions are not inverses.

95. Yes, for $f(x)=2x+4$ the inverse is $x=2y+4 \Rightarrow x-4=2y \Rightarrow \dfrac{x-4}{2}=y \Rightarrow f^{-1}(x)=\dfrac{1}{2}x-2$, which is equal to $g(x)$.

97. Since $(x_1,y_1)=(y_2,x_2)$, the x's and y's are switched, and the screen suggest that they are linear functions.

99. Reflect the graph across the line $y=x.$ See Figure 99.

101. Reflect the graph across the line $y=x.$ See Figure 101.

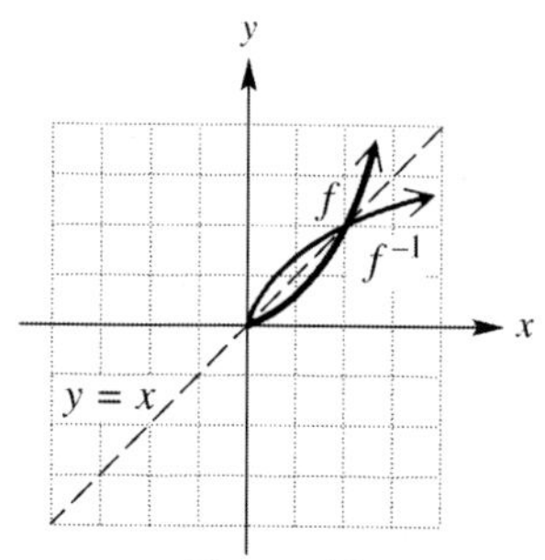

Figure 99

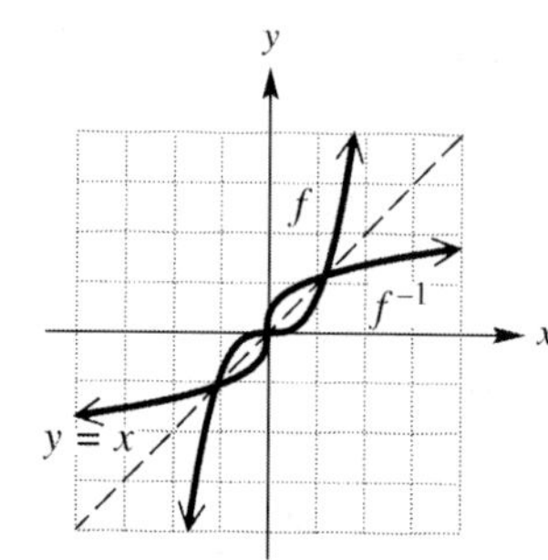

Figure 101

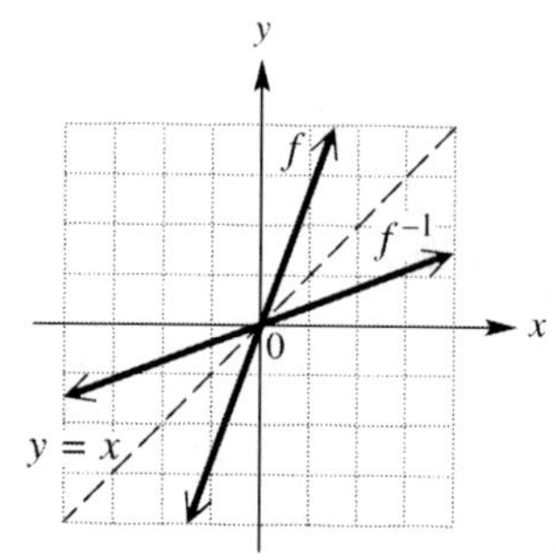

Figure 103

103. Reflect the graph across the line $y=x.$ See Figure 103.

105. It represents the cost in dollars of building 1000 cars.

107. With $m=a=\dfrac{a}{1}$, the inverse function will switch the x and y terms and the graph of the inverse is a reflection across the line $y=x$ which will have the slope $m=\dfrac{1}{a}$.

109. From the graph of the function in the given window, the function fails the horizontal line test and is not one-to-one. See Figure 109.

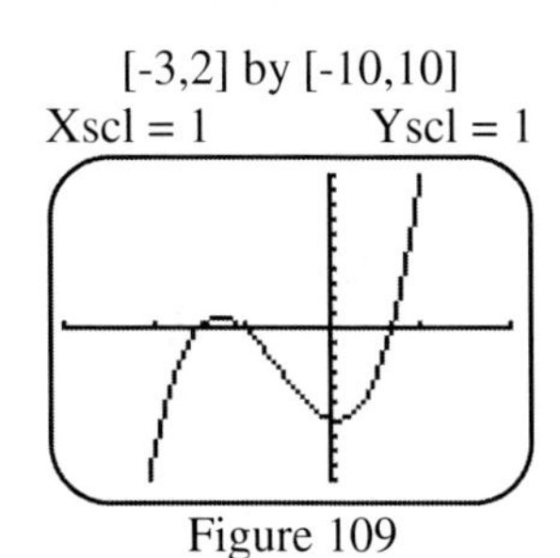

Figure 109

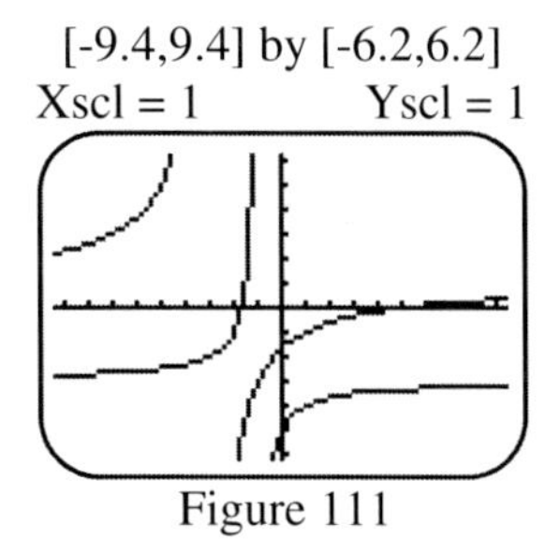

Figure 111

111. From the graph of the function in the given window, the function passes the horizontal line test and is one-to-one. The inverse is $x = \dfrac{y-5}{y+3} \Rightarrow xy + 3x = y - 5 \Rightarrow xy - y = -3x - 5 \Rightarrow y(x-1) = -3x - 5 \Rightarrow y = \dfrac{-3x-5}{x-1} \Rightarrow$

$f^{-1}(x) = \dfrac{-3x-5}{x-1}$. For both graphs, see Figure 111.

113. If $f(x) = 4x - 1$, then the inverse is $x = 4y - 1 \Rightarrow x + 1 = 4y \Rightarrow y = \dfrac{x+1}{4} \Rightarrow f^{-1}(x) = \dfrac{x+1}{4}$. Inputting the numbers given yields numbers that correspond to the letters: TREASURE HUNT IS ON.

115. Using the function $f(x) = x^3 + 1$ on the letters NO PROBLEM or the corresponding numbers 14, 15, 16, 18, 15, 2, 12, 5, and 13, yields the numbers 2745, 3376, 4097, 5833, 3376, 9, 1729, 126, and 2198. The inverse is $x = y^3 + 1 \Rightarrow x - 1 = y^3 \Rightarrow \sqrt[3]{x-1} = y \Rightarrow f^{-1}(x) = \sqrt[3]{x-1}$.

117. From the graph on a calculator, one possible answer that passes the horizontal line test is the domain $[0, \infty)$.

119. From the graph on a calculator, one possible answer that passes the horizontal line test is the domain $[6, \infty)$.

121. From the graph on a calculator, one possible answer that passes the horizontal line test is the domain $[0, \infty)$.

123. If $f(x) = -x^2 + 4,\ x \ge 0$, then the inverse is $x = -y^2 + 4,\ y \ge 0 \Rightarrow x - 4 = -y^2 \Rightarrow$ $4 - x = y^2 \Rightarrow y = \sqrt{4-x} \Rightarrow f^{-1}(x) = \sqrt{4-x}$.

125. If $f(x) = |x-6|,\ x \ge 6$, then the inverse is $x = |y-6|,\ y \ge 6 \Rightarrow x = y - 6 \Rightarrow x + 6 = y \Rightarrow$ $f^{-1}(x) = x + 6,\ x \ge 0$.

5.2: Exponential Functions

1. See Figure 1
3. See Figure 3

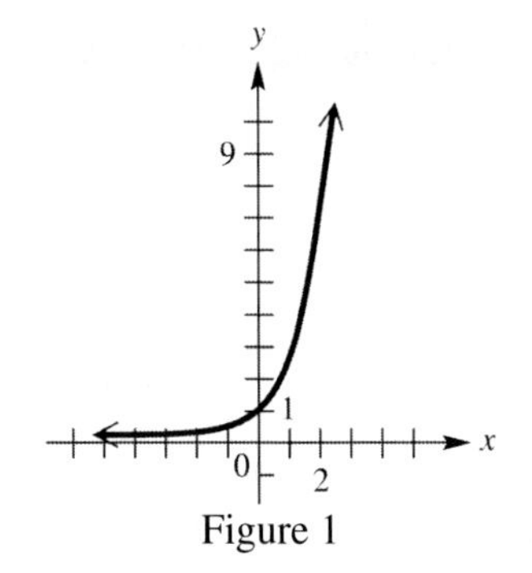

Figure 1

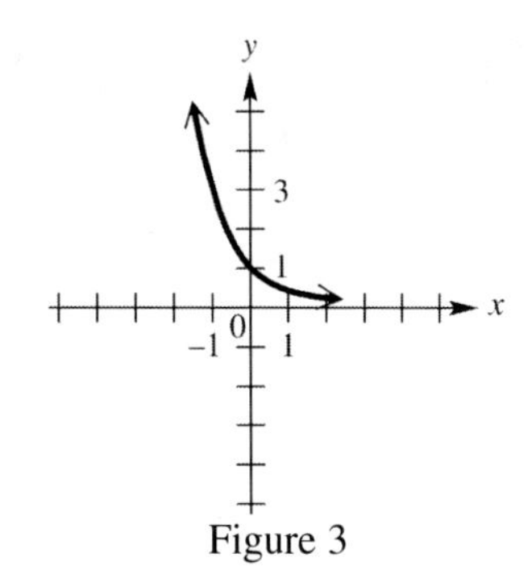

Figure 3

5. $4^x = 2 \Rightarrow (2^2)^x = 2^1 \Rightarrow 2^{2x} = 2^1 \Rightarrow 2x = 1 \Rightarrow x = \dfrac{1}{2}$.

7. $\left(\dfrac{1}{2}\right)^x = 4 \Rightarrow (2^{-1})^x = 2^2 \Rightarrow 2^{-x} = 2^2 \Rightarrow -x = 2 \Rightarrow x = -2$.

9. From the calculator, $2^{\sqrt{10}} \approx 8.952419619$.

11. From the calculator, $\left(\dfrac{1}{2}\right)^{\sqrt{2}} \approx 0.3752142272$.

13. From the calculator, $4.1^{-\sqrt{3}} \approx 0.0868214883$.

15. From the calculator, $\sqrt{7}^{\sqrt{7}} \approx 13.1207791$

17. From the calculator, the point $\left(\sqrt{10}, 8.9524196\right)$ lies on the graph of $y = 2^x$.

19. From the calculator, the point $(\sqrt{2}, 0.37521423)$ lies on the graph of $y = \left(\dfrac{1}{2}\right)^x$.

21. See Figure 21. All values can be input for x; therefore, the domain is $(-\infty, \infty)$. Only values where $f(x) > 0$ are possible solutions; therefore, the range is $(-\infty, 0)$. From the graph, the asymptote is $y = 0$.

23. See Figure 23. All values can be input for x; therefore, the domain is $(-\infty, \infty)$. Only values where $f(x) > 0$ are possible solutions; therefore, the range is $(0, \infty)$. From the graph, the asymptote is $y = 0$.

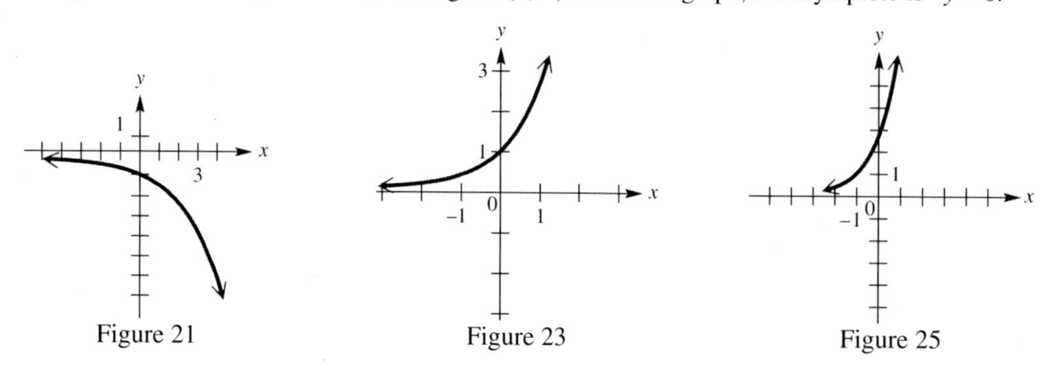

Figure 21 Figure 23 Figure 25

25. See Figure 25. All values can be input for x; therefore, the domain is $(-\infty, \infty)$. Only values where $f(x) > 0$ are possible solutions; therefore, the range is $(0, \infty)$. From the graph, the asymptote is $y = 0$.

27. See Figure 27. All values can be input for x; therefore, the domain is $(-\infty, \infty)$. Only values where $f(x) > 0$ are possible solutions; therefore, the range is $(0, \infty)$. From the graph, the asymptote is $y = 0$.

29. See Figure 29. All values can be input for x; therefore, the domain is $(-\infty, \infty)$. Only values where $f(x) > 0$ are possible solutions; therefore, the range is $(0, \infty)$. From the graph, the asymptote is $y = 0$.

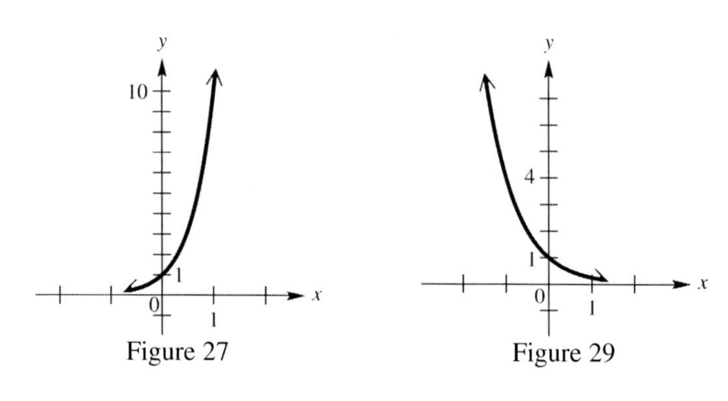

Figure 27 Figure 29

31. (a) Since the graph goes up to the right, $a > 1$.
    (b) From the graph, the domain is $(-\infty, \infty)$, the range is $(0, \infty)$, and the asymptote is $y = 0$.
    (c) Reflect f(x) across the x-axis. See Figure 31c.
    (d) From the graph, the domain is $(-\infty, \infty)$, the range is $(-\infty, 0)$, and the asymptote is $y = 0$.
    (e) Reflect f(x) across the y-axis. See Figure 31e.
    (f) From the graph, the domain is $(-\infty, \infty)$, the range is $(0, \infty)$, and the asymptote is $y = 0$.

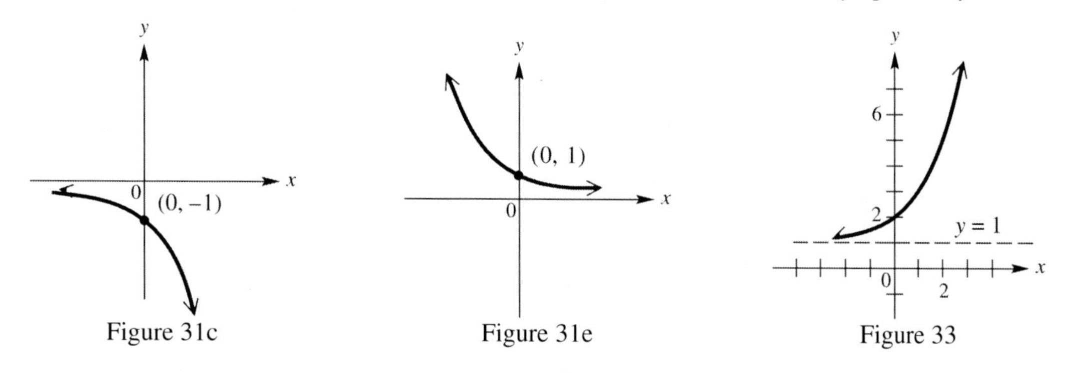

Figure 31c Figure 31e Figure 33

33. Shift the original graph (not shown) 1 unit upward. See Figure 33.
35. Shift the original graph (not shown) 1 unit left. See Figure 35.

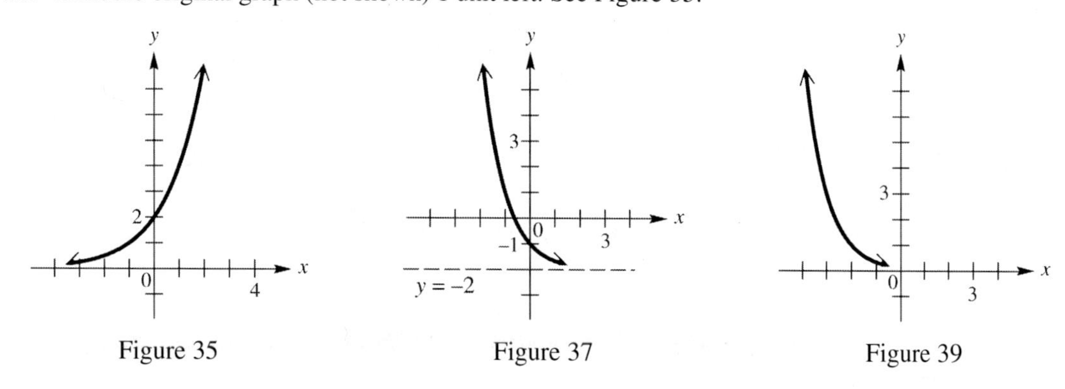

Figure 35 Figure 37 Figure 39

37. Shift the original graph (not shown) 2 units downward. See Figure 37.
38. Shift the original graph (not shown) 4 units upward. See Figure 38.
39. Shift the original graph (not shown) 2 units left. See Figure 39.
41. $2^{3-x} = 8 \Rightarrow 2^{3-x} = 2^3 \Rightarrow 3 - x = 3 \Rightarrow -x = 0 \Rightarrow x = 0.$

43. $12^{x-3}=1 \Rightarrow 12^{x-3}=12^0 \Rightarrow x-3=0 \Rightarrow x=3.$

45. $e^{4x-1}=(e^2)^e \Rightarrow e^{4x-1}=e^{2x} \Rightarrow 4x-1=2x \Rightarrow 2x=1 \Rightarrow x=\frac{1}{2}.$

47. $27^{4x}=9^{x+1} \Rightarrow (3^3)^{4x} \Rightarrow (3^2)^{x+1} \Rightarrow 3^{12x}=3^{2x+2} \Rightarrow 12x=2x+2 \Rightarrow\ 10x=2 \Rightarrow x=\frac{1}{5}.$

49. $4^{x-2}=2^{3x+3} \Rightarrow (2^2)^{x-2}=2^{3x+3} \Rightarrow 2^{2x-4}=2^{3x+3} \Rightarrow 2x-4=3x+3 \Rightarrow -7=x \Rightarrow x=-7.$

51. $(\sqrt{2})^{x+4}=4^x \Rightarrow (2^{1/2})^{x+4}=(2^2)^x \Rightarrow 2^{1/2x+2}=2^{2x} \Rightarrow \frac{1}{2}x+2=2x \Rightarrow 2=\frac{3}{2}x \Rightarrow 4=3x \Rightarrow x=\frac{4}{3}.$

53. $\left(\sqrt{2}\right)^{-2x}=\left(\frac{1}{2}\right)^{2x+3} \Rightarrow \left(2^{1/2}\right)^{-2x}=\left(2^{-1}\right)^{2x+3} \Rightarrow 2^{-x}=2^{-2x-3} \Rightarrow -x=-2x-3 \Rightarrow x=-3$

55. $(6)^{1-x}=\left(\frac{1}{36}\right)^{2x} \Rightarrow (6)^{1-x}=\left(6^{-2}\right)^{2x} \Rightarrow (6)^{1-x}=(6)^{-4x} \Rightarrow 1-x=-4x \Rightarrow 3x=-1 \Rightarrow x=-\frac{1}{3}$

57. (a) $2^{x+1}=8 \Rightarrow 2^{x+1}=2^3 \Rightarrow x+1=3 \Rightarrow x=2.$

(b) From a calculator graph of $y_1=2^{x+1}$ and $y_2=8,\ 2^{x+1}<8$ for the interval $(2,\infty).$

(c) From a calculator graph of $y_1=2^{x+1}$ and $y_2=8,\ 2^{x+1}<8$ for the interval $(-\infty,2).$

59. (a) $27^{4x}=9^{x+1} \Rightarrow (3^3)^{4x}=(3^2)^{x+1} \Rightarrow 3^{12x}=3^{2x+2} \Rightarrow 12x=2x+2 \Rightarrow 10x=2 \Rightarrow x=\frac{1}{5}.$

(b) From a calculator graph of $y_1=27^{4x}$ and $y_2=9^{x+1},\ 27^{4x}>9^{x+1}$ for the interval $\left(\frac{1}{5},\infty\right).$

(c) From a calculator graph of $y_1=27^{4x}$ and $y_2=9^{x+1},\ 27^{4x}<9$ for the interval $\left(-\infty,\frac{1}{5}\right).$

61. (a) $\left(\frac{1}{2}\right)^{-x}=\left(\frac{1}{4}\right)^{x+1} \Rightarrow (2^{-1})^{-x}=(2^{-2})^{x+1} \Rightarrow 2^x=2^{-2x-2} \Rightarrow x=-2x-2 \Rightarrow 3x=-2 \Rightarrow x=-\frac{2}{3}.$

(b) From a calculator graph of $=\varnothing$ and $y_2=\left(\frac{1}{4}\right)^{x+1},\ \left(\frac{1}{2}\right)^{-x}\geq\left(\frac{1}{4}\right)^{x+1}$ for the interval $\left[-\frac{2}{3},\infty\right).$

(c) From a calculator graph of $y_1=\left(\frac{1}{2}\right)^{-x}$ and $y_2=\left(\frac{1}{4}\right)^{x+1},\ \left(\frac{1}{2}\right)^{-x}\leq\left(\frac{1}{4}\right)^{x+1}$ for the interval $\left(-\infty,-\frac{2}{3}\right].$

63. (a) If $f(x)=27,$ then $27=a^3 \Rightarrow \sqrt[3]{27}=a \Rightarrow a=3.$ Therefore, $f(1)=3^1=3.$

(b) If $f(x)=27,$ then $27=a^3 \Rightarrow \sqrt[3]{27}=a \Rightarrow a=3.$ Therefore, $f(-1)=3^{-1}=\frac{1}{3}.$

(c) If $f(x)=27,$ then $27=a^3 \Rightarrow \sqrt[3]{27}=a \Rightarrow a=3.$ Therefore, $f(2)=3^2=9.$

(d) If $f(x)=27,$ then $27=a^3 \Rightarrow \sqrt[3]{27}=a \Rightarrow a=3.$ Therefore, $f(0)=3^0=1.$

65. If point (–3, 64) in on the graph of $f(x)=a^x,$ then $64=\frac{1}{a^3} \Rightarrow \frac{1}{64}=a^3 \Rightarrow \sqrt[3]{\frac{1}{64}} \Rightarrow a=\frac{1}{4}.$ The equation is $f(x)=\left(\frac{1}{4}\right)^x.$

67. $f(t)=\left(\frac{1}{3}\right)^{1-2t}=\left(\frac{1}{3}\right)^{1}\cdot\left(\frac{1}{3}\right)^{-2t}=\left(\frac{1}{3}\right)\left(\left(\frac{1}{3}\right)^{-2}\right)^{t}=\left(\frac{1}{3}\right)(3^2)^t\Rightarrow f(t)=\left(\frac{1}{3}\right)9^t.$

69. (a) Use the formula $A=p\left(1+\frac{r}{n}\right)^{nt}$ to find the amount in the account.

$$A=20000\left(1+\frac{0.03}{1}\right)^{(1)(4)}=20000(1.03)^4\Rightarrow A=\$22,510.18.$$

(b) Use the formula $A=p\left(1+\frac{r}{n}\right)^{nt}$ to find the amount in the account.

$$A=20000\left(1+\frac{0.03}{2}\right)^{(2)(4)}=20000(1.015)^8\Rightarrow A=\$22,529.85.$$

71. (a) Use the formula to find the $A=p\left(1+\frac{r}{n}\right)^{nt}$ amount in the account.

$$A=27500\left(1+\frac{0.0395}{365}\right)^{(365)(5)}=27500(1.000108219)^{1825}\Rightarrow A=\$33,504.34.$$

(b) Use the formula $A=Pe^{rt}$ to find the amount in the account.

$$A=27500\left(e^{(0.0395)(5)}\right)=27500e^{0.1975}\Rightarrow A=\$33,504.71.$$

73. Plan A: $A=40000\left(1+\frac{0.025}{4}\right)^{(4)(3)}=40000(1.00625)^{12}\Rightarrow A=\$43,105.30.$

Plan B: $A=40000e^{(0.024)(3)}=40000e^{(0.072)}\Rightarrow A=\$42,986.21$ Plan A is better by \$119.09.

75. Set $y_1=1000\left(1+\frac{0.05}{1}\right)^{x}\Rightarrow y_1=1000(1.05)^x$, and $y_2=1000\left(1+\frac{0.05}{12}\right)^{12x}$.

Graphing each on the same calculator screen shows the line y_2 slightly above line y_1. Also, using the table function for y_1, y_2 and y_2-y_1 yields the following differences: (1 year) – \$1.16, (2 years) – \$2.44, (5 years) – \$7.08, (10 years) – \$18.12, (20 years) –\$59.34, (30 years) – \$145.80, and (40 years) – \$318.43.

77. (a) See Figure 77a.

(b) Exponential would be better because the average rate of change between data points is not constant.

(c) See Figure 77c.

(d) $p(1500)=1013e^{-0.0001341(1500)}=828.4210207\Rightarrow p(1500)\approx 828$ mb.

$p(11,000)=1013e^{-0.0001341(11,000)}=231.7296764\Rightarrow p(11,000)\approx 232$ mb. $p(1500)$ is slightly lower than the actual, and $p(11,000)$ is slightly higher than the actual.

[-1000,11000] by [0,1200] Xscl = 1000 Yscl = 100

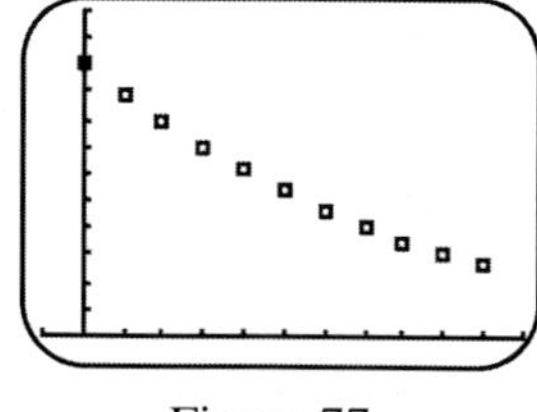

Figure 77a

[-1000,11000] by [0,1200] Xscl = 1000 Yscl = 100

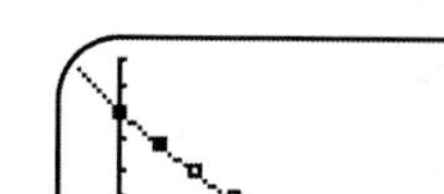

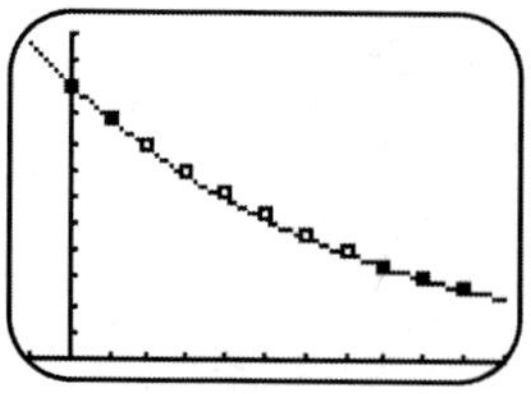

Figure 77c

[0,60] by [0,1.2] Xscl= 10 Yscl = 0.2

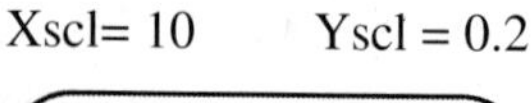

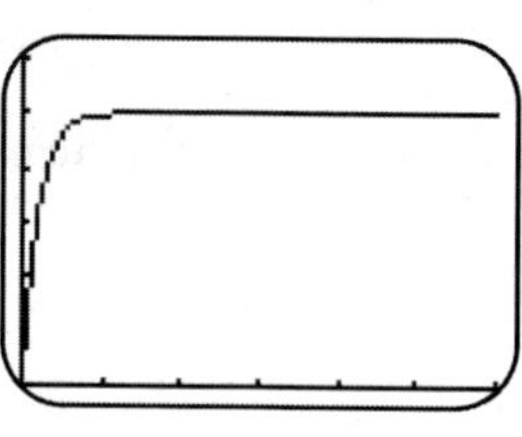

Figure 79

79. (a) $f(2)=1-e^{-5(2)}=1-0.367879441=0.632120559 \Rightarrow f(2)\approx 0.63$. There is a 63% chance that at least one car will enter the intersection during a 2-minute period.

    (b) See Figure 79. As time progresses, the probability increases and begins to approach 1. That is, it is almost certain that at least one car will enter the intersection during a 60-minute period.

81. With $a>0$, all inputs for x yield different solutions. The function is one-to-one and an inverse function exists.

82. See Figure 82.

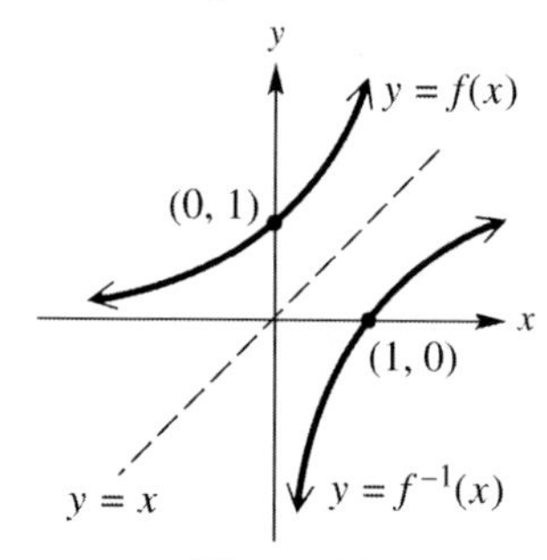

Figure 82

83. To find the inverse, switch the x and y, $x=a^y$.

84. If $a=10$ then $x=a^y$ will be $x=10^y$.

85. If $a=e$ then $x=a^y$ will be $x=e^y$.

86. If (p,q) is on f then switching the x and y means (q,p) will be on f^{-1}.

5.3: Logarithms and Their Properties

1. (a) By the definitions of logarithms, $2^4=16$ is equivalent to $\log_2 16=4$; it matches with C.

   (b) By the definitions of logarithms, $3^0=1$ is equivalent to $\log_3 1=0$; it matches with A.

   (c) By the definitions of logarithms, $10^{-1}=\frac{1}{10}=0.1$ is equivalent to $\log_{10} 0.1=-1$; it matches with E.

   (d) By the definitions of logarithms, $2^{1/2}=\sqrt{2}$ is equivalent to $\log_2\sqrt{2}=\frac{1}{2}$; it matches with B.

   (e) By the definitions of logarithms, $e^{-2}=\frac{1}{e^2}$ is equivalent to $\log_e\frac{1}{e^2}=-2$; it matches with F.

(f) By the definitions of logarithms, $\left(\frac{1}{2}\right)^{-3}=8$ is equivalent to $\log_{1/2} 8=-3$; it matches with D.

3. By the definitions of logarithms, $3^4=81$ is equivalent to $\log_3 81=4$.

5. By the definitions of logarithms, $\left(\frac{1}{2}\right)^{-4}=16$ is equivalent to $\log_{1/2} 16=-4$.

7. By the definitions of logarithms, $10^{-4}=0.0001$ is equivalent to $\log 0.0001=-4$.

9. By the definitions of logarithms and natural logarithms, $e^0=1$ is equivalent to $\ln 1=0$.

11. By the definitions of logarithms, $\log_6 36=2$ is equivalent to $6^2=36$.

13. By the definitions of logarithms, $\log_{\sqrt{3}} 81=8$ is equivalent to $\left(\sqrt{3}\right)^8=81$.

15. By the definitions of logarithms, $\log_{10} 0.001=-3$ is equivalent to $10^{-3}=0.001$.

17. By the definitions of logarithms, $\log\sqrt{10}=0.5$ is equivalent to $10^{0.5}=\sqrt{10}$.

19. By the definitions of logarithms, $\log_5 125=x \Rightarrow 5^x=125$. Since we know $5^3=125$, it follows that $x=3$.

21. By the definitions of logarithms, $\log_x 3^{12}=24 \Rightarrow x^{24}=3^{12} \Rightarrow \left(x^2\right)^{12}=3^{12}$. Since $x^2=3$, it follows that

$x=\sqrt{3}$.

23. By the definitions of logarithms, $\log_6 x=-3 \Rightarrow x=6^{-3} \Rightarrow x=\frac{1}{6^3} \Rightarrow x=\frac{1}{216}$.

25. By the definitions of logarithms, $\log_x 16=\frac{4}{3} \Rightarrow x^{4/3}=16 \Rightarrow \left(x^{4/3}\right)^{3/4}=16^{3/4} \Rightarrow x=\sqrt[4]{16^3} \Rightarrow x=8$.

27. By the definitions of logarithms, $\log_2(x+1)=3 \Rightarrow x+1=2^3 \Rightarrow x+1=8 \Rightarrow x=7$

29. By the definitions of logarithms, $\log_9 \frac{\sqrt[4]{27}}{3}=x \Rightarrow 9^x=\frac{\sqrt[4]{27}}{3} \Rightarrow \left(3^2\right)^x=\frac{3^{3/4}}{3} \Rightarrow 3^{2x}=3^{-1/4}$. Since $3^{2x}=3^{-1/4}$,

$2x=-\frac{1}{4} \Rightarrow x=-\frac{1}{8}$.

31. (a) Using the properties of logarithms, $3^{\log_3 7}=7$.

(b) Using the properties of logarithms, $4^{\log_4 9}=9$.

(c) Using the properties of logarithms, $12^{\log_{12} 4}=4$.

(d) Using the properties of logarithms, $a^{\log_a k}\,(k>0, a>0, a\neq 1)=k$.

33. (a) Using the properties of logarithms, $\log_3 1=0$.

(b) Using the properties of logarithms, $\log_4 1=0$.

(c) Using the properties of logarithms, $\log_{12} 1=0$.

(d) Using the properties of logarithms, $\log_a 1\,(a>0, a\neq 1)=0$.

35. By properties of logarithms, $\log 10^{1.5} = \log_{10} 10^{1.5} = 1.5$.

37. By properties of logarithms, $\log 10^{\sqrt{5}} = \log_{10} 10^{\sqrt{5}} = \sqrt{5}$.

39. By properties of logarithms, $\ln e^{2/3} = \frac{2}{3}$.

41. By properties of logarithms, $\ln e^{\pi} = \pi$.

43. By properties of logarithms, $\sqrt{7}\ln e^{\sqrt{7}} = \sqrt{7}\left(\sqrt{7}\right) = 7$.

45. From the calculator, $\log 43 \approx 1.633468456$.

47. From the calculator, $\log 0.783 \approx -1.062382379$.

49. From the calculator, $\log 28^3 = 3\log 28 = 3(1.447158031) \approx 4.341474094$.

51. From the calculator, $\ln 43 \approx 3.761200116$.

53. From the calculator, $\ln 0.783 \approx -0.244622583$.

55. From the calculator, $\ln 28^3 = 3\ln 28 = 3(3.33220451) \approx 9.996613531$.

57. Since $\text{pH} = -\log[H_3O^+]$ and grapefruit has $H_3O^+ = 6.3\times10^{-4}$, $\text{pH} = -\log(6.3\times10^{-4}) \approx 3.20066 \approx 3.2$.

59. Since $\text{pH} = -\log[H_3O^+]$ and crackers have $H_3O^+ = 3.9\times10^{-9}$, $\text{pH} = -\log(3.9\times10^{-9}) \approx 8.408935 \approx 8.4$.

61. Since $\text{pH} = -\log[H_3O^+]$ and soda pop has a $\text{pH}=2.7$,

$2.7 = -\log[H_3O^+] \Rightarrow -2.7 = \log[H_3O^+] \Rightarrow H_3O^+ = 10^{-2.7} \Rightarrow H_3O^+ \approx 0.001995 \approx 2\times10^{-3}$.

63. Since $\text{pH} = -\log[H_3O^+]$ and beer has a $\text{pH}=4.8$,

$4.8 = -\log[H_3O^+] \Rightarrow -4.8 = \log[H_3O^+] \Rightarrow H_3O^+ = 10^{-4.8} \Rightarrow H_3O^+ \approx 0.00001585 \approx 1.6\times10^{-5}$.

65. $\log_3 \frac{2}{5} = \log_3 2 - \log_3 5$.

67. $\log_2 \frac{6x}{y} = \log_2 6 + \log_2 x - \log_2 y$.

69. $\log_5 \frac{5\sqrt{7}}{3m} = \log_5 5 + \log_5 7^{1/2} - (\log_5 3 + \log_5 m) = 1 + \frac{1}{2}\log_5 7 - \log_5 3 - \log_5 m$.

71. $\log_4(2x+5y)$ cannot be rewritten.

73. $\log_k \frac{pq^2}{m} = \log_k p + \log_k q^2 - \log_k m = \log_k p + 2\log_k q - \log_k m$.

75. $\log_m \sqrt{\frac{r^3}{5z^5}} = \log_m \left(\frac{r^3}{5z^5}\right)^{1/2} = \frac{1}{2}[\log_m r^3 - (\log_m 5 + \log_m z^5)] = \frac{1}{2}(3\log_m r - \log_m 5 - 5\log_m z)$ or

$\frac{3}{2}\log_m r - \frac{1}{2}\log_m 5 - \frac{5}{2}\log_m z$

77. $\log_a x + \log_a y - \log_a m = \log_a \frac{xy}{m}$

79. $2\log_m a - 3\log_m b^2 = \log_m a^2 - \log_m \left(b^2\right)^3 = \log_m \frac{a^2}{b^6}$

81. $2\log_a (z-1) + \log_a (3z+2),\ z>0 \Rightarrow \log_a (z-1)^2 + \log_a (3z+2) = \log_a \left((z-1)^2 (3z+2)\right)$

83. $-\frac{2}{3}\log_5 5m^2 + \frac{1}{2}\log_5 25m^2 \Rightarrow \log_5 \left(5m^2\right)^{-2/3} + \log_5 \left(25m^2\right)^{1/2} \Rightarrow \log_5 \left(5^{-2/3} m^{-4/3}\right) + \log_5 5m \Rightarrow$

$\log_5 \left(5^{-2/3} m^{-4/3}\right)(5m) \Rightarrow \log_5 \left(5^{1/3} m^{-1/3}\right)$ or $\log_5 \sqrt[3]{\frac{5}{m}}$

85. $3\log x - 4\log y = \log x^3 - \log y^4 = \log \frac{x^3}{y^4}$

87. $\ln(a+b) + \ln a - \frac{1}{2}\ln 4 = \ln(a+b) + \ln a - \ln\sqrt{4} = \ln \frac{(a+b)a}{2}$

89. $\log_5 10 = \frac{\log 10}{\log 5} \approx 1.430676558$

91. $\log_{15} 5 = \frac{\log 5}{\log 15} \approx 0.5943161289$

93. $\log_{100} 83 = \frac{\log 83}{\log 100} \approx 0.9595390462$

95. $\log_{2.9} 7.5 = \frac{\log 7.5}{\log 2.9} \approx 1.892441722$

97. To get the graph of $y = -3^x + 7$, reflect the graph of $y = 3^x$ across the x-axis and shift 7 units upward.

98. See Figure 98.

[-5,5] by [-10,10]

Xscl = 1 Yscl = 1

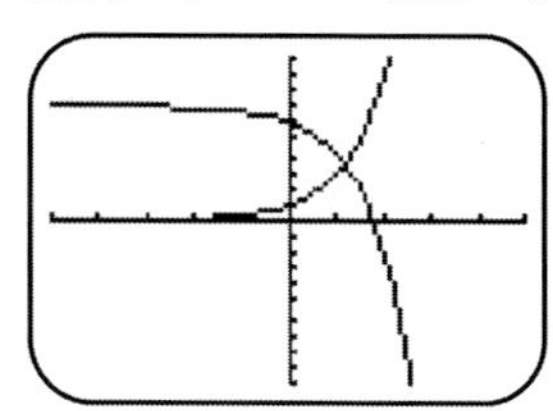

Figure 98

99. From the calculator's capabilities, the x-intercept of $y_2 = -3^x + 7$ is $x \approx 1.7712437492$.

100. If $0 = -3^x + 7$ then $3^x = 7$. Now using base 3 logarithm yields $\log_3 3^x = \log_3 7 \Rightarrow x = \log_3 7$.

101. $\log_3 7 = \frac{\log 7}{\log 3} = 1.77124374916$

102. The approximations are close enough to support the conclusion that the x-intercept is equal to $\log_3 7$.

103. $\ln\left|x+\sqrt{x^2+3}\right|+\ln\left|x-\sqrt{x^2+3}\right|=\ln 3 \Rightarrow \ln\left(\left|x+\sqrt{x^2+3}\right|\left|x-\sqrt{x^2+3}\right|\right)=\ln 3 \Rightarrow$

$\ln\left|\left(x^2+x\sqrt{x^2+3}-x\sqrt{x^2+3}-\left(\sqrt{x^2+3}\right)^2\right)\right|=\ln 3 \Rightarrow \ln\left|\left(x^2-x^2-3\right)\right|=\ln 3 \Rightarrow$

$\ln|-3|=\ln 3 \Rightarrow \ln 3=\ln 3$

105. $\frac{1}{3}\ln\left(\frac{x^2+1}{5}\right)-\frac{1}{3}\ln\left(\frac{x^2+4}{5}\right)=\ln\sqrt[3]{\frac{x^2+1}{x^2+4}} \Rightarrow \ln\left(\sqrt[3]{\frac{x^2+1}{5}}\right)-\ln\left(\sqrt[3]{\frac{x^2+4}{5}}\right)=\ln\sqrt[3]{\frac{x^2+1}{x^2+4}} \Rightarrow$

$\ln\left(\frac{\sqrt[3]{\frac{x^2+1}{5}}}{\sqrt[3]{\frac{x^2+4}{5}}}\right)=\ln\sqrt[3]{\frac{x^2+1}{x^2+4}} \Rightarrow \ln\sqrt[3]{\frac{x^2+1}{x^2+4}}=\ln\sqrt[3]{\frac{x^2+1}{x^2+4}}$

107. With a total of 100 individuals and 50 individuals of each species, $P_1=\frac{50}{100}=\frac{1}{2}$ and $P_2=\frac{50}{100}=\frac{1}{2}$. Now using the index of diversity, $H=-\left(\frac{1}{2}\log_2\left(\frac{1}{2}\right)+\frac{1}{2}\log_2\left(\frac{1}{2}\right)\right)=-\log_2\left(\frac{1}{2}\right)=-(-1) \Rightarrow H=1.$

109. (a) $S(100)=0.36\ln\left(1+\frac{100}{0.36}\right)\approx 2.0269\approx 2$

(b) $S(200)=0.36\ln\left(1+\frac{200}{0.36}\right)\approx 2.2758\approx 2$

(c) $S(150)=0.36\ln\left(1+\frac{150}{0.36}\right)\approx 2.1725\approx 2$

(d) $S(10)=0.36\ln\left(1+\frac{10}{0.36}\right)\approx 1.2095\approx 1$

Reviewing Basic Concepts (Sections 5.1—5.3)

1. No, because the x-values -2 and 2 both correspond to the y-value 4. In a one-to-one function, each y-value must correspond to exactly one x-value (and each x-value to exactly one y-value).

2. (a) Interchange the x- and y-values:

x	12	21	32	45
y	7	8	9	10

(b) For $f(x)=\frac{x+5}{4}$ the inverse is $x=\frac{y+5}{4} \Rightarrow 4x=y+5 \Rightarrow y=4x-5 \Rightarrow f^{-1}(x)=4x-5.$

3. Graph $f(x)=2x+3$ and reflect it across the line $y=x.$ See Figure 3.

4. See Figure 4.

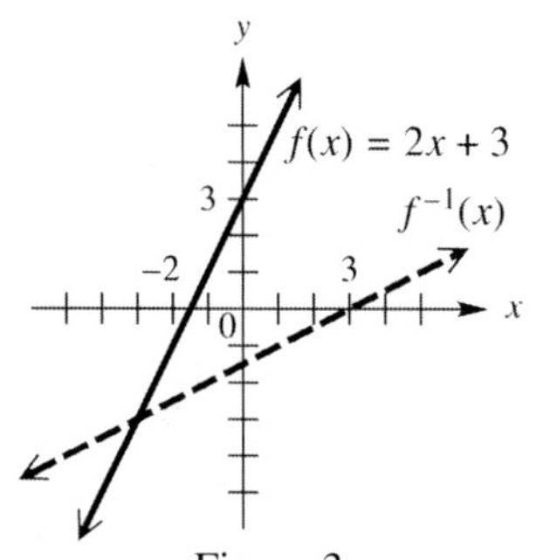

Figure 3

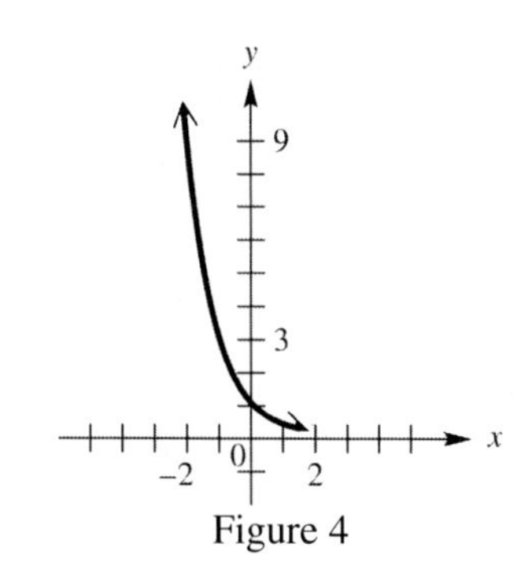

Figure 4

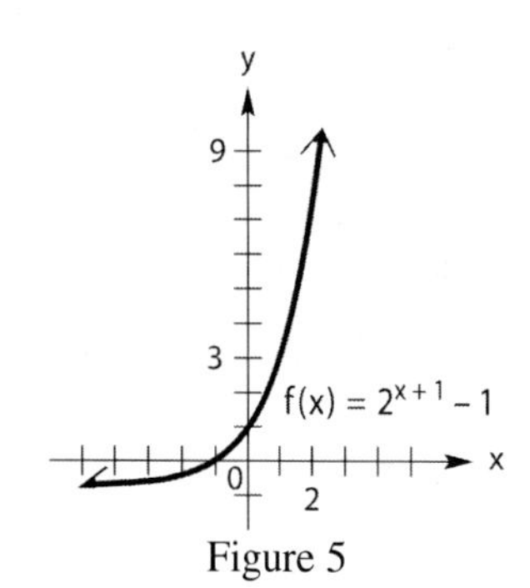

Figure 5

5. See Figure 5.
   (a) It is shifted 1 unit left and 1 unit downward.
   (b) Domain: $(-\infty, \infty)$ Range: $(-1, \infty)$
   (c) $y = -1$
   (d) The graph of the line passes the horizontal line test and is a one to one function.

6. $4^{2x} = 8 \Rightarrow \left(2^2\right)^{2x} = 2^3 \Rightarrow 2^{4x} = 2^3 \Rightarrow 4x = 3 \Rightarrow x = \frac{3}{4}.$

7. Using $A = P\left(1+\frac{r}{n}\right)^{nt}$, we get $A = 600\left(1+\frac{0.04}{4}\right)^{4(3)} = 600(1.01)^{12} = 676.10.$

   The interest earned is $\$676.10 - \$600 = \$76.10.$

8. (a) $\log\left(\frac{1}{\sqrt{10}}\right) = \log_{10} 10^{-1/2} = -\frac{1}{2}$

   (b) $2\ln e^{1.5} = \ln e^{2(1.5)} = \log_e e^3 = 3$

   (c) $\log_2 4 = 2$ because we know that $2^2 = 4.$

9. From the properties of logarithms, $\log\frac{3x^2}{5y} = \log 3 + \log x^2 - (\log 5 + \log y) = \log 3 + 2\log x - \log 5 - \log y.$

10. From the properties of logarithms, $\ln 4 + \ln x - 3\ln 2 = \ln 4 + \ln x - \ln 2^3 = \ln\frac{4x}{8} = \ln\frac{x}{2}.$

5.4: Logarithmic Functions

1. Reflect f across the line $y = x$ and interchange the x- and y-coordinates for each point. See Figure 1. From the graph of f^{-1}, Domain: $(0, \infty)$; Range: $(-\infty, \infty)$; the graph increases on its domain $(0, \infty)$; and the vertical asymptote is $x = 0$.

3. Reflect f across the line $y = x$ and interchange the x- and y-coordinates for each point. See Figure 3. From the graph of f^{-1}, Domain: $(0,\infty)$; Range: $(-\infty,\infty)$; the graph decreases on its domain $(0,\infty)$; and the vertical asymptote is $x = 0$.

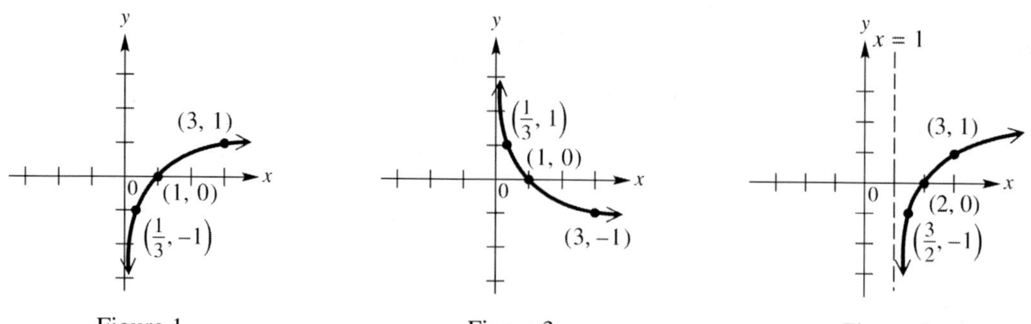

Figure 1 Figure 3 Figure 5

5. Reflect f across the line $y = x$ and interchange the x- and y-coordinates for each point. See Figure 5. From the graph of f^{-1}, Domain: $(1,\infty)$; Range: $(-\infty,\infty)$; the graph increases on its domain $(1,\infty)$; and the vertical asymptote is $x = 1$.

7. Logarithmic

9. Since the argument of a logarithm must be positive, $2x > 0 \Rightarrow x > 0$. The domain is $(0,\infty)$.

11. Since the argument of a logarithm must be positive, $(-x) > 0 \Rightarrow x < 0$. The domain is $(-\infty,0)$.

13. Since the argument of a natural logarithm must be positive, $(x^2 + 7) > 0$, this is true for all values of x, therefore the domain is $(-\infty,\infty)$.

15. Since the argument of a natural logarithm must be positive, $(-x^2 + 4) > 0$. Now set $(-x^2 + 4) = 0 \Rightarrow$ $x^2 - 4 = 0 \Rightarrow x^2 = 4 \Rightarrow x = -2,\ 2$. The endpoints –2 and 2 divide the number line into 3 intervals. The interval $(-\infty,-2)$ has a negative solution; the interval $(-2,2)$ has a positive solution; and the interval $(2,\infty)$ has a negative solution. Therefore, $(-x^2 + 4) > 0$ for the interval $(-2,2)$ and the domain is $y = x$

17. Since the argument of a natural logarithm must be positive, $\left(x^2 - 4x - 21\right) > 0$. Now set $x^2 - 4x - 21 = 0 \Rightarrow (x-7)(x+3) = 0 \Rightarrow x = -3,\ 7$. The endpoints -3 and 7 divide the number line into 3 intervals. The interval $(-\infty,-3)$ has a positive product; the interval $(-3,7)$ has a negative product; and the interval $(7,\infty)$ has a positive product. Therefore, $x^2 - 4x - 21 > 0$ for the interval $(-\infty,-3)\cup(7,\infty)$ and the domain is $(-\infty,-3)\cup(7,\infty)$.

19. Since the argument of a natural logarithm must be positive, $(x^3 - x) > 0$. Now set $x^3 - x = 0 \Rightarrow x(x^2 - 1) = 0 \Rightarrow x(x+1)(x-1) = 0 \Rightarrow x = -1,\ 0,\ 1$. The endpoints $-1, 0,$ and 1 divide the

number line into 4 intervals. The interval $(-\infty,-1)$ has a negative product; the interval $(-1,0)$ has a positive product; the interval $(0,1)$ has a negative product; and the interval $(1,\infty)$ has a positive product. Therefore, $x^3-x>0$ for the interval $(-1,0)\cup(1,\infty)$ and the domain is $(-1,0)\cup(1,\infty)$.

21. Since the argument of a natural logarithm must be positive, $\frac{x+3}{x-4}>0$. Now set both $x+3=0$ and $x-4=0\Rightarrow x=-3,\ 4$. The endpoints -3 and 4 divide the number line into 3 intervals. The interval $(-\infty,-3)$ has a positive product; the interval $(-3,4)$ has a negative product; and the interval $(4,\infty)$ has a positive product. Therefore, $\frac{x+3}{x-4}>0$ for the interval $(-\infty,-3)\cup(4,\infty)$ and the domain is $(-\infty,-3)\cup(4,\infty)$.

23. Since the argument of a natural logarithm must be positive, $|3x-7|>0$. Since this is true for all real x-values except when $3x-7=0$, we solve for $3x-7=0\Rightarrow 3x=7\Rightarrow x=\frac{7}{3}$. Therefore, $|3x-7|>0$ for the interval $\left(-\infty,\frac{7}{3}\right)\cup\left(\frac{7}{3},\infty\right)$, and the domain is $\left(-\infty,\frac{7}{3}\right)\cup\left(\frac{7}{3},\infty\right)$.

25. Shift the graph of $f(x)=\log_2 x$ upward 3 units to sketch the graph of $f(x)=(\log_2 x)+3$. See Figure 25.

27. Shift the graph of $f(x)=\log_2 x$ left 3 units and reflect all negative y-values across the x-axis to sketch the graph of $f(x)=|\log_2(x+3)|$. See Figure 27.

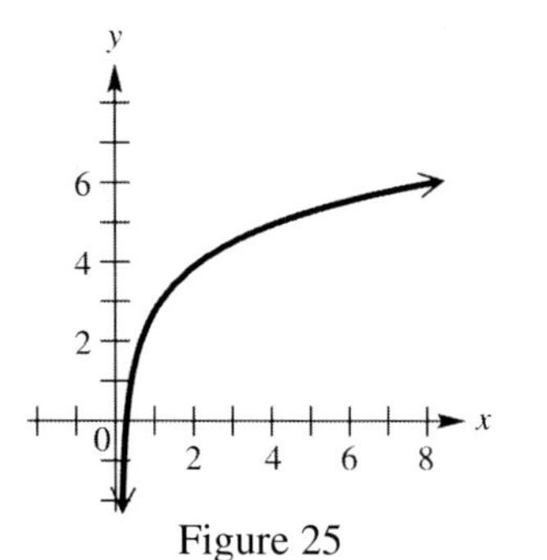

Figure 25

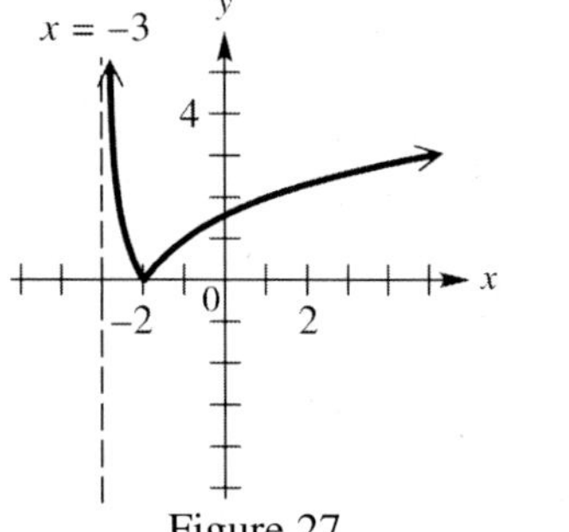

Figure 27

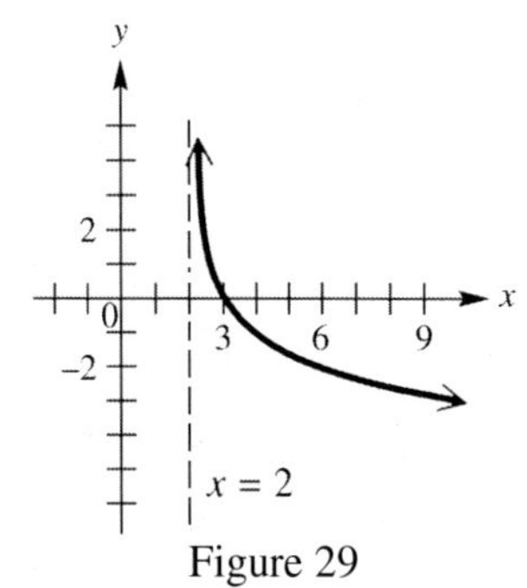

Figure 29

29. Shift the graph of $f(x)=\log_{1/2} x$ right 2 units to sketch the graph of $f(x)=\log_{1/2}(x-2)$. See Figure 29.

31. The graph of $y=e^x+3$ is the graph of $y=e^x$ shifted 3 units upward. The correct graph is B.

33. The graph of $y=e^{x+3}$ is similar to the graph of $y=e^x$ with a y-intercept of $(0,e^3)$. The correct graph is D.

35. The graph of $y=\ln x+3$ is the graph of $y=\ln x$ passing through $(1,-3)$. The correct graph is A.

37. The graph of $y=\ln(x-3)$ is the graph of $y=\ln x$ shifted 3 units right. The correct graph is C.

39. Graph a logarithmic function with base greater than 1 that has an asymptote at $x=0$ and passes through $(1,0)$ and $(5,1)$. See Figure 39.

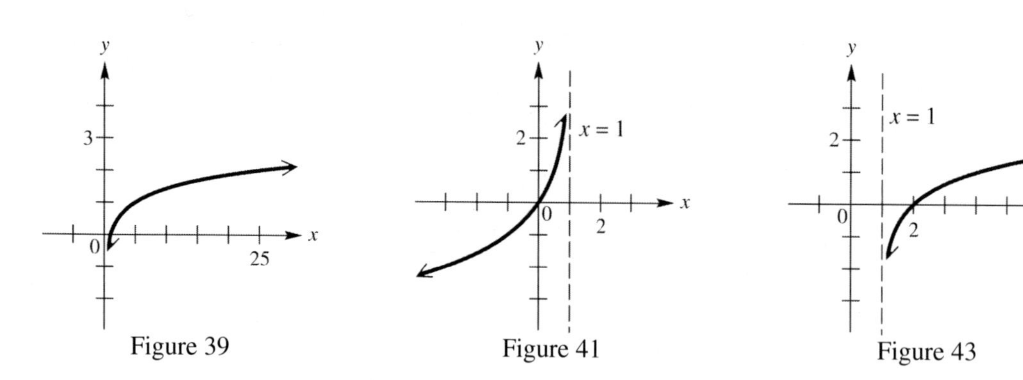

41. Graph a logarithmic function with base between 0 and 1 that is shifted 1 unit left then reflected across the *y*-axis, has an asymptote at $x=1$, and passes through $(0,0)$ and $(-1,-1)$. See Figure 41.

43. Graph a logarithmic function with base greater than 1 that is shifted 1 unit right, has an asymptote at $x=1$, and passes through $(2,0)$ and $(4,1)$. See Figure 43.

45. (a) The graph is shifted 4 units to the left.

    (b) See Figure 45.

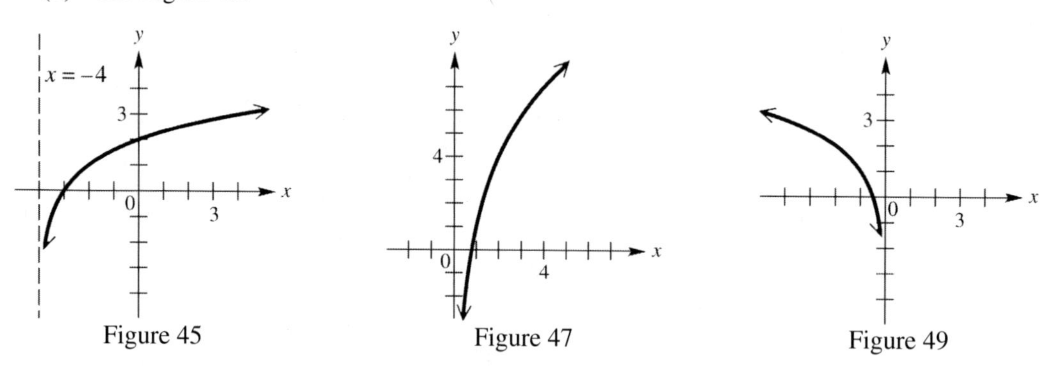

47. (a) The graph is stretched vertically by a factor of 3 and shifted 1 unit upward.

    (b) See Figure 47.

49. (a) The graph is reflected across the *y*-axis and shifted 1 unit upward.

    (b) See Figure 49.

51. The graphs are not the same because the domain of $y=\log x^2$ is $(-\infty,0)\cup(0,\infty)$ while the domain of $y=2\log x$ is $(0,\infty)$. The power rule does not apply if the argument is nonpositive.

53. (a) If $\log_9 27 = x$ then $9^x = 27 \Rightarrow \left(3^2\right)^x = 27 \Rightarrow 3^{2x} = 3^3 \Rightarrow 2x = 3 \Rightarrow x = \frac{3}{2}$.

    (b) By the change of base rule, $\log_9 27 = \frac{\log 27}{\log 9}$, then by calculator, $\frac{\log 27}{\log 9} \approx \frac{1.43136}{0.95424} = 1.5 = \frac{3}{2}$.

    (c) The point $\left(27, \frac{3}{2}\right)$ is on the graph of $y = \log_9 27$, which supports the answer in part a.

55. (a) If $\log_{16} \frac{1}{8} = x$ then $16^x = \frac{1}{8} \Rightarrow \left(2^4\right)^x = \frac{1}{2^3} \Rightarrow 2^{4x} = 2^{-3} \Rightarrow 4x = -3 \Rightarrow x = -\frac{3}{4}$.

(b) By the change of base rule, $\log_{16}\frac{1}{8} = \frac{\log\frac{1}{8}}{\log 16}$, then by calculator, $\frac{\log\frac{1}{8}}{\log 16} \approx \frac{-0.90309}{1.20412} = -0.75 = -\frac{3}{4}$.

(c) The point $\left(\frac{1}{8}, -\frac{3}{4}\right)$ is on the graph of $y = \log_{16}\frac{1}{8}$, which supports the answer in part a.

57. For the function $f(x) = 4^x - 3$, the inverse is found by interchanging the x- and y-values.

$x = 4^y - 3 \Rightarrow x + 3 = 4^y \Rightarrow \log_4(x+3) = \log_4 4^y \Rightarrow \log_4(x+3) = y$

Therefore, $f^{-1}(x) = \log_4(x+3)$. Graph f and f^{-1} in the same window. See Figure 57.

59. For the function $f(x) = -10^x + 4$, the inverse is found by interchanging the x- and y-values.

$x = -10^y + 4 \Rightarrow x - 4 = -10^y \Rightarrow -x + 4 = 10^y \Rightarrow \log(-x+4) = \log 10^y \Rightarrow \log(-x+4) = y$

Therefore, $f^{-1}(x) = \log(4-x)$. Graph f and f^{-1} in the same window. See Figure 59.

[-4.7,4.7] by [-3.1,3.1]
Xscl = 1 Yscl = 1

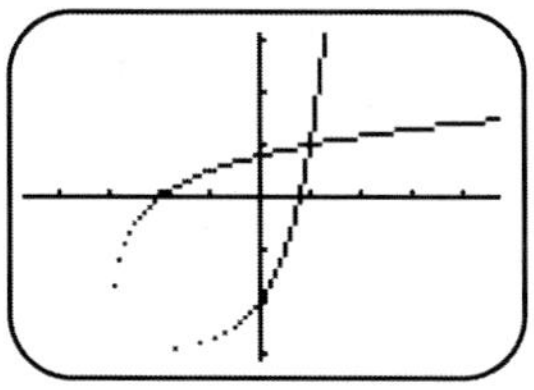

Figure 57

[-6,6.] by [-4.,4.]
Xscl = 1 Yscl = 1

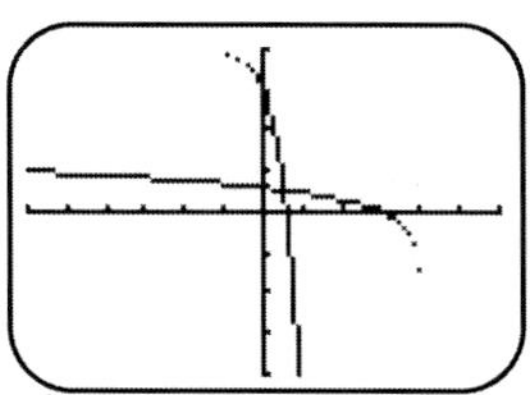

Figure 59

61. (a) First, find a. Using the given information, $2 = \log_a 3 \Rightarrow a^2 = 3 \Rightarrow a = \sqrt{3}$.

Then, $f\left(\frac{1}{9}\right) = \log_{\sqrt{3}}\left(\frac{1}{9}\right) = x \Rightarrow \left(\sqrt{3}\right)^x = \frac{1}{9} \Rightarrow 3^{1/2x} = 3^{-2} \Rightarrow \frac{1}{2}x = -2 \Rightarrow x = -4$.

(b) Since $a = \sqrt{3}$ from part a, $f(27) = \log_{\sqrt{3}} 27 = x \Rightarrow \left(\sqrt{3}\right)^x = 27 \Rightarrow 3^{\frac{1}{2}x} = 3^3 \Rightarrow \frac{1}{2}x = 3 \Rightarrow x = 6$.

(c) Since $a = \sqrt{3}$ from part a, find f^{-1} for the function $f(x) = \log_{\sqrt{3}} x$. The inverse is found by interchanging the x- and y-values: $x = \log_{\sqrt{3}} y \Rightarrow \left(\sqrt{3}\right)^x = y \Rightarrow f^{-1}(x) = \left(\sqrt{3}\right)^x$.

Now $f^{-1}(-2) = \left(\sqrt{3}\right)^{-2} = \frac{1}{3}$.

(d) Since $a = \sqrt{3}$ from part a, find f^{-1} for the function $f(x) = \log_{\sqrt{3}} x$. The inverse is found by interchanging the x- and y-values: $x = \log_{\sqrt{3}} y \Rightarrow \left(\sqrt{3}\right)^x = y \Rightarrow f^{-1}(x) = \left(\sqrt{3}\right)^x$.

Now, $f^{-1}(0) = \left(\sqrt{3}\right)^0 = 1$.

63. Graphing $y_1 = 2^{-x}$ and $y_2 = \log_{10} x$ in the same window yields the intersection point $(1.87, 0.27)$. Therefore, the solution is $x \approx 1.87$.

65. (a) The left side is a reflection of the right side with respect to the axis of the tower.

The graph of $f(-x)$ is the reflection of $f(x)$ with respect to the y-axis.

(b) Since the horizontal line on the top has one-half on each side of the y-axis, the x-coordinate on the right side will be $x=\frac{1}{2}(15.7488)\Rightarrow x=7.8744$. Using $f(7.8744)=-301\ln\left(\frac{7.8744}{207}\right)\approx 984$ feet.

(c) Graphing $y_1=-301\ln\left(\frac{x}{207}\right)$ and $y_2=500$ in the same window yields the approximate intersection point $(39.31, 500)$. Therefore, the height is approximately 39 feet.

67. (a) Using $x=9$, yields $f(9)=27+1.105\log(9+1)=27+1.105\log 10=28.105$ in.

(b) It tells us that at 99 miles from the eye of a typical hurricane, the barometric pressure is 29.21 inches.

5.5: Exponential and Logarithmic Equations and Inequalities

1. $3e^{2x}+1=5\Rightarrow 3e^{2x}=4\Rightarrow e^{2x}=\frac{4}{3}\Rightarrow 2x=\ln\frac{4}{3}\Rightarrow x=\frac{1}{2}\ln\frac{4}{3}$

3. $2\left(10^x\right)=14\Rightarrow 10^x=7\Rightarrow x=\log 7$

5. $\frac{1}{2}\log_2 x=\frac{3}{4}\Rightarrow \log_2 x=\frac{3}{2}\Rightarrow x=2^{3/2}$

7. $4\ln 3x=8\Rightarrow \ln 3x=2\Rightarrow 3x=e^2\Rightarrow x=\frac{1}{3}e^2$

9. (a) $3^x=7\Rightarrow \log 3^x=\log 7\Rightarrow x\log 3=\log 7\Rightarrow x=\frac{\log 7}{\log 3}$ (Other forms of the answer are possible.)

(b) From the calculator, the solution is $x\approx 1.771$.

11. (a) $\left(\frac{1}{2}\right)^x=5\Rightarrow \log\left(\frac{1}{2}\right)^x=\log 5\Rightarrow x\log\left(\frac{1}{2}\right)=\log 5\Rightarrow x=\frac{\log 5}{\log(1/2)}$

(Other forms of the answer are possible.)

(b) From the calculator, the solution is $x\approx -2.322$.

13. (a) $0.8^x=4\Rightarrow \log 0.8^x=\log 4\Rightarrow x\log 0.8=\log 4\Rightarrow x=\frac{\log 4}{\log 0.8}$ (Other forms of the answer are possible.)

(b) From the calculator, the solution is $x\approx -6.213$.

15. (a) $4^{x-1}=3^{2x}\Rightarrow \log 4^{x-1}=\log 3^{2x}\Rightarrow (x-1)\log 4=(2x)\log 3\Rightarrow x\log 4-\log 4-2x\log 3\Rightarrow$

$x\log 4-2x\log 3=\log 4\Rightarrow x(\log 4-2\log 3)=\log 4\Rightarrow x=\frac{\log 4}{\log 4-2\log 3}$

(Other forms of the answer are possible.)

(b) From the calculator, the solution is $x\approx -1.710$.

17. (a) $6^{x+1}=4^{2x-1}\Rightarrow \log 6^{x+1}=\log 4^{2x-1}\Rightarrow (x+1)\log 6=(2x-1)\log 4\Rightarrow x\log 6+\log 6=2x\log 4-\log 4\Rightarrow$

$x\log 6-2x\log 4=-\log 6-\log 4\Rightarrow x(\log 6-2\log 4)=-\log 6-\log 4\Rightarrow x=\frac{-\log 6-\log 4}{\log 6-2\log 4}$

(Other forms of the answer arepossible.)

(b) From the calculator, the solution is $x \approx 3.240$.

19. (a) No real number value for x can produce a negative solution. Therefore, the solution set is $\varnothing$.

21. (a) $e^{x-3} = 2^{3x} \Rightarrow \ln e^{x-3} = \ln 2^{3x} \Rightarrow (x-3)\ln e = 3x\ln 2 \Rightarrow x-3 = 3x\ln 2 \Rightarrow$

$x - 3x\ln 2 = 3 \Rightarrow x(1-3\ln 2) = 3 \Rightarrow x = \dfrac{3}{1-3\ln 2}$ (Other forms of the answer are possible.)

(b) From the calculator, the solution is $x \approx -2.779$.

23. (a) No real number value for x can produce a negative solution. Therefore, the solution set is $\varnothing$.

25. (a) $0.05(1.15)^x = 5 \Rightarrow \log 0.05 + \log(1.15)^x = \log 5 \Rightarrow x\log 1.15 = \log 5 - \log 0.05 \Rightarrow$

$x\log 1.15 = \log\dfrac{5}{0.05} \Rightarrow x\log 1.15 = \log 100 \Rightarrow x\log 1.15 = 2 \Rightarrow x = \dfrac{2}{\log 1.15}$

(Other forms of the answer are possible.)

(b) From the calculator, the solution is $x \approx 32.950$.

27. (a) $3(2)^{x-2} + 1 = 100 \Rightarrow 3(2)^{x-2} = 99 \Rightarrow \log 3 + \log 2^{x-2} = \log 99 \Rightarrow (x-2)\log 2 = \log 99 - \log 3 \Rightarrow$

$x\log 2 - 2\log 2 = \log 99 - \log 3 \Rightarrow x\log 2 = \log 99 - \log 3 + 2\log 2 \Rightarrow$

$x\log 2 = \log\left(\dfrac{99}{3}\right) + 2\log 2 \Rightarrow x = \dfrac{\log 33}{\log 2} + \dfrac{2\log 2}{\log 2} \Rightarrow x = 2 + \dfrac{\log 33}{\log 2}$

(Other forms of the answer are possible.)

(b) From the calculator, the solution is $x \approx 7.044$.

29. (a) $2(1.05)^x + 3 = 10 \Rightarrow 2(1.05)^x = 7 \Rightarrow \log 2 + \log 1.05^x = \log 7 \Rightarrow x\log 1.05 = \log 7 - \log 2 \Rightarrow$

$x\log 1.05 = \log\left(\dfrac{7}{2}\right) \Rightarrow x = \dfrac{\log 3.5}{\log 1.05}$. (Other forms of the answer are possible.)

(b) From the calculator, the solution is $x \approx 25.677$.

31. (a) $5(1.015)^{x-1980} = 8 \Rightarrow 1.015^{x-1980} = \dfrac{8}{5} \Rightarrow (x-1980)\log 1.015 = \log 1.6 \Rightarrow$

$x - 1980 = \dfrac{\log 1.6}{\log 1.015} \Rightarrow x = 1980 + \dfrac{\log 1.6}{\log 1.015}$ (Other forms of the answer are possible.)

(b) From the calculator, the solution is $x \approx 2011.568$.

33. $5\ln x = 10 \Rightarrow \ln x = \dfrac{10}{5} \Rightarrow \ln x = 2 \Rightarrow e^{\ln x} = e^2 \Rightarrow x = e^2$

35. $\ln(4x) = 1.5 \Rightarrow e^{\ln 4x} = e^{1.5} \Rightarrow 4x = e^{1.5} \Rightarrow x = \dfrac{e^{1.5}}{4}$

37. $\log(2-x) = 0.5 \Rightarrow 10^{\log(2-x)} = 10^{0.5} \Rightarrow 2 - x = \sqrt{10} \Rightarrow -x = \sqrt{10} - 2 \Rightarrow x = 2 - \sqrt{10}$

39. $\log_6(2x+4) = 2 \Rightarrow 6^{\log_6(2x+4)} = 6^2 \Rightarrow 2x + 4 = 36 \Rightarrow 2x = 32 \Rightarrow x = 16$

41. $\log_4(x^3 + 37) = 3 \Rightarrow 4^{\log 4(x^{33} - 37)} = 4^3 \Rightarrow x^3 + 37 = 64 \Rightarrow x^3 = 27 \Rightarrow \sqrt[3]{x^3} = \sqrt[3]{27} \Rightarrow x = 3$

43. $\ln x + \ln x^2 = 3 \Rightarrow \ln x(x^2) = 3 \Rightarrow \ln x^3 = 3 \Rightarrow e^{\ln x^3} = e^3 \Rightarrow x^3 = e^3 \Rightarrow \sqrt[3]{x^3} = \sqrt[3]{e^3} \Rightarrow x = e$

45. $2\ln(x-1)+30=34 \Rightarrow 2\ln(x-1)=4 \Rightarrow \ln(x-1)=2 \Rightarrow x-1=e^2 \Rightarrow x=e^2+1$

47. $5\log\left(x^2-1\right)+7=12 \Rightarrow 5\log\left(x^2-1\right)=5 \Rightarrow \log\left(x^2-1\right)=1 \Rightarrow x^2-1=10 \Rightarrow x^2=11 \Rightarrow x=\pm\sqrt{11}$

49. $3\log_2\left(3x^2+2\right)+1=2 \Rightarrow 3\log_2\left(3x^2+2\right)=1 \Rightarrow \log_2\left(3x^2+2\right)=\frac{1}{3} \Rightarrow 3x^2+2=2^{1/3} \Rightarrow x^2=\frac{2^{1/3}-2}{3} \Rightarrow x \text{ is } \varnothing$

51. $\log x+\log(x-21)=2 \Rightarrow \log x(x-21)=2 \Rightarrow \log(x^2-21x)=2 \Rightarrow 10^{\log(x^2-21x)}=10^2 \Rightarrow$

$x^2-21x=100 \Rightarrow x^2-21x-100=0 \Rightarrow (x-25)(x+4)=0 \Rightarrow x=-4,\ 25$

Since the argument of a logarithm can not be negative, the solution is $x=25$

53. $\ln(4x-2)-\ln 4=-\ln(x-2) \Rightarrow \ln(4x-2)+\ln(x-2)=\ln 4 \Rightarrow \ln(4x-2)(x-2)=\ln 4 \Rightarrow$

$\ln(4x^2-10x+4)=\ln 4 \Rightarrow e^{\ln(4x^2-10x+4)}=e^{\ln 4} \Rightarrow 4x^2-10x+4=4 \Rightarrow 4x^2-10x=0 \Rightarrow$

$(2x)(2x-5)=0 \Rightarrow x=0,\ \frac{5}{2}$

Since the argument of a logarithm cannot be zero, the solution is $x=2.5$.

55. $\log_5(x+2)+\log_5(x-2)=1 \Rightarrow \log_5(x+2)(x-2)=1 \Rightarrow \log_5(x^2-4)=1 \Rightarrow 5^{\log_5(x^2-4)}=5^1 \Rightarrow$

$x^2-4=5 \Rightarrow x^2=9 \Rightarrow x=-3,\ 3$

Since the argument of a logarithm cannot be negative, the solution is $x=3$.

57. $\log_7(4x)-\log_7(x+3)=\log_7 x \Rightarrow \log_7 4x=\log_7 x+\log_7(x+3) \Rightarrow \log_7 4x=\log_7 x(x+3) \Rightarrow$

$\log_7 4x=\log_7(x^2+3x) \Rightarrow 7^{\log 7 4x}=7^{\log 7(x^2+3x)} \Rightarrow 4x=x^2+3x \Rightarrow x^2-x=0 \Rightarrow x(x-1)=0 \Rightarrow x=0,\ 1$

Since the argument of a logarithm can not be zero, the solution is $x=1$.

59. $\ln e^x-2\ln e=\ln e^4 \Rightarrow x-2(1)=4 \Rightarrow x=6$

61. $\log x=\sqrt{\log x} \Rightarrow (\log x)^2=(\sqrt{\log x})^2 \Rightarrow (\log x)^2=\log x \Rightarrow (\log x)^2-\log x=0 \Rightarrow \log x(\log x-1)=0,$

then $\log x=0 \Rightarrow x=1$ or $\log x-1=0 \Rightarrow \log x=1 \Rightarrow x=10$. The solution are $x=1,\ 10$.

63. (a) Graph on a calculator $y_1=e^{-2\ln x}$ and $y_2=\frac{1}{16}$. The graphs intersect at $x=4$.

(b) From the graph, $e^{-2\ln x}<\frac{1}{16}$ for the interval $(4,\infty)$.

(c) From the graph, $e^{-2\ln x}>\frac{1}{16}$ for the interval $(0,4)$.

65. The statement is incorrect. We must reject any solution that is not in the domain of any logarithmic function in the equation.

67. If $1.5^{\log x}=e^5$ then $1.5^{\log x}-e^5=0$. From a calculator graph of $y_1=1.5^{\log x}-e^5$, the x-intercept or solution is $x\approx 19.106$.

69. $r=p-k\ln t \Rightarrow r-p=-k\ln t \Rightarrow \frac{r-p}{-k}=\ln t \Rightarrow \ln t=\frac{p-r}{k} \Rightarrow e^{\ln t}=e^{(p-r)/k} \Rightarrow t=e^{(p-r)/k}$

71. $T = T_0 + (T_1 - T_0)10^{-kt} \Rightarrow T - T_0 = (T_1 - T_0)10^{-kt} \Rightarrow \dfrac{T - T_0}{T_1 - T_0} = 10^{-kt} \Rightarrow$

$\log\left(\dfrac{T - T_1}{T_1 - T_0}\right) = \log 10^{-kt} \Rightarrow \log\left(\dfrac{T - T_1}{T_1 - T_0}\right) = -kt \Rightarrow t = -\dfrac{1}{k}\log\left(\dfrac{T - T_1}{T_1 - T_0}\right)$

73. $A = T_0 + Ce^{-kt} \Rightarrow A - T_0 = Ce^{-kt} \Rightarrow \dfrac{A - T_0}{C} = e^{-kt} \Rightarrow \ln\left(\dfrac{A - T_0}{C}\right) = \ln e^{-kt} \Rightarrow$

$-kt = \ln\left(\dfrac{A - T_0}{C}\right) \Rightarrow k = \dfrac{\ln\left(\dfrac{A - T_0}{C}\right)}{-t}$

75. $y = A + B(1 - e^{-Cx}) \Rightarrow y = A + B - Be^{-Cx} \Rightarrow y - A - B = -Be^{-Cx} \Rightarrow \dfrac{-y + A + B}{B} = e^{-Cx} \Rightarrow$

$\ln\left(\dfrac{A + B - y}{B}\right) = \ln e^{-Cx} \Rightarrow \ln\left(\dfrac{A + B - y}{B}\right) = -Cx \Rightarrow x = -\dfrac{\ln\left(\dfrac{A + B - y}{B}\right)}{C}$

77. $\log A = \log B - C\log x \Rightarrow \log A = \log B - \log x^C \Rightarrow \log A = \log \dfrac{B}{x^C} \Rightarrow 10^{\log A} = 10^{\log(B/x^C)} \Rightarrow A = \dfrac{B}{x^C}$

79. $A = P\left(1 + \dfrac{r}{n}\right)^{nt} \Rightarrow \dfrac{A}{P} = \left(1 + \dfrac{r}{n}\right)^{nt} \log\left(\dfrac{A}{P}\right) = \log\left(1 + \dfrac{r}{n}\right)^{nt} \Rightarrow \log\left(\dfrac{A}{P}\right) = nt\log\left(1 + \dfrac{r}{n}\right) \Rightarrow t = \dfrac{\log\left(\dfrac{A}{P}\right)}{n\log\left(1 + \dfrac{r}{n}\right)}$

81. $e^{2x} - 6e^x + 8 = 0 \Rightarrow (e^x - 4)(e^x - 2) = 0 \Rightarrow (e^x - 4) = 0 \text{ } or \text{ } (e^x - 2) = 0 \Rightarrow x = \{\ln 4, \ln 2\}$.

83. $2e^{2x} + e^x = 6 \Rightarrow 2e^{2x} + e^x - 6 = 0 \Rightarrow (2e^x - 3)(e^x + 2) = 0 \Rightarrow (2e^x - 3) = 0 \text{ } or \text{ } (e^x + 2) = 0 \Rightarrow x = \{\ln \dfrac{3}{2}\}$.

85. $\dfrac{1}{2}e^{2x} + e^x = 1 \Rightarrow \dfrac{1}{2}e^{2x} + e^x - 1 = 0.$ Let $a = e^x \Rightarrow \dfrac{1}{2}a^2 + a - 1 = 0$ and use the quadratic formula to solve:

$a = \dfrac{-1 \pm \sqrt{1^2 - (4)(\frac{1}{2})(-1)}}{(2)(\frac{1}{2})} = \dfrac{-1 \pm \sqrt{1 + 2}}{1} = -1 \pm \sqrt{3} \Rightarrow a = \sqrt{3} - 1 \Rightarrow \sqrt{3} - 1 = e^x \Rightarrow x = \ln\left(\sqrt{3} - 1\right)$

87. $3^{2x} + 35 = 12(3^x) \Rightarrow 3^{2x} - 12(3^x) + 35 = 0 \Rightarrow (3^x - 7)(3^x - 5) = 0 \Rightarrow (3^x - 7) \text{ } or \text{ } (3^x - 5) = 0 \Rightarrow$

$x = \{\log_3 7, \log_3 5\}$

89. $(\log_2 x)^2 + \log_2 x = 2 \Rightarrow (\log_2 x)^2 + \log_2 x - 2 = 0.$ Let $a = \log_2 x \Rightarrow a^2 + a - 2 = 0 \Rightarrow$

$(a + 2)(a - 1) = 0 \Rightarrow a = 1, -2.$ $1 = \log_2 x \Rightarrow x = 2 \text{ } or - 2 = \log_2 x \Rightarrow x = \dfrac{1}{4} \Rightarrow x = \{2, \dfrac{1}{4}\}$.

91. $(\ln x)^2 + 16 = 10\ln x \Rightarrow (\ln x)^2 - 10\ln x + 16 = 0.$ Let $a = \ln x \Rightarrow a^2 - 10a + 16 = 0 \Rightarrow$

$(a - 8)(a - 2) = 0 \Rightarrow a = 8, 2.$ $8 = \ln x \Rightarrow x = e^8 \text{ } or \text{ } 2 = \ln x \Rightarrow x = e^2 \Rightarrow x = \{e^8, e^2\}$

93. $-2e^x+5=0 \Rightarrow -2e^x=-5 \Rightarrow e^x=\frac{5}{2} \Rightarrow x=\ln\frac{5}{2}$. From the graph $f(x)<0$ when x is in the interval $\left(\ln\frac{5}{2},\infty\right)$ and $f(x)\geq 0$ when x is in the interval $\left(-\infty,\ln\frac{5}{2}\right]$.

95. $2\left(3^x\right)-18=0 \Rightarrow 2\left(3^x\right)=18 \Rightarrow 3^x=9 \Rightarrow x=2$. From the graph $f(x)<0$ when x is in the interval $(-\infty,2)$ and $f(x)\geq 0$ when x is in the interval $[2,\infty)$.

97. $3^{2x}-9^{x+1}=0 \Rightarrow 3^{2x}=9^{x+1} \Rightarrow 3^{2x}=3^{2x+2} \Rightarrow 2x=2x+2 \Rightarrow$ No solution. From the graph $f(x)<0$ when x is in the interval $(-\infty,\infty)$ and $f(x)\geq 0$ for no values of x.

99. $8-4\log_5(x)=0 \Rightarrow 8=4\log_5(x) \Rightarrow 2=\log_5(x) \Rightarrow x=25$. From the graph $f(x)<0$ when x is in the interval $(25,\infty)$ and $f(x)\geq 0$ for values of x in the interval $(0,25]$.

101. $\ln(x+2)=0 \Rightarrow x+2=1 \Rightarrow x=-1$. From the graph $f(x)<0$ when x is in the interval $(-2,-1)$ and $f(x)\geq 0$ for values of x in the interval $[-1,\infty)$.

103. $7-5\log x=0 \Rightarrow 5\log x=7 \Rightarrow \log x=\frac{7}{5} \Rightarrow x=10^{7/5}$. From the graph $f(x)<0$ when x is in the interval $\left(10^{7/5},\infty\right)$ and $f(x)\geq 0$ for values of x in the interval $\left(0,10^{7/5}\right]$.

105. For $x^2=2^x$, graph on a calculator $y_1=x^2-2^x$. From the calculator, the x-intercepts or solutions are -0.767, 2, and 4.

107. For $\log x=x^2-8x+14$, graph on a calculator $y_1=\log x-x^2+8x-14$, From the calculator the x-intercepts or solutions are 2.454 and 5.659.

109. For $e^x=\frac{1}{x+2}$, graph on a calculator $y_1=e^x-\frac{1}{x+2}$. From the calculator, the x-intercept or solution is -0.443.

111. $\log_2\sqrt{2x^2}-1=0.5 \Rightarrow \log_2\sqrt{2x^2}=1.5 \Rightarrow 2^{\log_2\sqrt{2x^2}}=2^{3/2} \Rightarrow \sqrt{2x^2}=\sqrt{2^3} \Rightarrow$
$2x^2=2^3 \Rightarrow 2x^2=8 \Rightarrow x^2=4 \Rightarrow x=-2,2.$

113. $\ln(\ln e^{-x})=\ln 3 \Rightarrow \ln(-x)=\ln 3 \Rightarrow -x=3 \Rightarrow x=-3$

115. If $y=2$, then $2=\frac{2-\log(100-x)}{0.42} \Rightarrow 2(0.42)=2-\log(100-x) \Rightarrow$
$0.84=2-\log(100-x) \Rightarrow -1.16=-\log(100-x) \Rightarrow 1.16=\log(100-x) \Rightarrow$
$10^{1.16}=10^{\log(100-x)} \Rightarrow 14.454398 \approx 100-x \Rightarrow -85.5 \approx -x \Rightarrow x \approx 85.5\%$

117. If $f(x)=33$, then $33=31.5+1.1\log(x+1) \Rightarrow 1.5=1.1\log(x+1) \Rightarrow \frac{1.5}{1.1}=\log(x+1) \Rightarrow$
$10^{1.5/1.1}=10^{\log(x+1)} \Rightarrow 23.1013 \approx x+1 \Rightarrow x \approx 22.1013 \Rightarrow x \approx 22$.

Reviewing Basic Concepts (Sections 5.4 and 5.5)

1. If $f(x) = 3^x$ and $f(x) \approx 10.98(1.14)^x$ then *f* and *g* are inverse functions, and their graphs are symmetric with respect to the line with equation $y = x$. The domain of *f* is the range of *g* and vice versa.

2. See Figure 2.

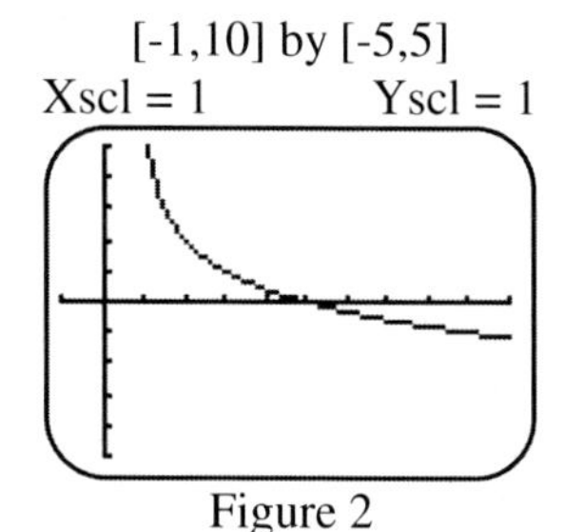

Figure 2

3. From the graph, the asymptote is $x = 5$ the *x*-intercept is (5, 0); and there is no *y*-intercept.

4. The graph of $f(x)$ is the same as the graph of $g(x)$, reflected across the *x*-axis, shifted 1 unit to the right, and shifted two units upward.

5. For $f(x) = 2 - \log_2(x-1)$, the inverse is determined by interchanging x and y:

   $x = 2 - \log_2(y-1) \Rightarrow x - 2 = -\log_2(y-1) \Rightarrow 2 - x = \log_2(y-1) \Rightarrow 2^{(2-x)} = 2^{\log(y-1)} \Rightarrow$

   $2^{(2-x)} = y - 1 \Rightarrow y = 1 + 2^{(2-x)} \Rightarrow f^{-1}(x) = 1 + 2^{2-x}$

6. $3^{2x-1} = 4^x \Rightarrow \log 3^{2x-1} = \log 4^x \Rightarrow (2x-1)\log 3 = x\log 4 \Rightarrow 2x\log 3 - \log 3 = x\log 4 \Rightarrow$

   $2x\log 3 - x\log 4 = \log 3 \Rightarrow x(2\log 3 - \log 4) = \log 3 \Rightarrow x = \dfrac{\log 3}{2\log 3 - \log 4} = \dfrac{\log 3}{\log \frac{9}{4}}$

7. $\ln 5x - \ln(x+2) = \ln 3 \Rightarrow \ln\left(\dfrac{5x}{x+2}\right) = \ln 3 \Rightarrow \dfrac{5x}{x+2} = 3 \Rightarrow 3(x+2) = 5x \Rightarrow 3x + 6 = 5x \Rightarrow 6 = 2x \Rightarrow x = 3$

8. $10^{5\log x} = 32 \Rightarrow \log 10^{5\log x} = \log 32 \Rightarrow 5\log x = \log 32 \Rightarrow \log x^5 = \log 32 \Rightarrow x^5 = 32 \Rightarrow x = 2$

9. $H = 1000(1 - e^{-kn}) \Rightarrow \dfrac{H}{1000} = 1 - e^{-kN} \Rightarrow 1 - \dfrac{H}{1000} = e^{-kN} \Rightarrow \ln\left(1 - \dfrac{H}{100}\right) = \ln e^{-kN} \Rightarrow$

   $\ln\left(1 - \dfrac{H}{1000}\right) = -kN \Rightarrow N = -\dfrac{1}{k}\ln\left(1 - \dfrac{H}{1000}\right)$

10. If $f(x) = 2300$, then $2300 = 280\ln(x+1) + 1925 \Rightarrow 375 = 280\ln(x+1) \Rightarrow$

    $e^{375/280} = e^{\ln(x+1)} \Rightarrow e^{375/280} = x + 1 \Rightarrow x = e^{375/280} - 1 \Rightarrow x \approx 2.8$ acres

5.6: Further Applications & Modeling with Exponential & Logarithmic Functions

1. Using the given function, $\frac{1}{3}A_0 = A_0e^{-0.0001216t} \Rightarrow \frac{1}{3} = e^{-0.0001216t} \Rightarrow \ln\frac{1}{3} = \ln e^{-0.0001216t} \Rightarrow$

   $\ln\frac{1}{3} = -0.0001216t \Rightarrow t = \frac{\ln\left(\frac{1}{3}\right)}{-0.0001216} \Rightarrow t \approx 9034.6 \approx 9{,}000$ years ago.

3. Using the given function, $0.15A_0 = A_0e^{-0.0001216t} \Rightarrow 0.15 = e^{-0.0001216t} \Rightarrow \ln 0.15 = \ln e^{-0.0001216t} \Rightarrow$

   $\ln 0.15 = -0.0001216t \Rightarrow t = \frac{\ln(0.15)}{-0.0001216} \Rightarrow t \approx 15601.3 \approx 16{,}000$ years old.

5. (a) With a half-life of 21.7 years and using the exponential decay function, our model is

   $\frac{1}{2}A_0 = A_0e^{-21.7k} \Rightarrow 0.5 = e^{-21.7k} \Rightarrow \ln 0.5 = \ln e^{-21.7k} \Rightarrow \ln 0.5 = -21.7k \Rightarrow k = \frac{\ln 0.5}{-21.7}$

   $k \approx 0.032$. The exponential decay model is $A(t) = A_0e^{-0.032t}$.

   (b) $400 = 500e^{-0.032t} \Rightarrow \frac{400}{500} = e^{-0.032t} \Rightarrow \ln 0.8 = \ln e^{-0.032t} \Rightarrow \ln 0.8 = -0.032t \Rightarrow \frac{\ln 0.8}{-0.032} = t \Rightarrow t \approx 6.97$ yr.

   (c) $A(10) = 500e^{-0.032(10)} \Rightarrow A(10) = 500e^{-0.32} \Rightarrow A(10) \approx 363$ grams.

7. (a) Since the amount remaining is less than half of the original 2 milligrams, the half-life is less than 200 days.

   (b) Using the exponential decay function, $A_0 = 2$, and the ordered pair $(100, 1.22)$ from the table, yields the formula: $1.22 = 2e^{-100k} \Rightarrow 0.61 = e^{-100k} \Rightarrow \ln 0.61 = \ln e^{-100k} \Rightarrow$

   $\ln 0.61 = -100k \Rightarrow k = \frac{\ln 0.61}{100} \Rightarrow k \approx .005$. The formula that models the data is $A = 2e^{-0.005t}$.

   (c) Graph $y_1 = 2e^{-0.005t}$ and $y_2 = 1$ on the calculator. The intersection of the graphs is the approximate half-life: 140 days.

9. (a) $R(x) = \log\frac{I}{I_o} \Rightarrow 7.4 = \log\frac{I}{I_o} \Rightarrow \frac{I}{I_0} = 10^{7.4} \Rightarrow I = 10^{7.4}I_o \approx 25{,}118{,}864I_0$

   (b) $R(x) = \log\frac{I}{I_o} \Rightarrow 6.3 = \log\frac{I}{I_o} \Rightarrow \frac{I}{I_0} = 10^{6.3} \Rightarrow I = 10^{6.3}I_o \approx 1{,}995{,}262I_0$

   (c) $\frac{10^{7.4}}{10^{6.3}} = 10^{1.1} \approx 12.6$ times more intense.

11. Use the equation to find the intensity of a star magnitude 1:

    $1 = 6 - \frac{5}{2}\log\frac{I}{I_0} \Rightarrow -5 = -\frac{5}{2}\log\frac{I}{I_0} \Rightarrow \log\frac{I}{I_0} \Rightarrow 10^2 = 10^{\log I/I_0} \Rightarrow 100 = \frac{I}{I_0} \Rightarrow I = 100I_0$. Now use the equation to find the intensity of a star magnitude 3:

$3 = 6 - \frac{5}{2}\log\frac{I}{I_0} \Rightarrow -3 = -\frac{5}{2}\log\frac{I}{I_0} \Rightarrow \frac{6}{5} = \log\frac{I}{I_0} \Rightarrow 10^{6/5} = 10^{\log I/I_0} \Rightarrow 10^{6/5} = \frac{I}{I_0} \Rightarrow I \approx 15.85I_0$. Comparing intensities of the magnitude 1 star to the magnitude 3 star yields $\frac{100I_0}{15.85I_0} \approx 6.31$. The magnitude 1 star is approximately 6.3 times more intense then the magnitude 3 star.

13. (a) First solve for k, using $f(x) = 60, T_0 = 20, t = 1C = 100 - 20 = 80; 60 = 20 + 80e^{-k(1)} \Rightarrow$ $40 = 80e^{-k} \Rightarrow 0.5 = e^{-k} \Rightarrow \ln 0.5 = \ln e^{-k} \Rightarrow \ln 0.5 = -k \Rightarrow k = -\ln 0.5 \Rightarrow k \approx 0.693$. The equation is $f(t) = 20 + 80x^{-0.693t}$.

    (b) $f(0.5) = 20 + 80e^{-0.693(0.5)} \Rightarrow f(0.5) \approx 76.6^\circ C$.

    (c) Solve for t, with $f(t) = 50, 50 = 20 + 80e^{-0.693t} \Rightarrow 30 = 80e^{-0.69et} \Rightarrow \frac{30}{80} = e^{-0.693t} \Rightarrow$ $\ln 0.375 = \ln e^{-0.693t} \Rightarrow \ln 0.375 = -0.693t \Rightarrow t = \frac{\ln 0.375}{-0.693} \Rightarrow t \approx 1.415$ or about 1 hour 25 minutes.

15. (a) Set $L_1 \to$ Year and $L_2 \to CFC\ 12(ppb)$ and use the exponential regression capabilities of the calculator to find the equation: $y \approx 0.72(1.041)^x$. Therefore the values are $C \approx 0.72$ and $a \approx 1.041$.

    (b) Use the equation from part a, to find $f(13): f(13) = 0.72(1.041)^{13} \Rightarrow f(13) \approx 1.21$.

17. (a) $A(3) = 1000(1.025)^3 = \$1076.89$

    (b) $A(10) = 1000(1.025)^{10} = \$1280.08$

    (c) $1900 = 1000(1.025)^x \Rightarrow 1.9 = 1.025^x \Rightarrow \ln(1.9) = x\ln 1.025 \Rightarrow x = \frac{\ln 1.9}{\ln 1.025} \approx 26$ years

19. (a) $B(1) = 3.5e^{0.02(1)} \approx 3.57$ million/mL

    (b) $B(6.5) = 3.5e^{0.02(6.5)} \approx 3.99$ million/mL

    (c) $3.5e^{0.02x} = 6 \Rightarrow e^{0.02x} = \frac{12}{7} \Rightarrow 0.02x = \ln\frac{12}{7} \Rightarrow x = \frac{\ln 12/7}{0.02} \approx 27$ hours

21. $3000 = 2500e^{0.0375t} \Rightarrow e^{0.0375t} = 1.2 \Rightarrow 0.0375t = \ln 1.2 \Rightarrow t = \frac{\ln 1.2}{0.0375} \approx 4.9$ years

23. $5000 = 2500e^{0.05t} \Rightarrow e^{0.05t} = 2 \Rightarrow 0.05t = \ln 2 \Rightarrow t = \frac{\ln 2}{0.05} \approx 13.9$ years

25. (a) The domain for this scenario would be $A \geq 1000$. The amount in the account must always be greater than or equal to the principal.

    (b) $T(1200) = 50\ln\frac{1200}{1000} \approx 9.1$ years

    (c) $50\ln\frac{A}{1000} = 23.5 \Rightarrow \ln\frac{A}{1000} = \frac{23.5}{50} \Rightarrow e^{23.5/50} = \frac{A}{1000} \Rightarrow A = 1000e^{23.5/50} \approx \$1600$

27. (a) Use the Compound Interest Formula

$$A = P\left(1+\frac{r}{n}\right)^{nt},\ 5000 = 1000\left(1+\frac{0.035}{4}\right)^{4t} \Rightarrow 5 = 1.00875^{4t} \Rightarrow \log 5 = \log 1.00875^{4t} \Rightarrow$$

$$t = \frac{\log 5}{4\log 1.00785} \Rightarrow \approx 46.2 \text{ years.}$$

(b) Use the Continuous Compound Interest Formula

$$A = Pe^{rt},\ 5000 = 1000e^{0.035t} \Rightarrow 5 = e^{0.035t} \Rightarrow \ln 5 = \ln e^{0.035t} \Rightarrow \ln 5 = 0.035t \Rightarrow$$

$$t = \frac{\ln 5}{0.035} \Rightarrow t \approx 46.0 \text{ years.}$$

29. Use the Compound Interest Formula

$$A = P\left(1+\frac{r}{n}\right)^{nt},\ 30000 = 27000\left(1+\frac{0.023}{4}\right)^{4t} \Rightarrow \frac{30000}{27000} = 1.00575^{4t} \Rightarrow \frac{10}{9} = 1.00575^{4t} \Rightarrow$$

$$\log\frac{10}{9} = \log 1.00575^{4t} \Rightarrow \log\frac{10}{9} = 4t \cdot \log 1.00575 \Rightarrow t = \frac{\log\frac{10}{9}}{4\log 1.00575} \Rightarrow t \approx 4.6 \text{ years.}$$

31. Use the Compound Interest Formula

$$A = P\left(1+\frac{r}{n}\right)^{nt},\ A = 60000\left(1+\frac{0.02}{4}\right)^{4(5)} \Rightarrow A = 60000(1.005)^{20} \Rightarrow A = \$66293.73$$

Now use the Continuous Compound Interest Formula $A = Pe^{rt}$, $A = 60000e^{0.018(5)} \Rightarrow A \approx \$65650.48$

The better investment is the 2% rate compounded quarterly; it will earn $643.27 more.

33. $R = \left(1+\dfrac{0.03}{4}\right)^{4} - 1 \Rightarrow R \approx 1.075^{4} - 1 \Rightarrow R \approx 0.030339$ or $R \approx 3.03\%$.

35. $P = 10{,}000\left(1+\dfrac{0.03}{2}\right)^{-2(5)} \Rightarrow P = 10{,}000(1.015)^{-10} \Rightarrow P = \$8616.67.$

37. $25000 = 30416\left(1+\dfrac{r}{1}\right)^{-1(5)} \Rightarrow \dfrac{25000}{30416} = (1+r)^{-5} \Rightarrow \left(\dfrac{25000}{30416}\right)^{-1/5} = 1+r \Rightarrow r \approx 0.03993$ or about 4%.

39. (a) Using $R = \dfrac{P}{\dfrac{1-(1+i)^{-n}}{i}}$, and $n = 12(4) = 48$, $i = \dfrac{0.075}{12} = 0.00625$ yields

$$R = \frac{8500}{\dfrac{1-(1+0.00625)^{-48}}{0.00625}} \Rightarrow R \approx \$205.52.$$

(b) The total interest paid will be $I = nR - P \Rightarrow I = 48(205.52) - 8500 \Rightarrow I \approx \$1364.96.$

41. (a) Using $R = \dfrac{P}{\dfrac{1-(1+i)^{-n}}{i}}$, and $n = 12(30) = 360$, $i = \dfrac{0.0725}{12}$ yields

$$R = \frac{125000}{\dfrac{1-(1+0.0725/12)^{-360}}{0.0725/12}} \Rightarrow R \approx \$852.72.$$

(b) The total interest paid will be $I = nR - P \Rightarrow I = 360(852.72) - 125000 \Rightarrow I \approx \$181{,}979.20.$

43. (a) First enter $y_1 = 1500\left(1+\frac{0.0575}{365}\right)^{365t}$ into the calculator, now using the table feature, the investment will triple when $y_1 = 4500$. From the table, $t \approx 19.1078 \Rightarrow t \approx 19$ years $+\ 0.1078(365)$ days $\Rightarrow t \approx 19$ years, 39 days. Analytically, for $y = 4500$, the solution is

$$4500 = 1500\left(1+\frac{0.0575}{365}\right)^{365t} \Rightarrow 3 = (1.000157534)^{365t} \Rightarrow \ln 3 = 365t \ln 1.000157534$$

$$\Rightarrow t = \frac{\ln 3}{365 \ln 1.000157534} \Rightarrow t \approx 19.1078 \text{ or 19 years and 39 days.}$$

(b) First enter $y_1 = 2000\left(1+\frac{0.08}{365}\right)^{365t}$ into the calculator, now using the table feature, the investment will triple when $y_1 = 2000$. From the table, $t \approx 11.455 \Rightarrow t \approx 11$ years $+\ 0.455(365)$ days $\Rightarrow t \approx 11$ years 166 days.

45. (a) $A(5) = 2,400,000e^{0.023(5)} \Rightarrow A(5) \approx 2,692,496 \Rightarrow A(5) \approx 2,700,000.$

(b) $A(10) = 2,400,000e^{0.023(5)} \Rightarrow A(10) \approx 3,020,640 \Rightarrow A(10) \approx 3,000,000.$

(c) $A(60) = 2,400,000e^{0.023(5)} \Rightarrow A(60) \approx 9,539,764 \Rightarrow A(60) \approx 9,500,000.$

47. (a) $42(2)^x = 400 \Rightarrow 2^x = \frac{200}{21} \Rightarrow x\ln 2 = \ln\frac{200}{21} \Rightarrow x = \frac{\ln 200/21}{\ln 2} \approx 3.25$, The result is in the third year.

(b) See Figure 47.

(c) $42(2)^x = 1000 \Rightarrow 2^x = \frac{500}{21} \Rightarrow x\ln 2 = \ln\frac{500}{21} \Rightarrow x = \frac{\ln 500/21}{\ln 2} \approx 4.57$, The result is in the fourth year.

[0, 5] by [350,450]
Xscl = 1 Yscl = 1

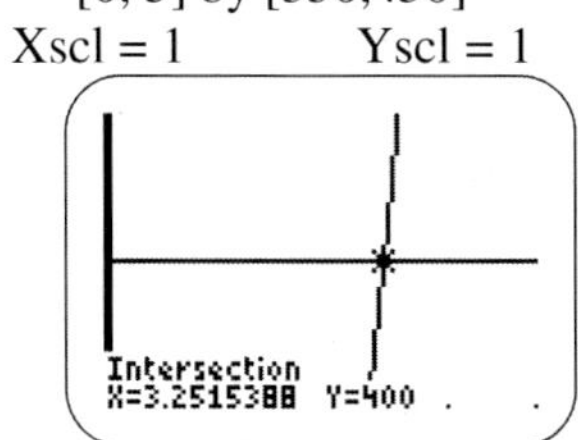

Figure 47

49. (a) Use the coordinate (2012, 17) to find the coefficient C: $17 = Ca^{(2012-2012)} \Rightarrow 17 = Ca^0 \Rightarrow 17 = C$,

Use (2060, 31) and $C = 17$ to find a: $31 = 17a^{(2060-2012)} \Rightarrow 31 = 17a^{48} \Rightarrow a^{48} = \frac{31}{17} \Rightarrow a = \sqrt[48]{\frac{31}{17}} \approx 1.0126$

The value of a is greater than or equal to one since the percentage is increasing.

(b) $P(2030) = 17(1.0126)^{2030-2012} = 17(1.0126)^{18} \approx 21.3\%$

(c) $17(1.0126)^{x-2012}=25\Rightarrow 1.0126^{x-2012}=\dfrac{25}{17}\Rightarrow (x-2012)\ln 1.0126=\ln\dfrac{25}{17}\Rightarrow$

$$x-2012=\frac{\ln\frac{25}{17}}{\ln 1.0126}\Rightarrow x=\frac{\ln\frac{25}{17}}{\ln 1.0126}+2012\approx 2042.8$$

51. $N(t)=100{,}000e^{rt}\Rightarrow 200{,}000=100{,}000e^{r(2)}\Rightarrow 2=e^{r(2)}\Rightarrow \ln 2=2r\Rightarrow r=\dfrac{\ln 2}{2}\Rightarrow r\approx 0.3466.$

$350{,}000=100{,}000e^{0.3466t}\Rightarrow 3.5=e^{0.3466t}\Rightarrow \ln 3.5=0.3466t\Rightarrow t=\dfrac{\ln 3.5}{0.3466}\Rightarrow t\approx 3.6$ hours.

53. Since x represents the number of years after 1980, we will use the coordinate (0, 30) to find the coefficient C: $30=Ca^0\Rightarrow 30=C$, The cost for each year after 1980 was 75% of the year prior and we have the coordinate (1, 22.5): $22.5=30a^{(1)}\Rightarrow 22.5=30a\Rightarrow a=\dfrac{22.5}{30}\Rightarrow a=0.75$

The function is $f(x)=30(0.75)^x$. Let $f(x)=1$ to find the year when the price was \$1.

$30(0.75)^x=1\Rightarrow 0.75^x=\dfrac{1}{30}\Rightarrow x\ln 0.75=\ln\dfrac{1}{30}\Rightarrow x=\dfrac{\ln\frac{1}{30}}{\ln 0.75}\approx 11.8$, Since x represents the number of years after 1980, the result is about 1992.

55. (a)

X	0	15	30	45	60	75	90	125
G(x)	7	21	57	111	136	158	164	178

(b) Since there are 261 people in the village the equation is $g(x)=261-f(x)$.

(c) Graph the data on the same calculator graph as $y_1=\dfrac{171}{1+18.6e^{-0.0747x}}$ and $y_2=18.3(1.024)^x$.

From the graph, y_1 is the better fit.

(d) Using $g(x)=261-f(x)$ and $g(x)=\dfrac{171}{1+18.6e^{-0.0747x}}$ yields $f(x)=261-\dfrac{171}{1+18.6e^{-0.0747x}}$.

57. (a) Use the capabilities of the exponential regression calculator and L_1 and L_2, to find the coefficients C and a: $C\approx 5.772,\ a\approx 1.277\Rightarrow f(x)=5.772(1.277)^{x-2008}$

(b) See Figure 57.

[2005,2015] by [2,20]

Xscl = 1 Yscl = 1

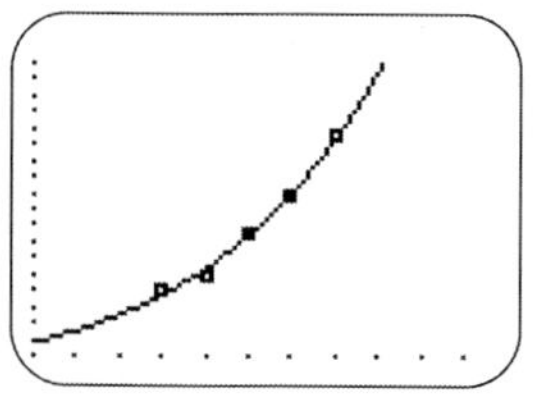

Figure 57

59. (a) $f(25) = \dfrac{9}{1+271e^{-0.122(25)}} \Rightarrow f(25) \approx 0.065$; and $f(65) = \dfrac{9}{1+271e^{-0.122(25)}} \Rightarrow f(65) \approx 0.82$. Among people age 25, 6.5% have some CHD, while among people age 65, 82% have some CHD.

(b) $0.50 = \dfrac{0.9}{1+271e^{-0.122x}} \Rightarrow 0.5(1+271e^{-0.122x}) = 0.9 \Rightarrow 1+271e^{-0.122x} = \dfrac{0.9}{0.5} \Rightarrow$

$$271e^{-0.122x} = \frac{0.9}{0.5} - 1 \Rightarrow e^{-0.122x} = \frac{\frac{0.9}{0.5}-1}{271} \Rightarrow \ln e^{-0.122x} = \ln\left(\frac{\frac{0.9}{0.5}-1}{271}\right) \Rightarrow$$

$$-0.122x = \ln\left(\frac{\frac{0.9}{0.5}-1}{271}\right) \Rightarrow x = \frac{\ln\left(\frac{\frac{0.9}{0.5}-1}{271}\right)}{-0.122} \Rightarrow x \approx 47.75.$$ The likelihood is 50% at about age 48.

Summary Exercises on Functions: Domains and Defining Equations and Composition

1. Choice A can be written as a function of x. $3x + 2y = 6 \Rightarrow y = f(x) = -\frac{3}{2}x + 3$

3. Choice C can be written as a function of x. $x^3 + y^3 = 5 \Rightarrow y = f(x) = \sqrt[3]{5 - x^3}$

5. Choice A can be written as a function of x. $x = \frac{2-y}{y+3} \Rightarrow y = f(x) = \frac{2-3x}{x+1}$

7. Choice D can be written as a function of x. $2x = \frac{1}{y^3} \Rightarrow y = f(x) = \sqrt[3]{\frac{1}{2x}}$

9. Choice C can be written as a function of x. $\frac{x}{4} - \frac{y}{9} = 0 \Rightarrow y = f(x) = \frac{9x}{4}$

11. Domain: $(-\infty, \infty)$

13. Domain: $(-\infty, \infty)$

15. Domain: $(-\infty, \infty)$ The domain is the set of all real numbers that make the denominator not equal to zero.

17. Domain: $(-\infty, -3) \cup (-3, 3) \cup (3, \infty)$ The domain is the set of all real numbers that make the denominator not equal to zero.

19. Domain: $(-4, 4)$ The domain is the set of all real numbers that make $16 - x^2 > 0$.

21. Domain: $(-\infty, -1] \cup [8, \infty)$ The domain is the set of all real numbers that make $x^2 - 7x - 8 \geq 0$.

Interval	Test Point	Value of x^2-7x-8	Sign of x^2-7x-8
$(-\infty,-1)$	-2	10	Positive
$(-1,8)$	0	-8	Negative
$(8,\infty)$	10	22	Positive

23. Domain: $(-\infty,\infty)$ The domain is the set of all real numbers that make the denominator not equal to zero.

25. Domain: $[1,\infty)$ The domain is the set of all real numbers that make $x^3-1\geq 0$.

27. Domain: $(-\infty,\infty)$ The domain is the set of all real numbers that make x^2+x+4 a real number.

29. Domain: $(-\infty,1)$ The domain is the set of all real numbers that make the denominator not equal to zero or $\frac{-1}{x^3-1}\geq 0$.

Interval	Test Point	Value of $\frac{-1}{x^3-1}$	Sign of $\frac{-1}{x^3-1}$
$(-\infty,1)$	0	1	Positive
$(1,\infty)$	2	$-\frac{1}{7}$	Negative

31. Domain: $(-\infty,\infty)$ The domain is the set of all real numbers that make $x^2+1>0$

33. Domain: $(-\infty,-2)\cup(-2,3)\cup(3,\infty)$ Since $\left(\frac{x+2}{x-3}\right)^2\geq 0$ for all real numbers, the domain of $f(x)$ is the set of all real numbers such that $\frac{x+2}{x-3}\neq 0$.

35. Domain: $(-\infty,0)\cup(0,\infty)$ The domain is the set of all real numbers such that $\frac{1}{x}$ is defined.

37. Domain: $(-\infty,\infty)$

39. Domain: $[-2,2]$ The domain is the set of all real numbers such that $16-x^4\geq 0$

Interval	Test Point	Value of $16-x^4$	Sign of $16-x^4$
$(-\infty, -2)$	–3	–65	Negative
$(-2, 2)$	0	16	Positive
$(2, \infty)$	3	–65	Negative

41. Domain: $(-\infty,-7]\cup(-4,3)\cup[9,\infty)$ The domain is the set of real numbers such that $\frac{x^2-2x-63}{x^2+x-12}\geq 0$.

$\frac{x^2-2x-63}{x^2+x-12}$ is not defined for $x^2+x-12=0\Rightarrow(x+4)(x-3)=0\Rightarrow\ x\neq -4$ or $x\neq 3$. Solve $\frac{x^2-2x-63}{x^2+x-12}=0$

to find the test intervals: $\frac{x^2-2x-63}{x^2+x-12}=0\Rightarrow\ x^2-2x-63=0\Rightarrow(x-9)(x+7)=0\Rightarrow\ x=9$ or $x=-7$.

Interval	Test Point	Value of $\frac{x^2-2x-63}{x^2+x-12}$	Sign
$(-\infty, -7)$	–10	$\frac{19}{26}$	Positive
$(-7, -4)$	–5	$-\frac{7}{2}$	Negative
$(-4, 3)$	0	$\frac{21}{4}$	Positive
$(3, 9)$	5	$-\frac{8}{3}$	Negative
$(9, \infty)$	10	$\frac{17}{98}$	Positive

43. Domain: $(-\infty,5]$ The domain is the set of real numbers such that $5-x\geq 0\Rightarrow 5\geq x$

45. Domain: $(-\infty,4)\cup(4,\infty)$ The domain is the set of real numbers such that

$\left|\frac{1}{4-x}\right|>0\Rightarrow\frac{1}{4-x}>0\Rightarrow 4-x>0\Rightarrow 4>x$ or $-\frac{1}{4-x}<0\Rightarrow -4+x<0\Rightarrow x<4$

47. Domain: $(-\infty,-5]\cup[5,\infty)$ The domain is the set of real numbers such that $\sqrt{x^2-25}$ is a real number. $x^2-25\geq 0\Rightarrow x^2\geq 25\Rightarrow x\geq 5$ or $x\leq -5$

49. Domain: $(-2,6)$ The domain is the set of real numbers such that $\frac{-3}{(x+2)(x-6)}>0$ and $(x+2)(x-6)\neq 0\Rightarrow$ $x\neq -2$ or $x\neq 6$.

Interval	Test Point	Value of $\frac{-3}{(x+2)(x-6)}$	Sign of $\frac{-3}{(x+2)(x-6)}$
$(-\infty, -2)$	-3	$-\frac{1}{3}$	Negative
$(-2, 6)$	0	$\frac{1}{4}$	Positive
$(6, \infty)$	7	$-\frac{1}{3}$	Negative

51. (a) $(f \circ g)(x) = f(g(x)) = f(5x+7) = -6(5x+7)+9 = -30x-42+9 = -30x-33$

The domain and range of both f and g are $(-\infty,\infty)$, so the domain of $f \circ g$ is $(-\infty,\infty)$.

(b) $(g \circ f)(x) = g(f(x)) = g(-6x+9) = 5(-6x+9)+7 = -30x+45+7 = -30x+52$

The domain of $g \circ f$ is $(-\infty,\infty)$.

53. (a) $(f \circ g)(x) = f(g(x)) = f(x+3) = \sqrt{x+3}$ The domain and range of g are $(-\infty,\infty)$, however, the domain and range of f are $[0,\infty)$. So, $x+3 \geq 0 \Rightarrow x \geq -3$. Therefore, the domain of $f \circ g$ is $[-3,\infty)$.

(b) $(g \circ f)(x) = g(f(x)) = g\left(\sqrt{x}\right) = \sqrt{x}+3$ The domain and range of g are $(-\infty,\infty)$, however, the domain and range of f are $[0,\infty)$.Therefore, the domain of $g \circ f$ is $[0,\infty)$.

55. (a) $(f \circ g)(x) = f(g(x)) = f(x^2+3x-1) = (x^2+3x-1)^3$

The domain and range of f and g are $(-\infty,\infty)$, so the domain of $f \circ g$ is $(-\infty,\infty)$.

(b) $(g \circ f)(x) = g(f(x)) = g(x^3) = \left(x^3\right)^2 + 3\left(x^3\right) - 1 = x^6 + 3x^3 - 1$

The domain and range of f and g are $(-\infty,\infty)$, so the domain of $g \circ f$ is $(-\infty,\infty)$.

57. (a) $(f \circ g)(x) = f(g(x)) = f(3x) = \sqrt{3x-1}$ The domain and range of g are $(-\infty,\infty)$, however, the domain and range of f are $[1,\infty)$. So, $3x-1 \geq 0 \Rightarrow x \geq \frac{1}{3}$. Therefore, the domain of $f \circ g$ is $\left[\frac{1}{3},\infty\right)$.

(b) $(g \circ f)(x) = g(f(x)) = g\left(\sqrt{x-1}\right) = 3\sqrt{x-1}$ The domain and range of g are $(-\infty,\infty)$, however, the range of f is $[0,\infty)$. So $x-1 \geq 0 \Rightarrow x \geq 1$. Therefore, the domain of $g \circ f$ is $[1,\infty)$.

59. (a) $(f \circ g)(x) = f(g(x)) = f(x+1) = \frac{2}{x+1}$ The domain and range of g are $(-\infty,\infty)$, however, the domain of f is $(-\infty,0) \cup (0,\infty)$. So, $x+1 \neq 0 \Rightarrow x \neq -1$. Therefore, the domain of $f \circ g$ is $(-\infty,-1) \cup (-1,\infty)$.

(b) $(g \circ f)(x) = g(f(x)) = g\left(\frac{2}{x}\right) = \frac{2}{x} + 1$ The domain and range of f is $(-\infty, 0) \cup (0, \infty)$, however, the domain and range of g are $(-\infty, \infty)$. So $x \neq 0$. Therefore, the domain of $g \circ f$ is $(-\infty, 0) \cup (0, \infty)$.

61. (a) $(f \circ g)(x) = f(g(x)) = f\left(-\frac{1}{x}\right) = \sqrt{-\frac{1}{x} + 2}$ The domain and range of g are $(-\infty, 0) \cup (0, \infty)$, however, the domain of f is $[-2, \infty)$. So, $-\frac{1}{x} + 2 \geq 0 \Rightarrow x < 0$ or $x \geq \frac{1}{2}$ (using test intervals). Therefore, the domain of $f \circ g$ is $(-\infty, 0) \cup \left[\frac{1}{2}, \infty\right)$.

(b) $(g \circ f)(x) = g(f(x)) = g\left(\sqrt{x+2}\right) = -\frac{1}{\sqrt{x+2}}$ The domain of f is $[-2, \infty)$ and its range is $(-\infty, \infty)$. The domain and range of g are $(-\infty, 0) \cup (0, \infty)$. So $x + 2 > 0 \Rightarrow x > -2$. Therefore, the domain of $g \circ f$ is $(-2, \infty)$.

63. (a) $(f \circ g)(x) = f(g(x)) = f\left(\frac{1}{x+5}\right) = \sqrt{\frac{1}{x+5}}$ The domain of g is $(-\infty, -5) \cup (-5, \infty)$, and the range of g is $(-\infty, 0) \cup (0, \infty)$. The domain of f is $[0, \infty)$. Therefore, the domain of $f \circ g$ is $(-5, \infty)$.

(b) $(g \circ f)(x) = g(f(x)) = g\left(\sqrt{x}\right) = \frac{1}{\sqrt{x}+5}$ The domain and range of f is $[0, \infty)$. The domain of g is $(-\infty, -5) \cup (-5, \infty)$. Therefore, the domain of $g \circ f$ is $[0, \infty)$.

65. (a) $(f \circ g)(x) = f(g(x)) = f\left(\frac{1}{x}\right) = \frac{1}{1/x - 2} = \frac{x}{1-2x}$ The domain and range of g are $(-\infty, 0) \cup (0, \infty)$. The domain of f is $(-\infty, -2) \cup (-2, \infty)$, and the range of f is $(-\infty, 0) \cup (0, \infty)$. So, $\frac{x}{1-2x} < 0 \Rightarrow x < 0$ or $0 < x < \frac{1}{2}$ or $x > \frac{1}{2}$ (using test intervals). Thus, $x \neq 0$ and $x \neq \frac{1}{2}$. Therefore, the domain of $f \circ g$ is $(-\infty, 0) \cup \left(0, \frac{1}{2}\right) \cup \left(\frac{1}{2}, \infty\right)$.

(b) $(g \circ f)(x) = g(f(x)) = g\left(\frac{1}{x-2}\right) = \frac{1}{1/(x-2)} = x - 2$ The domain and range of g are $(-\infty, 0) \cup (0, \infty)$. The domain of f is $(-\infty, -2) \cup (-2, \infty)$, and the range of f is $(-\infty, 0) \cup (0, \infty)$. Therefore, the domain of $g \circ f$ is $(-\infty, -2) \cup (-2, \infty)$.

67. (a) $(f \circ g)(x) = f(g(x)) = f\left(\sqrt{x}\right) = \log\left(\sqrt{x}\right)$ The domain of g is $(-\infty, \infty)$ and range of g is $[0, \infty)$. The domain of f is $(0, \infty)$, and the range of f is $(-\infty, \infty)$. So the domain of $f \circ g$ is $(0, \infty)$.

(b) $(g \circ f)(x) = g(f(x)) = g(\log x) = \left(\sqrt{\log x}\right)$ The domain of g is $(-\infty, \infty)$ and range of g is $[0, \infty)$. The domain of f is $(0, \infty)$, and the range of f is $(-\infty, \infty)$. So the domain of $g \circ f$ is $[1, \infty)$.

69. (a) $(f \circ g)(x) = f(g(x)) = f\left(\sqrt{x}\right) = e^{\sqrt{x}}$ The domain of g is $\left(-\infty,\infty\right)$ and range of g is $\left[0,\infty\right)$. The domain of f is $(-\infty,\infty)$, and the range of f is $(0,\infty)$. So the domain of $f \circ g$ is $\left[0,\infty\right)$.

(b) $(g \circ f)(x) = g(f(x)) = g\left(e^x\right) = \left(\sqrt{e^x}\right)$ The domain of g is $\left(-\infty,\infty\right)$ and range of g is $\left[0,\infty\right)$. The domain of f is $(-\infty,\infty)$, and the range of f is $(0,\infty)$. So the domain of $g \circ f$ is $\left(-\infty,\infty\right)$.

71. (a) $(f \circ g)(x) = f(g(x)) = f\left(\ln\sqrt{x}\right) = -\left(\ln\sqrt{x}\right)^2$ The domain g is $\left(0,\infty\right)$. The domain of f is $(-\infty,\infty)$ So the domain of $f \circ g$ is $\left(0,\infty\right)$.

(b) $(g \circ f)(x) = g(f(x)) = g\left(-x^2\right) = \ln\sqrt{-x^2}$ The function is undefined for negative values. So the domain of $g \circ f$ is $\varnothing$.

73. (a) $(f \circ g)(x) = f(g(x)) = f\left(\dfrac{x}{x-2}\right) = 5\left(\dfrac{x}{x-2}\right) - 2 = \dfrac{3x+4}{x-2}$ The domain g is $(-\infty,2)\cup\left(2,\infty\right)$. The domain of f is $(-\infty,\infty)$ So the domain of $f \circ g$ is $(-\infty,2)\cup\left(2,\infty\right)$.

(b) $(g \circ f)(x) = g(f(x)) = g\left(5x-2\right) = \dfrac{5x-2}{5x-2-2} = \dfrac{5x-2}{5x-4}$ The domain g is $(-\infty,2)\cup\left(2,\infty\right)$. The domain of f is $(-\infty,\infty)$ So the domain of $g \circ f$ is $\left(-\infty,\dfrac{4}{5}\right)\cup\left(\dfrac{4}{5},\infty\right)$.

Chapter 5 Review Exercises

1. The function is not one-to-one, the graph fails the horizontal line test.
2. The function is not one-to-one, the graph fails the horizontal line test.
3. The function is not one-to-one, the graph fails the horizontal line test.
4. The function is one-to-one, different x-values always produce different y-values.
5. The function is not one-to-one, the graph fails the horizontal line test.
6. The function is not one-to-one, the graph fails the horizontal line test.
7. All real numbers can be input for x, therefore the domain is $(-\infty,\infty)$.
8. All real numbers can be solutions to $f(x)$, therefore the range is $(-\infty,\infty)$.
9. Since f is one-to-one, it has an inverse.

10. For $f(x)=\sqrt[3]{2x-7}$ the inverse is

$$x=\sqrt[3]{2y-7}\Rightarrow x^3=2y-7\Rightarrow x^3+7=2y\Rightarrow y=\frac{x^3+7}{2}\Rightarrow f^{-1}(x)=\frac{x^3+7}{2}.$$

11. See Figure 11. The graphs are reflections across the line $y=x$.

[10,70] by [0,60]

Xscl = 10 Yscl = 10

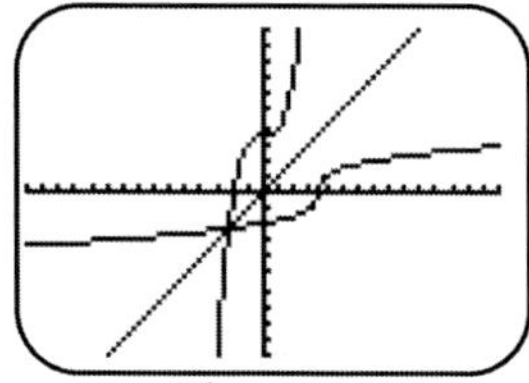

Figure 11

12. $(f\circ f^{-1})(x)=\sqrt[3]{2\left(\frac{x^3+7}{2}\right)-7}=\sqrt[3]{x^3}=x$ and $(f^{-1}\circ f)(x)=\frac{\left(\sqrt[3]{2x-7}\right)^3+7}{2}=\frac{2x-7+7}{2}=\frac{2x}{2}=x$.

13. $y=a^{x+2}$ is the graph of $y=a^x$ shifted 2 units left with a y-intercept of a^2. Therefore, it matches C.

14. $y=a^x+2$ is the graph of $y=a^x$ shifted 2 units upward with a y-intercept of 3. Therefore, it matches A.

15. $y=-a^x+2$ is the graph of $y=a^x$ reflected across the x-axis, then shifted 2 units upward, with a y-intercept of 1. Therefore, it matches D.

16. $y=a^{-x}+2$ is the graph of $y=a^x$ reflected across the y-axis, then shifted 2 units upward, with a y-intercept of 3. Therefore, it matches B.

17. Because the graph goes up to the left, the value of a is $0<x<1$.

18. All real numbers can be input for x, therefore the domain is $(-\infty,\infty)$.

19. From the graph, only values where $f(x)>0$, are possible solutions. Therefore, the range is $(0,\infty)$.

20. Since every real number a raised to the zero power equals 1, $f(0)=a^0=1$.

21. Reflect $f(x)$ across the line $y=x$. See Figure 21.

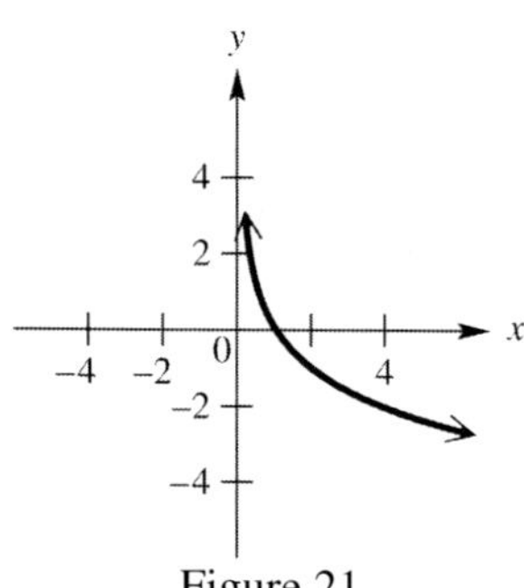

Figure 21

22. For $f(x) = a^x$, the inverse is, $x = a^y \Rightarrow y = \log_a x \Rightarrow f^{-1}(x) = \log_a x.$

23. See Figure 23. From the graph, the domain is $(-\infty, \infty)$; and the range is $(0, \infty)$.

24. See Figure 24. From the graph, the domain is $(-\infty, \infty)$; and the range is $(2, \infty)$.

25. See Figure 25. From the graph, the domain is $(-\infty, \infty)$; and the range is $(-\infty, 0)$.

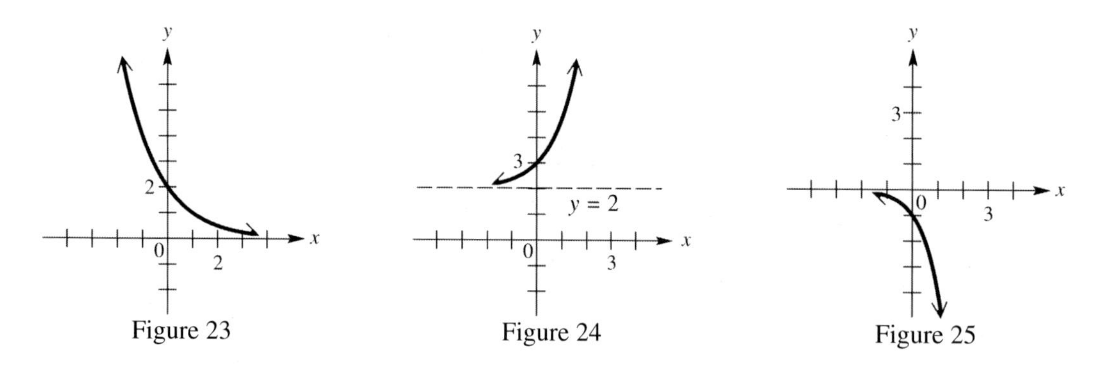

Figure 23 Figure 24 Figure 25

26. The graph of $f(x) = a^x$ with $0 < x < 1$ would have a range that decreases as the domain increases. Since the negative is on the exponent for $f(x) = a^{-x}$, the graph is reflected across the y-axis and is therefore now increasing as the domain increases. It is increasing on its domain.

27. (a) $\left(\frac{1}{8}\right)^{-2x} = 2^{x+3} \Rightarrow (2^{-3})^{-2x} = 2^{x+3} \Rightarrow 2^{6x} = 2^{x+3} \Rightarrow 6x = x + 3 \Rightarrow 5x = 3 \Rightarrow x = \frac{3}{5}$

(b) If $\left(\frac{1}{8}\right)^{-2x} \geq 2^{x+3}$, then $\left(\frac{1}{8}\right)^{-2x} - 2^{x+3} \geq 0$. Now graph $y_1 = \left(\frac{1}{8}\right)^{-2x} - 2^{x+3}$ on a calculator, use the answer from part (a), and from the graph the solution to $\left(\frac{1}{8}\right)^{-2x} - 2^{x+3} \geq 0$ is the interval $\left[\frac{3}{5}, \infty\right)$.

28. (a) $3^{-x} = \left(\frac{1}{27}\right)^{1-2x} \Rightarrow 3^{-x} = (3^{-3})^{1-2x} \Rightarrow 3^{-x} = 3^{-3+6x} \Rightarrow -x = -3 + 6x \Rightarrow 3 = 7x \Rightarrow x = \frac{3}{7}$

(b) If $3^{-x} < \left(\frac{1}{27}\right)^{1-2x}$, then $3^{-x} - \left(\frac{1}{27}\right)^{-2x} < 0$. Now graph $y_1 = 3^{-x} - \left(\frac{1}{27}\right)^{1-2x}$ on a calculator, use the answer from part (a), and from the graph the solution to $3^{-x} - \left(\frac{1}{27}\right)^{1-2x} < 0$ is the interval $\left(\frac{3}{7}, \infty\right)$.

29. (a) $0.5^{-x} = 0.25^{x+1} \Rightarrow 0.5^{-x} = (0.5^2)^{x+1} \Rightarrow 0.5^{-x} = 0.5^{2x+2} \Rightarrow -x = 2x+2 \Rightarrow -2 = 3x \Rightarrow x = -\frac{2}{3}$

(b) If $0.5^{-x} > 0.25^{x+1}$, then $0.5^{-x} - 0.25^{x+1} > 0$. Now graph $y_1 = 0.5^{-x} - 0.25^{x+1}$ on a calculator, use the answer from part (a), and from the graph the solution to $0.5^{-x} - 0.25^{x+1} > 0$ is the interval $\left(-\frac{2}{3}, \infty\right)$.

30. (a) $0.4x = 2.5^{1-x} \Rightarrow \log 0.4^x = \log 2.5^{1-x} \Rightarrow x \log 0.4 = (1-x)\log 2.5 \Rightarrow x\log 0.4 = \log 2.5 - x\log 2.5 \Rightarrow$ $x\log 0.4 + x\log 2.5 = \log 2.5 \Rightarrow x(\log 0.4 + \log 2.5) = \log 2.5 \Rightarrow x = \frac{\log 2.5}{\log 0.4 + \log 2.5}$, using a calculator, this is equal to $\frac{\log 2.5}{0}$, which is undefined. The solution is $\varnothing$.

(b) If $0.4^x < 2.5^{1-x}$ then $0.4x - 2.5^{1-x} < 0$. Now graph $y_1 = 0.4x - 2.5^{1-x}$ on a calculator, from the graph the solution to $0.4x - 2.5^{1-x} < 0$ is the interval $(-\infty, \infty)$.

31. We can find the intersection points of $y = x^2$ and $y = 2^x$ by using the intersection of graphs method and graph the equation $y_1 = x^2 - 2^x$ to finding the x-intercepts. Using the calculator the intercepts are $2, 4$, and -0.766664696. The missing point in common is $(-0.766664696, 0.58777475603)$.

32. Using the intersection of graphs method, graph $y = 3^x - \pi$ and find the x-intercept. From the calculator, the x-intercepts is $x = \{1.041978046\}$.

33. From the calculator, $\log 58.3 \approx 1.7656685547633 \approx 1.7657$.

34. From the calculator, $\log 0.00233 \approx -2.63264407897 \approx -2.6326$.

35. From the calculator, In $58.3 \approx 4.06560209336 \approx 4.0656$.

36. From the calculator, $\log_2 0.00233 = \frac{\log 0.00233}{\log 2} = -8.7455$.

37. By the definitions of logarithms, $\log_{13} 1 = x \Rightarrow 13^x = 1$ since we know $13^0 = 1$ it follows that $x = 0$.

38. By the properties of logarithms, $\ln e^{\sqrt{6}} = \log_e e^{\sqrt{6}} = \sqrt{6}.$

39. By the properties of logarithms, $\log_5 5^{12} = 12.$

40. By the properties of logarithms, $7^{\log_7 13} = 13.$

41. By the properties of logarithms, $3^{\log_3 5} = 5.$

42. By the change of base rule, $\log_4 9 = \dfrac{\log 9}{\log 4} \approx 1.58496250072 \approx 1.5850.$

43. The x-intercept is 1, and the graph is increasing. The correct graph is E.

44. Since $f(x) = \log_2(2x) = \log_2 + \log_2 x = 1 + \log_2 x$, the graph will be similar to that of $f(x) = \log_2 x$, but will have a vertical shift up 1. The x-intercept will be $\frac{1}{2}$, since $\log_2 2\left(\frac{1}{2}\right) = \log_2 1 = 0.$

The correct graph is D.

45. Since $f(x) = \log_2\left(\frac{1}{x}\right) = \log_2 x^{-1} = -\log_2 x$, the graph will be similar to that of $f(x) = \log_2 x$, but will be reflected across the x-axis. The x-intercept is 1. The correct graph is B.

46. Since $f(x) = \log_2\left(\frac{x}{2}\right) = \log_2\left(\frac{1}{2}x\right) = \log_2\left(\frac{1}{2}\right) + \log_2 x = -1 + \log_2 x$, the graph will be similar to that of $f(x) = \log_2 x$, but will have a vertical shift down 1 unit. The x-intercept will be 2, since $\log_2\left(\frac{2}{2}\right) = \log_2 1 = 0$. The correct graph is C.

47. With $f(x) = \log_2(x-1)$, the graph will be similar to that of $f(x) = \log_2 x$, but will be shifted 1 unit to the right; the x-intercept will be 2. The correct graph is F.

48. With $f(x) = \log_2(-x)$, the graph will be similar to that of $f(x) = \log_2 x$, but will be reflected across the y-axis. The x-intercept is -1, since $\log_2(-(-1)) = \log_2 1 = 0$. The correct graph is A.

49. See Figure 49. From the graph, the domain is $(0, \infty)$ and the range is $(-\infty, \infty)$.

50. See Figure 50. From the graph, the domain is $(2, \infty)$ and the range is $(-\infty, \infty)$.

51. See Figure 51. From the graph, the domain is $(-\infty, 0)$ and the range is $(-\infty, \infty)$.

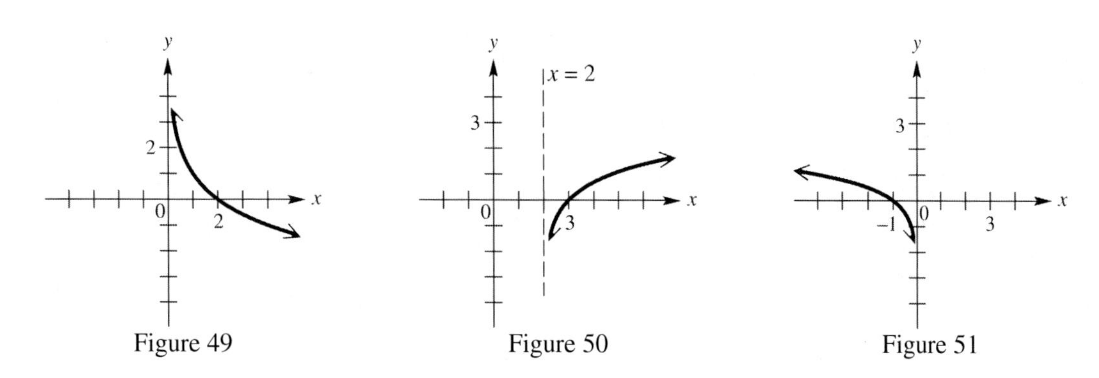

Figure 49 Figure 50 Figure 51

52. The functions in Exercises 49, 50, and 51 are the inverses, respectively, of those in Exercises 23, 24, and 25.

53. If the function has a point $(81,4)$, then $f(81)=4$. Therefore, $\log_a 81=4 \Rightarrow a^4=81 \Rightarrow a^4=3^4 \Rightarrow a=3$. The base is 3.

54. If the function has a point $\left(-4,\frac{1}{16}\right)$, then $f(-4)=\frac{1}{16}$. Therefore, $a^{-4}=\frac{1}{16} \Rightarrow a^{-4}=2^{-4} \Rightarrow a=2$. The base is 2.

55. $\log_3 \frac{mn}{5r}=\log_3 mn-\log_3 5r=\log_3 m+\log_3 n-(\log_3 5+\log_3 r)=\log_3 m+\log_3 n-\log_3 5-\log_3 r$

56. $\log_2 \frac{\sqrt{7}}{15}=\log_2 \sqrt{7}-\log_2 15=\log_2 7^{1/2}-\log_2 15=\frac{1}{2}\log_2 7-\log_2 15.$

57. $\log_5(x^2 y^4 \sqrt[5]{m^3 p})=\log_5 x^2+\log_5 y^4+\log_5(m^3 p)^{1/5}=2\log_5 x+4\log_5 y+\frac{1}{5}\log_5(m^3 p)=$

$2\log_5 x+4\log_5 y+\frac{1}{5}(\log_5 m^3+\log_5 p)=2\log_5 x+4\log_5 y+\frac{3}{5}\log_5 m+\frac{1}{5}\log_5 p$

58. The properties of logarithms do not apply to polynomials.

59. (a) $\log(x+3)+\log x=1 \Rightarrow \log(x+3)(x)=1 \Rightarrow 10^{\log(x+3)(x)}=10^1 \Rightarrow (x+3)(x)=10 \Rightarrow$
$x^2+3x-10=0 \Rightarrow (x+5)(x-2)=0 \Rightarrow x=-5,2.$

Since the argument of a logarithm cannot be negative, the solution is $x=2$.

(b) Graph $y_1=\log(x+3)+\log x$ and $y_2=1$ in the same window. From the graph, $y_1>y_2$ for the interval $(2,\infty)$.

60. (a) $\ln e^{\ln e} - \ln(x-4) = \ln 3 \Rightarrow \ln x - \ln(x-4) = \ln 3 \Rightarrow \ln\left(\frac{x}{x-4}\right) = \ln 3 \Rightarrow e^{\ln(x/x-4)} = e^{\ln x} \Rightarrow$

$\frac{x}{x-4} = 3 \Rightarrow x = 3x - 12 \Rightarrow -2x = -12 \Rightarrow x = 6$

(b) Graph $y_1 = \ln e^{\ln x} - \ln(x-4)$ and $y_2 = \ln 3$ in the same window. From the graph, $y_1 \le y_2$ for the interval $[6, \infty)$.

61. (a) $\ln e^{\ln 2} - \ln(x-1) = \ln 5 \Rightarrow \ln 2 - \ln(x-1) = \ln 5 \Rightarrow \ln\left(\frac{2}{x-1}\right) = \ln 5 \Rightarrow e^{\ln(2/x-1)} = e^{\ln 5} \Rightarrow$

$\frac{2}{x-1} = 5 \Rightarrow 2 = 5x - 5 \Rightarrow 5x = 7 \Rightarrow x = \frac{7}{5}$ or 1.4

(b) Graph $y_1 = \ln e^{\ln 2} - \ln(x-1)$ and $y_2 = \ln 5$ in the same window. From the graph, $y_1 \ge y_2$ for the interval $(1, 1.4]$.

62. (a) $8^x = 32 \Rightarrow (2^3)^x = 2^5 \Rightarrow 2^{3x} = 2^5 \Rightarrow 3x = 5 \Rightarrow x = \frac{5}{3}$

63. (a) $\frac{8}{27} = x^{-3} \Rightarrow \left(\frac{2}{3}\right)^3 = x^{-3} \Rightarrow \left(\frac{2}{3}\right)^3 = \left(\frac{1}{x}\right)^3 \Rightarrow \frac{2}{3} = \frac{1}{x} \Rightarrow 2x = 3 \Rightarrow x = \frac{3}{2}$

64. (a) $10^{2x-3} = 17 \Rightarrow \log 10^{2x-3} = \log 17 \Rightarrow (2x-3)\log 10 = \log 17$

$\Rightarrow 2x - 3 = \log 17 \Rightarrow 2x = 3 + \log 17 \Rightarrow x = \frac{3 + \log 17}{2}$

(b) From the calculator the solutions is $x \approx 2.115$.

65. (a) $e^{x+1} = 10 \Rightarrow \ln e^{x+1} = \ln 10 \Rightarrow (x+1)\ln e = \ln 10 \Rightarrow x+1 = \ln 10 \Rightarrow x = -1 + \ln 10$

(Other forms of the answer are possible.)

(b) From the calculator, the solutions is $x \approx 1.303$.

66. (a) $\log_{64} x = \frac{1}{3} \Rightarrow 64^{\log_{64} x} = 64^{1/3} \Rightarrow x = 64^{1/3} \Rightarrow x = 4$

67. (a) $\ln(6x) - \ln(x+1) = \ln 4 \Rightarrow \ln\frac{6x}{x+1} = \ln 4 \Rightarrow e^{\ln 6x/x+1} = e^{\ln 4} \Rightarrow \frac{6x}{x+1} = 4 \Rightarrow 4x + 4 = 6x \Rightarrow$

$4 = 2x \Rightarrow x = 2$

68. (a) $\log_{12}(2x)+\log_{12}(x-1)=1 \Rightarrow \log_{12}(2x)(x-1)=1 \Rightarrow \log_{12} 2x^2-2x=1 \Rightarrow$

$12^{\log_{12} 2x^2-2x}=12^1 \Rightarrow 2x^2-2x=12 \Rightarrow 2x^2-2x-12=0 \Rightarrow 2(x^2-x-6)=0 \Rightarrow$

$2(x-3)(x+2)=0 \Rightarrow x=-2, 3$. Since the argument of a logarithm cannot be negative, the solution is $x=3$.

69. (a) $\log_{16}\sqrt{x+1}=\frac{1}{4} \Rightarrow 16^{\log_{16}\sqrt{x+1}}=16^{1/4} \Rightarrow \sqrt{x+1}=16^{1/4} \Rightarrow x+1=(16^{1/4})^2 \Rightarrow$

$x+1=16^{1/2} \Rightarrow x+1=4 \Rightarrow x=3$

70. (a) $\ln x+3\ln 2=\ln\frac{2}{x} \Rightarrow \ln x+\ln 2^3=\ln\frac{2}{x} \Rightarrow \ln[(x)(2^3)]=\ln\frac{2}{3} \Rightarrow$

$\ln 8x=\ln\frac{2}{x} \Rightarrow e^{\ln 8x}=e^{\ln 2/x} \Rightarrow 8x=\frac{2}{x} \Rightarrow 8x^2=2 \Rightarrow x^2=\frac{2}{8} \Rightarrow x^2=\frac{1}{4} \Rightarrow x=-\frac{1}{2},\frac{1}{2}$

Since the argument of a natural logarithm can not be negative, the solution is $x=\frac{1}{2}$.

71. (a) $\ln[\ln(e^{-x})]=\ln 3 \Rightarrow \ln(-x)=\ln 3 \Rightarrow e^{\ln(-x)}=e^{\ln 3} \Rightarrow -x=3 \Rightarrow x=-3$

72. (a) $\ln e^x-\ln e^3=\ln e^5 \Rightarrow x-3=5 \Rightarrow x=8$

73. No real number value for x can produce a negative solution; therefore, $x=\varnothing$.

74. $N=a+b\ln\left(\frac{c}{d}\right) \Rightarrow N-a=b\ln\left(\frac{c}{d}\right) \Rightarrow \frac{N-a}{b}=\ln\left(\frac{c}{d}\right) \Rightarrow$

$e^{(N-a)/b}=e^{\ln(c/d)} \Rightarrow e^{(N-a)/b}=\frac{c}{d} \Rightarrow c=de^{(N-a)/b}$

75. $y=y_0e^{-kt} \Rightarrow \frac{y}{y_0}=e^{-kt} \Rightarrow \ln\left(\frac{y}{y_0}\right)=\ln e^{-kt} \Rightarrow \ln\left(\frac{y}{y_0}\right)=-kt \Rightarrow t=\frac{\ln(y/y_0)}{-k}$

76. Graph $y_1=\log_{10} x$ and $y_2=x-2$ in the same window. By the capabilities of the calculator, the x-coordinates of the intersections are $x \approx 0.010$ and $x \approx 2.376$.

77. Graph $y_1=2^{-x}$ and $y_2=\log_{10} x$ in the same window. By the capabilities of the calculator, the x-coordinate of the intersection is $x \approx 1.874$.

78. Graph $y_1=x^2-3$ and $y_2=\log x$ in the same window. By the capabilities of the calculator, the x-coordinates of the intersections are $x \approx 0.001$ and $x \approx 1.805$.

79. Use the Compound Interest Formula: $A = P\left(\frac{1+r}{n}\right)^{nt}$ with $A = \$4613$, $P = \$3500$, $t = 10$, and $n = 1$.

$$4613 = 3500\left(1+\frac{r}{1}\right)^{1(10)} \Rightarrow \frac{4613}{3500} = (1+r)^{10} \Rightarrow \left(\frac{4613}{3500}\right)^{0.1} = 1+r \Rightarrow r = \left(\frac{4613}{3500}\right)^{0.1} - 1 \Rightarrow$$

$r \approx 0.027996$. The annual interest rate, rounded to the nearest tenth, is 2.8% .

80. Use the Compound Interest Formula: $A = P\left(1+\frac{r}{n}\right)^{nt}$ with $A = \$58344$, $P = \$48000$, $r = 0.05$, and $n = 2$.

$$58,344 = 48,000\left(1+\frac{0.05}{2}\right)^{2t} \Rightarrow 58,344 = 48,000(1.025)^{2t} \Rightarrow \left(\frac{58,344}{48,000}\right) = 1.025^{2t} \Rightarrow 1.2155 = 1.025^{2t}$$

$$\Rightarrow \ln 1.2155 = \ln 1.025^{2t} \Rightarrow \ln 1.2155 = 2t \ln 1.025 \Rightarrow t = \frac{\ln 1.2155}{2\ln 1.025} \Rightarrow t \approx 3.951698 \text{ or about 4 years.}$$

81. For the first 8 years, use the Compound Interest Formula:

$A = P\left(1+\frac{r}{n}\right)^{nt}$ with $P = \$12000$, $t = 8$, $r = 0.05$, and $n = 1$.

$$A = 12,000\left(1+\frac{0.05}{1}\right)^{8(1)} \Rightarrow A = 12,000(1.05)^{8} \Rightarrow A \approx \$17,729.47$$

Now for the second 6 years, use the Compound Interest Formula:

$A = P\left(1+\frac{r}{n}\right)^{nt}$ with $P = \$17,729.47$, $t = 6$, $r = 0.06$, and $n = 1$.

$$A = 17,729.47\left(1+\frac{0.06}{1}\right)^{1(6)} \Rightarrow A = 17,729.47(1.06)^{6} \Rightarrow A \approx \$25,149.59$$

At the end of the 14 year period, $\$25,149.59$ would be in the account.

82. (a) Use the Compound Interest Formula: $A = P\left(1+\frac{r}{n}\right)^{nt}$ with $P = \$2000$, $r = .03$, $t = 5$, and $n = 4$.

$$A = 2000\left(1+\frac{0.03}{4}\right)^{4(5)} \Rightarrow A = 2000(1.0075)^{20} \Rightarrow A \approx \$2,322.37$$

(b) Use the Continuous Compound Interest Formula: $A = Pe^{rt}$, with $P = \$2000$, $r = 0.03$, and $t = 5$.

$A = 2000e^{(0.03)(.5)} \Rightarrow A \approx \$2323.67$.

(c) Use the Continuous Compound Interest Formula:

$A = Pe^{rt}$ with $A = \$6000$, $P = \$2000$, and $r = 0.03$.

$6000 = 2000e^{0.03t} \Rightarrow 3 = e^{0.03t} \Rightarrow \ln 3 = \ln e^{0.03t} \Rightarrow \ln 3 = 0.03t \Rightarrow t = \dfrac{\ln 3}{0.03} \Rightarrow t \approx 36.6204$

It would take about 36.6 years to triple.

83. (a) See Figure 83. Since L is increasing, heavier planes require longer runways.

(b) We can find the answer by solving $L(10)$ and $L(100)$. $L(10) = 3\log 10 = 3(1) = 3$ or 3000 feet; $L(100) = 3\log 100 = 3(2) = 6$ or 6000 feet. No, it does not increase by a factor of 10, but rather it increases by a factor of 2 to 6000 feet.

84. $2200 = 280\ln(x+1) + 1925 \Rightarrow 275 = 280(x+1) \Rightarrow \dfrac{275}{280} = \ln(x+1) \Rightarrow e^{275/280} = e^{\ln(x+1)} \Rightarrow$

$e^{275/280} = x+1 \Rightarrow x = e^{275/280} - 1 \Rightarrow x \approx 1.67$ acres.

85. (a) $P(0.5) = 0.04e^{-4(0.5)} \Rightarrow P(0.5) \approx 0.0054$ grams/liter

(b) $P(1) = 0.4e^{-4(1)} \Rightarrow P(1) \approx 0.00073$ grams/liter

(c) $P(2) = 0.04e^{-4(2)} \Rightarrow P(2) \approx 0.000013$ grams/liter

(d) $0.002 = 0.04e^{-4x} \Rightarrow 0.05 = e^{-4x} \Rightarrow \ln 0.05 = \ln e^{-4x} \Rightarrow \ln 0.05 = -4x \Rightarrow$

$x = \dfrac{\ln 0.05}{-4} \Rightarrow x \approx 0.75$ miles.

86. (a) $p(2) = 250 - 120(2.8)^{-0.5(2)} \Rightarrow p(2) \approx 207$

(b) $p(10) = 250 - 120(28)^{-0.5(10)} \Rightarrow p(10) \approx 249$

(c) See Figure 86. When $x = 2$, the graph supports the answer of 207 from part (a).

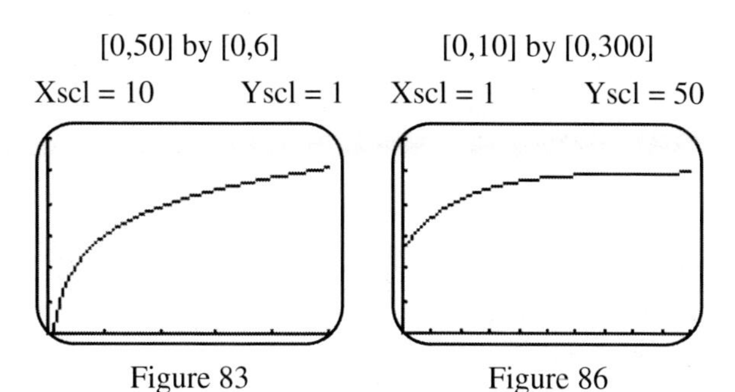

Figure 83 Figure 86

87. Find t when $v(t)=147$.

$$147=176\left(1-e^{-0.18t}\right)\Rightarrow\frac{147}{176}=1-e^{-0.18t}\Rightarrow e^{-0.18t}=1-\frac{147}{176}\Rightarrow e^{-0.18t}=\frac{29}{176}\Rightarrow$$

$$\ln e^{-0.18t}=\ln\left(\frac{29}{176}\right)\Rightarrow -0.18t=\ln\left(\frac{29}{176}\right)\Rightarrow t=\frac{\ln\left(\frac{29}{176}\right)}{-0.18}=t\approx 10.017712$$

It will take the skydiver about 10 seconds to attain the speed of 147 feet/second (100 mph)

Chapter 5 Test

1. (a) This function has a decreasing graph with a y-axis asymptote. This matches with B.

   (b) This function has an increasing graph with a y-intercept of 1. This matches with A.

   (c) This function has an increasing graph with a y-axis asymptote. This matches with C.

   (d) This function has a decreasing graph with a y-intercept of 1. This matches with D.

2. (a) Since the graph of $f(x)=3-4x$ passes the horizontal line test the function is one-to-one.

   By definition, a function has an inverse if and only if it is one to one, therefore this function has an inverse.

   (b) Since the graph of $f(x)=x^2-3x$ does not pass the horizontal line test the function is not one-to-one. By definition, a function has an inverse if and only if it is one to one, therefore this function does not have an inverse.

   (c) Since the graph of $f(x)=5(2)^x$ passes the horizontal line test the function is one-to-one. By definition, a function has an inverse if and only if it is one to one, therefore this function has an inverse.

3. (a) $f(x)=5x-7\Rightarrow y=5x-7\Rightarrow y+7=5x\Rightarrow\frac{y+7}{5}=x\Rightarrow f^{-1}(x)=\frac{x+7}{5}$

   (b) $f(x)=2\log x\Rightarrow y=2\log x\Rightarrow\frac{y}{2}=\log x\Rightarrow 10^{y/2}=x\Rightarrow f^{-1}(x)=10^{x/2}$

   (c) $f(x)=\frac{x+1}{x-2}\Rightarrow y=\frac{x+1}{x-2}\Rightarrow y(x-2)=x+1\Rightarrow xy-2y=x+1\Rightarrow xy-x=2y+1\Rightarrow$

   $x(y-1)=2y+1\Rightarrow x=\frac{2y+1}{y-1}\Rightarrow f^{-1}(x)=\frac{2x+1}{x-1}$

4. (a) See Figure 4.

   (b) From the graph, the domain is $(-\infty,\infty)$, and the range is $(-\infty, 8)$.

   (c) From the graph, it has a horizontal asymptote with equation $y=8$.

(d) Set $y=0$ to find the x-intercept. $0=-2^{x-1}+8 \Rightarrow 2^{x-1}=8 \Rightarrow 2^{x-1}=2^3 \Rightarrow x-1=3 \Rightarrow x=4$.

The x-intercept is (4, 0). Set $x=0$ to find the y-intercept. $y=-2^{0-1}+8 \Rightarrow y=-\frac{1}{2}+8 \Rightarrow y=7.5$.

The y-intercept is (0, 7.5).

(e) To find the inverse, interchange the x and y variables and solve for y. $x=-2^{y-1}+8 \Rightarrow$

$x-8=-2^{y-1} \Rightarrow 8-x=2^{y-1} \Rightarrow \log(8-x)=\log 2^{y-1} \Rightarrow \log(8-x)=(y-1)\log 2 \Rightarrow$

$\log(8-x)=y\log 2-1\log 2 \Rightarrow \log(8-x)+\log 2=y\log 2 \Rightarrow y=\dfrac{\log(8-x)+\log 2}{\log 2}$ or

$y=\dfrac{\log(8-x)}{\log 2}+1.$

[-10,10] by [-10,10]

Xscl = 1 Yscl = 1

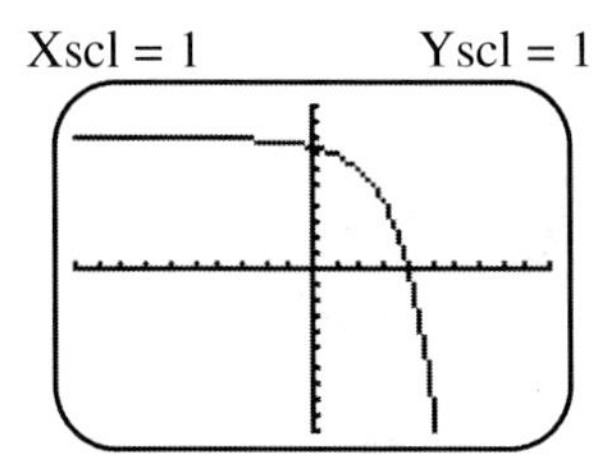

Figure 4

5. $(\frac{1}{8})^{2x+3}=16^{x+1} \Rightarrow (2^{-3})^{2x-3}=(2^4)^{x+1} \Rightarrow 2^{-6x+9}=2^{4x+4} \Rightarrow -6x+9=4x+4 \Rightarrow 5=10x \Rightarrow \quad x=\frac{1}{2}$ or $x=0.5$.

6. (a) Use the Compound Interest Formula $A=P\left(1+\frac{r}{n}\right)^{nt}$, with $P=\$10000$, $r=0.035$, $t=4$, and $n=4$.

$$A=10000\left(1+\frac{0.035}{4}\right)^{4(4)} \Rightarrow A\approx \$11,495.74.$$

(b) Use the Continuous Compound Interest Formula $A=Pe^{rt}$, with $P=\$10000$, $r=0.035$, and $t=4$.

$A=10,000e^{0.035(4)} \Rightarrow A\approx \$11,502.74.$

7. The expression, $\log_5 27$, is the exponent to which 5 must be raised in order to obtain 27. To find an approximation with a calculator, use the change-of-base rule: $\log_5 27=\dfrac{\log 27}{\log 5}$.

8. (a) From the calculator, $\log 45.6 \approx 1.659$.

(b) From the calculator, $\ln 470 \approx 6.153$.

(c) From the calculator, $\log_3 769 = \dfrac{\log 769}{\log 3} \approx 6.049.$

9. $\log\dfrac{m^3 n}{\sqrt{y}} = \log m^3 + \log n - \log\sqrt{y} = \log m^3 + \log n - \log y^{1/2} \Rightarrow 3\log m + \log n - \dfrac{1}{2}\log y.$

10. $2\log x + \dfrac{1}{2}\log y - 4\log z = \log x^2 + \log y^{1/2} - \log z^4 = \log\dfrac{x^2\sqrt{y}}{z^4}$

11. By the inverse property of logarithms $\ln e^y = y \Rightarrow \ln e^{2x} = 2x$

By the inverse property of logarithms $10^{\log y} = y \Rightarrow 10^{\log x^2} = x^2$

12. The domain of the function $f(x) = \ln(x)$ is $x > 0$. Therefore, the domain of the function $f(x) = \ln(2x+1)$ will be $2x+1 > 0 \Rightarrow 2x > -1 \Rightarrow x > -\dfrac{1}{2}$. The domain in interval notation is $\left(-\dfrac{1}{2}, \infty\right)$.

13. $A = Pe^{rt} \Rightarrow \dfrac{A}{P} = e^{rt} \Rightarrow \ln\dfrac{A}{P} = \ln e^{rt} \Rightarrow \ln A - \ln P = rt \Rightarrow t = \dfrac{\ln A - \ln P}{r}.$

14. (a) $\log_2 x + \log_2(x+2) = 3;\ x > 0 \Rightarrow \log_2[x(x+2)] = 3 \Rightarrow 2^{\log_2[x(x+2)]} = 2^3 \Rightarrow x(x+2) = 2^3 \Rightarrow$ $x^2 + 2x - 8 = 0 \Rightarrow (x+4)(x-2) = 0 \Rightarrow x = -4, 2$. Since the argument of a logarithm can not be negative, the solution cannot be the extraneous solution -4, so the solution is $x = 2$.

(b) Using the change-of-base rule on $y_1 = \log_2 x + \log_2(x+2) - 3$ yields $y_1 = \dfrac{\log x}{\log 2} + \dfrac{\log(x+2)}{\log 2} - 3.$

Graph the equation. See Figure 14. The x-intercept is $x = 2$.

(c) From the graph, $\log_2 x + \log_2(x+2) > 3$ for the interval $(2, \infty)$.

[-10,10] by [-10,10]

Xscl = 1 Yscl = 1

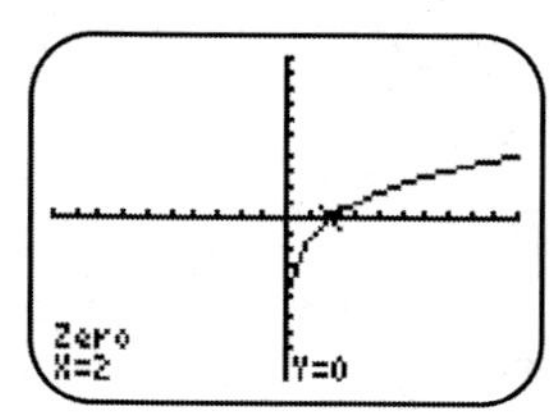

Figure 14

15. (a) $2e^{5x+2}=8 \Rightarrow e^{5x+2}=4 \Rightarrow \ln e^{5x+2}=\ln 4 \Rightarrow 5x+2=\ln 4 \Rightarrow 5x=\ln 4-2 \Rightarrow x=\dfrac{\ln 4-2}{5}.$

(Other forms of the answer are possible.)

(b) From the calculator, the solution is $x=-0.123$.

16. (a) $6^{2-x}=2^{3x+1} \Rightarrow \log 6^{2-x}=\log 2^{3x+1} \Rightarrow (2-x)\log 6=(3x+1)\log 2 \Rightarrow$

$2\log 6-x\log 6=3x\log 2+\log 2 \Rightarrow 2\log 6-\log 2=3x\log 2+x\log 6 \Rightarrow$

$\log\dfrac{6^2}{2}=x(\log(2^3)(6)) \Rightarrow \log 18=x(\log 48) \Rightarrow x=\dfrac{\log 18}{\log 48}.$

(b) From the calculator, the solution is $x=0.747$.

17. (a) $\log(\ln x)=1 \Rightarrow 10^{\log(\ln x)}=10^1 \Rightarrow \ln x=10 \Rightarrow e^{\ln x}=e^{10} \Rightarrow x=e^{10}.$

(b) From the calculator, the solution is $x=22,026.466$.

18. (a) The 2 in the equation is the original population; tripling this would yield $y=3\cdot 2$ The match is B.

(b) In this function, $y=3$; therefore, the match is D.

(c) In this function, $t=3$; therefore, the match is C.

(d) Since 4 month equals $\frac{1}{3}$ year, in this function $t=\frac{1}{3}$; therefore, the match is A.

19. (a) See Figure 19a. (b) See Figure 19b.

(c) See Figure 19c. (d) See Figure 19d.

Function (c) is the best at describing $A(t)$ because it starts at 350 and gradually decreases as time increases.

[0,10] by [0,500]
Xscl = 1 Yscl = 100

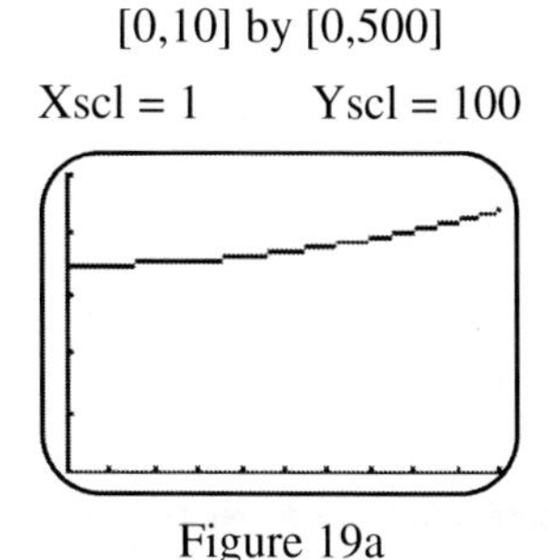

Figure 19a

[0,10] by [0,500]
Xscl = 1 Yscl = 100

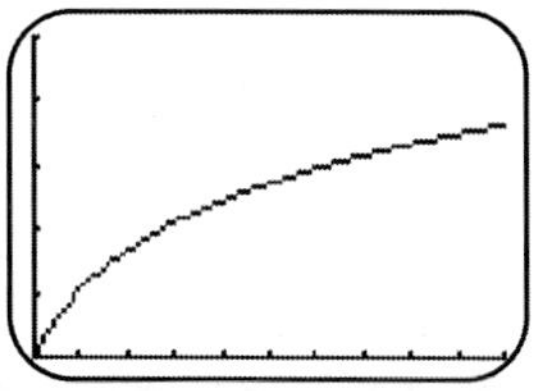

Figure 19b

[0,10] by [0,500]
Xscl= 1 Yscl = 100

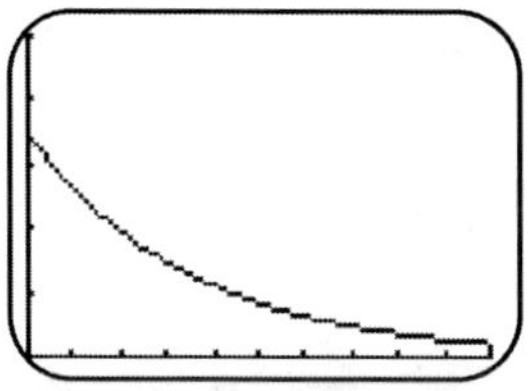

Figure 19c

[0,10] by [0,500]
Xscl = 1 Yscl= 100

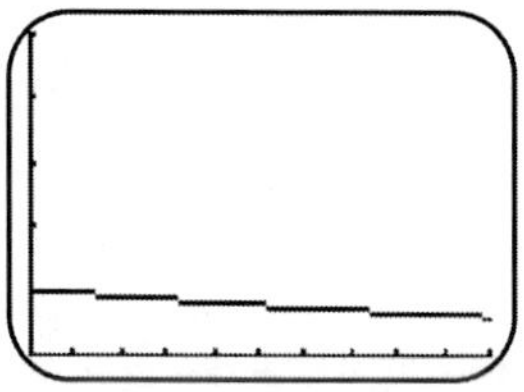

Figure 19d

20. (a) Using the exponential decay model $A(t)=A_0e^{-kt}$, we can solve for k with $A_0=2$ and $t=1600$.

$\frac{1}{2}(2)=2e^{-k(1600)} \Rightarrow 1=2e^{-k(1600)} \Rightarrow \frac{1}{2}=e^{-k(1600)} \Rightarrow \ln\frac{1}{2}=\ln e^{-k(1600)} \Rightarrow \ln\frac{1}{2}=-k(1600) \Rightarrow$

$k=\dfrac{\ln(5)}{-1600} \Rightarrow k\approx 0.000433.$ The model is $A(t)=2e^{-0.000433t}$.

(b) $A(9600) = 2e^{-0.000433(9600)} \Rightarrow A(9600) \approx 0.03$ grams.

(c) $0.5 = 2e^{-0.000433t} \Rightarrow 0.25 = e^{-0.000433t} \Rightarrow \ln 0.25 = \ln e^{-0.000433t} \Rightarrow$

$$\ln 0.25 = -0.000433\text{t} \Rightarrow \text{t} = \frac{\ln 0.25}{-0.000433} \Rightarrow t \approx 3200 \text{ years.}$$

Chapter 6: Systems and Matrices

6.1: Systems of Equations

1. From the graph, the population is greater in Jacksonville than in New Orleans from 2006 to 2009.

3. Approximately (2005, 1.26)

5. year, population (in millions)

7. From the graph, the solution is $(2,2)$. Using substitution, first solve [equation 1] for $x, x-y=0 \Rightarrow x=y$ [equation 3]. Substitute y for x in [equation 2]: $y+y=4 \Rightarrow 2y=4 \Rightarrow y=2$. Now substitute 2 for y in [equation 3]: $x=y \Rightarrow x=2$. The solution is $(2,2)$.

9. From the graph, the solution is $\left(\frac{1}{2},-2\right)$. Using substitution, first solve [equation 2] for x: $2x-3y=7 \Rightarrow 2x=3y+7 \Rightarrow x=\frac{3}{2}y+\frac{7}{2}$ [equation 3]. Substitute $\frac{3}{2}y+\frac{7}{2}$ for x in [equation 1]: $6\left(\frac{3}{2}y+\frac{7}{2}\right)+4y=-5 \Rightarrow 9y+21+4y=-5 \Rightarrow 13y=-26 \Rightarrow y=-2$. Now substitute -2 in for y in [equation 3]: $x=\frac{3}{2}(-2)+\frac{7}{2} \Rightarrow x=-3+\frac{7}{2} \Rightarrow x=\frac{1}{2}$. The solution is $\left(\frac{1}{2},-2\right)$.

11. Since $y=x$ in [equation 2], we can use substitution. Substitute x for y in [equation 1]: $6x-(x)=5 \Rightarrow 5x=5 \Rightarrow x=1$. Now substitute 1 for x in [equation 2]: $y=x \Rightarrow y=1$. The solution is $(1,1)$.

13. Using substitution, first solve [equation 1] for $x, x+2y=-1 \Rightarrow x=-2y-1$ [equation 3]. Substitute $-2y-1$ for x in [equation 2]: $2(-2y-1)+y=4 \Rightarrow -3y-2=4 \Rightarrow -3y=6 \Rightarrow y=-2$. Now substitute -2 for y in [equation 3]: $x=-2(-2)-1 \Rightarrow x=3$. The solution is $(3,-2)$.

15. Since $y=2x+3$ in [equation 1], we can use substitution. Substitute $2x+3$ for y in [equation 2]: $3x+4(2x+3)=78 \Rightarrow 11x+12=78 \Rightarrow 11x=66 \Rightarrow x=6$. Now substitute 6 for x in [equation 1]: $y=2(6)+3 \Rightarrow y=15$. The solution is $(6,15)$.

17. Using substitution, first solve [equation 1] for x: $3x-2y=12 \Rightarrow 3x=2y+12 \Rightarrow x=\frac{2}{3}y+4$ [equation 3]. Substitute $\frac{2}{3}y+4$ for x in [equation 2]: $5\left(\frac{2}{3}y+4\right)=4-2y \Rightarrow \frac{10}{3}y+2y=4-20 \Rightarrow \frac{16}{3}y=-16$ $\Rightarrow y=-3$. Now substitute -3 for y in [equation 3]: $x=\frac{2}{3}(-3)+4 \Rightarrow x=2$. The solution is $(2,-3)$.

19. Using substitution, first solve [equation 2] for y: $2x+y=5 \Rightarrow y=-2x+5$ [equation 3]. Substitute $-2x+5$ for y in [equation 1]: $4x-5(-2x+5)=-11 \Rightarrow 4x+10x-25=-11 \Rightarrow 14x=14 \Rightarrow x=1$. Now substitute 1 for y in [equation 3]: $y=-2(1)+5 \Rightarrow y=3$. The solution is $(1,3)$.

21. Using substitution, first solve [equation 2] for y: $9y = 31 + 2x \Rightarrow y = \frac{31+2x}{9}$ [equation 3]. Substitute

$\frac{31+2x}{9}$ for y in [equation 1]: $4x + 5\left(\frac{31+2x}{9}\right) = 7 \Rightarrow 9\left[4x + 5\left(\frac{31+2x}{9}\right) = 7\right] \Rightarrow$

$36x + 5(31+2x) = 63 \Rightarrow 36x + 155 + 10x = 63 \Rightarrow 46x = -92 \Rightarrow x = -2.$

Now substitute -2 for x in [equation 3]: $y = \frac{31+2(-2)}{9} \Rightarrow y = 3$. The solution is $(-2, 3)$.

23. Using substitution, first solve [equation 1] for x: $3x - 7y = 15 \Rightarrow 3x = 7y + 15 \Rightarrow x = \frac{7y+15}{3}$ [equation 3].

Substitute $\frac{7y+15}{3}$ for x in [equation 2]: $3\left(\frac{7y+15}{3}\right) + 7y = 15 \Rightarrow 7y + 15 + 7y = 15 \Rightarrow 14y = 0 \Rightarrow y = 0.$

Now substitute 0 for y in [equation 3]: $x = \frac{7(0)+15}{3} \Rightarrow x = 5$. The solution is $(5, 0)$.

25. Using substitution, first solve [equation 1] for x: $2x - 7y = 8 \Rightarrow 2x = 7y + 8 \Rightarrow x = \frac{7y+8}{2}$ [equation 3].

Substitute $\frac{7y+8}{2}$ for x in [equation 2]: $-3\left(\frac{7y+8}{2}\right) + \frac{21}{2}y = 5 \Rightarrow 2\left[-3\left(\frac{7y+8}{2}\right) + \frac{21}{2}y = 5\right] \Rightarrow$

$-3(7y+8) + 21y = 10 \Rightarrow -21y - 24 + 21y = 10 \Rightarrow -24 = 10.$

Since this is a false statement, the solution set is $\varnothing$.

27. Using substitution, first solve [equation 1] for x: $x - 2y = 4 \Rightarrow x = 4 + 2y$ [equation 3]. Substitute $x = 4 + 2y$ for x in [equation 2]: $-2(4+2y) + 4y = -8 \Rightarrow -8 - 4y + 4y = -8 \Rightarrow -8 = -8$. Since this is a true statement, the system is dependent and the solutions are $(4+2y, y)$ *or* $\left(x, \frac{x-4}{2}\right)$.

29. Multiply the first equation by 3 and add to eliminate the y-variable:

$$\begin{array}{rcl} 9x - 3y & = & -12 \\ x + 3y & = & 12 \\ \hline 10x & = & 0 \end{array}$$

$\Rightarrow x = 0$. Substitute 0 for x in the second equation: $(0) + 3y = 12 \Rightarrow y = 4$. The solution is $(0, 4)$.

31. Multiply the second equation by 2, and subtract to eliminate the x-variable:

$$\begin{array}{rcl} 4x + 3y & = & -1 \\ 4x + 10y & = & 6 \\ \hline -7y & = & -7 \end{array}$$

$\Rightarrow y = 1$. Substitute 1 for y in the first equation: $4x + 3(1) = -1 \Rightarrow 4x = -4 \Rightarrow x = -1.$ The solution is $(-1, 1)$.

33. Multiply the second equation by 4, and subtract to eliminate the x-variable:

$$\begin{array}{r} 12x \ -5y = \ 9 \\ 12x - 32y = -72 \\ \hline 27y = \ 81 \end{array}$$

$\Rightarrow y = 3$. Substitute 3 for y in the first equation: $12x - 5(3) = 9 \Rightarrow 12x = 24 \Rightarrow x = 2$.

The solution is $(2,3)$.

35. Multiply the first equation by 2 and add to eliminate both variables.

$$\begin{array}{r} 8x \ -2y = \ 18 \\ -8x \ +2y = -18 \\ \hline 0 = \ 0 \end{array}$$

$\Rightarrow$ infinite number of solutions. Therefore, the solutions have the following relationship: $4x - y = 9 \Rightarrow -y = -4x + 9 \Rightarrow y = 4x - 9$. The solution set is $\{(x, 4x-9)\}$ or $\left\{\left(\frac{y+9}{4}, y\right)\right\}$.

37. Multiply the first equation by 2 and add to eliminate both variables:

$$\begin{array}{r} 18x \ -10y = \ 2 \\ -18y \ +10y = \ 1 \\ \hline 0 \ = \ 3 \end{array}$$

$\Rightarrow \varnothing$

39. Multiply the first equation by 2 and subtract to eliminate both variables:\

$$\begin{array}{r} 6x \ +2y = \ 12 \\ 6y \ +2y = \ 1 \\ \hline 0 = \ 13 \end{array}$$

$\Rightarrow \varnothing$

41. First, multiplying both equations by 6 yields $3x + 2y = 48$ and $4x + 9y = 102$. Now multiply the first equation by 4, the second equation by 3, and subtract to eliminate the x-variable:

$$\begin{array}{r} 12x \ + \ 8y = 192 \\ 12x \ +27y = 306 \\ \hline -19y = -114 \end{array}$$

$\Rightarrow y = 6$. Substitute 6 for y in the first equation (multiplied by 6):

$3x + 2(6) = 48 \Rightarrow 3x = 36 \Rightarrow x = 12$. The solution is $(12,6)$.

43. Multiplying the first equation by 12 and the second by 6 yields $8x + 3y = 46$ and $x + 2y = 9$. Now multiply the second equation by 8, and subtract to eliminate the x-variable:

$$\begin{array}{r} 8x \ + \ 3y = \ 46 \\ 8x \ +16y = \ 72 \\ \hline -13y = -26 \end{array}$$

$\Rightarrow y = 2$. Substitute 2 for y in the second equation (multiplied by 6):

$x + 2(2) = 9 \Rightarrow x = 5$. The solution is $(5,2)$.

45. Graph each function in the same window. The intersection or solution is $(0.138, -4.762)$.

47. Graph each function in the same window. The intersection or solution is $(-8.708, -15.668)$.

49. An inconsistent system will conclude with no variables and a false statement, such as $0 = 1$. A system with dependent equations will conclude with no variables and a true statement, such as $0 = 0$.

From these equations, when we subtract, we will get the result $0 = 0$ when $k = -6$, and the result $0 = (-6 - k)$ when $k \neq -6$. Therefore, the system will have no solution when $k \neq -6$, and system will have infinitely many solutions when $k = -6$.

51. See Figure 51.

53. See Figure 53

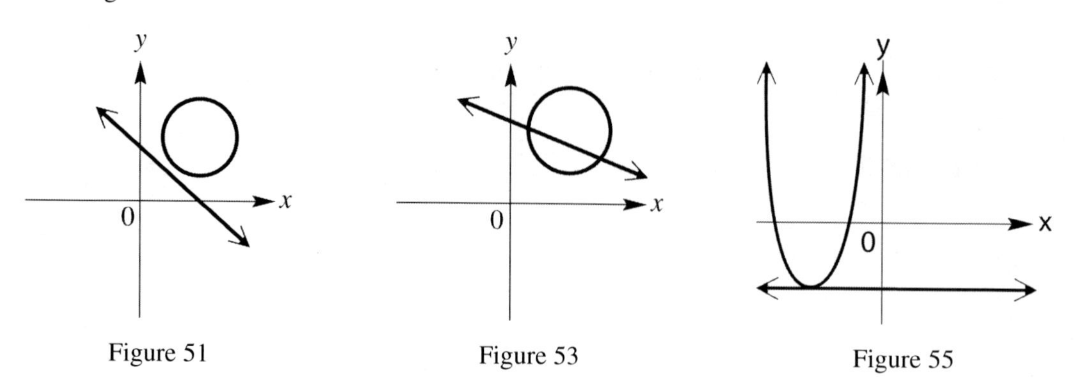

Figure 51 Figure 53 Figure 55

55. See Figure 55.

57. See Figure 57.

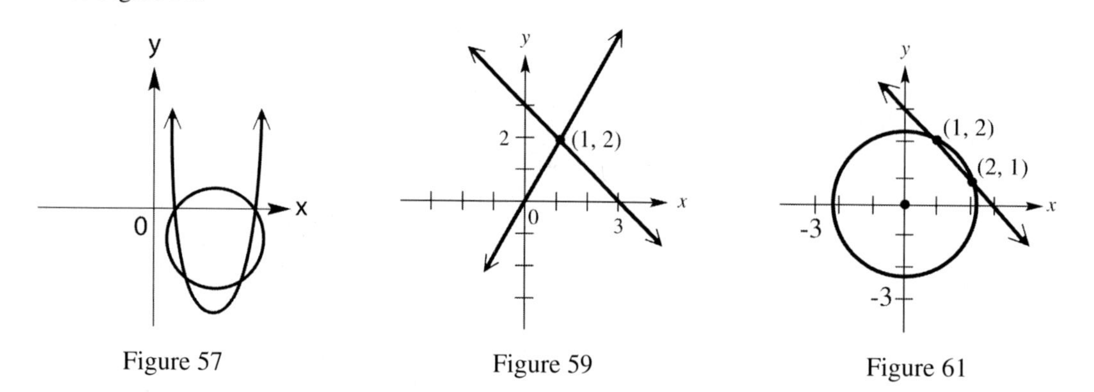

Figure 57 Figure 59 Figure 61

59. Check: $1+2=3 \Rightarrow 3=3,\quad 2(1)-2=0 \Rightarrow 0=0.$ See Figure 59.

61. Check: $1^2+2^2=5 \Rightarrow 5=5,\ 2=3-1 \Rightarrow 2=2,\ 2^2+1^2=5 \Rightarrow 5=5,\ 1=3-2 \Rightarrow 1=1.$

63. Check: $1=(-1)^2 \Rightarrow 1=1,\ (-1)^2+1^2=2,\ 1=1^2 \Rightarrow 1=1,\ 1^2+1^2=2 \Rightarrow 2=2.$

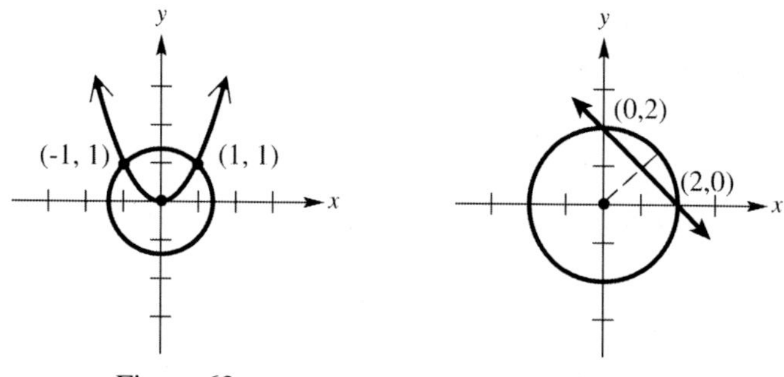

Figure 63 Figure 65

65. Check: $0^2+2^2=4 \Rightarrow 0+4=4 \Rightarrow 4=4,\ 0+2=2 \Rightarrow 2=2.$

$$2^2+0^2=4 \Rightarrow 4+0=4 \Rightarrow 4=4,\ 2+0=2 \Rightarrow 2=2.$$

67. Since [equation 1] is solved for y, we substitute $-x^2+2$ for y in [equation 2]:

$x-(-x^2+2)=0 \Rightarrow x^2+x-2 \Rightarrow (x+2)(x-1)=0 \Rightarrow x=-2$ or 1. Substituting these values into [equation 1] yields $y=-(-2)^2+2 \Rightarrow y=-4+2 \Rightarrow y=-2$ and $y=-(1)^2+2 \Rightarrow y=-1+2 \Rightarrow y=1$.

The solution set is $\{(-2,-2),(1,1)\}$.

69. First solve [equation 2] for x: $x-y=1 \Rightarrow x=y+1$ [equation 3]. Now substitute $y+1$ for x in [equation 1]: $(y+1)^2+y^2=5 \Rightarrow (y^2+2y+1)+y^2=5 \Rightarrow y^2+2y+1+y^2-5=0 \Rightarrow$

$2y^2+2y-4=0 \Rightarrow 2(y-1)(y+2)=0 \Rightarrow y=1$ or -2. Substituting these values into [equation 3]

yields $x=(1)+1 \Rightarrow x=2$ and $x=(-2)+1 \Rightarrow x=-1$. The solution set is $\{(-1,-2),(2,1)\}$.

71. First solve [equation 2] for x^2: $-x^2+y=-4 \Rightarrow x^2=y+4$ [equation 3]. Now substitute $y+4$ for x^2 in [equation 1]: $(y+4)+y^2=10 \Rightarrow y^2+y+4-10=0 \Rightarrow y^2+y-6=0 \Rightarrow (y-2)(y+3)=0 \Rightarrow y=2$ or -3. Substituting these values into [equation 3] yields $x^2=2+4 \Rightarrow x=\pm\sqrt{6}$ and $x^2=-3+4 \Rightarrow x=\pm 1$.

The solution set is $\{(-1,-3),(1,-3),(-\sqrt{6},2),(\sqrt{6},2)\}$.

73. Graph each function in the same window. The intersections or solutions are $(-0.79,0.62)$ and $(0.88,0.77)$.

75. Graph each function in the same window. The intersection or solution is $(0.06,2.88)$.

77. (a) Let x represent the amount spent in 2012, and let y represent the amount spent in 2011. The required system of equations is $x+y=1858$ and $x-y=226$.

(b) Adding the two equations results in $2x=2084 \Rightarrow x=1042$. From the first equation, $1042+y=1858 \Rightarrow y=816$. The solution set is $\{(1042,816)\}$.

(c) The amount spent in 2012 was \$1042 million and the amount spent in 2011 was \$816 million.

79. Let x be the selling price in 2010 and y be the selling price in 2012. Then, $0.7x=y$ and $x-y=195$. By substitution we have $x-0.7x=195 \Rightarrow 0.3x=195 \Rightarrow x=650$. Substitute $x = 650$ into $x-y=195 \Rightarrow$ $650-y=195 \Rightarrow y=455$. In 2010 the average price was \$650 and in 2012 the average price was \$455.

81. Let x = the width of the base; therefore, the length $2x$, and let y = the height of the box. Since the volume is 588 in^2, $2x^2y=588$ and so $y=\dfrac{588}{2x^2}=\dfrac{294}{x^2}$. Since the surface area is 448 in^2,

$2(2x)(x)+2(x)(y)+2(2x)(y)=448 \Rightarrow 4x^2+6xy=448$. Substituting for y in this equation yields

$4x^2+6x\cdot\dfrac{294}{x^2}=448$. Simplifying we get $4x^3-448x+1764=0$. By graphing the left side of the equation, the x-intercepts are approximately 5.17 and 7. See Figure 81. When $x \approx 5.17$, the dimensions are 5.17 by $2(5.17) \approx 10.34$ by $\dfrac{294}{(5.17)^2} \approx 11.00$ inches. When $x=7$, the dimensions are 7 by $2(7)=14$

by $\dfrac{294}{(7)^2}=6$ inches.

[0,10] by [-500,1500]
Xscl = 1 Yscl = 500

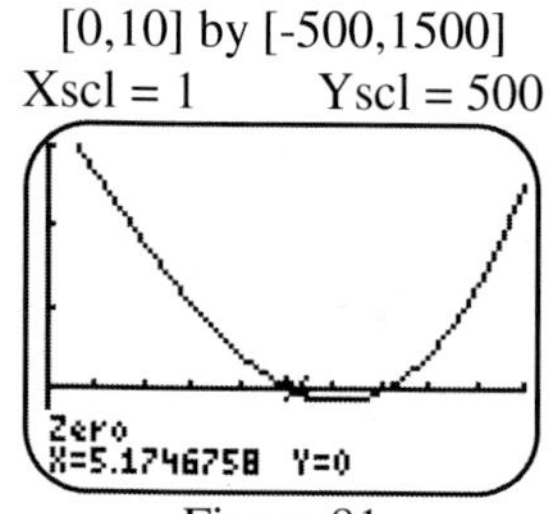

Figure 81

[0,10] by [0,10]
Xscl = 1 Yscl = 1

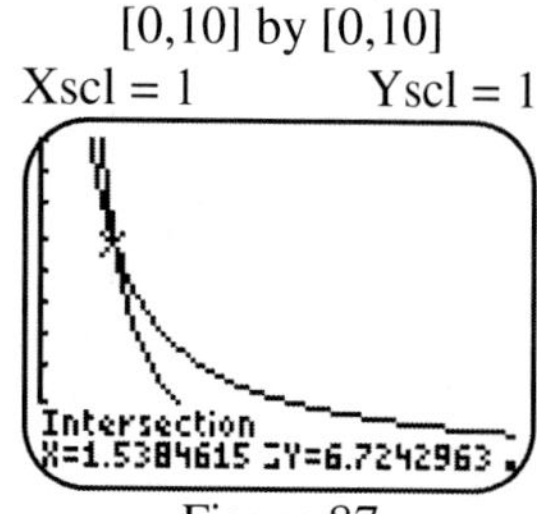

Figure 87

83. First let $H = 180$ in each equation and solve for y:

$$180 = 0.491x + 0.468y + 11.2 \Rightarrow 0.468y = 180 - 11.2 - 0.491x \Rightarrow y_1 = \frac{168.8 - 0.491x}{0.468}$$

$$180 = -0.981x + 1.872y + 26.4 \Rightarrow 1.872y = 180 - 26.4 - 0.981x \Rightarrow y_2 = \frac{153.6 + 0.981x}{1.872}.$$ Now graph each equation in the same window and the intersection is approximately $(177.1, 174.9)$. This means that if an athlete's maximum heart rate is 180 beats per minute (bpm), then it will be about 177 bpm after 5 seconds and 175 bpm after 10 seconds.

85. (a) Let x = the amount spent by Apple and y = the amount spent by Samsung. Then $x + y = 293$ and $x - y = 93$.

(b)
$$\begin{aligned} x + y &= 293 \\ x - y &= \ \ 93 \\ \hline 2x \quad &= 386 \Rightarrow x = 193. \end{aligned}$$

Then, $x - y = 93 \Rightarrow 193 - y = 93 \Rightarrow y = 100$ The solution is $(193, 100)$.

(c) From January to June 2012, Apple spent \$193 million and Samsung spent \$100 million.

87. If $V = \pi r^2 h$ and $V = 50$ then $50 = \pi r^2 h$; and $S.A = 2\pi rh$ and $S.A. = 65$ then $65 = 2\pi rh$. Now solve each for $h: 50 = \pi r^2 h \Rightarrow h_1 = \frac{50}{\pi r^2}$; and $65 = 2\pi rh \Rightarrow h_2 = \frac{65}{2\pi r}$. Graph each equation in the same window. See Figure 87. The intersection is approximately: (1.538, 6.724); therefore, $r \approx 1.538$ and $h \approx 6.724$.

89. Add the equations to eliminate the w_2-variable:

$$\begin{aligned} w_1 + \sqrt{2}w_2 &= 300 \\ \sqrt{3}w_1 - \sqrt{2}w_2 &= \ \ 0 \\ \hline (1+\sqrt{3})w_1 &= 300 \end{aligned} \qquad \Rightarrow w_1 = \frac{300}{(1+\sqrt{3})} \approx 109.8.$$

Now substitute 109.8 for w_1 in the first equation:

$$\left(\frac{300}{(1+\sqrt{3})}\right) + \sqrt{2}\,w_2 = 300 \Rightarrow \sqrt{2}w_2 = 300 - \frac{300}{(1+\sqrt{3})} \Rightarrow \sqrt{2}w_2 = \frac{300(1+\sqrt{3}) - 300}{1+\sqrt{3}} \Rightarrow$$

$$\sqrt{2}\,w_2 = \frac{300\sqrt{3}}{1+\sqrt{3}} \Rightarrow w_2 = \frac{300\sqrt{3}}{1+\sqrt{3}} \times \frac{1}{\sqrt{2}} = \frac{300\sqrt{3}}{\sqrt{2}+\sqrt{6}} \Rightarrow w_2 \approx 134.5.$$ The solution is (109.8, 134.5).

91. The total number of vehicles entering intersection A is $500+150=650$ vehicles per hour. The expression $x+y$ represents the number of vehicles leaving intersection A each hour. Therefore, we have $x+y=650$. The total number of vehicles leaving intersection B is $50+400=450$ vehicles per hour. There are 100 vehicles entering intersection B from the south and y vehicles entering intersection B from the west. Thus $y+100=450 \Rightarrow y=350$. Using substitution, substitute 350 in for y into the equation $x+y=650$: $x+350=650 \Rightarrow x=300$. At intersection A, a stoplight should allow for 300 vehicles per hour to travel south and 350 vehicles per hour to continue traveling east.

93. (a) $p=\frac{1}{10}q \Rightarrow p=\frac{1}{10}(15) \Rightarrow p=1.5$, The price will be \$1.50 for a supply of 15 units.

$p=15-\frac{2}{3}q \Rightarrow p=15-\frac{2}{3}(15) \Rightarrow p=5$, The price will be \$5 for a demand of 15 units.

(b) Since the two equations $p=\frac{1}{10}q$ and $p=15-\frac{2}{3}q$ are already solved for p we will set the two equations equal to each other and solve for q.

$\frac{1}{10}q=15-\frac{2}{3}q \Rightarrow 3q=450-20q \Rightarrow 23q=450 \Rightarrow q \approx 19.57 \approx 20$ units.

Substitute $q=20$ into $p=\frac{1}{10}(19.57) \Rightarrow p=1.96$. The equilibrium price is \$1.96 with a demand of about 20 units.

95. Find where $C=R, 20x+10{,}000=30x-11{,}000 \Rightarrow 21{,}000=10x \Rightarrow x=2{,}100$. Now substitute 2,100 in for x into the first equation: $C=20(2{,}100)+10{,}000=42{,}000+10{,}000 \Rightarrow C=52{,}000$. When $x=2{,}100$; $R=C=\$52{,}000$.

97. If $t=\frac{1}{x}$ and $u=\frac{1}{y}$, then $5\left(\frac{1}{x}\right)+15\left(\frac{1}{y}\right)=16 \Rightarrow 5t+15u=16$ and $5\left(\frac{1}{x}\right)+4\left(\frac{1}{y}\right)=5 \Rightarrow 5t+4u=5$. The equations are $5t+4u=16$ and $5t+4u=5$.

98. Subtract the two equations to eliminate the x-variable:

$$\begin{array}{l} 5t+15u=16 \\ \underline{5t+\ \ 4u=\ \ 5} \\ \quad 11u=11 \quad \Rightarrow u=1. \end{array}$$

Now substitute 1 for u in the first equation: $5t+15(1)=16 \Rightarrow 5t=1 \Rightarrow t=\frac{1}{5}$. Therefore, $u=1$ and $t=\frac{1}{5}$.

99. If $t=\frac{1}{x}$ and $t=\frac{1}{5}$, then $\frac{1}{x}=\frac{1}{5}$ and $x=5$. If $u=\frac{1}{y}$ and $u=1$, then $\frac{1}{y}=1 \Rightarrow y=1$.

100. $\frac{5}{x}+\frac{15}{y}=16 \Rightarrow xy\left(\frac{5}{x}+\frac{15}{y}=16\right) \Rightarrow 5y+15x=16xy \Rightarrow 5y-16xy=-15x \Rightarrow$

$y(5-16x)=-15x \Rightarrow y=\frac{-15x}{5-16x}.$

101. $\frac{5}{x}+\frac{4}{y}=5 \Rightarrow xy\left(\frac{5}{x}+\frac{4}{y}=5\right) \Rightarrow 5y+4x=5xy \Rightarrow 5y-5xy=-4x \Rightarrow y(5-5x)=-4x \Rightarrow y=\frac{-4x}{5-5x}.$

102. See Figure 102. The graphs intersect at (5, 1), which supports the indicated exercise.

[0,10] by [0,2]
Xscl = 1 Yscl = 0.5

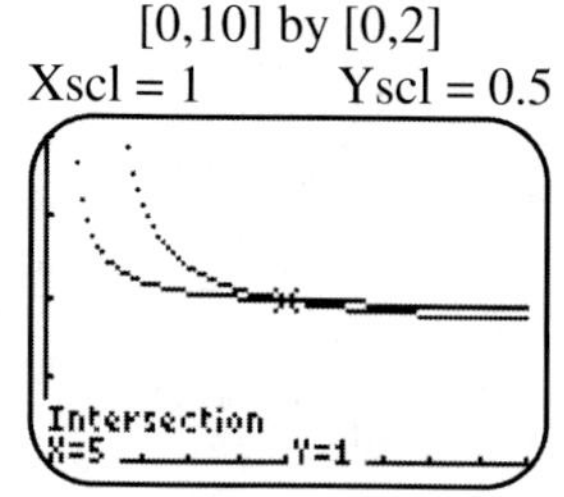

Figure 102

103. If we let $t=\frac{1}{x}$ and $u=\frac{1}{y}$, then our new equations are $2t+u=\frac{3}{2}$ and $3t-u=1$. Add these to eliminate the u-variable:

$$\begin{aligned} 2t+u&=\frac{3}{2} \\ 3t-u&=1 \\ \hline 5t\quad&=\frac{5}{2} \end{aligned}$$

$\Rightarrow t=\frac{1}{2}$. If $t=\frac{1}{x}$ and $t=\frac{1}{2}$, then $\frac{1}{x}=\frac{1}{2} \Rightarrow x=2$. Now substitute $t=\frac{1}{2}$ into $2t+u=\frac{3}{2}$, which yields $2\left(\frac{1}{2}\right)+u=\frac{3}{2} \Rightarrow u=\frac{1}{2}$. If $u=\frac{1}{y}$ and $u=\frac{1}{2}$, then $\frac{1}{y}=\frac{1}{2} \Rightarrow y=2$. The solution is (2,2)

105. If we let $t=\frac{1}{x}$ and $u=\frac{1}{y}$, then our new equations are $2t+3u=18$ and $4t-5u=-8$. Multiply the first equation by 2 and subtract to eliminate the t-variable.

$$\begin{aligned} 4t+6u&=36 \\ 4t-5u&=-8 \\ \hline 11u&=44 \end{aligned}$$

$\Rightarrow u=4$.

If $u=\frac{1}{y}$ and $u=4$, then $\frac{1}{y}=4 \Rightarrow 4y=1 \Rightarrow y=\frac{1}{4}$. Now substitute $u=4$ into $2t+3u=18$, which yields $2t+3(4)=18 \Rightarrow 2t=6 \Rightarrow t=3$. If $t=\frac{1}{x}$ and $t=3$, then $\frac{1}{x}=3 \Rightarrow 3x=1 \Rightarrow x=\frac{1}{3}$.

The solution is $\left(\frac{1}{3},\frac{1}{4}\right)$.

6.2: Solution of Linear Systems in Three Variables

1. For $2x+y-z=-1$ and $(-3,6,1): 2(-3)+(6)-1=-1 \Rightarrow -6+6-1=-1 \Rightarrow -1=-1.$

   For $x-y+3z=-6$ and $(-3,6,1): (-3)-6+3(1)=-6 \Rightarrow -3-6+3=-6 \Rightarrow -6=-6.$

   For $-4x+y+z=19$ and $(-3,6,1): -4(-3)+6+1=19 \Rightarrow 12+6+1=19 \Rightarrow 19=19.$

3. For $5x-y+2z=-0.4$ and $(-0.2,0.4,0.5)$: $5(-0.2)-(0.4)+2(0.5)=-0.4 \Rightarrow$

   $-1-0.4+1=0.4 \Rightarrow -0.4=-0.4.$

   For $x+4z=1.8$ and $(-0.2,0.4,0.5)$: $(-0.2)+4(0.5)=1.8 \Rightarrow -0.2+4(0.5)=1.8 \Rightarrow$

   $-0.2+2.0=1.8 \Rightarrow 1.8=1.8.$

   For $-3y+z=-0.7$ and $(-0.2,0.4,0.5)$: $-3(0.4)+(0.5)=-0.7 \Rightarrow -1.2+0.5=-0.7 \Rightarrow -0.7=-0.7.$

5. For $x-y+z=-2$ and $(-2,-1,3)$: $(-2)-(-1)+(3)=2 \Rightarrow -2+1+3=2 \Rightarrow 2=2.$

   For $3x-2y+z=-1$ and $(-2,-1,3)$: $3(-2)-2(-1)+(3)=-1 \Rightarrow -6+2+3=-1 \Rightarrow -1=-1.$

   For $x+y=-3$ and $(-2,-1,3)$: $(-2)+(-1)=-3 \Rightarrow -2+(-1)=-3 \Rightarrow -3=-3.$

7. First add [equation 1] and [equation 2] to eliminate z and produce [equation 4]. Second add [equation 2] and [equation 3] to eliminate y and z and produce [equation 5].

   $$\begin{array}{rrrrl} x+ & y+ & z= & 2 & [1] \\ 2x+ & y- & z= & 5 & [2] \\ \hline 3x+ & 2y & = & 7 & [4] \end{array} \qquad \begin{array}{rrrrl} 2x+ & y- & z= & 5 & [2] \\ x- & y+ & z= & -2 & [3] \\ \hline 3x & & = & 3 & [5] \end{array}$$

   Solve [equation 5] for $x: 3x=3 \Rightarrow x=1$. Substitute $x=1$ into [equation 4] to solve for y:

   $3(1)+2y=7 \Rightarrow 2y=4 \Rightarrow y=2$. Finally, substitute $x=1$ and $y=2$ into [equation 1] to solve for z:

   $(1)+(2)+z=2 \Rightarrow z=-1$. The solution is $(1,2,-1)$.

9. First, add [equation 1] and [equation 2] to eliminate y and produce [equation 4]. Second, multiply [equation 3] by 3 and add [equation 1] to eliminate y and produce[equation 5].

   $$\begin{array}{rrrrl} x+ & 3y+ & 4z= & 14 & [1] \\ 2x- & 3y+ & 2z= & 10 & [2] \\ \hline 3x & + & 6z= & 24 & [4] \end{array} \qquad \begin{array}{rrrrl} 9x- & 3y+ & 3z= & 27 & 3[3] \\ x+ & 3y+ & 4z= & 14 & [1] \\ \hline 10x & + & 7z= & 41 & [5] \end{array}$$

   Next multiply [equation 4] by 10 and add to [equation 5] multiplied by -3 to eliminate x and produce [equation 6]

   $$\begin{array}{rrrl} 30x & +60z= & 240 & 10[4] \\ -30x & -21z= & -123 & -3[5] \\ \hline & 39z= & 117 & [6] \end{array}$$

   Solve [equation 6] for $z: 39z=117 \Rightarrow z=3$. Substitute $z=3$ into [equation 4] to solve for x:

   $3x+6(3)=24 \Rightarrow 3x+6 \Rightarrow x=2$. Finally, substitute $x=2$ and $z=3$ into [equation 1] to solve for y:

   $(2)+3y+4(3)=14 \Rightarrow 3y=0 \Rightarrow y=0$. The solution is $(2,0,3)$.

11 First, multiply [equation 1] by 2 and add [equation 3] to eliminate all the variables.

$$\begin{array}{rl} 2x+4y+6z=16 & 2[1] \\ -2x-4y-6z=5 & [3] \\ \hline 0=21 & \end{array}$$

Since this is false, the solution set is $\varnothing$.

13. First, add [equation 1] and [equation 2] to eliminate z and produce [equation 4]. Second, multiply [equation 1] by 3 and add [equation 3] to eliminate z and produce [equation 5].

$$\begin{array}{rl} x+4y-z=6 & [1] \\ 2x-y+z=3 & [2] \\ \hline 3x+3y=9 & [4] \end{array} \qquad \begin{array}{rl} 3x+12y-3z=18 & 3[1] \\ 3x+2y+3z=16 & [2] \\ \hline 6x+14y=34 & [5] \end{array}$$

Next multiply [equation 4] by -2 and add to [equation 5] to eliminate x and produce [equation 6].

$$\begin{array}{rl} -6x-6y=-18 & -2[4] \\ 6x+14y=34 & [5] \\ \hline 8y=16 & [6] \end{array}$$

Solve [equation 6] for y: $y: 8y=16 \Rightarrow y=2$. Substitute $y=2$ into [equation 4] to solve for x: $3x+3(2)=9 \Rightarrow 3x=3 \Rightarrow x=1$. Finally, substitute $x=1$ and $y=2$ into [equation 1] to solve for z: $(1)+4(2)-z=6 \Rightarrow -z=-3 \Rightarrow z=3$. The solution is $(1,2,3)$.

15. First, multiply [equation 1] by -3 and add [equation 2] to eliminate y and produce [equation 4]. Second, multiply [equation 1] by 2 and add [equation 3] to eliminate y and produce [equation 5]

$$\begin{array}{rl} -15x-3y+9z=18 & -3[1] \\ 2x+3y+z=5 & [2] \\ \hline -13x+10z=23 & [4] \end{array} \qquad \begin{array}{rl} 10x+2y-6z=-12 & 2[1] \\ -3x-2y+4z=3 & [3] \\ \hline 7x-2z=-9 & [5] \end{array}$$

Next multiply [equation 5] by 5 and add [equation 4] to eliminate z and produce [equation 6]

$$\begin{array}{rl} 35x-10z=-45 & 5[5] \\ -13x+10z=23 & [4] \\ \hline 22x=-22 & [6] \end{array}$$

Solve [equation 6] for x: $22x=-22 \Rightarrow x=-1$. Substitute $x=-1$ into [equation 4] to solve for z: $-13(-1)+10z=23 \Rightarrow 10z=10 \Rightarrow z=1$. Finally, substitute $x=-1$ and $z=1$ into [equation 1] to solve for y: $5(-1)+y-3(1)=-6 \Rightarrow y=2$. The solution is $(-1,2,1)$.

17. First, add [equation 1] and [equation 3] to eliminate x and produce [equation 4]. Second, multiply [equation 3] by 3 and add [equation 2] to eliminate x and produce [equation 5].

$$\begin{array}{rl} x-3y-2z=-3 & [1] \\ -x-y+4z=3 & [3] \\ \hline -4y+2z=0 & [4] \end{array} \qquad \begin{array}{rl} -3x-3y+12z=9 & 3[3] \\ 3x+2y-z=12 & [2] \\ \hline -y+11z=21 & [5] \end{array}$$

Next multiply [equation 5] by -4 and add [equation 4] to eliminate y and produce [equation 6].

$$\begin{array}{rl} 4y-44z=-84 & -4[5] \\ -4y+2z=0 & [4] \\ \hline -42z=-84 & [6] \end{array}$$

Solve [equation 6] for z: $-42z=-84 \Rightarrow z=2$. Substitute $z=2$ into [equation 4] to solve for y: $-4y+2(2)=0 \Rightarrow -4y=-4 \Rightarrow y=1$. Finally, substitute $y=1$ and $z=2$ into [equation 1] to solve for x: $x-3(1)-2(2)=-3 \Rightarrow x=4$. The solution is $(4,1,2)$.

19. First eliminate x by multiplying [equation 1] by -2 and adding [equation 2].

$$\begin{array}{rl} -2x+4y-6z=-12 & -2\,[1] \\ 2x-y+2z=5 & [2] \\ \hline 3y-4z=-7 & \end{array}$$

Now solve this for y. $3y-4z=-7 \Rightarrow 3y=4z-7 \Rightarrow y=\dfrac{4z-7}{3}$. Now substitute this into [equation 2].

$$2x-\frac{4z-7}{3}+2z=5 \Rightarrow 3\left(2x-\frac{4z-7}{3}+2z=5\right) \Rightarrow 6x-4z+7+6z=15 \Rightarrow$$

$$6x+2z=8 \Rightarrow x=\frac{8-2z}{6} \Rightarrow x=\frac{4-z}{3}.$$ We have infinitely many solutions $\left(\dfrac{4-z}{3}, \dfrac{4z-7}{3}, z\right)$.

21. First eliminate x by adding [equation 1] to [equation 2] multiplied by -3:

$$\begin{array}{rl} 3x+4y-z=13 & [1] \\ -3x-3y-6z=-45 & -3\,[2] \\ \hline y-7z=-32 & \end{array}$$

Now solve this for y: $y-7z=-32 \Rightarrow y=7z-32$. Now substitute this into [equation 2].

$x+(7z-32)+2z=15 \Rightarrow x+9z=47 \Rightarrow x=47-9z.$

We have infinitely many solutions $(47-9z, 7z-32, z)$.

23. First, multiply [equation 1] by 3 and add [equation 2] to eliminate y and produce [equation 4], then simplify. Second, multiply [equation 3] by 3 and add [equation 2] to eliminate y and produce [equation 5] then simplify.

$$\begin{array}{rl} 24x-9y+18z=-6 & 3[1] \\ 4x+9y+4z=18 & [2] \\ \hline 28x+22z=12 & [4] \\ =14x+11z=6 & [4a] \end{array} \qquad \begin{array}{rl} 36x-9y+24z=-6 & 3[2] \\ 4x+9y+4z=18 & [2] \\ \hline 40x+28z=12 & [5] \\ =10x+7z=3 & [5a] \end{array}$$

Next multiply [equation 4a] by 5 and add [equation 5a] multiplied by -7 to eliminate x and produce [equation 6].

$$\begin{array}{rl} 70x+55z=30 & 5[4a] \\ -70x-49z=-21 & -7[5a] \\ \hline 6z=9 & [6] \end{array}$$

Solve [equation 6] for $z: 6z=9 \Rightarrow z=\dfrac{3}{2}$. Substitute $z=\dfrac{3}{2}$ into [equation 5a] to solve for x:

$10x+7\left(\dfrac{3}{2}\right)=3 \Rightarrow 2\left[10x+7\left(\dfrac{3}{2}\right)=3\right] \Rightarrow 20x+21=6 \Rightarrow 20x=-15 \Rightarrow x=-\dfrac{3}{4}$. Finally, substitute $x=-\frac{3}{4}$ and $z=\frac{3}{2}$ into [equation 2] to solve for y: $4\left(-\frac{3}{4}\right)+9y+4\left(\frac{3}{2}\right)=18\Rightarrow$

$-3+9y+6=18\Rightarrow 9y=15\Rightarrow y=\frac{15}{9}=\frac{5}{3}$. The solution is $\left(-\frac{3}{4},\frac{5}{3},\frac{3}{2}\right)$.

25. First, add [equation 1] and [equation 2] multiplied by -1 to eliminate x and produce [equation 4]. Second, add [equation 3] and [equation 4] to eliminate y and produce [equation 5].

$$\begin{array}{rrl} x-z= & 2 & [1] \\ -x-y= & 3 & -1[2] \\ \hline -y-z= & 5 & [4] \end{array} \qquad \begin{array}{rrl} y-\ \ z= & 1 & [3] \\ -y-\ \ z= & 5 & [4] \\ \hline -2z= & 6 & [5] \end{array}$$

Solve [equation 5] for $z: -2z=6\Rightarrow z=-3$. Substitute $z=-3$ into [equation 1] to solve for x:

$x-(-3)=2\Rightarrow x=-1$. Finally, substitute $z=-3$ into [equation 3] to solve for y:

$y-(-3)=1\Rightarrow y=-2$. The solution is $(-1,-2,-3)$.

27. First, multiply [equation 1] by 3 and add [equation 2] multiplied by -2 to eliminate y and produce [equation 4]. Second, add [equation 4] and [equation 3] multiplied by -9 to eliminate x.

$$\begin{array}{rrrrl} 9x+ & 6y- & 3z= & -3 & 3[1] \\ & -6y- & 2z= & -24 & -2[2] \\ \hline 9x & - & 5z= & -27 & [4] \end{array} \qquad \begin{array}{rrrl} 9x & - & 5z=-27 & [4] \\ -9x & + & 27z=\ \ 27 & [3] \\ \hline & & 22z=\ \ 0 & \Rightarrow z=0. \end{array}$$

Substitute $z=0$ to solve [equation 2] for y: $3y+0=12\Rightarrow y=4$. Finally, substitute $z=0$ into [equation 3] to solve for x: $x-3(0)=-3\Rightarrow x=-3$. The solution is $(-3,4,0)$.

29. First eliminate y by adding [equation 1] to [equation 2].

$$\begin{array}{rrrrr} x- & y+ & z & = & -6 \\ 4x+ & y+ & z & = & 7 \\ \hline 5x & & +2z & = & 1 \end{array}$$

Now solve this for x. $5x+2z=1\Rightarrow 5x=1-2x\Rightarrow x=\frac{1-2x}{5}$. Now substitute this into [equation 1].

$\frac{1-2z}{5}-y+z=-6\Rightarrow 5\left(\frac{1-2z}{5}-y+z=-6\right)\Rightarrow 1-2z-5y+5z=-30\Rightarrow -5y+3z=-31\Rightarrow$

$-5y=-3z-31\Rightarrow y=\frac{3z+31}{5}$. We have infinitely many solutions $\left(\frac{1-2z}{5},\frac{3z+31}{5},z\right)$.

31. First, multiply [equation 1] by -3 and add [equation 2] to eliminate x and produce [equation 4]. Second, add [equation 4] and [equation 3] to eliminate y.

$$\begin{array}{rrrrl} -6x- & 9y- & 12z= & -9 & -3[1] \\ 6x+ & 3y+ & 8z= & 6 & -2[2] \\ \hline & -6y- & 4z= & -3 & [4] \end{array} \qquad \begin{array}{rrrl} -6y- & 4z= & -3 & [4] \\ 6y- & 4z= & 1 & [3] \\ \hline & -8z= & -2 & \Rightarrow z=\frac{1}{4}. \end{array}$$

Substitute $z=\frac{1}{4}$ to solve [equation 3] for y: $6y-4\left(\frac{1}{4}\right)=1 \Rightarrow 6y=2 \Rightarrow y=\frac{1}{3}$. Finally, substitute $y=\frac{1}{3}$

and $z=\frac{1}{4}$ into [equation 1] to solve for x: $2x+3\left(\frac{1}{3}\right)+4\left(\frac{1}{4}\right)=3 \Rightarrow 2x=1 \Rightarrow x=\frac{1}{2}$.

The solution is $\left(\frac{1}{2},\frac{1}{3},\frac{1}{4}\right)$.

33. If we let $t=\frac{1}{x}$, $u=\frac{1}{y}$, and $v=\frac{1}{z}$, then our new equations are $t+u-v=\frac{1}{4}$ [1], $2t-u+3v=\frac{9}{4}$ [2], and $-t-2u+4v=1$ [3]. First, add [equation 1] and [equation 3] to eliminate t and produce [equation 4]. Second, add [equation 2] and [equation 3] multiplied by 2 to eliminate t and produce [equation 5].

$x^2-3x=4 \Rightarrow x^2-3x-4=0 \Rightarrow x=(x-4)(x+1)=0 \Rightarrow x=4,-1.$

$$\begin{array}{rl} t+u-v=\frac{1}{4} & [1] \\ -t-2u+4v=1 & [2] \\ \hline -u+3v=\frac{5}{4} & [4] \end{array} \qquad \begin{array}{rl} 2t-u+3v=\frac{4}{9} & [2] \\ -2t-4u+8v=2 & 2[3] \\ \hline -5u+11v=\frac{17}{4} & [5] \end{array}$$

Next multiply [equation 4] by -5 and add [equation 5] to eliminate u and produce [equation 6].

$$\begin{array}{rl} 5u-15v=-\frac{25}{4} & -5[4] \\ -5u+11v=\frac{17}{4} & [5] \\ \hline -4v=-\frac{8}{4} & [6] \end{array}$$

Solve [equation 6] for v. $-4v=-2 \Rightarrow v=-\frac{1}{2}$. If $v=-\frac{1}{2}$ and $v=\frac{1}{z}$, then $-\frac{1}{2}=\frac{1}{z} \Rightarrow z=-2$.

Substitute $v=\frac{1}{2}$ into [equation 4] to solve for u. $-u+3\left(\frac{1}{2}\right)=\frac{5}{4} \Rightarrow -u=-\frac{1}{4}$. If $u=\frac{1}{4}$ and $u=\frac{1}{y}$ then $\frac{1}{4}=\frac{1}{y} \Rightarrow y=4$. Finally, substitute $v=\frac{1}{2}$ and $u=\frac{1}{4}$ into [equation 3] to solve for t.

$-t-2\left(\frac{1}{4}\right)+4\left(\frac{1}{2}\right)=1 \Rightarrow -2t-1+4=2 \Rightarrow -2t=-1 \Rightarrow t=\frac{1}{2}$.

If $t=\frac{1}{2}$ and $t=\frac{1}{x}$ then $\frac{1}{2}=\frac{1}{x} \Rightarrow x=2$. The solution is $(2,4,2)$.

35. If we let $t=\frac{1}{x}$, $u=\frac{1}{y}$, and $v=\frac{1}{z}$ then our new equations are $2t-2u+v=-1$ [1], $4t+u-2v=-9$ [2], and $t+u-3v=-9$ [3]. First, add [equation 1] and [equation 2] multiplied by 2 to eliminate u and produce [equation 4]. Second, add [equation 1] and [equation 3] multiplied by 2 to eliminate u and produce [equation 5].

$$\begin{array}{rl} 2t - 2u + v = -1 & [1] \\ 8t + 2u - 4v = -18 & 2[2] \\ \hline 10t \quad - 3v = -19 & [4] \end{array} \qquad \begin{array}{rl} 2t - 2u + v = -1 & [1] \\ 8t + 2u - 6v = -18 & 2[3] \\ \hline 4t \quad - 5v = -19 & [5] \end{array}$$

Next multiply [equation 4] by 2 and add [equation 5] multiplied by -5 to eliminate t and produce [equation 6].

$$\begin{array}{rl} 20t - 6v = -38 & 2[2] \\ -20t + 25v = 95 & -5[5] \\ \hline 19v = 57 & [6] \Rightarrow v = 3 \end{array}$$

If $v = 3$ and $v = \frac{1}{z}$ then $3 = \frac{1}{z} \Rightarrow 3z = 1 \Rightarrow z = \frac{1}{3}$. Substitute $v = 3$ into [equation 5] to solve for t.

$4t + 5(3) = -19 \Rightarrow 4t = -4 \Rightarrow t = -1$. If $t = -1$ and $t = \frac{1}{x}$ then $-1 = \frac{1}{x} \Rightarrow -x = 1 \Rightarrow x = -1$. Finally, substitute $v = 3$ and $t = -1$ into [equation 1] to solve for u. $2(-1) - 2u + (3) = -1 \Rightarrow -2u = -2 \Rightarrow u = 1$.

If $u = 1$ and $= \frac{1}{y}$ then $1 = \frac{1}{y} \Rightarrow y = 1$. The solution is $\left(-1, 1, \frac{1}{3}\right)$.

37. Add [equation 1] and [equation 2] to eliminate z and produce [equation 4].

$$\begin{array}{rl} x - 4y + 2z = -2 & [1] \\ x + 2y - 2z = -3 & [2] \\ \hline 2x - 2y \quad = -5 & [4] \end{array}$$

Add [equation 4] to [equation 3] multiplied by -2.

$$\begin{array}{rl} 2x - 2y = -5 & [4] \\ -2x + 2y = -8 & -2[3] \\ \hline 0 = -13 & \end{array}$$

Since this is a false statement, the solution is $\varnothing$.

39. Subtract [equation 3] from [equation 1] to eliminate x and produce equation [4].

$$\begin{array}{rl} x + y + z = 0 & [1] \\ x + 3y + 3z = 5 & [3] \\ \hline -2y - 2z = -5 & [4] \end{array}$$

Subtract [equation 3] from [equation 2] to eliminate x and produce equation [5].

$$\begin{array}{rl} x - y - z = 3 & [2] \\ x + 3y + 3z = 5 & [3] \\ \hline -4y - 4z = -2 & [5] \end{array}$$

Multiply [equation 4] by 2 and subtract [equation 5].

$$\begin{array}{rl} -4y - 4z = -10 & 2[4] \\ -4y - 4z = -2 & [5] \\ \hline 0 = -8 & \end{array}$$

Since this is a false statement, the solution is $\varnothing$.

41. Add the first two equations.

$$\begin{array}{l} 2x - y + 2z = 6 \\ \underline{-x + y + \ \ z = 0} \\ x \qquad + 3z = 6 \end{array}$$

Add this equation to the third equation.

$$\begin{array}{l} x + 3z = \ \ 6 \\ \underline{-x - 3z = -6} \\ \qquad 0 = \ \ 0 \end{array}$$

Therefore, we have infinitely many solutions. $x + 3z = 6 \Rightarrow x = -3z + 6.$

Multiply the second equation by 2 and add to the first equation: \

$$\begin{array}{l} 2x - \ \ y + 2z = 6 \\ \underline{-2x + 2y + 2z = 0} \\ \qquad y + 4z = 6 \Rightarrow y = -4z + 6. \end{array}$$

We have infinitely many solutions $(-3z + 6, -4z + 6, z)$.

43. First, add [equation 2] and [equation 3] multiplied by -1 to eliminate y and produce [equation 4]. Second, add [equation 1] and [equation 3] multiplied by -20 to eliminate y.

$$\begin{array}{lll} 4x + 2y + 3z = & 280 & [2] \\ \underline{-3x - 2y - \ \ z = } & \underline{-180} & -1[3] \\ x \qquad + 2z = & 100 & [4] \end{array}$$

$$\begin{array}{lll} 25x + 40y + 20z = & 2200 & [1] \\ \underline{-60x - 40y - 20z =} & \underline{-3600} & -20[3] \\ -35x \qquad\qquad = & -1400 \Rightarrow x = 40 & \end{array}$$

Substitute $x = 40$ to solve [equation 4] for z. $40 + 2z = 100 \Rightarrow 2z = 60 \Rightarrow z = 30.$ Finally, substitute $x = 40$ and $z = 30$ into [equation 3] to solve for y. $3(40) + 2y + 30 = 180 \Rightarrow 2y = 30 \Rightarrow y = 15.$

The solution is $(40, 15, 30)$.

45. Let x = number of gallons of \$9 grade, y = number of gallons of \$3 grade, and z = number of gallons of \$4.50 grade. From the information the equations are $x + y + z = 300$ [1], $9x + 3y + 4.5z = 6(300)$ [2], and $z = 2y$ [3]. Substitute [equation 3] into both [equation 1] and [equation 2] to produce equations [4] and [5]. [3] into [1] produces $x + y + 2y = 300 \Rightarrow x + 3y = 300$ [4]. [3] into [2] produces $9x + 3y + 4.5(2y) = 1800 \Rightarrow 9x + 12y = 1800$ [5]. Now add [equation 5] to [equation 4] multiplied by -9.

$$\begin{array}{lll} 9x + 12y = & 1800 & [5] \\ \underline{-9x - 27y =} & \underline{-2700} & -9\ [4] \\ \qquad -15y = & -900 & \Rightarrow y = 60 \end{array}$$

Substitute $y = 60$, into [3]. z= 2(60) $\Rightarrow z = 120.$ Finally, substitute $y = 60$, and $z = 120$ into [1]. $x + 60 + 120 = 300 \Rightarrow x = 120.$ She should use 120 gallons of \$9 water, 60 gallons of \$3 water, and 120 gallons of \$4.50 water.

47. Let x = price of the senior ticket, y = price of adult ticket, and z = price of student ticket. From the information given the equations are $2x+y+2z=51$ [1], $y+5z=55$ [2], and $2x+2y+7z=75$ [3]. First solve [equation 2] for y and substitute into [equation 1] and create [equation 4].
$2x+55-5z+2z=51 \Rightarrow 2x-3z=-4$ Now substitute [equation 2] into [equation 3] and create [equation 5] $2x+2(55-5z)+7z=75 \Rightarrow 2x+110-10z+7z=75 \Rightarrow 2x-3z=-35$.

Subtracting [equation 4] from [equation 5] we will eliminate both the x and z variables.
$$\begin{array}{r} 2x-3z=-35 \\ \underline{2x-3z=-4} \\ 0=-31 \end{array}$$

This shows that there is no solution to this system of equations. The ticket pricing was inconsistent.

49. Let x = measure of the largest angle, y = measure of the medium angle, and z = measure of the smallest angle. From the information the equations are $a+b+c=180$ [1], $a=2b-55$ [2], and $c=b-25$ [3]. Substitute both [2] and [3] into [1]. $(2b-55)+b+(b-25)=180 \Rightarrow 4b-80=180 \Rightarrow 4b=260 \Rightarrow b=65$. Substitute $b=65$ into [3]. $c=65-25 \Rightarrow c=40$, and substitute $b=65$ into [2]. $a=2(65)-55 \Rightarrow a=130-55 \Rightarrow a=75$. The angles are 75°, 65°, and 40°.

51. Let x = the amount invested at 4%, y = the amount invested at 4.5%, and z = the amount invested at 2.5%. From the information the equations are $x+y+z=10000$ [1], $0.04x+0.025y+0.045z=415$ [2], and $y=2x$ [3]. Substitute [equation 3] into [equation 1] and [equation 3] into [equation 2] multiplied by 1000, to produce equations [4] and [5]. [3] into [1] $\Rightarrow x+2x+z=10000 \Rightarrow 3x+z=10000$ [4]. [3] into 1000[2] $\Rightarrow 40x+45(2x)+25z=415000 \Rightarrow 130x+25z=415000 \Rightarrow 26x+5z=83000$ [5]. Now multiply [equation 4] by -5 and add to [equation 5].

$$\begin{array}{lll} -15x-5y=-50000 & & -5[4] \\ \underline{26x+5y=\ \ 83000} & & [5] \\ 11x \qquad = \ 33000 & & \Rightarrow x=3000 \end{array}$$

Substitute x =3000, into [3]. $y=2(3000) \Rightarrow y=6000$. Finally, substitute x = 3000 and y = 6000 into [1]. $3000+6000+z=10000 \Rightarrow z=1000$. He invested \$3000 at 4%, \$6000 at 4.5%, and \$1000 at 2.5%.

53. Let x = the number of EZ models, y = the number of Compact models, and z = the number of Commercial models. From the information, the two equations are derived:
(weight) $10x+20y+60z=440$ [1] and (volume) $10x+8y+28z=248$ [2]. First, subtract [2] from [1].

$$\begin{array}{ll} 10x+20y+60z=440 & [1] \\ \underline{10x+\ 8y+28z=248} & [2] \end{array}$$

$$12y+32z=192 \qquad \Rightarrow 12y=-32z+192 \Rightarrow y=-\frac{8}{3}z+16.$$

Now substitute into [1]: $10x+20\left(-\frac{8}{3}z+16\right)+60z=440 \Rightarrow 3\left[10x+20\left(-\frac{8}{3}z+16\right)+60z=440\right] \Rightarrow$

$30x - 160z + 960 + 180z = 1320 \Rightarrow 30x = -20z + 360 \Rightarrow x = -\frac{2}{3}z + 12.$ From the equations for x and y, z must be a multiple of 3 for x and y to be whole numbers. There are three possibilities:

If z = 0, then $y = -\frac{8}{3}(0) + 16 = 16$ and $x = -\frac{2}{3}(0) + 12 = 12.$ Therefore, 12 EZ, 16 Comp., and 0 Comm.

If z = 3, then $y = -\frac{8}{3}(3) + 16 = 8$ and $x = -\frac{2}{3}(3) + 12 = 10.$ Therefore, 10 EZ, 8 Comp, and 3 Comm.

If z = 6, then $y = -\frac{8}{3}(6) + 16 = 0$ and $x = -\frac{2}{3}(6) + 12 = 8.$ Therefore, 8 EZ, 0 Comp., and 6 Comm.

If $z \geq 9,$ then y is less than zero; therefore, no solution is possible.

55. (a)

$$\begin{aligned} a + 20b + 2c &= 190 \\ a + 5b + 3c &= 320 \\ a + 40b + c &= 50 \end{aligned}$$

(b) Using technology, the solution is (30, –2, 100). So the equation is $P = 30 - 2A + 100S.$

(c) When A = 10 and S = 2500, P = 30 – 2(10) + 100(2.5) = $260,000.

57. Using the form $y = ax^2 + bx + c$ and the given points yields the following equations.

For $(-3, 0)$: $0 = a(-3)^2 + b(-3) + c \Rightarrow 9a - 3b + c = 0.$ [equation 1]

For $(0, -3)$: $-3 = a(0)^2 + b(0) + c \Rightarrow c = -3.$ [equation 2]

For $(1, 0)$: $0 = a(1)^2 + b(1) + c \Rightarrow a + b + c = 0.$ [equation 3]

First, substitute $c = -3$ into [equation 1] and [equation 3] and then add the two equations to eliminate b.

$$\begin{array}{lc} 9a - 3b - 3 = 0 & [1] \\ \underline{a + b - 3 = 0} & [3] \end{array} \Rightarrow \begin{array}{lc} 9a - 3b = 3 & [1] \\ \underline{3a + 3b = 9} & 3[3] \\ 12a \quad = 12 \Rightarrow a = 1 & \end{array}$$

Substitute $a = 1$ into [3]. $1 + b - 3 = 0 \Rightarrow b = 2$ The equation is $y = x^2 + 2x - 3.$

59. Using the form $y = ax^2 + bx + c$ and the given points yields the following equations.

For $(1.5, 6.25)$: $6.25 = a(1.5)^2 + b(1.5) + c \Rightarrow 2.25a + 1.5b + c = 6.25.$ [equation 1]

For $(0, -2)$: $-2 = a(0)^2 + b(0) + c \Rightarrow c = -2.$ [equation 2]

For $(-1.5, 3.25)$: $3.25 = a(-1.5)^2 + b(-1.5) + c \Rightarrow 2.25a - 1.5b + c = 3.25.$ [equation 3]

Substitute c = –2 into [1]. $2.25a + 1.5b - 2 = 6.25 \Rightarrow 2.25a + 1.5b = 8.25.$ [equation 4]

Substitute c = –2 into [3]. $2.25a - 1.5b - 2 = 3.25 \Rightarrow 2.25a - 1.5b = 5.25.$ [equation 5]

Now add [equation 4] and [equation 5]

$$\begin{array}{lc} 2.25a + 1.5b = 8.25 & [4] \\ \underline{2.25a - 1.5b = 5.25} & [5] \\ 4.5a \quad = 13.5 & \Rightarrow a = 3 \end{array}$$

Finally, substitute a = 3 and c = –2 into [1]. $6.25 = 2.25(3) + 1.5b + (-2) \Rightarrow b = 1.$

The equation is $y = 3x^2 + x - 2.$

61. Using the form $y = ax^2 + bx + c,$ and the given points yields the following equations.

For $(-1,4)$: $4 = a(-1)^2 + b(-1) + c \Rightarrow a - b + c = 4$. [equation 1]

For $(1,2)$: $2 = a(1)^2 + b(1) + c \Rightarrow a + b + c = 2$. [equation 2]

For $(3,8)$: $8 = a(3)^2 + b(3) + c \Rightarrow 9a + 3b + c = 8$. [equation 3]

First, add [equation 1] and [equation 2] multiplied to eliminate both b.

Second, add [equation 2] and [equation 3] multiplied by –3 to eliminate b and produce [equation 5].

$$\begin{array}{lll} a - b + c = 4 & [1] \\ a + b + c = 2 & [2] \\ \hline 2a \quad + 2c = 6 & [4] \end{array} \qquad \begin{array}{ll} -3a - 3b - 3c = -6 & -3[2] \\ 9a + 3b + \ \ c = \ \ 8 & [3] \\ \hline 6a \qquad\quad - 2c = 2 & [5] \end{array}$$

$$\begin{array}{ll} 2a + 2c = 6 & [4] \\ 6a - 2c = 2 & [5] \\ \hline 8a \qquad = 8 & \Rightarrow a = 1 \end{array}$$

Finally, substitute a = 1 into [5]. $6(1) - 2c = 2 \Rightarrow c = 2$.

Finally, substitute a = 1 and c = 2 into [1]. $(1) - b + (2) = 4 \Rightarrow -b = 1 \Rightarrow b = -1$.

The equation is $y = x^2 - x + 2$.

63. Using the form $y = ax^2 + bx + c$, and the given points yields the following equations.

For $(0,1)$: $1 = a(0)^2 + b(0) + c \Rightarrow c = 1$. [equation 1]

For $(1,0)$: $0 = a(1)^2 + b(1)^2 + c \Rightarrow a + b + c = 0$. [equation 2]

For $(2,-5)$: $-5 = a(2)^2 + b(2) + c \Rightarrow 4a + 2b + c = -5$. [equation 3]

Substitute c = 1 into [1]. $a + b + 1 = 0 \Rightarrow a + b = -1$. [equation 4]

Substitute c = 1 into [3]. $4a + 2b + 1 = -5 \Rightarrow 4a + 2b = -6$. [equation 5]

Now add [equation 4] and [equation 5] multiplied by -2.

$$\begin{array}{ll} -2a - 2b = \ \ 2 & -2[4] \\ \ \ 4a - 2b = -6 & [5] \\ \hline 2a \qquad = -4 & \Rightarrow a = -2 \end{array}$$

Finally, substitute a = -2 and c = 1 into [1]. $(-2) + b + (1) = 0 \Rightarrow b = 1$.

The equation is $y = -2x^2 + x + 1$.

65. Using the form $x^2 + y^2 + ax + by + c = 0$ and the given points yields the following equations.

For $(1,4)$: $(1)^2 + (4)^2 + a(1) + b(4) + c = 0 \Rightarrow 1 + 16 + a + 4b + c = 0 \Rightarrow a + 4b + c = -17$. [equation 1]

For $(5,2)$: $(5)^2 + (2)^2 + a(5) + b(2) + c = 0 \Rightarrow 25 + 4 + 5a + 2b + c = 0 \Rightarrow 5a + 2b + c = -29$. [equation 2]

For $(-3,-4)$: $(-3)^2 + (-4)^2 + a(-3) + b(-4) + c = 0 \Rightarrow 9 + 16 - 3a - 4b + c = 0 \Rightarrow -3a - 4b + c = -25$. [equation 3]

Now, add [equation 1] multiplied by -1 to [equation 2] to eliminate c and produce [equation 4].
Then, add [equation 3] multiplied by -1 to [equation 2] to eliminate c and produce [equation 5].

$$\begin{array}{lll} 5a+2b+c=-29 & [2] \\ \underline{-a-4b-c=\ \ 17} & -1[1] \\ 4a-2b \quad =-12 & [4] \end{array} \qquad \begin{array}{lll} 5a+2b+c=-29 & [2] \\ \underline{3a+4b-c=\ \ 25} & -1[3] \\ 8a+6b \quad =-4 & [5] \end{array}$$

Next multiply [equation 4] by 3 and add [equation 5] to eliminate b.

$$\begin{array}{lll} 12a-6b=-36 & 3[4] & \\ \underline{8a+6b=\ -4} & [5] & \\ 20a \quad =-40 & & \Rightarrow a=-2 \end{array}$$

Substitute $a=-2$ into [4]. $4(-2)-2b=-12 \Rightarrow -8-2b=-12 \Rightarrow -2b=-4 \Rightarrow b=2.$ Finally, substitute $a=-2$ and $b=2$ into [1]. $(-2)+4(2)+c=-17 \Rightarrow c=-23.$ The equation is $x^2+y^2-2x+2y-23=0.$

67. Using the form $x^2+y^2+ax+by+c=0$ and the given points yields the following equations.

For $(-1,3)$: $(-1)^2+(3)^2+a(-1)+b(3)+c=0 \Rightarrow 1+9-a+3b+c=0 \Rightarrow -a+3b+c=-10.$ [equation 1]

For $(6,2)$: $(6)^2+(2)^2+a(6)+b(2)+c=0 \Rightarrow 36+4+6a+2b+c=0 \Rightarrow 6a+2b+c=-40.$ [equation 2]

For $(-2,-4)$: $(-2)^2+(-4)^2+a(-2)+b(-4)+c=0 \Rightarrow 4+16-2a-4b+c=0 \Rightarrow -2a-4b+c=-20.$ [equation 3]

Now, subtract [equation 1] from [equation 2] to eliminate c and produce [equation 4].
Then, subtract [equation 3] from [equation 2] to eliminate c and produce [equation 5].

$$\begin{array}{lll} 6a+2b+c=-40 & [2] \\ \underline{a-3b-c=\ \ 10} & -1[1] \\ 7a-b \quad =-30 & [4] \end{array} \qquad \begin{array}{lll} 6a+2b+c=-40 & [2] \\ \underline{2a+4b-c=\ \ 20} & -1[3] \\ 8a+6b \quad =-20 & [5] \end{array}$$

Next multiply [equation 4] by 6 and add [equation 5] to eliminate b.

$$\begin{array}{lll} 42a-6b=-180 & 6[4] & \\ \underline{8a+6b=\ -20} & [5] & \\ 50a \quad =-200 & & \Rightarrow a=-4 \end{array}$$

Substitute $a=-4$ into [4]. $7(-4)-b=-30 \Rightarrow -b=-2 \Rightarrow b=2.$ Finally, substitute $a=-4$ and $b=2$ into [1]. $-(-4)+3(2)+c=-10 \Rightarrow c=-20.$ The equation is $x^2+y^2-4x+2y-20=0.$

69. Using the form $x^2+y^2+ax+by+c=0$ and the given points yields the following equations.

For $(2,1)$: $(2)^2+(1)^2+a(2)+b(1)+c=0 \Rightarrow 4+1+2a+b+c=0 \Rightarrow 2a+b+c=-5.$ [equation 1]

For $(-1,0)$: $(-1)^2+(0)^2+a(-1)+b(0)+c=0 \Rightarrow 1+0-a+0b+c=0 \Rightarrow -a+c=-1.$ [equation 2]

For $(3,3)$: $(3)^2+(3)^2+a(3)+b(3)+c=0 \Rightarrow 9+9+3a+3b+c=0 \Rightarrow 3a+3b+c=-18.$ [equation 3]

Now, multiply [equation 1] by –3 and add [equation 3] to eliminate b and produce [equation 4].

Next multiply [equation 2] by 2 and add [equation 4] to eliminate c.

$$\begin{array}{rl} -6a-3b-3c=\ \ 15 & -3\,[1] \\ 3a+3b+\ c=-18 & [3] \\ \hline -3a\qquad -2c=\ -3 & [4] \end{array} \qquad \begin{array}{rl} -2a+2c=-2 & 2\,[2] \\ -3a-2c=-3 & [4] \\ \hline -5a\qquad =-5 & \Rightarrow a=1. \end{array}$$

Substitute $a=1$ into [2]. $-(1)+c=-1 \Rightarrow c=0$. Finally, substitute $a=1$ and $c=0$ into [1].

$2(1)+b+0=-5 \Rightarrow b=-7$. The equation is $x^2+y^2+x-7y=0$.

71. Using the form $s(t)=at^2+bt+c$ and the given points yields the following equations.

For $(0,5)$: $5=a(0)^2+b(0)+c \Rightarrow c=5$.

For $(1,23)$: $23=a(1)^2+b(1)+c \Rightarrow a+b+c=23$. [equation 1]

For $(2,37)$: $37=a(2)^2+b(2)+c \Rightarrow 4a+2b+c=37$. [equation 2]

First, multiply [equation 1] by -2 and add [equation 2] to eliminate b and produce [equation 3].

$$\begin{array}{rl} -2a-2b-2c=-46 & -2\,[1] \\ 4a+2b+\ c=\ \ 37 & [2] \\ \hline 2a\qquad -c=\ -9 & [3] \end{array}$$

Substitute $c=5$ into [3]. $2a-(5)=-9 \Rightarrow 2a=-4 \Rightarrow a=-2$.

Finally, substitute $a=-2$ and $c=5$ into [1]. $(-2)+b+(5)=23 \Rightarrow b=20$.

The equation is $s(t)=-2t^2+20t+5$, and $s(8)=-2(8)^2+20(8)+5=-128+160+5=37$.

6.3: Solution of Linear Systems by Row Transformations

1. For $\begin{bmatrix} 2 & 4 \\ 4 & 7 \end{bmatrix}$, $\frac{1}{2}R_1 \to \begin{bmatrix} 1 & 2 \\ 4 & 7 \end{bmatrix}$.

3. For $\begin{bmatrix} 1 & 5 & 6 \\ -2 & 3 & -1 \\ 4 & 7 & 0 \end{bmatrix}$, $2R_1+R_2 \to \begin{bmatrix} 1 & 5 & 6 \\ 0 & 13 & 11 \\ 4 & 7 & 0 \end{bmatrix}$.

5. For $\begin{bmatrix} -3 & 1 & -4 \\ 2 & 1 & 3 \\ 10 & 5 & 2 \end{bmatrix}$, $-5R_2+R_3 \to \begin{bmatrix} -3 & 1 & -4 \\ 2 & 1 & 3 \\ 0 & 0 & -13 \end{bmatrix}$.

7. The augmented matrix is $\left[\begin{array}{cc|c} 2 & 3 & 11 \\ 1 & 2 & 8 \end{array}\right]$.

9. The augmented matrix is $\left[\begin{array}{cc|c} 1 & 5 & 6 \\ 1 & 0 & 3 \end{array}\right]$.

11. The augmented matrix is $\left[\begin{array}{ccc|c} 2 & 1 & 1 & 3 \\ 3 & -4 & 2 & -7 \\ 1 & 1 & 1 & 2 \end{array}\right]$.

13. The augmented matrix is $\left[\begin{array}{ccc|c} 1 & 1 & 0 & 2 \\ 0 & 2 & 1 & -4 \\ 0 & 0 & 1 & 2 \end{array}\right]$.

15. The system of equations is
$$\begin{aligned} 2x + y &= 1 \\ 3x - 2y &= -9. \end{aligned}$$

17. The system of equations is
$$\begin{aligned} x &= 2 \\ y &= 3 \\ z &= -2. \end{aligned}$$

19. The system of equations is
$$\begin{aligned} 3x + 2y + z &= 1 \\ 2y + 4z &= 22 \\ -x - 2y + 3z &= 15 \end{aligned}$$

21. The system can be written as $x + 2y = 3$ and $y = -1$. Substituting $y = -1$ into the first equation gives $x + 2(-1) = 3 \Rightarrow x = 5$. The solution is (5, –1).

23. The system can be written as $x - 5y = 6$ and $0 = 0$. The solution is $(5y + 6, y)$.

25. The system can be written as $x + y - z = 4$, $y - z = 2$, and $z = 1$. Substituting $z = 1$ into the second equation gives $y - (1) = 2 \Rightarrow y = 3$ Substituting $y = 3$ and $z = 1$ into the first equation gives $x + (3) - (1) = 4 \Rightarrow x = 2$. The solution is (2, 3, 1).

27. The system can be written as $x + 2y - z = 5$, $y - 2z = 1$, and $0 = 0$. Since $0 = 0$, there are an infinite number of solutions. The second equation gives $y = 1 + 2z$. Substituting this into the first equation gives $x + 2(1 + 2z) - z = 5 \Rightarrow x = 3 - 3z$. The solution can be written as $\{\,(3 - 3z, 1 + 2z, z) \mid z$ is a real number$\}$.

29. The system can be written as $x + 2y + z = -3$, $y - 3z = \frac{1}{2}$, and $0 = 4$. Since $0 = 4$ is false, there are no solutions.

31. Using the Row Echelon Method, the given system of equations yields the matrix $\left[\begin{array}{cc|c} 1 & 1 & 5 \\ 1 & -1 & -1 \end{array}\right]$, which by $-R_1 + R_2 \rightarrow \left[\begin{array}{cc|c} 1 & 1 & 5 \\ 0 & -2 & -6 \end{array}\right] \Rightarrow \left(-\frac{1}{2}R_2\right) \rightarrow \left[\begin{array}{cc|c} 1 & 1 & 5 \\ 0 & 1 & 3 \end{array}\right]$. From this matrix, we have the resulting equation $y = 3$. Now substitute $y = 3$ into the resulting R_1 equation. $x + (3) = 5 \Rightarrow x = 2$. The solution is $(2, 3)$.

33. Using the Row Echelon Method, the given system of equations yields the matrix $\left[\begin{array}{cc|c} 1 & 1 & -3 \\ 2 & -5 & -6 \end{array}\right]$, which by $-2R_1 + R_2 \rightarrow \left[\begin{array}{cc|c} 1 & 1 & -3 \\ 0 & -7 & 0 \end{array}\right] \Rightarrow \left(-\frac{1}{7}R_2\right) \rightarrow \left[\begin{array}{cc|c} 1 & 1 & -3 \\ 0 & 1 & 0 \end{array}\right]$. From this matrix, we have the resulting equation $y = 0$. Now substitute $y = 0$ into the resulting R_1 equation. $x + (0) = -3 \Rightarrow x = -3$.
The solution is $(-3, 0)$.

35. Using the Row Echelon Method, the given system of equations yields the matrix $\begin{bmatrix} 2 & -3 & | & 10 \\ 2 & 2 & | & 5 \end{bmatrix}$, which by

$-R_1 + R_2 \to \begin{bmatrix} 2 & -3 & | & 10 \\ 0 & 5 & | & -5 \end{bmatrix} \Rightarrow \left(\frac{1}{2}R_1\right)$ and $\left(\frac{1}{5}R_2\right) \to \begin{bmatrix} 1 & -\frac{3}{2} & | & 5 \\ 0 & 1 & | & -1 \end{bmatrix}$. From this matrix, we have the resulting

equation $y = -1$. Now substitute $y = -1$ into the resulting R_1 equation.

$x - \frac{3}{2}(-1) = 5 \Rightarrow 2x + 3 = 10 \Rightarrow 2x = 7 \Rightarrow x = \frac{7}{2}$. The solution is $\left(\frac{7}{2}, -1\right)$.

37. Using the Row Echelon Method, the given system of equations yields the matrix $\begin{bmatrix} 2 & -3 & | & 2 \\ 4 & -6 & | & 1 \end{bmatrix}$, which by

$-2R_1 + R_2 \to \begin{bmatrix} 2 & -3 & | & 2 \\ 0 & 0 & | & -3 \end{bmatrix}$. Since this matrix yields the equation $0 = -3$, which is false; therefore,

the solution is $\varnothing$.

39. Using the Row Echelon Method, the given system of equations yields the matrix $\begin{bmatrix} 6 & -3 & | & 1 \\ -12 & 6 & | & -2 \end{bmatrix}$, which by

$2R_1 + R_2 \to \begin{bmatrix} 6 & -3 & | & 1 \\ 0 & 0 & | & 0 \end{bmatrix}$. The matrix yields the equation $0 = 0$. Since this is true, there are ∞ solutions.

From the resulting R_1, the solutions have the relationship: $-3y = 1 - 6x \Rightarrow y = \frac{6x-1}{3}$; $\left[\text{or } x = \frac{3y+1}{6}\right]$.

The solution is $\left(x, \frac{6x-1}{3}\right)$ or $\left(\frac{3y+1}{6}, y\right)$.

41. Using the Row Echelon Method, the given system of equations yields the matrix $\begin{bmatrix} 1 & 1 & 0 & | & -1 \\ 0 & 1 & 1 & | & 4 \\ 1 & 0 & 1 & | & 1 \end{bmatrix}$, which by

$-R_1 + R_3 \to \begin{bmatrix} 1 & 1 & 0 & | & -1 \\ 0 & 1 & 1 & | & 4 \\ 0 & -1 & 1 & | & 2 \end{bmatrix} \Rightarrow R_2 + R_3 \to \begin{bmatrix} 1 & 1 & 0 & | & -1 \\ 0 & 1 & 1 & | & 4 \\ 0 & 0 & 2 & | & 6 \end{bmatrix} \Rightarrow \left(\frac{1}{2}R_3\right) \to \begin{bmatrix} 1 & 1 & 0 & | & -1 \\ 0 & 1 & 1 & | & 4 \\ 0 & 0 & 1 & | & 3 \end{bmatrix}$. From this matrix,

we have the resulting equation $z = 3$. Now use back-substitution. Substituting $z = 3$ into the resulting R_2 yields $y = 3 = 4 \Rightarrow y = 1$. Finally, substituting $y = 1$ and $z = 3$ into the resulting R_1 yields $x + 1 = -1 \Rightarrow x = -2$. The solution is $(-2, 1, 3)$.

43. Using the Row Echelon Method, the given system of equations yields the matrix $\left[\begin{array}{ccc|c} 1 & 1 & -1 & 6 \\ 2 & -1 & 1 & -9 \\ 1 & -2 & 3 & 1 \end{array}\right]$, which

by $(-2R_1+R_2)$ and $(-R_1+R_3)$ $\rightarrow \left[\begin{array}{ccc|c} 1 & 1 & -1 & 6 \\ 0 & -3 & 3 & -21 \\ 0 & -3 & 4 & -5 \end{array}\right] \Rightarrow -R_2+R_3 \rightarrow \left[\begin{array}{ccc|c} 1 & 1 & -1 & 6 \\ 0 & -3 & 3 & -21 \\ 0 & 0 & 1 & 16 \end{array}\right] \Rightarrow$

$\left(-\frac{1}{3}R_2\right) \rightarrow \left[\begin{array}{ccc|c} 1 & 1 & -1 & 6 \\ 0 & 1 & -1 & 7 \\ 0 & 0 & 1 & 16 \end{array}\right]$. From this matrix, we have the resulting equation $z=16$. Now use back

substitution, substituting $z=16$ into the resulting R_2. This yields $y-16=7 \Rightarrow y=23$. Finally, substituting $y=23$

and $z=16$ into the resulting R_1 yields $x+23-16=6 \Rightarrow x=-1$. The solution is $(-1,23,16)$.

45. Using the Row Echelon Method, the given system of equations yields the matrix $\left[\begin{array}{ccc|c} -1 & 1 & 0 & -1 \\ 0 & 1 & -1 & 6 \\ 1 & 0 & 1 & -1 \end{array}\right]$, which

by $R_1+R_3 \rightarrow \left[\begin{array}{ccc|c} -1 & 1 & 0 & -1 \\ 0 & 1 & -1 & 6 \\ 0 & 1 & 1 & -2 \end{array}\right] \Rightarrow -R_2+R_3 \rightarrow \left[\begin{array}{ccc|c} -1 & 1 & 0 & -1 \\ 0 & 1 & -1 & 6 \\ 0 & 0 & 2 & -8 \end{array}\right] \Rightarrow$

$(-R_1)$ and $\left(\frac{1}{2}R_2\right) \rightarrow \left[\begin{array}{ccc|c} 1 & -1 & 0 & 1 \\ 0 & 1 & -1 & 6 \\ 0 & 0 & 1 & -4 \end{array}\right]$. From this matrix, we have the resulting equation $z=-4$. Now use

back-substitution. Substituting $z=-4$ into the resulting R_2 yields $y+4=6 \Rightarrow y=2$. Finally, substituting $y=2$

and $z=-4$ into the resulting R_1 yields $-x+2=-1 \Rightarrow x=3$. The solution is $(3,2,-4)$.

47. Using the Row Echelon Method, the given system of equations yields the matrix $\left[\begin{array}{ccc|c} 2 & -1 & 3 & 0 \\ 1 & 2 & -1 & 5 \\ 0 & 2 & 1 & 1 \end{array}\right]$, which by

$R_1 \leftrightarrow R_2 \rightarrow \left[\begin{array}{ccc|c} 1 & 2 & -1 & 5 \\ 2 & -1 & 3 & 0 \\ 0 & 2 & 1 & 1 \end{array}\right] \Rightarrow -2R_1+R_2 \rightarrow \left[\begin{array}{ccc|c} 1 & 2 & -1 & 5 \\ 0 & -5 & 5 & -10 \\ 0 & 2 & 1 & -1 \end{array}\right] \Rightarrow 2R_2+5R_3 \rightarrow \left[\begin{array}{ccc|c} 1 & 2 & -1 & 5 \\ 0 & -5 & 5 & -10 \\ 0 & 0 & 15 & -15 \end{array}\right] \Rightarrow$

$\left(-\frac{1}{5}R_2\right)+\left(\frac{1}{15}R_3\right) \rightarrow \left[\begin{array}{ccc|c} 1 & 2 & -1 & 5 \\ 0 & 1 & -1 & 2 \\ 0 & 0 & 1 & -1 \end{array}\right]$. From this matrix, we have the resulting equation $z=-1$. Now use

back-substitution. Substituting $(x+1)(x-3) \Rightarrow 5x-3=A(x-3)+B(x+1)$. into the resulting R_2 yields

$y+1=2 \Rightarrow y=1$. Finally, substituting $y=1$ and $x=3 \Rightarrow 12=B(4) \Rightarrow B=3$. into the resulting R_1 yields

$x+2+1=5 \Rightarrow x=2$. The solution is $(2,1,-1)$.

49. Using the Row Echelon Method, the given system of equations yields the matrix $\begin{bmatrix} 1 & 1 & -2 & | & -6 \\ 1 & -1 & 1 & | & 4 \\ 2 & 0 & -1 & | & -1 \end{bmatrix}$, which

by $(R_1 - R_2) \to R_2$ $\begin{bmatrix} 1 & 1 & -2 & | & -6 \\ 0 & 2 & -3 & | & -10 \\ 2 & 0 & -1 & | & -1 \end{bmatrix} \Rightarrow (-2R_1 + R_3) \to R_3$ $\begin{bmatrix} 1 & 1 & -2 & | & -6 \\ 0 & 2 & -3 & | & -10 \\ 0 & -2 & 3 & | & 11 \end{bmatrix} \Rightarrow$

$(R_2 + R_3) \to R_3$ $\begin{bmatrix} 1 & 1 & -2 & | & -6 \\ 0 & 2 & -3 & | & -10 \\ 0 & 0 & 0 & | & 1 \end{bmatrix}$ From this matrix, we have the resulting equation $0 = 1$.

There is not solution to this system of equations.

51. Using the capabilities of the calculator, the Reduced Row Echelon Method of

$\left\| \begin{bmatrix} 0.07 & 0.23 & | & 9 \\ -1.25 & 0.33 & | & 2.4 \end{bmatrix} \right\|$ The solution is approximately $(7.785, 36.761)$.

53. Using the capabilities of the calculator, the Reduced Row Echelon Method of

$\begin{bmatrix} 2.1 & 0.5 & 1.7 & | & 4.9 \\ -2 & 1.5 & -1.7 & | & 3.1 \\ 5.8 & -4.6 & 0.8 & | & 9.3 \end{bmatrix} = \begin{bmatrix} 1 & 0 & 0 & | & 5.21127 \\ 0 & 1 & 0 & | & 3.73944 \\ 0 & 0 & 1 & | & -4.65493 \end{bmatrix}$. The solution is approximately $(5.211, 3.739, -4.655)$.

55. Using the capabilities of the calculator, the Reduced Row Echelon Method of

$\begin{bmatrix} 53 & 95 & 12 & | & 108 \\ 81 & -57 & -24 & | & -92 \\ -9 & 11 & -78 & | & 21 \end{bmatrix} = \begin{bmatrix} 1 & 0 & 0 & | & -0.24997 \\ 0 & 1 & 0 & | & 1.2838 \\ 0 & 0 & 1 & | & -0.05934 \end{bmatrix}$. The solution is approximately $(-0.250, 1.284, -0.059)$.

57. In both cases, we simply write the coefficients and do not write the variables. This is possible because we agree on the order in which the variables appear (descending degree).

59. The given system of equations yields the matrix, $\begin{bmatrix} 1 & -3 & 2 & | & 10 \\ 2 & -1 & -1 & | & 8 \end{bmatrix}$, which by

$-2R_1 + R_2 \to \begin{bmatrix} 1 & -3 & 2 & | & 10 \\ 0 & 5 & -5 & | & -12 \end{bmatrix} \Rightarrow \frac{1}{5}R_2 \begin{bmatrix} 1 & -3 & 2 & | & 10 \\ 0 & 1 & -1 & | & -\frac{12}{5} \end{bmatrix}$. Since we cannot eliminate more than one

coefficient there are an infinite number of solutions and they will be in the following relationship. From the

result of R_2, $y - z = -\dfrac{12}{5} \Rightarrow 5y - 5z = -12 \Rightarrow 5y = 5z - 12 \Rightarrow y = \dfrac{5z - 12}{5}$. Now substitute this into the

result of R_1: $x - 3\left(\dfrac{5z - 12}{5}\right) + 2z = 10 \Rightarrow 5x - 5z + 36 + 10z = 50 \Rightarrow 5x - 5z = 14 \Rightarrow$

$5x = 5z + 14 \Rightarrow x = \dfrac{5z + 14}{5}$. The solution is $\left(\dfrac{5z + 14}{5}, \dfrac{5z - 12}{5}, z\right)$.

61. The given system of equations yields the matrix $\left[\begin{array}{ccc|c} 1 & 2 & -1 & 0 \\ 3 & -1 & 1 & 6 \\ -2 & -4 & 2 & 0 \end{array}\right]$, which by

$(-3R_1+R_2)$ and $(2R_1+R_3) \to \left[\begin{array}{ccc|c} 1 & 2 & -1 & 0 \\ 0 & -7 & 4 & 6 \\ 0 & 0 & 0 & 0 \end{array}\right]$ since $-\frac{32}{11}z=-\frac{96}{11} \Rightarrow z=3.$ is $0=0,$ there are

infinitely many solutions and the solutions are in the following relationship: From the result of R_2,

$-7y+4z=6 \Rightarrow -7y=6-4z \Rightarrow y=\frac{4z-6}{7}$. Now substitute this into the result of R_1 :

$x+2\left(\frac{4z-6}{7}\right)-z=0 \Rightarrow 7x+8z-12-7z=0 \Rightarrow 7x+z=12 \Rightarrow 7x=12-z \Rightarrow x=\frac{12-z}{7}.$

The solution is $\left(\frac{12-z}{7}, \frac{4z-6}{7}, z\right)$.

63. The given system of equations yields the matrix $\left[\begin{array}{ccc|c} 1 & -2 & 1 & 5 \\ -2 & 4 & -2 & 2 \\ 2 & 1 & -1 & 2 \end{array}\right]$. By $(2R_1+R_2)$ and

$(-2R_1+R_3) \to \left[\begin{array}{ccc|c} 1 & -2 & 1 & 5 \\ 0 & 0 & 0 & 12 \\ 0 & 5 & -3 & -8 \end{array}\right]$. Since R_2 is $0=12,$ there is no solution or $\varnothing$.

65. The given system of equations yields the matrix $\left[\begin{array}{cccc|c} 1 & 3 & -2 & -1 & 9 \\ 4 & 1 & 1 & 2 & 2 \\ -3 & -1 & 1 & -1 & -5 \\ 1 & -1 & -3 & -2 & 2 \end{array}\right]$, which by $\left.\begin{array}{r} -4R_1+R_2 \\ 3R_1+R_3 \\ -R1+R_4 \end{array}\right\} \to$

$$\left[\begin{array}{cccc|c} 1 & 3 & -2 & -1 & 9 \\ 0 & -11 & 9 & 6 & -34 \\ 0 & 8 & -5 & -4 & 22 \\ 0 & -4 & -1 & -1 & -7 \end{array}\right] \Rightarrow \left.\begin{array}{r} R_3+2R_4 \\ 8R_2+11R_3 \end{array}\right\} \to \left[\begin{array}{cccc|c} 1 & 3 & -2 & -1 & 9 \\ 0 & -11 & 9 & 6 & -34 \\ 0 & 0 & 17 & 4 & -30 \\ 0 & 0 & -7 & -6 & 8 \end{array}\right] \Rightarrow$$

$$7R_3+17R_4 \to \left[\begin{array}{cccc|c} 1 & 3 & -2 & -1 & 9 \\ 0 & -11 & 9 & 6 & -34 \\ 0 & 0 & 17 & 4 & -30 \\ 0 & 0 & 0 & -74 & -74 \end{array}\right] \Rightarrow \left.\begin{array}{r} -\frac{1}{11}R_2 \\ -\frac{1}{17}R_3 \\ -\frac{1}{74}R_4 \end{array}\right\} \to \left[\begin{array}{cccc|c} 1 & 3 & -2 & -1 & 9 \\ 0 & 1 & \frac{-9}{11} & \frac{-6}{11} & \frac{34}{11} \\ 0 & 0 & 1 & \frac{4}{17} & -\frac{30}{17} \\ 0 & 0 & 0 & 1 & 1 \end{array}\right].$$ From this matrix, we

have the resulting equation $w=1$. Now use back-substitution. Substituting $w=1$ into the resulting R_3 yields

$z+\frac{4}{17}(1)=-\frac{30}{17} \Rightarrow z=-\frac{34}{17} \Rightarrow z=-2.$ Substituting $w=1$ and $z=-2$ into the resulting R_2 yields

$y-\frac{9}{11}(-2)-\frac{6}{11}(1)=\frac{34}{11}\Rightarrow y+\frac{18-6}{11}=\frac{34}{11}\Rightarrow y=\frac{34-12}{11}=\frac{22}{11}\Rightarrow y=2.$ Finally, substitute $y=2, z=-2,$ and $w=1$ into $R_1: \ x+3(2)-2(-2)-1=9\Rightarrow x+9=9\Rightarrow x=0.$

The solution is $(0, 2, -2, 1)$.

67. (a) Using the given equation model, the given system of equations and the capabilities of the calculator, the Reduced Row Echelon Method of: $\begin{bmatrix}1800 & 5000 & 1 & 1300\\ 3200 & 12000 & 1 & 5300\\ 4500 & 13000 & 1 & 6500\end{bmatrix}=\begin{bmatrix}1 & 0 & 0 & 0.5714286\\ 0 & 1 & 0 & 0.4571429\\ 0 & 0 & 1 & -2014.2857\end{bmatrix}$.

Therefore, the equation is $F=0.5714N+0.4571R-2014.$

(b) Let $N=3500$ and $R=12{,}500$. Then $F=0.5714(3500)+0.4571(12{,}500)-2014\Rightarrow \ F\approx 5699.65.$

This model predicts monthly food costs of approximately \$5700.

69. Let x represent the number of model A, and y the number of model B. Using the information yields

$$\begin{cases}2x+3y=34\\ 25x+30y=365\end{cases}\rightarrow\begin{bmatrix}2 & 3 & 34\\ 25 & 30 & 365\end{bmatrix}, \text{ which by } \left.\begin{matrix}\frac{1}{2}R_1\\ \frac{1}{5}R_2\end{matrix}\right\}\rightarrow\begin{bmatrix}1 & \frac{3}{2} & 17\\ 5 & 6 & 73\end{bmatrix}\Rightarrow -5R_1+R_2\rightarrow$$

$\begin{bmatrix}1 & \frac{3}{2} & 17\\ 0 & -\frac{3}{2} & -12\end{bmatrix}\Rightarrow -\frac{2}{3}R_2\rightarrow\begin{bmatrix}1 & \frac{3}{2} & 17\\ 0 & 1 & 8\end{bmatrix}$. From this matrix, we have the resulting equation $y=8$. Now substituting $y=8$ into R_1 yields $x+\frac{3}{2}(8)=7\Rightarrow x+12=17\Rightarrow x=5.$ The maximum number is Model A: 5 bicycles, and Model B: 8 bicycles.

71. Let $x=$ the amount of money at 2%, $y=$ the amount of money at 3%, and $z=$ the amount of money at 2.5%. Using the information yields $\begin{cases}x+y+z=12{,}500\\ y=1000+\frac{1}{2}x\Rightarrow -\frac{1}{2}x+y=1000.\\ 0.02x+0.03y+0.025z=305\end{cases}$ This system yields

$\begin{bmatrix}1 & 1 & 1 & 12{,}500\\ -0.5 & 1 & 0 & 1{,}000\\ 0.02 & 0.03 & 0.025 & 305\end{bmatrix}$, which by $\frac{1}{2}R_1+R_2$ and $-2R_1+100R_3$, $\begin{bmatrix}1 & 1 & 1 & 12{,}500\\ 0 & 1.5 & 0.5 & 7{,}250\\ 0 & 1 & 0.5 & 5{,}500\end{bmatrix}$ and

$$-\frac{2}{3}R_2+R_3\rightarrow\begin{bmatrix}1 & 1 & 1 & 12{,}500\\ 0 & 1.5 & 0.5 & 7{,}250\\ 0 & 0 & 1/6 & 4000/6\end{bmatrix}\Rightarrow 6R_3\rightarrow\begin{bmatrix}1 & 1 & 1 & 12{,}500\\ 0 & 1.5 & 0.5 & 7{,}250\\ 0 & 0 & 1 & 4000\end{bmatrix}$$

From this matrix, we have the resulting equation $z=4{,}000$ and $1.5y+0.5z=7{,}250$. Substitute $z=4000$ into $1.5y+0.5z=7{,}250\Rightarrow 1.5y+0.5(4000)=7250\Rightarrow y=3500$. Finally, substitute z=4000 and y =3500 into $R_1: x+y+z=12{,}500\Rightarrow x+3500+4000=12{,}500\Rightarrow x=5000$. The loans were: \$5,000 at 2%, \$3500 at 3%, and \$4000 at 2.5%.

73. (a) The three equations can be written as: $25=0^2+0+c, 260=2^2a+2b+c,$ and

$695=4^2a+4b+c\Rightarrow 16a+4b+c=695, 4a+2b+c=260, \textit{ and } c=25.$

Therefore, the matrix is: $\begin{bmatrix} 16 & 4 & 1 & | & 695 \\ 4 & 2 & 1 & | & 260 \\ 0 & 0 & 1 & | & 25 \end{bmatrix}$

(b) Using technology, $a = 25, b = 67.5, c = 25$. $f(x) = 25x^2 + 67.5x + 25$.

(c) See Figure 73

(d) For example, in 2014 the predicted sales is $f(6) = 25(6)^2 + 67.5(6) + 25 = 1330$ Answers may vary.

75. (a) Using $f(x) = ax^2 + bx + c$ and the data from the table yields:

From the point $(1990, 11)$: $11 = a(1990)^2 + b(1990) + c \Rightarrow 1990^2 a + |1990b + c = 11$.

From the point $(2010, 10)$: $10 = a(2010)^2 + b(2010) + c \Rightarrow 2010^2 a + 2010b + c = 10$.

From the point $(2030, 6)$: $6 = a(2030)^2 + b(2030) + c \Rightarrow 2030^2 a + 2030b + c = 6$.

(b) Now using the capabilities of the calculator, the Reduced Row Echelon Method of:

$$\begin{bmatrix} 1990^2 & 1990 & 1 & | & 11 \\ 2010^2 & 2010 & 1 & | & 10 \\ 2030^2 & 2030 & 1 & | & 6 \end{bmatrix} = \begin{bmatrix} 1 & 0 & 0 & | & -0.00375 \\ 0 & 1 & 0 & | & 14.95 \\ 0 & 0 & 1 & | & -14,889.125 \end{bmatrix}.$$

Therefore, the equation is $f(x) = -0.00375x^2 + 14.95x - 14889.125$.

(c) See Figure 75.

(d) Answers will vary, for example in 2015 the predicted ratio is $f(2015) \approx 9.3$.

[0,5] by [20,725]
Xscl = 1 Yscl = 5

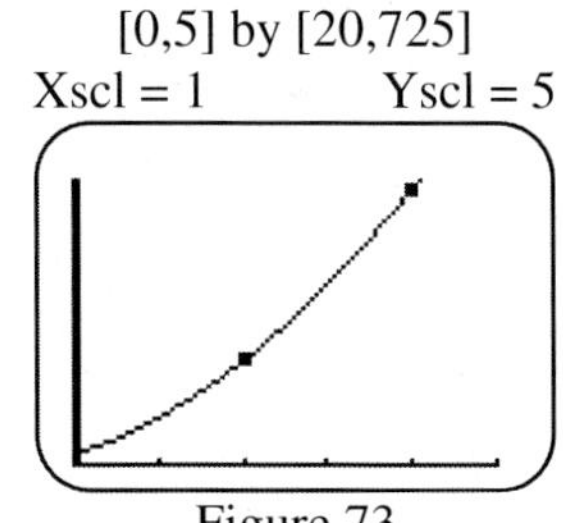

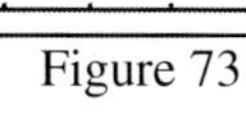

Figure 73

[1983,2035] by [5,12]
Xscl = 5 Yscl = 1

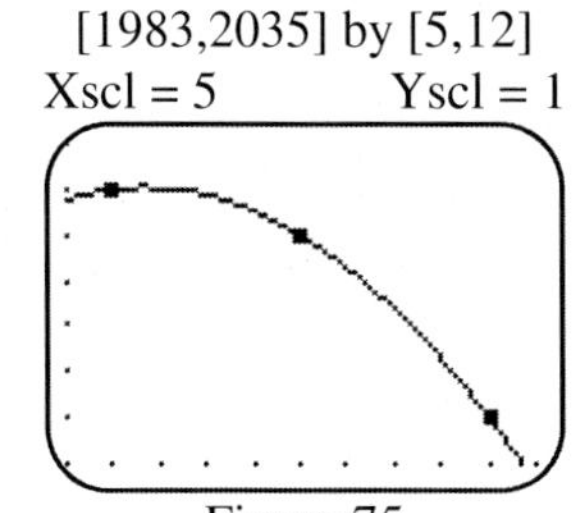

Figure 75

77. (a) At intersection A, incoming traffic is equal to $x + 5$. The outgoing traffic is given by $y + 7$. Therefore, $x + 5 = y + 7$. The incoming traffic at intersection B is $z + 6$ and the outgoing traffic is $x + 3$, so $z + 6 = x + 3$. Finally, at intersection C, the incoming flow is $y + 3$ and the outgoing flow is $z + 4$, so $y + 3 = z + 4$.

(b) The three equations are $\begin{cases} x+5=y+7 \Rightarrow x-y=2 \\ z+6=x+3 \Rightarrow x-z=3 \\ y+3=z+4 \Rightarrow y-z=1 \end{cases}$ which can be represented by $\left[\begin{array}{ccc|c} 1 & -1 & 0 & 2 \\ 1 & 0 & -1 & 3 \\ 0 & 1 & -1 & 1 \end{array}\right]$.

Then by $-R_1+R_2 \to \left[\begin{array}{ccc|c} 1 & -1 & 0 & 2 \\ 0 & 1 & -1 & 1 \\ 0 & 1 & -1 & 1 \end{array}\right] \Rightarrow -R_2+R_3 \to \left[\begin{array}{ccc|c} 1 & -1 & 0 & 2 \\ 0 & 1 & -1 & 1 \\ 0 & 0 & 0 & 0 \end{array}\right]$. Since in the result of

R_3 $0=0$, there are infinite solutions. They are $y-z=1 \Rightarrow y=z+1$ and $x-z=3 \Rightarrow x=z+3$. The solution is $(z+3,\ z+1,\ z)$, where $z \geq 0$.

(c) There are infinitely many solutions since some cars could be driving around the block continually.

79. (a) Using the given model and information from the table the equations are

$x=\frac{Dx}{D}=\frac{-114}{-114}=1,\ y=\frac{Dy}{D}=\frac{228}{-114}=-2,$

(b) The equations can be represented by $\left[\begin{array}{cccc|c} 1 & 871 & 11.5 & 3 & 239 \\ 1 & 847 & 12.2 & 2 & 234 \\ 1 & 685 & 10.6 & 5 & 192 \\ 1 & 969 & 14.2 & 1 & 343 \end{array}\right]$. Using the capabilities of the calculator, the Reduced Row Echelon Method produces the solutions $a \approx -715.457$, $b \approx 0.34756$, $c \approx 48.6585$, and $d \approx 30.71951$.

(c) The equation is $F=-715.457+0.34756A+48.6585P+30.71951W$.

(d) Input the given data into the equation.

$F=-715.457+0.34756(960)+48.6585(12.6)+30.71951(3) \Rightarrow F \approx 323.45623$. This is approximately 323, which is close to the actual value of 320.

Reviewing Basic Concepts (Section 6.1-6.3)

1. Multiply the first equation by 5, the second equation by 2, and subtract to eliminate the x-variable.

$10x-15y=90$

$\underline{10x+4y\ =14}$

$-19y=76 \Rightarrow y=-4$ Substitute (-4) for y in the second equation:

$10x+4(-4)=14 \Rightarrow 10x=30 \Rightarrow x=3$. The solution is $\{(3,-4)\}$.

2. The equations are: $2x+y=-4 \Rightarrow y_1=-2x-4$ and $-x+2y=2 \Rightarrow 2y=x+2 \Rightarrow y=\frac{x+2}{2} \Rightarrow$

$y_2=\frac{1}{2}x+1$. Graph y_1 and y_2 in the same window, the intersection is $\{(-2,0)\}$. See Figure 2.

3. Using substitution, first solve [equation 2] for x, $x+2y=2 \Rightarrow x=-2y+2$ [equation 3]. Substitute $(-2y+2)$ in for x, in [equation 1]: $5(-2y+2)+10y=10 \Rightarrow -10y+10+10y=10 \Rightarrow 10=10$. Since this is true, there is an infinite number of solutions given as $(-2y+2, y)$ *or* $\left(x, \dfrac{2-x}{2}\right)$.

4. Subtract [equation 2] from [equation 1] to eliminate both variables.

$$\begin{array}{l} x-y=6 \\ \underline{x+y=4} \\ \quad 0=2 \end{array}$$
$\Rightarrow$ Since this is false the solution is $\varnothing$.

5. First solve [equation 1] for y and substitute into [equation 2]: $6x+2y=10 \Rightarrow 2y=10-6x \Rightarrow y=5-3x$.

   Now substitute: $2x^2-3(-3x+5)=11 \Rightarrow 2x^2+9x-15=11 \Rightarrow 2x^2+9x-26=0 \Rightarrow$

   $(2x+13)(x-2)=0 \Rightarrow x=-6.5$ and $x = 2$. If $x = -6.5$, then by substitution :

   $y=-3(-6.5)+5 \Rightarrow y=19.5+5 \Rightarrow y=24.5$. If $x = 2$, then by substitution $y=-3(2)+5 \Rightarrow y=-1$.

   The solution is: $\{(2,-1),(-6.5,24.5)\}$. See Figure 5 for graphical support.

[-10,10] by [-10,10]
Xscl = 1 Yscl = 1

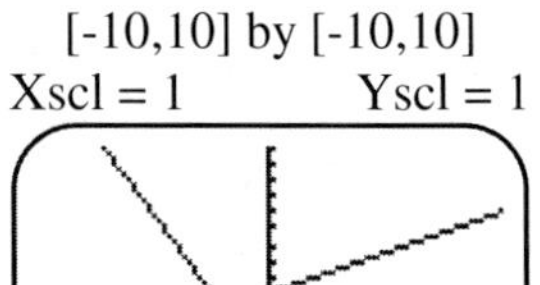

Figure 2

[-10,10] by [-10,10]
Xscl = 1 Yscl = 1

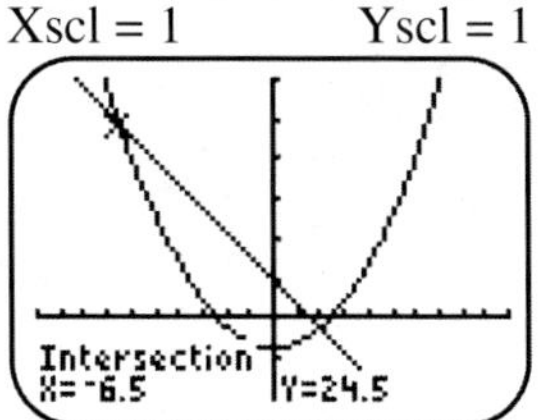

Figure 5

6. The equations can be represented by: $\left[\begin{array}{ccc|c} 1 & 1 & 1 & 1 \\ -1 & 1 & 1 & 5 \\ 0 & 1 & 2 & 5 \end{array}\right] . R_1+R_2 \rightarrow \left[\begin{array}{ccc|c} 1 & 1 & 1 & 1 \\ 0 & 2 & 2 & 6 \\ 0 & 1 & 2 & 5 \end{array}\right] \Rightarrow$

   $R_2-2R_1 \rightarrow \left[\begin{array}{ccc|c} 1 & 1 & 1 & 1 \\ 0 & 2 & 2 & 6 \\ 0 & 0 & -2 & -4 \end{array}\right]$.

   Now solve the resulting equation $R_3: -2z=-4 \Rightarrow z=2$. Substitute $z=2$ into the resulting equation R_2 : $2y+2(2)=6 \Rightarrow 2y=2 \Rightarrow y=1$. Finally, substitute $y=1$ and $z=2$ into R_1 :

   $x+(1)+(2)=1 \Rightarrow x=-2$. The solution is $\{(-2,1,2)\}$.

7. Using the Reduced Row Echelon method, the given system of equations yields the matrix, $\left[\begin{array}{ccc|c} 2 & 4 & 4 & 4 \\ 1 & 3 & 1 & 4 \\ -1 & 3 & 2 & -1 \end{array}\right]$.

   Using the capabilities of the calculator, the Reduced Row Echelon Method produces the solutions $x=2$, $y=1$, and $z=-1$.

8. Solve the augmented matrix $\begin{bmatrix} 2 & 1 & 2 & | & 10 \\ 1 & 0 & 2 & | & 5 \\ 1 & -2 & 2 & | & 1 \end{bmatrix}$, which by $\left.\begin{matrix} R_1 - 2R_2 \\ R_1 - 2R_3 \end{matrix}\right\} \rightarrow \begin{bmatrix} 2 & 1 & 2 & | & 10 \\ 0 & 1 & -2 & | & 0 \\ 0 & 5 & -2 & | & 8 \end{bmatrix} \Rightarrow$

$-5R_2 + R_3 \rightarrow \begin{bmatrix} 2 & 1 & 2 & | & 10 \\ 0 & 1 & -2 & | & 0 \\ 0 & 0 & 8 & | & 8 \end{bmatrix}$. Now solve the resulting equation $R_3: \ 8z = 8 \Rightarrow z = 1$. Substitute $z = 1$ into the resulting equation $R_2: \ y - 2(1) = 0 \Rightarrow y = 2$. Finally, substitute $y = 2$ and $z = 1$ into R_1: $2x + (2) + 2(1) = 10 \Rightarrow 2x = 6 \Rightarrow x = 3$. The solution is $(3, 2, 1)$.

9. Let x represent the number of LCD monitors and y represent the number of CRT monitors.

   Then from the information the equations are $\begin{matrix} x + y = 133 & [1] \\ x - y = 43 & [2] \end{matrix}$ Add [equation 1] and [equation 2]:

   $2x = 176 \Rightarrow x = 88$. Substitute x =88 into [1]: $88 + y = 143 \Rightarrow y = 45.$. There were approximately 88 million LCD monitors sold and 45 million CRT monitors sold in 2006.

10. Let x = amount at 2%, y = amount at 3%, and x = amount at 4%. Then from the information:

    $x + y + z = 5000$
    $0.02x + 0.03y + 0.04z = 165$
    $z = x + y \Rightarrow x + y - z = 0$

    Using the Reduced Row Echelon method, the given system of equations yields the matrix

    $\begin{bmatrix} 1 & 1 & 1 & | & 5000 \\ 0.02 & 0.03 & 0.04 & | & 165 \\ 1 & 1 & -1 & | & 0 \end{bmatrix}$.

    Using the capabilities of the calculator, the Reduced Row Echelon Method produces the solutions $x = 1000$, $y = 1500$, and $z = 2500$.

    There was \$1000 invested at 2%, \$1500 invested at 3%, and \$2500 invested at 4%.

6.4: Matrix Properties and Operations

1. Because the number of rows equals the number of columns, this is a 2×2 square matrix.
3. This is a 3×4 matrix.
5. Because there is only one column, this is a 2×1 column matrix.
7. Because the number of rows equals the number of columns, there is only one row and column, this is a 1×1 square matrix (one row and one column).
9. For $\begin{bmatrix} w & x \\ y & z \end{bmatrix} = \begin{bmatrix} 3 & 2 \\ -1 & 4 \end{bmatrix}$, $w = 3, x = 2, y = -1$, and $z = 4$.
11. For $\begin{bmatrix} 0 & 5 & x \\ -1 & 3 & y+2 \\ 4 & 1 & z \end{bmatrix} = \begin{bmatrix} 0 & w+3 & 6 \\ -1 & 3 & 0 \\ 4 & 1 & 8 \end{bmatrix}$, $w + 3 = 5 \Rightarrow w = 2$, $x = 6$, $y + 2 = 0 \Rightarrow y = -2$, and $z = 8$.

13. For $\begin{bmatrix} z & 4r & 8s \\ 6p & 2 & 5 \end{bmatrix} + \begin{bmatrix} -9 & 8r & 3 \\ 2 & 7 & 4 \end{bmatrix} = \begin{bmatrix} 2 & 36 & 27 \\ 20 & 7 & 12a \end{bmatrix}$,

$z-9=2 \Rightarrow z=11;\ 4r+8r=36 \Rightarrow r=3;\ 8s+3=27 \Rightarrow 8s=24 \Rightarrow s=3$

$6p+2=20 \Rightarrow 6p=18 \Rightarrow p=3;\ 5+4=12a \Rightarrow 9=12a \Rightarrow a=\dfrac{3}{4}$

15. The two matrices must have the same dimensions. To find the sum, add the corresponding entries. The sum will be a matrix with the same dimension.

17. $\begin{bmatrix} 6 & -9 & 2 \\ 4 & 1 & 3 \end{bmatrix} + \begin{bmatrix} -8 & 2 & 5 \\ 6 & -3 & 4 \end{bmatrix} = \begin{bmatrix} 6-8 & -9+2 & 2+5 \\ 4+6 & 1-3 & 3+4 \end{bmatrix} = \begin{bmatrix} -2 & -7 & 7 \\ 10 & -2 & 7 \end{bmatrix}$

19. $\begin{bmatrix} -6 & 8 \\ 0 & 0 \end{bmatrix} - \begin{bmatrix} 0 & 0 \\ -4 & -2 \end{bmatrix} = \begin{bmatrix} -6-0 & 8-0 \\ 0-(-4) & 0-(-2) \end{bmatrix} = \begin{bmatrix} -6 & 8 \\ 4 & 2 \end{bmatrix}$

21. $\begin{bmatrix} 6 & -2 \\ 5 & 4 \end{bmatrix} + \begin{bmatrix} -1 & 7 \\ 7 & -4 \end{bmatrix} = \begin{bmatrix} 6-1 & -2+7 \\ 5+7 & 4-4 \end{bmatrix} = \begin{bmatrix} 5 & 5 \\ 12 & 0 \end{bmatrix}$

23. $\begin{bmatrix} -8 & 4 & 0 \\ 2 & 5 & 0 \end{bmatrix} + \begin{bmatrix} 6 & 3 \\ 8 & 9 \end{bmatrix}$ Matrices must be the same size to add or subtract. We cannot add $(2\times3)+(2\times2)$.

25. $\begin{bmatrix} 9 & 4 & 1 & -2 \\ 5 & -6 & 3 & 4 \\ 2 & -5 & 1 & 2 \end{bmatrix} - \begin{bmatrix} -2 & 5 & 1 & 3 \\ 0 & 1 & 0 & 2 \\ -8 & 3 & 2 & 1 \end{bmatrix} + \begin{bmatrix} 2 & 4 & 0 & 3 \\ 4 & -5 & 1 & 6 \\ 2 & -3 & 0 & 8 \end{bmatrix} =$

$\begin{bmatrix} 9+2+2 & 4-5+4 & 1-1+0 & -2-3+3 \\ 5-0+4 & -6-1-5 & 3-0+1 & 4-2+6 \\ 2+8+2 & -5-3-3 & 1-2+0 & 2-1+8 \end{bmatrix} = \begin{bmatrix} 13 & 3 & 0 & -2 \\ 9 & -12 & 4 & 8 \\ 12 & -11 & -1 & 9 \end{bmatrix}$

27. $2\begin{bmatrix} 2 & -1 \\ 5 & 1 \\ 0 & 3 \end{bmatrix} + \begin{bmatrix} 5 & 0 \\ 7 & -3 \\ 1 & 1 \end{bmatrix} - \begin{bmatrix} 9 & -4 \\ 4 & 4 \\ 1 & 6 \end{bmatrix} = \begin{bmatrix} 4+5-9 & -2+0+4 \\ 1+7-4 & 2-3-4 \\ 0+1-1 & 6+1-6 \end{bmatrix} = \begin{bmatrix} 0 & 2 \\ 13 & -5 \\ 0 & 1 \end{bmatrix}$

29 $2\begin{bmatrix} 2 & -1 & -1 \\ -1 & 2 & -1 \\ -1 & -1 & 2 \end{bmatrix} + 3\begin{bmatrix} 1 & 2 & 3 \\ 2 & 1 & 3 \\ 2 & 3 & 1 \end{bmatrix} = \begin{bmatrix} 4+3 & -2+6 & -2+9 \\ -2+6 & 4+3 & -2+9 \\ -2+6 & -2+9 & 4+3 \end{bmatrix} = \begin{bmatrix} 7 & 4 & 7 \\ 4 & 7 & 7 \\ 4 & 7 & 7 \end{bmatrix}$.

31. $3\begin{bmatrix} 6 & -1 & 4 \\ 2 & 8 & -3 \\ -4 & 5 & 6 \end{bmatrix} + 5\begin{bmatrix} -2 & -8 & -6 \\ 4 & 1 & 3 \\ 2 & -1 & 5 \end{bmatrix} = \begin{bmatrix} 18 & -3 & 12 \\ 6 & 24 & -9 \\ -12 & 15 & 18 \end{bmatrix} + \begin{bmatrix} -10 & -40 & -30 \\ 20 & 5 & 15 \\ 10 & -5 & 25 \end{bmatrix} =$

$\begin{bmatrix} 18-10 & -3-40 & 12-30 \\ 6+20 & 24+5 & -9+15 \\ -12+10 & 15-5 & 18+25 \end{bmatrix} = \begin{bmatrix} 8 & -43 & -18 \\ 26 & 29 & 6 \\ -2 & 10 & 43 \end{bmatrix}$

33. $2[A] = 2\begin{bmatrix} -2 & 4 \\ 0 & 3 \end{bmatrix} = \begin{bmatrix} -4 & 8 \\ 0 & 6 \end{bmatrix}$

35. $2[A]-[B] = 2\begin{bmatrix} -2 & 4 \\ 0 & 3 \end{bmatrix} - \begin{bmatrix} -6 & 2 \\ 4 & 0 \end{bmatrix} = \begin{bmatrix} -4+6 & 8-2 \\ 0-4 & 6-0 \end{bmatrix} = \begin{bmatrix} 2 & 6 \\ -4 & 6 \end{bmatrix}$

37. $5[A]+0.5[B]=5\begin{bmatrix}-2 & 4\\ 0 & 3\end{bmatrix}+0.5\begin{bmatrix}-6 & 2\\ 4 & 0\end{bmatrix}=\begin{bmatrix}-10-3 & 20+1\\ 0+2 & 15+0\end{bmatrix}=\begin{bmatrix}-13 & 21\\ 2 & 15\end{bmatrix}$

39. $[A]=([A]+[B])-[B]=\begin{bmatrix}6 & 12 & 0\\ -10 & -4 & 11\end{bmatrix}-\begin{bmatrix}4 & 6 & -5\\ -6 & 3 & 2\end{bmatrix}=\begin{bmatrix}6-4 & 12-6 & 0+5\\ -10+6 & -4-3 & 11-2\end{bmatrix}=\begin{bmatrix}2 & 6 & 5\\ -4 & -7 & 9\end{bmatrix}$

41. A is 4×2 and B is 2×4. For *AB*, the number of columns of *A* is the same as the number of rows of *B*; therefore, *AB* has dimensions: $4(\text{rows of } A)\times4(\text{columns of } B)$ or *AB* is 4×4. For *BA*, the number of columns of *B* is the same as the number of rows of *A*; therefore, *BA* has dimensions: $2(\text{rows of } B)\times2(\text{columns of } A)$ or *BA* is 2×2.

43. A is 3×5 and B is 5×2. For *AB*, the number of columns of *A* is the same as the number of rows of B; therefore, *AB* has dimensions: $3(\text{rows of } A)\times2(\text{columns of } B)$ or AB is 3×2. For *BA*, the number of columns of *B* is not the same as the number of rows of *A*; therefore, *BA* is not defined.

45. A is 4×3 and B is 2×5. For *AB*, the number of columns of *A* is not the same as the number of rows of *B*; therefore, *AB* is not defined. For *BA*, the number of columns of *B* is not the same as the number of rows of *A*; therefore, *BA* is also not defined.

47. … the number of columns of *M* equal the number of rows of *N*.

49. A and B are both 2×2 so AB and BA are also both 2×2.

$$AB=\begin{bmatrix}1 & -1\\ 2 & 0\end{bmatrix}\begin{bmatrix}-2 & 3\\ 1 & 2\end{bmatrix}=\begin{bmatrix}-3 & 1\\ -4 & 6\end{bmatrix}; BA=\begin{bmatrix}-2 & 3\\ 1 & 2\end{bmatrix}\begin{bmatrix}1 & -1\\ 2 & 0\end{bmatrix}=\begin{bmatrix}4 & 2\\ 5 & -1\end{bmatrix}$$

51. Since both A and B are 2×3, the number of rows in B is not equal to the number of columns in A, so AB is undefined. Also, the number of rows in A is not equal to the number of columns in B so BA is undefined.

53. $AB=\begin{bmatrix}3 & -1\\ 1 & 0\\ -2 & -4\end{bmatrix}\begin{bmatrix}-2 & 5 & -3\\ 9 & -7 & 0\end{bmatrix}=$

$$\begin{bmatrix}3(-2)+(-1)(9) & 3(5)+(-1)(-7) & 3(-3)+(-1)(0)\\ 1(-2)+0(9) & 1(5)+0(-7) & 1(-3)+0(0)\\ -2(-2)+(-4)(9) & -2(5)+(-4)(-7) & -2(-3)+(-4)(0)\end{bmatrix}\Rightarrow AB=\begin{bmatrix}-15 & 22 & -9\\ -2 & 5 & -3\\ -32 & 18 & 6\end{bmatrix}$$

$$BA=\begin{bmatrix}-2 & 5 & -3\\ 9 & -7 & 0\end{bmatrix}\begin{bmatrix}3 & -1\\ 1 & 0\\ -2 & -4\end{bmatrix}=\begin{bmatrix}-2(3)+5(1)+(-3)(-2) & -2(-1)+5(0)+(-3)(-4)\\ 9(3)+(-7)(1)+0(-2) & 9(-1)+(-7)(0)+0(-4)\end{bmatrix}\Rightarrow BA=\begin{bmatrix}5 & 14\\ 20 & -9\end{bmatrix}$$

55. AB is undefined, we cannot multiply a 3×3 by a 2×3.

$$BA=\begin{bmatrix}-1 & 3 & -1\\ 7 & -7 & 1\end{bmatrix}\begin{bmatrix}1 & -1 & 0\\ 2 & -1 & 5\\ 6 & 1 & -4\end{bmatrix}=$$

$$\begin{bmatrix}-1(1)+3(2)+(-1)(6) & -1(-1)+3(-1)+(-1)(1) & -1(0)+3(5)+(-1)(-4)\\ 7(1)+(-7)(2)+1(6) & 7(-1)+-7(-1)+1(1) & 7(0)+(-7)(-5)+1(-4)\end{bmatrix}\Rightarrow BA=\begin{bmatrix}-1 & -3 & 19\\ -1 & 1 & -39\end{bmatrix}$$

57. $\begin{bmatrix}3 & -4 & 1\\5 & 0 & 2\end{bmatrix}\begin{bmatrix}-1\\4\\2\end{bmatrix}=\begin{bmatrix}3(1)-4(4)+1(2)\\5(-1)+0(4)+2(2)\end{bmatrix}=\begin{bmatrix}-3-16+2\\-5+0+4\end{bmatrix}=\begin{bmatrix}-17\\-1\end{bmatrix}$

59. $\begin{bmatrix}5 & 2\\-1 & 4\end{bmatrix}\begin{bmatrix}3 & -2\\1 & 0\end{bmatrix}=\begin{bmatrix}5(3)+2(1) & 5(-2)+2(0)\\-1(3)+4(1) & -1(-2)+4(0)\end{bmatrix}=\begin{bmatrix}15+2 & -10+0\\-3+4 & 2+0\end{bmatrix}=\begin{bmatrix}17 & -10\\1 & 2\end{bmatrix}$

61. $\begin{bmatrix}2 & 2 & -1\\3 & 0 & 1\end{bmatrix}\begin{bmatrix}0 & 2\\-1 & 4\\0 & 2\end{bmatrix}=\begin{bmatrix}2(0)+2(-1)+(-1)(0) & 2(2)+2(4)+(-1)(2)\\3(0)+0(-1)+1(0) & 3(2)+0(4)+1(2)\end{bmatrix}=\begin{bmatrix}0-2+0 & 4+8-2\\0+0+0 & 6+0+2\end{bmatrix}=\begin{bmatrix}-2 & 10\\0 & 8\end{bmatrix}$

63. $\begin{bmatrix}-2 & -3 & -4\\2 & -1 & 0\\4 & -2 & 3\end{bmatrix}\begin{bmatrix}0 & 1 & 4\\1 & 2 & -1\\3 & 2 & -2\end{bmatrix}=$

$\begin{bmatrix}-2(0)+(-3)(1)+(-4)(3) & -2(1)+(-3)(2)+(-4)(2) & -2(4)+(-3)(-1)+(-4)(-2)\\2(0)+(-1)(1)+0(3) & 2(1)+(-1)(2)+0(2) & 2(4)+(-1)(-1)+0(-2)\\4(0)+(-2)(1)+3(3) & 4(1)+(-2)(2)+3(2) & 4(4)+(-2)(-1)+3(-2)\end{bmatrix}=$

$\begin{bmatrix}0-3-12 & -2-6-8 & -8+3-8\\0-1+0 & 2-2+0 & 8+1+0\\0-2+9 & 4-4+6 & 16+2-6\end{bmatrix}=\begin{bmatrix}-15 & -16 & 3\\-1 & 0 & 9\\7 & 6 & 12\end{bmatrix}$

65. $\begin{bmatrix}-2 & 4 & 1\end{bmatrix}\begin{bmatrix}3 & -2 & 4\\2 & 1 & 0\\0 & -1 & 4\end{bmatrix}=\begin{bmatrix}-2(3)+4(2)+1(0) & -2(-2)+4(1)+1(-1) & -2(4)+4(0)+1(4)\end{bmatrix}=$
$\begin{bmatrix}-6+8+0 & 4+4-1 & -8+0+4\end{bmatrix}=[2 \quad 7 \quad -4]$

67. $\begin{bmatrix}p & q\\r & s\end{bmatrix}\begin{bmatrix}a & c\\b & d\end{bmatrix}=\begin{bmatrix}pa+qb & pc+qd\\ra+sb & rc+sd\end{bmatrix}$

69. $BA=\begin{bmatrix}5 & 1\\0 & -2\\3 & 7\end{bmatrix}\begin{bmatrix}4 & -2\\3 & 1\end{bmatrix}=\begin{bmatrix}5(4)+1(3) & 5(-2)+1(1)\\0(4)+(-2)(3) & 0(-2)+(-2)(1)\\3(4)+7(3) & 3(-2)+7(1)\end{bmatrix}=\begin{bmatrix}23 & -9\\-6 & -2\\33 & 1\end{bmatrix}$

71. $BC=\begin{bmatrix}5 & 1\\0 & -2\\3 & 7\end{bmatrix}\begin{bmatrix}-5 & 4 & 1\\0 & 3 & 6\end{bmatrix}=\begin{bmatrix}5(-5)+1(0) & 5(4)+1(3) & 5(1)+1(6)\\0(-5)-2(0) & 0(4)-2(3) & 0(1)-2(6)\\3(-5)+7(0) & 3(4)+7(3) & 3(1)+7(6)\end{bmatrix}=\begin{bmatrix}-25 & 23 & 11\\0 & -6 & -12\\-15 & 33 & 45\end{bmatrix}$

73. We cannot multiply A and B because we cannot multiply a (2×2) by a (3×2).

75. $A^2=\begin{bmatrix}4 & -2\\3 & 1\end{bmatrix}\begin{bmatrix}4 & -2\\3 & 1\end{bmatrix}=\begin{bmatrix}4(4)-2(3) & 4(-2)-2(1)\\3(4)+1(3) & 3(-2)+1(1)\end{bmatrix}=\begin{bmatrix}10 & -10\\15 & -5\end{bmatrix}$

77. $AB\neq BA; BC\neq CB; AC\neq CA.$. Matrix multiplication is not commutative.

79. (a) From the information the matrix is $\begin{bmatrix}50 & 100 & 30\\10 & 90 & 50\\60 & 120 & 40\end{bmatrix}$.

(b) The income matrix is $\begin{bmatrix} 12 \\ 10 \\ 15 \end{bmatrix}$.

(c) $\begin{bmatrix} 50 & 100 & 30 \\ 10 & 90 & 50 \\ 60 & 120 & 40 \end{bmatrix}\begin{bmatrix} 12 \\ 10 \\ 15 \end{bmatrix} = \begin{bmatrix} 50(12)+100(10)+30(15) \\ 10(12)+90(10)+50(15) \\ 60(12)+120(10)+40(15) \end{bmatrix} = \begin{bmatrix} 2050 \\ 1770 \\ 2520 \end{bmatrix}$

(d) $2050+1770+2520=\$6340$

81. (a) From R_2 the equation is $d_{n+1}=-0.05m_n+1.05d_n$ (hundredths). The deer population will grow at 5 %/year.

(b) Let $m_n=2000$ and $d_n=5000$ and multiply the given matrices:

$$\begin{bmatrix} m_{n+1} \\ d_{n+1} \end{bmatrix} = \begin{bmatrix} 0.51 & 0.4 \\ -0.05 & 1.05 \end{bmatrix}\begin{bmatrix} 2000 \\ 5000 \end{bmatrix} = \begin{bmatrix} 1020+2000 \\ -100+5250 \end{bmatrix} = \begin{bmatrix} 3020 \\ 5150 \end{bmatrix}$$

After 1 year there will be 3020 mountain lions and 515,000 deer.

Now multiply this 1 year matrix by the given matrix to find year 2.

$$\begin{bmatrix} m_{n+2} \\ d_{n+2} \end{bmatrix} = \begin{bmatrix} 0.51 & 0.4 \\ -0.05 & 1.05 \end{bmatrix}\begin{bmatrix} 3020 \\ 5150 \end{bmatrix} = \begin{bmatrix} 1540.2+2060 \\ -151+5407.5 \end{bmatrix} = \begin{bmatrix} 3600.2 \\ 5256.5 \end{bmatrix}$$

After 2 years there will be 3600 mountain lions and approximately 525,700 deer.

(c) Let $m_n.=4000$ and $d_n=5000$ and multiply the given matrices:

$$\begin{bmatrix} m_{n+1} \\ d_{n+1} \end{bmatrix} = \begin{bmatrix} 0.51 & 0.4 \\ -0.05 & 1.05 \end{bmatrix}\begin{bmatrix} 4000 \\ 5000 \end{bmatrix} = \begin{bmatrix} 2040+2000 \\ -200+5250 \end{bmatrix} = \begin{bmatrix} 4040 \\ 5050 \end{bmatrix}$$

After 1 year there will be 4040 mountain lions and 505,000 deer. Now multiply this 1 year matrix by the given matrix to find year 2. $\begin{bmatrix} m_{n+2} \\ d_{n+2} \end{bmatrix} = \begin{bmatrix} 0.51 & 0.4 \\ -0.05 & 1.05 \end{bmatrix}\begin{bmatrix} 4040 \\ 5050 \end{bmatrix} = \begin{bmatrix} 2060.4+2020 \\ -202+5302.5 \end{bmatrix} = \begin{bmatrix} 4080.4 \\ 5100.5 \end{bmatrix}$

After 2 years there will be 4080 mountain lions and approximately 510,050 deer. Mountain Lions: after $1\text{ year} = \dfrac{4040}{4000} = 1.01$; after 2 years $= \dfrac{4080}{4040} = 1.01$

Deer: after 1 year $= \dfrac{505{,}000}{500{,}000} = 1.01$; after 2 years $= \dfrac{510{,}050}{500{,}000} = 1.01$

83. $A+B=B+A$: $A+B=\begin{bmatrix} a_{11}+b_{11} & a_{12}+b_{12} \\ a_{21}+b_{21} & a_{22}+b_{22} \end{bmatrix} = B+A.$

85. $(AB)C=A(BC)$:

$$(AB)C = \begin{bmatrix} a_{11}b_{11}+a_{12}b_{21} & a_{11}b_{12}+a_{12}b_{22} \\ a_{21}b_{11}+a_{22}b_{21} & a_{21}b_{12}+a_{22}b_{22} \end{bmatrix}\begin{bmatrix} c_{11} & c_{12} \\ c_{21} & c_{22} \end{bmatrix} =$$

$$\begin{bmatrix} (a_{11}b_{11}c_{11}+a_{12}b_{21}c_{11})+(a_{11}b_{12}c_{21}+a_{12}b_{22}c_{21}) & (a_{11}b_{11}c_{12}+a_{12}b_{21}c_{12})+(a_{11}b_{12}c_{22}+a_{12}b_{22}c_{22}) \\ (a_{21}b_{11}c_{11}+a_{22}b_{21}c_{11})+(a_{21}b_{12}c_{21}+a_{22}b_{22}c_{21}) & (a_{21}b_{11}c_{12}+a_{22}b_{21}c_{12})+(a_{21}b_{12}c_{22}+a_{22}b_{22}c_{22}) \end{bmatrix}$$

$$A(BC) = \begin{bmatrix} a_{11} & a_{12} \\ a_{21} & a_{22} \end{bmatrix} \begin{bmatrix} b_{11}c_{11} + b_{12}c_{21} & b_{11}c_{12} + b_{12}c_{22} \\ b_{21}c_{11} + b_{22}c_{21} & b_{21}c_{12} + b_{22}c_{22} \end{bmatrix} =$$

$$\begin{bmatrix} (a_{11}b_{11}c_{11} + a_{11}b_{12}c_{21}) + (a_{12}b_{21}c_{11} + a_{12}b_{22}c_{21}) & (a_{11}b_{11}c_{12} + a_{11}b_{12}c_{22}) + (a_{12}b_{21}c_{12} + a_{12}b_{22}c_{22}) \\ (a_{21}b_{11}c_{11} + a_{21}b_{21}c_{21}) + (a_{22}b_{21}c_{11} + a_{22}b_{22}c_{21}) & (a_{21}b_{11}c_{12} + a_{21}b_{12}c_{22}) + (a_{22}b_{21}c_{12} + a_{22}b_{22}c_{22}) \end{bmatrix}$$

Therefore, $(\mathrm{AB})\mathrm{C} = \mathrm{A}(\mathrm{BC})$.

87. $c(A+B) = cA + cB$:

$$c(A+B) = \begin{bmatrix} c(a_{11} + b_{11}) & c(a_{12} + b_{12}) \\ c(a_{21} + b_{21}) & c(a_{22} + b_{22}) \end{bmatrix} = \begin{bmatrix} ca_{11} + cb_{11} & ca_{12} + cb_{12} \\ ca_{21} + cb_{21} & ca_{22} + cb_{22} \end{bmatrix} = cA + cB$$

89. $(cA)d = (cd)A$:

$$(cA)d = \begin{bmatrix} ca_{11} & ca_{12} \\ ca_{21} & ca_{22} \end{bmatrix} \bullet d = \begin{bmatrix} cda_{11} & cda_{12} \\ cda_{21} & cda_{22} \end{bmatrix} = (cd)A.$$

6.5: Determinants and Cramer's Rule

1. $\det \begin{bmatrix} -5 & 9 \\ 4 & -1 \end{bmatrix} = 5 - 36 = -31$

3. $\det \begin{bmatrix} -1 & -2 \\ 5 & 3 \end{bmatrix} = -3 - (-10) = 7$

5. $\det \begin{bmatrix} 9 & 3 \\ -3 & -1 \end{bmatrix} = -9 - (-9) = 0$

7. $\det \begin{bmatrix} 3 & 4 \\ 5 & -2 \end{bmatrix} = -6 - 20 = -26$

9. Use $A_{ij} = (-1)^{i+j} \cdot M_{ij}$, to find the cofactor of each element a.

   For a_{21}, find $A_{21} = (-1)^{2+1} \cdot \left(\det \begin{bmatrix} 0 & 1 \\ 2 & 1 \end{bmatrix} \right) = -1(0-2) = 2$

   For a_{22}, find $A_{22} = (-1)^{2+2} \cdot \left(\det \begin{bmatrix} -2 & 1 \\ 4 & 1 \end{bmatrix} \right) = 1(-2-4) = -6$

   For a_{23}, find $A_{23} = (-1)^{2+3} \cdot \left(\begin{bmatrix} -2 & 0 \\ 4 & 2 \end{bmatrix} \right) = -1(-4-0) = 4$

11. Use $A_{ij} = (-1)^{i+j} \cdot M_{ij}$, to find the cofactor of each element a.

   For a_{21}, find $A_{21} = (-1)^{2+1} \cdot \left(\det \begin{bmatrix} 2 & -1 \\ 4 & 1 \end{bmatrix} \right) = -1(2+4) = -6$

   For a_{22}, find $A_{22} = (-1)^{2+2} \cdot \left(\det \begin{bmatrix} 1 & -1 \\ 1 & 1 \end{bmatrix} \right) = 1(1-1) = 0$

For a_{23}, find $A_{23} = (-1)^{2+3} \cdot \left(\begin{bmatrix} 1 & 2 \\ -1 & -4 \end{bmatrix}\right) = -1(4+2) = -6$

13. Evaluate, expand by the second row. Therefore, det $= (-)(a_{21})(M_{21}) + (a_{22})(M_{22}) - (a_{23})(M_{23})$:

$$-2\left(\det\begin{bmatrix} -7 & 8 \\ 3 & 0 \end{bmatrix}\right) + 1\left(\det\begin{bmatrix} 4 & 8 \\ -6 & 0 \end{bmatrix}\right) - 3\left(\det\begin{bmatrix} 4 & -7 \\ -6 & 3 \end{bmatrix}\right) = -2(-24) + 1(48) - 3(-30) = 48 + 48 + 90 = 186.$$

15. Evaluate, expand by the first column. Therefore, det $= (a_{11})(M_{11}) - (a_{21})(M_{21}) + (a_{31})(M_{31})$:

$$1\left(\det\begin{bmatrix} 2 & -1 \\ 1 & 4 \end{bmatrix}\right) - (-1)\left(\det\begin{bmatrix} 2 & 0 \\ 1 & 4 \end{bmatrix}\right) + 0\left(\det\begin{bmatrix} 2 & 0 \\ 2 & -1 \end{bmatrix}\right) = 1(8+1) + 1(8-0) + 0 = 9 + 8 = 17.$$

17. Evaluate, expand by the first row. Therefore, det $= (a_{11})(M_{11}) - (a_{12})(M_{12}) + (a_{13})(M_{13})$:

$$10\left(\det\begin{bmatrix} 4 & 3 \\ 8 & 10 \end{bmatrix}\right) - 2\left(\det\begin{bmatrix} -1 & 3 \\ -3 & 10 \end{bmatrix}\right) + 1\left(\det\begin{bmatrix} -1 & 4 \\ -3 & 8 \end{bmatrix}\right) =$$

$10(40-24) - 2(-10+9) + 1(-8+12) = 160 + 2 + 4 = 166.$

19. Evaluate, expand by the second row will give a determinant of 0.

21. Evaluate, expand by the third column. Therefore, det $= (a_{13})(M_{13}) - (a_{23})(M_{23}) + (a_{33})(M_{33})$:

$$-1\left(\det\begin{bmatrix} 2 & 6 \\ -6 & -6 \end{bmatrix}\right) - 0\left(\det\begin{bmatrix} 3 & 3 \\ -6 & -6 \end{bmatrix}\right) + 2\left(\det\begin{bmatrix} 3 & 3 \\ 2 & 6 \end{bmatrix}\right) = -1(-12+36) - 0 + 2(18-6) = -24 + 24 = 0.$$

23. Evaluate, expand by the second row. Therefore, det $= (-)(a_{21})(M_{21}) + (a_{22})(M_{22}) - (a_{23})(M_{23})$:

$$-0.3\left(\det\begin{bmatrix} -0.8 & 0.6 \\ 4.1 & -2.8 \end{bmatrix}\right) + 0.9\left(\det\begin{bmatrix} 0.4 & -0.6 \\ 3.1 & -2.8 \end{bmatrix}\right) - 0.7\left(\det\begin{bmatrix} 0.4 & -0.8 \\ 3.1 & 4.1 \end{bmatrix}\right) =$$

$-0.3(2.24-2.46) + 0.9(-1.12-1.86) - 0.7(1.64+2.48) = .066 - 2.682 - 2.884 = -5.5.$

25. Evaluate, expand by the second row. Therefore, det $= (a_{31})(M_{31}) - (a_{32})(M_{32}) + (a_{33})(M_{33})$:

$$7(\det)\begin{bmatrix} -4 & 3 \\ 5 & -15 \end{bmatrix} + 9(\det)\begin{bmatrix} 17 & 3 \\ 11 & -15 \end{bmatrix} + 23(\det)\begin{bmatrix} 17 & -4 \\ 11 & 5 \end{bmatrix} =$$

$7(60-15) + 9(-255-33) + 23(85+44) = 690$

27. If $\det\begin{bmatrix} -0.5 & 2 \\ x & x \end{bmatrix} = 0$, then $-0.5x - 2x = 0 \Rightarrow -2.5x = 0 \Rightarrow x = 0$. The solution set is $\{0\}$.

29. If $\det\begin{bmatrix} 2x & 3 \\ 11 & x \end{bmatrix} = 6$, then $2x^2 - 11x = 6 \Rightarrow 2x^2 - 11x - 6 = 0 \Rightarrow x = (x-6)(2x+1) = 0 \Rightarrow x = 6, -\frac{1}{2}$. The solution set is $\left\{\frac{1}{2}, 6\right\}$.

31. Evaluate, expand by the second column. Therefore, det $= (a_{13})(M_{13}) - (a_{23})(M_{23}) + (a_{33})(M_{33})$:

$$0 - 1\left(\det\begin{bmatrix} 4 & 3 \\ -3 & x \end{bmatrix}\right) + (-1)\left(\det\begin{bmatrix} 4 & 3 \\ 2 & 0 \end{bmatrix}\right) = 5 - 1(4x+9) - 1(0-6) = 5 \Rightarrow -4x - 9 + 6 = 5 \Rightarrow$$

$-4x = 8 \Rightarrow x = -2$. The solution set is $\{-2\}$

33. Evaluate, expand by the second column. Therefore, $\det = (-)(a_{12})(M_{12}) + (a_{22})(M_{22}) - (a_{32})(M_{32})$:

$$-1\left(\det\begin{bmatrix}0 & x\\3 & 2\end{bmatrix}\right)+4\left(\det\begin{bmatrix}2x & -1\\3 & 2\end{bmatrix}\right)+0 = x \Rightarrow -1(0-3x)+4(4x+3) = x \Rightarrow$$

$3x+16x+12 = x \Rightarrow 18x = -12 \Rightarrow x = -\frac{12}{18}$. The solution set is $\left\{-\frac{2}{3}\right\}$.

35. Using the calculator, the determinant is 298.

37. Using the calculator, the determinant is -88.

39. With the given points find: $A = \frac{1}{2}\det\begin{bmatrix}0 & 0 & 1\\0 & 2 & 1\\1 & 4 & 1\end{bmatrix}$. Using the calculator the determinant is -2.

Therefore, $A = \left|\frac{1}{2}(-2)\right| \Rightarrow A = 1$ square units.

41. With the given points find $A = \frac{1}{2}\det\begin{bmatrix}2 & 5 & 1\\-1 & 3 & 1\\4 & 0 & 1\end{bmatrix}$. Using the calculator the determinant is 19. Therefore,

$A = \left|\frac{1}{2}(19)\right| \Rightarrow A = \frac{19}{2}$ square units.

43. With the given points find $A = \frac{1}{2}\det\begin{bmatrix}1 & 2 & 1\\4 & 3 & 1\\3 & 5 & 1\end{bmatrix}$. Using the calculator the determinant is 7. Therefore,

$A = \left|\frac{1}{2}(7)\right| \Rightarrow A = \frac{7}{2}$ square units.

45. If the three points form a triangle with no area ($D = 0$ using determinants), then the points must be collinear.

$D = \frac{1}{2}\det\begin{bmatrix}1 & -3 & 2\\3 & 11 & 1\\1 & 1 & 1\end{bmatrix} = 0$; The points are collinear.

47. If the three points form a triangle with no area ($D = 0$ using determinants), then the points must be collinear.

$D = \frac{1}{2}\det\begin{bmatrix}-2 & 4 & 2\\-5 & 4 & 3\\1 & 1 & 1\end{bmatrix} = 6 \neq 0$; The points are not collinear.

49. If the three points form a triangle with no area ($D = 0$ using determinants), then the points must be collinear.

$D = \frac{1}{2}\det\begin{bmatrix}4 & 6 & 12\\-1 & 0 & 4\\1 & 1 & 1\end{bmatrix} = 1 \neq 0$; The points are not collinear.

51. Since the second column is all zeros, by Determinant Theorem 1, the determinant is 0.

53. Use Determinant Theorem 6; multiplying column 1 by 2 and adding the result to column 3 yields the equivalent determinant, $\det\begin{bmatrix} 6 & 8 & 0 \\ -10 & 0 & 0 \\ 4 & 0 & 0 \end{bmatrix}$. Since column 3 has all zeros, the determinant is 0

(Determinant Theorem 1) and $\det\begin{bmatrix} 6 & 8 & -12 \\ -1 & 0 & 2 \\ 4 & 0 & -8 \end{bmatrix} = 0.$

55. Use Determinant Theorem 6; multiplying column 2 by 4 and adding the result to column 1, and multiplying column 2 by –4 and adding the result to column 3, yields the equivalent determinant $\det\begin{bmatrix} 0 & 1 & 0 \\ 2 & 0 & 1 \\ 8 & 2 & -4 \end{bmatrix}$. Now

expand this about row 1: $\det\begin{bmatrix} 0 & 1 & 0 \\ 2 & 0 & 1 \\ 8 & 2 & -4 \end{bmatrix} = 0(0-2)-1(-8-8)+0(4-0) = 16 \Rightarrow \det\begin{bmatrix} -4 & 1 & 4 \\ 2 & 0 & 1 \\ 0 & 2 & 4 \end{bmatrix} = 16.$

57. Find the determinants: $D = \det\begin{bmatrix} 1 & 1 \\ 2 & -1 \end{bmatrix} = -3,\ D_x = \det\begin{bmatrix} 4 & 1 \\ 2 & -1 \end{bmatrix} = -6,$ and $D_y = \det\begin{bmatrix} 1 & 4 \\ 2 & 2 \end{bmatrix} = -6.$

Then $x = \dfrac{Dx}{D} = \dfrac{-6}{-3} = 2$ and $y = \dfrac{D_y}{D} = \dfrac{-6}{-3} = 2$. The solution set is $\{(2,2)\}$.

59. Find the determinants: $D = \det\begin{bmatrix} 4 & 3 \\ 2 & 3 \end{bmatrix} = 6,\ D_x = \det\begin{bmatrix} -7 & 3 \\ -11 & 3 \end{bmatrix} = 12,$ and $D_y = \det\begin{bmatrix} 4 & -7 \\ 2 & -11 \end{bmatrix} = -30.$

Then $x = \dfrac{D_x}{D} = \dfrac{12}{2} = 2$ and $y = \dfrac{D_y}{D} = \dfrac{-30}{6} = -5$. The solution set is $\{(2,-5)\}$.

61. Find the determinant D: $D = \det\begin{bmatrix} 3 & 2 \\ 6 & 4 \end{bmatrix} = 0$; therefore, there are infinitely many solutions or no solutions.

Now use Row Echelon Method on the augmented matrix: $\left[\begin{array}{cc|c} 3 & 2 & 4 \\ 6 & 4 & 8 \end{array}\right],\ -2R_1 + R_2 \rightarrow \left[\begin{array}{cc|c} 3 & 2 & 4 \\ 0 & 0 & 0 \end{array}\right]$. Since

$0 = 0$, there are infinitely many solutions and using R_1, they are in the following

relationship: $3x + 2y = 4 \Rightarrow 3x = 4 - 2y \Rightarrow x = \dfrac{4-2y}{3}$. The solution is $\left(\dfrac{4-2y}{3}, y\right)$ or $\left(x, \dfrac{4-3x}{2}\right)$.

63. Find the determinants: $D = \det\begin{bmatrix} 2 & -3 \\ 1 & 5 \end{bmatrix} = 13,\ D_x = \det\begin{bmatrix} -5 & -3 \\ 17 & 5 \end{bmatrix} = 26,$ and $D = \det\begin{bmatrix} 2 & -5 \\ 1 & 17 \end{bmatrix} = 39.$

Then $x = \dfrac{D_x}{D} = \dfrac{26}{13} = 2$ and $y = \dfrac{D_y}{D} = \dfrac{39}{13} = 3$. The solution set is $\{(2,3)\}$.

65. Using your calculator, find the following determinants:

$$D = \det\begin{bmatrix} 4 & -1 & 3 \\ 3 & 1 & 1 \\ 2 & -1 & 4 \end{bmatrix} = 15,\ D_x = \det\begin{bmatrix} -3 & -1 & 3 \\ 0 & 1 & 1 \\ 0 & -1 & 4 \end{bmatrix} = -15,\ D_y = \det\begin{bmatrix} 4 & -3 & 3 \\ 3 & 0 & 1 \\ 2 & 0 & 4 \end{bmatrix} = 30, \text{ and}$$

$$D_z = \det\begin{bmatrix} 4 & -1 & -3 \\ 3 & 1 & 0 \\ 2 & -1 & 0 \end{bmatrix} = 15. \text{ Then } x = \frac{D_x}{D} = \frac{-15}{15} = -1,\ y = \frac{D_y}{D} = \frac{30}{15} = 2, \text{ and}$$

$z = \frac{D_z}{D} = \frac{15}{15} = 1$. The solution set is $\{(-1,2,1)\}$.

67. Using your calculator, find the following determinants:

$$D = \det\begin{bmatrix} 2 & -1 & 4 \\ 3 & 2 & -1 \\ 1 & 4 & 2 \end{bmatrix} = 63,\ D_x = \det\begin{bmatrix} -2 & -1 & 4 \\ -3 & 2 & -1 \\ 17 & 4 & 2 \end{bmatrix} = -189,\ D_y = \det\begin{bmatrix} 2 & -2 & 4 \\ 3 & -3 & -1 \\ 1 & 17 & 2 \end{bmatrix} = 252,$$

$$D_z = \det\begin{bmatrix} 2 & -1 & -2 \\ 3 & 2 & -3 \\ 1 & 4 & 17 \end{bmatrix} = 126. \text{ Then } x = \frac{D_x}{D} = \frac{-189}{63} = -3,\ y = \frac{D_y}{D} = \frac{252}{63} = 4, \text{ and } z = \frac{Dz}{D} = \frac{126}{63} = 2.$$

The solution set is $\{(-3,4,2)\}$.

69. Using your calculator, find the following determinants:

$$D = \det\begin{bmatrix} 5 & -1 & 0 \\ 3 & 0 & 2 \\ 0 & 4 & 3 \end{bmatrix} = -31,\ D_x = \det\begin{bmatrix} -4 & -1 & 0 \\ 4 & 0 & 2 \\ 22 & 4 & 3 \end{bmatrix} = 0,\ D_y = \det\begin{bmatrix} 5 & -4 & 0 \\ 3 & 4 & 2 \\ 0 & 22 & 3 \end{bmatrix} = -124, \text{ and}$$

$$D_z = \det\begin{bmatrix} 5 & -1 & -4 \\ 3 & 0 & 4 \\ 0 & 4 & 22 \end{bmatrix} = -62. \text{ Then } x = \frac{D_x}{D} = \frac{0}{-31} = 0,\ y = \frac{D_y}{D} = \frac{-124}{-31} = 4, \text{ and } z = \frac{Dz}{D} = \frac{-62}{-31} = 2.$$

The solution set is $\{(0,4,2)\}$.

71. Using your calculator, find the following determinant: $D = \det\begin{bmatrix} 2 & -1 & 3 \\ -2 & 1 & -3 \\ 5 & -1 & 1 \end{bmatrix} = 0$; therefore, there are infinitely many solutions or no solutions. If we add [equation 1] + [equation 2] the result is $0 \neq 3$; therefore, no solution or $\varnothing$.

73. Using your calculator, find the following determinants:

$$D = \det\begin{bmatrix} 3 & 2 & 0 & -1 \\ 2 & 0 & 1 & 2 \\ 1 & 2 & -1 & 0 \\ 2 & -1 & 1 & 1 \end{bmatrix} = -9;\ D_x = \det\begin{bmatrix} 0 & 2 & 0 & -1 \\ 5 & 0 & 1 & 2 \\ -2 & 2 & -1 & 0 \\ 2 & -1 & 1 & 1 \end{bmatrix} = 9$$

$$D_y = \det\begin{bmatrix} 3 & 0 & 0 & -1 \\ 2 & 5 & 1 & 2 \\ 1 & -2 & -1 & 0 \\ 2 & 2 & 1 & 1 \end{bmatrix} = -18;\ D_z = \det\begin{bmatrix} 3 & 2 & 0 & -1 \\ 2 & 0 & 5 & 2 \\ 1 & 2 & -2 & 0 \\ 2 & -1 & 2 & 1 \end{bmatrix} = -45;\ D_w = \det\begin{bmatrix} 3 & 2 & 0 & 0 \\ 2 & 0 & 1 & 5 \\ 1 & 2 & -1 & -2 \\ 2 & -1 & 1 & 2 \end{bmatrix} = -9$$

Then $x = \frac{D_x}{D} = \frac{9}{-9} = -1;\ y = \frac{D_y}{D} = \frac{-18}{-9} = 2;\ z = \frac{D_z}{D} = \frac{-45}{-9} = 5;\ w = \frac{D_w}{D} = \frac{-9}{-9} = 1.$

The solution set is $\{(-1, 2, 5, 1)\}$.

75. If $D = 0$, Cramer's rule cannot be applied because there is no unique solution. There are either no solutions or infinitely many solutions.

6.6: Solution of Linear Systems by Matrix Inverses

1. Yes, $AB = BA = \begin{bmatrix} 1 & 0 \\ 0 & 1 \end{bmatrix}$.

$$\begin{bmatrix} 5 & 7 \\ 2 & 3 \end{bmatrix}\begin{bmatrix} 3 & -7 \\ -2 & 5 \end{bmatrix} = \begin{bmatrix} 15-14 & -35+35 \\ 6-6 & -14+15 \end{bmatrix} = \begin{bmatrix} 1 & 0 \\ 0 & 1 \end{bmatrix};$$

$$\begin{bmatrix} 3 & -7 \\ -2 & 5 \end{bmatrix}\begin{bmatrix} 5 & 7 \\ 2 & 3 \end{bmatrix} = \begin{bmatrix} 15-14 & 21-21 \\ -10+10 & -14+15 \end{bmatrix} = \begin{bmatrix} 1 & 0 \\ 0 & 1 \end{bmatrix}$$

3. No, $AB \neq \begin{bmatrix} 1 & 0 \\ 0 & 1 \end{bmatrix}$.

$$\begin{bmatrix} -1 & 2 \\ 3 & -5 \end{bmatrix}\begin{bmatrix} -5 & -2 \\ -3 & -1 \end{bmatrix} = \begin{bmatrix} 5-6 & 2-2 \\ -15+15 & -6+5 \end{bmatrix} = \begin{bmatrix} -1 & 0 \\ 0 & -1 \end{bmatrix}$$

5. No, $AB \neq \begin{bmatrix} 1 & 0 & 0 \\ 0 & 1 & 0 \\ 0 & 0 & 1 \end{bmatrix}$. Use your calculator to multiply the matrices. $\begin{bmatrix} 0 & 1 & 0 \\ 0 & 0 & -2 \\ 1 & -1 & 0 \end{bmatrix}\begin{bmatrix} 1 & 0 & 1 \\ 1 & 0 & 0 \\ 0 & -1 & 0 \end{bmatrix} = \begin{bmatrix} 1 & 0 & 0 \\ 0 & 2 & 0 \\ 0 & 0 & 1 \end{bmatrix}$

7. Yes. Using the calculator, multiply the matrices. $AB = \begin{bmatrix} 1 & 0 & 0 \\ 0 & 1 & 0 \\ 0 & 0 & 1 \end{bmatrix} = BA$

9. Using $A^{-1} = \frac{1}{\det A}\begin{bmatrix} d & -b \\ -c & a \end{bmatrix}$ to find the inverse yields $A^{-1} = \frac{1}{15-14}\begin{bmatrix} 5 & -7 \\ -2 & 3 \end{bmatrix} = \begin{bmatrix} 5 & -7 \\ -2 & 3 \end{bmatrix}$.

11. Using $A^{-1} = \frac{1}{\det A}\begin{bmatrix} d & -b \\ -c & a \end{bmatrix}$ to find the inverse yields

$$A^{-1} = \frac{1}{-4+6}\begin{bmatrix} 4 & 2 \\ -3 & -1 \end{bmatrix} = \left(\frac{1}{2}\right)\begin{bmatrix} 4 & 2 \\ -3 & -1 \end{bmatrix} = \begin{bmatrix} 2 & 1 \\ -\frac{3}{2} & -\frac{1}{2} \end{bmatrix}.$$

13. The $\det A = -12-(-12) = 0$; therefore, No inverse.

15. Using $A^{-1} = \frac{1}{\det A}\begin{bmatrix} d & -b \\ -c & a \end{bmatrix}$ to find the inverse yields

$$A^{-1} = \frac{1}{0.06-0.1}\begin{bmatrix} 0.1 & -0.2 \\ -0.5 & 0.6 \end{bmatrix} = (-25)\begin{bmatrix} 0.1 & -0.2 \\ -0.5 & 0.6 \end{bmatrix} = \begin{bmatrix} -2.5 & 5 \\ 12.5 & -15 \end{bmatrix}.$$

17. $$A\,|\,I_3 = \left[\begin{array}{ccc|ccc} 0 & 0 & 1 & 1 & 0 & 0 \\ 1 & 0 & 0 & 0 & 1 & 0 \\ 0 & 1 & 0 & 0 & 0 & 1 \end{array}\right] \begin{array}{l} R_2 \to \\ R_3 \to \\ R_1 \to \end{array} \left[\begin{array}{ccc|ccc} 1 & 0 & 0 & 0 & 1 & 0 \\ 0 & 1 & 0 & 0 & 0 & 1 \\ 0 & 0 & 1 & 1 & 0 & 0 \end{array}\right]; \; A^{-1} = \begin{bmatrix} 0 & 1 & 0 \\ 0 & 0 & 1 \\ 1 & 0 & 0 \end{bmatrix}$$

19. $$A\,|\,I_3 = \left[\begin{array}{ccc|ccc} 1 & 0 & 1 & 1 & 0 & 0 \\ 2 & 1 & 3 & 0 & 1 & 0 \\ -1 & 1 & 1 & 0 & 0 & 1 \end{array}\right] \begin{array}{l} \\ R_2 - 2R_1 \to \\ R_3 + R_1 \to \end{array} \left[\begin{array}{ccc|ccc} 1 & 0 & 1 & 1 & 0 & 0 \\ 0 & 1 & 1 & -2 & 1 & 0 \\ 0 & 1 & 2 & 1 & 0 & 1 \end{array}\right] \begin{array}{l} \\ \\ R_3 - R_2 \to \end{array} \left[\begin{array}{ccc|ccc} 1 & 0 & 0 & 1 & 0 & 0 \\ 0 & 1 & 1 & -2 & 1 & 0 \\ 0 & 0 & 1 & 3 & -1 & 1 \end{array}\right]$$

$$\begin{array}{l} R_1 - R_3 \to \\ R_2 - R_3 \to \\ \end{array} \left[\begin{array}{ccc|ccc} 1 & 0 & 0 & -2 & 1 & -1 \\ 0 & 1 & 0 & -5 & 2 & -1 \\ 0 & 0 & 1 & 3 & -1 & 1 \end{array}\right]; \; A^{-1} = \begin{bmatrix} -2 & 1 & -1 \\ -5 & 2 & -1 \\ 3 & -1 & 1 \end{bmatrix}$$

21. The inverse will not exist if its determinant is equal to 0.

23. Solve using the calculator. Since $\det A = -1$, the inverse exists and $A^{-1} = \begin{bmatrix} 1 & 0 & 0 \\ 0 & -1 & 0 \\ -1 & 0 & 1 \end{bmatrix}$.

25. Solve using the calculator. Since $\det A = -1$, the inverse exists and $A^{-1} = \begin{bmatrix} 15 & 4 & -5 \\ -12 & -3 & 4 \\ -4 & -1 & 1 \end{bmatrix}$.

27. Solve using the calculator. Since $\det A = 0.036$, the inverse exists and $A^{-1} = \begin{bmatrix} -\frac{10}{3} & \frac{5}{9} & -\frac{10}{9} \\ \frac{20}{3} & \frac{5}{9} & \frac{80}{9} \\ -5 & \frac{5}{6} & -\frac{20}{3} \end{bmatrix}$.

29. Solve using the calculator. Since $\det A = 0$, there is No inverse.

31. Solve using the calculator. Since $\det A \approx 9.207$, the inverse exists and

$$A^{-1} = \begin{bmatrix} 0.0543058761 & -0.054358761 \\ 1.846399787 & 0.153600213 \end{bmatrix}.$$

33. Solve using the calculator. Since $\det A = 0.001$ the inverse exists and $A^{-1} = \begin{bmatrix} -20 & 10 & -10 \\ -50 & 20 & -10 \\ 30 & -10 & 10 \end{bmatrix}$.

35. Using the matrix inverse method, put the system into the proper matrix form: $\begin{bmatrix} 2 & -1 \\ 3 & 1 \end{bmatrix}\begin{bmatrix} x \\ y \end{bmatrix} = \begin{bmatrix} -8 \\ -2 \end{bmatrix} \Rightarrow$

$\begin{bmatrix} x \\ y \end{bmatrix} = \begin{bmatrix} 2 & -1 \\ 3 & 2 \end{bmatrix}^{-1}\begin{bmatrix} -8 \\ -2 \end{bmatrix} \Rightarrow \frac{1}{5}\begin{bmatrix} 1 & 1 \\ -3 & 2 \end{bmatrix}\begin{bmatrix} -8 \\ -2 \end{bmatrix} \Rightarrow \begin{bmatrix} \frac{1}{5} & \frac{1}{5} \\ -\frac{3}{5} & \frac{2}{5} \end{bmatrix}\begin{bmatrix} -8 \\ -2 \end{bmatrix} \Rightarrow \begin{bmatrix} -\frac{8}{5} - \frac{2}{5} \\ \frac{24}{5} - \frac{4}{5} \end{bmatrix} = \begin{bmatrix} -2 \\ 4 \end{bmatrix}$. The solution set is $\{(-2,4)\}$.

37. Using the matrix inverse method, put the system into the proper matrix form: $\begin{bmatrix} 2 & 3 \\ 3 & 4 \end{bmatrix}\begin{bmatrix} x \\ y \end{bmatrix} = \begin{bmatrix} -10 \\ -12 \end{bmatrix} \Rightarrow$

$\begin{bmatrix} x \\ y \end{bmatrix} = \begin{bmatrix} 2 & 3 \\ 3 & 3 \end{bmatrix}^{-1}\begin{bmatrix} -10 \\ -12 \end{bmatrix} \Rightarrow -1\begin{bmatrix} 4 & -3 \\ -3 & 2 \end{bmatrix}\begin{bmatrix} -10 \\ -12 \end{bmatrix} \Rightarrow \begin{bmatrix} -4 & 3 \\ 3 & -2 \end{bmatrix}\begin{bmatrix} -10 \\ -12 \end{bmatrix} \Rightarrow \begin{bmatrix} 40-36 \\ -30+24 \end{bmatrix} = \begin{bmatrix} 4 \\ -6 \end{bmatrix}$.

The solution set is $\{(4,-6)\}$.

39. Using the matrix inverse method, put the system into the proper matrix form: $\begin{bmatrix} 2 & -5 \\ 2 & -5 \end{bmatrix}\begin{bmatrix} x \\ y \end{bmatrix} = \begin{bmatrix} 10 \\ 15 \end{bmatrix} \Rightarrow$

$\begin{bmatrix} x \\ y \end{bmatrix} = \begin{bmatrix} 2 & -5 \\ 2 & -5 \end{bmatrix}^{-1}\begin{bmatrix} 10 \\ 15 \end{bmatrix} \Rightarrow$ The matrix $\begin{bmatrix} 2 & -5 \\ 2 & -5 \end{bmatrix}^{-1}$ does not exist so there is no solution to this system.

41. Solve on the calculator using the matrix inverse method: $\begin{bmatrix} x \\ y \\ z \end{bmatrix} = \begin{bmatrix} 2 & 0 & 4 \\ 3 & 1 & 5 \\ -1 & 1 & -2 \end{bmatrix}^{-1} \bullet \begin{bmatrix} 14 \\ 19 \\ -7 \end{bmatrix} = \begin{bmatrix} 3 \\ 0 \\ 2 \end{bmatrix}$; therefore, the solution set is $\{(3,0,2)\}$.

43. Solve on the calculator using the matrix inverse method: $\begin{bmatrix} x \\ y \\ z \end{bmatrix} = \begin{bmatrix} 1 & 3 & 1 \\ 1 & -2 & 3 \\ 2 & -3 & -1 \end{bmatrix}^{-1} \bullet \begin{bmatrix} 2 \\ -3 \\ 34 \end{bmatrix} = \begin{bmatrix} 12 \\ -\frac{15}{11} \\ -\frac{65}{11} \end{bmatrix}$; therefore, the solution set is $\left\{\left(12, -\frac{15}{11}, -\frac{65}{11}\right)\right\}$.

45. Solve on the calculator using the matrix inverse method: $\begin{bmatrix} x \\ y \\ z \\ w \end{bmatrix} = \begin{bmatrix} 1 & 3 & -2 & -1 \\ 4 & 1 & 1 & 2 \\ -3 & -1 & 1 & -1 \\ 1 & -1 & -3 & -2 \end{bmatrix}^{-1} \bullet \begin{bmatrix} 9 \\ 2 \\ -5 \\ 2 \end{bmatrix} = \begin{bmatrix} 0 \\ 2 \\ -2 \\ 1 \end{bmatrix}$;

therefore, the solution set is $\{(0,2,-2,1)\}$.

47. Solve on the calculator using the matrix inverse method: $\begin{bmatrix} x \\ y \end{bmatrix} = \begin{bmatrix} 1 & -\sqrt{2} \\ 0.75 & 1 \end{bmatrix}^{-1} \bullet \begin{bmatrix} 2.6 \\ -7 \end{bmatrix} = \begin{bmatrix} -3.542308934 \\ -4.343268299 \end{bmatrix}$;

therefore the solution set is $\{(-3.542308934, -4.343268299)\}$.

49. Solve on the calculator using the matrix inverse method: $\begin{bmatrix} x \\ y \\ z \end{bmatrix} = \begin{bmatrix} \pi & e & \sqrt{2} \\ e & \pi & \sqrt{2} \\ \sqrt{2} & e & \pi \end{bmatrix}^{-1} \bullet \begin{bmatrix} 1 \\ 2 \\ 3 \end{bmatrix} = \begin{bmatrix} -.9704156969 \\ 1.391914631 \\ .1874077432 \end{bmatrix}$;

therefore, the solution set is $\{(-0.9704156959, 1.391914631, 0.1874077432)\}$.

51. If $P(x) = ax^3 + bx^2 + cx + d$, then the ordered pair $(-1, 14)$ yields the equation $-a + b - c + d = 14$, the ordered pair $(1.5, 1.5)$ yields the equation $3.375a + 2.25b + 1.5c + d = 1.5$, the ordered pair $(2, -1)$ yields the equation $8a + 4b + 2c + d = -1$, and the ordered pair $(3, -18)$ yields the equation $27a + 9b + 3c + d = -18$. Now solve on the calculator using the matrix inverse method:

$\begin{bmatrix} a \\ b \\ c \\ d \end{bmatrix} = \begin{bmatrix} -1 & 1 & -1 & 1 \\ 3.375 & 2.25 & 1.5 & 1 \\ 8 & 4 & 2 & 1 \\ 27 & 9 & 3 & 1 \end{bmatrix}^{-1} \bullet \begin{bmatrix} 14 \\ 1.5 \\ -1 \\ -18 \end{bmatrix} = \begin{bmatrix} -2 \\ 5 \\ -4 \\ 3 \end{bmatrix}$; therefore, the equation is $P(x) = -2x^3 + 5x^2 - 4x + 3$.

53. If $P(x) = ax^4 + bx^3 + cx^2 + dx + e$, then the ordered pair (–2, 13) yields the equation $16a - 8b + 4c - 2d + e = 13$, the ordered pair (–1, 2) yields the equation $a - b + c - d + e = 2$, the ordered pair (0, –1) yields the equation $e = -1$, the ordered pair (1, 4) yields the equation $a + b + c + d + e = 4$, and the ordered pair (2, 41) yields the equation $16a + 8b + 4c + 2d + e = 41$. Now solve on the calculator using the matrix inverse method: $\begin{bmatrix} a \\ b \\ c \\ d \\ e \end{bmatrix} = \begin{bmatrix} 16 & -8 & 4 & -2 & 1 \\ 1 & -1 & 1 & -1 & 1 \\ 0 & 0 & 0 & 0 & 1 \\ 1 & 1 & 1 & 1 & 1 \\ 16 & 8 & 4 & 2 & 1 \end{bmatrix}^{-1} \bullet \begin{bmatrix} 13 \\ 2 \\ -1 \\ 4 \\ 41 \end{bmatrix} = \begin{bmatrix} 1 \\ 2 \\ 3 \\ -1 \\ -1 \end{bmatrix}$; therefore, the equation is $P(x) = x^4 + 2x^3 + 3x^2 - x - 1$.

55. Let x be the number of CDs of type A, let y be the number of CDs of type B, and let z be the number of CDs of type C. Then

$$\begin{aligned} 2x + 3y + 4z &= 120.91 \\ x + 4y &= 62.95 \\ 2x + y + 3z &= 79.94 \end{aligned}$$

Now solve on the calculator using the matrix inverse method: $\begin{bmatrix} x \\ y \\ z \end{bmatrix} = \begin{bmatrix} 2 & 3 & 4 \\ 1 & 4 & 0 \\ 2 & 1 & 3 \end{bmatrix}^{-1} \bullet \begin{bmatrix} 120.91 \\ 62.95 \\ 79.94 \end{bmatrix} = \begin{bmatrix} 10.99 \\ 12.99 \\ 14.99 \end{bmatrix}$.

The cost of a Type A CD is \$10.99, the cost a Type B CD is \$12.99, and the cost of a Type C CD is \$14.99.

57. (a) Using the model $T = aA + bI + c$ and the data from the table, the equations are

$$\begin{aligned} 113a + 308b + c &= 10{,}170 \\ 133a + 622b + c &= 15{,}305 \\ 155a + 1937b + c &= 21{,}289 \end{aligned}$$

(b) Solve on the calculator using the matrix inverse method:

$$\begin{bmatrix} a \\ b \\ c \end{bmatrix} = \begin{bmatrix} 113 & 308 & 1 \\ 133 & 622 & 1 \\ 155 & 1937 & 1 \end{bmatrix}^{-1} \bullet \begin{bmatrix} 10{,}170 \\ 15{,}305 \\ 21{,}289 \end{bmatrix} = \begin{bmatrix} 251.3175021 \\ 0.3460189769 \\ -0.18335.45158 \end{bmatrix}$$ The formula is $T \approx 251A + 0.346I - 18{,}300.$

(c) $T \approx 251(118) + 0.346(311) - 18{,}300 \Rightarrow T \approx 11{,}426.$ This is quite close to the actual value of 11,314.

59. (a) Using the model $P = a + bS + cC$ and the data from the table, the equations are

$$a + 1500b + 8c = 122$$
$$a + 2000b + 5c = 130$$
$$a + 2200b + 10c = 158$$

(b) Solve on the calculator using the matrix inverse method $\begin{bmatrix} a \\ b \\ c \end{bmatrix} = \begin{bmatrix} 1 & 1500 & 8 \\ 1 & 2000 & 5 \\ 1 & 2200 & 10 \end{bmatrix}^{-1} \bullet \begin{bmatrix} 122 \\ 130 \\ 158 \end{bmatrix} = \begin{bmatrix} 30 \\ 0.04 \\ 4 \end{bmatrix}.$

The formula is $G \approx 30 + 0.04S + 4C.$

The selling price is $p = 30 + 0.04(1800) + 4(7) \Rightarrow p = 130$ or \$130,000.

61. If $A = (A^{-1})^{-1}$, then $A = \begin{bmatrix} 5 & -9 \\ -1 & 2 \end{bmatrix}^{-1} = \begin{bmatrix} 2 & 9 \\ 1 & 5 \end{bmatrix}.$

63. If $A = (A^{-1})^{-1}$, then $A = \begin{bmatrix} \frac{2}{3} & -\frac{1}{3} & 0 \\ \frac{1}{3} & -\frac{5}{3} & 1 \\ \frac{1}{3} & -\frac{1}{3} & 0 \end{bmatrix}^{-1} = \begin{bmatrix} 1 & 0 & 1 \\ -1 & 0 & 2 \\ -2 & 1 & 3 \end{bmatrix}.$

65. This is a shortened method. First form the augmented matrix: $[A/I] = \left[\begin{array}{ccc|ccc} a & 0 & 0 & 1 & 0 & 0 \\ 0 & b & 0 & 0 & 1 & 0 \\ 0 & 0 & c & 0 & 0 & 1 \end{array}\right].$ Since a, b, and c are all non-zero $\frac{1}{a}$, $\frac{1}{b}$, and $\frac{1}{c}$ all exist. Use these values and solve for the inverse.

$$\left.\begin{array}{l} \frac{1}{a}R_1 \\ \frac{1}{b}R_2 \\ \frac{1}{c}R_3 \end{array}\right\} \rightarrow \left[\begin{array}{ccc|ccc} 1 & 0 & 0 & \frac{1}{a} & 0 & 0 \\ 0 & 1 & 0 & 0 & \frac{1}{b} & 0 \\ 0 & 0 & 1 & 0 & 0 & \frac{1}{c} \end{array}\right].$$ Therefore, $A^{-1} = \begin{bmatrix} \frac{1}{a} & 0 & 0 \\ 0 & \frac{1}{b} & 0 \\ 0 & 0 & \frac{1}{c} \end{bmatrix}.$

Reviewing Basic Concepts (Sections 6.4—6.6)

1. $A - B = \begin{bmatrix} -5-0 & 4+2 \\ 2-3 & -1+4 \end{bmatrix} = \begin{bmatrix} -5 & 6 \\ -1 & 3 \end{bmatrix}.$

2. $-3B = \begin{bmatrix} -3(0) & -3(-2) \\ -3(3) & -3(-4) \end{bmatrix} = \begin{bmatrix} 0 & 6 \\ -9 & 12 \end{bmatrix}$

3. $A^2 = \begin{bmatrix} -5 & 4 \\ 2 & -1 \end{bmatrix}\begin{bmatrix} -5 & 4 \\ 2 & -1 \end{bmatrix} = \begin{bmatrix} -5(-5)+4(2) & -5(4)+4(-1) \\ 2(-5)-1(2) & 2(4)-1(-1) \end{bmatrix} = \begin{bmatrix} 33 & -24 \\ -12 & 9 \end{bmatrix}.$

4. Using the calculator, $CD=\begin{bmatrix}1 & 3 & -3\\0 & 6 & 0\\4 & 2 & 2\end{bmatrix}$.

5. $\det A=-5(-1)-2(4)=5-8=-3$.

6. Evaluate, expand by the first column. Therefore, $=(a_{11})(M_{11})-(a_{21})(M_{21})+(a_{31})(M_{31})$:

$$2\left(\det\begin{bmatrix}1 & 0\\-1 & 4\end{bmatrix}\right)-(-2)\left(\det\begin{bmatrix}-3 & 1\\-1 & 4\end{bmatrix}\right)+0\left(\det\begin{bmatrix}-3 & 1\\1 & 0\end{bmatrix}\right)=2(4+0)+2(-12+1)+0(0-1)=$$

$2(4)+2(-11)=8-22=-14$.

7. Using $A^{-1}=\dfrac{1}{\det A}\begin{bmatrix}d & -b\\-c & a\end{bmatrix}$ to find the inverse yields $A^{-1}=\dfrac{1}{5-8}\begin{bmatrix}-1 & -4\\-2 & -5\end{bmatrix}=\begin{bmatrix}\frac{1}{3} & \frac{4}{3}\\\frac{2}{3} & \frac{5}{3}\end{bmatrix}$.

8. Solve using the calculator, $C^{-1}=\begin{bmatrix}-\frac{2}{7} & -\frac{11}{4} & \frac{1}{14}\\-\frac{4}{7} & -\frac{4}{7} & \frac{1}{7}\\-\frac{1}{7} & -\frac{1}{7} & \frac{2}{7}\end{bmatrix}$.

9. The equations are $\dfrac{\sqrt{3}}{2}(w_1+w_2)=100\Rightarrow\dfrac{\sqrt{3}}{2}w_1+\dfrac{\sqrt{3}}{2}w_2=100$ and $w_1-w_2=0$. Now we find the determinants: $D=\det\begin{bmatrix}\frac{\sqrt{3}}{2} & \frac{\sqrt{3}}{2}\\1 & -1\end{bmatrix}=-\sqrt{3}$, $D_{w_1}=\det\begin{bmatrix}100 & \frac{\sqrt{3}}{2}\\0 & -1\end{bmatrix}=-100$, and $D_{w_2}=\det\begin{bmatrix}\frac{\sqrt{3}}{2} & 100\\1 & 0\end{bmatrix}=-100$.

Then $w_1=\dfrac{D_{w1}}{D}=\dfrac{-100}{-\sqrt{3}}=\dfrac{100}{\sqrt{3}}\cdot\dfrac{\sqrt{3}}{\sqrt{3}}=\dfrac{100\sqrt{3}}{3}\approx 57.7$ and $w_2=\dfrac{D_{w1}}{D}=\dfrac{-100}{-\sqrt{3}}=\dfrac{100}{\sqrt{3}}\cdot\dfrac{\sqrt{3}}{\sqrt{3}}=\dfrac{100\sqrt{3}}{3}\approx 57.7$.

Both w_1 and w_2 are approximately 57.7 pounds.

10. $\begin{bmatrix}x\\y\\z\end{bmatrix}=\begin{bmatrix}2 & 1 & 2\\0 & 1 & 2\\1 & -2 & 2\end{bmatrix}^{-1}\bullet\begin{bmatrix}10\\4\\1\end{bmatrix}=\begin{bmatrix}3\\2\\1\end{bmatrix}$. The solution is $(3,2,1)$.

6.7: Systems of Inequalities and Linear Programming

1. See Figure 1.
3. See Figure 3.

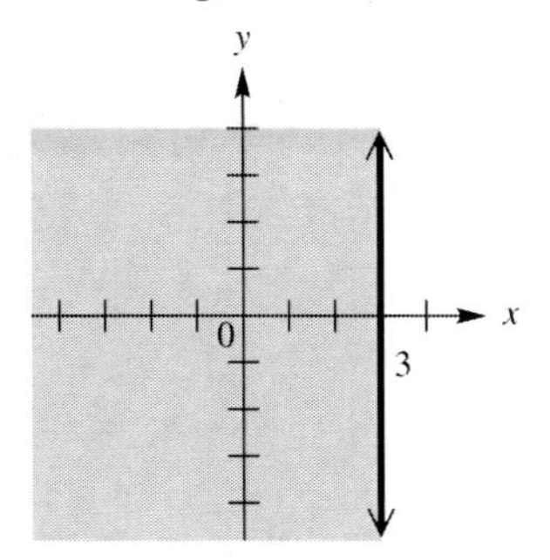

Figure 1

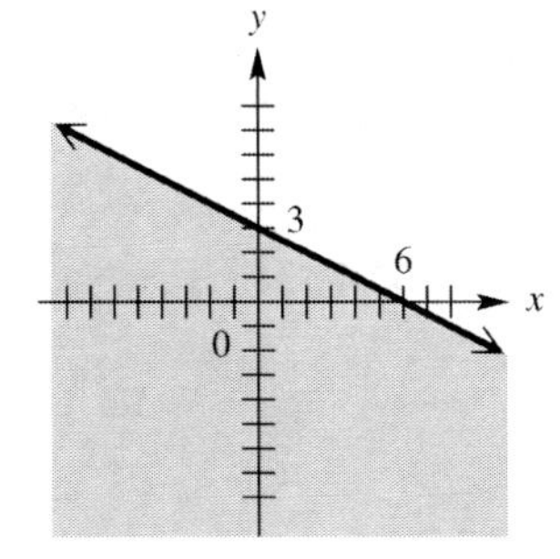

Figure 3

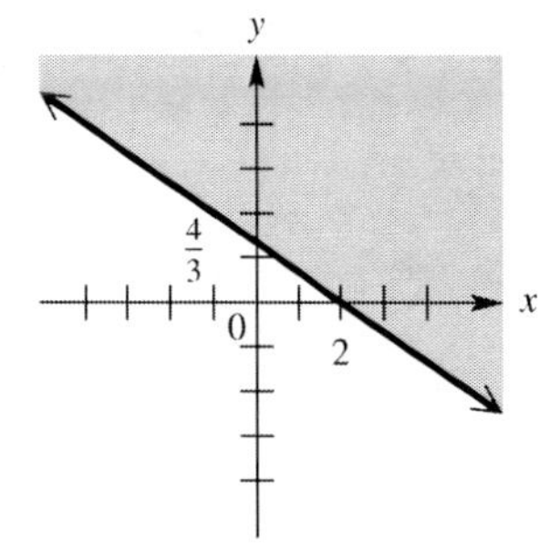

Figure 5

5. See Figure 5.
7. See Figure 7.
9. See Figure 9

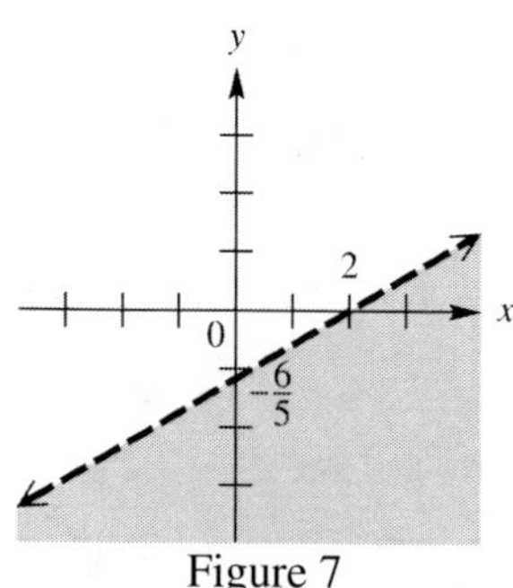

Figure 7

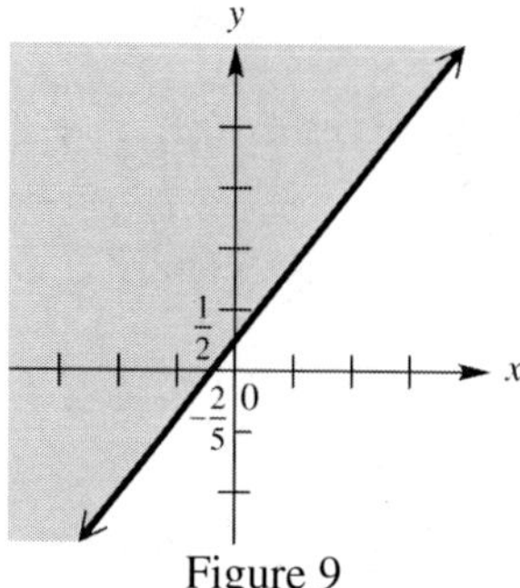

Figure 9

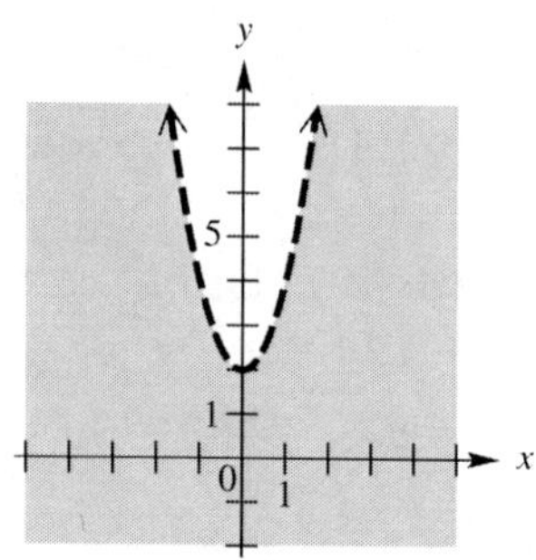

Figure 11

11. See Figure 11.
13. See Figure 13.
15. See Figure 15

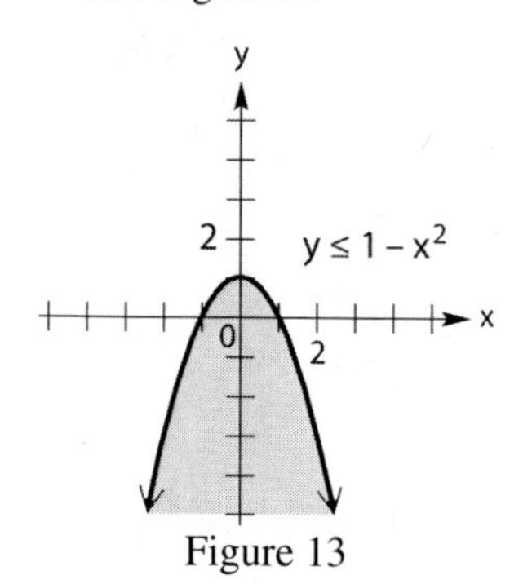

Figure 13

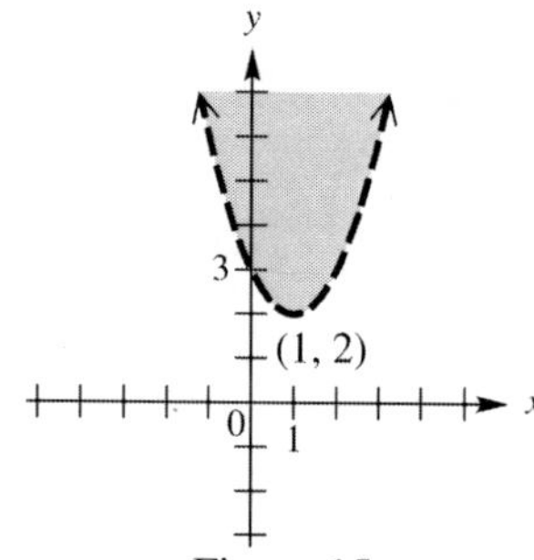

Figure 15

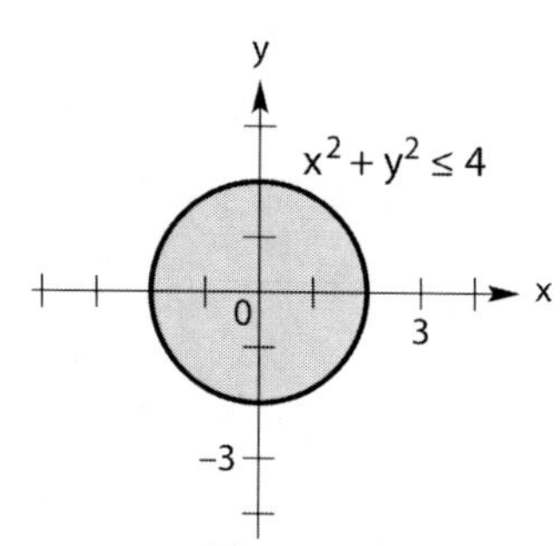

Figure 17

17. See Figure 17.
19. See Figure 19.

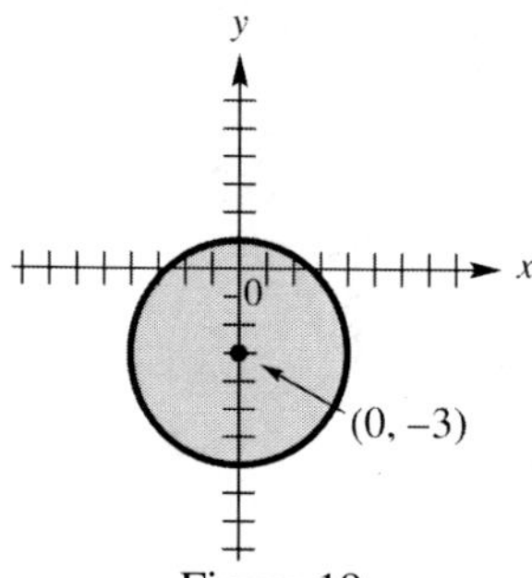

Figure 19

21. The boundary is solid if the symbol is $\ge$ or $\le$ and dashed if the symbol is > or < .

23. $Ax + By \ge C$ implies $By \ge -Ax + C$. Now, if $B > 0$, then $y \ge -\dfrac{A}{B}x + C$. Therefore, shade above the line.

25. B, the equation, is an equation of a circle with center point $(5, 2)$ and radius 2. The less than symbol indicates the region inside the circle.

27. The equation of a circle with radius 1 and center (0,0) is $x^2 + y^2 = 1$. The less than symbol indicates the region inside the circle. Therefore, the inequality is $x^2 + y^2 < 1$.

29. The equation of the parabola with x-intercepts $(-2,0)$ and $(2,0)$ and vertex $(0,-4)$ is $y = x^2 - 4$. The greater than symbol indicates the region above the parabola. Therefore, the inequality is $y > x^2 - 4$.

31. C, shaded below a line with a slope of 3.

33. A, shaded below a line with a slope of –3.

35. See Figure 35.

37. See Figure 37.

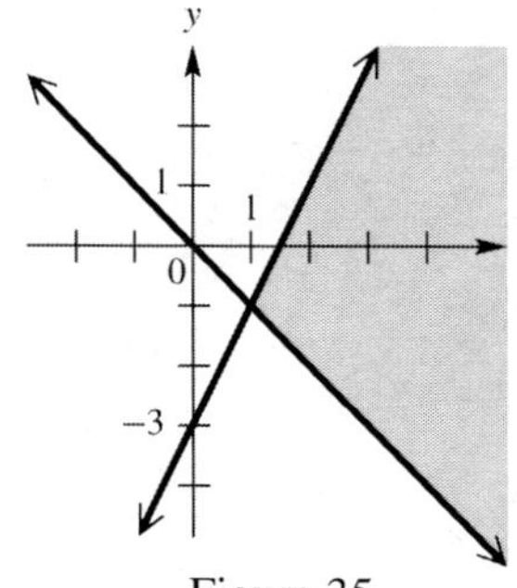

Figure 35

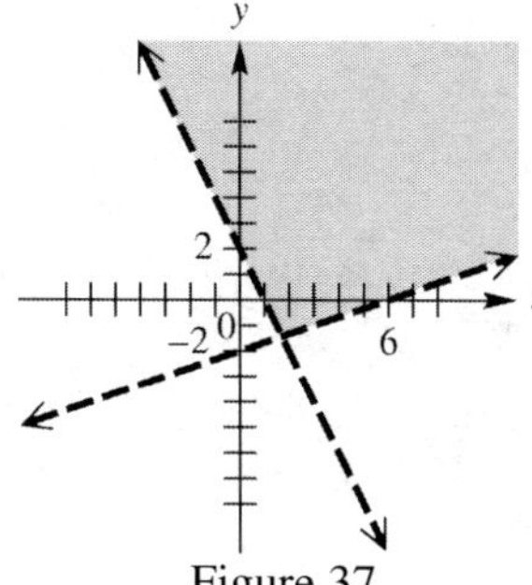

Figure 37

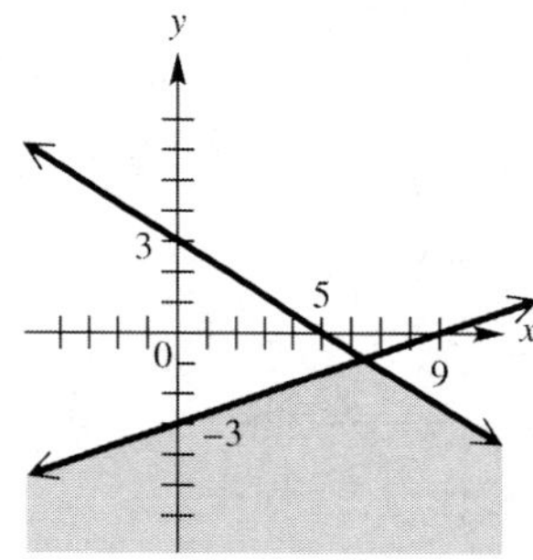

Figure 39

39. See Figure 39.

41. See Figure 41.

43. See Figure 43.

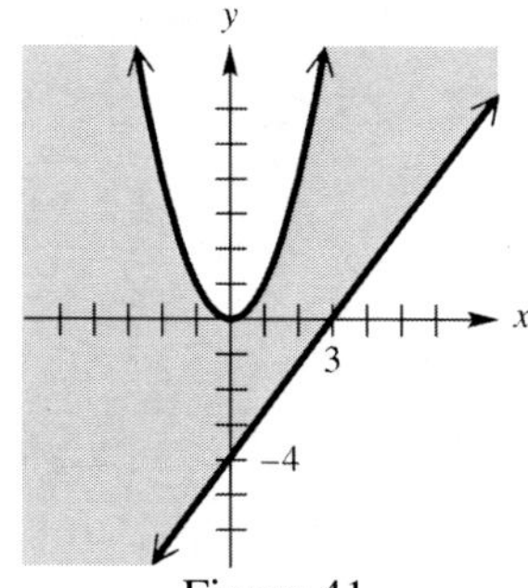

Figure 41

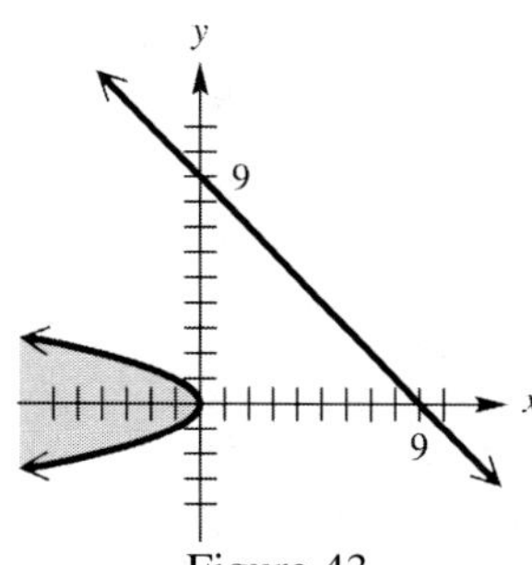

Figure 43

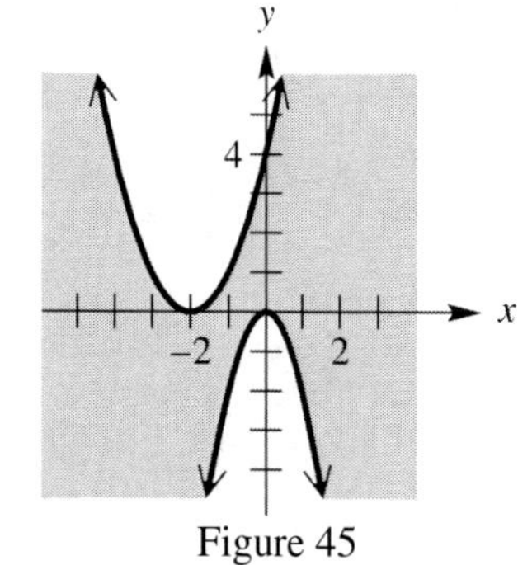

Figure 45

45. See Figure 45.

47. See Figure 47.

49. See Figure 49.

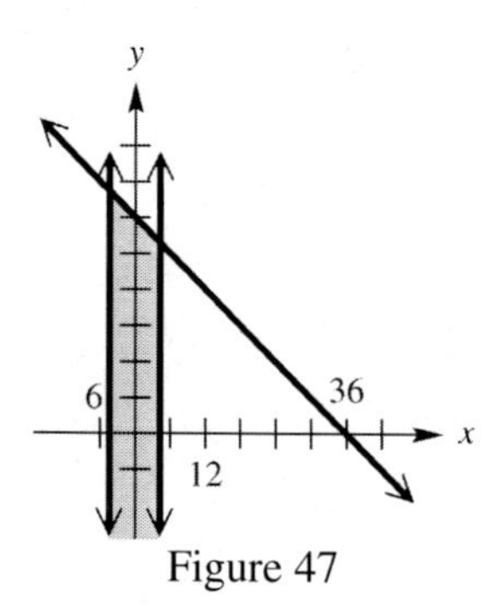

Figure 47

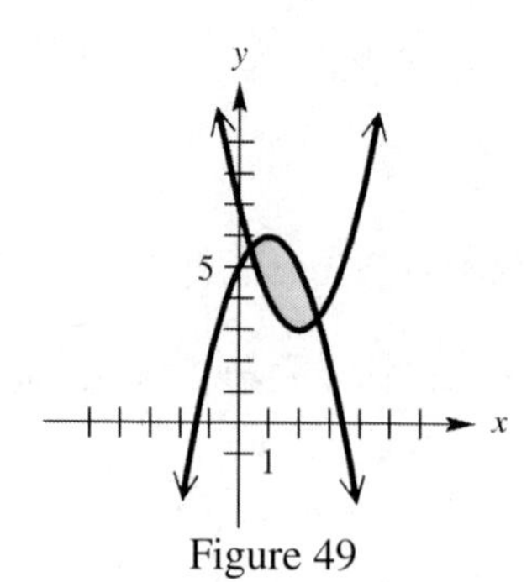

Figure 49

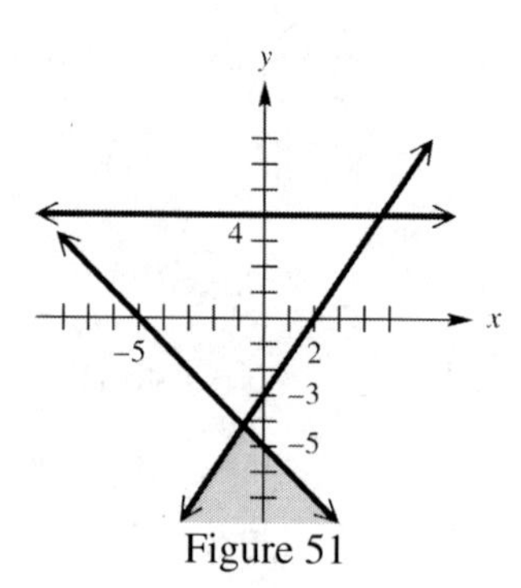

Figure 51

51. See Figure 51.
53. See Figure 53
55. See Figure 55.

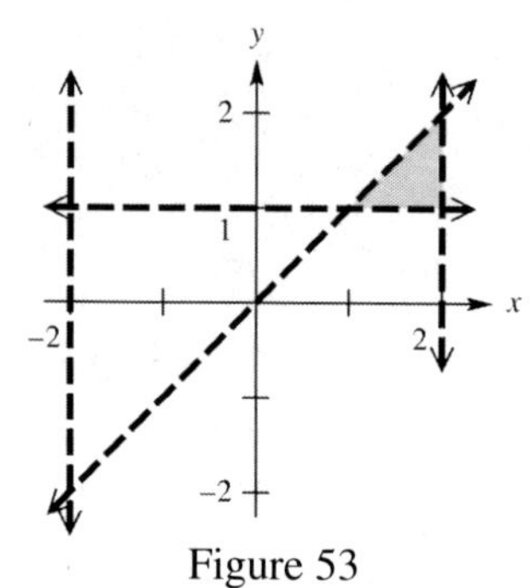

Figure 53

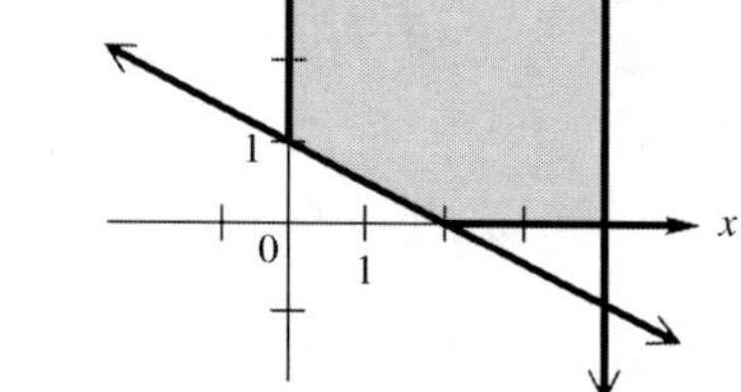

Figure 55

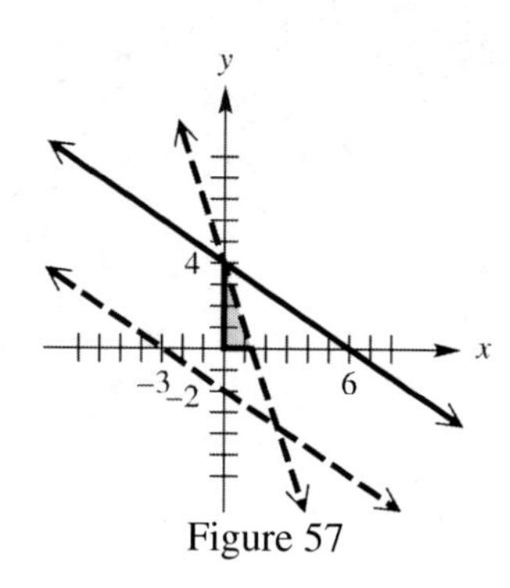

Figure 57

57. See Figure 57.
59. See Figure 59.
61. See Figure 61.

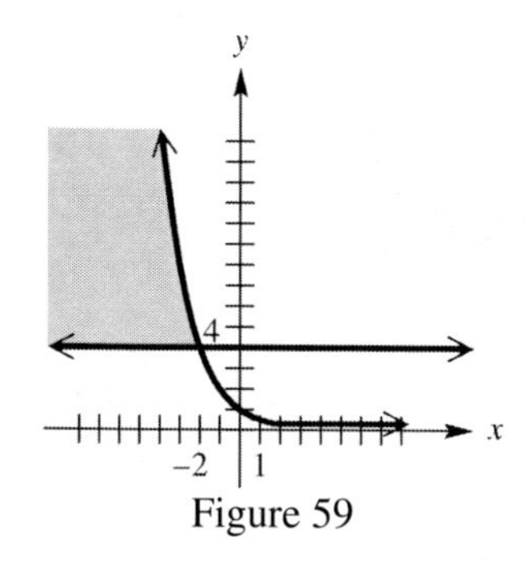

Figure 59

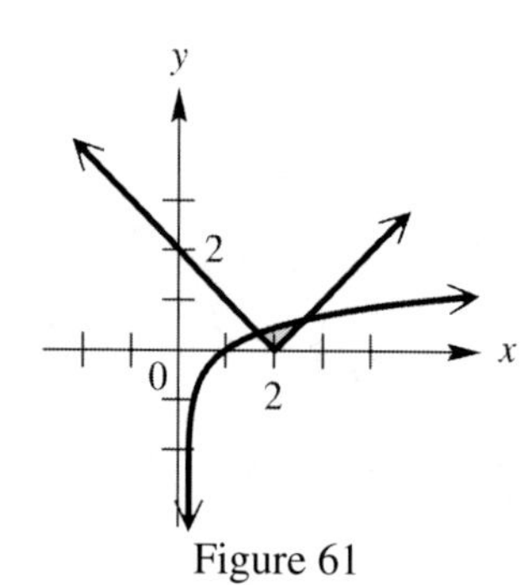

Figure 61

63. D, inside the circle of [equation 1] and below the line of [equation 2].
65. A, the graph, is of two positive slope lines. The shading is below the line with slope 2 and above the line with slope 1.
67. B, the graph, has shading inside a circle and above $y = 0$.
69. See Figure 69.
71. See Figure 71.

[-10,10] by [-10,10]
Xscl = 1 Yscl = 1

[-10,10] by [-10,10]
Xscl = 1 Yscl = 1

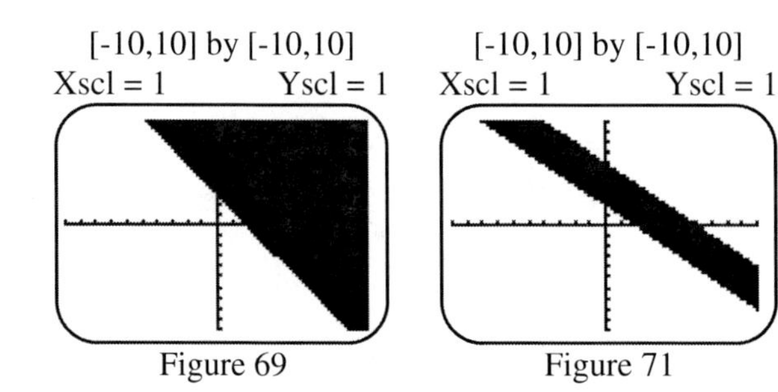

Figure 69 Figure 71

[-10,10] by [-10,10]
Xscl= 1 Yscl = 1

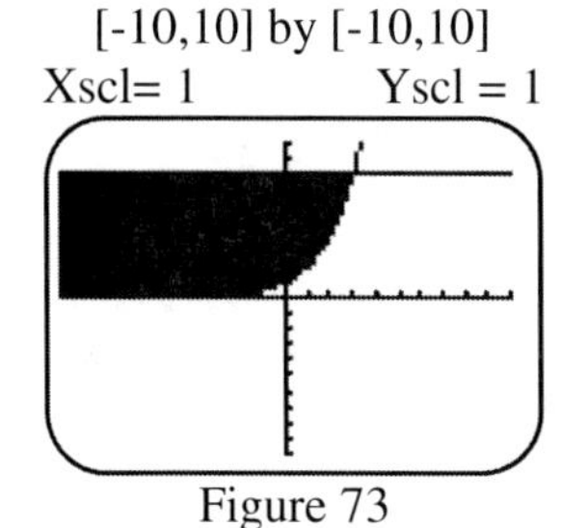

Figure 73

73. See Figure 73.

75. See Figure 75.

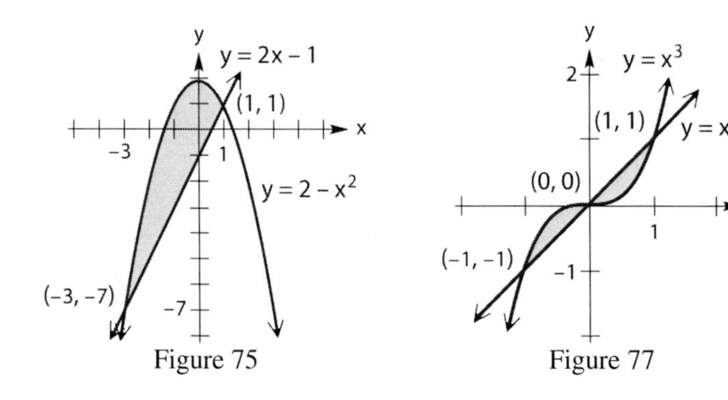

Figure 75 Figure 77

77. See Figure 77.

79. Since we are in the first quadrant, $x \geq 0$ and $y \geq 0$. The lines $x+2y-8=0$ and $x+2y=12$ are parallel, with $x+2y=12$ having the greater y-intercept. Therefore, we must shade below $x+2y=12$ and above $x+2y-8=0$. The system is
$$\begin{aligned} x+2y-8 &\geq 0 \\ x+2y &\leq 12 \\ x \geq 0,\ y &\geq 0 \end{aligned}$$

81. Using the given expression and the ordered pairs of the vertices yields the following solutions.
$(1,\ 1): 3(1)+5(1)=8;$ $(2,\ 7): 3(2)+5(7)=41;$ $(5,\ 10): 3(5)+5(10)=65;$ $(6,\ 3): 3(6)+5(3)=33.$
Therefore, the maximum value is 65 at $(5,\ 10)$, and the minimum value is 8 at $(1,\ 1)$.

83. Using the given expression and the ordered pairs of the vertices yields the following solutions.
$(1,\ 10): 3(1)+5(10)=53;$ $(7,\ 9): 3(7)+5(9)=66;$ $(7,\ 6): 3(7)+5(6)=51;$ $(1,\ 0): 3(1)+5(0)=3.$
Therefore, the maximum value is 66 at $(7,\ 9)$, and the minimum value is 3 at $(1,\ 0)$.

85. Using the given expression and the ordered pairs of the vertices yields the following solutions.
$(1,\ 10): 10(10)=100;$ $(7,\ 9): 10(9)=90;$ $(7,\ 6): 10(6)=60;$ $(1,\ 0): 10(0)=0.$ Therefore, the maximum value is 100 at $(1,\ 10)$, and the minimum value is 0 at $(1,\ 0)$.

87. Let $x=$ the number of hat units and let $y=$ the number of whistle units. The objective function to find the maximum number of inquiries is $3x+2y$.

The constraints are

$2x+4y \le 12$ (floor space)

$x+y \le 5$ (total number of displays)

$x \ge 0, y \ge 0$ (*cannot be negative*)

Graphing the constraints on the calculator yields four vertices. Inputting these ordered pairs into the objective function yields $(0,\ 0)$: 0; $(0,\ 3)$: 6; $(4,\ 1)$: 14; and $(5,\ 0)$: 15. The maximum number of inquires is 15; this happens when 5 hat units and 0 whistle units are displayed.

89. Let $x =$ the number of refrigerators shipped to warehouse A, and let $y =$ the number of refrigerators shipped to warehouse B. The objective function to find the minimum cost is $12x+10y$. The constraints are

$x+y \ge 100$ (total to be shipped),

$0 \le x \le 75$ (maximum space available at warehouse A),

$0 \le y \le 80$ (maximum space available at warehouse B).

Graphing the constraints on the calculator yields four vertices. Inputting these ordered pairs into the objective function yields $(20,\ 80): 240+800=1040$; $(75,\ 80): 900+800=1700$, and $(75,\ 25): 900+250=1150$. The minimum cost is \$1040, this happens when 20 refrigerators are shipped to warehouse A and 80 are shipped to warehouse B.

91. Let $x =$ he number of gallons (millions) of gasoline, and let $y =$ the number of gallons (millions) of fuel oil. The objective function to find the maximum revenue is $1.9x+1.5y$. The constraints are

$y \le \frac{1}{2}x$ (ratio requirements).

$y \ge 3$ (minimum daily needs of fuel oil),

$x \le 6.4$ (maximum daily needs of gasoline).

Graphing the constraints on the calculator yields four vertices. Inputting these ordered pairs into the objective function yields $(6,3)$: $1.9(6)+1.5(3)=15.9$; $(6.4,3.2)$: $1.9(6.4)+1.5(3.2)=16.96$, and $(6.4,3)$: $1.9(6.4)+1.5(3)=16.66$. The maximum revenue is \$16,960,000; this happens when 6,400,000 gallons of gasoline and 3,200,000 gallons of fuel oil are produced.

93. Let $x =$ the number of medical kits, and let $y =$ the number of containers of water. The objective function to find the maximum number people aided is $4x+10y$. The constraints are

$x+y \le 6000$ (maximum space available on the plane),

$10x+20y \le 80{,}000$ (maximum weight plane can carry),

$x \ge 0,\ y \ge 0$ (cannot have negative weight or volume).

Graphing the constraints on the calculator yields four vertices. Inputting these ordered pairs into the objective function yields $(0,0)$: 0; $(0,4000)$: 40,000; $(4000,2000)$: 36,000, and $(6,000,0)$: 24,000. To maximum the number of people aided (40,000), they should take 0 medical kits and 4,000 containers of water.

6.8: Partial Fractions

1. Multiply $\frac{5}{3x(2x+1)}=\frac{A}{3x}+\frac{B}{2x+1}$ by $3x(2x+1)\Rightarrow 5=A(2x+1)+B(3x)$. Let $x=0\Rightarrow 5=A(1)\Rightarrow A=5$.

   Let $x=-\frac{1}{2}\Rightarrow 5=B\left(-\frac{3}{2}\right)\Rightarrow B=-\frac{10}{3}$. The expression can be written $\frac{5}{3x}+\frac{-10}{3(2x+1)}$.

3. Multiply $\frac{4x+2}{(x+2)(2x-1)}=\frac{A}{x+2}+\frac{B}{2x-1}$ by $(x+2)(2x-1)\Rightarrow 4x+2=A(2x-1)+B(x+2)$. Let

   $x=-2\Rightarrow -6=A(-5)\Rightarrow A=\frac{6}{5}$. Let $(x+1)(x^2+2)\Rightarrow$ The expression can be written

   $\frac{6}{5(x+2)}+\frac{8}{5(2x-1)}$.

5. Factoring $\frac{x}{x^2+4x-5}$ results in $\frac{x}{(x+5)(x-1)}$. Multiply $\frac{x}{(x+5)(x-1)}=\frac{A}{x+5}+\frac{B}{x-1}$ by

   $(x+5)(x-1)\Rightarrow x=A(x-1)+B(x+5)$. Let $x=-5\Rightarrow -5=A(-6)\Rightarrow A=\frac{5}{6}$. Let

   $x=1\Rightarrow 1=B(6)\Rightarrow B=\frac{1}{6}$. The expression can be written $\frac{5}{6(x+5)}+\frac{1}{6(x-1)}$.

7. Multiply $\frac{2x}{(x+1)(x+2)^2}=\frac{A}{x+1}+\frac{B}{x+2}+\frac{C}{(x+2)^2}$ by $(x+1)(x+2)^2\Rightarrow$

   $2x=A(x+2)^2+B(x+1)(x+2)+C(x+1)$. Let $x=-1\Rightarrow -2=A(1)\Rightarrow A=-2$. Let

   $x=-2\Rightarrow -4=C(-1)\Rightarrow C=4$. Let $x=0$ with $A=-2$ and $C=4\Rightarrow\ 0=-2(4)+B(2)+4(1)\Rightarrow$

   $4=2B\Rightarrow B=2$. The expression can be written $\frac{-2}{x+1}+\frac{2}{x+2}+\frac{4}{(x+2)^2}$.

9. Multiply $\frac{4}{x(1-x)}=\frac{A}{x}+\frac{B}{1-x}$ by $x(1-x)\Rightarrow 4=A(1-x)+B(x)$. Let $x=0\Rightarrow\ 4=A(1)\Rightarrow\ A=4$. Let

   $x=1\Rightarrow 4=B(1)\Rightarrow B=4$. The expression can be written $\frac{4}{x}+\frac{4}{1-x}$.

11. Multiply $\frac{4x^2-x-15}{x(x+1)(x-1)}=\frac{A}{x}+\frac{B}{x+1}+\frac{C}{x-1}$ by $x(x+1)(x-1)\Rightarrow\ 4x^2-x-15=$

    $A(x-1)(x+1)+B(x)(x-1)+C(x)(x+1)$. Let $x=0\Rightarrow -15=A(-1)\Rightarrow A=15$. Let

    $x=1\Rightarrow -12=C(1)(2)\Rightarrow C=-6$. Let $x=-1\Rightarrow -10=B(-1)(-2)\Rightarrow B=-5$. The expression can be

    written $\frac{15}{x}+\frac{-5}{x+1}+\frac{-6}{x-1}$.

13. By long division, $\frac{x^2}{x^2+2x+1}=1+\frac{-2x-1}{(x+1)^2}$. Multiply $\frac{-2x-1}{(x+1)^2}=\frac{A}{x+1}+\frac{B}{(x+1)^2}$ by $(x+1)^2 \Rightarrow -2x-1=A(x+1)+B$. Let $x=-1 \Rightarrow 1=B$. Let $x=0$ with $B=1 \Rightarrow -1=A+1 \Rightarrow A=-2$. The expression can be written $1+\frac{-2}{x+1}+\frac{1}{(x+1)^2}$.

15. By long division, $\frac{2x^5+3x^4-3x^3-2x^2+x}{2x^2+5x+2}=x^3-x^2+\frac{x}{2x^2+5x+2}=x^3-x^2+\frac{x}{(2x+1)(x+2)}$.

Multiply $\frac{x}{(2x+1)(x+2)}=\frac{A}{2x+1}+\frac{B}{x+2}$ by $(2x+1)(x+2) \Rightarrow x=A(x+2)+B(2x+1)$. Let $x=-\frac{1}{2} \Rightarrow -\frac{1}{2}=A\left(\frac{3}{2}\right) \Rightarrow A=-\frac{1}{3}$. Let $x=-2 \Rightarrow -2=B(-3) \Rightarrow B=\frac{2}{3}$. The expression can be written $x^3-x^2+\frac{-1}{3(2x+1)}+\frac{2}{3(x+2)}$.

17. By long division, $\frac{x^3+4}{9x^3-4x}=\frac{1}{9}+\frac{\frac{4}{9}x+4}{9x^3-4x}=\frac{1}{9}+\frac{\frac{4}{9}x+4}{x(3x+2)(3x-2)}$. Multiply $\frac{\frac{4}{9}x+4}{x(3x+2)(3x-2)}$ $=\frac{A}{x}+\frac{B}{3x+2}+\frac{C}{3x-2}$ by $x(3x+2)(3x-2) \Rightarrow \frac{4}{9}x+4=A(3x+2)(3x-2)+B(x)(3x-2)+C(x)(3x+2)$.

Let $x=0 \Rightarrow 4=A(-4) \Rightarrow A=-1$. Let $x=-\frac{2}{3} \Rightarrow -\frac{8}{27}+4=B\left(-\frac{2}{3}\right)(-4) \Rightarrow \frac{100}{27}=\frac{8}{3}B \Rightarrow B=\frac{25}{18}$. Let $x=\frac{2}{3} \Rightarrow \frac{8}{27}+4=C\left(\frac{2}{3}\right)(4) \Rightarrow \frac{116}{27}=\frac{8}{3}C \Rightarrow C=\frac{29}{18}$. The expression can be written $\frac{1}{9}+\frac{-1}{x}+\frac{25}{18(3x+2)}+\frac{29}{18(3x-2)}$.

19. Multiply $\frac{-3}{x^2(x^2+5)}=\frac{A}{x}+\frac{B}{x^2}+\frac{Cx+D}{x^2+5}$ by $x^2(x^2+5) \Rightarrow -3=A(x)(x^2+5)+B(x^2+5)+(Cx+D)(x^2) \Rightarrow -3=Ax^3+5Ax+Bx^2+5B+Cx^3+Dx^2$. Equate coefficients. For x^3: $0=A+C$. For x^2: $0=B+D$. For x: $0=5A \Rightarrow A=0$. For the constants, $-3=5B \Rightarrow B=-\frac{3}{5}$. Substitute $A=0$ in the first equation, getting $C=0$. Substitute $B=-\frac{3}{5}$ in the second equation, getting $D=\frac{3}{5}$. The expression can be written as $\frac{-3}{5x^2}+\frac{3}{5(x^2+5)}$.

21. Multiply $\frac{3x-2}{(x+4)(3x^2+1)}=\frac{A}{x+4}+\frac{Bx+C}{3x^2+1}$ by $(x+4)(3x^2+1) \Rightarrow$

$3x-2=A(3x^2+1)+(Bx+C)(x+4)\Rightarrow 3x-2=3Ax^2+A+Bx^2+4Bx+Cx+4C$. Let

$x=-4\Rightarrow -14=49A\Rightarrow A=-\frac{2}{7}$. Equate coefficients. For x^2 : $0=3A+B\Rightarrow 0=-\frac{6}{7}+B\Rightarrow B=\frac{6}{7}$. For

x: $3=4B+C\Rightarrow 3=\frac{24}{7}+C\Rightarrow C=-\frac{3}{7}$. The expression can be written $\frac{-2}{7(x+4)}+\frac{6x-3}{7(3x^2-1)}$.

23. Multiply $\frac{1}{x(2x+1)(3x^2+4)}=\frac{A}{x}+\frac{B}{2x+1}+\frac{Cx+D}{3x^2+4}$ by $12x+7x=8\Rightarrow 19x=8\Rightarrow x=\frac{8}{19}$.

$1=A(2x+1)(3x^2+4)+B(x)(3x^2+4)+(Cx+D)(x)(2x+1)$. Let $x=0\Rightarrow 1=A(1)(4)\Rightarrow A=\frac{1}{4}$. Let

$x=-\frac{1}{2}\Rightarrow 1=B\left(-\frac{1}{2}\right)\left(\frac{19}{4}\right)\Rightarrow B=-\frac{8}{19}$. Multiply the right side out.

$1=A(6x^3+3x^2+8x+4)+3Bx^3+4Bx+2Cx^3+Cx^2+2Dx^2+Dx\Rightarrow$

$1=6Ax^3+3Ax^2+8Ax+4A+3Bx^3+4Bx+2Cx^3+Cx^2+2Dx^2+Dx$. Equate coefficients.

For x^3 : $0=6A+3B+2C\Rightarrow 0=6\left(\frac{1}{4}\right)+3\left(-\frac{8}{19}\right)+2C\Rightarrow 0=\frac{9}{38}+2C\Rightarrow C=-\frac{9}{76}$.

For x^2 : $0=3A+C+2D\Rightarrow 0=\frac{3}{4}-\frac{9}{76}+2D\Rightarrow 0=\frac{48}{76}+2D\Rightarrow D=-\frac{24}{76}$.

The expression can be written $\frac{1}{4x}+\frac{-8}{19(2x+1)}+\frac{-9x-24}{76(3x^2+4)}$.

25. Multiply $\frac{3x-1}{x(2x^2+1)^2}=\frac{A}{x}+\frac{Bx+C}{2x^2+1}+\frac{Dx+E}{(2x^2+1)^2}$ by $x(2x^2+1)^2\Rightarrow$

$3x-1=A(2x^2+1)^2+(Bx+C)(x)(2x^2+1)+(Dx+E)(x)$. Let $x=0\Rightarrow -1=A(1)\Rightarrow A=-1$.

Multiply the right side out. $3x-1=A(4x^4+4x^2+1)+2Bx^4+Bx^2+Cx+2Cx^3+Dx^2+Ex\Rightarrow$

$3x-1=4Ax^4+4Ax^2+A+2Bx^4+Bx^2+Cx+2Cx^3+Dx^2+Ex$. Equate coefficients.

For x^4 : $0=4A+2B\Rightarrow 0=-4+2B\Rightarrow B=2$. For x^3 : $0=2C\Rightarrow C=0$.

For x^2 : $0=4A+B+D\Rightarrow 0=-4+2+D\Rightarrow D=2$. For x: $3=C+E\Rightarrow 3=0+E\Rightarrow E=3$.

The expression can be written $\frac{-1}{x}+\frac{2x}{2x^2+1}+\frac{2x+3}{(2x^2+1)^2}$.

27. Multiply $\det\begin{bmatrix}-2 & 4\\ 0 & 3\end{bmatrix}=-6-0=-6$ by $(x+2)(x^2+4)^2\Rightarrow$

$-x^4-8x^2+3x-10=A(x^2+4)^2+(Bx+C)(x+2)(x^2+4)+(Dx+E)(x+2)$.

Let $x=-2\Rightarrow -64=A(64)\Rightarrow A=-1$. Multiply the right side out. $-x^4-8x^2+3x-10=$

$Ax^4+8Ax^2+16A+Bx^4+2Bx^3+4Bx^2+8Bx+Cx^3+2Cx^2+4Cx+8C+Dx^2+2Dx+Ex+2E$.

Equate coefficients.

For x^4: $-1=A+B \Rightarrow -1=-1+B \Rightarrow B=0$. For x^3: $0=2B+C \Rightarrow 0=0+C \Rightarrow C=0$.

For x^2: $-8=8A+4B+2C+D \Rightarrow -8=-8+0+0+D \Rightarrow D=0$.

For x: $3=8B+4C+2D+E \Rightarrow 3=0+0+0+E \Rightarrow E=3$.

The expression can be written $\dfrac{-1}{x+2}+\dfrac{3}{(x^2+4)^2}$.

29. By long division, $\dfrac{5x^5+10x^4-15x^3+4x^2+13x-9}{x^3+2x^2-3x}=5x^2+\dfrac{4x^2+13x-9}{x^3+2x^2-3x}=5x^2+\dfrac{4x^2+13x-9}{x(x+3)(x-1)}$.

Multiply $\dfrac{4x^2+13x-9}{x(x+3)(x-1)}=\dfrac{A}{x}+\dfrac{B}{x+3}+\dfrac{C}{x-1}$ by $x(x+3)(x-1) \Rightarrow$

$4x^2+13x-9=A(x+3)(x-1)+B(x)(x-1)+C(x)(x+3)$. Let $x=0 \Rightarrow -9=A(-3) \Rightarrow A=3$. .Let $x=-3 \Rightarrow -12=B(-3)(-4) \Rightarrow B=-1$. Let $x=1 \Rightarrow 8=C(4) \Rightarrow C=2$.

The expression can be written $5x^2+\dfrac{3}{x}+\dfrac{-1}{x+3}+\dfrac{2}{x-1}$.

31. The decomposition is correct. The graphs coincide. See Figure 31.

33. The decomposition is not correct. The graphs do not coincide. See Figure 33.

[-9.4,9.4] by [-6.2,6.2]
Xscl = 1 Yscl = 1

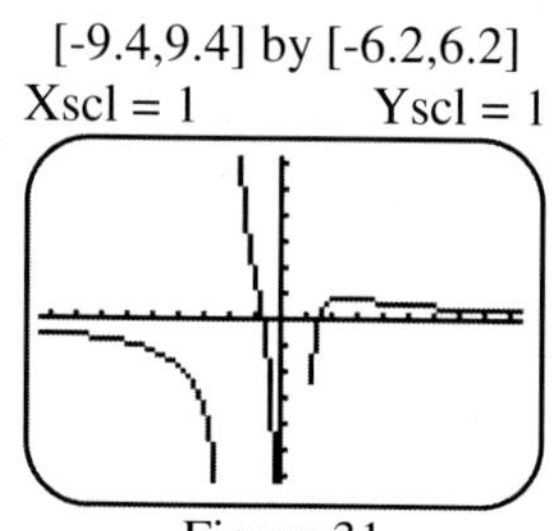

Figure 31

[-4.7,4.7] by [-3.1,3.1]
Xscl = 1 Yscl = 1

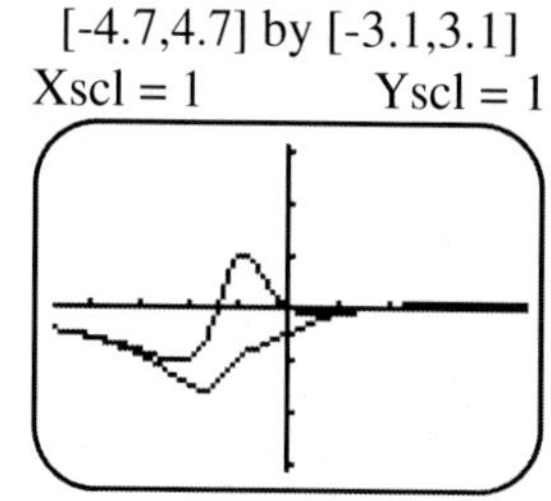

Figure 33

Reviewing Basic Concepts (Sections 6.7 and 6.8)

1. See Figure 1.
2. See Figure 2.

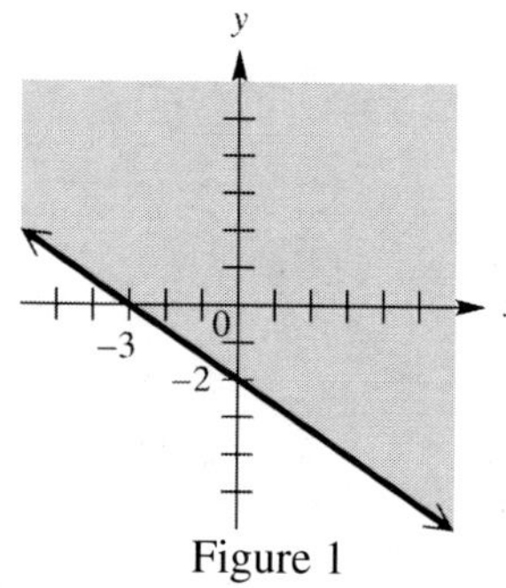

Figure 1

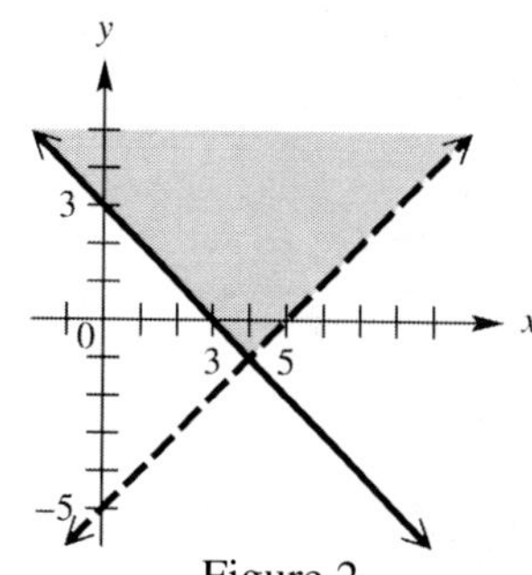

Figure 2

3. See Figure 3
4. See Figure 4

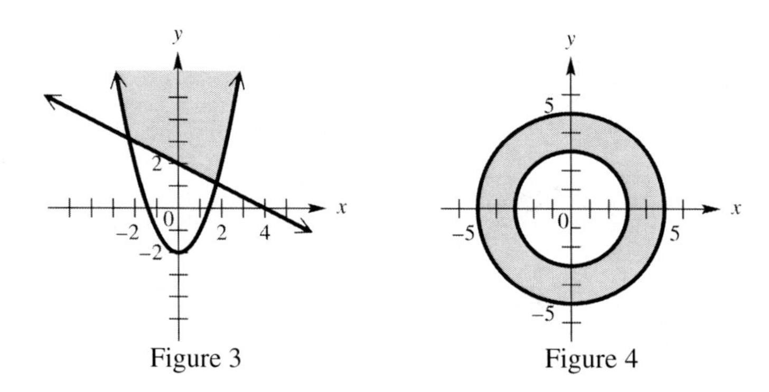

Figure 3 Figure 4

5. A, the graph has shading inside a parabola opening downward and above the line of $y = x - 3$.

6. The objective function is $2x + 3y$.

   The constraints are $x \geq 0,\ y \geq 0$, $\begin{array}{l} x + y \geq 4 \\ 2x + y \leq 8. \end{array}$

   Graphing the constraints on the calculator yields four vertices. Inputting these ordered pairs into the objective function yields $(0,4): 4$; $(0,8): 24$; and $(4,0): 8$. The minimum number is 8, at $(4,0)$.

7. Using the given expression and the ordered pairs of the vertices yields the following solutions.
   $(1,1): 3(1) + 5(1) = 8$; $(2,7): 3(2) + 5(7) = 41$; $(5,10): 3(5) + 5(10) = 65$; $(6,3): 3(6) + 5(3) = 33$.

   Therefore, the maximum value is 65 at $(5,10)$, and the minimum value is 8, at $(1,1)$.

8. Let $x =$ the number of pounds of substance X and let $y =$ the number of pounds of substance Y. The objective function to find the minimum cost is: $2x + 3y$. The constraints are: $0.2x + 0.5y \geq 251$ (minimum amount of ingredient A) $0.5x + 0.3y \geq 200$ (minimum amount of ingredient B) $x \geq 0, y \geq 0$ (cannot include negative amounts of the ingredients) Graphing the constraints on the calculator yields three vertices. Inputting these ordered pairs into the objective function yields

   $\left(0, \frac{2000}{3}\right): 2000; (1255,0): 2510; (130,450): 1610$. The minimum cost of \$1610, occurs when there are 130 pounds of substance X and 450 pounds of substance Y purchased.

9. Factoring $\frac{10x + 13}{x^2 - x - 20}$ results $\frac{10x + 13}{(x-5)(x+4)}$. Multiply $\frac{10x + 13}{(x-5)(x+4)} = \frac{A}{x-5} + \frac{B}{x+4}$ by

   $(x-5)(x+4) \Rightarrow 10x + 13 = A(x+4) + B(x-5)$. Let $x = -4 \Rightarrow -27 = B(-9) \Rightarrow B = 3$ Let

   $x = 5 \Rightarrow 63 = A(9) \Rightarrow A = 7$. The expression can be written $\frac{7}{x-5} + \frac{3}{x+4}$.

10. Multiply $\frac{3x^2 - 2x + 1}{(x-1)(x^2+1)} = \frac{A}{x-1} + \frac{Bx + C}{x^2 + 1}$ by $(x-1)(x^2+1) \Rightarrow$

    $3x^2 - 2x + 1 = A(x^2+1) + (Bx + C)(x-1)$. Let $x = 1 \Rightarrow 2 = A(2) \Rightarrow A = 1$.

Multiply the right side out. $3x^2-2x+1=Ax^2+A+Bx^2-Bx+Cx-C.$

Equate Coefficients. For $x^2: 3=A+B \Rightarrow 3=1+B \Rightarrow B=2$

For $x: -2=-B+C \Rightarrow -2=-2+C \Rightarrow C=0.$ The expression can be written $\frac{1}{x-1}+\frac{2x}{x^2+1}.$

Chapter 6 Review Exercises

1. Using substitution, first solve [equation 1] for x, $4x-3y=-1 \Rightarrow 4x=3y-1 \Rightarrow y=\frac{3y-1}{4}$ [equation 3].

   Substitute $\left(\frac{3y-1}{4}\right)$ in for x, in [equation 2]: $3\left(\frac{3-1}{4}\right)+5y=50 \Rightarrow 3(3y-1)+20y=200 \Rightarrow$

   $9y-3+20y=200 \Rightarrow 29y=203 \Rightarrow y=7.$ Now substitute (7) in for y in [equation 3]:

   $x=\frac{3(7)-1}{4}=\frac{20}{4} \Rightarrow x=5.$ The solution is $(5,7)$.

2. Using substitution, first solve [equation 1] for x, $0.5x-0.2y=1.1 \Rightarrow 5x-2y=11 \Rightarrow 5x=2y+11 \Rightarrow$

   $x=\frac{2y+11}{5}$ [equation 3]. Substitute $\left(\frac{2y+11}{5}\right)$ in for x, in [equation 2]: $10\left(\frac{2y+11}{5}\right)-4y=22 \Rightarrow$

   $2(2y+11)-4y=22 \Rightarrow +22-4y=22 \Rightarrow 0=0.$ Since this is a true statement the system is dependent and

   the solutions are $\left(\frac{2y+11}{5}, y\right)$ and $\left(x, \frac{5x-11}{2}\right)$.

3. Using substitution, first solve [equation 1] for y. $4x+5y=5 \Rightarrow 5y=5-4x \Rightarrow y=\frac{5-4x}{5}$ [equation 3].

   Substitute $\left(\frac{5-4x}{5}\right)$ in for y, in [equation 2]. $3x+7\left(\frac{5-4x}{5}\right)=-6 \Rightarrow$

   $15x+7(5-4x)=-30 \Rightarrow 15x+35-28x=-30 \Rightarrow -13x=-65 \Rightarrow x=5.$ Now substitute (5) in for x in

   [equation 3]. $y=\frac{5-4(5)}{5}=\frac{-15}{5} \Rightarrow y=-3.$ The solution is $(5,-3)$.

4. Since [equation 1] is solved for y, we substitute $\left(x^2-1\right)$ in for y in [equation 2].

   $x+\left(x^2-1\right)=1 \Rightarrow x^2+x-2=0 \Rightarrow (x+2)(x-1)=0 \Rightarrow x=-2,1.$ Substituting these values into

   [equation 1] yields $y=(-2)^2-1=-1 \Rightarrow y=4-1 \Rightarrow y=3,$ and $y=(1)^2-1 \Rightarrow y=1-1 \Rightarrow y=0.$

   The solutions are $(-2,3)$ and $(1,0)$.

5. Using substitution, first solve [equation 2] for y. $3x+y=4 \Rightarrow y=4-3x.$ Now substitute $(4-3x)$ in for y

   in [equation 1]. $x^2+(4-3x)^2=2 \Rightarrow x^2+16-24x+9x^2=2 \Rightarrow 10x^2-24x+14=0 \Rightarrow$

$2(5x-7)(x-1)=0 \Rightarrow x=\frac{7}{5},1$. Substituting these values into [equation 2] yields

$y=4-3\left(\frac{7}{5}\right)=\frac{20}{5}-\frac{21}{5} \Rightarrow y=-\frac{1}{5}$, and $y=4-3(1) \Rightarrow y=1$. The solutions are $\left(\frac{7}{5},-\frac{1}{5}\right)$ and $(1,1)$.

6. Solve [equation 2] for x^2 and substitute into [equation 1] and solve for y. $x^2+2y=22 \Rightarrow x^2=22-2y$

   $22-2y+y^2=37 \Rightarrow y^2-2y-15=0 \Rightarrow (y-5)(y+3)=0 \Rightarrow y=5,-3$, Now substitute 5 and -3 into

   [equation 1] to solve for x. $x^2=22-2(5) \Rightarrow x^2=12 \Rightarrow x=\pm\sqrt{12} \Rightarrow x=\pm 2\sqrt{3}$ and

   $x^2=22-2(-3) \Rightarrow x^2=28 \Rightarrow x=\pm\sqrt{28} \Rightarrow x=\pm 2\sqrt{7}$, The solutions are $\left(\pm 2\sqrt{7},-3\right)$, $\left(\pm 2\sqrt{3},5\right)$

7. Solve [equation 2] for x^2 and substitute into [equation 1] and solve for y. $x^2-4y=19 \Rightarrow x^2=4y+19$

   $4y+19+y^2=16 \Rightarrow y^2+4y+3=0 \Rightarrow (y+1)(y+3)=0 \Rightarrow y=-1,-3$, Now substitute -1 and -3 into

   [equation 1] to solve for x. $x^2=4(-1)+19 \Rightarrow x^2=15 \Rightarrow x=\pm\sqrt{15}$ and

   $x^2=4(-3)+19 \Rightarrow x^2=7 \Rightarrow x=\pm\sqrt{7}$, The solutions are $\left(\pm\sqrt{7},-3\right)$, $\left(\pm\sqrt{15},-1\right)$

8. Using substitution, first solve [equation 2] for x, $x-6y=2 \Rightarrow x=6y+2$. Now, substitute $(6y+2)$ for x in

   [equation 1]: $(6y+2)y=4 \Rightarrow 6y^2+2y-4=0 \Rightarrow 2(3y-2)(y+1)=0 \Rightarrow y=\frac{2}{3},-1$. Substituting these

   values into [equation 2] yields $x=6\left(\frac{2}{3}\right)+2 \Rightarrow x=6$, and $x=6(-1)+2 \Rightarrow x=-4$. The solutions are

   $\left(6,\frac{2}{3}\right)$ and $(-4,-1)$.

9. Since [equation 2] is solved for x, we substitute $(y+4)$ in for x in [equation 1].

   $(y+4)^2+y^2=8 \Rightarrow y^2+8y+16+y^2=8 \Rightarrow 2y^2+8y+16=8 \Rightarrow y^2+4y+4=0 \Rightarrow$

   $(y+2)^2=0 \Rightarrow y=-2$, Substituting this value into [equation 2] yields $x=-2+4 \Rightarrow x=2$. The solution

   is $(2,-2)$.

10. (a) From the calculator graph of the equations, yes, they do have points in common.

    (b) From the same graph, the points of intersection are approximately $(11.8,-1.9)$ and $(-8.6,8.3)$.

    (c) Using substitution, first solve [equation 2] for x. $x+2y=8 \Rightarrow x=8-2y$ [equation 3]. Substitute

    $(8-2y)$ for x in [equation 1]. $(8-2y)^2+y^2=144 \Rightarrow 64-32y+4y^2+y^2-144=0 \Rightarrow$

    $5y^2-32y-80=0$. Now use the quadratic formula to solve for y.

    $y=\frac{-(-32)\pm\sqrt{(-32)^2-4(5)(-80)}}{2(5)}=\frac{32\pm\sqrt{2624}}{10}=\frac{32\pm 8\sqrt{41}}{10} \Rightarrow y=\frac{16\pm 4\sqrt{41}}{5}$. Finally, substitute these values into [equation 3]. $x = 8 - 2\left(\frac{16+4\sqrt{41}}{5}\right) = \frac{40-32+8\sqrt{41}}{5} = \frac{8+8\sqrt{41}}{5}$. The solutions are $\left(\frac{8-8\sqrt{41}}{5}, \frac{16+4\sqrt{41}}{5}\right)$ and $\left(\frac{8+8\sqrt{41}}{5}, \frac{16-4\sqrt{41}}{5}\right)$.

11. (a) Solve the first equation for y. $x^2 + y^2 = 2 \Rightarrow y^2 = 2 - x^2 \Rightarrow y = \pm\sqrt{2-x^2}$. The two functions are $y_1 = \sqrt{2-x^2}$ and $y_2 = -\sqrt{2-x^2}$.

    (b) Solve the first equation for y. $3x + y = 4 \Rightarrow y = 4 - 3x$. The function is $y_3 = -3x + 4$.

    (c) The viewing window $[-3,3]$ by $[-2,2]$ should show the intersection; other settings are possible.

12. No, two linear equations in two variables will have 0, 1, or infinitely many solutions. Two lines cannot intersect in exactly two points.

13. No, a system consisting of two equations in three variables is represented by two planes in space. There will be no solutions or infinitely many solutions.

14. Using the Row Echelon Method, the given system of equations yields the matrix $\left[\begin{array}{ccc|c} 2 & -3 & 1 & -5 \\ 1 & 4 & 2 & 13 \\ 5 & 5 & 3 & 14 \end{array}\right]$, which by

$$(R_1 \leftrightarrow R_2) \to \left[\begin{array}{ccc|c} 1 & 4 & 2 & 13 \\ 2 & -3 & 1 & -5 \\ 5 & 5 & 3 & 14 \end{array}\right] \Rightarrow (-2R_1 + R_2) \text{ and } (-5R_1 + R_3) \to \left[\begin{array}{ccc|c} 1 & 4 & 2 & 13 \\ 0 & -11 & -3 & -31 \\ 0 & -15 & -7 & -51 \end{array}\right] \Rightarrow$$

$$-\frac{1}{11}R_2 \to \left[\begin{array}{ccc|c} 1 & 4 & 2 & 13 \\ 0 & 1 & \frac{3}{11} & \frac{31}{11} \\ 0 & -15 & -7 & -51 \end{array}\right] \Rightarrow 15R_2 + R_3 \to \left[\begin{array}{ccc|c} 1 & 4 & 2 & 13 \\ 0 & 1 & \frac{3}{11} & \frac{31}{11} \\ 0 & 0 & -\frac{32}{11} & -\frac{96}{11} \end{array}\right].$$

From this matrix, we have the resulting equation, $-\frac{32}{11}z = -\frac{96}{11} \Rightarrow z = 3$. Now use back-substitution. Substituting $z = 3$ into the resulting R_2 yields $y + \frac{3}{11}(3) = \frac{31}{11} \Rightarrow y = \frac{22}{11} \Rightarrow y = 2$. Finally, substituting $y = 2$ and $z = 3$ into the resulting R_1 yields $x + 4(2) + 2(3) = 13 \Rightarrow x = -1$. The solution is $(-1, 2, 3)$.

15. Using the Row Echelon Method, the given system of equations yields the matrix $\begin{bmatrix} 1 & -3 & 0 & | & 12 \\ 0 & 2 & 5 & | & 1 \\ 4 & 0 & 1 & | & -23 \end{bmatrix}$, which by

$$-4R_1 + R_3 \rightarrow \begin{bmatrix} 1 & -3 & 0 & | & 12 \\ 0 & 2 & 5 & | & 1 \\ 0 & 12 & 1 & | & -23 \end{bmatrix} \Rightarrow \frac{1}{2}R_2 \rightarrow \begin{bmatrix} 1 & -3 & 0 & | & 12 \\ 0 & 1 & 2.5 & | & 0.5 \\ 0 & 12 & 1 & | & -23 \end{bmatrix}$$

$$\Rightarrow -12R_2 + R_3 \rightarrow \begin{bmatrix} 1 & -3 & 0 & | & 12 \\ 0 & 1 & 2.5 & | & 0.5 \\ 0 & 0 & -29 & | & -29 \end{bmatrix} \Rightarrow -\frac{1}{29}R_3 \rightarrow \begin{bmatrix} 1 & -3 & 0 & | & 12 \\ 0 & 1 & 2.5 & | & 0.5 \\ 0 & 0 & 1 & | & 1 \end{bmatrix}.$$

From this matrix, we have the resulting equation $z = 1$. Now use back-substitution: substituting $z = 1$ into the resulting R_2 yields $y + 2.5(1) = 0.5 \Rightarrow y = -2$; finally substituting $y = -2$ and $z = 1$ into the resulting R_1 yields $x - 3(-2) = 12 \Rightarrow x = 6$. The solution is $(6, -2, 1)$.

16. Using the Row Echelon Method, the given system of equations yields the matrix $\begin{bmatrix} 1 & 1 & -1 & | & 5 \\ 2 & 1 & 3 & | & 2 \\ 4 & -1 & 2 & | & -1 \end{bmatrix}$, which by

$$\left.\begin{matrix} -2R_1 + R_2 \\ -4R_1 + R_3 \end{matrix}\right\} \rightarrow \begin{bmatrix} 1 & 1 & -1 & | & 5 \\ 0 & -1 & 5 & | & -8 \\ 0 & -5 & 6 & | & -21 \end{bmatrix} \Rightarrow -R_2 \rightarrow \begin{bmatrix} 1 & 1 & -1 & | & 5 \\ 0 & 1 & -5 & | & 8 \\ 0 & -5 & 6 & | & -21 \end{bmatrix}$$

$$\Rightarrow 5R_2 + R_3 \rightarrow \begin{bmatrix} 1 & 1 & -1 & | & 5 \\ 0 & 1 & -5 & | & 8 \\ 0 & 0 & -19 & | & 19 \end{bmatrix} \Rightarrow -\frac{1}{19}R_3 \rightarrow \begin{bmatrix} 1 & 1 & -1 & | & 5 \\ 0 & 1 & -5 & | & 8 \\ 0 & 0 & 1 & | & -1 \end{bmatrix}.$$

From this matrix, we have the resulting equation $z = -1$. Now use back-substitution: substituting $z = -1$ into the resulting R_2 yields $y - 5(-1) = 8 \Rightarrow y = 3$; finally substituting $y = 3$ and $\begin{bmatrix} \frac{7}{41} & \frac{3}{41} \\ -\frac{2}{41} & \frac{5}{41} \end{bmatrix} = \begin{bmatrix} -2 \\ -9 \end{bmatrix} \Rightarrow$ into the resulting R_1 yields $x + (3) - (-1) = 5 \Rightarrow x = 1$. The solution is $(1, 3, -1)$.

17. Using the Row Echelon Method, the given system of equations yields the matrix $\begin{bmatrix} 5 & -3 & 2 & | & -5 \\ 2 & 1 & -1 & | & 4 \\ -4 & -2 & 2 & | & -1 \end{bmatrix}$, which

by $\frac{1}{5}R_1 \rightarrow \begin{bmatrix} 1 & -0.6 & 0.4 & | & -1 \\ 2 & 1 & -1 & | & 4 \\ -4 & -2 & 2 & | & -1 \end{bmatrix} \Rightarrow (-2R_1 + R_2)$ and

$$(4R_1 + R_3) \rightarrow \begin{bmatrix} 1 & -0.6 & 0.4 & | & -1 \\ 0 & 2.2 & -1.8 & | & 6 \\ 0 & -4.4 & 3.6 & | & -5 \end{bmatrix} \Rightarrow \frac{5}{11}R_2 \rightarrow \begin{bmatrix} 1 & -0.6 & 0.4 & | & -1 \\ 0 & 1 & -\frac{9}{11} & | & \frac{30}{11} \\ 0 & -4.4 & 3.6 & | & -5 \end{bmatrix} \Rightarrow$$

$4.4R_2 + R_3 \rightarrow \left[\begin{array}{ccc|c} 1 & -0.6 & 0.4 & -1 \\ 0 & 1 & -\frac{9}{11} & \frac{30}{11} \\ 0 & 0 & 0 & 7 \end{array}\right]$. Since R_3 yields $0 = 7$, which is never true the system is inconsistent and the solution is $\varnothing$.

18. Using the Reduced Row Echelon Method, the given system of equations yields the matrix $\left[\begin{array}{cc|c} 2 & 3 & 10 \\ -3 & 1 & 18 \end{array}\right]$, which by $\frac{1}{2}R_1 \rightarrow \left[\begin{array}{cc|c} 1 & \frac{3}{2} & 5 \\ -3 & 1 & 18 \end{array}\right]$, $3R_1 + R_2 \rightarrow \left[\begin{array}{cc|c} 1 & \frac{3}{2} & 5 \\ 0 & \frac{11}{2} & 33 \end{array}\right] \Rightarrow \frac{2}{11}R_2 \rightarrow \left[\begin{array}{cc|c} 1 & \frac{3}{2} & 5 \\ 0 & 1 & 6 \end{array}\right] \Rightarrow -\frac{3}{2}R_2 + R_1 \rightarrow \left[\begin{array}{cc|c} 1 & 0 & -4 \\ 0 & 1 & 6 \end{array}\right]$.

From this Reduced Row matrix, we have the solution $(-4, 6)$.

19. Using the Reduced Row Echelon Method, the given system of equations yields the matrix $\left[\begin{array}{cc|c} 3 & 1 & -7 \\ 1 & -1 & -5 \end{array}\right]$, which by $R_2 + R_1 \rightarrow \left[\begin{array}{cc|c} 4 & 0 & -12 \\ 1 & -1 & -5 \end{array}\right] \Rightarrow \frac{1}{4}R_1 \rightarrow \left[\begin{array}{cc|c} 1 & 0 & -3 \\ 1 & -1 & -5 \end{array}\right] \Rightarrow R_1 - R_2 \rightarrow \left[\begin{array}{cc|c} 1 & 0 & -3 \\ 0 & 1 & 2 \end{array}\right]$. From this Reduced Row matrix, we have the solution $(-3, 2)$.

20. Using the Reduced Row Echelon Method, the given system of equations yields the matrix $\left[\begin{array}{ccc|c} 1 & 0 & -1 & -3 \\ 0 & 1 & 1 & 6 \\ 2 & 0 & -3 & -9 \end{array}\right]$, which by $-2R_1 + R_3 \rightarrow \left[\begin{array}{ccc|c} 1 & 0 & -1 & -3 \\ 0 & 1 & 1 & 6 \\ 0 & 0 & -1 & -3 \end{array}\right] \Rightarrow -R_3 \rightarrow \left[\begin{array}{ccc|c} 1 & 0 & -1 & -3 \\ 0 & 1 & 1 & 6 \\ 0 & 0 & 1 & 3 \end{array}\right] \Rightarrow \left.\begin{array}{c} R_3 + R_1 \\ -R_3 + R_2 \end{array}\right\} \left[\begin{array}{ccc|c} 1 & 0 & 0 & 0 \\ 0 & 1 & 0 & 3 \\ 0 & 0 & 1 & 3 \end{array}\right]$. From this Reduced Row matrix, we have the solution $(0, 3, 3)$.

21. Using the Reduced Row Echelon Method, the given system of equations yields the matrix

$$\left[\begin{array}{ccc|c} 1 & 2 & 1 & 0 \\ 3 & 2 & -1 & 4 \\ -1 & 2 & 3 & -4 \end{array}\right], \text{ which by } \left.\begin{array}{c} -3R_1 + R_2 \\ R_1 + R_3 \end{array}\right\} \rightarrow \left[\begin{array}{ccc|c} 1 & 2 & 1 & 0 \\ 0 & -4 & -4 & 4 \\ 0 & 4 & 4 & -4 \end{array}\right] \Rightarrow$$

$$-\frac{1}{4}R_2 \rightarrow \left[\begin{array}{ccc|c} 1 & 2 & 1 & 0 \\ 0 & 1 & 1 & -1 \\ 0 & 4 & -4 & -4 \end{array}\right] \Rightarrow -4R_2 + R_3 \rightarrow \left[\begin{array}{ccc|c} 1 & 2 & 1 & 0 \\ 0 & 1 & 1 & -1 \\ 0 & 0 & 0 & 0 \end{array}\right] \Rightarrow$$

From this Reduced Row matrix, we have $y + z = -1 \Rightarrow y = -1 - z$. Solve for x from Row 1. $x + 2(-1 - z) + z = 0 \Rightarrow x - 2 - z = 0 \Rightarrow x = z + 2$. The system is dependent and the solution is given by $\{(z + 2, -z - 1, z)\}$.

22. $\begin{bmatrix} -5 & 4 & 9 \\ 2 & -1 & -2 \end{bmatrix} + \begin{bmatrix} 1 & -2 & 7 \\ 4 & -5 & -5 \end{bmatrix} = \begin{bmatrix} -5+1 & 4-2 & 9+7 \\ 2+4 & -1-5 & -2-5 \end{bmatrix} = \begin{bmatrix} -4 & 2 & 16 \\ 6 & -6 & -7 \end{bmatrix}$.

23. $\begin{bmatrix}3\\2\\5\end{bmatrix}-\begin{bmatrix}8\\-4\\6\end{bmatrix}+\begin{bmatrix}1\\0\\2\end{bmatrix}=\begin{bmatrix}3-8+1\\2+4+0\\5-6+2\end{bmatrix}=\begin{bmatrix}-4\\6\\1\end{bmatrix}.$

24. $\begin{bmatrix}2&5&8\\1&9&2\end{bmatrix}-\begin{bmatrix}3&4\\7&1\end{bmatrix}$. We cannot subtract matrices of unlike size $(2\times3)-(2\times3)$, the solution is $\varnothing$.

25. $3\begin{bmatrix}2&4\\-1&4\end{bmatrix}-2\begin{bmatrix}5&8\\2&-2\end{bmatrix}=\begin{bmatrix}6&12\\-3&12\end{bmatrix}-\begin{bmatrix}10&16\\4&-4\end{bmatrix}=\begin{bmatrix}6-10&12-16\\-3-4&12+4\end{bmatrix}=\begin{bmatrix}-4&-4\\-7&16\end{bmatrix}.$

26. $-1\begin{bmatrix}3&-5&2\\1&7&-4\end{bmatrix}+5\begin{bmatrix}0&2\\-1&3\end{bmatrix}$. We cannot add matrices of unlike size $(2\times3)+(2\times2)$, the solution is $\varnothing$.

27. $10\begin{bmatrix}2x&y\\5y&6x\end{bmatrix}+2\begin{bmatrix}-3x&6y\\2y&5x\end{bmatrix}=\begin{bmatrix}20x&10y\\50y&60x\end{bmatrix}+\begin{bmatrix}-6x&12y\\4y&10x\end{bmatrix}=\begin{bmatrix}20x-6x&10y+12y\\50y+4y&60x+10x\end{bmatrix}=\begin{bmatrix}14x&22y\\54y&70x\end{bmatrix}$

28. The sum of two $m\times n$ matrices A and B is founded by adding corresponding elements.

29. $\begin{bmatrix}-8&6\\5&2\end{bmatrix}\begin{bmatrix}3&-1\\7&2\end{bmatrix}=\begin{bmatrix}-8(3)+6(7)&-8(-1)+6(2)\\5(3)+2(7)&5(-1)+2(2)\end{bmatrix}=\begin{bmatrix}-24+42&8+12\\15+14&-5+4\end{bmatrix}=\begin{bmatrix}18&20\\29&-1\end{bmatrix}.$

30. $\begin{bmatrix}3&2&-1\\4&0&6\end{bmatrix}\begin{bmatrix}-2&0\\0&2\\3&1\end{bmatrix}=\begin{bmatrix}3(-2)+2(0)-1(3)&3(0)+2(2)-1(1)\\4(-2)+0(0)+6(3)&4(0)+0(2)+6(1)\end{bmatrix}=\begin{bmatrix}-9&3\\10&6\end{bmatrix}.$

31. $\begin{bmatrix}1&-2&4&2\\0&1&-1&8\end{bmatrix}\begin{bmatrix}-1\\2\\0\\1\end{bmatrix}=\begin{bmatrix}1(-1)-2(2)+4(0)+2(1)\\0(-1)+1(2)-1(0)+8(1)\end{bmatrix}=\begin{bmatrix}-3\\10\end{bmatrix}.$

32. $\begin{bmatrix}1&2&5\\-3&4&7\\0&2&-1\end{bmatrix}\begin{bmatrix}4&2&3\\10&-5&6\end{bmatrix}=$ We cannot multiply matrices of size $(3\times3)\times(2\times3)$, the solution is $\varnothing$.

33. $\begin{bmatrix}4&2&3\\10&-5&6\end{bmatrix}\begin{bmatrix}1&2&5\\-3&4&7\\0&2&-1\end{bmatrix}=\begin{bmatrix}4-6+0&8+8+6&20+14-3\\10+15+0&20-20+12&50-35-6\end{bmatrix}=\begin{bmatrix}-2&22&31\\25&12&9\end{bmatrix}.$

34. $\begin{bmatrix}3&-1&0\end{bmatrix}\begin{bmatrix}1&3&2\\2&-4&0\\5&7&3\end{bmatrix}=\begin{bmatrix}3-2+0&9+4+0&6+0+0\end{bmatrix}=\begin{bmatrix}1&13&6\end{bmatrix}.$

35. Yes; $AB=\begin{bmatrix}3&2\\13&9\end{bmatrix}\begin{bmatrix}9&-2\\-13&3\end{bmatrix}=\begin{bmatrix}27-26&-6+6\\117-117&-26+27\end{bmatrix}=\begin{bmatrix}1&0\\0&1\end{bmatrix}.$

$BA=\begin{bmatrix}9&-2\\-13&3\end{bmatrix}\begin{bmatrix}3&2\\13&9\end{bmatrix}=\begin{bmatrix}27-26&18-18\\-39+39&-26+27\end{bmatrix}=\begin{bmatrix}1&0\\0&1\end{bmatrix}.$

36. Yes; $AB=\begin{bmatrix}1&0\\2&-3\end{bmatrix}\begin{bmatrix}1&0\\\frac{2}{3}&-\frac{1}{3}\end{bmatrix}=\begin{bmatrix}1+0&0+0\\2-2&0+1\end{bmatrix}=\begin{bmatrix}1&0\\0&1\end{bmatrix}$ $BA=\begin{bmatrix}1&1\\\frac{2}{3}&-\frac{1}{3}\end{bmatrix}\begin{bmatrix}1&0\\2&-3\end{bmatrix}\begin{bmatrix}1+0&0+0\\\frac{2}{3}-\frac{2}{3}&0+1\end{bmatrix}=\begin{bmatrix}1&0\\0&1\end{bmatrix}.$

37. No; $AB=\begin{bmatrix}2&0&6\\0&1&0\\1&0&1\end{bmatrix}\begin{bmatrix}-1&0&\frac{3}{2}\\0&1&0\\\frac{1}{4}&0&-1\end{bmatrix}=\begin{bmatrix}-2+0+\frac{3}{2}&0+0+0&3+0-6\\0+0+0&0+1+0&0+0+0\\-1+0+\frac{1}{4}&0+0+0&\frac{3}{2}+0-1\end{bmatrix}=\begin{bmatrix}-\frac{1}{2}&0&-3\\0&1&0\\-\frac{3}{4}&0&\frac{1}{2}\end{bmatrix}$.

38. Yes; $AB=\begin{bmatrix}1&0&2\\0&2&4\\0&0&1\end{bmatrix}\begin{bmatrix}1&0&-2\\0&\frac{1}{2}&-2\\0&0&1\end{bmatrix}=\begin{bmatrix}1+0+0&0+0+0&-2+0+2\\0+0+0&0+1+0&0-4+4\\0+0+0&0+0+0&0+0+1\end{bmatrix}=\begin{bmatrix}1&0&0\\0&1&0\\0&0&1\end{bmatrix}$.

$BA=\begin{bmatrix}1&0&-2\\0&\frac{1}{2}&-2\\0&0&1\end{bmatrix}\begin{bmatrix}1&0&2\\0&2&4\\0&0&1\end{bmatrix}=\begin{bmatrix}1+0+0&0+0+0&2+0-2\\0+0+0&0+1+0&0+2-2\\0+0+0&0+0+0&0+0+1\end{bmatrix}=\begin{bmatrix}1&0&0\\0&1&0\\0&0&1\end{bmatrix}$.

39. $\det A=30-30=0$. Since the determinant is equal to 0, A^{-1} does not exist.

40. Using $A^{-1}=\frac{1}{\det A}\begin{bmatrix}d&-b\\-c&a\end{bmatrix}$ to find the inverse yields $A^{-1}=\begin{bmatrix}-4&2\\0&3\end{bmatrix}^{-1}=-\frac{1}{12}\begin{bmatrix}3&-2\\0&-4\end{bmatrix}=\begin{bmatrix}-\frac{1}{4}&\frac{1}{6}\\0&\frac{1}{3}\end{bmatrix}$.

41. Using $A^{-1}=\frac{1}{\det A}\begin{bmatrix}d&-b\\-c&a\end{bmatrix}$ to find the inverse yields $A^{-1}=\begin{bmatrix}2&0\\-1&5\end{bmatrix}^{-1}=\frac{1}{10}\begin{bmatrix}5&0\\1&2\end{bmatrix}=\begin{bmatrix}\frac{1}{2}&0\\\frac{1}{10}&\frac{1}{5}\end{bmatrix}$.

42. Solve using the calculator, $A^{-1}=\begin{bmatrix}\frac{1}{4}&\frac{1}{2}&\frac{1}{2}\\\frac{1}{4}&-\frac{1}{2}&\frac{1}{2}\\\frac{1}{8}&-\frac{1}{4}&-\frac{1}{4}\end{bmatrix}=\begin{bmatrix}0.25&0.5&0.5\\0.25&-0.5&0.5\\0.125&-0.25&-0.25\end{bmatrix}$.

43. Solve using the calculator $A^{-1}=\begin{bmatrix}\frac{2}{3}&0&-\frac{1}{3}\\\frac{1}{3}&0&-\frac{2}{3}\\-\frac{2}{3}&1&\frac{1}{3}\end{bmatrix}$.

44. Solve using the calculator, the determinant of $A=0$; therefore, there is no inverse.

45. Using the matrix inverse method, put the system into the proper matrix form:

$\begin{bmatrix}1&1\\2&3\end{bmatrix}\begin{bmatrix}x\\y\end{bmatrix}=\begin{bmatrix}4\\10\end{bmatrix}\Rightarrow\begin{bmatrix}x\\y\end{bmatrix}=\begin{bmatrix}1&1\\2&3\end{bmatrix}^{-1}\begin{bmatrix}4\\10\end{bmatrix}\Rightarrow\begin{bmatrix}3&-1\\-2&1\end{bmatrix}\begin{bmatrix}4\\10\end{bmatrix}\Rightarrow\begin{bmatrix}12-10\\-8+10\end{bmatrix}=\begin{bmatrix}2\\2\end{bmatrix}$. The solution is $(2,2)$.

46. Using the matrix inverse method, put the system into the proper matrix form:

$\begin{bmatrix}5&-3\\2&7\end{bmatrix}\begin{bmatrix}x\\y\end{bmatrix}=\begin{bmatrix}-2\\-9\end{bmatrix}\Rightarrow\begin{bmatrix}x\\y\end{bmatrix}=\begin{bmatrix}5&-3\\2&7\end{bmatrix}^{-1}\begin{bmatrix}-2\\-9\end{bmatrix}\Rightarrow\begin{bmatrix}\frac{7}{41}&\frac{3}{41}\\-\frac{2}{41}&\frac{5}{41}\end{bmatrix}=\begin{bmatrix}-2\\-9\end{bmatrix}\Rightarrow\begin{bmatrix}-\frac{14}{41}-\frac{27}{41}\\\frac{4}{41}-\frac{45}{41}\end{bmatrix}=\begin{bmatrix}-1\\-1\end{bmatrix}$.

The solution is $(-1,-1)$.

47. Using the matrix inverse method, put the system into the proper matrix form:

$\begin{bmatrix}2&1\\3&-2\end{bmatrix}\begin{bmatrix}x\\y\end{bmatrix}=\begin{bmatrix}5\\4\end{bmatrix}\Rightarrow\begin{bmatrix}x\\y\end{bmatrix}=\begin{bmatrix}2&1\\3&-2\end{bmatrix}^{-1}\begin{bmatrix}5\\4\end{bmatrix}\Rightarrow\begin{bmatrix}\frac{2}{7}&\frac{1}{7}\\\frac{3}{7}&-\frac{2}{7}\end{bmatrix}\begin{bmatrix}5\\4\end{bmatrix}\Rightarrow\begin{bmatrix}\frac{10}{7}+\frac{4}{7}\\\frac{15}{7}-\frac{8}{7}\end{bmatrix}=\begin{bmatrix}2\\1\end{bmatrix}$. The solution is $(2,1)$.

48. Using the matrix inverse method, put the system into the proper matrix form:

$$\begin{bmatrix}1 & -2\\3 & 1\end{bmatrix}\begin{bmatrix}x\\y\end{bmatrix}=\begin{bmatrix}7\\7\end{bmatrix}\Rightarrow\begin{bmatrix}x\\y\end{bmatrix}=\begin{bmatrix}1 & -2\\3 & 1\end{bmatrix}^{-1}\begin{bmatrix}7\\7\end{bmatrix}\Rightarrow\begin{bmatrix}\frac{1}{7} & \frac{2}{7}\\-\frac{3}{7} & \frac{1}{7}\end{bmatrix}\begin{bmatrix}7\\7\end{bmatrix}\Rightarrow\begin{bmatrix}\frac{7}{7}+\frac{14}{7}\\-\frac{21}{7}+\frac{7}{7}\end{bmatrix}=\begin{bmatrix}3\\-2\end{bmatrix}.$$ The solution is $(3,-2)$.

49. Solve on the calculator using the matrix inverse method:

$$\begin{bmatrix}x\\y\\z\end{bmatrix}=\begin{bmatrix}1 & 2 & 0\\0 & 3 & -1\\1 & 2 & -1\end{bmatrix}^{-1}\cdot\begin{bmatrix}-1\\-5\\-3\end{bmatrix}=\begin{bmatrix}1\\-1\\2\end{bmatrix}$$; therefore, the solution is $(1,-1,2)$.

50. The determinant of the coefficient matrix is $\det\begin{bmatrix}3 & -2 & 4\\4 & 1 & -5\\-6 & 4 & -8\end{bmatrix}=0$; therefore, infinitely many solutions or no solutions. Now use Row Echelon Method on the augmented matrix:

$$\left[\begin{array}{ccc|c}3 & -2 & 4 & 1\\4 & 1 & -5 & 2\\-6 & 4 & -8 & -2\end{array}\right]\Rightarrow 2R_1+R_3\rightarrow\left[\begin{array}{ccc|c}3 & -2 & 4 & 1\\4 & 1 & -5 & 2\\0 & 0 & 0 & 0\end{array}\right].$$ Since $0=0$, the equation has an infinite number of solutions. Continue to use Row Echelon Method to find these solutions: $R_1+2R_2\rightarrow\left[\begin{array}{ccc|c}3 & -2 & 4 & 1\\11 & 0 & -6 & 5\\0 & 0 & 0 & 0\end{array}\right].$

Now solve for R_2 for x: $11x-6z=5\Rightarrow 11x=6z+5\Rightarrow x=\frac{6z+5}{11}$. Now substitute $x=\frac{6z+5}{11}$ into R_1 and solve for y:

$$3\left(\frac{6z+5}{11}\right)-2y+4z=1\Rightarrow 18z+15-22y+44z=11\Rightarrow -22y=-62z-4\Rightarrow y=\frac{62z+4}{22}=\frac{31z+2}{11}.$$

The solution is $\left(\frac{6z+5}{11},\frac{31z+2}{11},z\right)$. Other forms are possible.

51. Solve on the calculator using the matrix inverse method:

$$\begin{bmatrix}x\\y\\z\end{bmatrix}=\begin{bmatrix}1 & 1 & 1\\2 & -1 & 0\\0 & 3 & 1\end{bmatrix}^{-1}\cdot\begin{bmatrix}1\\-2\\2\end{bmatrix}=\begin{bmatrix}-1\\0\\2\end{bmatrix}$$; therefore, the solution is $(-1,0,2)$.

52. Solve on the calculator using the matrix inverse method:

$$\begin{bmatrix}x\\y\\z\end{bmatrix}=\begin{bmatrix}1 & 0 & 0\\0 & 1 & 1\\2 & 0 & -3\end{bmatrix}^{-1}\cdot\begin{bmatrix}-3\\6\\-9\end{bmatrix}=\begin{bmatrix}-3\\5\\1\end{bmatrix}$$; therefore, the solution is $(-3,5,1)$.

53. Solve on the calculator using the matrix inverse method:

$$\begin{bmatrix}x\\y\\z\end{bmatrix}=\begin{bmatrix}2 & -4 & 4\\1 & -3 & 2\\1 & -1 & 2\end{bmatrix}^{-1}\cdot\begin{bmatrix}0\\-3\\1\end{bmatrix}\Rightarrow\varnothing.$$ Since the inverse of the 3×3 matrix does not exist this system is inconsistent.

54. One solution to the solution set $\{(4-y, y)\}$ is $\{(4-1,1)\} \Rightarrow \{(3,1)\}$. Answers may vary.

55. $\det\begin{bmatrix} -1 & 8 \\ 2 & 9 \end{bmatrix} = -9-16 = -25.$

56. $\det\begin{bmatrix} -2 & 4 \\ 0 & 3 \end{bmatrix} = -6-0 = -6.$

57. Evaluate, expand by the second column. Therefore, $\det = (-)(a_{12})(M_{12}) + (a_{22})(M_{22}) - (a_{32})(M_{32})$:

$$-4\left(\det\begin{bmatrix} 3 & 2 \\ -1 & 3 \end{bmatrix}\right) + 0 - 0 = -4(9+2) = -4(11) = -44.$$

58. Evaluate, expand by the second column. Therefore, $\det = (-)(a_{12})(M_{12}) + (a_{22})(M_{22}) - (a_{32})(M_{32})$:

$$-2\left(\det\begin{bmatrix} 4 & 3 \\ 5 & 2 \end{bmatrix}\right) + 0 - (-1)\left(\det\begin{bmatrix} -1 & 3 \\ 4 & 3 \end{bmatrix}\right) = -2(8-15) + 1(-3-12) = -2(-7) + 1(-15) = 14 - 15 = -1.$$

59. If $\det\begin{bmatrix} -3 & 2 \\ 1 & x \end{bmatrix} = 5$, then $-3x - 2 = 5 \Rightarrow -3x = 7 \Rightarrow x = \frac{7}{3}$. The solution set is $\left\{-\frac{7}{3}\right\}$.

60. If $\det\begin{bmatrix} 3x & 7 \\ -x & 4 \end{bmatrix} = 8$, then $12x + 7x = 8 \Rightarrow 19x = 8 \Rightarrow x = \frac{8}{19}$. The solution set is $\left\{\frac{8}{19}\right\}$.

61. Evaluate, expand by the third column. Therefore, $\det = (a_{13})(M_{13}) - (a_{23})(M_{23}) + (a_{33})(M_{33})$:

$0 - (-1)\left(\det\begin{bmatrix} 2 & 5 \\ 0 & 2 \end{bmatrix}\right) + 0 = 4 \Rightarrow 4 = 4$. Since this is always true, all real numbers can be input for x.

62. Evaluate, expand by the first row. Therefore, $\det = (a_{11})(M_{11}) - (a_{12})(M_{12}) + (a_{13})(M_{13})$:

$$6x\left(\det\begin{bmatrix} 5 & 3 \\ 2 & -1 \end{bmatrix}\right) - 2\left(\det\begin{bmatrix} 1 & 3 \\ x & -1 \end{bmatrix}\right) + 0 = 2x \Rightarrow 6x(-5-6) - 2(-1-3x) = 2x \Rightarrow$$

$-66x + 2 + 6x = 2x \Rightarrow -60x + 2 = 2x \Rightarrow -62x = -2 \Rightarrow x = \frac{-2}{-62}$. The solution set is $\left\{\frac{1}{31}\right\}$.

63. (a) $D = \det\begin{bmatrix} 3 & -1 \\ 2 & 1 \end{bmatrix} = 5.$

(b) $D_x = \det\begin{bmatrix} 28 & -1 \\ 2 & 1 \end{bmatrix} = 30.$

(c) $D_y = \det\begin{bmatrix} 3 & 28 \\ 2 & 2 \end{bmatrix} = -50.$

(d) $x = \frac{D_x}{D} = \frac{30}{5} = 6;\ y = \frac{D_y}{D} = \frac{-50}{5} = -10$. The solution is $(6, -10)$.

64. (a) $A =$ the coefficient matrix: $A = \begin{bmatrix} 3 & -1 \\ 2 & 1 \end{bmatrix}$.

(b) $B =$ the answer matrix: $B = \begin{bmatrix} 28 \\ 2 \end{bmatrix}$.

(c) To solve for x and y, multiply A^{-1} by B: $\begin{bmatrix} x \\ y \end{bmatrix} = A^{-1}B = \begin{bmatrix} 6 \\ -10 \end{bmatrix}$; therefore, the solution is $(6,-10)$.

65. If $D=0$, there would be division by 0, which is undefined. The system will have no solutions or infinitely many solutions.

66. Find the determinants $D = \det\begin{bmatrix} 3 & 1 \\ 5 & 4 \end{bmatrix} = 7$, $D_x = \det\begin{bmatrix} -1 & 1 \\ 10 & 4 \end{bmatrix} = -14$, and $D_y = \det\begin{bmatrix} 3 & -1 \\ 5 & 10 \end{bmatrix} = 35$. Then

$x = \frac{D_x}{D} = \frac{-14}{7} = -2$ and $y = \frac{D_y}{D} = \frac{35}{7} = 5$. The solution is $(-2,5)$.

67. Find the determinants $D = \det\begin{bmatrix} 3 & 7 \\ 5 & -1 \end{bmatrix} = -38$, $D_x = \det\begin{bmatrix} 2 & 7 \\ -22 & -1 \end{bmatrix} = 152$, and $D_y = \det\begin{bmatrix} 3 & 2 \\ 5 & -22 \end{bmatrix} = -76$.

Then $x = \frac{D_x}{D} = \frac{152}{-38} = -4$ and $y = \frac{D_y}{D} = \frac{-76}{-38} = 2$. The solution is $(-4,2)$.

68. Find the determinants $D = \det\begin{bmatrix} 2 & -5 \\ 3 & 4 \end{bmatrix} = 23$, $D_x = \det\begin{bmatrix} 8 & -5 \\ 10 & 4 \end{bmatrix} = 82$, and $D_y = \det\begin{bmatrix} 2 & 8 \\ 3 & 10 \end{bmatrix} = -4$. Then

$x = \frac{D_x}{D} = \frac{82}{23}$ and $y = \frac{D_y}{D} = \frac{-4}{23}$. The solution is $\left(\frac{82}{23}, -\frac{4}{23}\right)$.

69. Using your calculator, find the following determinants:

$D = \det\begin{bmatrix} 3 & 2 & 1 \\ 4 & -1 & 3 \\ 1 & 3 & -1 \end{bmatrix} = 3$, $D_x = \det\begin{bmatrix} 2 & 2 & 1 \\ -16 & -1 & 3 \\ 12 & 3 & -1 \end{bmatrix} = -12$, $D_y = \det\begin{bmatrix} 3 & 2 & 1 \\ 4 & -16 & 3 \\ 1 & 12 & -1 \end{bmatrix} = 18$, and

$D_z = \det\begin{bmatrix} 3 & 2 & 2 \\ 4 & -1 & -16 \\ 1 & 3 & 12 \end{bmatrix} = 6$ Then $x = \frac{D_x}{D} = \frac{-12}{3} = -4$, $y = \frac{D_y}{D} = \frac{18}{3} = 6$, and $z = \frac{D_z}{D} = \frac{6}{3} = 2$.

The solution is $(-4,6,2)$.

70. The determinant of the coefficient matrix is $\det\begin{bmatrix} 5 & -2 & -1 \\ -5 & 2 & 1 \\ 1 & -4 & -2 \end{bmatrix} = 0$; therefore, infinitely many or no solutions. Now use Row Echelon Method on the augmented matrix: $\left.\begin{matrix} R_3 \to R_1 \\ R_2 \to R_3 \end{matrix}\right\} \to \left[\begin{array}{ccc|c} 1 & -4 & -2 & 0 \\ 5 & -2 & -1 & 8 \\ -5 & 2 & 1 & -8 \end{array}\right] \Rightarrow$

Since $R_2 + R_3$ produces $0=0$, there are infinitely many solutions and they are dependent solutions.

Continue to use Row Echelon Method to find these solutions: $R_1 + 2R_3 \to \left[\begin{array}{ccc|c} 1 & -4 & -2 & 0 \\ 5 & -2 & -1 & 8 \\ -9 & 0 & 0 & -16 \end{array}\right]$.

Solve for x using R_3: $-9x=-16 \Rightarrow x=\frac{16}{9}$. Now substitute $x=\frac{16}{9}$ into R_1 and solve for y:

$$\frac{16}{9}-4y-2z=0 \Rightarrow 16-36y-18z=0 \Rightarrow -36y=18z-16 \Rightarrow y=\frac{18z-16}{-36}=\frac{8-9z}{18}.$$

The solution is $\left(\frac{16}{9},\frac{8-9z}{18},z\right)$.

71. Using your calculator, find the following determinants:

$$D=\det\begin{bmatrix}-1 & 3 & -4\\ 2 & 4 & 1\\ 3 & 0 & -1\end{bmatrix}=67;\ D_x=\det\begin{bmatrix}2 & 3 & -4\\ 3 & 4 & 1\\ 9 & 0 & -1\end{bmatrix}=172;\ D_y=\det\begin{bmatrix}-1 & 2 & -4\\ 2 & 3 & 1\\ 3 & 9 & -1\end{bmatrix}=-14;$$

$$D_z=\det\begin{bmatrix}-1 & 3 & 2\\ 2 & 4 & 3\\ 3 & 0 & 9\end{bmatrix}=-87 \text{ Then } x=\frac{D_x}{D}=\frac{172}{67},\ y=\frac{D_y}{D}=\frac{-14}{67}, \text{ and } z=\frac{D_z}{D}=\frac{-87}{67}.$$

The solution is $\left(\frac{172}{67},-\frac{14}{67},-\frac{87}{67}\right)$.

72. Let x = the amount of rice in cups and y = the amount of soybeans in cups. Then from the information the system of equations is $15x+22.5y=9.5$ [equation 1] and $810x+270y=324$ [equation 2]. Multiply [equation 1] by -12 and add [equation 2] to eliminate y:

$$\begin{aligned}-180x-270y&=-114\\ 810x+270y&=\ 324\\ \hline 630x\qquad\quad &=\ 210 \Rightarrow x=\frac{210}{630}=\frac{1}{3}.\end{aligned}$$

Now substitute $x=\frac{1}{3}$ into [equation 1] and solve for y:

$$15\left(\frac{1}{3}\right)+22.5y=9.5 \Rightarrow 5+22.5y=9.5 \Rightarrow 2.5y=4.5 \Rightarrow y=0.20 \text{ or } \frac{1}{5}.$$

The meal should include $\frac{1}{3}$ cup of rice and $\frac{1}{5}$ cup of soybeans.

73. Let x = the number of CD's and y = the number of plastic holders. From the information the system of equations is obtained: $x+y=100$ [equation 1] and $0.40x+0.30y=38.00$ [equation 2]. Multiply [equation 1] by -30 and add [equation 2] multiplied by 100 to eliminate y:

$$\begin{aligned}-30x-30y&=-3000\\ 40x+30y&=\ 3800\\ \hline 10x\qquad\quad &=\ 800 \Rightarrow x=80.\end{aligned}$$

Now substitute $x=80$ into [equation 1] and solve for y:

$80+y=100 \Rightarrow y=20$. They should send 80 CD's and 20 holders.

74. Let x = the number pounds of \$4.60 tea, y = the number of pounds of \$5.75 tea, and z = the number of pounds of \$6.50 tea. From the information, the system of equations is obtained: $x+y+z=20$ [equation 1],

$4.6x + 5.75y + 6.5z = 20(5.25) = 105$ and $x = y + z \Rightarrow x - y - z = 0$. Now create the coefficient matrix and solve on the calculator using the matrix inverse method:

$$A = \begin{bmatrix} 1 & 1 & 1 \\ 4.6 & 5.75 & 6.5 \\ 1 & -1 & -1 \end{bmatrix}. \text{ Therefore, } \begin{bmatrix} x \\ y \\ z \end{bmatrix} = \begin{bmatrix} 1 & 1 & 1 \\ 4.6 & 5.75 & 6.5 \\ 1 & -1 & -1 \end{bmatrix}^{-1} \cdot \begin{bmatrix} 20 \\ 105 \\ 0 \end{bmatrix} = \begin{bmatrix} 10 \\ 8 \\ 2 \end{bmatrix}.$$

They should use 10 pounds of \$4.60 tea, 8 pounds of \$5.75 tea, and 2 pounds of \$6.50 tea.

75. Let x = the amount of 5% solution (ml). y = the amount of 15% solution (ml), and z = the amount of 10% solution (ml). Then from the information, the system of equations is $x + y + z = 20$ [equation 1], $0.05x + 0.15y + 0.10z = 0.08(20) = 1.6$, and $x = y + z + 2 \Rightarrow x - y - z = 2$. Now create the coefficient matrix and solve on the calculator using the matrix inverse method:

$$A = \begin{bmatrix} 1 & 1 & 1 \\ 5 & 15 & 10 \\ 1 & -1 & -1 \end{bmatrix}; \text{ therefore, } \begin{bmatrix} x \\ y \\ z \end{bmatrix} = \begin{bmatrix} 1 & 1 & 1 \\ 5 & 15 & 10 \\ 1 & -1 & -1 \end{bmatrix}^{-1} \cdot \begin{bmatrix} 20 \\ 160 \\ 2 \end{bmatrix} = \begin{bmatrix} 11 \\ 3 \\ 6 \end{bmatrix}.$$

They should use 11 ml of 5% solution, 3 ml of 15% solution, and 6 ml of 10% solution.

76. (a) Using $P = a + bA + cW$ and the data from the table yields,

$$a + 39b + 142c = 113$$
$$a + 53b + 181c = 138$$
$$a + 65b + 191c = 152$$

Now using the capabilities of the calculator and the Reduced Row Echelon Method:

$$\left[\begin{array}{ccc|c} 1 & 39 & 142 & 113 \\ 1 & 53 & 181 & 138 \\ 1 & 65 & 191 & 152 \end{array}\right] = \left[\begin{array}{ccc|c} 1 & 0 & 0 & 32.780488 \\ 0 & 1 & 0 & 0.9024390 \\ 0 & 0 & 1 & 0.3170732 \end{array}\right].$$

Therefore, the equation is $P \approx 32.78 + 0.9024A + 0.3171W$.

(b) Using $A = 55$ and $W = 175$ yields $P \approx 32.78 + 0.9024(55) + 0.3171(175) \Rightarrow P \approx 138$.

77. Using the equation for a polynomial of degree 3 and given points yields:

$(-2,1)$: $-8a + 4b - 2c + d = 1$

$(-1,6)$: $-a + b - c + d = 6$

$(2,9)$: $8a + 4b + 2c + d = 9$

$(3,26)$: $27a + 9b + 3c + d = 26$

The equations can be represented by $\left[\begin{array}{cccc|c} -8 & 4 & -2 & 1 & 1 \\ -1 & 1 & -1 & 1 & 6 \\ 8 & 4 & 2 & 1 & 9 \\ 27 & 9 & 3 & 1 & 26 \end{array}\right]$.

Using the capabilities of the calculator, the Reduced Row Echelon Method produces the solutions: $a = 1,\ b = 0,\ c = -2,\ d = 5$. The equation is $P(x) = x^3 - 2x + 5$.

78. Using $f(x)=ax^2+bx+c$ and the data from the graph yields:

From the point $(-6,4)$: $36a-6b+c=4$.

From the point $(-4,-2)$: $16a-4b+c=-2$.

From the point $(2,4)$: $4a+2b+c=4$.

Now using your calculator, find the following determinants:

$$D=\det\begin{bmatrix}36&-6&1\\16&-4&1\\4&2&1\end{bmatrix}=-96;\ D_a=\det\begin{bmatrix}4&-6&1\\-2&-4&1\\4&2&1\end{bmatrix}=-48;\ D_b=\det\begin{bmatrix}36&4&1\\16&-2&1\\4&4&1\end{bmatrix}=-192;$$

$$D_c=\det\begin{bmatrix}36&-6&4\\16&-4&-2\\4&2&4\end{bmatrix}=192 \text{ Then } a=\frac{D_a}{D}=\frac{-48}{-96}=\frac{1}{2},\ b=\frac{D_b}{D}=\frac{-192}{-96}=2, \text{ and}$$

$c=\frac{D_c}{D}=\frac{192}{-96}=-2$. The equation is $P(x)=\frac{1}{2}x^2+2x-2$.

79. See Figure 79.

80. See Figure 80.

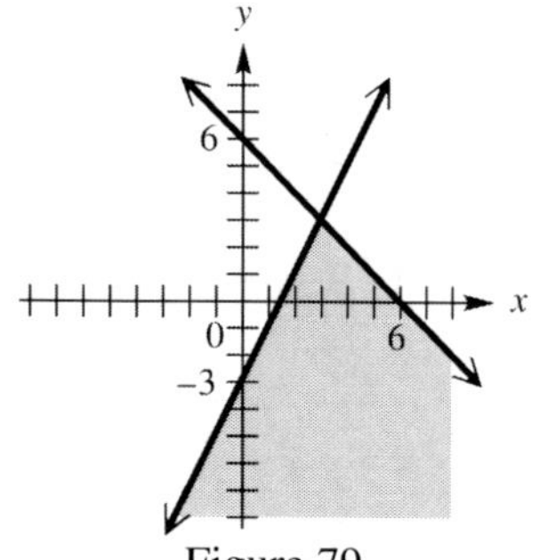

Figure 79

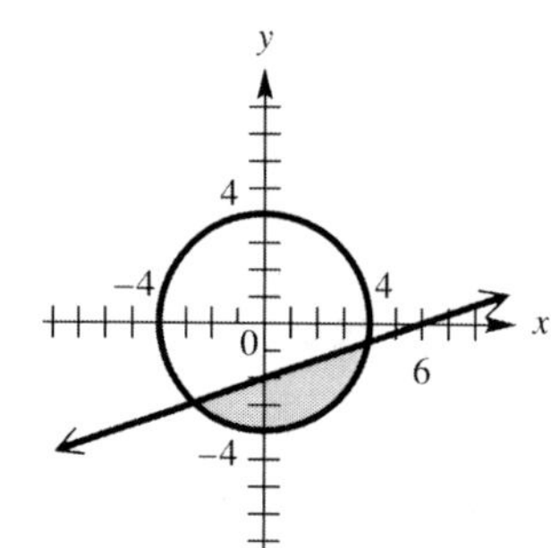

Figure 80

81. Graphing $x\geq 0,\ y\geq 0,\ 3x+2y\leq 12$ and $5x+y\geq 5$ in the same window yields four vertices. Using the given expression and the ordered pairs of the vertices yields the following solutions:

$(0,5)$: $2(0)+4(5)=20$

$(0,6)$: $2(0)+4(6)=24$

$(1,0)$: $2(1)+4(0)=2$

$(4,0)$: $2(4)+4(0)=8$

Therefore, the maximum value is 24 at (0,6).

82. Graphing $x\geq 0,\ y\geq 0,\ x+y\leq 50,\ 2x+y\geq 20$ and $x+2y\geq 30$ in the same window yields five vertices. Using the given expression and the ordered pairs of the vertices yields the following solutions:

$(0,50):\ 4(0)+2(50)=100$

$(0,20):\ 4(0)+2(20)=40$

$\left(\frac{10}{3},\frac{40}{3}\right):\ 4\left(\frac{10}{3}\right)+2\left(\frac{40}{3}\right)=40$

$(30.0):\ 4(30)+2(0)=120$

$(50,0):\ 4(50)+2(0)=200$

Therefore, the minimum value is 40 at (0,20) and $\left(\frac{10}{3},\frac{40}{3}\right)$

Note: values on the line $2x+y=20$ also give minimums of 40 between the two points.

83. Let x = the number radios produced daily and let y = the number of Blu-ray players produced daily.

The objective function to find the maximum profit is $15x+35y$.

The constraints are

$5\le x\le 25$ (radio production restrictions)

$0\le y\le 30$ (Blu-ray player maximum production)

$x\le y$ (radio production less than or equal to Blu-ray player production)

Graphing the constraints on the calculator yields four vertices. Inputting these ordered pairs into the objective function yields (5,30): 1125, (5,5): 250, (25,30): 1425, and (25,25): 1250.

The maximum profit is $1425, this happens when 25 radios and 30 Blu-ray players are manufactured.

84. Factoring $\frac{5x-2}{x^2-4}$ results in $\frac{5x-2}{(x-2)(x+2)}$. Multiply $\frac{5x-2}{(x-2)(x+2)}=\frac{A}{x-2}+\frac{B}{x+2}$ by

$(x-2)(x+2)\Rightarrow 5x-2=A(x+2)+B(x-2)$. Let $x=-2\Rightarrow -12=B(-4)\Rightarrow B=3$.

Let $x=2\Rightarrow 8=A(4)\Rightarrow A=2$. The expression can be written $\frac{2}{x-2}+\frac{3}{x+2}$.

85. Factoring $\frac{x+2}{x^3+2x^2+x}$ results in $\frac{x+2}{x(x+1)^2}$. Multiply $\frac{x+2}{x^3-x^2+4x}=\frac{A}{x}+\frac{B}{x+1}+\frac{C}{(x+1)^2}$ by

$x(x+1)^2\Rightarrow x+2=A(x+1)^2+Bx(x+1)+Cx$. Let $x=0\Rightarrow 2=A(1)\Rightarrow A=2$. Let

$x=-1\Rightarrow 1=C(-1)\Rightarrow C=-1$. Let $x=1,\ A=2,\ C=-1\Rightarrow$

$3=2(4)+B(1)(2)-1\Rightarrow 3=8=2B-1\Rightarrow B=-2$

The expression can be written $\frac{2}{x}+\frac{-2}{x+1}+\frac{-1}{(x+1)^2}$ or $\frac{2}{x}-\frac{2}{x+1}-\frac{1}{(x+1)^2}$.

86. Factoring $\frac{x+2}{x^3-x^2+4x}$ results in $\frac{x+2}{x(x^2-x+4)}$. Multiply $\frac{x+2}{x(x^2-x+4)}=\frac{A}{x}+\frac{Bx+C}{x^2-x+4}$ by

$x(x^2-x+4)\Rightarrow\ x+2=A(x^2-x+4)+(Bx+C)(x)\Rightarrow x+2=(A+B)x^2+(C-A)x+4A.$

By equating coefficients $A+B=0$, $C-A=1$, and $4A=2 \Rightarrow A=\frac{1}{2}$. Since $A+B=0$ and

$A=\frac{1}{2}$, $B=-\frac{1}{2}$. Since $C-A=1$ and $A=\frac{1}{2}$, $C=\frac{3}{2}$.

The expression can be written $\frac{\frac{1}{2}}{x}+\frac{-\frac{1}{2}x+\frac{3}{2}}{x^2-x+4}=\frac{1}{2x}+\frac{-x+3}{2\left(x^2-x+4\right)}$.

87. Factoring $\frac{6x^2-x-3}{x^3-x}=\frac{6x^2-x-3}{x\left(x^2-1\right)}=\frac{6x^2-x-3}{x(x-1)(x+1)}$. Multiply $\frac{6x^2-x-3}{x(x-1)(x+1)}=\frac{A}{x}+\frac{B}{x-1}+\frac{C}{x+1}$ by

$x(x-1)(x+1) \Rightarrow 6x^2-x-3=A(x-1)(x+1)+(B)(x)(x+1)+(C)x(x-1)$.

Let $x=0 \Rightarrow -3=-A \Rightarrow A=3$. Let $x=1 \Rightarrow 2=2B \Rightarrow B=1$. Let $x=-1 \Rightarrow 4=2C \Rightarrow C=2$.

The expression can be written $\frac{3}{x}+\frac{1}{x-1}+\frac{2}{x+1}$.

Chapter 6 Test

1. (a) The first is the equation of a circle; the second is the equation of a line.

   (b) A circle and a line could intersect in 0, 1, or 2 points.

   (c) First solve the second equation for y: $2x-y=0 \Rightarrow y=2x$. Now substitute $y=2x$ into

   Equation 1: $x^2+(2x)^2=5 \Rightarrow x^2+4x^2=5 \Rightarrow 5x^2=5 \Rightarrow x^2=1 \Rightarrow x=\pm 1$

   Now substitute these values into $y=2x$: $y=2(1) \Rightarrow y=2$, and $y=2(-1) \Rightarrow y=-2$

   The solutions are $(-1,-2)$ and $(1,2)$.

   (d) Graph on a calculator shows two points of intersection. See Figure 1.

[-5,5] by [-5,5]
Xscl = 1 Yscl = 1

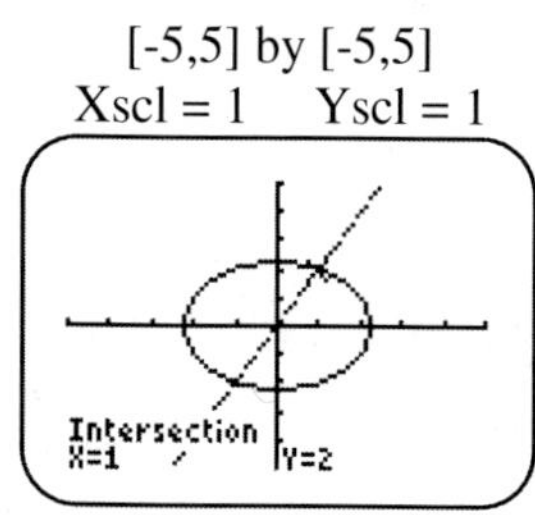

[-5,5] by [-5,5]
Xscl = 1 Yscl = 1

Intersection
X=-1 Y=-2

Figure 1

2. (a) First solve [equation 1] for x and substitute the result into [equation 2]. $x-2y=1 \Rightarrow x=1+2y$

   $2(1+2y)+y=7 \Rightarrow 2+4y+y=7 \Rightarrow 5y=5 \Rightarrow y=1$. Now substitute this value into [equation 1]

   $x=1+2(1) \Rightarrow x=3$, The solution set is $\{(3,1)\}$.

   (b) First solve [equation 1] for y and substitute the result into [equation 2]. $3x-y=1 \Rightarrow y=3x-1$

$-6x+2(3x-1)=-2 \Rightarrow -6x+6x+-2=-2 \Rightarrow -2=-2$. This implies that we have an infinite number of solutions. We will show the solution set as $\{(x, 3x-1)\}$ or $\left\{\left(\frac{y+1}{3}, y\right)\right\}$.

3. (a) Multiply [equation 2] by -3 and add it to [equation 1] to eliminate the x variable.

$$3x-4y=7 \qquad [1]$$
$$-3x-9y=\frac{-27}{2} \qquad -3[2]$$
$$-13y=-\frac{13}{2} \Rightarrow y=\frac{1}{2}$$

Now substitute this y–value into [equation 1] to solve for x. $x+3\left(\frac{1}{2}\right)=\frac{9}{2} \Rightarrow x=\frac{6}{2}=3$.

The solution set is $\left\{\left(3, \frac{1}{2}\right)\right\}$.

(b) Multiply [equation 1] by 2 and add it to [equation 2] to eliminate the x variable.

$$2x-2y=10 \qquad 2[1]$$
$$-2x+2y=1 \qquad [2]$$
$$0=11 \Rightarrow \text{No solution}$$

We have found that there is no solution to this system of equations. The solution set is $\varnothing$.

4. First solve [equation 1] for x^2 and substitute the result into [equation 2]. $x^2-y=5 \Rightarrow x^2=y+5$

$y+5+y^2=11 \Rightarrow y^2+y-6=0 \Rightarrow (y+3)(y-2)=0 \Rightarrow y=-3, 2$, Now substitute these values into [equation 1] to find the values of x: $x^2=(-3)+5 \Rightarrow x^2=2 \Rightarrow x=\pm\sqrt{2}$, and

$x^2=(2)+5 \Rightarrow x^2=7 \Rightarrow x=\pm\sqrt{7}$, The solution set is $\left\{\left(-\sqrt{2},-3\right),\left(\sqrt{2},-3\right),\left(-\sqrt{7},2\right),\left(\sqrt{7},2\right)\right\}$.

5. (a) Using the Row Echelon Method, the given system of equations yields the matrix,

$$\left[\begin{array}{ccc|c} 2 & 1 & 1 & 3 \\ 1 & 2 & -1 & 3 \\ 3 & -1 & 1 & 5 \end{array}\right], \text{ which by } (R1 \leftrightarrow R2) \rightarrow \left[\begin{array}{ccc|c} 1 & 2 & -1 & 3 \\ 2 & 1 & 1 & 3 \\ 3 & -1 & 1 & 5 \end{array}\right] \Rightarrow (-2R_1+R_2) \text{ and}$$

$$\Rightarrow (-3R_1+R_3) \rightarrow \left[\begin{array}{ccc|c} 1 & 2 & -1 & 3 \\ 0 & -3 & 3 & -3 \\ 0 & -7 & 4 & -4 \end{array}\right] \Rightarrow (7R_2-3R_3) \rightarrow \left[\begin{array}{ccc|c} 1 & 2 & -1 & 3 \\ 0 & -3 & 3 & -3 \\ 0 & 0 & 9 & -9 \end{array}\right].$$ From this matrix, we have

the resulting equation, $9z=-9 \Rightarrow z=-1$. Now use back-substitution: $z=-1$ substituting into the resulting $x=2:-12=C(4) \Rightarrow C=-3.$ yields, $-3y+3(-1)=-3 \Rightarrow -3y=0 \Rightarrow y=0;$ finally substituting $y=0$ and $z=-1$ into the resulting R_1 yields, $x+2(0)-(-1)=3 \Rightarrow x=2.$

The solution is $\{(2, 0, -1)\}$.

(b) First subtract [equation 1] from [equation 3] to eliminate the x variable and create [equation 4]. Then add –2 multiplied by [equation 1] to [equation 2] to eliminate the x variable and create [equation 5].

$$\begin{array}{rl} x+2y+2z=5 & [3] \\ -x-y+\ \ z=-1 & -1[1] \\ \hline y+\ \ 3z=4 & [4] \end{array} \qquad \begin{array}{rl} 2x+3y+\ \ z=6 & [2] \\ -2x-2y+2z=-2 & -2[1] \\ \hline y+\ \ 3z=4 & [5] \end{array}$$

Note that equations [4] and [5] are the same and this implies that we have an infinite number of solutions. Solve [equation 4] for y and substitute into [equation 1] to solve for x:

$y+3z=4 \Rightarrow y=4-3z$ then $x+(4-3z)-z=1 \Rightarrow x=4z-3$. The solution set is

$\{(4z-3, 4-3z, z)\}$.

6. (a) $3\begin{bmatrix} 2 & 3 \\ 1 & -4 \\ 5 & 9 \end{bmatrix} - \begin{bmatrix} -2 & 6 \\ 3 & -1 \\ 0 & 8 \end{bmatrix} = \begin{bmatrix} 6 & 9 \\ 3 & -12 \\ 15 & 27 \end{bmatrix} - \begin{bmatrix} -2 & 6 \\ 3 & -1 \\ 0 & 8 \end{bmatrix} = \begin{bmatrix} 6+2 & 9-6 \\ 3-3 & -12+1 \\ 15-0 & 27-8 \end{bmatrix} = \begin{bmatrix} 8 & 3 \\ 0 & -11 \\ 15 & 19 \end{bmatrix}.$

(b) Cannot add $(1\times 2)+(1\times 2)+(2\times 2)$; therefore, the solution set is $\varnothing$.

(c) $\begin{bmatrix} 2 & 1 & -3 \\ 4 & 0 & 5 \end{bmatrix}\begin{bmatrix} 1 & 3 \\ 2 & 4 \\ 3 & -2 \end{bmatrix} = \begin{bmatrix} 2+2-9 & 6+4+6 \\ 4+0+15 & 12+0-10 \end{bmatrix} = \begin{bmatrix} -5 & 16 \\ 19 & 2 \end{bmatrix}.$

7. (a) AB can be found; it will be $n\times n$

(b) BA can be found; it will be $n\times n$

(c) $AB = BA$ is not necessarily true, since matrix multiplication is not commutative.

(d) Since the number of rows of the first matrix is not equal to the number of columns of the second matrix, AC cannot be found.

Since the number of rows of the first matrix is equal to the number of columns of the second matrix, CA can be found; it will be $m\times n$.

8. (a) $\det\begin{bmatrix} 4 & 9 \\ -5 & -11 \end{bmatrix} = -44+45 = 1$

(b) Evaluate, expand by the second column. Therefore, $\det = (-)(a_{12})(M_{12})+(a_{22})(M_{22})-(a_{32})(M_{32})$:

$$0\left(\det\begin{bmatrix} -1 & 9 \\ 12 & -3 \end{bmatrix}\right)+7\left(\det\begin{bmatrix} 2 & 8 \\ 12 & -3 \end{bmatrix}\right)-5\left(\det\begin{bmatrix} 2 & 8 \\ -1 & 9 \end{bmatrix}\right) = 0+7(-6-96)-5(18+8) \Rightarrow$$

$7(-102)-5(26) = -844$

9. Find the determinants: $D = \det\begin{bmatrix} 2 & -3 \\ 4 & 5 \end{bmatrix} = 22$, $D_x = \det\begin{bmatrix} -33 & -3 \\ 11 & 5 \end{bmatrix} = -132$, and $D_y = \det\begin{bmatrix} 2 & -33 \\ 4 & 11 \end{bmatrix} = 154$.

Then $x = \dfrac{D_x}{D} = \dfrac{-132}{22} = -6$ and $y = \dfrac{D_y}{D} = \dfrac{154}{22} = 7$. The solution set is $\{(-6, 7)\}$.

10. (a) For the system: $A=\begin{bmatrix}1 & 1 & -1\\ 2 & -3 & -1\\ 1 & 2 & 2\end{bmatrix}$, $X=\begin{bmatrix}x\\ y\\ z\end{bmatrix}$, and $B=\begin{bmatrix}-4\\ 5\\ 3\end{bmatrix}$.

(b) Solve using the calculator, $A^{-1}=\begin{bmatrix}\frac{1}{4} & \frac{1}{4} & \frac{1}{4}\\ \frac{5}{16} & -\frac{3}{16} & \frac{1}{16}\\ -\frac{7}{16} & \frac{1}{16} & \frac{5}{16}\end{bmatrix}$.

(c) $\begin{bmatrix}x\\ y\\ z\end{bmatrix}=\begin{bmatrix}1 & 1 & -1\\ 2 & -3 & -1\\ 1 & 2 & 2\end{bmatrix}^{-1}\begin{bmatrix}-4\\ 5\\ 3\end{bmatrix}=\begin{bmatrix}1\\ -2\\ 3\end{bmatrix}$. The solution is $\{(1,-2,3)\}$.

(d) For the new system: $A=\begin{bmatrix}0.5 & 1 & 1\\ 2 & -3 & -1\\ 1 & 2 & 2\end{bmatrix}$ and by the calculator, $\det A=0$. Since the determinant equals zero, there is no inverse and the matrix inverse method cannot be used.

11. (a) Using $f(x)=ax^2+bx+c$ and the data from the table yields

$(0,1.3)$: $0^2a+0b+c=1.3$

$(100,2.5)$: $100^2a+100b+c=2.5$

$(200,8.9)$: $200^2a+200b+c=8.9$

Now using the capabilities of the calculator and the Reduced Row Echelon Method:

$$\left[\begin{array}{ccc|c}0 & 0 & 1 & 1.3\\ 100^2 & 100 & 1 & 2.5\\ 200^2 & 200 & 1 & 8.9\end{array}\right]=\left[\begin{array}{ccc|c}1 & 0 & 0 & 0.000264\\ 0 & 1 & 0 & -0.014\\ 0 & 0 & 1 & 1.3\end{array}\right].$$

Therefore, the equation is $f(x)=0.00026x^2-0.014x+1.3$.

(b) Using the functions of the calculator we see that that the population will reach 8 billion in about 2040. See Figure 11.

[-5,250] by [0,15]
Xscl = 5 Yscl = 1

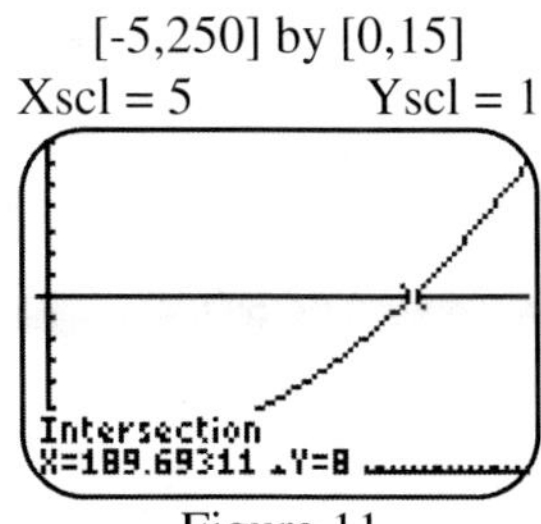

Figure 11

12. B, the graph is shaded above the line; therefore, $y>2-x$ and outside the parabola; therefore, $y<x^2-5$.

13. Let x = the number type X cabinets and let y = the number of type Y cabinets.

The objective function to find the maximum storage space is $8x+12y$. The constraints are,

$100x + 200y \leq 1400$ (cost of cabinets), $6x + 8y \leq 72$ (floor space), and $x \geq 0$ and $y \geq 0$ (cannot have a negative number of cabinets). Graphing the constraints on the calculator yields four vertices. Inputting these ordered pairs into the objective function yields,

$(0,0):\ 8(0) + 12(0) = 0$

$(12,0):\ 8(12) + 12(0) = 96$

$(8,3):\ 8(8) + 12(3) = 100$

$(0.7):\ 8(0) + 12(7) = 84$

The maximum storage is 100 cubic feet, when there are 8 cabinets of type X and 3 cabinets of type Y.

14. Factoring $\dfrac{7x-1}{x^2-x-6}$ results in $\dfrac{7x-1}{(x-3)(x+2)}$. Multiply $\dfrac{7x-1}{(x-3)(x+2)} = \dfrac{A}{x-3} + \dfrac{B}{x+2}$ by

$(x-3)(x+2) \Rightarrow 7x - 1 = A(x+2) + B(x-3).$

Let $x = -2 \Rightarrow -15 = B(-5) \Rightarrow B = 3$. Let $x = 3 \Rightarrow 20 = A(5) \Rightarrow A = 4$.

The expression can be written $\dfrac{4}{x-3} + \dfrac{3}{x+2}$.

15. The expression $\dfrac{x^2-11x+6}{(x+2)(x-2)^2}$ is factored. Multiply $\dfrac{x^2-11x+6}{(x+2)(x-2)^2} = \dfrac{A}{x+2} + \dfrac{B}{x-2} + \dfrac{C}{(x-2)^2}$ by

$(x+2)(x-2)^2 \Rightarrow x^2 - 11x + 6 = A(x-2)^2 + B(x-2)(x+2) + C(x+2).$

Let $x = -2 \Rightarrow 32 = A(16) \Rightarrow A = 2$. Let $x = 2:\ -12 = C(4) \Rightarrow C = -3$.

Let $x = 0,\ A = 2,\ C = -3:\ 6 = 4(2) - 4B + 2(-3) \Rightarrow$

$6 = 8 - 4B - 6 \Rightarrow 6 = 2 - 4B \Rightarrow 4 = -4B \Rightarrow B = -1.$

The expression can be written $\dfrac{2}{x+2} + \dfrac{-1}{x-2} + \dfrac{-3}{(x-2)^2}$.

Chapter 7: Analytic Geometry and Nonlinear Systems

7.1: Circles and Parabolas

1. E. Since $x=2y^2$. is equivalent to $y^2=4\left(\frac{1}{8}\right)x$, this is a parabola that opens to the right $(c>0)$.

3. H. Since $x^2=-3y$ is equivalent to $x^2=4\left(-\frac{3}{4}\right)y$, this is a parabola that opens downward $(c<0)$.

5. F. This is the equation of a circle centered at the origin with radius $\sqrt{5}$.

7. D. This is the equation of a circle centered at the point $(-3,4)$ with radius $\sqrt{25}=5$.

9. Here $h=1$, $k=4$ and $r^2=3^2=9$. The equation is $(x-1)^2+(y-4)^2=9$.

11. A circle that is centered at the origin with $r^2=1^2=1$ has equation $x^2+y^2=1$.

13. Here $h=\frac{2}{3}$, $k=-\frac{4}{5}$ and $r^2=\left(\frac{3}{7}\right)^2=\frac{9}{49}$. The equation is $\left(x-\frac{2}{3}\right)^2+\left(y+\frac{4}{5}\right)^2=\frac{9}{49}$.

15. The radius is the distance between $(-1,2)$ and $(2,6)$: $r=\sqrt{(2-(-1))^2+(6-2)^2}=\sqrt{9+16}=5$. Here

    $h=-1$, $k=2$ and $r^2=5^2=25$. The equation is $(x+1)^2+(y-2)^2=25$.

17. If the center is $(-3,-2)$, the circle must touch the x-axis at the point $(-3,0)$. The radius is 2. Here

    $h=-3$, $k=-2$ and $r^2=2^2=4$. The equation is $(x+3)^2+(y+2)^2=4$.

19. The equation is that of a circle with center (3, 3) and radius 0. That is, the graph is the point (3, 3).

21. Midpoint: $\left(\frac{5+(-1)}{2},\frac{-9+3}{2}\right)=(2,-3)\Rightarrow$ The center of the circle is (2,-3).

    Distance: $d=\sqrt{(2-5)^2+(-3-(-9))^2}=\sqrt{9+36}=\sqrt{45}\Rightarrow$ The radius of the circle is $\sqrt{45}$ units. The

    equation of the circle is $(x-2)^2+(y+3)^2=45$.

23. Midpoint: $\left(\frac{-5+(1)}{2},\frac{-7+1}{2}\right)=(-2,-3)\Rightarrow$ The center of the circle is (-2,-3).

    Distance: $d=\sqrt{(-2-1)^2+(-3-(1))^2}=\sqrt{9+16}=\sqrt{25}=5\Rightarrow$ The radius of the circle is 5 units. The

    equation of the circle is $(x+2)^2+(y+3)^2=25$.

25. Midpoint: $\left(\frac{-5+(5)}{2},\frac{0+0}{2}\right)=(0,0)\Rightarrow$ The center of the circle is (0,0).

    Distance: $d=\sqrt{(0-(-5))^2+(0-0)^2}=\sqrt{25}=5\Rightarrow$ The radius of the circle is 5 units.

    The equation of the circle is $(x)^2+(y)^2=25$.

27. In a circle, the radius is the distance from the center to any point on the circle.

29. This is the equation of a circle centered at the origin with radius $\sqrt{4}=2$. See Figure 29. From the figure, the domain is $[-2,2]$, and the range is $[-2,2]$.

31. This is the equation of a circle centered at the origin with radius $\sqrt{0}=0$. The graph is only the point (0, 0). See Figure 31. From the figure, the domain is $\{0\}$, and the range is $\{0\}$.

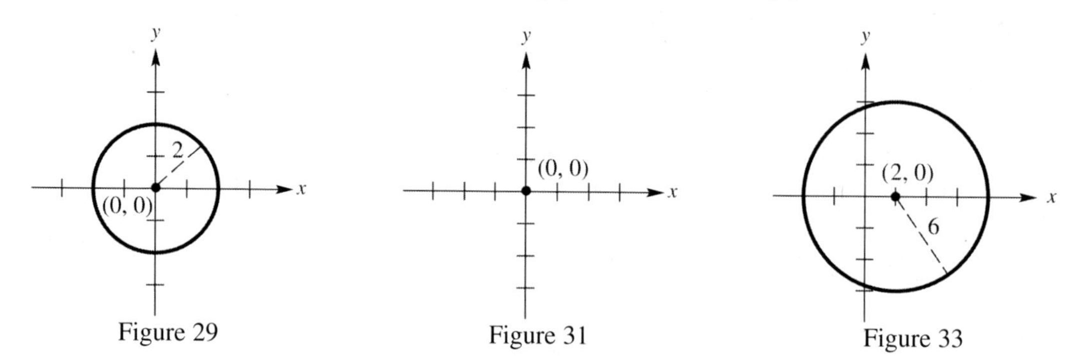

Figure 29 Figure 31 Figure 33

33. This is the equation of a circle centered at $(2,0)$ with radius $\sqrt{36}=6$. See Figure 33. From the figure, the domain is $[-4,8]$, and the range is $[-6,6]$.

35. This is the equation of a circle centered at $(5,-4)$ with radius $\sqrt{49}=7$. See Figure 35. From the figure, the domain is $[-2,12]$, and the range is $[-11,3]$.

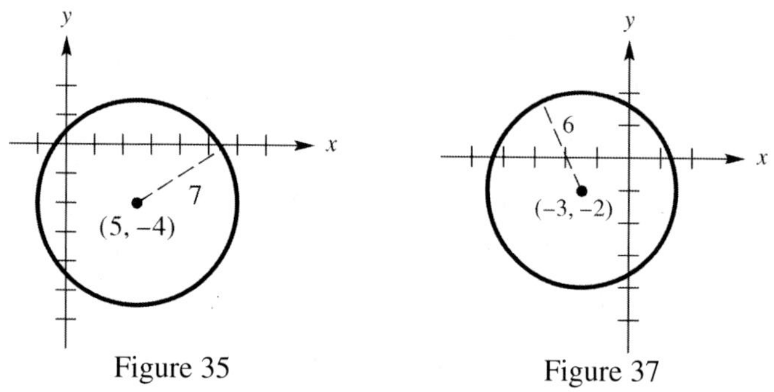

Figure 35 Figure 37

37. This is the equation of a circle centered at $(-3,-2)$ with radius $\sqrt{36}=6$. See Figure 37. From the figure, the domain is $[-9,3]$, and the range is $[-8,4]$.

39. $x^2+(y-2)^2+10=9 \Rightarrow x^2+(y-2)^2=-1$. This is the equation of a circle centered at $(0,2)$ with radius $\sqrt{-1}$. No such graph exists. The domain is $\varnothing$ and the range is $\varnothing$.

41. $x^2+y^2=81 \Rightarrow y^2=81-x^2 \Rightarrow y=\pm\sqrt{81-x^2}$ Graph $y_1=\sqrt{81-x^2}$ and $y_2=\sqrt{81-x^2}$ as shown in Figure 41. From the figure, the domain is $[-9,9]$, and the range is $[-9,9]$.

43. $(x-3)^2+(y-2)^2=25 \Rightarrow (y-2)^2=25-(x-3)^2 \Rightarrow y-2=\pm\sqrt{25-(x-3)^2} \Rightarrow y=2\pm\sqrt{25-(x-3)^2}$.

Graph $y_1=2-\sqrt{25-(x-3)^2}$ and $y_2=2+\sqrt{25-(x-3)^2}$ as shown in Figure 43. From the figure, the domain is $[-2,8]$, and the range is $[-3,7]$.

[-14.1,14.1] by [-9.3,9.3]
Xscl = 1 Yscl = 1

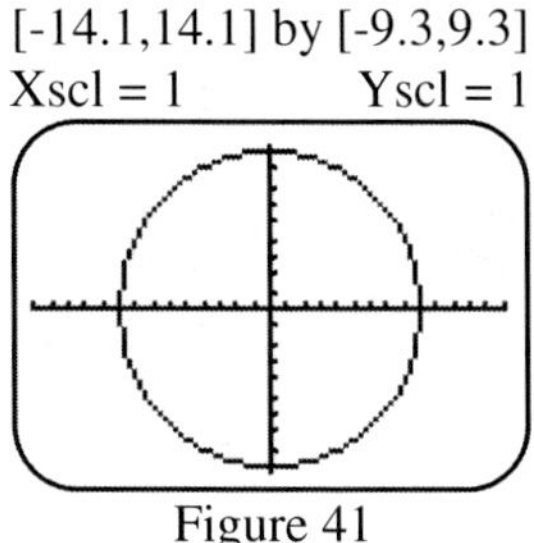

Figure 41

[-9.4,9.4] by [-4.2,8.2]
Xscl = 1 Yscl = 1

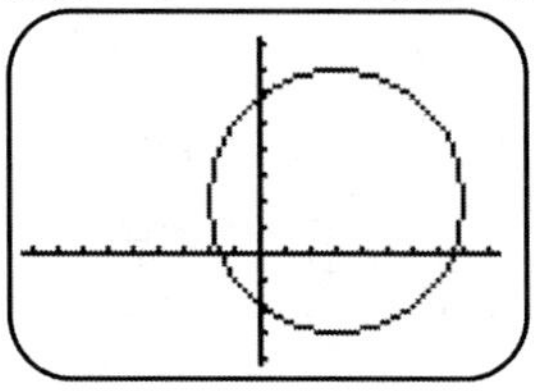

Figure 43

45. $x^2+6x+y^2+8y+9=0 \Rightarrow (x^2+6x+9)+(y^2+8y+16)=-9+9+16 \Rightarrow (x+3)^2+(y+4)^2=16$. The graph is a circle with center $(-3,-4)$, and radius $r=4$.

47. $x^2-4x+y^2+12y=-4 \Rightarrow (x^2-4x+4)+(y^2+12y+36)=-4+4+36 \Rightarrow (x-2)^2+(y+6)^2=36$.

The graph is a circle with center $(2,-6)$, and radius $r=6$.

49. $4x^2+4x+4y^2-16y-19=0 \Rightarrow 4\left(x^2+x+\frac{1}{4}\right)+4\left(y^2-4y+4\right)=19+1+16 \Rightarrow$

$4\left(x+\frac{1}{2}\right)^2+4(y-2)^2=36 \Rightarrow \left(x+\frac{1}{2}\right)^2+(y-2)^2=9$. The graph is a circle with center $\left(-\frac{1}{2},2\right)$, and radius $r=3$.

51. $x^2+2x+y^2-6y+14=0 \Rightarrow (x^2+2x+1)+(y^2-6y+9)=-14+1+9 \Rightarrow (x+1)^2+(y-3)^2=-4$. The graph does not exist since the value for the radius is not a real number.

53. $x^2-2x+y^2+4y=0 \Rightarrow (x^2-2x+1)+(y^2+4y+4)=1+4 \Rightarrow (x-1)^2+(y+2)^2=5$. The graph is a circle with center $(1,-2)$ and radius $r=\sqrt{5}$.

55. $b^2=\frac{1}{2}$. $(x+2)^2+y^2=\frac{4}{9}$. The graph is a circle with center $(-2,0)$, and radius $r=\frac{2}{3}$.

57. D. Since $(x-4)^2=y+2$ is equivalent to $(x-4)^2=4\left(\frac{1}{4}\right)(y+2)$, the parabola has vertex $(4,-2)$, and it opens upward $(c>0)$.

59. C. Since $y+2=-(x-4)^2$ is equivalent to $(x-4)^2=4\left(-\frac{1}{4}\right)(y+2)$, the parabola has vertex $(4,-2)$, and it opens downward $(c<0)$.

61. F. Since $(y-4)^2 = x+2$ is equivalent to $(y-4)^2 = 4\left(\frac{1}{4}\right)(x+2)$, the parabola has vertex $(-2,4)$, and it opens to the right $(c>0)$.

63. E. Since $x+2 = -(y-4)^2$ is equivalent to $(y-4)^2 = 4\left(-\frac{1}{4}\right)(x+2)$, the parabola has vertex $(-2,4)$, and it opens to the left $(c<0)$.

65. (a) If both coordinates of the vertex are negative, the vertex is in quadrant III.
    (b) If the first coordinate of the vertex is negative and the second is positive, the vertex is in quadrant II.
    (c) If the first coordinate of the vertex is positive and the second is negative, the vertex is in quadrant IV.
    (d) If both coordinates of the vertex are positive, the vertex is in quadrant I.

67. Since $x^2 = 16y$ is equivalent to $x^2 = 4(4)y$, the equation is in the form $x^2 = 4cy$ with $c = 4$. The focus is $(0,4)$, and the equation of the directrix is $y = -4$. The axis is $x = 0$, or the y-axis.

69. Since $x^2 = -\frac{1}{2}y$ is equivalent to $x^2 = 4\left(-\frac{1}{8}\right)y$, the equation is in the form $x^2 = 4cy$ with $c = -\frac{1}{8}$. The focus is $\left(0, -\frac{1}{8}\right)$, and the equation of the directrix is $y = \frac{1}{8}$. The axis is $x = 0$, or the y-axis.

71. Since $y^2 = \frac{1}{16}x$ is equivalent to $y^2 = 4\left(\frac{1}{64}\right)x$, the equation is in the form $y^2 = 4cx$ with $c = \frac{1}{64}$. The focus is $\left(\frac{1}{64}, 0\right)$, and the equation of the directrix is $x = -\frac{1}{64}$. The axis is $y = 0$, or the x-axis.

73. Since $y^2 = -16x$ is equivalent to $y^2 = 4(-4)x$, the equation is in the form $y^2 = 4cx$ with $c = -4$. The focus is $(-4,0)$. and the equation of the directrix is $x = 4$. The axis is $y = 0$, or the x-axis.

75. If the vertex is $(0,0)$ and the focus is $(0,-2)$, then the parabola opens downward and $c = -2$. The equation is $x^2 = 4cy \Rightarrow x^2 = -8y$.

77. If the vertex is $(0,0)$ and the focus is $\left(-\frac{1}{2}, 0\right)$, then the parabola opens to the left and $c = -\frac{1}{2}$. The equation is $y^2 = 4cx \Rightarrow y^2 = -2x$.

79. If the vertex is $(0,0)$ and the parabola opens to the right, the equation is in the form $y^2 = 4cx$. Find the value or c by using the fact that the parabola passes through $(2, -2, \sqrt{2})$. Thus, $(-2, \sqrt{2})^2 = 4c(2) \Rightarrow c = 1$. The equation is $y^2 = 4cx \Rightarrow y^2 = 4x$.

81. If the vertex is (0, 0) and the parabola opens downward, the equation is in the form $x^2 = 4cy$. Find the value of c by using the fact that the parabola passes through $\left(\sqrt{10}, -5\right)$. Thus, $\left(\sqrt{10}\right)^2 = 4c(-5) \Rightarrow c = -\frac{1}{2}$. The equation is $x^2 = 4cy \Rightarrow x^2 = -2y$.

83. If the vertex is (0, 0) and the parabola has y-axis symmetry, the equation is in the form $x^2 = 4cy$. Find the value of c by using the fact that the parabola passes through $(2, -4)$. Thus, $(2)^2 = 4c(-4) \Rightarrow c = -\frac{1}{4}$. The equation is $x^2 = 4cy \Rightarrow x^2 = -y$.

85. If the focus is (0,2) and the vertex is (0,1), the parabola opens upward and $c = 1$. Substituting in $(x-h)^2 = 4c(y-k)$, we get $(x-0)^2 = 4(1)(y-1)$ or $x^2 = 4(y-1)$.

87. If the focus is (0,0) and the directrix has equation $x = -2$, the vertex is (-1,0) and $c = 1$. The parabola opens to the right. Substituting in $(y-k)^2 = 4c(x-h)$, we get $(y-0)^2 = 4(1)(x-(-1))$ or $y^2 = 4(x+1)$.

89. If the focus is (-1,3) and the directrix has equation $y = 7$, the vertex is (-1,5) and $c = -2$. The parabola opens downward. Substituting in $(y-k)^2 = 4c(x-h)$, we get $(x+1)^2 = 4(-2)(y-5)$ or $(x+1)^2 = -8(y-5)$.

91. Since the parabola has a horizontal axis, the equation is in the form $(y-k)^2 = 4c(x-h)$. Find the value of c by using the fact that the parabola passes through (-4,0) and the vertex is (-2,3). Substituting $x = -4$, $y = 0$, $h = -2$, and $k = 3$ yields $(0-3)^2 = 4c(-4(-2)) \Rightarrow c = -\frac{9}{8}$. The equation is $(y-3)^2 = -\frac{9}{2}(x+2)$.

93. The equation $y = (x+3)^2 - 4$ can be written as $(x+3)^2 = 4\left(\frac{1}{4}\right)(y+4)$. The vertex is $(-3, -4)$. The vertical axis has equation $x = -3$, and the parabola opens upward. See Figure 93. From the figure, the domain is $(-\infty, \infty)$, and the range is $[-4, \infty)$.

95. The equation $y = -2(x+3)^2 + 2$ can be written as $(x+3)^2 = 4\left(-\frac{1}{8}\right)(y-2)$. The vertex is (-3,2). The vertical axis has equation $x = -3$, and the parabola opens downward. See Figure 95. From the figure, the domain is $(-\infty, \infty)$, and the range is $(-\infty, 2]$.

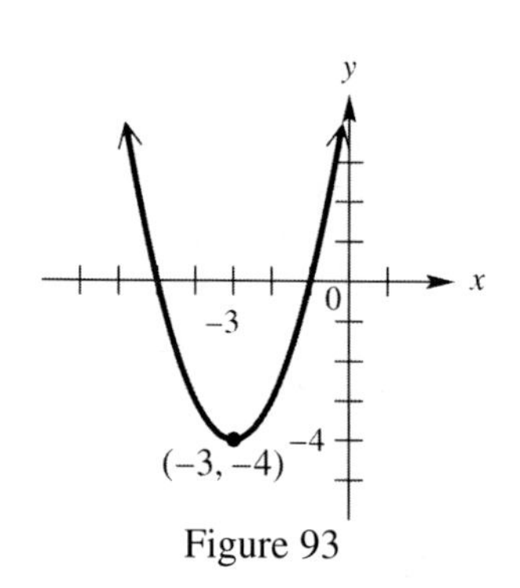

Figure 93

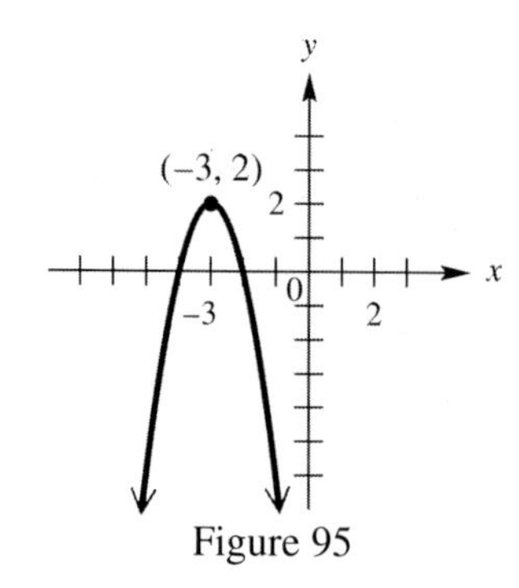

Figure 95

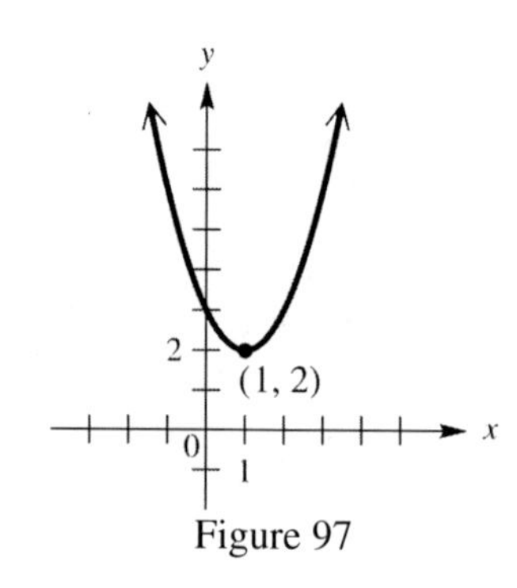

Figure 97

97. Rewrite the equation: $y = x^2 - 2x + 3 \Rightarrow y - 3 + 1 = x^2 - 2x + 1 \Rightarrow y - 2 = (x-1)^2$. The equation $y - 2 = (x-1)^2$ can be written as $(x-1)^2 = 4\left(\frac{1}{4}\right)(y-2)$. The vertex is (1,2).The vertical axis has equation $x = 1$, and the parabola opens upward. See Figure 97. From the figure, the domain is $(-\infty, \infty)$, and the range is $[2, \infty)$.

99. Rewrite the equation: $y = 2x^2 - 4x + 5 \Rightarrow y - 5 + 2 = 2\left(x^2 - 2x + 1\right) \Rightarrow y - 3 = 2\left(x-1\right)^2$. The equation $y - 3 = 2\left(x-1\right)^2$ can be written as $\left(x-1\right)^2 = 4\left(\frac{1}{8}\right)(y-3)$. The vertex is (1,3). The vertical axis has equation $x = 1$, and the parabola opens upward. See Figure 99. From the figure, the domain is $\left(-\infty, \infty\right)$, and the range is $\left[3, \infty\right)$.

101. The equation $x = y^2 + 2$ can be written as $(y-0)^2 = 4\left(\frac{1}{4}\right)(x-2)$. The vertex is (2,0). The horizontal axis has equation $y = 0$, and the parabola opens to the right. See Figure 101. From the figure, the domain is $\left[2, \infty\right)$ and the range is $(-\infty, \infty)$.

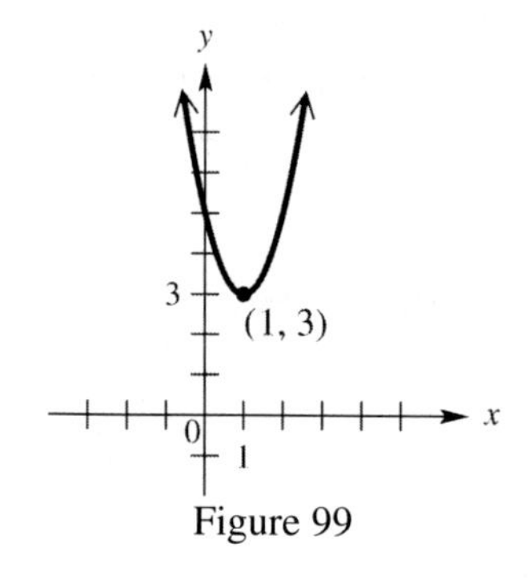

Figure 99

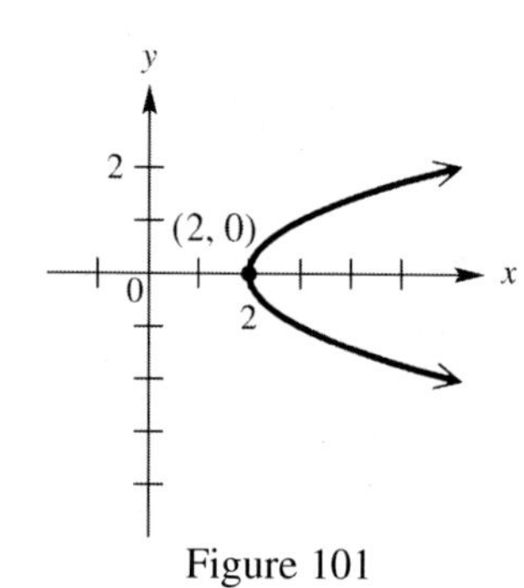

Figure 101

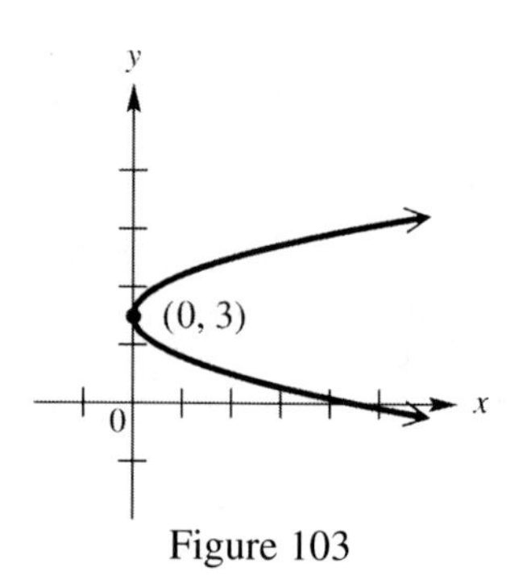

Figure 103

103. The equation $x = \left(y-3\right)^2$ can be written as $\left(y-3\right)^2 = 4\left(\frac{1}{4}\right)\left(x-0\right)$. The vertex is (0,3). The horizontal axis has equation $y = 3$, and the parabola opens to the right. See Figure 103. From the figure, the domain is $\left[0, \infty\right)$, and the range is $(-\infty, \infty)$.

105. The equation $x=(y-4)^2+2$ can be written as $(y-4)^2=4\left(\frac{1}{4}\right)(x-2)$. The vertex is (2,4). The horizontal axis has equation $y=4$, and the parabola opens to the right. See Figure 105. From the figure, the domain is $[2,\infty)$, and the range is $(-\infty,\infty)$.

107. Rewrite the equation: $x=\frac{2}{3}y^2-4y+8 \Rightarrow \frac{3}{2}x=y^2-6y+12 \Rightarrow \frac{3}{2}x-12+9=y^2-6y+9 \Rightarrow$ $\frac{3}{2}x-3=(y-3)^2 \Rightarrow \frac{3}{2}(x-2)=(y-3)^2$. The equation $\frac{3}{2}(x-2)=(y-3)^2$ can be written $(y-3)^2=4\left(\frac{3}{8}\right)(x-2)$. The vertex is $(2,3)$. The horizontal axis has equation $y=3$ and the parabola opens to the right. See Figure 107. From the figure, the domain is $[2,\infty)$ and the range is $(-\infty,\infty)$.

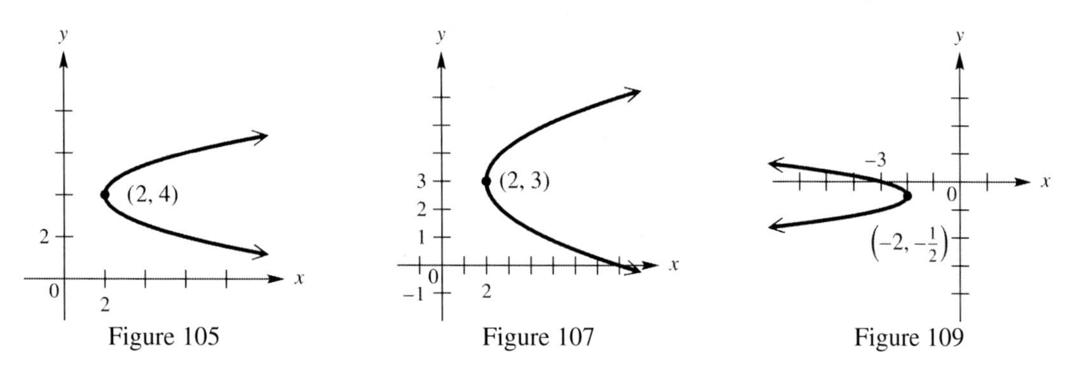

Figure 105 Figure 107 Figure 109

109. Rewrite the equation: $x=-4y^2-4y-3 \Rightarrow x+3-1=-4\left(y^2+y+\frac{1}{4}\right) \Rightarrow x+2=-4\left(y+\frac{1}{2}\right)^2$. The equation $x+2=-4\left(y+\frac{1}{2}\right)^2$ can be written $\left(y+\frac{1}{2}\right)^2=4\left(-\frac{1}{16}\right)(x+2)$. The vertex is $\left(-2,-\frac{1}{2}\right)$. The horizontal axis has equation $y=-\frac{1}{2}$ and the parabola opens to the left. See Figure 109. From the figure, the domain is $(-\infty,-2]$ and the range is $(-\infty,\infty)$.

111. Rewrite the equation: $x=2y^2-4y+6 \Rightarrow x-6+2=2(y^2-2y+1) \Rightarrow x-4=2(y-1)^2$. The equation $x-4=2(y-1)^2$ can be written $(y-1)^2=4\left(\frac{1}{8}\right)(x-4)$. The vertex is $(4,1)$. The horizontal axis has equation $y=1$ and the parabola opens to the right. See Figure 111. From the figure, the domain is $[4,\infty)$ and the range is $(-\infty,\infty)$.

113. Rewrite the equation: $2x=y^2-2y+9 \Rightarrow 2x-9+1=y^2-2y+1 \Rightarrow 2(x-4)=(y-1)^2$. The equation $2(x-4)=(y-1)^2$ can be written $(y-1)^2=4\left(\frac{1}{2}\right)(x-4)$. The vertex is $(4, 1)$. The horizontal axis has

equation $y=1$ and the parabola opens to the right. See Figure 113. From the figure, the domain is $[4,\infty)$ and the range is $(-\infty,\infty)$.

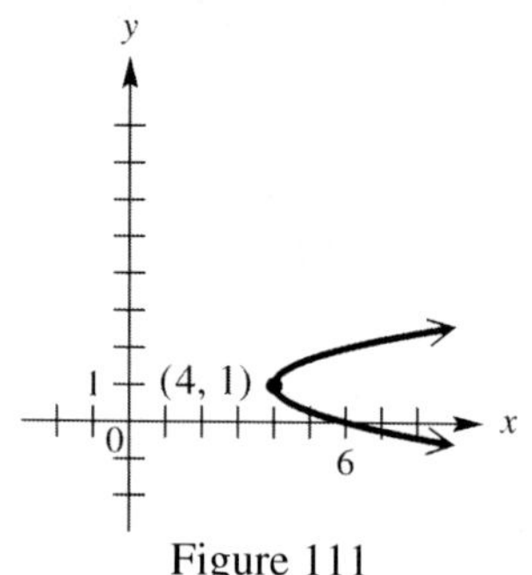

Figure 111

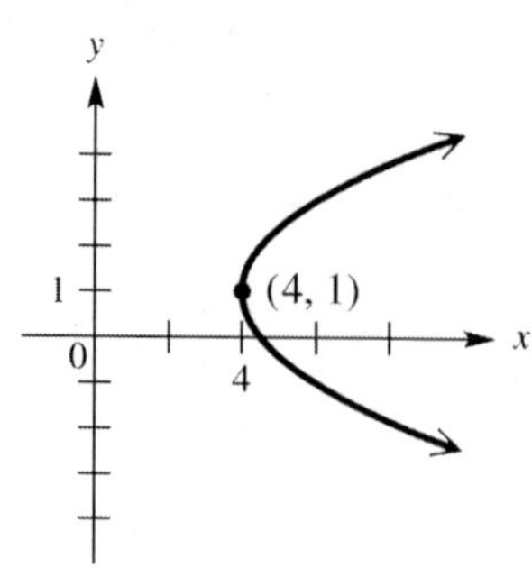

Figure 113

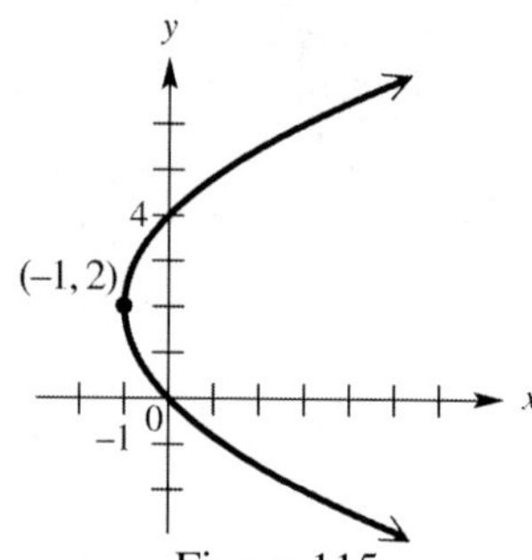

Figure 115

115. Rewrite the equation: $y^2-4y+4=4x+4 \Rightarrow (y-2)^2=4(x+1)$. The equation $(y-2)^2=4(x+1)$ can be written $(y-2)^2=4(1)(x+1)$. The vertex is $(-1, 2)$. The horizontal axis has equation $y=2$ and the parabola opens to the right. See Figure 115. From the figure, the domain is $[-1,\infty)$ and the range is $(-\infty,\infty)$.

117. Since the directrix has equation $x=-c$, a point on the directrix has the form $(-c, y)$. Let (x, y) be a point on the parabola. By definition, the distance from the focus $(c,0)$ to point (x, y) on the parabola, must be equal to the distance from point $(-c, y)$ on the directrix to point (x, y) on the parabola. That is

$$\sqrt{(x-c)^2+(y-0)^2}=\sqrt{(x+c)^2+(y-y)^2} \Rightarrow$$

$$(x-c)^2+y^2=(x+c)^2 \Rightarrow x^2-2xc+c^2+y^2=x^2+2xc+c^2 \Rightarrow -2xc+y^2=2xc \Rightarrow y^2=4xc.$$

119. (a) For Mars, $y=\frac{19}{11}x-\frac{12.6}{3872}x^2$. For the moon, $y=\frac{19}{11}x-\frac{5.2}{3872}x^2$. Graph $y_1=(19/11)x-(12.6/3872)x^2$ and $y_2=(19/11)x-(5.2/3872)x^2$ as shown in Figure 119.

(b) From the graph, the ball thrown on Mars reaches a maximum height of $y\approx 229$ and the ball thrown on the moon reaches a maximum height of $y\approx 555$.

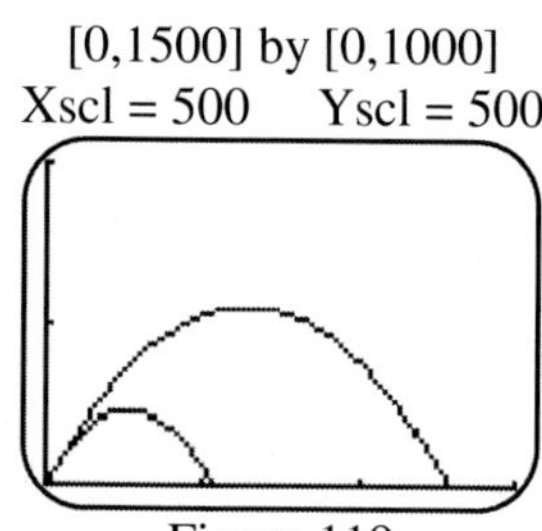

Figure 119

121. $y=-\frac{5\times10^{-9}}{2(10^7)}(0.4)^2=-4\times10^{-17}$; the alpha particle is deflected 4×10^{-17} meter downward.

123. Let the vertex of the parabola be (0, 12). The equation of the parabola is of the form $(x-h)^2 = 4c(y-k)$.

By substitution, the equation is $(x-0)^2 = 4c(y-12) \Rightarrow x^2 = 4c(y-12)$. Since the parabola passes through the point (6, 0), the value of c can be found by substitution:

$6^2 = 4c(0-12) \Rightarrow 36 = 4c(-12) \Rightarrow -3 = 4c \Rightarrow c = -\frac{3}{4}$. The equation is $x^2 = -3(y-12)$. Noting that the y-coordinate 9 feet up is 9, half the width can be found by substitution: $x^2 = -3(9-12) \Rightarrow x^2 = 9 \Rightarrow x = 3$; The width is 6 feet.

7.2: Ellipses and Hyperbolas

1. G. This is an ellipse with $a^2 = 16$, $b^2 = 4$, and $c = \sqrt{16-4} = \sqrt{12} = 2\sqrt{3}$. The Foci are $\left(0, \pm 2\sqrt{3}\right)$.

3. F. This is a hyperbola centered at (0, 0) with a horizontal transverse axis.

5. E. Since $h = -2$ and $k = 4$, this is an ellipse centered at $(-2, 4)$.

7. D. Since $h = -2$ and $k = 4$, this is a hyperbola centered at $(-2, 4)$.

9. A circle can be interpreted as an ellipse whose foci have the same coordinates. The "coinciding foci" give the center of the circle.

11. $\frac{x^2}{9} + \frac{y^2}{4} = 1 \Rightarrow a = 3$ and $b = 2$. $a^2 - b^2 = 3^2 - 2^2 = 5 = c^2 \Rightarrow c = \sqrt{5}$. The foci are $(\pm\sqrt{5}, 0)$. The endpoints of the major axis (vertices) are $(\pm 3, 0)$ so the domain is $[-3, 3]$. The endpoints of the minor axis are $(0, \pm 2)$, so the range is $[-2, 2]$. The ellipse is graphed in Figure 11.

13. $9x^2 + 6y^2 = 54 \Rightarrow \frac{x^2}{6} + \frac{y^2}{9} = 1 \Rightarrow a = 3$ and $b = \sqrt{6}$. $a^2 - b^2 = 9 - 6 = 3 = c^2 \Rightarrow c = \sqrt{3}$. The foci are $\left(0, \pm\sqrt{3}\right)$. The endpoints of the major axis (vertices) are $(0, \pm 3)$ so the range is $[-3, 3]$. The endpoints of the minor axis are $\left(\pm\sqrt{6}, 0\right)$ so the domain is $\left[-\sqrt{6}, \sqrt{6}\right]$. The ellipse is graphed in Figure 13.

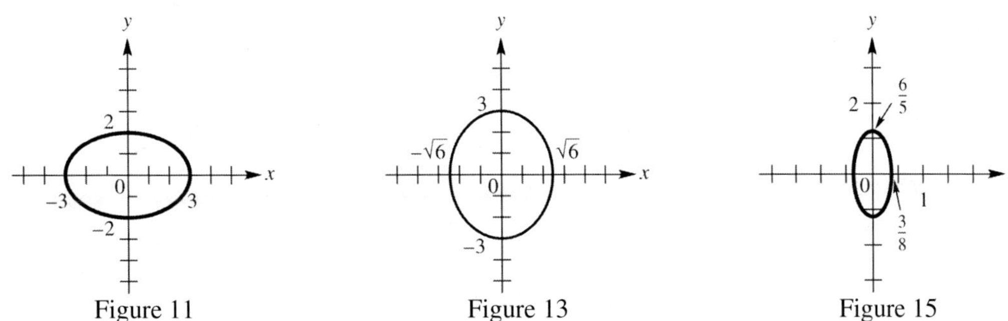

Figure 11 Figure 13 Figure 15

15. $\frac{25y^2}{36} + \frac{64x^2}{9} = 1 \Rightarrow \frac{y^2}{\frac{36}{25}} + \frac{x^2}{\frac{9}{64}} = 1 \Rightarrow a = \sqrt{\frac{36}{25}} = \frac{6}{5}$ and $b = \sqrt{\frac{9}{64}} = \frac{3}{8}$

The endpoints of the major axis (vertices) are $\left(0, \pm\frac{6}{5}\right)$ so the range is $\left[-\frac{6}{5}, \frac{6}{5}\right]$. The endpoints of the minor axis are $\left(\pm\frac{3}{8}, 0\right)$ so the domain is $\left[-\frac{3}{8}, \frac{3}{8}\right]$. See Figure 15.

17. The ellipse is centered at $(1, -3)$. The major axis is vertical and has length $2a = 10$. The length of the minor axis is $2b = 6$ The graph is shown in Figure 17. The domain is $[-2, 4]$ and the range is $[-8, 2]$.

19. The ellipse is centered at $(2, 1)$. The major axis is horizontal and has length $2a = 8$. The length of the minor axis is $2b = 6$. The graph is shown in Figure 19. The domain is $[-2, 6]$ and the range is $[-2, 4]$.

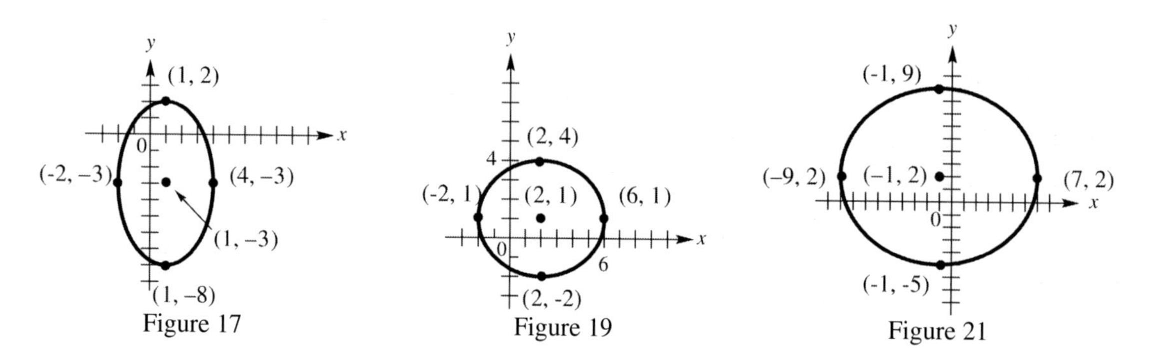

Figure 17 Figure 19 Figure 21

21. The ellipse is centered at $(-1, 2)$. The major axis is horizontal and has length $2a = 16$. The length of the minor axis is $2b = 14$. The graph is shown in Figure 21. The domain is $[-9, 7]$ and the range is $[-5, 9]$.

23. The ellipse is centered between the foci at $(0, 0)$. The major axis is horizontal with $a = 4$. Since the foci are $(\pm 2, 0)$, we know that $c = 2$. Since $c^2 = a^2 - b^2$, the value of b can be found by substitution:

$b^2 = a^2 - c^2 = 4^2 - 2^2 = 16 - 4 = 12 \Rightarrow b = \sqrt{12}$. The equation is $\frac{x^2}{16} + \frac{y^2}{12} = 1$.

25. The ellipse is centered between the foci at $(0, 0)$. The major axis is vertical with $a = 2\sqrt{2}$. Since the foci are $(0, \pm 2)$, we know that $c = 2$. Since $c^2 = a^2 - b^2$, the value of b can be found by substitution:

$b^2 = a^2 - c^2 = \left(2\sqrt{2}\right)^2 - 2^2 = 8 - 4 = 4 \Rightarrow b = 2$. The equation is $\frac{x^2}{4} + \frac{y^2}{8} = 1$.

27. The ellipse is centered between the endpoint of the major axis of $(0, 0)$. The major axis is horizontal with $a = 4$ and the minor axis is vertical with $b = 2$. The equation is $\frac{x^2}{16} + \frac{y^2}{4} = 1$.

29. The ellipse is centered between the endpoints of the major axis at $(0, 0)$. The major axis is horizontal with $a = 6$. Since $c^2 = a^2 - b^2$, the value of b can be found by substitution:

$b^2 = a^2 - c^2 = 6^2 - 4^2 = 36 - 16 = 20 \Rightarrow b = \sqrt{20}$. The equation is $\frac{x^2}{36} + \frac{y^2}{20} = 1$.

31. Since the center is $(3,-2)$, we know that $h=3$ and $k=-2$. Since $c^2=a^2-b^2$, the value of b can be found by substitution: $b^2=a^2-c^2=5^2-3^2=25-9=16 \Rightarrow b=4$. The major axis is vertical so the equation is $\frac{(x-3)^2}{16}+\frac{(y+2)^2}{25}=1$.

33. The ellipse is centered between the foci at $(0,0)$. The major axis is vertical with $a=3$. Since the foci are $(0,\pm 2)$, we know that $c=2$. Since $c^2=a^2-b^2$, the value of b can be found by substitution: $b^2=a^2-c^2=3^2-2^2=9-4=5 \Rightarrow b=\sqrt{5}$. The equation is $\frac{y^2}{9}+\frac{x^2}{5}=1$.

35. Since the center is $(5,2)$, we know that $h=5$ and $k=2$. Since the minor axis is horizontal and has length 8, $b=4$. Since $c^2=a^2-b^2$ and $c=3$, the value of a can be found by substitution: $3^2+4^2=a^2 \Rightarrow a^2=25 \Rightarrow a=5$. The equation is $\frac{(x-5)^2}{25}+\frac{(y-2)^2}{16}=1$.

37. The ellipse is centered between the vertices at $(4,5)$ and $a=\frac{9-1}{2}=4$. The major axis is vertical. Since the minor axis has length 6, $b=3$. The equation is $\frac{(x-4)^2}{9}+\frac{(y-5)^2}{16}=1$.

39. $9x^2+18x+4y^2-8y-23=0 \Rightarrow 9(x^2+2x)+4(y^2-2y)=23 \Rightarrow 9(x^2+2x+1)+4(y^2-2y+1)=$

$23+9+4 \Rightarrow 9(x+1)^2+4(y-1)^2=36 \Rightarrow \frac{(x+1)^2}{4}+\frac{(y-1)^2}{9}=1$

The center is $(-1,1)$. The vertices are $(-1,1-3),(-1,1+3)$ or $(-1,-2),(-1,4)$.

41. $4x^2+8x+y^2+2y+1=0 \Rightarrow 4(x^2+2x)+(y^2+2y)=-1 \Rightarrow 4(x^2+2x+1)+(y^2+2y+1)=$

$-1+4+1 \Rightarrow 4(x+1)^2+(y+1)^2=4 \Rightarrow \frac{(x+1)^2}{1}+\frac{(y+1)^2}{4}=1$

The center is $(-1,-1)$. The vertices are $(-1,-1-2)$, $(-1,-1+2)$ or $(-1,-3),(-1,1)$.

43. $4x^2+16x+5y^2-10y+1=0 \Rightarrow 4(x^2+4x)+5(y^2-2y)=-1 \Rightarrow 4(x^2+4x+4)+5(y^2-2y+1)=$

$-1+16+5 \Rightarrow 4(x+2)^2+5(y-1)^2=20 \Rightarrow \frac{(x+2)^2}{5}+\frac{(y-1)^2}{4}=1$

The center is $(-2,1)$. The vertices are $(-2-\sqrt{5},1),(-2+\sqrt{5},1)$.

45. $16x^2-16x+4y^2+12y=51 \Rightarrow 16(x^2-x)+4(y^2+3y)=51 \Rightarrow 16\left(x^2-x+\frac{1}{4}\right)+4\left(y^2+3y+\frac{9}{4}\right)=$

$51+4+9 \Rightarrow 16\left(x-\frac{1}{2}\right)^2+4\left(y+\frac{3}{2}\right)^2=64 \Rightarrow \frac{\left(x-\frac{1}{2}\right)^2}{4}+\frac{\left(y+\frac{3}{2}\right)^2}{16}=1$

The center is $\left(\frac{1}{2},-\frac{3}{2}\right)$. The vertices are $\left(\frac{1}{2},-\frac{3}{2}-4\right),\left(\frac{1}{2},-\frac{3}{2}+4\right)$ or $\left(\frac{1}{2},-\frac{11}{2}\right),\left(\frac{1}{2},\frac{5}{2}\right)$.

47. The transverse axis is horizontal with $a=4$ and $b=3$. The asymptotes are $y=\pm\frac{3}{4}x$. See Figure 47. The domain is $(-\infty,-4)\cup[4,\infty)$ and the range is $(-\infty,\infty)$.

49. $49y^2-36x^2=1764\Rightarrow\frac{y^2}{36}-\frac{x^2}{49}=1$ The transverse axis is vertical with $a=6$ and $b=7$. The asymptotes are $y=\pm\frac{6}{7}x$. See Figure 49. The domain is $(-\infty,\infty)$ and the range is $(-\infty,-6]\cup[6,\infty)$.

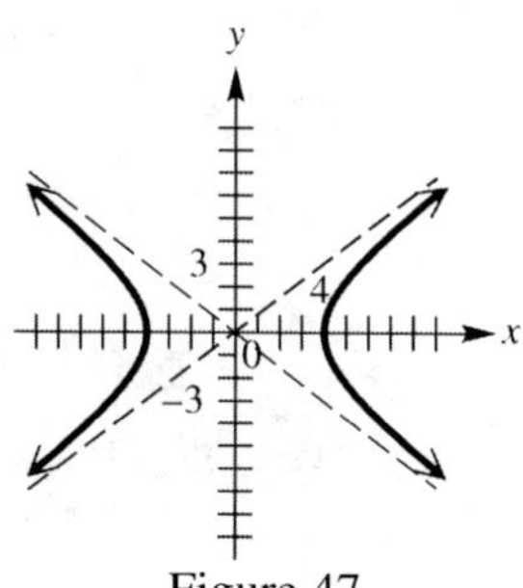

Figure 47

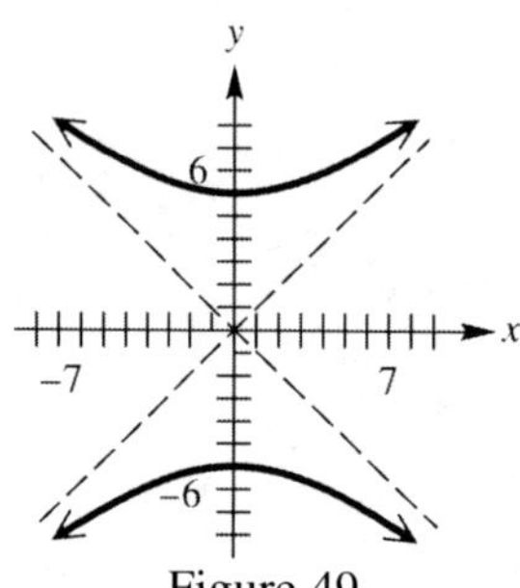

Figure 49

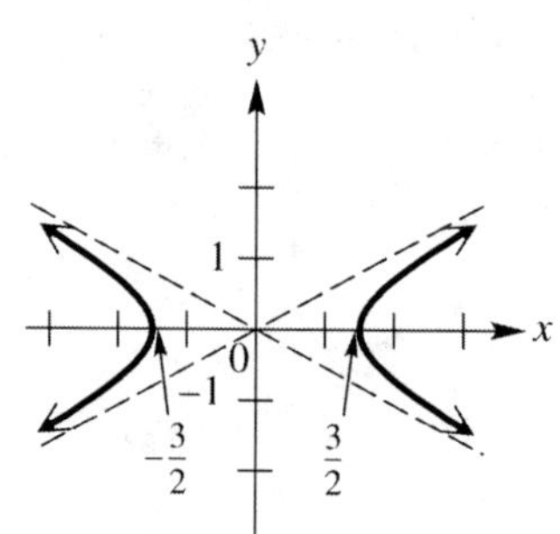

Figure 51

51. $\frac{4x^2}{9}-\frac{25y^2}{16}=1\Rightarrow\frac{x^2}{\frac{9}{4}}-\frac{y^2}{\frac{16}{25}}=1$. The transverse axis is horizontal with $a=\frac{3}{2}$ and $b=\frac{4}{5}$. The asymptotes are $y=\pm\frac{18}{15}x$. See Figure 51. The domain is $\left(-\infty,-\frac{3}{2}\right]\cup\left[\frac{3}{2},\infty\right)$ and the range is $(-\infty,\infty)$.

53. $9x^2-4y^2=1\Rightarrow\frac{x^2}{\frac{1}{9}}-\frac{y^2}{\frac{1}{4}}=1$. The transverse axis is horizontal with $a=\frac{1}{3}$ and $b=\frac{1}{2}$. The asymptotes are $y=\pm\frac{3}{2}x$. See Figure 53. The domain is $\left(-\infty,-\frac{1}{3}\right]\cup\left[\frac{1}{3},\infty\right)$ and the range is $(-\infty,\infty)$.

55. The center is $(1,-3)$ and the transverse axis is horizontal with $a=3$ and $b=5$. See Figure 55. The domain is $(-\infty,-2]\cup[4,\infty)$ and the range is $(-\infty,\infty)$.

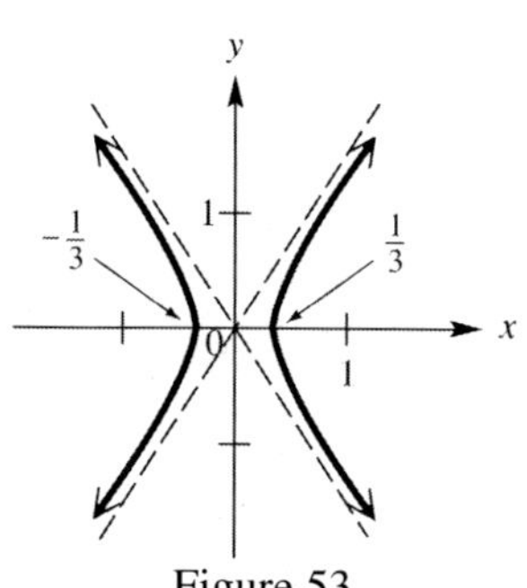

Figure 53

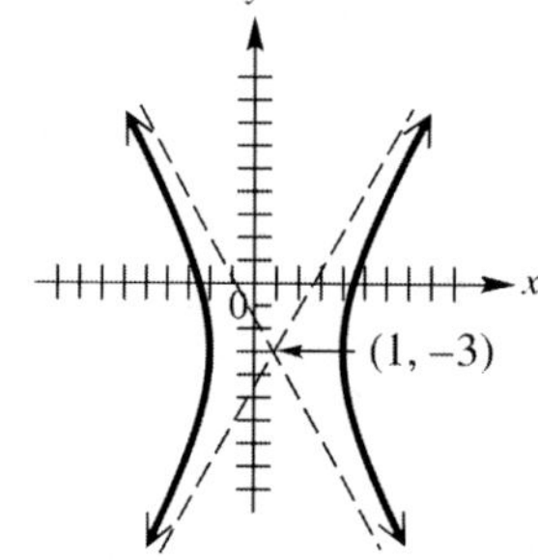

Figure 55

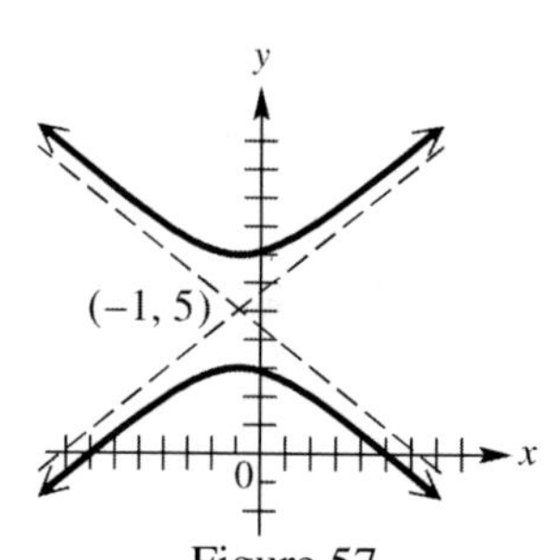

Figure 57

57. The center is $(-1,5)$ and the transverse axis is vertical with $a=2$ and $b=3$. See Figure 57. The domain is $(-\infty,\infty)$ and the range is $(-\infty,3]\cup[7,\infty)$.

59. $16(x+5)^2-(y-3)^2=1 \Rightarrow \dfrac{(x+5)^2}{\frac{1}{16}}-\dfrac{(y-3)^2}{1}=1$ The center is (–5, 3) and the transverse axis is horizontal with $a=\dfrac{1}{4}$ and $b=1$. See Figure 59. The domain is $\left(-\infty,-5-\dfrac{1}{4}\right]\cup\left[-5+\dfrac{1}{4},\infty\right)$ or $\left(-\infty,-\dfrac{21}{4}\right]\cup\left[-\dfrac{19}{4},\infty\right)$ and the range is $(-\infty,\infty)$.

61. $9(x-2)^2-4(y+1)^2=36 \Rightarrow \dfrac{(x-2)^2}{4}-\dfrac{(y+1)^2}{9}=1$. The center is (2, –1) and the transverse axis is vertical with $a=2$ and $b=3$. See Figure 61. The domain is $(-\infty,0]\cup[4,\infty)$ and the range is $(-\infty,\infty)$.

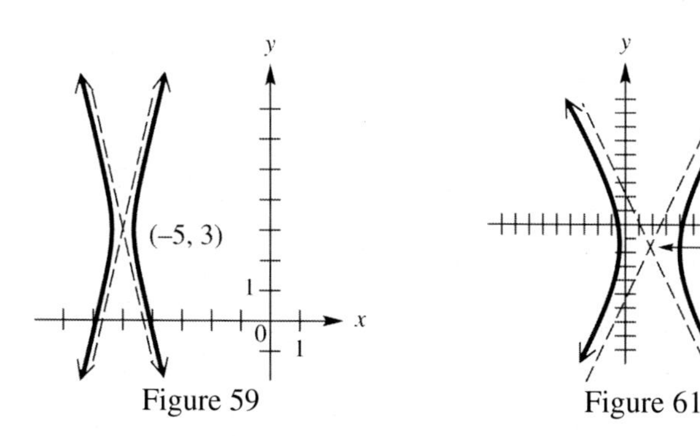

Figure 59 Figure 61

63. The hyperbola has a horizontal transverse axis with $c=4$. The x-intercepts coincide with the vertices, so $a=3$. The center is located between the foci at (0, 0). Since $c^2=a^2+b^2$, the value of b can be found by substitution: $b^2=c^2-a^2=4^2-3^2=16-9=7 \Rightarrow b=\sqrt{7}$. The equation is $\dfrac{x^2}{9}-\dfrac{y^2}{7}=1$.

65. The asymptotes intersect at the origin so the center is (0, 0). The hyperbola has a vertical transverse axis and the y-intercepts coincide with the vertices so $a=3$. From the asymptotes, $\dfrac{a}{b}=\dfrac{3}{5}$ with $a=3 \Rightarrow b=5$. The equation is $\dfrac{y^2}{9}-\dfrac{x^2}{25}=1$.

67. The center is located between the vertices at (0, 0). The hyperbola has a vertical transverse axis with $a=6$. From the asymptotes, $\dfrac{a}{b}=\dfrac{1}{2}$ with $a=6 \Rightarrow b=12$. The equation is $\dfrac{y^2}{36}-\dfrac{x^2}{144}=1$.

69. The center is located between the vertices at (0, 0). The hyperbola has a horizontal transverse axis with $a=3$. The equation is of the form $\dfrac{x^2}{a^2}-\dfrac{y^2}{b^2}=1$. By substitution using the point (6, 1),

$\frac{6^2}{3^2}-\frac{1^2}{b^2}=1\Rightarrow\frac{36}{9}-1=\frac{1}{b^2}\Rightarrow 3=\frac{1}{b^2}\Rightarrow b^2=\frac{1}{3}$. The equation is $\frac{x^2}{3^2}-\frac{y^2}{\frac{1}{3}}=1$ or $\frac{x^2}{9}-3y^2=1$.

71. The center is located between the foci at (0, 0). The hyperbola has a vertical transverse axis with $c=\sqrt{13}$. From the asymptotes, $\frac{a}{b}=5$. Also, from $c^2=a^2+b^2\Rightarrow a^2+b^2=13$. Solving these equations simultaneously results in $a^2=\frac{25}{2}$ and $b^2=\frac{1}{2}$. The equation is $\frac{y^2}{\frac{25}{2}}-\frac{x^2}{\frac{1}{2}}=1$ or $\frac{2y^2}{25}-2x^2=1$.

73. The center is located between the vertices at (4, 3). The hyperbola has a vertical transverse axis with $a=2$. From the asymptotes, $\frac{a}{b}=7$ with $a=2\Rightarrow b=\frac{2}{7}$. The equation is $\frac{(y-3)^2}{4}-\frac{(x-4)^2}{\frac{4}{49}}=1$ or $\frac{(y-3)^2}{4}-\frac{49(x-4)^2}{4}=1$.

75. With center (1, −2) and vertex (3, −2), we know the hyperbola has a horizontal transverse axis with $a=2$. With center (1, −2) and focus (4, −2), we know $c=3$. Since $c^2=a^2+b^2$, the value of b can be found by substitution: $b^2=c^2-a^2=3^2-2^2=9-4=5\Rightarrow b=\sqrt{5}$. The equation is $\frac{(x-1)^2}{4}-\frac{(y+2)^2}{5}=1$.

77. $x^2-2x-y^2+2y=4\Rightarrow(x^2-2x+1)-(y^2-2y+1)=4+1-1\Rightarrow(x-1)^2-(y-1)^2=4\Rightarrow$ $\frac{(x-1)^2}{4}-\frac{(y-1)^2}{4}=1$. The center is (1, 1). The vertices are (1−2, 1), (1+2, 1) or (−1, 1), (3, 1).\

79. $3y^2+24y-2x^2+12x+24=0\Rightarrow 3(y^2+8y)-2(x^2-6x)=-24\Rightarrow$

$3(y^2+8y+16)-2(x^2-6x+9)=-24+48-18\Rightarrow 3(y+4)^2-2(x-3)^2=6\Rightarrow\frac{(y+4)^2}{2}-\frac{(x-3)^2}{3}=1$.

The center is $(3,4)$. The vertices are $(3,-4-\sqrt{2})$, $(3,-4+\sqrt{2})$.

81. $x^2-6x-2y^2+7=0\Rightarrow(x^2-6x+9)-2y^2=-7+9\Rightarrow(x-3)^2-2(y-0)^2=2\Rightarrow$ $\frac{(x-3)^2}{2}-\frac{(y-0)^2}{1}=1$. The center is $(3,0)$. The vertices are $(3-\sqrt{2},0)$, $(3+\sqrt{2},0)$.

83. $4y^2+32y-5x^2-10x+39=0\Rightarrow 4(y^2+8y)-5(x^2+2x)=-39\Rightarrow$

$4(y^2+8y+16)-5(x^2+2x+1)=-39+64-5\Rightarrow 4(y+4)^2-5(x+1)^2=20\Rightarrow\frac{(y+4)^2}{5}-\frac{(x+1)^2}{4}=1$.

The center is $(-1,-4)$. The vertices are $(-1,-4-\sqrt{5})$, $(-1,-4+\sqrt{5})$.

85. If the focus is (0,2) and the vertex is (0,1), the parabola opens upward and $c=1$. Substituting in $(x-h)^2=4c(y-k)$, we get $(x-0)^2=4(1)(y-1)$ or $x^2=4(y-1)$.

86. If the focus is (-1,2) and the vertex is (3,2), the parabola opens to the left and $c = -4$. Substituting in $(y-k)^2 = 4c(x-h)$, we get $(y-2)^2 = -16(x-3)$.

87. If the focus is (0,0) and the directrix has equation $x = -2$, the vertex is (-1,0) and $c = 1$. The parabola opens to the right. Substituting in $(y-k)^2 = 4c(x-h)$, we get $(y-0)^2 = 4(1)(x-(-1))$ or $y^2 = 4(x+1)$.

88. If the focus is (2,1) and the directrix has equation $x = -1$, the vertex is $\left(\frac{1}{2}, 1\right)$ and $c = \frac{3}{2}$. The parabola opens to the right. Substituting in $(y-k)^2 = 4c(x-h)$, we get $(y-1)^2 = 4\left(\frac{3}{2}\right)\left(x - \frac{1}{2}\right)$ or $(y-1)^2 = 6\left(x - \frac{1}{2}\right)$.

89. If the focus is (-1,3) and the directrix has equation $y = 7$, the vertex is (-1,5) and $c = -2$. The parabola opens downward. Substituting in $(y-k)^2 = 4c(x-h)$, we get $(x+1)^2 = 4(-2)(y-5)$ or $(x+1)^2 = -8(y-5)$.

90. If the focus is (1,2) and the directrix has equation $y = 4$, the vertex is (1,3) and $c = -1$. The parabola opens downward. Substituting in $(x-h)^2 = 4c(y-k)$, we get $(x-1)^2 = 4(-1)(y-3)$ or $(x-1)^2 = -4(y-3)$.

91. The patient and the emitter are 12 units apart. These positions represent the foci of the ellipse so $c = 6$. With the minor axis measuring 16 units, $b = 8$. Since $c^2 = a^2 - b^2$, the value of a can be found by substitution: $a^2 = b^2 + c^2 = 6^2 + 8^2 = 36 + 64 = 100 \Rightarrow a = 10$. The equation is $\frac{x^2}{100} + \frac{y^2}{64} = 1$.

93. A major axis measuring 620 feet indicates that $a = 310$. A minor axis measuring 513 feet indicates that $b = 256.5$. Then $5c^2 = a^2 - b^2 = 310^2 - 256.5^2 \Rightarrow c = \sqrt{310^2 - 256.5^2} \approx 174.1$. The distance between the foci is $2c \approx 2(174.1) \approx 348.2$ feet.

95. Using a vertical major axis, $a = 15$. The minor axis has length 20, so $b = 10$. The equation is $\frac{y^2}{225} + \frac{x^2}{100} = 1$. Assuming the truck drives exactly in the middle of the road, we want to find y when $x = 6$. $\frac{y^2}{225} + \frac{6^2}{100} = 1 \Rightarrow \frac{y^2}{225} = 1 - \frac{36}{100} \Rightarrow y^2 = 225\left(1 - \frac{36}{100}\right) \Rightarrow y = \sqrt{225\left(1 - \frac{36}{100}\right)} = 12$. The truck must be just under 12 feet high to pass through.

97. (a) Since $c = \sqrt{a^2 - b^2} = \sqrt{4465^2 - 4462^2} \approx 163.6$, one focus is located at the point $(163.6, 0)$. The graph representing Earth is a circle with radius 3960 with center (163.6, 0). The equation for Earth is $(x-163.6)^2 + y^2 = 3690^2$. To graph this equation, solve for y and graph two parts.
$(x-163.6)^2 + y^2 = 3690^2 \Rightarrow y^2 = 3690^2 - (x-163.6)^2 \Rightarrow y = \pm\sqrt{3690^2 - (x-163.6)^2}$ To graph the

ellipse, solve for y and graph two parts.

$$\frac{x^2}{4465^2}+\frac{y^2}{4462^2}=1 \Rightarrow \frac{y^2}{4462^2}=1-\frac{x^2}{4465^2} \Rightarrow y^2=4462^2\left(1-\frac{x^2}{4465^2}\right) \Rightarrow y=\pm\sqrt{4462^2\left(1-\frac{x^2}{4465^2}\right)}$$

The graphs are shown in Figure 97.

(b) The minimum distance is $4465-(3960+163.6)\approx 341$ miles. The maximum distance is $4465-(3960+163.6)\approx 669$ miles.

[-6750,6750] by [-4500,4500]

Xscl = 1000 Yscl = 1000

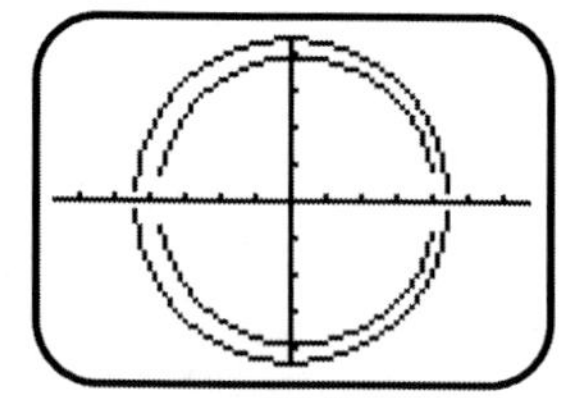

Figure 97

99. (a) Find a and b in the equation $\frac{x^2}{a^2}-\frac{y^2}{b^2}=1.$ Because the equations of the asymptotes of a hyperbola with horizontal transverse axes are $y=\pm\frac{b}{a}x,$ and the given asymptotes are $y=\pm x,$ it follows that $\frac{a}{b}=1$ or $a=b.$ Since the line $y=x$ intersects the x-axis at a 45° angle, the triangle shown in the third quadrant is a 45°- 45°- 90° right triangle and both legs must have length d. Then, by the Pythagorean theorem, $c^2=d^2+d^2=2d^2.$ That gives $c=d\sqrt{2}.$ Also, for a hyperbola $c^2=a^2+b^2,$ and since $a=b,$ $c^2=a^2+a^2=2a^2.$ That gives $c=a\sqrt{2}.$ From these two equations, $a\sqrt{2}=d\sqrt{2}$ and so $a=d.$ That is, $a=b=d=5\times10^{-14}.$ Thus the equation of the trajectory of A, where $x>0,$ is given by $\frac{x^2}{\left(5\times10^{-14}\right)^2}-\frac{y^2}{\left(5\times10^{-14}\right)^2}=1.$ Solving for x yields $x^2-y^2=\left(5\times10^{-14}\right)^2 \Rightarrow$ $x^2=y^2+2.5\times10^{-27} \Rightarrow x=\sqrt{y^2+2.5\times10^{-27}}.$ This equation represents the right half of the hyperbola, as shown in the textbook.

(b) Since $a=5\times10^{-14},$ the distance from the origin to the vertex is $5\times10^{-14}.$ The distance from N to the origin can be found using the Pythagorean theorem. Let h represent this distance, then $h^2=d^2+d^2.$ That is, $h^2=\left(5\times10^{-14}\right)^2+\left(5\times10^{-14}\right)^2 \Rightarrow h^2=5\times10^{-27} \Rightarrow h\approx7\times10^{-14}.$ The minimum distance between the centers of the alpha particle and the gold nucleus is $5\times10^{-14}+7\times10^{-14}\approx1.2\times10^{-13}.$

101. Let (x,y) be any point on the ellipse and start with the distance formula.

$\sqrt{(x+3)^2+y^2}+\sqrt{(x-3)^2+y^2}=10$	Given Equation.
$\sqrt{(x+3)^2+y^2}=10-\sqrt{(x-3)^2+y^2}$	Subtract $\sqrt{(x-3)^2+y^2}.$

$(x+3)^2 + y^2 = 100 - 20\sqrt{(x-3)^2 + y^2} + (x-3)^2 + y^2$ Square each side.

$(x+3)^2 - (x-3)^2 - 100 = -20\sqrt{(x-3)^2 + y^2}$ Subtract $(x-3)^2$ and 100.

$x^2 + 6x + 9 - x^2 + 6x - 9 - 100 = -20\sqrt{(x-3)^2 + y^2}$ Expand the binomials.

$12x - 100 = -20\sqrt{(x-3)^2 + y^2}$ Simplify.

$25 - 3x = 5\sqrt{(x-3)^2 + y^2}$ Divide each side by -4.

$625 - 150x + 9x^2 = 25\left(x^2 - 6x + 9 + y^2\right)$ Square each side.

$625 - 150x + 9x^2 = 25x^2 - 150x + 225 + 25y^2$ Multiply the right side.

$-16x^2 - 25y^2 = -400$ Simplify.

$\frac{-16x^2}{-400} + \frac{-25y^2}{-400} = \frac{-400}{-400}$ Divide each side by

$x = \ln(t-1) \Rightarrow t = e^x + 1.$

$\frac{x^2}{25} + \frac{y^2}{16} = 1$ Simplify.

Reviewing Basic Concepts (Sections 7.1 and 7.2)

1. (a) The circle is defined in B.
   (b) The parabola is defined in D.
   (c) The ellipse is defined in A.
   (d) The hyperbola is defined in C.

2. $12x^2 - 4y^2 = 48 \Rightarrow \frac{x^2}{4} - \frac{y^2}{12} = 1$. The transverse axis is horizontal with $a = 2$ and $b = \sqrt{12}$. The asymptotes are $y = \pm\frac{\sqrt{12}}{2}x$. See Figure 2.

3. Rewrite the equation. $y = 2x^2 + 3x - 1 \Rightarrow y + 1 + \frac{9}{8} = 2\left(x^2 + \frac{3}{2}x + \frac{9}{16}\right) \Rightarrow y + \frac{17}{8} = 2\left(x + \frac{3}{4}\right)^2$. The equation $y + \frac{17}{8} = 2\left(x + \frac{3}{4}\right)^2$ can be written $y + \frac{17}{8} = 4\left(\frac{1}{2}\right)\left(x + \frac{3}{4}\right)^2$. The vertex is $\left(-\frac{3}{4}, \frac{17}{8}\right)$.

   See Figure 3.

4. $x^2+y^2-2x+2y-2=0 \Rightarrow \left(x^2-2x+1\right)+\left(y^2+2y+1\right)=2+1+1 \Rightarrow \left(x-1\right)^2+\left(y+1\right)^2=4.$ The graph is a circle with center $(1,-1)$ and radius $\left(-4+\frac{1}{16},1\right) \Rightarrow \left(-\frac{63}{16},-1\right)$ See Figure 4.

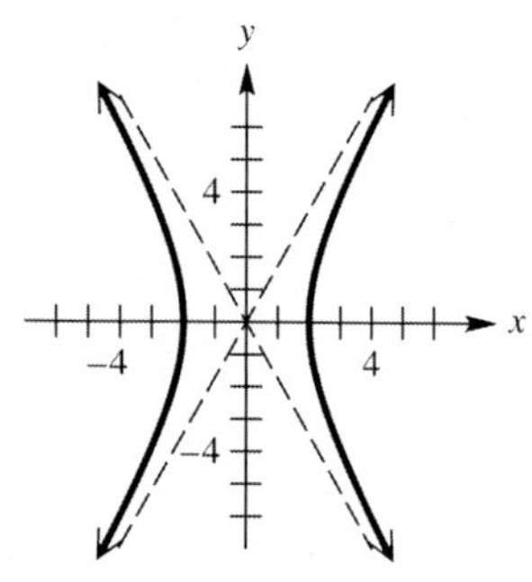

Figure 2

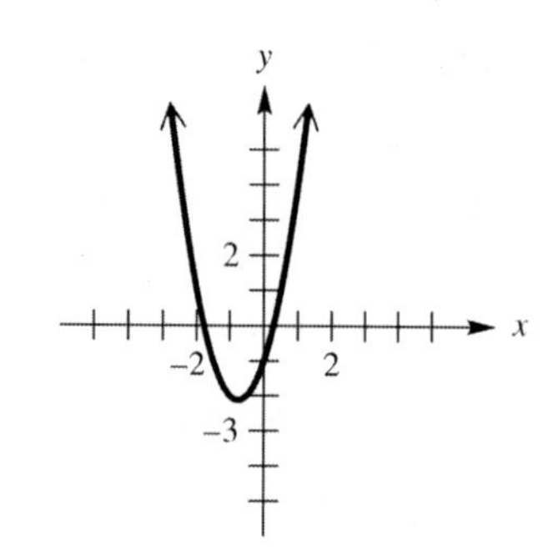

Figure 3

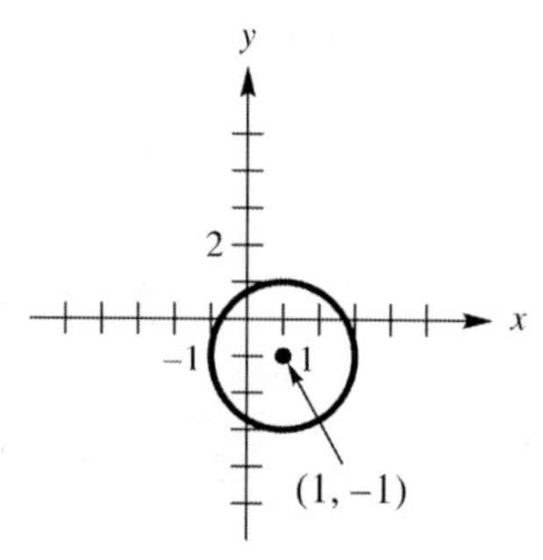

Figure 4

5. $4x^2+9y^2=36 \Rightarrow \frac{x^2}{9}+\frac{y^2}{4}=1 \Rightarrow a=3$ and $b=2$. The ellipse is graphed in Figure 5.

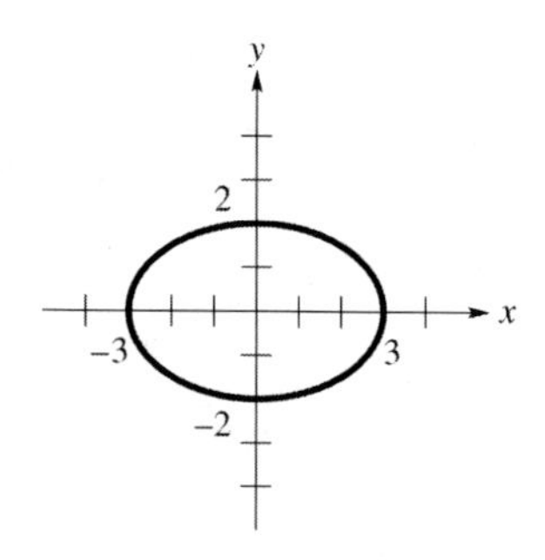

Figure 5

6. If $c<a$, it is an ellipse. If $c>a$, it is a hyperbola.

7. A circle with center $(2,-1)$ and radius 3 has equation $\left(x-2\right)^2+\left(y+1\right)^2=9$.

8. The ellipse is centered between the foci at $(0,0)$. The major axis is horizontal with $a=6$. Since the foci are $(\pm 4,0)$, we know that $c=4$. Since $c^2=a^2-b^2$, the value of b can be found by substitution: $b^2=a^2-c^2=6^2-4^2=36-16=20 \Rightarrow b=\sqrt{20}$. The equation is $\frac{x^2}{36}+\frac{y^2}{20}=1$.

9. The hyperbola has a vertical transverse axis with $c=4$. The vertices are $(0,\pm 2)$ so $a=2$. The center is located between the foci at $(0,0)$. Since $c^2=a^2+b^2$, the value of b can be found by substitution: $b^2=c^2-a^2=4^2-2^2=16-4=12 \Rightarrow b=\sqrt{12}$. The equation is $\frac{y^2}{4}-\frac{x^2}{12}=1$.

10. If the vertex is $(0,0)$ and the focus is $\left(0,\frac{1}{2}\right)$, then the parabola opens upward and $c=\frac{1}{2}$. The equation is

$x^2=4cy \Rightarrow x^2=2y.$

7.3: The Conic Sections and Nonlinear Systems

1. $x^2+y^2=144 \Rightarrow (x-0)^2+(y-0)^2=12^2$ The graph of this equation is a circle with center $(0,0)$ and radius 12. Also, note in our original equation, the x^2- and y^2-terms have the same positive coefficient.

3. $y=2x^2+3x-4 \Rightarrow y=2\left(x^2+\frac{3}{2}x\right)-4 \Rightarrow y=2\left(x^2+\frac{3}{2}x+\frac{9}{16}-\frac{9}{16}\right)-4 \Rightarrow y=2\left(x+\frac{3}{4}\right)^2+2\left(-\frac{9}{16}\right)-4$

   $\Rightarrow y=2\left(x+\frac{3}{4}\right)^2-\frac{9}{8}-4 \Rightarrow y=2\left(x+\frac{3}{4}\right)^2-\frac{41}{8} \Rightarrow y-\left(-\frac{41}{8}\right)=2\left[x-\left(-\frac{3}{4}\right)\right]^2$ The graph of this equation is a parabola opening upwards with a vertex of $\left(-\frac{3}{4},-\frac{41}{8}\right)$. Also, note our original equation has an x^2- term, but no y^2-term.

5. $x-1=-3(y-4)^2$ The graph of this equation is a parabola opening to the left with a vertex of $(1,4)$. Also, note when expanded, our original equation has a y^2-term, but no x^2-term.

7. $\frac{x^2}{49}+\frac{y^2}{100}=1 \Rightarrow \frac{x^2}{7^2}+\frac{y^2}{10^2}=1$ The graph of this equation is an ellipse centered at the origin and x-intercepts of 7 and –7, and y-intercepts of 10 and –10. Also, note in our original equation, the x^2- and y^2-terms both have different positive coefficients.

9. $\frac{x^2}{4}-\frac{y^2}{16}=1 \Rightarrow \frac{x^2}{2^2}-\frac{y^2}{4^2}=1$ The graph of this equation is a hyperbola centered at the origin with x-intercepts of 2 and –2, and asymptotes of $y=\pm\frac{4}{2}x=\pm 2x.$ Also, note in our original equation, the x^2- and y^2-terms have coefficient that are opposite in sign.

11. $\frac{x^2}{25}-\frac{y^2}{25}=1 \Rightarrow \frac{x^2}{5^2}-\frac{y^2}{5^2}=1$ The graph of this equation is a hyperbola centered at the origin with x-intercepts of 5 and –5, and asymptotes of $y=\pm\frac{5}{5}x=\pm x.$ Also, note in our original equation, the x^2- and y^2-terms have coefficients that are opposite in sign.

13. $\frac{x^2}{4}=1-\frac{y^2}{9}\Rightarrow\frac{x^2}{4}+\frac{y^2}{9}=1\Rightarrow\frac{(x-0)^2}{2^2}+\frac{(y-0)^2}{3^2}=1$ The equation is of the form $\frac{(x-h)^2}{b^2}+\frac{(y-k)^2}{a^2}=1$

with $a=3,\ b=2,\ h=0,$ and $k=0,$ so the graph of the given equation is an ellipse.

15. $\frac{(x+3)^2}{16}+\frac{(y-2)^2}{16}=1\Rightarrow(x+3)^2+(y-2)^2=16\Rightarrow[x-(-3)]^2+(y-2)^2=4^2$ The equation is of the form $(x-h)^2+(y-k)^2=r^2$ with $r=4,\ h=-3,$ and $k=2,$ so the graph of the given equation is a circle.

17. $x^2-6x+y=0\Rightarrow y=-x^2+6x\Rightarrow y=-\left(x^2-6x+9-9\right)\Rightarrow y=-(x-3)^2+9\Rightarrow y-9=-(x-3)^2$

The equation is of the form $y-k=a(x-h)^2$ with $a=-1,\ h=3,$ and $k=9,$ so the graph of the given equation is a parabola.

19. $4(x-3)^2+3(y+4)^2=0\Rightarrow\frac{4(x-3)^2}{12}+\frac{3(y+4)^2}{12}=0\Rightarrow\frac{(x-3)^2}{3}+\frac{[y-(-4)]^2}{4}=0$

The graph is the point $(3,-4)$.

21. $x-4y^2-8y=0\Rightarrow x=4y^2+8y\Rightarrow x=4\left(y^2+2y+1-1\right)\Rightarrow x=4(y+1)^2-4\Rightarrow x-(-4)=4[y-(-1)]^2$

The equation is of the form $x-h=a(y-k)^2$ with $a=4,\ h=-4,$ and $k=-1,$ so the graph of the given equation is a parabola.

23. $6x^2-12x+6y^2-18y+25=0\Rightarrow 6\left(x^2-2x+1-1\right)+6\left(y^2-3y+\frac{9}{4}-\frac{9}{4}\right)=-25\Rightarrow$

$6\left(x^2-2x+1\right)-6+6\left(y^2-3y+\frac{9}{4}\right)-\frac{27}{2}=-25\Rightarrow 6(x-1)^2+6\left(y-\frac{3}{2}\right)^2=-25+6+\frac{27}{2}$

$6(x-1)^2+6\left(y-\frac{3}{2}\right)^2=-\frac{50}{2}+\frac{12}{2}+\frac{27}{2}\Rightarrow 6(x-1)^2+6\left(y-\frac{3}{2}\right)^2=-\frac{11}{2}\Rightarrow(x-1)^2+\left(y-\frac{3}{2}\right)^2=-\frac{11}{12}$

A sum of squares can never be negative. This equation has no graph.

25. $x^2=4y-8\Rightarrow x^2=4(y-2)\Rightarrow y-2=\frac{1}{4}(x-0)^2$ The equation is of the form $y-k=a(x-h)^2$ with $a=\frac{1}{4},\ h=0,$ and $k=2,$ so the graph of the given equation is a parabola with vertex (0, 2) and vertical axis $x=0$ (the y-axis). Use the vertex and axis and plot a few additional points.

27. $x^2=25+y^2\Rightarrow x^2-y^2=25\Rightarrow\frac{x^2}{25}-\frac{y^2}{25}=1\Rightarrow\frac{(x-0)^2}{5^2}-\frac{(y-0)^2}{5^2}=1$ The equation is of the form

$\frac{(x-h)^2}{a^2}-\frac{(y-k)^2}{b^2}=1$ with $a=5,\ b=5,\ h=0,$ and $k=0,$ so the graph of the given equation is a hyperbola with center (0, 0), vertices (–5, 0) and (5, 0), and asymptotes $y=\pm x$.

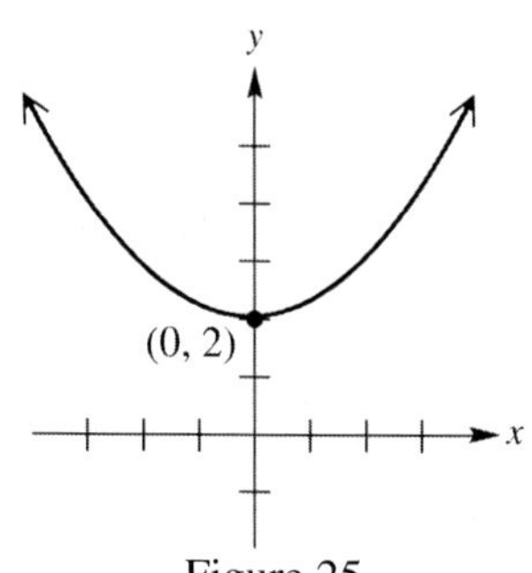

Figure 25

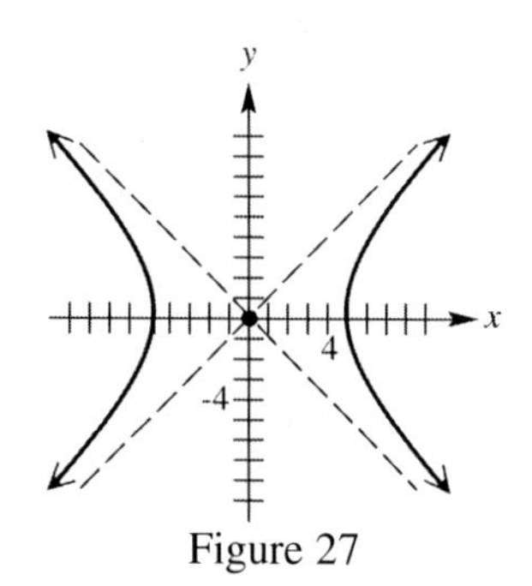

Figure 27

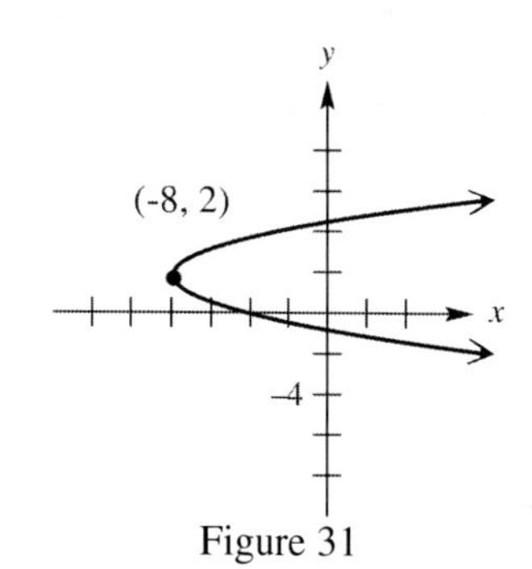

Figure 31

29. $\frac{x^2}{4}+\frac{y^2}{4}=-1 \Rightarrow x^2+y^2=-4$ A sum of squares can never be negative. This equation has no graph.

31. $y^2-4y=x+4 \Rightarrow y^2-4y+4-4=x+4 \Rightarrow (y-2)^2-4=x+4 \Rightarrow x+8=(y-2)^2 \Rightarrow$

$x-(-8)=(y-2)^2$ The equation is of the form $x-h=a(y-k)^2$ with $a=1$, $h=-8$, and $k=2$, so the graph of the given equation is a parabola with vertex (−8, 2) and horizontal axis $y = 2$.

33. $3x^2+6x+3y^2-12y=12 \Rightarrow x^2+2x+y^2-4y=4 \Rightarrow (x^2+2x+1-1)+(y^2-4y+4-4)=4$

$(x+1)^2-1+(y-2)^2-4=4 \Rightarrow (x-1)^2+(y-2)^2=4+1+4 \Rightarrow (x+1)^2+(y-2)^2=9$

$\Rightarrow [x-(-1)]^2+(y-2)^2=3^2$ The equation is of the form $(x-h)^2+(y-k)^2=r^2$ with

$r=3$, $h=-1$, and $k=2$, so the graph of the given equation is a circle with center (−1, 2) and radius 3.

35. $4x^2-8x+9y^2-36y=-4 \Rightarrow 4(x^2-2x+1-1)+9(y^2-4y+4-4)=-4 \Rightarrow$

$4(x-1)^2-4+9(y-2)^2-36=-4 \Rightarrow 4(x-1)^2+9(y-2)^2=36 \Rightarrow$

$\frac{4(x-1)^2}{36}+\frac{9(y-2)^2}{36}=1 \Rightarrow \frac{(x-1)^2}{9}+\frac{(y-2)^2}{4}=1 \Rightarrow \frac{(x-1)^2}{3^2}+\frac{(y-2)^2}{2^2}=1$

The equation is of the form $\frac{(x-h)^2}{a^2}+\frac{(y-k)^2}{b^2}=1$ with $a=3$, $b=2$, $h=1$, and $k=2$, so the graph of the given equation is an ellipse with center (1, 2) and vertices (−2, 2), (4, 2), (1, 0) and (1, 4).

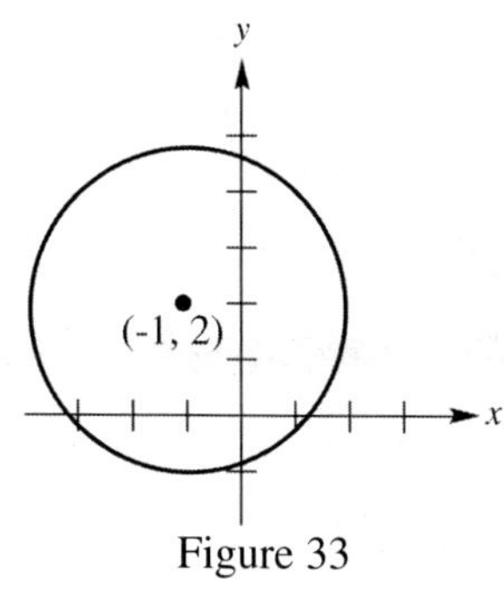

Figure 33

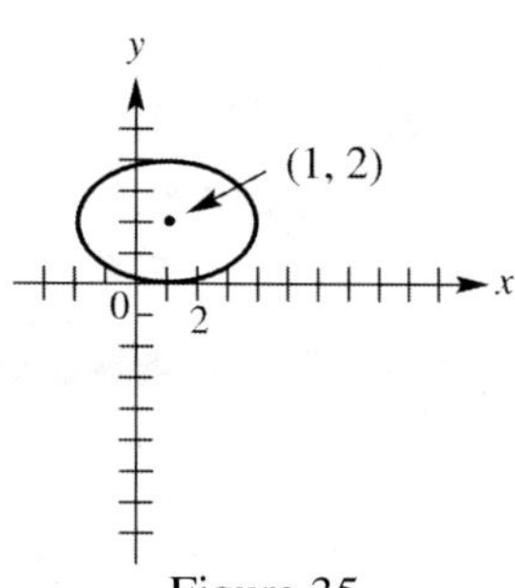

Figure 35

37. Since the sum of the distances from two points (foci) is a constant, the conic section is an ellipse.

39. Since the ratio of the distance from a point to (3, 0) and the distance from a point to the line $x = \frac{4}{3}$ is 1.5, the eccentricity is greater than 1. The conic section is a hyperbola.

41. $12x^2 + 9y^2 = 36 \Rightarrow \frac{x^2}{3} + \frac{y^2}{4} = 1 \Rightarrow a = 2,\ b = \sqrt{3},$ and $c = \sqrt{4-3} = 1;\ d = \frac{c}{a} = \frac{1}{2}.$

43. $x^2 - y^2 = 4 \Rightarrow \frac{x^2}{4} - \frac{y^2}{4} = 1 \Rightarrow a = 2,\ b = 2,$ and $c = \sqrt{4+4} = \sqrt{8};\ e = \frac{c}{a} = \frac{\sqrt{8}}{2} = \sqrt{2}.$

45. $4x^2 + 7y^2 = 28 \Rightarrow \frac{x^2}{7} + \frac{y^2}{4} = 1 \Rightarrow a = \sqrt{7},\ b = 2,$ and $c = \sqrt{7-4} = \sqrt{3};\ e = \frac{c}{a} = \frac{\sqrt{3}}{\sqrt{7}} = \frac{\sqrt{21}}{7}.$

47. $x^2 - 9y^2 = 18 \Rightarrow \frac{x^2}{18} - \frac{y^2}{2} = 1 \Rightarrow a = \sqrt{18},\ b = \sqrt{2},$ and $c = \sqrt{18+2} = \sqrt{20};\ e = \frac{c}{a} = \frac{\sqrt{20}}{\sqrt{18}} = \frac{\sqrt{10}}{3}.$

49. Since $e = 1$, the conic is a parabola. With center $(0,0)$ and focus $(0,8)$, the equation is $x^2 = 4cy \Rightarrow x^2 = 32y.$

51. Since $0 < e < 1$, the conic is an ellipse with $c = 3$. Now $\frac{c}{a} = e \Rightarrow \frac{3}{a} = \frac{1}{2} \Rightarrow a = 6.$ For an ellipse, $b^2 = a^2 - c^2 = 36 - 9 = 27.$ The equation is $\frac{x^2}{36} + \frac{y^2}{27} = 1.$

53. Since $e > 1$, the conic is a hyperbola with $a = 6$. Now $\frac{c}{a} = e \Rightarrow \frac{c}{6} = 2 \Rightarrow c = 12.$ For a hyperbola, $b^2 = c^2 - a^2 = 144 - 36 = 108.$ The equation is $\frac{x^2}{36} - \frac{y^2}{108} = 1.$

55. Since $e = 1$, the conic is a parabola. With center $(0,0)$ and focus $(0,-1)$, the equation is $x^2 = 4cy \Rightarrow x^2 = -4y.$

57. Since $0 < e < 1$, the conic is an ellipse with $a = 3$. Now $\frac{c}{a} = e \Rightarrow \frac{c}{3} = \frac{4}{5} \Rightarrow c = \frac{12}{5}.$ For an ellipse, $b^2 = a^2 - c^2 = 9 - \frac{144}{25} = \frac{81}{25}.$ The equation is $\frac{x^2}{\frac{81}{25}} + \frac{y^2}{9} = 1$ or $\frac{25x^2}{81} + \frac{y^2}{9} = 1.$

59. From the graph, the coordinates of P (a point on the graph) are (–3, 8), the coordinates of F (a focus) are (3, 0), the equation of L (the directrix) is $x = 27$. By the distance formula, the distance from P to F is

$$\sqrt{(x_2 - x_1)^2 + (y_2 - y_1)^2} = \sqrt{[3-(-3)]^2 + (0-8)^2} = \sqrt{6^2 + (-8)^2} = \sqrt{36+64} = \sqrt{100} = 10$$

The distance from a point to a line is defined as the perpendicular distance, so the distance from P to L is $|27 - (-3)| = 30.$ Thus, $e = \frac{\text{Distance of } P \text{ from } F}{\text{Distance of } P \text{ from } L} = \frac{10}{30} = \frac{1}{3}.$

61. From the graph, we see that $F=\left(\sqrt{2},0\right)$ and L is the vertical line $x=-\sqrt{2}$. Choose $(0,0)$, the vertex of the parabola, as P. Distance of P from $F=\sqrt{2}$, and distance of P from $L=\sqrt{2}$. Thus, we have $e=\frac{\text{Distance of } P \text{ from } F}{\text{Distance of } P \text{ from } L}=\frac{\sqrt{2}}{\sqrt{2}}=1.$

63. From the graph, we see that $P = (9, -7.5)$, $F = (9, 0)$ and L is the vertical line $x = 4$. Distance of P from $F=7.5$, and distance of P from $L=5$. Thus, $e=\frac{\text{Distance of } P \text{ from } F}{\text{Distance of } P \text{ from } L}=\frac{7.5}{5}=1.5.$

65. Add the equations to eliminate the y^2:

$$\begin{aligned} x^2+y^2&=10\\ 2x^2-y^2&=17\\ \hline 3x^2\quad\;&=27\Rightarrow x^2=9\Rightarrow x=\pm 3. \end{aligned}$$

Substituting these values into [equation 1] yields $(-3)^2+y^2=10\Rightarrow y^2=1\Rightarrow y=\pm 1$ and $(3)^2+y^2=10\Rightarrow y^2=1\Rightarrow y=\pm 1$. The solution set is $\{(-3,1),(-3,-1),(3,1),(3,-1)\}$.

67. Multiply [equation 1] by 3 and subtract to eliminate the x^2:

$$\begin{aligned} 3x^2+6y^2&=27\\ 3x^2-4y^2&=27\\ \hline 10y^2&=0\Rightarrow 10y^2=0\Rightarrow y=0. \end{aligned}$$

Substituting 0 into [equation 1] for y, yields $x^2+2(0)^2=9\Rightarrow x^2=9\Rightarrow x=\pm 3$. The solution set is $\{(3,0),(-3,0)\}$.

69. Multiply [equation 1] by 3 and [equation 2] by 2 and subtract to eliminate both x^2 and y^2:

$$\begin{aligned} 6x^2+6y^2&=60\\ 6x^2+6y^2&=60\\ \hline 0\quad&=0 \end{aligned}$$

$\Rightarrow$ infinite number of solutions. The solutions have the following relationship:

$2x^2+2y^2=20\Rightarrow 2y^2=20-2x^2\Rightarrow y^2=10-x^2\Rightarrow y=\pm\sqrt{10-x^2}$. The solution set is $\left\{\left(x,-\sqrt{10-x^2}\right),\left(x,\sqrt{10-x^2}\right)\right\}$.

71. First solve [equation 2] for x: $x-y=-2\Rightarrow x=y-2$ [equation 3]. Now substitute $y-2$ for x in [equation 1]: $3(y-2)^2+2y^2=5\Rightarrow 3\left(y^2-4y+4\right)+2y^2=5\Rightarrow 3y^2-12y+12+2y^2-5=0\Rightarrow$

$5y^2-12y+7=0\Rightarrow(5y-7)(y-1)=0\Rightarrow y=1$ or $\frac{7}{5}$. Substituting these values into [equation 3] yields $x=(1)-2\Rightarrow x=-1$ and $x=\left(\frac{7}{5}\right)-2\Rightarrow x=-\frac{3}{5}$. The solution set is $\left\{\left(-\frac{3}{5},\frac{7}{5}\right),(-1,1)\right\}$.

73. Add the equations to eliminate the y^2 :

$$\begin{array}{l} x^2 + y^2 = 8 \\ \underline{x^2 - y^2 = 0} \\ 2x^2 \qquad = 8 \Rightarrow x^2 = 4 \Rightarrow x = \pm 2. \end{array}$$

Substituting these values into [equation 2] yields $(-2)^2 - y^2 = 0 \Rightarrow y^2 = 4 \Rightarrow y = \pm 2$ and $(2)^2 - y^2 = 0 \Rightarrow y^2 = 4 \Rightarrow y = \pm 2$ The solution set is $\{(-2,-2),(-2,2),(2,-2),(2,2)\}$.

75. Subtract the equations to eliminate both x^2 and y^2

$$\begin{array}{l} x^2 + xy + y^2 = 3 \\ \underline{x^2 - xy + y^2 = 1} \\ \quad 2xy \qquad = 2 \Rightarrow y = \dfrac{1}{x} \end{array}$$

Substituting this result into [equation 2] yields $x^2 - x\left(\frac{1}{x}\right) + \left(\frac{1}{x}\right)^2 = 1 \Rightarrow x^2 - 1 + \frac{1}{x^2} = 1 \Rightarrow$

$x^4 - x^2 + 1 = x^2 \Rightarrow x^4 - 2x^2 + 1 = 0 \Rightarrow \left(x^2 - 1\right)\left(x^2 - 1\right) = 0 \Rightarrow x = \pm 1$. Substitute these values into

$y = \frac{1}{x}$ yields $y = \frac{1}{-1} = -1$ and $y = \frac{1}{1} = 1$ The solution set is $\{(1,1),(-1,-1)\}$.

77. Add the equations to eliminate both xy and y^2

$$\begin{array}{l} x^2 \quad - xy + y^2 = 5 \\ \underline{2x^2 + xy - y^2 = 10} \\ 3x^2 \qquad\qquad = 15 \Rightarrow x^2 = 5 \Rightarrow x = \pm\sqrt{5} \end{array}$$

Substituting these results into [equation 1] yields $\left(-\sqrt{5}\right)^2 - \left(-\sqrt{5}\right)y + y^2 = 5 \Rightarrow 5 + \sqrt{5}y + y^2 = 5 \Rightarrow$

$y^2 + \sqrt{5}y = 0 \Rightarrow y\left(y + \sqrt{5}\right) = 0 \Rightarrow y = 0, -\sqrt{5}$ and $\left(\sqrt{5}\right)^2 - \left(\sqrt{5}\right)y + y^2 = 5 \Rightarrow 5 - \sqrt{5}y + y^2 = 5 \Rightarrow$

$y^2 - \sqrt{5}y = 0 \Rightarrow y\left(y - \sqrt{5}\right) = 0 \Rightarrow y = 0, \sqrt{5}$.

The solution set is $\left\{\left(\sqrt{5},0\right),\left(-\sqrt{5},0\right),\left(\sqrt{5},\sqrt{5}\right),\left(-\sqrt{5},-\sqrt{5}\right)\right\}$.

79. First solve [equation 3] for y and substitute into [equation 2]:

$$\begin{array}{l} 2x^2 + y^2 + 3z^2 = 3 \\ 2x + y - z = 1 \\ x + y \qquad = 0 \end{array}$$

$x + y = 0 \Rightarrow y = -x$: $2x - x - z = 1 \Rightarrow x - z = 1 \Rightarrow z = x - 1$, Substituting $y = -x$ and $z = x - 1$ into [equation 1] yields: $2x^2 + (-x)^2 + 3(x-1)^2 = 3 \Rightarrow 2x^2 + x^2 + 3x^2 - 6x + 3 = 3 \Rightarrow 6x^2 - 6x = 0 \Rightarrow$

$6x(x-1) = 0 \Rightarrow x = 0,1$. Substitute these values to find y and z. $y = -(0) = 0$, $z = 0 - 1 = -1$ and $y = -(1) = -1$, $z = 1 - 1 = 0$, The solution set is $\{(0,0,-1),(1,-1,0)\}$.

81. See Figure 81.

83. See Figure 83.

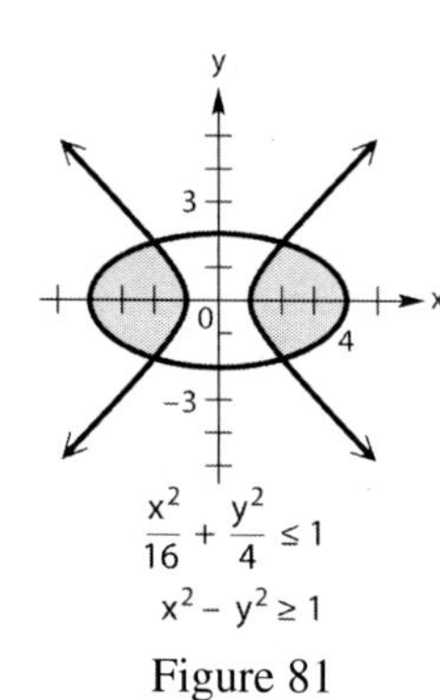

Figure 81

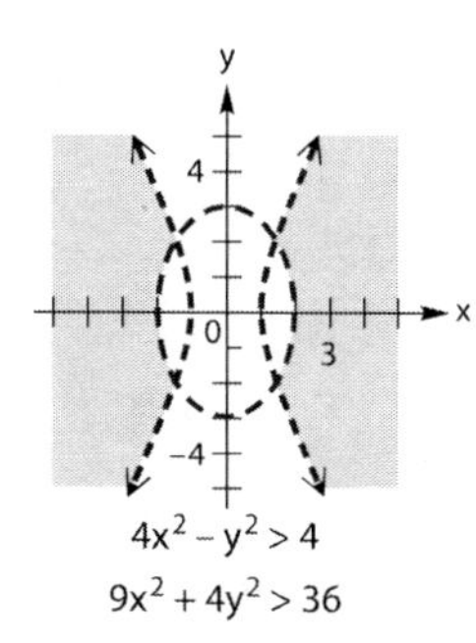

Figure 83

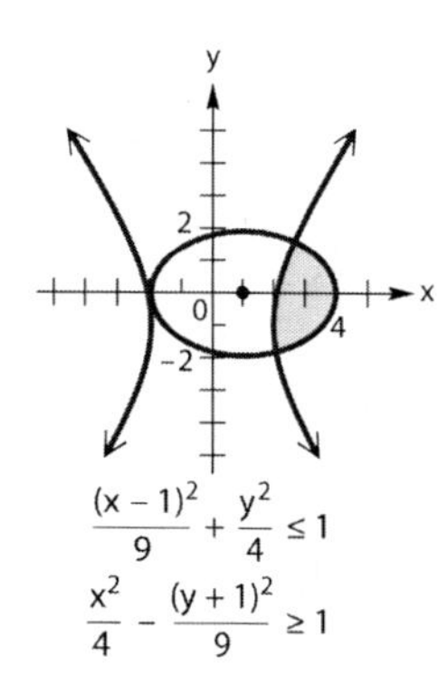

Figure 85

85. See Figure 85.

87. For an ellipse, $c = \sqrt{a^2 - b^2} = \sqrt{5013 - 4970} = \sqrt{43}$. The eccentricity is $e = \frac{c}{a} = \frac{\sqrt{43}}{\sqrt{5013}} \approx .093$.

89. (a) Earth orbits every $365 \cdot 24 \cdot 60 \cdot 60 = 31{,}536{,}000$ seconds. Thus $\frac{y^2}{16} - \frac{x^2}{9} = 1$ The maximum velocity of Earth is $v_{\max} = \frac{2\pi(1.496 \times 10^8)}{31{,}536{,}000}\sqrt{\frac{1+.0167}{1-.0167}} \approx 30.3$ km per sec. The maximum velocity of Earth is $v_{\min} = \frac{2\pi(1.496 \times 10^8)}{31{,}536{,}000}\sqrt{\frac{1+.0167}{1-.0167}} \approx 29.3$ km per sec.

(b) The minimum and maximum velocities are equal. Therefore, the planet's velocity is constant.

(c) A planet is at its maximum and minimum distances from a focus when it is located at the vertices of the ellipse. Thus the minimum and maximum velocities of a planet will occur at the vertices of the elliptical orbit, which are $a + c$ for the minimum and $a - c$ for the maximum.

91. Here $a + c = 94.6$ and $a - c = 91.4$. Solving these equations simultaneously results in $a = 93$ and $c = 1.6$. The eccentricity is $e = \frac{c}{a} = \frac{1.6}{93} \approx 0.0172$.

7.4: Parametric Equations

1.

t	-2	-1	0	1	2
x	-3	-1	1	3	5
y	-4	-3	-2	-1	0

3.

t	-2	-1	0	1	2
x	-1	0	1	2	3
y	3	0	-1	0	3

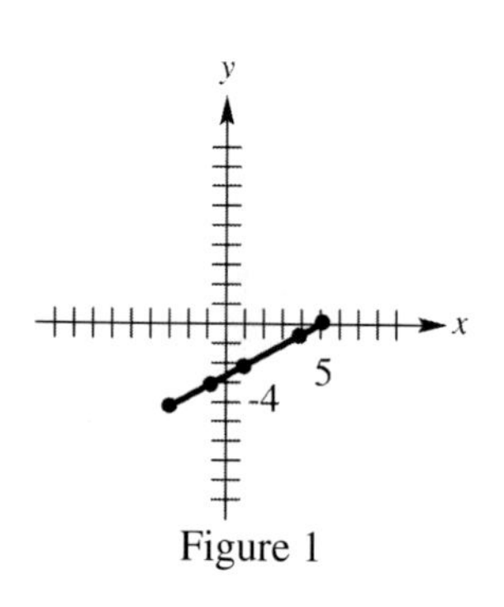

Figure 1

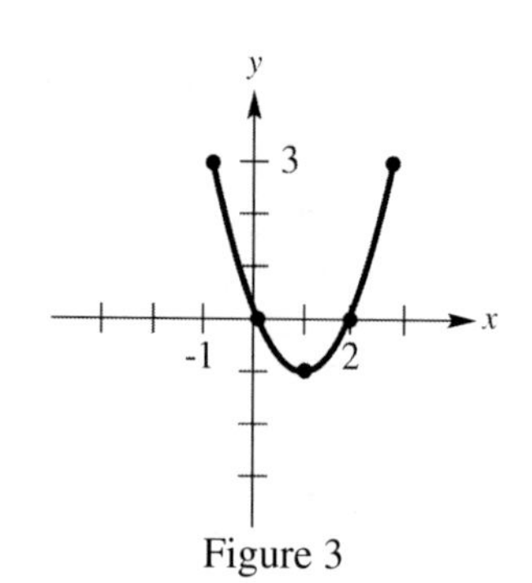

Figure 3

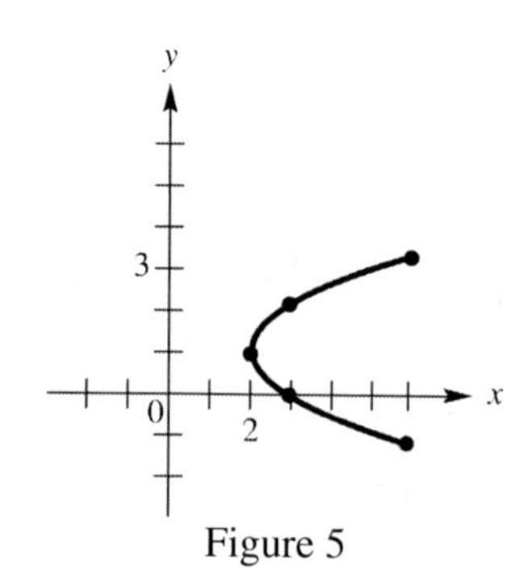

Figure 5

5.

t	-2	-1	0	1	2
x	6	3	2	3	6
y	3	2	1	0	1

7. See Figure 7. From the first equation, $x=2t \Rightarrow t=\frac{1}{2}x$. By substitution in the second equation, $y=\frac{1}{2}x+1$.

   When t is in $[-2,3]$, the range of $x=2t$ is x in $[-4,6]$.

9. See Figure 9. From the first equation, $x=\sqrt{t} \Rightarrow t=x^2$. By substitution in the second equation,

   $y=3x^2-4$. When t is in $[0,4]$, the range of $x=\sqrt{t}$ is x in $[0,2]$.

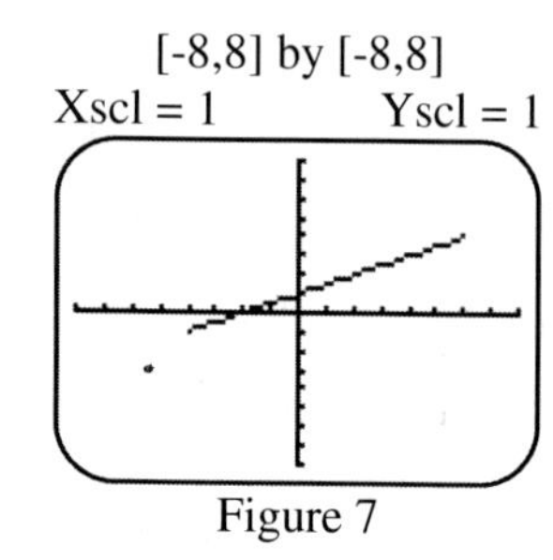

Figure 7

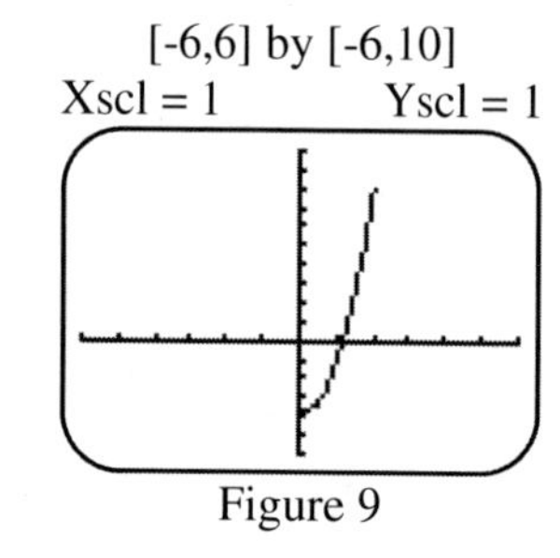

Figure 9

11. See Figure 11. From the first equation, $x=t^3+1 \Rightarrow t^3=x-1$. By substitution in the second equation,

    $y=x-2$. When t is in $[-3,3]$, the range of $x=t^3+1$ is x in $[-26,28]$.

13. See Figure 13. From the second equation, $y=\sqrt{3t-1} \Rightarrow t=\frac{y^2+1}{3}$. By substitution in the second equation,

    $x=2^{(y^2+1)/3}$. When t is in $\left[\frac{1}{3},4\right]$, the range of $y=\sqrt{3t=-1}$ is y in $[0,\sqrt{11}]$.

[-30,30] by [-30,30]
Xscl = 2 Yscl = 2

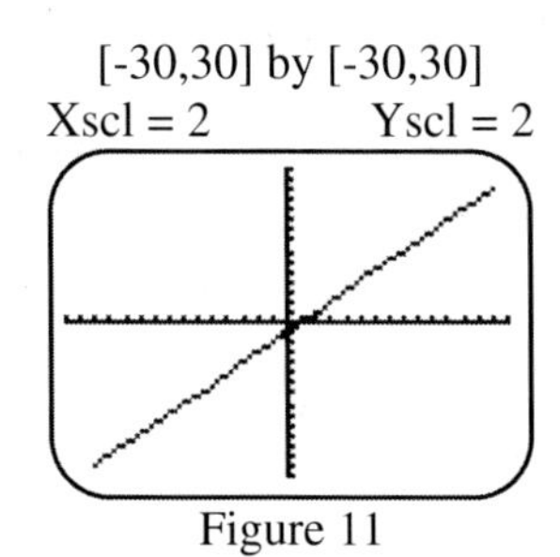

Figure 11

[-2,30] by [-2,10]
Xscl = 1 Yscl = 1

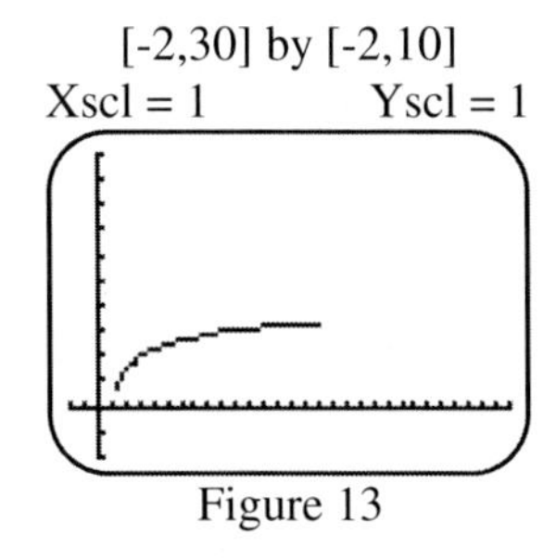

Figure 13

[-6,6] by [-4,4]
Xscl= 1 Yscl = 1

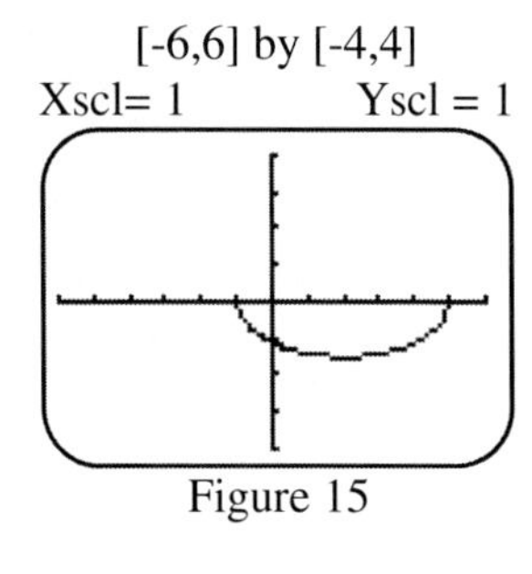

Figure 15

[-6,6] by [-4,4]
Xscl = 1 Yscl= 1

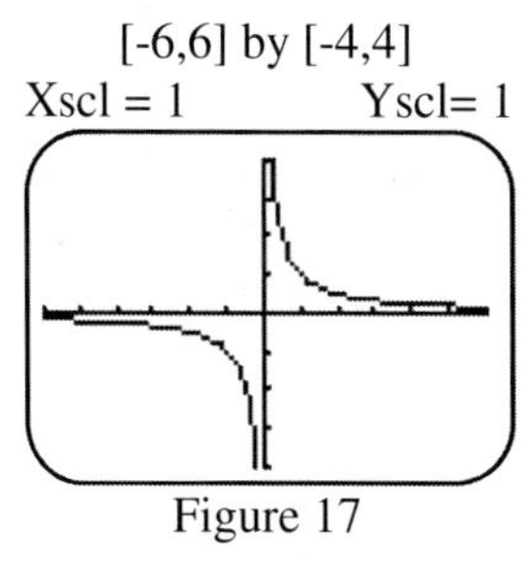

Figure 17

15. See Figure 15. From the first equation, $x = t + 2 \Rightarrow t = x - 2$. By substitution in the second equation, $y = -\frac{1}{2}\sqrt{9-(x-2)^2}$. When t is in $[-3,3]$, the range of $x = t + 2$ is x in $[-1,5]$.

17. See Figure 17. From the first equation, $x = t \Rightarrow t = x$. By substitution in the second equation, $y = \frac{1}{x}$. When t is in $(-\infty,0)\cup(0,\infty)$, the range of $x = t$ is x in $(-\infty,0)\cup(0,\infty)$.

19. From the first equation, $x = 3t \Rightarrow t = \frac{1}{3}x$. By substitution in the second equation, $y = \frac{1}{3}x - 1$. When t is in $(-\infty,\infty)$, the range of $x = 3t$ is x in $(-\infty,\infty)$.

21. From the second equation, $y = t + 1 \Rightarrow t = y - 1$. By substitution in the first equation, $x = 3(y-1)^2$. When t is in $(-\infty,\infty)$, the range of $y = t + 1$ is y in $(-\infty,\infty)$.

23. From the second equation, $y = 4t^3 \Rightarrow t = \sqrt[3]{\frac{y}{4}}$. By substitution in the first equation, $x = 3\left(\frac{y}{4}\right)^{2/3}$. When t is in $(-\infty,\infty)$, the range of $y = 4t^3$ is y in $(-\infty,\infty)$.

25. From the first equation, $x = t \Rightarrow t = x$. By substitution in the second equation, $y = \sqrt{x^2+2}$. When t is in $(-\infty,\infty)$, the range of $x = t$ is x in $(-\infty,\infty)$.

27. From the first equation, $x = e^t \Rightarrow t = \ln x$. By substitution in the second equation, $y = \frac{1}{x}$. When t is in $(-\infty,\infty)$, the range of $x = e^t$ is x in $(0,\infty)$.

29. From the first equation, $x = \frac{1}{\sqrt{t+2}} \Rightarrow t = \frac{1}{x^2} - 2$. By substitution in the second equation, $y = 1 - 2x^2$. When t is in $(-2,\infty)$, the range of $x = \frac{1}{\sqrt{t+2}}$ is x in $(0,\infty)$.

31. From the first equation, $x = t + 2 \Rightarrow t = x - 2$. By substitution in the second equation, $y = \frac{1}{x}$. When t has the restriction $t \neq 2$, the range of $x = t + 2$ has the restriction $x \neq 0$.

33. From the first equation, $x = t^2 \Rightarrow t = \sqrt{x}$. By substitution in the second equation, $y = \ln x$. When t is in $(0, \infty)$, the range of $x = \ln t$ is x in $(0, \infty)$.

35. For $x = \frac{1}{2}t$, $y = 2\left(\frac{1}{2}t\right) + 3 \Rightarrow y = t + 3$. For $x = \frac{t+3}{2}$, $y = 2\left(\frac{t+3}{2}\right) + 3 \Rightarrow y = t + 6$.

37. For $x = \frac{1}{3}t$, $y\sqrt{3\left(\frac{1}{3}t\right) + 2} \Rightarrow y = \sqrt{t+2}$ for t in $[-2, \infty)$. For $x = \frac{t-2}{3}$, $y = \sqrt{3\left(\frac{t-2}{3}\right) + 2} \Rightarrow y = \sqrt{t}$ for t in $[0, \infty)$.

39. For $x = t^3 + 1$, $t^3 + 1 = y^3 + 1 \Rightarrow y = t$. For $x = t$, $t = y^3 + 1 \Rightarrow y = \sqrt[3]{t-1}$.

41. For $x = \sqrt{t+1}$, $\sqrt{t+1} = \sqrt{y+1} \Rightarrow y = t$ for t in $[-1, \infty)$. For $x = \sqrt{t}$, $\sqrt{t} = \sqrt{y+1} \Rightarrow y + 1 = t \Rightarrow y = t - 1$ for t in $[0, \infty)$.

43. (a) Find t when $y = 0$. $400 \cdot \frac{\sqrt{2}}{2}t - 16t^2 = 0 \Rightarrow 16t^2 - 200\sqrt{2}t = 0 \Rightarrow t(16t - 200\sqrt{2}) = 0 \Rightarrow$

$$16t - 200\sqrt{2} = 0 \Rightarrow 16t = 200\sqrt{2} \Rightarrow t = \frac{200\sqrt{2}}{16} \approx 17.7 \text{ seconds.}$$

(b) Find x when $t \approx 17.7$. $x = 400 \cdot \frac{\sqrt{2}}{2}(17.7) \approx 5000$ feet.

(c) Find y when $x \approx 8.85$ (half the total time). $y = 400 \cdot \frac{\sqrt{2}}{2}(8.85) - 16(8.85)^2 \approx 1250$ feet.

45. See Figure 45. From the first equation, $x = 60\,t \Rightarrow t = \frac{x}{60}$. By substitution in the second equation,

$$y = 80\left(\frac{x}{60}\right) - 16\left(\frac{x}{60}\right)^2 \Rightarrow y = \frac{4}{3}x - \frac{x^2}{225}.$$

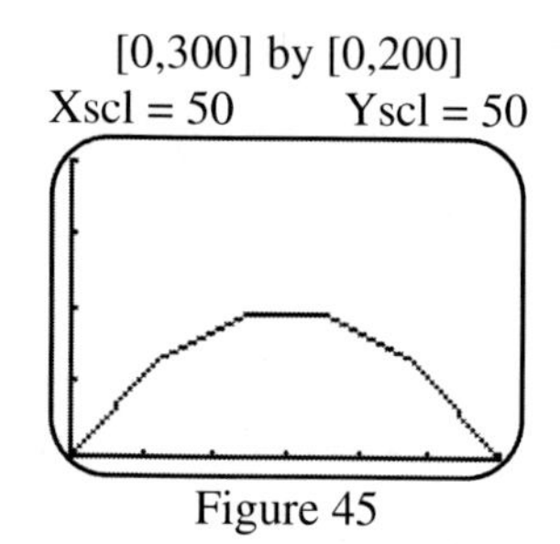

Figure 45

47. From the first equation, $x = v_0 \frac{\sqrt{2}}{2} t \Rightarrow t = \frac{2x}{v_0\sqrt{2}}$. By substitution in the second equation,

$$y = v_0 \frac{\sqrt{2}}{2}\left(\frac{2x}{v_0\sqrt{2}}\right) - 16\left(\frac{2x}{v_0\sqrt{2}}\right)^2 \Rightarrow y = x - \frac{32}{(v_0)^2}x^2.$$

49. A line through (x_1, y_1) with slope m is given by $y - y_1 = m(x - x_1)$. For $x = t$, $y - y_1 = m(t - x_1)$.

For $t = x - x_1$, $y - y_1 = mt \Rightarrow y = mt + y_1$. Many answers are possible.

Reviewing Basic Concepts (Sections 7.3 and 7.4)

1. $3x^2 + y^2 - 6x + 6y = 0 \Rightarrow 3(x^2 - 2x + 1) + (y^2 + 6y + 9) = 0 + 3 + 9 \Rightarrow$

$3(x-1)^2 + (y+3)^2 = 12 \Rightarrow \frac{(x-1)^2}{4} + \frac{(y+3)^2}{12} = 1$; ellipse centered at $(1, -3)$.

2. $y^2 - 2x^2 + 8y - 8x - 4 = 0 \Rightarrow (y^2 + 8y + 16) - 2(x^2 + 4x + 4) = 4 + 16 - 8 \Rightarrow$

$(y+4)^2 - 2(x+2)^2 = 12 \Rightarrow \frac{(y+4)^2}{12} - \frac{(x+2)^2}{6} = 1$; hyperbola centered at $(-4, -2)$.

3. $3y^2 + 12y + 5x = 3 \Rightarrow 3(y^2 + 4y + 4) = -5x + 3 + 12 \Rightarrow 3(y+2)^2 = -5(x-3) \Rightarrow (y+2)^2 = \frac{5}{3}(x-3)$;

parabola with center $(3, -2)$ that opens to the left.

4. $x^2 + 25y^2 = 25 \Rightarrow \frac{x^2}{25} + \frac{y^2}{1} = 1 \Rightarrow a = 5$, $b = 1$, and $c = \sqrt{25-1} = \sqrt{24}$. $e = \frac{c}{a} = \frac{\sqrt{24}}{5} = \frac{2\sqrt{6}}{5}$.

5. $8y^2 - 4x^2 = 8 \Rightarrow \frac{y^2}{1} - \frac{x^2}{2} = 1 \Rightarrow a = 1$, $b = \sqrt{2}$, and $c = \sqrt{1+2} = \sqrt{3}$. $e = \frac{c}{a} = \frac{\sqrt{3}}{1} = \sqrt{3}$.

6. $3x^2 + 4y^2 = 108 \Rightarrow \frac{x^2}{36} + \frac{y^2}{27} = 1 \Rightarrow a = 6$, $b = \sqrt{27}$, and $c = \sqrt{36-27} = \sqrt{9}$. $e = \frac{c}{a} = \frac{3}{6} = \frac{1}{2}$.

7. Since $e = 1$, the conic is a parabola. With center $(0,0)$ and focus $(-2,0)$, the equation is

$y^2 = 4cx \Rightarrow y^2 = -8x$.

8. The ellipse is centered between the foci at $(0,0)$. The major axis is horizontal with $a=5$. Since the foci are $(\pm 3,0)$, we know that $c=3$. Since $c^2=a^2-b^2$, the value of b can be found by substitution.

   $b^2=a^2-c^2=5^2-3^2=25-9=16 \Rightarrow b=4$. The equation is $\frac{x^2}{25}+\frac{y^2}{16}=1$.

9. The hyperbola has a vertical transverse axis with $c=5$. The vertices are $(0,\pm 4)$, so $a=4$. The center is located between the foci at $(0,0)$. Since $c^2=a^2+b^2$, the value of b can be found by substitution.

   $b^2=c^2-a^2=5^2-4^2=25-16=9 \Rightarrow b=3$. The equation is $\frac{y^2}{16}-\frac{x^2}{9}=1$.

10. The center is located between the vertices at (0, 0). The hyperbola has a horizontal transverse axis with $a=3$. From the asymptotes, $\frac{b}{a}=\frac{2}{3}$ with. $a=3 \Rightarrow b=2$. The equation is $\frac{x^2}{9}-\frac{y^2}{4}=1$.

11. Add 2 multiplied by [equation 1] to equation 2 to eliminate x^2:

$$
\begin{array}{r}
4x^2+2y^2=18 \\
-4x^2+3y^2=27 \\
\hline
5y^2=45 \Rightarrow y^2=9 \Rightarrow y=\pm 3
\end{array}
$$

   Substitute these values into [equation 1] to solve for x: $2x^2+(3)^2=9 \Rightarrow 2x^2=0 \Rightarrow x=0$, $2x^2+(-3)^2=9 \Rightarrow 2x^2=0 \Rightarrow x=0$, The solution set is $\{(0,-3),(0,3)\}$.

12. See Figure 12.

13. Using a vertical minor axis, $b=9$. The major axis has length 30, so $a=15$. The equation is $\frac{x^2}{225}+\frac{y^2}{81}=1$. When $x=6$, $\frac{6^2}{225}+\frac{y^2}{81}=1 \Rightarrow \frac{y^2}{81}=1-\frac{36}{225} \Rightarrow y^2=81\left(1-\frac{36}{225}\right) \Rightarrow$

   $y=\sqrt{81\left(1-\frac{36}{225}\right)} \approx 8.25$ feet.

14. (a) See Figure 14.

(b) From the first equation, $x = 2t \Rightarrow t = \frac{x}{2}$. By substitution in the second equation, $y = \sqrt{\frac{x^2}{4} + 1}$.

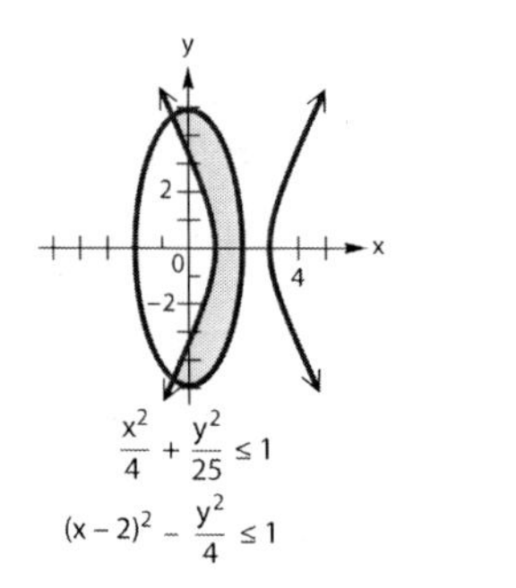

Figure 12

Figure 14

Chapter 7 Review Exercises

1. Here $h = -2,\ k = 3,$ and $r^2 = 5^2 = 25$. The equation is $(x+2)^2 + (y-3)^2 = 25$. See Figure 1. From the figure, the domain is $[-7,3]$, and the range is $[-2,8]$.

2. Here $h = \sqrt{5},\ k = -\sqrt{7},$ and $r^2 = \left(\sqrt{3}\right)^2 = 3$. The equation is $\left(x - \sqrt{5}\right)^2 + \left(y + \sqrt{7}\right)^2 = 3$. See Figure 2. From the figure, the domain is $\left[\sqrt{5} - \sqrt{3}, \sqrt{5} + \sqrt{3}\right]$, and the range is $\left[-\sqrt{7} - \sqrt{3}, -\sqrt{7} + \sqrt{3}\right]$.

3. The radius is the distance between $(-8,1)$ and $(0,16)$. $r = \sqrt{(0-(-8))^2 + (16-1)^2} = \sqrt{64+225} = 17$. Here $h = -8,\ k = 1,$ and $r^2 = 17^2 = 289$. The equation is $(x+8)^2 + (y-1)^2 = 289$. See Figure 3. From the figure, the domain is $[-25,9]$, and the range is $[-16,18]$.

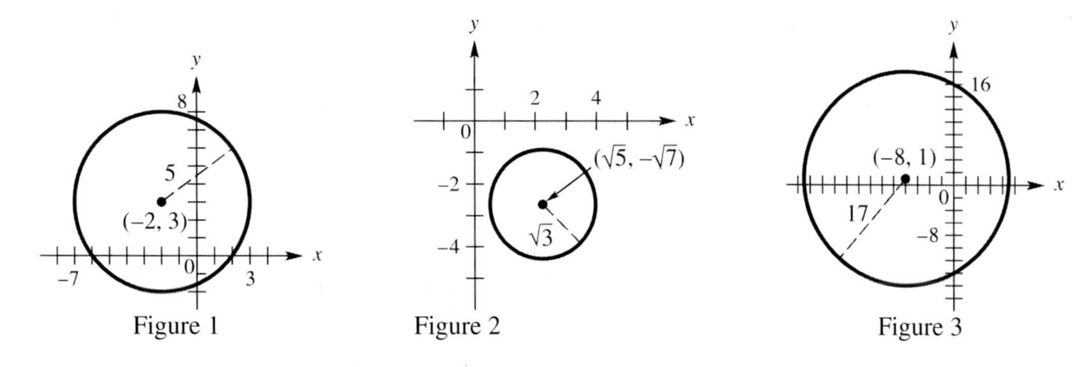

Figure 1 Figure 2 Figure 3

4. If the center is $(3,-6)$, the circle must touch the x-axis at the point $(3,0)$. The radius is 6. Here $h = 3$, $k = -6,$ and $r^2 = 6^2 = 36$. The equation is $(x-3)^2 + (y+6)^2 = 36$. See Figure 4. From the figure, the domain is $[-3,9]$, and the range is $[-12,0]$.

5. $x^2-4x+y^2+6y+12=0 \Rightarrow (x^2-4x+4)+\left(y^2+6y+9\right)=-12+4+9 \Rightarrow (x-2)^2+(y+3)^2=1.$

The circle has center $(2,-3)$, and radius $r=1$.

6. $x^2-6x+y^2-10y+30=0 \Rightarrow (x^2-6x+9)+\left(y^2-10y+25\right)=-30+9+25 \Rightarrow (x-3)^2+(y-5)^2=4.$

The circle has center $(3,5)$, and radius $r=2$.

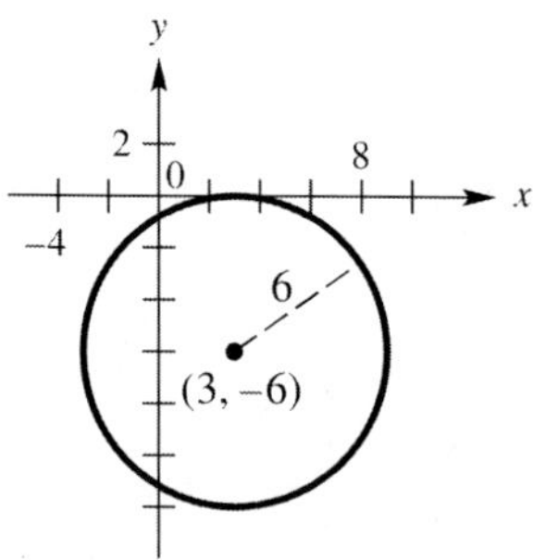

Figure 4

7. $2x^2+14x+2y^2+6y=-2 \Rightarrow x^2+7x+y^2+3y=-1 \Rightarrow$

$$\left(x^2+7x+\frac{49}{4}\right)+\left(y^2+3y+\frac{9}{4}\right)=-1+\frac{49}{4}+\frac{9}{4} \Rightarrow \left(x+\frac{7}{2}\right)^2+\left(y+\frac{3}{2}\right)^2=\frac{54}{4}.$$

The circle has center $\left(-\frac{7}{2},-\frac{3}{2}\right)$, and radius $r=\frac{\sqrt{54}}{2}=\frac{3\sqrt{6}}{2}$.

8. $3x^2+3y^2+33x-15y=0 \Rightarrow x^2+11x+y^2-5=0 \Rightarrow$

$$\left(x^2+11x+\frac{121}{4}\right)+\left(y^2-5y+\frac{25}{4}\right)=0+\frac{121}{4}+\frac{25}{4} \Rightarrow \left(x+\frac{11}{2}\right)^2+\left(y-\frac{5}{2}\right)^2=\frac{146}{4}.$$

The circle has center $\left(-\frac{11}{2},\frac{5}{2}\right)$, and radius $r=\frac{\sqrt{146}}{2}$.

9. The equation is that of a circle with center $(4,5)$, and radius 0. That is, the graph is the point $(4,5)$.

10. Since $y^2=-\frac{2}{3}x$ is equivalent to $y^2=4\left(-\frac{1}{6}\right)x$, the equation is in the form $y^2=4cx$ with $c=-\frac{1}{6}$. The focus is $\left(-\frac{1}{6},0\right)$, and the equation of the directrix is $x=\frac{1}{6}$. The axis is $y=0$, or the x-axis. The graph is shown in Figure 10. From the figure, the domain is $(-\infty,0]$, and the range is $(-\infty,\infty)$.

11. Since $y^2=2x$ is equivalent to $y^2=4\left(\frac{1}{2}\right)x$, the equation is in the form $y^2=4cx$ with $c=\frac{1}{2}$. The focus is $\left(\frac{1}{2},0\right)$, and the equation of the directrix is $x=-\frac{1}{2}$. The axis is $y=0$, or the x-axis. The graph is shown in Figure 11. From the figure, the domain is $[0,\infty)$, and the range is $(-\infty,\infty)$.

12. Since $3x^2 - y = 0$ is equivalent to $x^2 = 4\left(\frac{1}{12}\right)y$, the equation is in the form $x^2 = 4cy$ with $c = \frac{1}{12}$. The focus is $\left(0, \frac{1}{12}\right)$, and the equation of the directrix is $y = -\frac{1}{12}$. The axis is $x = 0$, or the y-axis. The graph is shown in Figure 12. From the figure, the domain is $(-\infty, \infty)$, and the range is $[0, \infty)$.

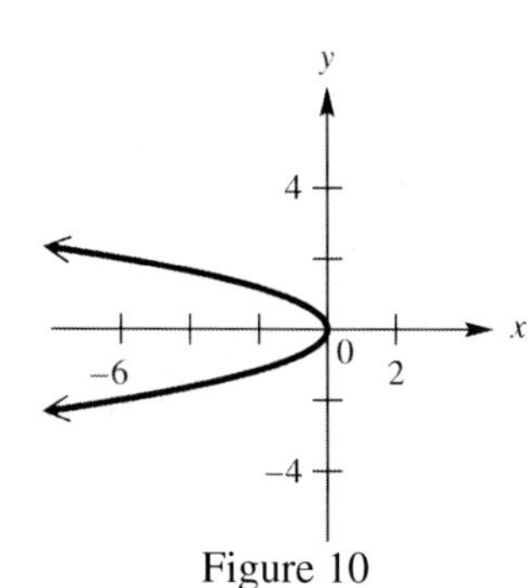

Figure 10

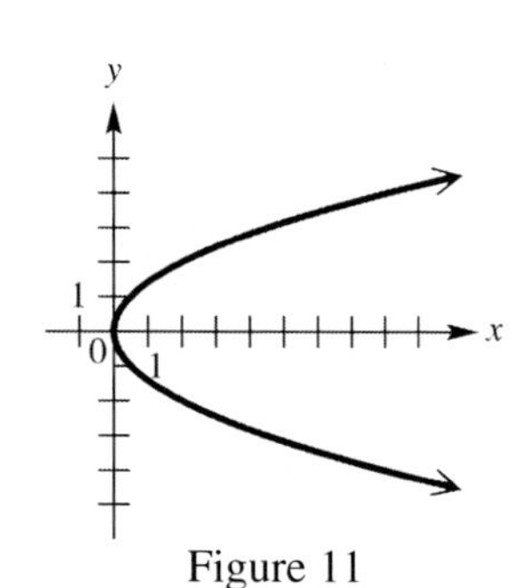

Figure 11

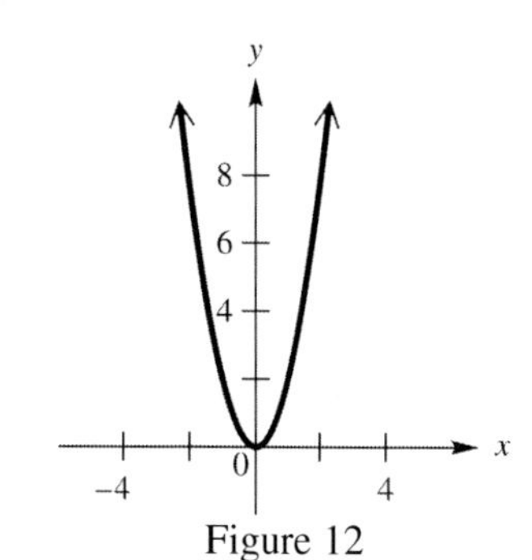

Figure 12

13. Since $x^2 + 2y = 0$ is equivalent to $x^2 = 4\left(-\frac{1}{2}\right)y$, the equation is in the form $x^2 = 4cy$ with $c = -\frac{1}{2}$. The focus is $\left(0, -\frac{1}{2}\right)$, and the equation of the directrix is $y = \frac{1}{2}$. The axis is $x = 0$, or the y-axis. The graph is shown in Figure 13. From the figure, the domain is $(-\infty, \infty)$, and the range is $(-\infty, 0]$.

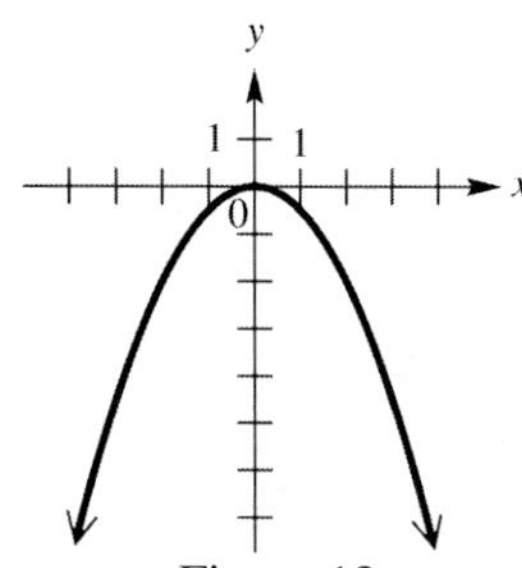

Figure 13

14. If the vertex is $(0,0)$, and the focus is $(4,0)$, then the parabola opens to the right and $c = 4$. The equation is $y^2 = 4cx \Rightarrow y^2 = 16x$.

15. If the vertex is $(0,0)$, and the parabola opens to the right, the equation is in the form $y^2 = 4cx$. Find the value of c by using the fact that the parabola passes through $(2,5)$. Thus, $(5)^2 = 4c(2) \Rightarrow c = \frac{25}{8}$. The equation is $y^2 = 4cx \Rightarrow y^2 = \frac{25}{2}x$.

16. If the vertex is $(0,0)$, and the parabola opens downward, the equation is in the form $x^2 = 4cy$. Find the value of c by using the fact that the parabola passes through $(3,-4)$. Thus, $(3)^2 = 4c(-4) \Rightarrow c = -\frac{9}{16}$. The equation is $x^2 = 4cy \Rightarrow x^2 = -\frac{9}{4}y$.

17. If the vertex is $(0,0)$, and the focus is $(0,-3)$, then the parabola opens downward and $c = -3$. The equation is $x^2 = 4cy \Rightarrow x^2 = -12y$.

18. If the equation has the x-term squared, it has a vertical axis, and opens up if the coefficient of x^2 is positive or down if the coefficient is negative. If the y-term is squared, it has a horizontal axis, and opens to the right if the coefficient of y^2 is positive or to the left if the coefficient is negative.

19. If the focus is $(2,6)$, and the vertex is t$(-5,6)$, the parabola opens to the right and $c = 7$. Substituting in $(y-k)^2 = 4c(x-h)$, we get $(y-6)^2 = -28(x+5)$.

20. If the focus is $(4,5)$, and the vertex is $(4,3)$, the parabola opens upward and $c = 2$. Substituting in $(x-h)^2 = 4c(y-k)$, we get $(x-4)^2 = 8(y-3)$.

21. $\frac{x^2}{5} + \frac{y^2}{9} = 1 \Rightarrow a = 3$ and $b = \sqrt{5}$. The endpoints of the major axis (vertices) are $(0,\pm 3)$, so the range is $[-3,3]$, The endpoints of the minor axis are $(\pm\sqrt{5},0)$, so the domain is $[-\sqrt{5},\sqrt{5}]$. See Figure 21.

22. $\frac{x^2}{16} + \frac{y^2}{4} = 1 \Rightarrow a = 4$ and $b = 2$. The endpoints of the major axis (vertices) are $(\pm 4,0)$, so the domain is $[-4,4]$. The endpoints of the minor axis are $(0,\pm 2)$, so the range is $[-2,2]$. See Figure 22.

23. The transverse axis is horizontal with $a = 8$ and $b = 6$. The asymptotes are $y = \pm\frac{3}{4}x$. See Figure 23. The domain is $(-\infty,-8]\cup[8,\infty)$, and the range is $(-\infty,\infty)$. The vertices are $(-8,0)$ and $(8,0)$.

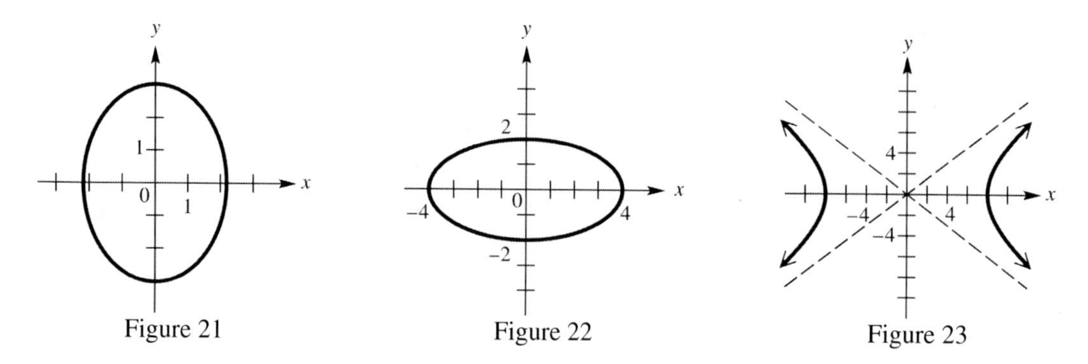

Figure 21 Figure 22 Figure 23

24. The transverse axis is vertical with $a = 5$ and $b = 3$. The asymptotes are $y = \pm\frac{5}{3}x$. See Figure 24. The domain is $(-\infty,\infty)$, and the range is $(-\infty,-5]\cup[5,\infty)$. The vertices are $(0,-5)$ and $(0,5)$.

25. The ellipse is centered at $(3,-1)$. The major axis is horizontal and has length $2a=4$, so the vertices are $(1,-1)$ and $(5,-1)$. The length of the minor axis is $2b=2$. The graph is shown in Figure 25. The domain is $[1,5]$, and the range is $[-2,0]$.

26. The ellipse is centered at $(2,-3)$. The major axis is horizontal and has length $2a=6$, so the vertices are $(-1,-3)$ and $(5,-3)$. The length of the minor axis is $2b=4$. The graph is shown in Figure 26. The domain is $[-1,5]$ and the range is $[-5,-1]$.

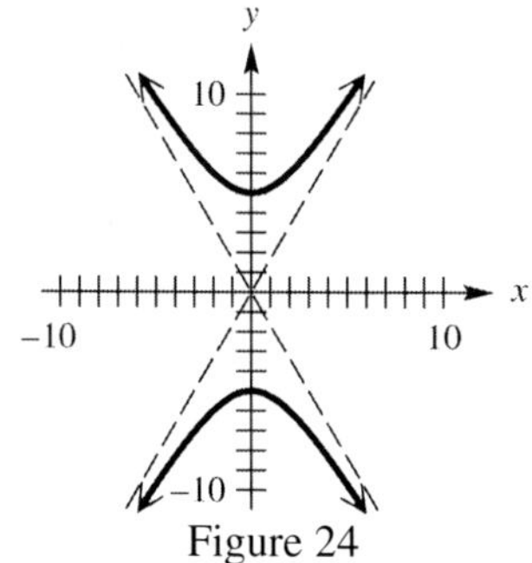

Figure 24

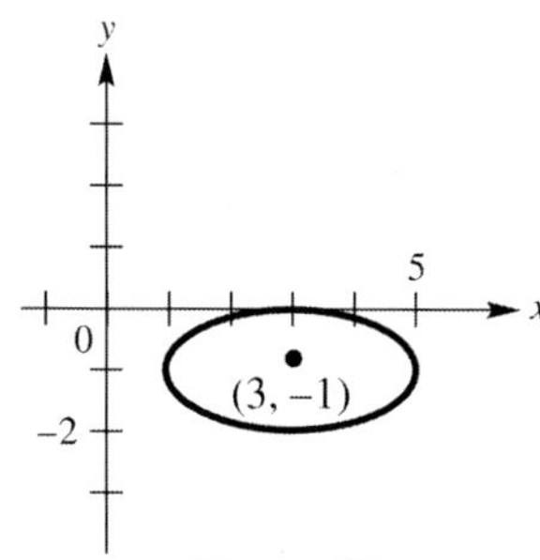

Figure 25

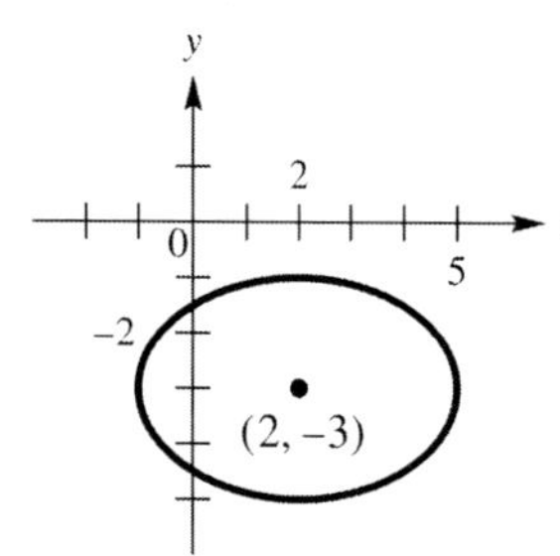

Figure 26

27. The center is $(-3,-2)$ and the transverse axis is vertical with $a=2$ and $b=3$. See Figure 27. The domain is $(-\infty,\infty)$ and the range is $(-\infty,-4]\cup[0,\infty)$. The vertices are $(-3,-4)$ and $(-3,0)$.

28. The center is $(-1,2)$ and the transverse axis is horizontal with $a=4$ and $b=2$. See Figure 28. The domain is $(-\infty,-5]\cup[3,\infty)$ and the range is $(-\infty,\infty)$. The vertices are $(-5,2)$ and $(3,2)$.

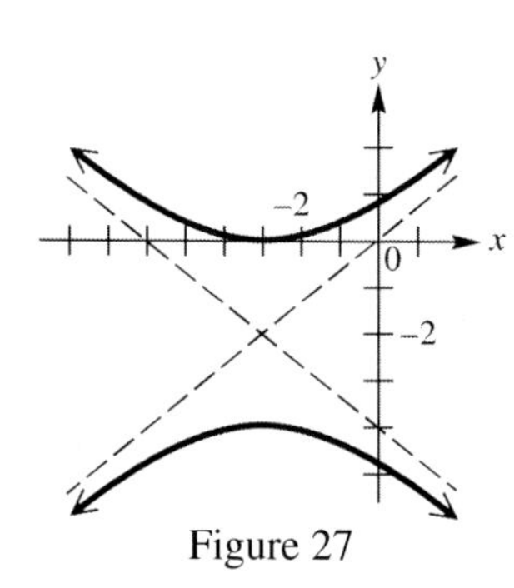

Figure 27

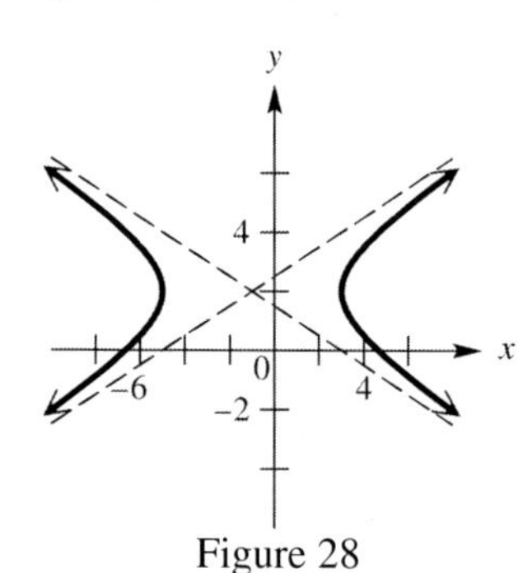

Figure 28

29. The major axis is vertical with $a=4$. Since one focus is $(0,2)$, we know that $c=2$. Since $c^2=a^2-b^2$, the value of b can be found by substitution: $b^2=a^2-c^2=4^2-2^2=16-4=12 \Rightarrow b=\sqrt{12}$. The equation is $\dfrac{x^2}{12}+\dfrac{y^2}{16}=1$.

30. The major axis is horizontal with $a=6$. Since one focus is $(-2,0)$, we know that $c=2$. Since $c^2=a^2-b^2$, the value of b can be found by substitution:

$b^2=a^2-c^2=6^2-2^2=36-4=32 \Rightarrow b=\sqrt{32}$. The equation is $\dfrac{x^2}{36}+\dfrac{y^2}{32}=1$.

31. The hyperbola has a vertical transverse axis with $c = 5$. The y-intercepts coincide with the vertices so $a = 4$. Since $c^2 = a^2 + b^2$, the value of b can be found by substitution:

$b^2 = c^2 - a^2 = 5^2 - 4^2 = 25 - 16 = 9 \Rightarrow b = 3$. The equation is $\frac{y^2}{16} - \frac{x^2}{9} = 1$.

32. The hyperbola has a vertical transverse axis. The y-intercept coincides with a vertex so $a = 2$. The equation is of the form $\frac{y^2}{a^2} - \frac{x^2}{b^2} = 1$. By substitution using the point (2, 3):

$\frac{3^2}{2^2} - \frac{2^2}{b^2} = 1 \Rightarrow \frac{9}{4} - 1 = \frac{4}{b^2} \Rightarrow \frac{5}{4} = \frac{4}{b^2} \Rightarrow 5b^2 = 16 \Rightarrow b^2 = \frac{16}{5}$. The equation is $\frac{y^2}{4} - \frac{x^2}{\frac{16}{5}} = 1$ or $\frac{y^2}{4} - \frac{5x^2}{16} = 1$.

33. Since $0 < e < 1$, the conic is an ellipse with $c = 3$. Now $\frac{c}{a} = e \Rightarrow \frac{3}{a} = \frac{2}{3} \Rightarrow a = \frac{9}{2}$. For an ellipse, $b^2 = a^2 - c^2 = \frac{81}{4} - 9 = \frac{45}{4}$. The equation is $\frac{x^2}{\frac{45}{4}} + \frac{y^2}{\frac{81}{4}} = 1$ or $\frac{4x^2}{45} + \frac{4y^2}{81} = 1$.

34. Since $e > 1$, the conic is a hyperbola with $c = 5$. Now $\frac{c}{a} = e \Rightarrow \frac{5}{a} = \frac{5}{2} \Rightarrow a = 2$. For a hyperbola, $b^2 = c^2 - a^2 = 25 - 4 = 21$. The equation is $\frac{x^2}{4} - \frac{y^2}{21} = 1$.

35. (a) $x^2 + y^2 + 2x + 6y - 15 = 0 \Rightarrow x^2 + 2x + 1 + y^2 + 6y + 9 = 15 + 9 + 1 \Rightarrow (x+1)^2 + (y+3)^2 = 25$; The center is $(-1, -3)$.

(b) The radius is $r = \sqrt{25} = 5$.

(c) $(x+1)^2 + (y+3)^2 = 25 \Rightarrow (y+3)^2 = 25 - (x+1)^2 \Rightarrow y + 3 = \pm\sqrt{25 - (x+1)^2} \Rightarrow$ $y = -3 \pm \sqrt{25 - (x+1)^2}$; Graph $y = -3 - \sqrt{25 - (x+1)^2}$ and $y = -3 + \sqrt{25 - (x+1)^2}$.

36. D. $4x^2 + y^2 = 36 \Rightarrow \frac{x^2}{9} + \frac{y^2}{36} = 1$; This is an ellipse with major axis on the y-axis.

37. E. $x = 2y^2 + 3 \Rightarrow 4\left(\frac{1}{8}\right)(x-3) = (y-0)^2$; This is a parabola that opens to the right.

38. A. $(x-1)^2 + (y+2)^2 = 36$; This is a circle with center $(1, -2)$ and radius 6.

39. C. $\frac{x^2}{36} + \frac{y^2}{9} = 1$; This is an ellipse with major axis on the x-axis.

40. B. $(y-1)^2 - (x-2)^2 = 36 \Rightarrow \frac{(y-1)^2}{36} - \frac{(x-2)^2}{36} = 1$; This is a hyperbola with center $(2, 1)$.

41. F. $y^2 = 36 + 4x^2 \Rightarrow y^2 - 4x^2 = 36 \Rightarrow \frac{y^2}{36} - \frac{x^2}{9} = 1$; This is a hyperbola with transverse axis on the y-axis.

42. $4x^2+8x+25y^2-250y=-529 \Rightarrow 4(x^2+2x)+25(y^2-10y)=-529 \Rightarrow$

$4(x^2+2x+1)+25(y^2-10y+25)=-529+4+625 \Rightarrow 4(x+1)^2+25(y-5)^2=100 \Rightarrow$

$\frac{(x+1)^2}{25}+\frac{(y-5)^2}{4}=1$; The center is $(-1,5)$. The vertices are $(-1-5,5), (-1+5,5)$ or $(-6,5),(4,5)$.

43. $5x^2+20x+2y^2-8y=-18 \Rightarrow 5(x^2+4x)+2(y^2-4y)=-18 \Rightarrow$

$5(x^2+4x+4)+2(y^2-4y+4)=-18+20+8 \Rightarrow 5(x+2)^2+2(y-2)^2=10 \Rightarrow$

$\frac{(x+2)^2}{2}+\frac{(y-2)^2}{5}=1$; The center is $(-2,2)$. The vertices are $\left(-2, 2-\sqrt{5}\right), \left(-2, 2+\sqrt{5}\right)$.

44. $x^2+4x-4y^2+24y=36 \Rightarrow (x^2+4x+4)-4(y^2-6y+9)=36+4-36 \Rightarrow$

$(x+2)^2-4(y-3)^2=4 \Rightarrow \frac{(x+2)^2}{4}-\frac{(y-3)^2}{1}=1$. The center is $(-2,3)$. The vertices are

$(-2-2,3), (-2+2,0)$ or $(-4,3), (0,3)$.

45. $4y^2+8y-3x^2+6x=11 \Rightarrow 4(y^2+2y+1)-3(x^2-2x+1)=11+4-3 \Rightarrow$

$4(y+1)^2-3(x-1)^2=12 \Rightarrow \frac{(y+1)^2}{3}-\frac{(x-1)^2}{4}=1$. The center is $(1,-1)$.

The vertices are $\left(1,-1-\sqrt{3}\right), \left(1,-1+\sqrt{3}\right)$.

46. $9x^2+25y^2=225 \Rightarrow \frac{x^2}{25}+\frac{y^2}{9}=1 \Rightarrow a=5, b=3,$ and $c=\sqrt{25-9}=4; e=\frac{c}{a}=\frac{4}{5}$

47. $4x^2+9y^2=36 \Rightarrow \frac{x^2}{9}+\frac{y^2}{4}=1 \Rightarrow a=3, b=2,$ and $c=\sqrt{9-4}=\sqrt{5};\ e=\frac{c}{a}=\frac{\sqrt{5}}{3}$

48. $9x^2-y^2=9 \Rightarrow \frac{x^2}{1}-\frac{y^2}{9}=1 \Rightarrow a=1, b=3,$ and $c=\sqrt{1+9}=\sqrt{10};\ e=\frac{c}{a}=\frac{\sqrt{10}}{1}=\sqrt{10}$

49. The parabola opens to the right so the equation has the form $(y-k)^2=4c(x-h)$. With vertex $(-3,2)$ and y-intercept (0, 5), $(5,-2)^2=4c(0+3) \Rightarrow 9=12c \Rightarrow c=\frac{3}{4}$. The equation is $(y-2)^2=3(x+3)$. This can also be written $x=\frac{1}{3}(y-2)^2-3$.

50. The center is located between the foci at (0, 0). The hyperbola has a vertical transverse axis with $c=12$. From the asymptotes, $\frac{a}{b}=1$. Also $c^2=a^2+b^2 \Rightarrow a^2+b^2=144$. Solving these equations simultaneously results in $a^2=72$ and $b^2=72$. The equation is $\frac{y^2}{72}-\frac{x^2}{72}=1$.

51. The foci are (0, 0) and (4, 0), so the center is (2, 0) and $c=2$. The sum of the distances is 8 so

$2a=8 \Rightarrow a=4$. For an ellipse, $b^2=a^2-c^2=16-4=12$. The equation is $\dfrac{(x-2)^2}{16}+\dfrac{y^2}{12}=1$.

52. The foci are (0, 0) and (0, 4), so the center is (0, 2) and $c=2$. The difference of the distances is 2 so

$2a=2 \Rightarrow a=1$. For a hyperbola, $b^2=c^2-a^2=4-1=3$. The equation is $\dfrac{(y-2)^2}{1}-\dfrac{x^2}{3}=1$.

53. Add the two equations to eliminate the y^2:

$$\begin{aligned} 2x^2 - y^2 &= 8 \\ 4x^2 + y^2 &= 16 \\ \hline 6x^2 \quad &= 24 \Rightarrow x^2 = 4 \Rightarrow x = \pm 2 \end{aligned}$$

Substitute these values into [equation 2] to find the value of y: $4(2)^2+y^2=16 \Rightarrow y^2=0 \Rightarrow y=0$,

$4(-2)^2+y^2=16 \Rightarrow y^2=0 \Rightarrow y=0$, The solution set is $\{(2,0),(-2,0)\}$.

54. See Figure 54.

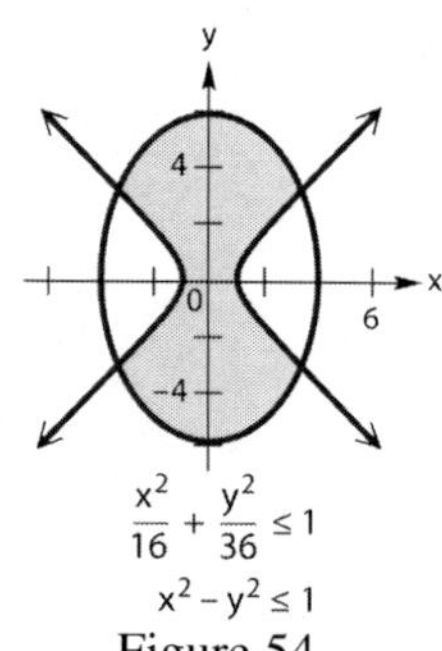

Figure 54

55. See Figure 55.

56. See Figure 56.

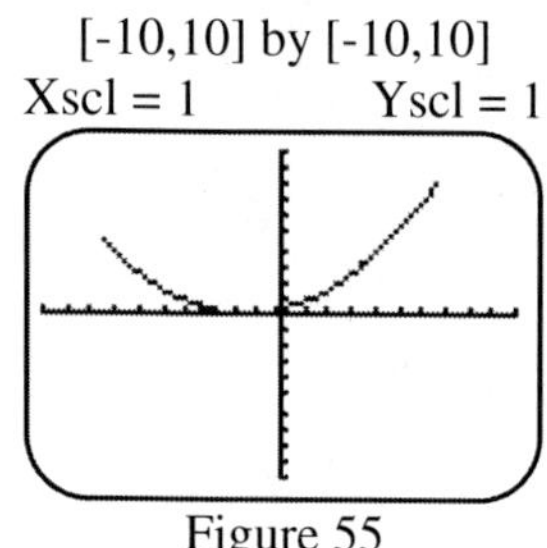

Figure 55

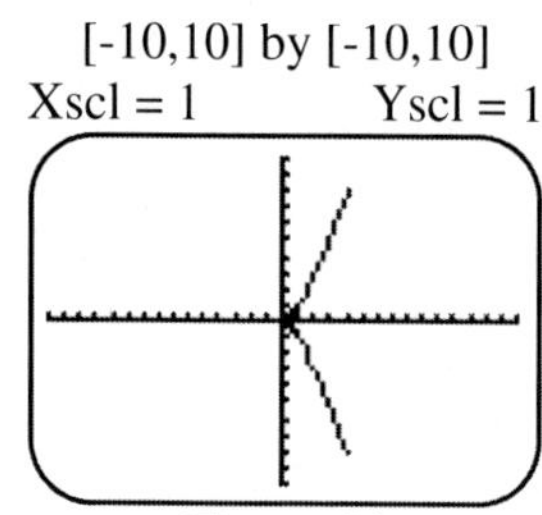

Figure 56

57. From the 1st equation, $x=\sqrt{t-1} \Rightarrow t=x^2+1$. By substitution in the 2nd equation $y=\sqrt{x^2+1}$. This is equivalent to $y^2-x^2=1$. When t is in $[1,\infty)$, the range of $x=\sqrt{t-1}$ is x in $[0,\infty)$.

58. From the 1st equation, $x=3t+2 \Rightarrow t=\dfrac{x-2}{3}$. By substitution in the 2nd equation $y=\dfrac{x-2}{3}-1$. This is equivalent to $x-3y=5$. When t is in $[-5, 5]$, the range of $x=3t+2$ is x in $[-13,17]$.

59. Since the major axis has length 134.5 million miles, $2a = 134.5 \Rightarrow a = 67.25$. From the given eccentricity, $\frac{c}{a} = 0.006775 \Rightarrow \frac{c}{67.25} = 0.006775 \Rightarrow c = 67.25(0.006775) = 0.4456$ million miles. The smallest distance is $67.25 - 0.4456 \approx 66.8$ million miles and the largest distance is $67.25 + 0.4456 \approx 67.7$ million miles.

60. Since the smallest distance between the comet and the sun is 89 million miles, $a - c = 89$. The given eccentricity provides the equation $\frac{c}{a} = 0.964$. Solving these two equations simultaneously gives $a \approx 2472.222$ and $c \approx 2383.222$. Then $b^2 = a^2 - c^2 \Rightarrow b^2 = 2472.222 - 2383.222 \Rightarrow b^2 \approx 432{,}135$. The equation is $\frac{x^2}{a^2} + \frac{y^2}{b^2} = 1 \Rightarrow \frac{x^2}{6{,}111{,}882} + \frac{y^2}{432{,}135} = 1$.

61. The value of $\frac{k}{\sqrt{D}}$ is $\frac{2.82 \times 10^7}{\sqrt{42.5 \times 10^6}} \approx 4326$. Since $2090 < 4326$, $V < \frac{k}{\sqrt{D}}$. The trajectory is elliptic.

62. The velocity must be more than 4326 m per sec. The minimum increase is $4326 - 2090 \approx 2236$ m per sec.

63. The required increase in velocity is less when D is larger.

64. $Ax^2 + Cy^2 + Dx + Ey + F = 0 \Rightarrow A\left(x^2 + \frac{D}{A}x\right) + C\left(y^2 + \frac{E}{C}y\right) = -F \Rightarrow$

$$A\left(x^2 + \frac{D}{A}x + \frac{D^2}{4A^2}\right) + C\left(y^2 + \frac{E}{C}y + \frac{E^2}{4C^2}\right) = -F + \frac{D^2}{4A} + \frac{E^2}{4C} \Rightarrow A\left(x + \frac{D}{2A}\right)^2 + C\left(y + \frac{E}{2C}\right)^2$$

$$= \frac{CD^2 + AE^2 - 4ACF}{4AC} \Rightarrow \frac{\left(x + \frac{D}{2A}\right)^2}{\frac{CD^2 + AE^2 - 4ACF}{4A^2C}} + \frac{\left(y + \frac{E}{2C}\right)^2}{\frac{CD^2 + AE^2 - 4ACF}{4AC^2}} = 1$$

The center is $\left(-\frac{D}{2A}, -\frac{E}{2C}\right)$.

Chapter 7 Test

1. (a) B. This is a hyperbola with center $(-3, -2)$.

   (b) A. This is a circle with center (3, 2) and radius 4.

   (c) D. This is a circle with center $(-3, 2)$ and radius 4.

   (d) E. This is a parabola that opens downward.

   (e) F. This is a parabola that opens to the right.

   (f) C. This is an ellipse with center $(-3, -2)$.

2. $y^2 = \frac{1}{8}x \Rightarrow (y - 0)^2 = 4\left(\frac{1}{32}\right)(x - 0)$; this is a parabola with vertex (0, 0) that opens to the right. Since $c = \frac{1}{32}$, the focus is located at $\left(\frac{1}{32}, 0\right)$ and the equation of the directrix is $x = -\frac{1}{32}$.

3. See Figure 3. This is the graph of a function with domain $[-6, 6]$ and range $[-1, 0]$.

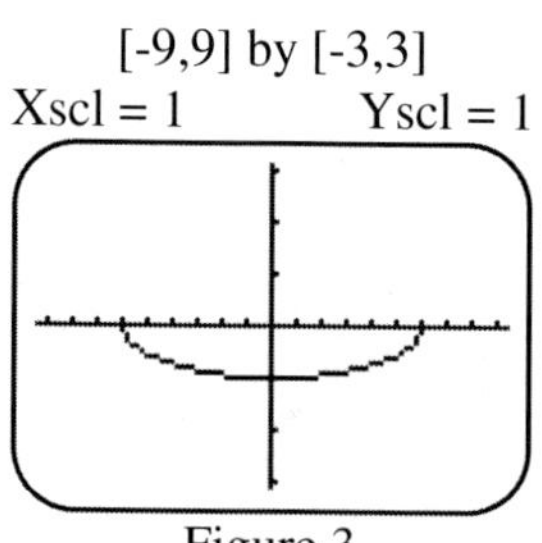

Figure 3

4. $\frac{x^2}{25} - \frac{y^2}{49} = 1 \Rightarrow \frac{y^2}{49} = \frac{x^2}{25} - 1 \Rightarrow y^2 = 49\left(\frac{x^2}{25} - 1\right) \Rightarrow y = \pm\sqrt{49\left(\frac{x^2}{25} - 1\right)} \Rightarrow \ y = \pm 7\sqrt{\frac{x^2}{25} - 1}$ The equations are $y_1 = 7\sqrt{\frac{x^2}{25} - 1}$ and $y_2 = -7\sqrt{\frac{x^2}{25} - 1}$.

5. This is a hyperbola with vertical transverse axis. Here $a = 2$ and $b = 3$. The asymptotes are $y = \pm\frac{2}{3}x$. See Figure 5. Since $c = \sqrt{a^2 + b^2} = \sqrt{2^2 + 3^2} = \sqrt{13}$, the foci are $(0, -\sqrt{13})$ and $(0, \sqrt{13})$. The center is (0, 0). The vertices are $(0, -2)$ and (0, 2).

6. $x^2 + 4y^2 + 2x - 16y + 17 = 0 \Rightarrow x^2 + 2x + 1 + 4(y^2 - 4y + 4) = -17 + 1 + 16 \Rightarrow \ (x+1)^2 + 4(y-2)^2 = 0$; The only point that satisfies this equation is $(-1, 2)$. See Figure 6.

7. $y^2 - 8y - 2x + 22 = 0 \Rightarrow y^2 - 8y + 16 = 2x - 22 + 16 \Rightarrow (y-4)^2 = 2(x-3) \Rightarrow (y-4)^2 = 4\left(\frac{1}{2}\right)(x-3)$; This is a parabola with vertex (3, 4) and focus (3.5, 4). See Figure 7.

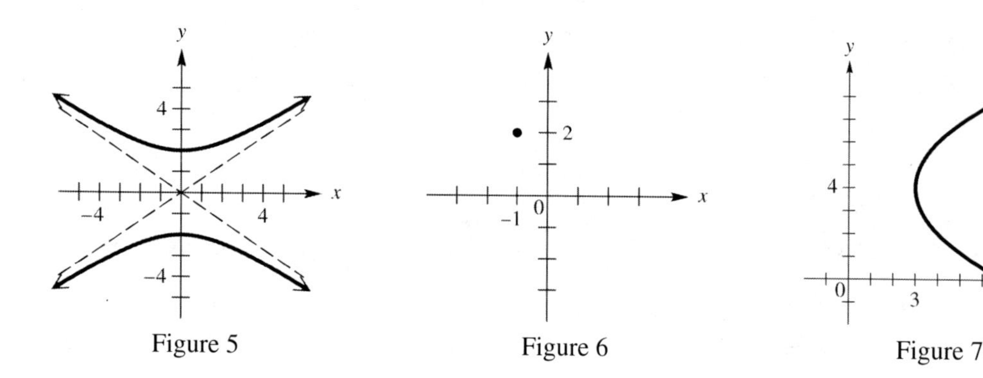

Figure 5 Figure 6 Figure 7

8. $x^2 + (y-4)^2 = 9 \Rightarrow (x-0)^2 + (y-4)^2 = 9$; This is a circle with center (0, 4) and radius 3. See Figure 8.

9. This is an ellipse with horizontal major axis. Here $a = 7$ and $b = 4$ so $c = \sqrt{7^2 - 4^2} = \sqrt{33}$. The center is $(3, -1)$. The vertices are $(3-7, -1)$ and $(3+7, -1)$ or $(-4, -1)$ and $(10, -1)$. The foci are $\left(3 + \sqrt{33}, -1\right)$ and $\left(3 - \sqrt{33}, -1\right)$. See Figure 9.

10. From the 2nd equation $y = t - 1 \Rightarrow t = y + 1$. Substituting in the 1st equation yields $x = 4(y+1)^2 - 4$. This equation can be written $(y+1)^2 = 4\left(\frac{1}{16}\right)(x+4)$. This is a parabola that opens to the right. The vertex is $(-4, -1)$ and the focus is $\left(-4 + \frac{1}{16}, -1\right) \Rightarrow \left(-\frac{63}{16}, -1\right)$. See Figure 10.

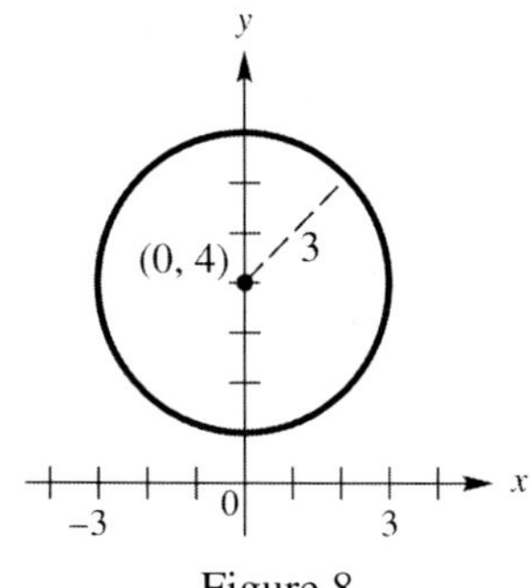

Figure 8

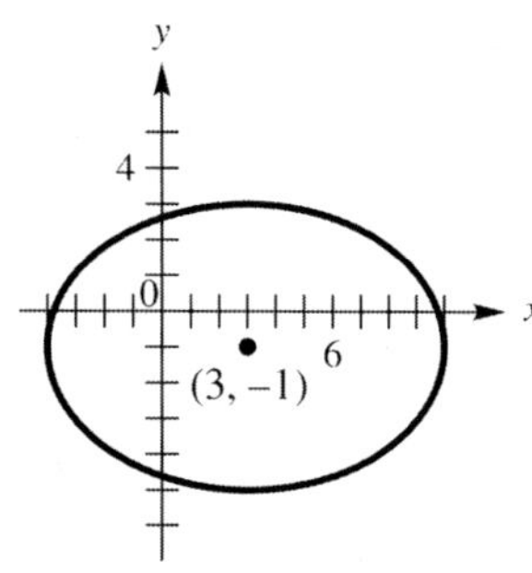

Figure 9

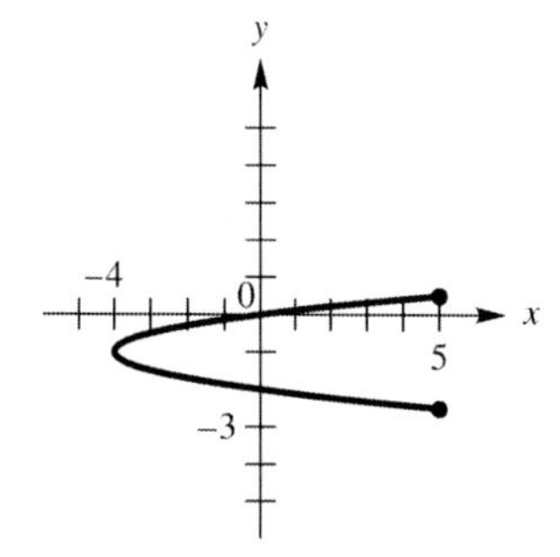

Figure 10

11. (a) Since $e = 1$, the conic is a parabola. With center (0, 0) and focus $(0, -2)$, we know that $c = -2$. The equation is $x^2 = 4cy \Rightarrow x^2 = -8y \Rightarrow y = -\frac{1}{8}x^2$.

    (b) Since $0 < e < 1$, the conic is an ellipse with $a = 3$. Now $\frac{c}{a} = e \Rightarrow \frac{c}{3} = \frac{5}{6} \Rightarrow c = \frac{15}{6}$. For an ellipse, $b^2 = a^2 - c^2 = 9 - \frac{225}{36} = \frac{11}{4}$. The equation is $\frac{x^2}{\frac{11}{4}} + \frac{y^2}{9} = 1$ or $\frac{4x^2}{11} + \frac{y^2}{9} = 1$.

12. Using a vertical minor axis, $b = 12$. The major axis has length 40 so $a = 20$. The equation is $\frac{x^2}{400} + \frac{y^2}{144} = 1$. When $x = 10$, $\frac{10^2}{400} + \frac{y^2}{144} = 1 \Rightarrow \frac{y^2}{144} = 1 - \frac{1}{4} \Rightarrow y^2 = 144\left(\frac{3}{4}\right) \Rightarrow y = 12\sqrt{\left(\frac{3}{4}\right)} \approx 10.39\,ft$.

13. Add 4 multiplied by [equation 1] and 3 multiplied by [equation 2] to eliminate x^2:

$$\begin{array}{r} 12x^2 + 8y^2 = 20 \\ -12x^2 + 9y^2 = -3 \\ \hline 17y^2 = 17 \Rightarrow y^2 = 1 \Rightarrow y = \pm 1 \end{array}$$

Substitute these values into [equation 1] to find the values of x: $3(1)^2 + 2y^2 = 5 \Rightarrow 2y^2 = 2 \Rightarrow y = \pm 1$, $3(-1)^2 + 2y^2 = 5 \Rightarrow 2y^2 = 2 \Rightarrow y = \pm 1$, The solution set is $\{(1,1), (1,-1), (-1,1), (-1,-1)\}$.

14. See Figure 14.

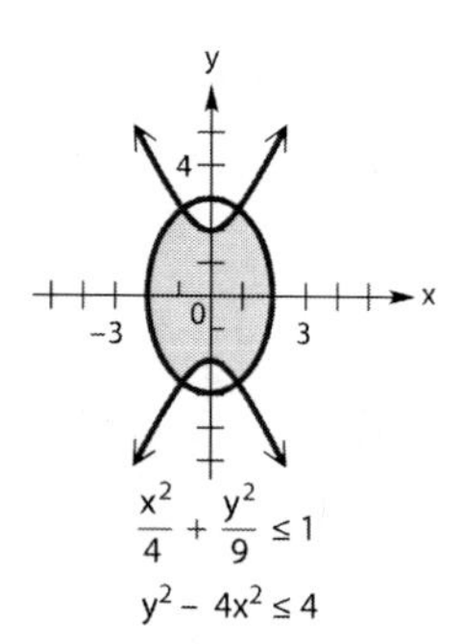

Figure 14

15. See Figure 15.

[-5,5] by [0,10]
Xscl = 1 Yscl = 1

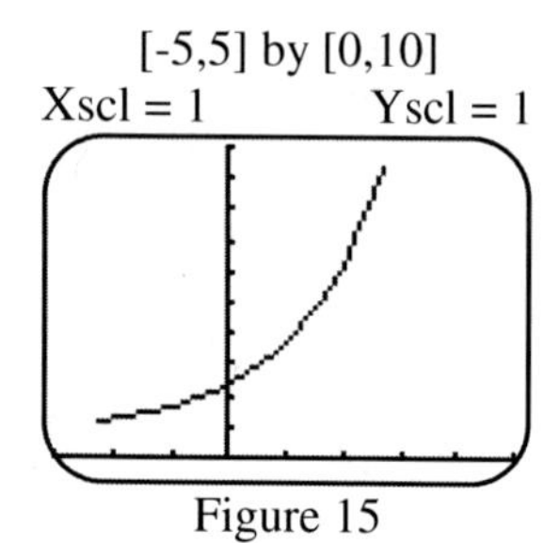
Figure 15

16. $x = \dfrac{1}{t+3} \Rightarrow t = \dfrac{1}{x} - 3.$ Substituting this in $y = t + 3 \Rightarrow y = \dfrac{1}{x}, x \neq 0.$

Chapter 8: Trigonometric Functions and Applications

8.1: Angles and Their Measures

1. An angle of 360° has an equivalent radian measure of 2π.

3. The least positive angle coterminal with -180° has degree measure 180°.

5. The formula relating $r, \theta, and\ s$ is $s = r\theta$.

7. (a) $c + 30^\circ = 90^\circ \Rightarrow c = 60^\circ$

   (b) $s + 30^\circ = 180^\circ \Rightarrow s = 150^\circ$

9. (a) $c + 45^\circ = 90^\circ \Rightarrow c = 45^\circ$

   (b) $s + 45^\circ = 180^\circ \Rightarrow s = 135^\circ$

11. (a) Since $90^\circ = \frac{\pi}{2}$ radians, $c + \frac{\pi}{4} = \frac{\pi}{2} \Rightarrow c = \frac{2\pi}{4} - \frac{\pi}{4} \Rightarrow c = \frac{\pi}{4}$.

    (b) Since $180^\circ = \pi$ radians, $s + \frac{\pi}{4} = \pi \Rightarrow s = \frac{4\pi}{4} - \frac{\pi}{4} \Rightarrow s = \frac{3\pi}{4}$.

13. (a) Since a complete revolution is 360°, then $\frac{180^\circ}{360^\circ} = \frac{1}{2}$ of a revolution.

    (b) Since a complete revolution is 360°, then $\frac{40^\circ}{360^\circ} = \frac{1}{9}$ of a revolution.

    (c) Since a complete revolution is 360°, then $\frac{1^\circ}{360^\circ} = \frac{1}{180}$ of a revolution.

15. (a) $c + x^\circ = 90^\circ \Rightarrow c = (90 - x)^\circ$

    (b) $s + x^\circ = 180^\circ \Rightarrow s = (180 - x)^\circ$

17. Since $360^\circ \div 12^\circ = 30^\circ$, every 5 minute section is 30°. Therefore 5 o'clock equals $5 \cdot 30^\circ = 150^\circ$.

19. Since $360^\circ \div 60^\circ = 6^\circ$, every 1 minute section is 6°. Therefore 3:15 equals 1 minute $+\left(\frac{1}{4} \cdot 6\,\text{minutes}\right)$

    or $1.25 \cdot 6^\circ = 7^\circ 30'$

21. $7x + 11x = 180 \Rightarrow 18x = 180 \Rightarrow x = 10$, therefore the angles are: $7(10) = 70^\circ$ and $11(10) = 110^\circ$.

23. $(5k + 5) + (3k + 5) = 90 \Rightarrow 8k + 10 = 90 \Rightarrow 8k = 80 \Rightarrow k = 10$, therefore the angles are: $5(10) + 5 = 55^\circ$ and $3(10) + 5 = 35^\circ$.

25. $(6x - 4) + (8x - 12) = 180 \Rightarrow 14x - 16 = 180 \Rightarrow 14x = 196 \Rightarrow x = 14$, therefore the angles are: $6(14) - 4 = 80^\circ$ and $8(14) - 12 = 100^\circ$.

27. $62^\circ 18' + 21^\circ 41' = 83^\circ 59'$

29. $71^\circ 18' - 47^\circ 29' = 70^\circ 78' - 47^\circ 29' = 23^\circ 49'$

31. $90^\circ - 72^\circ 58' 11'' = 89^\circ 59' 60'' - 72^\circ 58' 11'' = 17^\circ 1' 49''$

33. $20^\circ 54' = \left(20\frac{54}{60}\right)^\circ = 20.9^\circ$

35. $91°35'54" = 91°35'\frac{54}{60} = 91°35.9' = \left(91\frac{35.9}{60}\right)° = 91.598°$

37. $31.4296° = 31° + .4296(60') = 31°25.776' = 31°25'.776(60") = 31°25'47"$

39. $89.9004° = 89° + .9004(60') = 89°54.024' = 89°54'.024(60") = 89°54'1"$

41. See Figure 41. The coterminal angles are: $75° + 360° = 435°$ and $75° - 360° = -285°$, are in quadrant I.

43. See Figure 43. The coterminal angles are: $174° + 360° = 534°$ and $174° - 360° = -186°$, are in quadrant II.

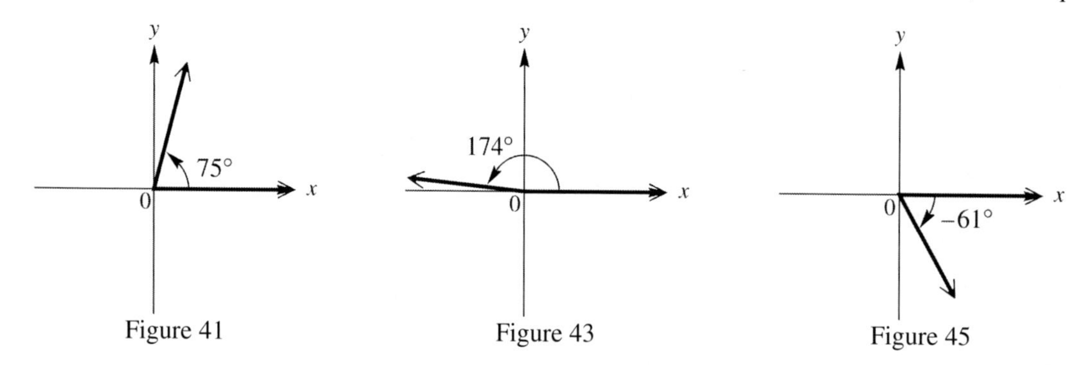

Figure 41 Figure 43 Figure 45

45. See Figure 45. The coterminal angles are: $-61° + 360° = 299°$ and $-61° - 360° = -421°$ are in quadrant IV.

47. The smallest positive degree measure of the coterminal angle is: $-40° + 360° = 320°$.

49. The smallest positive degree measure of the coterminal angle is: $450° - 360° = 90°$.

51. The smallest positive radian measure of the coterminal angle is: $-\frac{\pi}{4} + 2\pi = -\frac{\pi}{4} + \frac{8\pi}{4} = \frac{7\pi}{4}$.

53. The smallest positive radian measure of the coterminal angle is: $-\frac{3\pi}{2} + 2\pi = -\frac{3\pi}{2} + \frac{4\pi}{2} = \frac{\pi}{2}$.

55. All degree measure coterminal angles will be multiples of $360°$, therefore: $30° + n \cdot 360°$.

57. All degree measure coterminal angles will be multiples of $360°$, therefore: $-90° + n \cdot 360°$.

59. All radian measure coterminal angles will be multiples of 2π, therefore: $\frac{\pi}{4} + n \cdot 2\pi \Rightarrow \frac{\pi}{4} + 2n\pi$.

61. All radian measure coterminal angles will be multiples of 2π, therefore: $-\frac{3\pi}{4} + n \cdot 2\pi \Rightarrow -\frac{3\pi}{4} + 2n\pi$.

63. $60° \cdot \frac{\pi}{180} = \frac{\pi}{3}$ radians.

65. $150° \cdot \frac{\pi}{180} = \frac{5\pi}{6}$ radians.

67. $-45° \cdot \frac{\pi}{180} = -\frac{\pi}{4}$ radians.

69. $\frac{\pi}{3} \cdot \frac{180}{\pi} = 60°$

71. $\frac{7\pi}{4} \cdot \frac{180}{\pi} = 315°$

73. $\dfrac{11\pi}{6}\cdot\dfrac{180}{\pi}=330°$

75. $39°\dfrac{\pi}{180}=\dfrac{39}{180}\cdot\pi=.68$

77. $139°10'=\left(139\dfrac{10}{60}\right)°=139.167°\Rightarrow 139.167°\cdot\dfrac{\pi}{180}=\dfrac{139.167}{180}\cdot\pi=2.43$

79. $64.29°\cdot\dfrac{\pi}{180}=\dfrac{64.29}{180}\cdot\pi=1.12$

81. $2\cdot\dfrac{180}{\pi}=\dfrac{360}{\pi}=114.592°=114°+.592(60')=114°35'$

83. $1.74\cdot\dfrac{180}{\pi}=\dfrac{313.2}{\pi}=99.695°=99°+.695(60')=99°42'$

85. $-1.3\cdot\dfrac{180}{\pi}=\dfrac{-234}{\pi}=-74.485°=-74°+.485(60')=-74°29'$

87. Going around the circle counterclockwise and starting at $0°=0$ radians; $30°=\dfrac{\pi}{6}$ radians; $45°=\dfrac{\pi}{4}$ radians; $60°=\dfrac{\pi}{3}$ radians; $90°=\dfrac{\pi}{2}$ radians; $120°=\dfrac{2\pi}{3}$ radians; $135°=\dfrac{3\pi}{4}$ radians; $150°=\dfrac{5\pi}{6}$ radians; $180°=\pi$ radians; $210°=\dfrac{7\pi}{6}$ radians; $225°=\dfrac{5\pi}{4}$ radians; $240°=\dfrac{4\pi}{3}$ radians; $270°=\dfrac{3\pi}{2}$ radians; $300°=\dfrac{5\pi}{3}$ radians; $315°=\dfrac{7\pi}{3}$ radians; and $330°=\dfrac{11\pi}{6}$ radians.

89. Since arc length is found by the formula $s=r\theta,$ the arc length is: $s=4\cdot\dfrac{\pi}{2}\Rightarrow s=2\pi.$

91. Since arc length is found by the formula $s=r\theta,$ the radius is: $6\pi=r\cdot\dfrac{3\pi}{4}\Rightarrow r=6\pi\cdot\dfrac{4}{3\pi}\Rightarrow r=8.$

93. Since arc length is found by the formula $s=r\theta,$ the central angle is: $3=3\theta\Rightarrow\theta=\dfrac{3}{3}\Rightarrow\theta=1.$

95. Since arc length is found by the formula $s=r\theta,$ the arc length is: $s=12.3\left(\dfrac{2\pi}{3}\right)\Rightarrow s=8.2\pi\Rightarrow s\approx 25.76$ cm.

97. Since $60°=60\cdot\dfrac{\pi}{180}=\dfrac{\pi}{3}$ radians and arc length is found by the formula $s=r\theta,$ the arc length is: $s=4.82\left(\dfrac{\pi}{3}\right)\Rightarrow s\approx 1.607\pi\Rightarrow s\approx 5.05$ m.

99. Since $44°-33°=11°,\ 11°\cdot\dfrac{\pi}{180}\approx .06111\pi$ radians, and arc length is found by the formula $s=r\theta,$ the arc length is: $s=6400(.06111\pi)\Rightarrow s\approx 391.1\pi\Rightarrow s\approx 1200$ km.

101. Since $41+12=53^{\circ}, 53^{\circ}\dfrac{\pi}{180}\approx 0.29444\pi$ radians, and arc length is found by the formula $s=r\theta,$ the arc length is: $s=6400(.29444\pi)\Rightarrow s\approx 1884.42\pi \Rightarrow s\approx 5900$ km.

103. Using $w=\dfrac{\theta}{t}$ yields: $w=\dfrac{\frac{3\pi}{4}}{8}=\dfrac{3\pi}{4}\cdot\dfrac{1}{8}\Rightarrow w=\dfrac{3\pi}{32}$ radians per second.

105. Using $w=\dfrac{\theta}{t}$ yields: $\dfrac{5\pi}{27}=\dfrac{\frac{2x}{9}}{t}\Rightarrow 5\pi t=\dfrac{27}{1}\cdot\dfrac{2\pi}{9}\Rightarrow 5\pi t=6\pi\Rightarrow t=\dfrac{6\pi}{5\pi}\Rightarrow t=\dfrac{6}{5}$ minutes.

107. Using $s=rwt$ yields: $s=6\left(\dfrac{\pi}{3}\right)(9)\Rightarrow s=18\pi$ cm.

109. Using $s=rwt$ yields: $6\pi=2\left(\dfrac{\pi}{4}\right)t\Rightarrow 6\pi=\dfrac{\pi}{2}t\Rightarrow 6\pi\,.\,\dfrac{2}{\pi}=t\Rightarrow t=12$ sec.

111. (a) The weight will rise the same distance as the arc length of the rotation. Therefore since arc length is found by the formula $s=r\theta$, the arc length is : $s=9.27\left(71+\dfrac{50}{60}\right)\cdot\dfrac{\pi}{180}\Rightarrow 9.27(71.83)\dfrac{\pi}{180}\Rightarrow s\approx 11.6$

(b) The weight will rise the same distance as the arc length of the rotation. Therefore since arc length is found by the formula $s=r\theta$, the angle is: $6=9.27(\theta)\Rightarrow\theta=.647$ Changing this to degree measure yields: $.647\cdot\dfrac{180}{\pi}=\dfrac{116.505}{\pi}\approx 37.085°\approx 37°+.085(60')\approx 37°5'$.

113. First find the arc length that the smaller wheel rotates, arc length is found by the formula $s=r\theta$ the arc length is: $s=5.23\left(60\cdot\dfrac{\pi}{180}\right)\Rightarrow s\approx 1.743\Rightarrow s\approx 5.477\,\text{cm}$ Now the larger wheel travels the same arc length, therefore the angle of rotation is: $5.477\approx 8.16\theta\Rightarrow\theta\approx .671$ Changing this to degree measure yields: $0.671\cdot\dfrac{180}{\pi}=\dfrac{120.78}{\pi}\approx 38.5°$

115. First find the arc length that the pedal wheel rotates, arc length is found by the formula $s=r\theta$ the arc length is: $s=4.72\left(180\cdot\dfrac{\pi}{180}\right)\Rightarrow s\approx 4.72\pi\Rightarrow s\approx 14.828$ in. Now the smallest wheel travels the same arc length, therefore the angle of rotation is: $14.828=1.38\theta\Rightarrow\theta\approx 10.745$. Finally, using this radian rotation calculate the distance traveled with a wheel radius of 13.6 inches: $s\approx 13.6(10.745)\Rightarrow s\approx 146$ in.

117. Using the area of a sector formula $A=\dfrac{1}{2}r^2\theta$ yields: $A=\dfrac{1}{2}(29.2)^2\left(\dfrac{5\pi}{6}\right)\Rightarrow A\approx 1120\,\text{m}^2$.

119. Using the area of a sector formula $A=\dfrac{1}{2}r^2\theta$ yields: $A=\dfrac{1}{2}(12.7)^2\left(81°.\dfrac{\pi}{180}\right)\Rightarrow A\approx 114\,\text{cm}^2$.

121. First find using $s = r\theta$ yields: $2\pi = 6\theta \Rightarrow \theta = \frac{\pi}{3}$. Now using the area of a sector formula $A = \frac{1}{2}r^2\theta$ yields:

$$A = \frac{1}{2}(6)^2\left(\frac{\pi}{3}\right) \Rightarrow A \approx \frac{36\pi}{6} = 6\pi.$$

123. Since the area scanned in 1 second would yield: $\theta = \frac{1}{48} \cdot 2\pi = \frac{\pi}{24}$ radians, we can solve for area using $A = \frac{1}{2}r^2\theta$. Therefore, $A = \frac{1}{2}(240)^2\left(\frac{\pi}{24}\right) \Rightarrow A \approx 3800\,\text{mi}^2$.

125. To find the portion cleaned we will find the 10 inch radius sector and subtract the inside 3 inch radius sector. The 10 in radius is: $A = \frac{1}{2}10^2\left(95° \cdot \frac{\pi}{180}\right) \Rightarrow A \approx 82.9\text{ in}^2$. The 3 in radius is:

$A = \frac{1}{2}3^2\left(95° \cdot \frac{\pi}{180}\right) \Rightarrow A \approx 7.5\text{ in}^2$. Subtracting these yields: $82.9 - 7.5 \approx 75.4\text{in}^2$.

127. (a) Each sector is: $360° \div 26 = 13.85°$.

(b) Using the area of a sector formula $A = \frac{1}{2}r^2\theta$ and $r = \frac{50}{2} \Rightarrow r = 25$ yields:

$$A = \frac{1}{2}25^2\left(13.85° \cdot \frac{\pi}{180}\right) \Rightarrow A \approx 76\,\text{m}^2.$$

129. If the diameter is 26, then the radius is 13, therefore 15 radians per second of rotation would have a linear speed of: $v = rw \Rightarrow v = 13.15 = 195$ inches per second which is equal to $195 \div 12 = 16\frac{1}{4}$ feet per second. This is equal to: $(16.25 \div 5280) \cdot 60 \cdot 60 \approx 11.1$ mph.

131. First we need to find inches: $15\,\text{mph} = 15.5280.12 = 950,400$ inches. Next we need to find θ using the arc length formula and a radius of : $r = \frac{2.25}{2} = 1.125 : 950,400 = 1.125\theta \Rightarrow \theta = 844,800$. Now use the angular speed formula: $w = \frac{844,800}{(60)(60)} \Rightarrow w = 234.67$ radians per second.

133. (a) The radian angle formed by 1 day is: $\theta = \frac{1}{365}(2\pi) \Rightarrow \theta = \frac{2\pi}{365}$ radians per day.

(b) With $\theta = \frac{2\pi}{365}$ per day and $t = 24$ hours per day, the angular speed is: $V = \frac{\theta}{t} \Rightarrow v = \frac{\frac{2\pi}{365}}{24} \Rightarrow v = \frac{\pi}{4380}$ radians per hour.

(c) With $r = 93,000,000$ and $\theta = \frac{\pi}{4380}$ radians per hour, the linear speed is:

$$v = rw \Rightarrow v = 93,000,000\left(\frac{\pi}{4380}\right) \Rightarrow v = 66,700\text{ mph}.$$

135. With a radius of $r=\frac{10}{2}=5$, a θ of 5000 revolutions per minute, and a time of 60 seconds in a minute, the angular speed is: $w=\frac{\theta}{t}\Rightarrow w=\frac{5000(2\pi)}{60}\Rightarrow w=\frac{500\pi}{3}$ radians per second and the linear speed is is: $v=rw\Rightarrow v=5\left(\frac{500\pi}{3}\right)\Rightarrow v=\frac{2500\pi}{3}$ inches per second.

137. First find the radian measure of $7°12'$: $\theta=\frac{7(60)+12}{360(60)}.2\pi\Rightarrow\theta\approx.1257$ radians. Next use $\theta=.12566$ and the arc length formula to find the radius: $496=r(.12566)\Rightarrow r=3947$ miles. Now we can find the circumference: $C=2r\pi\Rightarrow C=2(3947)\pi\Rightarrow C\approx 24{,}800$ miles.

8.2: Trigonometric Functions and Fundamental Identities

1. $x=5, y=12\Rightarrow r=\sqrt{x^2+y^2}=\sqrt{5^2+(12)^2}=\sqrt{169}=13$,

$\sin\theta=\frac{y}{r}=\frac{12}{13};\cos\theta=\frac{x}{r}=\frac{5}{13};\tan\theta=\frac{y}{x}=\frac{12}{5};\ \csc\theta=\frac{r}{y}=\frac{13}{12};\sec\theta=\frac{r}{x}=\frac{13}{5};\cot\theta=\frac{y}{x}=\frac{5}{12}.$

3. They are the same for each trigonometric function. They are coterminal angles.

5. $x=5, y=-12\Rightarrow r=\sqrt{x^2+y^2}=\sqrt{5^2+(-12)^2}=\sqrt{169}=13$, See Figure 5.

$\sin\theta=\frac{y}{r}=\frac{-12}{13}=-\frac{12}{13};\cos\theta=\frac{x}{r}=\frac{5}{13};\tan\theta=\frac{y}{x}=\frac{-12}{5}=-\frac{12}{5}$

$\csc\theta=\frac{r}{y}=\frac{13}{-12}=-\frac{13}{12};\sec\theta=\frac{r}{x}=\frac{13}{5};\cot\theta=\frac{5}{-12}=-\frac{5}{12}.$

7. $x=-3, y=4\Rightarrow r=\sqrt{x^2+y^2}=\sqrt{(-3)^2+(4)^2}=\sqrt{25}=5$ See Figure 7.

$\sin\theta=\frac{y}{r}=\frac{4}{5};\cos\theta=\frac{x}{r}=\frac{-3}{5}=-\frac{3}{5};\tan\theta=\frac{y}{x}=\frac{4}{-3}=-\frac{4}{3}$

$\csc\theta=\frac{r}{y}=\frac{5}{4};\sec\theta=\frac{r}{x}=\frac{5}{-3}=-\frac{5}{3};\cot\theta=\frac{-3}{4}=-\frac{3}{4}.$

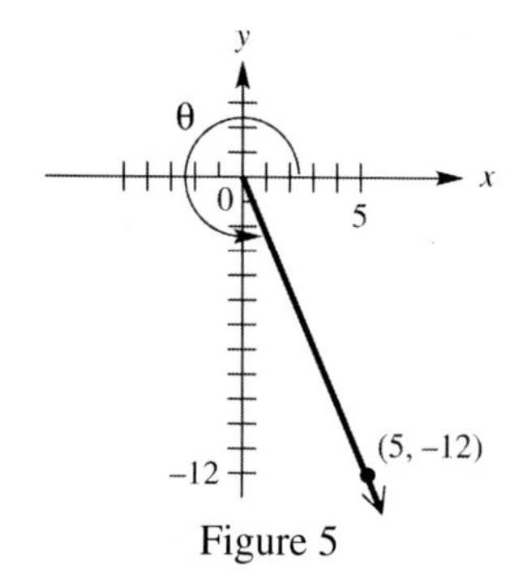

Figure 5

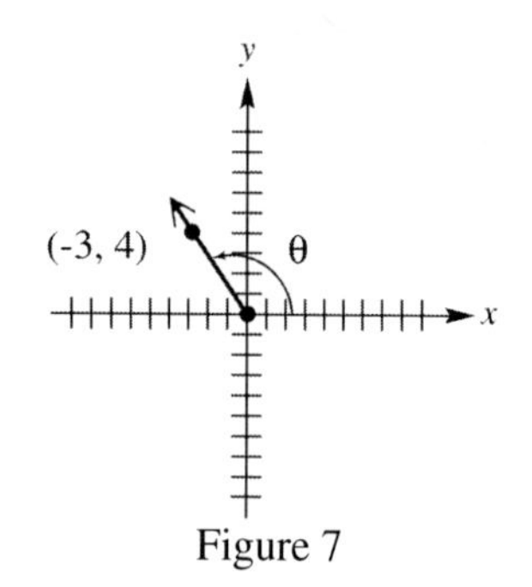

Figure 7

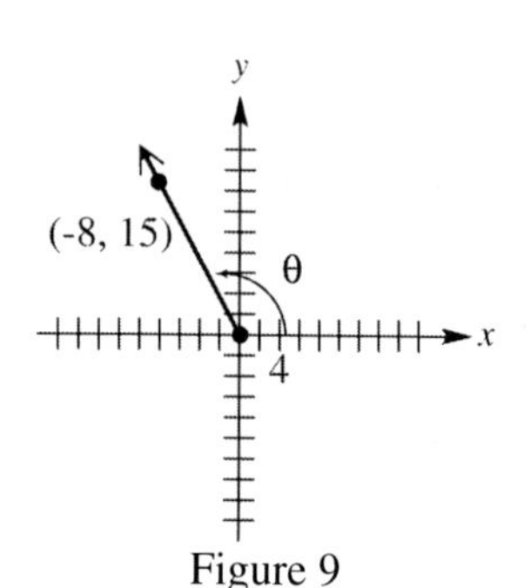

Figure 9

9. $x=-8, y=15\Rightarrow r=\sqrt{x^2+y^2}=\sqrt{(-8)^2+(15)^2}=\sqrt{289}=17$ See Figure 9.

$\sin\theta = \frac{y}{r} = \frac{15}{17}; \cos\theta = \frac{x}{r} = \frac{-8}{17} = -\frac{8}{17}; \tan\theta = \frac{y}{x} = \frac{15}{-8} = -\frac{15}{8}$

$\csc\theta = \frac{r}{y} = \frac{17}{15}; \sec\theta = \frac{r}{x} = \frac{17}{-8} = -\frac{17}{8}; \cot\theta = \frac{-8}{15} = -\frac{8}{15}.$

11. $x = 7, y = -24 \Rightarrow r = \sqrt{x^2 + y^2} = \sqrt{(7)^2 + (-24)^2} = \sqrt{625} = 25$ See Figure 11.

$\sin\theta = \frac{y}{r} = \frac{-24}{25} = -\frac{24}{25}; \cos\theta = \frac{x}{r} = \frac{7}{25}; \tan\theta = \frac{y}{x} = \frac{-24}{7} = -\frac{24}{7}$

$\csc\theta = \frac{r}{y} = \frac{25}{-24} = -\frac{25}{24}; \sec\theta = \frac{r}{x} = \frac{25}{7}; \cot\theta = \frac{-7}{24} = -\frac{7}{24}.$

13. $x = 0, y = 2 \Rightarrow r = \sqrt{x^2 + y^2} = \sqrt{(0)^2 + (2)^2} = \sqrt{4} = 2$ See Figure 13.

$\sin\theta = \frac{y}{r} = \frac{2}{2} = 1; \cos\theta = \frac{x}{r} = \frac{0}{2} = 0; \tan\theta = \frac{y}{x} = \frac{2}{0} \Rightarrow \textit{undefined}.$

$\csc\theta = \frac{r}{y} = \frac{2}{2} = 1; \sec\theta = \frac{r}{x} = \frac{2}{0} \Rightarrow \textit{undefined};; \cot\theta = \frac{0}{2} = 0.$

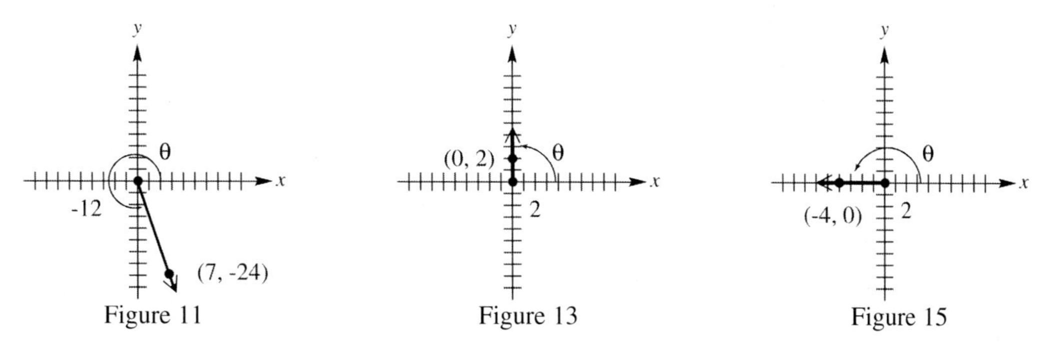

Figure 11 Figure 13 Figure 15

15. $x = -4, y = 0 \Rightarrow r = \sqrt{x^2 + y^2} = \sqrt{(-4)^2 + (0)^2} = \sqrt{16} = 4$ See Figure 15.

$\sin\theta = \frac{y}{r} = \frac{0}{4} = 0; \cos\theta = \frac{x}{r} = \frac{-4}{4} = -1; \tan\theta = \frac{y}{x} = \frac{0}{-4} = 0$

$\csc\theta = \frac{r}{y} = \frac{4}{0} \Rightarrow \textit{undefined}; \sec\theta = \frac{r}{x} = \frac{4}{-4} = -1; \cot\theta = \frac{-4}{0} \Rightarrow \textit{undefined}.$

17. $x = 0, y = -4 \Rightarrow r = \sqrt{x^2 + y^2} = \sqrt{(0)^2 + (-4)^2} = \sqrt{16} = 4$ See Figure 17.

$\sin\theta = \frac{y}{r} = \frac{-4}{4} = -1; \cos\theta = \frac{x}{r} = \frac{0}{4} = 0; \tan\theta = \frac{y}{x} = \frac{-4}{0} \Rightarrow \textit{undefined}.$

$\csc\theta = \frac{r}{y} = \frac{4}{-4} = -1; \sec\theta = \frac{r}{x} = \frac{4}{0} \Rightarrow \textit{undefined};; \cot\theta = \frac{0}{-4} = 0.$

19. $x = 1, y = \sqrt{3} \Rightarrow r = \sqrt{x^2 + y^2} = \sqrt{(1)^2 + (\sqrt{3})^2} = \sqrt{4} = 2$ See Figure 19.

$\sin\theta = \frac{y}{r} = \frac{\sqrt{3}}{2}; \cos\theta = \frac{x}{r} = \frac{1}{2}; \tan\theta = \frac{y}{x} = \frac{\sqrt{3}}{1} = \sqrt{3}$

$\csc\theta = \frac{r}{y} = \frac{2}{\sqrt{3}} = \frac{2}{\sqrt{3}} \cdot \frac{\sqrt{3}}{\sqrt{3}} = \frac{2\sqrt{3}}{3}; \sec\theta = \frac{r}{x} = \frac{2}{1} = 2; \cot\theta = \frac{1}{\sqrt{3}} = \frac{1}{\sqrt{3}} \cdot \frac{\sqrt{3}}{\sqrt{3}} = \frac{\sqrt{3}}{3}.$

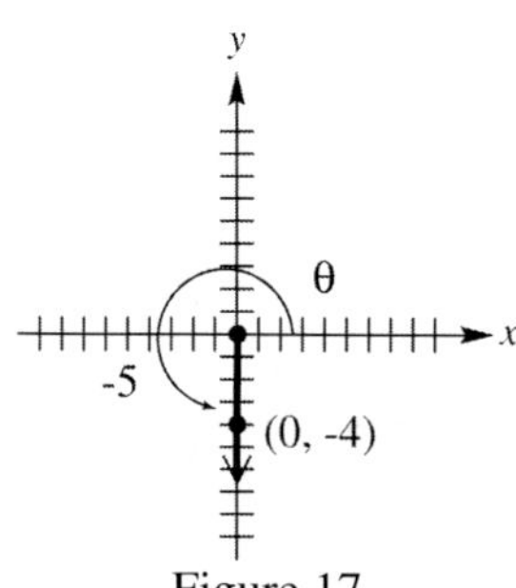

Figure 17

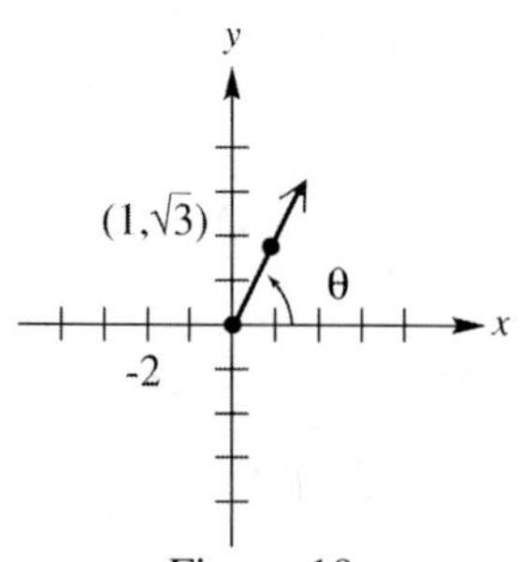

Figure 19

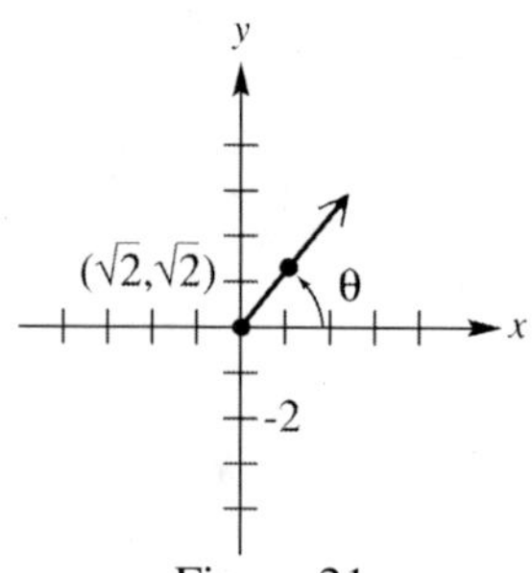

Figure 21

21. $x=-\sqrt{2}, y=-\sqrt{2} \Rightarrow r=\sqrt{x^2+y^2}=\sqrt{(-\sqrt{2})^2+(-\sqrt{2})^2}=\sqrt{4}=2$ See Figure 21.

$$\sin\theta=\frac{y}{r}=\frac{\sqrt{2}}{2};\cos\theta=\frac{x}{r}=\frac{\sqrt{2}}{2}=;\tan\theta=\frac{y}{x}=\frac{\sqrt{2}}{\sqrt{2}}=1$$

$$\csc\theta=\frac{r}{y}=\frac{2}{\sqrt{2}}=\frac{2}{\sqrt{2}}\cdot\frac{\sqrt{2}}{\sqrt{2}}=\sqrt{2};\sec\theta=\frac{r}{x}=\frac{2}{\sqrt{2}}=\frac{2}{\sqrt{2}}\cdot\frac{\sqrt{2}}{\sqrt{2}}=\sqrt{2};\cot\theta=\frac{\sqrt{2}}{\sqrt{2}}=1.$$

23. $x=-2\sqrt{3}, y=-2 \Rightarrow r=\sqrt{x^2+y^2}=\sqrt{(-2\sqrt{3})^2+(-2)^2}=\sqrt{16}=4$ See Figure 23.

$$\sin\theta=\frac{y}{r}=\frac{-2}{4}=-\frac{1}{2};\cos\theta=\frac{x}{r}=\frac{-2\sqrt{3}}{4}=-\frac{\sqrt{3}}{2};\tan\theta=\frac{y}{x}=\frac{-2}{-2\sqrt{3}}=\frac{1}{\sqrt{3}}=\frac{\sqrt{3}}{3}$$

$$\csc\theta=\frac{r}{y}=\frac{4}{-2}=-2;\sec\theta=\frac{r}{x}=\frac{4}{-2\sqrt{3}}=-\frac{2}{\sqrt{3}}=-\frac{2\sqrt{3}}{3};\cot\theta=\frac{-2\sqrt{3}}{-2}=\sqrt{3}.$$

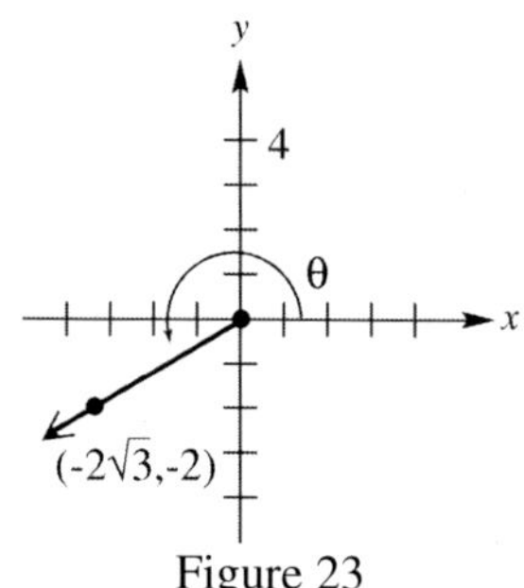

Figure 23

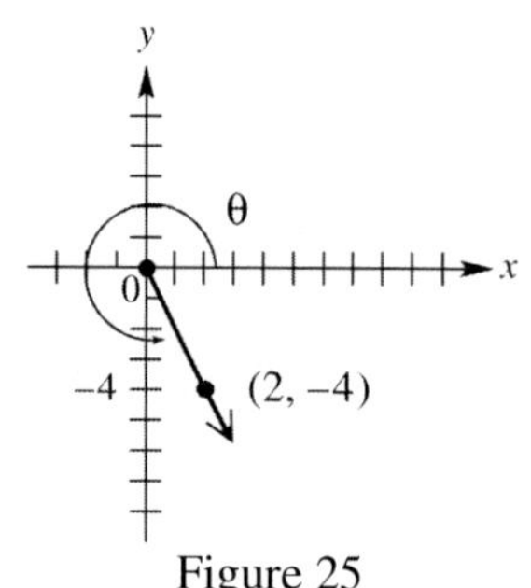

Figure 25

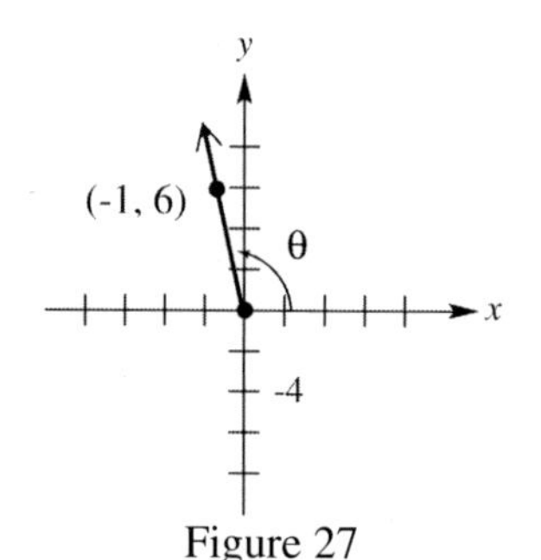

Figure 27

25. Let $x=2, y=-4 \Rightarrow r=\sqrt{x^2+y^2}=\sqrt{(2)^2+(-4)^2}=\sqrt{20}=2\sqrt{5}$ See Figure 25.

$$\sin\theta=\frac{y}{r}=\frac{-4}{2\sqrt{5}}=-\frac{2}{\sqrt{5}}=-\frac{2\sqrt{5}}{5};\cos\theta=\frac{x}{r}=\frac{2}{2\sqrt{5}}=\frac{1}{\sqrt{5}}=\frac{\sqrt{5}}{5};\tan\theta=\frac{y}{x}=\frac{-4}{2}=-2$$

$$\csc\theta=\frac{r}{y}=\frac{2\sqrt{5}}{-4}=-\frac{\sqrt{5}}{2};\sec\theta=\frac{r}{x}=\frac{2\sqrt{5}}{2}=\sqrt{5};\cot\theta=\frac{2}{-4}=-\frac{1}{2}.$$

27. Let $x=-1, y=6 \Rightarrow r=\sqrt{x^2+y^2}=\sqrt{(-1)^2+(6)^2}=\sqrt{37}$ See Figure 27.

$$\sin\theta=\frac{y}{r}=\frac{6}{\sqrt{37}}=\frac{6\sqrt{37}}{37};\cos\theta=\frac{x}{r}=\frac{-1}{\sqrt{37}}=-\frac{\sqrt{37}}{37};\tan\theta=\frac{y}{x}=\frac{6}{-1}=-6$$

$$\csc\theta=\frac{r}{y}=\frac{\sqrt{37}}{6};\sec\theta=\frac{r}{x}=\frac{\sqrt{37}}{-1}=-\sqrt{37};\cot\theta=\frac{-1}{6}=-\frac{1}{6}.$$

29. Let $x=-7, y=-4 \Rightarrow r=\sqrt{x^2+y^2}=\sqrt{(-7)^2+(-4)^2}=\sqrt{65}$ See Figure 29.

$\sin\theta=\frac{y}{r}=\frac{-4}{\sqrt{65}}=-\frac{4\sqrt{65}}{65};\cos\theta=\frac{x}{r}=\frac{-7}{\sqrt{65}}=-\frac{7\sqrt{65}}{65};\tan\theta=\frac{y}{x}=\frac{-4}{-7}=\frac{4}{7}$

$\csc\theta=\frac{r}{y}=\frac{\sqrt{65}}{-4}=-\frac{\sqrt{65}}{4};\sec\theta=\frac{r}{x}=\frac{\sqrt{65}}{-7}=-\frac{\sqrt{65}}{7};\cot\theta=\frac{-7}{-4}=\frac{7}{4}.$

31. Let $x=2, y=-2 \Rightarrow r=\sqrt{x^2+y^2}=\sqrt{(2)^2+(-2)^2}=\sqrt{8}=2\sqrt{2}$. See Figure 31.

$\sin\theta=\frac{y}{r}=\frac{-2}{2\sqrt{2}}=-\frac{1}{\sqrt{2}}=-\frac{\sqrt{2}}{2};\cos\theta=\frac{x}{r}=\frac{2}{2\sqrt{2}}=\frac{1}{\sqrt{2}}=\frac{\sqrt{2}}{2};\tan\theta=\frac{y}{x}=\frac{-2}{2}=-1$

$\csc\theta=\frac{r}{y}=\frac{2\sqrt{2}}{-2}=-\sqrt{2};\sec\theta=\frac{r}{x}=\frac{2\sqrt{2}}{2}=\sqrt{2};\cot\theta=\frac{2}{-2}=-1.$

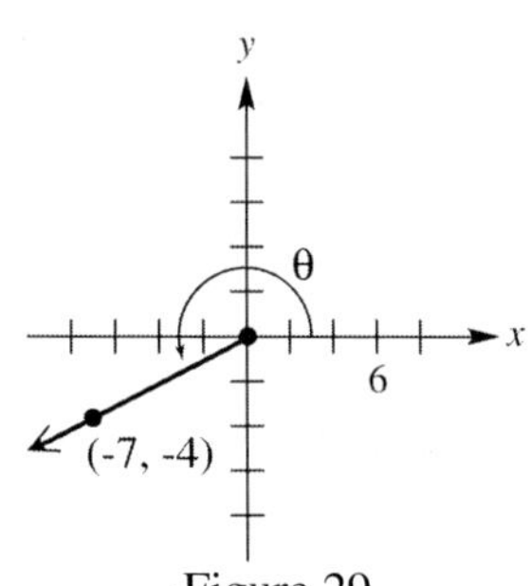

Figure 29

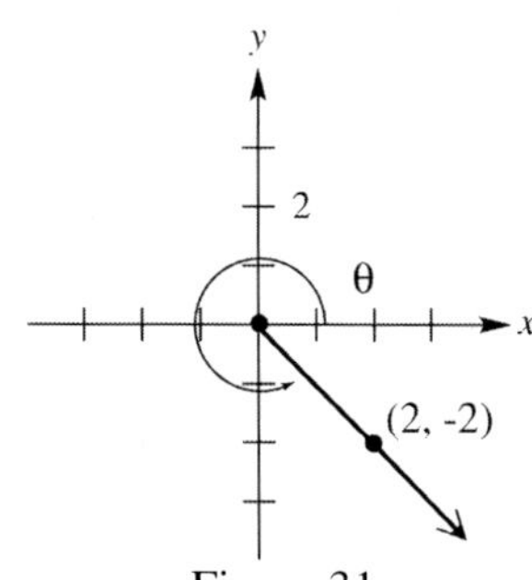

Figure 31

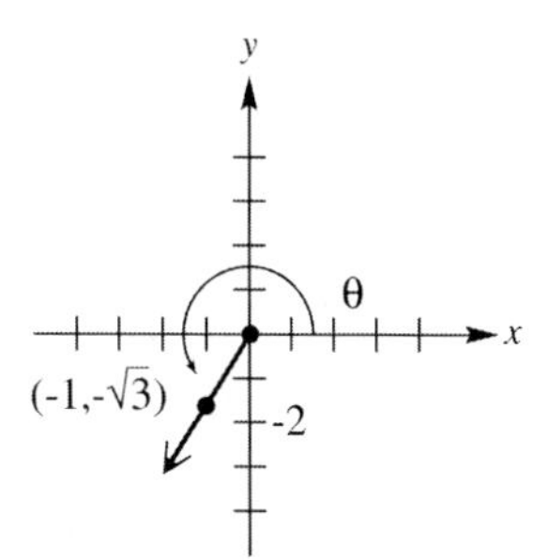

Figure 33

33. Let $x=-1, y=-\sqrt{3} \Rightarrow r=\sqrt{x^2+y^2}=\sqrt{(-1)^2+(-\sqrt{3})^2}=\sqrt{4}=2$. See Figure 33.

$\sin\theta=\frac{y}{r}=\frac{-\sqrt{3}}{2}=-\frac{\sqrt{3}}{2};\cos\theta=\frac{x}{r}=\frac{-1}{2}=-\frac{1}{2};\tan\theta=\frac{y}{x}=\frac{-\sqrt{3}}{-1}=\sqrt{3}$

$\csc\theta=\frac{r}{y}=\frac{2}{-\sqrt{3}}=-\frac{2\sqrt{3}}{3};\sec\theta=\frac{r}{x}=\frac{2}{-1}=-2;\cot\theta=\frac{-1}{-\sqrt{3}}=\frac{\sqrt{3}}{3}.$

Use the Figure below for Exercises 35-51.

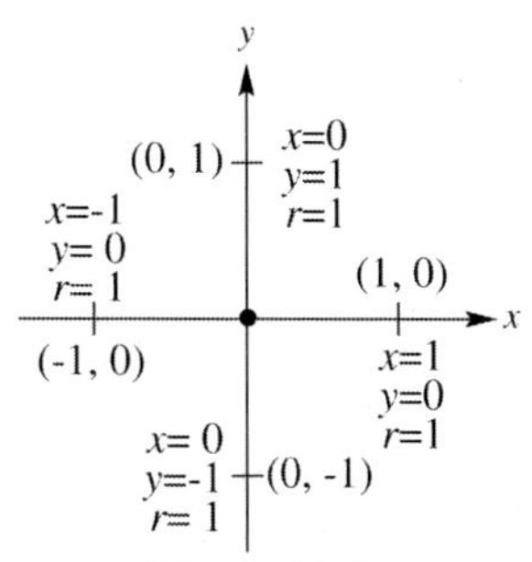

Figure 35-51

35. $\cos 90^\circ=\frac{x}{r}=\frac{0}{1}=0$

37. $\tan 180^\circ=\frac{y}{x}=\frac{0}{-1}=0$

39. $\sec 180^\circ=\frac{r}{x}=\frac{1}{-1}=-1$

41. The quadrantal angle $\theta = -270^\circ$ is coterminal with $-270^\circ + 360^\circ = 90^\circ$. $\sin(-270)^\circ = \sin 90^\circ = \frac{y}{r} = \frac{1}{1} = 1$

43. The quadrantal angle $\theta = 540^\circ$ is coterminal with $540^\circ - 360^\circ = 180^\circ$.

$\cot 540^\circ = \cot 180^\circ = \frac{x}{y} = \frac{-1}{0} \Rightarrow$ undefined.

45. The quadrantal angle $\theta = -450^\circ$ is coterminal with $720^\circ - 450^\circ = 270^\circ$.

$\csc(-450)^\circ = \csc 270^\circ = \frac{r}{y} = \frac{-1}{1} = -1$

47. The quadrantal angle $\theta = 1800^\circ$ is coterminal with $1800^\circ - 5(360)^\circ = 0^\circ$.

$\sin(1800)^\circ = \sin 0^\circ = \frac{y}{r} = \frac{0}{1} = 0$

49. The quadrantal angle $\theta = 1800^\circ$ is coterminal with $1800^\circ - 5(360)^\circ = 0^\circ$.

$\csc(1800)^\circ = \csc 0^\circ = \frac{r}{y} = \frac{1}{0} \Rightarrow$ undefined.

51. The quadrantal angle $\theta = 1800^\circ$ is coterminal with $1800^\circ - 5(360)^\circ = 0^\circ$.

$\sec(1800)^\circ = \sec 0^\circ = \frac{r}{x} = \frac{1}{1} = 1$

53. $\sec\theta = \frac{1}{\cos\theta} = \frac{1}{\frac{2}{3}} = \frac{3}{2}$

55. $\csc\theta = \frac{1}{\sin\theta} = \frac{1}{-\frac{3}{7}} = -\frac{7}{3}$

57. $\cot\theta = \frac{1}{\tan\theta} = \frac{1}{5}$

59. $\cos\theta = \frac{1}{\sec\theta} = \frac{1}{-\frac{5}{2}} = -\frac{2}{5}$

61. $\sin\theta = \frac{1}{\csc\theta} = \frac{1}{\sqrt{2}} = \frac{\sqrt{2}}{2}$

63. $\tan\theta = \frac{1}{\cot\theta} = \frac{1}{-2.5} = -0.4$

65. This angle is a quadrantal angle whose terminal side lies either on the positive part of the y-axis or the negative part of the y-axis. Any point on these terminal sides would have the form $(0, k)$, where k is a real number, $k \neq 0$. $\cos\left[(2n+1)\cdot 90^\circ\right] = \frac{x}{r} = \frac{0}{\sqrt{0^2 + k^2}} = \frac{0}{\sqrt{k^2}} = 0.$

67. This angle is a quadrantal angle whose terminal side lies either on the positive part of the x-axis or the negative part of the x-axis. Any point on these terminal sides would have the form $(k, 0)$, where k is a real number, $k \neq 0$. $\cos\left[(2n+1)\cdot 180^\circ\right] = \frac{x}{r} = \frac{-k}{\sqrt{0^2 + k^2}} = -1$

69. Since $\sin\theta > 0, \csc > 0.$ The functions are greater than 0 (positive) in quadrants I and II.

71. Since $\sin\theta > 0$ in quadrants I and II and $\cos\theta > 0$ in quadrants I and IV. Both conditions are met only in quadrant I.

73. Since $\tan\theta < 0$ in quadrants II and IV and $\cos\theta < 0$ in quadrants II and III. Both conditions are met only in quadrant II.

75. Since $\sec\theta > 0$ in quadrants I and IV and $\csc\theta > 0$ in quadrants I and II. Both conditions are met only in quadrant I.

77. Since $\sec\theta < 0$ in quadrants II and III and $\csc\theta < 0$ in quadrants III and IV. Both conditions are met only in quadrant III.

79. Since $\sin\theta < 0 \Rightarrow \csc\theta < 0$ in quadrants III and IV. Both functions are less than 0 in quadrants III and IV.

81. The answers to 71 and 75 are the same because functions in exercise 71 are the reciprocals of the function in exercise 75.

83. Impossible because the range of $\sin\theta$ is [-1,1].

85. Possible because the range of $\cos\theta$ is [-1,1].

87. Possible because the range of $\tan\theta$ is $(-\infty,\infty)$

89. Impossible because the range of $\sec\theta$ is $(-\infty,-1]\cup[-1,\infty)$.

91. Possible because the range of $\csc\theta$ is $(-\infty,-1]\cup[-1,\infty)$.

93. Possible because the range of $\cot\theta$ is $(-\infty,\infty)$.

95. Possible because the range of $\sin\theta$ is $[-1,1]$ and the range of $\csc\theta$ is $(-\infty,-1]\cup[-1,\infty)$.

97. Impossible because the range of $\cos\theta$ is [-1,1].

99. $\tan\theta = \dfrac{15}{-8}$ with θ in quadrant II, $\tan\theta = \dfrac{y}{x} \Rightarrow x = -8, y = 15.$

$$x^2 + y^2 = r^2 \Rightarrow (-8)^2 + (15)^2 = r^2 \Rightarrow r^2 = 289 \Rightarrow r = 17.$$

$$\sin\theta = \frac{y}{r} = \frac{15}{17}, \cos\theta = \frac{x}{r} = \frac{-8}{17} = -\frac{8}{17}, \tan\theta = \frac{y}{x} = \frac{15}{-8} = -\frac{15}{8}$$

$$\csc\theta = \frac{r}{y} = \frac{17}{15}, \sec\theta = \frac{r}{x} = \frac{17}{-8} = -\frac{17}{8}, \cot\theta = \frac{x}{y} = \frac{-8}{15} = -\frac{8}{15}$$

101. $\sin\theta = \dfrac{\sqrt{5}}{7}$ with θ in quadrant I, $\sin\theta = \dfrac{y}{r} \Rightarrow y = \sqrt{5}, r = 7.$

$$x^2 + y^2 = r^2 \Rightarrow x^2 + \left(\sqrt{5}\right)^2 = (7)^2 \Rightarrow x^2 = 44 \Rightarrow x = \pm 2\sqrt{11}.\ \theta \text{ is in quadrant I, let } x = 2\sqrt{11}.$$

$$\sin\theta = \frac{y}{r} = \frac{\sqrt{5}}{7}, \cos\theta = \frac{x}{r} = \frac{2\sqrt{11}}{7}, \tan\theta = \frac{y}{x} = \frac{\sqrt{5}}{2\sqrt{11}} = \frac{\sqrt{55}}{22}$$

$$\csc\theta = \frac{r}{y} = \frac{7}{\sqrt{5}} = \frac{7\sqrt{5}}{5}, \sec\theta = \frac{r}{x} = \frac{7}{2\sqrt{11}} = \frac{7\sqrt{11}}{22}, \cot\theta = \frac{x}{y} = \frac{2\sqrt{11}}{\sqrt{5}} = \frac{2\sqrt{55}}{5}$$

103. $\cot\theta = \frac{\sqrt{3}}{8}$ with θ in quadrant I, $\cot\theta = \frac{x}{y} \Rightarrow x = \sqrt{3}, y = 8.$

$x^2 + y^2 = r^2 \Rightarrow \left(\sqrt{3}\right)^2 + (8)^2 = r^2 \Rightarrow r^2 = 67 \Rightarrow r = \sqrt{67}.$

$\sin\theta = \frac{y}{r} = \frac{8}{\sqrt{67}} = \frac{8\sqrt{67}}{67}, \cos\theta = \frac{x}{r} = \frac{\sqrt{3}}{\sqrt{67}} = \frac{\sqrt{201}}{67}, \tan\theta = \frac{y}{x} = \frac{8}{\sqrt{3}} = \frac{8\sqrt{3}}{3}$

$\csc\theta = \frac{r}{y} = \frac{\sqrt{67}}{8}, \sec\theta = \frac{r}{x} = \frac{\sqrt{67}}{\sqrt{3}} = \frac{\sqrt{201}}{3}, \cot\theta = \frac{x}{y} = \frac{\sqrt{3}}{8}$

105. $\sin\theta = \frac{\sqrt{2}}{6}$ with $\cos\theta < 0$, θ in quadrant II, $\sin\theta = \frac{y}{r} \Rightarrow y = \sqrt{2}, r = 6.$

$x^2 + y^2 = r^2 \Rightarrow x^2 + \left(\sqrt{2}\right)^2 = (6)^2 \Rightarrow x^2 = 34 \Rightarrow x = \pm\sqrt{34}.$ θ is in quadrant II, let $x = -\sqrt{34}.$

$\sin\theta = \frac{y}{r} = \frac{\sqrt{2}}{6}, \cos\theta = \frac{x}{r} = -\frac{\sqrt{34}}{6}, \tan\theta = \frac{y}{x} = -\frac{\sqrt{2}}{\sqrt{34}} = -\frac{\sqrt{68}}{34} = -\frac{\sqrt{17}}{17}$

$\csc\theta = \frac{r}{y} = \frac{6}{\sqrt{2}} = 3\sqrt{2}, \sec\theta = \frac{r}{x} = -\frac{6}{\sqrt{34}} = -\frac{6\sqrt{34}}{34} = -\frac{3\sqrt{34}}{17}, \cot\theta = \frac{x}{y} = -\frac{\sqrt{34}}{\sqrt{2}} = -\frac{\sqrt{68}}{2} = -\sqrt{17}$

107. $\sec\theta = -4$ with $\sin\theta > 0$, θ in quadrant II, $\sec\theta = \frac{r}{x} \Rightarrow x = -1, r = 4.$

$x^2 + y^2 = r^2 \Rightarrow (-1)^2 + y^2 = (4)^2 \Rightarrow y^2 = 15 \Rightarrow y = \pm\sqrt{15}.$ θ is in quadrant II, let $y = \sqrt{15}.$

$\sin\theta = \frac{y}{r} = \frac{\sqrt{15}}{4}, \cos\theta = \frac{x}{r} = -\frac{1}{4}, \tan\theta = \frac{y}{x} = -\frac{\sqrt{15}}{1} = -\sqrt{15}$

$\csc\theta = \frac{r}{y} = \frac{4}{\sqrt{15}} = \frac{4\sqrt{15}}{15}, \sec\theta = \frac{r}{x} = -4 = -\frac{6\sqrt{34}}{34} = -\frac{3\sqrt{34}}{17}, \cot\theta = \frac{x}{y} = -\frac{1}{\sqrt{15}} = -\frac{\sqrt{15}}{15}$

109. Use the Pythagorean Identity $\sin^2\theta + \cos^2\theta = 1$ to write $\cos\theta, \sin^2\theta + \cos^2\theta = 1 \Rightarrow \cos^2 = 1 - \sin^2\theta \Rightarrow$

$\cos\theta = \pm\sqrt{1-\sin^2\theta}$. Since θ is acute, it is in quadrant I and $\cos\theta = \sqrt{1-\sin^2\theta}$.

111. First use the Pythagorean Identity $1 + \cot^2\theta = \csc^2\theta$ to write $\csc\theta, 1 + \cot^2\theta = \csc^2\theta \Rightarrow$

$\csc\theta = \pm\sqrt{1+\cot^2\theta}$ since θ is in quadrant III, $\csc\theta = -\sqrt{1+\cot^2\theta}$. Substituting this into the Reciprocal

Identity $\sin\theta = \frac{1}{\csc\theta}$ yields: $\sin\theta = -\frac{1}{\sqrt{1+\cot^2\theta}}$ or $-\frac{\sqrt{1+\cot^2\theta}}{1+\cot^2\theta}$.

113. Use the Pythagorean Identity $\sin^2\theta + \cos^2\theta = 1$ to write $\cos\theta, \sin^2\theta + \cos^2\theta = 1 \cos^2 = 1 - \sin^2\theta \Rightarrow$

$\cos\theta = \pm\sqrt{1-\sin^2\theta}$. Since θ is in quadrant I or IV, $\cos\theta = \sqrt{1-\sin^2\theta}$. Now substituting this into the

Quotient Identity $\tan\theta = \frac{\sin\theta}{\cos\theta}$ yields $\tan\theta = \frac{\sin\theta}{\sqrt{1-\sin^2\theta}}$ or $\frac{\sin\theta\sqrt{1-\sin^2\theta}}{1-\sin^2\theta}$.

115. Dividing $x^2 + y^2 = r^2$ by y^2 yields: $\frac{x^2}{y^2} + \frac{y^2}{y^2} = \frac{r^2}{y^2}$ which is equal to:

$\cot^2\theta + 1 = \csc^2\theta \Rightarrow 1 + \cot^2\theta = \csc^2\theta$.

117. False $\sin\theta + \cos\theta \neq 1$ for all values of θ. For example, if $\theta = 30°$, then

$\sin 30° + \cos 30° = \frac{1}{2} + \frac{\sqrt{3}}{2} = \frac{1+\sqrt{3}}{2} \neq 1$.

119. (a) The slope of the line is .03, so when x increases by 100 feet, y increases by 3 feet. Thus the highway has a grade of 3%.

(b) Using the point from part (a), $(100, 3)$ we first solve for r and then find $\sin\theta$:

$r = \sqrt{100^2 + 3^2} \Rightarrow r = \sqrt{10,009}$, then $\sin\theta = \frac{y}{r} = \frac{-6}{10,009}$. Now use the formula $R = W(\sin\theta)$, to find

grade resistance $R = 25,000\left(\frac{3}{\sqrt{10,009}}\right) \approx 750$ pounds.

Reviewing Basic Concepts (Sections 8.1 - 8.2)

1. (a) complement: $c + 35° = 90° \Rightarrow c = 55°$ supplement: $s + 35° = 180° \Rightarrow s = 145°$

(b) complement, since $90° = \frac{\pi}{2}$ radians, $c + \frac{\pi}{4} = \frac{\pi}{2} \Rightarrow c = \frac{2\pi}{4} - \frac{\pi}{4} \Rightarrow c = \frac{\pi}{4}$.

supplement, since $180° = \pi$ radians, $s + \frac{\pi}{4} = \pi \Rightarrow s = \frac{4\pi}{4} - \frac{\pi}{4} \Rightarrow s = \frac{3\pi}{4}$.

2. $32.25° = 32° + .25°(60)' = 32°15'0"$.

3. $59°35'30" = \left(59 + \frac{35}{60} + \frac{30}{(60)(60)}\right)° = 59.591\overline{6}°$. .

4. (a) The smallest positive degree measure of the coterminal angle is: $560° - 360° = 200°$.

(b) The smallest positive radian measure of the coterminal angle is: $-\frac{2\pi}{3} + 2\pi = -\frac{2\pi}{3} + \frac{6\pi}{3} = \frac{4\pi}{3}$.

5. (a) $240° \cdot \frac{\pi}{180} = \frac{4\pi}{3}$ (b) $\frac{3\pi}{4} \cdot \frac{180}{\pi} = 135°$

6. (a) Since $120° = 120° \cdot \frac{\pi}{180} = \frac{2\pi}{3}$ and arc length is found by the formula $s = r\theta$, , the arc length is:

$s = 3\left(\frac{2\pi}{3}\right) \Rightarrow s = 2\pi$ cm.

(b) Using the area of a sector formula $A = \frac{1}{2}r^2\theta$ yields: $A = \frac{1}{2}(3)^2\left(\frac{2\pi}{3}\right) \Rightarrow A = 3\pi$ cm^2.

7. 1. $x = -2, y = 5 \Rightarrow r = \sqrt{x^2 + y^2} = \sqrt{(-2)^2 + (5)^2} = \sqrt{29}$

$\sin\theta=\frac{y}{r}=\frac{5}{\sqrt{29}}=\frac{5\sqrt{29}}{29};\cos\theta=\frac{x}{r}=\frac{-2}{\sqrt{29}}=-\frac{2\sqrt{29}}{29};\tan\theta=\frac{y}{x}=\frac{5}{-2}=-\frac{5}{2}$

$\csc\theta=\frac{r}{y}=\frac{\sqrt{29}}{5};\sec\theta=\frac{r}{x}=\frac{\sqrt{29}}{-2}=-\frac{\sqrt{29}}{2};\cot\theta=\frac{x}{y}=-\frac{2}{5}.$

8. The angle of $270^\circ \Rightarrow x=0, y=-1$. $r=\sqrt{x^2+y^2}=\sqrt{0^2+(-1)^2}=\sqrt{1}=1$,.

   $\sin\theta=\frac{y}{r}=\frac{-1}{1}=-1;\cos\theta=\frac{x}{r}=\frac{0}{1}=0;\tan\theta=\frac{y}{x}=\frac{-1}{0}\Rightarrow \textit{undefined}.$

   $\csc\theta=\frac{r}{y}=\frac{1}{-1}=-1;\sec\theta=\frac{r}{x}=\frac{1}{0}\Rightarrow \textit{undefined};\cot\theta=\frac{x}{y}=\frac{0}{-1}=0.$

9. (a) Impossible because the range of $\cos\theta$ is [-1,1].

   (b) Possible because the range of $\tan\theta$ is $(-\infty,\infty)$

   (c) Possible because the range of $\csc\theta$ is $(-\infty,-1]\cup[-1,\infty)$.

   (d) Impossible because the range of $\sec\theta$ is $(-\infty,-1]\cup[-1,\infty)$.

10. $\sin\theta=-\frac{2}{3}$ with θ in quadrant III, $\sin\theta=\frac{y}{r}\Rightarrow y=-2, r=3.$

    $x^2+y^2=r^2\Rightarrow(x)^2+(-2)^2=(3)^2\Rightarrow x^2=5\Rightarrow x=\pm\sqrt{5}.$ θ is in quadrant III, let $x=-\sqrt{5}.$

    $\sin\theta=\frac{y}{r}=-\frac{2}{3},\ \cos\theta=\frac{x}{r}=\frac{-\sqrt{5}}{3},\ \tan\theta=\frac{y}{x}=\frac{-2}{-\sqrt{5}}=\frac{2\sqrt{5}}{5}$

    $\csc\theta=\frac{r}{y}=\frac{3}{-2}=-\frac{3}{2},\ \sec\theta=\frac{r}{x}=\frac{3}{-\sqrt{5}}=-\frac{3\sqrt{5}}{5},\ \cot\theta=\frac{x}{y}=\frac{-\sqrt{5}}{-2}=\frac{\sqrt{5}}{2}$

8.3: Right Triangles and Evaluating Trigonometric Functions

1. $\sin A=\frac{\text{opp}}{\text{hyp}}=\frac{21}{29}$ $\cos A=\frac{\text{adj}}{\text{hyp}}=\frac{20}{29}$ $\tan A=\frac{\text{opp}}{\text{adj}}=\frac{21}{20}$

   $\csc A=\frac{\text{hyp}}{\text{opp}}=\frac{29}{21}$ $\sec A=\frac{\text{hyp}}{\text{adj}}=\frac{29}{20}$ $\cot A=\frac{\text{adj}}{\text{opp}}=\frac{20}{21}$

3. $\sin A=\frac{\text{opp}}{\text{hyp}}=\frac{n}{p}$ $\cos A=\frac{\text{adj}}{\text{hyp}}=\frac{m}{p}$ $\tan A=\frac{\text{opp}}{\text{adj}}=\frac{n}{m}$

   $\csc A=\frac{\text{hyp}}{\text{opp}}=\frac{p}{n}$ $\sec A=\frac{\text{hyp}}{\text{adj}}=\frac{p}{m}$ $\cot A=\frac{\text{adj}}{\text{opp}}=\frac{m}{n}$

5. Using the 30° - 60° right triangle rules, for a 30° angle the hypotenuse $=2$, the opposite side $=1$, and the adjacent side $=\sqrt{3}$, therefore $\tan 30^\circ=\frac{1}{\sqrt{3}}=\frac{\sqrt{3}}{3}$ and $\cot 30^\circ=\frac{\sqrt{3}}{1}=\sqrt{3}$.

7. Using the 30° - 60° right triangle rules, for a 60° angle the hypotenuse $=2$, the opposite side $=\sqrt{3}$, and the adjacent side $=1$, therefore $\sin 60^\circ=\frac{\sqrt{3}}{2};\cot 60^\circ=\frac{1}{\sqrt{3}}=\frac{\sqrt{3}}{3}$; and $\csc 60^\circ=\frac{2}{\sqrt{3}}=\frac{2\sqrt{3}}{3}$.

9. First find the reference angle: $(180° - 135°) = 45°$. Using $45°$-$45°$ right triangle rules, for a $45°$ angle the hypotenuse $= \sqrt{2}$, the opposite side $= 1$, and the adjacent side $= 1$. Since $\theta = 135°$ is in quadrant II, x is negative and cosine, secant, tangent, and cotangent are negative, therefore $\tan 135° = -\tan 45° = -\frac{1}{1} = -1$; and $\cot 135° = -\cot 45° = -\frac{1}{1} = -1$.

10. First find the reference angle: $(180° - 150°) = 30°$. Using $30°$-$60°$ right triangle rules, for a $30°$ angle the hypotenuse $= 2$, the opposite side $= 1$, and the adjacent side $= \sqrt{3}$. Since $\theta = 150°$ is in quadrant II, x is negative and cosine, secant, tangent, and cotangent are negative, therefore $\sin 150° = \sin 30° = \frac{1}{2}$; $\cot 150° = -\cot 30° = -\frac{\sqrt{3}}{1} = -\sqrt{3}$; and $\sec 150° = -\sec 30° = -\frac{2}{\sqrt{3}} = -\frac{2\sqrt{3}}{3}$.

11. First find the reference angle: $(210° - 180°) = 30°$. Using $30°$-$60°$ right triangle rules, for a $30°$ angle the hypotenuse $= 2$, the opposite side $= 1$, and the adjacent side $= \sqrt{3}$. Since $\theta = 210°$ is in quadrant III, x is negative and sine, cosine, secant, and cotangent are negative, therefore $\cos 210° = -\cos 30° = -\frac{\sqrt{3}}{2}$; and $\sec 210° = -\sec 30° = -\frac{2}{\sqrt{3}} = -\frac{2\sqrt{3}}{3}$.

13. (a) Using the $30°$ - $60°$ right triangle rules, for a $30°$ angle the hypotenuse $= 2$, the opposite side $= 1$, and the adjacent side $= \sqrt{3}$, therefore $\tan 30° = \frac{1}{\sqrt{3}} = \frac{\sqrt{3}}{3}$.

 (b) Using the calculator, $\tan 30° = \frac{\sqrt{3}}{3} \approx .5773502692$

15. (a) Using the $30°$ - $60°$ right triangle rules, for a $30°$ angle the hypotenuse $= 2$, the opposite side $= 1$ ¸and the adjacent side $= \sqrt{3}$, therefore $\sin 30° = \frac{1}{2}$.

 (b) $\sin 30° = \frac{1}{2}$ is a rational number.

17. (a) Using the $30°$ - $60°$ right triangle rules, for a $30°$ angle the hypotenuse $= 2$, the opposite side $= 1$ ¸and the adjacent side $= \sqrt{3}$, therefore $\sec 30° = \frac{2}{\sqrt{3}} = \frac{2\sqrt{3}}{3}$.

 (b) Using the calculator, $\sec 30° = (\cos 30°)^{-1} = \frac{2\sqrt{3}}{3} \approx 1.154700538$

19. (a) Using the $45°$ - $45°$ right triangle rules, for a $45°$ angle the hypotenuse $= \sqrt{2}$, the opposite side $= 1$, and the adjacent side $= 1$, therefore $\csc 45° = \frac{\sqrt{2}}{1} = \sqrt{2}$.

 (b) Using the calculator, $\csc 45° = (\sin 45°)^{-1} = \sqrt{2} \approx 1.414213562$

21. (a) Using the $45°$ - $45°$ right triangle rules, for a $45°$ angle the hypotenuse $= \sqrt{2}$, the opposite side $= 1$, and the adjacent side $= 1$, therefore $\cos 45° = \frac{1}{\sqrt{2}} = \frac{\sqrt{2}}{2}$.

(b) Using the calculator, $\cos 45° = \frac{\sqrt{2}}{2} \approx .7071067812$

23. (a) Because $\theta = \frac{\pi}{3} = 60°$ we use the $30°$ - $60°$ right triangle rules, for a $60°$ angle the hypotenuse $= 2$, the opposite side $= \sqrt{3}$, and the adjacent side $= 1$, therefore $\sin \frac{\pi}{3} = \sin 60° = \frac{\sqrt{3}}{2}$.

(b) $\sin \frac{\pi}{3} = \sin 60° = \frac{\sqrt{3}}{2} = .8660254038$

25. (a) Because $\theta = \frac{\pi}{3} = 60°$ we use the $30°$ - $60°$ right triangle rules, for a $60°$ angle the hypotenuse $= 2$, the opposite side $= \sqrt{3}$, and the adjacent side $= 1$, therefore $\tan \frac{\pi}{3} = \tan 60° = \frac{\sqrt{3}}{1} = \sqrt{3}$.

(b) $\tan \frac{\pi}{3} = \tan 60° = \frac{\sqrt{3}}{1} = \sqrt{3} = 1.732050808$

27. (a) Because $\theta = \frac{\pi}{6} = 30°$ we use the $30°$ - $60°$ right triangle rules, for a $30°$ angle the hypotenuse $= 2$, the opposite side $= 1$, and the adjacent side $= \sqrt{3}$, therefore $\csc \frac{\pi}{6} = \csc 30° = \frac{2}{1} = 2$.

(b) $\csc \frac{\pi}{6} = \csc 30° = 2$ is a rational number.

29. (a) Because $\theta = \frac{\pi}{3} = 60°$ we use the $30°$ - $60°$ right triangle rules, for a $60°$ angle the hypotenuse $= 2$, the opposite side $= \sqrt{3}$, and the adjacent side $= 1$, therefore $\csc \frac{\pi}{3} = \csc 60° = \frac{2}{\sqrt{3}} = \frac{2\sqrt{3}}{3}$.

(b) $\csc \frac{\pi}{3} = \csc 60° = \left(\sin \frac{\pi}{3} \right)^{-1} = \left(\sin 60° \right)^{-1} = \frac{2\sqrt{3}}{3} = 1.154700538$

31. The exact value of $\sin 45°$ is $\frac{\sqrt{2}}{2}$ The decimal value given is an approximation.

33. $\cot 73° = \tan(90° - 73°) = \tan 17°$.

35. $\sin 38° = \cos(90° - 38°) = \cos 52°$.

37. $\tan 25°43' = \cot(90° - 25°43') = \cot(89°60' - 25°43') = \cot 64°17'$

39. $\cos \frac{\pi}{5} = \sin\left(\frac{\pi}{2} - \frac{\pi}{5} \right) = \sin\left(\frac{5\pi}{10} - \frac{2\pi}{10} \right) = \sin \frac{3\pi}{10}$

41. $\tan 0.5 = \cot\left(\frac{\pi}{2} - 0.5 \right)$

43. $\cos 1 = \sin\left(\frac{\pi}{2} - 1\right)$

45. If $\theta = 98°$ then $\theta' = 180° - 98° = 82°$

47. If $\theta = 230°$ then $\theta' = 230° - 180° = 50°$

49. If $\theta = -135°$ then $\theta' = 180° + (-135°) = 45°$

51. If $\theta = 750°$ then $\theta' = 750° - 720° = 30°$

53. If $\theta = \frac{4\pi}{3}$ then $\theta' = \frac{4\pi}{3} - \pi = \frac{4\pi}{3} - \frac{3\pi}{3} = \frac{\pi}{3}$.

55. If $\theta = -\frac{4\pi}{3} = 2\pi - \frac{4\pi}{3} = \frac{6\pi}{3} - \frac{4\pi}{3} = \frac{2\pi}{3}$ then $\theta' = \pi - \frac{2\pi}{3} = \frac{3\pi}{3} - \frac{2\pi}{3} = \frac{\pi}{3}$

57. It is easy to find one-half of 2, which is 1. This is, then, the measure of the side opposite the $30°$ angle, and the ratios are easily found. Yes, any positive number could have been used.

59. First find the reference angle: $(360° - 300°) = 60°$. Using $30°$-$60°$ right triangle rules, for a $60°$ angle the hypotenuse $= 2$, the opposite side $= \sqrt{3}$, and the adjacent side $= 1$. Since $\theta = 300°$ is in quadrant IV, *y* is negative and sine, tangent, cotangent, and cosecant are negative, therefore:

$\sin 300° = -\sin 60° = -\frac{\sqrt{3}}{2}$ $\qquad$ $\csc 300° = -\csc 60° = -\frac{2}{\sqrt{3}} = -\frac{2\sqrt{3}}{3}$

$\cos 300° = \cos 60° = \frac{1}{2}$ $\qquad$ $\sec 300° = \sec 60° = \frac{2}{1} = 2$

$\tan 300° = -\tan 60° = -\frac{\sqrt{3}}{1} = -\sqrt{3}$ $\qquad$ $\cot 300° = -\cot 60° = -\frac{1}{\sqrt{3}} = -\frac{\sqrt{3}}{3}$

61. First find the reference angle: $(405° - 360°) = 45°$. Using the $45°$-$45°$ right triangle rules, for a $45°$ angle the hypotenuse $= \sqrt{2}$, the opposite side $= 1$, and the adjacent $= 1$. Since $\theta = 405°$ is in quadrant I, all functions are positive, therefore:

$\sin 405° = \sin 45° = \frac{1}{\sqrt{2}} = \frac{\sqrt{2}}{2}$ $\qquad$ $\csc 405° = \csc 45° = \frac{\sqrt{2}}{1} = \sqrt{2}$

$\cos 405° = \cos 45° = \frac{1}{\sqrt{2}} = \frac{\sqrt{2}}{2}$ $\qquad$ $\sec 405° = \sec 45° = \frac{\sqrt{2}}{1} = \sqrt{2}$

$\tan 405° = \tan 45° = \frac{1}{1} = 1$ $\qquad$ $\cot 405° = \cot 45° = \frac{1}{1} = 1$

63. First find the reference angle: $\left(2\pi - \frac{11\pi}{6}\right) = \frac{\pi}{6}$. Since $\frac{\pi}{6}$ radians is equal to $30°$ we use the $30°$-$60°$ right triangle rules, for a $30°$ angle the hypotenuse $= 2$, the opposite side $= 1$, and the adjacent side $= \sqrt{3}$. Since $\theta = \frac{11\pi}{6}$ is in quadrant IV, *y* is negative and sine, tangent, cotangent, and cosecant are negative, therefore:

$\sin\frac{11\pi}{6} = -\sin\frac{\pi}{6} = -\frac{1}{2}$ $\qquad$ $\csc\frac{11\pi}{6} = -\csc\frac{\pi}{6} = \frac{2}{1} = -2$

$\cos\frac{11\pi}{6} = \cos\frac{\pi}{6} = \frac{\sqrt{3}}{2}$ $\sec\frac{11\pi}{6} = \sec\frac{\pi}{6} = \frac{2}{\sqrt{3}} = \frac{2\sqrt{3}}{3}$

$\tan\frac{11\pi}{6} = -\tan\frac{\pi}{6} = -\frac{1}{\sqrt{3}} = -\frac{\sqrt{3}}{3}$ $\cot\frac{11\pi}{6} = -\cot\frac{\pi}{6} = -\frac{\sqrt{3}}{1} = -\sqrt{3}$

65. First find the reference angle: $\left(2\pi - \frac{7\pi}{4}\right) = \frac{\pi}{4}$ Since $\frac{\pi}{4}$ radians is equal to $45°$ we use the $45°$-$45°$ right triangle rules, for a $45°$ angle the hypotenuse $= \sqrt{2}$, the opposite side $= 1$, and the adjacent $= 1$. Since $\theta = -\frac{7\pi}{4}$ is in quadrant I, all functions are positive, therefore:

$\sin-\frac{7\pi}{4} = \sin\frac{\pi}{4} = -\frac{1}{\sqrt{2}} = \frac{\sqrt{2}}{2}$ $\csc-\frac{7\pi}{4} = \csc\frac{\pi}{4} = -\frac{\sqrt{2}}{1} = \sqrt{2}$

$\cos-\frac{7\pi}{4} = \cos\frac{\pi}{4} = \frac{1}{\sqrt{2}} = \frac{\sqrt{2}}{2}$ $\sec-\frac{7\pi}{4} = \sec\frac{\pi}{4} = \frac{\sqrt{2}}{1} = \sqrt{2}$

$\tan-\frac{7\pi}{4} - \tan\frac{\pi}{4} = \frac{1}{1} = 1$ $\cot-\frac{7\pi}{4} = \cot\frac{\pi}{4} = \frac{1}{1} = 1$

67. First find the reference angle: $\left(4\pi - \frac{19\pi}{6}\right) = \frac{5\pi}{6}$, then $\theta = \pi - \frac{5\pi}{6} = \frac{\pi}{6}$. Since $\frac{\pi}{6}$ radians is equal to 30° we use the 30°-60° right triangle rules, for a 30° angle the hypotenuse $= 2$, the opposite side $= 1$, and the adjacent side $= \sqrt{3}$. Since $\theta = -\frac{19\pi}{6}$ is in quadrant II, x is negative and cosine, secant, tangent, and cotangent are negative, therefore:

$\sin-\frac{19\pi}{6} = \sin\frac{\pi}{6} = \frac{1}{2}$ $\csc-\frac{19\pi}{6} = \csc = \frac{\pi}{6} = \frac{2}{1} = 2$

$\cos-\frac{19\pi}{6} = -\cos = \frac{\pi}{6} = -\frac{\sqrt{3}}{2}$ $\sec-\frac{19\pi}{6} = -\sec = \frac{\pi}{6} = -\frac{2\sqrt{3}}{3}$

$\tan-\frac{19\pi}{6} = -\tan\frac{\pi}{6} = -\frac{1}{\sqrt{3}} = -\frac{\sqrt{3}}{3}$ $\cot-\frac{19\pi}{6} = -\cot\frac{\pi}{6} = -\frac{\sqrt{3}}{1}$

69. Calculate in degree mode: $\tan 29^{\circ} \approx .5543090515$

71. Calculate in degree mode: $\cot 41°24' = \cot 41\frac{24}{60}° = \cot 41.4° = (\tan 41.4°)^{-1} \approx 1.134277349$

73. Calculate in degree mode: $\sec 183°48' = \sec 145\frac{48}{60}° = \sec 183.8° = (\cos 183.8°)^{-1} \approx -1.002203376$

75. Calculate in degree mode: $\tan(-80°6') = \tan\left(-80\frac{6}{60}°\right) \approx \tan(-80.1°) \approx -5.729741647$

77. Calculate in radian mode: $\sin 2.5 \approx .5984721441$

79. Calculate in radian mode: $\tan 5 \approx -3.3805155006$

81. (a) Find the reference angle: $\theta' = \left(\frac{7\pi}{6} - \pi\right) = \frac{\pi}{6}$. Since $\theta = \frac{7\pi}{6}$ is in quadrant III,sine is negative and the reference angle function is: $-\sin\frac{\pi}{6}$

(b) Since $\frac{\pi}{6}$ radians is equal to 30° we use the 30°-60° right triangle rules, for a $30°$ angle the hypotenuse $=2$, the opposite side $=1$, and the adjacent side $=\sqrt{3}$, therefore $-\sin\frac{\pi}{6} = -\frac{1}{2}$.

(c) Calculate in radian mode: $\sin\frac{7\pi}{6} = -0.5$, which is equal to $-\sin\frac{\pi}{6} = -\frac{1}{2}$.

83. (a) Find the reference angle: $\cos\frac{\pi}{3} = \frac{1}{2}$. Since $\theta = \frac{3\pi}{4}$ is in quadrant II, tangent is negative and the reference angle function is: $-\tan\frac{\pi}{4}$.

(b) Since $\frac{\pi}{4}$ radians is equal to $45°$ we use the $45°$-$45°$ right triangle rules, for a $45°$ angle the hypotenuse $=\sqrt{2}$, the opposite side $=1$, and the adjacent $=1$.therefore $-\tan\frac{\pi}{4} = -\frac{1}{1} = -1$.

(c) Calculate in radian mode: $\tan\frac{3\pi}{4} = -1$, which is equal to $-\tan\frac{\pi}{4} = -\frac{1}{1} = -1$.

85. (a) Find the reference angle: $\theta' = \left(\frac{7\pi}{6} - \pi\right) = \frac{\pi}{6}$. Since $\sin A = \frac{opp}{hyp} = \frac{60}{61}$ is in quadrant III,cosine is negative and the reference angle function is: $-\cos\frac{\pi}{6}$.

(b) Since $\frac{\pi}{6}$ radians is equal to $30°$.we use the $30°$-$60°$ right triangle rules, for a 30° angle the hypotenuse $=2$, the opposite side $=1$, and the adjacent side $=\sqrt{3}$, therefore $-\cos\frac{\pi}{6} = -\frac{\sqrt{3}}{2}$.

(c) Calculate in radian mode: $\cos\frac{7\pi}{6} = -.8660254038$, which is equal to $-\cos\frac{\pi}{6} = -\frac{\sqrt{3}}{2}$.

87. Since $\sin\theta$ is positive in quadrant I and II, values for $\cot 315° = -\cot 45° = -\frac{1}{1} = -1$ can only be in these quadrants.In quadrant I, from the $30°$-$60°$ right triangle, $\sin 30° = \frac{1}{2}$, therefore $30°$ is one value. In quadrant II, use the reference angle $30°$ to find the other: $180° - 30° = 150°$. Therefore $\theta = 30°$ or $150°$.

89. Since $\tan\theta$ is negative in quadrant II and IV, values for θ can only be in these quadrants. From the $30°$-$60°$ right triangle, $\tan 60° = \sqrt{3}$, therefore $60°$ is the reference angle. In quadrant II, use the reference angle $60°$ to find the first value: $180° - 60° = 120°$. In quadrant IV, use the reference angle to find the other value: $360° - 60° = 300°$. Therefore, $\theta = 120°$ or $300°$.

91. Since $\cot\theta$ is negative in quadrant II and IV, values for θ can only be in these quadrants. From the $30°-60°$ right triangle, $\cot 60° = \dfrac{\sqrt{3}}{3}$, therefore $60°$ is the reference angle. In quadrant II, use the reference angle $60°$ to find the first value: $360°-60°$. In quadrant IV, use the reference angle $60°$ to find the other value: $360°-60° = 300°$. Therefore, $\theta = 120°$ or $300°$

93. Since $\cos\theta$ is positive in quadrant I and IV, values for θ can only be in these quadrants. In quadrant I, we use the inverse trigonometric function (degree mode) of our calculator to find the angle: $\cos^{-1}.68716510 \Rightarrow \theta = 46.59388121°$.In quadrant IV, use the found reference angle from quadrant I to find $\theta: 360° - 46.59388121° = 313.4061188°$ Therefore, $\theta \approx 46.59388121°$ or $B = 90.00° - 61.00° \Rightarrow B = 29.00'$.

95. Since $\sin\theta$ is positive in quadrant I and II, values for θ can only be in these quadrants. In quadrant I, we use the inverse trigonometric function (degree mode) of our calculator to find the angle: $\sin^{-1}.41298643 \Rightarrow \theta = 24.39257624°$ .In quadrant I1, use the found reference angle from quadrant Ito find θ: 24.39257624°= 155.6074238°. Therefore, $\theta \approx 24.39257624$ or $155.6074238°$.

97. Since $\tan\theta$ is positive in quadrant I and III, values for θ can only be in these quadrants. In quadrant I, we use the inverse trigonometric function (degree mode) of our calculator to find the angle: $\tan^{-1}.876292035 \Rightarrow \theta = 41.24818261°$ .In quadrant III, use the found reference angle from quadrant Ito find $\theta: 180° + 41.24818261° = 221.2481826°$ .Therefore, $\theta \approx 41.24818261°$ or $221.2481826°$.

99. Since $\tan\theta$ is positive in quadrant I and III, values for θ can only be in these quadrants. In quadrant I, we use the inverse trigonometric function (radian mode) of our calculator to find the angle: $\tan^{-1}.21264138 \Rightarrow \theta = .2095206607°$. In quadrant III, use the found reference angle from quadrant I to find $\theta: \pi + .2095206607 = 3.351113314$ radians. Therefore, $\theta \approx .2095206607$ or 3.351113314 radians.

101. Since $\cot\theta$ is positive in quadrant I and III, values for θ can only be in these quadrants.In quadrant I, we use the inverse trigonometric function (radian mode) of our calculator to find the angle: Since $\cot\theta = (\tan\theta)^{-1} \Rightarrow (\tan\theta)^{-1} - .29949853 \Rightarrow \tan\theta = (.29949853)^{-1} \Rightarrow \theta = \tan^{-1}(.29949853)^{-1} \Rightarrow$ $\theta \approx 1.2799966$ radians .In quadrant III, use the found reference angle from quadrant I to find $\theta \approx 1.2799966$ radians .Therefore, $\theta \approx 1.27979966$ or 4.42139314 radians.

103. Using the point in the given example, $\sqrt{(x_1)^2 + (y_1)^2} = r$ would yield: $r \approx 6.9032258$. This represents the distance from the point (x_1, y_1) to the origin.

104. Using the point in the given example and degree mode yields:

$$\tan^{-1}\left(\frac{y_1}{x_1}\right) = \tan^{-1}\left(\frac{5.9783689}{3.4516129}\right) \approx 59.9999° \approx 60°.$$

105. Using the point in the given example and degree mode yields:

$$\sin^{-1}\left(\frac{y_1}{r}\right) = \sin^{-1}\left(\frac{5.9783689}{6.9032258}\right) \approx 59.9999° \approx 60°.$$

106. Using the point in the given example and degree mode yields: $\cos^{-1}\left(\dfrac{x_1}{r}\right) = \cos^{-1}\left(\dfrac{3.4516129}{6.9032258}\right) \approx 60°$.

107. It is a measure (approximately $60°$, for the example) of the angle formed by the positive x-axis and the ray $y=\sqrt{3}x,\ x\geq 0.$

108. ...make a conjecture: The slope of a line passing through the origin is equal to the tangent of the angle it forms with the positive x-axis.

109. Solve for θ using the formula and given values: $400=5000(\sin\theta)\Rightarrow \sin\theta=\dfrac{400}{5000}\Rightarrow\theta=\sin^{-1}\dfrac{2}{25}\approx 4.6°.$

111. Solve for c_2 using the formula and given values:

$$\frac{3\times10^8}{c_2}=\frac{\sin 46°}{\sin 31°}\Rightarrow c_2=\frac{(3\times10^8)\sin 31°}{\sin 46°}\approx 214{,}796{,}154\approx 2\times10^8 \text{ min/sec}.$$

113. Solve for θ using the formula and given values:

$$\frac{3\times10^8}{1.5\times10^8}=\frac{\sin 40°}{\sin\theta_2}\Rightarrow(3\times10^8)\sin\theta_2=(1.5\times10^8)\sin 40°\Rightarrow\sin\theta_2=\frac{(1.5\times10^8)\sin 28°}{3\times10^8}\Rightarrow$$

$$\theta_2=\sin^{-1}\left(\frac{(1.5\times10^8)\sin 40°}{3\times10^8}\right)\Rightarrow\theta_2\approx 18.747°\approx 19°.$$

115. Solve for using the formula and given values:

$$\frac{3\times10^8}{2.254\times10^8}=\frac{\sin 90°}{\sin\theta_2}\Rightarrow(3\times10^8)\sin\theta_2=(1.254\times10^8)\sin 90°\Rightarrow\sin\theta_2=\frac{(2.254\times10^8)\sin 90°}{3\times10^8}\Rightarrow$$

$$\theta_2=\sin^{-1}\frac{(2.254\times10^8)\sin 90°}{3\times10^8}\Rightarrow\theta_2\approx 48.706°\approx 48.7°.$$

117. (a) First change 55 and 30 mph to feet per second: $\dfrac{55\text{ mi}}{1\text{ hr}}\cdot\dfrac{5280\text{ ft}}{1\text{ mi}}\cdot\dfrac{1\text{ hr}}{3600\text{ sec}}=80\dfrac{2}{3}\text{ ft/sec}$ and

$\dfrac{30\text{ mi}}{1\text{ hr}}\cdot\dfrac{5280\text{ ft}}{1\text{ mi}}\cdot\dfrac{1\text{ hr}}{3600\text{ sec}}=44\text{ft/sec}.$ Now solve for D using the formula, given, and found

values: $D=\dfrac{1.05\left(\left(80\frac{2}{3}\right)^2-(44)^2\right)}{64.4(.4+.02+\sin 3.5°)}\approx 154.9303\Rightarrow D\approx 155$ feet.

(b) Solve for D using the formula and given values:

$$D=\frac{1.05\left(\left(80\frac{2}{3}\right)^2-(44)^2\right)}{64.4(.4+.02+\sin(-2°))}\approx 193.5313\Rightarrow D\approx 194 \text{ feet}.$$

(c) As the grade decreases from uphill to downhill, the braking distance increases, which corresponds to driving experience.

8.4: Applications of Right Triangles

1. To find $B: B=90°-36°20'\Rightarrow B=89°60'-36°20'\Rightarrow B=53°40'$

To find a: $\sin A = \dfrac{a}{c} \Rightarrow \sin 36°20' = \dfrac{a}{964} \Rightarrow a = 964(\sin 36°20') \approx 571.1526 \Rightarrow a \approx 571$ m.

To find b: $\cos A = \dfrac{b}{c} \Rightarrow \cos 36°20' = \dfrac{a}{964} \Rightarrow b = 964(\cos 36°20') \approx 776.5827 \Rightarrow b \approx 777$ m.

3. To find M: $M = 90° - 51.2° \Rightarrow M = 38.8°$.

To find n: $\tan N = \dfrac{n}{m} \Rightarrow \tan 51.2° = \dfrac{n}{124} = n = 124(\tan 51.2°) \approx 154.2249 \Rightarrow n \approx 154$ m.

To find p: $\cos N = \dfrac{n}{m} \Rightarrow \cos 51.2° = \dfrac{124}{p} \Rightarrow p = \dfrac{124}{\cos 51.2°} \approx 197.8922 \Rightarrow c \approx 198$ m.

5. To find A: $A = 90° - 42.0892° \Rightarrow A = 47.9108°$.

To find a: $\tan B = \dfrac{b}{a} \Rightarrow a = \dfrac{b}{\tan B} = \dfrac{56.851}{\tan 42.0892°} \approx 62.942095 \Rightarrow a \approx 62.942$ cm.

To find c: $\sin B = \dfrac{b}{c} \Rightarrow c = \dfrac{b}{\sin B} = \dfrac{56.851}{\sin 42.0892°} \approx 84.81594 \Rightarrow c \approx 84.816$ cm.

7. To find A: $\tan A = \dfrac{a}{b} \Rightarrow \tan A = \dfrac{7.1}{9.7} \Rightarrow A = \tan^{-1} \dfrac{7.1}{9.7} \Rightarrow A \approx 36.2026° \Rightarrow A \approx 36°$. To find B:

$B = 90° - 36° \Rightarrow B = 54°$. To find c: $\sin A = \dfrac{a}{c} \Rightarrow \sin 36° = \dfrac{7.1}{c} \Rightarrow c = \dfrac{7.1}{\sin 36°} \approx 12.07924 \Rightarrow c \approx 12$ ft.

9. To find A: $\cos A = \dfrac{b}{c} \Rightarrow \cos A = \dfrac{7.3}{11} \Rightarrow A = \cos^{-1} \dfrac{7.3}{11} \Rightarrow A \approx 48.42220° \Rightarrow A \approx 48°$.

To find B: $B = 90° - 48° \Rightarrow B = 42°$.

To find a: $\sin A = \dfrac{a}{c} \Rightarrow \sin 48° = \dfrac{a}{11} \Rightarrow a = 11(\sin 48°) \approx 8.174593 \Rightarrow a \approx 8.2$ ft.

11. To find B: $B = 90.00° - 28.00° \Rightarrow B = 62.00°$.

To find a: $\sin A = \dfrac{a}{c} \Rightarrow \sin 28.00° = \dfrac{a}{17.4} \Rightarrow a = 17.4(\sin 28.00°) \approx 8.1688 \Rightarrow a \approx 8.17$ ft.

To find b: $\cos A = \dfrac{b}{c} \Rightarrow \cos 28.00° = \dfrac{b}{17.4} \Rightarrow b = 17.4(\cos 28.00°) \approx 15.3633 \Rightarrow b \approx 15.4$ ft.

13. To find A: $A = 90.00° - 73.00° \Rightarrow A = 17.00°$.

To find a: $\tan B = \dfrac{b}{a} \Rightarrow \tan 73.00° = \dfrac{128}{a} \Rightarrow a = \dfrac{128}{\tan 73.00°} \approx 39.1335 \Rightarrow a \approx 39.1$ in.

To find c: $\sin B = \dfrac{b}{c} \Rightarrow \sin 73.00° = \dfrac{128}{c} \Rightarrow c = \dfrac{128}{\sin 73.00°} \approx 133.849 \Rightarrow c \approx 134$ in.

15. To find c, use Pythagorean Theorem:

$c^2 = a^2 + b^2 \Rightarrow c^2 = (76.4)^2 + (39.3)^2 \Rightarrow c^2 = 7381.45 \Rightarrow c \approx 85.9154 \Rightarrow c \approx 85.9$ yd.

To find A: $\tan A = \dfrac{a}{b} \Rightarrow \tan A = \dfrac{76.4}{39.3} \Rightarrow A = \tan^{-1}\left(\dfrac{76.4}{39.3}\right) \approx 62.7788° \Rightarrow A \approx 62°46'44'' \Rightarrow A \approx 62°50'$.

To find B: $B = 90°00 - 62°50' \Rightarrow B = 89°60' - 62°50' \Rightarrow B = 27°10'$.

17. The other acute angle requires the least work to find, simply subtract the given angle from $90°$.

19. Because x and y are known, use tangent to find: $\tan\theta = \frac{y}{x} \Rightarrow \tan\theta = \frac{3.68}{4.6} \Rightarrow \theta = \tan^{-1}\left(\frac{3.68}{4.6}\right) \approx 38.6598°$.

21. Because AD and BC are parallel, angle DAB is congruent to angle ABC, as they are alternate interior angles of the transversal AB. (A theorem of elementary geometry assures us of this.)

23. It is measured clockwise from the north.

25. $\sin 43°50' = \frac{d}{13.5} \Rightarrow d = 13.5(\sin 43°50') \Rightarrow d \approx 9.34959996 \Rightarrow d \approx 9.35$ m.

27. $\tan 23.4° = \frac{5.75}{x} \Rightarrow x = \frac{5.75}{\tan 23.4°} \approx 13.2875 \Rightarrow x \approx 13.3$ ft.

29. First find the distance between the buildings, using the angle of depression and $h_1 = 30.0$ feet, since a point on the building horizontally across the street is, like the window, 30 feet from the ground. Find d:

$\tan 20.0 = \frac{30.0}{d} \Rightarrow d = \frac{30.0}{\tan 20.0°} \approx 82.424$. Now use the found distance and the angle elevation to find h_2 (the height from the horizontal point on the building to the top of the building):

$\tan 50° = \frac{h_2}{d} \Rightarrow \tan 50° = \frac{h_2}{82.424} \Rightarrow h_2 = 82.424(\tan 50°) \approx 98.229$. Finally, the height of the building is:

$h_1 + h_2 = 30 + 98.229 = 128.229 \approx 128$ feet.

31. First find the complementary angle to the angle of depression:

$\tan C = \frac{12.2}{5.93} \Rightarrow C = \tan^{-1}\left(\frac{12.02}{5.93}\right) \Rightarrow C \approx 63.74°$.

Therefore, the angle of depression is: $90° - 63.74° = 26.26° \approx 26.3°$or $26°20'$.

33. To find A: $\tan A = \frac{a}{b} \Rightarrow \tan A = \frac{1.0837}{1.4923} \Rightarrow A = \tan^{-1}\frac{1.0837}{1.4923} \Rightarrow A \approx 35.9869° \Rightarrow A \approx 35.987°$ or $35°59'10''$.To find B: $B = 90.0° - 35.987° \Rightarrow B = 54.013°$ or $54°00'50''$.

35. Let be the base of the smaller right triangle, then: $\tan 21°10' = \frac{x}{135+y} \Rightarrow x = (135+y)\tan 21°10'$ and

$\tan 35°30' = \frac{x}{y} \Rightarrow x = y(\tan 35°30')$. Now $(135+y)\tan 21°10' = y(\tan 35°30') \Rightarrow$

$135(\tan 21°10') = y(\tan 35°30) - y(\tan 21°10') \Rightarrow 135(\tan 21°10') = y(\tan 35°30' - \tan 21°10') \Rightarrow$

$y = \frac{135(\tan 21°10')}{\tan 35°30' - \tan 21°10'} \Rightarrow y \approx 160.30258$. Use the small triangle to solve for x: $x = y(\tan 35°30') \Rightarrow$

$x = 160.30258(\tan 35°30') \Rightarrow x \approx 114.343$. The height of the pyramid is approximately: 114 feet.

37. Let h= the height of the house and x = the height of the antenna, then using the angles of elevation yields the following equations: $\tan 18°10' = \frac{h}{28.0} \Rightarrow h = 28.0\ (\tan\ 18°10')$. and

$\tan 27°10' = \frac{h+x}{28.0} = h + x = 28.0(\tan 27°10')$. Using these equations we can use substitution for *h* and solve for *x*: $28.0(\tan 18°10') + x = 28.0(\tan 27°10') \Rightarrow x = 28.0(\tan 27°10') - 28.0(\tan 18°10') \Rightarrow x \approx 5.18157$.

The height of the antenna is approximately 5.18 meters.

39. First solve for the hypotenuse of the top triangle: $\cos 30°50' = \frac{198.4}{c} \Rightarrow c = \frac{198.4}{\cos 30°50'} \Rightarrow c \approx 231.05719$.

Now use this value and the smaller angle of the bottom triangle $\cot 30° = \frac{3}{\sqrt{3}} = \frac{3\sqrt{3}}{3} = \sqrt{3}$ solve for *x*:

$\sin 21°30' = \frac{x}{c} \Rightarrow \sin 21°30' = \frac{x}{231.05719} \Rightarrow x = 231.05719(\sin 21°30') \Rightarrow x \approx 84.7$ m.

41. The angle between the two ships is $90°[180 - (28°10' + 61°50')]$; the first ship sails 96 miles (4×24) and the second ship sails 112 miles (4×28). Using Pythagorean Theorem we can find *c*, the distance between them: $c^2 = 96^2 + 112^2 \Rightarrow c^2 = 21,760 \Rightarrow c \approx 17.51$. The ships are approximately 148 miles apart.

43. From exercise 42, use angle $A = 53°40'$ and solve for *d*, the distance from point *A* to the transmitter: $\sin A = \cos A.$. The distance is approximately 1.48 miles.

45. The angle at the top of the triangle is $90°((27° + (180° - 117°))$, therefore a right triangle is formed. Now use Pythagorean Theorem to solve for *x*, the distance between starting and ending points: $c^2 = 50^2 = 140^2 \Rightarrow c^2 = 22,100 \Rightarrow c \approx 148.66$. The distance is approximately 150 kilometers.

47. (a) A right triangle is formed, since the two acute angles at the bottom of the triangle add up to $90°: [A(180° - 129°25') + B(39°25') = 90°]$. Now using angle $B = 39°25'$, of the triangle, solve for distance AC, $\sin 39°25' = \frac{x}{15} \Rightarrow x = 15(\sin 39°35') \Rightarrow x \approx 9.524$ miles

(b) Use Pythagorean Theorem to find distance BC. $9.524^2 + b^2 = 15^2 \Rightarrow b = \sqrt{15^2 - 9.524^2} \Rightarrow b \approx 11.59$

49. Let the length of RS be *x*: $\tan 32°10' = \frac{x}{53.1} \Rightarrow x = 53.1(\tan 32°10') \Rightarrow x \approx 33.3957 \approx 33.4$ m.

51. First solve for *h*, the height of the searchlight beam: $\tan 30° = \frac{h}{1000} \Rightarrow h = 1000(\tan 30°) \Rightarrow h \approx 577.35$. The cloud ceiling is searchlight beam height plus observer height, therefore: $6 + 577.35 = 583.35 \approx 583$ feet.

53. (a) First solve for β and then for *d*, using the given formula and information:

$\beta \approx \frac{57.3S}{R} \Rightarrow \beta \approx \frac{57.3(336)}{600} \Rightarrow \beta \approx 32.088°$. Therefore: $d = 600\left(1 - \cos\left(\frac{32.088}{2}\right)°\right) \Rightarrow$

$d \approx 23.3702 \approx 23.4$ feet.

(b) First solve for β and then for d, using the given formula and information:

$$\beta \approx \frac{57.3S}{R} \Rightarrow \beta \approx \frac{57.3(485)}{600} \Rightarrow \beta \approx 46.3175^\circ.$$

$$d = 600\left(1-\cos\left(\frac{46.3175}{2}\right)^\circ\right) \Rightarrow d \approx 48.34877 \approx 48.3 \text{ ft.}$$

(c) The faster the speed, the more land that needs to be cleared on the inside of the curve.

55. (a) $\tan\theta = \frac{y}{x}$

(b) $\tan\theta = \frac{y}{x} \Rightarrow x(\tan\theta) = y \Rightarrow x = \frac{y}{\tan\theta}$

57. Find $a: \cos 60^\circ = \frac{a}{24} \Rightarrow a = 24(\cos 60^\circ) = 24\left(\frac{1}{2}\right) = 12$

Find $b: \sin 60^\circ = \frac{b}{24} \Rightarrow b = 24(\sin 60^\circ) = 24\left(\frac{\sqrt{3}}{2}\right) = 12\sqrt{3}$

Using $b = 12\sqrt{3}$, find $c: \sin 45^\circ = \frac{12\sqrt{3}}{c} \Rightarrow c = \frac{12\sqrt{3}}{\cos 45^\circ} = \frac{12\sqrt{3}}{\frac{\sqrt{2}}{2}} = 12\sqrt{3}\cdot\frac{2}{\sqrt{2}}\cdot\frac{\sqrt{2}}{\sqrt{2}} = 12\sqrt{6}$

Using $c = 12\sqrt{6}$, find $d: \sin 45^\circ = \frac{d}{12\sqrt{6}} \Rightarrow d = 12\sqrt{6}(\sin 45^\circ) = 12\sqrt{6}\left(\frac{\sqrt{2}}{2}\right) = 6\sqrt{12} = 6\left(2\sqrt{3}\right) = 12\sqrt{3}$.

59. Find $a: \sin 60^\circ = \frac{7}{a} \Rightarrow a = \frac{7}{\sin 60^\circ} = \frac{7}{\frac{\sqrt{3}}{2}} = 7\cdot\frac{2}{\sqrt{3}}\cdot\frac{\sqrt{3}}{\sqrt{3}} = \frac{14\sqrt{3}}{3}$ Since the triangle is a $45^\circ - 45^\circ$ right triangle, $a = n$ and $n = \frac{14\sqrt{3}}{3}$

Using $a = \frac{14\sqrt{3}}{3}$, find $m: \cos 60^\circ = \frac{m}{\frac{14\sqrt{3}}{3}} \Rightarrow m = \frac{14\sqrt{3}}{3}(\cos 60^\circ) = \frac{14\sqrt{3}}{3}\cdot\frac{1}{2} = \frac{7\sqrt{3}}{3}$

Using $n = \frac{14\sqrt{3}}{3}$, find $q: \cos 45^\circ = \frac{\frac{14\sqrt{3}}{3}}{q} \Rightarrow q = \frac{\frac{14\sqrt{3}}{3}}{\cos 45^\circ} = \frac{\frac{14\sqrt{3}}{3}}{\frac{\sqrt{2}}{2}} = \frac{14\sqrt{3}}{3}\cdot\frac{2}{\sqrt{2}}\cdot\frac{\sqrt{2}}{\sqrt{2}} = \frac{28\sqrt{6}}{6} = \frac{14\sqrt{6}}{3}$

61. First, bisect the upper angle, which will also be the height (h) of the original triangle. Now solve for h using the smaller right triangle: $\sin 60^\circ = \frac{h}{s} \Rightarrow h = s(\sin 60^\circ) \Rightarrow h = \frac{s\sqrt{3}}{2}$. Finally find the area of the original triangle: $A = \frac{1}{2}bh = \frac{1}{2}\cdot s\cdot\frac{s\sqrt{3}}{2} \Rightarrow A = \frac{s^2\sqrt{3}}{4}$..

Reviewing Basic Concepts (Sections 8.3 - 8.4)

1. $\sin A = \frac{\text{opp}}{\text{hyp}} = \frac{18}{17}$ $\cos A = \frac{\text{adj}}{\text{hyp}} = \frac{15}{17}$ $\tan A = \frac{\text{opp}}{\text{adj}} = \frac{8}{15}$

   $\csc A = \frac{\text{hpy}}{\text{opp}} = \frac{17}{8}$ $\sec A = \frac{\text{hyp}}{\text{adj}} = \frac{17}{15}$ $\cot A = \frac{\text{adj}}{\text{opp}} = \frac{15}{8}$

2. By the $30°\text{-}60°$ right triangle,

   $\sin 30° = \frac{1}{2}$ $\cos 30° = \frac{\sqrt{3}}{2}$ $\tan 30° = \frac{\sqrt{3}}{3}$

   $\csc 30° = \frac{2}{1} = 2$ $\sec 30° = \frac{2}{\sqrt{3}} = \frac{2\sqrt{3}}{3}$ $\cot 30° = \frac{3}{\sqrt{3}} = \frac{3\sqrt{3}}{3} = \sqrt{3}$

   By the $45°\text{-}45°$ right triangle,

   $\sin 45° = \frac{\sqrt{2}}{2}$ $\cos 45° = \frac{\sqrt{2}}{2}$ $\tan 45° = \frac{\sqrt{2}}{\sqrt{2}} = 1$

   $\csc 45° = \frac{2}{\sqrt{2}} = \frac{2\sqrt{2}}{2} = \sqrt{2}$ $\sec 45° = \frac{2}{\sqrt{2}} = \frac{2\sqrt{2}}{\sqrt{2}} = \sqrt{2}$ $\cot 45° = \frac{\sqrt{2}}{\sqrt{2}} = 1$

   By the $30°\text{-}60°$ right triangle,

   $\sin 60° = \frac{\sqrt{3}}{2}$ $\cos 60° = \frac{1}{2}$ $\tan 60° = \frac{\sqrt{3}}{1} = \sqrt{3}$

   $\csc 30° = \frac{2}{\sqrt{3}} = \frac{2\sqrt{3}}{3}$ $\sec 60° = \frac{2}{1} = 2$ $\cot 60° = \frac{1}{\sqrt{3}} = \frac{\sqrt{3}}{3}$

3. (a) $\sin 27° = \cos(90° - 27°) = \cos 63°$

   (b) $\tan \frac{\pi}{5} = \cot\left(\frac{\pi}{2} - \frac{\pi}{5}\right) = \cot\left(\frac{5\pi}{10} - \frac{2\pi}{10}\right) = \cot \frac{3\pi}{10}$

4. (a) If $\theta = 100°$ then $\theta' = 180° - 100° = 80°$

   (b) If $\theta = -365° = 720° - 365° = 355°$ then $\theta' = 360° - 355° = 5°$

   (c) If $\theta = \frac{8\pi}{3} - 2\pi = \frac{8\pi}{3} - \frac{6\pi}{3} = \frac{2\pi}{3}$ then $\theta° = \pi - \frac{2\pi}{3} = \frac{3\pi}{3} - \frac{2\pi}{3} = \frac{\pi}{3}$

5. First find the reference angle: $(360° - 315°) = 45°$. Using the $45° - 45°$ right triangle rules, for a $45°$ angle the hypotenuse $= \sqrt{2}$, the opposite side $= 1$, and the adjacent $= 1$. Since $\theta = 315°$ is in quadrant IV, y is negative and sine, tangent, cotangent, and cosecant are negative, therefore:

   $\sin 315° = -\sin 45° = -\frac{1}{\sqrt{2}} = -\frac{\sqrt{2}}{2}$ $\csc 315° = -\csc 45° = -\frac{\sqrt{2}}{1} = -\sqrt{2}$

   $\cos 315° = \cos 45° = \frac{1}{\sqrt{2}} = \frac{\sqrt{2}}{2}$ $\sec 315° = \sec 45° = \frac{\sqrt{2}}{1} = \sqrt{2}$

$\tan 315° = -\tan 45° = -\frac{1}{1} = -1$ $\cot 315° = -\cot 45° = -\frac{1}{1} = -1$

6. (a) Using the calculator, $\sin 46°30' \approx .725374371$

   (b) Using the calculator, $\tan(-100°) \approx 5.67128182$

   (c) Using the calculator, $\csc 4 = (\sin 4)^{-1} \approx -1.321348709$

7. Since $\tan\theta$ is negative in quadrant II and IV, values for θ can only be in these quadrants. From the 30°- 60° right triangle, $\tan 30° = \frac{\sqrt{3}}{3}$, therefore 30° is the reference angle. In quadrant II, use the reference angle 30° to find the first value: $180° - 30° = 150°$. In quadrant IV, use the reference angle to find the other value: $360° - 30° = 330°$. Therefore, $\theta = 150°$ or $330°$

8. Since $(2,-5)$, is positive in quadrant I and II, values for θ can only be in these quadrants. In quadrant I, we use the inverse trigonometric function (radian mode) of our calculator to find the angle: $\sin^{-1} .68163876 \Rightarrow \theta = .75$. In quadrant II, use the found reference angle from quadrant I to find $\theta: \pi - .75 \approx 2.391592654$. Therefore, $\theta \approx .75$ or 2.391592654.

9. Solve for h, height using the angle of elevation and given shadow length:

   $\tan 35.3° = \frac{h}{48} \Rightarrow h = 48\tan 35.3° \Rightarrow h \approx 33.986 \approx 34$ feet.

10. Let x be the base of the smaller right triangle and h the height of Mt. Kilimanjaro, then:

   $\tan 13.7° = \frac{h}{x} \Rightarrow x = \frac{h}{\tan 13.7°}$ and $\tan 10.4° = \frac{h}{x+5}$. Now substitute the second equation into the first equation: $\tan 10.4° = \frac{h}{\frac{h}{\tan 13.7°}+5}$, now multiply by $\frac{\tan 13.7}{\tan 13.7}$ and solve for h:

   $$\tan 10.4° = \frac{h}{\frac{h}{\tan 13.7°}+5} \cdot \frac{\tan 13.7°}{\tan 13.7°} \Rightarrow \tan 10.4° = \frac{h(\tan 13.7°)}{h+5(\tan 13.7°)} \Rightarrow$$

   $$\tan 10.4°(h+5(\tan 13.7°)) = h(\tan 13.7°) \Rightarrow h(\tan 10.4°) + 5(\tan 13.7°)(\tan 10.4°) = h(\tan 13.7°) \Rightarrow$$

   $$h(\tan 10.4°) - h(\tan 13.7°) = -5(\tan 13.7°)(\tan 10.4°) \Rightarrow$$

   $$h(\tan 10.4° - \tan 13.7°) = -5(\tan 13.7°)(\tan 10.4°) \Rightarrow h = \frac{-5(\tan 13.7°)(\tan 10.4°)}{\tan 10.4° - \tan 13.7°} \Rightarrow h \approx 3.713588 \text{ miles}$$

   Finally convert to feet: $h \approx \frac{3.713588 \text{ mi}}{1} \cdot \frac{5280 \text{ ft}}{1 \text{ mi}} \approx 19{,}607.7$

   The mountain is approximately 19,600 feet high.

8.5: The Circular Functions

1. From the properties of $30°-60°$ right triangles, an angle of $s=\frac{7\pi}{6}$ radians intersects the unit circle at the point $\left(-\frac{\sqrt{3}}{2},-\frac{1}{2}\right)$. Since $\sin s=y, \sin\frac{7\pi}{6}=-\frac{1}{2}$.

3. From the properties of $45°-45°$ right triangles, an angle of $s=\frac{3\pi}{4}$ radians intersects the unit circle at the point $\left(-\frac{\sqrt{2}}{2},\frac{\sqrt{2}}{2}\right)$. Since $\tan s=\frac{y}{x}, \tan\frac{3\pi}{4}=\frac{\sqrt{2}}{2}\div\left(-\frac{\sqrt{2}}{2}\right)=-1$.

5. From the properties of $30°-60°$ right triangles, an angle of $s=\frac{2\pi}{3}$ radians intersects the unit circle at the point $\left(-\frac{1}{2},\frac{\sqrt{3}}{2}\right)$. Since $\sec s=\frac{1}{x}, \sec\frac{2\pi}{3}=\frac{1}{-\frac{1}{2}}=-2$.

7. From the properties of $30°-60°$ right triangles, an angle of $s=\frac{5\pi}{6}$ radians intersects the unit circle at the point $\left(-\frac{\sqrt{3}}{2},\frac{1}{2}\right)$. Since $\cot s=\frac{x}{y}, \cot\frac{5\pi}{6}=\frac{-\frac{\sqrt{3}}{2}}{\frac{1}{2}}=-\frac{\sqrt{3}}{2}\cdot\frac{2}{1}=-\sqrt{3}$.

9. First find a coterminal angle: $\theta'=2\pi-\frac{5\pi}{6}\Rightarrow\theta'=\frac{7\pi}{6}$. From the properties of $30°-60°$ right triangles, an angle of $s=\frac{7\pi}{6}$ radians intersects the unit circle at the point $\left(-\frac{\sqrt{3}}{2},-\frac{1}{2}\right)$. Since $\sin s=y, \sin\frac{7\pi}{6}=-\frac{1}{2}$.

11. First find a coterminal angle: $\theta'=\frac{23\pi}{6}\pi-2\pi\ \Rightarrow\ \theta'=\frac{11\pi}{6}$. From the properties of $30°-60°$ right triangles, $s=\frac{11\pi}{6}$ radians intersects the unit circle at $\left(\frac{\sqrt{3}}{2},-\frac{1}{2}\right)$. Since $\sec s=\frac{1}{x}, \sec\frac{11\pi}{6}=\frac{1}{\frac{\sqrt{3}}{2}}=\frac{2\sqrt{3}}{3}$.

13. An angle of $s=\frac{\pi}{2}$ radians intersects the unit circle at the point $(0,1)$. Therefore,

$\sin s=y, \sin\frac{\pi}{2}=1$ $\quad\cos s=x, \cos\frac{\pi}{2}=0$ $\quad\tan s=\frac{y}{x}, \tan\frac{\pi}{2}=\frac{1}{0}=\text{undefined}$ $\quad\csc=\frac{1}{y}, \csc\frac{\pi}{2}=\frac{1}{1}=1$

$\sec s=\frac{1}{x}, \sec\frac{\pi}{2}=\frac{1}{0}=\text{undefined}$ $\quad\cot s=\frac{x}{y}, \cot\frac{\pi}{2}=\frac{0}{1}=0$

15. From the properties of $45°-45°$ right triangles, an angle of $s=-\frac{\pi}{4}$ radians intersects the unit circle at the point $\left(\frac{\sqrt{2}}{2},-\frac{\sqrt{2}}{2}\right)$. Therefore, $\sin s=y, \sin\frac{\pi}{4}=-\frac{\sqrt{2}}{2}$ $\quad\csc=\frac{1}{y}, \csc\frac{\pi}{4}=\frac{1}{-\frac{\sqrt{2}}{2}}=-\frac{2}{\sqrt{2}}=-\frac{2\sqrt{2}}{2}=-\sqrt{2}$

$\cos s=x, \cos-\frac{\pi}{4}=\frac{\sqrt{2}}{2}$ $\quad\sec s=\frac{1}{x}, \sec-\frac{\pi}{4}=\frac{1}{\frac{\sqrt{2}}{2}}=\frac{2}{\sqrt{2}}=\frac{2\sqrt{2}}{2}=\sqrt{2}$

$\tan s=\frac{y}{x}, \tan-\frac{\pi}{4}=\frac{-\frac{\sqrt{2}}{2}}{\frac{\sqrt{2}}{2}}=-\frac{\sqrt{2}}{2}\cdot\frac{2}{\sqrt{2}}=-1$ $\quad\cot s=\frac{x}{y}, \cot-\frac{\pi}{4}=\frac{\frac{\sqrt{2}}{2}}{-\frac{\sqrt{2}}{2}}=\frac{2}{\sqrt{2}}\cdot\left(-\frac{2}{\sqrt{2}}\right)=-1$

17. An angle of $s=\pi$ radians intersects the unit circle at the point $(-1,0)$. Therefore, $\sin s=y, \sin\pi=0$

$\cos s = x, \cos\pi = -1$ $\tan s = \frac{y}{x}, \tan\pi = \frac{0}{-1} = 0$

$\csc = \frac{1}{y}, \csc\pi = \frac{1}{0} = \text{undefined}$ $\sec s = \frac{1}{x}, \sec\pi = \frac{1}{-1} = -1$ $\cot s = \frac{x}{y}, \cot\pi = \frac{-1}{0} = \text{undefined}$

19. From the properties of $30°-60°$ right triangles, an angle of $s = -\frac{\pi}{3}$ radians intersects the unit circle at the point $\left(\frac{1}{2}, -\frac{\sqrt{3}}{2}\right)$. Therefore, $\sin s = y, \sin -\frac{\pi}{3} = -\frac{\sqrt{3}}{2}$ $\csc s = \frac{1}{y}, \csc -\frac{\pi}{3} = \frac{1}{-\frac{\sqrt{3}}{2}} = -\frac{2}{\sqrt{3}} = -\frac{2\sqrt{3}}{2}$

$\cos s = x, \cos -\frac{\pi}{3} = \frac{1}{2}$ $\sec s = \frac{1}{x}, \sec -\frac{\pi}{3} = \frac{1}{\frac{1}{2}} = 2$ $\tan s = \frac{y}{x}, \tan -\frac{\pi}{3} = \frac{-\frac{\sqrt{3}}{2}}{\frac{1}{2}} = -\frac{\sqrt{3}}{2} \cdot \left(\frac{2}{1}\right) = -\sqrt{3}$

$\cot s = \frac{x}{y}, \cot -\frac{\pi}{3} = \frac{\frac{1}{2}}{-\frac{\sqrt{3}}{2}} = \frac{1}{2} \cdot \left(-\frac{2}{\sqrt{3}}\right) = -\frac{1}{\sqrt{3}} = -\frac{\sqrt{3}}{3}$

21. An angle of $s = \frac{\pi}{2}$ radians intersects the unit circle at the point $(0, -1)$. Therefore,

$\sin s = y, \sin -\frac{\pi}{2} = -1$ $\cos s = x, \cos -\frac{\pi}{2} = 0$ $\tan s = \frac{y}{x}, \tan -\frac{\pi}{2} = \frac{-1}{0} = \text{undefined}$

$\csc s = \frac{1}{y}, \csc -\frac{\pi}{2} = \frac{1}{-1} = -1$ $\sec s = \frac{1}{x}, \sec -\frac{\pi}{2} = \frac{1}{0} = \text{undefined}$ $\cot s = \frac{x}{y}, \cot -\frac{\pi}{2} = \frac{0}{-1} = 0$

23. An angle of $s = 2\pi$ radians intersects the unit circle at the point $(1, 0)$. Therefore,

$\sin s = y, \sin 2\pi = 0$ $\cos s = x, \cos 2\pi = 1$ $\tan s = \frac{y}{x}, \tan 2\pi = \frac{0}{1} = 0$

$\csc s = \frac{1}{y}, \csc 2\pi = \frac{1}{0} = \text{undefined}$ $\sec s = \frac{1}{x}, \sec 2\pi = \frac{1}{1} = 1$ $\cot s = \frac{x}{y}, \cot 2\pi = \frac{1}{0} = \text{undefined}$

25. From the properties of $30°-60°$ right triangles, an angle of $s = \frac{\pi}{6}$ radians intersects the unit circle at the point $\left(\frac{\sqrt{3}}{2}, \frac{1}{2}\right)$. Therefore, $\sin s = y, \sin \frac{\pi}{6} = \frac{1}{2}$ $\csc s = \frac{1}{y}, \csc \frac{\pi}{6} = \frac{1}{\frac{1}{2}} = 2$

$\cos s = x, \cos \frac{\pi}{6} = \frac{\sqrt{3}}{2}$ $\sec s = \frac{1}{x}, \sec \frac{\pi}{6} = \frac{1}{\frac{\sqrt{3}}{2}} = \frac{2}{\sqrt{3}} = \frac{2\sqrt{3}}{3}$ $\tan s = \frac{y}{x}, \tan \frac{\pi}{6} = \frac{\frac{1}{2}}{\frac{\sqrt{3}}{2}} = \frac{1}{2} \cdot \frac{2}{\sqrt{3}} = \frac{1}{\sqrt{3}} = \frac{\sqrt{3}}{3}$

$\cot s = \frac{x}{y}, \cot \frac{\pi}{6} = \frac{\frac{\sqrt{3}}{2}}{\frac{1}{2}} = \frac{\sqrt{3}}{2} \cdot \frac{2}{1} = \sqrt{3}$

27. From the properties of $45°-45°$ right triangles, an angle of $s = -\frac{7\pi}{4}$ radians intersects the unit circle at the point $\left(\frac{\sqrt{2}}{2}, \frac{\sqrt{2}}{2}\right)$. Therefore, $\sin s = y, \sin \frac{7\pi}{4} = \frac{\sqrt{2}}{2}$ $\csc = \frac{1}{y}, \csc \frac{5\pi}{4} = \frac{1}{\frac{\sqrt{2}}{2}} = \frac{2}{\sqrt{2}} = \frac{2\sqrt{2}}{2} = \sqrt{2}$

$\cos s = x, \cos \frac{7\pi}{4} = \frac{\sqrt{2}}{2}$ $\sec s = \frac{1}{x}, \sec \frac{7\pi}{4} = \frac{1}{\frac{\sqrt{2}}{2}} = \frac{2}{\sqrt{2}} = -\frac{2\sqrt{2}}{2} = \sqrt{2}$

$\tan s = \frac{y}{x}, \tan \frac{7\pi}{4} = \frac{\frac{\sqrt{2}}{2}}{\frac{\sqrt{2}}{2}} = 1$ $\cot s = \frac{x}{y}, \cot \frac{7\pi}{4} = \frac{\frac{\sqrt{2}}{2}}{\frac{\sqrt{2}}{2}} = 1$

29. Using the calculator (radian mode), $\sin .6109 \approx .5736049112 \approx 0.5736$

31. Using the calculator (radian mode), $\cos(-1.1519) \approx .4067524531 \approx 0.4068$

33. Using the calculator (radian mode), $\tan 4.0203 \approx 1.206484913 \approx 1.2065$

35. Using the calculator (radian mode), $\csc(-9.4946) = [\sin(-9.4946)]^{-1} \approx 14.33376901 \approx 14.3338$

37. Using the calculator (radian mode), $\sec 2.8440 = [\cos 2.8440]^{-1} \approx -1.045975716 \approx -1.0460$

39. Using the calculator (radian mode), $\cot 6.0301 = [\tan 6.0301]^{-1} \approx -3.866512664 \approx -3.8665$

41. If $s = \theta$ intersects the unit circle at $\left(\frac{\sqrt{2}}{2}, \frac{\sqrt{2}}{2}\right)$ then:

$\sin s = y, \sin\theta = \frac{\sqrt{2}}{2}$ $\quad$ $\csc s = \frac{1}{y}, \csc\theta = \frac{1}{\frac{\sqrt{2}}{2}} = \frac{2}{\sqrt{2}} = \sqrt{2}$

$\cos s = x, \cos\theta = \frac{\sqrt{2}}{2}$ $\quad$ $\sec s = \frac{1}{x}, \sec\theta = \frac{1}{\frac{\sqrt{2}}{2}} = \frac{2}{\sqrt{2}} = \frac{2\sqrt{2}}{2} = \sqrt{2}$

$\tan s = \frac{y}{x}, \tan\theta = \frac{\frac{\sqrt{2}}{2}}{\frac{\sqrt{2}}{2}} = 1$ $\quad$ $\cot s = \frac{x}{y}, \cot\theta = \frac{\frac{\sqrt{2}}{2}}{\frac{\sqrt{2}}{2}} = 1$

43. If $s = \theta$ intersects the unit circle at $\left(\frac{5}{13}, -\frac{12}{13}\right)$ then:

$\sin s = y, \sin\theta = -\frac{12}{13}$ $\quad$ $\cos s = x, \cos\theta = \frac{5}{13}$ $\quad$ $\tan s = \frac{y}{x}, \tan\theta = \frac{-\frac{12}{13}}{\frac{5}{13}} = -\frac{12}{13}\cdot\frac{13}{5} = -\frac{12}{5}$

$\csc s = \frac{1}{y}, \csc\theta = \frac{1}{-\frac{12}{13}} = -\frac{13}{12}$ $\quad$ $\sec s = \frac{1}{x}, \sec\theta = \frac{1}{\frac{5}{13}} = \frac{13}{5}$ $\quad$ $\cot s = \frac{x}{y}, \cot\theta = \frac{\frac{5}{15}}{-\frac{12}{13}} = \frac{5}{13}\cdot\left(-\frac{13}{12}\right) = -\frac{5}{12}$

45. Since on a unit circle $\cos\theta = x$ from the figure, $\cos 0.8 = 0.7$

47. Since on a unit circle $\sin\theta = y$ from the figure, $\sin 2 = 0.9$

49. Since on a unit circle $\sin\theta = y$ from the figure, $\sin 3.8 = -0.6$

51. If on a unit circle $\cos\theta = x$ then $x = -.65$. Therefore, from the figure, when $x = -.65, \theta \approx 2.3$ or 4

53. If on a unit circle $\sin\theta = y$ then $y = 0.7$. Therefore, from the figure, when $y = 0.7, \theta \approx 0.8$

55. Using the calculator (radian mode), if $\tan s = .21264138$, then $s = \tan^{-1} .21264138 \approx .2095$

57. Using the calculator (radian mode), if $\sin s = .99184065$, then $s = \sin^{-1} .99184065 \approx 1.4430$

59. Using the calculator (radian mode), if $\cot s = .09637041$, then $\tan s = .09637041^{-1}$ and

$s = \tan^{-1}(.09637041^{-1}) \approx 1.4747$

61. First find θ . Since this is a unit circle, $\sin\theta = y$, Therefore, $\sin\theta = .83305413$ and

$\theta = \sin^{-1} .83305413 \Rightarrow \theta \approx .98460588$. Now with $r = 1$ use the arc length formula

$s = r\theta \Rightarrow s \approx (1)(.98460588) \Rightarrow s \approx .9846$

63. If $\cos s = x$ then $x = \cos 2.5 \Rightarrow x = -.80144$; if $\sin s = y$ then $y = \sin 2.5 \Rightarrow y = .59847$.

Therefore, $(x, y) \approx (-0.8011, 0.5985)$.

65. First find positive s, $s = 4\pi - 7.4 \Rightarrow s \approx 5.16637$ If $\cos s = x$; then $x = \cos 5.16637 \Rightarrow x = .438547$; if $\sin s = y$ then $y = \sin 5.16637 \Rightarrow y = -.898708$. Therefore, $(x, y) \approx (.4385, -.8987)$.

67. If $s = 51$, then $\cos 51 \approx .7422$ and $\sin 5.1 \approx .6702$. Since cosine is positive, x is positive, and since sine is positive y is positive, therefore an angle of s radians lies in quadrant I.

69. If $s = 65$, then $\cos 65 \approx -.52245$ and $\sin 65 \approx .82683$. Since cosine is negative, x is negative, and since sine is positive y is positive, therefore an angle of s radians lies in quadrant II.

71. For parts (a)-(f) refer to Figure 82 and $\theta = 60^\circ$.

(a) $\cos\theta = OQ = \cos 60^\circ = \dfrac{1}{2}$,

(b) $\sin\theta = PQ = \sin 60^\circ = \dfrac{\sqrt{3}}{2}$

(c) $\tan\theta = VR = \tan 60^\circ = \sqrt{3}$

(d) $\sec\theta = OV = \sec 60^\circ = 2$

(e) $\csc\theta = OU = \csc 60^\circ = \dfrac{2}{\sqrt{3}} = \dfrac{2\sqrt{3}}{3}$

(f) $\cot\theta = US = \cot 60^\circ = \dfrac{1}{\sqrt{3}} = \dfrac{\sqrt{3}}{3}$

73. $f(7) = 1.95\cos\left[\dfrac{\pi}{6}(7 - 6.6)\right] + 12.15 \Rightarrow 1.95\cos(.20943951) + 12.15 \Rightarrow$

$1.95(.9781475601) + 12.15 \Rightarrow f(7) \approx 14$ hours. On July 1, there are about 14 hours of daylight.

75. (a) For January, $x = 0$ therefore $t = 60 - 30\cos\dfrac{0\pi}{6} \Rightarrow t = 60 - 30(\cos 0) \Rightarrow t = 60 - 30(1) \Rightarrow t = 30^\circ$

(b) For April, $x = 3$ therefore $t = 60 - 30\cos\dfrac{3\pi}{6} \Rightarrow t = 60 - 30\left(\cos\dfrac{\pi}{2}\right) \Rightarrow t = 60 - 30(0) \Rightarrow t = 60^\circ$

(c) For May, $x = 4$ therefore $t = 60 - 30\cos\dfrac{4\pi}{6} \Rightarrow t = 60 - 30\left(\cos\dfrac{2\pi}{3}\right) \Rightarrow t = 60 - 30\left(-\dfrac{1}{2}\right) \Rightarrow t = 75^\circ$

(d) For June, $x = 5$ therefore

$$t = 60 - 30\left(\cos\frac{5\pi}{6}\right) \Rightarrow t = 60 - 30\left(-\frac{\sqrt{3}}{2}\right) \Rightarrow t = 60 + 15\sqrt{3} \Rightarrow t \approx 60 + 25.98 \Rightarrow t \approx 86^\circ$$

(e) For August, $\frac{2\pi}{\left(\frac{1}{2}\right)} = 4\pi$ therefore

$$t = 60 - 30\left(\cos\frac{7\pi}{6}\right) \Rightarrow t = 60 - 30\left(-\frac{\sqrt{3}}{2}\right) \Rightarrow t = 60 + 15\sqrt{3} \Rightarrow t \approx 60 + 25.98 \Rightarrow t \approx 86°$$

(f) For October, $x = 9$ therefore $t = 60 - 30\cos\frac{9\pi}{6} \Rightarrow t = 60 - 30\left(\cos\frac{3\pi}{2}\right) \Rightarrow t = 60 - 30(0) \Rightarrow t \approx 60°$

77. Using the given formula and $F(t) = 1$ which corresponds to a full moon, yields:

$1 = \frac{1}{2}(1 - \cos t) \Rightarrow 2 = 1 - \cos t \Rightarrow 1 = -\cos t \Rightarrow \cos t = -1$. Therefore, the full moon phase angles will be

$t = \pm\pi, \pm 3\pi, \pm 5\pi \cdots$ Now use $F(t) = 0$. Which corresponds to a new moon:

$0 = \frac{1}{2}(1 - \cos t) \Rightarrow 0 = 1 - \cos t \Rightarrow -1 = -\cos t \Rightarrow \cos t = 1$. Therefore, the new moon phase angles will

be $t = 0, \pm 2\pi, \pm 4\pi, \ldots.$

8.6: Graphs of the Sine and Cosine Functions

1. Since $\sin 0 = 0$, $\sin\frac{\pi}{2} = 1$, the amplitude is 1, and the period is $\frac{2\pi}{(1)} = 2\pi$, the graph is G.

3. Since $-\sin 0 = 0$, $-\sin\frac{\pi}{2} = -1$, the amplitude is 1, and the period is $\frac{2\pi}{(1)} = 2\pi$, the graph is E.

5. Since $\sin 2(0) = 0$, $\sin 2\left(\frac{\pi}{4}\right) = \sin\frac{\pi}{2} = 1$, the amplitude is 1, and the period is $\frac{2\pi}{(2)} = \pi$, the graph is B.

7. Since $2\sin 0 = 2(0) = 0$, $2\sin\frac{\pi}{2} = 2(1) = 2$, the amplitude is 2, and the period is $\frac{2\pi}{(1)} = 2\pi$, the graph is F.

9. The graph of $y = \sin\left(x - \frac{\pi}{4}\right)$ is the graph of $y = \sin x$ translated $\frac{\pi}{4}$ units to the right, therefore the graph is D.

11. The graph of $y = \cos\left(x - \frac{\pi}{4}\right)$ is the graph of $y = \cos x$ translated $\frac{\pi}{4}$ units to the right, therefore the graph is H.

13. The graph of $y = 1 + \sin x$ is the graph of $y = \sin x$ translated 1 unit upward, therefore the graph is B.

15. The graph of $y = 1 + \cos x$ is the graph of $y = \cos x$ translated 1 unit upward, therefore the graph is F.

17. The equation $y = 3\sin(2x - 4)$ has an amplitude 3; a period of $\frac{2\pi}{(2)} = \pi$; and a phase shift $\frac{4}{2} = 2$, therefore B.

19. The equation $y = 4\sin(3x - 2)$ has an amplitude 4; a period of $\frac{2\pi}{(3)}$; and a phase shift $\frac{2}{3}$ therefore C.

21. See Figure 21. From the equation, the amplitude is 2.

23. See Figure 23. From the equation, the amplitude is $\frac{2}{3}$.

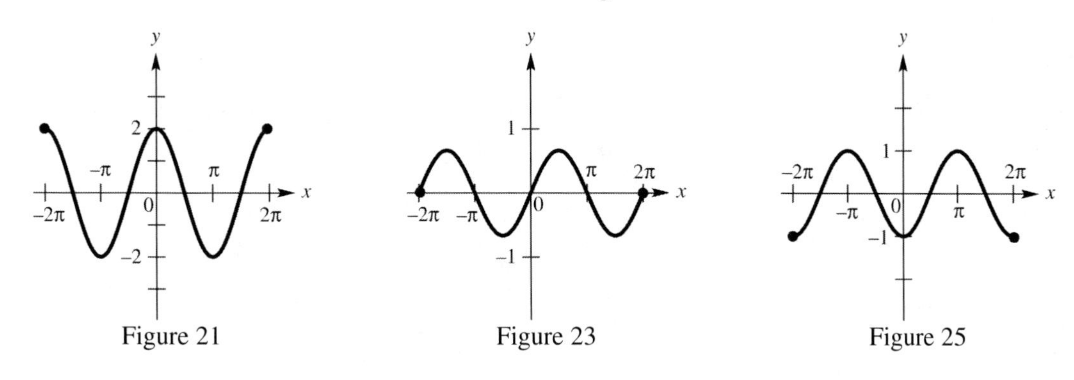

Figure 21 Figure 23 Figure 25

25. See Figure 25. From the equation, the amplitude is 1.
27. See Figure 27. From the equation, the amplitude is 2.
29. See Figure 29. From the equation, the period is and the amplitude is 1.

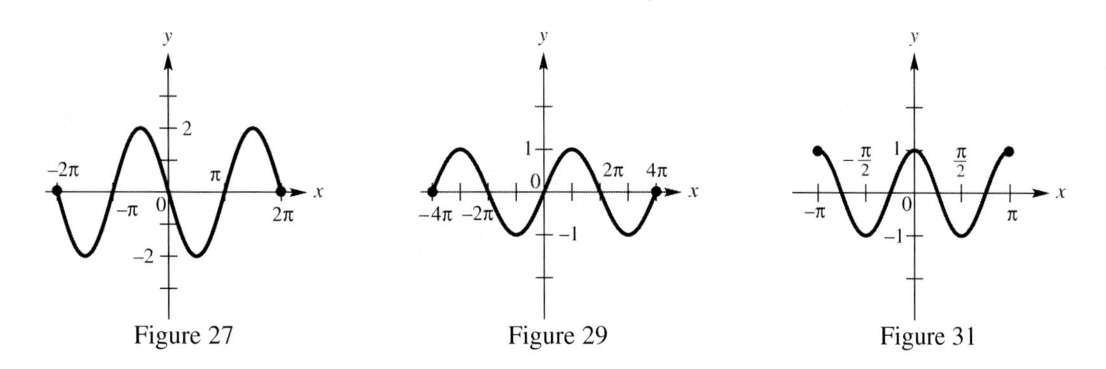

Figure 27 Figure 29 Figure 31

31. See Figure 31. From the equation, the period is π and the amplitude is 1.
33. See Figure 33. From the equation, the period is 8π and the amplitude is 2.
35. See Figure 35. From the equation, the period is $\frac{2\pi}{3}$ and the amplitude is 2.

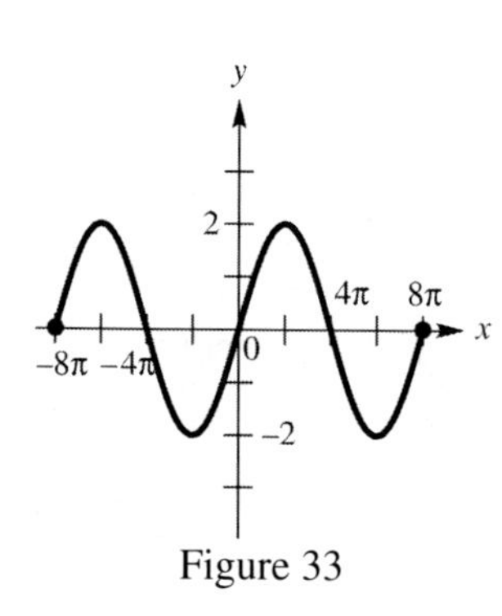

Figure 33

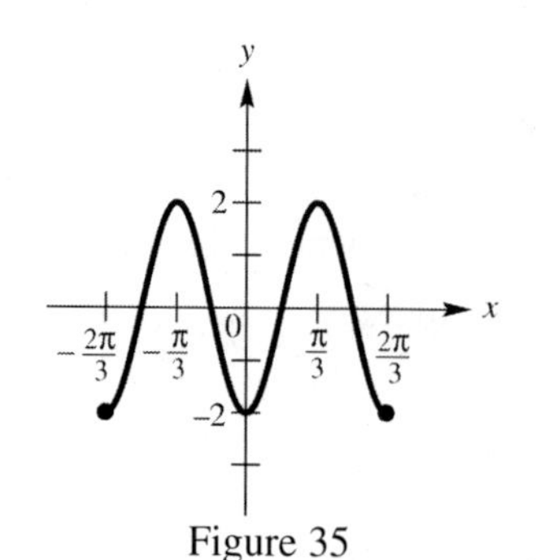

Figure 35

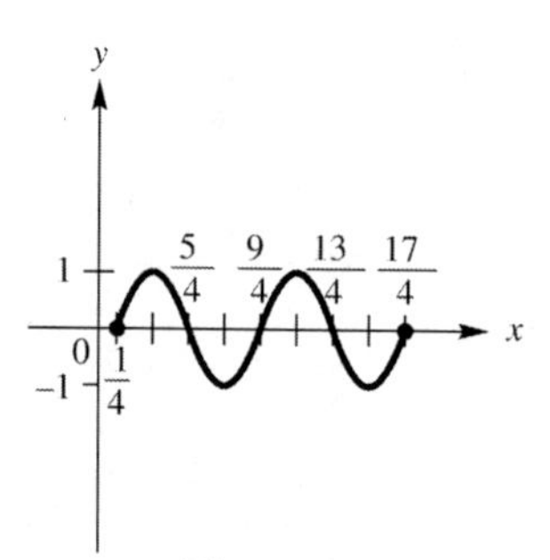

Figure 37

37. See Figure 37. From the equation, the phase shift is $\frac{\pi}{4}$.

38. See Figure 38. From the equation, the phase shift is $\frac{\pi}{3}$.

39. See Figure 39. From the equation, the phase shift is $\frac{\pi}{3}$.

41. See Figure 41. From the equation , the phase shift is $-\pi$.

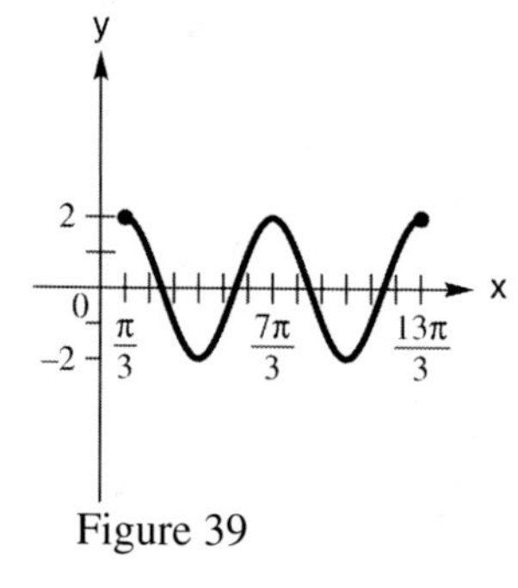

Figure 39

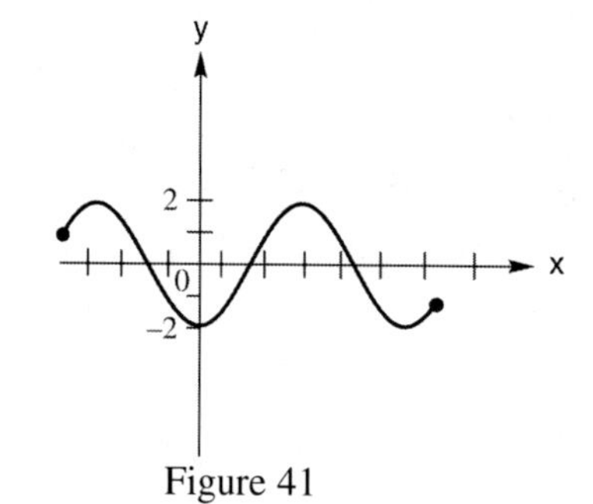

Figure 41

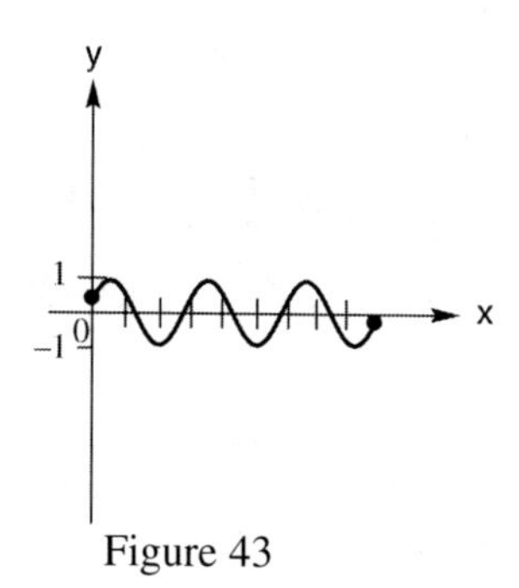

Figure 43

43. See Figure 43. Note that $y=\sin\left(2x+\frac{\pi}{4}\right)$ can be written as $y=\sin\left(2\left(x+\frac{\pi}{8}\right)\right)$ and the phase shift is $-\frac{\pi}{8}$.

45. (a) From the equation, the amplitude is: $|-4|=4$.

(b) From the equation, the period is: $\theta=\frac{2\pi}{2}=\pi$

(c) Since $y=-4\sin(2x-\pi)\Rightarrow y=-4\sin\left[2\left(x-\frac{\pi}{2}\right)\right]$, the phase shift is: $\frac{\pi}{2}$.

(d) No vertical translation.

(e) The range is: $[-4,4]$. See Figure 45.

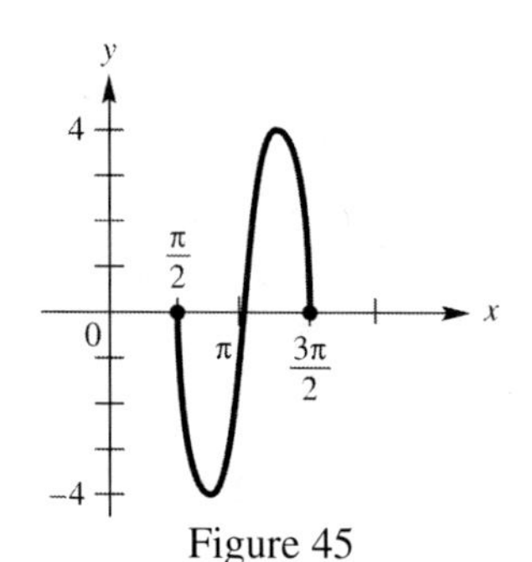

Figure 45

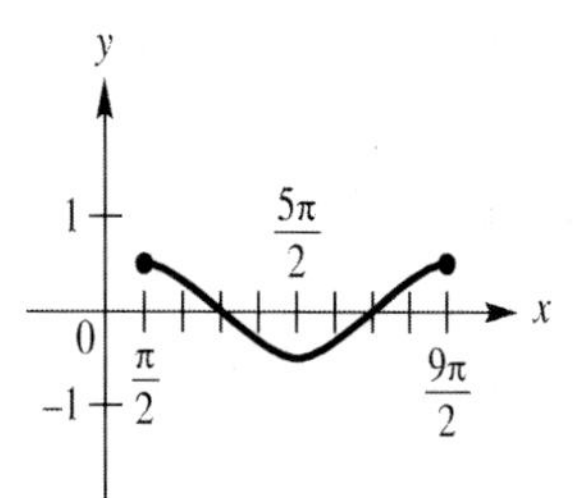

Figure 47

47. (a) From the equation, the amplitude is: $\frac{1}{2}$.

(b) From the equation, the period is: $\frac{2\pi}{\frac{1}{2}} = 4\pi$.

(c) Since $y = \frac{1}{2}\cos\left(\frac{1}{2}x - \frac{\pi}{4}\right) \Rightarrow y = \frac{1}{2}\cos\left[\frac{1}{2}\left(x - \frac{\pi}{2}\right)\right]$, the phase shift is: $\frac{\pi}{2}$.

(d) No vertical translation.

(e) The range is: $\left[-\frac{1}{2}, \frac{1}{2}\right]$. See Figure 47.

49. (a) From the equation, the amplitude is: $\left|-\frac{2}{3}\right| = \frac{2}{3}$.

(b) From the equation, the period is: $\frac{2\pi}{\left(\frac{3}{4}\right)} = \frac{8\pi}{3}$.

(c) From the equation, there is no phase shift.

(d) From the equation, the vertical translation is: upward 1 unit.

(e) The range is: $\left[-\frac{2}{3}, \frac{2}{3}\right]$ translated upward 1 unit, therefore, $\left[\frac{1}{3}, \frac{5}{3}\right]$. See Figure 49.

51. (a) From the equation, the amplitude is: $|-2| = 2$.

(b) From the equation, the period is: $\frac{2\pi}{\left(\frac{1}{2}\right)} = 4\pi$.

(c) From the equation, there is no phase shift.

(d) From the equation, the vertical translation is: upward 1 unit.

(e) The range is: $[-2, 2]$ translated upward 1 unit, therefore, $[-1, 3]$. See Figure 51.

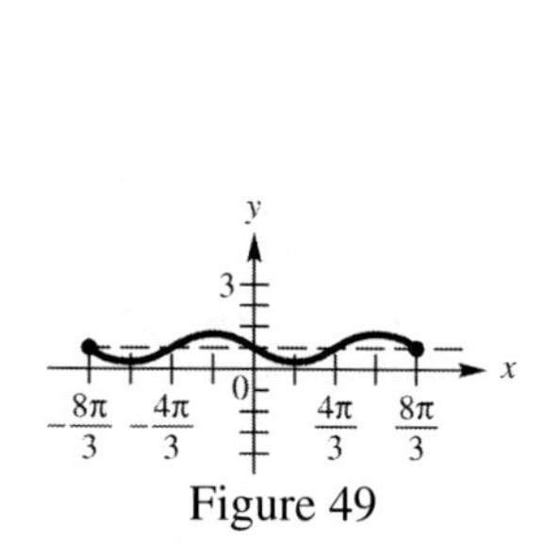

Figure 49

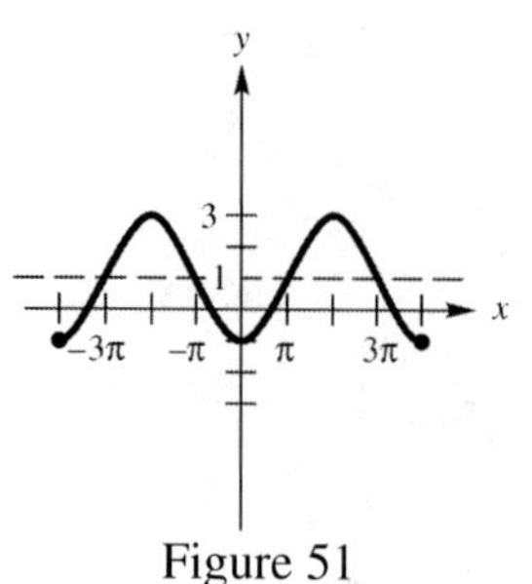

Figure 51

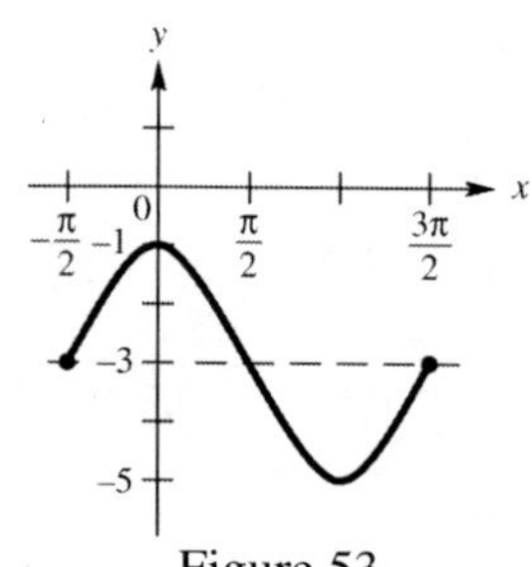

Figure 53

53. (a) From the equation, the amplitude is: 2.

(b) From the equation, the period is: $\frac{2\pi}{(1)} = 2\pi.$

(c) From the equation, the phase shift is: $-\frac{\pi}{2}.$

(d) From the equation, the vertical translation is: downward 3 unit.

(e) The range is: $[-2,2]$ translated downward 3 units, therefore, $[-5,-1]$. See Figure 53.

55. (a) From the equation, the amplitude is:1.

(b) From the equation, the period is: $\frac{2\pi}{(2)} = \pi.$

(c) From the equation, the phase shift is: $-\frac{\pi}{4}.$

(d) From the equation, the vertical translation is: upward $\frac{1}{2}$ unit.

(e) The range is: $[-1,1]$ translated upward $\frac{1}{2}$ unit, therefore, $\left[-\frac{1}{2},\frac{3}{2}\right]$. See Figure 55.

57. (a) From the equation, the amplitude is: 2.

(b) From the equation, the period is: $\cos T = 2\pi$

(c) From the equation, the phase shift is: π.

(d) From the equation, there is no vertical translation.

(e) The range is: $[-2,2]$. See Figure 57.

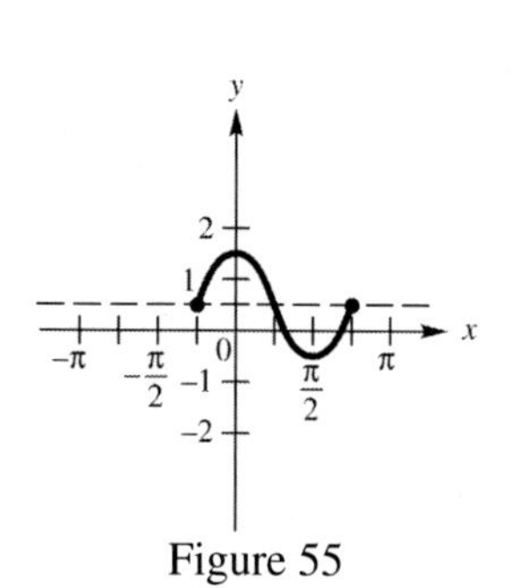

Figure 55

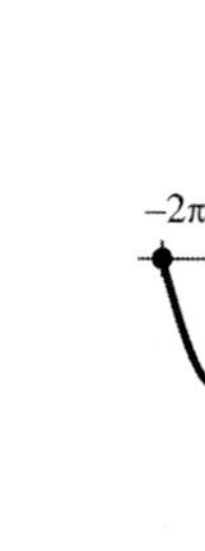
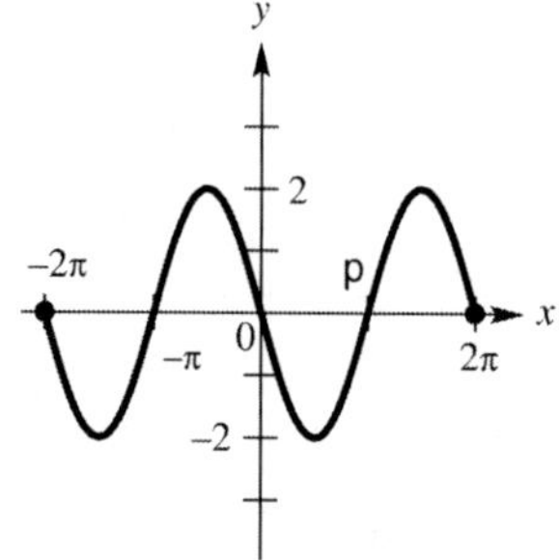

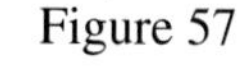
Figure 57

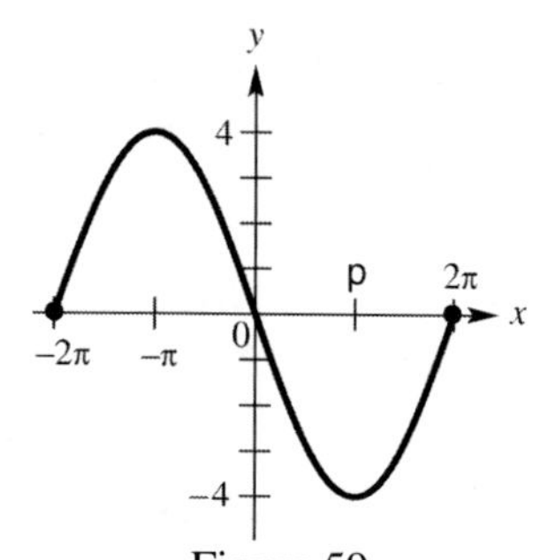

Figure 59

59. (a) From the equation, the amplitude is: 4.

(b) From the equation, the period is: $\dfrac{2\pi}{\left(\dfrac{1}{2}\right)} = 4\pi.$

(c) Since $y = 4\cos\left(\dfrac{1}{2}x + \dfrac{\pi}{2}\right) \Rightarrow y = 4\cos\left[\dfrac{1}{2}(x+\pi)\right]$, the phase shift is: $-\pi$.

(d) From the equation, there is no vertical translation.

(e) The range is: $[-4, 4]$. See Figure 59.

61. (a) From the equation, the amplitude is: $|-1| = 1$.

(b) From the equation, the period is: $\dfrac{2\pi}{(3)}$.

(c) Since $y = 2 - \sin\left(3x - \dfrac{\pi}{5}\right) \Rightarrow y = 2 - \sin\left[3\left(x - \dfrac{\pi}{15}\right)\right]$, the phase shift is: $\dfrac{\pi}{15}$.

(d) From the equation, the vertical translation is: upward 2 units.

(e) The range is: $[-1, 1]$. translated upward 2 units, therefore, $[1, 3]$. See Figure 61.

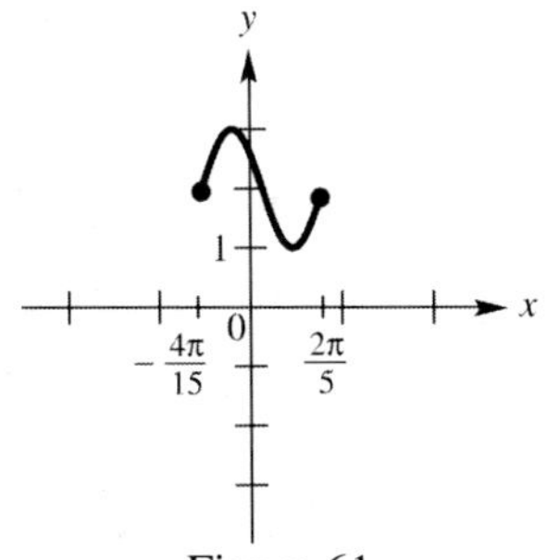

Figure 61

63. The amplitude is $\dfrac{1}{2}[2-(-2)] = \dfrac{1}{2}(4) = 2$, so $a = 2$. One complete cycle of the graph is achieved in π units, so the period $\pi = \dfrac{2\pi}{b} \Rightarrow b = 2$. Comparing the given graph with the general and cosine curves, we see that this graph is a cosine curve. Substituting $a = 2$ and $b = 2$, the function is $y = 2\cos 2x$.

65. The amplitude is $\dfrac{1}{2}[3-(-3)] = \dfrac{1}{2}(6) = 3$, so $a = 3$. One-half of a complete cycle of the graph is achieved in 2π units, so the period $2 \cdot 2\pi = 4\pi \Rightarrow 4\pi = \dfrac{2\pi}{b} \Rightarrow b = \dfrac{1}{2}$. Comparing the given graph with the general and cosine curves, we see that this graph is a reflection of the cosine curve. Thus $a = -3$.

Substituting $a = -3$ *and* $b = \dfrac{1}{2}$, the function is $y = -3\cos\dfrac{1}{2}x$.

67. The amplitude is $\frac{1}{2}[3-(-3)]=\frac{1}{2}(6)=3$, so $a=3$. One complete cycle of the graph is achieved in $\frac{\pi}{2}$ units, so the period $\frac{\pi}{2}=\frac{2\pi}{b}\Rightarrow b=4$. Comparing the given graph with the general and cosine curves, we see that this graph is a sine curve. Substituting $a=3$ *and* $b=4$, the function is $y=3\sin 4x$.

69. This is the sine curve shifted down one unit, so the equation is $y=-1+\sin x$.

71. The maximum is at $\left(\frac{\pi}{3},1\right)$, so the cosine curve has been shifted $\frac{\pi}{3}$ units to the right. Thus, the equation is $y=\cos\left(x-\frac{\pi}{3}\right)$.

73. (a) From the graph, the maximum is 40°F and the minimum is –40°F.

(b) From the graph, the amplitude is 40 and the period is 12. Since there is no vertical translation, the amplitude will help us find the maximum and minimum average temperature values of 40° and –40° and the period represents the 12 months of the calendar year. The monthly average temperatures vary by 80°F over a 12-month period.

(c) The x-intercepts represents the two months when the average temperature is 0°F.

75. (a) From the graph, the maximum is 87°F in July, and the minimum is 62°F in January or late December.

(b) Since a minimum temperature 50° would increase the range of temperatures, the amplitude would increase. See Figure 75.

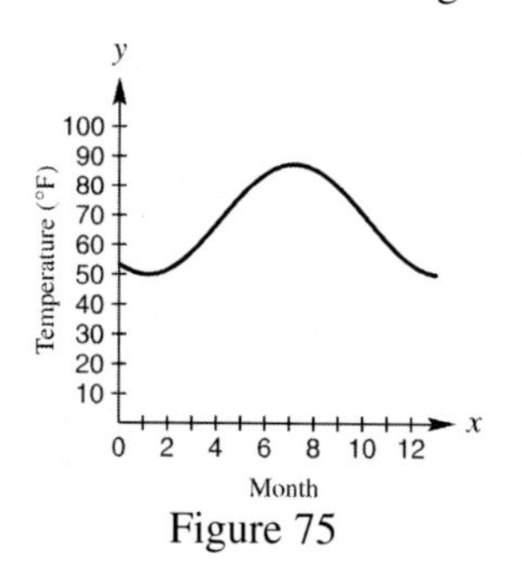

Figure 75

77. The graph repeats each day, so the period is 24 hours.

79. (a) From the equation, the amplitude is 34;the period is $\frac{\pi}{6}=\frac{2\pi}{b}\Rightarrow b=12$, and the phase shift is 4.3.

(b) May corresponds to $x=5$, $f(5)=34\sin\left[\frac{\pi}{6}(5-4.3)\right]\approx 12.2^\circ F$

December corresponds to $x=12$, $f(12)=34\sin\left[\frac{\pi}{6}(12-4.3)\right]\approx -26.4^\circ F$

(c) Since half of the months have average temperatures above zero and half have average temperatures below zero, we would conjecture that the average yearly temperature is $0^\circ F$.

81. (a) We predict the average yearly temperature by finding the mean of the average monthly temperatures:

$$\frac{51+55+63+67+77+86+90+84+71+59+52}{12}=\frac{845}{12}=70.4°\text{F}.$$

This is very close to the actual value of $70.4°\text{F}$.

(b) See Figure 81b.

(c) Since amplitude $a\approx\frac{90-51}{2}\Rightarrow c=70.5;$ the period is 12 months or $b=\frac{2\pi}{(12)}=\frac{\pi}{6};$ the vertical translation is $c=\frac{51+90}{2}\Rightarrow c=70.5;$ and the phase shift can be found by using the minimum temperature value. Since the coldest month is January, when $x=1$, $b=(1-d)$ must equal $(-\pi+2\pi n)$, where n is an integer, since the cosine function is a minimum at these values. Letting $n=0$, we can solve for d: $\frac{\pi}{6}(1-d)=-x\Rightarrow 1-d=-6\Rightarrow d=7$. The table shows that temperatures are actually a little warmer after July than before, so we choose for a better approximation. Now using *a, b, c,* and *d* yields: $f(x)=19.5\cos\left[\frac{\pi}{6}(x-7.2)\right]+70.5.$

(d) See Figure 81d. The function gives an excellent model for the given data.

(e) Set the calculator to the nearest hundredth and the regression equation is, see Figure 81e.

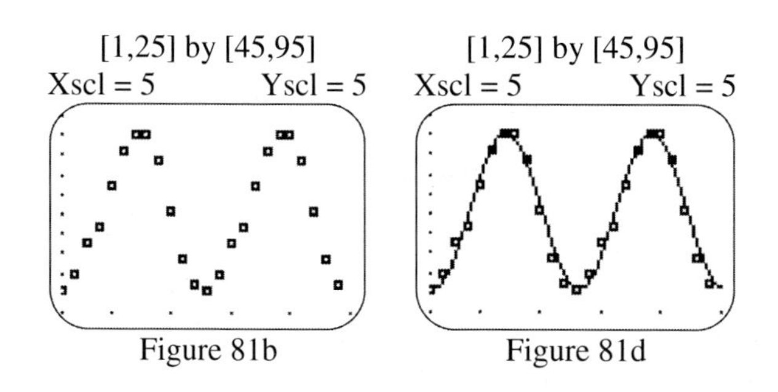

Figure 81b Figure 81d

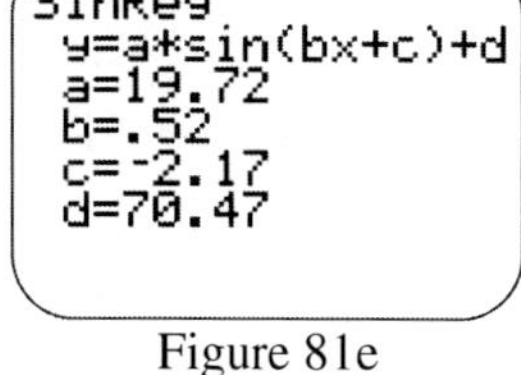

Figure 81e

83. (a) See Figure 83a.

(b) The maximum monthly average temperature is 68° and the minimum is 40°. The midpoint of these values is $\frac{1}{2}(68+40)=50\Rightarrow d=50$. Half the difference between the temperatures is $\frac{1}{2}(68-40)=14\Rightarrow a=14$. Since the temperatures cycle every 12 months, $b=\frac{2\pi}{12}=\frac{\pi}{6}$. The maximum of the $y = \sin x$ graph occurs when $x = \frac{\pi}{2}$ while the maximum in the table occurs when

$x = 7$. Thus $\frac{\pi}{6}(7 - c) = \frac{\pi}{2} \Rightarrow 7 - c = 3 \Rightarrow c = 4$. The function is $f(x) = 14\sin\left(\frac{\pi}{6}(x-4)\right) + 50$.

(c) See Figure 83c.

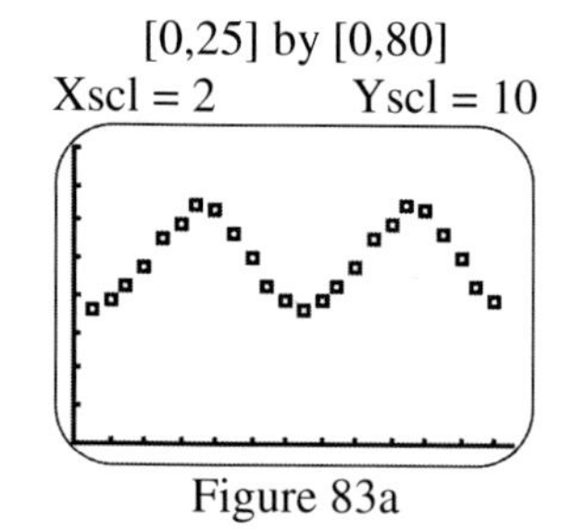

Figure 83a

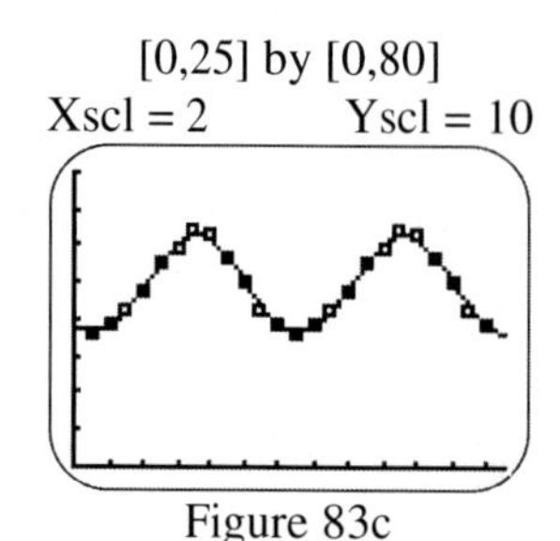

Figure 83c

85. (a) The maximum monthly average temperature is 92° and the minimum is 58°. The midpoint of these values is $\frac{1}{2}(92 + 58) = 75 \Rightarrow d = 50$. Half the difference between the temperatures is $\frac{1}{2}(92 - 58) = 17 \Rightarrow a = 17$. Since the temperatures cycle every 12 months, $b = \frac{2\pi}{12} = \frac{\pi}{6}$. The maximum of the $y = \cos x$ graph occurs when $x = 0$ while the maximum in the table occurs when $x = 7$. Thus $\frac{\pi}{6}(7 - c) = 0 \Rightarrow 7 - c = 0 \Rightarrow c = 7$. The function is $f(x) = 17\cos\left(\frac{\pi}{6}(x-7)\right) + 75$.

(b) Yes, different values of c are possible of the form $c = 7 + 12n$ where n is an integer.

87. (a) The maximum current speed is $18bk$ and the minimum is $-18bk$.. The midpoint of these values is $\frac{1}{2}(18 + (-18)) = 0 \Rightarrow d = 0$. Half the difference between the speeds is $\frac{1}{2}(18 - (-18)) = 18 \Rightarrow a = 18$. Since the tides cycle every 12.4 hours, $b = \frac{2\pi}{12.4} = \frac{\pi}{6.2}$. The maximum of the $y = \cos x$ graph occurs when $x = 0$ while the maximum in the table occurs when $x = 9.8$. Thus $\frac{\pi}{6.2}(9.8 - c) = 0 \Rightarrow 9.8 - c = 0 \Rightarrow c = 9.8$. The function is $f(x) = 18\cos\left(\frac{\pi}{6.2}(x-9.8)\right)$.

(b) The canal contains the most water at 3.7 and 16.1 hours after midnight (high tides). At these points water is rushing out of the canal at 18bk. The canal contains the least water at 9.8 and 22.2 hours after midnight (low tides). At these points water is rushing into the canal at 18bk. See Figure 87.

89. (a) See Figure 89.

(b) If $x = 1970$ corresponds to 1970 then the equation is:

$c(x) = .4(x - 1970)^2 + .6(x - 1970) + 330 + 7.5\ \sin\left[2\pi(x - 1970)\right]$.

[0,24] by [-20,20]
Xscl = 4 Yscl = 5

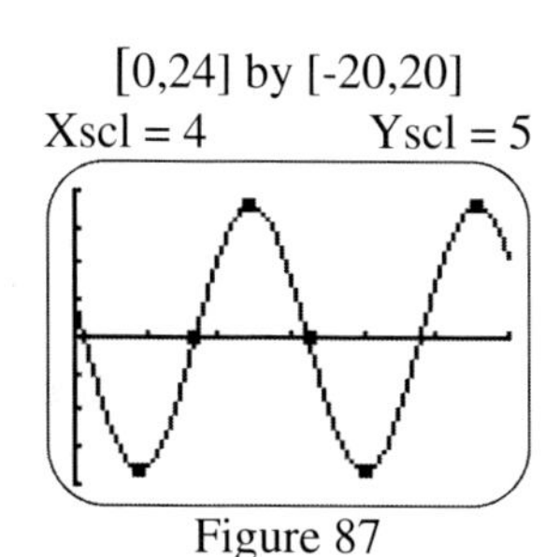

Figure 87

[5,25] by [320,380]
Xscl = 5 Yscl= 10

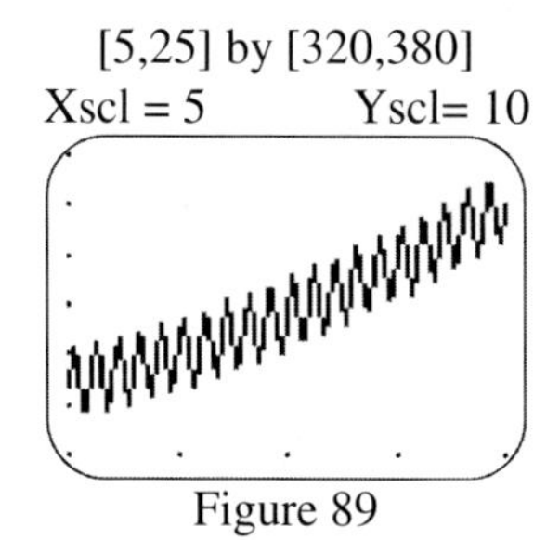

Figure 89

Reviewing Basic Concepts (Sections 8.5 - 8.6)

1. (a) An angle of $s=-2\pi$ radians intersects the unit circle at the point $(1,0)$.

   (b) From the properties of $45°-45°$ right triangles, an angle of $s=\dfrac{5\pi}{4}$ radians intersects the unit circle at the point $\left(-\dfrac{\sqrt{2}}{2},-\dfrac{\sqrt{2}}{2}\right)$.

   (c) An angle of $s=\dfrac{5\pi}{2}$ radians intersects the unit circle at the point $(0,1)$.

2. An angle of $s=-\dfrac{5\pi}{2}$ radians intersects the unit circle at the point $(0,-1)$. Therefore, since:

$\sin s=y,\ \sin-\dfrac{5\pi}{2}=-1$ $\cos s=x,\ \cos-\dfrac{5\pi}{2}=0$ $\tan s=\dfrac{y}{x},\tan-\dfrac{5\pi}{2}=\dfrac{-1}{0}=\text{undefined}$

$\csc s=\dfrac{1}{y},\csc-\dfrac{5\pi}{2}=\dfrac{1}{-1}=-1$ $\sec s=\dfrac{1}{x},\sec-\dfrac{5\pi}{2}=\dfrac{1}{0}=\text{undefined}$ $\cot s=\dfrac{x}{y},\cot-\dfrac{5\pi}{2}=\dfrac{0}{-1}=0$

3. (a) From the properties of $30°-60°$ right triangles, an angle of $s=\dfrac{7\pi}{6}$ radians intersects the unit circle at the point $\left(-\dfrac{\sqrt{3}}{2},-\dfrac{1}{2}\right)$. Therefore, since:

$\sin s=y,\sin\dfrac{7\pi}{6}=-\dfrac{1}{2}$ $\csc s=\dfrac{1}{y},\csc\dfrac{7\pi}{6}=\dfrac{1}{-\frac{1}{2}}=-2$

$\cos s=x,\cos\dfrac{7\pi}{6}=-\dfrac{\sqrt{3}}{2}$ $\sec s=\dfrac{1}{x},\sec\dfrac{7\pi}{6}=\dfrac{1}{-\frac{\sqrt{3}}{2}}=-\dfrac{2}{\sqrt{3}}=-\dfrac{2\sqrt{3}}{3}$

$\tan s=\dfrac{y}{x},\tan\dfrac{7\pi}{6}=\dfrac{-\frac{1}{2}}{-\frac{\sqrt{3}}{2}}=-\dfrac{1}{2}\cdot\left(-\dfrac{2}{\sqrt{3}}\right)=\dfrac{\sqrt{3}}{3}$ $\cot s=\dfrac{x}{y},\cot\dfrac{7\pi}{6}=\dfrac{-\frac{\sqrt{3}}{2}}{-\frac{1}{2}}=-\dfrac{\sqrt{3}}{2}\cdot\left(-\dfrac{2}{1}\right)=\sqrt{3}$

   (b) From the properties of $30°-60°$ right triangles, an angle of $s=-\dfrac{2\pi}{3}$ radians intersects the unit circle at the point $\left(-\dfrac{1}{2},-\dfrac{\sqrt{3}}{2}\right)$. Therefore, since:

$\sin s=y,\ \sin-\dfrac{2\pi}{3}=-\dfrac{\sqrt{3}}{2}$ $\csc s=\dfrac{1}{y},\ \csc-\dfrac{2\pi}{3}=\dfrac{1}{-\frac{\sqrt{3}}{2}}=-\dfrac{2}{\sqrt{3}}=-\dfrac{2\sqrt{3}}{3}$

$\cos s = x,\ \cos -\frac{2\pi}{3} = -\frac{1}{2}$ $\qquad$ $\sec s = \frac{1}{x},\ \sec -\frac{2\pi}{3} = \frac{1}{-\frac{1}{2}} = -2$

$\tan s = \frac{y}{x}, \tan -\frac{2\pi}{3} = \frac{-\frac{\sqrt{3}}{2}}{-\frac{1}{2}} = -\frac{\sqrt{3}}{2} \cdot -\frac{2}{1} = \sqrt{3}$ $\qquad$ $\cot s = \frac{x}{y}, \cot -\frac{2\pi}{3} = \frac{-\frac{1}{2}}{-\frac{\sqrt{3}}{2}} = -\frac{1}{2} \cdot -\frac{2}{\sqrt{3}} = \frac{1}{\sqrt{3}} = \frac{\sqrt{3}}{3}$

4. Using the calculator (radian mode),

   $\sin 2.25 \approx .7780731969$; $\cos 2.25 \approx -.6281736227$;

   $\tan 2.25 \approx -1.238627616$; $\csc 2.25 \approx 1.285226125$; $\sec 2.25 \approx -1.591916572$; and $\cot 2.25 \approx -.8073451511$.

5. Since cosine is positive in quadrants I and IV and sine is negative in quadrants II and IV, then the point corresponding to 100 is in quadrant IV.

6. From the equation, the amplitude is $|-1| = 1$ and the period is $\frac{2\pi}{(1)} = 2\pi$. See Figure 6.

7. From the equation, the amplitude is 3, the period is $\frac{2\pi}{(\pi)} = 2$, and the phase shift is $-\frac{\pi}{\pi} = -1$. See Figure 7.

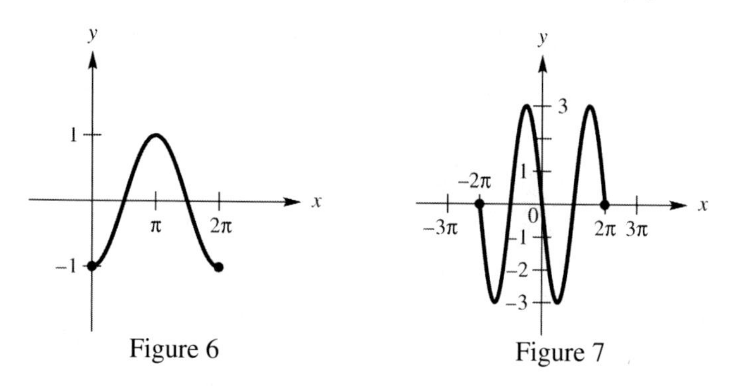

Figure 6 $\qquad$ Figure 7

8. (a) Since the amplitude is 6.5 and the vertical shift is 12.4, the maximum number of daylight hours is $6.5 + 12.4 = 18.9$ hours and the minimum is $-6.5 + 12.4 = 5.9$ hours.

   (b) The amplitude represents half the difference in the daylight hours between the longest and shortest days. The period represents 12 months or one year.

8.7 Graphs of the Other Circular Functions

1. The basic cosecant graph reflected across the x-axis, therefore graph B.

3. The basic tangent graph reflected across the x-axis, therefore graph E.

5. The basic tangent graph translated $\frac{\pi}{4}$ units to the right, therefore graph D.

7. True, $x = \frac{\pi}{2}$ is the smallest positive asymptote of the tangent function.

9. False, secant values are undefined when $x = \frac{\pi}{2} + \pi k$ while cosecant values are undefined when $x = \pi k$.

11. $\sec(-x) = \dfrac{1}{\cos(-x)} = \dfrac{1}{\cos x} = \sec x$

13. $\tan(-x) = \dfrac{\sin(-x)}{\cos(-x)} = \dfrac{-\sin x}{\cos x} = -\tan x$

15. (a) From the equation, the period is: $\dfrac{2\pi}{\left(\dfrac{1}{2}\right)} = 4\pi$.

    (b) From the equation, there is no phase shift.

    (c) From the equation, the range is $(-\infty,-2]\cup[2,\infty)$.

17. (a) From the equation, the period is: $\dfrac{2\pi}{(1)} = 2\pi$.

    (b) From the equation, the phase shift is: $-\dfrac{\pi}{2}$ units to the right.

    (c) From the equation, the range is $(-\infty,-2]\cup[2,\infty)$.

19. (a) From the equation, the period is: $\dfrac{\pi}{\left(\dfrac{1}{3}\right)} = 3\pi$ (The phase shift for the basic cotangent function is π.)

    (b) From the equation, the phase shift is: $\left(\dfrac{\pi}{2}\text{ units to the right}\right)$

    (c) The range of all cotangent functions is $(-\infty,\infty)$.

21. (a) From the equation, the period is: $\dfrac{2\pi}{(2)} = \pi$.

    (b) From the equation, the phase shift is: $-\dfrac{\pi}{2}$. $\left(\dfrac{\pi}{2}\text{ units to the left}\right)$

    (c) From the equation, the range is $\left(-\infty,-\dfrac{1}{2}\right]\cup\left[\dfrac{1}{2},\infty\right)$.

23. (a) From the equation, the period is: $\dfrac{\pi}{(1)} = \pi$. (The phase shift for the basic tangent function is π.)

    (b) From the equation, the phase shift is: $-\dfrac{\pi}{4}$.

    (c) The range of all tangent functions is $(-\infty,\infty)$.

25. Graph a secant function with a period of 2π a range of $(-\infty,-1]\cup[1,\infty)$ and no phase shift. See Figure 25.

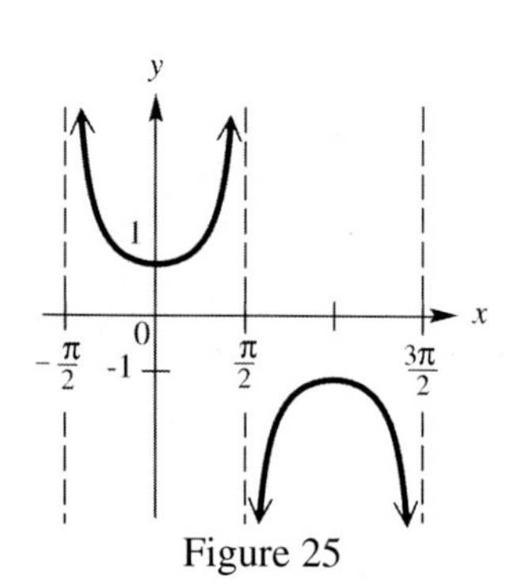

Figure 25

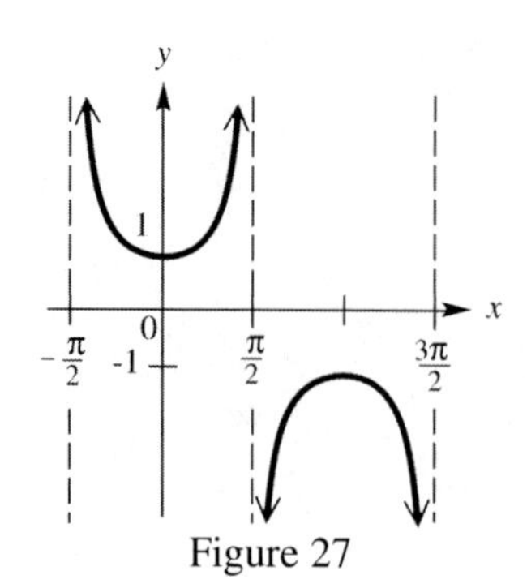

Figure 27

27. Graph a secant function with a period of 2π a range of $(-\infty,-1]\cup[1,\infty)$ and a phase shift of 2π to the right. See Figure 27.

29. Graph a secant function with a period of $\frac{\pi}{\left(\frac{1}{4}\right)}=8\pi$ a range of $(-\infty,-3]\cup[3,\infty)$ and no phase shift. See Figure 29.

31. Graph a cosecant function with a period of $\frac{2\pi}{(1)}=2\pi,$ a range $\left(-\infty,-\frac{1}{2}\right]\cup\left[\frac{1}{2},\infty\right),$ of reflected across the x-axis, and a phase shift $\frac{\pi}{2}$ units to the left. See Figure 31.

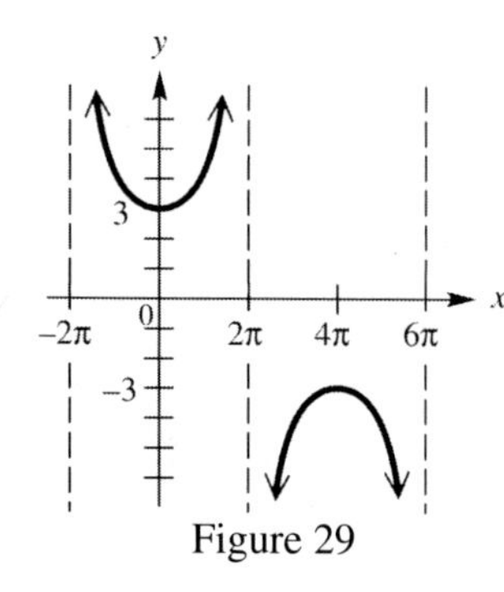

Figure 29

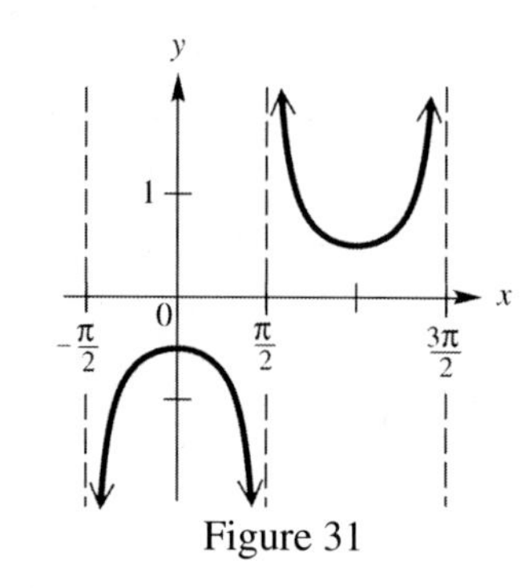

Figure 31

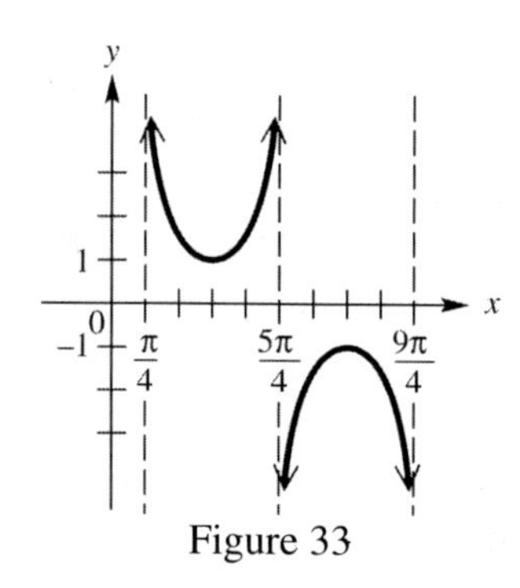

Figure 33

33. Graph a cosecant function with a period of $\frac{2\pi}{(1)}=2\pi,$ and a phase shift $\frac{\pi}{4}$ units to the right. See Figure 33.

35. Graph a secant function with a period of $\frac{2\pi}{(1)}=2\pi,$ and a phase shift $\frac{\pi}{4}$ units to the left. See Figure 35.

37. Since $y=\sec\left(\frac{1}{2}x+\frac{\pi}{3}\right)\Rightarrow y=\sec\frac{1}{2}\left(x+\frac{2\pi}{3}\right),$ graph a secant function with a period $\frac{2\pi}{\left(\frac{1}{2}\right)}=4\pi,$ and a phase shift $\frac{2\pi}{3}$ units to the left. See Figure 37.

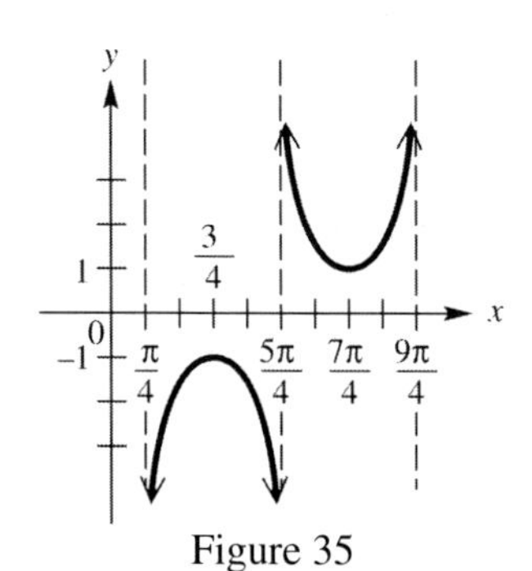

Figure 35

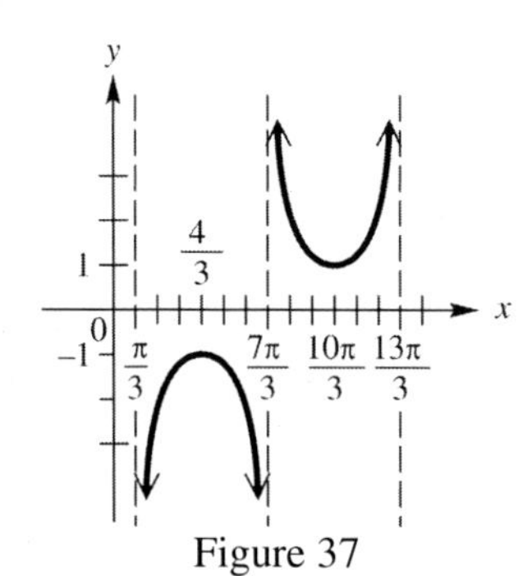

Figure 37

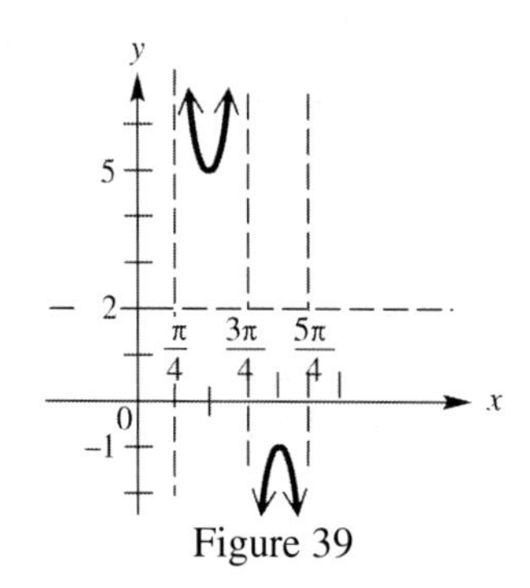

Figure 39

39. Since $y = 2 + 3\sec(2x - \pi) \Rightarrow y = 2 + 3\sec 2\left(x - \frac{\pi}{2}\right)$, graph a secant function with a period $\frac{2\pi}{(2)} = \pi$ a vertical shift 2 units upward therefore a range of $(-\infty, -1] \cup [5, \infty)$, and a phase shift $\frac{\pi}{2}$ units to the right. See Figure 39.

41. Graph a cosecant function with a period $\frac{2\pi}{(1)} = 2\pi$, reflected across the x-axis, a vertical shift 1 unit upward therefore a range of $\left(-\infty, \frac{1}{2}\right] \cup \left[\frac{3}{2}, \infty\right)$, and a phase shift $\frac{3\pi}{4}$ units to the right. See Figure 41.

43. Graph a tangent function with a period of π, and no phase shift. See Figure 43.

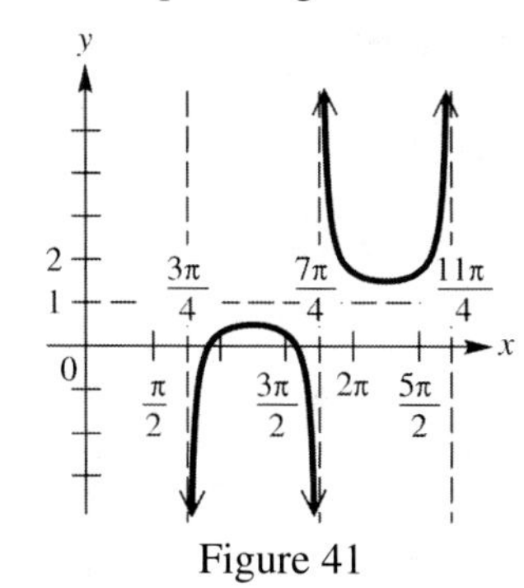

Figure 41

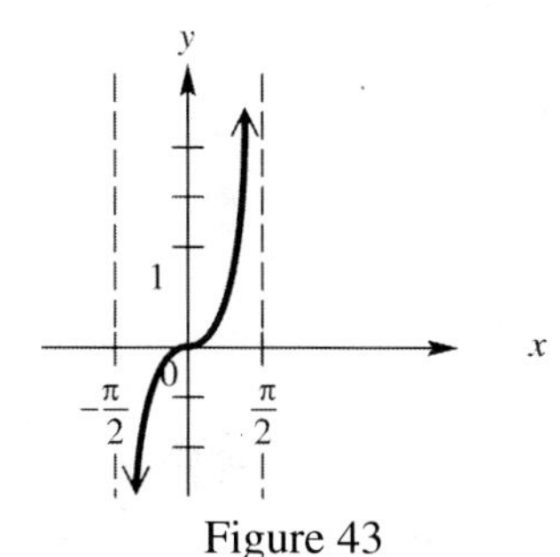

Figure 43

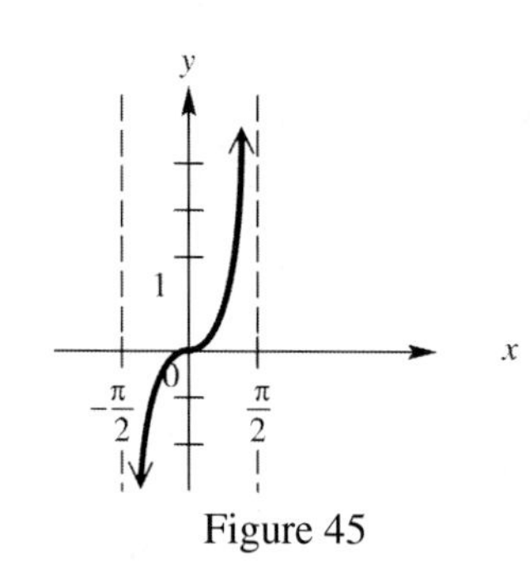

Figure 45

45. Graph a tangent function with a period of π, and a phase shift π units to the right. See Figure 45.

47. Graph a tangent function with a period $\frac{\pi}{(4)}$. See Figure 47.

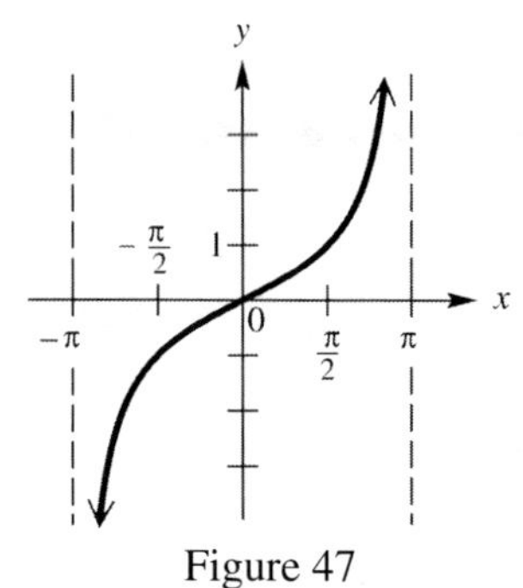

Figure 47

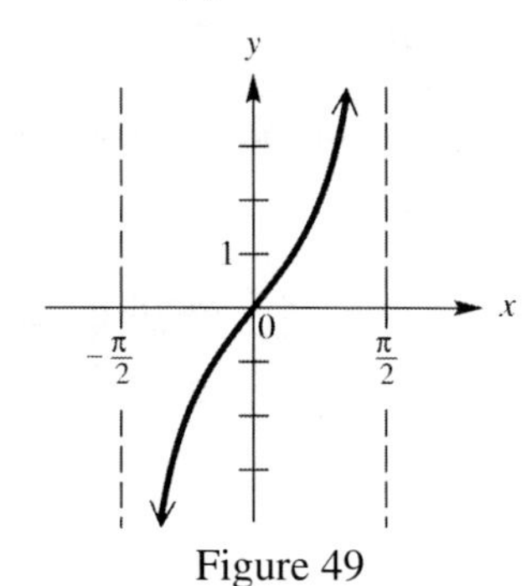

Figure 49

49. Graph a tangent function with a period $\frac{\pi}{(1)} = \pi$; and passing through the points $(0,0)$ (midpoint), $\left(-\frac{\pi}{4},-2\right)$ and $\left(\frac{\pi}{4},2\right)$. See Figure 49.

51. Graph a tangent function with a period $\frac{\pi}{\left(\frac{1}{4}\right)} = 4\pi$; and passing through the points $(0,0)$ (midpoint), $(-\pi,-2)$,and $(\pi,2)$. See Figure 51.

53. Graph a cotangent function with a period $\frac{\pi}{(3)}$; and passing through the points $\left(\frac{\pi}{6},0\right)$ (midpoint), $\left(\frac{\pi}{12},1\right)$, and $\left(\frac{\pi}{4},-1\right)$. See Figure 53.

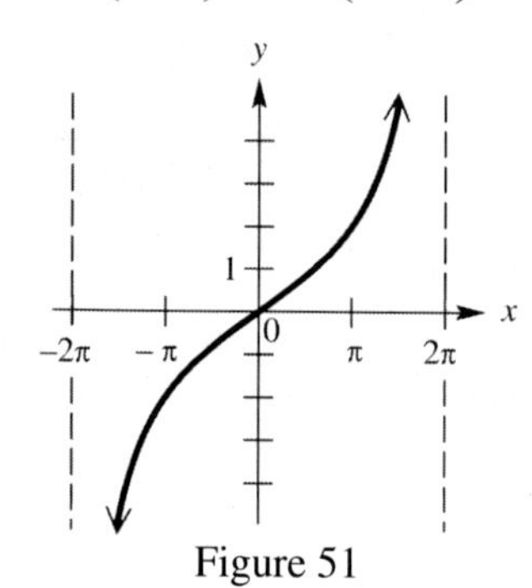

Figure 51

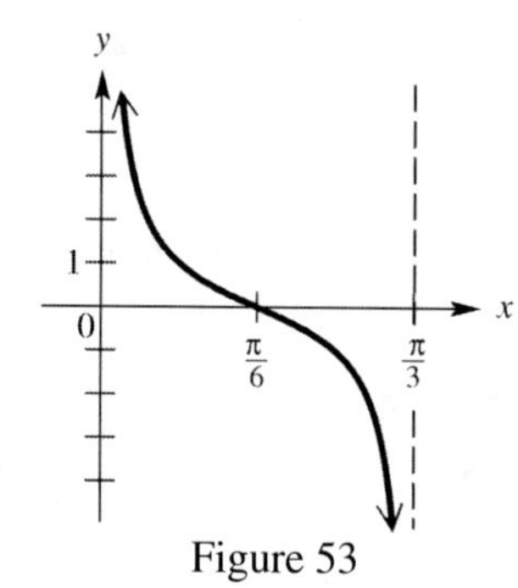

Figure 53

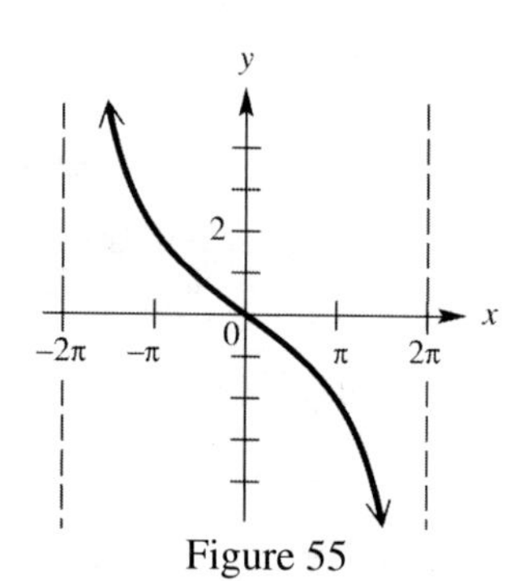

Figure 55

55. Graph a tangent function with a period $\frac{\pi}{\left(\frac{1}{4}\right)} = 4\pi$, is reflected across the x-axis, and passes through the points $(0,0)$ (midpoint), $(-\pi,2)$, and $(\pi,-2)$. See Figure 55.

57. Graph a cotangent function with a period $\frac{\pi}{(4)}$; and passing through the points $\left(\frac{\pi}{8},0\right)$ (midpoint), $\left(\frac{\pi}{16},\frac{1}{2}\right)$, and $\left(\frac{3\pi}{16},-\frac{1}{2}\right)$. See Figure 57.

59. Since $y = \tan(2x-\pi) \Rightarrow y = \tan 2\left(x-\frac{\pi}{2}\right)$, graph a tangent function, over a two period interval, with a period $\frac{\pi}{(2)}$ and a phase shift of $\frac{\pi}{2}$ units to the right. See Figure 59.

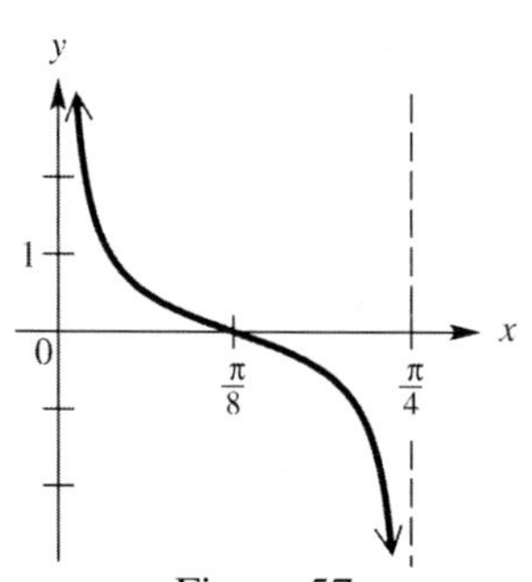

Figure 57

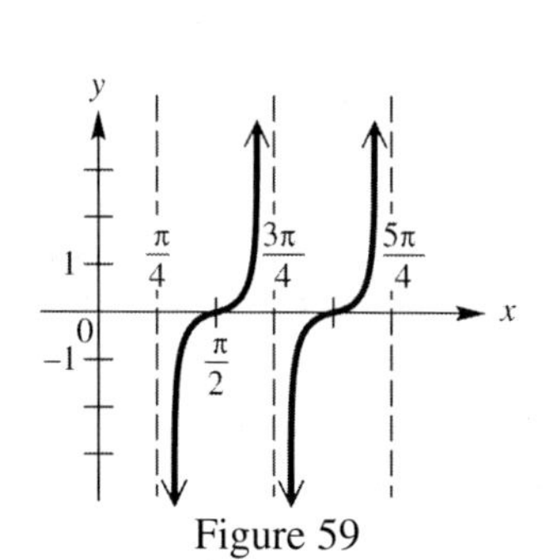

Figure 59

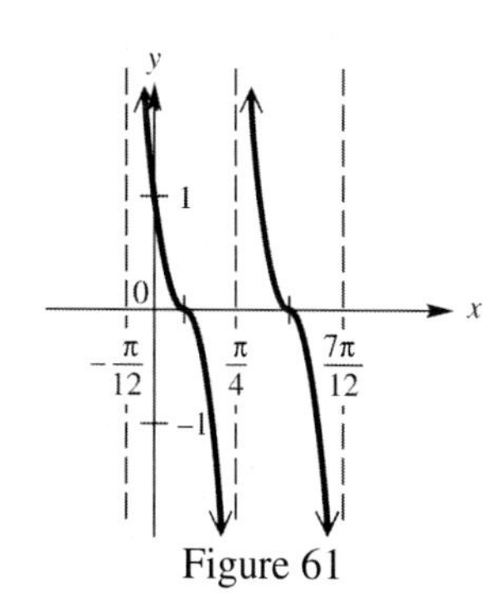

Figure 61

61. Since $y = \cot\left(3\pi + \frac{x}{4}\right) \Rightarrow y = \cot 3\left(x + \frac{\pi}{12}\right)$, graph a cotangent function, over a two period interval, with a period $\frac{\pi}{(3)}$ and a phase shift of $\frac{\pi}{12}$ units to the left. See Figure 61.

63. Graph a tangent function with a period $\frac{\pi}{(1)} = \pi$ and vertical shift 1 unit upward. See Figure 63.

65. Graph a cotangent function with a period $\frac{\pi}{(1)} = \pi$, is reflected across the x-axis, and vertical shift 1 unit upward. See Figure 65.

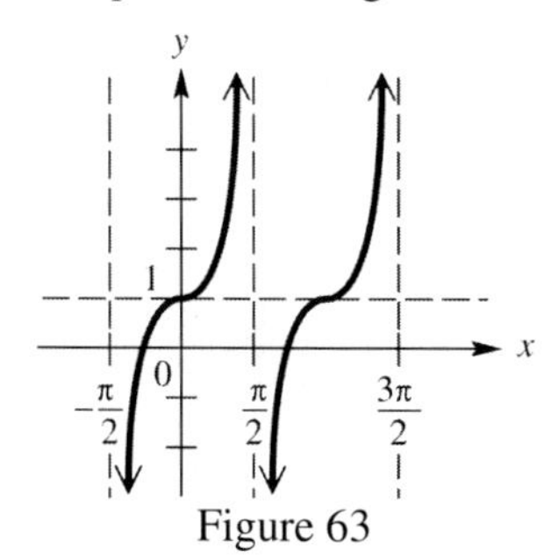

Figure 63

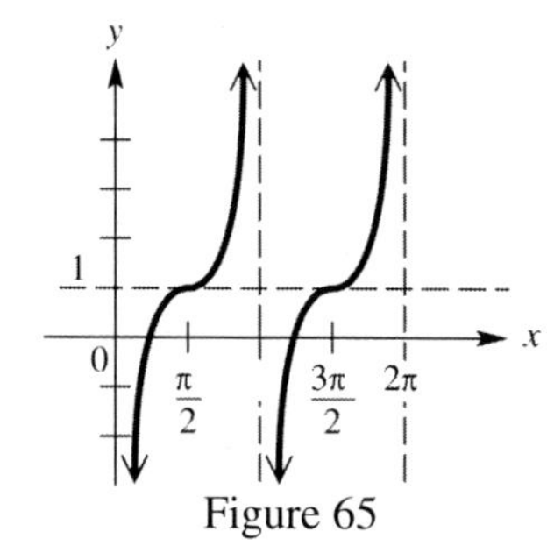

Figure 65

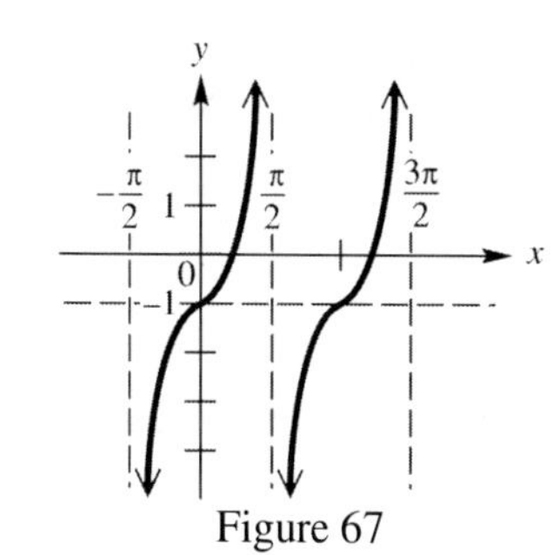

Figure 67

67. Graph a tangent function with a period $\frac{\pi}{(1)} = \pi$, and vertical shift 1 unit downward, and passes through the points $(0,-1)$ (midpoint), $\left(-\frac{\pi}{4},-3\right)$ and $\left(\frac{\pi}{4},1\right)$ See Figure 67.

69. Since $y = -1 + \frac{1}{2}\cot(2x - 3\pi) \Rightarrow y = -1 + \frac{1}{2}\cot 2\left(x - \frac{3\pi}{2}\right)$, graph a cotangent function, over a two period interval, with a period $\frac{\pi}{(2)}$, a phase shift of $\frac{3\pi}{2}$ units to the right, has a vertical shift 1 unit downward, and passes through the points $\left(\frac{3\pi}{4},-1\right)$ (midpoint), $\left(\frac{5\pi}{8},-\frac{1}{2}\right)$ and $\left(\frac{7\pi}{8},-\frac{3}{2}\right)$. See Figure 69.

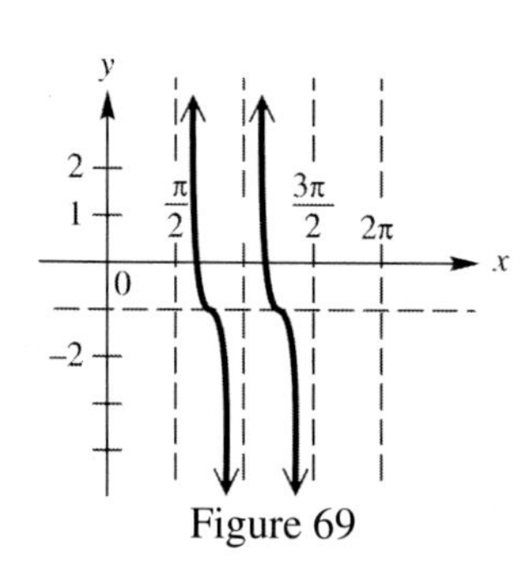

Figure 69

71. Since the asymptotes are at $-\frac{\pi}{2}, \frac{\pi}{2}$, and $\frac{3\pi}{2}$, this is a tangent function of the form $y = a\tan x$. The graph passes through the point $\left(\frac{\pi}{4}, -2\right)$. Substituting these values into the equation above gives

    $y = a\tan x \Rightarrow -2 = a\tan\frac{\pi}{4} \Rightarrow -2 = a\cdot 1 \Rightarrow -2 = a$ Thus, the equation of the graph is $y = -2\tan x$.

73. Since the asymptotes are at $0, \frac{\pi}{3}$, and $\frac{2\pi}{3}$, this is a tangent function of the form $y = a\tan bx$. The period of the function is $\frac{\pi}{b} = \frac{\pi}{3} \Rightarrow b = 3$. The graph passes through the point $\left(\frac{\pi}{12}, 1\right)$. Substituting these values into the equation above gives $y = a\cot bx \Rightarrow 1 = a\cot\left(3\cdot\frac{\pi}{12}\right) \Rightarrow 1 = a\cot\frac{\pi}{4} \Rightarrow 1 = a.$

    Thus, the equation of the graph is $y = \cot 3x.$

75. Since the asymptotes are at multiples of π, this is a cotangent function of the form $y = a\cot x$. The graph passes through the point $\left(\frac{\pi}{4}, -3\right)$. Substituting these values into the equation above gives

    $y = a\cot x \Rightarrow -3 = a\cot\frac{\pi}{4} \Rightarrow -3 = a\cdot 1 \Rightarrow -3 = a$ Thus, the equation of the graph is $y = -3\cot x.$

77. Since the graph crosses the y-axis at (0,1), this is a secant graph with $a = 1$. The period is $\left|-\frac{\pi}{4} - \frac{\pi}{4}\right| = \frac{\pi}{2}$.

    Thus, $b = \frac{2\pi}{\frac{\pi}{2}} \Rightarrow b = 4$. The equation of the graph is $y = \sec 4x.$

79. This is the graph of $y = \csc x$ translated two units down. Thus, the equation of the graph is $y = -2 + \csc x$.

81. (a) $d = 4\tan(2\pi\cdot 0) \Rightarrow d = 0$ m.

    (b) $d = 4\tan(2\pi\cdot 0.2) \Rightarrow d \approx 12.3$ m.

    (c) $d = 4\tan(2\pi\cdot 0.2) \Rightarrow d \approx -12.3m.$

    (d) $d = 4\tan(2\pi\cdot 0.2) \Rightarrow d \approx 12.3m.$

(e) The value $t = .25$ yields: $d = 4\tan(2\pi)(.25) \Rightarrow d = 4\tan\frac{\pi}{2}$. Since $\tan\frac{\pi}{2}$ is undefined, .25 is a meaningless value for t.

8.8 Harmonic Motion

1. (a) If $s(0) = 2$, then we can solve for a: $2 = \cos(w \cdot 0) \Rightarrow 2 = a(1) \Rightarrow a = 2$. Now, since the period is .5 seconds, $\frac{2\pi}{w} = \frac{1}{2} \Rightarrow w = 4\pi$. Thus $s(t) = a\cos(wt) \Rightarrow s(t) = 2\cos(4\pi t)$.

   (b) $s(1) = 2\cos(4\pi(1)) \Rightarrow s(1) = 2$. The weight is neither moving upward nor downward. At $t = 1$ the motion of the weight is changing from up to down. The calculator graph supports this.

3. (a) If $s(0) = -3$, then we can solve for a: $-3 = a\cos(w \cdot 0) \Rightarrow -3 = a(1) \Rightarrow a = -3$. Now, since the period is .8 seconds, $\frac{2\pi}{w} = \frac{8}{10} \Rightarrow 8w = 2.5\pi$. Thus $s(t) = a\cos(wt) \Rightarrow s(t) = -3\cos(2.5\pi t)$.

   (b) $s(1) = -3\cos(2.5\pi(1)) \Rightarrow s(1) = 0$. The spring is at its natural length one second after the weight is released. The weight is moving upward. The calculator graph supports this.

5. Using $F = 27.5$, note that $b = w = 2\pi(27.5) = 55\pi$, therefore $s(t) = a\cos(55\pi t)$. Now since $s(0) = .21$,we can solve for a: $.21 = a\cos(55\pi(0)) \Rightarrow .21 = a(1) \Rightarrow a = .21$. The equation is: $s(t) = .21\cos(55\pi t)$.

   See Figure 5.

7. Using $F = 110$ note that $b = w = 2\pi(55) = 110\pi$, therefore $s(t) = a\cos(110\pi t)$. Now since $s(0) = .14$, we can solve for a: $.14 = a\cos(110\pi(0)) \Rightarrow .14 = a(1) \Rightarrow a = .14$. The equation is: $s(t) = .14\cos(110\pi t)$.

   See Figure 7.

[0,0.05] by [-0.3,0.3]
Xscl = 0.01 Yscl = 0.1

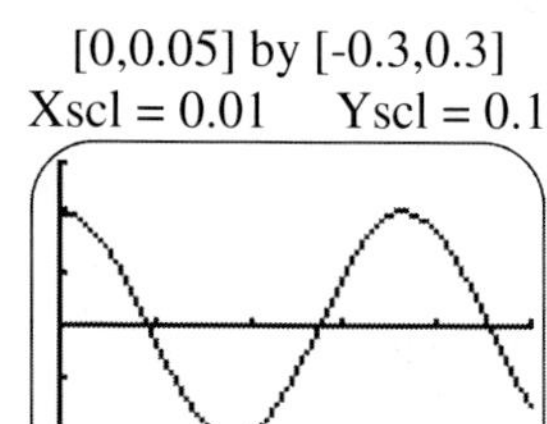

Figure 5

[0,0.05] by [-0.3,0.3]
Xscl = 0.01 Yscl = 0.1

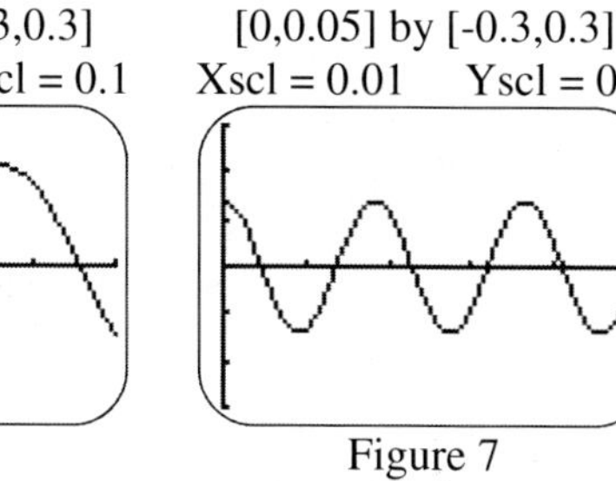

Figure 7

9. (a) Using the simple harmonic motion equation $s(t) = a\sin wt$ with $a = 2$ (radius 2) and $w = 2$ (2 radians/sec) yields the equation: $s(t) = 2\sin 2t$ Therefore, from this equation, the amplitude is $|a| = |2| = 2$, the period is $\frac{2\pi}{2} = \pi$, and the frequency is $\frac{w}{2\pi} = \frac{2}{2\pi} = \frac{1}{\pi}$.

(b) Using the simple harmonic motion equation $s(t) = a\sin wt$ with $a = 2$ (radius 2) and $w = 4$ (4 radians/sec) yields the equation: $s(t) = 2\sin 4t$. Therefore, from this equation, the amplitude is $|a| = |2| = 2$, the period is $\frac{2\pi}{4} = \frac{\pi}{2}$, and the frequency is $\frac{w}{2\pi} = \frac{4}{2\pi} = \frac{2}{\pi}$.

11. Using $P = 1$ and $P = 2\pi\sqrt{\frac{L}{32}}$ (from problem 10), we can solve for L:

$$1 = 2\pi\sqrt{\frac{L}{32}} \Rightarrow \frac{1}{2\pi} = \sqrt{\frac{L}{32}} \Rightarrow 2\pi = \sqrt{\frac{32}{L}} \Rightarrow (2\pi)^2 = \frac{32}{L} \Rightarrow 4\pi^2 L = 32 \Rightarrow L = \frac{32}{4\pi^2} \Rightarrow L = \frac{8}{\pi^2}$$

13. (a) Using the formula , $s(t) = a\sin\sqrt{\frac{k}{m}}t$, as the simple harmonic motion formula gives us $w = \sqrt{\frac{k}{m}}$, therefore $w = \sqrt{\frac{2}{1}} \Rightarrow w = \sqrt{2}$. With a spring stretch of $\frac{1}{2}$ foot the amplitude is: $\frac{1}{2}$; the period is: $\frac{2\pi}{3} = \frac{2\pi}{\sqrt{2}} = \frac{2\sqrt{2}\pi}{2} = \sqrt{2}\pi$; and the frequency is: $\frac{w}{2\pi} = \frac{\sqrt{2}}{2\pi}$.

(b) Using the answers from part (a) and the simple harmonic motion formula $s(t) = a\sin wt$ yields: $s(t) = \frac{1}{2}\sin\sqrt{2}t$.

15. (a) From the equation, $a = -4$ and $w = 10$, therefore the maximum height is the amplitude $|-4| = 4$ in.

(b) From the equation, $a = -4$ and $w = 10$, therefore frequency $\frac{w}{2\pi} = \frac{10}{2\pi} = \frac{5}{\pi}$ cycles/second; and the period is: $\frac{2\pi}{w} = \frac{2\pi}{10} = \frac{\pi}{5}$ seconds.

(c) If $s(t) = 4$, solve $4 = -4\cos 10t \Rightarrow -1 = \cos 10t \Rightarrow 10t = \pi \Rightarrow t = \frac{\pi}{10}$. The weight first reaches its maximum height after $\frac{\pi}{10}$ seconds.

(d) $s(1.466) = -4\cos(10(1.466)) \approx 2$. After 1.466 seconds, the weight is about 2 inches above the equilibrium position.

17. (a) If the spring is pulled down 2 inches, then $a = -2$ Now use the given period to find w:
$\frac{2\pi}{w} = \frac{1}{3} \Rightarrow w = 6\pi$ Therefore the equation is $s(t) = -2\cos 6\pi t$

(b) The frequency is: $\frac{w}{2\pi} = \frac{6\pi}{2\pi} = 3$ cycles/second.

19. (a) From the equation the amplitude is $a = 2$, the spring is compressed 2 inches.

(b) From the equation $w = 2\pi$ therefore the frequency is $\frac{w}{2\pi} = \frac{2\pi}{2\pi} = 1$ cycle per second.

(c) Graph the equation on a calculator, using the capabilities of the calculator $D(t) = \frac{1}{e}$ inches when $t \approx 0.2053$ sec.

Reviewing Basic Concepts (Sections 8.7 – 8.8)

1. Graph the cosecant function with a period of 2π and a phase shift of $\frac{\pi}{4}$. The domain of the function is $\left\{x \mid x \neq \frac{\pi}{4} + n\pi, \text{ where } n \text{ is an integer}\right\}$ and the range is $(-\infty, -1] \cup [-1, \infty)$. See Figure 1.

2. Graph the secant function with a period of 2π and a phase shift of $-\frac{\pi}{4}$. The domain of the function is $\left\{x \mid x \neq \frac{\pi}{4} + n\pi, \text{ where } n \text{ is an integer}\right\}$ and the range is $(-\infty, -1] \cup [-1, \infty)$. See Figure 2.

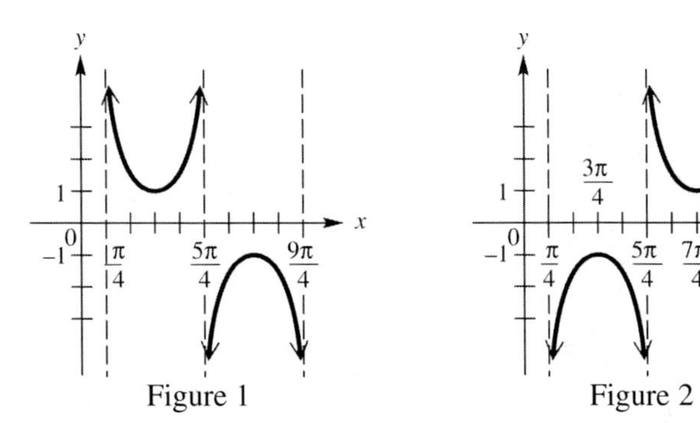

Figure 1 Figure 2

3. Graph the tangent function with a period of π. The domain of the function is $\left\{x \mid x \neq \frac{\pi}{2} + n\pi, \text{ where } n \text{ is an integer}\right\}$ and the range is $(-\infty, \infty)$. See Figure 3.

4. Graph the cotangent function with a period of π. The domain of the function is $\{x \mid x \neq n\pi, \text{ where } n \text{ is an integer}\}$ and the range is $(-\infty, \infty)$. See Figure 4.

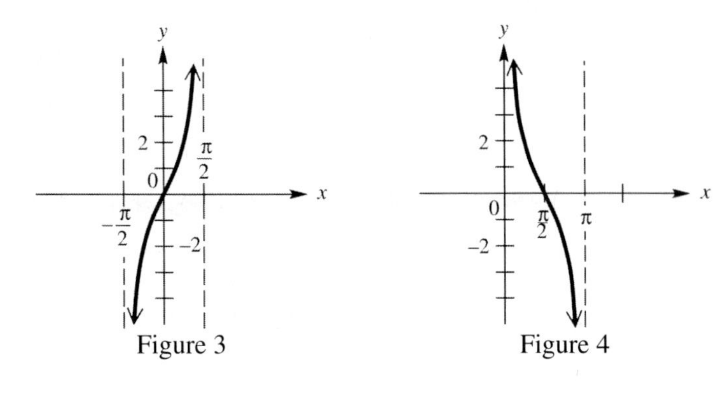

Figure 3 Figure 4

5. (a) From the equation, $a = -4$, therefore the amplitude is $|-4| = 4$ The maximum height is 4 inches.

(b) Solve for t: $4 = -4\cos 8\pi t \Rightarrow -1 = \cos 8\pi t \Rightarrow 8\pi t = \pi \Rightarrow t = \frac{\pi}{8\pi} \Rightarrow t = \frac{1}{8}$ seconds.

(c) The period is: $\frac{2\pi}{w} = \frac{2\pi}{8\pi} = \frac{1}{4}$; and the frequency is: $\frac{w}{2\pi} = \frac{8\pi}{2\pi} = 4$ cycles/second.

Chapter 8 Review

1. The smallest positive degree measure of the coterminal angle is: $-174° + 360° = 186°$.
2. All degree measure coterminal angles will be multiples of $360°$, therefore: $270° + n \cdot 360°$.
3. First find the radians per second: $\theta = \frac{320(2\pi)}{60} \Rightarrow \theta = \frac{32\pi}{3}$ radians per second. Now multiply by $\frac{2}{3}$ seconds

   and by $\frac{180}{\pi}$ to change to degrees: $\theta = \frac{32\pi}{3} \cdot \frac{2}{3} \cdot \frac{180}{\pi} \Rightarrow \theta = 1280°$.
4. First find degree rotated per second: $\theta = \frac{650(360°)}{60} \Rightarrow \theta = 3900°$. Now multiply by 2.4 seconds:

   $3900.2.4 = 9360°$.
5. 1 radian $= 1 \cdot \frac{180}{\pi} = 57.3°$ therefore 1 radian $\approx 57.3° > 1°$.
6. (a) Change to degrees and determine the quadrant: $3 \cdot \frac{180}{\pi} = 171.9°$, therefore quadrant II.

   (b) Change to degrees and determine the quadrant: $4 \cdot \frac{180}{\pi} = 292.2°$, therefore quadrant III.

   (c) Change to degrees and determine the quadrant: $-2 \cdot \frac{180}{\pi} = 114.6°$, therefore quadrant III.

   (d) Change to degrees and determine the quadrant: $7 \cdot \frac{180}{\pi} = 401.1° - 360° = 41.1°$, therefore quadrant I.
7. $120° \cdot \frac{\pi}{180} = \frac{2\pi}{3}$
8. $800° \cdot \frac{\pi}{180} = \frac{40\pi}{9}$
9. $\frac{5\pi}{4} \cdot \frac{180}{\pi} = 225°$
10. $-\frac{6\pi}{5} \cdot \frac{\pi}{180} = -216°$
11. Using the arc length formula, 1 rotation has a length: $s = r\theta \Rightarrow s = 2(2\pi) \Rightarrow s = 4\pi$ inches. Then 20 minutes

    equals: $\frac{20}{60} \cdot 4\pi = \frac{4\pi}{3}$ inches.
12. Using the arc length formula, 1 rotation has a length: $s = r\theta \Rightarrow s = 2(2\pi) \Rightarrow s = 4\pi$ inches. Then 3 hours

    equals: $3.4\pi = 12\pi$ inches.

13. Since arc length is found by the formula $s = r\theta$, the radius is: $s = 15.2 \cdot \frac{3\pi}{4} \Rightarrow s = 35.8$ cm.

14. Using the area of a sector formula $A = \frac{1}{2}r^2\theta$ yields: $A = \frac{1}{2}(28.69)^2\left(\frac{7\pi}{4}\right) \Rightarrow A \approx 2263 \text{ in}^2$.

15. Because the central angle is very small, the arc length is approximately equal to the length of the inscribed chord. Using the arc length formula yields: $s = 2000\left(1 + \frac{10}{60}\right)\left(\frac{\pi}{180}\right) \approx 41$ yards.

16. Since arc length is found by the formula $s = r\theta$, the central angle is: $4 = 8\theta \Rightarrow \theta = \frac{4}{8} \Rightarrow \theta = \frac{1}{2}$ radians. Now using the area of a sector formula $A = \frac{1}{2}r^2\theta$ yields: $A = \frac{1}{2}(8)^2\left(\frac{1}{2}\right) \Rightarrow A = 16 \text{ units}^2$.

17. First find the value of r: $r = \sqrt{(-3)^2 + (-3)^2} \Rightarrow r = \sqrt{18.} = 3\sqrt{2}$. Now each of the following is:

$\sin\theta = \frac{y}{r} = \frac{-3}{3\sqrt{2}} = -\frac{\sqrt{2}}{2}$ $\cos\theta = \frac{x}{r} = \frac{-3}{3\sqrt{2}} = -\frac{\sqrt{2}}{2}$ $\tan\theta = \frac{y}{x} = \frac{-3}{-3} = 1$

$\csc\theta = \frac{r}{y} = \frac{3\sqrt{2}}{-3} = -\sqrt{2}$ $\sec\theta = \frac{r}{x} = \frac{3\sqrt{2}}{-3} = -\sqrt{2}$ $\cot\theta = \frac{x}{y} = \frac{-3}{-3} = 1$

18. First find the value of r: $r = \sqrt{(1)^2 + (-\sqrt{3})^2} \Rightarrow r = \sqrt{4.} = 2$. Now each of the following is:

$\sin\theta = \frac{y}{r} = \frac{-\sqrt{3}}{2} = -\frac{\sqrt{3}}{2}$ $\cos\theta = \frac{x}{r} = \frac{1}{2}$ $\tan\theta = \frac{y}{x} = \frac{-\sqrt{3}}{1} = -\sqrt{3}$

$\csc\theta = \frac{r}{y} = \frac{2}{-\sqrt{3}} = -\frac{2\sqrt{3}}{3}$ $\sec\theta = \frac{r}{x} = \frac{2}{1} = 2$ $\cot\theta = \frac{x}{y} = \frac{1}{-\sqrt{3}} = -\frac{\sqrt{3}}{3}$

19. Using the calculator:

$\sin 180° = 0$ $\cos 180° = -1$ $\tan 180° = 0$

$\csc 180°$ is undefined $\sec 180° = -1$ $\cot 180°$ is undefined

20. First find the value of r: $r = \sqrt{(3)^2 + (-4)^2} \Rightarrow r = \sqrt{25} = 5$. Now each of the following is:

$\sin\theta = \frac{y}{r} = \frac{-4}{5} = -\frac{4}{5}$ $\cos\theta = \frac{x}{r} = \frac{3}{5}$ $\tan\theta = \frac{y}{x} = \frac{-4}{3} = -\frac{4}{3}$

$\csc\theta = \frac{r}{y} = \frac{5}{-4} = -\frac{5}{4}$ $\sec\theta = \frac{r}{x} = \frac{5}{3}$ $\cot\theta = \frac{x}{y} = \frac{3}{-4} = -\frac{3}{4}$

21. First find the value of r: $r = \sqrt{(9)^2 + (-2)^2} \Rightarrow r = \sqrt{85}$. Now each of the following is:

$\sin\theta = \frac{y}{r} = \frac{-2}{\sqrt{85}} = -\frac{2\sqrt{85}}{85}$ $\cos\theta = \frac{x}{r} = \frac{9}{\sqrt{85}} = \frac{9\sqrt{85}}{85}$ $\tan\theta = \frac{y}{x} = \frac{-2}{9} = -\frac{2}{9}$

$\csc\theta = \frac{r}{y} = \frac{\sqrt{85}}{-2} = -\frac{\sqrt{85}}{2}$ $\sec\theta = \frac{r}{x} = \frac{\sqrt{85}}{9}$ $\cot\theta = \frac{x}{y} = \frac{9}{-2} = -\frac{9}{2}$

22. First find the value of r: $r = \sqrt{(-2\sqrt{2})^2 + (2\sqrt{2})^2} \Rightarrow r = \sqrt{16.} = 4$. Now each of the following is:

$\sin\theta = \frac{y}{r} = \frac{2\sqrt{2}}{4} = \frac{\sqrt{2}}{2}$ $\cos\theta = \frac{x}{r} = \frac{-2\sqrt{2}}{4} = -\frac{\sqrt{2}}{2}$ $\tan\theta = \frac{y}{x} = \frac{2\sqrt{2}}{-2\sqrt{2}} = -1$

$\csc\theta = \frac{r}{y} = \frac{4}{2\sqrt{2}} = \sqrt{2}$ $\sec\theta = \frac{r}{x} = \frac{4}{-2\sqrt{2}} = -\sqrt{2}$ $\cot\theta = \frac{x}{y} = \frac{-2\sqrt{2}}{2\sqrt{2}} = -1$

23. If the terminal side of a quadrantal angle lies along the y-axis, a point on the terminal side would be of the form $(0, k)$ where k is a real number, $k \neq 0$

$\sin\theta = \frac{y}{r} = \frac{k}{r}$ $\cos\theta = \frac{x}{r} = \frac{0}{r} = 0$ $\tan\theta = \frac{y}{x} = \frac{k}{0}$ is undefined

$\csc\theta = \frac{r}{y} = \frac{r}{k}$ $\sec\theta = \frac{r}{x} = \frac{r}{0}$ is undefined $\cot\theta = \frac{x}{y} = \frac{0}{k} = 0$

24. Since the range of $\sec\theta$ is: $\sec\theta \leq -1$ or $\sec\theta \geq 1, \sec\theta = -\frac{2}{3}$ is impossible.

25. Since the range of $\tan\theta$ is: $-\infty \leq \tan\theta \leq \infty, \tan\theta = 1.4$ is possible.

26. Since $\sin\theta = \frac{\sqrt{3}}{5}, y = \sqrt{3}$ and $r = 5$. Now use Pythagorean Theorem to solve for x: $r^2 = x^2 + y^2 \Rightarrow$

$25 = x^2 + 3 \Rightarrow x^2 = 22 \Rightarrow x = \pm\sqrt{22}$. Since $\cos\theta < 0, x = -\sqrt{22}.$ Therefore,

$\sin\theta = \frac{\sqrt{3}}{5}$ $\cos\theta = \frac{x}{r} = \frac{-\sqrt{22}}{5} = -\frac{\sqrt{22}}{5}$ $\tan\theta = \frac{y}{x} = \frac{\sqrt{3}}{-\sqrt{22}} = -\frac{\sqrt{66}}{22}$

$\csc\theta = \frac{r}{y} = \frac{5}{\sqrt{3}} = \frac{5\sqrt{3}}{3}$ $\sec\theta = \frac{r}{x} = \frac{5}{-\sqrt{22}} = -\frac{5\sqrt{22}}{22}$ $\cot\theta = \frac{x}{y} = \frac{-\sqrt{22}}{\sqrt{3}} = -\frac{\sqrt{66}}{3}$

27. Since $\cos\theta = \frac{\sqrt{3}}{5}, x = -5$ and $r = 8$. Now use Pythagorean Theorem to solve for x:

$r^2 = x^2 + y^2 \Rightarrow 64 = 25 + y^2 \Rightarrow y^2 = 39 \Rightarrow y = \pm\sqrt{39}.$ Since θ is in quadrant III, $y = -\sqrt{39}.$ Therefore,

$\sin\theta = \frac{y}{r} = \frac{-\sqrt{39}}{8} = -\frac{\sqrt{39}}{8}$ $\cos\theta = -\frac{5}{8}$ $\tan\theta = \frac{y}{x} = \frac{-\sqrt{39}}{-5} = \frac{\sqrt{39}}{5}$

$\csc\theta = \frac{r}{y} = \frac{8}{-\sqrt{39}} = -\frac{8\sqrt{39}}{39}$ $\sec\theta = \frac{r}{x} = \frac{8}{-5} = -\frac{8}{5}$ $\cot\theta = \frac{x}{y} = \frac{-5}{-\sqrt{39}} = \frac{5\sqrt{39}}{39}$

28. The sine function is negative in quadrant III and IV. The cosine function is positive in quadrants I and IV. Therefore, since $\sin\theta < 0$ and $\cos > 0$, θ must be in quadrant IV and in quadrant IV $\tan\theta$ is negative.

29. $\sin A = \frac{\text{opp}}{\text{hyp}} = \frac{40}{58} = \frac{20}{29}$ $\cos A = \frac{\text{adj}}{\text{hyp}} = \frac{42}{58} = \frac{21}{29}$ $\tan A = \frac{\text{opp}}{\text{adj}} = \frac{40}{42} = \frac{20}{21}$

$\csc A = \frac{\text{hyp}}{\text{opp}} = \frac{58}{40} = \frac{29}{20}$ $\sec A = \frac{\text{hyp}}{\text{adj}} = \frac{58}{42} = \frac{29}{21}$ $\cot A = \frac{\text{adj}}{\text{opp}} = \frac{42}{40} = \frac{21}{20}$

30. $\sin A = \frac{\text{opp}}{\text{hyp}} = \frac{60}{61}$ $\cos A = \frac{\text{adj}}{\text{hyp}} = \frac{11}{61}$ $\tan A = \frac{\text{opp}}{\text{adj}} = \frac{60}{11}$

$\csc A = \frac{\text{hyp}}{\text{opp}} = \frac{61}{60}$ $\sec A = \frac{\text{hyp}}{\text{adj}} = \frac{61}{11}$ $\cot A = \frac{\text{adj}}{\text{opp}} = \frac{11}{60}$

31. First find the reference angle: $(360° - 300°) = 60°$. Using the $30°$-$60°$ right triangle rules, for a $60°$ angle the hypotenuse $= 2$, the opposite side $= \sqrt{3}$ and the adjacent side=1.Since $\theta = 300°$ is in quadrant IV, y is negative and sine, tangent, cotangent, and cosecant are negative, therefore:

$\sin 300° = -\sin 60° = -\frac{\sqrt{3}}{2}$ $\csc 300° = -\csc 60° = -\frac{2}{\sqrt{3}} = -\frac{2\sqrt{3}}{3}$

$\cos 300° = \cos 60° = \frac{1}{2}$ $\sec 300° = \sec 60° = \frac{2}{1} = 2$

$\tan 300° = -\tan 60° = -\frac{\sqrt{3}}{1} = -\sqrt{3}$ $\cot 300° = -\cot 60° = -\frac{1}{\sqrt{3}} = -\frac{\sqrt{3}}{3}.$

32. First find the reference angle: $-225°$ is coterminal with $-225° + 360° = 135°$ and the reference angle is: $360° - 135° = 45°$. Using the $45°$-$45°$ right triangle rules, for a $45°$ angle the hypotenuse $= \sqrt{2}$, the opposite side=1, and the adjacent=1. Since $\theta = -225°$ is in quadrant II, x is negative and cosine, tangent, cotangent, and secant are negative, therefore:

$\sin(-225°) = \sin 45° = \frac{1}{\sqrt{2}} = \frac{\sqrt{2}}{2}$ $\csc(-225°) = \csc 45° = \frac{\sqrt{2}}{1} = \sqrt{2}$

$\cos(-225°) = -\cos 45° = -\frac{1}{\sqrt{2}} = -\frac{\sqrt{2}}{2}$ $\sec(-225°) = -\sec 45° = -\frac{\sqrt{2}}{1} = -\sqrt{2}$

$\tan(-225°) = -\tan 45° = -\frac{1}{1} = -1$ $\cot(-225°) = -\cot 45° = -\frac{1}{1} = -1$

33. First find the reference angle: $-390°$ is co-terminal with $-390° + 720° = 300°$ and the reference angle is: $360° - 330° = 30°$. Using the $30° - 60°$ right triangle rules, for a $30°$ angle the hypotenuse $= 2$, the opposite side=1, and the adjacent $= \sqrt{3}$. Since $\theta = -390°$ is in quadrant IV, y is negative and sine, tangent, cotangent, and cosecant are negative, therefore:

$\sin(-390°) = -\sin 30° = -\frac{1}{2}$ $\csc(-390°) = -\csc 30° = -\frac{20}{1} = -2$

$\cos(-390°) = \cos 30° = -\frac{\sqrt{3}}{2}$ $\sec(-390°) = \sec 30° = -\frac{2}{\sqrt{3}} = \frac{2\sqrt{3}}{3}$

$\tan(-390°) = -\tan 30° = -\frac{1}{\sqrt{3}} = -\frac{\sqrt{3}}{3}$ $\cot(-390°) = -\cot 30° = -\frac{\sqrt{3}}{1} = -\sqrt{3}$

34. $\sin 72°30' = \sin\left(72 + \frac{30}{60}\right)^{\circ} = \sin 72° \approx .95371695$

35. $\sec 222°30' = \sec\left(222 + \frac{30}{60}\right)° = \sec 222.5° = (\cos 222.5°)^{-1} \approx -1.3563417$

36. $\cot 305.6° = (\tan 305.6°)^{-1} \approx -.7159268$

37. $\tan 11.7689° \approx .20834446$

38. If $\theta = 135°, \theta' = 45°$, if $\theta = 45°, \theta' = 45°$, if $\theta = 300°, \theta' = 60°$, and if $\theta = 140°, \theta' = 40°$. Of these reference angles, 40° is the only one which is not a special angle, so D, tan 140°, is the only one which cannot be determined exactly.

39. $\theta = \sin^{-1} .82584121 \approx 55.673870° \approx 55.7°$

40. $\cot\theta = 1.1249386 \Rightarrow \frac{1}{\tan\theta} = 1.1249386 \Rightarrow \tan\theta = (1.1249386)^{-1} \Rightarrow \theta = \tan^{-1}(1.1249386)^{-1} \approx$

$41.635092° \approx 41.6°$

41. To find B: $B = 90° - 58°30' \Rightarrow B = 89°60' - 58°30' \Rightarrow B = 31°30'$.

To find a: $\sin A = \frac{a}{c} \Rightarrow \sin 58°30' = \frac{a}{748} \Rightarrow a = 748(\sin 58°30') \approx 637.7748 \Rightarrow a \approx 638.$

To find b: $\cos A = \frac{b}{c} \Rightarrow \cos 58°30' = \frac{b}{748} \Rightarrow b = 748(\cos 58°30') \approx 390.8289 \Rightarrow b \approx 391$

42. To find B: $B = 90° - 39.72° \Rightarrow B = 50.28°$.

To find a: $\tan A = \frac{a}{c} \Rightarrow \tan 39.72° = \frac{a}{38.97} \Rightarrow a = 38.97(\tan 39.72°) \approx 32.3765 \Rightarrow a \approx 32.38$

To find c: $\cos A = \frac{b}{c} \Rightarrow \cos 39.72° = \frac{38.97}{c} \Rightarrow \mathrm{c} = \frac{38.97}{\cos 39.72°} \approx 50.6646 \Rightarrow \mathrm{c} \approx 50.66$

43. Draw a picture of these points (not shown). Since $344° - 254° = 90°$, points A, B, and C form a right triangle with angle C the right angle and angle. $B = 42°((254° - 180°) - 32°)$.

Now set $d =$ the distance between point A and point B, now use the sine function to solve for d:

$\sin 42° = \frac{780}{d} \Rightarrow d = \frac{780}{\sin 42°} \Rightarrow d \approx 1165.692$. The distance from A to B is approximately 1200 meters.

44. Draw a picture of these points (not shown). A right triangle is formed since the angle where the ship turns is $90°(35° + 55°$ by alternate interior angles$)$. Solve by Pythagorean theorem:

$d^2 = 80^2 + 74^2 \Rightarrow d^2 = 11{,}876 \Rightarrow d \approx 108.97706$. The ship is approximately 110 km from the pier.

45. A right triangle is formed with a bottom angle of $36°(360° - 324°)$, and the adjacent side to this 36° angle is 110 mph $(2 \cdot 55)$. Let x be the distance between the cars and solve for x using cosine:

$\cos 36° = \frac{110}{x} \Rightarrow x = \frac{110}{\cos 36°} \Rightarrow x \approx 135.96748$. The cars are approximately 140 miles apart.

46. Two right triangles are formed. Let x be the distance the boat travels, and y be the distance from shore at the second observation. First, solve for y using the larger right triangle: $\angle 1 = 27°$ (by alternate interior angles)

$\Rightarrow \tan 27° = \dfrac{150}{x+y} \Rightarrow x + y = \dfrac{150}{\tan 27°} \Rightarrow y = \dfrac{150}{\tan 27°} - x.$ Next, solve for y using the smaller right triangle:

$< 2 = 39°$ (by alternate interior angles) $\Rightarrow \tan 39° = \dfrac{150}{y} \Rightarrow y = \dfrac{150}{\tan 39°}.$ Finally, since each equation equals

y, set them equal to each other: $\dfrac{150}{\tan 27°} - x = \dfrac{150}{\tan 39°} \Rightarrow x = \dfrac{150}{\tan 27°} \Rightarrow x \approx 109.157.$

The boat travels about 109 feet.

47. Let $h =$ height of the tower, now use the tangent function to solve for h:

$\tan 38°20' = \dfrac{h}{93.2} \Rightarrow h = 93.2 \tan 38°20' \approx 73.69300534 \Rightarrow h \approx 73.7$ feet.

48. The angle at the point on the ground is 29.5° (by geometry, alternate interior angles are equal). Now set $h =$ height of the tower and use the tangent function to solve for h:

$\tan 29.5° = \dfrac{h}{36.0} \Rightarrow h = 36.0 \tan 29.5° \Rightarrow h \approx 20.36782 \Rightarrow h \approx 20.4$ meters.

49. Let x be the length of the base of the smaller right triangle. First, solve for x using the larger triangle:

$\tan 29° = \dfrac{h}{392 + x} \Rightarrow 392 + x = \dfrac{h}{\tan 29°} \Rightarrow x = \dfrac{h}{\tan 29°} - 392.$ Next, solve for x using the smaller triangle:

$\tan 49° = \dfrac{h}{x} \Rightarrow x = \dfrac{h}{\tan 49°}.$ Finally, since each equation is equal to x, set them equal to each other:

$\dfrac{h}{\tan 29°} - 392 = \dfrac{h}{\tan 49°} \Rightarrow \dfrac{h}{\tan 29°} - \dfrac{h}{\tan 49°} = 392 \Rightarrow h \tan 49° - h \tan 29° = 392 \tan 29° \tan 49° \Rightarrow$

$h(\tan 49° - \tan 29°) = 392 \tan 29° \tan 49° \Rightarrow h = \dfrac{392 \tan 29° \tan 49°}{\tan 49° - \tan 29°} \Rightarrow h \approx 419.3585 \Rightarrow h \approx 419.$

50. (a) $\sin\theta = \dfrac{x_Q - x_P}{d} \Rightarrow x_Q = x_P + d\sin\theta.$ Similarly, $\cos\theta = \dfrac{y_Q - y_P}{d} \Rightarrow y_Q = y_P + d\cos\theta.$

(b) Using the given information yields:

$x_Q = x_P + d\sin\theta \Rightarrow 123.62 + 193.86 \sin 17°19'22'' = 123.62 + 193.86 \sin 17.3228° \Rightarrow x_Q \approx 181.34$

$y_Q = y_P + d\cos\theta \Rightarrow 337.95 + 193.86 \cos 17°19'22'' = 337.95 + 193.86 \cos 17.3228° \Rightarrow y_Q \approx 523.02$

The coordinates of Q are $(181.34, 523.02)$.

51. From the properties 30°-60° of right triangles, an angle of $s = \dfrac{2\pi}{3}$ radians intersects the unit circle at the point $\left(-\dfrac{1}{2}, \dfrac{\sqrt{3}}{2}\right)$. Since $\cos s = x, \cos\dfrac{2\pi}{3} = -\dfrac{1}{2}.$

52. From the properties $30°\text{-}60°$ of right triangles, an angle of $s=-\frac{11\pi}{6}$ radians intersects the unit circle at the point $\left(\frac{\sqrt{3}}{2},\frac{1}{2}\right)$. Since $\sin s=y, \sin\left(-\frac{11\pi}{6}\right)=\frac{1}{2}$.

53. First a coterminal angle: $4\pi-\frac{7\pi}{3}=\frac{5\pi}{3}\Rightarrow\theta'=-\frac{\pi}{3},$ From the properties of $30°\text{-}60°$ right triangles, an angle of $s=-\frac{\pi}{3}$ radians intersects the unit circle at $\left(\frac{1}{2},-\frac{\sqrt{3}}{2}\right)$. Since $\tan s=\frac{y}{x},\ \tan-\frac{\pi}{3}=-\frac{\sqrt{3}}{2}\cdot\frac{2}{1}=-\sqrt{3}.$

54. From the properties $45°\text{-}45°$ of right triangles, an angle of $s=\frac{5\pi}{4}$ radians intersects the unit circle at the point $\left(-\frac{\sqrt{2}}{2},-\frac{\sqrt{2}}{2}\right)$. Since $\cot s=\frac{x}{y}, \cot\left(\frac{5\pi}{4}\right)=\frac{-\frac{\sqrt{2}}{2}}{-\frac{\sqrt{2}}{2}}=1$.

55. First find $\theta:\theta'=2\pi-\frac{11\pi}{6}\Rightarrow\theta'=\frac{\pi}{6}$. From the properties of $30°\text{-}60°$ right triangles, an angle of $s=\frac{\pi}{6}$ radians intersects the unit circle at the point $\left(\frac{\sqrt{3}}{2},\frac{1}{2}\right)$. Since $\csc s=\frac{1}{y},\ \csc\frac{\pi}{6}=\frac{1}{\frac{1}{2}}=2$

56. First a coterminal angle: $4\pi-3\pi=\pi\Rightarrow\theta'=\pi,$ an angle of $s=\pi$ radians intersects the unit circle at $(-1,0)$. Since $\sec s=\frac{1}{x},\ \sec 3\pi=\frac{1}{-1}=-1.$

57. Using the calculator (radian mode), $\cos(-.2443)\approx.97030688\approx.9703$

58. Using the calculator (radian mode), $\cot 3.0543=[\tan 3.0543]^{-1}\approx-11.426605\approx-11.4266$.

59. From the properties of $30°\text{-}60°$ right triangles, an angle of $s=\frac{\pi}{3}$ radians intersects the unit circle at the point $\left(\frac{1}{2},\frac{\sqrt{3}}{2}\right)$. Since $\cos s=\frac{1}{2}\Rightarrow s=\frac{\pi}{3}$

60. Using the calculator (radian mode), if $\sin s=.49244294,$ then $s=\sin^{-1}.49244294\approx.51489440\approx.5149$

61. First find θ. Since this is a unit circle, $\sin\theta=y,$ therefore $\sin\theta=-.5250622$ and $\theta=\sin^{-1}-.5250622\Rightarrow$ $\theta\approx-.5528.$ Now with $r=1,$ use the arc length formula $s=|r\theta|\Rightarrow s\approx|(1)(-.5528)|\Rightarrow s\approx.5528.$ The length of the shortest arc of the circle from $(1,0)$ to $(.85106383, -.5250622)$ is .5528.

62. B, from the equation, the amplitude is 4 and the period is $\frac{2\pi}{2}=\pi.$

63. The amplitude is 2, the period is $\frac{2\pi}{(1)}=2\pi,$ and there is no vertical translation or phase shift. See Figure 63.

64. The amplitude (since a tangent function) is not applicable, the period is $\frac{\pi}{(3)}$, and there is no vertical translation or phase shift. See Figure 64.

65. The amplitude is $\frac{1}{2}$, the period is $\frac{2\pi}{(3)}$, and there is no vertical translation or phase shift. See Figure 65.

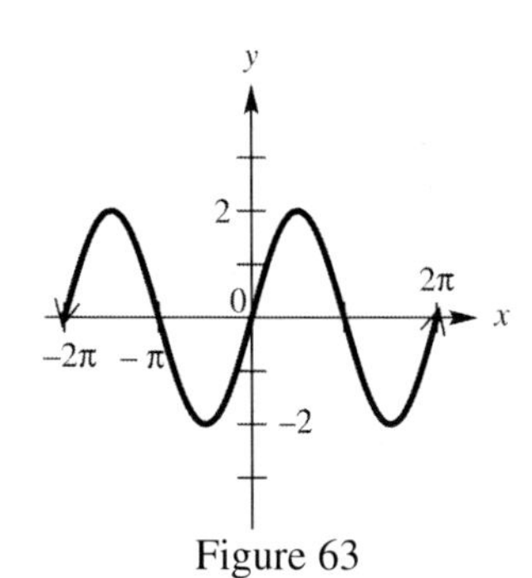

Figure 63

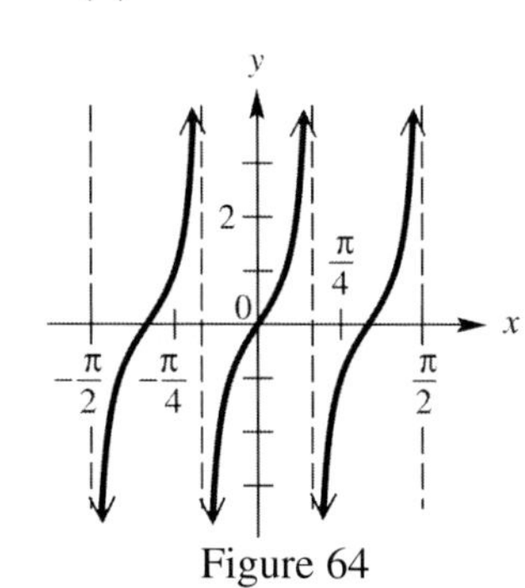

Figure 64

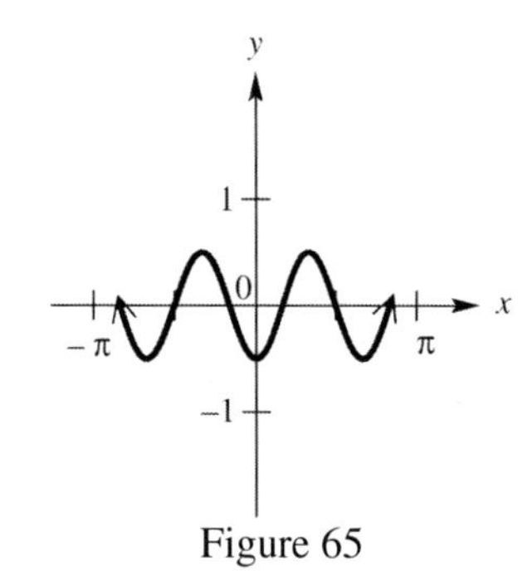

Figure 65

66. The amplitude is 2, the period is $\frac{2\pi}{(5)}$, and there is no vertical translation or phase shift. See Figure 66.

67. The amplitude is 2, the period is $\frac{2\pi}{\left(\frac{1}{4}\right)} = 8\pi$, the vertical shift is 1 unit upward, and there is no phase shift. See Figure 67.

68. The amplitude is $\frac{1}{4}$, the period is $\frac{2\pi}{\left(\frac{2}{3}\right)} = 3\pi$, the vertical shift is 3 units upward, and there is no phase shift. See Figure 68.

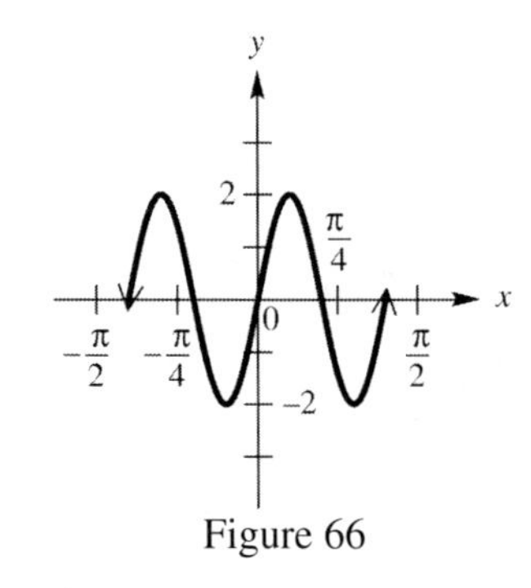

Figure 66

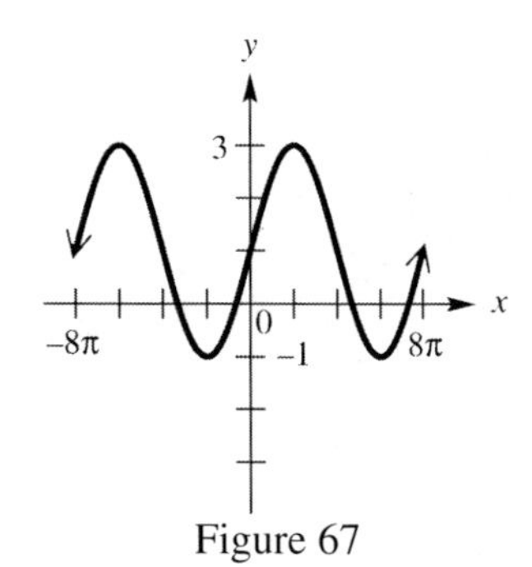

Figure 67

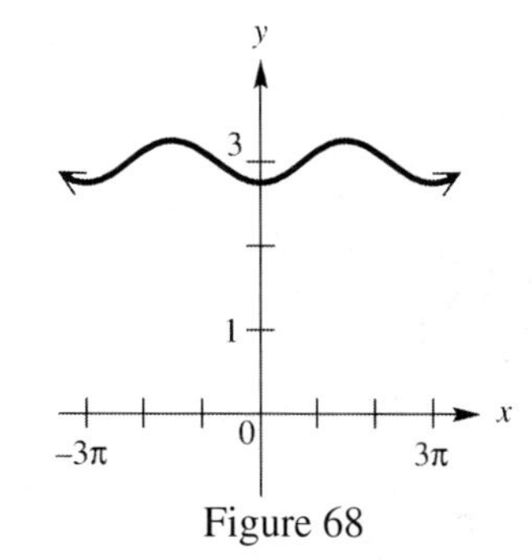

Figure 68

69. The amplitude is 3, the period is $\frac{2\pi}{(1)} = 2\pi$, there is no vertical shift, and the phase shift is $-\frac{\pi}{2}$. See Figure 69.

70. The amplitude is $|-1| = 1$, the period is $\frac{2\pi}{(1)} = 2\pi$, there is no vertical shift, and the phase shift is $\frac{3\pi}{4}$. See Figure 70.

71. Since $y=\frac{1}{2}\csc\left(2x-\frac{\pi}{4}\right) \Rightarrow y=\frac{1}{2}\csc 2\left(x-\frac{\pi}{8}\right)$ and the equation, the amplitude is not applicable (cosecant function), the period is $\frac{2\pi}{(2)}=\pi$, there is no vertical shift, and the phase shift is $\frac{\pi}{8}$. See Figure 71.

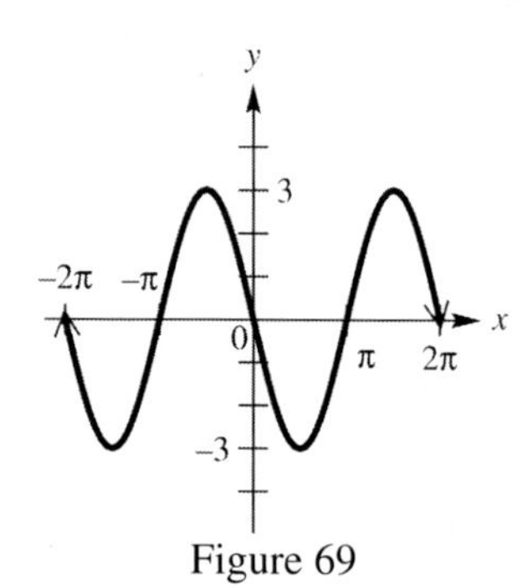

Figure 69

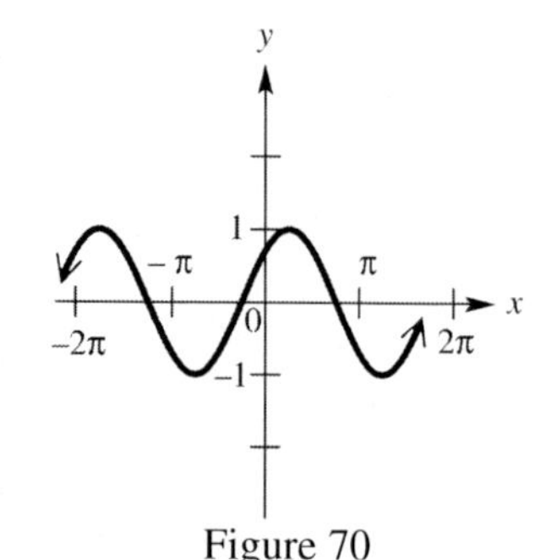

Figure 70

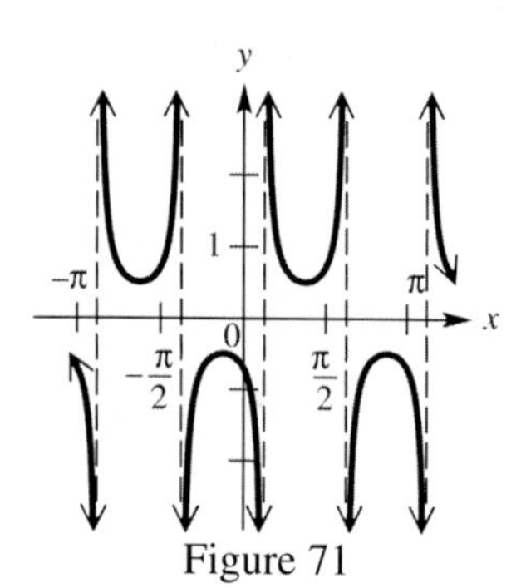

Figure 71

72. There is no amplitude, the period is $\frac{\pi}{\left(\frac{1}{2}\right)}=2\pi$, and there is no vertical translation or phase shift.

See Figure 72.

73. There is no amplitude, the period is $\frac{2\pi}{(2)}=\pi$, and there is no vertical translation or phase shift. See Figure 73.

74. Since $y=\csc(2x-\pi)=\csc 2\left(x-\frac{\pi}{2}\right)$ and now from the equation, the amplitude is not applicable, (cosecant function) the period is $\frac{2\pi}{(2)}=\pi$, there is no vertical shift, and the phase shift is $\frac{\pi}{2}$. See Figure 74.

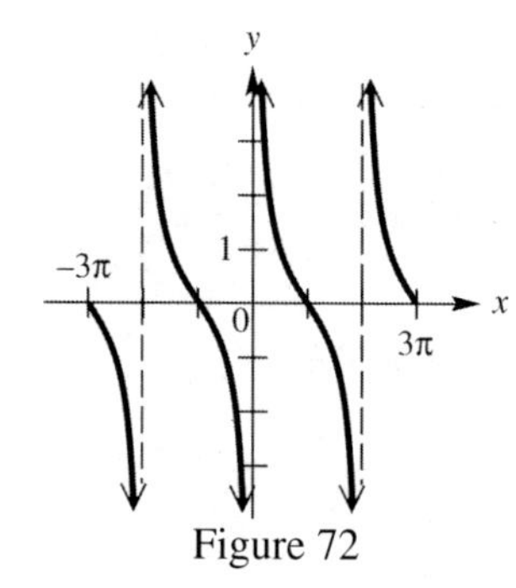

Figure 72

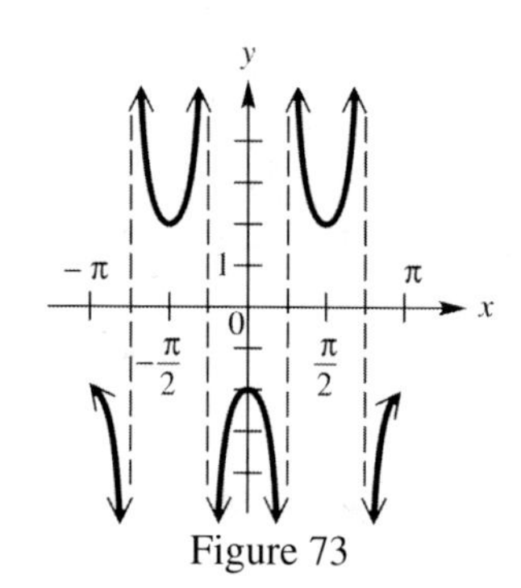

Figure 73

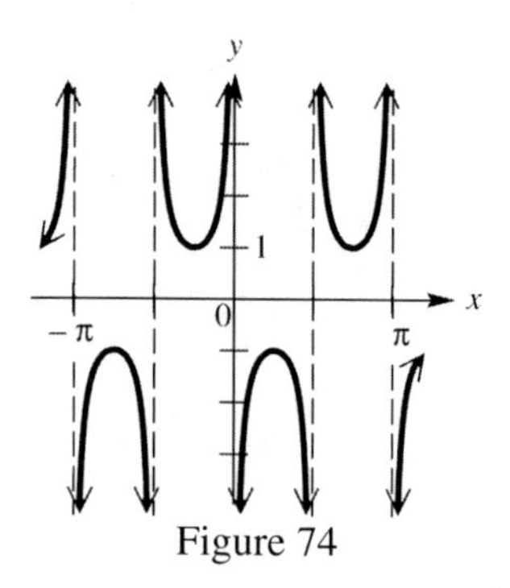

Figure 74

75. Graph a cosine function with an amplitude of 3 and a period of $\frac{2\pi}{(2)}=\pi$. See Figure 75.

76. Graph a cotangent function with a period of $\frac{\pi}{(3)}$, and passing through the points $\left(\frac{\pi}{6},0\right)$ (midpoint), $\left(\frac{\pi}{12},0\right)$, and $\left(\frac{\pi}{4},-\frac{1}{2}\right)$. See Figure 76.

77. Graph a cosine function with amplitude 1, period $\frac{2\pi}{(1)} = 2\pi$, and phase shift $\frac{\pi}{4}$ units to the right.

See Figure 77.

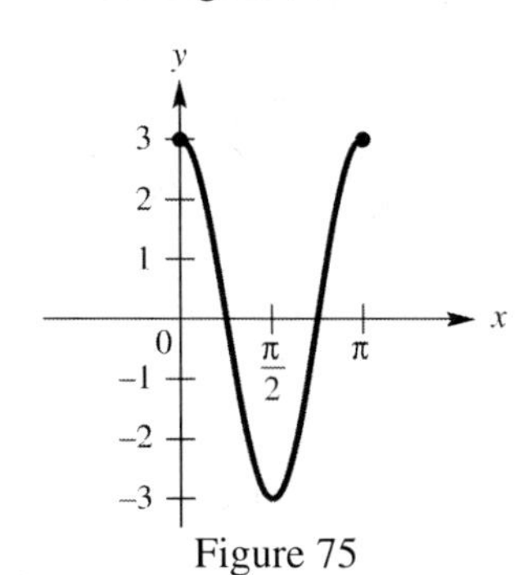

Figure 75

Figure 76

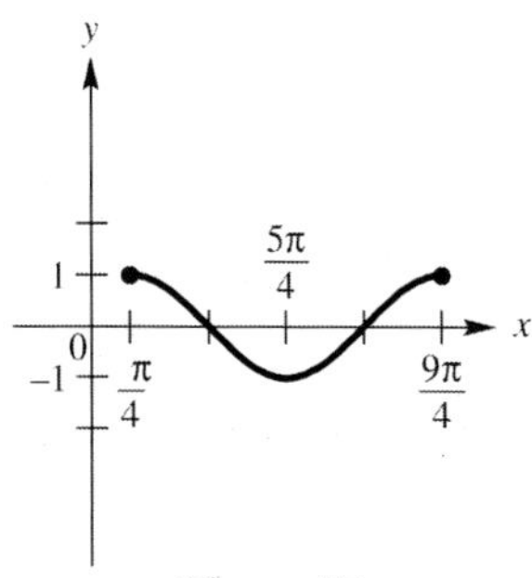

Figure 77

78. Graph a tangent function with a period of $\frac{\pi}{(1)} = \pi$, a phase shift $\frac{\pi}{2}$ units to the right, and passing through the points $\left(\frac{\pi}{2},0\right)$ (midpoint), $\left(\frac{\pi}{4},-1\right)$, and $\left(\frac{3\pi}{4},1\right)$. See Figure 78.

79. Graph a cosine function with amplitude 2, period $\frac{2\pi}{(3)}$, and vertical translation 1 unit upward. See Figure 79.

80. Graph a sine function with a amplitude of $|-3| = 3$, a period of $\frac{2\pi}{(2)} = \pi$, and a vertical translation 1 unit downward. See Figure 80.

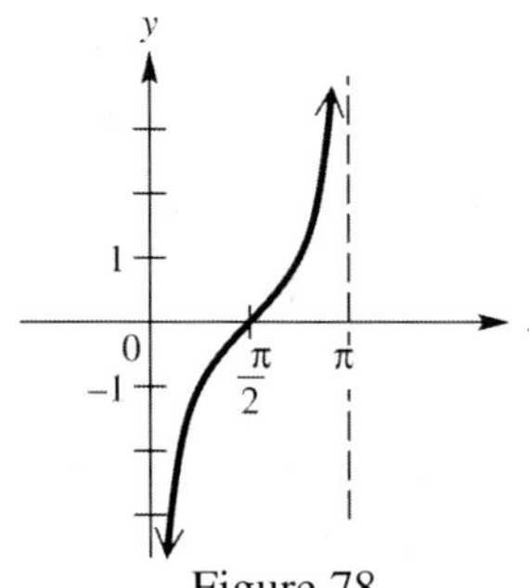

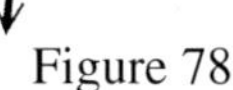
Figure 78

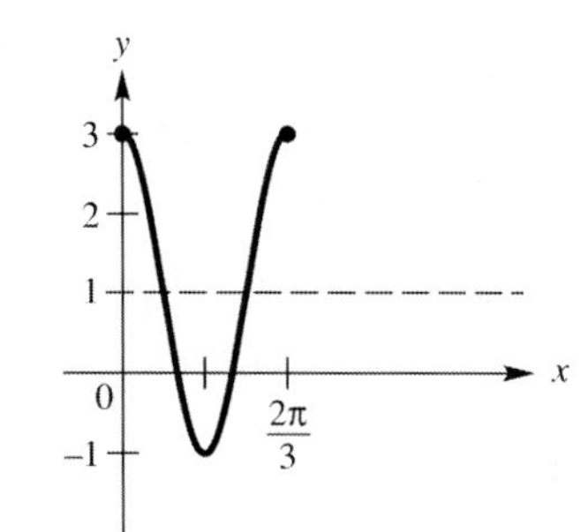

Figure 79

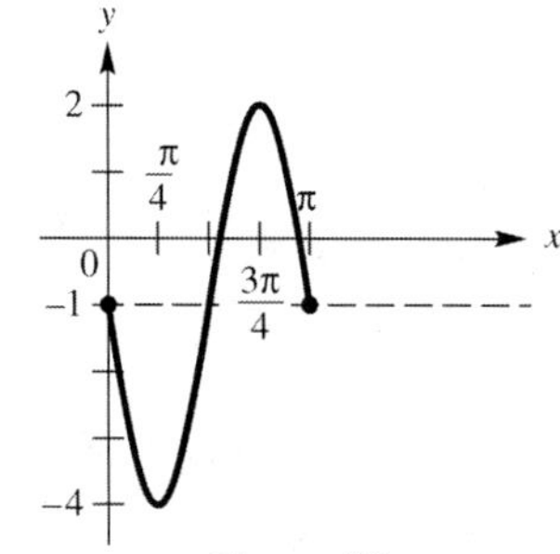

Figure 80

81. Since the period is π, the function is either $\tan x$ or $\cot x$. With an x-intercept $n\pi$, the function is tangent.

82. Since the period is 2π, the function is not $\tan x$ or $\cot x$. With the graph passing through $(0,0)$, the function is sine.

83. Since the period is 2π, the function is not $\tan x$ or $\cot x$. With the graph passing through $\left(\frac{\pi}{2},0\right)$, the function is cosine.

84. Since the period is 2π, the function is not $\tan x$ or $\cot x$. With the graph having the domain: $\{x|x \neq n\pi,$ where n is an integer$\}$ the function is cosecant.

85. Since the period is the function is either $\tan x$ or $\cot x$. With the graph decreasing over the interval $(0, \pi)$, the function is cotangent.

86. Since the period is 2π, the function is not $\tan x$ or $\cot x$. With the graph having asymptotes: $x = (2n+1)\frac{\pi}{2}$, where n is an integer, the function is secant.

87. Since the graph shows an amplitude of 3, a period of $\pi = \frac{2\pi}{(2)}$, and passes through $\left(\frac{\pi}{4}, 0\right)$ which means it has a phase shift of $\frac{\pi}{4}$, the equation is $y = 3\sin\left[2\left(x - \frac{\pi}{4}\right)\right]$. (Other answers are possible)

88. Since the graph shows an amplitude of 4, a period of $4\pi = \frac{2\pi}{\left(\frac{1}{2}\right)}$, and passes through $(0,0)$ which means it has no phase shift, the equation is $y = 4\sin\frac{1}{2}x$. (Other answers are possible)

89. Since the graph shows an amplitude of $\frac{1}{3}$, a period of $4\pi = \frac{2\pi}{\left(\frac{\pi}{2}\right)}$, and passes through $(0,0)$ which means it has no phase shift, the equation is $y = \frac{1}{3}\sin\frac{\pi}{2}x$. (Other answers are possible)

90. Since the graph shows an amplitude of π, a period of $2 = \frac{2\pi}{(\pi)}$, and passes through $\left(\frac{1}{2}, 0\right)$ which means it has a phase shift of $\frac{1}{2}$, the equation is $y = \pi\sin\left[\pi\left(x - \frac{1}{2}\right)\right]$. (Other answers are possible)

91. (a) Let January correspond to $x = 1$, Februrary to $x = 2, \ldots\ldots$, and December of the 2nd year to $x = 24$. The data appear to follow the pattern of a translated sine graph. See Figure 91a.

(b) Using the model $f(x) = a\sin\left[b(x-d)\right] + c$, we can find the constants: The maximum average monthly temperature is 72° F, and the minimum is 22° F. Let the amplitude a be $\frac{72-22}{2} = \frac{50}{2}$, so $a = 25$ .Since the period is $12 = \frac{2\pi}{b}$, $b = \frac{\pi}{6}$. The data are centered vertically around the line $y = \frac{75+25}{2} = 50$, therefore $c = 50$. The minimum temperature occurs in January. Thus, when $x = 1$, $b(x-d) = -\frac{\pi}{2}$ since the sine function is minimum at $-\frac{\pi}{2}$. Using this and solving for d yields: $\frac{\pi}{6}(1-d) = -\frac{\pi}{2} \Rightarrow d = 4$. Since the months before January are slightly colder than the months following January the value of d

can be slightly adjusted to give a better visual fit, Trying $d = 4.2$ gives a better fit. Using these constant values yields the equation: $f(x) = 25\sin\left[\frac{\pi}{6}(x-4.2)\right] + 50.$

(c) The constant values a controls the amplitude, b the period, c the vertical shift, and d the phase shift.

(d) See Figure 91d. The function gives an excellent model.

(e) The regression function gives $y = 25.77\sin(0.52x - 2.19) + 50.57$: see Figure 91e.

[1,25] by [20,80]
Xscl = 5 Yscl = 10

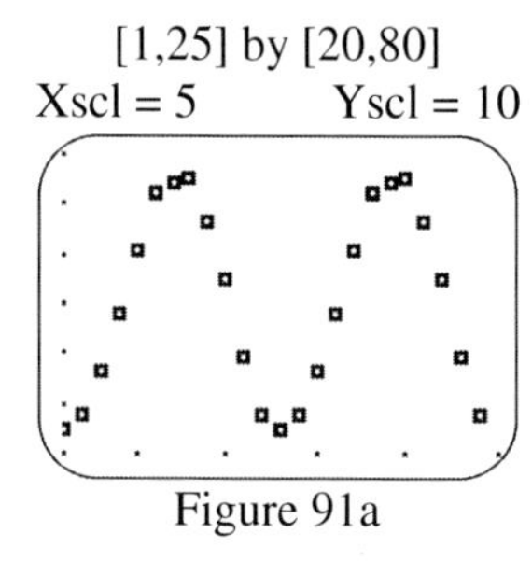

Figure 91a

[1,25] by [20,80]
Xscl = 5 Yscl = 10

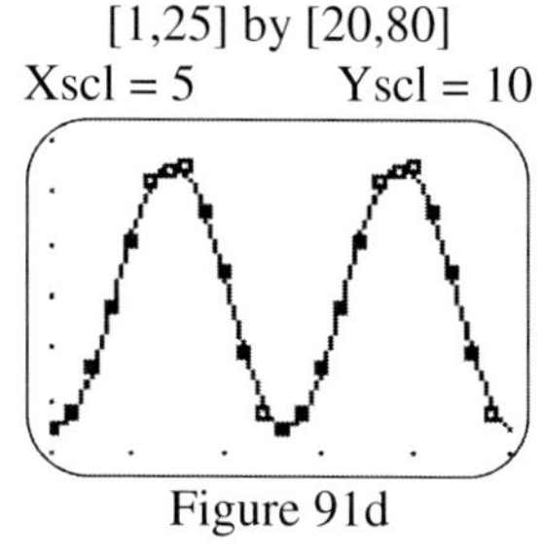

Figure 91d

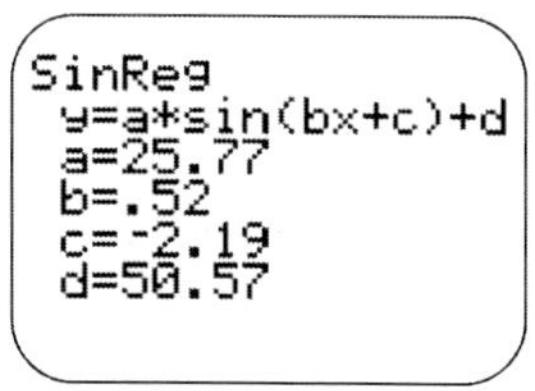

Figure 91e

92. (a) If the shorter leg of the right triangle has length $h_2 - h_2$, then $\cot\theta = \frac{d}{h_2 - h_1} \Rightarrow d = (h_2 - h_1)\cot\theta.$

(b) With the given values the equation is: $d = (55-5)cod\theta \Rightarrow d = 50\cot\theta.$ Since the period is π, and the graph wanted is for the interval $0 < \theta \le \frac{\pi}{2}$, the graph will be the left half of a cotangent function. The asymptote is the line $\theta = 0$ and when $\theta = \frac{\pi}{4}, d = 50(1) = 50.$ See Figure 92.

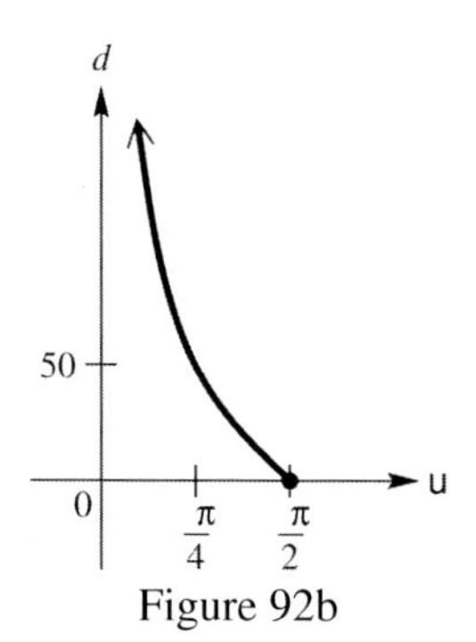

Figure 92b

93. (a) $\csc\theta = \frac{d}{h} \Rightarrow d = h\csc\theta$

(b) If $d = 2h$ and $\csc\theta = \frac{d}{h}$, then $\csc\theta = \frac{2h}{h} \Rightarrow \csc\theta = 2.$ If $\csc\theta = 2$, then $\sin\theta = \frac{1}{2}$, and $\theta = \sin^{-1}\left(\frac{1}{2}\right) = \frac{\pi}{6}.$ The value of d is double that of h when the sun is 30° above the horizon.

(c) $\csc\frac{\pi}{2} = 1$ and. $\csc\frac{\pi}{3} = 1.15.$ When the sun is lower in the sky, $\theta = \frac{\pi}{3}$, sunlight is filtered by more atmosphere. There is less ultraviolet light reaching the earth's surface, and therefore, there is less likelihood of becoming sunburned. In the case of $\frac{\pi}{3}$, sunlight passes through 15% more atmosphere.

94. (a) January 1 of the base year would be $t = 0$, which yields:

$$P(0) = 7(1 - \cos 2\pi(0))(0+10) + 100e^{2(0)} = 7(1-1)(10) + 100(1) \Rightarrow P(0) = 100$$

(b) July 1 of the base year would be $t = 5$, which yields:

$$P(.5) = 7(1 - \cos 2\pi(.5))(.5+10) + 100e^{2(.5)} = 7(1-(-1))(10.5) + 100e^{1} \Rightarrow P(.5) \approx 258$$

(c) January 1 of the following year would be $t = 1$, which yields:

$$P(1) = 7(1 - \cos 2\pi(1))(1+10) + 100e^{2(1)} = 7(1-1)(11) + 100e^{3} \Rightarrow P(1) \approx 122$$

(d) July 1 of the following year would be $t = 1.5$, which yields:

$$P(1.5) = 7(1 - \cos 2\pi(1.5))(1.5+10) + 100e^{2(1.5)} = 7(1-(-1))(11.5) + 100e^{3} \Rightarrow P(1) \approx 296$$

95. From the equation, the amplitude is 4, the period is $\frac{2\pi}{(\pi)} = 2$, and the frequency is $\frac{(\pi)}{2\pi} = \frac{1}{2}$.

96. From the equation, the amplitude is 3, the period is $\frac{2\pi}{(2)} = \pi$, and the frequency is $\frac{(2)}{2\pi} = \frac{1}{\pi}$.

97. It represents the number of oscillations in one second. The position from the initial point after 1.5 seconds is given as $4\sin\left(\frac{3\pi}{2}\right) = -4 \Rightarrow 4$ inches below the initial point; after 2 seconds, $4\sin(2\pi) = 0 \Rightarrow$ at the intial point; after 3.25 seconds, $4\sin\left(\frac{13\pi}{4}\right) = -2\sqrt{2} \Rightarrow 2\sqrt{2}$ inches below the initial point.

98. The period is the time to complete one cycle. The amplitude is the maximum distance (on either side) from the initial point.

Chapter 8 Test:

1. The smallest positive degree measure of the coterminal angle is: $-157° + 360° = 203°$.

2. First find the revolutions per second: $\frac{450 \text{ rev}}{1\text{min}} \cdot \frac{1\text{min}}{60\text{sec}} = 7.5$ revolutions per second. Now find the degrees: $7.5(360°) = 2700°$. The point on the edge of the tire moves $2700°$ in one second.

3. (a) $120° \cdot \frac{\pi}{180} = \frac{2\pi}{3}$ radians.

(b) $\frac{9\pi}{10} \cdot \frac{180}{\pi} = 162°$

4. (a) Since arc length is found by the formula $s = r^{\theta}$, the central angle is: $200 = 150\theta \Rightarrow \theta = \frac{200}{150} \Rightarrow \theta = \frac{4}{3}$.

(b) Using the area of a sector formula $A = \frac{1}{2}r^2\theta$ yields: $A = \frac{1}{2}(150)^2\left(\frac{4}{3}\right) \Rightarrow A = 15{,}000 \text{ km}^2$

5. Since $36° = 36 \cdot \frac{\pi}{180} = \frac{\pi}{5}$ radians and arc length is found by the formula $s = r\theta$, the arc length is:

$s = 12\left(\frac{\pi}{5}\right) \Rightarrow s = \frac{12}{5}\pi \Rightarrow s \approx 7.54 \text{ in}$

6. (a) The smallest positive degree measure of the coterminal angle is: $-150° + 360° = 210°$.

(b) First find the reference angle: $(210° - 180°) = 30°$. Using the $30°$-$60°$ right triangle rules, for a $30°$ angle the hypotenuse $= 2$, the opposite side $= 1$, and the adjacent $= \sqrt{3}$. Since $\theta = -150°$ is in quadrant III, both x and y are negative and the point is: $\left(-\sqrt{3}, -1\right)$.

(c) $\sin(-150°) = -\sin 30° = -\frac{1}{2}$ $\csc(-150°) = -\csc 30° = -\frac{2}{1} = -2$

$\cos(-150°) = -\cos 30° = -\frac{\sqrt{3}}{2}$ $\sec(-150°) = -\sec 30° = -\frac{2}{\sqrt{3}} = -\frac{2\sqrt{3}}{3}$

$\tan(-150°) = \tan 30° = \frac{1}{\sqrt{3}} = \frac{\sqrt{3}}{3}$ $\cot(-150°) = \cot 30° = -\frac{\sqrt{3}}{1} = \sqrt{3}$

(d) $-150° = -\frac{150°}{1} \cdot \frac{\pi}{180°} = -\frac{5\pi}{6}$

7. If $\sin s = .82584121$, then $s = \sin^{-1} .82584121 \approx .97169234$.

8. If $\cos\theta < 0$, then θ is in quadrant II or III. If $\cot\theta > 0$, is in quadrant I or III. Therefore, θ terminates in quadrant III.

9. With the coordinate $(2, -5)$, we will have a radius of $r = \sqrt{(2)^2 + (-5)^2} = \sqrt{4 + 25} \Rightarrow r = \sqrt{29}$. Therefore:

$\sin\theta = \frac{y}{r} = \frac{-5}{\sqrt{29}} = -\frac{5\sqrt{29}}{29}$ $\cos\theta = \frac{x}{r} = \frac{2}{\sqrt{29}} = \frac{2\sqrt{29}}{29}$ $\tan\theta = \frac{y}{x} = \frac{-5}{2} = -\frac{5}{2}$.

10. Since $\cos\theta = \frac{4}{5}$, then $x = 4$ and $r = 5$. Now use the pythagorean theorem to find y

$5^2 = 4^2 + y^2 \Rightarrow 25 = 16 + y^2 \Rightarrow y^2 = 9 \Rightarrow y = \pm 3$. Since θ is in quadrant IV, $y = -3$.

Therefore: $\sin\theta = \frac{y}{r} = \frac{-3}{5} = -\frac{3}{5}$ $\tan\theta = \frac{y}{x} = \frac{-3}{4} = -\frac{3}{4}$

$\csc\theta = \frac{r}{y} = \frac{5}{-3} = -\frac{5}{3}$ $\sec\theta = \frac{r}{x} = \frac{5}{4} = \frac{5}{4}$ $\cot\theta = \frac{x}{y} = \frac{4}{-3} = -\frac{4}{3}$

11. Use knowledge of $(30°$-$60°)$ and $(45°$-$45°)$ right triangles to find each:

To find w, use $\sin 30° = \frac{4}{w} \Rightarrow w = \frac{4}{\sin 30°} = \frac{4}{\frac{1}{2}} = 8$.

To find x, use $\tan 45° = \frac{4}{x} \Rightarrow x = \frac{4}{\tan 45°} = \frac{4}{1} = 4$.

To find y, use $\cos 30° = \frac{y}{w} = \frac{y}{8} \Rightarrow y = 8\cos 30° = 8 \cdot \frac{\sqrt{3}}{2} = 4\sqrt{3}$.

To find z, use $\sin 45° = \frac{4}{z} \Rightarrow z = \frac{4}{\sin 45°} = \frac{4}{\frac{1}{\sqrt{2}}} = 4\sqrt{2}$.

12. First, $\cot(-750°)$ is coterminal with $(3 \cdot 360° - 750°) = \cot 330°$. Now find the reference angle: $\cot(360° - 330°) = \cot 30°$. Finally, since $\cot(-750°)$ is terminal in quadrant IV, we will find $-\cot 30°$ and using the $(30° - 60°)$ right triangle: $-\cot 30° = -\sqrt{3}$.

13. (a) $\sin 78°21' = \sin\left(78 + \frac{21}{60}\right)° = \sin 78.35° \approx .97939940$

    (b) $\tan 11.7689° \approx .20834446$

    (c) $\sec 58.9041° = \frac{1}{\cos 58.9041°} = (\cos 58.9041°)^{-1} \approx 1.9362132$

14. To find $B: B = 90° - 58°30' \Rightarrow B = 89°60' - 58°30' \Rightarrow B = 31°30'$.

    To find $a: \sin A = \frac{a}{c} \Rightarrow \sin 58°30' = \frac{a}{748} \Rightarrow a = 748(\sin 58°30') \approx 637.7748 \Rightarrow a \approx 638$.

    To find $b: \cos A = \frac{b}{c} \Rightarrow \cos 58°30' = \frac{b}{748} \Rightarrow b = 748(\cos 58°30') \approx 390.8289 \Rightarrow b \approx 391$

15. Let x be the height of the flag pole. Now use the tangent function to solve for x:

    $\tan 32°10' = \frac{x}{24.7} \Rightarrow x = 24.7 \tan 32°10' \approx 15.5$ feet.

16. Draw a picture of the ships travels (not shown). Angle ABX is 55° (using geometry—alternate interior angles are equal) and angle ABC is $55° + 35° = 90°$, therefore triangle ABC is a right triangle and we can use Pythagorean theorem to find $AC: AC^2 = 80^2 + 74^2 \Rightarrow AC = \sqrt{80^2 + 74^2} \approx 108.977 \Rightarrow AC \approx 110$ km.

17. From the equation, the amplitude is 2, the period is $\frac{2\pi}{(1)} = 2\pi$, the vertical translation is 1 unit downward, and the phase shift is π units to the left. See Figure 17.

18. From the equation, the amplitude is $|-1| = 1$ the period is $\frac{2\pi}{(2)} = \pi$, and there is no vertical translation or phase shift, and the graph is reflected across the x-axis. See Figure 18.

19. From the equation, the amplitude (since a tangent function) is not applicable, the period is $\frac{\pi}{(1)} = \pi$, and the phase shift is $\frac{\pi}{2}$ units to the right. See Figure 19.

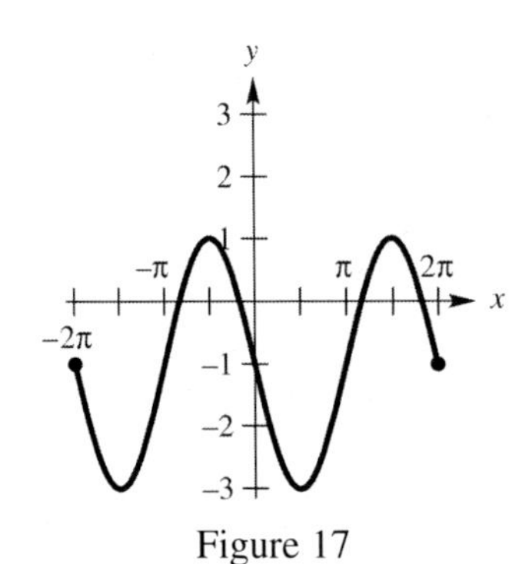

Figure 17

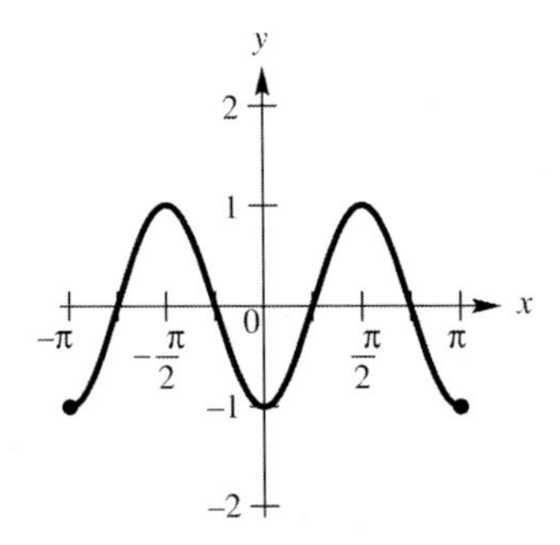

Figure 18

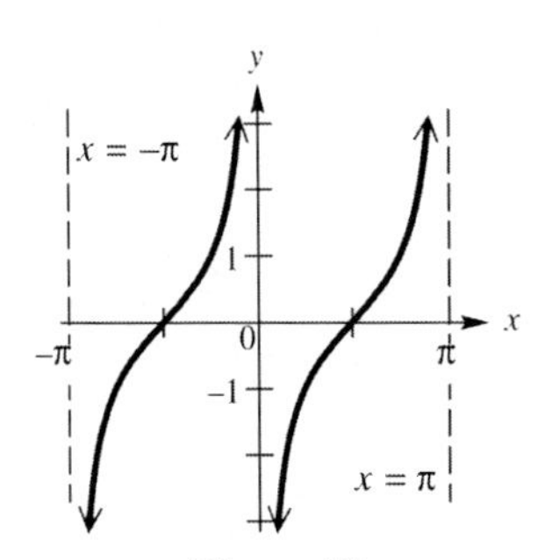

Figure 19

20. Graph the secant function with a period of 2π and shifted up 1 unit. See Figure 20.

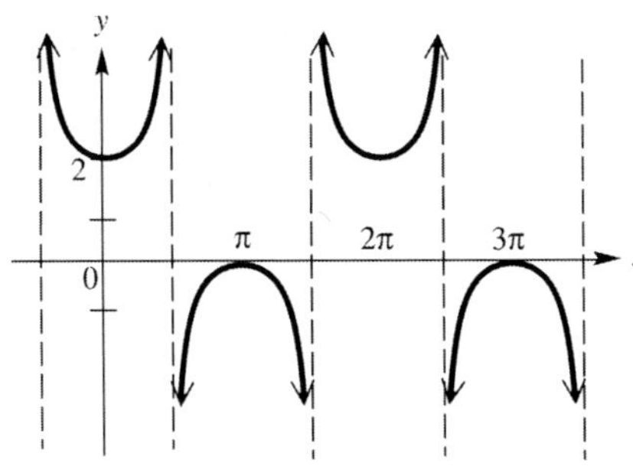

Figure 20

21. The graph has an amplitude of 2 therefore the value of a is 2. The graph repeats itself twice in the interval of 2π therefore the value of b is 2. The shape of the graph shows that we are graphing the sine function. Therefore, the equation of the graph is $y = 2\sin 2x$.

22. (a) Convert the times to decimals hours and enter the values in your calculator as described. Now make a scatterplot of the data on the calculator. See Figure 22a.

(b) The minimum sunset time is 4:35 or 4.58; the maximum sunset time is 7:33 or 7.55. Using these values, the amplitude is $\frac{7.55-4.58}{2}. = 1.48$ or $1:29$. In the equation $y = a\sin b(x-d)+c$, the amplitude is a, therefore $a = 1.48$.

(c) The times repeat every 12 months, so the period is $\frac{2\pi}{(12)} = \frac{\pi}{6} \approx .52$. Therefore $b \approx .52$.

(d) We want the cycle to start with March 21, which corresponds to 3.68, therefore the phase shift is $\frac{2\pi}{(12)} = \frac{\pi}{6}d \approx 3.68$.

(e) The vertical shift is the average of the maximum and minimum sunset times,

so $c = \frac{4.58+7.55}{2} = 6.065$. Therefore $c \approx 6.07$.

(f) Using $y = a\sin b(x-d)+c$ yields: $y = 1.48\sin\left[.52(x-3.68)\right]+6.07 \Rightarrow y = 1.48\sin(.52x-1.91)+6.07$.

See Figure 22f. (Use radian mode)

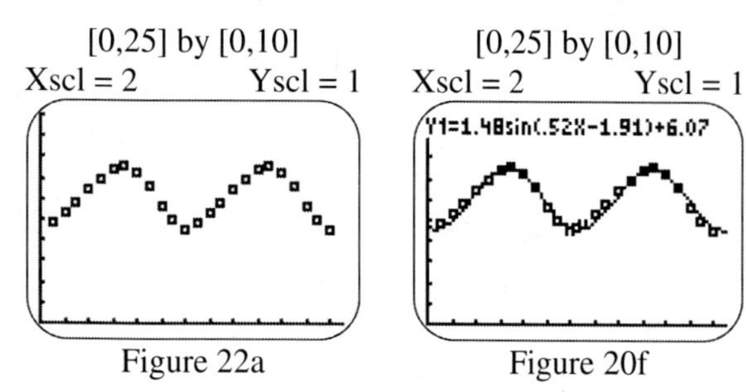

Figure 22a Figure 20f

23. Because the amplitude is $|-3| = 3$, the maximum height the weight rises above an equilibrium position of $y = 0$ is 3 inches.

24. If $s(t) = 3$, then $3 = -3\cos 2\pi t \Rightarrow -1 = \cos 2\pi t$. Since $\cos\theta = -1$ when $\theta = \pi$, then $2\pi t = \pi \Rightarrow 2t = 1 \Rightarrow t = \frac{1}{2}$. Therefore it reaches the maximum height at $\frac{1}{2}$ seconds.

Chapter 9: Trigonometric Identities and Equations

9.1: Trigonometric Identities

1. Since by the negative-number identities $\sin x = -\sin(-x)$, the function is: Odd.

3. Since by the negative-number identities $\tan x = -\tan(-x)$, the function is: Odd.

5. Since by the negative-number identities $\sec x = \sec(-x)$, the function is: Even.

7. By a negative-number identity, if $\cos(-\theta) = \cos\theta$, then $\cos(-4.38) = \cos\ 4.38$.

9. By a negative-number identity, if $\sin(-\theta) = -\sin\theta$, then $\sin(-0.5) = -\sin\ 0.5$.

11. By a negative-number identity, if $\tan(-\theta) = -\tan\theta$, then $\tan\left(-\frac{\pi}{7}\right) = -\tan\frac{\pi}{7}$.

13 By a quotient identity, $\frac{\cos x}{\sin x} = \cot x$, therefore B.

15. By a negative-number identity, $\cos(-x) = \cos x$, therefore E.

17. By a Pythagorean identity, $1 = \sin^2 x + \cos^2 x$, therefore A.

19. Using a Pythagorean identity and then a quotient identity, $\sec^2 x - 1 = \tan^2 x = \frac{\sin^2 x}{\cos^2 x}$, therefore A.

21. Using a Pythagorean identity, $1 + \sin^2 x = (\csc^2 x - \cot^2 x) + \sin^2 x$, therefore D.

23. The correct identity is $1 + \cot^2 x = \csc^2 x$; the function must have the argument "x" or θ, t, etc.

25. $\sin\theta$ in terms of $\cot\theta$: $\sin\theta = \frac{1}{\csc\theta} = \frac{1}{\pm\sqrt{1+\cot^2\theta}} = \pm\frac{\sqrt{1+\cot^2\theta}}{1+\cot^2\theta}$.

$P = 16k\left[1 - \sin^2(2\pi t)\right]$ in terms of $\sec\theta$:

$$\sin\theta = \cos\theta \cdot \frac{\sin\theta}{\cos\theta} = \cos\theta \cdot \tan\theta = \frac{1}{\sec\theta} \cdot (\pm\sqrt{\sec^2\theta - 1}) = \pm\frac{\sqrt{\sec^2\theta - 1}}{\sec\theta}.$$

27. $\tan\theta$ in terms of $\sin\theta$: $\tan\theta = \frac{\sin\theta}{\cos\theta} = \frac{\sin\theta}{\pm\sqrt{1-\sin^2\theta}} = \pm\frac{\sin\theta\sqrt{1-\sin^2\theta}}{1-\sin^2\theta}$.

$\tan\theta$ in terms of $\cos\theta$: $\tan\theta = \frac{\sin\theta}{\cos\theta} = \pm\frac{\sqrt{1-\cos^2\theta}}{\cos\theta}$.

$\tan\theta$ in terms of $\sec\theta$: $\tan\theta = \pm\sqrt{\sec^2\theta - 1}$.

$\tan\theta$ in terms of $\csc\theta$: $\tan\theta = \frac{1}{\cot\theta} = \frac{1}{\pm\sqrt{\csc^2\theta - 1}} = \pm\frac{\sqrt{\csc^2\theta - 1}}{\csc^2\theta - 1}$.

29. $\sec\theta$ in terms of $\sin\theta$: $\sec\theta = \frac{1}{\cos\theta} = \frac{1}{\pm\sqrt{1-\sin^2\theta}} = \pm\frac{\sqrt{1-\sin^2\theta}}{1-\sin^2\theta}$.

$\sec\theta$ in terms of $\tan\theta$: $\sec\theta = \pm\sqrt{\tan^2\theta + 1}$.

$\sec\theta$ in terms of $\cot\theta$: $\sec\theta = \pm\sqrt{\tan^2\theta+1} = \pm\sqrt{\frac{1}{\cot^2\theta}+1} = \pm\sqrt{\frac{1+\cot^2\theta}{\cot^2\theta}} = \pm\frac{\sqrt{1+\cot^2\theta}}{\cot\theta}$.

$\sec\theta$ in terms of $\csc\theta$: $\sec\theta = \frac{1}{\cos\theta} = \frac{1}{\pm\sqrt{1-\sin^2\theta}} = \frac{1}{\pm\sqrt{1-\frac{1}{\csc^2\theta}}} = \frac{1}{\pm\sqrt{\frac{\csc^2\theta-1}{\csc^2\theta}}} =$

$\frac{\pm\sqrt{\csc^2\theta}}{\sqrt{\csc^2\theta-1}} = \frac{\pm\csc\theta}{\sqrt{\csc^2\theta-1}} = \pm\frac{\csc\theta\sqrt{\csc^2\theta-1}}{\csc^2\theta-1}$.

31. $\tan\theta\cos\theta = \frac{\sin\theta}{\cos\theta}.\frac{\cos\theta}{1} = \sin\theta.$

33. $\frac{\sin\beta\tan\beta}{\cos\beta} = \frac{\sin\beta}{\cos\beta}\cdot\tan\beta = \tan\beta\bullet\tan\beta = \tan^2\beta.$

35. $\sec^2 x - 1 = (\tan^2 x+1)-1 = \tan^2 x.$

37. $\frac{\sin^2 x}{\cos^2 x} + \sin x\csc x = \tan^2 x + \frac{\sin x}{1}\cdot\frac{1}{\sin x} = \tan^2 x + 1 = \sec^2 x.$

39. $\cot\theta\sin\theta = \frac{\cos\theta}{\sin\theta}\cdot\sin\theta = \cos\theta.$

41. $\cos\theta\csc\theta = \cos\theta\cdot\frac{1}{\sin\theta} = \frac{\cos\theta}{\sin\theta}$, which simplifies to $\cot\theta$.

43. $\frac{\cot^2\theta}{\csc^2\theta} = \frac{\frac{\cos^2\theta}{\sin^2\theta}}{\frac{1}{\sin^2\theta}} = \frac{\cos^2\theta}{\sin^2\theta}\cdot\frac{\sin^2\theta}{1} = \cos^2\theta.$

45. $\sin^2\theta+\cos^2\theta = 1 \Rightarrow 1-\cos^2\theta = \sin^2\theta$

47. $\frac{1}{\sec^2\theta-1} = \frac{1}{\frac{1}{\cos^2\theta}-1} = \frac{1}{\frac{1-\cos^2\theta}{\cos^2\theta}} = \frac{\cos^2\theta}{\sin^2\theta} = \cot^2\theta.$

49. $\sin^2\theta(\csc^2\theta-1) = \sin^2\theta\cdot\cot^2\theta = \sin^2\theta\cdot\frac{\cos^2\theta}{\sin^2\theta} = \cos^2\theta.$

51. $(1-\cos\theta)(1+\sec\theta) = (1-\cos\theta)\left(1+\frac{1}{\cos}\right) = 1+\frac{1}{\cos\theta}-\cos\theta-1 = \frac{1}{\cos\theta}-\cos\theta,$

which simplifies to $\sec\theta-\cos\theta$.

53. It is in terms of sine and cosine, but simplifies to $\frac{\cos^2\theta-\sin^2\theta}{\sin\theta\cos\theta} = \frac{\cos^2\theta}{\sin\theta\cos\theta} - \frac{\sin^2\theta}{\sin\theta\cos\theta} =$

$\frac{\cos\theta}{\sin\theta} - \frac{\sin\theta}{\cos\theta} = \cot\theta - \tan\theta.$

55. $\sec\theta-\cos\theta = \frac{1}{\cos\theta} - \frac{\cos^2\theta}{\cos\theta} = \frac{1-\cos^2\theta}{\cos\theta} = \frac{\sin^2\theta}{\cos\theta}.$

57. $\sin\theta(\csc\theta - \sin\theta) = \sin\theta\left(\dfrac{1}{\sin\theta} - \sin\theta\right) = 1 - \sin^2\theta = \cos^2\theta.$

59. $\csc\theta\sec\theta\tan\theta = \dfrac{1}{\sin\theta}\dfrac{1}{\cos\theta}\dfrac{\sin\theta}{\cos\theta} = \dfrac{1}{\cos^2\theta} = \sec^2\theta.$

61. $\cot\theta + \dfrac{1}{\cot\theta} = \dfrac{\cos\theta}{\sin\theta} + \tan\theta = \dfrac{\cos\theta}{\sin\theta} + \dfrac{\sin\theta}{\cos\theta} = \dfrac{\cos^2\theta}{\sin\theta\cos\theta} + \dfrac{\sin^2\theta}{\sin\theta\cos\theta} = \dfrac{\sin^2\theta + \cos^2\theta}{\sin\theta\cos\theta} =$

$\dfrac{1}{\sin\theta\cos\theta}$, or $\csc\theta\sec\theta$.

63. $\tan s(\cot s + \csc s) = \tan s\left(\dfrac{1}{\tan s} + \dfrac{1}{\sin s}\right) = 1 + \dfrac{\tan s}{\sin s} = 1 + \left(\dfrac{\sin s}{\cos s} \div \sin s\right) = 1 + \left(\dfrac{\sin s}{\cos s} \cdot \dfrac{1}{\sin s}\right) =$

$1 + \dfrac{1}{\cos s} = 1 + \sec s.$

65. $\dfrac{1}{\csc^2\theta} + \dfrac{1}{\sec^2\theta} = \sin^2\theta + \cos^2\theta = 1.$

67. $\dfrac{\cos x}{\sec x} + \dfrac{\sin x}{\csc x} = \left(\dfrac{\cos x}{1} \div \dfrac{1}{\cos x}\right) + \left(\dfrac{\sin x}{1} \div \dfrac{1}{\sin x}\right) = \left(\dfrac{\cos x}{1} . \dfrac{\cos x}{1}\right) + \left(\dfrac{\sin x}{1} \cdot \dfrac{\sin x}{1}\right) = \sin^2 x + \cos^2 x = 1.$

69. $(1 + \sin t)^2 + \cos^2 t = 1 + 2\sin t + \sin^2 t + \cos^2 t = 1 + 2\sin t + 1 = 2 + 2\sin t.$

71. $\dfrac{1}{1 + \cos x} - \dfrac{1}{1 - \cos x} = \dfrac{1(1 - \cos x)}{(1 + \cos x)(1 - \cos x)} - \dfrac{1(1 + \cos x)}{(1 + \cos x)(1 - \cos x)} = \dfrac{1 - \cos x - 1 - \cos x}{1 - \cos^2 x} =$

$\dfrac{-2\cos x}{\sin^2 x} = -\dfrac{2\cos x}{\sin^2 x}$, or $-2\cot x\csc x$.

73. $\dfrac{\cot\theta}{\csc\theta} = \dfrac{\frac{\cos\theta}{\sin\theta}}{\frac{1}{\sin\theta}} = \dfrac{\cos\theta}{\sin\theta} \cdot \dfrac{\sin\theta}{1} = \dfrac{\cos\theta}{1} = \cos\theta.$

75. $\cos^2\theta(\tan^2\theta + 1) = \cos^2\theta(\sec^2\theta) = \cos^2\theta\left(\dfrac{1}{\cos^2\theta}\right) = 1.$

77. $\dfrac{\tan^2\gamma + 1}{\sec\gamma} = \dfrac{\sec^2\gamma}{\sec\gamma} = \sec\gamma.$

79. $\dfrac{1 - \sin^2\beta}{\cos\beta} = \dfrac{\cos^2\beta}{\cos\beta} = \cos\beta$

81. $\dfrac{1 - \cos x}{1 + \cos x} = \dfrac{1 - \cos x}{1 + \cos x} \cdot \dfrac{1 - \cos x}{1 - \cos x} = \dfrac{1 - 2\cos x + \cos^2 x}{1 - \cos^2 x} = \dfrac{1 - 2\cos x + \cos^2 x}{\sin^2 x}$. Work with the right side of the equation. $(\cot x - \csc x)^2 = \left(\dfrac{\cos x}{\sin x} - \dfrac{1}{\sin x}\right)^2 = \left(\dfrac{\cos x - 1}{\sin x}\right)^2 = \dfrac{\cos^2 x - 2\cos x - 1}{\sin^2 x.}$. Therefore,

$\dfrac{1 - \cos x}{1 + \cos x} = (\cot x - \csc x)^2.$

83. $$\frac{\cot\alpha+1}{\cot\alpha-1}=\frac{\frac{\cos\alpha}{\sin\alpha}+1}{\frac{\cos\alpha}{\sin\alpha}-1}=\frac{\frac{\cos\alpha+\sin\alpha}{\sin\alpha}}{\frac{\cos\alpha-\sin\alpha}{\sin\alpha}}=\frac{\cos\alpha+\sin\alpha}{\cos\alpha-\sin\alpha}=\frac{\cos\alpha+\sin\alpha}{\cos\alpha-\sin\alpha}\cdot\frac{\frac{1}{\cos\alpha}}{\frac{1}{\cos\alpha}}$$

$$\frac{\frac{\cos\alpha}{\cos\alpha}+\frac{\sin\alpha}{\cos\alpha}}{\frac{\cos\alpha}{\cos\alpha}-\frac{\sin\alpha}{\cos\alpha}}=\frac{1+\tan\alpha}{1-\tan\alpha}.$$

85. $\sin^2\alpha+\tan^2\alpha+\cos^2\alpha=\sin^2\alpha+\cos^2\alpha+\tan^2\alpha=1+\tan^2\alpha=\sec^2\alpha.$

87. $$\frac{\sin^2\gamma}{\cos\gamma}=\frac{1-\cos^2\gamma}{\cos\gamma}=\frac{1}{\cos\gamma}-\frac{\cos^2\gamma}{\cos\gamma}=\sec\gamma-\cos\gamma.$$

89. $$\frac{\cos\theta}{\sin\theta\cot\theta}=\frac{\cos\theta}{\sin\theta\cdot\frac{\cos\theta}{\sin\theta}}=\frac{\cos\theta}{\cos\theta}=1.$$

91. $\tan^2\gamma\sin^2\gamma=\left(\sec^2\gamma-1\right)\left(1-\cos^2\gamma\right)=\sec^2\gamma-\sec^2\gamma\cos^2\gamma-1+\cos^2\gamma=$

$$\sec^2\gamma-1+\cos^2\gamma-\sec^2\gamma\cos^2\gamma=\tan^2\gamma+\cos^2\gamma-\frac{1}{\cos^2\gamma}\bullet\cos^2\gamma=\tan^2\gamma+\cos^2\gamma-1.$$

93. $$\frac{(\sec\theta-\tan\theta)^2+1}{\sec\theta\csc\theta-\tan\theta\csc\theta}=\frac{\sec^2\theta-2\sec\theta\tan\theta+\tan^2\theta+1}{\csc\theta(\sec\theta-\tan\theta)}=\frac{\sec^2\theta-2\sec\theta\tan\theta+\sec^2\theta}{\csc\theta(\sec\theta-\tan\theta)}=$$

$$\frac{2\sec^2\theta-2\sec\theta\tan\theta}{\csc\theta(\sec\theta-\tan\theta)}=\frac{2\sec\theta(\sec\theta-\tan\theta)}{\csc\theta(\sec\theta-\tan\theta)}=\frac{2\sec\theta}{\csc\theta}=2\left(\frac{1}{\cos\theta}\cdot\frac{\sin\theta}{1}\right)=2\tan\theta.$$

95. $$\frac{1}{\tan\alpha-\sec\alpha}+\frac{1}{\tan\alpha+\sec\alpha}=\frac{(\tan\alpha+\sec\alpha)+(\tan\alpha-\sec\alpha)}{\tan^2\alpha-\sec^2\alpha}=\frac{2\tan\alpha}{\tan^2\alpha-\left(\tan^2+1\right)}=\frac{2\tan\alpha}{-1}=-2\tan\alpha.$$

97. $\sec^4 x-\sec^2 x=\sec^2 x\left(\sec^2 x-1\right)=\sec^2 x\cdot\tan^2 x$. Simplify the right side of the equation.

$\tan^4 x+\tan^2 x=\tan^2 x(\tan^2 x+1)=\tan^2 x\cdot\sec^2 x.$ Therefore, $\sec^4 x-\sec^2 x=\tan^4 x+\tan^2 x.$

99. $$\frac{\sec^4 s-\tan^4 s}{\sec^2 s+\tan^2 s}=\frac{\left(\sec^2 s-\tan^2 s\right)\left(\sec^2 s+\tan^2 s\right)}{\sec^2 s+\tan^2 s}=\sec^2 s-\tan^2 s.$$

101. $$\frac{\tan t-\cot t}{\tan t+\cot t}=\frac{\tan t-\frac{1}{\tan t}}{\tan t+\frac{1}{\tan t}}=\frac{\tan t-\frac{1}{\tan t}}{\tan t+\frac{1}{\tan t}}\cdot\frac{\tan t}{\tan t}=\frac{\tan^2 t-1}{\tan^2 t+1}=\frac{\tan^2 t-1}{\sec^2 t}$$

103. $$\sin^2\alpha\sec^2\alpha+\sin^2\alpha\csc^2\alpha=\sin^2\alpha\left(\sec^2\alpha+\csc^2\alpha\right)=\sin^2\alpha\left(\frac{1}{\cos^2\alpha}+\frac{1}{\sin^2\alpha}\right)=$$

$\tan^2\alpha+1=\sec^2\alpha.$

105. $\sec^2\theta - 2\sec\theta\tan\theta + \tan^2\theta = \dfrac{1}{\cos^2\theta} - 2\cdot\dfrac{1}{\cos\theta}\cdot\dfrac{\sin\theta}{\cos\theta} + \dfrac{\sin^2\theta}{\cos^2\theta} = \dfrac{1-2\sin\theta+\sin^2\theta}{\cos^2\theta}$

$= \dfrac{(1-\sin\theta)^2}{1-\sin^2\theta} = \dfrac{(1-\sin\theta)^2}{(1-\sin\theta)(1+\sin\theta)} = \dfrac{1-\sin\theta}{1+\sin\theta}.$

107. $\dfrac{1+\sin\theta}{1-\sin\theta} - \dfrac{1-\sin\theta}{1+\sin\theta} = \dfrac{(1+\sin\theta)(1+\sin\theta)-(1-\sin\theta)(1-\sin\theta)}{(1-\sin\theta)(1+\sin\theta)} = \dfrac{1+2\sin\theta+\sin^2\theta-1+2\sin\theta-\sin^2\theta}{1-\sin^2\theta}$

$= \dfrac{4\sin\theta}{\cos^2\theta} = \dfrac{4\sin\theta}{\cos\theta}\cdot\dfrac{1}{\cos\theta} = 4\tan\theta\sec\theta.$

109. $(2\sin x+\cos x)^2 + (2\cos x-\sin x)^2 = \left(4\sin^2 x+4\sin x\cos x+\cos^2 x\right)+\left(4\cos^2 x-4\sin x\cos x+\sin^2 x\right)=$

$4\left(\sin^2 x+\cos^2 x\right)+\left(\cos^2 x+\sin^2 x\right)=4+1=5.$

111. $\sec x-\cos x+\csc x-\sin x-\sin x\tan x = \dfrac{1}{\cos x}-\cos x+\dfrac{1}{\sin x}-\sin x-\sin x\left(\dfrac{\sin x}{\cos x}\right)=$

$\left(\dfrac{1}{\cos x}-\cos x\right)+\left(\dfrac{1}{\sin x}-\sin x\right)-\dfrac{\sin^2 x}{\cos x}=\dfrac{1-\cos^2 x}{\cos x}+\dfrac{1-\sin^2 x}{\sin x}-\dfrac{\sin^2 x}{\cos x}=$

$\left(\dfrac{1-\cos^2 x}{\cos x}-\dfrac{\sin^2 x}{\cos x}\right)+\dfrac{1-\sin^2 x}{\sin x}=\dfrac{1-\cos^2 x-\sin^2 x}{\cos x}+\dfrac{\cos^2 x}{\sin x}=\dfrac{1-\left(\cos^2 x+\sin^2 x\right)}{\cos x}+\dfrac{\cos^2 x}{\sin x}=$

$\dfrac{1-1}{\cos x}+\cos x\cdot\dfrac{\cos x}{\sin x}=\cos x\cot x.$

113. (a) Using an identity yields $I = k\left(1-\sin^2\theta\right)$.

(b) The function reaches its maximum value of k when $\cos^2\theta = 1$. This happens when $\theta = 0$.

115. (a) See Figure 115a. The total mechanical energy E is always 2. The spring has maximum potential energy when it is fully stretched but not moving. The spring has maximum kinetic energy when it is not stretched but is moving fastest.

(b) Let $Y_1 = P(t)$, $Y_2 = k(t)$, $Y_3 = E(t) = 2$ for all inputs; see Figure 115b. The spring is stretched the most (has the greatest potential energy) when $t = 0.25,\ 0.5,\ 0.75,$ etc. At these times kinetic energy is 0.

(c) $E(t)=P(t)+k(t)=2\cos^2(4\pi t)+2\sin^2(4\pi t)=2\left(\cos^2(4\pi t)+\sin^2(4\pi t)\right)=2(1)=2.$

[0,0.5] by [-1,3]
Xscl =0.25 Yscl = 1

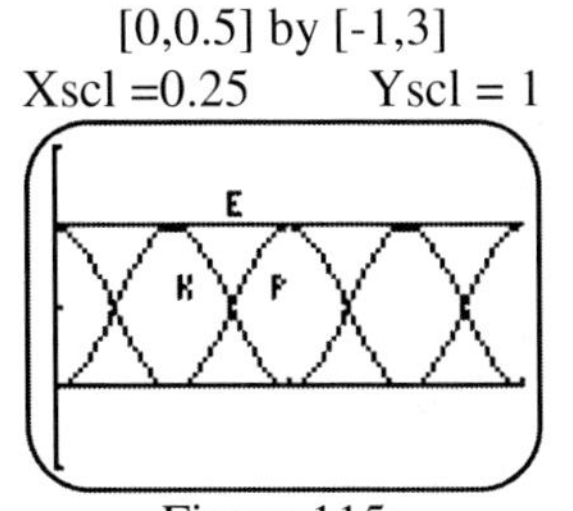

Figure 115a

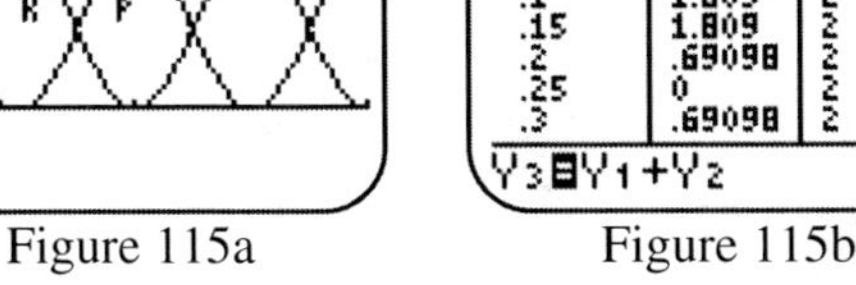

Figure 115b

9.2: Sum and Difference Identities

1. By cosine of a sum, $\cos(x+y)=\cos x\cos y-\sin x\sin y$, therefore F.

3. By sine of a sum, $\sin(x+y)=\sin x\cos y+\cos x\sin y$, therefore C.

5. $\sin\frac{\pi}{12}=\sin\left(\frac{\pi}{3}-\frac{\pi}{4}\right)=\sin\frac{\pi}{3}\cos\frac{\pi}{4}-\cos\frac{\pi}{3}\sin\frac{\pi}{4}=\frac{\sqrt{3}}{2}\cdot\frac{\sqrt{2}}{2}-\frac{1}{2}\cdot\frac{\sqrt{2}}{2}=\frac{\sqrt{6}-\sqrt{2}}{4}$.

7. $\sin\left(-\frac{5\pi}{12}\right)=-\sin\left(\frac{5\pi}{12}\right)=-\sin\left(\frac{\pi}{6}+\frac{\pi}{4}\right)=-\left(\sin\frac{\pi}{6}\cos\frac{\pi}{4}+\cos\frac{\pi}{6}\sin\frac{\pi}{4}\right)=$
$-\left(\frac{1}{2}\cdot\frac{\sqrt{2}}{2}+\frac{\sqrt{3}}{2}\cdot\frac{\sqrt{2}}{2}\right)=\frac{-\sqrt{2}-\sqrt{6}}{4}$.

9. $\sin\frac{13\pi}{12}=\sin\left(\frac{5\pi}{4}-\frac{\pi}{6}\right)=\sin\frac{5\pi}{4}\cos\frac{\pi}{6}-\cos\frac{5\pi}{4}\sin\frac{\pi}{6}=\frac{-\sqrt{2}}{2}\cdot\frac{\sqrt{3}}{2}-\frac{-\sqrt{2}}{2}\cdot\frac{1}{2}=\frac{-\sqrt{6}+\sqrt{2}}{4}$.

11. $\cos 75^\circ=\cos(45^\circ+30^\circ)=\cos 45^\circ\cos 30^\circ-\sin 45^\circ\sin 30^\circ=\frac{\sqrt{2}}{2}\cdot\frac{\sqrt{3}}{2}-\frac{\sqrt{2}}{2}\cdot\frac{1}{2}=\frac{\sqrt{6}-\sqrt{2}}{4}$.

13. $\tan 105^\circ=\tan(60^\circ+45^\circ)=\frac{\tan 60^\circ+\tan 45^\circ}{1-\tan 60^\circ\tan 45^\circ}=\frac{\sqrt{3}+1}{1-\sqrt{3}}\cdot\frac{1+\sqrt{3}}{1+\sqrt{3}}=\frac{2\sqrt{3}+4}{-2}=-\sqrt{3}-2$.

15. $\cos(-15^\circ)=\cos 15^\circ=\cos(60^\circ-45^\circ)=\cos 60^\circ\cos 45^\circ+\sin 60^\circ\sin 45^\circ=\frac{1}{2}\cdot\frac{\sqrt{2}}{2}+\frac{\sqrt{3}}{2}\cdot\frac{\sqrt{2}}{2}=\frac{\sqrt{2}+\sqrt{6}}{4}$.

17. $\cos\frac{\pi}{3}\cos\frac{2\pi}{3}-\sin\frac{\pi}{3}\sin\frac{2\pi}{3}=\cos\left(\frac{\pi}{3}+\frac{2\pi}{3}\right)=\cos\pi=-1$.

19. $\sin 76^\circ\cos 31^\circ+\cos 76^\circ\sin 31^\circ=\sin(76^\circ-31^\circ)=\sin 45^\circ=\frac{\sqrt{2}}{2}$.

21. $\frac{\tan 80^\circ+\tan 55^\circ}{1-\tan 80^\circ\tan 55^\circ}=\tan(80^\circ+55^\circ)=\tan 135^\circ=-1$.

23. $\sin(180^\circ-x)=\sin 180^\circ\cos x-\cos 180^\circ\sin x=(0\cdot\cos x)-(-1)\sin x=0-(-\sin x)=\sin x$.

25. $\cos(180^\circ+x)=\cos 180^\circ\cos x-\sin 180^\circ\sin x=(-1)\cos x-(0\cdot\sin x)=-\cos x$.

27. $\sin(x-90^\circ)=\sin x\cos 90^\circ-\cos x\sin 90^\circ=(\sin x\cdot 0)-(\cos x\cdot 1)=-\cos x$.

29. $\tan(180^\circ-x)=\frac{\tan 180^\circ-\tan x}{1+\tan 180^\circ\tan x}=\frac{0-\tan x}{1+(0\cdot\tan x)}=\frac{-\tan x}{1}=-\tan x$.

31. $\cos\left(\frac{\pi}{2}-x\right)=\cos\frac{\pi}{2}\cos x+\sin\frac{\pi}{2}\sin x=(0)(\cos x)+(1)(\sin x)=\sin x$.

33. $\cos\left(\frac{3\pi}{2}+x\right)=\cos\frac{3\pi}{2}\cos x-\sin\frac{3\pi}{2}\sin x=(0)(\cos x)-(-1)(\sin x)=\sin x$.

35. $\sin(\pi+x)=\sin\pi\cos x+\cos\pi\sin x=(0)(\cos x)+(-1)(\sin x)=-\sin x$.

37. $\cos(135^\circ-x)=\cos 135^\circ\cos x+\sin 135^\circ\sin x=-\frac{\sqrt{2}}{2}\cdot\cos x+\frac{\sqrt{2}}{2}\sin x=\frac{\sqrt{2}(\sin x-\cos x)}{2}$.

39. $\tan(45°+x)=\frac{\tan 45°+\tan x}{1-\tan 45°\tan x}=\frac{1+\tan x}{1-(1)\tan x}=\frac{1+\tan x}{1-\tan x}.$

41. $\tan(\pi-x)=\frac{\tan\pi-\tan x}{1+\tan\pi\tan x}=\frac{0-\tan x}{1+(0)\tan x}=-\tan x.$

43. $\tan\left(\frac{5\pi}{4}-x\right)=\frac{\tan\frac{5\pi}{4}-\tan x}{1+\tan\frac{5\pi}{4}\tan x}=\frac{1-\tan x}{1+\tan x}.$

45. $\cos(2\pi-x)=\cos 2\pi\cos x+\sin 2\pi\sin x=\cos x+0\cdot\sin x=\cos x.$

47. If $0<A<\frac{\pi}{2}$ and $0<B<\frac{\pi}{2}$, then A and B are both found in quadrant I. Next, using $\cos A=\frac{3}{5}$ and $\sin B=\frac{5}{13}$, find $\sin A$ and $\cos B$.

$$\sin^2 A=1-\cos^2 A=1-\left(\frac{3}{5}\right)^2=1-\frac{9}{25}=\frac{16}{25}\Rightarrow\sin A=\pm\sqrt{\frac{16}{25}}\Rightarrow\sin A=\frac{4}{5};$$

$$\cos^2 B=1-\sin^2 B=1-\left(\frac{5}{13}\right)^2=1-\left(\frac{25}{169}\right)=\frac{144}{169}\Rightarrow\cos B=\pm\sqrt{\frac{144}{169}}\Rightarrow\cos B=\frac{12}{13}.$$

Now use $\cos A=\frac{3}{5}$, $\sin B=\frac{5}{13}$, $\sin A=\frac{4}{5}$, and $\cos B=\frac{12}{13}$ to find

(a) $\sin(A+B)=\sin A\cos B+\cos A\sin B=\frac{4}{5}\cdot\frac{12}{13}+\frac{3}{5}\cdot\frac{5}{13}=\frac{48+15}{65}\Rightarrow\sin(A+B)=\frac{63}{65}.$

(b) $\sin(A-B)=\sin A\cos B-\cos A\sin B=\frac{4}{5}\cdot\frac{12}{13}-\frac{3}{5}\cdot\frac{5}{13}=\frac{48-15}{65}\Rightarrow\sin(A-B)=\frac{33}{65}.$

(c) First find $\tan A$ and $\tan B$: $\tan A=\frac{\sin A}{\cos A}=\frac{\frac{4}{5}}{\frac{3}{5}}=\frac{4}{5}\cdot\frac{5}{3}=\frac{4}{3}$; $\tan B=\frac{\sin B}{\cos B}=\frac{\frac{5}{13}}{\frac{12}{13}}=\frac{5}{13}\cdot\frac{13}{12}=\frac{5}{12}.$

Now solve $\tan(A+B)=\frac{\tan A+\tan B}{1-\tan A\tan B}=\frac{\frac{4}{3}+\frac{5}{12}}{1-\frac{4}{3}\cdot\frac{5}{12}}\cdot\frac{36}{36}=\frac{48+15}{36-20}\Rightarrow\tan(A+B)=\frac{63}{16}.$

(d) $\tan(A-B)=\frac{\tan A-\tan B}{1+\tan A\tan B}=\frac{\frac{4}{3}-\frac{5}{12}}{1+\frac{4}{3}\cdot\frac{5}{12}}\cdot\frac{36}{36}=\frac{48-15}{36+20}\Rightarrow\tan(A-B)=\frac{33}{56}.$

(e) Since $\sin(A+B)>0$ and $\tan(A+B)>0$, $A+B$ is found in quadrant I.

(f) Since $\sin(A-B)>0$ and $\tan(A-B)>0$, $A-B$ is found in quadrant I.

49. If $\pi<A<\frac{3\pi}{2}$ and $\pi<B<\frac{3\pi}{2}$, then A and B are both found in quadrant III. Next, using $\cos A=-\frac{8}{17}$ and $\cos B=-\frac{3}{5}$, find $\sin A$ and $\sin B$.

$$\sin^2 A=1-\cos^2 A=1-\left(-\frac{8}{17}\right)^2=1-\frac{64}{289}=\frac{225}{289}\Rightarrow\sin A=\pm\sqrt{\frac{225}{289}}\Rightarrow\sin A=\pm\frac{15}{17}.$$

Since A is found in quadrant III, $\sin A = -\frac{15}{17}$.

$\sin^2 B = 1 - \cos^2 B = 1 - \left(-\frac{3}{5}\right)^2 = 1 - \frac{9}{25} = \frac{16}{25} \Rightarrow \sin B = \pm\sqrt{\frac{16}{25}} \Rightarrow \sin B = \pm\frac{4}{5}$. Since $B$is in quadrant III,

$\sin B = -\frac{4}{5}$. Now use $\cos A = -\frac{8}{17}$, $\cos B = -\frac{3}{5}$, $\sin A = -\frac{15}{17}$, and $\sin B = -\frac{4}{5}$ to find:

(a) $\sin(A+B) = \sin A\cos B + \cos A\sin B = \left(-\frac{15}{17}\right)\left(-\frac{3}{5}\right) + \left(-\frac{8}{17}\right)\left(-\frac{4}{5}\right) = \frac{45+32}{85} \Rightarrow \sin(A+B) = \frac{77}{85}$.

(b) $\sin(A-B) = \sin A\cos B - \cos A\sin B = \left(-\frac{15}{17}\right)\left(-\frac{3}{5}\right) - \left(-\frac{8}{17}\right)\left(-\frac{4}{5}\right) = \frac{45-32}{85} \Rightarrow \sin(A-B) = \frac{13}{85}$.

(c) First find $\tan A$ and $\tan B$.

$$\tan A = \frac{\sin A}{\cos A} = \frac{-\frac{15}{17}}{-\frac{8}{17}} = \left(-\frac{15}{17}\right)\left(-\frac{17}{8}\right) = \frac{15}{8};\quad \tan B = \frac{\sin B}{\cos B} = \frac{-\frac{4}{5}}{-\frac{3}{5}} = \left(-\frac{4}{5}\right)\left(-\frac{5}{3}\right) = \frac{4}{3}.$$

Now solve $\tan(A+B) = \frac{\tan A + \tan B}{1 - \tan A\tan B} = \frac{\frac{15}{8} + \frac{4}{3}}{1 - \frac{15}{8}\cdot\frac{4}{3}}\cdot\frac{24}{24} = \frac{45+32}{24-60} \Rightarrow \tan(A+B) = -\frac{77}{36}$.

(d) $\tan(A-B) = \frac{\tan A - \tan B}{1 + \tan A\tan B} = \frac{\frac{15}{8} - \frac{4}{3}}{1 + \frac{15}{8}\cdot\frac{4}{3}}\cdot\frac{24}{24} = \frac{45-32}{24+60} \Rightarrow \tan(A-B) = \frac{13}{84}$.

(e) Since $\sin(A+B) > 0$ and $\tan(A+B) < 0$, $A+B$ is found in quadrant II.

(f) Since $\sin(A-B) > 0$ and $\tan(A-B) > 0$, $A-B$ is found in quadrant I.

51. $\sin(x+x) = \sin(2x) = \sin x\cos x + \sin x\cos x = 2\sin x\cos x$.

53. $\sin(x+y) + \sin(x-y) = \sin x\cos y + \cos x\sin y + \sin x\cos y - \cos x\sin y = 2\sin x\cos y$.

55. $\frac{\cos(A-B)}{\cos A\sin B} = \frac{\cos A\cos B + \sin A\sin B}{\cos A\sin B} = \frac{\cos A\cos B}{\cos A\sin B} + \frac{\sin A\sin B}{\cos A\sin B} = \frac{\cos B}{\sin B} + \frac{\sin A}{\cos A} =$

$\cot B + \tan A = \tan A + \cot B$.

57. $\frac{\sin(A-B)}{\sin B} + \frac{\cos(A-B)}{\cos B} = \frac{\sin A\cos B - \cos A\sin B}{\sin B} + \frac{\cos A\cos B + \sin A\sin B}{\cos B} =$

$\frac{\sin A\cos^2 B + \sin A\sin^2 B}{\sin B\cos B} = \frac{\sin A\left(\cos^2 B + \sin^2 B\right)}{\sin B\cos B} = \frac{\sin A}{\sin B\cos B}$.

59. $\frac{\tan x - \tan y}{\tan x + \tan y} = \frac{\frac{\sin x}{\cos x} - \frac{\sin y}{\cos y}}{\frac{\sin x}{\cos x} + \frac{\sin y}{\cos y}} = \frac{\frac{\sin x\cos y - \cos x\sin y}{\cos x\cos y}}{\frac{\sin x\cos y + \cos x\sin y}{\cos x\cos y}} = \frac{\sin x\cos y - \cos x\sin y}{\sin x\cos y + \cos x\sin y} = \frac{\sin(x-y)}{\sin(x+y)}$.

61. Since there are 60 cycles per second, the number of cycles in 0.05 seconds is given by (0.05 sec.)(60 cycles/sec.) = 3 cycles.

63. *(a)* $F = \frac{0.6(170)\sin(30° + 90°)}{\sin 12°} \approx 424.8659171 \approx 425$ pounds.

(b) Using the sine of a sum identity yields $F = \frac{0.6W\sin(\theta + 90°)}{\sin 12°} \Rightarrow$

$$F = \frac{0.6W}{\sin 12^\circ}(\sin\theta\cos 90^\circ + \cos\theta\sin 90^\circ) \approx 2.8858W(\sin\theta(0)+\cos\theta(1)) \approx 2.9W\cos\theta.$$

(c) The function F is at a maximum when $\sin(\theta+90^\circ)=1$. $\sin^{-1}1=\theta+90^\circ \Rightarrow 90^\circ=\theta+90^\circ \Rightarrow \theta=0^\circ$.

65. (a) Using the given values, $P=\frac{a}{r}\cos\left[\frac{2\pi r}{\lambda}-ct\right]=\frac{0.4}{10}\cos\left[\frac{20\pi}{4.9}-1026t\right]$. Graph this on $0\le t\le 10$. See Figure 65a. At this distance, the pressure P is oscillating.

(b) Using the given values, $P=\frac{a}{r}\cos\left[\frac{2\pi r}{\lambda}-ct\right]=\frac{3}{r}\cos\left[\frac{2\pi r}{4.9}-10{,}260\right]$. Graph this on $0\le t\le 20$. See Figure 65a. At this distance, the pressure P is oscillating. See Figure 65b. As the radius increases, the pressure P oscillates, and the amplitude decreases as r increases.

(c) If $r=n\lambda$, $P=\frac{a}{r}\cos\left[\frac{2\pi r}{\lambda}-ct\right]=\frac{a}{n\lambda}\cos\left[\frac{2\pi n\lambda}{\lambda}-ct\right]=\frac{a}{n\lambda}\cos[2\pi n-ct]=$

$$\frac{a}{2n\lambda}[\cos(2\pi n)\cos(ct)+\sin(2\pi n)\sin(ct)]=\frac{a}{n\lambda}[(1)\cos(ct)+(0)\sin(ct)]=\frac{a}{n\lambda}\cos(ct).$$

[0,0.05] by [-0.05,0.05] [0,20] by [-2,2]

Xscl =0.01 Yscl = 1

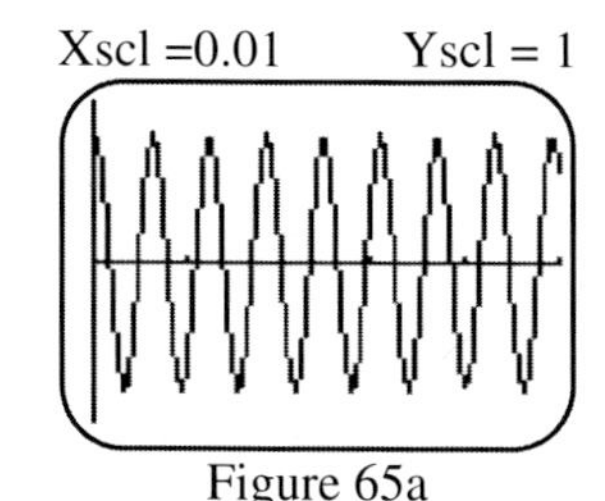

Figure 65a

Xscl = 1 Yscl = 1

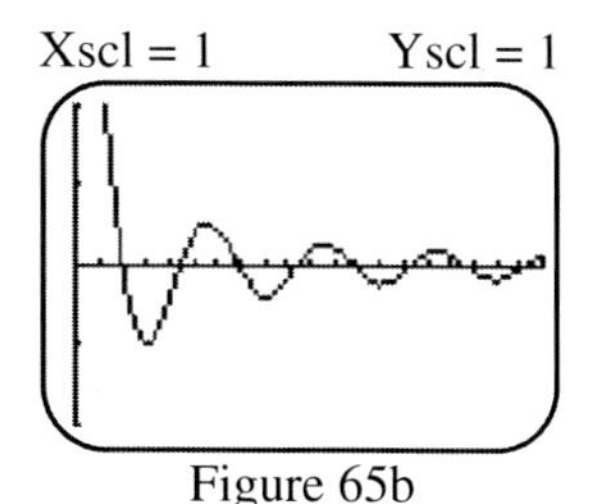

Figure 65b

Reviewing Basic Concepts (Sections 9.1 – 9.2)

1. $\frac{\csc x}{\cot x}-\frac{\cot x}{\csc x}=\frac{1}{\sin x}\cdot\frac{\sin x}{\cos x}-\frac{\cos x}{\sin x}\cdot\frac{\sin x}{1}=\frac{1}{\cos x}-\cos x=\frac{1-\cos^2 x}{\cos x}=\frac{\sin^2 x}{\cos x}.$

2. $\tan\left(-\frac{\pi}{12}\right)=\tan\left(\frac{\pi}{4}-\frac{\pi}{3}\right)=\frac{\tan\left(\frac{\pi}{4}\right)-\tan\left(\frac{\pi}{3}\right)}{1+\tan\left(\frac{\pi}{4}\right)\tan\left(\frac{\pi}{3}\right)}=\frac{1-\sqrt{3}}{1+1(\sqrt{3})}=\frac{1-\sqrt{3}}{1+\sqrt{3}}\cdot\frac{1-\sqrt{3}}{1-\sqrt{3}}=\frac{4-2\sqrt{3}}{-2}$ or $\sqrt{3}-2$.

3. $\cos 18^\circ\cos 108^\circ+\sin 18^\circ\sin 108^\circ=\cos(18^\circ-108^\circ)=\cos(-90^\circ)=\cos 90^\circ=0.$

4. $\sin\left(x-\frac{\pi}{4}\right)=\sin x\cos\frac{\pi}{4}-\cos x\sin\frac{\pi}{4}=\sin x\left(\frac{\sqrt{2}}{2}\right)-\cos x\left(\frac{\sqrt{2}}{2}\right)=\frac{\sqrt{2}}{2}(\sin x-\cos x).$

5. If A is found in quadrant II and $\sin A=\frac{2}{3}$, then $y=2$, $r=3$, and, by the Pythagorean theorem, $x^2+2^2=3^2 \Rightarrow x^2+4=9 \Rightarrow x^2=5 \Rightarrow x=\pm\sqrt{5}$. Because A is found in quadrant II, $x=-\sqrt{5}$; therefore, $\sin A=\frac{2}{3}$, $\cos A=-\frac{\sqrt{5}}{3}$, and $\tan A=-\frac{2}{\sqrt{5}}$.

If B is in quadrant III and $\cos B=-\frac{1}{2}$, then $x=-1$, $r=2$, and, by the Pythagorean theorem, $(-1)^2+y^2=2^2 \Rightarrow 1+y^2=4 \Rightarrow y^2=3 \Rightarrow y=\pm\sqrt{3}$. Because B is found in quadrant III, $y=-\sqrt{3}$;

therefore, $\sin B=-\frac{\sqrt{3}}{2}$, $\cos B=-\frac{1}{2}$, and $\tan B=\frac{\sqrt{3}}{1}=\sqrt{3}$. Use these values to find each compound value. $\sin(A+B)=\sin A\cos B+\cos A\sin B=\frac{2}{3}\left(-\frac{1}{2}\right)+\left(-\frac{\sqrt{5}}{3}\right)\left(-\frac{\sqrt{3}}{2}\right)=-\frac{2}{6}+\frac{\sqrt{15}}{6}\Rightarrow$

$\sin(A+B)=\frac{-2+\sqrt{15}}{6}$.

$\cos(A-B)=\cos A\cos B+\sin A\sin B=\left(-\frac{\sqrt{5}}{3}\right)\left(-\frac{1}{2}\right)+\frac{2}{3}\left(-\frac{\sqrt{3}}{2}\right)=-\frac{\sqrt{5}}{6}-\frac{2\sqrt{3}}{6}\Rightarrow$

$\cos(A-B)=\frac{\sqrt{5}-2\sqrt{3}}{6}$. $\tan(A-B)=\frac{\tan A-\tan B}{1+\tan A\tan B}=\frac{-\frac{2}{\sqrt{5}}-\sqrt{3}}{1+\left(-\frac{2}{\sqrt{5}}\right)\cdot\left(\sqrt{3}\right)}\cdot\frac{\sqrt{5}}{\sqrt{5}}\Rightarrow\tan(A-B)=\frac{-2-\sqrt{15}}{\sqrt{5}-2\sqrt{3}}$.

6. $\csc^2\theta-\cot^2\theta=1+\cot^2\theta-\cot^2\theta=1$.

7. $\frac{\sin t}{1-\cos t}=\frac{\sin t}{1-\cos t}\cdot\frac{1+\cos t}{1+\cos t}=\frac{\sin t(1+\cos t)}{1-\cos^2 t}=\frac{\sin t(1+\cos t)}{\sin^2 t}=\frac{1+\cos t}{\sin t}$.

8. $\frac{\cot A-\tan A}{\csc A\sec A}=\frac{\frac{\cos A}{\sin A}-\frac{\sin A}{\cos A}}{\frac{1}{\sin A}\cdot\frac{1}{\cos A}}\cdot\frac{\sin A\cos A}{\sin A\cos A}=\frac{\cos^2 A-\sin^2 A}{1}=\cos^2 A-\sin^2 A$.

9. $\frac{\sin(x-y)}{\sin x\sin y}=\frac{\sin x\cos y-\cos x\sin y}{\sin x\sin y}=\frac{\sin x\cos y}{\sin x\sin y}-\frac{\cos x\sin y}{\sin x\sin y}=\frac{\cos y}{\sin y}-\frac{\cos x}{\sin x}=$
$\cot y-\cot x$.

10. $V=20\sin\left(\frac{\pi t}{4}-\frac{\pi}{2}\right)=20\left[\sin\frac{\pi t}{4}\cos\frac{\pi}{2}-\cos\frac{\pi t}{4}\sin\frac{\pi}{2}\right]=20\left[\sin\frac{\pi t}{4}(0)-\cos\frac{\pi t}{4}(1)\right]=-20\cos\frac{\pi t}{4}$.

9.3: Further Identities

1. First find $\cos\theta$. If $\sin\theta=\frac{2}{5}$, then $\cos^2\theta=1-\sin^2\theta=1-\left(\frac{2}{5}\right)^2=1-\frac{4}{25}\Rightarrow\cos\theta=\pm\frac{\sqrt{21}}{5}$, since for $\cos\theta<0$, $\cos\theta=\frac{\sqrt{21}}{5}$. Now use $\sin\theta=\frac{2}{5}$ and $\cos\theta=-\frac{\sqrt{21}}{5}$ to find:

   (a) $\sin 2\theta=2\sin\theta\cos\theta=2\left(\frac{2}{5}\right)\left(-\frac{\sqrt{21}}{5}\right)=-\frac{4\sqrt{21}}{25}$.

   (b) $\cos 2\theta=\cos^2\theta-\sin^2\theta=\left(-\frac{\sqrt{21}}{5}\right)^2-\left(\frac{2}{5}\right)^2=\frac{21}{25}-\frac{4}{25}=\frac{17}{25}$.

3. First find $\sin\theta$ and $\cos\theta$. If $\tan\theta=2$, then $\sec^2\theta=1+\tan^2\theta=1+2^2=1+4=5\Rightarrow\sec\theta=\pm\sqrt{5}$.

   Since $\cos\theta>0$, and θ is in quadrant I, $\sec\theta=\sqrt{5}$. Now use this to find $\cos\theta$. $\cos\theta=\frac{1}{\sec\theta}=\frac{1}{\sqrt{5}}=\frac{\sqrt{5}}{5}$.

   Then use this to find $\sin x$. $\sin^2\theta=1-\cos^2\theta=1-\left(\frac{\sqrt{5}}{5}\right)^2=1-\frac{5}{25}=\frac{20}{25}=\frac{4}{5}\Rightarrow\sin\theta=\pm\sqrt{\frac{4}{5}}=\pm\frac{2\sqrt{5}}{5}$.

$\sin\theta = \frac{2\sqrt{5}}{5}$ (θ is in quadrant I). Use $\sin x = \frac{2\sqrt{5}}{5}$ and $\cos x = \frac{\sqrt{5}}{5}$ to find:

(a) $\sin 2\theta = 2\sin\theta\cos\theta = 2\left(\frac{2\sqrt{5}}{5}\right)\left(\frac{\sqrt{5}}{5}\right) = \frac{20}{25} = \frac{4}{5}.$

(b) $\cos 2\theta = \cos^2\theta - \sin^2\theta = \left(\frac{\sqrt{5}}{5}\right)^2 - \left(\frac{2\sqrt{5}}{5}\right)^2 = \frac{5}{25} - \frac{20}{25} = -\frac{15}{25} = -\frac{3}{5}.$

5. First find $\cos\theta$. I $\sin\theta = -\frac{\sqrt{5}}{7}$, then $\cos^2\theta = 1 - \sin^2\theta = 1 - \left(-\frac{\sqrt{5}}{7}\right)^2 = 1 - \frac{5}{49} = \frac{44}{49} \Rightarrow$

$\cos\theta = \pm\sqrt{\frac{44}{49}} = \pm\frac{2\sqrt{11}}{7}$. Since $\cos\theta > 0$, $\cos\theta = \frac{2\sqrt{11}}{7}$. Now use $\sin\theta = -\frac{\sqrt{5}}{7}$ and $\cos\theta = \frac{2\sqrt{11}}{7}$ to find:

(a) $\sin 2\theta = 2\sin\theta\cos\theta = 2\left(-\frac{\sqrt{5}}{7}\right)\left(\frac{2\sqrt{11}}{7}\right) = -\frac{4\sqrt{55}}{49}.$

(b) $\cos 2\theta = 1 - 2\sin^2\theta = 1 - 2\left(-\frac{\sqrt{5}}{7}\right)^2 = 1 - 2\left(\frac{5}{49}\right) = \frac{39}{49}.$

7. $\cos^2 15° - \sin^2 15° = \cos[2(15°)] = \cos 30° = \frac{\sqrt{3}}{2}.$

9. $1 - 2\sin^2 15° = \cos[2(15°)] = \cos 30° = \frac{\sqrt{3}}{2}.$

11. $2\cos^2 67.5° - 1 = \cos[2(67.5°)] = \cos 135° = -\frac{\sqrt{2}}{2}.$

13. $\frac{\tan 51°}{1 - \tan^2 51°} = \frac{1}{2} \cdot \frac{2\tan 51°}{1 - \tan^2 51°} = \frac{1}{2}\tan[2 \cdot 51°] = \frac{1}{2}\tan 102°.$

15. $\frac{1}{4} - \frac{1}{2}\sin^2 47.1° = \frac{1}{4}\left[4\left(\frac{1}{4} - \frac{1}{2}\sin^2 47.1°\right)\right] = \frac{1}{4}\left[1 - 2\sin^2 47.1°\right] = \frac{1}{4}\cos[2(47.1°)] = \frac{1}{4}\cos 94.2°.$

17. $4\sin 15°\cos 15° = 2\sin 2(15°) = 2\sin 30° = 2 \cdot \frac{1}{2} = 1.$

19. When graphing the equation on the calculator, the graph looks like that of $\cos 2x$. Verify this result.

$\cos^4 x - \sin^4 x = (\cos^2 x - \sin^2 x)(\cos^2 x + \sin^2 x) = (\cos^2 x - \sin^2 x)(1) = \cos 2x.$

21. $\cos 3x = \cos(2x + x) = \cos 2x\cos x - \sin 2x\sin x = (2\cos^2 x - 1)\cos x - (2\sin x\cos x)\sin x =$

$2\cos^3 x - \cos x - 2\sin^2 x\cos x = 2\cos^3 x - \cos x - 2(1 - \cos^2 x)\cos x =$

$2\cos^3 x - \cos x - 2\cos x + 2\cos^3 x = 4\cos^3 x - 3\cos x.$

The graph of each equation supports that they are equal.

23. $\tan 4x = \tan 2(2x) = \dfrac{2\tan 2x}{1-\tan^2 2x} = \dfrac{2\left(\frac{2\tan x}{1-\tan^2 x}\right)}{1-\left(\frac{2\tan x}{1-\tan^2 x}\right)^2}\cdot\dfrac{\left(1-\tan^2 x\right)^2}{\left(1-\tan^2 x\right)^2} = \dfrac{4\tan x(1-\tan^2 x)}{(1-\tan^2 x)^2-4\tan^2 x} =$

$\dfrac{4\tan x-4\tan^3 x}{1-2\tan^2 x+\tan^4 x-4\tan^2 x} = \dfrac{4\tan x-4\tan^3 x}{1-6\tan^2 x+\tan^4 x}$. The graph of each equation supports that they are equal.

25. Since $\sin\dfrac{\pi}{12}$ is found in quadrant I, the sine value will be positive; therefore:

$\sin\dfrac{\pi}{12} = \sin\left(\dfrac{\pi/6}{2}\right) = \sqrt{\dfrac{1-\cos\frac{\pi}{6}}{2}} = \sqrt{\dfrac{1-\frac{\sqrt{3}}{2}}{2}\cdot\dfrac{2}{2}} = \sqrt{\dfrac{2-\sqrt{3}}{4}} = \dfrac{\sqrt{2-\sqrt{3}}}{2}.$

27. $\tan\left(-\dfrac{\pi}{8}\right) = -\tan\left(\dfrac{\pi}{8}\right) = -\tan\left(\dfrac{\pi/4}{2}\right) = -\left(\dfrac{1-\cos\frac{\pi}{4}}{\sin\frac{\pi}{4}}\right) = -\dfrac{1-\frac{\sqrt{2}}{2}}{\frac{\sqrt{2}}{2}}\cdot\dfrac{2}{2} = -\dfrac{2-\sqrt{2}}{\sqrt{2}}\cdot\dfrac{\sqrt{2}}{\sqrt{2}} = -\dfrac{2\sqrt{2}-2}{2} = 1-\sqrt{2}.$ An

Alternate method yields $-\sqrt{3-2\sqrt{2}}$.

29. Since $\sin 67.5°$ is found in quadrant I, the sine value will be positive, therefore:

$\sin 67.5° = \sin\left(\dfrac{135°}{2}\right) = \sqrt{\dfrac{1-\cos 135°}{2}} = \sqrt{\dfrac{1-\left(-\frac{\sqrt{2}}{2}\right)}{2}\cdot\dfrac{2}{2}} = \sqrt{\dfrac{2+\sqrt{2}}{4}} = \dfrac{\sqrt{2+\sqrt{2}}}{2}.$

31. If $\cos x = \dfrac{1}{4}$ and $0 < x < \dfrac{\pi}{2}$, then x is found in quadrant I. With x in quadrant I, the cosine value will be positive; therefore: $\cos\dfrac{x}{2} = \sqrt{\dfrac{1+\cos x}{2}} = \sqrt{\dfrac{1+\frac{1}{4}}{2}\cdot\dfrac{4}{4}} = \sqrt{\dfrac{4+1}{8}} = \dfrac{\sqrt{5}}{\sqrt{8}}\cdot\dfrac{\sqrt{2}}{\sqrt{2}} = \dfrac{\sqrt{10}}{4}.$

33. If $\sin x = \dfrac{3}{5}$ and $\dfrac{\pi}{2} < x < \pi$, then x is found in quadrant II. With x in quadrant II, the cosine value will be negative; therefore: $\cos x = -\sqrt{1-\sin^2 x} = -\sqrt{1-\dfrac{9}{25}} = -\sqrt{\dfrac{16}{25}} = -\dfrac{4}{5}$. Now use $\sin x = \dfrac{3}{5}$ and $\cos x = -\dfrac{4}{5}$ to find $\tan\dfrac{x}{2}$. $\tan\dfrac{x}{2} = \dfrac{1-\cos\theta}{\sin\theta} = \dfrac{1-\left(-\frac{4}{5}\right)}{\frac{3}{5}}\cdot\dfrac{5}{5} = \dfrac{5+4}{3} = 3.$

35. If $\tan x = \dfrac{\sqrt{7}}{3}$ and $\pi < x < \dfrac{3\pi}{2}$, then x is found in quadrant III. With x in quadrant III, both sine and secant values are negative; therefore: $\sec x = -\sqrt{1+\tan^2 x} = -\sqrt{1+\dfrac{7}{9}} = -\sqrt{\dfrac{16}{9}} = -\dfrac{4}{3}$. Therefore $\cos x = -\dfrac{3}{4}$, and

$\sin x = -\sqrt{1-\cos^2 x} = -\sqrt{1-\dfrac{9}{16}} = -\sqrt{\dfrac{7}{16}} = -\dfrac{\sqrt{7}}{4}$. Now use $\sin x = -\dfrac{\sqrt{7}}{4}$ and $\cos x = -\dfrac{3}{4}$ to find $\tan\dfrac{x}{2}$.

$\tan\dfrac{x}{2} = \dfrac{1-\cos x}{\sin x} = \dfrac{1-\left(-\frac{3}{4}\right)}{\frac{-\sqrt{7}}{4}}\cdot\dfrac{4}{4} = \dfrac{4+3}{-\sqrt{7}}\cdot\dfrac{\sqrt{7}}{\sqrt{7}} = \dfrac{7\sqrt{7}}{-7} = -\sqrt{7}.$

37. If $\tan x = 2$ and $0 < x < \frac{\pi}{2}$, then x is found in quadrant I. With x in quadrant I, then $\sec x > 0$.

$\sec^2 x = \tan^2 x + 1 \Rightarrow \sec^2 x = 2^2 + 1 = 5 \Rightarrow \sec x = \sqrt{5} \Rightarrow \cos x = \frac{1}{\sqrt{5}} = \frac{\sqrt{5}}{5}$. Since

$0 < x < \frac{\pi}{2} \Rightarrow 0 < \frac{x}{2} < \frac{\pi}{4} \Rightarrow \frac{x}{2}$ is in quadrant I and $\frac{x}{2} > 0$. $\sin\frac{x}{2} = \sqrt{\frac{1-\cos x}{2}} = \sqrt{\frac{1-\frac{\sqrt{5}}{5}}{2}} = \frac{\sqrt{50-10\sqrt{5}}}{10}$.

39. Since $\frac{\pi}{2} < x < \pi \Rightarrow \cos x < 0$. $\cos x = \sqrt{\frac{1+\cos 2x}{2}} \Rightarrow \sqrt{\frac{1+\left(-\frac{5}{12}\right)}{2}} = -\sqrt{\frac{7}{24}} = -\frac{\sqrt{42}}{12}$

41. (a) The function $\tan\frac{\pi}{2}$ is undefined, so it cannot be used.

(b) $\tan\left(\frac{\pi}{2} + x\right) = \frac{\sin(\frac{\pi}{2} + x)}{\cos(\frac{\pi}{2} + x)}$.

(c) $\tan\left(\frac{\pi}{2} + x\right) = \frac{\sin\frac{\pi}{2}\cos x + \cos\frac{\pi}{2}\sin x}{\cos\frac{\pi}{2}\cos x - \sin\frac{\pi}{2}\sin x} = \frac{1\cdot\cos x + 0\cdot\sin x}{0\cdot\cos x - 1\cdot\sin x} = \frac{\cos x}{-\sin x} = -\cot x$.

43. $\sqrt{\frac{1-\cos 40^\circ}{2}} = \sin\frac{40^\circ}{2} = \sin 20^\circ$.

45. $\sqrt{\frac{1-\cos 147^\circ}{1+\cos 147^\circ}} = \tan\frac{147^\circ}{2} = \tan 73.5^\circ$.

47. $\frac{1-\cos 59.74^\circ}{\sin 59.74^\circ} = \tan\frac{59.74^\circ}{2} = \tan 29.87^\circ$.

49. $\frac{2\cos 2\alpha}{\sin 2\alpha} = \frac{2(\cos^2\alpha - \sin^2\alpha)}{2\sin\alpha\cos\alpha} = \frac{\cos^2\alpha}{\sin\alpha\cos\alpha} - \frac{\sin^2\alpha}{\sin\alpha\cos\alpha} = \frac{\cos\alpha}{\sin\alpha} - \frac{\sin\alpha}{\cos\alpha} = \cot\alpha - \tan\alpha$.

51. $\sec^2\frac{x}{2} = \left(\frac{1}{\cos\frac{x}{2}}\right)^2 = \left(\pm\frac{1}{\sqrt{\frac{1+\cos x}{2}}}\right)^2 = \frac{1}{\frac{1+\cos x}{2}} = \frac{2}{1+\cos x}$.

53. Working the right side of the equation.

$$\frac{2-\sec^2\theta}{\sec^2\theta} = \frac{2-\frac{1}{\cos^2\theta}}{\frac{1}{\cos^2\theta}} = \frac{2-\frac{1}{\cos^2\theta}}{\frac{1}{\cos^2\theta}}\cdot\frac{\cos^2\theta}{\cos^2\theta} = 2\cos^2\theta - 1 = \cos 2\theta.$$

55. If $\tan s + \cot s = \frac{\sin s}{\cos s} + \frac{\cos s}{\sin s} = \frac{\sin^2 s + \cos^2 s}{\sin s\cos s} = \frac{1}{\sin s\cos s}$, and if $2\csc 2s = 2\left(\frac{1}{2\sin 2s}\right) =$

$2\left(\frac{1}{2\sin s\cos s}\right) = \frac{1}{\sin s\cos s}$, then $\tan s + \cot s = 2\csc 2s$.

57. Working the left side of the equation. $\sin^2\frac{x}{2}=\left(\pm\sqrt{\frac{1-\cos x}{2}}\right)^2=\frac{1-\cos x}{2}$.

Working the right side of the equation. $\frac{\tan x-\sin x}{2\tan x}=\frac{\frac{\sin x}{\cos x}-\sin x}{2\cdot\frac{\sin x}{\cos x}}=\frac{\frac{\sin x}{\cos x}-\sin x}{2\cdot\frac{\sin x}{\cos x}}\cdot\frac{\cos x}{\cos x}=$

$\frac{\sin x-\cos x\sin x}{2\sin x}=\frac{\sin x(1-\cos x)}{2\sin x}=\frac{1-\cos x}{2}$.

59. Working the right side of the equation.

$$\frac{1-\tan^2 x}{1+\tan^2 x}=\frac{1-\frac{\sin^2 x}{\cos^2 x}}{1+\frac{\sin^2 x}{\cos^2 x}}=\frac{1-\frac{\sin^2 x}{\cos^2 x}}{1+\frac{\sin^2 x}{\cos^2 x}}\cdot\frac{\cos^2 x}{\cos^2 x}=\frac{\cos^2 x-\sin^2 x}{\cos^2 x+\sin^2 x}=\frac{\cos^2 x-\sin^2 x}{1}=\cos 2x.$$

61. $\sin 2\alpha\cos 2\alpha=\sin 2\alpha\left(1-2\sin^2\alpha\right)=\sin 2\alpha-2\sin 2\alpha\sin^2\alpha=$

$\sin 2\alpha-2(2\sin\alpha\cos\alpha)\sin^2=\sin 2\alpha-4\sin^3\alpha\cos\alpha.$

63. $\sin 4\alpha=2\sin 2\alpha\cos 2\alpha=2(2\sin\alpha\cos\alpha)\cos 2\alpha=4\sin\alpha\cos\alpha\cos 2\alpha.$

65. $2\sin 58^\circ\cos 102^\circ=2\left[\frac{1}{2}[\sin(58^\circ+102^\circ)+\sin(58^\circ-102^\circ)]\right]=\sin 160^\circ+\sin(-44^\circ)=\sin 160^\circ-\sin 44^\circ.$

67. $2\cos 85^\circ\sin 140^\circ=\frac{1}{2}[2[\sin(85^\circ+140^\circ)-\sin(85^\circ-140^\circ)]]=\sin 225^\circ-\sin(-55^\circ)=\sin 225^\circ+\sin 55^\circ.$

69. $\cos 4x-\cos 2x=-2\sin\left(\frac{4x+2x}{2}\right)\sin\left(\frac{4x-2x}{2}\right)=-2\sin 3x\sin x.$

71. $\sin 25^\circ+\sin(-48)^\circ=2\sin\left(\frac{25^\circ+(-48^\circ)}{2}\right)\cos\left(\frac{25^\circ-(-48^\circ)}{2}\right)=2\sin(-11.5^\circ)\cos 36.5^\circ=$

$-2\sin 11.5^\circ\cos 36.5^\circ.$

73. $\cos 4x+\cos 8x=2\cos\left(\frac{4x+8x}{2}\right)\cos\left(\frac{4x-8x}{2}\right)=2\cos 6x\cos(-2x)=2\cos 6x\cos 2x.$

75. (a) Since R is the radius of the circle, the dashed line has length $R-b$, so $\cos\frac{\theta}{2}=\frac{R-b}{R}$.

(b) $\tan\frac{\theta}{4}=\frac{1-\cos\frac{\theta}{2}}{\sin\frac{\theta}{2}}=\frac{1-\frac{R-b}{R}}{\frac{50}{R}}=\frac{R-(R-b)}{50}=\frac{b}{50}$.

(c) If $b=12$, then $\tan\frac{\theta}{4}=\frac{b}{50}$ yields $\tan\frac{\theta}{4}=\frac{12}{50}$. Now solve for θ.

$\frac{\theta}{4}=\tan^{-1}\left(\frac{12}{50}\right)\Rightarrow\theta=4\tan^{-1}\left(\frac{12}{50}\right)\approx 53.98^\circ\Rightarrow\theta\approx 54^\circ.$

77. (a) Graph $W=VI=[163\sin(120\pi t)][1.23\sin(120\pi t)]=200.49\sin^2(120\pi t)$. See Figure 77.

(b) The minimum wattage is 0 watts and occurs whenever $\sin(120\pi t)=0$. The maximum wattage will occur when $\sin(120\pi t)=1$. This would be $200.49(1)=200.49$ watts.

(c) Use the identity, $\sin^2 A = \frac{1}{2}(1-\cos 2A)$.

$$W = 200.49\sin^2(120\pi t) = 200.49\left[\frac{1}{2}(1-\cos(240\pi t))\right] = -100.245\cos(240\pi t)+100.245.$$

From this equation, $a=-100.245$, $w=240\pi$, and $c=100.245$.

(d) Graphing $(120\pi t)$ and $W=-100.245\cos(240\pi t)+100.245$ in the same window shows that both equations have the same graph.

(e) Graph W and $y=100.245$ together. The cosine (or sine) graph of $y=\tan\ x$ appears to be vertically centered about this line. An estimate for the average wattage consumed is 100.245 watts. The light bulb would be rated at about 100 watts.

[0,0.5] by [-1,6]
Xscl = 0.1 Yscl = 1

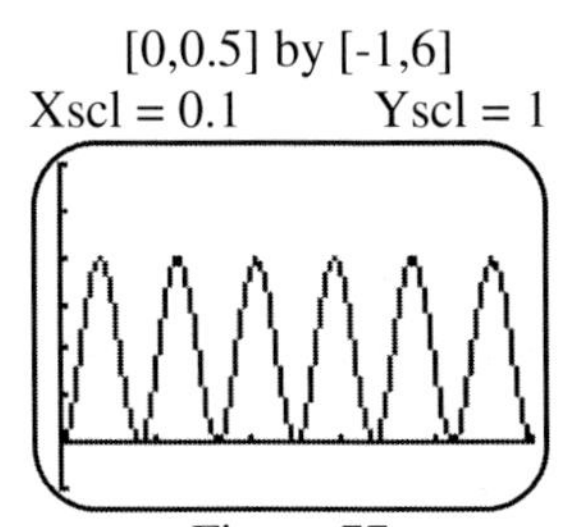

Figure 77

79. $\tan\theta+\cot\theta = \frac{\sin\theta}{\cos\theta}+\frac{\cos\theta}{\sin\theta} = \frac{\sin^2\theta}{\sin\theta\cos\theta}+\frac{\cos^2\theta}{\sin\theta\cos\theta} = \frac{\sin^2\theta+\cos^2\theta}{\sin\theta\cos\theta} =$

$\frac{1}{\sin\theta\cos\theta} = \frac{1}{\cos\theta}\cdot\frac{1}{\sin\theta} = \sec\theta\csc\theta.$

80. $\csc\theta\cos^2\theta+\sin\theta = \frac{1}{\sin\theta}\cdot\cos^2\theta+\sin\theta = \frac{\cos^2\theta}{\sin\theta}+\sin\theta = \frac{\cos^2\theta+\sin^2\theta}{\sin\theta} = \frac{1}{\sin\theta} = \csc\theta.$

81. $\tan\frac{x}{2} = \frac{1-\cos x}{\sin x} = \frac{1}{\sin x}-\frac{\cos x}{\sin x} = \csc x-\cot x.$

82. $\sec(\pi-x) = \frac{1}{\cos(\pi-x)} = \frac{1}{\cos\pi\cos x+\sin\pi\sin x} = \frac{1}{-1\cdot\cos x+0\cdot\sin x} = \frac{1}{-\cos x} = -\sec x.$

83. $\frac{\sin t}{1+\cos t} = \tan\frac{t}{2} = \frac{1-\cos t}{\sin t}.$

84. $\frac{1-\sin t}{\cos t} = \frac{1}{\cos t}-\frac{\sin t}{\cos t} = \sec t-\tan t = (\sec t-\tan t)\cdot\frac{\sec t+\tan t}{\sec t+\tan t} = \frac{\sec^2 t-\tan^2 t}{\sec t+\tan t} = \frac{1}{\sec t+\tan t}.$

85. $\sin 2\theta = 2\sin\theta\cos\theta = 2\sin\theta\cos\theta\cdot\frac{\cos\theta}{\cos\theta} = \frac{2\sin\theta\cos^2\theta}{\cos\theta} = 2\cdot\frac{\sin\theta}{\cos\theta}\cdot\cos^2\theta = 2\tan\theta\cos^2\theta =$

$2\cdot\tan\theta\cdot\frac{1}{\sec^2\theta} = \frac{2\tan\theta}{\sec^2\theta} = \frac{2\tan\theta}{1+\tan^2\theta}.$

86. $\frac{2}{1+\cos\ x}-\tan^2\frac{x}{2} = \frac{2}{1-\cos x}-\frac{\sin^2\frac{x}{2}}{\cos^2\frac{x}{2}} = \frac{2}{1+\cos x}-\left(\sqrt{\frac{1-\cos x}{1+\cos x}}\right)^2 = \frac{2}{1+\cos x}-\frac{1-\cos x}{1+\cos x} = \frac{1+\cos x}{1+\cos x} = 1.$

87. $\cot\theta - \tan\theta = \frac{\cos\theta}{\sin\theta} - \frac{\sin\theta}{\cos\theta} = \frac{\cos^2\theta}{\sin\theta\cos\theta} - \frac{\sin^2\theta}{\sin\theta\cos\theta} = \frac{\cos^2\theta - \sin^2\theta}{\sin\theta\cos\theta} = \frac{\cos^2\theta - \left(1-\cos^2\theta\right)}{\sin\theta\cos\theta} = \frac{2\cos^2\theta - 1}{\sin\theta\cos\theta}.$

88. $1 - \tan^2\frac{\theta}{2} = 1 - \left(\sqrt{\frac{1-\cos\theta}{1+\cos\theta}}\right)^2 = 1 - \frac{1-\cos\theta}{1+\cos\theta} = \frac{1+\cos\theta}{1+\cos\theta} - \frac{1-\cos\theta}{1+\cos\theta} = \frac{2\cos\theta}{1+\cos\theta}.$

89. $\frac{\sin\theta + \tan\theta}{1+\cos\theta} = \frac{\frac{\sin\theta}{1} + \frac{\sin\theta}{\cos\theta}}{1+\cos\theta} = \frac{\frac{\cos\theta\sin\theta + \sin\theta}{\cos\theta}}{1+\cos\theta} = \frac{\frac{\sin\theta(\cos\theta+1)}{\cos\theta}}{1+\cos\theta} = \frac{\sin\theta}{\cos\theta} = \tan\theta.$

90. $\csc^4 x - \cot^4 x = \left(\csc^2\theta + \cot^2\theta\right)\left(\csc^2 - \cot^2\theta\right) = \left(\csc^2\theta + \cot^2\theta\right)\left(1 + \cot^2\theta - \cot^2\theta\right) =$

$\left(\csc^2\theta + \cot^2\theta\right)(1) = \left(\csc^2\theta + \cot^2\theta\right) = \frac{1}{\sin^2\theta} + \frac{\cos^2 x}{\sin^2 x} = \frac{1+\cos^2 x}{\sin x} = \frac{1+\cos^2 x}{1-\cos^2 x}.$

91. $\frac{\tan^2 t + 1}{\tan t \csc^2 t} = \frac{\sec^2 t}{\tan t \csc^2 t} = \frac{\sec^2 t}{\csc^2 t} \cdot \frac{1}{\tan t} = (\tan^2 t) \cdot \frac{1}{\tan t} = \tan t.$

92. $\frac{\sin s}{1+\cos s} + \frac{1+\cos s}{\sin s} = \frac{\sin^2 s}{\sin s(1+\cos s)} + \frac{1 + 2\cos s + \cos^2 s}{\sin s(1+\cos s)} = \frac{\sin^2 s + \cos^2 s + 2\cos s + 1}{\sin s(1+\cos s)} =$

$\frac{1 + 2\cos s + 1}{\sin s(1+\cos s)} = \frac{2 + 2\cos s}{\sin s(1+\cos s)} = \frac{2(1+\cos s)}{\sin s(1+\cos s)} = \frac{2}{\sin s} = 2 \cdot \frac{1}{\sin s} = 2\csc s.$

93. $\tan 4\theta = \tan 2(2\theta) = \frac{2\tan 2\theta}{1 - \tan^2 2\theta} = \frac{2\tan 2\theta}{1 - (\sec^2\theta - 1)} = \frac{2\tan 2\theta}{2 - \sec^2\theta}.$

94. $\tan\left(\frac{x}{2} + \frac{\pi}{4}\right) = \frac{\tan\frac{x}{2} + \tan\frac{\pi}{4}}{1 - \tan\frac{x}{2}\tan\frac{\pi}{4}} = \frac{\frac{\sin x}{1+\cos x} + 1}{1 - \frac{\sin x}{1+\cos x}(1)} = \frac{\frac{\sin x + 1 + \cos x}{1+\cos x}}{\frac{1+\cos x - \sin x}{1+\cos x}} = \frac{1 + \cos x + \sin x}{1 + \cos x - \sin x} =$

$\frac{1+\cos x+\sin x}{1+\cos x-\sin x} \cdot \frac{\cos x}{\cos x} = \frac{\cos x + \cos^2 x + \cos x \sin x}{(1+\cos x - \sin x)\cos x} = \frac{\cos x + 1 - \sin^2 x + \cos x \sin x}{(1+\cos x - \sin x)\cos x} =$

$\frac{\cos x + (1+\sin x)(1-\sin x) + \cos x \sin x}{(1+\cos x - \sin x)\cos x} = \frac{\cos x(1+\sin x) + (1+\sin x)(1-\sin x)}{(1+\cos x - \sin x)\cos x} =$

$\frac{(1+\sin x)[\cos x + (1-\sin x)]}{(1+\cos x - \sin x)\cos x} = \frac{1+\sin x}{\cos x} = \frac{1}{\cos x} + \frac{\sin x}{\cos x} = \sec x + \tan x.$

95. $\frac{\cot s - \tan s}{\cos s + \sin s} = \frac{\frac{\cos s}{\sin s} - \frac{\sin s}{\cos s}}{\cos s + \sin s} = \frac{\frac{\cos s}{\sin s} - \frac{\sin s}{\cos s}}{\cos s + \sin s} \cdot \frac{\sin s \cos s}{\sin s \cos s} = \frac{\cos^2 s - \sin^2 s}{(\cos s + \sin s)\sin s \cos s} =$

$\frac{(\cos s + \sin s)(\cos s - \sin s)}{(\cos s + \sin s)\sin s \cos s} = \frac{\cos s - \sin s}{\sin s \cos s}$

96. $\frac{\tan\theta - \cot\theta}{\tan\theta + \cot\theta} = \frac{\frac{\sin\theta}{\cos\theta} - \frac{\cos\theta}{\sin\theta}}{\frac{\sin\theta}{\cos\theta} + \frac{\cos\theta}{\sin\theta}} = \frac{\frac{\sin\theta}{\cos\theta} - \frac{\cos\theta}{\sin\theta}}{\frac{\sin\theta}{\cos\theta} + \frac{\cos\theta}{\sin\theta}} \cdot \frac{\cos\theta\sin\theta}{\cos\theta\sin\theta} = \frac{\sin^2\theta - \cos^2\theta}{\sin^2\theta + \cos^2\theta} = \frac{\sin^2\theta - \cos^2\theta}{1} =$

$\sin^2\theta - \cos^2\theta = \left(1 - \cos^2\theta\right) - \cos^2\theta = 1 - 2\cos^2\theta$

97. $\dfrac{\tan(x+y)-\tan y}{1+\tan(x+y)\tan y}=\dfrac{\dfrac{\tan x+\tan y}{1-\tan x\tan y}-\tan y}{1+\dfrac{\tan x+\tan y}{1-\tan x\tan y}\cdot\tan y}=\dfrac{\dfrac{\tan x+\tan y}{1-\tan x\tan y}-\tan y}{1+\dfrac{\tan x+\tan y}{1-\tan x\tan y}\cdot\tan y}\cdot\dfrac{1-\tan x\tan y}{1-\tan x\tan y}=$

$\dfrac{\tan x+\tan y-\tan y(1-\tan x\tan y)}{1-\tan x\tan y+(\tan x+\tan y)\tan y}=\dfrac{\tan x+\tan x\tan^2 y}{1-\tan x\tan y+\tan x\tan y+\tan^2 y}=\dfrac{\tan x(1+\tan^2 y)}{1+\tan^2 y}=\tan x$

98. $2\cos^2\dfrac{x}{2}\tan x=2\left(\pm\sqrt{\dfrac{1+\cos x}{2}}\right)^2\cdot\dfrac{\sin x}{\cos x}=2\cdot\dfrac{1+\cos x}{2}\cdot\dfrac{\sin x}{\cos x}=(1+\cos x)\cdot\dfrac{\sin x}{\cos x}=$

$\dfrac{\sin x}{\cos x}+\sin x=\tan x+\sin x$

99. $\dfrac{\cos^4 x-\sin^4 x}{\cos^2 x}=\dfrac{(\cos^2 x-\sin^2 x)(\cos^2 x+\sin^2 x)}{\cos^2 x}=\dfrac{\cos^2 x-\sin^2 x}{\cos^2 x}=\dfrac{\cos^2 x}{\cos^2 x}-\dfrac{\sin^2 x}{\cos^2 x}=1-\tan^2 x$

100. $\dfrac{\csc t+1}{\csc t-1}=\dfrac{\dfrac{1}{\sin t}+1}{\dfrac{1}{\sin t}-1}=\dfrac{\dfrac{1}{\sin t}+1}{\dfrac{1}{\sin t}-1}\cdot\dfrac{\sin t}{\sin t}=\dfrac{1+\sin t}{1-\sin t}=\dfrac{1+\sin t}{1-\sin t}\cdot\dfrac{1+\sin t}{1+\sin t}=\dfrac{(1+\sin t)^2}{1-\sin^2 t}=\dfrac{(1+\sin t)^2}{\cos^2 t}=$

$\left(\dfrac{1+\sin t}{\cos t}\right)^2=\left(\dfrac{1}{\cos t}+\dfrac{\sin t}{\cos t}\right)^2=(\sec t+\tan t)^2$

101. $\dfrac{2(\sin x-\sin^3 x)}{\cos x}=\dfrac{2\sin x(1-\sin^2 x)}{\cos x}=\dfrac{2\sin x\cos^2 x}{\cos x}=2\sin x\cos x=\sin 2x$

102. $\dfrac{1}{2}\cot\dfrac{x}{2}-\dfrac{1}{2}\tan\dfrac{x}{2}=\dfrac{1}{2}\cdot\dfrac{1}{\tan\frac{x}{2}}-\dfrac{1}{2}\tan\dfrac{x}{2}=\dfrac{1}{2}\cdot\dfrac{1}{\dfrac{\sin x}{1+\cos x}}-\dfrac{1}{2}\cdot\dfrac{1-\cos x}{\sin x}=\dfrac{1+\cos x}{2\sin x}-\dfrac{1-\cos x}{2\sin x}=$

$\dfrac{1+\cos x-(1-\cos x)}{2\sin x}=\dfrac{2\cos x}{2\sin x}=\dfrac{\cos x}{\sin x}=\cot x$

103. $\dfrac{\sin(x+y)}{\cos(x-y)}=\dfrac{\sin x\cos y+\cos x\sin y}{\cos x\cos y+\sin x\sin y}=\dfrac{\cos x\cos y+\cos x\sin y}{\cos x\cos y+\sin x\sin y}\cdot\dfrac{\dfrac{1}{\sin x\ \sin y}}{\dfrac{1}{\sin x\ \sin y}}=\dfrac{\dfrac{\cos y}{\sin y}+\dfrac{\cos x}{\sin x}}{\dfrac{\cos x\ \cos y}{\sin x\ \sin y}+1}=$

$\dfrac{\cot y+\cot x}{\cot x\cot y+1}=\dfrac{\cot x+\cot y}{1+\cot x\cot y}.$

104. $\dfrac{1}{\sec t-1}+\dfrac{1}{\sec t+1}=\dfrac{\sec t+1}{\sec^2 t-1}+\dfrac{\sec t-1}{\sec^2 t-1}=\dfrac{\sec t+1}{\tan^2 t}+\dfrac{\sec t-1}{\tan^2 t}=\dfrac{2\sec t}{\tan^2 t}=2\sec t\cdot\dfrac{1}{\tan^2 t}=$

$2\sec t\cdot\cot^2 t=2\cdot\dfrac{1}{\cos t}\cdot\dfrac{\cos^2 t}{\sin^2 t}=\dfrac{2\cos t}{\sin^2 t}=2\cdot\dfrac{\cos t}{\sin t}\cdot\dfrac{1}{\sin t}=2\cot t\csc t.$

9.4: The Inverse Circular Functions

1. one-to-one
3. $\cos y$
5. π

7. (a) The domain of $y = \sin^{-1} x$ is the range of $y = \sin x$, therefore the domain is: $[-1,1]$

(b) The range of $y = \sin^{-1} x$ is the domain of $y = \sin x$, therefore the range is: $\left[-\frac{\pi}{2}, \frac{\pi}{2}\right]$

(c) For this function, as x increases, y increases. Therefore, it in an increasing function.

(d) Arcsin (-2) is not defined since -2 is not in the domain.

9. (a) The domain of $y = \tan^{-1} x$ is the range of $y = \tan\ x$, therefore the domain is: $(-\infty, \infty)$

(b) The range of $y = \tan^{-1} x$ is the domain of $y = \tan\ x$, therefore the range is: $\left(-\frac{\pi}{2}, \frac{\pi}{2}\right)$

(c) For this function, as x increases, y increases. Therefore, it in an increasing function.

(d) No, since the domain is $(-\infty, \infty)$ arctan x is defined for all numbers.

11. $y = \tan^{-1} 1 \Rightarrow 1 = \tan y \Rightarrow y = \frac{\pi}{4} \left(range: \left(-\frac{\pi}{2}, \frac{\pi}{2}\right)\right)$

13. $y = \cos^{-1}(-1) \Rightarrow -1 = \cos y \Rightarrow y = \pi \left(range: [0, \pi]\right)$

15. $y = \sin^{-1}(-1) \Rightarrow -1 = \sin y \Rightarrow y = -\frac{\pi}{2} \left(range: \left[-\frac{\pi}{2}, \frac{\pi}{2}\right]\right)$

17. $y = \arctan\ 0 \Rightarrow 0 = \tan y \Rightarrow y = 0 \left(range: \left(-\frac{\pi}{2}, \frac{\pi}{2}\right)\right)$

19. $y = \arccos\ 0 \Rightarrow 0 = \cos y \Rightarrow y = \frac{\pi}{2} \left(range: [0, \pi]\right)$

21. $y = \sin^{-1}\left(\frac{\sqrt{2}}{2}\right) \Rightarrow \frac{\sqrt{2}}{2} = \sin y \Rightarrow y = \frac{\pi}{4} \left(range: \left[-\frac{\pi}{2}, \frac{\pi}{2}\right]\right)$

23. $y = arc\cos\left(-\frac{\sqrt{3}}{2}\right) \Rightarrow -\frac{\sqrt{3}}{2} = \cos y \Rightarrow y = \frac{5\pi}{6} \left(range: [0, \pi]\right)$

25. $y = \cot^{-1}(-1) \Rightarrow y = \tan^{-1}\left(\frac{1}{-1}\right) + \pi \Rightarrow y = \tan^{-1}(-1) + \pi \Rightarrow (-1 = \tan y) + \pi \Rightarrow$

$y = -\frac{\pi}{4} + \pi \Rightarrow y = \frac{3\pi}{4} \left(range: (0, \pi)\right).$

27. $y = \csc^{-1}(-2) \Rightarrow y = \sin^{-1}\left(\frac{1}{-2}\right) \Rightarrow y = \sin^{-1}\left(-\frac{1}{2}\right) \Rightarrow -\frac{1}{2} = \sin y \Rightarrow y = -\frac{\pi}{6} \left(range: \left[-\frac{\pi}{2}, 0\right) \cup \left(0, \frac{\pi}{2}\right]\right)$

29. $y = arc\sec \frac{2\sqrt{3}}{3} \Rightarrow y = \cos^{-1}\left(\frac{1}{\frac{2\sqrt{3}}{3}}\right) \Rightarrow y = \cos^{-1}\left(\frac{3}{2\sqrt{3}}\right) \Rightarrow y = \cos^{-1}\frac{\sqrt{3}}{2} \Rightarrow$

$\frac{\sqrt{3}}{2} = \cos y \Rightarrow y = \frac{\pi}{6} \left(range: \left[0, \frac{\pi}{2}\right) \cup \left(\frac{\pi}{2}, \pi\right]\right)$

31. $y=\tan^{-1}\left(\sqrt{3}\right)\Rightarrow\sqrt{3}=\tan y\Rightarrow y=\frac{\pi}{3}\left(range:\left(-\frac{\pi}{2},\frac{\pi}{2}\right)\right)$

33. $y=\csc^{-1}(2)\Rightarrow y=\sin^{-1}\left(\frac{1}{2}\right)\Rightarrow\frac{1}{2}=\sin y\Rightarrow y=\frac{\pi}{6}\left(range:\left[-\frac{\pi}{2},0\right)\cup\left(0,\frac{\pi}{2}\right]\right)$

35. $\theta=\arctan(-1)\Rightarrow-1=\tan\theta\Rightarrow\theta=-45^\circ\ \left(range:(-90^\circ,90^\circ)\right)$

37. $\theta=\arcsin\left(-\frac{\sqrt{3}}{2}\right)\Rightarrow-\frac{\sqrt{3}}{2}=\sin\theta\Rightarrow\theta=-60^\circ\ \left(range:\left[-90^\circ,90^\circ\right]\right)$

39. $\theta=\cot^{-1}\left(-\frac{\sqrt{3}}{3}\right)\Rightarrow\theta=\tan^{-1}\left(\frac{1}{-\frac{\sqrt{3}}{3}}\right)+180^\circ\Rightarrow\theta=\tan^{-1}\left(-\frac{3}{\sqrt{3}}\right)+180^\circ\Rightarrow$

$\theta=\tan^{-1}(-\sqrt{3})+180^\circ\Rightarrow(-\sqrt{3}=\tan\theta)+180^\circ\Rightarrow\theta=-60^\circ+180^\circ\Rightarrow\theta=120^\circ\ \left(range:\left(0^\circ,180^\circ\right)\right)$

41. $\theta=\csc^{-1}(-2)\Rightarrow\theta=\sin^{-1}\left(\frac{1}{-2}\right)\Rightarrow\theta=\sin^{-1}\left(-\frac{1}{2}\right)\Rightarrow\left(-\frac{1}{2}=\sin\theta\right)\Rightarrow\theta=-30^\circ$

$\left(range:\left[90^\circ,0^\circ\right)\cup\left(0^\circ,90^\circ\right]\right)$

43. $\theta=\sin^{-1}(2)\Rightarrow2=\sin\theta\Rightarrow\theta$ does not exist (range of $\sin\theta:[-1,1]$)

45. $\theta=\sec^{-1}\left(-\frac{1}{2}\right)\Rightarrow\theta=\cos^{-1}(-2)\Rightarrow-2=\cos\theta\Rightarrow\theta$ does not exist (range of $\cos\theta:[-1,1]$)

47. $\theta=\sin^{-1}(-.13349122)\Rightarrow\theta\approx-7.6713835^\circ$

49. $\theta=arc\cos\ (-.39876459)\Rightarrow\theta\approx113.500970^\circ$

51. $\theta=\csc^{-1}(1.9422833)\Rightarrow\theta=\sin^{-1}\frac{1}{1.9422833}\Rightarrow\theta=\sin^{-1}(1.9422833)^{-1}\Rightarrow\theta\approx30.987961^\circ$

53. $y=\arctan1.1111111\Rightarrow y\approx0.83798122$

55. $y=\cot^{-1}(-.92170128)\Rightarrow y=\left(\tan^{-1}\frac{1}{-.92170128}\right)+\pi\Rightarrow y=(\tan^{-1}(-.92170128)^{-1})+\pi\Rightarrow$

$y\approx-.8261201193+\pi\Rightarrow y\approx2.315472534$

57. $y=\arcsin.92837781\Rightarrow y\approx1.1900238$

59. Graph $y=\cot^{-1}x$, see Figure 59

61. Graph $y=sec^{-1}x$, see Figure 61.

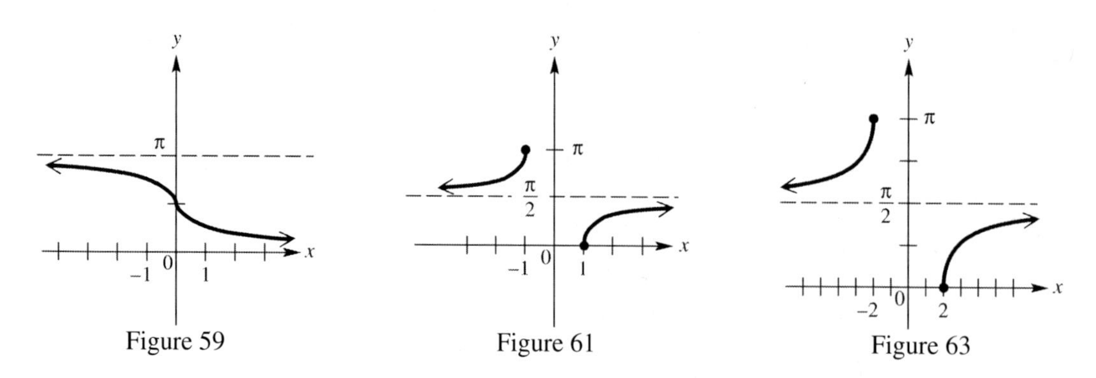

Figure 59 Figure 61 Figure 63

63. Graph $y = arc\sec\frac{1}{2}x\left(like\ the\ graph\ of\ y = \sec^{-1} x, with\ a\ domain\left(-\infty,2\right]\cup\left[2,\infty\right)\right)$ See Figure 63.

65. The domain of $y = \tan^{-1} x$ is $(-\infty,\infty)$ Therefore 1.003 is in this domain and $\tan^{-1} 1.003$ can be found.

67. Since $\sin(\sin^{-1} x) = x$ when $-1 < x < 1, \sin\left(\sin^{-1}\frac{1}{2}\right) = \frac{1}{2}$

69. First find $\sin\frac{4\pi}{3} = -\frac{\sqrt{3}}{2}$ now solve for $\sin^{-1}\left(-\frac{\sqrt{3}}{2}\right) \Rightarrow -\frac{\sqrt{3}}{2} = \sin\theta \Rightarrow \theta = -\frac{\pi}{3}$

71. First find $\cos\frac{3\pi}{2} = 0$ now solve for $\cos^{-1} 0 \Rightarrow 0 = \cos\theta \Rightarrow \theta = \frac{\pi}{2}$.

73. Since $\sin^2\frac{x}{2} = \left(\pm\sqrt{\frac{1-\cos x}{2}}\right)^2 = \frac{1-\cos x}{2}$ for all values of x $\tan(\tan^{-1} 5) = 5$

75. Since $\sec(\sec^{-1} x) = x$ when $x \le -1$ or $x \ge 1, \sec(\sec^{-1} 2) = 2$.

77. First find $\tan\frac{5\pi}{6} = -\frac{1}{\sqrt{3}} = \frac{-\sqrt{3}}{3}$ now solve for $\tan^{-1}\left(\frac{-\sqrt{3}}{3}\right) \Rightarrow \frac{-\sqrt{3}}{3} = \tan\theta \Rightarrow \theta = -\frac{\pi}{6}$.

79. Since $\cos\theta = \frac{3}{4}$ is in quadrant I. Sketch and label a triangle in quadrant I $x = 3$ and $r = 4$. Now use Pythagorean Theorem to find *y*: $3^2 + y^2 = 4^2 \Rightarrow y^2 = 16 - 9 \Rightarrow y^2 = 7 \Rightarrow y = \sqrt{7}$. Therefore, $\tan\theta = \frac{\sqrt{7}}{3}$ and $\tan\left(\arccos\frac{3}{4}\right) = \frac{\sqrt{7}}{3}$.

81. Since $\tan\theta = -2$ then θ is in quadrant IV. Sketch and label a triangle in quadrant IV, with $x = 1$ and $y = -1$ Now use Pythagorean theorem to find *r*: $1^2 + (-2)^2 = r^2 \Rightarrow r^2 = 1 + 4 \Rightarrow r^2 = 5 \Rightarrow r = \sqrt{5}$ Therefore, $\cos\theta = \frac{1}{\sqrt{5}} = \frac{\sqrt{5}}{5}$ and $\cos(\tan^{-1}(-2)) = \frac{\sqrt{5}}{5}$.

83. Since $\tan\theta = \frac{12}{5}$ then θ is in quadrant I. Sketch and label a triangle in quadrant I, with $x = 5$ and $y = 12$ Use Pythagorean theorem to find *r*: $5^2 + 12^2 = r^2 \Rightarrow r^2 = 25 + 144 \Rightarrow r^2 = 169 \Rightarrow r = 13$ Now $\sin 2\theta = 2\sin\theta\cos\theta$ therefore, $\sin 2\theta = 2\left(\frac{12}{13}\right)\left(\frac{5}{13}\right) = \frac{120}{169}$ and $\sin\left(2\tan^{-1}\frac{12}{5}\right) = \frac{120}{169}$.

85. Since $\tan\theta = \frac{4}{3}$, then θ is in quadrant I. Sketch and label a triangle in quadrant I, with $= \frac{\sqrt{u^2+5}}{u}$ and $y = 4$. Use Pythagorean theorem to find r: $3^2 + 4^2 = r^2 \Rightarrow r^2 = 9 + 16 \Rightarrow r^2 = 25 \Rightarrow r = 5$. Now $\cos 2\theta = \cos^2\theta - \sin^2\theta$, therefore, $\cos 2\theta = \left(\frac{3}{5}\right)^2 - \left(\frac{4}{5}\right)^2 = -\frac{7}{25}$ and $\cos\left(2\arctan\frac{4}{3}\right) = -\frac{7}{25}$.

87. Since $\cos\theta = \frac{1}{5}$, then θ is in quadrant I. Sketch and label a triangle in quadrant I, with $x = 1$ and $r = 5$. Use Pythagorean theorem to find y: $1^2 + y^2 = 5^2 \Rightarrow y^2 = 25 - 1 \Rightarrow y^2 = 24 \Rightarrow y = \sqrt{24} = 2\sqrt{6}$. Now $\sin 2\theta = 2\sin\theta\cos\theta$, therefore, $\sin 2\theta = 2\left(\frac{2\sqrt{6}}{5}\right)\left(\frac{1}{5}\right) = \frac{4\sqrt{6}}{25}$ and $\sin\left(2\cos^{-1}\frac{1}{5}\right) = \frac{4\sqrt{6}}{25}$.

89. Since $\sin\theta = \frac{3}{5}$, then θ is in quadrant I. Sketch and label a triangle in quadrant I, with $y = 3$ and $r = 5$.

Use Pythagorean theorem to find x: $x^2 + 3^2 = 5^2 \Rightarrow x^2 = 25 - 9 \Rightarrow x^2 = 16 \Rightarrow x = 4$. Since $\cos\beta = \frac{12}{13}$, then β is in quadrant I. Use Pythagorean theorem to find y:

$12^2 + y^2 = 13^2 \Rightarrow y^2 = 169 - 144 \Rightarrow y^2 = 25 \Rightarrow y = 5$. Now $\cos(\theta - \beta) = \cos\theta\cos\beta + \sin\theta\sin\beta$, therefore,

$\cos(\theta + \beta) = \frac{4}{5}\cdot\frac{12}{13} + \frac{3}{5}\cdot\frac{5}{13} = \frac{48}{65} + \frac{15}{65} = \frac{63}{65}$ and $\cos\left(\sin^{-1}\frac{3}{5} - \cos^{-1}\frac{12}{13}\right) = \frac{63}{65}$

91. Use $\tan\theta = \frac{3}{4}$ and $\tan\beta = \frac{12}{5}$ in $\tan(\theta + \beta) = \frac{\tan\theta + \tan\beta}{1 - \tan\theta\tan\beta}$, therefore, $\tan x = -2 + \sqrt{2}$ and

$$\tan\left(\tan^{-1}\frac{3}{4} + \tan^{-1}\frac{12}{5}\right) = \frac{\frac{3}{4} + \frac{12}{5}}{1 - \left(\frac{3}{4}\cdot\frac{12}{5}\right)} - \frac{63}{16}$$

93. Since $\tan\theta = \frac{5}{12}$, then θ is in quadrant I. Sketch and label a triangle in quadrant I, with $x = 12$ and $y = 5$.

Use Pythagorean theorem to find r: $12^2 + 5^2 = r^2 \Rightarrow r^2 = 144 + 25 \Rightarrow r^2 = 169 \Rightarrow r = 13$. Since $\tan\beta = \frac{3}{4}$, then β is in quadrant I. Sketch and label a triangle in quadrant I, with $x = 4$ and $y = 3$.

Use Pythagorean theorem to find y: $4^2 + 3^2 = r^2 \Rightarrow r^2 = 16 + 9 \Rightarrow r^2 = 25 \Rightarrow r = 5$.

Now $\cos(\theta - \beta) = \cos\theta\cos\beta + \sin\theta\sin\beta$, therefore, $\cos(\theta + \beta) = \frac{12}{13}\cdot\frac{4}{5} + \frac{5}{13}\cdot\frac{3}{5} = \frac{48}{65} + \frac{15}{65} = \frac{63}{65}$ and

$\cos\left(\tan^{-1}\frac{5}{12} - \tan^{-1}\frac{3}{4}\right) = \frac{63}{65}$

95. Since $\sin\theta = \frac{1}{2}$, then θ is in quadrant I. Sketch and label a triangle in quadrant I, with $y = 1$ and $r = 2$.

Use Pythagorean theorem to find x: $x^2+1^2=2^2 \Rightarrow x^2=4-1 \Rightarrow x^2=3 \Rightarrow x=\sqrt{3}$. Since $\tan\beta=(-3)$, then β is in quadrant IV. Sketch and label a triangle in quadrant IV, with $x=1$ and $y=-3$. Use Pythagorean theorem to find r: $1^2+(-3)^2=r^2 \Rightarrow r^2=1+9 \Rightarrow r^2=10 \Rightarrow y=\sqrt{10}$. Now

$\sin(\theta+\beta)=\sin\theta\cos\beta+\cos\theta\sin\beta$, therefore

$\sin(\theta+\beta)=\frac{1}{2}\cdot\frac{1}{\sqrt{10}}+\frac{\sqrt{3}}{2}\cdot\frac{-3}{\sqrt{10}}=\frac{1}{2\sqrt{10}}+\frac{-3\sqrt{3}}{2\sqrt{10}}=\frac{\sqrt{10}-3\sqrt{30}}{20}$ and $\sin\left(\sin^{-1}\frac{1}{2}-\tan^{-1}(-3)\right)=\frac{\sqrt{10}-3\sqrt{30}}{20}$

97. $\cos(\tan^{-1}5)\approx 0.894427191$

99. $\tan(\arcsin .12251014)\approx 0.1234399811$

101. $\sec(\cos^{-1}u)=\frac{1}{\cos(\cos^{-1}u)}=\frac{1}{u}$

103. Since $\cos\theta=u$, then θ is in quadrant I. Sketch and label a triangle in quadrant I,with $x=u$ and $r=1$.

Use Pythagorean theorem to find y: $u^2+y^2=1^2 \Rightarrow y^2=1-u^2 \Rightarrow y=\sqrt{1-u^2}$.

Now $\sin\theta=\frac{\sqrt{1-u^2}}{1}=\sqrt{1-u^2}$ and $\sin(\arccos u)=\sqrt{1-u^2}$.

105. Since $\sin\theta=u$, then θ is in quadrant I. Sketch and label a triangle in quadrant I, with $y=u$ and $r=1$.

Use Pythagorean theorem to find x: $x^2+u^2=1^2 \Rightarrow x^2=1-u^2 \Rightarrow x=\sqrt{1-u^2}$.

Now $\cot\theta=\frac{\sqrt{1-u^2}}{u}$ and $\cot(\arcsin u)=\frac{\sqrt{1-u^2}}{u}$.

107. Since $\sec\theta=\frac{u}{2}$, then θ is in quadrant I. Sketch and label a triangle in quadrant I, $x=2$ and $r=u$.

Use Pythagorean theorem to find y: $2^2+y^2=u^2 \Rightarrow y^2=u^2-4 \Rightarrow y=\sqrt{u^2-4}$.

Now $\sin\theta=\frac{\sqrt{u^2-1}}{u}$ and $\sin\left(\sec^{-1}\frac{u}{2}\right)=\frac{\sqrt{u^2-4}}{u}$.

109. Since $\theta\frac{u}{\sqrt{u^2+2}}$, then θ is in quadrant I. Sketch and label a triangle in quadrant I. Use $\sqrt{u^2+2}$ and

Pythagorean theorem to find $x.x^2u^2=\left(\sqrt{u^2+2}\right)^2 \Rightarrow x^2=u^2+2-u^2 \Rightarrow x^2=2 \Rightarrow x=\sqrt{2}$. Now

$\tan\theta=\frac{u}{\sqrt{2}}=\frac{u\sqrt{2}}{2}$ and $\tan\left(\sin^{-1}\frac{u}{\sqrt{u^2+2}}\right)=\frac{u\sqrt{2}}{2}$.

111. Since $\cos\theta=\frac{\sqrt{4-u^2}}{u}$, then θ is in quadrant I. Sketch and label a triangle in quadrant I. Use $x=\sqrt{4-u^2}$

and Pythagorean theorem to find $r.\left(\sqrt{4-u^2}\right)^2+u^2=r^2 \Rightarrow r^2=4-u^2+u^2 \Rightarrow r^2=4 \Rightarrow r=2$. Now

$\sec\theta=\frac{2}{\sqrt{4-u^2}}=\frac{2\sqrt{4-u^2}}{4-u^2}$ and $\sec\left(arc\cot\frac{\sqrt{4-u^2}}{u}\right)=\frac{2\sqrt{4-u^2}}{4-u^2}$.

113. (a) $\theta = \arcsin\left(\sqrt{\dfrac{v^2}{2v^2+64(0)}}\right) = arcsin\left(\sqrt{\dfrac{v^2}{2v^2}}\right) = arcsin\sqrt{\dfrac{1}{4}} = \arcsin\dfrac{1}{\sqrt{2}} = 45°$

(b) $\theta = \arcsin\left(\sqrt{\dfrac{v^2}{2v^2+64(6)}}\right) = \arcsin\left(\sqrt{\dfrac{v^2}{2v^2+384}}\right)$. As $v \to \infty$, $\sqrt{\dfrac{v^2}{2v^2+384}} \to \sqrt{\dfrac{1}{2}}$, therefore

$\theta = \arcsin\sqrt{\dfrac{1}{2}} = \arcsin\dfrac{1}{\sqrt{2}} = 45^\circ$. The equation of the asymptote is $\theta = 45^\circ = \cos 45^\circ, x = y$.

115. (a) For $3: \theta = \tan^{-1}\left(\dfrac{3(3)}{(3)^2+4}\right) = \tan^{-1}\dfrac{9}{13} \approx 34.70 \approx 35^\circ$

(b) For $6: \theta = \tan^{-1}\left(\dfrac{3(6)}{(6)^2+4}\right) = \tan^{-1}\dfrac{18}{40} = \tan^{-1}\dfrac{9}{20} \approx 24.23 \approx 24^\circ$

(c) For $9: \theta = \tan^{-1}\left(\dfrac{3(9)}{(9)^2+4}\right) = \tan^{-1}\dfrac{27}{85} \approx 17.62 \approx 18^\circ$

(d) Sketch and label the two triangles formed above the sight line. From this sketch $(\theta + a) = \dfrac{3+1}{x} = \dfrac{4}{x}$

and $\tan a = \dfrac{1}{x}$. Now use these values in the tangent of a sum identity

$$\tan(\theta+a) = \frac{\tan\theta + \tan a}{1-\tan\theta\tan a} : \frac{4}{x} = \frac{\tan\theta + \frac{1}{2}}{1-(\tan\theta)\left(\frac{1}{x}\right)} \Rightarrow \frac{4}{x} = \frac{x\tan\theta+1}{x-\tan\theta} \Rightarrow 4(x-\tan\theta+1) =$$

$$4x - 4\tan\theta = x^2\tan\theta + x \Rightarrow 4x - x = x^2\tan\theta + 4\tan\theta \Rightarrow 3x = \tan\theta(x^2+4) \Rightarrow$$

$$\tan\theta = \frac{3x}{x^2+4} \Rightarrow \theta = \tan^{-1}\left(\frac{3x}{x^2+4}\right)$$

(e) Graph $y_1 = \tan^{-1}\left(\dfrac{3x}{x^2+4}\right)$, Figure 115. The maximum value occurs when $x \approx 2$ feet.

[-4,4] by [-1,1]
Xscl = 1 Yscl = 1

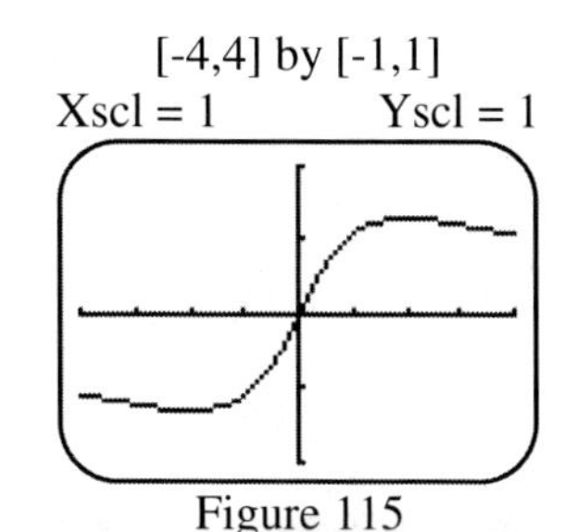

Figure 115

117. Let be the angle to the right of θ, and let β be the angle to the left of θ. Then $\beta + \theta + a = \pi$ and

$\theta = \pi - a - \beta$. Now from the picture $\tan a = \dfrac{150}{x}$, therefore $a = \tan^{-1}\left(\dfrac{105}{x}\right)$: and $\tan\beta = \dfrac{75}{100-x}$, therefore $\beta=\tan^{-1}\left(\frac{75}{100-x}\right)$. As a result: $\theta=\pi-\tan^{-1}\left(\frac{150}{x}\right)-\tan^{-1}\left(\frac{75}{100-x}\right)=$

$\pi-\arctan\left(\frac{150}{x}\right)-\arctan\left(\frac{75}{100-x}\right)$.

Reviewing Basic Concepts (Sections 9.3 - 9.4)

1. Since $\frac{\pi}{2}<x<\pi$ and $\cos 2x=-\frac{5}{12}$, x is found in quadrant II. Now solve for x:

$\cos 2x=1-2\sin^2 x\Rightarrow -\frac{5}{12}==1-2\sin^2 x\Rightarrow 2\sin^2 x=\frac{17}{12}\Rightarrow \sin^2 x=\frac{17}{24}\Rightarrow \sin x=\frac{\sqrt{17}}{\sqrt{24}}$

(positive because in quadrant II) $\Rightarrow \sin x=\frac{\sqrt{17}}{2\sqrt{6}}.\frac{\sqrt{6}}{\sqrt{6}}\Rightarrow \sin x=\frac{\sqrt{102}}{12}$. Sketch and label a triangle in quadrant II, with $y=\sqrt{102}$ and $r=12$. Now use Pythagorean theorem to find θ. Therefore, since x is in quadrant II, $\tan x=-\frac{\sqrt{102}}{\sqrt{42}}=-\frac{\sqrt{102}}{\sqrt{42}}.\frac{\sqrt{42}}{\sqrt{42}}=-\frac{\sqrt{4284}}{42}=-\frac{6\sqrt{119}}{42}\Rightarrow \tan x=-\frac{\sqrt{119}}{7}$.

2. Since $\sin\theta=-\frac{1}{3}$ and θ is in quadrant III. Sketch and label a triangle in quadrant III, with $y=-1$ and $r=3$.

Now use pythagorean theorem to find $x: x^2+(-1)^2=3^2\Rightarrow x^2=9-1\Rightarrow x^2=8\Rightarrow x=\sqrt{8}=2\sqrt{2}$,

Since x is in quadrant III, $x=-2\sqrt{2}$. Therefore, $\cos\theta=-\frac{2\sqrt{2}}{3}$. Now find:

$\sin 2\theta=2\sin\theta\cos\theta=2\left(-\frac{1}{2}\right)\left(-\frac{2\sqrt{2}}{3}\right)=\frac{4\sqrt{2}}{9}$;

$\cos 2\theta=2\cos^2\theta-1=2\left(-\frac{2\sqrt{2}}{3}\right)^2-1=2\left(\frac{8}{9}\right)-1=\frac{7}{9}$: $\tan 2\theta=\frac{\sin 2\theta}{\cos 2\theta}=\frac{\frac{4\sqrt{2}}{9}}{\frac{7}{9}}=\frac{4\sqrt{2}}{7}$.

3. $\sin 75^\circ=\sin\frac{150^\circ}{2}=\sqrt{\frac{1-\cos 150^\circ}{2}}=\sqrt{\frac{1-\left(-\frac{v3}{2}\right)}{2}.\frac{2}{2}}=\frac{\sqrt{2+\sqrt{3}}}{2}$ or:

$\sin 75^\circ=\sin(30^\circ+45^\circ)=\sin 30^\circ\cos 45^\circ+\cos 30^\circ\sin 45^\circ=\frac{1}{2}.\frac{\sqrt{2}}{2}+\frac{\sqrt{3}}{2}.\frac{\sqrt{2}}{2}=\frac{\sqrt{2}}{4}+\frac{\sqrt{6}}{4}=\frac{\sqrt{2}+\sqrt{6}}{4}$

4. $2\sin 25^\circ\cos 150^\circ=2\left[\frac{1}{2}\left[\sin(25^\circ+150^\circ)+\sin(25^\circ-150^\circ)\right]\right]=\sin 175^\circ+\sin(-125^\circ)=\sin 175^\circ-\sin 125^\circ$

5. (a) If the left side equals $\sin^2\frac{x}{2}=\left(\pm\sqrt{\frac{1-\cos x}{2}}\right)^2=\frac{1-\cos x}{2}$, and the right side equals:

$\frac{\tan x-\sin x}{2\tan x}=\frac{\frac{\sin x}{\cos x}-\sin x}{2\frac{\sin x}{\cos x}}=\frac{\frac{\sin x}{\cos x}-\sin x}{2\frac{\sin x}{\cos x}}\cdot\frac{\cos x}{\cos x}=\frac{\sin x-\sin x\cos x}{2\sin x}=\frac{\sin x(1-\cos x)}{2\sin x}=\frac{1-\cos x}{2}$, then

$\sin^2\frac{x}{2}=\frac{\tan x-\sin x}{2\tan x}$.

(b) If the left side equals: $\frac{\sin 2x}{2\sin x}=\frac{2\sin x\cos x}{2\sin x}=\cos x$, and the right side equals:

$\cos^2\frac{x}{2}-\sin^2\frac{x}{2}=\left(\pm\sqrt{\frac{1+\cos x}{2}}\right)^2-\left(\pm\sqrt{\frac{1-\cos x}{2}}\right)^2=\frac{1+\cos x}{2}-\frac{1-\cos x}{2}=$

$\frac{1+\cos x-(1-\cos x)}{2}=\frac{2\cos x}{2}=\cos x$, then $\frac{\sin 2x}{2\sin x}=\cos^2\frac{x}{2}-\sin^2\frac{x}{2}$.

6. (a) $y=\arccos\frac{\sqrt{3}}{2}\Rightarrow\frac{\sqrt{3}}{2}=\cos y\Rightarrow y=\frac{\pi}{6}$

(b) $y=\sin^{-1}\left(-\frac{\sqrt{2}}{2}\right)\Rightarrow-\frac{\sqrt{2}}{2}=\sin y\Rightarrow y=-\frac{\pi}{4}$

7. (a) $\theta=\arccos .5\Rightarrow .5=\cos\theta\Rightarrow\theta=60°$

(b) $\theta=\cot^{-1}(-1)\Rightarrow-1=\cot\theta\Rightarrow\theta=135°$

8. Graph $y=2\csc^{-1}x$ (like the graph of $y=\csc^{-1}x$, with a range $2\left(-\frac{\pi}{2},\frac{\pi}{2}\right)=(-\pi,\pi)$), see Figure 8.

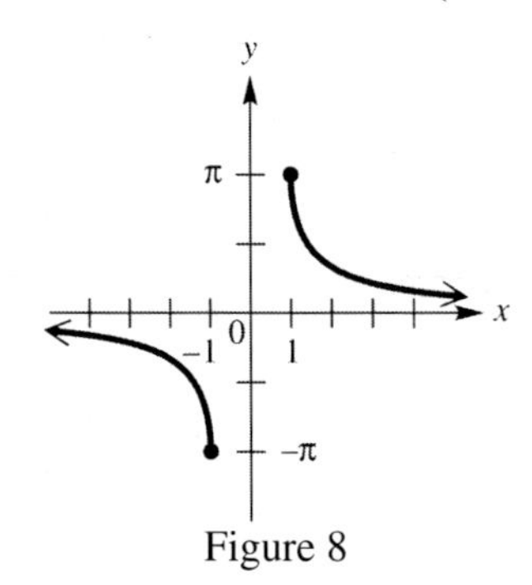

Figure 8

9. (a) Since $\sin\theta=-\frac{2}{3}$, then θ is in quadrant IV. Sketch and label a triangle in quadrant IV, with $y=-2$ and $r=3$. Now use Pythagorean theorem to find $x: x^2+(-2)^2=3^2\Rightarrow x^2=9-4\Rightarrow x^2=5\Rightarrow x=\sqrt{5}$.

Therefore, $\cot\theta=\frac{\sqrt{5}}{-2}=-\frac{\sqrt{5}}{2}$ and $\cot\left(\arcsin\left(-\frac{2}{3}\right)\right)=-\frac{\sqrt{5}}{2}$.

(b) Since $\tan\theta=\frac{5}{12}$, then θ is in quadrant I. Sketch and label a triangle in quadrant I, with $x=12$ and $y=5$. Use Pythagorean theorem to find $r: 12^2+5^2=r^2=144+25\Rightarrow r^2=169\Rightarrow r=13$. Since

$\sin\beta=\frac{3}{5}$, then β is in quadrant I. Sketch and label a triangle in quadrant I, with $y=3$ and $r=5$. Use Pythagorean theorem to find $x: x^2+3^2=5^2 \Rightarrow r^2=25-9 \Rightarrow x^2=16 \Rightarrow x=4$. Now $\cos(\theta-\beta)=\cos\theta\cos\beta+\sin\theta\sin\beta$, therefore, $\cos(\theta+\beta)=\frac{12}{13}\cdot\frac{4}{5}+\frac{5}{13}\cdot\frac{3}{5}=\frac{48}{65}+\frac{15}{65}=\frac{63}{65}$ and $\cos\left(\tan^{-1}\frac{5}{12}-\sin^{-1}\frac{3}{5}\right)=\frac{63}{65}$.

10. Since $\cot\theta=u$, then θ is in quadrant I. Sketch and label a triangle in quadrant I, with $x=u$ and $y=1$. Use Pythagorean theorem to find $r: u^2+1^2=r^2 \Rightarrow r^2=u^2+1 \Rightarrow r=\sqrt{u^2+1}$. Now $\sin\theta=\frac{1}{\sqrt{u^2+1}}=\frac{1}{\sqrt{u^2+1}}\cdot\frac{\sqrt{u^2+1}}{\sqrt{u^2+1}}=\frac{\sqrt{u^2+1}}{u^2+1}$ and $\sin(\operatorname{arccot} u)=\frac{\sqrt{u^2+1}}{u^2+1}$.

9.5: Trigonometric Equations and Inequalities (I)

1. $2\cos x+1=0 \Rightarrow 2\cos x=-1 \Rightarrow \cos x=-\frac{1}{2} \Rightarrow x=\frac{2\pi}{3},\frac{4\pi}{3}$. The solution set is $\left\{\frac{2\pi}{3},\frac{4\pi}{3}\right\}$.

3. $5\sin x-6=0 \Rightarrow 5\sin x=6 \Rightarrow \sin x=\frac{6}{5} \Rightarrow x$ does not exist. The solution set is $\varnothing$.

5. $2\tan x+1=-1 \Rightarrow 2\tan x=-2 \Rightarrow \tan x=-1 \Rightarrow x=\frac{3\pi}{4},\frac{7\pi}{4}$. The solution set is $\left\{\frac{3\pi}{4},\frac{7\pi}{4}\right\}$.

7. $2\cos x+5=6 \Rightarrow 2\cos x=1 \Rightarrow \cos x=\frac{1}{2} \Rightarrow x=\frac{\pi}{3},\frac{5\pi}{3}$. The solution set is $\left\{\frac{\pi}{3},\frac{5\pi}{3}\right\}$.

9. $2\csc x+4=\csc x+6 \Rightarrow \csc x=2 \Rightarrow \sin x=\frac{1}{2} \Rightarrow x=\frac{\pi}{6},\frac{5\pi}{6}$. The solution set is $\left\{\frac{\pi}{6},\frac{5\pi}{6}\right\}$.

11. From the factored equation $(\cot x-1)(\sqrt{3}\cot x+1)=0$, either $\cot x-1=0 \Rightarrow \cot x=1 \Rightarrow x=\frac{\pi}{4},\frac{5\pi}{4}$, or $\sqrt{3}\cot x+1=0 \Rightarrow \sqrt{3}\cot x=-1 \Rightarrow \cot x=-\frac{1}{\sqrt{3}} \Rightarrow \cot x=-\frac{\sqrt{3}}{3} \Rightarrow x=\frac{2\pi}{3},\frac{5\pi}{3}$.

    The solution set is $\left\{\frac{\pi}{4},\frac{2\pi}{3},\frac{5\pi}{4},\frac{5\pi}{3}\right\}$.

13. First set the equation equal to zero and factor. $\cos x\cot x=\cos x \Rightarrow \cos x\cot x-\cos x=0 \Rightarrow$ $\cos x(\cot x-1)=0$. Now from this factored equation either $\cos x=0 \Rightarrow x=\frac{\pi}{2},\frac{3\pi}{2}$, or $\cot x-1=0 \Rightarrow \cot x=1 \Rightarrow x=\frac{\pi}{4},\frac{5\pi}{4}$. The solution set is $\left\{\frac{\pi}{4},\frac{\pi}{2},\frac{5\pi}{4},\frac{3\pi}{2}\right\}$.

15. First set the equation equal to zero and factor. $\sin^2 x-2\sin x+1=0 \Rightarrow (\sin x-1)^2=0$. Now, from this factored equation, $\sin x-1=0 \Rightarrow \sin x=1 \Rightarrow x=\frac{\pi}{2}$. The solution set is $\left\{\frac{\pi}{2}\right\}$.

17. If $4(1+\sin x)(1-\sin x)=3$, then $1-\sin^2 x=\frac{3}{4}\Rightarrow\cos^2 x=\frac{3}{4}\Rightarrow\cos x=\pm\frac{\sqrt{3}}{2}\Rightarrow x=\left\{\frac{\pi}{6},\frac{5\pi}{6},\frac{7\pi}{6},\frac{11\pi}{6}\right\}$.

The solution set is $\left\{\frac{\pi}{6},\frac{5\pi}{6},\frac{7\pi}{6},\frac{11\pi}{6}\right\}$.

19. First set the equation equal to zero and factor. $\tan x+1=\sqrt{3}+\sqrt{3}\cot x\Rightarrow\tan x+1-\sqrt{3}-\frac{\sqrt{3}}{\tan x}=0\Rightarrow$

$\tan^2 x+\tan\left(1-\sqrt{3}\right)-\sqrt{3}=0\Rightarrow\left(\tan x+1\right)\left(\tan x-\sqrt{3}\right)=0$. Now from this factored equation, either

$\tan x+1=0\Rightarrow\tan x=-1\Rightarrow x=\frac{3\pi}{4},\frac{7\pi}{4}$, or $\tan x-\sqrt{3}=0\Rightarrow\ \tan x=\sqrt{3}\Rightarrow\ x=\frac{\pi}{3},\frac{4\pi}{3}$.

The solution set is $\left\{\frac{\pi}{3},\frac{3\pi}{4},\frac{4\pi}{3},\frac{7\pi}{4}\right\}$.

21. First set the equation equal to zero and factor. $2\sin x-1=\csc x\Rightarrow 2\sin x-1-\frac{1}{\sin x}=0\Rightarrow$

$2\sin^2 x-\sin x-1=0\Rightarrow(2\sin x+1)(\sin x-1)=0$. From this factored equation, either

$2\sin x+1=0\Rightarrow 2\sin x=-1\Rightarrow\sin x=-\frac{1}{2}\Rightarrow x=\frac{7\pi}{6},\frac{11\pi}{6}$, or $\sin x-1=0\Rightarrow\sin x=1\Rightarrow x=\frac{\pi}{2}$.

The solution set is $\left\{x=\frac{\pi}{2},\frac{7\pi}{6},\frac{11\pi}{6}\right\}$.

23. $\cos^2 x-\sin^2 x=1\Rightarrow 1-\sin^2 x-\sin^2 x=1\Rightarrow 1-2\sin^2 x=1\Rightarrow\ -2\sin^2 x=0\Rightarrow$

$\sin^2 x=0\Rightarrow\sin x=0\Rightarrow x=0,\pi$. The solution set is $\{0,\pi\}$.

25. (a) First simplify. $-2\cos x+1=0\Rightarrow -2\cos x=-1\Rightarrow\cos x=\frac{1}{2}$. Now the reference angle is $\frac{\pi}{3}$ and x is found in quadrant I and IV; therefore, $f(x)=0$ when $x=\frac{\pi}{3},\frac{5\pi}{3}$.

(b) Graph $y_1=-2\cos x+1$ on the calculator. From this graph, $f(x)>0$ when the graph is above the x-axis; therefore, it is true for the interval $\left(\frac{\pi}{3},\frac{5\pi}{3}\right)$.

(c) Graph $y_1=-2\cos x+1$ on the calculator. From this graph, $f(x)<0$ when the graph is below the x-axis; therefore, it is true for the interval $\left[0,\frac{\pi}{3}\right)\cup\left(\frac{5\pi}{3},2\pi\right)$.

27. (a) First simplify. $\tan^2 x-3=0\Rightarrow\tan^2 x=3\Rightarrow\tan x=\pm\sqrt{3}$. Now the reference angle is $\frac{\pi}{3}$ and x is found in all quadrants; therefore, $f(x)=0$ when $x=\left\{\frac{\pi}{3},\frac{2\pi}{3},\frac{4\pi}{3},\frac{5\pi}{3}\right\}$.

(b) Graph $y_1 = \tan^2 x - 3$ on the calculator. From this graph, the asymptotes are $\frac{\pi}{2}$ and $\frac{3\pi}{2}$, and $f(x) > 0$ when the graph is above the x-axis; therefore, it is true for the interval $\left(\frac{\pi}{3},\frac{\pi}{2}\right)\cup\left(\frac{\pi}{2},\frac{2\pi}{3}\right)\cup\left(\frac{4\pi}{3},\frac{3\pi}{2}\right)\cup\left(\frac{5\pi}{3},2\pi\right)$.

(c) Graph $y_1 = \tan^2 x - 3$ on the calculator. From this graph, $f(x) < 0$ when the graph is below the x-axis; therefore, it is true for the interval $\left[0,\frac{\pi}{3}\right)\cup\left(\frac{2\pi}{3},\frac{4\pi}{3}\right)\cup\left(\frac{3\pi}{2},\frac{5\pi}{3}\right)$.

29. (a) First, set the equation equal to zero, and factor. $2\cos^2 x - \sqrt{3}\cos x = 0 \Rightarrow \cos x\left(2\cos x - \sqrt{3}\right) = 0$. Now from this factored equation, either $\cos x = 0 \Rightarrow x = \frac{\pi}{2},\frac{3\pi}{2}$, or $2\cos x - \sqrt{3} = 0 \Rightarrow$ $2\cos x = \sqrt{3} \Rightarrow \cos x = \frac{\sqrt{3}}{2}$; therefore, the reference angle is $\frac{\pi}{6}$. Since x is found in quadrant I and IV, and $x = \frac{\pi}{6},\frac{11\pi}{6}$, the solution for $f(x) = 0$ is true when $x = \left\{\frac{\pi}{6},\frac{\pi}{2},\frac{3\pi}{2},\frac{11\pi}{6}\right\}$.

(b) Graph $y_1 = 2\cos^2 x - \sqrt{3}\cos x$ on the calculator. From this graph, $f(x) > 0$ when the graph is above the x-axis; therefore, it is true for the interval $\left(0,\frac{\pi}{6}\right)\cup\left(\frac{\pi}{2},\frac{3\pi}{2}\right)\cup\left(\frac{11\pi}{6},2\pi\right)$.

(c) Graph $y_1 = 2\cos^2 x - \sqrt{3}\cos x$ on the calculator. From this graph, $f(x) < 0$ when the graph is below the x-axis; therefore, it is true for the interval $\left(\frac{\pi}{6},\frac{\pi}{2}\right)\cup\left(\frac{3\pi}{2},\frac{11\pi}{6}\right)$.

31. (a) First, set the equation equal to zero, and factor. $\sin^2 x\cos x - \cos x = 0 \Rightarrow \cos x\left(\sin^2 x - 1\right) = 0$. Now, from this factored equation, either $\cos x = 0 \Rightarrow x = \frac{x}{2},\frac{3\pi}{2}$, or $\sin^2 x - 1 = 0 \Rightarrow$ $\sin^2 x = 1 \Rightarrow \sin x = \pm 1 \Rightarrow x = \frac{\pi}{2},\frac{3\pi}{2}$; therefore, $f(x) = 0$ when $x = \left\{\frac{\pi}{2},\frac{3\pi}{2}\right\}$.

(b) Graph $y_1 = \sin^2 x\cos x - \cos x$ on the calculator. From this graph, $f(x) > 0$ when the graph is above the x-axis; therefore, it is true for the interval $\left(\frac{\pi}{2},\frac{3\pi}{2}\right)$.

(c) Graph $y_1 = \sin^2 x\cos x - \cos x$ on the calculator. From this graph, $f(x) < 0$ when the graph is below the x-axis; therefore, it is true for the interval $\left[0,\frac{\pi}{2}\right)\cup\left(\frac{3\pi}{2},2\pi\right)$.

33. $\tan\theta + \cot\theta = 0 \Rightarrow \tan\theta + \frac{1}{\tan\theta} = 0 \Rightarrow \tan^2\theta + 1 = 0 \Rightarrow \tan^2\theta = -1 \Rightarrow \tan\theta$ does not exist, The solution set is $\varnothing$.

35. $2\tan^2\theta\sin\theta-\tan^2\theta=0\Rightarrow\tan^2\theta(2\sin\theta-1)=0\Rightarrow\tan^2\theta=0$ or $2\sin\theta-1=0$;

$\tan^2\theta=0\Rightarrow\tan\theta=0$ or $2\sin\theta-1=0\Rightarrow2\sin\theta=1\Rightarrow\sin\theta=\frac{1}{2}$, Over the interval $\left[0^\circ,360^\circ\right)$, the equation $\tan\theta=0$ has two solutions. These are 0° and 180°. In the same interval, the equation $\sin\theta=\frac{1}{2}$ has two solutions, the angles in quadrants I and II that have a reference angle is 30°. These are 30° and 150°. The solution set is $\left\{0^\circ,30^\circ,150^\circ,180^\circ\right\}$.

37. $\sec^2\theta\tan\theta=2\tan\theta\Rightarrow\sec^2\theta\tan\theta-2\tan\theta=0\Rightarrow\tan\theta\left(\sec^2\theta-2\right)=0\Rightarrow\tan\theta=0$ or $\sec^2\theta-2=0$;

$\sec^2\theta-2=0\Rightarrow\sec^2\theta=2\Rightarrow\sec\theta=\pm\sqrt{2}$, Over the interval $\left[0^\circ,360^\circ\right)$, the equation $\tan\theta=0$ has two solutions. These are 0° and 180°. In the same interval, the equation $\sec\theta=\sqrt{2}$ has two solutions, the angles in quadrants I and IV that have a reference angle is 45°. These are 45° and 315°. The equation $\sec\theta=-\sqrt{2}$ has two solutions, the angles in quadrants II and III that have a reference angle of 45°. These are 135° and 225° The solution set is $\left\{0^\circ,45^\circ,135^\circ,180^\circ,225^\circ,315^\circ\right\}$.

39. $9\sin^2\theta-6\sin\theta=1\Rightarrow9\sin^2\theta-6\sin\theta-1=0$, Using the quadratic formula with $a=9$, $b=-6$, and $c=-1$. $\sin\theta=\dfrac{6\pm\sqrt{36-4(9)(-1)}}{2(9)}=\dfrac{6\pm\sqrt{36+36}}{18}=\dfrac{6\pm\sqrt{72}}{18}=\dfrac{6\pm6\sqrt{2}}{18}=\dfrac{1\pm\sqrt{2}}{3}$; The quadrant I angle is $\sin^{-1}\left(\dfrac{1+\sqrt{2}}{3}\right)\approx53.6^\circ$. The quadrant II angle is approximately $180^\circ-53.6^\circ=126.4^\circ$. The other solution is $\sin^{-1}\left(\dfrac{1-\sqrt{2}}{3}\right)\approx-7.9^\circ$. Since this angle is not in the interval $\left[0^\circ,360^\circ\right)$ we will use the reference angle 7.9° to find angles in Quadrant III and IV. The quadrant III angle is approximately $180^\circ+7.9^\circ=187.9^\circ$. The quadrant IV angle is approximately $360^\circ-7.9^\circ=352.1^\circ$. The solution set is $\left\{53.6^\circ,126.4^\circ,187.9^\circ,352.1^\circ\right\}$.

41. $\tan^2\theta+4\tan\theta+2=0$, Using the quadratic formula with $a=1$, $b=4$, and $c=2$.

$\tan\theta=\dfrac{-4\pm\sqrt{16-4(1)(2)}}{2(1)}=\dfrac{-4\pm\sqrt{16-8}}{2}=\dfrac{-4\pm\sqrt{8}}{2}=\dfrac{-4\pm2\sqrt{2}}{2}=-2\pm\sqrt{2}$; The quadrant II angle is $\tan^{-1}\left(-2+\sqrt{2}\right)\approx-30.4^\circ$. Since this angle is not in the interval $\left[0^\circ,360^\circ\right)$ we will use the reference angle 30.4° to find angles in Quadrant II and IV. The quadrant II angle is approximately $180^\circ - 30.4^\circ = 149.6^\circ$. The quadrant IV angle is approximately $360^\circ - 30.4^\circ = 329.6^\circ$. The other solution is $\tan^{-1}\left(-2-\sqrt{2}\right) \approx -73.7^\circ$. Since this angle is not in the interval $\left[0^\circ, 360^\circ\right)$ we will use the reference angle 73.7° to find angles in Quadrant II and IV. The quadrant II angle is approximately $180^\circ - 73.7^\circ = 106.3^\circ$. The quadrant IV angle is approximately $360^\circ - 73.7^\circ = 286.3^\circ$. The solution set is $\left\{106.3^\circ, 149.6^\circ, 286.3^\circ, 329.6^\circ\right\}$.

43. $\sin^2\theta - 2\sin\theta + 3 = 0$, Using the quadratic formula with $a = 1$, $b = -2$, and $c = 3$

$$\sin\theta = \frac{2 \pm \sqrt{4-4(1)(3)}}{2(1)} = \frac{2 \pm \sqrt{4-12}}{2} = \frac{2 \pm \sqrt{-8}}{2} = \frac{2 \pm 2i\sqrt{2}}{2} = 1 \pm i\sqrt{2}$$; Since the result is not a real number, the equation has no real solutions. The solution set is the empty set.

45. $\cot\theta + 2\csc\theta = 3 \Rightarrow \dfrac{\cos\theta}{\sin\theta} + \dfrac{2}{\sin\theta} = 3 \Rightarrow \cos\theta + 2 = 3\sin\theta \Rightarrow (\cos\theta + 2)^2 = (3\sin\theta)^2 \Rightarrow$

$\cos^2\theta + 4\cos\theta + 4 = 9\sin^2\theta \Rightarrow \cos^2\theta + 4\cos\theta + 4 = 9\left(1 - \cos^2\theta\right) \Rightarrow$

$\cos^2\theta + 4\cos\theta + 4 = 9 - 9\cos^2\theta \Rightarrow 10\cos^2\theta + 4\cos\theta - 5 = 0$

We use the quadratic formula with $a = 10$, $b = 4$, and $c = -5$.

$$\cos\theta = \frac{-4 \pm \sqrt{4^2 - 4(10)(-5)}}{2(10)} = \frac{-4 \pm \sqrt{16+200}}{20} = \frac{-4 \pm \sqrt{216}}{20} = \frac{-4 \pm 6\sqrt{6}}{20} = \frac{-2 \pm 3\sqrt{6}}{10}$$

Since $\cos\theta = \dfrac{-2+3\sqrt{6}}{10} > 0$ (and less than 1), we will obtain two angles. One angle will be in quadrant I and the other will be in quadrant IV. Using a calculator, if $\cos\theta = \dfrac{-2+3\sqrt{6}}{10} \approx 0.53484692$, the quadrant I angle will be approximately 57.7°. The quadrant IV angle will be approximately $360° - 57.7° = 302.3°$.

Since $\cos\theta = \dfrac{-2-3\sqrt{6}}{10} < 0$ (and greater than –1), we will obtain two angles. One angle will be in quadrant II and the other will be in quadrant III. Using a calculator, if $\cos\theta = \dfrac{-2-3\sqrt{6}}{10} \approx -0.93484692$, the quadrant II angle will be approximately 159.2°. The reference angle is 180° – 159.2° = 20.8°. Thus, the quadrant III angle will be approximately 180° + 20.8° = 200.8°. Since the solution was found by squaring both sides of an equation, we must check that each proposed solution is a solution of the original equation. 302.3° and 200.8° do not satisfy our original equation. Thus, they are not elements of the solution set.

Solution set: {57.7°, 159.2°}

47. $\tan^3 x = 3\tan x \Rightarrow \tan^3 - 3\tan x = 0$.

48. First factor. $\tan^3 x - 3\tan x = 0 \Rightarrow \tan x(\tan^2 x - 3) = 0$. Now, from this factored equation, either

$\tan x = 0 \Rightarrow x = 0, \pi,$ or $\tan^2 x - 3 = 0 \Rightarrow \tan^2 x = 3 \Rightarrow$ $\tan x = \pm\sqrt{3} \Rightarrow x = \frac{\pi}{3}, \frac{2\pi}{3}, \frac{4\pi}{3}, \frac{5\pi}{3}$; therefore, the

solution set is $\left\{0, \frac{\pi}{3}, \frac{2\pi}{3}, \pi, \frac{4\pi}{3}, \frac{5\pi}{3}\right\}$.

49. $\frac{\tan^2 x}{\tan x} = \frac{3\tan x}{\tan x} \Rightarrow \tan^2 x = 3 \Rightarrow \tan x = \pm\sqrt{3} \Rightarrow x = \frac{\pi}{3}, \frac{2\pi}{3}, \frac{4\pi}{3}, \frac{5\pi}{3}$.

50. The answers do not agree. The solutions 0 and π were lost when dividing by $\tan x$.

51. First, put the equation in quadratic form, and factor.

$3\sin^2 x - \sin x = 2 \Rightarrow 3\sin^2 x - \sin x - 2 = 0 \Rightarrow (3\sin x + 2)(\sin x - 1) = 0$. From this factored form, either

$3\sin x + 2 = 0 \Rightarrow 3\sin x = -2 \Rightarrow \sin x = -\frac{2}{3}$, or $\sin x - 1 = 0 \Rightarrow \sin x = 1$. Now find x:

(i) For $\sin x = 1$: $x = \sin^{-1}(1) \Rightarrow 1 = \sin x \Rightarrow x = \frac{\pi}{2}$.

(ii) For $\sin x = -\frac{2}{3}$: find the reference angle (calculate in radian mode). $\sin^{-1}\left(\frac{2}{3}\right) \approx 0.7297$.

Since x is in quadrants III and IV, $x = \pi + 0.7297 \approx 3.87$, and $x = 2\pi - 0.7297 \approx 5.55$; therefore , $f(x) = 0$,

when $x = \left\{\frac{\pi}{2}, 3.87, 5.55\right\}$.

53. First, put the equation in quadratic form, and factor by the quadratic formula.

$\tan x = \frac{-4 \pm \sqrt{(4)^2 - 4(1)(2)}}{2(1)} = \frac{-4 \pm \sqrt{8}}{2} = \frac{-4 \pm 2\sqrt{2}}{2} \Rightarrow \tan x = -2 \pm \sqrt{2}$. Now find x:

(i) For $\tan x = -2 + \sqrt{2}$, find the reference angle (calculate in radian mode). $\tan^{-1}\left(-2 + \sqrt{2}\right) \approx 0.5299$.

Since x is found in quadrants II and IV, $x = \pi - 0.5299 \approx 2.61$, and $x = 2\pi - 0.5299 \approx 5.75$.

(ii) For $\tan x = -2 - \sqrt{2}$, find the reference angle (calculate in radian mode). $\tan^{-1}\left(-2 - \sqrt{2}\right) \approx 1.2859$.

Since x is found in quadrants II and IV, $x = \pi - 1.2859 \approx 1.86$, and $x = 2\pi - 1.2859 \approx 5.00$.

Therefore, $f(x) = 0$ when $x = \{1.86, 2.61, 5.00, 5.75\}$.

55. First, set the equation equal to zero, and factor by the quadratic formula:

$2\cos^2 x + 2\cos x = 1 \Rightarrow 2\cos^2 x + 2\cos x - 1 = 0 \Rightarrow \cos x = \frac{-2 \pm \sqrt{(2)^2 - 4(2)(-1)}}{2(2)} =$

$\frac{-2 \pm \sqrt{12}}{4} = \frac{-2 \pm 2\sqrt{3}}{4} \Rightarrow \cos x = \frac{-1 \pm \sqrt{3}}{2}$. Now find x:

(i) For $x = \frac{-1 + \sqrt{3}}{2}$, find the reference angle (calculate in radian mode): $\cos^{-1}\left(\frac{-1 \pm \sqrt{3}}{2}\right) \approx 1.196$. Since x is found in quadrants I and IV, $x \approx 1.20$, and $x = 2\pi - 1.196 = 5.09$.

(ii) For $x=\frac{-1-\sqrt{3}}{2}$, find the reference angle (calculate in radian mode): $\cos^{-1}\left(\frac{-1-\sqrt{3}}{2}\right)=\varnothing$, because $\frac{-1-\sqrt{3}}{2}$ is outside the range of $\cos x$.

Therefore, $f(x)=0$ when $x=\{1.20, 5.09\}$.

57. First, set the equation equal to zero, then use identities to put it into the quadratic form and factor. $\sec^2\theta=2\tan\theta+4 \Rightarrow \tan^2\theta+1-2\tan\theta-4=0 \Rightarrow \tan^2\theta-2\tan\theta-3=0 \Rightarrow (\tan\theta-3)(\tan\theta+1)=0$. From this factored form, either $\tan\theta-3=0 \Rightarrow \tan\theta=3$, or $\tan\theta+1=0 \Rightarrow \tan\theta=-1$; therefore, $\tan x=-1, 3$.
Now find x:

(i) For $\tan x=3$, find the reference angle (calculate in degree mode): $\cos^{-1}3 \approx 71.57^\circ$. Since x is found in quadrants I and III, $x \approx 71.57^\circ$ and $x=180^\circ+71.57^\circ=251.57^\circ$.

(ii) For $\tan x=-1$, find the reference angle (calculate in degree mode): $\tan^{-1}(-1)=45^\circ$. Since x is found in quadrants II and IV, $x=180^\circ-45^\circ=135^\circ$, and $x=360^\circ-45^\circ=315^\circ$.

Therefore, $f(x)=0$ when $x=\{71.6^\circ, 135^\circ, 251.6^\circ, 315^\circ\}$.

59. Use identities to get the equation in quadratic form, and factor by the quadratic formula:
$2\sin\theta=1-2\cos\theta \Rightarrow (2\sin\theta)^2=(1-2\cos\theta)^2 \Rightarrow 4\sin^2\theta=1-4\cos\theta+4\cos^2\theta \Rightarrow$

$4(1-\cos^2\theta)=1-4\cos\theta+4\cos^2\theta \Rightarrow 4-4\cos^2\theta=1-4\cos\theta+4\cos^2\theta \Rightarrow$

$0=8\cos^2\theta-4\cos\theta-3 \Rightarrow \cos\theta=\frac{4\pm\sqrt{(-4)^2-4(8)(-3)}}{2(8)}=\frac{4\pm\sqrt{112}}{16}=\frac{4\pm4\sqrt{7}}{16} \Rightarrow \cos\theta=\frac{1\pm\sqrt{7}}{4}$.

Now find x:

(i) For $\cos x=\frac{1+\sqrt{7}}{4}$, find the reference angle (calculate in degree mode). $\cos\theta=\frac{1+\sqrt{7}}{4} \Rightarrow$
$\cos\theta \approx 0.91143783 \Rightarrow \theta=\cos^{-1}0.91143783 \approx 24.30^\circ$. Since x is found in quadrants I and IV, $x \approx 24.30^\circ$, and $x=360^\circ-24.30=335.70^\circ$.

(ii) For $\cos\theta=\frac{1-\sqrt{7}}{4}$, find the reference angle (calculate in degree mode). $\cos\theta=\frac{1-\sqrt{7}}{4} \Rightarrow$
$\cos\theta \approx -0.41143783 \Rightarrow \theta=\cos^{-1}(-0.41143783) \approx 65.70^\circ$. Since x is found in quadrants II and III, $x=180^\circ-65.70^\circ \approx 114.30^\circ$, and $x=180^\circ+65.70^\circ \approx 245.70^\circ$.

Finally, since the equation was square, we will need to check each solution.
Check 24.30°. $2\sin 24.30^\circ \approx 0.82$, and $1-2\cos 24.30^\circ \approx -0.82$; this is not a solution.
Check 114.30°. $2\sin 114.30^\circ \approx 1.82$, and $1-2\cos 114.30^\circ \approx 1.82$; this is a solution.
Check 245.70°. $2\sin 245.70^\circ \approx -1.82$, and $1-2\cos 114.30^\circ \approx 1.82$; this is not a solution.
Check 335.70°. $2\sin 335.70^\circ \approx -0.82$, and $1-2\cos 335.70^\circ \approx -0.82$; this is a solution.
Therefore, $f(x)=0$ when $x=\{114.3^\circ, 335.7^\circ\}$.

61. $\cos\theta+1=0 \Rightarrow \cos\theta=-1 \Rightarrow \theta=180^\circ$ in the interval $[0^\circ, 360^\circ)$. The solution set is $\{180^\circ+360^\circ n$, where n is any integer$\}$.

63. $3\csc x - 2\sqrt{3} = 0 \Rightarrow 3\csc x = 2\sqrt{3} \Rightarrow \csc x = \frac{2\sqrt{3}}{3} \Rightarrow x = \frac{\pi}{3}, \frac{2\pi}{3}$ in the interval $[0, 2\pi)$.

The solution set is $\left\{\frac{\pi}{3} + 2n\pi, \frac{2\pi}{3} + 2n\pi, \text{ where } n \text{ is any integer}\right\}$.

65. $6\sin^2\theta + \sin\theta = 1 \Rightarrow 6\sin^2\theta + \sin\theta - 1 = 0 \Rightarrow (3\sin\theta - 1)(2\sin\theta + 1) = 0 \Rightarrow$
$\sin\theta = \frac{1}{3} \Rightarrow \theta \approx 19.5°$ or $\theta \approx 180° - 19.5° = 160.5°$ or $\sin\theta = -\frac{1}{2} \Rightarrow \theta = 210°$ or $\theta = 330°$

The solution set is {19.5° + 360°n,160.5° + 360°n, 210° + 360°n, 330° + 360°n, where n is any integer}.

67. $2\cos^2 x + \cos x - 1 = 0 \Rightarrow (2\cos x - 1)(\cos x + 1) = 0 \Rightarrow 2\cos x - 1 = 0 \Rightarrow$

$\cos x = \frac{1}{2}$ or $\cos x + 1 = 0 \Rightarrow \cos x = -1$, Over the interval $[0, 2\pi)$, the equation $\cos x = \frac{1}{2}$ has two solutions. The angles in quadrants I and IV that have a reference angle of $\frac{\pi}{3}$ are $\frac{\pi}{3}$ and $\frac{5\pi}{3}$. In the same interval, $\cos x = -1$ when the angle is π. Thus, the solution set is $\left\{\frac{\pi}{3} + 2n\pi, \pi + 2n\pi,\right.$ and $\left.\frac{5\pi}{3} + 2n\pi, \text{ where } n \text{ is any integer}\right\}$.

69. $\sin\theta\cos\theta - \sin\theta = 0 \Rightarrow \sin\theta(\cos\theta - 1) = 0 \Rightarrow \sin\theta = 0 \Rightarrow \theta = 0°$ or $\theta = 180°$ or $\cos\theta = 1 \Rightarrow \theta = 0°$
The solution set is {180°n, where n is any integer}.

71. $\sin x(3\sin x - 1) = 1 \Rightarrow 3\sin^2 x - \sin x - 1 = 0$, We use the quadratic formula with $a = 3$, $b = -1$, and $c = -1$.
$\sin x = \frac{-(-1) \pm \sqrt{(-1)^2 - 4(3)(-1)}}{2(3)} = \frac{1 \pm \sqrt{1+12}}{6} = \frac{1 \pm \sqrt{13}}{6}$, Since $\sin x = \frac{1+\sqrt{13}}{6} > 0$ (and less than 1), we will obtain two angles. One angle will be in quadrant I and the other will be in quadrant II. Using a calculator, if $\sin x = \frac{1+\sqrt{13}}{6} \approx 0.76759188$, the quadrant I angle will be approximately 0.8751.

The quadrant II angle will be approximately $\pi - 0.8751 \approx 2.2665$. Since $\sin x = \frac{1-\sqrt{13}}{6} < 0$ (and greater than −1), we will obtain two angles. One angle will be in quadrant III and the other will be in quadrant IV. Using a calculator, if $\sin x = \frac{1-\sqrt{13}}{6} \approx -0.43425855$, then $x \approx -0.4492$. Since this solution is not in the interval $[0, 2\pi)$, we must use it as a reference angle to find angles in the interval. Our reference angle will be 0.4492. The angle in quadrant III will be approximately $\pi + 0.4492 \approx 3.5908$. The angle in quadrant IV will be approximately $2\pi - 0.4492 \approx 5.8340$. Thus, the solution set is { $0.8751 + 2n\pi$, $2.2665 + 2n\pi$, $3.5908 + 2n\pi$, and $5.8340 + 2n\pi$, where n is any integer}.

73. $5+5\tan^2\theta=6\sec\theta \Rightarrow 5\left(1+\tan^2\theta\right)=6\sec\theta \Rightarrow 5\sec^2\theta=6\sec\theta \Rightarrow 5\sec^2\theta-6\sec\theta=0 \Rightarrow$

$\sec\theta(5\sec\theta-6)=0 \Rightarrow \sec\theta=0$ or $5\sec\theta-6=0 \Rightarrow \sec\theta=\frac{6}{5}$, $\sec\theta=0$ is an impossible value since the secant function must be either ≥ 1 or ≤ -1. Since $\sec\theta=\frac{6}{5}>1$, we will obtain two angles. One angle will be in quadrant I and the other will be in quadrant IV. Using a calculator, if $\sec\theta=\frac{6}{5}=1.2$, the quadrant I angle will be approximately $33.6°$. The quadrant IV angle will be approximately $360°-33.6°=326.4°$. Thus, the solution set is { $33.6°+360°n$ and $326.4°+360°n$, where n is any integer}.

75. $\frac{2\tan\theta}{3-\tan^2\theta}=1 \Rightarrow 2\tan\theta=3-\tan^2\theta \Rightarrow \tan^2\theta+2\tan\theta-3=0 \Rightarrow (\tan\theta-1)(\tan\theta+3)=0 \Rightarrow$

$\tan\theta-1=0 \Rightarrow \tan\theta=1$ or $\tan\theta+3=0 \Rightarrow \tan\theta=-3$, Over the interval [0°, 360°), the equation $\tan\theta=1$ has two solutions 45° and 225°. Over the same interval, the equation $\tan\theta=-3$ has two solutions that are approximately $-71.6°+180°=108.4°$ and $-71.6°+360°=288.4°$.

Thus, the solutions are $45°+360°n$, $108.4°+360°n$, $225°+360°n$ and $288.4°+360°n$, where n is any integer. Since the period of the tangent function is $180°$, the solution set can also be written as { $45°+180°n$ and $108.4°+180°n$, where n is any integer}.

77. The equation $2x-1=0$ has one solution. The equation $2\sin x-1=0$ has an infinite number of solutions because the sine function is periodic.

79. Graph $y=2\sin x-1+2\cos x$ (not shown). The x-intercepts are $\{1.99, 5.86\}$.

81. Graph $y=2\cos^3 x+\sin x+1$ (not shown). The x-intercepts are {2.68, 4.46, 4.71}.

83. Graph $y=\ln x-\cos x$ (not shown). The x-intercept is {1.30}.

85. (a) $14=\frac{35}{3}+\frac{7}{3}\sin\frac{2x\pi}{365} \Rightarrow \frac{7}{3}=\frac{7}{3}\sin\frac{2x\pi}{365} \Rightarrow 1=\sin\frac{2x\pi}{365} \Rightarrow \frac{2x\pi}{365}=\frac{\pi}{2} \Rightarrow 4\pi x=365\pi \Rightarrow x\approx 91.25.$

Fourteen hours of daylight will occur approximately 91.3 days after March 21, or on June 20.

(b) The minimum will occur when $\sin\frac{2\pi x}{365}=-1$; therefore, $\frac{2\pi x}{365}=\frac{3\pi}{2} \Rightarrow 4\pi x=1095\pi \Rightarrow x\approx 273.75.$

The minimum hours of daylight will occur approximately 273.8 days after March 21, December 19.

(c) $10=\frac{35}{3}+\frac{7}{3}\sin\frac{2\pi x}{365} \Rightarrow 30=35+7\sin\frac{2\pi x}{365} \Rightarrow -\frac{5}{7}=\sin\frac{2\pi x}{365}$. Now, for $\sin\frac{2\pi x}{365}$, find the reference angle(calculate in radian mode): $\frac{2\pi x}{365}=\sin^{-1}\frac{5}{7} \Rightarrow \frac{2\pi x}{365}\approx 0.7956$. Since $\sin\frac{2\pi x}{365}=-\frac{5}{7}$, $\frac{2\pi x}{365}$ is found in quadrants III and IV, and $\frac{2\pi x}{365}\approx \pi+0.7956 \Rightarrow \frac{2\pi x}{365}\approx 3.9372 \Rightarrow x\approx 228.7$, or

$\frac{2\pi x}{365} \approx 5.4876 \Rightarrow x \approx 318.8$. There will be about 10 hours of daylight twice, and they are approximately 228.7 days after March 21, or on Nov 4, and 318.8 days after March 21, or on Feb 2.

87. First, put the equation in quadratic form, and then factor by the quadratic formula:

$$0.342(80)\cos\theta + 2\cos^2\theta = \frac{16(80)^2}{(60)^2} \Rightarrow 27.36\cos\theta + 2\cos^2\theta = \frac{256}{9} \Rightarrow 2\cos^2\theta + 27.36\cos\theta - \frac{256}{9} = 0 \Rightarrow$$

$$\cos\theta = \frac{-27.36 \pm \sqrt{748.5696 - 4(2)(-\frac{256}{9})}}{2(2)} = \frac{-27.36 \pm 31.243}{4}.$$ Therefore,

$$\cos\theta = \frac{-27.36 + 31.243}{4} = 0.97075, \text{ or } \cos\theta = \frac{-27.36 - 31.243}{4} = -14.65.$$

Since the range of $\cos\theta$ is $-1 < \cos\theta < 1$, the only solution is $\theta = \cos^{-1} 0.97075 \approx 13.892^\circ \approx 14^\circ$.

89. Let $s(t) = \frac{2+\sqrt{3}}{2}$, and find $\sin t$. $\sin t + 2\cos t = \frac{2+\sqrt{3}}{2} \Rightarrow 2\sin t + 4\cos t = 2 + \sqrt{3} \Rightarrow$

$$4\cos t = \left(2+\sqrt{3}\right) - 2\sin t \Rightarrow (4\cos t)^2 = \left(\left(2+\sqrt{3}\right) - 2\sin t\right)^2 \Rightarrow$$

$$16\cos^2 t = \left(7+4\sqrt{3}\right) - 4\left(2+\sqrt{3}\right)\sin t + 4\sin^2 t \Rightarrow 16\left(1-\sin^2 t\right) = \left(7+4\sqrt{3}\right) - 4\left(2+\sqrt{3}\right)\sin t + 4\sin^2 t \Rightarrow$$

$$0 = 20\sin^2 t - 4\left(2+\sqrt{3}\right)\sin t + \left(-9+4\sqrt{3}\right) \Rightarrow 0 = \left(2\sin t - \sqrt{3}\right)\left(10\sin t - 4 + 3\sqrt{3}\right).$$ Now, when

$2\sin t - \sqrt{3} = 0 \Rightarrow 2\sin t = \sqrt{3} \Rightarrow \sin t = \frac{\sqrt{3}}{2}$, and when

$$10\sin t - 4 + 3\sqrt{3} = 0 \Rightarrow 10\sin t = 4 - 3\sqrt{3} \Rightarrow \sin t = \frac{4-3\sqrt{3}}{10}.$$

To find a solution, choose $\sin t = \frac{\sqrt{3}}{2}$, which yields $t = \frac{\pi}{3}$.

9.6: Trigonometric Equations and Inequalities (II)

1. Dividing each by 2 yields $x = \frac{\pi}{3}, \pi, \frac{4\pi}{3}$.

3. Use identities to solve for x. $\sin\frac{x}{2} = \cos\frac{x}{2} \Rightarrow \left(\sin\frac{x}{2}\right)^2 = \left(\cos\frac{x}{2}\right)^2 \Rightarrow$

$\frac{1-\cos x}{2} = \frac{1+\cos x}{2} \Rightarrow 1-\cos x = 1+\cos x \Rightarrow -2\cos x = 0 \Rightarrow x = \frac{\pi}{2}, \frac{3\pi}{2}$. Since both sides of the equation have been squared, we must check each solution. Check $x = \frac{\pi}{2}$. $\sin\frac{\left(\frac{\pi}{2}\right)}{2} = \sin\frac{\pi}{4} = \frac{\sqrt{2}}{2}$, and $\cos\frac{\frac{\pi}{2}}{2} = \cos\frac{\pi}{4} = \frac{\sqrt{2}}{2}$; therefore, $x = \frac{\pi}{2}$ is a solution. Check $x = \frac{3\pi}{2}$. $\sin\frac{\left(\frac{3\pi}{2}\right)}{2} = \sin\frac{3\pi}{4} = \frac{\sqrt{2}}{2}$, and $\cos\frac{\left(\frac{3\pi}{2}\right)}{4} = -\frac{\sqrt{2}}{2}$; therefore, $x = \frac{3\pi}{2}$ is not a solution. The solution is $x = \frac{\pi}{2}$.

5. Use identities to solve for x. $\sin^2\left(\frac{x}{2}\right)-1=0 \Rightarrow \left(\frac{\sqrt{1-\cos x}}{2}\right)^2-1=0 \Rightarrow \frac{1-\cos x}{2}-1=0 \Rightarrow$

$1-\cos x-2=0 \Rightarrow -\cos x=1 \Rightarrow \cos x=-1 \Rightarrow x=\pi.$

7. Use identities to solve for x. $\sin 2x=2\cos^2 x \Rightarrow 2\sin x\cos x=2\cos^2 x \Rightarrow$

$2\sin x\cos x-2\cos^2 x=0 \Rightarrow 2\cos x(\sin x-\cos x)=0.$ Now, either $2\cos x=0 \Rightarrow \cos x=0 \Rightarrow x=\frac{\pi}{2},\frac{3\pi}{2},$

$\sin x-\cos x=0 \Rightarrow \sin x=\cos x \Rightarrow x=\frac{\pi}{4},\frac{5\pi}{4}.$ Therefore, $x=\frac{\pi}{4},\frac{\pi}{2},\frac{5\pi}{4},\frac{3\pi}{2}.$

9. Use identities to solve for x. $\cos x-1=\cos 2x \Rightarrow \cos x-1=2\cos^2 x-1 \Rightarrow \cos x-2\cos^2 x=0 \Rightarrow$

$\cos x(1-2\cos x)=0.$ Now, either $\cos x=0 \Rightarrow x=\frac{\pi}{2},\frac{3\pi}{2},$ or $1-2\cos x=0 \Rightarrow$

$-2\cos x=-1 \Rightarrow \cos x=\frac{1}{2} \Rightarrow x=\frac{\pi}{3},\frac{5\pi}{3}.$ Therefore, $x=\frac{\pi}{3},\frac{\pi}{2},\frac{3\pi}{2},\frac{5\pi}{3}.$

11. Use identities to solve for x: $\sin 2x-\cos x=0 \Rightarrow 2\sin x\cos x-\cos x=0 \Rightarrow \cos x(2\sin x-1)=0.$

Now, either $\cos x=0 \Rightarrow x=\frac{\pi}{2},\frac{3\pi}{2}$ or $2\sin x-1=0 \Rightarrow$

$2\sin x=1 \Rightarrow \sin x=\frac{1}{2} \Rightarrow x=\frac{\pi}{6},\frac{5\pi}{6}.$ Therefore, $x=\frac{\pi}{2},\frac{\pi}{6},\frac{5\pi}{6},\frac{3\pi}{2}.$

13. (a) If the required interval for x is $0 \le x < 2\pi$, then $0 \le 2x < 4\pi$. Therefore, if $\cos 2x=\frac{\sqrt{3}}{2}$, then

$2x=\frac{\pi}{6},\frac{11\pi}{6},\frac{13\pi}{6},\frac{23\pi}{6}.$ Now, dividing each solution by 2 (or multiplying by $\frac{1}{2}$) yields

$x=\frac{\pi}{12},\frac{11\pi}{12},\frac{13\pi}{12},\frac{23\pi}{12}.$

(b) Using a calculator, the graph of $\cos 2x=\frac{\sqrt{3}}{2}$ is above the graph of $y=\frac{\sqrt{3}}{2}$ for the interval:

$\left[0,\frac{\pi}{12}\right)\cup\left(\frac{11\pi}{12},\frac{13\pi}{12}\right)\cup\left(\frac{23\pi}{12},2\pi\right).$

15. (a) If the required interval for x is $0 \le x < 2\pi$, then $0 \le 3x < 6\pi$. Therefore, if $\sin 3x=-1$, then

$3x=\frac{3\pi}{2},\frac{7\pi}{2},\frac{11\pi}{2}.$ Now, dividing each solution by 3 (or multiplying by $\frac{1}{3}$) yields $x=\frac{\pi}{2},\frac{7\pi}{6},\frac{11\pi}{6}.$

(b) Using a calculator, the graph of $\sin 3x=-1$ is never below the graph of $y=-1$; therefore, the interval is $\varnothing$.

17. (a) First, solve for $\cos 2x$. $\sqrt{2}\cos 2x = -1 \Rightarrow \cos 2x = -\frac{1}{\sqrt{2}}$. If the required interval for x is $0 \le x < 2\pi$, then $0 \le 2x < 4\pi$. Therefore, if $\cos 2x = -\frac{1}{\sqrt{2}}$, then $2x = \frac{3\pi}{4}, \frac{5\pi}{4}, \frac{11\pi}{4}, \frac{13\pi}{4}$. Now dividing each solution by 2 (or multiplying by $\frac{1}{2}$) yields $x = \frac{3\pi}{8}, \frac{5\pi}{8}, \frac{11\pi}{8}, \frac{13\pi}{8}$.

(b) Using a calculator, the graph of $\sqrt{2}\cos 2x = -1$ intersects or is below the graph of $y = -1$ for the interval: $\left[\frac{3\pi}{8}, \frac{5\pi}{8}\right] \cup \left[\frac{11\pi}{8}, \frac{13\pi}{8}\right]$.

19. (a) First solve for $\sin\frac{x}{2}$. $\sin\frac{x}{2} = \sqrt{2} - \sin\frac{x}{2} \Rightarrow 2\sin\frac{x}{2} = \sqrt{2} \Rightarrow \sin\frac{x}{2} = \frac{\sqrt{2}}{2}$. If the required interval for x is $0 \le x < 2\pi$, then $0 \le \frac{x}{2} < \pi$. Therefore, if $\sin\frac{x}{2} = \frac{\sqrt{2}}{2}$, then $\frac{x}{2} = \frac{\pi}{4}, \frac{3\pi}{4}$.

Now multiplying each solution by 2 yields $x = \frac{\pi}{2}, \frac{3\pi}{2}$.

(b) Using a calculator, the graph of $y = \sin\frac{x}{2}$ is above the graph of $y = \sqrt{2} - \sin\frac{x}{2}$ for the interval: $\left(\frac{\pi}{2}, \frac{3\pi}{2}\right)$.

21. $\sqrt{2}\sin 3x - 1 = 0 \Rightarrow \sqrt{2}\sin 3x = 1 \Rightarrow \sin 3x = \frac{1}{\sqrt{2}} \Rightarrow \sin 3x = \frac{\sqrt{2}}{2}$, In quadrant I and II, sine is positive. Thus,

$3x = \frac{\pi}{4} + 2n\pi, \frac{3\pi}{4} + 2n\pi \Rightarrow x = \frac{\pi}{12} + \frac{2n\pi}{3}, \frac{\pi}{4} + \frac{2n\pi}{3}$

Solution set: $\left\{\frac{\pi}{12} + \frac{2n\pi}{3}, \frac{\pi}{4} + \frac{2n\pi}{3}\text{, where } n \text{ is any integer}\right\}$

23. $\cos\frac{\theta}{2} = 1 \Rightarrow \frac{\theta}{2} = 0° + 360°n \Rightarrow \theta = 720°n$. Solution set: $\{720°n$, where n is any integer$\}$

25. $2\sqrt{3}\sin\frac{x}{2} = 3 \Rightarrow \sin\frac{x}{2} = \frac{3}{2\sqrt{3}} \Rightarrow \sin\frac{x}{2} = \frac{3\sqrt{3}}{6} \Rightarrow \sin\frac{x}{2} = \frac{\sqrt{3}}{2}$

Since $0 \le \theta < 2\pi$, $0° \le \frac{x}{2} < \pi$. Thus, $\frac{x}{2} = \frac{\pi}{3} + 2n\pi, \frac{2\pi}{3} + 2n\pi \Rightarrow x = \frac{2\pi}{3} + 4n\pi, \frac{4\pi}{3} + 4n\pi$

Solution set: $\left\{\frac{2\pi}{3} + 4n\pi, \frac{4\pi}{3} + 4n\pi\text{, where } n \text{ is any integer}\right\}$

27. $2\sin\theta = 2\cos 2\theta \Rightarrow \sin\theta = \cos 2\theta \Rightarrow \sin\theta = 1 - 2\sin^2\theta \Rightarrow 2\sin^2\theta + \sin\theta - 1 = 0 \Rightarrow$
$(2\sin\theta - 1)(\sin\theta + 1) = 0 \Rightarrow 2\sin\theta - 1 = 0$ or $\sin\theta + 1 = 0$

Over the interval $[0°, 360°)$, we have $2\sin\theta - 1 = 0 \Rightarrow 2\sin\theta = 1 \Rightarrow \sin\theta = \frac{1}{2} \Rightarrow \theta = 30°$ or $150°$

$\sin\theta + 1 = 0 \Rightarrow \sin\theta = -1 \Rightarrow \theta = 270°$, Solution set: $\{30° + 360°n,\ 150° + 360°n,$
$270° + 360°n$, where n is any integer$\}$

29. $1 - \sin x = \cos 2x \Rightarrow 1 - \sin x = 1 - 2\sin^2 x \Rightarrow 2\sin^2 x - \sin x = 0 \Rightarrow \sin x(2\sin x - 1) = 0$

Over the interval $[0,\ 2\pi)$, we have $\sin x = 0 \Rightarrow x = 0$ or π.

$2\sin x - 1 = 0 \Rightarrow \sin x = \frac{1}{2} \Rightarrow x = \frac{\pi}{6}$ or $\frac{5\pi}{6}$

Solution set: $\left\{n\pi,\ \frac{\pi}{6} + 2n\pi,\ \frac{5\pi}{6} + 2n\pi,\right.$ where n is any integer$\}$

31. $3\csc^2 \frac{x}{2} = 2\sec x \Rightarrow \frac{3}{\sin^2 \frac{x}{2}} = \frac{2}{\cos x} \Rightarrow \sin^2 \frac{x}{2} = \frac{3}{2}\cos x \Rightarrow \frac{1 - \cos x}{2} = \frac{3}{2}\cos x \Rightarrow$

$1 - \cos x = 3\cos x \Rightarrow 1 = 4\cos x \Rightarrow \frac{1}{4} = \cos x$

Over the interval $[0,\ 2\pi)$, we have $\cos x = \frac{1}{4} \Rightarrow x = 1.3181$ or $x = 4.9651$

Solution set: $\{1.3181 + 2n\pi,\ 4.9651 + 2n\pi,$ where n is any integer$\}$

33. $2 - \sin 2\theta = 4\sin 2\theta \Rightarrow 2 = 5\sin 2\theta \Rightarrow \sin 2\theta = \frac{2}{5} \Rightarrow \sin 2\theta = 0.4$ Since $0 \le \theta < 360°$, $0° \le 2\theta < 720°$.

In quadrant I and II, sine is positive. $\sin 2\theta = 0.4 \Rightarrow 2\theta = 23.6°, 156.4°, 383.6°, 516.4°$

Thus, $\theta = 11.8°, 78.2°, 191.8°, 258.2°$.

Solution set: $\{11.8° + 360°n,\ 78.2° + 360°n,\ 191.8° + 360°n,\ 258.2° + 360°n,$ where n is any integer$\}$ or

$\{11.8° + 180°n,\ 78.2° + 180°n,$ where n is any integer$\}$

35. $2\cos^2 2\theta = 1 - \cos 2\theta \Rightarrow 2\cos^2 2\theta + \cos 2\theta - 1 = 0 \Rightarrow (2\cos 2\theta - 1)(\cos 2\theta + 1) = 0$

Since $0 \le \theta < 360° \Rightarrow 0° \le 2\theta < 720°$, we have $2\cos 2\theta - 1 = 0 \Rightarrow 2\cos 2\theta = 1 \Rightarrow \cos 2\theta = \frac{1}{2}$. Thus,

$2\theta = 60°,\ 300°,\ 420°,\ 660° \Rightarrow \theta = 30°,\ 150°,\ 210°,\ 330°$ or $\cos 2\theta + 1 = 0 \Rightarrow \cos 2\theta = -1$

$2\theta = 180°,\ 540° \Rightarrow \theta = 90°,\ 270°$

Solution set: $\{30° + 360°n,\ 90° + 360°n,\ 150° + 360°n,\ 210° + 360°n,\ 270° + 360°n,$

$330° + 360°n,$ where n is any integer$\}$ or $\{30° + 180°n,\ 90° + 180°n,\ 150° + 180°n,$ where n is any integer$\}$

37. Graph $y = \sin x + \sin 3x - \cos x$. Using the capabilities of the calculator, the x-intercepts are

$0.262, 1.309, 1.571, 3.403, 4.451,$ and 4.712.

39. Graph $y = \cos 2x + \cos x$. Using the capabilities of the calculator, the x-intercepts are

$1.047, 3.142,$ and 5.236.

41. Graph $y = \cos\frac{x}{2} - 2\sin 2x$. Using the capabilities of the calculator, the x-intercepts are

$0.259, 1.372, 3.142, 4.911,$ and 6.024.

43. Since $1+\tan^2 x = \sec^2 x$, we will express the equation in terms of secant. $\sec 2x + \tan 2x = 2 \Rightarrow$

$\tan 2x = 2 - \sec 2x \Rightarrow \tan^2 2x = 4 - 4\sec x + \sec^2 2x \Rightarrow \sec^2 2x - 1 = 4 - 4\sec x + \sec^2 2x \Rightarrow$

$0 = 5 - 4\sec 2x \Rightarrow \sec 2x = \frac{5}{4} \Rightarrow \frac{1}{\cos 2x} = \frac{5}{4} \Rightarrow \cos 2x = \frac{4}{5}$. Multiply the inequality $0 \le x < 2\pi$ by 2 to find

the interval for 2x is $[0, 4\pi)$. Using a calculator and the fact that cosine is positive in quadrants I and IV, we

get $2x \approx 0.64350111, 5.6396842, 6.9266864, 11.922870.$. Dividing by 2

$x \approx 0.321750555, 2.819421, 3.463342, 5.961435$. Check answers for extraneous solutions:

$\sec(2(0.32175055)) + \tan(2(0.32175055)) = 2$, $\sec(2(3.463342)) + \tan(2(3.463342)) = 2$,

$\sec(2(5.961435)) + \tan(2(5.961435)) \ne 2$, $\sec(2(2.8198421)) + \tan(2(2.8198421)) \ne 2$

Therefore, $x \approx 0.32175055, 3.463342$.

45. Since $\sin^2 x + \cos^2 x = 1 \Rightarrow \cos^2 x = 1 - \sin^2 x$, we will use this as a substitution in the equation below.

$\cos 2x = 1 - \sin 2x \Rightarrow \cos^2 2x = 1 - 2\sin 2x + \sin^2 2x \Rightarrow 1 - \sin^2 2x = 1 - 2\sin 2x + \sin^2 2x \Rightarrow$

$2\sin^2 2x - 2\sin 2x = 0 \Rightarrow 2\sin 2x(\sin 2x - 1) = 0 \Rightarrow 2\sin 2x = 0$ *or* $\sin 2x - 1 = 0$. Multiply the inequality

$0 \le x < 2\pi$ by 2 to find the interval for 2x is $[0, 4\pi)$. We know that $x = 0, \frac{\pi}{2}, \pi, 2\pi, \frac{5\pi}{2}$. Dividing by 2.

$x = 0, \frac{\pi}{4}, \frac{\pi}{2}, \pi, \frac{5\pi}{4}$. Check answers for extraneous solutions: $\cos(2\cdot 0) = 1 - \sin(2\cdot 0) \Rightarrow 1 = 1$,

$\cos\left(2\cdot\frac{\pi}{4}\right) = 1 - \sin\left(2\cdot\frac{\pi}{4}\right) \Rightarrow 0 = 0$, $\cos\left(2\cdot\frac{\pi}{2}\right) = 1 - \sin\left(2\cdot\frac{\pi}{2}\right) \Rightarrow -1 \ne 0$, $\cos(2\cdot\pi) = 1 - \sin(2\cdot\pi) \Rightarrow 1 = 1$,

$\cos\left(2\cdot\frac{5\pi}{4}\right) = 1 - \sin\left(2\cdot\frac{5\pi}{4}\right) \Rightarrow 0 = 0$. Therefore, $x = 0, \frac{\pi}{4}, \pi, \frac{5\pi}{4}$.

47. The function $\frac{\tan 2\theta}{2} \ne \tan\theta$, because the 2 in 2θ is not a factor of the numerator. It is a factor in the

argument of the tangent function.

49. (a) If $e = 0$, then $0 = 20\sin\left(\frac{\pi t}{4} - \frac{\pi}{2}\right)$. Since $\arcsin 0 = 0$, we can solve the following equation for t.

$\frac{\pi t}{4} - \frac{\pi}{2} = 0 \Rightarrow \frac{\pi t}{4} = \frac{\pi}{2} \Rightarrow 2\pi t = 4\pi \Rightarrow t = 2$ seconds.

(b) If $e = 10\sqrt{3}$, then $10\sqrt{3} = 20\sin\left(\frac{\pi t}{4} - \frac{\pi}{2}\right) \Rightarrow \frac{\sqrt{3}}{2} = \sin\left(\frac{\pi t}{4} - \frac{\pi}{2}\right)$. Since $\arcsin\frac{\sqrt{3}}{2} = \frac{\pi}{2}$, we can solve the following equation for t. $\frac{\pi t}{4} - \frac{\pi}{2} = \frac{\pi}{3} \Rightarrow \frac{\pi t}{4} = \frac{\pi}{3} + \frac{\pi}{2} \Rightarrow \frac{\pi t}{4} = \frac{10\pi}{12} \Rightarrow 12\pi t = 40\pi \Rightarrow t = \frac{10}{3}$ seconds.

51. (a) For $x = t$, graph $P(t) = 0.004\sin\left(2\pi(261.63)t + \frac{\pi}{7}\right)$. See Figure 51.

(b) $0 = 0.004\sin\left[2\pi(261.63)t + \frac{\pi}{7}\right] \Rightarrow 0 = \sin(1643.87t + 0.45)$. Since $\sin(1643.87t + 0.45) = 0$, $1643.87t + 0.45 = n\pi$, where n is an integer, and $t = \frac{n\pi - 0.45}{1643.87}$. If $n = 0$, then $t \approx -0.000274$. If $n = 1$, then $t \approx 0.00164$. If $n = 2$, then $t \approx 0.00355$. If $n = 3$, then $t \approx 0.00546$. The only solutions for $t \in [0, 0.005]$ are $t = 0.00164$, 0.00355.

(c) Graph $y = \sin(1643.87t + 0.45)$ on the calculator. The x-intercepts are $t \approx 0.00164$ and 0.00355; therefore, from the graph, $P < 0$ for the interval $(0.00164, 0.00355)$.

(d) The inequality $P < 0$ implies that there is a decrease in pressure, so an eardrum would be vibrating outward.

[0,0.005] by [-0.005,0.005]
Xscl = 0.001 Yscl = 1

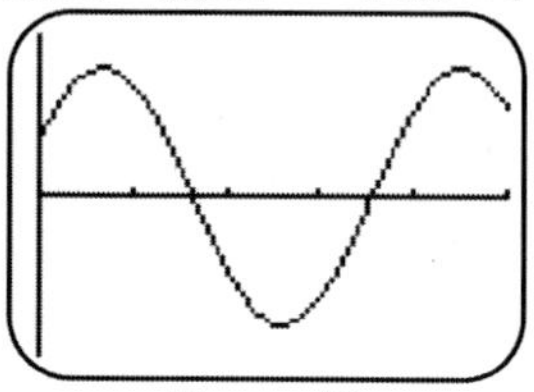

Figure 51

[0.15,1.15] by [-0.01,0.01]
Xscl =0.05 Yscl =0.01

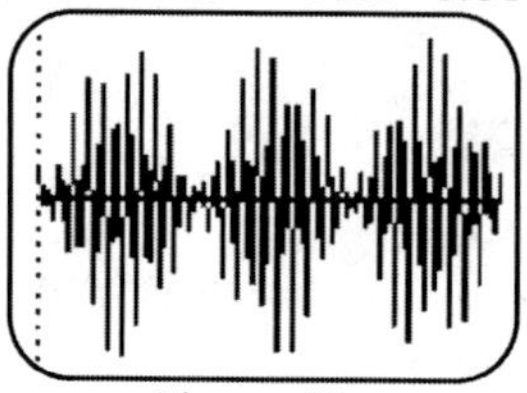

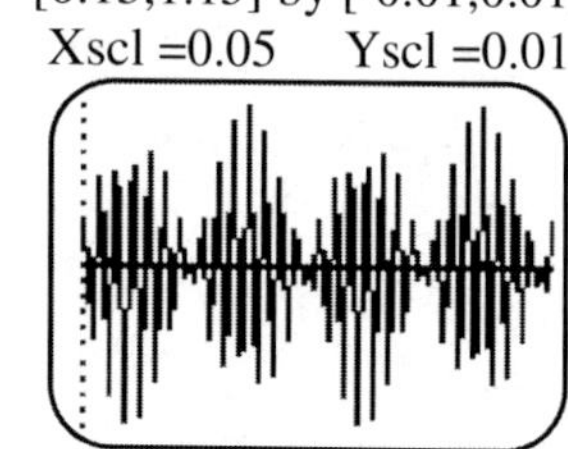

Figure 53a

[0.15,1.15] by [-0.01,0.01]
Xscl =0.05 Yscl =0.01

Figure 53b

53. (a) For $x = t$, graph $P(t) = 0.005\sin 440\pi t + 0.005\sin 446\pi t$. See Figure 53a. There are 3 beats a second.

(b) For $x = t$, graph $P(t) = 0.005\sin 440\pi t + 0.005\sin 432\pi t$. See Figure 53b. There are 4 beats a second.

(c) The number of beats is equal to the absolute value of the difference in the frequencies of the two tones.

Reviewing Basic Concepts (Sections 9.5 - 9.6)

1. If the required interval for x is $0 \le x < 2\pi$, then $0 \le 2x < 4\pi$. Therefore, if $\cos 2x = \frac{\sqrt{3}}{2}$, then $2x = \frac{\pi}{6}, \frac{11\pi}{6}, \frac{13\pi}{6}, \frac{23\pi}{6}$. Now, dividing each solution by 2 (or multiplying by ½) yields $x = \frac{\pi}{12}, \frac{11\pi}{12}, \frac{13\pi}{12}, \frac{23\pi}{12}$.

2. First solve for $\sin x$. $2\sin x + 1 = 0 \Rightarrow 2\sin x = -1 \Rightarrow \sin x = -\frac{1}{2}$. If $\sin x = -\frac{1}{2}$, then $x = \frac{7\pi}{6}, \frac{11\pi}{6}$.

3. Using the factored equation, either $\tan x - 1 = 0 \Rightarrow \tan x = 1 \Rightarrow x = \frac{\pi}{4}, \frac{5\pi}{4}, \ldots$ or

$\cos x - 1 = 0 \Rightarrow \cos x = 1 \Rightarrow x = 0, 2\pi, ..$ Therefore, $x = 2n\pi, \frac{\pi}{4} + n\pi$ where n is any integer.

4. Solve for x. $2\cos^2 x = \sqrt{3}\cos x \Rightarrow 2\cos^2 x - \sqrt{3}\cos x = 0 \Rightarrow \cos x(2\cos x - \sqrt{3}) = 0$. Now, either

$\cos x = 0 \Rightarrow x = \frac{\pi}{2}, \frac{3\pi}{2}, \ldots$ or $2\cos x - \sqrt{3} = 0 \Rightarrow 2\cos x = \sqrt{3} \Rightarrow \cos x = \frac{\sqrt{3}}{2} \Rightarrow x = \frac{\pi}{6}, \frac{11\pi}{6}, \ldots$ Therefore,

$x = \frac{\pi}{6} + 2n\pi, \frac{\pi}{2} + n\pi, \frac{11\pi}{6} + 2n\pi,$ where n is any integer.

5. First, set the equation equal to zero and factor by the quadratic formula.

$3\cot^2\theta - 3\cot\theta - 1 = 0 \Rightarrow \cot\theta = \frac{3 \pm \sqrt{(-3)^2 - 4(3)(-1)}}{2(3)} = \frac{3 \pm \sqrt{21}}{6}$. If $\cot\theta = \frac{3 \pm \sqrt{21}}{6}$, then

$\tan\theta = \frac{6}{3 \pm \sqrt{21}}$. Now find θ.

(i) For $\tan\theta = \frac{6}{3 + \sqrt{21}}$: Find the reference angle (calculate in degree mode). $\tan\theta = \frac{6}{3 + \sqrt{21}} \Rightarrow$

$\tan\theta \approx 0.79128785 \Rightarrow \theta = \tan^{-1} 0.79128785 \approx 38.4°$. Since θ is found in quadrants I and III,

$\theta \approx 38.4°$, and $\theta = 180° + 38.4 = 218.4°$.

(ii) For $\tan\theta = \frac{6}{3 - \sqrt{21}}$: Find the reference angle (calculate in degree mode). $\tan\theta = \frac{6}{3 - \sqrt{21}} \Rightarrow$

$\tan\theta \approx -3.79128785 \Rightarrow \theta = \tan^{-1}(-3.79128785) \approx 75.2°$. Since θ is found in quadrants II and IV,

$\theta = 180° - 75.2° \approx 104.8°$, and $\theta = 360° - 75.2° \approx 284.8°$.

Therefore, $\theta = 38.4°, 104.8°, 218.4°, 284.8°$.

6. First, factor by the quadratic formula. $4\cos^2\theta + 4\cos\theta - 1 = 0 \Rightarrow$

$\cos\theta = \frac{-4 \pm \sqrt{(4)^2 - 4(4)(-1)}}{2(4)} = \frac{-4 \pm \sqrt{32}}{8} = = \frac{-4 \pm 4\sqrt{2}}{8} \Rightarrow \cos\theta = \frac{-1 \pm \sqrt{2}}{2}$. Now find x.

(i) For $\cos\theta = \frac{-1 + \sqrt{2}}{2}$: find the reference angle (calculate in degree mode): $\cos\theta = \frac{-1 + \sqrt{2}}{2} \Rightarrow$

$\cos\theta \approx 0.20710678 \Rightarrow \theta = \cos^{-1}(0.20710678) \approx 78.0°$. Since θ is found in quadrants I and IV,

$\theta \approx 78.0°$, and $\theta = 360° - 78.0° = 282.0°$.

(ii) For $\cos\theta = \frac{-1 - \sqrt{2}}{2}$, find the reference angle (calculate in degree mode). $\cos\theta = \frac{-1 - \sqrt{2}}{2} \Rightarrow$

$\cos\theta =\approx -1.207$. This value is outside the range of $\cos\theta$; therefore, $\varnothing$.

7. Use identities to solve for θ. $2\sin\theta - 1 = \csc\theta \Rightarrow 2\sin\theta - 1 = \frac{1}{\sin\theta} \Rightarrow 2\sin^2\theta - \sin\theta = 1 \Rightarrow$

$2\sin^2\theta - \sin\theta - 1 = 0 \Rightarrow (2\sin\theta + 1)(\sin\theta - 1) = 0$. Now, either $2\sin\theta + 1 = 0 \Rightarrow$

$2\sin\theta = -1 \Rightarrow \sin\theta = -\dfrac{1}{2} \Rightarrow \theta = 210°, 330°$, or $\sin\theta - 1 = 0 \Rightarrow \sin\theta = 1 \Rightarrow \theta = 90°$.

Therefore, $\theta = 90°, 210°, 330°$.

8. If $\sec^2\dfrac{\theta}{2} = 2$, then $\cos^2\dfrac{\theta}{2} = \dfrac{1}{2}$. If the required interval for θ is $0 \le \theta < 360°$, then $0 \le \dfrac{\theta}{2} < 180°$. If $\cos\dfrac{\theta}{2} = \pm\dfrac{1}{\sqrt{2}} = \pm\dfrac{\sqrt{2}}{2}$, then $\dfrac{\theta}{2} = 45°, 135°$. Multiplying these solutions by 2 yields $\theta = 90°, 270°$.

9. Graph $y = x^2 + \sin x - x^3 - \cos x$. Using the capabilities of the calculator, the x-intercepts are $x \approx 0.68058878, 1.4158828$.

10. Graph $y_1 = x^3 - \cos^2 x$ and $y_2 = \dfrac{1}{2}x - 1$ in the same screen. Using the capabilities of the calculator, the intersection of these two graphs is $x \approx 0, 0.37600772$.

Chapter 9 Review Exercises

1. If $f(-x) = -f(x)$, then the function is an odd function, and (x, y) yields $(-x, -y)$. The following functions are odd: sine, tangent, cotangent, and cosecant.
2. If $f(-x) = f(x)$, then the function is an even function, and (x, y) yields $(-x, y)$. The following functions are even: cosine and secant.
3. Since cosine is an even function, $\cos(-x) = \cos x$, and $\cos(-3) = \cos 3$.
4. Since sine is an odd function, $\sin(-x) = -\sin x$, and $\sin(-3) = -\sin 3$.
5. Since tangent is an odd function, $\tan(-x) = -\tan x$, and $\tan(-3) = -\tan 3$.
6. Since secant is an even function $\sec(-x) = \sec x$, and $\sec(-3) = \sec 3$.
7. Since cosecant is an odd function, $\csc(-x) = -\csc x$, and $\csc(-3) = -\csc 3$.
8. Since cotangent is an odd function, $\cot(-x) = -\cot x$, and $\cot(-3) = -\cot 3$.
9. By reciprocal identity, $\sec x = \dfrac{1}{\cos x}$, therefore B.
10. By quotient identity, $\tan x = \dfrac{\sin x}{\cos x}$, therefore C.
11. By quotient identity, $\cot x = \dfrac{\cos x}{\sin x}$, therefore F.
12. By Pythagorean identity, $\tan^2 x + 1 = \sec^2 x = \dfrac{1}{\cos^2 x}$, therefore E.
13. By reciprocal identity, $\tan^2 x = \dfrac{1}{\cot^2 x}$, therefore D.

14. By reciprocal identity, $\csc x = \dfrac{1}{\sin x}$, therefore A.

15. $\dfrac{\cot\theta}{\sec\theta} = \dfrac{\frac{\cos\theta}{\sin\theta}}{\frac{1}{\cos\theta}} = \dfrac{\cos\theta}{\sin\theta}\cdot\dfrac{\cos\theta}{1} = \dfrac{\cos^2\theta}{\sin\theta}$.

16. $\tan^2\theta(1+\cot^2\theta) = \dfrac{\sin^2\theta}{\cos^2\theta}(\csc^2\theta) = \dfrac{\sin^2\theta}{\cos^2\theta}\left(\dfrac{1}{\sin^2\theta}\right) = \dfrac{1}{\cos^2\theta}$.

17. $\csc\theta + \cot\theta = \dfrac{1}{\sin\theta} + \dfrac{\cos\theta}{\sin\theta} = \dfrac{1+\cos\theta}{\sin\theta}$.

18. Since $x = \dfrac{3}{5}$, and x is in quadrant IV, $\sin x = -\sqrt{1-\cos^2 x}$; therefore,

$\sin x = -\sqrt{1-\left(\dfrac{3}{5}\right)^2} = -\sqrt{1-\dfrac{9}{25}} = -\sqrt{\dfrac{16}{25}} = -\dfrac{4}{5}$. Now use these values to solve for

$\tan x = \dfrac{\sin x}{\cos x} = \dfrac{-\frac{4}{5}}{\frac{3}{5}} = -\dfrac{4}{3}$, $\sec x = \dfrac{1}{\cos x} = \dfrac{5}{3}$, $\csc x = \dfrac{1}{\sin x} = -\dfrac{5}{4}$, $\cot x = -\dfrac{3}{4}$.

19. Since $x = -\dfrac{5}{4}$, and, since $\dfrac{\pi}{2} < x < \pi$, x is in quadrant II, $\sec x = -\sqrt{1+\tan^2 x} = -\sqrt{1+\dfrac{25}{16}} = -\sqrt{\dfrac{41}{16}} = -\dfrac{\sqrt{41}}{4}$.

Now, use these to find the other functions.

$\cos x = \dfrac{1}{\sec x} = -\dfrac{4}{\sqrt{41}} = -\dfrac{4\sqrt{41}}{41}$; $\cot x = \dfrac{1}{\tan x} = -\dfrac{4}{5}$; $\sin x = \sqrt{1-\cos^2 x} =$

$\sqrt{1-\left(\dfrac{4\sqrt{41}}{41}\right)^2} = \sqrt{1-\dfrac{656}{1681}} = \sqrt{\dfrac{1025}{1681}} = \dfrac{5\sqrt{41}}{41}$; $\csc x = \dfrac{1}{\sin x} = \dfrac{41}{5\sqrt{41}} \bullet \dfrac{\sqrt{41}}{\sqrt{41}} = \dfrac{41\sqrt{41}}{5(41)} = \dfrac{\sqrt{41}}{5}$.

20. E, by a sum identity, $\cos 210° = \cos(150° + 60°) = \cos 150°\cos 60° - \sin 150°\sin 60°$.

21. B, by a cofunctional identity, $\sin 35° = \cos(90° - 35°) = \cos 55°$.

22. J, by a cofunctional identity, $\tan 35° = \tan(90° - (-35°)) = \cot 125°$.

23. A, by a negative number identity, $-\sin 35° = \sin(-35°)$.

24. I, by a negative number identity, $\cos 35° = \cos(-35°)$.

25. C, by a half-number identity, $\cos 75° = \cos\dfrac{150°}{2} = \sqrt{\dfrac{1+\cos 150°}{2}}$.

26. H, by a sum identity, $\sin 75° = \sin 15°\cos 60° + \cos 15°\sin 60°$.

27. D, by a double number identity, $\sin 300° = 2\sin 150°\cos 150°$.

28. G, by a double number identity, $\cos 300° = \cos(2\cdot 150°) = \cos^2 150° - \sin^2 150°$.

29. F, by a cofunctional identity, $\tan(-55°) = \cot(-35°)$.

30. $\sin^2 x - \sin^2 y = (1-\cos^2 x) - (1-\cos^2 y) = -\cos^2 x + \cos^2 y = \cos^2 y - \cos^2 x$.

31. $2\cos^3 x-\cos x=\cos x(2\cos^2 x-1)=\dfrac{1}{\sec x}(\cos 2x)=\dfrac{\cos^2 x-\sin^2 x}{\sec x}.$

32. The right side of the equation equals $\dfrac{\sin 2x+\sin x}{\cos 2x-\cos x}=\dfrac{2\sin x\cos x+\sin x}{2\cos^2 x-1-\cos x}=\dfrac{\sin x(2\cos x+1)}{(2\cos x+1)(\cos x-1)}=\dfrac{\sin x}{\cos x-1}.$

The left side of the equation equals $-\cot\dfrac{x}{2}=\dfrac{1}{\tan\frac{x}{2}}=-\dfrac{1}{\frac{1-\cos x}{\sin x}}=\dfrac{-\sin x}{1-\cos x}=\dfrac{\sin x}{\cos x-1}$. Since both sides of the equation equal the same expression, $-\cot\dfrac{x}{2}=\dfrac{\sin 2x+\sin x}{\cos 2x-\cos x}.$

33. The left side of the equation equals $\dfrac{\sin^2 x}{2-2\cos x}=\dfrac{1-\cos^2 x}{2(1-\cos x)}=\dfrac{(1-\cos x)(1+\cos x)}{2(1-\cos x)}=\dfrac{1+\cos x}{2}$. The right side of the equation equals $\cos^2\dfrac{x}{2}=\left(\pm\sqrt{\dfrac{1+\cos x}{2}}\right)^2=\dfrac{1+\cos x}{2}$. Since both sides of the equation equal the same expression, $\dfrac{\sin^2 x}{2-2\cos x}=\cos^2\dfrac{x}{2}.$

34. $\dfrac{\sin 2x}{\sin x}=\dfrac{2\sin x\cos x}{\sin x}=2\cos x=\dfrac{2}{\sec x}.$

35. The right side of the equation equals $\cos A-\dfrac{\tan A}{\csc A}=\cos A-\dfrac{\frac{\sin A}{\cos A}}{\frac{1}{\sin A}}=\cos A-\dfrac{\sin^2 A}{\cos A}=$ $\dfrac{\cos^2 A-\sin^2 A}{\cos A}=\dfrac{\cos 2A}{\cos A}$. The left side of the equation equals $2\cos A-\sec A=2\cos A-\dfrac{1}{\cos A}$ $=\dfrac{2\cos^2 A-1}{\cos A}=\dfrac{\cos 2A}{\cos A}$. Since both sides of the equation equal the same expression, $2\cos A-\sec A=\cos A-\dfrac{\tan A}{\csc A}.$

36. $\dfrac{2\tan B}{\sin 2B}=\dfrac{\frac{2\sin B}{\cos B}}{2\sin B\cos B}=\dfrac{2\sin B}{\cos B}\cdot\dfrac{1}{2\sin B\cos B}=\dfrac{1}{\cos^2 B}=\sec^2 B.$

37. $2\tan\alpha\csc 2\alpha=\dfrac{2\sin\alpha}{\cos\alpha}\cdot\dfrac{1}{\sin 2\alpha}=\dfrac{2\sin\alpha}{\cos\alpha(2\sin\alpha\cos\alpha)}=\dfrac{1}{\cos^2\alpha}=\sec^2\alpha=1+\tan^2\alpha.$

38. $\cot\dfrac{t}{2}=\dfrac{1}{\tan\frac{t}{2}}=\dfrac{1}{\frac{1-\cos t}{\sin t}}=\dfrac{\sin t}{1-\cos t}.$

39. $\dfrac{2\cot x}{\tan 2x}=\dfrac{2\cos x}{\sin x}\div\dfrac{\sin 2x}{\cos 2x}=\dfrac{2\cos x}{\sin x}\cdot\dfrac{1-2\sin^2 x}{2\sin x\cos x}=\dfrac{1-2\sin^2 x}{\sin^2 x}=\dfrac{1}{\sin^2 x}-\dfrac{2\sin^2 x}{\sin^2 x}=\csc^2 x-2.$

40. $\tan\theta\sin 2\theta=\dfrac{\sin\theta}{\cos\theta}(2\sin\theta\cos\theta)=2\sin^2\theta=2(1-\cos^2\theta)=2-2\cos^2\theta.$

41. $2\tan x\csc 2x-\tan^2 x=\tan x(2\csc 2x-\tan x)=\tan x\left(\dfrac{2}{\sin 2x}-\dfrac{\sin x}{\cos x}\right)=$

$$\tan x\left(\frac{2}{2\sin x\cos x}-\frac{\sin x}{\cos x}\right)=\tan x\left(\frac{2-2\sin^2 x}{2\sin x\cos x}\right)=\tan x\left(\frac{2(1-\sin^2 x)}{2\sin x\cos x}\right)=$$

$$\frac{\sin x}{\cos x}.\left(\frac{\cos^2 x}{\sin x\cos x}\right)=\frac{\sin x}{\cos x}.\frac{\cos x}{\sin x}=1.$$

42. $y=\sin^{-1}\frac{\sqrt{2}}{2}\Rightarrow\frac{\sqrt{2}}{2}=\sin y\Rightarrow y=\frac{\pi}{4}.\ \left(\text{range: }\left[-\frac{\pi}{2},\frac{\pi}{2}\right]\right)$

43. $y=\arccos\left(-\frac{1}{2}\right)\Rightarrow-\frac{1}{2}=\cos y\Rightarrow y=\frac{2\pi}{3}.\ \left(\text{range: }[0,\pi]\right)$

44. $y=\arctan\frac{\sqrt{3}}{3}\Rightarrow\frac{\sqrt{3}}{3}=\tan y\Rightarrow y=\frac{\pi}{6}.\ \left(\text{range: }\left[-\frac{\pi}{2},\frac{\pi}{2}\right]\right)$

45. $y=\sec^{-1}(-2)\Rightarrow y=\cos^{-1}\left(-\frac{1}{2}\right)\Rightarrow-\frac{1}{2}=\cos y\Rightarrow y=\frac{2\pi}{3}.\ \left(\text{range: }\left[0,\frac{\pi}{2}\right)\cup\left(\frac{\pi}{2},\pi\right]\right)$

46. $y=\operatorname{arccsc}\frac{2\sqrt{3}}{3}\Rightarrow y=\arcsin\frac{3}{2\sqrt{3}}\Rightarrow y=\arcsin\frac{\sqrt{3}}{2}\Rightarrow\frac{\sqrt{3}}{2}=\sin y\Rightarrow y=\frac{\pi}{3}.\ \left(\text{range: }\left[-\frac{\pi}{2},0\right)\cup\left(0,\frac{\pi}{2}\right]\right)$

47. $y=\cot^{-1}(-1)\Rightarrow y=\tan^{-1}\left(\frac{1}{-1}\right)+\pi\Rightarrow y=\tan^{-1}(-1)+\pi\Rightarrow(-1=\tan y)+\pi\Rightarrow$

$y=-\frac{\pi}{4}+\pi\Rightarrow y=\frac{3\pi}{4}.\ \ (\text{range: }(0,\pi))$

48. $\theta=\arccos\frac{1}{2}\Rightarrow\frac{1}{2}=\cos\theta\Rightarrow\theta=60^\circ.\ \left(\text{range: }[0^\circ,180^\circ]\right)$

49. $\theta=\arcsin\left(-\frac{\sqrt{3}}{2}\right)\Rightarrow-\frac{\sqrt{3}}{2}=\sin\theta\Rightarrow\theta=-60^\circ.\ \left(\text{range: }[-90^\circ,90^\circ]\right)$

50. $\theta=\tan^{-1}(0)\Rightarrow 0=\tan\theta\Rightarrow\theta=0^\circ.\ \left(\text{range: }[-90^\circ,90^\circ]\right)$

51. $\theta=\arcsin(-0.656059029)\Rightarrow\theta\approx-41^\circ.$

52. $\theta=\arccos(0.7095707365)\Rightarrow\theta\approx 44.8^\circ.$

53. $\theta=\arctan(-0.1227845609)\Rightarrow\theta\approx-7^\circ.$

54. $\theta=\cot^{-1}(4.70463109)\Rightarrow\theta=\tan^{-1}\frac{1}{4.704630109}\Rightarrow\theta=\tan^{-1}(4.704630109)^{-1}\Rightarrow\theta\approx 12^\circ.$

55. $\theta=\sec^{-1}(28.65370835)\Rightarrow\theta=\cos^{-1}\frac{1}{28.65370835}\Rightarrow\theta=\cos^{-1}(28.65370835)^{-1}\Rightarrow\theta\approx 88^\circ.$

56. $\theta=\csc^{-1}(19.10732261)\Rightarrow\theta=\sin^{-1}\frac{1}{19.10732261}\Rightarrow\theta=\sin^{-1}(19.10732261)^{-1}\Rightarrow\theta\approx 3^\circ.$

57. The value -3 is in the domain of the inverse tangent function but not in the domain of the inverse sine function.

58. The function $\cos x$ is defined for every real number, but $\arccos x$ is defined only on the interval $[-1,1]$. The function $\arccos(\cos x) = x$ only for x in the interval $[0, \pi]$.

59. Since $\sin(\sin^{-1} x) = x$ when $-1 < x < 1$, $\sin\left(\sin^{-1}\frac{1}{2}\right) = \frac{1}{2}$.

60. Since $\cos\theta = \frac{3}{4}$, θ is in quadrant I. Sketch and label a triangle in quadrant I, with $x = 3$ and $r = 4$. Now use the Pythagorean theorem to find y. $3^2 + y^2 = 4^2 \Rightarrow y^2 = 16 - 9 \Rightarrow y^2 = 7 \Rightarrow y = \sqrt{7}$.

Therefore, $\sin\theta = \frac{\sqrt{7}}{4}$, and $\sin\left(\cos^{-1}\frac{3}{4}\right) = \frac{\sqrt{7}}{4}$.

61. Since $\tan\theta = \frac{3}{1}$, θ is in quadrant I. Sketch and label a triangle in quadrant I, with $x = 1$ and $y = 3$. Use the Pythagorean theorem to find r. $1^2 + 3^2 = r^2 \Rightarrow r^2 = 1 + 9 \Rightarrow r^2 = 10 \Rightarrow r = \sqrt{10}$.

Therefore, $\cos\theta = \frac{1}{\sqrt{10}} = \frac{\sqrt{10}}{10}$, and $\cos(\arctan 3) = \frac{\sqrt{10}}{10}$.

62. Since $\theta = -\frac{1}{3}$, θ is in quadrant IV. Sketch and label a triangle in quadrant I, with $y = -1$ and $r = 3$. Use the Pythagorean theorem to find x. $x^2 + (-1)^2 = 3^2 \Rightarrow x^2 = 9 - 1 \Rightarrow x^2 = 8 \Rightarrow x = \sqrt{8} = 2\sqrt{2}$. Now,

$\cos 2\theta = \cos^2\theta - \sin^2\theta$, thus $\cos 2\theta = \left(\frac{2\sqrt{2}}{3}\right)^2 - \left(-\frac{1}{3}\right)^2 = \frac{8}{9} - \frac{1}{9} = \frac{7}{9}$, and $\cos\left(2\sin^{-1}\frac{1}{3}\right) = \frac{7}{9}$,

Finally, if $\cos\left(2\sin^{-1}\left(-\frac{1}{3}\right)\right) = \frac{7}{9}$, then $\sec\left(2\sin^{-1}\left(-\frac{1}{3}\right)\right) = \frac{9}{7}$.

63. Since $\cos = \frac{3\pi}{2} = 0$, we can solve $\cos^{-1}(0) \Rightarrow 0 = \cos y \Rightarrow y = \frac{\pi}{2}$. Therefore, $\cos^{-1}\left(\cos\frac{3\pi}{2}\right) = \frac{\pi}{2}$.

64. Since $\sin\theta = \frac{3}{5}$, θ is in quadrant I. Sketch and label a triangle in quadrant I, with $y = 3$ and $r = 5$. Use the Pythagorean theorem to find x. $x^2 + 3^2 = 5^2 \Rightarrow x^2 = 25 - 9 \Rightarrow x^2 = 16 \Rightarrow x = 4$. Since $\cos\beta = \frac{5}{7}$, then β is in quadrant I. Sketch and label a triangle in quadrant I, with $x = 5$ and $r = 7$. Use the Pythagorean theorem to find y. $5^2 + y^2 = 7^2 \Rightarrow y^2 = 49 - 25 \Rightarrow y^2 = 24 \Rightarrow y = \sqrt{24} = 2\sqrt{6}$. Now use $\tan\theta = \frac{3}{4}$ and $\tan\beta = \frac{2\sqrt{6}}{5}$ in $\tan(\theta + \beta) = \frac{\tan\theta + \tan\beta}{1 - \tan\theta\tan\beta}$; therefore, $\frac{\frac{3}{4} + \frac{2\sqrt{6}}{5}}{1 - \left(\frac{3}{4}\cdot\frac{2\sqrt{6}}{5}\right)} = \frac{\frac{3}{4} + \frac{2\sqrt{6}}{5}}{1 - \left(\frac{3}{4}\cdot\frac{2\sqrt{6}}{5}\right)}\cdot\frac{20}{20} = \frac{15 + 8\sqrt{6}}{20 - 6\sqrt{6}}$

$= \frac{15 + 8\sqrt{6}}{20 - 6\sqrt{6}}\cdot\frac{20 + 6\sqrt{6}}{20 + 6\sqrt{6}} = \frac{588 + 250\sqrt{6}}{184} = \frac{294 + 125\sqrt{6}}{92}$. Therefore, $\tan\left(\sin^{-1}\frac{3}{5} + \cos^{-1}\frac{5}{7}\right) = \frac{294 + 125\sqrt{6}}{92}$.

65. Since $\theta = u$, θ is in quadrant I. Sketch and label a triangle in quadrant I, with $x = 1$ and $y = u$. Use the Pythagorean theorem to find *r*. $1^2 + u^2 = r^2 \Rightarrow r^2 = 1 + u^2 \Rightarrow r = \sqrt{u^2+1}$. Now $\sin\theta = \dfrac{u}{\sqrt{u^2+1}} = \dfrac{u\sqrt{u^2+1}}{u^2+1}$, and $\sin\left(\tan^{-1} u\right) = \dfrac{u\sqrt{u^2+1}}{u^2+1}$.

66. Since $\tan\theta = \dfrac{u}{\sqrt{1-u^2}}$, θ is in quadrant I. Sketch and label a triangle in quadrant I, $x = \sqrt{1-u^2}$ and $y = u$. Use the Pythagorean theorem to find *r*. $\left(\sqrt{1-u^2}\right)^2 + u^2 = r^2 \Rightarrow r^2 = 1 - u^2 + u^2 \Rightarrow r = 1$. Now $\cos\theta = \dfrac{\sqrt{1-u^2}}{1} = \sqrt{1-u^2}$, and $\cos\left(\arctan\dfrac{u}{\sqrt{1-u^2}}\right) = \sqrt{1-u^2}$.

67. Since $\cos\theta = \dfrac{u}{\sqrt{u^2+1}}$, θ is in quadrant I. Sketch and label a triangle in quadrant I, with $x = u$, and $r = \sqrt{u^2+1}$. Use the Pythagorean theorem to find *y*. $u^2 + y^2 = \left(\sqrt{u^2+1}\right)^2 \Rightarrow y^2 = u^2 + 1 - u^2 \Rightarrow y = 1$. Now $\tan\theta = \dfrac{1}{u}$ and $\tan(\arccos u) = \dfrac{1}{u}$.

68. $\sin^2 x = 1 \Rightarrow \sin x = \pm 1 \Rightarrow x = \dfrac{\pi}{2}, \dfrac{3\pi}{2}$. The solution set is $x = \left\{\dfrac{\pi}{2}, \dfrac{3\pi}{2}\right\}$.

69. $2\tan x - 1 = 0 \Rightarrow 2\tan x = 1 \Rightarrow \tan x = \dfrac{1}{2}$. Now, for $x = \dfrac{1}{2}$, find the reference angle (calculate in radian mode) $x = \tan^{-1}\left(\dfrac{1}{2}\right) \approx 0.463647609$. Since x is found in quadrants I and III, $x \approx 0.463647609$, and $x = \pi + 0.463647609 \approx 3.605240263$. The solution set is $x = \{0.463647609, 3.605240263\}$.

70. First, factor and solve for $\sin x$. $3\sin^2 x - 5\sin x + 2 = 0 \Rightarrow (3\sin x - 2)(\sin x - 1) = 0$. Now, either $3\sin x - 2 = 0 \Rightarrow 3\sin x = 2 \Rightarrow \sin x = \dfrac{2}{3}$, or $\sin x - 1 = 0 \Rightarrow \sin x = 1$. If $\sin x = \dfrac{2}{3}$, find the reference angle (calculate in radian mode) $x = \sin^{-1}\left(\dfrac{2}{3}\right) \approx 0.729726562$. Since x is found in quadrants I and II, $x \approx 0.729726562$, and $x = \pi - 0.729726562 \approx 2.411864997$. If $\sin x = 1$, then $x = \dfrac{\pi}{2}$. The solution is $x = 0.729726562, \dfrac{\pi}{2}, 2.411864997$.

71. Use identities to find $\tan x$. $\tan x = \cot x \Rightarrow \tan x = \dfrac{1}{\tan x} \Rightarrow \tan^2 x = 1 \Rightarrow \tan x = \pm 1$. If $\tan x = \pm 1$, then the solution is $x = \dfrac{\pi}{4}, \dfrac{3\pi}{4}, \dfrac{5\pi}{4}, \dfrac{7\pi}{4}$.

72. First set equal to zero, factor, and solve for $\cot x$. $5\cot^2 x + 3\cot x = 2 \Rightarrow 5\cot^2 x + 3\cot x - 2 = 0 \Rightarrow$

$(5\cot x - 2)(\cot x + 1) = 0$. Now, either $5\cot x - 2 = 0 \Rightarrow 5\cot x = 2 \Rightarrow \cot x = \frac{2}{5}$, or

$\cot x + 1 = 0 \Rightarrow \cot x = -1$. If $\cot x = \frac{2}{5}$, then $\tan x = \frac{5}{2}$. Now find the reference angle in radian mode.

$x = \tan^{-1}\left(\frac{5}{2}\right) \approx 1.19028995$. Since x is found in quadrants I and III, $x \approx 1.19028995$, and

$x = \pi + 1.19028995 \approx 4.331882603$. If $\cot x = -1$, then $x = \frac{3\pi}{4}, \frac{7\pi}{4}$.

The solutions are $x \approx 1.19028995, \frac{3\pi}{4}, 4.331882603, \frac{7\pi}{4}$.

73. Use identities to find $\cos x$. $\sec\frac{x}{2} = \cos\frac{x}{2} \Rightarrow \frac{1}{\cos\frac{x}{2}} = \cos\frac{x}{2} \Rightarrow 1 = \cos^2\frac{x}{2} \Rightarrow 1 = \left(\pm\sqrt{\frac{1+\cos x}{2}}\right)^2 \Rightarrow$

$1 = \frac{1+\cos x}{2} \Rightarrow 2 = 1 + \cos x \Rightarrow \cos x = 1$. If $\cos x = 1$, then the solution is $x = 0$.

74. Use identities and factor. $\sin 2x = \cos 2x + 1 \Rightarrow \sin 2x - \cos 2x - 1 = 0 \Rightarrow$

$2\sin x\cos x - (2\cos^2 x - 1) - 1 = 0 \Rightarrow 2\sin x\cos x - 2\cos^2 x = 0 \Rightarrow 2\cos x(\sin x - \cos x) = 0$. Now, either

$2\cos x = 0 \Rightarrow \cos x = 0 \Rightarrow x = \frac{\pi}{2}, \frac{3\pi}{2}$, or $\sin x - \cos x = 0 \Rightarrow \sin x = \cos x \Rightarrow$

$\cos^2 x = \sin^2 x \Rightarrow \cos^2 x = 1 - \cos^2 x \Rightarrow 2\cos^2 x = 1 \Rightarrow \cos^2 x = \frac{1}{2} \Rightarrow \cos x = \pm\frac{1}{\sqrt{2}} \Rightarrow \; x = \frac{\pi}{4}, \frac{3\pi}{4}, \frac{5\pi}{4}, \frac{7\pi}{4}$.

Since we squared the equation to solve it, we need to check the second set of answers. The functions $\cos x$ and $\sin x$ have different signs in quadrants II and IV, so only $\frac{\pi}{4}$ and $\frac{5\pi}{4}$ will check.

The solution is $x = \frac{\pi}{4}, \frac{3\pi}{4}, \frac{5\pi}{4}, \frac{7\pi}{4}$.

75. First solve for $\sin 2x$. $2\sin 2x = 1 \Rightarrow \sin 2x = \frac{1}{2}$. If the required interval for x is $0 \le x < 2\pi$, then

$0 \le 2x < 4\pi$. Therefore, if $2x = \frac{1}{2}$, then $2x = \frac{\pi}{6}, \frac{5\pi}{6}, \frac{13\pi}{6}, \frac{17\pi}{6}$. Now, dividing each solution by 2

(or multiplying by ½) yields $x = \frac{\pi}{12}, \frac{5\pi}{12}, \frac{13\pi}{12}, \frac{17\pi}{12}$.

76. Use identities and factor. $\sin 2x + \sin 4x = 0 \Rightarrow \sin 2x + 2\sin 2x\cos 2x = 0 \Rightarrow \; \sin 2x(1 + 2\cos 2x) = 0$. If the required interval for x is $0 \le x < 2\pi$, then $0 \le 2x < 4\pi$. Now, either $\sin 2x = 0 \Rightarrow 2x = 0, \pi, 2\pi, 3\pi$, or

$1+2\cos 2x=0\Rightarrow 2\cos 2x=-1\Rightarrow 1+\cos 2x=-\frac{1}{2}\Rightarrow\ 2x=\frac{2\pi}{3},\frac{4\pi}{3},\frac{8\pi}{3},\frac{10\pi}{3}$. Divide each solution by 2 (or multiply by ½) to get: $x=0,\frac{\pi}{3},\frac{\pi}{2},\frac{2\pi}{3},\pi,\frac{4\pi}{3},\frac{3\pi}{2},\frac{5\pi}{3}$.

77. Use identities and solve for x. $\cos x-\cos 2x=2\cos x\Rightarrow -\cos x-\left(2\cos^2 x-1\right)=0\Rightarrow$ $-\cos x-2\cos^2 x+1=0\Rightarrow 2\cos^2 x+\cos x-1=0\Rightarrow\left(2\cos x-1\right)\left(\cos x+1\right)=0$. Now, either $2\cos x-1=0\Rightarrow 2\cos x=1\Rightarrow \cos x=\frac{1}{2}\Rightarrow x=\frac{\pi}{3},\frac{5\pi}{3}$, or $\cos x+1=0\Rightarrow\ \cos x=-1\Rightarrow x=\pi$.

The solution is $x=\frac{\pi}{3},\pi,\frac{5\pi}{3}$.

78. If the required interval for x is $0\le x<2\pi$, then $0\le 2x<4\pi$. Therefore, if $\tan 2x=\sqrt{3}$, then $2x=\frac{\pi}{3},\frac{4\pi}{3},\frac{7\pi}{3},\frac{10\pi}{3}$. Now, dividing each solution by 2 (or multiplying by ½) yields $x=\frac{\pi}{6},\frac{2\pi}{3},\frac{7\pi}{6},\frac{5\pi}{3}$.

Since we are looking for all solutions we can generalize the solution set as follows:

$$\left\{\frac{\pi}{6}+n\pi,\frac{2\pi}{3}+n\pi,\text{ where } n \text{ is any integer}\right\}$$

79. Factor to solve for $\cos\frac{x}{2}$. $\cos^2\frac{x}{2}-2\cos\frac{x}{2}+1=0\Rightarrow\left(\cos\frac{x}{2}-1\right)\left(\cos\frac{x}{2}-1\right)=0$. If the required interval for x is $0\le x<2\pi$, then $0\le\frac{x}{2}<\pi$. If $\cos\frac{x}{2}-1=0\Rightarrow\cos\frac{x}{2}=1\Rightarrow\frac{x}{2}=0$, then multiplying this by 2 yields the solution $x=0$. Since we are looking for all solutions we can generalize the solution set as follows: $\left\{2n\pi,\text{ where } n \text{ is any integer}\right\}$

80. Graph $y=\sin 2x-\cos 2x-1$. From the calculator, the x intercepts are $\frac{\pi}{4},\frac{\pi}{2},\frac{5\pi}{4},\frac{3\pi}{2}$.

(a) For $\sin 2x>\cos 2x+1$, the graph is above the x-axis for the interval $\left(\frac{\pi}{4},\frac{\pi}{2}\right)\cup\left(\frac{5\pi}{4},\frac{3\pi}{2}\right)$.

(b) For $2x<\cos 2x+1$, the graph is below the x-axis for the interval $\left[0,\frac{\pi}{4}\right)\cup\left(\frac{\pi}{2},\frac{5\pi}{4}\right)\cup\left(\frac{3\pi}{2},2\pi\right)$.

81. (a) Let α be the angle to the left of θ, then $\tan\alpha=\frac{5}{x}$ and $\alpha=\arctan\frac{5}{x}$. Now,

$$\tan(\alpha+\theta)=\frac{5+10}{x}\Rightarrow\alpha+\theta=\arctan\frac{15}{x}\Rightarrow\ \theta=\arctan\frac{15}{x}-\alpha\Rightarrow\theta=\arctan\frac{15}{x}-\arctan\frac{5}{x}.$$

(b) Graph $f(x)=\arctan\left(\frac{15}{x}\right)-\arctan\left(\frac{5}{x}\right)$. See Figure 81. The maximum occurs at $x\approx 8.6602567$.

[0,20] by [-1,1] [0,0.01] by [-0.006,0.006]

Xscl = 1 Yscl = 1 Xscl = 0.001 Yscl = 0.001

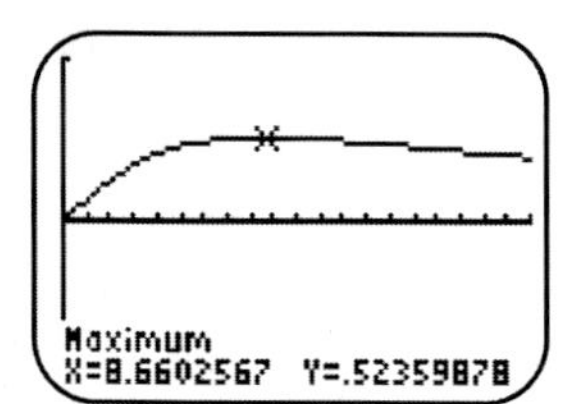

Figure 81

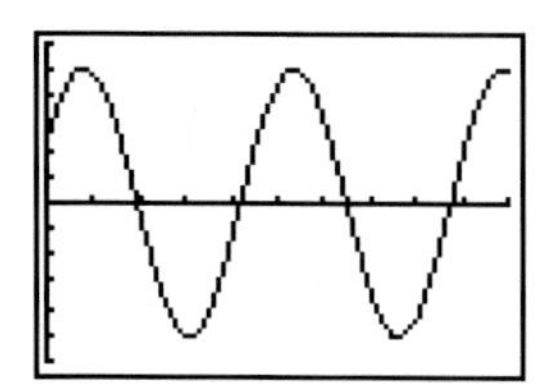
Figure 82

82. (a) Make sure your calculator is in radian mode. Let $A_1 = 0.0012$, $\phi_1 = 0.052$, $A_2 = 0.004$, and $\phi_2 = 0.61$.

$$A = \sqrt{(0.0012\cos .052 + 0.004\cos 0.61)^2 + (0.0012\sin 0.052 + 0.004\sin 0.61)^2} \approx 0.00506$$

$$\phi = \arctan\left(\frac{0.0012\sin 0.052 + 0.004\sin 0.61}{0.0012\cos 0.052 + 0.004\cos 0.61}\right) \approx 0.484$$

If $f = 220$, then $P = A\sin(2\pi ft + \phi)$ becomes $P = 0.00506\sin(440\pi t + 0.484)$.

(b) The two graphs are the same. See Figure 82.

83. $40 = 100\sin(2\pi \cdot 60)t \Rightarrow 0.4 = \sin 120\pi t \Rightarrow 120\pi t = \sin^{-1} 0.4 \Rightarrow t = \dfrac{\sin^{-1} 0.4}{120\pi} \Rightarrow t \approx 0.00109 \approx 0.001$ seconds.

84. $50 = 100\sin(2\pi \cdot 120)t \Rightarrow 0.5 = \sin 240\pi t \Rightarrow 240\pi t = \sin^{-1} 0.5 \Rightarrow t = \dfrac{\sin^{-1} 0.5}{240\pi} \Rightarrow t \approx 0.000694 \approx 0.0007$

seconds.

85. For $0.752 = \dfrac{\sin\theta_1}{\sin\theta_2}$, if $\theta_2 = 90°$, then $0.752 = \dfrac{\sin\theta_1}{\sin 90°} \Rightarrow 0.752 = \dfrac{\sin\theta_1}{1} \Rightarrow 0.752 = \sin\theta_1 \Rightarrow$

$\theta_1 = \sin^{-1} 0.752 \Rightarrow \theta_1 \approx 48.8°$.

86. If $\theta_1 > 48.8°$, then $\theta_2 > 90°$ and the light beam will stay completely under the water. The beam will be reflected at the surface of the water.

Chapter 9 Test

1. If $\sin y = -\dfrac{3}{5}$, and $\pi < y < \dfrac{3\pi}{2}$, then y is in quadrant III. Sketch and label a triangle in quadrant III, with $y = -3$ and $r = 5$. Use the Pythagorean theorem to find x.

$x^2 + (-3)^2 = 5^2 \Rightarrow x^2 = 25 - 9 \Rightarrow x^2 = 16 \Rightarrow x = \pm 4$. Because x is found in quadrant III, $x = -4$ Therefore,

$\cos y = -\dfrac{4}{5}$. If $\cos x = -\dfrac{4}{5}$ and $\dfrac{\pi}{2} < x < \pi$, then x is in quadrant II. Sketch and label a triangle in quadrant II, with $x = -4$ and $r = 5$. Use the Pythagorean theorem to find y.

$(-4)^2 + y^2 = 5^2 \Rightarrow y^2 = 25 - 16 \Rightarrow y^2 = 9 \Rightarrow y = \pm 3$ Therefore, $\sin x = \dfrac{3}{5}$. Using these values,

$$\sin(x + y) = \sin x\cos y + \cos x\sin y \Rightarrow \left(\frac{3}{5}\right)\left(-\frac{4}{5}\right) + \left(-\frac{4}{5}\right)\left(-\frac{3}{5}\right) = -\frac{12}{25} + \frac{12}{25} = 0.$$

2. If $\sin y = -\frac{3}{5}$, and $\pi < y < \frac{3\pi}{2}$, then y is in quadrant III. Sketch and label a triangle in quadrant III, with $y = -3$ and $r = 5$. Use the Pythagorean theorem to find x.

   $x^2 + (-3)^2 = 5^2 \Rightarrow x^2 = 25 - 9 \Rightarrow x^2 = 16 \Rightarrow x = \pm 4$. Because x is found in quadrant III, $x = -4$. Therefore, $\cos y = -\frac{4}{5}$. If $\cos x = -\frac{4}{5}$ and $\frac{\pi}{2} < x < \pi$, then x is in quadrant II. Sketch and label a triangle in quadrant II, with $x = -4$ and $r = 5$. Use the Pythagorean theorem to find y.

   $(-4)^2 + y^2 = 5^2 \Rightarrow y^2 = 25 - 16 \Rightarrow y^2 = 9 \Rightarrow y = \pm 3$ Therefore, $\sin x = \frac{3}{5}$. Using these values,

   $\cos(x - y) = \cos x \cos y + \sin x \sin y \Rightarrow \left(-\frac{4}{5}\right)\left(-\frac{4}{5}\right) + \left(\frac{3}{5}\right)\left(-\frac{3}{5}\right) = \frac{16}{25} - \frac{9}{25} = \frac{7}{25}$.

3. If $\sin y = -\frac{3}{5}$, and $\pi < y < \frac{3\pi}{2}$, then y is in quadrant III. Sketch and label a triangle in quadrant III, with $y = -3$ and $r = 5$. Use the Pythagorean theorem to find x.

   $x^2 + (-3)^2 = 5^2 \Rightarrow x^2 = 25 - 9 \Rightarrow x^2 = 16 \Rightarrow x = \pm 4$. Because x is found in quadrant III, $x = -4$ Therefore, $\cos y = -\frac{4}{5}$. If $\cos x = -\frac{4}{5}$ and $\frac{\pi}{2} < x < \pi$, then x is in quadrant II. Using these values

   $$\tan \frac{y}{2} = \frac{1 - \cos y}{\sin y} \Rightarrow \frac{1 - \left(-\frac{4}{5}\right)}{-\frac{3}{5}} = \frac{\frac{9}{5}}{-\frac{3}{5}} = -3$$

4. If $\cos x = -\frac{4}{5}$ and $\frac{\pi}{2} < x < \pi$, then x is in quadrant II. Sketch and label a triangle in quadrant II, with $x = -4$ and $r = 5$. Use the Pythagorean theorem to find y.

   $(-4)^2 + y^2 = 5^2 \Rightarrow y^2 = 25 - 16 \Rightarrow y^2 = 9 \Rightarrow y = \pm 3$ Therefore, $\sin x = \frac{3}{5}$.

   Using these values $\cos 2x = 2\cos^2 x - 1 \Rightarrow 2\left(-\frac{4}{5}\right)^2 - 1 = \frac{32}{25} - \frac{25}{25} = \frac{7}{25}$.

5. $\tan^2 x - \sec^2 x = \frac{\sin^2 x}{\cos^2 x} - \frac{1}{\cos^2 x} = \frac{\sin^2 x - 1}{\cos^2 x} = -\frac{\cos^2 x}{\cos^2 x} = -1$.

6. Graph $y = \sec x - \sin x \tan x$ on the calculator. The graph looks like $\cos x$, so the identity is $\sec x - \sin x \tan x = \cos x$. Verify this: $\sec x - \sin x \tan x = \frac{1}{\cos x} - \frac{\sin x}{1} \cdot \frac{\sin x}{\cos x} = \frac{1 - \sin^2 x}{\cos x} = \frac{\cos^2 x}{\cos x} = \cos x$.

   Therefore, $\sec x - \sin x \tan x = \cos x$.

7. $\sec^2 B = \frac{1}{\cos^2 B} = \frac{1}{1 - \sin^2 B}$.

8. $\dfrac{\cot A-\tan A}{\csc A\sec A}=\dfrac{\frac{\cos A}{\sin A}-\frac{\sin A}{\cos A}}{\frac{1}{\sin A}\cdot\frac{1}{\cos A}}=\dfrac{\cos^2 A-\sin^2 A}{\sin A\cos A}\cdot\dfrac{\sin A\cos A}{1}=\cos^2 A-\sin^2 A=\cos 2A.$

9. $\dfrac{\cos x+1}{\sin x+\tan x}=\dfrac{\cos x+1}{\sin x+\dfrac{\sin x}{\cos x}}=\dfrac{\cos x+1}{\dfrac{\sin x\cos x+\sin x}{\cos x}}=\dfrac{\cos x(\cos x+1)}{\sin x(\cos x+1)}=\dfrac{\cos x}{\sin x}=\cot x.$

10. $\dfrac{\sec^2 x-1}{1+\tan^2 x}=\dfrac{\sec^2 x-1}{\sec^2 x}=\dfrac{\sec^2 x}{\sec^2 x}-\dfrac{1}{\sec^2 x}=1-\cos^2 x=\sin^2 x.$

11. $\cos(270°-\theta)=\cos 270°\cos\theta+\sin 270°\sin\theta=0\cdot\cos\theta+(-1)\sin\theta=-\sin\theta.$

12. $\sin(\pi+\theta)=\sin\pi\cos\theta+\cos\pi\sin\theta=0\cdot\cos\theta+(-1)\sin\theta=-\sin\theta.$

13. (a) See Figure 13.

(b) The domain is $[-1,1]$, the same as $f(x)=\sin^{-1}x$. The range is that of $f(x)=\sin^{-1}x$, multiplied by -2; therefore, it is $[-\pi,\pi]$.

(c) The domain of $f(x)=-2\sin^{-1}x$ is $[-1,1]$, and 2 is not in this interval. (No number has sine value 2.)

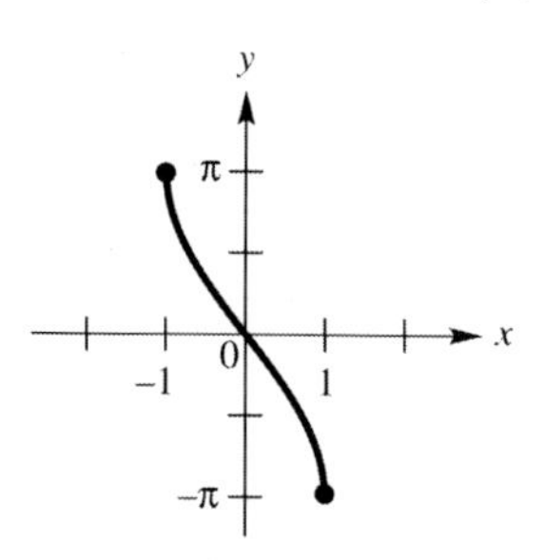

Figure 13

14. $y=\arccos\left(-\dfrac{1}{2}\right)\Rightarrow -\dfrac{1}{2}=\cos y\Rightarrow y=\dfrac{2\pi}{3}.$

15. $y=\tan^{-1}(0)\Rightarrow 0=\tan y\Rightarrow y=0.$

16. $y=\csc^{-1}\dfrac{2\sqrt{3}}{3}\Rightarrow\dfrac{2\sqrt{3}}{3}=\csc y\Rightarrow\sin y=\dfrac{3}{2\sqrt{3}}\Rightarrow\sin y=\dfrac{\sqrt{3}}{2}\Rightarrow y=\dfrac{\pi}{3}.$

17. Since $\sin\left(\dfrac{5\pi}{6}\right)=\dfrac{1}{2}$, $y=\sin^{-1}\left(\dfrac{1}{2}\right)\Rightarrow\dfrac{1}{2}=\sin y\Rightarrow y=\dfrac{\pi}{6}.$

18. Since $\sin\theta=\dfrac{2}{3}$, θ is in quadrant I. Sketch and label a triangle in quadrant I, with $y=2$ and $r=3$. Use the Pythagorean theorem to find x. $x^2+2^2=3^2\Rightarrow x^2=9-4\Rightarrow x^2=5\Rightarrow x=\pm\sqrt{5}$. Therefore, $\cos\theta=\dfrac{\sqrt{5}}{3}$, and $\cos\left(\arcsin\dfrac{2}{3}\right)=\dfrac{\sqrt{5}}{3}.$

19. Since $\theta=\dfrac{1}{3}$, θ is in quadrant I. Sketch and label a triangle in quadrant I, with $x=1$ and $r=3$. Use the Pythagorean theorem to find y. $1^2+y^2=3^2\Rightarrow y^2=9-1\Rightarrow y^2=8\Rightarrow y=\pm\sqrt{8}=\pm2\sqrt{2}.$

Now, $\sin 2\theta = 2\sin\theta\cos\theta$; therefore, $\sin 2\theta = 2\left(\frac{2\sqrt{2}}{3}\right)\left(\frac{1}{3}\right) = \frac{4\sqrt{2}}{9}$, and $\sin\left(2\cos^{-1}\frac{1}{3}\right) = \frac{4\sqrt{2}}{9}$.

20. $\sec(\cos^{-1} u) = \frac{1}{\cos(\cos^{-1} u)} = \frac{1}{u}$.

21. Since $\theta = \frac{u}{1}$, θ is in quadrant I. Sketch and label a triangle in quadrant I. Use $r = 1$, $y = u$, and the Pythagorean theorem to find x. $x^2 + u^2 = 1^2 \Rightarrow x^2 = 1 - u^2 \Rightarrow x = \sqrt{1-u^2}$. Now, $\tan\theta = \frac{u}{\sqrt{1-u^2}} = \frac{u\sqrt{1-u^2}}{1-u^2}$, and $\tan(\arcsin u) = \frac{u\sqrt{1-u^2}}{1-u^2}$.

22. Use identities to solve for θ. $\sin^2\theta = \cos^2\theta + 1 \Rightarrow 1 - \cos^2\theta = \cos^2\theta + 1 \Rightarrow 0 = 2\cos^2\theta \Rightarrow$ $\cos\theta = 0 \Rightarrow \theta = \frac{\pi}{2}, \frac{3\pi}{2}$.

23. Use identities and factor. $\csc^2\theta - 2\cot\theta = 4 \Rightarrow 1 + \cot^2\theta - 2\cot\theta = 4 \Rightarrow \cot^2\theta - 2\cot\theta - 3 = 0 \Rightarrow$ $(\cot\theta - 3)(\cot\theta + 1) = 0$. Now, either $\cot\theta - 3 = 0 \Rightarrow \cot\theta = 3$, or $\cot\theta + 1 = 0 \Rightarrow \cot\theta = -1$. If $\cot\theta = 3$, then $\tan\theta = \frac{1}{3} \Rightarrow \theta = \tan^{-1}\left(\frac{1}{3}\right)$. We find the reference angle in degree mode.

$\tan^{-1}\left(\frac{1}{3}\right) \approx 18.4°$. Since θ is in quadrants I and III, $\theta \approx 18.4°$, and $\theta \approx 180° + 18.4° \approx 198.4°$.

If $\cot\theta = -1 \Rightarrow \theta = 135°, 315°$, the solution set is $\theta \approx \{18.4°, 135°, 198.4°, 315°\}$.

24. Use identities and factor. $\cos x = \cos 2x \Rightarrow \cos x = 2\cos^2 x - 1 \Rightarrow 0 = 2\cos^2 x - \cos x - 1 \Rightarrow$ $(2\cos x + 1)(\cos x - 1) = 0$. Now, either $2\cos x + 1 = 0 \Rightarrow 2\cos x = -1 \Rightarrow \cos x = -\frac{1}{2} \Rightarrow x = \frac{2\pi}{3}, \frac{4\pi}{3}$, or $\cos x - 1 = 0 \Rightarrow \cos x = 1 \Rightarrow x = 0$. Therefore, the solution set is $x = \left\{0, \frac{2\pi}{3}, \frac{4\pi}{3}\right\}$.

25. If the required interval for θ is $0° \le \theta < 360°$, then $0° \le \frac{\theta}{2} < 180°$.

Now solve for $\frac{\theta}{2}$. $2\sqrt{3}\sin\frac{\theta}{2} = 3 \Rightarrow \sin\frac{\theta}{2} = \frac{3}{2\sqrt{3}} \Rightarrow \sin\frac{\theta}{2} = \frac{\sqrt{3}}{2} \Rightarrow \frac{\theta}{2} = 60°, 120°$. Finally, multiplying the solutions by 2 yields the solution $\theta = 120°, 240°$.

26. Graph $y = 2\sin x - 1$. Using the capabilities of the calculator, the x-intercepts are $\frac{\pi}{6}$ and $\frac{5\pi}{6}$. From the graph, $2\sin x - 1 \le 0$, or the graph of the equation is below the x-axis for the interval $\left[0, \frac{\pi}{6}\right] \cup \left[\frac{5\pi}{6}, 2\pi\right)$.

27. (a) $2\sin x + 1 = 2 \Rightarrow 2\sin x = -1 \Rightarrow \sin x = -\frac{1}{2}$ then $x = \frac{7\pi}{6}, \frac{11\pi}{6}$. Since we are looking for all solutions we can generalize the solution set as follows: $\left\{\frac{7\pi}{6} + 2n\pi, \frac{11\pi}{6} + 2n\pi, \text{ where } n \text{ is any integer}\right\}$

(b) $\tan x + \sec x = 1 \Rightarrow \dfrac{\sin x}{\cos x} + \dfrac{1}{\cos x} = 1 \Rightarrow \sin x + 1 = \cos x \Rightarrow (\sin x + 1)^2 = \cos^2 x \Rightarrow$

$(\sin x + 1)^2 = 1 - \sin^2 x \Rightarrow \sin^2 x + 2\sin x + 1 = 1 - \sin^2 x \Rightarrow 2\sin^2 x + 2\sin x = 0 \Rightarrow \sin x(\sin x + 1) = 0 \Rightarrow$

$\sin x = 0$ or $\sin x = -1$. If $\sin x = -1$ then $\cos x = 0$ and $\sec x$ and $\tan x$ is undefined. If

$\sin x = 0$ then x *is* $0, \pi, 2\pi$. Since we are looking for all solutions we can generalize the solution set as

follows: $\{n\pi$, where n is any integer$\}$

(c) $\cos^2 x - \sin^2 x = \dfrac{1}{2} \Rightarrow (1 - \sin^2 x) - \sin^2 x = \dfrac{1}{2} \Rightarrow -2\sin^2 x = -\dfrac{1}{2} \Rightarrow \sin^2 x = \dfrac{1}{4} \Rightarrow \sin x = \pm\dfrac{1}{2}$ then

$x = \dfrac{\pi}{6}, \dfrac{5\pi}{6}, \dfrac{7\pi}{6}, \dfrac{11\pi}{6}$. Since we are looking for all solutions we can generalize the solution set as

follows: $\left\{\dfrac{\pi}{6} + n\pi, \dfrac{5\pi}{6} + n\pi \text{ where } n \text{ is any integer}\right\}$

28. (a) Since t will repeat every 12 months, the domain for 1 year will be $[0, 11]$. The function $T(x)$ is limited by the range of cosine; at the minimum range of (-1), $50 + 50\cos(-1) = 0$, and at the maximum range of (1), $50 + 50\cos(1) = 100$; therefore, the range of the function is $[0, 100]$ in hundreds.

(b) See Figure 28.

(c) From the graph, the maximum of 10,000 animals occurs at 0 or 12 months (July). The minimum of 0 animals occurs at 6 months (January).

(d) $T(3) = 50 + 50\cos\left(\dfrac{(3)\pi}{6}\right) = 50 + 50(0) = 50$ hundreds $= 5000$. There are 5000 animals at 3 months and at 9 months. This is supported by the graph.

(e) $75 = 50 + 50\cos\left(\dfrac{\pi}{6}t\right) = 25 \Rightarrow 50\cos\left(\dfrac{\pi}{6}t\right) \Rightarrow \dfrac{1}{2} = \cos\left(\dfrac{\pi}{6}t\right) \Rightarrow \dfrac{\pi}{6}t = \dfrac{\pi}{3},$

or $\dfrac{\pi}{6}t = \dfrac{5\pi}{3} \Rightarrow t = 2, 10$. At 2 months (September) and at 10 months (May), there will be 7500 animals.

(f) $T = 50 + 50\cos\left(\dfrac{\pi}{6}t\right) \Rightarrow T - 50 = 50\cos\left(\dfrac{\pi}{6}t\right) \Rightarrow \dfrac{T-50}{50} = \cos\left(\dfrac{\pi}{6}t\right) \Rightarrow$

$\dfrac{\pi}{6}t = \arccos\left(\dfrac{T-50}{50}\right) \Rightarrow t = \dfrac{6}{\pi}\arccos\left(\dfrac{T-50}{50}\right).$

[0,12] by [0,110]

Xscl = 1 Yscl = 10

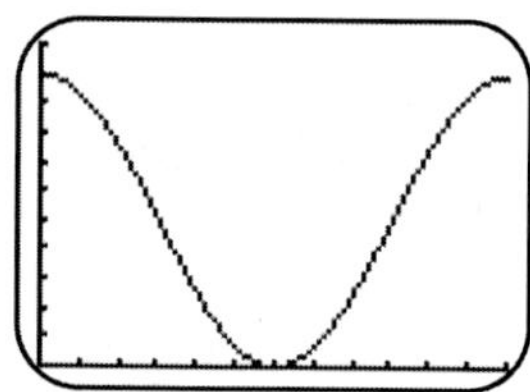

Figure 28

Chapter 10: Applications of Trigonometry and Vectors

10.1 The Law of Sines

1. The proportion shown in C is not valid because it cannot be written in the form of the Law of Sines.
3. The given information is AAS which does not result in the ambiguous case.
5. The given information is SSS which does not result in the ambiguous case.
7. The given information is SSA which results in the ambiguous case.
9. The given information is SSA which results in the ambiguous case.
11. Angle C has measure $180° - 60° - 75° = 45°$. Now, using the Law of Sines,

$$\frac{a}{\sin 60°} = \frac{\sqrt{2}}{\sin 45°} \Rightarrow \text{a} = \frac{\sqrt{2}\sin 60°}{\sin 45°} \Rightarrow \text{a} = \frac{\sqrt{2}\cdot\frac{\sqrt{3}}{2}}{\frac{\sqrt{2}}{2}} \Rightarrow \text{a} = \frac{\frac{\sqrt{6}}{2}}{\frac{\sqrt{2}}{2}} \Rightarrow \text{a} = \frac{\sqrt{6}}{2}\cdot\frac{2}{\sqrt{2}} = \sqrt{3}.$$

13. Angle A has measure $180° - 120° - 15° = 45°$. Now, using the Law of Sines,

$$\frac{a}{\sin 45°} = \frac{5}{\sin 120°} \Rightarrow a = \frac{5\sin 45°}{\sin 120°} \Rightarrow a = \frac{5\cdot\frac{\sqrt{2}}{2}}{\frac{\sqrt{3}}{2}} \Rightarrow a = \frac{\frac{\sqrt{6}}{2}}{\frac{3}{2}} \Rightarrow a = \frac{5\sqrt{6}}{3}.$$

15. Angle C has measure $180° - 37° - 48° = 95°$. Now, using the Law of Sines,

$$\frac{a}{\sin 37°} = \frac{18}{\sin 95°} \Rightarrow a = \frac{18\sin 37°}{\sin 95°} \Rightarrow a \approx 10.87;\ \frac{b}{\sin 48°} = \frac{18}{\sin 95°} \Rightarrow b = \frac{18\sin 48°}{\sin 95°} \Rightarrow b \approx 13.43.$$

The solutions are $C = 95°,\ b \approx 13$ m, $a \approx 11$ m.

17. Angle B has measure $180° - 115.5° - 27.2° = 37.3°$. Now, using the Law of Sines,

$$\frac{a}{\sin 27.2°} = \frac{76.0}{\sin 115.5°} \Rightarrow a = \frac{76.0\sin 27.2°}{\sin 115.5°} \Rightarrow a \approx 38.49;$$

$$\frac{b}{\sin 37.3°} = \frac{76.0}{\sin 115.5°} \Rightarrow b = \frac{76.0\sin 37.3°}{\sin 115.5°} \Rightarrow b \approx 51.03.$$ The solutions are $B = 37.3°,\ a \approx 38.5$ ft, $b \approx 51.0$ ft.

19. Angle B has measure $180° - 37° - 95° = 48°$. Now, using the Law of Sines,

$$\frac{a}{\sin 37°} = \frac{18}{\sin 95°} \Rightarrow a = \frac{18\sin 37°}{\sin 95°} \Rightarrow a \approx 10.87;$$

$$\frac{b}{\sin 48°} = \frac{18}{\sin 95°} \Rightarrow b = \frac{18\sin 48°}{\sin 95°} \Rightarrow b \approx 13.43.$$ The solutions are $B = 48°,\ a \approx 11$ m, $b \approx 13$ m.

21. Angle A has measure $180° - 74.08° - 69.38° = 36.54°$. Now, using the Law of Sines,

$$\frac{a}{\sin 36.54°} = \frac{45.38}{\sin 74.08°} \Rightarrow a = \frac{45.38\sin 36.54°}{\sin 74.08°} \Rightarrow a \approx 28.096;$$

$$\frac{b}{\sin 69.38°} = \frac{45.38}{\sin 74.08°} \Rightarrow b = \frac{45.38\sin 69.38°}{\sin 74.08°} \Rightarrow b \approx 44.167.$$

The solutions are $A = 36.54°,\ a \approx 28.10$ m, $b \approx 44.17$ m.

23. Angle A has measure $180° - 38°40' - 91°40' = 49°40'$. Now, using the Law of Sines,

$$\frac{b}{\sin 38°40'} = \frac{19.7}{\sin 49°40'} \Rightarrow b = \frac{19.7 \sin 38°40'}{\sin 49°40'} \Rightarrow b \approx 16.146;$$

$$\frac{c}{\sin 91°40'} = \frac{19.7}{\sin 49°40'} \Rightarrow c = \frac{19.7 \sin 91°40'}{\sin 49°40'} \Rightarrow c \approx 25.832.$$

The solutions are $A = 49°40'$, $b \approx 16.1$ cm, $c \approx 25.8$ cm.

25. Angle C has measure $180° - 35.3° - 52.8° = 91.9°$. Now, using the Law of Sines,

$$\frac{a}{\sin 35.3°} = \frac{675}{\sin 52.8°} \Rightarrow a = \frac{675 \sin 35.3°}{\sin 52.8°} \Rightarrow a \approx 489.69;$$

$$\frac{c}{\sin 91.9°} = \frac{675}{\sin 52.8°} \Rightarrow c = \frac{675 \sin 91.9°}{\sin 52.8°} \Rightarrow c \approx 846.96.$$ The solutions are $C = 91.9°$, $a \approx 490$ ft, $c \approx 847$ ft.

27. Angle B has measure $180° - 39.70° - 30.35° = 109.95° \approx 110.0°$. Now, using the Law of Sines,

$$\frac{a}{\sin 39.70°} = \frac{39.74}{\sin 110.0°} \Rightarrow a = \frac{39.74 \sin 39.70°}{\sin 110.0°} \Rightarrow a \approx 27.01;$$

$$\frac{c}{\sin 30.35°} = \frac{39.74}{\sin 110.0°} \Rightarrow c = \frac{39.74 \sin 30.35°}{\sin 110.0°} \Rightarrow c \approx 21.37.$$

The solutions are $B = 110.0°$, $a \approx 27.01$ m, $c \approx 21.37$ m.

29. Angle A has measure $180° - 42.88° - 102.40° = 34.72°$. Now, using the Law of Sines,

$$\frac{a}{\sin 34.72°} = \frac{3974}{\sin 42.88°} \Rightarrow a = \frac{3974 \sin 34.72°}{\sin 42.88°} \Rightarrow a \approx 3326;$$

$$\frac{c}{\sin 102.40°} = \frac{3974}{\sin 42.88°} \Rightarrow c = \frac{3974 \sin 102.40°}{\sin 42.88°} \Rightarrow c \approx 5704.$$

The solutions are $A = 34.72°$, $a \approx 3326$ ft, $c \approx 5704$ ft.

31. Three given angles do not determine a unique triangle. Data set A does not determine a unique triangle.

33. (a) Consider the point (3, 0) which is on the x-axis and 4 units below the given point. If the length of the line segment is $4 < h < 5$, two triangles can be drawn. One triangle would intersect the x-axis to the left of (3, 0) and to the right of (0, 0), and the other would intersect the x-axis to the right of (3, 0).

(b) A line segment with length $h = 4$ would form a single right triangle. Also, if $h \geq 5$, a single triangle would be formed with the line segment intersecting the positive x-axis to the right of (3, 0).

(c) If $h < 4$, no triangle could be drawn because the line segment would not intersect the positive x-axis.

35. Using the Law of Sines, $\frac{\sin A}{31} = \frac{\sin 48°}{26} \Rightarrow \sin A = \frac{31 \sin 48°}{26} \Rightarrow A = \sin^{-1}\left(\frac{31 \sin 48°}{26}\right) \approx 62.4°$.

Since $180° - 62.4° = 117.6°$ and $48° + 117.6° < 180°$, there are two triangles.

37. Using the Law of Sines, $\frac{\sin B}{61} = \frac{\sin 58°}{50} \Rightarrow \sin B = \frac{61 \sin 58°}{50} \Rightarrow B = \sin^{-1}\left(\frac{61 \sin 58°}{50}\right) \Rightarrow B \approx \sin^{-1}(1.03)$.

Since 1.03 is not in the domain of the inverse sine function, no such angle exists.
Thus no such triangle exists.

39. Using the Law of Sines, $\dfrac{\sin B}{41.5}=\dfrac{\sin 29.7°}{27.2}\Rightarrow \sin B=\dfrac{41.5\sin 29.7°}{27.2}\Rightarrow B=\sin^{-1}\left(\dfrac{41.5\sin 29.7°}{27.2}\right)$

$\approx 49.1°$ or $180-49.1=130.9°$

If $B=49.1,\ C=180°-29.7°-49.1°=101.2°$. If $B=130.9°,\ C=180°-29.7°-130.9°=19.4°$.

The solutions are $B_1=49.1°,\ C_1=101.2°$ or $B_2=130.9°,\ C_2=19.4°$.

41. Using the Law of Sines,

$\dfrac{\sin A}{859}=\dfrac{\sin 74.3°}{783}\Rightarrow \sin A=\dfrac{859\sin 74.3°}{783}\Rightarrow A=\sin^{-1}\left(\dfrac{859\sin 74.3°}{783}\right)\Rightarrow A\approx \sin^{-1}(1.056)$. Since 1.056 is not in the domain of the inverse sine function, no such angle exists. Thus no such triangle exists.

43. Using the Law of Sines, $\dfrac{\sin B}{5.432}=\dfrac{\sin 142.13°}{7.297}\Rightarrow \sin B$

$=\dfrac{5.432\sin 142.13°}{7.297}\Rightarrow B=\sin^{-1}\left(\dfrac{5.432\sin 142.13°}{7.297}\right)\approx 27.19°$ Since

$180°-27.19°=152.81$ and $142.13°+152.81°>180°$ there is only one triangle.

$C=180°-142.13°-27.19°=10.68°$. The solutions are . $B=27.19°,\ c=10.68°$.

45. Using the Law of Sines,

$\dfrac{\sin B}{8.14}=\dfrac{\sin 42.5°}{15.6}\Rightarrow \sin B=\dfrac{8.14\sin 42.5°}{15.6}\Rightarrow B=\sin^{-1}\left(\dfrac{8.14\sin 42.5°}{15.6}\right)\approx 20.6°$.

Since $180°-42.5°-20.6°=159.4°$ and $42.5°+159.4°>180°$, there is one triangle.

$C=180°-42.5°-20.6°=116.9°.\ \dfrac{c}{\sin 116.9°}=\dfrac{15.6}{\sin 42.5°}\Rightarrow c=\dfrac{15.6\sin 116.9°}{\sin 42.5°}\approx 20.6.$

The solutions are $B=20.6°,\ C=116.9°,\ c=20.6$ ft.

47. Using the Law of Sines,

$\dfrac{\sin C}{145}=\dfrac{\sin 72.2°}{78.3}\Rightarrow \sin C=\dfrac{145\sin 72.2°}{78.3}\Rightarrow C=\sin^{-1}\left(\dfrac{145\sin 72.2°}{78.3}\right)\approx c\approx \sin^{-1}(1.76).$

Since 1.76 is not in the domain of the inverse sine function, no such angle exists.

Thus no such triangle exists.

49. Using the Law of Sines,

$\dfrac{\sin B}{11.8}=\dfrac{\sin 38°40'}{9.72}\Rightarrow \sin B=\dfrac{11.8\sin 38°40'}{9.72}\Rightarrow B=\sin^{-1}\left(\dfrac{11.8\sin 38°.40'}{9.72}\right)\approx 49°20'$. There are two angles:

$B_1=49°20'$ and $B_2=180-49°20'=130°40$(since $38°40'+130°40'<180°$)

$C_1=180°-38°40'-49°20'=92°00'$ and $C_2=180°-38°40'-130°40'=10°40'$

$\dfrac{c_1}{\sin 92°00'}=\dfrac{9.72}{\sin 38°40'}\Rightarrow c_1=\dfrac{9.72\sin 92°00'}{\sin 38°40'}\approx 15.5\quad \dfrac{c_2}{\sin 10°40'}=\dfrac{9.72}{\sin 38°40'}\Rightarrow c_2=\dfrac{9.72\sin 10°40'}{\sin 38°40'}\approx 2.88.$

The solutions are $B_1=49°20',\ C_1=92°00',\ c_1=15.5$ km or $B_2=130°40',\ C_2=10°40',\ c_2=2.88$ km

51. Using the Law of Sines,

$$\frac{\sin A}{7540} = \frac{\sin 32°50'}{5180} \Rightarrow \sin A = \frac{7540 \sin 32°50'}{5180} \Rightarrow A = \sin^{-1}\left(\frac{7540 \sin 32°50'}{5180}\right) \approx 52°10'.$$

There are two angles: $A_1 = 52°10'$ and $A_2 = 180° - 52°10' = 127°50'$ (since $32°50' + 127°50' < 180°$)

$C_1 = 180° - 32°50' - 52°10' = 95°00'$ and $C_2 = 180° - 32°50' - 127°50' = 19°20'$

$$\frac{c_1}{\sin 95°00'} = \frac{5180}{\sin 32°50'} \Rightarrow c_1 = \frac{5180 \sin 95°00'}{\sin 32°50'} \approx 9520$$

$$\frac{c_2}{\sin 19°20'} = \frac{5180}{\sin 32°50'} \Rightarrow c_2 = \frac{5180 \sin 19°20'}{\sin 32°50'} \approx 3160.$$ The solutions are

$A_1 = 52°10'$, $C_1 = 95°00'$, $c_1 = 9520$ cm or $A_2 = 127°50'$, $C_2 = 19°20'$, $c_2 = 3160$ cm.

53. The Pythagorean theorem only applies to right triangles.

55. If we are given only three sides, then any equation from the Law of Sines will contain two unknowns. At least one angle must be given.

57. Angle A has measure $180° = 112°10' - 15°20' = 52°20'$. Now, using the Law of Sines,

$$\frac{AB}{\sin 15°20'} = \frac{354}{\sin 52°30'} \Rightarrow AB = \frac{354 \sin 15°20'}{\sin 52°30'} \Rightarrow AB \approx 118\text{m}.$$

59. Triangle ABC has the following measurements:

$A = 90° - 47.7° = 42.3°, B = 302.5° - 270° - 270° = 32.5°, C = 180° - 42.3° - 32.5° = 105.2°$ and $c = 3.46$.

Now, using the Law of Sines, $\frac{b}{\sin 32.5°} = \frac{3.46}{\sin 105.2°} \Rightarrow b = \frac{3.46 \sin 32.5°}{\sin 105.2°} \Rightarrow b \approx 1.93\text{mi}.$

61. Using the Law of Sines, $\frac{x}{\sin 54.8°} = \frac{12}{\sin 70.4°} \Rightarrow x = \frac{12 \sin 54.8°}{\sin 70.4°} \Rightarrow x \approx 10.4$ in.

63. Note that the distance between the centers of the small gear and, middle-size gear is $2.7 + 1.6 = 4.3$ and the distance between the centers of the small gear and large gear is $3.6 + 1.6 = 5.2$. Use the Law of Sines to find the angle α on the large gear. $\frac{\sin \alpha}{4.3} = \frac{\sin 38°}{5.2} \Rightarrow \sin \alpha = \frac{4.3 \sin 38°}{5.2} \Rightarrow \alpha \approx \sin^{-1}\left(\frac{4.3 \sin 38°}{5.2}\right) \approx 30.6°.$

It follows that $\theta = 180° - 38° - 30.6° = 111.4° \approx 111°$.

65. Let A, B, and C be the ship's initial position, new position, and position of the lighthouse, respectively. Triangle ABC has the following measurements:

$A = 180° - 37° = 143°$, $B = 25°$, $C = 180° - 143° - 25° = 12°$, and $c = 2.5$.

Now, using the Law of Sines, $\frac{a}{\sin 143°} = \frac{2.5}{\sin 12°} \Rightarrow a = \frac{2.5 \sin 143°}{\sin 12°} \Rightarrow a \approx 7.2$ mi and

$$\frac{b}{\sin 25°} = \frac{2.5}{\sin 12°} \Rightarrow a = \frac{2.5 \sin 25°}{\sin 12°} \Rightarrow a \approx 5.1 \text{ mi}.$$

67. Triangle ABC has the following measurements:

$A = 90° - 22.4° + 45° = 112.6°,\ B = 45° - 10.6° = 34.4°,$
$C = 180° - 112.6° - 34.4° = 33°,\ c = 25.5$

Now, using the Law of Sines, $\dfrac{b}{\sin 34.4°} = \dfrac{25.5}{\sin 33°} = \dfrac{25.5}{\sin 33°} \Rightarrow b = \dfrac{25.5 \sin 34.4°}{\sin 33°} \Rightarrow b \approx 26.5$ km.

69. Let R be the position of the rocket. Triangle T_1RT_2 has the following measurements:

$T_1 = 28.1°,\ T_2 = 180° - 79.5° = 100.5°,\ R = 180° - 28.1° - 100.5° = 51.4°$, and $r = 1.73$.

Now, using the Law of Sines, $\dfrac{t_2}{\sin 100.5°} = \dfrac{1.73}{\sin 51.4°} \Rightarrow t_2 = \dfrac{1.73 \sin 100.5°}{\sin 51.4°} \Rightarrow t_2 \approx 2.18$ km.

71. Use the Law of Sines to find angle W.

$$\frac{\sin W}{11.2} = \frac{\sin 25.5°}{28.6} \Rightarrow \sin W = \frac{11.2 \sin 25.5°}{28.6} \Rightarrow W = \sin^{-1}\left(\frac{11.2 \sin 25.5°}{28.6}\right) \approx 9.7°.$$

Thus, $P = 180° - 25.5° - 9.7° = 144.8°$. Now use the Law of Sines to find the required distance.

$$\frac{p}{\sin 144.8°} = \frac{11.2}{\sin 9.7°} \Rightarrow p = \frac{11.2 \sin 144.8°}{\sin 9.7°} \Rightarrow p \approx 38.3 \text{ cm}.$$

73. $\dfrac{\sin C}{2\sqrt{5}} = \dfrac{\sin 30°}{\sqrt{5}} \Rightarrow \sin C = \dfrac{2\sqrt{5} \sin 30°}{\sqrt{5}} = 1$. Thus $C = 90°$. This is a right triangle.

75. The longest side must be opposite the largest angle. Thus, *B* must be larger than *A*, which is impossible because *A* and *B* cannot both be obtuse.

77. Let *B*, *M*, and *D* be the positions of Bochum, the moon, and Donaueschingen, respectively. Triangle *BMD* has the following measurements:

$B = 52.6997°,\ D = 180° - 52.7430° = 127.2570°,$
$M = 180° - 52.6997° - 127.2570° = 0.433,\ m = 398$

Now, using the Law of Sines, $\dfrac{d}{\sin 127.2570°} = \dfrac{398}{\sin 0433°} \Rightarrow a = \dfrac{398 \sin 127.2570°}{\sin .0433°} \Rightarrow a \approx 419{,}000$ km.

This compares favorably to the actual value.

79. (a) Note: this solution utilizes the area formula introduced in section 10.2. First note that

$\dfrac{R}{\sin C} = \dfrac{r}{\sin A} \Rightarrow r = \dfrac{R \sin A}{\sin C} \Rightarrow r = \dfrac{R \sin A}{\sin(A+B)}$. Substituting in the area formula gives

$$A = 10\left(\frac{1}{2} Rr \sin B\right) = 5R\left(\frac{R \sin A}{\sin(A+B)}\right)\sin B = \left[5 \frac{\sin A \sin B}{\sin(A+B)} R^2\right].$$

(b) $A = \left[5 \dfrac{\sin 18° \sin 36°}{\sin(18° + 36°)} R^2\right] \approx 1.12257R^2$.

(c) (i) The stripe has height $10 \div 13 \approx .76923$ in. and area of $11.4(.76923) \approx 8.77\text{in}^2$.

(ii) The 50 stars have area $A = 50(1.12257)(.308)^2 \approx 5.32\text{in}^2$.

(iii) Red occupies the greatest area on the flag.

80. On the given interval, the graph is increasing.

81. If $B < A$, and $\sin B < \sin A$ because $y \sin x$ is increasing on $\left[0, \frac{\pi}{2}\right]$.

82. $\frac{b}{\sin B} = \frac{a}{\sin A} \Rightarrow b = \frac{a \sin B}{\sin A}$.

83. $b = \frac{a \sin B}{\sin A} = a \cdot \frac{\sin B}{\sin A} \cdot \text{Since} \frac{\sin B}{\sin A} < 1, b = a \cdot \frac{\sin B}{\sin A} < a \cdot 1 = a,$ so $b < a$.

84. If $B < A$, then $b < a$,but $b > a$ in triangle ABC

10.2 The Law of Cosines and Area Formulas

1. (a) Two sides and the included angle are given. This is form SAS.
   (b) SAS should be solved using the Law of Cosines.
3. (a) Two sides and a non-included angle are given. This is form SSA.
   (b) SSA should be solved using the Law of Sines.
5. (a) Two angles and the included side are given. This is form ASA.
   (b) ASA should be solved using the Law of Sines.
7. (a) Two angles and the included side are given. This is form ASA.
   (b) ASA should be solved using the Law of Sines

9. $a^2 = 3^2 + 8^2 - 2(3)(8)\cos 60° \Rightarrow a^2 = 73 - 48(.5) \Rightarrow a^2 = 73 - 24 \Rightarrow a^2 = 49 \Rightarrow a = 7.$

11 $1^2 = 1^2 + \left(\sqrt{3}\right)^2 - 2(1)\left(\sqrt{3}\right)\cos\theta \Rightarrow 1 = 4 - 2\sqrt{3}\cos\theta \Rightarrow \cos\theta = \frac{-3}{-2\sqrt{3}} \Rightarrow \cos\theta = \frac{\sqrt{3}}{2} \Rightarrow \theta = 30°$

13. Use the Law of Cosines to find a: $a^2 = 4^2 + 6^2 - 2(4)(6)\cos 61° \Rightarrow a = \sqrt{52 - 48\cos 61°} \approx 5.36.$

Use the Law of Sines to find B: $\frac{\sin B}{4} = \frac{\sin 61°}{5.36} \Rightarrow \sin B = \frac{4\sin 61°}{5{,}36} \Rightarrow B = \sin^{-1}\left(\frac{4\sin 61°}{5.36}\right) \approx 40.7°.$

Then $C = 180° - 61° - 40.7° = 78.3°$. The solutions are $a \approx 5.4,\ B \approx 40.7°,\ C \approx 78.3°.$

15. Use the Law of Cosines to find A: $4^2 = 8^2 + 10^2 - 2(8)(10)\cos A \Rightarrow A = \cos^{-1}\left(\frac{16-164}{-160}\right) \approx 22.33°.$ Use the Law of Sines to find B: $\frac{\sin B}{10} = \frac{\sin 22.33°}{4} \Rightarrow \sin B = \frac{10\sin 22.33°}{4} \Rightarrow B = \sin^{-1}\left(\frac{10\sin 22.33°}{4}\right) \approx 71.8°$

Since angle B is obtuse, $B = 180° - 71.8° = 108.2°.$ The value of C is $180° - 22.3° - 108.2° = 49.5°$.

The solutions are $A \approx 22.3°, B \approx 108.2°,\ C \approx 49.5°.$

17. Use the Law of Cosines to find A: $5^2 = 7^2 + 9^2 - 2(7)(9)\cos A \Rightarrow A = \cos^{-1}\left(\frac{25-130}{-126}\right) \approx 33.55°.$ Use the Law of Sines to find B: $\frac{\sin B}{7} = \frac{\sin 33.55°}{5} \Rightarrow \sin B = \frac{7\sin 33.55°}{5} \Rightarrow B = \sin^{-1}\left(\frac{7\sin 33.55°}{5}\right) \approx 50.7°$. The value of C is $180° - 33.6° - 50.7° = 95.7°$ The solutions are $A \approx 33.6°,\ B \approx 50.7°,\ C \approx 95.7°.$

19. Use the Law of Cosines to find c:

$c^2 = 5.71^2 + 4.21^2 - 2(5.71)(4.21)\cos 28.3° \Rightarrow c = \sqrt{50.3282 - 48.0782\cos 28.3°} \approx 2.83$ Use the Law of Sines to find A: $\frac{\sin A}{4.21} = \frac{\sin 28.3°}{28.3} \Rightarrow \sin A = \frac{4.21\sin 28.3°}{28.3} \Rightarrow A = \sin^{-1}\left(\frac{4.21\sin 28.3°}{28.3}\right) \approx 44.9°.$

Then $B = 180° - 28.3° - 44.9° = 106.8°$. The solutions are $c \approx 2.83$ in, $A \approx 44.9°,\ B \approx 106.8°$.

21. Use the Law of Cosines to find c:

$$c^2 = 8.94^2 + 7.23^2 - 2(8.94)(7.23)\cos 45.6° \Rightarrow c = \sqrt{132.1965 - 129.274\cos 45.6°} \approx 6.46$$

Use the Law of Sines to find A:

$$\frac{\sin A}{7.23} = \frac{\sin 45.6°}{6.46} \Rightarrow \sin A = \frac{7.23\sin 45.6°}{6.46} \Rightarrow A = \sin^{-1}\left(\frac{7.23\sin 45.6°}{6.46}\right) \approx 53.1°.$$

Then $B = 180° - 45.6° - 53.1° = 81.3°$. The solutions are $c \approx 6.46$ m, $A \approx 53.1°,\ B \approx 81.3°$.

23. Use the Law of Cosines to find A:

$$9.3^2 = 5.7^2 + 8.2^2 - 2(5.7)(8.2)\cos A \Rightarrow A = cos^{-1}\left(\frac{86.49 - 99.73}{-93.48}\right) \approx 81.86°.$$

Use the Law of Sines to find B:

$$\frac{\sin B}{5.7} = \frac{\sin 81.86°}{9.3} \Rightarrow \sin B = \frac{5.7\sin 81.86°}{9.3} \Rightarrow B = \sin^{-1}\left(\frac{5.7\sin 81.86°}{9.3}\right) \approx 37.35°.$$

The value of C is $180° - 82° - 37° = 61°$. The solutions are $A \approx 82°,\ B \approx 37°,\ C \approx 61°$.

25. Use the Law of Cosines to find A:

$$42.9^2 = 37.6^2 + 62.7^2 - 2(37.6)(62.7)\cos A \Rightarrow A = \cos^{-1}\left(\frac{1840.41 - 5345.05}{-4715.04}\right) \approx 41.99°.$$

Use the Law of Sines to find B:

$$\frac{\sin B}{37.6} = \frac{\sin 41.99°}{42.9} \Rightarrow \sin B = \frac{37.6\sin 41.99°}{42.9} \Rightarrow B = \sin^{-1}\left(\frac{37.6\sin 41.99°}{42.9}\right) \approx 35.90°$$

$C \approx 180° - 42°00' - 35°50' = 102°10'$ The solutions are $A \approx 42°00', B \approx 35°50' C \approx 102°10'$.

27. Use the Law of Cosines to find A:

$$965^2 = 1240^2 + 876^2 - 2(1240)(876)\cos A \Rightarrow A = \cos^{-1}\left(\frac{931,225 - 2,304,976}{-2,172,480}\right) \approx 50.78°.$$

Use the Law of Sines to find B:

$$\frac{\sin B}{876} = \frac{\sin 50.78°}{965} \Rightarrow \sin B = \frac{876\sin 50.78°}{965} \Rightarrow B = \sin^{-1}\left(\frac{876\sin 50.78°}{965}\right) \approx 44.69°$$

$C \approx 180° - 50°50' - 44°40' = 80°30'$. The solutions are $A \approx 50°50',\ B \approx 44°40'\ C \approx 84°30'$.

29. Use the Law of Cosines to find a:

$$a^2 = 143^2 + 89.6^2 - 2(143)(89.6)\cos 80°40 \Rightarrow a = \sqrt{28,477.16 - 25,625\cos 80°40'} \approx 155.95.$$

Use the Law of Sines to find B:

$$\frac{\sin B}{143} = \frac{\sin 80°40'}{155.95} \Rightarrow \sin B = \frac{143 \sin 80°40'}{155.95} \Rightarrow B = \sin^{-1}\left(\frac{143 \sin 80°40'}{155.95}\right) \approx 64°50'$$

Then $C = 180° - 80°40' - 64°50' = 34°30'$ The solutions are $a \approx 156\,\text{cm}$, $B \approx 64°50'$, $C \approx 34°30'$.

31. Use the Law of Cosines to find b:

$$b^2 = 8.919^2 + 6.427^2 - 2(8919)(6.427)\cos 74.80° \Rightarrow b = \sqrt{120.85489 - 114.644826 \cos 74.80°} \approx 9.529.$$

Use the Law of Sines to find A:

$$\frac{\sin A}{8919} = \frac{\sin 74.80°}{9.529} \Rightarrow \sin A = \frac{8.919 \sin 74.80°}{9.529} \Rightarrow A = \sin^{-1}\left(\frac{8919 \sin 74.80°}{9.529}\right) \approx 64.59°.$$

Then $C = 180° - 74.80° - 64.59° = 40.61°$. The solutions are $b \approx 9.529$ in, $A \approx 64.59°$, $C \approx 40.61°$.

33. Use the Law of Cosines to find a:

$$a^2 = 6.28^2 + 12.2^2 - 2(6.28)(12.2)\cos 112.8° \Rightarrow a = \sqrt{188.2784 - 153.232 \cos 112.8°} \approx 15.7.$$

Use the Law of Sines to find B:

$$\frac{\sin B}{6.28} = \frac{\sin 112.8°}{15.7} \Rightarrow \sin B = \frac{6.25 \sin 112.8°}{15.7} \Rightarrow B = \sin^{-1}\left(\frac{6.28 \sin 112.8°}{15.7}\right) \approx 21.6°.$$

Then $C = 180° - 112.8° - 21.6° = 45.6°$. The solutions are $a \approx 15.7\,\text{m}$, $B \approx 21.6°$, $C \approx 45.6°$.

35. Use the Law of Cosines to find A: $3.0^2 = 5.0^2 + 6.0^2 - 2(5.0)(6.0)\cos A \Rightarrow A = \cos^{-1}\left(\frac{9-61}{-60}\right) \approx 30°.$

Use the Law of Sines to find B: $\frac{\sin B}{5.0} = \frac{\sin 30°}{3.0} \Rightarrow \sin B = \frac{5.0 \sin 30°}{3.0} \Rightarrow B = \sin^{-1}\left(\frac{5.0 \sin 30°}{3.0}\right) \approx 56°.$

The value of C is $180° - 30° - 56° = 94°$. The solutions are $A \approx 30°$, $B \approx 56°$, $C \approx 94°$.

37. Case 1. First use the triangle sum formula to find angle A: $179°60' - 50°52' - 28°37' = 101°30'$

Use the Law of Sines: $\frac{59.49}{\sin 100°31'} = \frac{b}{\sin 50°52'} \Rightarrow \sin 100°31' b = 59.49 \cdot \sin 50°52' \Rightarrow b = 46.93$

$$\frac{59.49}{\sin 100°31'} = \frac{c}{\sin 28°31'} \Rightarrow \sin 100°31' c = 59.49 \cdot \sin 28°31' \Rightarrow c = 28.98$$

39. Case 3. First use the Law of Cosines to find b:

$$b^2 = (3.961)^2 + (5.308)^2 - 2(3.961)(5.308)\cos 58°12' = 21.7063 \Rightarrow b = 4.659$$

Use the Law of Sines tofind A: $\frac{3.961}{\sin A} = \frac{4.659}{\sin 58°12'} \Rightarrow 4.659 \sin A = 3.961 \cdot \sin 58°12' \Rightarrow A = 46°16'$

Finally, use the triangle sum formula to find angle C: $179°60' - 46°16' - 58°12' = 75°32'$

41. Case 4. First use the Law of Cosines to find angle C.

$$(65.88)^2 = (51.41)^2 + (37.29)^2 - 2(51.41)(37.29)\cos C \Rightarrow C = 94°35'$$

Use the Law of Sines to find Angle A.

$$\frac{51.41}{\sin A} = \frac{65.88}{\sin 94^\circ 35'} \Rightarrow 65.88 \sin A = 51.41 \cdot \sin 94^\circ 35' \Rightarrow A = 51^\circ 04'$$

Finally, use the triangle sum formula to find angle B: $179^\circ 60' - 51^\circ 4' - 94^\circ 35' = 34^\circ 21'$

43. Case 2. Use the Law of Sines to find angle B:

$$\frac{7.031}{\sin 41^\circ 12'} = \frac{9.947}{\sin B} \Rightarrow 7.031 \sin B = 9.947 \cdot \sin 41^\circ 12 \Rightarrow B = 68^\circ 43'$$

There are two triangles $B_1 = 68^\circ 43'$ *and* $B_2 = 180 - 68^\circ 43' = 111^\circ 17'$. Since

$41^\circ 12' + 111^\circ 17' < 180^\circ, C_1 = 180 - 41^\circ 12' - 68^\circ 43' = 70^\circ 05'$ and $C_2 = 180 - 41^\circ 12' - 111^\circ 17' = 27^\circ 31'$

Use the Law of Sines to find c_1 *and* c_2:

$$\frac{c_1}{\sin 70^\circ 05'} = \frac{7.031}{\sin 41^\circ 12'} \Rightarrow c_1 \sin 41^\circ 12' = 7.031 \cdot \sin 70^\circ 05' \Rightarrow c_1 = 10.04$$

$$\frac{c_2}{\sin 27^\circ 31'} = \frac{7.031}{\sin 41^\circ 12'} \Rightarrow c_2 \sin 41^\circ 12' = 7.031 \cdot \sin 27^\circ 31 \Rightarrow c_2 = 4.933$$

45. Case 2. Use the Law of Sines to find angle A:

$$\frac{34.22}{\sin 14^\circ 19'} = \frac{27.16}{\sin A} \Rightarrow 34.22 \sin A = 27.16 \cdot \sin 14^\circ 19' \Rightarrow A = 11^\circ 19'$$

There is one triangle since $180^\circ - 11^\circ 19' + 14^\circ 19' > 180^\circ$. Angle B $= 180^\circ - 11^\circ 19' - 14^\circ 19' = 154^\circ 22'$

Use the Law of Sines to find b:

$$\frac{34.22}{\sin 14^\circ 19'} = \frac{b}{\sin 154^\circ 22'} \Rightarrow b \sin 14^\circ 19' = 34.22 \cdot \sin 154^\circ 22' \Rightarrow b = 59.87$$

47. Case 2. Use the Law of Sines to find angle A:

$\frac{2634}{\sin A} = \frac{2220}{\sin 18^\circ 37'} \Rightarrow 2220 \sin A = 2634 \cdot \sin 18^\circ 37' \Rightarrow \sin A > 1.$. There is no triangle that exists.

49. The absolute value of $\cos\theta$ will be greater than 1. The calculator will give an error message (or a complex number) when using the inverse cosine function.

51. Use the Law of Cosines to find c:

$$c^2 = 350^2 + 286^2 - 2(350)(286)\cos 46.3^\circ \Rightarrow c = \sqrt{204{,}296 - 200{,}200 \cos 46.3^\circ} \approx 257 \text{ m}$$

53. Let the vertical line passing through airport C intersect the line from A to B at the point D. Now, using the right triangle CDB we can find the measure of angle B. The angle of this right triangle at position C is given by $180^\circ - 128^\circ 40' = 51^\circ 20'$. Thus $B = 180^\circ - 90^\circ - 51^\circ 20' = 38^\circ 40'$. Use the Law of Cosines on ABC to find b: $b^2 = 450^2 + 359^2 - 2(450)(359)\cos 38^\circ 40' \Rightarrow d = \sqrt{331{,}323{,}100 \cos 38^\circ 40'} \approx 281$ km

55. Let A, B, and C be the ship's initial position, new position and position of the rock, respectively. Triangle ABC has the following measurements:

$A = 90° - 45°20' = 44°40'$, $B = 308°40' - 270° = 38°40'$, $C = 180° - 44°40' - 38°40' = 96°40'°$, and

$c = 15.2$. Now, using the Law of Sines, $\frac{a}{\sin 44°40} = \frac{15.2}{\sin 96°40'} \Rightarrow a = \frac{15.2 \sin 44°40'}{\sin 96°40'} \Rightarrow a \approx 10.8$ mi.

57. Let *A*, *B*, and *C* be the positions of the airplane, battleship and submarine, respectively. Triangle *ABC* has the following measurements: $A = 24°10' - 17°30' = 6°40'$, $B = 17°30'$, $C = 180° - 6°40' - 70°30' = 155°50'°$,

and $c = 5120$. Now, using the Law of Sines, $\frac{a}{\sin 6°40'} = \frac{5120}{\sin 155°50'} \Rightarrow a = \frac{5120 \sin 6°40'}{\sin 155°50'} \Rightarrow a \approx 1450$ ft

59. Use the Law of Cosines to find the distance *d*:

$d^2 = 10^2 + 10^2 - 2(10)(10)\cos 128° \Rightarrow d = \sqrt{200 - 200\cos 128°} \approx 18$ ft.

61. Use the Law of Cosines to find the distance *c*: $c^2 = 3800^2 + 2900^2 - 2(3800)(2900)\cos 110° \Rightarrow$

$c = \sqrt{22,850,000 - 22,040,000\cos 110°} \approx 5500$ m

63. The distances traveled by the sound are $3(344) = 1032$ m and $6(344) = 2064$ m. Use the Law of Cosines to find the distance *d*:

$d^2 = 1032^2 + 2064^2 - 2(1032)(2064)\cos 42.2° \Rightarrow d = \sqrt{5,325,120 - 4,260,096\cos 42.2°} \approx 1473$ m

65. Use the Law of Cosines to find the angle of the triangle located at position *A*:

$9^2 = 17^2 + 21^2 - 2(17)(21)\cos A \Rightarrow A = \cos^{-1}\left(\frac{81 - 730}{-714}\right) \approx 25°$. The bearing is $325° + 25° = 350°$.

67. If the satellite travels one orbit in 2 hours, it is traveling at a rate of $360 \div 120 = 3°$ per minute. At 12:03 P.M. the angle formed at the center of Earth is 9°. Noting that the sides of the triangle shown are 6400 km and $6400 + 1600 = 8000$ km, use the Law of Cosines to find the distance *d*:

$d^2 = 8000^2 + 6400^2 - 2(8000)(6400)\cos 9° \Rightarrow d = \sqrt{104,960,000 - 102,400,000\cos 9°} \approx 2000$ km

69. Use the Law of Cosines to find angle θ

$57.8^2 = 25.9^2 + 32.5^2 - 2(25.9)(32.5)\cos\theta \Rightarrow \theta = \cos^{-1}\left(\frac{3340.84 - 1727.06}{-1683.5}\right) \approx 163.5°$.

71. Use the Law of Cosines to find *x*: $x^2 = 25^2 + 25^2 - 2(25)(25)\cos 52° \Rightarrow x = \sqrt{1250 - 1250\cos 52°} \approx 22$ ft

73. Since *A* is obtuse, $90° < A < 180°$. The cosine of a quadrant II angle is negative.

74. In $a^2 = b^2 + c^2 - 2ab\cos A$, $\cos A$ is negative, so $a^2 = b^2 + c^2 +$ (a positive quantity). Thus, $a^2 > b^2 + c^2$.

75. $b^2 + c^2 > b^2$ and $b^2 + c^2 > c^2$. If $a^2 > b^2 + c^2$, then $a^2 > b^2$ and $a^2 > c^2$ from which $a > b$ and $a > c$ because *a*, *b*, and *c* are nonnegative.

76. Because *A* is obtuse, it is the largest angle, so the longest side should be *a*, not *c*.

77. $A = \frac{1}{2}(13.6)(10.1)\sin 42.5° \approx 46.4$ m^2

79. $s = \frac{1}{2}(12 + 16 + 25) = 26.5; A = \sqrt{26.5(26.5 - 12)(26.5 - 16)(26.5 - 25)} \approx 78$ m^2

81. $s=\frac{1}{2}(76.3+109+98.8)=142.05;\ \ A=\sqrt{142.05(142.05-76.3)(142.05-109)(142.05-98.8)}\approx 3650\text{ ft}^2$

83. $s=\frac{1}{2}(25.4+38.2+19.8)=41.7; A=\sqrt{41.7(41.7-25.4)(41.7-38.2)(41.7-19.8)}\approx 228\text{ yd}^2$

85. $s=\frac{1}{2}(75+68+85)=114; A=\sqrt{114(114-75)(114-68)(114-85)}\approx 2435\,\text{m}^2$ The number of cans needed is $2435\div 75\approx 32.5$ or 33 cans.

87. $A=\frac{1}{2}(15.2)(16.1)\sin 125°\approx 100\text{ m}^2$

89. $s=\frac{1}{2}(9+10+17)=18; A=\sqrt{18(18-9)(18-10)(18-17)}=36\text{ m}^2$ and $9+10+17=36$

91. (a) Using the law of sines, we have $\frac{\sin C}{c}=\frac{\sin A}{a}\Rightarrow\frac{\sin C}{15}=\frac{\sin 60°}{13}\Rightarrow\sin C=\frac{15\sin 60°}{13}=$

$\frac{15}{13}\cdot\frac{\sqrt{3}}{2}\approx 0.99926008$. There are two angles C between 0° and 180° that satisfy the condition. Since $\sin C\approx 0.99926008$, to the nearest tenth value of C is $C_1=87.8°$. Supplementary angles have the same sine value, so another possible value of C is $B_2=180°-87.8°=92.2°$.

(b) By the law of cosines, we have $\cos C=\frac{a^2+b^2-c^2}{2ab}\Rightarrow\cos C=\frac{13^2+7^2-15^2}{2(13)(7)}=$

$\frac{-7}{182}\approx -0.03846154\Rightarrow C\approx 92.2°$

(c) With the law of cosines, we are required to find the inverse cosine of a negative number; therefore; we know angle C is greater than 90°.

10.3 Vectors and their Applications

1. The vector pairs that have the same direction and length are **m** & **p** and **n** & **r**.

3. m = 2t, p = 2t, or $t=\frac{1}{2}$m and $t=\frac{1}{2}$p; also m = 1p and n = 1r or p = 1m and r = 1n

5. See Figure 5.

7. See Figure 7.

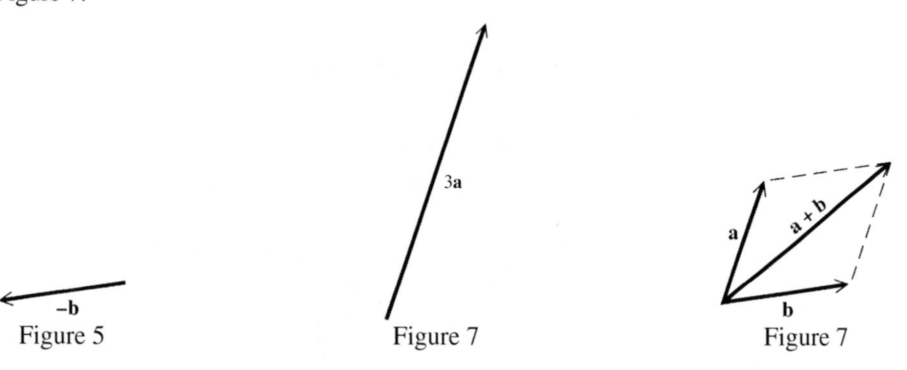

Figure 5 Figure 7 Figure 7

9. See Figure 9.

11. See Figure 11.

13. See Figure 13.

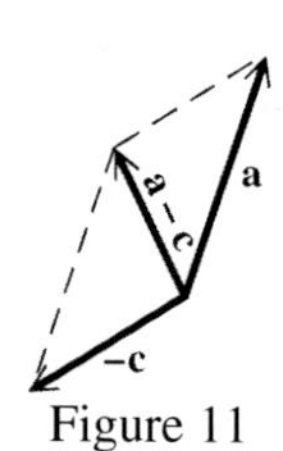

Figure 11

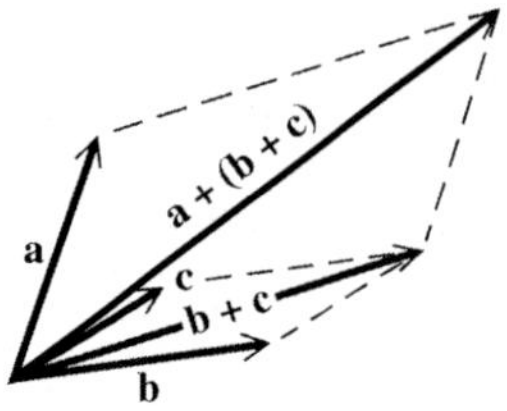

Figure 13

15. See Figure 15.

17. (a) $u + v = \langle -8,8\rangle + \langle 4,8\rangle = \langle -8+4, 8+8\rangle = \langle -4,16\rangle$

(b) $u - v = \langle -8,8\rangle - \langle 4,8\rangle = \langle -8-4, 8-8\rangle = \langle -12,0\rangle$

(c) $-u = -\langle -8,8\rangle = \langle 8,-8\rangle$

19. (a) $u + v = \langle 4,8\rangle + \langle 4,-8\rangle = \langle 4+4, 8+(-8)\rangle = \langle 8,0\rangle$

(b) $u - v = \langle 4,8\rangle - \langle 4,-8\rangle = \langle 4-4, 8-(-8)\rangle = \langle 0,16\rangle$

(c) $-u = -\langle 4,8\rangle = \langle -4,-8\rangle$

21. (a) $u + v = \langle -8,4\rangle + \langle 8,8\rangle = \langle -8+8, 4+8\rangle = \langle 0,12\rangle$

(b) $u - v = \langle -8,4\rangle - \langle 8,8\rangle = \langle -8-8, 4-8\rangle = \langle -16,-4\rangle$

(c) $-u = -\langle -8,4\rangle = \langle 8,-4\rangle$

23. (a) $2u = 2(2i) = 4i$

(b) $2u + 3v = 2(2i) + 3(i+j) = 4i + 3i + 3j = 7i + 3j$

(c) $v - 3u = i + j - 3(2i) = i + j - 6i = -5i + j$

25. (a) $2u = 2\langle -1,2\rangle = \langle -2,4\rangle$

(b) $2u + 3v = 2\langle -1,2\rangle + 3\langle 3,0\rangle = \langle -2,4\rangle + \langle 9,0\rangle = \langle -2,+9, 4+0\rangle = \langle 7,4\rangle$

(c) $v - 3u = \langle 3,0\rangle - 3\langle -1,2\rangle = \langle 3,0\rangle + \langle 3,-6\rangle = \langle 3+3, 0+(-6)\rangle = \langle 6,-6\rangle$

27. $u + v = \langle -2,5\rangle + \langle 4,3\rangle = \langle -2+4, 5+3\rangle = \langle 2,8\rangle$

29. $v - u = \langle 4,3\rangle - \langle -2,5\rangle = \langle 4-(-2), 3-5\rangle = \langle 6,-2\rangle$

31. $-5v = -5\langle 4,3\rangle = \langle -20,-15\rangle$

33. $v = \langle 6\cos 30^\circ, 6\sin 30^\circ\rangle = \left\langle 6\cdot\frac{\sqrt{3}}{2}, 6\cdot\frac{1}{2}\right\rangle = \langle 3\sqrt{3}, 3\rangle$

35. $v = \langle 9\cos 225^\circ, 9\sin 225^\circ \rangle = \left\langle 9\cdot\left(-\frac{\sqrt{2}}{2}\right), 9\cdot\left(-\frac{\sqrt{2}}{2}\right)\right\rangle = \left\langle -\frac{9\sqrt{2}}{2}, -\frac{9\sqrt{2}}{2}\right\rangle$

37. $v = \langle 4\cos 40^\circ, 4\sin 40^\circ \rangle \approx \langle 3.06, 2.57\rangle$

39. $v = \langle 5\cos(-35^\circ), 5\sin(-35^\circ)\rangle \approx \langle 4.10, -2.87\rangle$

41. The adjacent angle in the parallelogram is supplementary to 40°, thus it has measure 140°.

$|v|^2 = 40^2 + 60^2 - 2(40)(60)\cos 140° \approx 8877.0133 \Rightarrow |v| = \sqrt{8877.0133} \approx 94.2$ lb

43. The adjacent angle in the parallelogram is supplementary to 110°, thus it has measure 70°.

$|v|^2 = 15^2 + 25^2 - 2(15)(25)\cos 70° \approx 593.4849 \Rightarrow |v| = \sqrt{593.4849} \approx 24.4$ lb

45. $\langle -5, 8\rangle = -5\mathbf{i} + 8\mathbf{j}$

47. $\langle 2, 0\rangle = 2\mathbf{i} + 0\mathbf{j} = 2\mathbf{i}$

49. $(8\cos 45°)\mathbf{i} + (8\sin 45°)\mathbf{j} = \left(8\cdot\frac{\sqrt{2}}{2}\right)\mathbf{i} + \left(8\cdot\frac{\sqrt{2}}{2}\right)\mathbf{j} = 4\sqrt{2}\mathbf{i} + 4\sqrt{2}\mathbf{j}$

51. $(.6\cos 115°)\mathbf{i} + (.6\sin 115°)\mathbf{j} = -.25\mathbf{i} + .54\mathbf{j}$

53. $|\langle 1,1\rangle| = \sqrt{1^2+1^2} = \sqrt{1+1} = \sqrt{2}$; direction angle: $\tan^{-1}(1) = 45°$

55. $|\langle 8\sqrt{2}, -8\sqrt{2}\rangle| = \sqrt{(8\sqrt{2})^2 + (-8\sqrt{2})^2} = \sqrt{128+128} = \sqrt{256} = 16$; direction angle: $\tan^{-1}\left(\frac{-8\sqrt{2}}{8\sqrt{2}}\right) = 315°$

57. $|\langle 15, -8\rangle| = \sqrt{15^2 + (-8)^2} = \sqrt{225+64} = \sqrt{289} = 17$; direction angle: $\tan^{-1}\left(\frac{-8}{15}\right) = 331.9°$

59. $|\langle -6, 0\rangle| = \sqrt{(-6)^2 + 0^2} = \sqrt{36+0} = \sqrt{36} = 6$; direction angle: $\tan^{-1}\left(\frac{0}{-6}\right) = 180°$

61. $\langle 6,-1\rangle\cdot\langle 2,5\rangle = (6)(2) + (-1)(5) = 12 - 5 = 7$

63. $\langle 2,-3\rangle\cdot\langle 6,5\rangle = (2)(6) + (-3)(5) = 12 - 15 = -3$

65. $4\mathbf{i}\cdot(5\mathbf{i}\text{-}9\mathbf{j}) = (4)(5) + (0)(-9) = 20 + 0 = 20$

67. $\cos\theta = \frac{\mathbf{u}\cdot\mathbf{v}}{|\mathbf{u}||\mathbf{v}|} = \frac{\langle 2,1\rangle\cdot\langle -3,1\rangle}{|\langle 2,1\rangle||\langle -3,1\rangle|} = \frac{(2)(-3)+(1)(1)}{\sqrt{4+1}\sqrt{9+1}} = \frac{-5}{\sqrt{5}\sqrt{10}} = \frac{-5}{\sqrt{50}} = \frac{-5}{5\sqrt{2}} = -\frac{1}{\sqrt{2}} \Rightarrow \theta = \cos^{-1}\left(-\frac{1}{\sqrt{2}}\right) = 135°$

69. $\cos\theta = \frac{\mathbf{u}\cdot\mathbf{v}}{|\mathbf{u}||\mathbf{v}|} = \frac{\langle 1,2\rangle\cdot\langle -6,3\rangle}{|\langle 1,2\rangle||\langle -6,3\rangle|} = \frac{(1)(-6)+(2)(3)}{\sqrt{1+4}\sqrt{36+9}} = \frac{0}{\sqrt{5}\sqrt{45}} = 0 \Rightarrow \theta = \cos^{-1}0 = 90°$

71. $\cos\theta = \frac{\mathbf{u}\cdot\mathbf{v}}{|\mathbf{u}||\mathbf{v}|} = \frac{(\mathbf{i}+7\mathbf{j})\cdot(\mathbf{i}+\mathbf{j})}{|\mathbf{i}+7\mathbf{j}||\mathbf{i}+\mathbf{j}|} = \frac{(1)(1)+(7)(1)}{\sqrt{1+49}\sqrt{1+1}} = \frac{8}{\sqrt{50}\sqrt{2}} = \frac{8}{\sqrt{100}} = \frac{8}{10} = \frac{4}{5} \Rightarrow \theta = \cos^{-1}\left(\frac{4}{5}\right) \approx 36.87°$

73. $\cos\theta = \frac{\mathbf{u}\cdot\mathbf{v}}{|\mathbf{u}||\mathbf{v}|} = \frac{(\mathbf{i}+\mathbf{j})\cdot(3\mathbf{i}+4\mathbf{j})}{|\mathbf{i}+\mathbf{j}||3\mathbf{i}+4\mathbf{j}|} = \frac{(1)(3)+(1)(4)}{\sqrt{1+1}\sqrt{9+16}} = \frac{7}{5\sqrt{2}} = \frac{7\sqrt{2}}{10} \Rightarrow \theta = \cos^{-1}\left(\frac{7\sqrt{2}}{10}\right) \approx 8.13°$

75. $(3\mathbf{u})\cdot\mathbf{v}=3\langle -2,1\rangle\cdot\langle 3,4\rangle=\langle -6,3\rangle\cdot\langle 3,4\rangle=(-6)(3)+(3)(4)=-18+12=-6$

77. $\mathbf{u}\cdot\mathbf{v}\text{-}\mathbf{u}\cdot\mathbf{w}=\langle -2,1\rangle\cdot\langle 3,4\rangle-\langle -2,1\rangle\cdot\langle -5,12\rangle=[(-2)(3)+(1)(4)][(-2)(-5)+(1)(12)]=-24$

79. $\langle 1,2\rangle\cdot\langle -6,3\rangle=(1)(-6)+(2)(3)=-6+6=0\Rightarrow$ orthogonal

81. $\langle 1,0\rangle\cdot\langle\sqrt{2},0\rangle=(1)(\sqrt{2})+(0)(0)=\sqrt{2}+0=\sqrt{2}\Rightarrow$ not orthogonal

83. $(\sqrt{5}\mathbf{i}-2\mathbf{j})\cdot(-5\mathbf{i}+2\sqrt{5}\mathbf{j})=(\sqrt{5})(-5)+(-2)(2\sqrt{5})=-5\sqrt{5}-4\sqrt{5}=-9\sqrt{5}\Rightarrow$ not orthogonal

85. By completing the parallelogram with the resultant vector as the diagonal, the triangle formed by two sides of the parallelogram and the resultant has a side of measure 176 and angles with the following measures: $41°10'$, $78°50'-41°10'=37°40'$, and $180°-41°10'-37°40'=101°10'$. Using the Law of Sines,

$$\frac{F_2}{\sin 41°10'}=\frac{176}{\sin 37°40'}\Rightarrow F_2=\frac{176\sin 41°10'}{\sin 37°40'}\approx 190\text{ lb}$$

$$\frac{R}{\sin 101°10'}=\frac{176}{\sin 37°40'}\Rightarrow F_2=\frac{176\sin 101°10'}{\sin 37°40'}\approx 283\text{ lb}$$

87. $\theta=\sin^{-1}\left(\frac{25}{80}\right)\approx 18°$

89. $\sin 2.3°=\frac{F}{60}\Rightarrow F=60\sin 2.3°\approx 2.4\text{ tons}$

91. $\theta=\sin^{-1}\left(\frac{18}{60}\right)\approx 17.5°$

93. The weight of the crate is $\sin 46°20'=\frac{F}{89.6}\Rightarrow F=89.6\sin 46°20'\approx 64.8\text{ lb}$.

The tension on the other rope is $\cos 46°20'=\frac{F}{89.6}\Rightarrow F=89.6\cos 46°20'\approx 61.9\text{ lb}$.

95. Let *A*, *B*, and *C* be the ship's initial position, turning position, and final position, respectively. Triangle *ABC* has the following measurements: $B=90°+34.0°=124.0°$, $a=4.6$, and $c=10.4$.

Now, $b^2=10.4^2+4.6^2-2(10.4)(4.6)\cos 124.0°\Rightarrow b=\sqrt{129.32-95.68\cos 124.0°}\approx 13.5$ mi

$$\frac{\sin A}{4.6}=\frac{\sin 124.0°}{13.5}\Rightarrow \sin A=\frac{4.6\sin 124.0°}{13.5}\Rightarrow A=\sin^{-1}\left(\frac{4.6\sin 124.0°}{13.5}\right)\approx 16.4°.$$

The ship is 13.5 miles from its starting point on a bearing of $34.0°+16.4°=50.4°$.

97. Let *A*, *B*, and *C* be the ship's initial position, turning position, and final position, respectively. Triangle *ABC* has the following measurements:

$B=360°-[317°-(189°-180°)]=52°, a=47.8, \text{and } c=18.5$

Now, $b^2=47.8^2+18.5^2-2(47.8)(18.5)\cos 52°\Rightarrow b=\sqrt{2627.09-1768.6\cos 52°}\approx 39.2\text{ mi}$

99. The speed of the boat in still water, 20, is the hypotenuse of a right triangle with the shorter leg representing the speed of the current and the longer leg representing the actual speed of the boat. The smallest angle of the right triangle is $90° - 80° = 10°$. The actual speed of the boat is $20\cos 10° \approx 19.7$ mph. The speed of the current is $20\sin 10° \approx 3.5$ mph.

101. If the wind is from the direction of 114°, the bearing of the wind is $114° + 180° = 294°$. The angle between the wind and the flight of the jet is $(360° - 294°) + [180° - (360° - 223°)] = 119°$. If b is the missing side of the triangle, $b^2 = 450^2 + 39^2 - 2(450)(39)\cos 119° \Rightarrow b = \sqrt{204{,}021 - 35{,}100\cos 119°} \approx 470$ mph. Then $\frac{\sin A}{39} = \frac{\sin 119°}{470} \Rightarrow \sin A = \frac{39\sin 119°}{470} \Rightarrow A = \sin^{-1}\left(\frac{39\sin 119°}{470}\right) \approx 4.2°$. The ground speed of the jet is 470 mph, and the bearing is $233° + 4.2° = 237°$.

103. The triangle formed is a right triangle with one leg representing the wind speed of 42 mph, the hypotenuse representing airspeed, and the other leg representing the groundspeed. The angle between the airspeed and groundspeed is $90° - 74.9° = 15.1°$. Let g represent the hypotenuse and let a represent the unknown leg. The groundspeed is given by $\sin 15.1° = \frac{42}{g} \Rightarrow g = \frac{42}{\sin 15.1°} \approx 161$ mph. The airspeed is given by

$\tan 15.1° = \frac{42}{a} \Rightarrow g = \frac{42}{\tan 15.1°} \approx 156$ mph.

105. Let A, B, and C represent the angle between the flight path and the desired bearing, the angle between the wind vector and the original flight path, and the angle between the wind vector and the desired bearing. Triangle ABC has the following measurements: $C = 64°30'$, $a = 35$, and $c = 190$. Using the Law of Sines,

$$\frac{\sin A}{35} = \frac{\sin 91.3°}{190} \Rightarrow \sin A = \frac{35\sin 64°30'}{190} \Rightarrow A = \sin^{-1}\left(\frac{35\sin 64°30'}{190}\right) \approx 9°30'.$$

So $B = 180° - 9°30' - 64°30' = 106°$. Use the Law of Sines to find b:

$$\frac{b}{\sin 106°} = \frac{190}{\sin 64°30'} \Rightarrow b = \frac{190\sin 106°}{\sin 64°30'} \approx 202 \text{ mph}.$$ The bearing is $r = \frac{3}{1+\cos\theta} \Rightarrow 64\ 64°30' + 9°30' = 74°$

and the ground speed is 202 mph.

107. Let A, B, and C represent the angle between the required flight path and due north, the angle between the wind vector and due north, and the angle between the wind vector and the required flight path, respectively. Triangle ABC has the following measurements: $C = 360° - 328° = 32°, a = 11$, and $c = 400 \div 2.5 = 160$

Using the Law of Sines, $\frac{\sin A}{11} = \frac{\sin 32°}{160} \Rightarrow \sin A = \frac{11\sin 32°}{160} \Rightarrow A = \sin^{-1}\left(\frac{11\sin 32°}{160}\right) \approx 2°$

So $B = 180° - 2° - 32° = 146°$. Use the Law of Sines to find b:

$$\frac{b}{\sin 146°} = \frac{160}{\sin 32°} \Rightarrow b = \frac{160\sin 146°}{\sin 32°} \approx 170 \text{ mph}.$$

The bearing is $360° - 2° = 358°$ and the ground speed is 170 mph.

109. If the wind is from the direction of 245°, the bearing of the wind is $245° - 180° = 65°$. The angle between the wind and the flight of the plane is $65° + (180° + 174°) = 71°$. If b is the missing side of the triangle,

$b^2 = 240^2 + 30^2 - 2(240)(30)\cos 71° \Rightarrow b = \sqrt{58500 - 14400\cos 71°} \approx 230$ km per hr.

Then, $\dfrac{\sin A}{30} = \dfrac{\sin 71°}{230} \Rightarrow \sin A = \dfrac{30\sin 71°}{230} \Rightarrow A = \sin^{-1}\left(\dfrac{30\sin 71°}{230}\right) \approx 7°$.

The ground speed is 230 km per hr and the bearing is $174° - 7° = 167°$.

111. (a) $|\mathbf{R}| = \sqrt{1^2 + (-2)^2} = \sqrt{5} \approx 2.2; |\mathbf{A}| = \sqrt{.5^2 + 1^2} = \sqrt{1.25} \approx 1.1$ About 2.2 inches of rain fell. The area of the opening is about 1.1 in^2.

(b) $V = |\mathbf{R} \cdot \mathbf{A}| = |(1)(.5) + (-2)(1)| = |-1.5| = 1.5$; The volume of rain was 1.5 in^3.

(c) To collect the maximum amount of rain, **R** and **A** should be parallel and point in opposite directions.

Reviewing Basic Concepts (Sections 10.1—10.3)

1. Angle B has measure $180° - 44° - 62° = 74°$. Now, using the Law of Sines,

$\dfrac{c}{\sin 62°} = \dfrac{12}{\sin 44°} \Rightarrow \dfrac{12\sin 62°}{\sin 44°} \Rightarrow c \approx 15.3;\ \dfrac{b}{\sin 74°} = \dfrac{12}{\sin 44°} \Rightarrow b = \dfrac{12\sin 74°}{\sin 44°} \Rightarrow b \approx 16.6$. The solutions are $B = 74°,\ b \approx 16.6,\ c = 15.3$.

2. There are two solutions. Using the Law of Sines,

$\dfrac{\sin B}{8} = \dfrac{\sin 32°}{6} \Rightarrow \sin B = \dfrac{8\sin 32°}{6} \Rightarrow B = \sin^{-1}\left(\dfrac{8\sin 32°}{6}\right) \approx 45.0°$. There are two angles: $B_1 = 45.0°$ and $B_2 = 180° - 45.0° = 135.0°(\text{since } 32° + 135.0° < 180°)$. $c_1 = 180° - 32° - 45.0° = 103°$ and

$c_2 = 180° - 32° - 135° = 13°$ $\dfrac{c_1}{\sin 103°} = \dfrac{6}{\sin 32°} \Rightarrow c_1 = \dfrac{6\sin 103°}{\sin 32°} \approx 11.0$

$\dfrac{c_2}{\sin 13°} = \dfrac{6}{\sin 32°} \Rightarrow c_2 = \dfrac{6\sin 13°}{\sin 32°} \approx 2.5$

The solutions are $B_1 = 45.0°, c_1 = 103°, c_1 = 11.0$ or $B_2 = 135.0°, c_2 = 13°, c_2 = 2.5$.

3. Using the Law of Sines, $\dfrac{\sin A}{7} = \dfrac{\sin 41°}{12} \Rightarrow \sin A = \dfrac{7\sin 41°}{12} \Rightarrow A = \sin^{-1}\left(\dfrac{7\sin 41°}{12}\right) \approx 22.5°$. Since $180° - 22.5° = 157.5°$ and $41° + 157.5° > 180°$, there is one triangle.

$B = 180° - 41° - 22.5° = 116.5° \cdot \dfrac{b}{\sin 116.5°} = \dfrac{12}{\sin 41°} \Rightarrow b = \dfrac{12\sin 116.5°}{\sin 41°} \approx 16.4$. The solutions are $A = 22.5°, B = 116.5°, b = 16.4$.

4. (a) Use the Law of Cosines to find b:

$b^2 = 8.1^2 + 8.3^2 - 2(8.1)(8.3)\cos 51° \Rightarrow b = \sqrt{134.5 - 134.46\cos 51°} \approx 7.063$

Use the Law of Sines: $\dfrac{\sin A}{8.1}=\dfrac{\sin 51^\circ}{7.063}\Rightarrow \sin A=\dfrac{8.1\sin 51^\circ}{7.063}\Rightarrow A=\sin^{-1}\left(\dfrac{8.1\sin 51^\circ}{7.063}\right)\approx 63.0^\circ$

Then $C=180^\circ-51^\circ-63.0^\circ=66.0^\circ$. The solutions are $b\approx 7.1,\ A\approx 63.0^\circ,\ C\approx 66.0^\circ$.

(b) Use the Law of Cosines to find A: $14^2=8^2+9^2-2(8)(9)\cos A\Rightarrow A=\cos^{-1}\left(\dfrac{196-145}{-144}\right)\approx 110.7^\circ$

Use the Law of Sines to find B:

$\dfrac{\sin B}{9}=\dfrac{\sin 110.7^\circ}{14}\Rightarrow \sin B=\dfrac{9\sin 110.7^\circ}{14}\Rightarrow B=\sin^{-1}\left(\dfrac{9\sin 110.7^\circ}{14}\right)\approx 37.0^\circ$.

The value of C is $180^\circ-110.7^\circ-37.0^\circ=32.3^\circ$. The solutions are $A\approx 110.7^\circ,\ B\approx 37.0^\circ,\ C\approx 32.3^\circ$.

5. $A=\dfrac{1}{2}(4.5)(5.2)\sin 55^\circ\approx 9.6$

6. $s=\dfrac{1}{2}(6+7+9)=11; A=\sqrt{11(11-6)(11-7)(11-9)}\approx 21$

7. (a) $2\mathbf{v}+\mathbf{u}=2(2\mathbf{i}-\mathbf{j})+(-3\mathbf{i}+2\mathbf{j})=4\mathbf{i}-2\mathbf{j}+(-3\mathbf{i}+2\mathbf{j})=(4\mathbf{i}-3\mathbf{i})+(-2\mathbf{j}+2\mathbf{j})=\mathbf{i}$

(b) $2\mathbf{v}=2(2\mathbf{i}-\mathbf{j})=4\mathbf{i}-2\mathbf{j}$

(c) $\mathbf{v}-3\mathbf{u}=(2\mathbf{i}-\mathbf{j})-3(-3\mathbf{i}+2\mathbf{j})=2\mathbf{i}-\mathbf{j}+9\mathbf{i}-6\mathbf{j})=(2\mathbf{i}+9\mathbf{i})+(-\mathbf{j}-6\mathbf{j})=11\mathbf{i}-7\mathbf{j}$

8. $\langle 3,-2\rangle\cdot\langle -1,3\rangle=(3)(-1)+(-2)(3)=-3-6=-9$

$\cos\theta=\dfrac{\mathbf{u}\cdot\mathbf{v}}{|\mathbf{u}|\,|\mathbf{v}|}=\dfrac{-9}{\sqrt{9+4}\sqrt{1+9}}=\dfrac{-9}{\sqrt{13}\sqrt{10}}=\dfrac{-9}{\sqrt{130}}\Rightarrow\theta=\cos^{-1}\left(-\dfrac{9}{\sqrt{130}}\right)\approx 142.1^\circ$

9. The adjacent angle in the parallelogram is supplementary to 52°, thus it has measure 128°.

$|\mathbf{v}|^2=100^2+130^2-2(100)(130)\cos 128^\circ\approx 42{,}907.19836\Rightarrow|\mathbf{v}|=\sqrt{42{,}907.19836}\approx 207$ lb

10. Angle B is supplementary to 52° and has measure $180^\circ-57^\circ=123^\circ$. The other angle of the triangle, C, has measure $180^\circ-52^\circ-123^\circ=5^\circ$. Using the Law of Sines, $\dfrac{a}{\sin 52^\circ}=\dfrac{950}{\sin 5^\circ}\Rightarrow a=\dfrac{950\sin 52^\circ}{\sin 5^\circ}\Rightarrow a\approx 8589$. If a vertical line from the plane to the ground intersects the ground at D, then triangle CBD is a right triangle.

Thus, $\sin 57^\circ=\dfrac{h}{8589}\Rightarrow h=8589\sin 57^\circ\approx 7200$ ft.

10.4 Trigonometric (Polar) Form of Complex Numbers

1. The modulus of a complex number represents the magnitude (length) of the vector representing it in the complex plane.
3. See Figure 3.
5. See Figure 5.

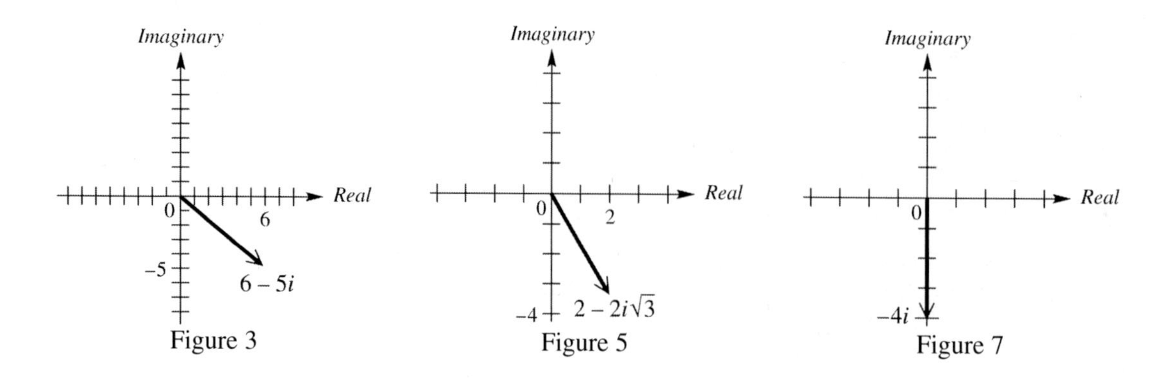

Figure 3 Figure 5 Figure 7

7. See Figure 7.
9. See Figure 9.

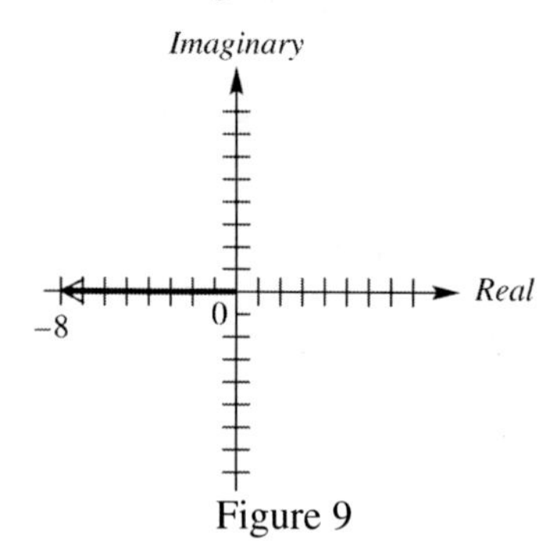

Figure 9

Figure 10

11. Since the vector has terminal point $(1,-4)$, the rectangular form is $1-4i$.
13. The imaginary part must be 0.
15. $a=0$
17. $(4-3i)+(-1+2i)=(4-1)+(-3+2)i=3-i$
19. $(-3+0i)+(0+3i)=(-3+0)+(0+3)i=-3+3i$
21. $(2+6i)+(0-2i)=(2+0)+(6-2)i=2+4i$
23. $(4-2i)+(5+0i)=(4+5)+(-2+0)i=9-2i$
25. $r=\sqrt{1^2+1^2}=\sqrt{2}$
27. $r=\sqrt{12^2+(-5)^2}=\sqrt{169}=13$
29. $r=\sqrt{(-6)^2+0^2}=\sqrt{36}=6$
31. $r=\sqrt{2^2+(-3)^2}=\sqrt{13}$
33. $2(\cos 45°+i\sin 45°)=2\left(\frac{\sqrt{2}}{2}+i\frac{\sqrt{2}}{2}\right)=\sqrt{2}+i\sqrt{2}$
35. $10 \text{ cis } 90°=10(\cos 90°+i\sin 90°)=10(0+1i)=10i$

37. $4(\cos 240° \_ i\sin 240°) = 4\left(-\frac{1}{2} - i\frac{\sqrt{3}}{2}\right) = -2 - 2i\sqrt{3}$

39. $\left(\cos\frac{\pi}{6} + i\sin\frac{\pi}{6}\right) = \left(\frac{\sqrt{3}}{2} + i\frac{1}{2}\right) = \frac{\sqrt{3}}{2} + \frac{1}{2}i$

41. $5\text{ cis}\left(-\frac{\pi}{6}\right) = 5\left[\cos\left(-\frac{\pi}{6}\right) + i\sin\left(-\frac{\pi}{6}\right)\right] = 5\left(\frac{\sqrt{3}}{2} - i\frac{1}{2}\right) = \frac{5\sqrt{3}}{2} - \frac{5}{2}i$

43. $\sqrt{2}\text{cis}\pi = \sqrt{2}(\cos\pi + i\sin\pi) = \sqrt{2}(-1 - 0i) = -\sqrt{2}$

45. $r = \sqrt{3^2 + (-3)^2} = \sqrt{18} = 3\sqrt{2}; \tan\theta = \frac{y}{x} \Rightarrow \tan\theta = \frac{-3}{3} \Rightarrow \tan\theta = -1 \Rightarrow \theta = \tan^{-1}(-1) = -45°$ Since θ is in quadrant IV and $0° \le \theta \le 360°$, $\theta = 36 - 45 = 315°$. Therefore $3 - 3i = 3\sqrt{2}\left[\cos(315°) + i\sin(315°)\right]$.

47. $r = \sqrt{1^2 + \left(\sqrt{3}\right)^2} = \sqrt{4} = 2; \tan\theta = \frac{y}{x} \Rightarrow \tan\theta = \frac{\sqrt{3}}{1} \Rightarrow \tan\theta = \sqrt{3} \Rightarrow \theta = \tan^{-1}\left(\sqrt{3}\right) = 60°$

Since θ is in quadrant I, $\theta = 60°$. Therefore $1 + i\sqrt{3} = 2\left(\cos 60° + i\sin 60°\right)$.

49. $r = \sqrt{0^2 + (-2)^2} = \sqrt{4} = 2; \tan\theta = \frac{y}{x} \Rightarrow \tan\theta = \frac{-2}{0} \Rightarrow \tan\theta$ is undefined $\Rightarrow \theta = 90°$. Since θ is on the negative y-axis and $0° \le \theta \le 360°$, $\theta = 90° + 180° = 270°$. Therefore $-2i = 2\left[\cos(270°) + i\sin(270°)\right]$.

51. $r = \sqrt{\left(4\sqrt{3}\right)^2 + 4^2} = \sqrt{64} = 8; \tan\theta = \frac{y}{x} \Rightarrow \tan\theta = \frac{1}{\sqrt{3}} \Rightarrow \tan\theta = \frac{1}{\sqrt{3}} \Rightarrow \theta = \tan^{-1}\left(\frac{1}{\sqrt{3}}\right) = \frac{\pi}{6}$.

Since θ is in quadrant I and $0 \le \theta \le 2\pi$, $\theta = \frac{\pi}{6}$. Therefore $4\sqrt{3} + 4i = 8\left(\cos\frac{\pi}{6} + i\sin\frac{\pi}{6}\right)$.

53. $r = \sqrt{\left(-\sqrt{2}\right)^2 + \left(\sqrt{2}\right)^2} = \sqrt{4} = 2; \tan\theta = \frac{y}{x} \Rightarrow \tan\theta = -\frac{\sqrt{2}}{\sqrt{2}} \Rightarrow \tan\theta = -1 \Rightarrow \theta = \tan^{-1}(-1) = -\frac{\pi}{4}$

Since θ is in quadrant II and $0 \le \theta \le 2\pi$, $\theta = -\frac{\pi}{4} + \pi = \frac{3\pi}{4}$. Therefore $-\sqrt{2} + i\sqrt{2} = 2\left(\cos\frac{3\pi}{4} + i\sin\frac{3\pi}{4}\right)$.

55. $r = \sqrt{(-4)^2\, 0^2} = \sqrt{16} = 4; \tan\theta = \frac{y}{x} \Rightarrow \tan\theta = \frac{0}{-4} \Rightarrow \tan\theta = 0 \Rightarrow \theta = \tan^{-1}(0) = 0$

Since θ is on the negative x-axis and $0 \le \theta \le 2\pi$, $\theta = 0 + \pi = \pi$. Therefore $-4 = 4(\cos\pi + i\sin\pi)$.

57. $[3(\cos 60° + i\sin 60°)][2(\cos 90° + i\sin 90°)] = (3\cdot 2)[\cos(60° + 90°) + i\sin(60° + 90°)]$

$= 6(\cos 150° + i\sin 150°) = 6\left(-\frac{\sqrt{3}}{2} + \frac{1}{2}i\right) = -3\sqrt{3} + 3i$

59. $[2(\cos 45° + i\sin 45°)][2(\cos 225° + i\sin 225°)] = (2\cdot 2)[\cos(45° + 225°) + i\sin(45° + 225°)]$

$= 4(\cos 270° + i\sin 270°) = 4(0 - 1i) = -4i$

61. $\left[5\text{cis}\frac{\pi}{2}\right]\left[3\text{cis}\frac{\pi}{2}\right]=\left[5\left(\cos\frac{\pi}{2}+i\sin\frac{\pi}{2}\right)\right]\left[3\left(\cos\frac{\pi}{4}+i\sin\frac{\pi}{4}\right)\right]$

$=(5\cdot 3)\left[\cos\left(\frac{\pi}{2}+\frac{\pi}{4}\right)\right]+i\sin\left(\frac{\pi}{2}+\frac{\pi}{4}\right)=15\left(\cos\frac{3\pi}{4}+i\sin\frac{3\pi}{4}\right)=15\left(-\frac{\sqrt{2}}{2}+\frac{\sqrt{2}}{2}i\right)=-\frac{15\sqrt{2}}{2}+\frac{15\sqrt{2}}{2}i$

63. $\left(\sqrt{3}\text{cis}\frac{\pi}{4}\right)\left(\sqrt{3}\text{cis}\frac{5\pi}{4}\right)=\left[\sqrt{3}\left(\cos\frac{\pi}{4}+i\sin\frac{\pi}{4}\right)\right]\left[\sqrt{3}\left(\cos\frac{5\pi}{4}+i\sin\frac{5\pi}{4}\right)\right]$

$=(\sqrt{3}\cdot\sqrt{3})\left[\cos\left(\frac{\pi}{4}+\frac{5\pi}{4}\right)+i\sin\left(\frac{\pi}{4}+\frac{5\pi}{4}\right)\right]=3\left(\cos\frac{3\pi}{2}+i\sin\frac{3\pi}{2}\right)=3(0+1i)=-3i$

65. $\frac{4(\cos 120°+i\sin 120°)}{2(\cos 150°+i\sin 150°)}=\frac{4}{2}[\cos(120°-150°)+i\sin(120°-150°)]$

$=2[\cos(-30°)+i\sin(-30°)]=2\left(\frac{\sqrt{3}}{2}-\frac{1}{2}i\right)=\sqrt{3}-i$

67. $\frac{10(\cos\frac{5\pi}{4}+i\sin\frac{5\pi}{4})}{5(\cos\frac{\pi}{4}+i\sin\frac{\pi}{4})}=\frac{10}{5}\left[\cos(\frac{5\pi}{4}-\frac{\pi}{4})+i\sin(\frac{5\pi}{4}-\frac{\pi}{4})\right]=2(\cos\pi+i\sin\pi)=2(-1-0i)=-2$

69. $\frac{3cis\frac{61\pi}{36}}{9cis\frac{13\pi}{36}}=\frac{3(\cos\frac{61\pi}{36}+i\sin\frac{61\pi}{36})}{9(\cos\frac{13\pi}{36}+i\sin\frac{13\pi}{36})}=\frac{3}{9}[\cos(\frac{61\pi}{36}-\frac{13\pi}{36})+i\sin(\frac{61\pi}{36}-\frac{13\pi}{36})]=$

$\frac{1}{3}\left(\cos\frac{4\pi}{3}+\sin\frac{4\pi}{3}\right)=\frac{1}{3}\left(-\frac{1}{2}-\frac{\sqrt{3}}{2}i\right)=-\frac{1}{6}-\frac{\sqrt{3}}{6}i$

71. $\frac{8}{\sqrt{3}+i}$ numerator: $8 = 8 + 0i$ and $r=\sqrt{8^2+0^2}=8$ $\theta=0°$ since $\cos 0° = 1$ and $\sin 0° = 0$, so $8 =$

8 cis $0°$; denominator: $\sqrt{3}+i$ and $r=\sqrt{\left(\sqrt{3}\right)^2+1^2}=\sqrt{3+1}=\sqrt{4}=2$; $\tan\theta=\frac{1}{\sqrt{3}}=\frac{\sqrt{3}}{3}$

Since x and y are both positive, θ is in quadrant I, so $\theta = 30°$. Thus $\sqrt{3}+i=2\,\text{cis}\,30°$.

$\frac{8}{\sqrt{3}+i}=\frac{8\,\text{cis}\,0°}{2\,\text{cis}\,30°}=\frac{8}{2}\text{cis}\left(0-30°\right)=4\left[\cos\left(-30°\right)+i\sin\left(-30°\right)\right]=4\left(\frac{\sqrt{3}}{2}-\frac{1}{2}i\right)=2\sqrt{3}-2i$

73. $\frac{-i}{1+i}$ numerator: $-i=0-i$ and $r=\sqrt{0^2+\left(-1\right)^2}=\sqrt{0+1}=\sqrt{1}=1$, $\theta=270°$ since $\cos 270°=0$ and

$\sin 270°=-1$, so $-i=1\,\text{cis}270°$.; denominator: $1+i$ $r=\sqrt{1^2+1^2}=\sqrt{1+1}=\sqrt{2}$ and $\tan\theta=\frac{y}{x}=\frac{1}{1}=1$

Since x and y are both positive, θ is in quadrant I, so $\theta=45°$. Thus, $1+i=\sqrt{2}$ cis $45°$

$\frac{-i}{1+i}=\frac{\text{cis }270°}{\sqrt{2}\text{ cis }45°}=\frac{1}{\sqrt{2}}\text{cis}\left(270°-45°\right)=\frac{\sqrt{2}}{2}\text{ cis }225°=\frac{\sqrt{2}}{2}\left(\cos 225°+i\sin 225°\right)=$

$$\frac{\sqrt{2}}{2}\left(-\frac{\sqrt{2}}{2}-i\cdot\frac{\sqrt{2}}{2}\right)=-\frac{1}{2}-\frac{1}{2}i$$

75. $\dfrac{2\sqrt{6}-2i\sqrt{2}}{\sqrt{2}-i\sqrt{6}}$ numerator: $2\sqrt{6}-2i\sqrt{2}$ and $r=\sqrt{\left(2\sqrt{6}\right)^2+\left(-2\sqrt{2}\right)^2}=\sqrt{24+8}=\sqrt{32}=4\sqrt{2}$

$\tan\theta=\dfrac{-2\sqrt{2}}{2\sqrt{6}}=-\dfrac{1}{\sqrt{3}}=-\dfrac{\sqrt{3}}{3}$. Since x is positive and y is negative, θ is in quadrant IV, so $\theta=-30°$.

Thus, $2\sqrt{6}-2i\sqrt{2}=4\sqrt{2}\operatorname{cis}(-30°)$. denominator: $\sqrt{2}-i\sqrt{6}$ and

$r=\sqrt{\left(\sqrt{2}\right)^2+\left(-\sqrt{6}\right)^2}=\sqrt{2+6}=\sqrt{8}=2\sqrt{2}$, $\tan\theta=\dfrac{-\sqrt{6}}{\sqrt{2}}=-\sqrt{3}$

Since x is positive and y is negative, θ is in quadrant IV, so $\theta=-60°$. Thus, $\sqrt{2}-i\sqrt{6}=2\sqrt{2}\operatorname{cis}(-60°)$

$$\frac{2\sqrt{6}-2i\sqrt{2}}{\sqrt{2}-i\sqrt{6}}=\frac{4\sqrt{2}\operatorname{cis}(-30°)}{2\sqrt{2}\operatorname{cis}(-60°)}=\frac{4\sqrt{2}}{2\sqrt{2}}\operatorname{cis}\left[-30°-(-60°)\right]=2\operatorname{cis}30°=2(\cos 30°+i\sin 30°)=$$

$$2\left(\frac{\sqrt{3}}{2}+i\frac{1}{2}\right)=\sqrt{3}+i$$

77. (a) $|a+bi|=\sqrt{a^2+b^2}$ and $|a-bi|=\sqrt{a^2+(-b^2)}=\sqrt{a^2+b^2}$

(b) $Z_1^2-1=(a+bi)^2-1=a^2+2abi-b^2-1=a^2-b^2-1+2abi$

$Z_2^2-1=(a-bi)^2-1=a^2-2abi-b^2-1=a^2-b^2-1-2abi$

(c) If $z_1=a+bi$ and $z_2=a-bi$, then z_1^2-1 and z_2^2-1 are also conjugates with the same modulus.

Therefore, if z_1 is in the Julia set, so is z_2. Thus *(a, b)* in the Julia set implies $(a,-b)$ is also in the set.

(d) Yes, see part (e) below.

(e) $z_1^2-1=(a+bi)^2-1=a^2+2abi-b^2-1=a^2-b^2-1+2abi$

$z_2^2-1=(-a+bi)^2-1=a^2-2abi-b^2-1=a^2-b^2-1-2abi$. If $z_1=a+bi$ and

$z_2=-a+bi$, then z_1^2-1 and z_2^2-1 are also conjugates with the same modulus.

Therefore, if z_1 is in the Julia set, so is z_2. Thus (a,b) in the Julia set implies $(a,-b)$ is also in the set.

79. First find the trigonometric form of $Z=6+3i: r=\sqrt{6^2+3^2}=\sqrt{45}\approx 6.71$;

$\tan\theta=\dfrac{3}{6}\Rightarrow\tan\theta=\dfrac{1}{2}\Rightarrow\theta\tan^{-1}\dfrac{1}{2}\Rightarrow\theta\approx 26.6°$. Therefore $Z\approx 6.71(\cos 26.6°+i\sin 26.6°)$.

$$I=\frac{E}{Z}=\frac{8(\cos 20°+i\sin 20°)}{6.71(\cos 26.57°+i\sin 26.57°)}=\frac{8}{6.71}[\cos(20°-26.57°)+i\sin(20°-26.57°)]\approx 1.18-.14i$$

81. Noting that $Z=\dfrac{1}{\dfrac{1}{Z_1}+\dfrac{1}{Z_2}}=\dfrac{1}{\dfrac{Z_2+Z_1}{Z_1Z_2}}=\dfrac{Z_1Z_2}{Z_1+Z_2}$, we find the following:

First find the trigonometric form of $Z_1 = 50 + 25i$: $r = \sqrt{50^2 + 25^2} = \sqrt{3125}$;

$\tan\theta = \dfrac{25}{50} \Rightarrow = \tan^{-1}\dfrac{1}{2} \Rightarrow \theta \approx 25.565°$. The value is $\sqrt{3125}(\cos 26.565° + i\sin 26.565°)$.

Next find the trigonometric form of $Z_2 = 60 + 20i$: $r = \sqrt{60^2 + 20^2} = \sqrt{4000}$;

$\tan\theta = \dfrac{20}{60} \Rightarrow\Rightarrow \theta = \tan^{-1}\dfrac{1}{3} \Rightarrow \theta \approx 18.435°$. The value is $\sqrt{4000}(\cos 18.435° + i\sin 18.435°)$. Finally find

the trigonometric form of $Z_1 + Z_2 = 110 + 45i : r = \sqrt{110^2 + 45^2} = \sqrt{14,125}$;

$\tan\theta = \dfrac{45}{110} \Rightarrow\Rightarrow \theta = \tan^{-1}\dfrac{9}{22} \Rightarrow \theta \approx 22.249°$. The value is $\sqrt{14,125}(\cos 22.249° + i\sin 22.249°)$.

$$Z = \frac{[\sqrt{3125}(\cos 26.565° + i\sin 26.565°)][\sqrt{4000}(\cos 18.435° + i\sin 18.435°)]}{\sqrt{14,125}(\cos 22.249° + i\sin 22.249°)} =$$

$$\frac{\sqrt{3125}\cdot\sqrt{4000}}{\sqrt{14,125}}[(\cos(26.565° + 18.435° - 22.249°) + i\sin(26.565° + 18.435° - 22.249°)] \approx 27.43 + 11.5i$$

83. To square a complex number in trigonometric form, square the modulus r and double the argument θ.

85. (a) If the real part equals 0, then $\cos\theta = 0 \Rightarrow \theta = \dfrac{\pi}{2}$.

(b) If the imaginary part equals 0, then $\sin\theta = 0 \Rightarrow \theta = \pi$.

10.5 Powers and Roots of Complex Numbers

1. $[3(\cos 30° + i\sin 30°)]^3 = 3^3[\cos(3.30°) + i\sin(3.30°)] = 27(\cos 90° + i\sin 90°) = 27(0 + 1i) = 27i$

3. $\left(\cos\dfrac{\pi}{4} + i\sin\dfrac{\pi}{4}\right)^8 = \cos\left(8\cdot\dfrac{\pi}{4}\right) + i\sin\left(8\cdot\dfrac{\pi}{4}\right) = \cos 2\pi + i\sin 2\pi = 1 + 0i = 1$

5. $\left[2\left(\cos\dfrac{2\pi}{3} + i\sin\dfrac{2\pi}{3}\right)\right]^3 = 2^3\left[\cos\left(3.\dfrac{2\pi}{3}\right) + i\sin\left(3.\dfrac{2\pi}{3}\right)\right] = 8(\cos 2\pi + i\sin 2\pi) = 8(1 + 0i) = 8$

7. $[3cis100°]^3 = [3(\cos 100° + i\sin 100°)]^3 = 3^3[\cos(3.100°) + i\sin(3.100°)] =$

$27(\cos 300° + i\sin 300°) = 27\left(\dfrac{1}{2} - \dfrac{\sqrt{3}}{2}i\right) = \dfrac{27}{2} - \dfrac{27\sqrt{3}}{2}i$

9. First find the trigonometric form of $\sqrt{3} + i$; $r = \sqrt{(\sqrt{3})^2 + 1^2} = \sqrt{4} = 2$;

$\tan\theta = \dfrac{y}{x} \Rightarrow \tan\theta = \dfrac{1}{\sqrt{3}} \Rightarrow\Rightarrow \theta = \tan^{-1}\dfrac{1}{\sqrt{3}} = 30°$ $\sqrt{3} + i = 2(\cos 30° + i\sin 30°)$.

$(\sqrt{3} + i)^3 = [2(\cos 30° + i\sin 30°)]^3 = 2^3[\cos(3.30°) + i\sin(3.30°)] = 8(\cos 90° + i\sin 90°) + 8(0 + 1i) = 8i$

11. First find the trigonometric form of $1 + i\sqrt{3}$; $r = \sqrt{1^2 + (\sqrt{3})^2} = \sqrt{4} = 2$;

$\tan\theta = \frac{y}{x} \Rightarrow \tan\theta = \frac{\sqrt{3}}{1} \Rightarrow\Rightarrow \theta = \tan^{-1}\sqrt{3} = 60°$. Therefore

$1+i\sqrt{3} = 2(\cos 60° + i\sin 60°)$. $(1+i\sqrt{3})^4 = [2(\cos 60° + i\sin 60°)]^4 = 2^4[\cos(4\cdot 60°) + i\sin(4\cdot 60°)] =$

$16(\cos 240° + i\sin 240°) = 16\left(-\frac{1}{2} - \frac{\sqrt{3}}{2}i\right) = -8 - 8i\sqrt{3}$

13. First find the trigonometric form of $-\frac{\sqrt{2}}{2} + \frac{\sqrt{2}}{2}i; r = \sqrt{\left(-\frac{\sqrt{2}}{2}\right)^2 + \left(\frac{\sqrt{2}}{2}\right)^2} = \sqrt{1} = 1;$

$\tan\theta = \frac{y}{x} \Rightarrow \tan\theta = \left[\frac{\sqrt{2}}{2} \div \left(-\frac{\sqrt{2}}{2}\right)\right] \Rightarrow \tan\theta = -1 \Rightarrow \theta = \tan^{-1}1 = -45°$. Since θ is in quadrant II,

$\theta = -45° + 180° = 135°$. Therefore $-\frac{\sqrt{2}}{2} + \frac{\sqrt{2}}{2}i = \cos 135° + i\sin 135°$.

$\left(-\frac{\sqrt{2}}{2} + \frac{\sqrt{2}}{2}i\right)^4 = (\cos 135° + i\sin 135°)^4 = \cos(4.135°) + i\sin(4.135°) = \cos 540° + i\sin 540° = -1$

15. First find the trigonometric form of $1-i; r = \sqrt{1^2 + (-1)^2} = \sqrt{2};$

$\tan\theta = \frac{y}{x} \Rightarrow \tan\theta = \frac{-1}{1} \Rightarrow\Rightarrow \theta = \tan^{-1}(-1) = -45°$. Therefore $1 - i = \sqrt{2}[\cos(-45°) + i\sin(-45°)]$.

$(1-i)^6 = [\sqrt{2}(\cos(-45°) + i\sin(-45°)]^6 = (\sqrt{2})^6\{\cos[6.(-45°)] + i\sin[6.(-45°)]\}$

$= 8[\cos(-270°) + i\sin(-270°)] = 8(0 + 1i) = 8i$

17. First find the trigonometric form of $-2-2i; r = \sqrt{(-2)^2 + (-2)^2} = \sqrt{8} = 2\sqrt{2};$

$\tan\theta = \frac{y}{x} \Rightarrow \tan\theta = \frac{-2}{-2} \Rightarrow\Rightarrow \theta = \tan^{-1}1 = 225°$. Therefore, $-2 - 2i = 2\sqrt{2}(\cos 225° + i\sin 225°)$.

$(-2-2i)^5 = [2\sqrt{2}(\cos 225° + i\sin 225°)]^5 = (2\sqrt{2})^5[\cos(5.225°) + i\sin(5.225°)] =$

$128\sqrt{2}(\cos 1125° + i\sin 1125°) = 128\sqrt{2}\left(\frac{\sqrt{2}}{2} + \frac{\sqrt{2}}{2}i\right) = 128 + 128i$

19. Note that $1 = (\cos 0° + i\sin 0°)$. Here $r = 1$, so $\sqrt[3]{r} = 1$, $n = 3$, and $k = 0$, 1, or 2. $\alpha = \frac{0° + 360°\cdot k}{3}$: For $k = 0$, $\alpha = 0°$. For $k = 1$, $\alpha = 120°$. For $k = 2$, $\alpha = 240°$. The cube roots are $\cos 0° + i\sin 0°, \cos 120° + i\sin 120°$, and $\cos 240° + i\sin 240°$. See Figure 19.

21. Here $r = 8$ so $\sqrt[3]{r} = 2$, $n = 3$, and $k = 0, 1$, or 2. $\alpha = \frac{60° + 360°\cdot k}{3}$: For $k = 0$, $\alpha = 20°$. For $k = 1$, $\alpha = 140°$. For $k = 2$, $\alpha = 260°$. The cube roots are $2(\cos 20° + i\sin 20°), 2(\cos 140° + i\sin 140°)$, and $2(\cos 260° + i\sin 260°)$. See Figure 21.

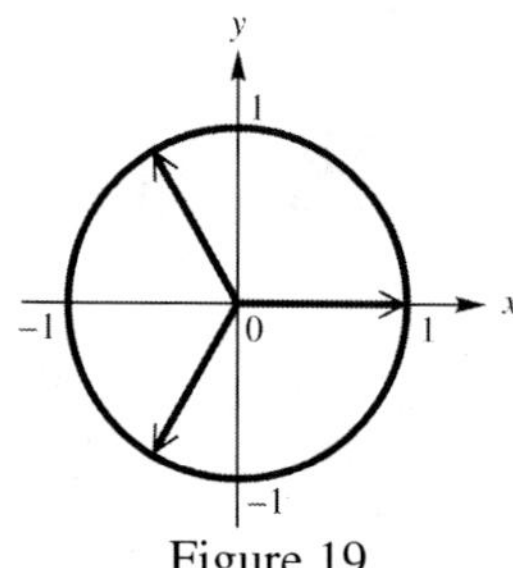

Figure 19

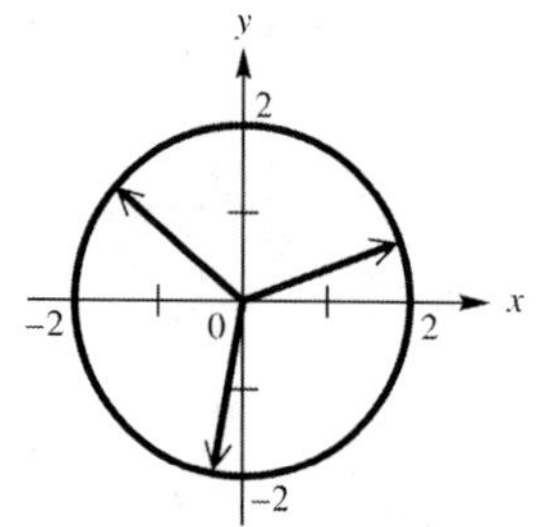

Figure 21

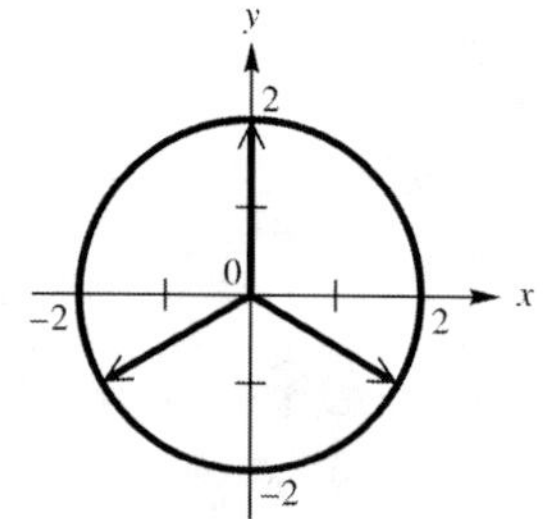

Figure 23

23. Note that $-8i = 8(\cos 270° + i \sin 270°)$. Here $r = 8$, so $\sqrt[3]{r} = 2$, $n = 3$, and $k = 0, 1,$ or 2. For

$\alpha = \dfrac{270° + 360° \cdot k}{3}$: For $k = 0$, $\alpha = 90°$. For $k = 1, \alpha = 210°$. For $k = 2$, $\alpha = 330°$. The cube roots

are $2(\cos 90° + i \sin 90°)$, $2(\cos 210° + i \sin 210°)$, and $2(\cos 330° + i \sin 330°)$. See Figure 23.

25. Note that $-64 = 64(\cos 180° + i \sin 180°)$. Here $r = 64$, so $\sqrt[3]{r} = 4$, $n = 3$, and $k = 0, 1,$ or 2.

$\alpha = \dfrac{180° + 360° \cdot k}{3}$: For $k = 0$, $\alpha = 60°$. For $k = 1$, $\alpha = 180°$. For $k = 2$, $\alpha = 300°$. The cube roots are

$4(\cos 60° + i \sin 60°), 4(\cos 180° + i \sin 180°)$, and $4(\cos 300° + i \sin 300°)$. See Figure 25.

27. Note that $1 + i\sqrt{3} = 2(\cos 60° + i \sin 60°)$. Here $r = 2$, so $\sqrt[3]{r} = \sqrt[3]{2}$, $n = 3$, and $k = 0,1$, or 2.

$\alpha = \dfrac{60° + 360° \cdot k}{3}$: For $k = 0$, $\alpha = 20°$. For $k = 1$, $\alpha = 140°$. For $k = 2$, $\alpha = 260°$. The cube roots are

$\sqrt[3]{2}(\cos 20° + i \sin 20°), \sqrt[3]{2}(\cos 140° + i \sin 140°)$, and $\sqrt[3]{2}(\cos 260° + i \sin 260°)$. See Figure 27.

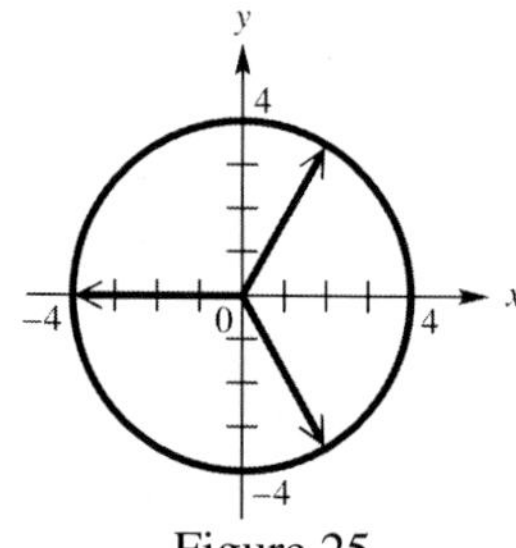

Figure 25

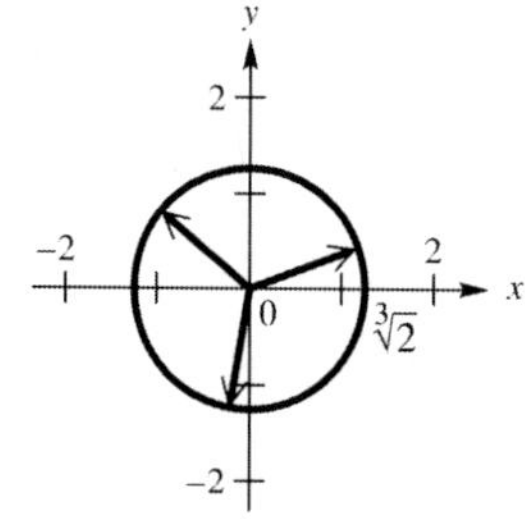

Figure 27

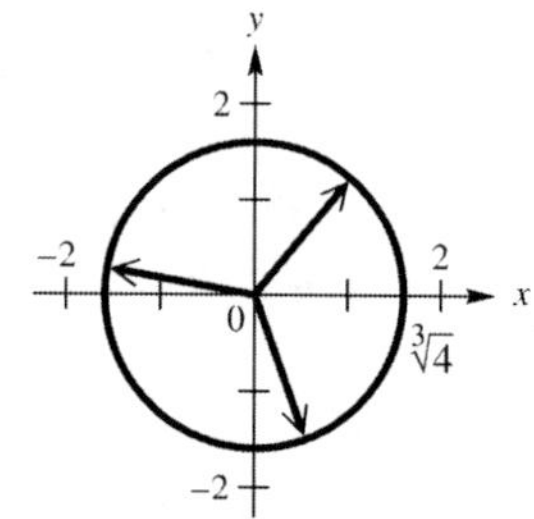

Figure 29

29. Note that $-2\sqrt{3} + 2i = 4(\cos 150° + i \sin 150°)$. Here $r = 4$, so $\sqrt[3]{r} = \sqrt[3]{4}$, $n = 3$, and

$k = 0, 1, 2$. $\alpha = \dfrac{150° + 360° \cdot k}{3}$: For $k = 0$, $\alpha = 50°$. For $k = 1$, $\alpha = 170°$. For $k = 2$, $\alpha = 290°$. The cube

roots are $\sqrt[3]{4}(\cos 50° + i \sin 50°)$, $\sqrt[3]{4}(\cos 170° + i \sin 170°)$, and $\sqrt[3]{4}(\cos 290° + i \sin 290°)$. See Figure 29.

31. The complex number is $4(\cos 120° + i \sin 120°)$. $r = 4$, so $\sqrt{r} = 2$, $n = 2$, and $k = 0$ or 1.

$\alpha = \dfrac{120° + 360° \cdot k}{2}$: For $k = 0$, $\alpha = 60°$. For $k = 1$, $\alpha = 240°$. The square roots are

$2(\cos 60° + i \sin 60°) = 1 + i\sqrt{3}$ and $2(\cos 240° + i \sin 240°) = -1 - i\sqrt{3}$.

33. The complex number is $\cos 180° + i\sin 180°$ Here $r = 1$ so, $\sqrt[3]{r} = 1$, $n = 3$, and $k = 0, 1,$ or 2. $\alpha = \frac{180° + 360° \cdot k}{3}$: For $k = 0$, $\alpha = 60°$. For $k = 1$, $\alpha = 180°$. For $k = 2$, $\alpha = 300°$. The cube roots are $\cos 60° + i\sin 60° = \frac{1}{2} + \frac{\sqrt{3}}{2}i$, $\cos 180° + i\sin 180° = -1$, and $\cos 300° + i\sin 300° = \frac{1}{2} - \frac{\sqrt{3}}{2}i.$

35. Note that $i = \cos 90° + i\sin 90°$. Here $r = 1$, so $\sqrt{r} = 1$, $n = 2$, and $k = 0$ or 1. $\alpha = \frac{90° + 360° \cdot k}{2}$: For $k = 0$, $\alpha = 45°$. For $k = 1$, $\alpha = 225°$. The square roots are $\cos 45° + i\sin 45° = \frac{\sqrt{2}}{2} + \frac{\sqrt{2}}{2}i$ and $\cos 225° + i\sin 225° = -\frac{\sqrt{2}}{2} - \frac{\sqrt{2}}{2}i.$

37. Note that $64i = 64(\cos 90° + i\sin 90°)$. Here $r = 64$, So $\sqrt[3]{r} = 4$, $n = 3$, and $k = 0, 1,$ or 2. $\alpha = \frac{90° + 360° \cdot k}{3}$: For $k = 0$, $\alpha = 30°$. For $k = 1$, $\alpha = 150°$. For $k = 2$, $\alpha = 270°$. The cube roots are $4(\cos 30° + i\sin 30°) = 2\sqrt{3} + 2i$, $4(\cos 150° + i\sin 150°) = -2\sqrt{3} + 2i$ and $4(\cos 270° + i\sin 270°) = -4i.$

39. Note that $81 = 81(\cos 0° + i\sin 0°)$. Here $r = 81$, so $\sqrt[4]{r} = 3$, $n = 4$, and $k = 0,1,2,$ or 3. $\alpha = \frac{0° + 360° \cdot k}{4}$: For $k = 0$, $\alpha = 0°$. For $k = 1$, $\alpha = 90°$. For $k = 2$, $\alpha = 180°$. For $k = 3$, $\alpha = 270°$. The fourth roots are $3(\cos 0° + i\sin 0°) = 3$, $3(\cos 90° + i\sin 90°) = 3i$, $3(\cos 180° + i\sin 180°) = -3$, and $3(\cos 270° + i\sin 270°) = -3i.$

41. (a) Note that $1 = \cos 0° + i\sin 0°$. Here $r = 1$, so $\sqrt[4]{r} = 1$, $n = 4$, and $k = 0,1,2,$ or 3 $\alpha = \frac{0° + 360° k}{4}$: For $k = 0$, $\alpha = 0°$. For $k = 1$, $\alpha = 90°$. For $k = 2$, $\alpha = 180°$. For $k = 3$, $\alpha = 270°$. The fourth roots are $\cos 0° + i\sin 0° = 1$, $\cos 90° + i\sin 90° = i \cos 180° + i\sin 180° = -1$ and $\cos 270° + i\sin 270° = -i.$ See Figure 41a.

(b) Note that $1 = \cos 0° + i\sin 0°$. Here $r = 1$, so $\sqrt[6]{r} = 1$, $n = 6$, and $k = 0,1,2,3,4$ or 5. $\alpha = \frac{0° + 360° \cdot k}{6}$: For $k = 0$, $\alpha = 0°$. For $k = 1$, $\alpha = 60°$. For $k = 2$, $\alpha = 120°$. For $k = 3$, $\alpha = 180°$. For $k = 4$, $\alpha = 240°$. For $k = 5$, $\alpha = 300°$. The sixth roots are

$\cos 0° + i\sin 0° = 1$, $\cos 60° + i\sin 60° = \frac{1}{2} + \frac{\sqrt{3}}{2}i$, $\cos 120° + i\sin 120° = -\frac{1}{2} + \frac{\sqrt{3}}{2}i$,

$\cos 180° + i\sin 180° = -1$, $\cos 240° + i\sin 240° = -\frac{1}{2} - \frac{\sqrt{3}}{2}i$ $\cos 300° + i\sin 300° = \frac{1}{2} - \frac{\sqrt{3}}{2}i.$

See Figure 41b

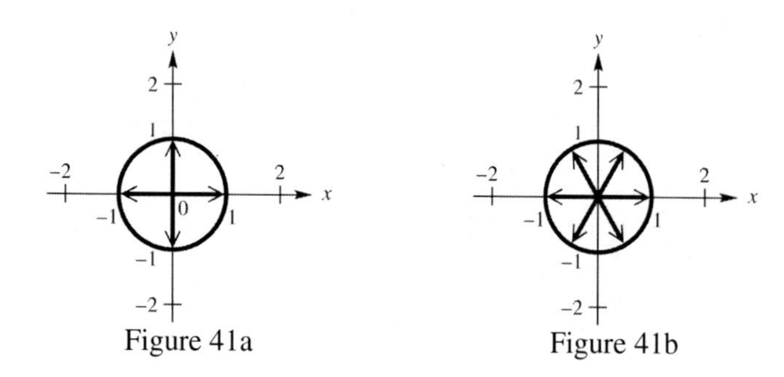

Figure 41a Figure 41b

43. The argument for a positive real number is $\theta = 0°$. By the nth root theorem with $k = 0$, $\alpha = 0°$. Thus, one nth root has an argument of $0°$ and it must be real.

45. $x^3 + 8 = (x+2)(x^2 - 2x + 4)$

46. $x + 2 = 0 \Rightarrow x = -2$

47. $x^2 - 2x + 4 = 0 \Rightarrow x^2 - 2x + 1 = -4 + 1 \Rightarrow (x-1)^2 = -3 \Rightarrow x - 1 = \pm\sqrt{-3} \Rightarrow x = 1 \pm i\sqrt{3}$

48. Note that $8(\cos 180° + i \sin 180°)$. Here $r = 8$, so $\sqrt[3]{r} = 2$, $n = 3$, and $k = 0, 1$, or 2. $\alpha = \dfrac{180° + 360° \cdot k}{3}$: For $k = 0$, $\alpha = 60°$. For $k = 1$, $\alpha = 180°$. For $k = 2$, $\alpha = 300°$. The cube roots are $2(\cos 60° + i \sin 60°), 2(\cos 180° + i \sin 180°)$, and $2(\cos 300° + i \sin 300°)$

49. $2(\cos 60° + i \sin 60°) = 1 + i\sqrt{3}, 2(\cos 180° + i \sin 180°) = -2$, and $2(\cos 300° + i \sin 300°) = 1 - i\sqrt{3}$.

50. The results are the same.

51. $x^4 + 1 = 0 \Rightarrow x^4 = -1$. Find the fourth roots of -1. Note that $-1 = \cos 180° + i \sin 180°$. Here $r = 1$, so $\sqrt[4]{r} = 1$, $n = 4$, and $k = 0, 1, 2$, or 3 $\alpha = \dfrac{180° + 360° \cdot k}{4}$: For $k = 0$, $\alpha = 45°$. For $k = 1$, $\alpha = 135°$. For $k = 2$, $\alpha = 225°$. For $k = 3$, $\alpha = 315°$. The fourth roots are $\cos 45° + i \sin 45°, \cos 135° + i \sin 135°, \cos 225° + i \sin 225°$, and $\cos 315° + i \sin 315°$.

53. $x^5 - i = 0 \Rightarrow x^5 = i$. Find the fifth roots of i. Note that $i = \cos 90° + i \sin 90°$. Here $r = 1$, so $\sqrt[5]{r} = 1$, $n = 5$, and $k = 0, 1, 2, 3$, or 4. $\alpha = \dfrac{90° + 360° \cdot k}{5}$: For $k = 0$, $\alpha = 18°$. For $k = 1$, $\alpha = 90°$. For $k = 2$, $\alpha = 162°$. For $k = 3$, $\alpha = 234°$. For $k = 4$, $\alpha = 306°$. The fifth roots are $\cos 18° + i \sin 18°, \cos 90° + i \sin 90°, \cos 162° + i \sin 162°, \cos 234° + i \sin 234°$, and $\cos 306° + i \sin 306°$.

55. $x^3 + 1 = 0 \Rightarrow x^3 = -1$. Find the cube roots of -1. Note that $-1 = \cos 180° + i \sin 180°$. Here $r = 1$, so $\sqrt[3]{r} = 1$, $n = 3$, and $k = 0, 1$, or 2. $\alpha = \dfrac{180° + 360° \cdot k}{3}$: For $k = 0$, $\alpha = 60°$. For $k = 1$, $\alpha = 180°$. For $k = 2$, $\alpha = 300°$. The fourth roots are $\cos 60° + i \sin 60°$, $\cos 180° + i \sin 180°$, and $\cos 300° + i \sin 300°$.

57. $x^3 - 8 = 0 \Rightarrow x^3 = 8$. Find the cube roots of 8. Note that $8 = 8(\cos 0° + i\sin 0°)$. Here $r = 8$, so $\sqrt[3]{r} = 2$, $n = 3$, and $k = 0, 1$, or 2. $\alpha = \dfrac{0° + 360° \cdot k}{3}$. For $k = 0$, $\alpha = 0°$. For $k = 1$, $\alpha = 120°$. For $k = 2$, $\alpha = 240°$. The cube roots are $2(\cos 0° + i\sin 0°)$, $2(\cos 120° + i\sin 120°)$, and $2(\cos 240° + i\sin 240°)$.

59. (a) Using a calculator yields: $f(z_1) = \dfrac{2(i)^3 + 1}{3(i)^2} = -\dfrac{1}{3} + \dfrac{2}{3}i; f(z_2) = \dfrac{2(\frac{1}{3} + \frac{2}{3}i)^3 + 1}{3(-\frac{1}{3} + \frac{2}{3}i)^2} = -\dfrac{131}{225} + \dfrac{208}{225}i$

The value is approaching w_2 and the pixel should be colored blue.

(b) Using a calculator yields:

$$f(z_1) = \frac{2(2+i)^3 + 1}{3(2+i)^2} = -\frac{103}{75} + \frac{46}{75}i; f(z_2) = \frac{2(\frac{103}{75} + \frac{46}{75}i)^3 + 1}{3(-\frac{103}{75} + \frac{46}{75}i)^2} \approx 1.01 + .23i$$

The value is approaching w_2 and the pixel should be colored red.

(c) Using a calculator yields: $f(z_1) = \dfrac{2(-1-i)^3 + 1}{3(-1-i)^2} = -\dfrac{2}{3} - \dfrac{5}{6}i; f(z_2) = \dfrac{2(-\frac{2}{3} - \frac{5}{6}i)^3 + 1}{3(-\frac{2}{3} + \frac{5}{6}i)^2} \approx -.51 - .84i$

The value is approaching w_3 and the pixel should be colored yellow.

Reviewing Basic Concepts (Sections 10.4 and 10.5)

1. $2(\cos 60° + i\sin 60°) = 2\left(\dfrac{1}{2} + \dfrac{\sqrt{3}}{2}\right) = 1 + i\sqrt{3}.$

2. $|3 - 4i| = \sqrt{3^2 + (-4)^2} = \sqrt{25} = 5$

3. $r = \sqrt{(-\sqrt{2})^2 + (\sqrt{2})^2} = \sqrt{4} = 2; \tan\theta = \dfrac{y}{x} \Rightarrow \tan\theta = \dfrac{\sqrt{2}}{-\sqrt{2}} \Rightarrow \tan\theta = -1 \Rightarrow \theta = \tan^{-1}(-1) = -45°$

Since θ is in quadrant II, $\theta = -45° + 180° = 135°$. Therefore $-\sqrt{2} + i\sqrt{2} = 2(\cos 135° + i\sin 135°)$.

4. $[4(\cos 135° + i\sin 135°)][2(\cos 45° + i\sin 45°)] = (4 \cdot 2)[\cos(135° + 45°) + i\sin(135° + 45°)]$

$= 8(\cos 180° + i\sin 180°) = 8(-1 + 0i) = -8$

5. $\dfrac{4(\cos 135° + i\sin 135°)}{2(\cos 45° + i\sin 45°)} = \dfrac{4}{2}[\cos(135° - 45°) + i\sin(135° - 45°)] = 2(\cos 90° + i\sin 45°) = 2(0 + i) = 2i$

6. $[4\text{cis}17°]^3 = [4(\cos 17° + i\sin 17°)]^3 = 4^3[\cos(3 \cdot 17°) + i\sin(3 \cdot 17°)] = 64(\cos 51° + i\sin 51°) = 64cis51°$

7. First find the trigonometric form of $2 - 2i$; $r = \sqrt{2^2 + 2^2} = \sqrt{8} = 2\sqrt{2}$;

$\tan\theta = \dfrac{y}{x} \Rightarrow \tan\theta = \dfrac{-2}{2} \Rightarrow \Rightarrow \theta = \tan^{-1}(-1) = -45°$. Therefore $2 - 2i = 2\sqrt{2};[\cos(-45°) + i\sin(-45°)]$.

$(2 - 2i) = \{2\sqrt{2};[\cos(-45°) + i\sin(-45°)]\}^4 = (2\sqrt{2})^4\{\cos[4 \cdot (-45°)] + i\sin[4 \cdot (-45°)]\}$

$= 64\left[\cos(-180°) + i\sin(180°)\right] = 64(-1+0i) = -64$

8. Note that $-64 = 64(\cos 180° + i\sin 180°)$. Here $r = 64$, so $\sqrt[3]{r} = 4$, $n = 3$, and $k = 0, 1$, or 2.

   $\alpha = \dfrac{180° + 360° \cdot k}{3}$: For $k = 0$, $\alpha = 60°$. For $k = 1$, $\alpha = 180°$. For $k = 2$, $\alpha = 300°$. The cube roots

   are $4(\cos 60° + i\sin 60°) = 2 + 2i\sqrt{3}$, $4(\cos 180° + i\sin 180°) = -4$, and $4(\cos 300° + i\sin 300°) = 2 - 2i\sqrt{3}$.

9. Note that $2i = 2(\cos 90° + i\sin 90°)$. Here $r = 2$, so $\sqrt{r} = \sqrt{2}$, $n = 2$, and $k = 0$ or $\alpha = \dfrac{90° + 360° \cdot k}{2}$:

   For $k = 0$, $\alpha = 45°$. For $k = 1$, $\alpha = 225°$. The square roots are $\sqrt{2}(\cos 45° + i\sin 45°) = 1 + i$

   and $\sqrt{2}(\cos 225° + i\sin 225°) = -1 - i$.

10. Find the cube roots of -1. Note that $-1 = \cos 180° + i\sin 180°$. Here $r = 1$, so $\sqrt[3]{r} = 1$, $n = 3$, and $k = 0, 1$, or

    2. $\alpha = \dfrac{180° + 360° \cdot k}{3}$: For $k = 0$, $\alpha = 60°$. For $k = 1$, $\alpha = 180°$. For $k = 2$, $\alpha = 300°$. The cube roots

    are $\cos 60° + i\sin 60°$, $\cos 180° + i\sin 180°$, and $\cos 300° + i\sin 300°$.

10.6 Polar Equations and Graphs

1. (a) The point is in quadrant II.

   (b) The point is in quadrant I.

   (c) The point is in quadrant IV.

   (d) The point is in quadrant III.

3. See Figure 3-13. Two other forms are $(1, 45° + 360°) = (1, 405°)$ or $(-1, 45° + 180°) = (-1, 225°)$.

5. See Figure 3-13. Two other forms are $(-2, 135° + 360°) = (-2, 495°)$ or $(2, 135° + 180°) = (2, 315°)$.

7. See Figure 3-13. Two other forms are $(5, -60° + 360°) = (5, 300°)$ or $(-5, -60° + 180°) = (-5, 120°)$.

9. See Figure 3-13. Two other forms are $(-3, -210° + 360°) = (-3, 150°)$ or $(3, -210° + 180°) = (3, -30°)$.

11. See Figure 3-13. Two other forms are $\left(3, \dfrac{5\pi}{3} + 2\pi\right) = \left(3, \dfrac{11\pi}{3}\right)$ or $\left(-3, \dfrac{5\pi}{3} - \pi°\right) = \left(-3, \dfrac{2\pi}{3}\right)$.

13. See Figure 3-13. Two other forms are $(2, 0 + \pi) = (2, \pi)$ or $(-2, 0 + 2\pi) = (-2, 2\pi)$.

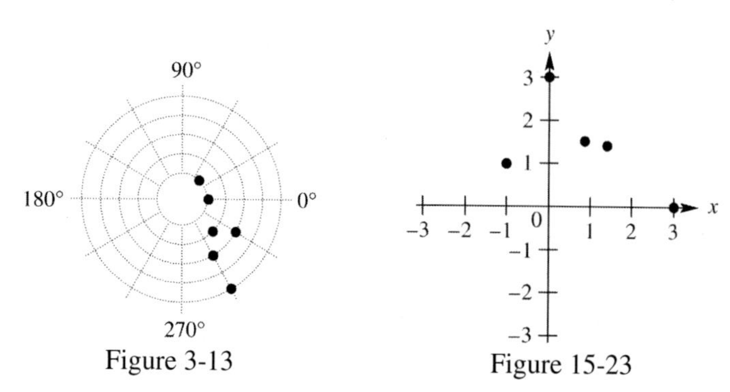

Figure 3-13

Figure 15-23

15. See Figure 15-23. $r=\sqrt{(-1)^2+1^2}=\sqrt{2}$; $\theta=\tan^{-1}\left(\dfrac{1}{-1}\right)=-45°$. Since the point is in QII, $\left(\sqrt{2},135°\right)$. Two forms are $\left(\sqrt{2},135°\right)$ and $\left(-\sqrt{2},315°\right)$.

17. See Figure 15-23. $r=\sqrt{0^2+3^2}=3$; $\theta=\tan^{-1}\left(\dfrac{3}{0}\right)\Rightarrow$ undefined $\Rightarrow\theta=90°$ (positive y-axis). Two forms are $(3,90°)$ and $(-3,270°)$.

19. See Figure 15-23. $r=\sqrt{\left(\sqrt{2}\right)^2+\left(\sqrt{2}\right)^2}=2$; $\theta=\tan^{-1}\left(\dfrac{\sqrt{2}}{\sqrt{2}}\right)=45°$. Two forms are $(2,4.5°)$. and $(-2,225°)$.

21. See Figure 15-23. $r=\sqrt{\left(\dfrac{\sqrt{3}}{2}\right)^2+\left(\dfrac{3}{2}\right)^2}=\sqrt{3}$; $\theta=\tan^{-1}\left(\dfrac{3}{2}\div\dfrac{\sqrt{3}}{2}\right)=60°$. Two forms are $\left(\sqrt{3},60°\right)$ and $\left(-\sqrt{3},240°\right)$.

23. See Figure 15-23. $r=\sqrt{3^2+0^2}=3$; $\theta=\tan^{-1}\left(\dfrac{0}{3}\right)=0°$ (positive x-axis). Two forms are $(3,0°)$ and $(-3,180°)$.

25. $x-y=4\Rightarrow r\cos\theta-r\sin\theta=4\Rightarrow r(\cos\theta-\sin\theta)=4\Rightarrow r=\dfrac{4}{\cos\theta-\sin\theta}$. See Figure 25.

27. $x^2+y^2=16\Rightarrow(r\cos\theta)^2-(r\sin\theta)^2=16\Rightarrow r^2\left(\cos^2\theta+\sin^2\theta\right)=16\Rightarrow r=4, r=-4$. See Figure 27.

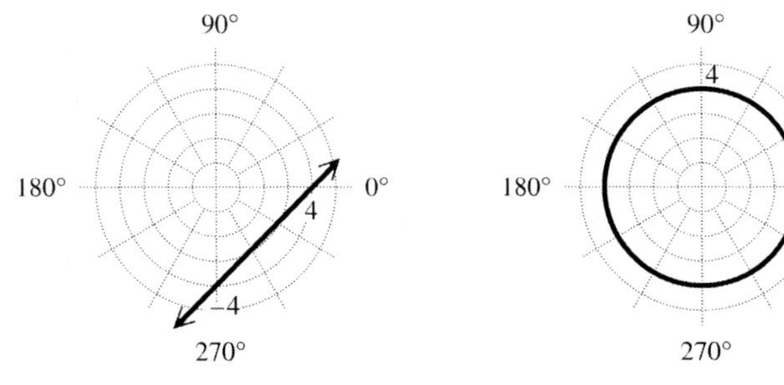

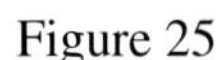
Figure 25

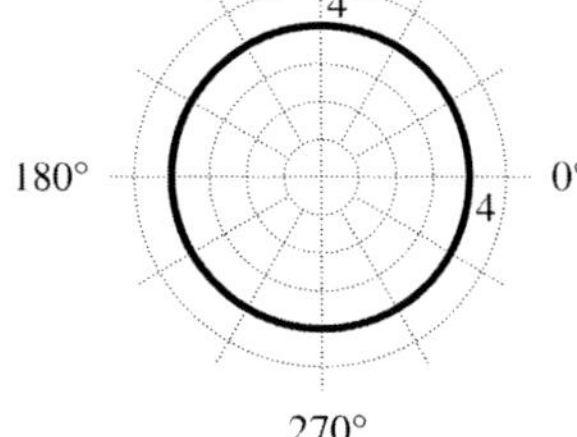

Figure 27

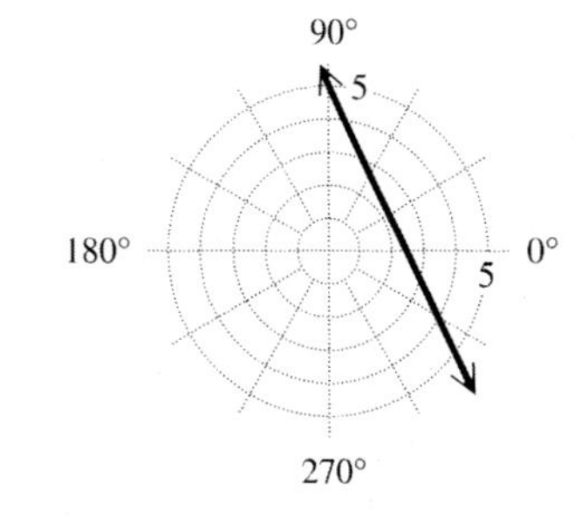

Figure 29

29. $2x+y=5\Rightarrow 2r\cos\theta+r\sin\theta=5\Rightarrow r(2\cos\theta+\sin\theta)=5\Rightarrow r=\dfrac{5}{2\cos\theta+\sin\theta}$. See Figure 29.

31. This is a circle of radius 3. Graph C.

33. This is a rose with four petals. Graph A.

35. The graph is a cardioid. See Figure 35.

37. The graph is a limaçon. See Figure 37.

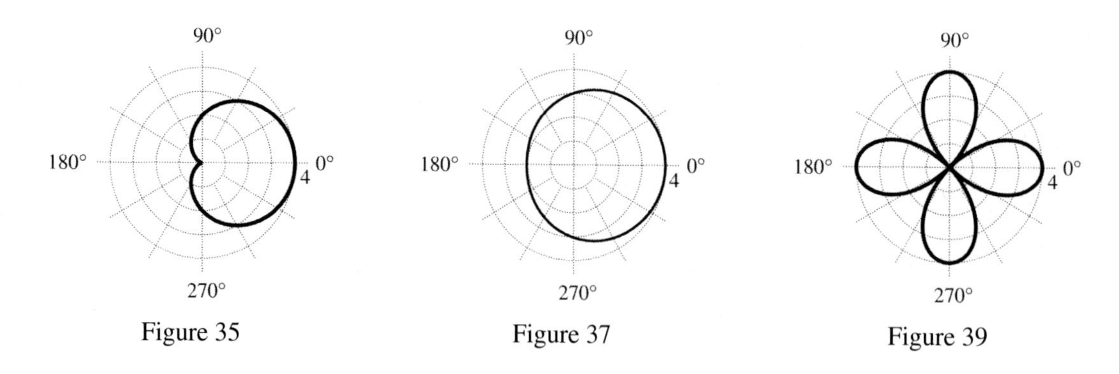

Figure 35 Figure 37 Figure 39

39. The graph is a four-leaved rose. See Figure 39.

41. The graph is a lemniscate. See Figure 41.

43. The graph is a cardioid. See Figure 43

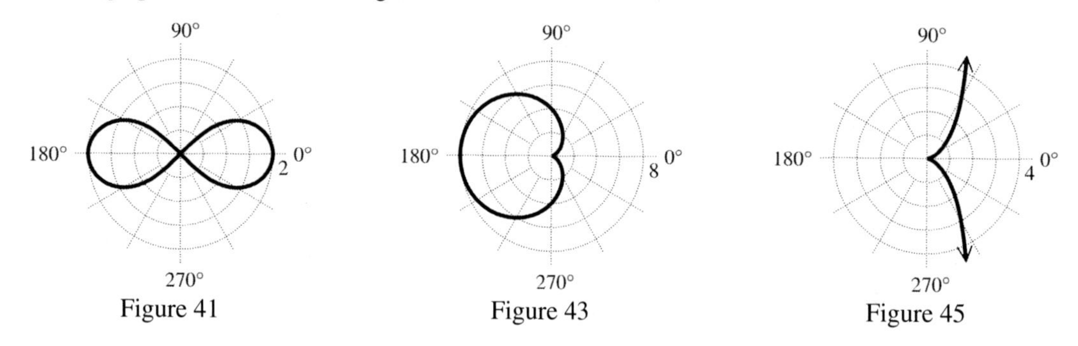

Figure 41 Figure 43 Figure 45

45. The graph is a cissoid. See Figure 45.

47. To graph $(r,\theta), r<0,$ you could locate $\theta,$ add $180°$ to it, and move $|r|$ units along the terminal ray of $\theta+180°$ in standard position.

49. The angle must be quadrantal. That is, it must be coterminal with $0°, 90°, 180°,$ or $270°.$

51. The graph would be reflected across the line $\theta=\frac{\pi}{2}$ (the *y*-axis).

53. The value of *a* determines the length of the petals, and the value of *n* determines the number of petals. If *n* is odd, there are *n* petals. If *n* is even, there are 2*n* petals.

55. $r=2\sin\theta \Rightarrow r^2=2r\sin\theta \Rightarrow x^2+y^2=2y \Rightarrow x^2+y^2-2y+1=1 \Rightarrow x^2+(y-1)^2=1.$ See Figure 55.

57. $r=\dfrac{2}{1-\cos\theta} \Rightarrow r-r\cos\theta=2 \Rightarrow \sqrt{x^2+y^2}-x-2 \Rightarrow \sqrt{x^2+y^2}=x+2 \Rightarrow x^2+y^2=(x+2)^2$

$\Rightarrow x^2+y^2=x^2+4x+4 \Rightarrow y^2=4x+4 \Rightarrow y^2=4(x+1).$ See Figure 57.

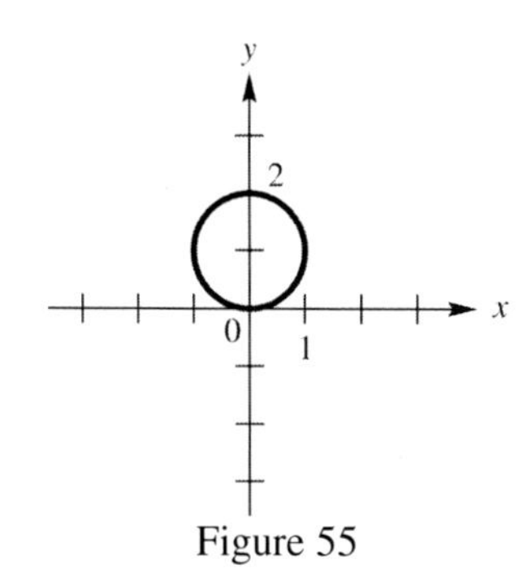

Figure 55

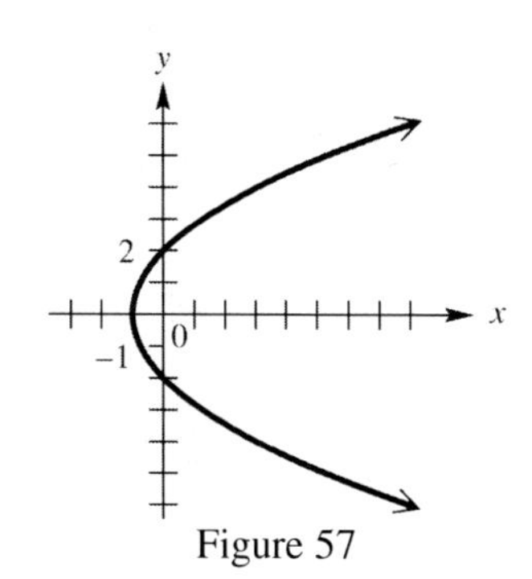

Figure 57

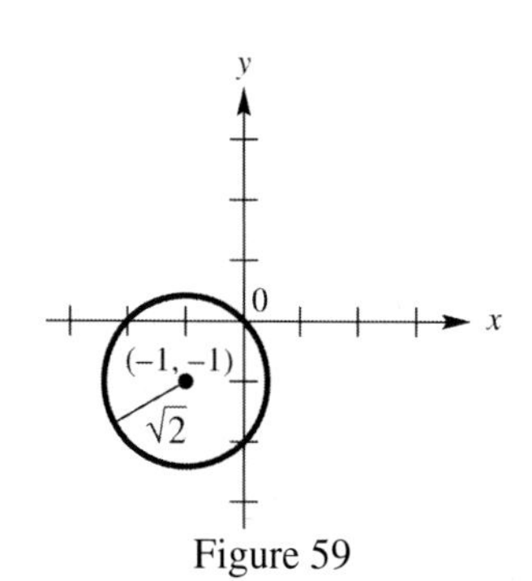

Figure 59

59. $r=-2\cos\theta-2\sin\theta \Rightarrow r^2=-2r\cos\theta-2r\sin\theta \Rightarrow\ x^2+y^2=-2x-2y \Rightarrow x^2+y^2+2x+2y=0 \Rightarrow$

$x^2+2x+1+y^2+2y+1=2 \Rightarrow (x+1)^2+(y+1)^2=2.$ See Figure 59.

61. $r=2\sec\theta \Rightarrow r=\dfrac{2}{\cos\theta}=2 \Rightarrow x=2.$ See Figure 61.

63. $r=\dfrac{2}{\cos\theta+\sin\theta} \Rightarrow r\cos\theta+r\sin\theta=2 \Rightarrow x+y=2.$ See Figure 63.

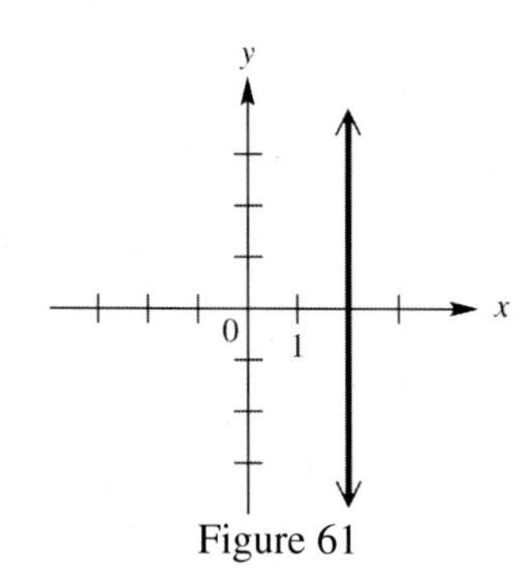

Figure 61

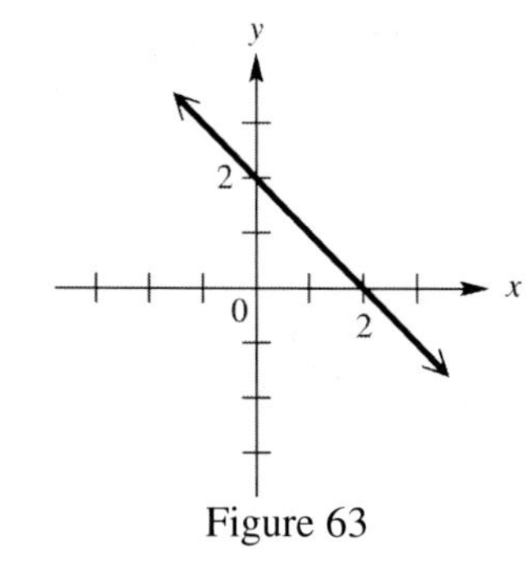

Figure 63

Figure 63

65. See Figure 65.

67. See Figure 67.

[-15,15] by [-15,15]
Xscl = 1 Yscl = 1

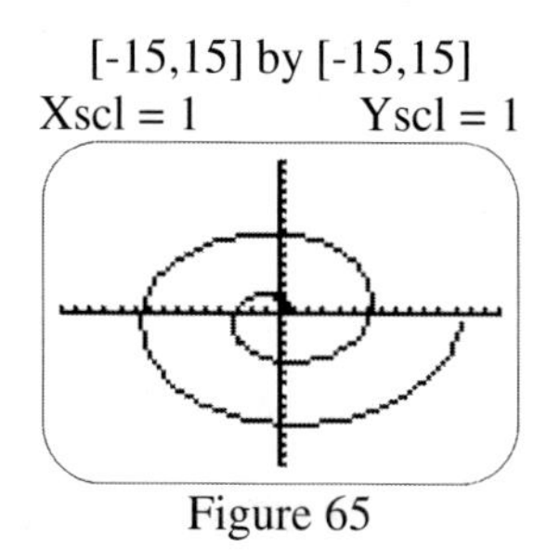

Figure 65

[-20,20] by [-20,20]
Xscl = 1 Yscl = 5

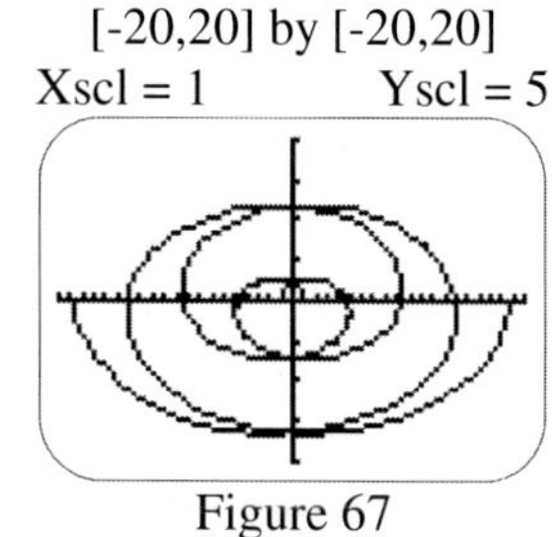

Figure 67

[-2.4,2.4] by [-1.6,1.6]
Xscl = 0.2 Yscl = 0.1

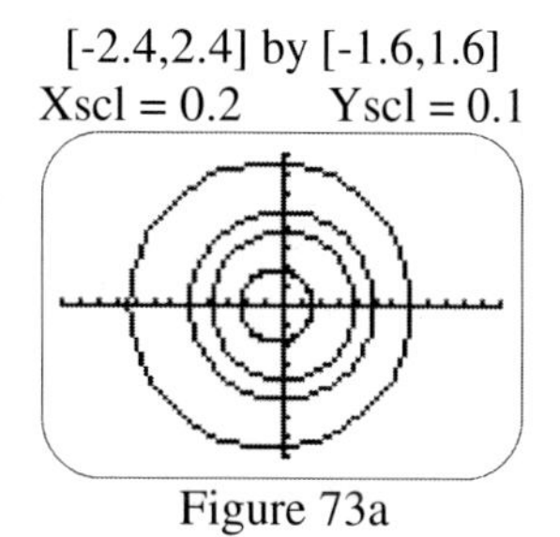

Figure 73a

[-60,60] by [-40,40]
Xscl = 10 Yscl = 10

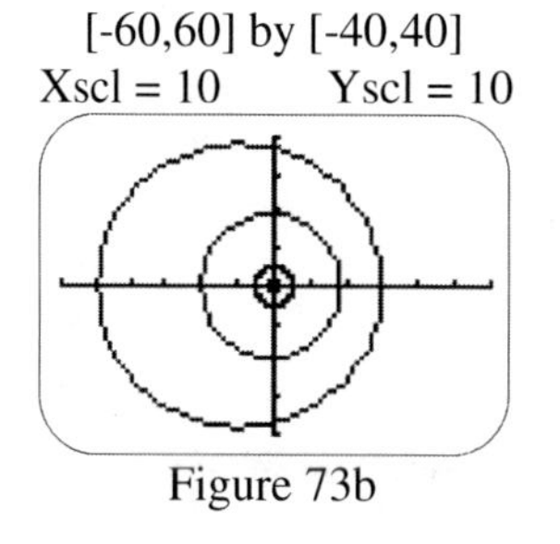

Figure 73b

69. $4\sin\theta=1+2\sin\theta \Rightarrow 2\sin\theta=1 \Rightarrow \sin\theta=\dfrac{1}{2} \Rightarrow \theta=\sin^{-1}\dfrac{1}{2} \Rightarrow \theta=\dfrac{\pi}{6}$ or $\dfrac{5\pi}{6}$ Then

$r=4\sin\theta \Rightarrow r=4\sin\dfrac{\pi}{6} \Rightarrow r=4\left(\dfrac{1}{2}\right) \Rightarrow r=2.$. The intersection points are $\left(2,\dfrac{\pi}{6}\right)$ and $\left(2\dfrac{5\pi}{6}\right)$.

71. $2+\sin\theta = 2+\cos\theta \Rightarrow \sin\theta = \cos\theta \Rightarrow \tan\theta = 1 \Rightarrow \theta = \frac{\pi}{4}$ or $\frac{5\pi}{4}$.

When $\theta = \frac{\pi}{4}, r = 2+\sin\theta \Rightarrow r = 2+\sin\frac{\pi}{4} \Rightarrow r = 2+\left(\frac{\sqrt{2}}{2}\right) \Rightarrow r = \frac{4+\sqrt{2}}{2}$.

When $\theta = \frac{5\pi}{4}, r = 2+\sin\theta \Rightarrow r = 2+\sin\frac{5\pi}{4} \Rightarrow r = 2+\left(\frac{\sqrt{2}}{2}\right) \Rightarrow r = \frac{4-\sqrt{2}}{2}$.

The intersection points are $\left(\frac{4+\sqrt{2}}{2}, \frac{\pi}{4}\right)$ and $\left(\frac{4-\sqrt{2}}{2}, \frac{5\pi}{4}\right)$.

73. (a) See Figure 73a.

(b) See Figure 73b. Earth is closest to the sun.

(c) By graphing the orbits of Neptune and Pluto (not shown), we see that Pluto is not always farthest from the sun.

10.7 More Parametric Equations

1. At $t = 2,\ x = 3(2)+6 = 12$ and $y = -2(2)+4 = 0$. The correct choice is C.

3. At $t = 5,\ x = 5$ and $y = 5^2 = 25$. The correct choice is A.

5. Note that $x = 3\sin t \sin \Rightarrow x^2 = 9\sin^2$ and $y = 3\cos t \Rightarrow y^2 = 9\cos^2 t$. Thus

$x^2+y^2 = 9\sin^2 t + 9\cos^2 t \Rightarrow x^2+y^2 = 9(\sin^2 t+\cos^2 t) \Rightarrow x^2+y^2 = 9$. This is a circle centered at (0, 0) with radius 3.

7. Here $y+x = 2\sin^2 t + 2\cos^2 t \Rightarrow y+x = 2(\sin^2 t+\cos^2 t) \Rightarrow y+x = 2 \Rightarrow y = 2-x$. This is a line segment connecting the points (2, 0) and (0, 2).

9. Note that $x = 3\tan t \Rightarrow x^2 = 9\tan^2 t$ and $y = 2\sec t \Rightarrow y^2 = 4\sec^2 t$. Thus

$9y^2 - 4x^2 = 36\sec^2 t - 36\tan^2 t \Rightarrow \frac{9y^2}{36} - \frac{4x^2}{36} = \sec^2 t - \tan^2 t \Rightarrow \frac{y^2}{4} - \frac{x^2}{9} = 1$. This is the upper branch of a hyperbola.

11. (a) $x = t+2,\ y = t^2$, for t in [–1, 1] See Figure 11.

t	$x = t+2$	$y = t^2$
-1	$-1+2=1$	$(-1)^2 = 1$
0	$0+2=2$	$0^2 = 0$
1	$1+2=3$	$1^2 = 1$

(b) $x-2 = t$, therefore $y = (x-2)^2$ or $y = x^2-4x+4$. Since t is in [–1, 1], x is in [–1 + 2, 1 + 2] or [1, 3].

13. (a) $x = \sqrt{t},\ y = 3t-4$, for t in [0, 4]. See Figure 13.

t	$x=\sqrt{t}$	$y=3t-4$
0	$\sqrt{0}=0$	$3(0)-4=-4$
1	$\sqrt{1}=1$	$3(1)-4=-1$
2	$\sqrt{2}\approx 1.4$	$3(2)-4=2$
3	$\sqrt{3}\approx 1.7$	$3(3)-4=5$
4	$\sqrt{4}=2$	$3(4)-4=8$

(b) $x=\sqrt{t}$, $y=3t-4$ Since $x=\sqrt{t}\Rightarrow x^2=t$, we have $y=3x^2-4$. Since t is in [0, 4], x is in $[\sqrt{0},\sqrt{4}]$ or [0, 2].

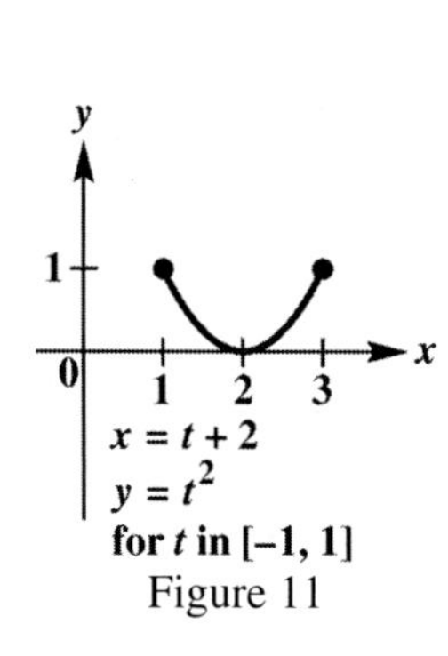

Figure 11

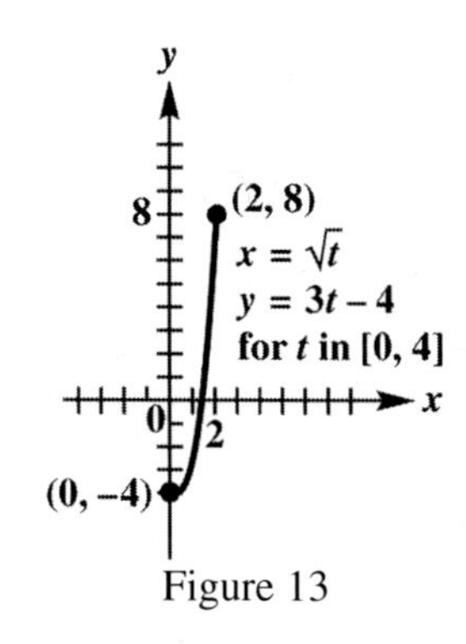

Figure 13

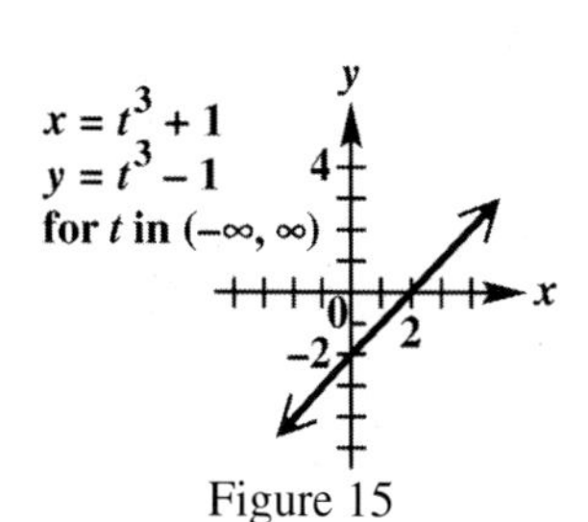

Figure 15

15. (a) $x=t^3+1$, $y=t^3-1$, for t in $(-\infty,\infty)$. See Figure 15.

t	$x=t^3+1$	$y=t^3-1$
−2	$(-2)^3+1=-7$	$(-2)^3-1=-9$
−1	$(-1)^3+1=0$	$(-1)^3-1=-2$
0	$0^3+1=1$	$0^3-1=-1$
1	$1^3+1=2$	$1^3-1=0$
2	$2^3+1=9$	$2^3-1=7$
3	$3^3+1=28$	$3^3-1=26$

(b) Since $x=t^3+1$, we have $x-1=t^3$. Since $y=t^3-1$, we have $y=(x-1)-1=x-2$. Since t is in $(-\infty,\infty)$, x is in $(-\infty,\infty)$.

17. (a) $x=t+2$, $y=\dfrac{1}{t+2}$, for $t\neq 2$.. See Figure 17.

(b) Since $x=t+2$ and $y=\dfrac{1}{t+2}$, we have $y=\dfrac{1}{x}$. Since $t\neq -2$, $x\neq -2+2$, $x\neq 0$. Therefore, x is in $(-\infty, 0)\cup(0,\infty)$.

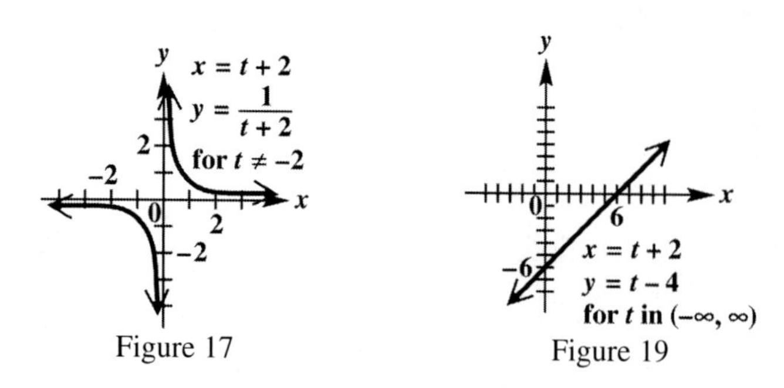

Figure 17 Figure 19

19. (a) $x = t + 2, y = t - 4$, for t in $(-\infty, \infty)$. See Figure 19.

(b) Since $x = t + 2$, we have $t = x - 2$. Since $y = t - 4$, we have $y = (x - 2) - 4 = x - 6$. Since t is in $(-\infty, \infty)$, x is in $(-\infty, \infty)$.

21. (a) The graph traces a circle of radius 3 once. See Figure 21a.

(b) The graph traces a circle of radius 3 twice. See Figure 21b.

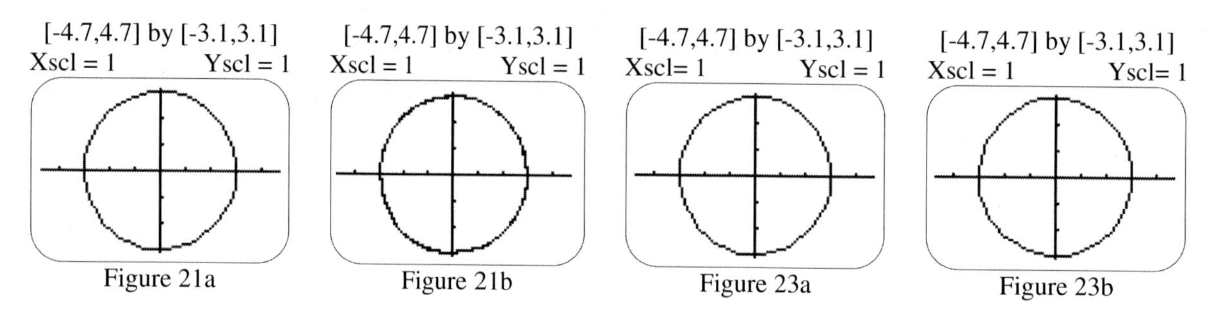

Figure 21a Figure 21b Figure 23a Figure 23b

23. (a) The graph traces a circle of radius 3 once counterclockwise starting at (3, 0). See Figure 23a.

(b) The graph traces a circle of radius 3 once clockwise starting at (0, 3). See Figure 23b.

25. $xy = \sin t \csc t \Rightarrow xy = 1 \Rightarrow y = \frac{1}{x}$. For t in the interval $(0, \pi)$, $x = \sin t$ takes on the values in $(0,1)$.

See Figure 25.

27. $x = 2 + \sin t \Rightarrow x - 2 = \sin t \Rightarrow (x-2)^2 = \sin^2 t$ and $y = 1 + \cos t \Rightarrow y - 1 = \cos t \Rightarrow (y-1)^2 = \cos^2 t$.

Thus $(x-2)^2 + (y-1)^2 = \sin^2 t + \cos^2 t \Rightarrow (x-2)^2 + (y-1)^2 = 1$. For

in the interval $[0, 2\pi]$, $x = 2 + \sin t$ takes on values in $[1,3]$. See Figure 27.

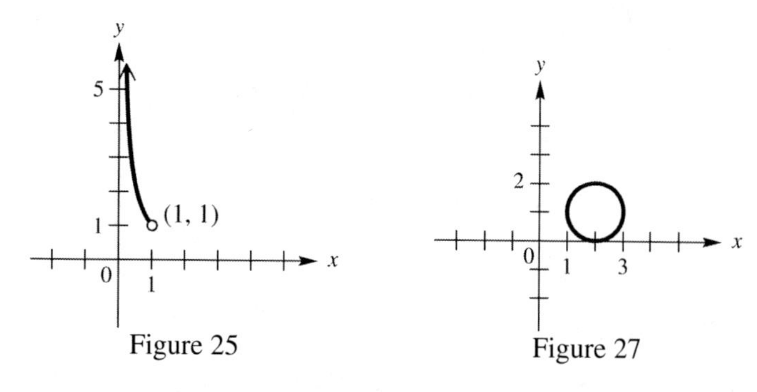

Figure 25 Figure 27

29. See Figure 29.

31. See Figure 31.

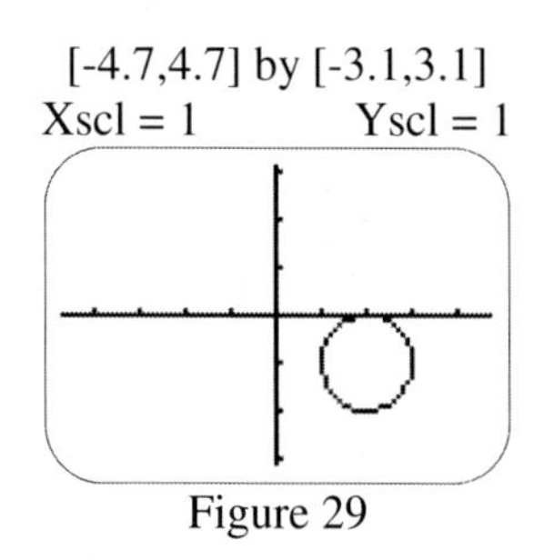

Figure 29

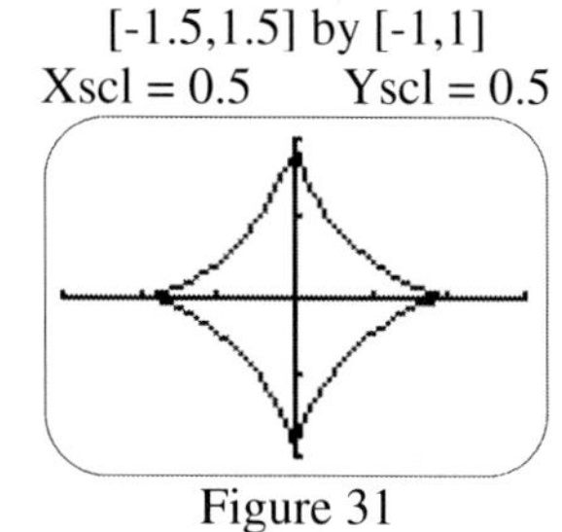

Figure 31

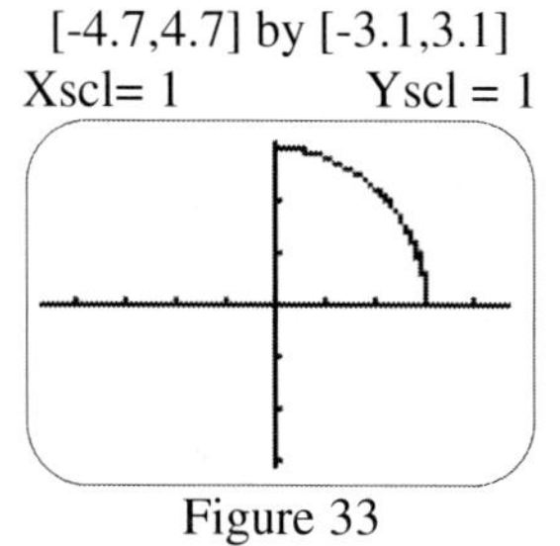

Figure 33

33. See Figure 33.

35. See Figure 35.

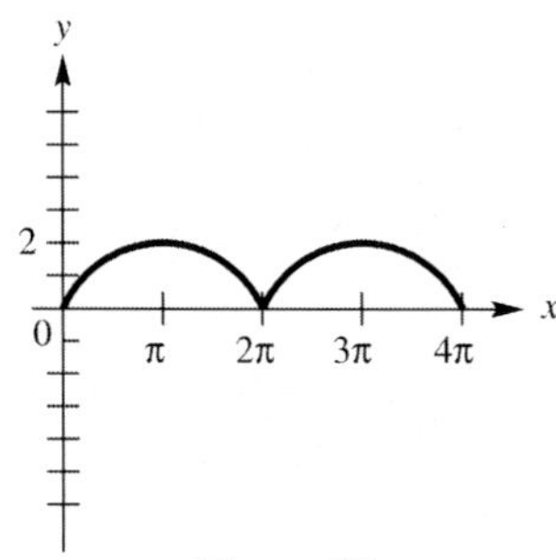

Figure 35

37. The letter graphed is F. See Figure 37.

39. The letter graphed is D. See Figure 39.

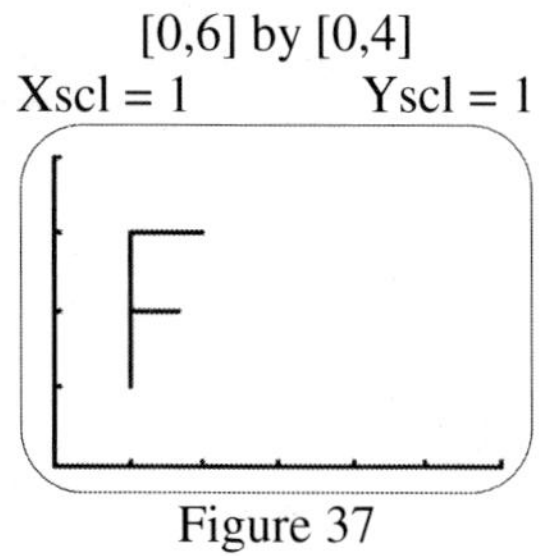

Figure 37

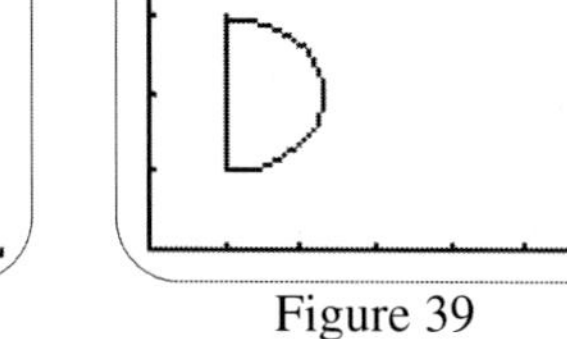

Figure 39

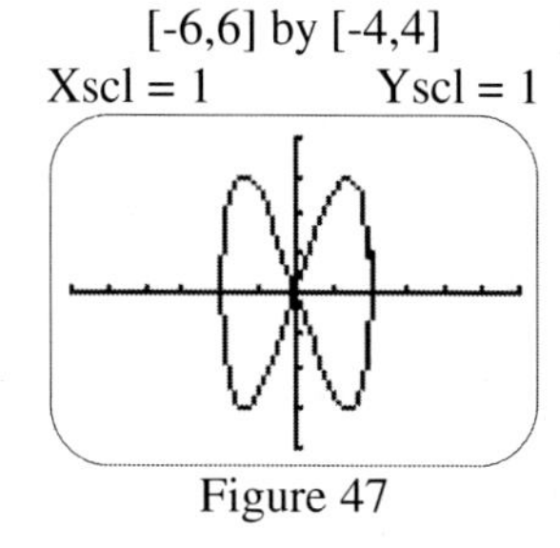

Figure 47

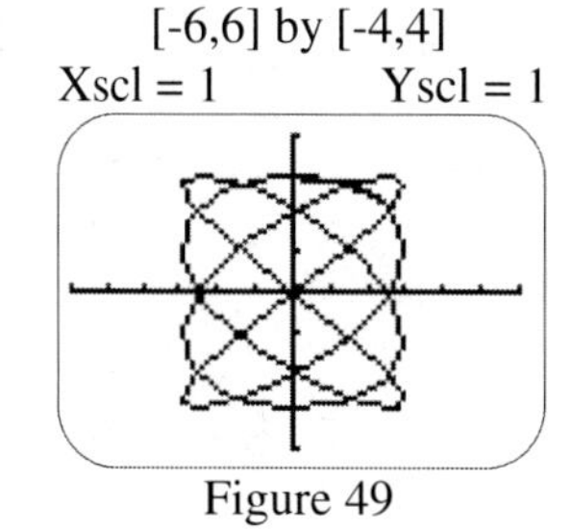

Figure 49

41. Answers may vary. One possibility is: $x_1 = 0, y_1 = 2t, x_2 = t, y_2 = 0$ for $0 \le t \le 1$.

43. Answers may vary. One possibility is: $x_1 = \sin t, y_1 = \cos t, x_2 = 0, y_2 = t - 2$ for $0 \le t \le \pi$.

45. Answers may vary.

47. See Figure 47.

49. See Figure 49.

51. (a) $x = (48\cos 60°)t \Rightarrow x = 48\left(\frac{1}{2}\right)t \Rightarrow x = 24t$

$$y = (48\sin 60°)t - 16t^2 + 0 \Rightarrow y = 48\left(\frac{\sqrt{3}}{2}\right)t - 16t^2 \Rightarrow y = -16t^2 + 24\sqrt{3}t$$

(b) $x = 24t \Rightarrow t = \frac{x}{24}$, thus $y = -16t^2 + 24\sqrt{3}t \Rightarrow y = -16\left(\frac{x}{24}\right)^2 + 24\sqrt{3}\left(\frac{x}{24}\right) \Rightarrow y = -\frac{1}{36}x^2 + \sqrt{3}x$

(c) Solve the equation $-16t^2 + 24\sqrt{3}t = 0$ to determine when the rocket is at ground level. The solutions are 0 and 2.6 seconds. Therefore the rocket is in flight for about 2.6 seconds. The horizontal distance traveled is $x = 24t \Rightarrow x = 24(2.6) \approx 62$ feet.

53. (a) $x = (88\cos 20°)t, \; y = -16t^2 + (88\sin 20°)t + 2$,

(b) $x = (88\cos 20°)t \Rightarrow t = \frac{x}{88\cos 20°}$, thus

$$y = 2 - 16\left(\frac{x}{88\cos 20°}\right)^2 + (88\sin 20°)\left(\frac{x}{88\cos 20°}\right) \Rightarrow y = 2 - \frac{x^2}{484\cos^2 20°} + (\tan 20°)x$$

(c) Use the quadratic formula to solve the equation $y = -16t^2 + (88\sin 20°)t + 2$ to determine when the ball is at ground level. The solutions are $-.064$ and 1.945 seconds. Therefore the ball is in flight for about 1.9 seconds. The horizontal distance traveled is $x = (88\cos 20°)(1.945) \approx 161$ feet.

55. $x = (88\cos 45°)t = x = 88\left(\frac{\sqrt{2}}{2}\right)t \Rightarrow x = 44\sqrt{2}t$

$y = (88\cos 45°)t - 2.66t^2 + 0 \Rightarrow y = 88\left(\frac{\sqrt{2}}{2}\right)t - 2.66t^2 \Rightarrow -2.66t^2 + 44\sqrt{2}t$. Solve the equation $-2.66t^2 + 44\sqrt{2}t = 0$ to determine when the ball is at ground level. The solutions are 0 and 23.393 seconds. The horizontal distance traveled is $x = 44\sqrt{2}t \Rightarrow x = 24\sqrt{2}(23.393) \approx 1456$ feet.

57. (a) See Figure 57.

(b) $(88\cos\theta)t = 82.69265063t \Rightarrow \cos\theta = \frac{82.69265063}{88} \Rightarrow \cos^{-1}\left(\frac{82.69265063}{88}\right) \approx 20.0°$

(c) $x = (88\cos 20.0°)t$ and $y = -16t^2 + (88\sin 20.0°)t$

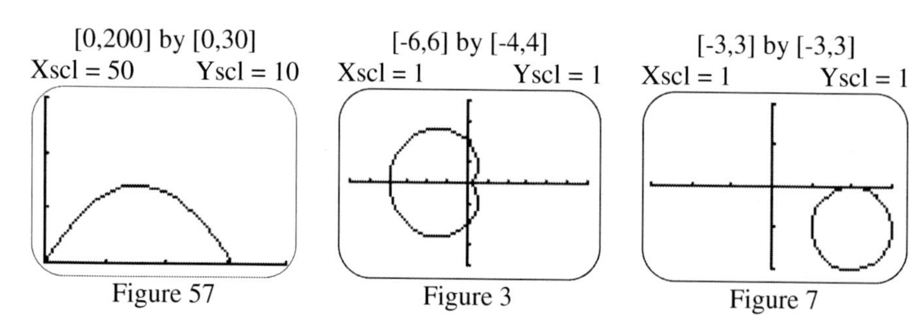

Figure 57 Figure 3 Figure 7

59. Since $x = r\cos\theta$ and $y = r\sin\theta$, by substitution, $x = a\theta\cos\theta$ and $y = a\theta\sin\theta$ for θ in $(-\infty, \infty)$

Reviewing Basic Concepts (Sections 10.6 and 10.7)

1. The point is in quadrant IV.

2. $r=\sqrt{(-2)^2+2^2}=\sqrt{8}=2\sqrt{2};\ \theta=\tan^{-1}\left(\frac{2}{-2}\right)=-45°$. Since the point is in QII $\theta=135°$. Two forms are $\left(2\sqrt{2},135°\right)$ and $\left(-2\sqrt{2},-45°\right)$.

3. The graph is a cardioid. See Figure 3

4. $r=2\cos\theta \Rightarrow r^2=2r\cos\theta \Rightarrow x^2+y^2=2x \Rightarrow x^2-2x+1+y^2=1 \Rightarrow (x-1)^2+y^2=1$

5. $x+y=6 \Rightarrow r\cos\theta+r\sin\theta=6 \Rightarrow r(\cos\theta+\sin\theta)=6 \Rightarrow r=\dfrac{6}{\cos\theta+\sin\theta}$

6. Note that $x=2\cos t \Rightarrow x^2=4\cos^2 t$ and $y=4\sin t \Rightarrow y^2=16\sin 2t$. Thus

$$4x^2+y^2=16\cos^2 t+16\sin^2 t \Rightarrow 4x^2+y^2=16\left(\cos^2 t+\sin^2 t\right) \Rightarrow 4x^2+y^2=16 \Rightarrow \frac{x^2}{4}+\frac{y^2}{16}=1.$$

7. See Figure 7.

[-6,6] by [-4,4]
Xscl = 1 Yscl = 1

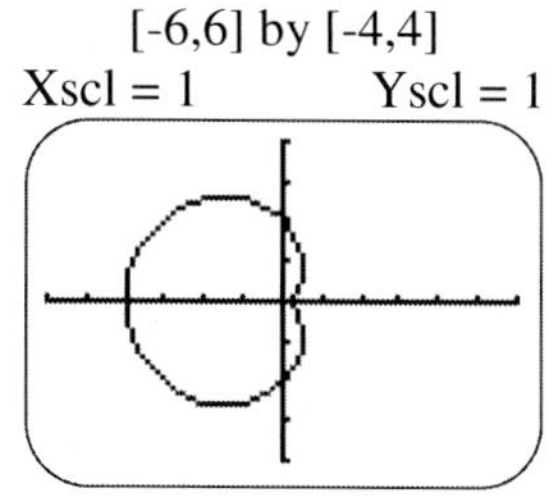

Figure 3

[-3,3] by [-3,3]
Xscl = 1 Yscl = 1

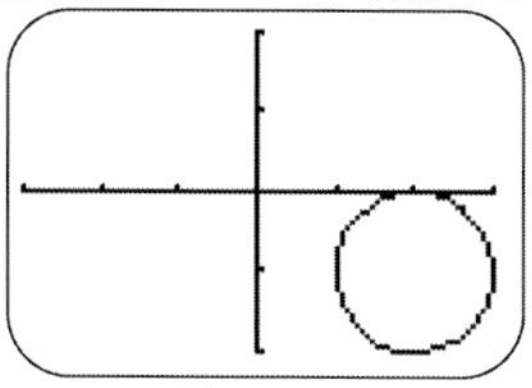

Figure 7

8. $x=(88\cos 45°)t \Rightarrow x=88\left(\frac{\sqrt{2}}{2}\right)t \Rightarrow x=44\sqrt{2}t$

$y=(88\cos 45°)t-16t^2+50 \Rightarrow y=88\left(\frac{\sqrt{2}}{2}\right)t-16t^2+50 \Rightarrow y=50-16t^2+88\sqrt{2}t$. Use the quadratic formula to solve the equation $50-16t^2+44\sqrt{2}t=0$ to determine when the ball is at ground level. The solutions are $-.683$ and 4.5725 seconds. The horizontal distance traveled is $x=44\sqrt{2}(4.5725)\approx 285$ feet.

Chapter 10 Review

1. Using the Law of Sines, $\dfrac{b}{\sin 39°30'}=\dfrac{96.3}{\sin 74°10'} \Rightarrow b=\dfrac{96.3\sin 39°30'}{\sin 74°10'} \Rightarrow b\approx 63.7$ m.

2. Using the Law of Sines, $\dfrac{b}{\sin 25.0°}=\dfrac{165}{\sin 100.2°} \Rightarrow b=\dfrac{165\sin 25.0°}{\sin 100.2°} \Rightarrow b\approx 70.9$ m.

3. Using the Law of Cosines,

$$86.14^2=253.2^2+241.9^2-2(253.2)(241.9)\cos A \Rightarrow A=\cos^{-1}\left(\frac{7420.0996-122,625.85}{-122,498.16}\right)\approx 19.87°$$

4. Using the Law of Cosines,

$19.7^2 = 14.8^2 + 31.8^2 - 2(14.8)(31.8)\cos B \Rightarrow B = \cos^{-1}\left(\frac{388.09 - 1230.28}{-941.28}\right) \approx 26.5°$.

5. Using the Law of Sines,

$\frac{\sin B}{69.8} = \frac{\sin 129°40'}{127} \Rightarrow \sin B = \frac{69.8 \sin 129°40'}{127} \Rightarrow B = \sin^{-1}\left(\frac{69.8 \sin 129°40'}{127}\right) \approx 25°00'$.

6. Using the Law of Sines,

$\frac{\sin A}{340} = \frac{\sin 39°50'}{268} \Rightarrow \sin A = \frac{340 \sin 39°50'}{268} \Rightarrow A = \sin^{-1}\left(\frac{340 \sin 39°50'}{268}\right) \approx 54°20'$.

Since $180° - 54°20' = 125°40'$ and $39°50' + 125°40' < 180°$, the other possible angle is . $180° - 54°20' = 125°40'$

7. Using the Law of Cosines,

$b^2 = 127^2 + 69.8^2 - 2(127)(69.8)\cos 120.7° \Rightarrow b = \sqrt{21{,}001.04 - 17{,}729 \cos 120.7°} \approx 173$ ft.

8. Using the Law of Cosines,

$a^2 = 184^2 + 192^2 - 2(184)(192)\cos 46.2° \Rightarrow a = \sqrt{70{,}720 - 70.656 \cos 46.2°} \approx 148$ cm

9. $A = \frac{1}{2}(840.6)(715.9)\sin(149.3°) \approx 153600 \text{ m}^2$.

10. $A = \frac{1}{2}(6.90)(10.2)\sin 35°10' \approx 20.3 \text{ ft}^2$

11. $s = \frac{1}{2}(.913 + .816 + .582) = 1.1555$; $A = \sqrt{1.1555(1.1555 - .913)(1.1555 - .816)(1.1555 - .582)} \approx .234 \text{ km}^2$

12. $s = \frac{1}{2}(43 + 32 + 51) = 63; A = \sqrt{63(63 - 43)(63 - 32)(63 - 51)} \approx 680 \text{ m}^2$

13. First note that $C = 180° - 47°20' - 24°50' = 107°50'$. Now, using the Law of Sines,

$\frac{a}{\sin 24°50'} = \frac{8.4}{\sin 107°50'} \Rightarrow a = \frac{8.4 \sin 24°50'}{\sin 107°50'} \Rightarrow a \approx 3.71$ mi. If a perpendicular line segment joining *C* to the the ground intersects the ground at *D*, we can use right triangle *BDC* to find the distance *d* to the balloon:

$\sin 47°20' = \frac{d}{3.71} \Rightarrow d = 3.71 \sin 47°20' \approx 2.7$ mi.

14. Triangle *XYZ* has the following measurements: $X = 48°, Z = 36°, Y = 180° - 48° - 36° = 96°$, and $y = 10$.

Using the Law of Sines, $\frac{x}{\sin 48°} = \frac{10}{\sin 96°} \Rightarrow x = \frac{10 \sin 48°}{\sin 96°} \Rightarrow x \approx 7\text{km}$.

15. Let *C* represent the position of the illegal transmitter. Triangle *ABC* has the following measurements: $A = 90° - 48° = 42°, B = 302° - 270° = 32°, C = 180° - 42° - 32° = 160°$, and $c = 3.46$. Using the Law of Sines, $\frac{b}{\sin 32°} = \frac{3.46}{\sin 106°} \Rightarrow b = \frac{3.46 \sin 32°}{\sin 106°} \Rightarrow b \approx 1.91$ mi.

16. First note that $A = 180° - 58.4° - 27.9° = 93.7°$. Now, using the Law of Sines,

$\frac{c}{\sin 27.9°} = \frac{125}{\sin 93.7°} \Rightarrow c = \frac{125 \sin 27.9°}{\sin 93.7°} \Rightarrow c \approx 58.6$ ft.

17. Let *A, B,* and *C* represent the lower end of the brace, the lower end of the pole, and the upper end of the pole, respectively. Then $A = 180° - 115° - 22° = 43°, B = 115°, C = 22°$, and $a = 8.0$.

    Using the Law of Cosines, $\frac{b}{\sin 115°} = \frac{8.0}{\sin 43°} \Rightarrow b = \frac{8.0 \sin 115°}{\sin 43°} \Rightarrow b \approx 11$ ft.

18. Using the Law of Cosines,

    $c^2 = 15^2 + 12.2^2 - 2(15)(12.2)\cos 70.3° \Rightarrow c = \sqrt{373/.84 - 366 \cos 70.3°} \approx 15.8$ ft.

19. Using the Law of Cosines, $150^2 = 102^2 + 135^2 - 2(102)(135)\cos C \Rightarrow \cos^{-1}\left(\frac{22,500 - 28,629}{-27,540}\right) \approx 77.1°$.

20. This in not an ambiguous case because we are not given SSA.

21. This triangle cannot exist because a+b=c.

22. (a) If A and B lie on a horizontal line with C positioned above the line, then C lies $10 \sin 30° = 5$ units above the line. Thus there will be exactly one value if b = 5 or if $b \geq 10$.

    (b) There will be two possible values if $5 < b < 10$.

    (c) There will be no value if $b < 5$.

23. If $C = 90°$, the value of $2ab\cos\theta$ is 0. The Law of Cosines becomes the Pythagorean theorem.

24. If the cosine value is positive, the angle is acute. If the cosine value is negative, the angle is obtuse.

25. See Figure 25.

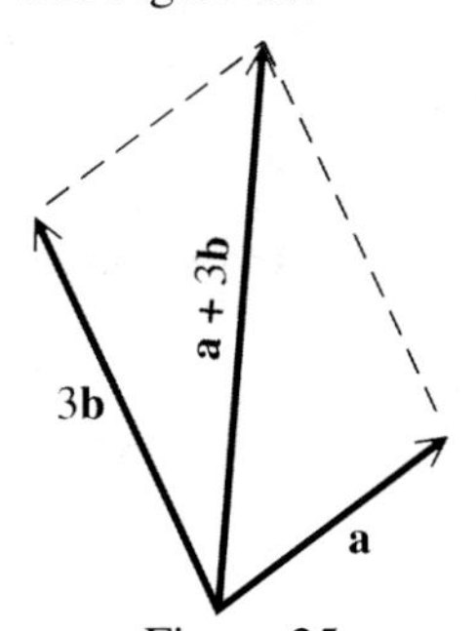

Figure 25

26. $|\langle 21, -20\rangle| = \sqrt{21^2 + (-20)^2} = \sqrt{441 + 400} = \sqrt{841} = 29$; direction angle: $\tan^{-1}\left(\frac{-20}{21}\right) \approx 316.4°$.

27. $|\langle -9, 12\rangle| = \sqrt{(-9)^2 + 12^2} = \sqrt{81 + 144} = \sqrt{225} = 15$; direction angle: $\tan^{-1}\left(\frac{12}{-9}\right) \approx 126.9°$.

28. $\langle 50\cos 45°, 50\sin 45°\rangle = \left\langle 50 \cdot \frac{\sqrt{2}}{2}, 50 \cdot \frac{\sqrt{2}}{2}\right\rangle = \langle 25\sqrt{2}, 25\sqrt{2}\rangle$.

29. $\langle 69.2\cos 75°, 69.2\sin 75°\rangle \approx \langle 17.9, 66.8\rangle$.

30. $\langle 964\cos 154°20', 964\sin 154°20'\rangle \approx \langle -869, 418\rangle$.

31. (a) $\langle 6,2\rangle \cdot \langle 3.-2\rangle = (6)(3)+(2)(-2) = 18-4 = 14$.

(b) $\cos\theta = \dfrac{u\cdot v}{|u||v|} = \dfrac{\langle 6,2\rangle\cdot\langle 3,-2\rangle}{|\langle 6,2\rangle||\langle 3,-2\rangle|} = \dfrac{(6)(3)+(2)(-2)}{\sqrt{36+4}\sqrt{9+4}} = \dfrac{14}{\sqrt{40}\sqrt{13}} = \dfrac{14}{\sqrt{520}} \Rightarrow$

$\theta = \cos^{-1}\left(\dfrac{14}{\sqrt{520}}\right) = 52.13°$.

32. (a) $\langle 2\sqrt{3},2\rangle\cdot\langle 5,5\sqrt{3}\rangle = (2\sqrt{3})(5)+(2)(5\sqrt{3}) = 10\sqrt{3}+10\sqrt{3}+10\sqrt{3} = 20\sqrt{3}$.

(b) $\cos\theta = \dfrac{u\cdot v}{|u||v|} = \dfrac{\langle 2\sqrt{3},2\rangle\cdot\langle 5,5\sqrt{3}\rangle}{|\langle 2\sqrt{3},2\rangle||\langle 5m5\sqrt{3}\rangle} = \dfrac{(2\sqrt{3})(5)+(2)(5\sqrt{3})}{\sqrt{12+4}\sqrt{25+75}} = \dfrac{20\sqrt{3}}{\sqrt{16}\sqrt{100}} = \dfrac{20\sqrt{3}}{40} = \dfrac{\sqrt{3}}{2} \Rightarrow$

$\theta = \cos^{-1}\left(\dfrac{\sqrt{3}}{2}\right) = 30°$.

33. The vectors are orthogonal because $u\cdot v = \langle 5,-1\rangle\cdot\langle -2,-10\rangle = (5)(-2)+(-1)(-10) = -10+10 = 0$.

34. The adjacent angle in the parallelogram is supplementary to $10°+15° = 25°$, thus it has measure $155°$.

$|v|^2 = 12^2+18^2-2(12)(18)\cos 155° \approx 859.524964 \Rightarrow |v| = \sqrt{859.524964} \approx 29$ lb.

35. The adjacent angle in the parallelogram is supplementary to $45°$, thus it has measure $135°$.

$|v|^2 = 1000^2+2000^2-2(1000)(2000)\cos 135° \approx 7{,}828{,}427.125 \Rightarrow |v| = \sqrt{7{,}828{,}427.125} \approx 2798$ The resultant is about 2800 newtons. The required angle can be found using the Law of Cosines:

$2000^2 = 1000^2+2798^2-2(1000)(2798)\cos A \Rightarrow \cos^{-1}\left(\dfrac{4{,}000{,}000-8{,}828{,}804}{-5{,}596{,}000}\right) \approx 30.4°$.

36. Let A represent the plan's initial position. Let B represent the point at which the vector representing the desired flight path $(310°\text{ bearing})$ intersects the vector representing the wind. Triangle ABC has the following measurements: $C = (360°-310°)+(121°-180°) = 82°$, $a = 37$, and $c = 520$. Using the Law of Sines, $\dfrac{\sin A}{37} = \dfrac{\sin 82°}{520} \Rightarrow \sin A = \dfrac{37\sin 82°}{520} \Rightarrow A = \sin^{-1}\dfrac{(37\sin 82°)}{520} \approx 4°$. The bearing the plane should take is $310°-4° = 306°$. Noting that $B = 180°-82°-4° = 94°$, $\dfrac{b}{\sin 94°} = \dfrac{520}{\sin 82°} \Rightarrow b = \dfrac{520\sin 94°}{\sin 82°} \Rightarrow b \approx 524$. The plane's actual speed is about 524 mph.

37. The angle between the current and the due north swim path is $180°-12° = 168°$. If b is the missing side of the triangle, $b^2 = 3.2^2+5.1^2-2(3.2)(5.1)\cos 168° \Rightarrow b = \sqrt{36.25-32.64\cos 168°} \approx 8.3$ mph. Then $\dfrac{\sin A}{5.1} = \dfrac{\sin 168°}{8.3} \Rightarrow \sin A = \dfrac{5.1\sin 168°}{8.3} \Rightarrow A = \sin^{-1}\left(\dfrac{5.1\sin 168°}{8.3}\right) \approx 7°20'$. The swimmer's actual speed is 8.3 mph, and the resulting bearing is $7°20'$.

38. (a) The speed of the wind is $|\mathbf{v}| = \sqrt{6^2+8^2} = \sqrt{36+64} = \sqrt{100} = 10$ mph.

(b) $3\mathbf{v} = 3(6\mathbf{i}+8\mathbf{j}) = 18\mathbf{i}+24\mathbf{j}$; This represents a 30 mph wind in the direction of v.

(c) $\mathbf{u}$ represents a southeast wind with speed $|\mathbf{u}| = \sqrt{(-8)^2+8^2} = \sqrt{64+64} = \sqrt{128} \approx 11{,}3$ mph.

39. See Figure 39.

40. See Figure 40.

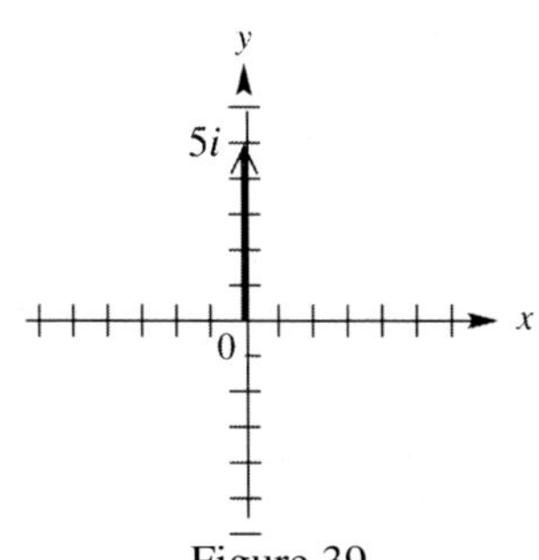

Figure 39

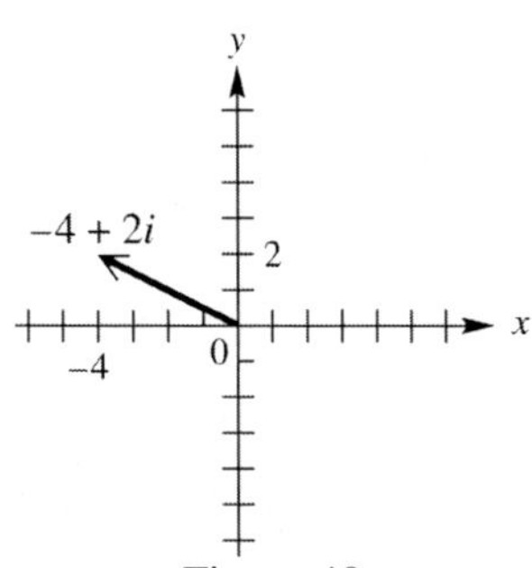

Figure 40

41. $(7+3i)+(-2+i) = (7+(-2))+(3+1)i = 5+4i$.

42. $(2-4i)+(5+i) = (2+5)+(-4+1)i = 7-3i$.

43. $r = \sqrt{(-2)^2+2^2} = \sqrt{8} = 2\sqrt{2}$; $\tan\theta = \dfrac{2}{-2} \Rightarrow \tan\theta = -1 \Rightarrow \theta = \tan^{-1}(-1) = -45°$. Since θ is in quadrant II,

$\theta = -45° + 180° = 135°$. Therefore $-2+2i = 2\sqrt{2}\text{cis}135°$.

44. $3(\cos 90° + i\sin 90°) = 3(0+1i) = 3i$.

45. $2\text{cis}225° = 2(\cos 225° + i\sin 225°) = 2\left(-\dfrac{\sqrt{2}}{2} - \dfrac{\sqrt{2}}{2}i\right) = -\sqrt{2} - \sqrt{2i}$.

46. $r = \sqrt{(-4)^2+(4\sqrt{3})^2} = \sqrt{64} = 8$; $\tan\theta = \dfrac{4\sqrt{3}}{-4} \Rightarrow \theta = -60°$. Since θ is in quadrant II, $\theta = -60° + 180° = 120°$.

Therefore $-4+4i\sqrt{3} = 8\text{cis}120°$.

47. $[5(\cos 90° + i\sin 90°)][6(\cos 180° + i\sin 180°)] = (5.6)[\cos(90°+180°) + i\sin(90°+180°)]$

$= 30(\cos 270° + i\sin 270°) = 30(0-i) = -30i.$

48. $(3\text{cis}135°)(2\text{cis}105°) = [3(\cos 135° + i\sin 135°)][2(\cos 105° + i\sin 105°)]$

$= (3.2)[\cos(135°+105°) + i\sin(135°+105°)] = 6(\cos 240° + i\sin 240°) = 6\left(-\dfrac{1}{2} - \dfrac{\sqrt{3}}{2}i\right) = -3 - 3\sqrt{3i}.$

49. $\dfrac{2(\cos 60° + i\sin 60°)}{8(\cos 300° + i\sin 300°)} = \dfrac{2}{8}[\cos(60°-300°)] + i\sin(60°-300°)$

$\dfrac{1}{4}[\cos(-240°) + i\sin(-240°)] = \dfrac{1}{4}\left(-\dfrac{1}{2} + \dfrac{\sqrt{3}}{2}i\right) = -\dfrac{1}{8} + \dfrac{\sqrt{3}}{8}i$.

50. $\dfrac{4\text{cis}270°}{2\text{cis}90°} = \dfrac{4(\cos 270° + i\sin 270°)}{2(\cos 90°) + i\sin 90°} = \dfrac{4}{2}\left[\cos(270° - 90°) + i\sin(270° - 90°)\right]$

$= 2(\cos 180° + i\sin 180°) = 2(-1 + 0i) = -2.$

51. First find the trigonometric form of $\sqrt{3} + i$; $r = \sqrt{(\sqrt{3})^2 + 1^2} = \sqrt{4} = 2$;

$\tan\theta = \dfrac{y}{x} \Rightarrow \tan\theta = \dfrac{1}{\sqrt{3}} \Rightarrow\Rightarrow \theta = \tan^{-1}\dfrac{1}{\sqrt{3}} = 30°$. Therefore $\sqrt{3} + i = 2(\cos 30° + i\sin 30°)$

$(\sqrt{3} + i)^3 = \left[2(\cos 30° + i\sin 30°)\right]^3 = 2^3\left[\cos(3\cdot 30°) + i\sin(3\cdot 30°)\right] = 8(\cos 90° + i\sin 90°) = 8(0 + 1i) = 8i.$

52. First find the trigonometric form of $2 - 2i$; $r = \sqrt{2^2 + (-2)^2} = \sqrt{8} = 2\sqrt{2}$;

$\tan\theta = \dfrac{y}{x} \Rightarrow \tan\theta = \dfrac{-2}{2} \Rightarrow\Rightarrow \theta = \tan^{-1}(-1) = 315°$. Therefore $2 - 2i = 2\sqrt{2}(\cos 315° + i\sin 315°)$.

$(2 - 2i)^5 = \left[2\sqrt{2}(\cos 315° + i\sin 315°)\right]^5 = (2\sqrt{2})^5\left[\cos(5.315°) + i\sin(5\cdot 315)\right] =$

$128\sqrt{2}(\cos 1575° + i\sin 1575°) = 128\sqrt{2}\left(-\dfrac{\sqrt{2}}{2} + \dfrac{\sqrt{2}}{2}i\right) = -128 + 128i$

53. $(\cos 100° + i\sin 100°)^6 = \cos(6.100°) + i\sin(6\cdot 100°) = \cos 600° + i\sin 600° = -\dfrac{1}{2} - \dfrac{\sqrt{3}}{2}i$.

54. $(\text{cis}20°)^3 = (\cos 20° + i\sin 20°)^3 = \cos(3.20°) + i\sin(3.20°) = \cos 60° + i\sin 60° = \dfrac{1}{2} + \dfrac{\sqrt{3}}{2}i$

55. Note that $-27i = 27(\cos 270° + i\sin 270°)$. Here $r = 27$, so $\sqrt[3]{r} = 3$, $n = 3$ and $k = 0, 1,$ or 2.

$\alpha = \dfrac{270° + 360°\cdot k}{3}$: For $k = 0$, $\alpha = 90°$. For $k = 1$, $\alpha = 210°$. For $k = 2$, $\alpha = 330°$. The cube roots are

3cis90°, 3cis210°, and 3cis330°. See Figure 55.

56. Note that $16i = 16(\cos 90° + i\sin 90°)$. Here r = 16 so $\sqrt[4]{r} = 2$, $n = 4$ and $k = 0,1,2,$or 3.

$\alpha = \dfrac{90° + 360°\cdot k}{4}$: For $k = 0$, $\alpha = 22.5°$. For $k = 1$, $\alpha = 112.5°$. For $k = 2$, $\alpha = 202.5°$. For

$k = 3$, $\alpha = 292.5°$. The fourth roots are $2cis22.5°$, 2cis112.5°, 2cis202.5°, and 2cis 292.5°. See Figure 56.

57. Note that $32 = 32(\cos 0° + i\sin 0°)$.

Here $r = 32$ so $\sqrt[5]{r} = 2$, $n = 5$ and $k = 0.1.2.3.$or 4. $\alpha = \dfrac{0° + 360\cdot k}{5}$: For $k = 0$, $\alpha = 0°$.

For $k = 1$, $\alpha = 72°$. For $k = 2$, $\alpha = 144°$. For $k = 3, \alpha = 216°$. For $k = 4$, $\alpha = 288°$. The fifth roots are

2cis0°, 2cis72°, 2cis144°, 2cis216°, and 2cis288°. See Figure 57.

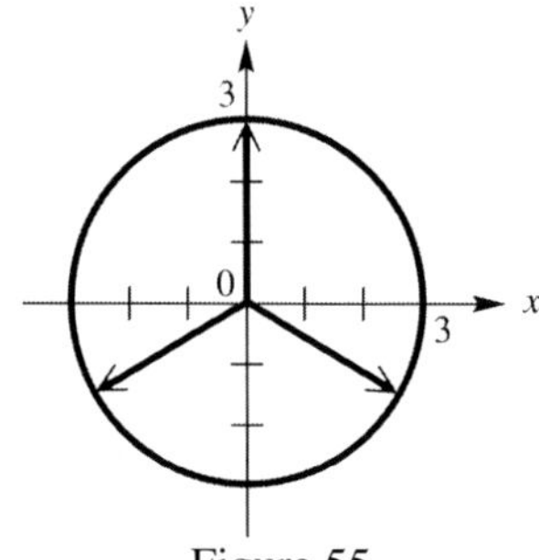

Figure 55

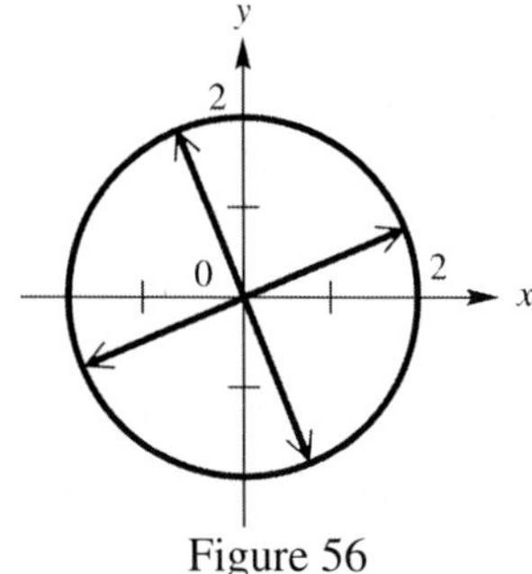

Figure 56

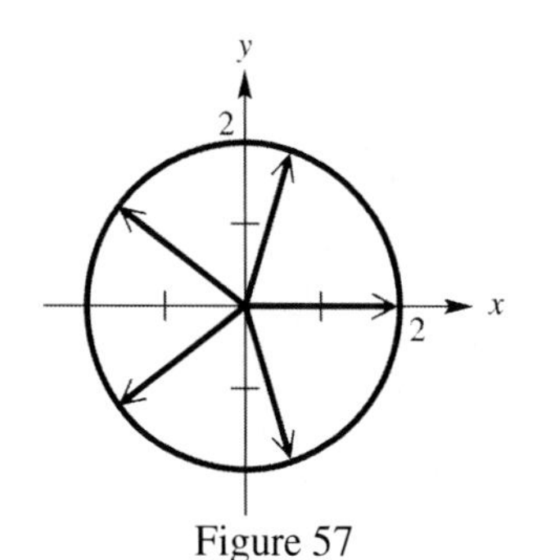

Figure 57

58. $x^4 + i = 0 \Rightarrow x^4 = -i$. Find the fourth roots of –i. Note that $-i = \cos 270° + i \sin 270°$.

Here $r = 1$ so $\sqrt[4]{r} = 1$, $n = 4$ and $k = 0,1,2$, or 3. $\alpha = \dfrac{270° + 360° \cdot k}{4}$: For $k = 0$, $\alpha = 67.5°$.

For $k = 1$, $\alpha = 157.5°$. For $k = 2$, $\alpha = 247.5°$. For $k = 3$, $\alpha = 337.5°$. The four roots are

cis67.5°, cis157.7°, cis247.5°, and cis337.5°.

59. $x = r\cos\theta = 12\cos 225 = 12\left(-\dfrac{\sqrt{2}}{2}\right) = -6\sqrt{2}; y = r\sin\theta = 12\sin 225° = 12\left(-\dfrac{\sqrt{2}}{2}\right) = -6\sqrt{2}$

The rectangular coordinates are $\left(-6\sqrt{2}, -6\sqrt{2}\right)$.

60. $x = r\cos\theta = -8\cos\left(-\dfrac{\pi}{3}\right) = -8\left(\dfrac{1}{2}\right) = -4; y = r\sin\theta = -8\sin\left(-\dfrac{\pi}{3}\right) = -8\left(-\dfrac{\sqrt{3}}{2}\right) = 4\sqrt{3}$

The rectangular coordinates are $\left(-4, 4\sqrt{3}\right)$.

61. $r = \sqrt{(-6)^2 + 6^2} = \sqrt{72} = 6\sqrt{2}; \theta = \tan^{-1}\left(\dfrac{6}{-6}\right) = -45°$. Since the point is in QII, $\theta = 135°$.

One possible form is $\left(6\sqrt{2}, 135°\right)$.

62. $r = \sqrt{0^2 + (-5)^2} = 5$; $\theta = \tan^{-1}\left(\dfrac{-5}{0}\right) \Rightarrow$ undefined $\Rightarrow \theta = 270°$ (negative y-axis).

One possible form is $(5, 270°)$.

63. See Figure 63.

64. See Figure 64.

65. See Figure 65.

[-6,6] by [-4,4]
Xscl = 1 Yscl = 1

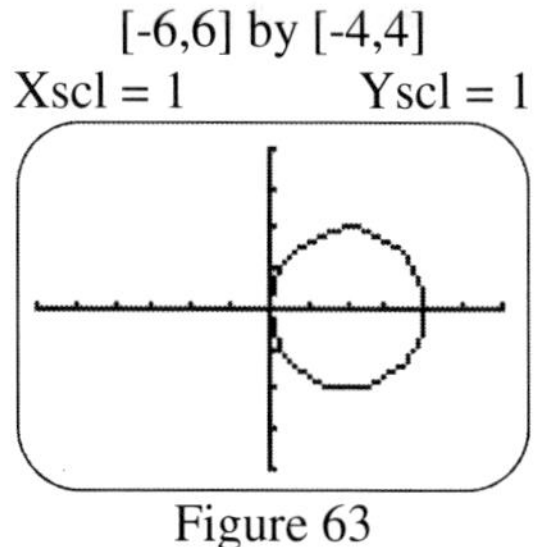
Figure 63

[-6,6] by [-4,4]
Xscl = 1 Yscl = 1

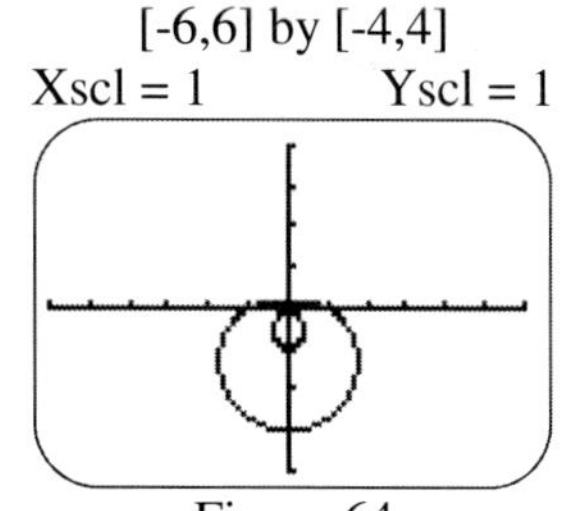
Figure 64

[-6,6] by [-4,4]
Xscl= 1 Yscl = 1

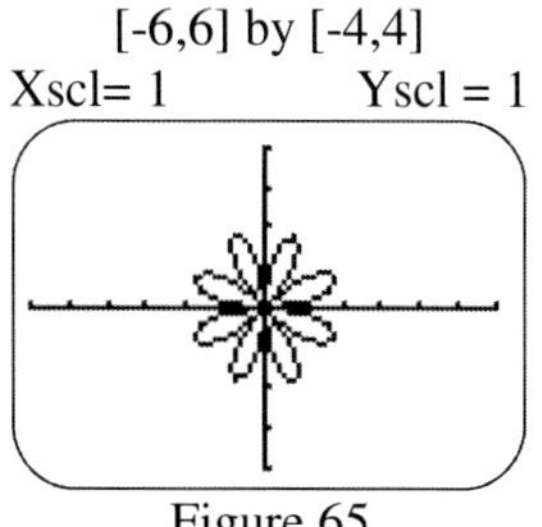
Figure 65

66. See Figure 66.

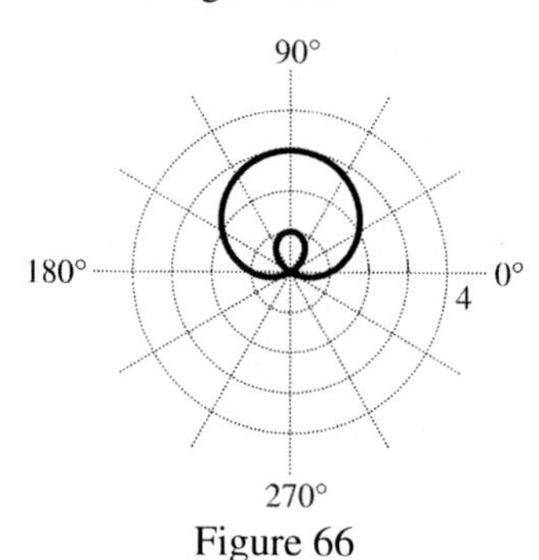

Figure 66

67. $r = \dfrac{3}{1+\cos\theta} \Rightarrow r + r\cos\theta = 3 + \sqrt{x^2+y^2} + x = 3 \Rightarrow \sqrt{x^2+y^2} = 3 - x \Rightarrow x^2 + y^2 = (3-x)^2 \Rightarrow x^2 + y^2$

$= 9 - 6x + x^2 \Rightarrow y^2 = -6x + 9 \Rightarrow y^2 = -6\left(x - \dfrac{3}{2}\right)$

68. $r = \dfrac{4}{2\sin\theta - \cos\theta} \Rightarrow 2r\sin\theta - r\cos\theta = 4 \Rightarrow 2y - x = 4$

69. $r = \sin\theta + \cos\theta \Rightarrow r^2 = r\sin\theta + r\cos\theta \Rightarrow x^2 + y^2 = y + x \Rightarrow x^2 + y^2 - x - y = 0 \Rightarrow$

$x^2 - x + \dfrac{1}{4} + y^2 - y + \dfrac{1}{4} = \dfrac{1}{4} + \dfrac{1}{4} \Rightarrow \left(x - \dfrac{1}{2}\right)^2 + \left(y - \dfrac{1}{2}\right)^2 = \dfrac{1}{2}$

70. $r = 2 \Rightarrow r^2 = 4 \Rightarrow x^2 + y^2 = 4$

71. $x = -3 \Rightarrow r\cos\theta = -3 \Rightarrow r = \dfrac{-3}{\cos\theta} \Rightarrow r = -\dfrac{3}{\cos\theta}$

72. $y = x \Rightarrow r\sin\theta = r\cos\theta \Rightarrow \dfrac{r\sin\theta}{r\cos\theta} = 1 \Rightarrow \tan\theta = 1$

73. $y = x^2 \Rightarrow r\sin\theta = r^2\cos^2\theta \Rightarrow \dfrac{r\sin\theta}{r\cos^2\theta} = r \Rightarrow r = \tan\theta.\dfrac{1}{\cos\theta} \Rightarrow r = \dfrac{\tan\theta}{\cos\theta}$

74. $x = y^2 \Rightarrow r\cos\theta = r^2\sin^2\theta \Rightarrow \dfrac{r\cos\theta}{r\sin^2\theta} = r \Rightarrow r = \cot\theta.\dfrac{1}{\sin\theta} \Rightarrow r = \dfrac{\cot\theta}{\sin\theta}$

75. Note that $x = \cos 2t \Rightarrow x = \cos^2 t - \sin^2 t$ and $y = \sin t \Rightarrow y^2 = \sin^2 t$. Thus $x + 2y^2 = \cos^2 t - \sin^2 t + 2\sin^2 t \Rightarrow x + 2y^2 = \cos^2 t + \sin^2 t \Rightarrow x + 2y^2 = 1 \Rightarrow 2y^2 + x - 1 = 0$. This can also be written $y^2 = -\dfrac{1}{2}(x-1)$. For t in the interval $(-\pi, \pi)$, $x = \cos 2t$ takes on the values in. $[-1, 1]$.

76. Note that $x = 5\tan t \Rightarrow x^2 = 25\tan^2 t$ and $y = 3\sec t \Rightarrow y^2 = 9\sec^2 t$.

Thus $25y^2 - 9x^2 = 225\sec^2 t - 225\tan^2 t \Rightarrow \frac{25y^2}{225} - \frac{9x^2}{225} = \sec^2 t - \tan^2 t \Rightarrow \frac{y^2}{9} - \frac{x^2}{25} = 1$. This can also be written $y = 3\sqrt{1+\frac{x^2}{25}}$. For t in the interval $\left(-\frac{\pi}{2}, \frac{\pi}{2,}\right)$, $x = 5$ takes on the values in $(-\infty, \infty)$.

77. See Figure 77.

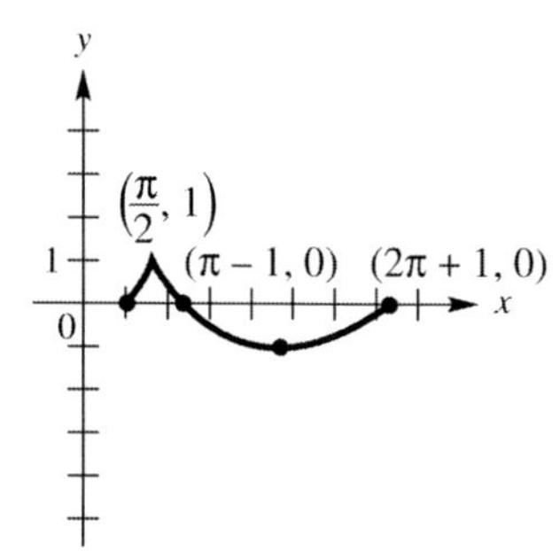

Figure 77

78. $x = (118\cos 27^\circ)t;\ y = (118\sin 27^\circ)t - 16t^2 + 3.2 \Rightarrow y = 3.2 - 16t^2 + (118\sin 27^\circ)t$

Use the quadratic formula to solve the equation $y = 3.2 - 16t^2 + (118\sin 27^\circ)t = 0$ to determine when the ball is at ground level. The solutions are $-.059$ and 3.4069 seconds. The horizontal distance traveled is $x = (1118\cos 27^\circ)(3.4069) \approx 360$ feet.

Chapter 10 Test

1. Find C, given $A = 25.2°$, $a = 6.92$ yd, $b = 4.82$ yd.

Use the law of sines to first find the measure of angle B.

$$\frac{\sin 25.2°}{6.92} = \frac{\sin B}{4.82} \Rightarrow \sin B = \frac{4.82\sin 25.2°}{6.92} \Rightarrow B = \sin^{-1}\left(\frac{4.82\sin 25.2°}{6.92}\right) \approx 17.3°$$

Use the fact that the angles of a triangle sum to 180° to find the measure of angle C.

$C = 180° - A - B = 180° - 25.2° - 17.3° = 137.5°$

2. Find c, given $C = 118°$, $b = 131$ km, $a = 75.0$ km.

Using the law of cosines to find the length of c.

$$c^2 = a^2 + b^2 - 2ab\cos C \Rightarrow c^2 = 75.0^2 + 131^2 - 2(75.0)(131)\cos 118° \approx 32011.12 \Rightarrow c \approx 178.9 \text{ km}$$

c is approximately 179 km. (rounded to two significant digits)

3. Find B, given $a = 17.3$ ft, $b = 22.6$ ft, $c = 29.8$ ft.

Using the law of cosines, find the measure of angle B.

$$b^2 = a^2 + c^2 - 2ac\cos B \Rightarrow \cos B = \frac{a^2 + c^2 - b^2}{2ac} = \frac{17.3^2 + 29.8^2 - 22.6^2}{2(17.3)(29.8)} \approx 0.65617605 \Rightarrow B \approx 49.0°$$

4. $a = 14$, $b = 30$, $c = 40$ We can use Heron's formula to find the area.

$s = \frac{1}{2}(a+b+c) = \frac{1}{2}(14+30+40) = 42$

$\mathcal{A} = \sqrt{s(s-a)(s-b)(s-c)} = \sqrt{42(42-14)(42-30)(42-40)} = \sqrt{42 \cdot 28 \cdot 12 \cdot 2} = \sqrt{28,224} = 168$ sq units

5. This is SAS, so we can use the formula $\mathcal{A} = \frac{1}{2}zy\sin X$. $\mathcal{A} = \frac{1}{2} \cdot 6 \cdot 12 \sin 30° = \frac{1}{2} \cdot 6 \cdot 12 \cdot \frac{1}{2} = 18$ sq units

6. Since $B > 90°$, b must be the longest side of the triangle.

(a) $b > 10$

(b) none

(c) $b \le 10$

7. $A = 60°, b = 30 \text{ m}, c = 45 \text{ m}$ This is SAS, so use the law of cosines to find a:

$a^2 = b^2 + c^2 - 2bc\cos A \Rightarrow a^2 = 30^2 + 45^2 - 2 \cdot 30 \cdot 45\cos 60° = 1575 \Rightarrow a = 15\sqrt{7} \approx 40$ m

Now use the law of sines to find B: $\frac{\sin B}{b} = \frac{\sin A}{a} \Rightarrow \frac{\sin B}{30} = \frac{\sin 60°}{15\sqrt{7}} \Rightarrow \sin B = \frac{30\sin 60°}{15\sqrt{7}} \Rightarrow B \approx 41°$

$C = 180° - A - B = 180° - 60° - 41° = 79°$

8. $b = 1075$ in., $c = 785$ in., $C = 38°30'$ We can use the law of sines.

$\frac{\sin B}{b} = \frac{\sin C}{c} \Rightarrow \frac{\sin B}{1075} = \frac{\sin 38°30'}{785} \Rightarrow \sin B = \frac{1075\sin 38°30'}{785} \Rightarrow B_1 \approx 58.5° = 58°30'$ or
$B_2 = 180° - 58°30' = 121°30'$

Solving separately for triangles A_1B_1C and A_2B_2C, we have the following.

$A_1B_1C: A_1 = 180° - B_1 - C = 180° - 58°30' - 38°30' = 83°00'$

$\frac{a_1}{\sin A_1} = \frac{b}{\sin B_1} \Rightarrow \frac{a_1}{\sin 83°} = \frac{1075}{\sin 58°30'} \Rightarrow a_1 = \frac{1075\sin 83°}{\sin 58°30'} \approx 1250$ in. (three significant digits) A_2B_2C:

$A_2 = 180° - B_2 - C = 180° - 121°30' - 38°30' = 20°00'$

$\frac{a_2}{\sin A_2} = \frac{b}{\sin B_2} \Rightarrow \frac{a_2}{\sin 20°} = \frac{1075}{\sin 121°30'} \Rightarrow a_2 = \frac{1075\sin 20°}{\sin 121°30'} \approx 431$ in. (rounded to three significant digits)

9. magnitude: $|\mathbf{v}| = \sqrt{(-6)^2 + 8^2} = \sqrt{36+64} = \sqrt{100} = 10$

angle: $\tan\theta = \frac{y}{x} \Rightarrow \tan\theta = \frac{8}{-6} = -\frac{4}{3} \approx -1.33333333 \Rightarrow \theta \approx -53.1° \Rightarrow \theta = -53.1° + 180° = 126.9°$

(θ lies in quadrant II) The magnitude $|\mathbf{v}|$ is 10 and $\theta = 126.9°$.

10. (a) $\mathbf{u} + \mathbf{v} = \langle -1, 3\rangle + \langle 2, -6\rangle = \langle -1+2, 3+(-6)\rangle = \langle 1, -3\rangle$

(b) $-3\mathbf{v} = -3\langle 2, -6\rangle = \langle -3\cdot 2, -3(-6)\rangle = \langle -6, 18\rangle$

(c) $\mathbf{u}\cdot\mathbf{v} = \langle -1, 3\rangle\cdot\langle 2, -6\rangle = -1(2)+3(-6) = -2-18 = -20$

(d) $|\mathbf{u}| = \sqrt{(-1)^2+3^2} = \sqrt{1+9} = \sqrt{10}$

11. $\mathbf{u}\cdot\mathbf{v} = \langle 4, 3\rangle\cdot\langle 1, 5\rangle = 4(1)+3(5) = 19$, $|\mathbf{u}| = \sqrt{4^2+3^2} = \sqrt{25} = 5$, $|\mathbf{v}| = \sqrt{1^2+5^2} = \sqrt{26}$

$\cos\theta = \dfrac{\mathbf{u}\cdot\mathbf{v}}{|\mathbf{u}||\mathbf{v}|} \Rightarrow \cos\theta = \dfrac{19}{5\sqrt{26}} \Rightarrow \theta = \cos^{-1}\left(\dfrac{19}{5\sqrt{26}}\right) \approx 41.8°$

12. Given $A = 45°, B = 30°$ and $AB = 4.2$ mi, first find the measure of angle C.

$C = 180° - 45° - 30° = 105°$

Use this information and the law of sines to find AC. $\dfrac{AC}{\sin 45°} = \dfrac{4.2}{\sin 105°} \Rightarrow$

$AC = \dfrac{4.2\sin 45°}{\sin 105°} \approx 3.075$ mi Drop a perpendicular line from C to segment AB.

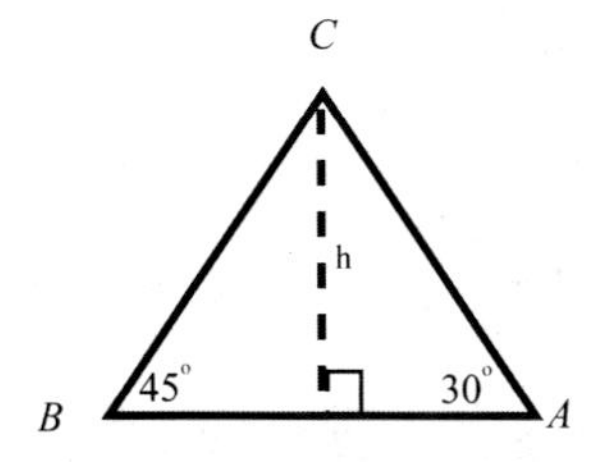

Thus, $\sin 30° = \dfrac{h}{3.075} \Rightarrow h \approx 3.075\sin 30° \approx 1.5$ mi. The balloon is 2.7 miles off the ground.

13. horizontal: $x = |\mathbf{v}|\cos\theta = 569\cos 127.5° \approx -346$ and vertical: $y = |\mathbf{v}|\sin\theta = 569\sin 127.5° \approx 451$

The vector is $\langle -346, 451\rangle$.

14. Consider the figure below.

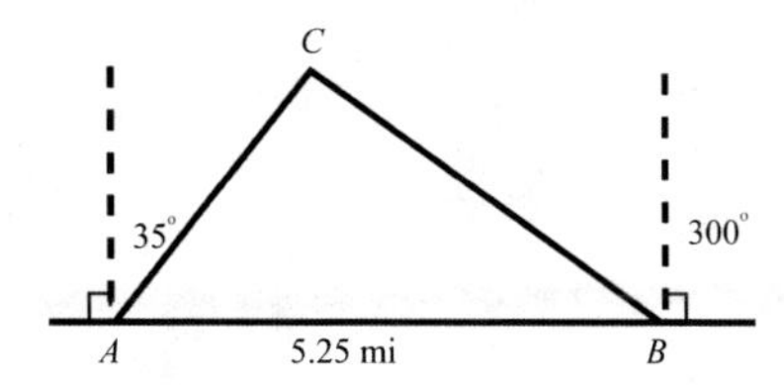

Since the bearing is 35° from A, angle A in ABC must be 90° – 35° = 55°. Since the bearing is 300° from B, angle B in ABC must be 300° – 270° = 30°. The angles of a triangle sum to 180°, so

$C = 180° - A - B = 180° - 55° - 30° = 95°$. Using the law of sines, we have

the distance from A to the transmitter is 2.64 miles. (rounded to two significant digits)

15. Since $m\angle DAC = 8.0°$, $m\angle CAB = 90° - 8.0° = 82.0°$. $m\angle B = 66°$, so $m\angle C = 180° - 82° - 66° = 32°$. Now use the law of sines to find AC: $\frac{AC}{\sin B} = \frac{AB}{\sin C} \Rightarrow \frac{AC}{\sin 66°} = \frac{8.0}{\sin 32°} \Rightarrow AC = \frac{8.0 \sin 66°}{\sin 32°} \approx 13.8 \approx 14$ m

16. AX is the airspeed vector. Since the plane is flying 630 miles due north in 3 hours, the ground speed is 210 mph. The measure of angle ACX is $180° - 42° = 138°$.

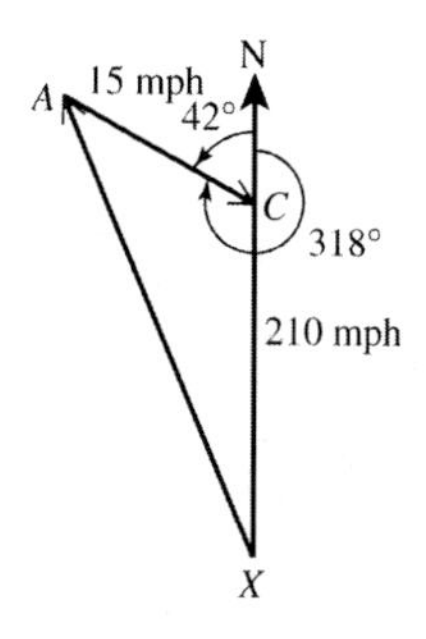

$|\mathbf{AX}|^2 = 15^2 + 210^2 - 2(15)(210)\cos 138° = 49006.8124 \Rightarrow |\mathbf{AX}| \approx 221.3748 \approx 220$ mph (two significant digits). Using the law of sines to find the measure of angle X, we have

$$\frac{\sin X}{15} = \frac{\sin 138°}{220} \Rightarrow \sin X = \frac{15 \sin 138°}{220} \Rightarrow X = \sin^{-1}\left(\frac{15 \sin 138°}{220}\right) \approx 2.6°$$

The plane's bearing is $360° - 2.6° = 357.4° \approx 357°$.

17. $|\mathbf{AC}| = 16.0$ lb; $|\mathbf{BA}| = 50.0$ lb

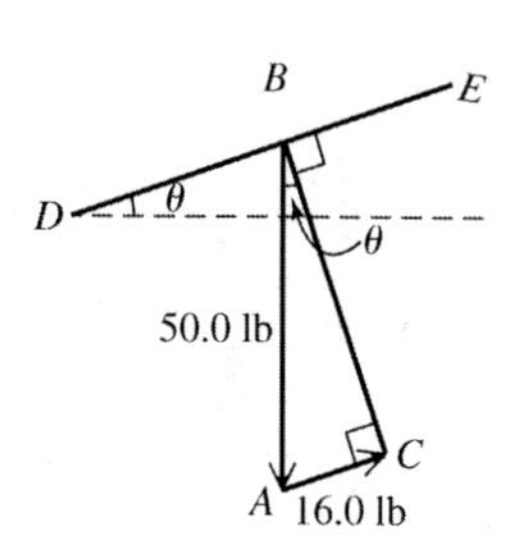

$$\sin\theta = \frac{16.0}{50.0} \Rightarrow \theta = \sin^{-1}\left(\frac{16.0}{50.0}\right) \approx 18.7°$$

18. $w + z = (2 - 4i) + (5 + i) = 7 - 3i$

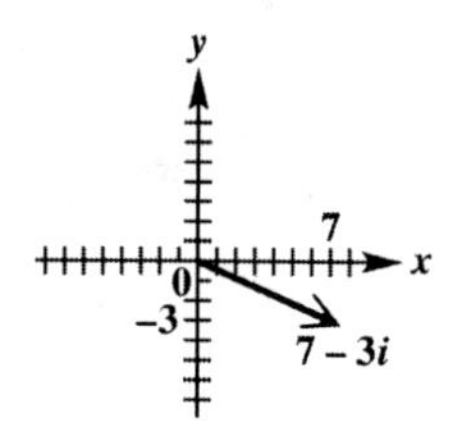

19. (a) $i^{15} = i^{12+3} = i^{12} \cdot i^3 = \left(i^4\right)^3 \cdot i^3 = 1(-i) = -i$

(b) $(1+i)^2 = (1+i)(1+i) = 1+i+i+i^2 = 1+2i+(-1) = 2i$

20. (a) $3i \Rightarrow r = \sqrt{0^2+3^2} = \sqrt{0+9} = \sqrt{9} = 3$

The point $(0,3)$ is on the positive y-axis, so, $\theta = 90°$. Thus, $3i = 3(\cos 90° + i\sin 90°)$.

(b) $1+2i \Rightarrow r = \sqrt{1^2+2^2} = \sqrt{1+4} = \sqrt{5}$

Since θ is in quadrant I, $\theta = \tan^{-1}\left(\frac{2}{1}\right) = \tan^{-1} 2 \approx 63.43°$. Thus, $1+2i = \sqrt{5}(\cos 63.43° + i\sin 63.43°)$.

(c) $-1-\sqrt{3}i \Rightarrow r = \sqrt{(-1)^2+(-\sqrt{3})^2} = \sqrt{1+3} = \sqrt{4} = 2$ Since θ is in quadrant III,

$\theta = \tan^{-1}\left(\frac{-\sqrt{3}}{-1}\right) = \tan^{-1}\sqrt{3} = 240°$. Thus, $-1-\sqrt{3}i = 2(\cos 240° + i\sin 240°)$

21. (a) $3(\cos 30° + i\sin 30°) = 3\left(\frac{\sqrt{3}}{2} + \frac{1}{2}i\right) = \frac{3\sqrt{3}}{2} + \frac{3}{2}i$

(b) $4 \text{ cis } 40° \approx 3.06 + 2.57i$

(c) $3(\cos 90° + i \sin 90°) = 3(0 + 1 \cdot i) = 0 + 3i = 3i$

22. (a) $wz = 8 \cdot 2[\cos(40° + 10°) + i\sin(40° + 10°)] = 16(\cos 50° + i\sin 50°)$

(b) $\frac{w}{z} = \frac{8}{2}[\cos(40° - 10°) + i\sin(40° - 10°)] = 4(\cos 30° + i\sin 30°) = 4\left(\frac{\sqrt{3}}{2} + \frac{1}{2}i\right) = 2\sqrt{3} + 2i$

(c) $z^3 = [2(\cos 10° + i\sin 10°)]^3 = 2^3(\cos 3 \cdot 10° + i\sin 3 \cdot 10°) = 8(\cos 30° + i\sin 30°) =$

$8\left(\frac{\sqrt{3}}{2} + \frac{1}{2}i\right) = 4\sqrt{3} + 4i$

23. Since $r^4(\cos 4\alpha + i\sin 4\alpha) = 16(\cos 270° + i\sin 270°)$, then we have $r^4 = 16 \Rightarrow r = 2$ and

$4\alpha = 270° + 360° \cdot k \Rightarrow \alpha = \frac{270° + 360° \cdot k}{4} = 67.5° + 90° \cdot k$, k any integer. If $k = 0$, then $\alpha = 67.5°$.

If $k = 1$, then $\alpha = 157.5°$, If $k = 2$, then $\alpha = 247.5°$., If $k = 3$, then $\alpha = 337.5°$.

The fourth roots of $-16i$ are $2(\cos 67.5° + i\sin 67.5)$, $2(\cos 157.5° + i\sin 157.5°)$, $2(\cos 247.5° + i\sin 247.5°)$,

and $2(\cos 337.5° + i\sin 337.5°)$.

24. Answers may vary.

(a) $(0, 5) \Rightarrow r = \sqrt{0^2 + 5^2} = \sqrt{0 + 25} = \sqrt{25} = 5$

The point (0, 5) is on the positive y-axis. Thus, $\theta = 90°$. One possibility is (5, 90°). Alternatively, if $\theta = 90° - 360° = -270°$, a second possibility is (5, –270°).

(b) $(-2, -2) \Rightarrow r = \sqrt{(-2)^2 + (-2)^2} = \sqrt{4+4} = \sqrt{8} = 2\sqrt{2}$

Since θ is in quadrant III, $\theta = \tan^{-1}\left(\frac{-2}{-2}\right) = \tan^{-1} 1 = 225°$. One possibility is $\left(2\sqrt{2}, 225°\right)$.

Alternatively, if $\theta = 225° - 360° = -135°$, a second possibility is $\left(2\sqrt{2}, -135°\right)$.

25. (a) $(3, 315°) \Rightarrow x = r\cos\theta \Rightarrow x = 3\cos 315° = 3 \cdot \frac{\sqrt{2}}{2} = \frac{3\sqrt{2}}{2}$ and $y = r\sin\theta \Rightarrow$

$y = 3\sin 315° = 3\left(-\frac{\sqrt{2}}{2}\right) = \frac{-3\sqrt{2}}{2}$ The rectangular coordinates are $\left(\frac{3\sqrt{2}}{2}, \frac{-3\sqrt{2}}{2}\right)$.

(b) $(-4, 90°) \Rightarrow x = r\cos\theta \Rightarrow x = -4\cos 90° = 0$ and $y = r\sin\theta \Rightarrow y = -4\sin 90° = -4$

The rectangular coordinates are $(0, -4)$.

26. $r = 1 - \cos\theta$ is a cardioid.

θ	0°	30°	45°	60°	90°	135°
$r = 1 - \cos\theta$	0	0.1	0.3	0.5	1	1.7
θ	180°	225°	270°	315°	360°	
$r = 1 - \cos\theta$	2	1.7	1	0.3	0	

$r = 1 - \cos\theta$

27. $r = 3\cos 3\theta$ is a three-leaved rose.

θ	0°	30°	45°	60°	90°	120°	135°	150°	180°
$r = 3\cos 3\theta$	3	0	–2.1	–3	0	3	2.1	0	–3

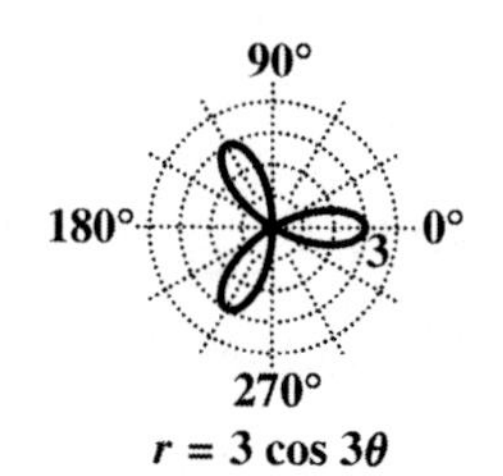

$r = 3\cos 3\theta$

Graph is retraced in the interval (180°, 360°).

28. (a) Since $r = \dfrac{4}{2\sin\theta - \cos\theta} = \dfrac{4}{-1\cdot\cos\theta + 2\sin\theta}$, we can use the general form for the polar equation of a line, $r = \dfrac{c}{a\cos\theta + b\sin\theta}$, with $a = -1$, $b = 2$, and $c = 4$, we have $-x + 2y = 4$ or $x - 2y = -4$. The graph is a line with intercepts (–4, 0) and (0, 2). See Figure 28a

(b) $r = 6$ represents the equation of a circle centered at the origin with radius 6, namely $x^2 + y^2 = 36$. See Figure 28b.

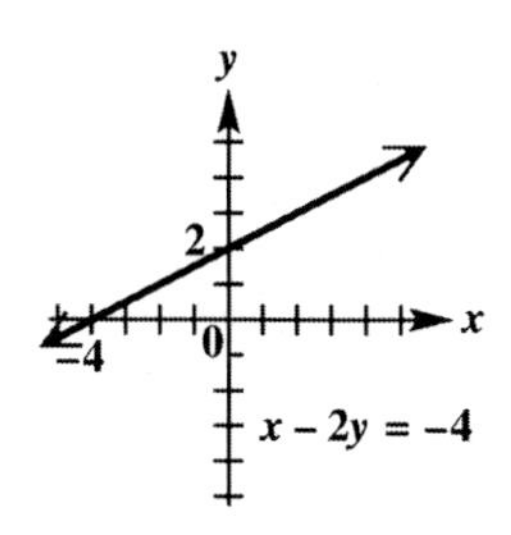

Figure 28a

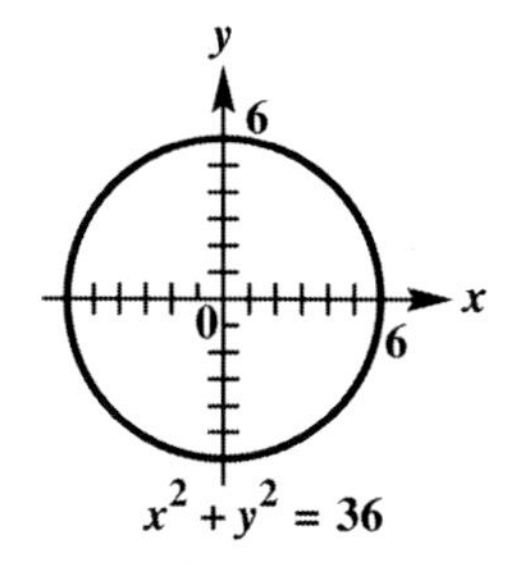

Figure 28b

29. $x = 2t - 1$, $y = t^2$ for t in [–2, 3] See Figure 29.

t	x	Y
–2	–5	4
–1	–3	1
0	–1	0
1	1	1
2	3	4
3	5	9

Since $x = 2t - 1 \Rightarrow t = \dfrac{x+1}{2}$ and $y = t^2$, we have $y = \left(\dfrac{x+1}{2}\right)^2 = \dfrac{1}{4}(x+1)^2$, where x is in [–5, 5]

30. $x = 2\cos 2t$, $y = 2\sin 2t$ for t in $[0, 2\pi]$. See Figure30.

T	0	$\frac{\pi}{8}$	$\frac{\pi}{4}$	$\frac{3\pi}{8}$	$\frac{\pi}{2}$	$\frac{5\pi}{8}$
X	2	$\sqrt{2}$	0	$-\sqrt{2}$	–2	$-\sqrt{2}$
Y	0	$\sqrt{2}$	2	$\sqrt{2}$	0	$-\sqrt{2}$

T	$\frac{3\pi}{4}$	π	$\frac{5\pi}{4}$	$\frac{3\pi}{2}$	$\frac{7\pi}{4}$	2π
X	0	2	0	–2	0	2
Y	–2	0	2	0	–2	0

Since $x = 2\cos 2t \Rightarrow \cos 2t = \frac{x}{2}$, $y = 2\sin 2t \Rightarrow \sin 2t = \frac{y}{2}$, and $\cos^2(2t) + \sin^2(2t) = 1$, we have

$$\left(\frac{x}{2}\right)^2 + \left(\frac{y}{2}\right)^2 = 1 \Rightarrow \frac{x^2}{4} + \frac{y^2}{4} = 1 \Rightarrow x^2 + y^2 = 4,$$ where x is in $[-1, 1]$.

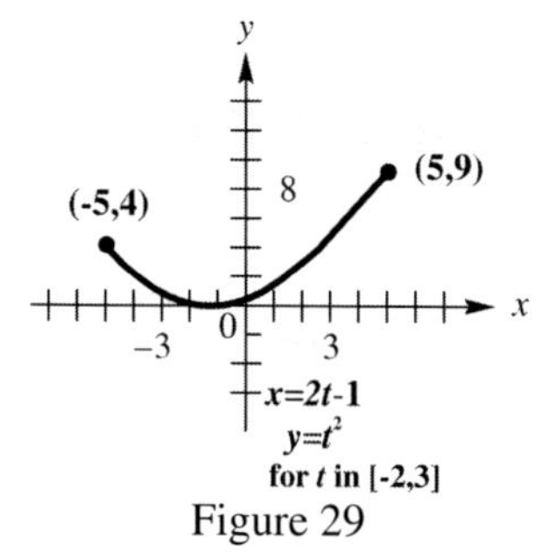

Figure 29

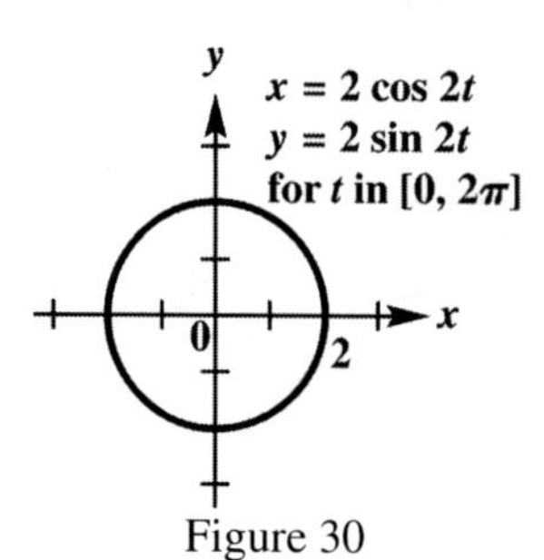

Figure 30

Chapter 11: Further Topics in Algebra

11.1: Sequences and Series

1. $a_1 = 4(1)+10 = 14;\ a_2 = 4(2)+10 = 18;\ a_3 = 4(3)+10 = 22;\ a_4 = 4(4)+10 = 26;\ a_5 = 4(5)+10 = 30$

   The first five terms of the sequence are 14, 18, 22, 26, and 30.

3. $a_1 = 2^{1-1} = 2^0 = 1;\ a_2 = 2^{2-1} = 2^1 = 2;\ a_3 = 2^{3-1} = 2^2 = 4;\ a_4 = 2^{4-1} = 2^3 = 8;\ a_5 = 2^{5-1} = 2^4 = 16$

   The first five terms of the sequence are 1, 2, 4, 8, and 16.

5. $a_1 = \left(\frac{1}{3}\right)^1 (1-1) = \left(\frac{1}{3}\right)(0) = 0;\ a_2 = \left(\frac{1}{3}\right)^2 (2-1) = \left(\frac{1}{9}\right)(1) = \frac{1}{9};$

   $a_3 = \left(\frac{1}{3}\right)^3 (3-1) = \left(\frac{1}{27}\right)(2) = \frac{2}{27};\ a_4 = \left(\frac{1}{3}\right)^4 (4-1) = \left(\frac{1}{81}\right)(3) = \frac{1}{27};\ a_5 = \left(\frac{1}{3}\right)^5 (5-1) = \left(\frac{1}{243}\right)(4) = \frac{4}{243}$

   The first five terms of the sequence are $0,\ \frac{1}{9},\ \frac{2}{27},\ \frac{1}{27},$ and $\frac{4}{243}$.

7. $a_1 = (-1)^1[(2)(1)] = -2;\ a_2 = (-1)^2[(2)(2)] = 4;\ a_3 = (-1)^3[(2)(3)] = -6;$

   $a_4 = (-1)^4[(2)(4)] = 8;\ a_5 = (-1)^5[(2)(5)] = -10$

   The first five terms of the sequence are $-2,\ 4,\ -6,\ 8,$ and -10.

9. $a_1 = \frac{4(1)-1}{1^2+2} = \frac{3}{3} = 1;\ a_2 = \frac{4(2)-1}{2^2+2} = \frac{7}{6};\ a_3 = \frac{4(3)-1}{3^2+2} = \frac{11}{11} = 1;$

   $a_4 = \frac{4(4)-1}{4^2+2} = \frac{15}{3} = \frac{5}{6};\ a_5 = \frac{4(5)-1}{5^2+2} = \frac{19}{27}$ The first five terms of the sequence are $1,\ \frac{7}{6},\ 1,\ \frac{5}{6},$ and $\frac{19}{27}$.

11. The terms of a sequence are numbered according to their position in the sequence; a_1 is the first term, a_2 is the second term, etc. So a_n will be the nth term in the sequence for any positive integer value of n.

13. The sequence has 7 terms; it is finite.

15. The sequence has 4 terms; it is finite.

17. The sequence has no last term; it is infinite.

19. The sequence is defined for integer values of n from 1 through 10; it is finite.

21. $a_1 = -2;\ a_2 = a_1 + 3 = -2+3 = 1;\ a_3 = a_2 + 3 = 1+3 = 4;\ a_4 = a_3 + 3 = 4+3 = 7$

    The first four terms of the sequence are $-2,\ 1,\ 4,$ and 7.

23. $a_1 = 1;\ a_2 = 1;\ a_3 = a_2 + a_1 = 1+1 = 2;\ a_4 = a_3 + a_2 = 2+1 = 3.$

    The first four terms of the sequence are 1, 1, 2, and 3.

25. $a_1 = 5;\ a_2 = 3(2) + 3a_1 = 6 + 3(5) = 21;\ a_3 = 3(3) + 3a_2 = 9 + 3(21) = 72;$
    $a_4 = 3(4) + 3a_3 = 12 + 3(72) = 228.$

    The first four terms of the sequence are 5, 21, 72, and 228.

27. $a_1 = 2;\ a_2 = 3;\ a_3 = a_2 \cdot a_1 = 3(2) = 6;\ a_4 = a_3 \cdot a_2 = 6(3) = 18$

The first four terms of the sequence are 2, 3, 6, and 18.

29. $\sum_{i=1}^{5} (2i+1) = 3+5+7+9+11 = 35$

31. $\sum_{j=1}^{4} \frac{1}{j} = \frac{1}{1} + \frac{1}{2} + \frac{1}{3} + \frac{1}{4} = \frac{25}{12}$

33. $\sum_{i=1}^{4} i^i = 1^1 + 2^2 + 3^3 + 4^4 = 288$

35. $\sum_{k=1}^{6} (-1)^k \bullet k = (-1)^1(1) + (-1)^2(2) + (-1)^3(3) + (-1)^4(4) + (-1)^5(5) + (-1)^6(6) = 3$

37. $\sum_{i=2}^{5} (6-3i) = (6-6) + (6-9) + (6-12) + (6-15) = -18$

39. $\sum_{i=-2}^{3} 2(3)^i = 2(3)^{-2} + 2(3)^{-1} + 2(3)^0 + 2(3)^1 + 2(3)^2 + 2(3)^3 = \frac{728}{9}$

41. $\sum_{i=-1}^{5} (i^2 - 2i) = \left[(-1)^2 - 2(-1)\right] + \left[0^2 - 2(0)\right] + \left[1^2 - 2(1)\right] + \left[2^2 - 2(2)\right] + \left[3^2 - 2(3)\right] +$

$\left[4^2 - 2(4)\right] + \left[5^2 - 2(5)\right] = 28$

43. $\sum_{i=1}^{5} (3^i - 4) = (3^1 - 4) + (3^2 - 4) + (3^3 - 4) + (3^4 - 4) + (3^5 - 4) = 343$

45. $\sum_{i=1}^{5} x_i = (-2) + (-1) + (0) + (1) + (2) = -2 - 1 + 0 + 1 + 2$

47. $\sum_{i=1}^{5} (2x_i + 3) = \left[2(-2)+3\right] + \left[2(-1)+3\right] + \left[2(0)+3\right] + \left[2(1)+3\right] + \left[2(2)+3\right] = -1+1+3+5+7$

49. $\sum_{i=1}^{3} (3x_i - x_i^2) = \left[3(-2) - (-2)^2\right] + \left[3(-1) - (-1)^2\right] + \left[3(0) - (0)^2\right] = -10 - 4 + 0$

51. $\sum_{i=2}^{5} \frac{x_i + 1}{x_i + 2} = \frac{-1+1}{-1+2} + \frac{0+1}{0+2} + \frac{1+1}{1+2} + \frac{2+1}{2+2} = 0 + \frac{1}{2} + \frac{2}{3} + \frac{3}{4}$

53. $\sum_{i=1}^{4} f(x_i)\Delta x = \left[4(0)-7\right](0.5) + \left[4(2)-7\right](0.5) + \left[4(4)-7\right](0.5) + \left[4(6)-7\right](0.5) = -0.35 + 0.5 + 4.5 + 8.5$

55. $\sum_{i=1}^{4} f(x_i)\Delta x = \left[2(0)^2\right](0.5) + \left[2(2)^2\right](0.5) + \left[2(4)^2\right](0.5) + \left[2(6)^2\right](0.5) = 0 + 4 + 16 + 36$

57. $\sum_{i=1}^{4} f(x_i)\Delta x = \left(\frac{-2}{0+1}\right)(0.5) + \left(\frac{-2}{2+1}\right)(0.5) + \left(\frac{-2}{4+1}\right)(0.5) + \left(\frac{-2}{6+1}\right)(0.5) = -1 - \frac{1}{3} - \frac{1}{5} - \frac{1}{7}$

59. $\sum_{i=1}^{100} 6 = (100)(6) = 600$

61. $\sum_{i=1}^{15} i^2 = \frac{(15)(15+1)[2(15)+1]}{6} = 1240$

63. $\sum_{i=1}^{5} (5i+3) = 5\left[\frac{(5)(5+1)}{2}\right] + 5(3) = 90$

65. $\sum_{i=1}^{5} (4i^2 - 2i + 6) = 4\left[\frac{(5)(5+1)[2(5)+1]}{6}\right] - 2\left[\frac{5(5+1)}{2}\right] + 5(6) = 220$

67. $\sum_{i=1}^{4} (3i^3 + 2i - 4) = 3\left[\frac{4^2(4+1)^2}{4}\right] + 2\left[\frac{(4)(4+1)}{2}\right] + 4(-4) = 304$

69. $\sum_{i=1}^{60} (i^3 - 2i^2) = \frac{60^2(60+1)^2}{4} - 2\left[\frac{(60)(60+1)[2(60)+1]}{6}\right] = 3,201,280$

71. $\sum_{i=1}^{77} (i^2 + 52i + 672) = \left[\frac{(77)(77+1)[2(77)+1]}{6}\right] + 52\left[\frac{77(77+1)}{2}\right] + 77(672) = 363,055$

73. A single fraction in the sum $\frac{2}{5(1)} + \frac{2}{5(2)} + \frac{2}{5(3)} + ... + \frac{2}{5(100)}$ can be shown as $\frac{2}{5(i)}$ where i begins at 1 and ends at 100. This sum can be shown as $\sum_{i=1}^{100} \frac{2}{5i}$.

75. A single fraction in the sum $1 + \frac{1}{2} + \frac{1}{3} + \frac{1}{4} + ... + \frac{1}{9}$ can be shown as $\frac{1}{i}$ where i begins at 1 and ends at 9. This sum can be shown as $\sum_{i=1}^{9} \frac{1}{i}$.

77. The graphing calculator graph of the first ten terms of this sequence is shown in Figure 77. From this graph, the terms of the sequence appear to converge to the value $\frac{1}{2}$ which, in fact, they do.

79. The graphing calculator graph of the first ten terms of this sequence is shown in Figure 79. From this graph, the terms of the sequence appear to diverge and, in fact, they do.

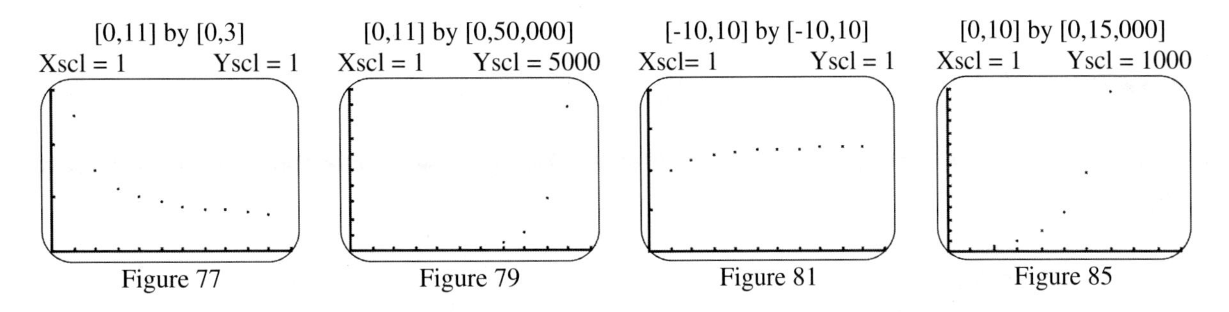

Figure 77 Figure 79 Figure 81 Figure 85

81. The graphing calculator graph of the first ten terms of this sequence is shown in Figure 81. From this graph, the terms of the sequence appear to converge to the value somewhat greater than 2.5. In fact, they converge to $e \approx 2.71828$.

83. The sum of the first six terms of the series is

$$\frac{1}{1^4}+\frac{1}{2^4}+\frac{1}{3^4}+\frac{1}{4^4}+\frac{1}{5^4}+\frac{1}{6^4}=1+\frac{1}{16}+\frac{1}{81}+\frac{1}{256}+\frac{1}{625}+\frac{1}{1296}\approx 1.081123534\Rightarrow$$

$$\frac{\pi^4}{90}\approx 1.081123534\Rightarrow \pi^4\approx 97.30111806\Rightarrow \pi\approx 3.140721718.$$

This approximation of π is accurate to three decimal places when rounded to 3.141.

85. (a) The number of bacteria doubles every 40 minutes, so $N_{j+1}=2N_j$ for $j\geq 1$.

(b) Two hours = 120 minutes $\Rightarrow 120=40(j-1)\Rightarrow 3=j-1\Rightarrow j=4$.

$N_1=230; N_2=2(230)=460; N_3=2(460)=920; N_4=2(920)=1840$. If there are initially 230 bacteria, then there will be 1840 bacteria after two hours.

(c) The graph of the first seven terms of this sequence is shown in Figure 85.

(d) From the graph, the growth rate is seen to be very rapid. Doubling the number of bacteria at equal intervals of time produces an exponential growth rate.

87. (a) $1+1+\frac{1^2}{2!}+\frac{1^3}{3!}+\frac{1^4}{4!}+\frac{1^5}{5!}+\frac{1^6}{6!}+\frac{1^7}{7!}\approx 2.718254;\ e\approx 2.718282$. Eight terms of the series gives an estimate that matches to four decimal places.

(b) $1+(-1)+\frac{(-1)^2}{2!}+\frac{(-1)^3}{3!}+\frac{(-1)^4}{4!}+\frac{(-1)^5}{5!}+\frac{(-1)^6}{6!}+\frac{(-1)^7}{7!}\approx 0.367857;\ e\approx 0.367879$. Eight terms of the series gives an estimate that matches to four decimal places.

(c) $1+\frac{1}{2}+\frac{\left(\frac{1}{2}\right)^2}{2!}+\frac{\left(\frac{1}{2}\right)^3}{3!}+\frac{\left(\frac{1}{2}\right)^4}{4!}+\frac{\left(\frac{1}{2}\right)^5}{5!}+\frac{\left(\frac{1}{2}\right)^6}{6!}+\frac{\left(\frac{1}{2}\right)^7}{7!}\approx 1.648721;\ \sqrt{e}=e^{1/2}\approx 1.648721$. Eight terms of the series gives an estimate that matches to six decimal places.

11.2: Arithmetic Sequences and Series

1. $5-2=3;\ 8-5=3;\ 11-8=3$. The common difference for this arithmetic sequence is 3.

3. $-2-3=-5;\ -7-(-2)=-5;\ -12-(-7)=-5$. The common difference for this arithmetic sequence is -5.

5. $(2x+5y)-(x+3y)=x+2y; (3x+7y)-(2x+5y)=x+2y$. The common difference for this arithmetic sequence is $x+2y$.

7. 8; $8+6=14;\ 14+6=20;\ 20+6=26;\ 26+6=32$. The first five terms of this arithmetic sequence are 8, 14, 20, 26, and 32.

9. 5; $5+(-2)=3;\ 3+(-2)=1;\ 1+(-2)=-1;\ (-1)+(-2)=-3$. The first five terms of this arithmetic sequence 5, 3, 1, -1, and -3.

11. $a_3=10;\ a_1=10-2(-2)=14;\ a_2=10-(-2)=12;\ a_4=10+(-2)=8;\ a_5=8+(-2)=6$. The first five terms of this arithmetic sequence are 14, 12, 10, 8, and 6.

13. $a_8 = 5+(8-1)(2) = 19;\ a_n = 5+(n-1)(2) = 2n+3$. The eighth term of the sequence is 19. The nth term of the sequence is $2n+3$.

14. $a_8 = -3+(8-1)(-4) = -31;\ a_n = -3+(n-1)(-4) = -4n+1$ The eighth term of the sequence is -31 The nth term of the sequence is $-4n+1$.

15. $a_3 = 2;\ d = 1 \Rightarrow a_1 = 2-2(1) = 0;\ a_8 = 0+(8-1)(1) = 7;\ a_n = 0+(n-1)(1)$. The eighth term of the sequence is 7. The nth term of the sequence is $n-1$.

17. $a_1 = 8;\ a_6 = 6 \Rightarrow d = 6-8 = -2;\ a_8 = 8+(8-1)(-2) = -6;\ a_n = 8+(n-1)(-2) = -2n+10$. The eighth term of the sequence is -6. The nth term of the sequence is $-2n+10$.

19. $a_{10} = 6;\ a_{12} = 15 \Rightarrow 2d = 15-6 = 9;\ d = 4.5;\ a_1 = 6-(9)(4.5) = -3.45;$

$a_8 = -34.5+(8-1)(4.5) = -3;\ a_n = -34.5+(n-1)(4.5) = 4.5n-39$. The eighth term of the sequence is -3 The nth term of the sequence is $4.5n-39$.

21. $a_1 = x;\ a_2 = x+3 \Rightarrow d = (x+3)-x = 3;\ a_8 = x+(8-1)(3) = x+21;\ a_n = x+(n-1)(3) = x+3n-3$.

The eighth term of the sequence is $x+21$. The nth term of the sequence is $x+3n-3$.

23. $a_3 = \pi+2\sqrt{e}; a_4 = \pi+3\sqrt{e} \Rightarrow d = \left(\pi+3\sqrt{3}\right)-\left(\pi+2\sqrt{e}\right) \Rightarrow a_1 = \pi.\ a_8 = \pi+7\sqrt{e}\ \ a_n = \pi+(n-1)\sqrt{e}$

The eighth term is $\pi+7\sqrt{e}$ and the nth term is $\pi+(n-1)\sqrt{e}$.

25. $a_5 = 27;\ a_{15} = 87 \Rightarrow 87 = 27+10d \Rightarrow 10d = 60 \Rightarrow d = 6;\ 27 = a_1+(5-1)(6) \Rightarrow a_1 = 27-24 = 3$.

The first term of the sequence is 3.

27. $a_5 = -3;\ a_{18} = -29 \Rightarrow -29 = -3+13d \Rightarrow 13d = -26 \Rightarrow d = -2;$

$-29 = a_1+(18-1)(-2) \Rightarrow a_1 = -29+34 = 5$. The first term of the sequence is 5.

29. $S_3 = 75;\ a_3 = 22 \Rightarrow 75 = \frac{3}{2}[a_1+(22)] \Rightarrow a_1 = 50-22 \Rightarrow a_1 = 28$. The first term of the sequence is 28.

31. $S_{16} = -160;\ a_{16} = -25 \Rightarrow -160 = \frac{16}{2}[a_1+(-25)] \Rightarrow 8a_1 = -160+200 \Rightarrow a_1 = 5$. The first term of the sequence is 5.

33. $a_8 = -160;\ d = 3 \Rightarrow a_{10} = 8+(10-1)(3) = 35 \Rightarrow S_{10} = \frac{10}{2}(8+35) = 215$. The sum of the first ten terms of the sequence is 215.

35. $a_3 = 5;\ a_4 = 8 \Rightarrow d = 8-5 = 3 \Rightarrow 5 = a_1+(3-1)(3) \Rightarrow a_1 = -1 \Rightarrow a_{10} = -1+(10-1)(3) = 26 \Rightarrow$

$S_{10} = \frac{10}{2}(-1+26) = 125$. The sum of the first ten terms of the sequence is 125.

37. $a_1 = 5;\ d = 9-5 = 4 \Rightarrow a_{10} = 5+(10-1)(4) = 41 \Rightarrow S_{10} = \frac{10}{2}(5+41) = 230$. The sum of the first ten terms of the sequence is 230.

39. $a_1 = 10;\ a_{10} = 5.5 \Rightarrow S_{10} = \frac{10}{2}(10+55) = 77.5$. The sum of the first ten terms of the sequence is 77.5.

41. $S_{20} = 1090;\ a_{20} = 102 \Rightarrow 1090 = \frac{20}{2}(a_1 + 102) \Rightarrow 109 = a_1 + 102 \Rightarrow a_1 = 7 \Rightarrow 102 = 7 + (20-1)d \Rightarrow$ $19d = 95 \Rightarrow d = 5$. The first term of the sequence is 7 and the common difference is 5.

43. $S_{12} = -108;\ a_{12} = -19 \Rightarrow -108 = \frac{12}{2}[a_1 + (-19)] \Rightarrow 6a_1 = -108 + 114 = 6 \Rightarrow$

$a_1 = 1 \Rightarrow -19 = 1 + (12-1)d \Rightarrow 11d = -20 \Rightarrow d = -\frac{20}{11}$. The first term of the sequence is 1 and the common difference is $-\frac{20}{11}$.

45. From the graph, $a_1 = -2, a_2 = -1 \Rightarrow d = -1-(-2) = 1 \Rightarrow a_n = -2 + (n-1) = n-3$. Also from the graph, $D:\{1,2,3,4,5,6\};\ R:\{-2,-1,0,1,2,3\}$. The nth term of the sequence is $n-3$. The domain of the sequence is the set $\{1,2,3,4,5,6\}$ and the range of the sequence is set $\{-2,-1,0,1,2,3\}$.

47. From the graph, $a_1 = 2.5,\ a_2 = 2 \Rightarrow d = 2 - 2.5 = -0.5 \Rightarrow a_n = 2.5 + (n-1)(-0.5) = -0.5n + 3$. Also from the graph, $D:\{1,2,3,4,5,6\};\ R:\{0.5,1,1.5,2,2.5\}$. The nth term of the sequence is $-0.5n+3$. The domain of the sequence is the set $\{1,2,3,4,5,6\}$ and the range of the sequence is set $\{0.5,1,1.5,2,2.5\}$.

49. From the graph, $a_1 = 10,\ a_2 = -10 \Rightarrow d = -10 - 10 = -20 \Rightarrow a_n = 10 + (n-1)(-20) = -20n + 30$. Also from the graph, $D:\{1,2,3,4,5\};\ R:\{-70,-50,-30,-10,10\}$. The nth term of the sequence is $-20n+30$. The domain of the sequence is the set $\{1,2,3,4,5\}$ and the range of the sequence is set $\{-70,-50,-30,-10,10\}$.

51. $a_1 = 3,\ a_8 = 17;\ 3+5+7+9+11+13+15+17 = S_8 = \frac{8}{2}(3+17) = 80$. The sum of the series is 80.

53. $a_1 = 1;\ a_{50} = 50;\ 1+2+3+4+\cdots+50 = S_{50} = \frac{50}{2}(1+50) = 1275$. The sum of the series is 1275.

55. $a_1 = -7;\ d = -4-(-7) = 3;\ 101 = -7 + (n-1)(3) \Rightarrow 3n = 111 \Rightarrow n = 37;$

$-7 + (-4) + (-1) + 2 + 5 + \ldots + 98 + 101 = S_{37} = \frac{37}{2}(-7+101) = 1739$. The sum of the series is 1739.

57. $a_1 = 5(1) = 5;\ a_{40} = 5(40) = 200;\ S_{40} = \frac{40}{20}(5+200) = 4100$. The sum of the series is 4100.

59. $a_1 = 1+4 = 5;\ a_3 = 3+4 = 7;\ \sum_{i=1}^{3}(i+4) = S_3 = \frac{3}{2}(5+7) = 18$. The sum is 18.

61. $a_1 = 2(1)+3 = 5; a_{10} = 2(10)+3 = 23;\ \sum_{j=1}^{10}(2j+3) = S_{10} = \frac{10}{2}(5+23) = 140$. The sum is 140.

63. $a_1 = -5-8(1) = -13;\ a_{12} = -5-8(12) = -101;\ \sum_{i=1}^{12}(-5-8i) = S_{12} = \frac{12}{2}[-13+(-101)] = -684.$

The sum is -684.

65. $a_1 = 1;\ a_{1000} = 1000;\ \sum_{i=1}^{1000} i = S_{1000} = \frac{1000}{2}(1+1000) = 500{,}500.$ The sum is 500,500.

67. $a_n = 4.2n+9.73$. Using the sequence feature of a graphing calculator, we obtain $s_{10} = 328.3$.

69. $a_n = \sqrt{8}n+\sqrt{3}$. Using the sequence feature of a graphing calculator, we obtain $s_{10} = 172.884$.

71. $a_1 = 51;\ d = 1;\ 71 = 51+(n-1)(1) \Rightarrow n = 71-50 = 21;\ \sum_{i=51}^{71} i = s_{21} = \frac{21}{2}(51+71) = 1281.$ The sum of all the integers from 51 to 71 is 1281.

73. $a_1 = 1;\ a_{12} = 12;\ s_{12} = \frac{12}{2}(1+12) = 78;$ chimes per 24 hours $= 2(78) = 156$. Chimes per 30 days $= (30)(156) = 4680$. The clock will chime 4680 times in a month of 30 days.

75. $a_1 = 49{,}000,\ d = 580,\ n = 11;\ a_{11} = 49{,}000+(11-1)(580) = 54{,}800.$ The population five years from now will be 54,800.

77. $a_1 = 18;\ a_{31} = 28;\ s_{31} = \frac{31}{2}(18+28) = 713.$ A total of 713 inches of material will be needed.

79. (a) $a_1 = 98.2,\ a_3 = 109.8 \Rightarrow 109.8 = 98.2+(3-1)d \Rightarrow 2d = 11.6 \Rightarrow d = 5.8.$ The common difference of the arithmetic sequence describing the child's height would be 5.8 centimeters.

(b) $a_6 = 98.2+(6-1)(5.8) = 127.2.$ We would expect the child's height to be 127.2 centimeters at age 8.

81. $d_{n+1}-d_n = (a_{n+1}+c\bullet b_{n+1})-(a_n+c\bullet b_n) = (a_{n+1}-a_n)+c(b_{n+1}-b_n) = j+ck,$ where *j* is the common difference for the sequence $a_1,\ a_2,\ a_3,\ \ldots$ and *k* is the common difference for the sequence $b_1, b_2, b_3, \ldots$ Thus, since $d_{n+1}-d_n$ is a constant, $d_1,\ d_2,\ d_3,\ \ldots$ is also an arithmetic sequence.

11.3: Geometric Sequences and Series

1. $a_1 = \frac{5}{3};\ a_2 = \left(\frac{5}{3}\right)(3) = 5;\ a_3 = (5)(3) = 15;\ a_4 = (15)(3) = 45.$ The first four terms of the sequence are $\frac{5}{3}, 5, 15, 45.$

3. $a_4 = 5,\ a_5 = 10 \Rightarrow r = \frac{10}{5} = 2;\ a_3 = \frac{5}{2};\ a_2 = \frac{\frac{5}{2}}{2} = \frac{5}{4};\ a_1 = \frac{\frac{5}{4}}{2} = \frac{5}{8}.$ The first five terms of the sequence are $\frac{5}{8}, \frac{5}{4}, \frac{5}{2}, 5, 10.$

5. $a_5 = 5(-2)^{5-1} = (5)(16) = 80; a_n = 5(-2)^{n-1}$. The fifth term of the sequence is 80 and the nth term of the sequence is $5(-2)^{n-1}$.

7. $a_2 = -4, r = -3 \Rightarrow a_1 = \frac{4}{3}; a_5 = \left(\frac{4}{3}\right)(-3)^{5-1} = 108; a_n = \left(\frac{4}{3}\right)(-3)^{n-1}$. The fifth term of the sequence is 108 and the nth term of the sequence is $\left(\frac{4}{3}\right)(-3)^{n-1}$.

9. $a_4 = 243, r = -3 \Rightarrow a_1 = \frac{243}{(-3)^{4-1}} = -9; a_5 = (-9)(-3)^{5-1} = -729; a_n = (-9)(-3)^{n-1}$. The fifth term of the sequence is -729 and the nth term of the sequence is $(-9)(-3)^{n-1}$.

11. $r = \frac{-12}{-4} = \frac{-36}{-12} = \frac{-108}{-36} = 3; a_5 = (-4)(3)^{5-1} = -324; a_n = (-4)(3)^{n-1}$. The fifth term of the sequence is -324 and the nth term of the sequence is $(-4)(3)^{n-1}$.

13. $r = \frac{2}{\frac{4}{5}} = \frac{5}{2} = \frac{\frac{25}{2}}{5} = \frac{5}{2}; a_5 = \left(\frac{4}{5}\right)\left(\frac{5}{2}\right)^{5-1} = \frac{125}{4}; a_n = \left(\frac{4}{5}\right)\left(\frac{5}{2}\right)^{n-1}$. The fifth term of the sequence is $\frac{125}{4}$ and the nth term of the sequence is $\left(\frac{4}{5}\right)\left(\frac{5}{2}\right)^{n-1}$.

15. $r = \frac{-5}{10} = \frac{\frac{5}{2}}{-5} = \frac{-\frac{5}{4}}{\frac{5}{2}} = -\frac{1}{2}; a_5 = (10)\left(-\frac{1}{2}\right)^{5-1} = \frac{5}{8}; a_n = 10\left(-\frac{1}{2}\right)^{n-1}$. The fifth term of the sequence is $\frac{5}{8}$ and the nth term of the sequence is $10\left(-\frac{1}{2}\right)^{n-1}$.

17. $a_3 = 5, a_8 = \frac{1}{625} \Rightarrow r^{7-2} = \frac{\frac{1}{625}}{5} = \frac{1}{3125} \Rightarrow r = \frac{1}{5}; a_1 = \frac{5}{\left(\frac{1}{5}\right)^2} = 125$. The first term of the sequence is 125 and the common ratio is $\frac{1}{5}$.

19. $a_4 = -\frac{1}{4}, a_9 = -\frac{1}{128} \Rightarrow r^{8-3} = \frac{-\frac{1}{128}}{-\frac{1}{4}} = \frac{1}{32} \Rightarrow r = \frac{1}{2}; a_1 = \frac{-\frac{1}{4}}{\left(\frac{1}{2}\right)^3} = -2$. The first term of the sequence is -2 and the common ratio is $\frac{1}{2}$.

21. $r = \frac{8}{2} = \frac{32}{8} = \frac{128}{32} = 4; S_5 = \frac{2(1-4^5)}{1-4} = 682$. The sum of the first five terms of the sequence is 682.

23. $r = \frac{-9}{18} = \frac{\frac{9}{2}}{-9} = \frac{-\frac{9}{4}}{\frac{9}{2}} = -\frac{1}{2}; S_5 = \frac{18\left[1-\left(-\frac{1}{2}\right)^5\right]}{1-\left(-\frac{1}{2}\right)} = \frac{99}{8}$. The sum of the first five terms of the sequence is $\frac{99}{8}$.

25. $S_5 = \frac{(8.423)\left[1-(2.859)^5\right]}{1-2.859} \approx 860.95$. Rounded to the nearest hundredth, the sum of the first five terms of the sequence is 860.95.

27. $a_1 = 3; r = 3; S_5 = \frac{3(1-3^5)}{1-3} = 363$. The sum is 363.

29. $a_1 = (48)\left(\frac{1}{2}\right) = 24; r = \frac{1}{2}; S_6 = \frac{24\left[1-\left(\frac{1}{2}\right)^6\right]}{1-\frac{1}{2}} = \frac{189}{4}$. The sum is $\frac{189}{4}$.

31. $a_1 = (-2)^4 = 16; r = 2; n = 10-3 = 7; S_7 = \frac{16\left(1-(-2)^7\right)}{1-(-2)} = 688$. The sum is 688.

33. $a_1 = -2; r = 2; S_8 = \frac{(-2)\left[1-(2)^8\right]}{1-(2)} = -510$, In this case the index begins at 2 so we will subtract the value of a_1 from the sum. $-510-(-2) = -508$. The sum is -508.

35. $a_1 = 5; r = 2; S_6 = \frac{(5)\left[1-(2)^6\right]}{1-(2)} = 315$, The sum is 315.

37. $a_1 = -1; r = \frac{1}{2}; S_4 = \frac{(-1)\left[1-\left(\frac{1}{2}\right)^4\right]}{1-\left(\frac{1}{2}\right)} = \frac{-\frac{15}{16}}{\frac{1}{2}} = -\frac{15}{8}$, The sum is $-\frac{15}{8}$.

39. The sum of the terms of an infinite geometric sequence exists if the absolute value of the common ratio is less than 1.

41. $a_1 = 0.8, r = 0.1 \Rightarrow S_\infty = \frac{0.8}{1-0.1} = \frac{8}{9}$. The sum of the geometric series is $\frac{8}{9}$.

43. $a_1 = 0.45, r = 0.01 \Rightarrow S_\infty = \frac{0.45}{1-0.01} = \frac{5}{11}$. The sum of the geometric series is $\frac{5}{11}$.

45. $a_1 = 0.378, r = 0.001 \Rightarrow S_\infty = \frac{0.378}{1-0.001} = \frac{14}{37}$. The sum of the geometric series is $\frac{14}{37}$.

47. $r = \frac{24}{12} = \frac{48}{24} = \frac{96}{48} = 2; |r| > 1 \Rightarrow$ the sum will not converge. Because the common ratio, 2, is greater than 1, the sum of the terms of the geometric sequence does not converge.

49. $r=\frac{-24}{-48}=\frac{-12}{-24}=\frac{-6}{-12}=\frac{1}{2}; |r|<1 \Rightarrow$ the sum would converge. The common ratio is $\frac{1}{2}$ and the sum of the terms of the geometric sequence would converge.

51. $r=\frac{2}{16}=\frac{\frac{1}{4}}{2}=\frac{\frac{1}{32}}{\frac{1}{4}}=\frac{1}{8}; S_\infty=\frac{16}{1-\frac{1}{8}}=\frac{128}{7}$. The sum is $\frac{128}{7}$.

53. $r=\frac{10}{100}=\frac{1}{10}; S_\infty=\frac{100}{1-\frac{1}{10}}=\frac{1000}{9}$. The sum is $\frac{1000}{9}$.

55. $r=\frac{\frac{2}{3}}{\frac{4}{3}}=\frac{\frac{1}{3}}{\frac{2}{3}}=\frac{1}{2}; S_\infty=\frac{\frac{4}{3}}{1-\frac{1}{2}}=\frac{8}{3}$. The sum is $\frac{8}{3}$.

57. $a_1=3, r=\frac{1}{4} \Rightarrow S_\infty=\frac{3}{1-\frac{1}{4}}=4$. The sum is 4.

59. $a_1=0.3, r=0.3 \Rightarrow S_\infty=\frac{0.3}{1-0.3}=\frac{3}{7}$. The sum is $\frac{3}{7}$.

61. Since $5^{-k}=\left(\frac{1}{5}\right)^k$ we have $a_1=\frac{1}{5}, r=\frac{1}{5} \Rightarrow S_\infty=\frac{\frac{1}{5}}{1-\frac{1}{5}}=\frac{1}{4}$. The sum is $\frac{1}{4}$.

63. $a_1=\frac{1}{5}, r=-\frac{1}{2} \Rightarrow S_\infty=\frac{\frac{1}{5}}{1-\left(-\frac{1}{2}\right)}=\frac{2}{15}$. The sum is $\frac{2}{15}$.

65. See Figure 65. Rounded to the nearest thousandth, the sum is -97.739.

67. See Figure 67. Rounded to the nearest thousandth, the sum is 0.212.

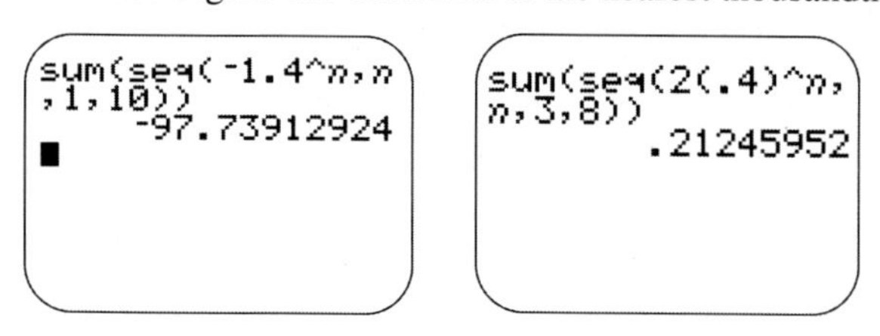

Figure 65 Figure 67

69. $S=1000\left[\frac{(1+0.04)^9-1}{0.04}\right]\approx 10,582.80$. Rounded to the nearest penny, the future value of the annuity is \$10,582.80.

71. $S=2430\left[\frac{(1+0.025)^{10}}{0.025}\right]\approx 27,244.22$. Rounded to the nearest penny, the future value of the annuity is \$27,224.22.

73. (a) $a_1 = 1276(0.916)^1 \approx 1169; r = 0.916$. Rounded to the nearest whole number, the first term of the sequence is 1169. The common ratio is 0.916.

(b) $a_{10} = 1276(0.916)^{10} \approx 531; a_{20} = 1276(0.916)^{20} \approx 221$. Rounded to the nearest whole number, the tenth term of the sequence is 531 and the twentieth term of the sequence is 221. This means that a person 10 years from retirement should have saved 531% of his or her annual income and a person 20 years from retirement should have saved 221% of his or her annual income.

75. (a) $a_2 = 2a_1; a_3 = 2a_2 = 4a_1; ...a_n = a_1(20)^{n-1}$. The nth term of the geometric series is $a_1(2)^{n-1}$.

(b) $a_1 = 100; a_n > 1{,}000{,}000 \Rightarrow (2)^{n-1} > 10{,}000 \Rightarrow n-1 > \dfrac{4}{\log(2)} \Rightarrow n > 14.25 = 15.$

(c) $n = 4$ time = $(15-1)$ 40 minutes $= 560$ minutes $= 9$ hours 20 minutes. After 9 hours and 20 minutes, the number of bacteria will exceed one million.

77. $a_1 = 0.8, r = 0.8 \Rightarrow a_9 = (0.8)(0.8)^{9-1} \approx 0.134 = 13.4\%$. After nine drainings and replacements with water, the strength of the mixture will be approximately 13.4%.

79. $a_1 = (100{,}000)(1-0.2) = 80{,}000; r = 0.8 \Rightarrow a_6 = 80{,}000(0.8)^{6-1} = 26{,}214.40...$ The value of the machine at the end of six years will be $26,214.40.

81. $a_1 = 100, r = 1.5 \Rightarrow a_n = 100(1.5)^{n-1}; \ a_4 = 100(1.5)^{4-1} = 337.5$ In the fourth generation there will be about 338 flies.

83. $a_1 = 40, r = 0.8 \Rightarrow S_\infty = \dfrac{40}{1-0.8} = 200$. The total length of arc through which the pendulum will swing is 200 centimeters.

85. $a_1 = 2, r = 2 \Rightarrow S_5 = \dfrac{2(1-2^5)}{1-2} = 62; \ S_{10} = \dfrac{2(1-2^{10})}{1-2} = 2046$. Going back 5 generations, a person has 62 ancestors and going back 10 generations, a person has 2046 ancestors

87. Option 1: Arithmetic sequences with $a_1 = 5000$ and $d = 10{,}000 \Rightarrow$

$$S_{30} = \frac{30}{2}\left[2(5000) + (30-1)(10{,}000)\right] = 4{,}500{,}000.$$

Option 2: Geometric sequences with $a_1 = 0.01$ and $r = 2 \Rightarrow S_{30} = \dfrac{0.01(1-2^{30})}{1-2} = 10{,}737{,}418.23$.

Option 1 pays a total of $4,500,000 while Option 2 pays a total of $10,737,418.23. Choose Option 2.

89. $a_1 = 2, r = \dfrac{1}{2} \Rightarrow a_8 = 2\left(\dfrac{1}{2}\right)^{8-1} = \dfrac{1}{64}$. The eighth such triangle will have sides of length $\dfrac{1}{64}$.

Reviewing Basic Concepts (Sections 11.1—11.3)

1. $a_1 = (-1)^{1-1}(4 \cdot 1) = 4; a_2 = (-1)^{2-1}(4 \cdot 2) = -8; a_3 = (-1)^{3-1}(4 \cdot 3) = 12; a_4 = (-1)^{4-1}(4 \cdot 4);$ $a_5 = (-1)^{5-1}(4 \cdot 5) = 20$. The first five terms of the sequence are $4, -8, 12, -16, 20$.

2. $a_1 = 3\cdot 1+1=4;\ a_5 = 3\cdot 5+1=16;\ s_5 = \frac{5}{2}(4+16)=50$. The series sum is 50.

3. From Figure 3, it can be seen that as n increases, the terms of the sequence converge to 1.

[0,11] by [-2,2]
Xscl = 1 Yscl = 1

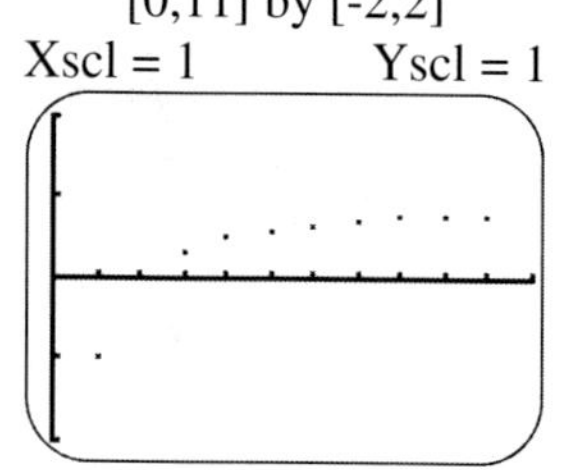

Figure 3

4. $a_1 = 8,\ d=-2 \Rightarrow a_2 = 8-2=-6,\ a_3 = 6-2=4,\ a_4 = 4-2=2,\ a_5 = 2-2=0$. The first five terms of the sequence are 8, 6, 4, 2, 0.

5. $a_5 = 5,\ a_8 = 17 \Rightarrow 3d = 17-5 \Rightarrow d=4 \Rightarrow a_1 = 5-4\cdot 4=-11$. The first term of the sequence is -11.

6. $a_{10} = 2+5(10-1)=47;\ S_{10} = \frac{10}{2}(2+47)=245$. The sum of the first ten terms of the sequence is 245.

7. $a_3 = (-2)(-3)^{3-1} = -18;\ a_n = (-2)(-3)^{n-1}$. .The third term of the sequence is -18 and the nth term of the sequence is $(-2)(-3)^{n-1}$.

8. $r = \frac{3}{5} = \frac{\frac{9}{5}}{3} = \frac{3}{5};\ \frac{244}{625} = 5\left(\frac{3}{5}\right)^{n-1} \Rightarrow \left(\frac{3}{5}\right)^{n-1} = \frac{243}{3125} \Rightarrow n-1=5 \Rightarrow n=6.$

$$S_6 = \frac{5\left[1-\left(\frac{3}{5}\right)^6\right]}{1-\frac{3}{5}} = \frac{7448}{625}$$. The series is geometric with common ratio $\frac{3}{5}$. The series sum is $\frac{7448}{625}$.

9. $a_1 = 3\left(\frac{2}{3}\right) = 2.\ S_\infty = \frac{2}{1-\frac{2}{3}} = 6$. The series sum is 6.

10. $S = 500\left[\frac{(1+0.035)^{13}-1}{0.035}\right] \approx 8056.52$. Rounded to the nearest penny, the future value of the annuity is $8,056.52.

11.4: Counting Theory

1. $4! = 4\cdot 3\cdot 2\cdot 1 = 24$

3. $(4-2)! = 2! = 2\cdot 1 = 2$

5. $\frac{6!}{5!} = \frac{6\cdot 5\cdot 4\cdot 3\cdot 2\cdot 1}{5\cdot 4\cdot 3\cdot 2\cdot 1} = 6$

7. $\frac{8!}{6!} = \frac{8\cdot 7\cdot 6\cdot 5\cdot 4\cdot 3\cdot 2\cdot 1}{6\cdot 5\cdot 4\cdot 3\cdot 2\cdot 1} = 8\cdot 7 = 56$

9. $3! \cdot 4 = 3 \cdot 2 \cdot 1 \cdot (4) = 24$

11. Divide the common factor $6 \cdot 5 \cdot 4 \cdot 3 \cdot 2 \cdot 1$ from both the numerator and denominator to begin. Then multiply the factors 7 and 8 which remain in the denominator to obtain 56.

13. $P(7,7) = \frac{7!}{(7-7)!} = \frac{7!}{0!} = 7 \cdot 6 \cdot 5 \cdot 4 \cdot 3 \cdot 2 \cdot 1 = 5040$. The value of $P(7,7)$ is 50,400

15. $P(9,2) = \frac{9!}{(9-2)!} = \frac{9!}{7!} = 9 \cdot 8 = 72$. The value of $P(9,2)$ is 72.

17. $P(5,1) = \frac{5!}{(5-1)!} = \frac{5!}{4!} = 5$. The value of $P(5,4)$ is 5.

19. $C(4,2) = \frac{4!}{(4-2)!2!} = \frac{4!}{2!2!} = \frac{4 \cdot 3}{2 \cdot 1} = 6$. The value of $C(4,2)$ is 6.

21. $C(6,0) = \frac{6!}{(6-0)!0!} = \frac{6!}{6!} = 1$. The value of $C(6,0)$ is 1.

23. $C(12,4) = \frac{12!}{(12-4)!4!} = \frac{12!}{8!4!} = \frac{12 \cdot 11 \cdot 10 \cdot 9}{4 \cdot 3 \cdot 2 \cdot 1} = 495$. The value of $C(12,4)$ is 495.

25. ${}_{20}P_5 = 20 \ {}_nP_r \ 5 = 1{,}860{,}480$ The value of ${}_{20}P_5$ is $1{,}860{,}480$

27. ${}_{15}P_8 = 15_n P_r 8 = 259{,}459{,}200$. The value of ${}_{15}P_8$ is $259{,}459{,}200$.

29. ${}_{20}C_5 = 20_n C_r 5 = 15{,}504$. The value of ${}_{20}C_5$ is $15{,}504$.

31. ${}_{15}C_8 = 15_n C_r 8 = 6435$. The value of ${}_{15}C_8$ is 6435.

33. (a) A telephone number involves a permutation of digits because order matters.
    (b) A social security number involves a permutation of digits because order matters.
    (c) A hand of cards in poker involves a combination of cards because order does not matter.
    (d) A committee of politicians involves a combination of persons because order does not matter.
    (e) The "combination" on a combination lock involves a permutation of numbers because order matters.
    (f) A lottery choice of six numbers where the order does not matter involves a combination of numbers.
    (g) An automobile license plate involves a permutation of characters because order matters.

35. $5 \times 3 \times 2 = 30$. 30 different types of homes are available.

37. (a) $2 \times 25 \times 24 \times 23 = 27{,}600$. 27,600 radio station calls can be made.
    (b) $2 \times 26 \times 26 \times 26 = 35{,}152$. 35,152 radio station calls can be made.
    (c) $2 \times 24 \times 24 \times 1 = 1104$. 1104 radio station calls can be made.

39. $3 \times 5 = 15$. 15 first and middle name combinations are possible.

41. (a) $26 \times 26 \times 26 \times 10 \times 10 \times 10 = 17{,}576{,}000$. 17,576,000 different license plates are possible.
    (b) $10 \times 10 \times 10 \times 26 \times 26 \times 26 = 17{,}576{,}000$. 17,576,000 additional plates are possible.
    (c) $26 \times 10 \times 10 \times 10 \times 26 \times 26 \times 26 = 456{,}976{,}000$. The new scheme provides 456,976,000 plates.

43. $6_n P_r 6 = 720$. 720 arrangements of the people are possible.

45. $6_n P_r 3 = 120$. 120 course schedules are possible.

47. $15_n P_r 3 = 2730$. 2730 slates of 3 officers are possible.

49. $5_n P_r 5 = 120$; $10_n P_r 5 = 30,240$. With 5 players, 120 assignments are possible and with 10 players, 30,240 assignments are possible.

51. $30_n P_r 4 = 27,405$. 27,405 different groups are possible.

53. $6_n C_r 3 = 20$. 20 different garnished hamburgers are possible .

55. $15_n C_r 2 = 105$; $15_n C_r 4 = 1365$. 105 samples of 2 may be drawn and 1365 samples of 4 may be drawn.

57. $8_n C_r 2 = 28$. 28 samples of 2 may be drawn where both marbles are blue.

59. (a) $9_n C_r 3 = 84$. 84 delegations are possible.

(b) $5_n C_r 3 = 10$. 10 delegations could have all liberals.

(c) $(5_n C_r 2)(4_n C_r 1) = (10)(4) = 40$. 40 delegations could have 2 liberals and 1 conservative.

(d) $8_n C_r 2 = 28$. 28 delegations are possible that include the mayor.

61. $8_n P_r 4 = 1680$. 1680 course schedules are possible.

63. $6_n C_r 4 = 15$. 15different soups can be made.

65. $12_n P_r 11 = 479,001,600$. 479,001,600 different seatings are possible.

67. (a) $8_n C_r 5 = 56$. 56 committees of all men may be chosen.

(b) $11_n C_r 5 = 462$. 462 committees of all women may be chosen.

(c) $(8_n C_r 3)(11_n C_r 2) = (56)(55) = 3080$. 3080 committees of 3 men and 2 women may be chosen.

(d) $(8_n C_r 5)(11_n C_r 0) + (8_n C_r 4)(11_n C_r 1) + (8_n C_r 3)(11_n C_r 2) + (8_n C_r 2)(11_n C_r 3) =$ $(56)(1) + (70)(11) + (56)(55) + (28)(165) = 8526$. 8526 committees with no more than 3 women are possible.

69. $2^{12} = 4096$. 4096 codes are possible.

71. $(10 \times 10 \times 10)(10 \times 10 \times 10) = 1,000,000$. 1,000,000 different combinations are possible.

73. Circular ring $\Rightarrow 1$ key to start; then, $3_n P_r 3 = 6$ ways to add the remaining 3. The keys can be put on in 6 distinguishable ways.

75. $P(n, n-1) = \dfrac{n!}{[n-(n-1)]!} = \dfrac{n!}{1!} = n!$; $P(n,n) = \dfrac{n!}{(n-n)!} = \dfrac{n!}{0!} = n!$. Thus, $P(n, n-1) = P(n,n)$.

77. $P(n,0) = \dfrac{n!}{(n-0)!} = \dfrac{n!}{n!} = 1$. Thus, $P(n,0) = 1$.

79. $C(n,0) = \dfrac{n!}{(n-0)!0!} = \dfrac{n!}{n!} = 1$. Thus, $C(n,0) = 1$.

81. $C(n,n-1)=\dfrac{n!}{[n-(n-1)]!(n-1)!}=\dfrac{n!}{1!(n-1)!}=\dfrac{n(n-1)!}{(n-1)!}=n.$ Thus, $C(n,n-1)=n.$

11.5: The Binomial Theorem

1. $\dfrac{6!}{3!3!}=\dfrac{6\cdot5\cdot4\cdot3\cdot2\cdot1}{3\cdot2\cdot1\cdot3\cdot2\cdot1}=20$ The expression is equal to 20.

3. $\dfrac{7!}{3!4!}=\dfrac{7\cdot6\cdot5\cdot4\cdot3\cdot2\cdot1}{3\cdot2\cdot1\cdot4\cdot3\cdot2\cdot1}=35.$ The expression is equal to 35.

5. $\begin{pmatrix}8\\3\end{pmatrix}=C(8,3)=\dfrac{8!}{(8-3)!3!}=\dfrac{8!}{5!3!}=56$ The expression is equal to 56.

7. $\begin{pmatrix}10\\8\end{pmatrix}=C(10,8)=\dfrac{10!}{(10-8)!10!}=\dfrac{10!}{2!10!}=45.$ The expression is equal to 45.

9. $\begin{pmatrix}13\\13\end{pmatrix}=C(13,13)=\dfrac{13!}{(13-13)!13!}=\dfrac{13!}{13!}=1.$ The expression is equal to 1. 11.

$\begin{pmatrix}8\\3\end{pmatrix}=C(8,3)=\dfrac{8!}{(8-3)!3!}=\dfrac{8!}{5!3!}=56$. The expression is equal to 56.

13. $\begin{pmatrix}100\\2\end{pmatrix}=C(100,2)=\dfrac{100!}{(100-2)!2!}=\dfrac{100!}{98!2!}=4950$. The expression is equal to 4950.

15. $\begin{pmatrix}5\\0\end{pmatrix}=C(5,0)=\dfrac{5!}{(5-0)!0!}=\dfrac{5!}{5!}=1$. The expression is equal to 1.

17. $\begin{pmatrix}n\\n-1\end{pmatrix}=\dfrac{n!}{(n-1)!(n-(n-1))!}=\dfrac{n!}{(n-1)!(1)!}=\dfrac{n(n-1)!}{(n-1)!}=n.$ The expression is equal to n.

19. The expansion of $(x+y)^8$ has a term where x is raised to each of the following powers: 8, 7, 6, 5, 4, 3, 2, 1, and 0. Thus, there are 9 terms.

21. $(2x)^4=16x^4;(3y)^4=81y^4$. In the expansion of $(2x)^4$, $16x^4$ is the first term and $81y^4$ is the last term.

23. The binomial expansion for $(x+y)^6$ is given by:

$$(x+y)^6=\begin{pmatrix}6\\0\end{pmatrix}x^6+\begin{pmatrix}6\\1\end{pmatrix}x^5y+\begin{pmatrix}6\\2\end{pmatrix}x^4y^2+\begin{pmatrix}6\\3\end{pmatrix}x^3y^3+\begin{pmatrix}6\\4\end{pmatrix}x^2y^4+\begin{pmatrix}6\\5\end{pmatrix}xy^5+\begin{pmatrix}6\\6\end{pmatrix}y^6=$$

$$x^6+6x^5y+15x^4y^2+20x^3y^3+15x^2y^4+6xy^5+y^6.$$

25. The binomial expansion for $(p-q)^5$ is given by:

$$(p-q)^5=\begin{pmatrix}5\\0\end{pmatrix}p^5+\begin{pmatrix}5\\1\end{pmatrix}p^4(-q)+\begin{pmatrix}5\\2\end{pmatrix}p^3(-q)^2+\begin{pmatrix}5\\3\end{pmatrix}p^2(-q)^3+\begin{pmatrix}5\\4\end{pmatrix}p(-q)^4+\begin{pmatrix}5\\5\end{pmatrix}(-q)^5=$$

$$p^5-5p^4q+10p^3q^2-10p^2q^3+5pq^4-q^5.$$

27. The binomial expansion for $\left(r^2+s\right)^5$ is given by:

$$\left(r^2+s\right)^5=\binom{5}{0}\left(r^2\right)^5+\binom{5}{1}\left(r^2\right)^4 s+\binom{5}{2}\left(r^2\right)^3 s^2+\binom{5}{3}\left(r^2\right)^2 s^3+\binom{5}{4}\left(r^2\right)s^4+\binom{5}{5}s^5=$$

$$r^{10}+5r^8s+10r^6s^2+10r^4s^3+5r^2s^4+s^5.$$

29. The binomial expansion for $(p+2q)^4$ is given by:

$$(p+2q)^4=\binom{4}{0}p^4+\binom{4}{1}p^3(2q)+\binom{4}{2}p^2(2q)^2+\binom{4}{3}p(2q)^3+\binom{4}{4}(2q)^4=$$

$$p^4+8p^3q+24p^2q^2+32pq^3+16q^4.$$

31. The binomial expansion for $(7p+2q)^4$ is given by:

$$(7p+2q)^4=\binom{4}{0}(7p)^4+\binom{4}{1}(7p)^3(2q)+\binom{4}{2}(7p)^2(2q)^2+\binom{4}{3}(7p)(2q)^3+\binom{4}{4}(2q)^4=$$

$$2401p^4+2744p^3q+1176p^2q^2+224pq^3+16q^4.$$

33. The binomial expansion for $(3x-2y)^6$ is given by:

$$(3x-2y)^6=\binom{6}{0}(3x)^6+\binom{6}{1}(3x)^5(-2y)+\binom{6}{2}(3x)^4(-2y)^2+\binom{6}{3}(3x)^3(-2y)^3+\binom{6}{4}(3x)^2(-2y)^4+$$

$$\binom{6}{5}(3x)(-2y)^5+\binom{6}{6}(-2y)^6=729x^6-2916x^5y+4860x^4y^2-4320x^3y^3+2160x^2y^4-576xy^5+64y^6.$$

35. The binomial expansion for $\left(\frac{m}{2}-1\right)^6$ is given by:

$$\left(\frac{m}{2}-1\right)^6=\binom{6}{0}\left(\frac{m}{2}\right)^6+\binom{6}{1}\left(\frac{m}{2}\right)^5(-1)+\binom{6}{2}\left(\frac{m}{2}\right)^4(-1)^2+\binom{6}{3}\left(\frac{m}{2}\right)^3(-1)^3+\binom{6}{4}\left(\frac{m}{2}\right)^2(-1)^4+$$

$$\binom{6}{5}\left(\frac{m}{2}\right)(-1)^5+\binom{6}{6}(-1)^6=\frac{m^6}{64}-\frac{3m^5}{16}+\frac{15m^4}{16}-\frac{5m^3}{2}+\frac{15m^2}{4}-3m+1.$$

37. The binomial expansion for $\left(\sqrt{2}r+\frac{1}{m}\right)^4$ is given by:

$$\left(\sqrt{2r}+\frac{1}{m}\right)^4=\binom{4}{0}\left(\sqrt{2}r\right)^4+\binom{4}{1}\left(\sqrt{2}r\right)^3\left(\frac{1}{m}\right)+\binom{4}{2}\left(\sqrt{2}r\right)^2\left(\frac{1}{m}\right)^2+\binom{4}{3}\left(\sqrt{2}r\right)\left(\frac{1}{m}\right)^3+\binom{4}{4}\left(\frac{1}{m}\right)^4=$$

$$4r^4+\frac{8\sqrt{2}r^3}{m}+\frac{4\sqrt{2}r^2}{m^2}+\frac{4\sqrt{2}r}{m^3}+\frac{1}{m^4}.$$

39. $\binom{8}{5}(4h)^3(-j)^5=-3584h^3j^5$ The sixth term of the binomial expansion of $(4h-j)^8$ is $-3584n^3j^5$.

41. $\binom{22}{14}(a^2)(b^{14}) = 319,770a^{16}b^{14}$ The fifteenth term of the binomial expansion of $(a^2+b)^{22}$ is $319,770a^{16}b^{14}$.

43. $\binom{20}{14}(x)^6(-y^3)^{14} = 38,760x^6y^{42}$ The fifteenth term of the binomial expansion of $(x-y^3)^{20}$ is $38,760x^6y^{42}$.

45. $\binom{8}{4}(3x^7)(2y^3)^4 = 90,720x^{28}y^{12}$ The middle term of the binomial expansion of $(3x^7+2y^3)^8$ is the fifth term which is $90,720x^{28}y^{12}$.

47. $n-4=7$ and $n-7=4 \Rightarrow n=11$. The coefficients of the fifth and eighth terms in the expansion of $(x+y)^n$ are the same for n=11.

49. $10! = 3,628,800$; $\sqrt{2\pi(10)}(10^{10})(e^{-10}) \approx 3,598,695.619$. The exact value of 10! is 3,628,800 and the Stirling's formula approximation of 10! is about 3,598,695.619.

50. $\frac{3,628,800 - 3,598,695.619}{3,628,800} \approx 0.00830 = 0.830\%$. The percentage error in the Stirling's formula approximation of 10! is about 0.830%.

51. $12! = 479,001,600$; $\sqrt{2\pi(12)}(12^{12})(e^{-12}) \approx 475,687,486.5$;

$\frac{479,001,600 - 475,687,486.5}{479,001,600} \approx 0.00692 = 0.692\%$. The exact value of 12! is 479,001,600 and the Stirling's formula approximation of 12! is about 475,687,486.5 which has a percent error of about 0.692%.

52. $13! = 6,227,020,800$; $\sqrt{2\pi(13)}(13^{13})(e^{-13}) \approx 6,187,239,475$;

$\frac{6,227,020,800 - 6,187,239,475}{6,227,020,800} \approx 0.00639 = 0.639\%$. The exact value of 13! Is 6,227,020,800 and the Stirling's formula approximation of 13! is about 6,187,239,475 which has a percent error of about 0.639%. Based on this series of exercises, it appears that the percent error in the Stirling's formula approximation of *n!* decreases as *n* increases.

53. $(1.02)^{-3} = (1+0.02)^{-3} \approx 1 + (-3)(0.02) + \frac{(-3)(-3-1)}{2!}(0.02)^2 +$

$\frac{(-3)(-3-1)(-3-2)}{3!}(0.02)^3 \approx 0.942$. To the nearest thousandth, $(1.02)^{-3}$ is 0.942.

55. $(1.01)^{3/2}=(1+0.01)^{3/2}\approx 1+\binom{3}{2}(0.01)+\frac{\left(\frac{3}{2}\right)\left(\frac{3}{2}-1\right)}{2!}(0.01)^2+\frac{\left(\frac{3}{2}\right)\left(\frac{3}{2}-2\right)}{3!}(0.01)^3\approx 1.015$. To the nearest thousandth, $(1.01)^{3/2}$ is 1.015.

Reviewing Basic Concepts (Sections 11.4 and 11.5)

1. $4!=24$. There are 24 different arrangements.

2. $P(7,3)=\frac{7!}{(7-3)!}=\frac{7!}{4!}=210.$

3 $C(10,4)=\frac{10!}{(10-4)!4!}=\frac{10!}{6!4!}=210$

4. ${}_6C_2\cdot{}_5C_2\cdot{}_3C_1=450$. There are 450 different arrangements for this basketball team.

5. $9\cdot4\cdot2=72$. There are 72 different homes available.

6. To expand $(x+y)^n$ use row *n+1* of Pascal's Triangle.

7. Given $(a+2b)^4$. We will let $x = a$ and $y = 2b$ in the binomial theorem. Using the 5th row of Pascal's Triangle we find the coefficients 1,4,6,4,1

$\Rightarrow (a+2b)^4=(a)^4+4(a)^3(2b)+6(a)^2(2b)^2+4(a)(2b)^3+(2b)^4$

8. $\binom{6}{2}(x)^4(-2y)^2=60x^4y^2$. The third term of the binomial $(x-2y)^6$ is $60x^4y^2$.

11.6: Mathematical Induction

1. The domain of the variable must be all positive integers (natural numbers).

3. Since $2^1=2(1)$ and $2^2=2(2)$, the statement is not true for $n=1$ and $n=2$.

5. $3+6+9+\cdots+3n=\frac{3n(n+1)}{2}$

(i) Show that the statement is true for $n=1$: $3(1)=\frac{3(1)(2)}{2}\Rightarrow 3=3.$

(ii) Assume that S_k is true: $3+6+9+\cdots+3k=\frac{3k(k+1)}{2}$. Show that S_{k+1} is true: $3+6+\cdots+3(k+1)=\frac{3(k+1)(k+2)}{2}$. Add $3(k+1)$ to each side of

$S_k:\ 3+6+9+\cdots+3k+3(k+1)=\frac{3k(k+1)}{2}+3(k+1)=\frac{3k(k+1)+6(k+1)}{2}=\frac{(k+1)(3k+6)}{2}=\frac{3(k+1)(k+2)}{2}.$

Since S_k implies S_{k+1}, the statement is true for every positive integer n.

7. $5+10+15+\cdots+5n=\dfrac{5n(n+1)}{2}$

(i) Show that the statement is true for $n=1$: $5(1)=\dfrac{5(1)(2)}{2}\Rightarrow 5=5$.

(ii) Assume that S_k is true: $5+10+15+\cdots+5k=\dfrac{5k(k+1)}{2}$. Show that S_{k+1} is true:

$5+10+\cdots+5(k+1)=\dfrac{5(k+1)(k+2)}{2}$. Add $5(k+1)$ to each side of S_k :

$$5+10+15+\cdots+5k+5(k+1)=\frac{5k(k+1)}{2}+5(k+1)=$$

$$\frac{5k(k+1)+10(k+1)}{2}=\frac{(k+1)(5k+10)}{2}=\frac{5(k+1)(k+2)}{2}.$$

Since S_k implies S_{k+1}, the statement is true for every positive integer n.

9. $3+3^2+3^3+\cdots+3^n=\dfrac{3(3^n-1)}{2}$

(i) Show that the statement is true for $n=1$: $3^1=\dfrac{3(3^1-1)}{2}\Rightarrow 3=3$.

(ii) Assume that S_k is true: $3+3^2+3^3+\cdots+3^k=\dfrac{3(3^k-1)}{2}$. Show that S_{k+1} is true:

$3+3^2+3^3+\cdots+3^{k+1}=\dfrac{3(3^{k+1}-1)}{2}$ Add 3^{k+1} to each side of S_k :

$$3+3^2+3^3+\cdots 3^k+3^{k+1}=\frac{3\left(3^k-1\right)}{2}+3^{k+1}=\frac{3\left(3^k-1\right)+2\left(3^{k+1}\right)}{2}=$$

$$\frac{3^{k+1}-3+2(3^{k+1})}{2}=\frac{3(3^{k+1})-3}{2}=\frac{3(3^{k+1}-1)}{2}$$

Since S_k implies S_{k+1}, the statement is true for every positive integer n.

11. $1^3+2^3+3^3+\cdots+n^3=\dfrac{n^2(n+1)^2}{4}$

(i) Show that the statement is true for $n=1$: $1^3=\dfrac{1^2(1+1)^2}{4}\Rightarrow 1=1$.

(ii) Assume that S_k is true: $1^3+2^3+3^3+\cdots+k^3=\dfrac{k^2(k+1)^2}{4}$. Show that S_{k+1} is true:

$1^3+2^3+3^3+\cdots+(k+1)^3=\dfrac{(k+1)^2(k+2)^2}{4}$. Add $(k+1)^3$ to each side of

$$S_k: 1^3+2^3+3^3+\cdots+k^3+(k+1)^3=\frac{k^2(k+1)^2}{4}+(k+1)^3=$$

$$\frac{k^2(k+1)^2+4(k+1)^3}{4}=\frac{(k+1)^2(k^2+4k+4)}{4}=\frac{(k+1)^2(k+2)^2}{4}$$

Since S_k implies S_{k+1}, the statement is true for every positive integer n.

13. $\frac{1}{1\cdot 2}+\frac{1}{2\cdot 3}+\cdots+\frac{1}{n(n+1)}=\frac{n}{n+1}$

(i) Show that the statement is true for $n=1$: $\frac{1}{1(1+1)}=\frac{1}{1+1}\Rightarrow\frac{1}{2}=\frac{1}{2}$.

(ii) Assume that S_k is true: $\frac{1}{1\cdot 2}+\frac{1}{2\cdot 3}+\cdots+\frac{1}{k(k+1)}=\frac{k}{k+1}$. Show that S_{k+1} is true:

$\frac{1}{1\cdot 2}+\frac{1}{2\cdot 3}+\cdots+\frac{1}{(k+1)(k+2)}=\frac{k+1}{k+2}$. Add $\frac{1}{(k+1)(k+2)}$ to each side of

$$S_k: \frac{1}{1\cdot 2}+\frac{1}{2\cdot 3}+\cdots+\frac{1}{(k+1)(k+2)}=\frac{k}{k+1}+\frac{1}{(k+1)(k+2)}=$$

$$\frac{k(k+2)+1}{(k+1)(k+2)}=\frac{k^2+2k+1}{(k+1)(k+2)}=\frac{(k+1)(k+1)}{(k+1)(k+2)}=\frac{k+1}{(k+2)}.$$

Since S_k implies S_{k+1}, the statement is true for every positive integer n.

15. $\frac{4}{5}+\frac{4}{5^2}+\frac{4}{5^3}+\cdots+\frac{4}{5^n}=1-\frac{1}{5^n}$

(i) Show that the statement is true for $n=1$: $\frac{4}{5^1}=1-\frac{1}{5^1}\Rightarrow\frac{4}{5}=\frac{4}{5}$.

(ii) Assume that S_k is true: $\frac{4}{5}+\frac{4}{5^2}+\frac{4}{5^3}+\cdots+\frac{4}{5^k}=1-\frac{1}{5^k}$.

Show that S_{k+1} is true: $\frac{4}{5}+\frac{4}{5^2}+\cdots+\frac{4}{5^{k+1}}=1-\frac{1}{5^{k+1}}$. Add $\frac{4}{5^{k+1}}$ to each side of S_k :

$$\frac{4}{5}+\frac{4}{5^2}+\frac{4}{5^3}+\cdots+\frac{4}{5^k}+\frac{4}{5^{k+1}}=1-\frac{1}{5^k}+\frac{4}{5^{k+1}}=1-\frac{1}{5^k}\cdot\frac{5}{5}+\frac{4}{5^{k+1}}=1-\frac{5}{5^{k+1}}+\frac{4}{5^{k+1}}=1-\frac{1}{5^{k+1}}.$$

Since S_k implies S_{k+1}, the statement is true for every positive integer n.

17. $\frac{1}{1\cdot 4}+\frac{1}{4\cdot 7}+\cdots+\frac{1}{(3n-2)(3n+1)}=\frac{n}{3n+1}$

(i) Show that the statement is true for $n=1$: $\frac{1}{1\cdot 4}=\frac{1}{3(1)+1}\Rightarrow\frac{1}{4}=\frac{1}{4}$.

(ii) Assume that S_k is true: $\frac{1}{1\cdot 4}+\cdots+\frac{1}{(3k-2)(3k+1)}=\frac{k}{3k+1}$.

Show that S_{k+1} is true: $\frac{1}{1\cdot 4}+\cdots+\frac{1}{[3(k+1)-2][3(k+1)+1]}=\frac{k+1}{3(k+1)+1}$. Add

$\frac{1}{[3(k+1)-2][3(k+1)+1]}$ to each side of S_k : $\frac{1}{1\cdot 4}+\cdots+\frac{1}{[3(k+1)-2][3(k+1)+1]}=$

$\frac{k}{3k+1}+\frac{1}{[3(k+1)-2][3(k+1)+1]}=\frac{k}{3k+1}+\frac{1}{(3k+1)(3k+4)}=\frac{k(3k+4)+1}{(3k+1)(3k+4)}=$

$\frac{3k^2+4k+1}{(3k+1)(3k+4)}=\frac{(3k+1)(k+1)}{(3k+1)(3k+4)}=\frac{k+1}{3k+4}=\frac{k+1}{3(k+1)+1}$.

Since S_k implies S_{k+1}, the statement is true for every positive integer n.

19. When $n=1$, $3^1<6(1)\Rightarrow 3<6$. When $n=2$, $3^2<6(2)\Rightarrow 9<12$. When $n=3$, $3^3>6(3)\Rightarrow 27>18$. For all $n\geq 3$, $3^n>6n$. The only values are 1 and 2.

21. When $n=1$, $2^1>1^2\Rightarrow 2>1$. When $n=2$, $2^2=2^2\Rightarrow 4=4$. When $n=3$, $2^3<3^2\Rightarrow 8<9$. When $n=4$, $2^4=4^2\Rightarrow 16=16$. For all $n\geq 5$, $2^n>n^2$. The only values are 2, 3, and 4.

23. $(a^m)^n=a^{mn}$

(i) Show that the statement is true for $n=1$: $(a^m)^1=a^{m\cdot 1}\Rightarrow a^m=a^m$.

(ii) Assume that S_k is true: $(a^m)^k=a^{mk}$.

Show that S_{k+1} is true: $(a^m)^{k+1}=a^{m(k+1)}$.

Multiply each side of S_k by a^m : $(a^m)^k\cdot(a^m)^1=a^{mk}\cdot a^m\Rightarrow(a^m)^{k+1}=a^{mk+m}\Rightarrow$

$(a^m)^{k+1}=a^{m(k+1)}$.

Since S_k implies S_{k+1}, the statement is true for every positive integer n.

25. $2^n>2n$, if $n\geq 3$

(i) Show that the statement is true for $n=3$: $2^3>2(3)\Rightarrow 8>6$.

(ii) Assume that S_k is true: $2^k>2k$.

Show that S_{k+1} is true: $2^{k+1}>2(k+1)$. Multiply each side of S_k by 2:

$2^k\cdot 2>2k\cdot 2\Rightarrow 2^{k+1}>2(2k)$. Because $2k>k+1$ for all $k>1$, we have $2(2k)>2(k+1)$; therefore, $2^{k+1}>2(k+1)$.

Since S_k implies S_{k+1}, the statement is true for every positive integer $n\geq 3$.

27. $a^n>1$, if $a>1$

(i) Show that the statement is true for $n=1$: $a^1>1\Rightarrow a>1$, which is true by the given restriction.

(ii) Assume that S_k is true: $a^k > 1$. Show that S_{k+1} is true: $a^{k+1} > 1$.

Multiply each side of S_k by a: $a^k \cdot a > 1 \cdot a \Rightarrow a^{k+1} > a$. Because $a > 1$, we may substitute 1 for a in the expression. That is, $a^{k+1} > 1$.

Since S_k implies S_{k+1}, the statement is true for every positive integer n.

29. $a^n < a^{n-1}$, if $0 < a < 1$

(i) Show that the statement is true for $n = 1$: $a^1 < a^0 \Rightarrow a < 1$, which is true by the given restriction.

(ii) Assume that S_k is true: $a^k < a^{k-1}$. Show that S_{k+1} is true: $a^{k+1} < a^k$.

Multiply each side of S_k by a: $a^k \cdot a < a^{k-1} \cdot a \Rightarrow a^{k+1} < a^k$.

Since S_k implies S_{k+1}, the statement is true for every positive integer n.

31. $n! > 2^n$, if $n \geq 4$

(i) Show that the statement is true for $n = 4$: $4! > 2^4 \Rightarrow 24 > 16$.

(ii) Assume that S_k is true: $k! > 2^k$. Show that S_{k+1} is true: $(k+1)! > 2^{k+1}$.

Multiply each side of S_k by $k+1$: $(k+1)k! > 2^k(k+1) \Rightarrow (k+1)! > 2^k(k+1)$.

Because $(k+1) > 2$ for all $k \geq 4$, we may substitute 2 for $(k+1)$ in the expression.

That is, $(k+1)! > 2^k(2)$ or $(k+1)! > 2^{k+1}$.

Since S_k implies S_{k+1}, the statement is true for every positive integer $n \geq 4$.

33. The number of handshakes is $\dfrac{n^2 - n}{2}$ if $n \geq 2$.

(i) Show that the statement is true for $n = 2$. The number of handshakes for 2 people is $\dfrac{2^2 - 2}{2} = \dfrac{2}{2} = 1$, which is true.

(ii) Assume that S_k is true: the number of handshakes for k people is $\dfrac{k^2 - k}{2}$.

Show that S_{k+1} is true: the number of handshakes for $k+1$ people is

$\dfrac{(k+1)^2 - (k+1)}{2} = \dfrac{k^2 + 2k + 1 - k - 1}{2} = \dfrac{k^2 + k}{2}$. When a person joins a group of k people, each person must shake hands with the new person. Since there are a total of k people that will shake hands with the new person, the total number of handshakes for $k+1$ people is

$$\frac{k^2 - k}{2} + k = \frac{k^2 - k + 2k}{2} = \frac{k^2 + k}{2}.$$

Since S_k, implies S_{k+1}, the statement is true for every positive integer $n \geq 2$.

35. The first figure has perimeter $P = 3$. When a new figure is generated, each side of the previous figure increases in length by a factor of $\frac{4}{3}$. Thus, the second figure has perimeter $P = 3\left(\frac{4}{3}\right)$, the third figure has perimeter $P = 3\left(\frac{4}{3}\right)^2$, and so on. In general, the nth figure has perimeter $P = 3\left(\frac{4}{3}\right)^{n-1}$.

37. With 1 ring, 1 move is required. With 2 rings, 3 moves are required. Note that $3 = 2 + 1$. With 3 rings, 7 moves are required. Note that $7 = 2^2 + 2 + 1$. With n rings, $2^{n-1} + 2^{n-2} + \cdots + 2^1 + 1 = 2^n - 1$ moves are required.

(i) Show that the statement is true for $n-1$. The number of moves for 1 ring is $2^1 - 1 = 1$, which is true.

(ii) Assume that S_k is true: the number of moves for k rings is $2^k - 1$. Show that S_{k+1} is true: the number of moves for $k+1$ rings is $2^{k+1} - 1$. Assume $k+1$ rings are on the first peg. Since S_k is true, the top k rings can be moved to the second peg in $2^k - 1$ moves. Now move the bottom ring to the third peg. Since S_k is true, move the k rings from the second peg on top of the ring on the third peg in $2^k - 1$ moves. The total number of moves is $\left(2^k - 1\right) + 1 + \left(2^k - 1\right) = 2 \cdot 2^k - 1 = 2^{k+1} - 1$.

Since S_k implies S_{k+1}, the statement is true for every positive integer n.

11.7: Probability

1. Since the coin has a head on each side, the sample space is $S = \{H\}$.

3. Since each of the three coins can be either heads or tails, the sample space is
$S = \{(H,H,H),(H,H,T),(H,T,H),(H,T,T),(T,H,H),(T,H,T),(T,T,H),(T,T,T)\}$.

5. On each spin, the spinner may land on 1, 2, or 3. The sample space is
$S = \{(1,1),(1,2),(1,3),(2,1),(2,2),(2,3),(3,1),(3,2),(3,3)\}$.

7. (a) The event $E_1 = \{H\}$. The probability of the event, $P(E_1) = 1$.

(b) The event $E_2 = \varnothing$. The probability of the event, $P(E_2) = 0$.

9. (a) The event $E_1 = \{(1,1),(2,2),(3,3)\}$. The probability of the event, $P(E_1) = \frac{3}{9} = \frac{1}{3}$.

(b) The event $E_2 = \{(1,1),(1,3),(2,1),(2,3),(3,1)(3,3)\}$. The probability of the event, $P(E_2) = \frac{6}{9} = \frac{2}{3}$.

(c) The event $E_3 = \{(2,1),(2,3)\}$. The probability of the event, $P(E_3) = \frac{2}{9}$.

11. All probability values must be greater than or equal to 0 and less than or equal to 1. Since $\frac{6}{5} > 1$, it cannot be a probability.

13. (a) The probability of drawing a yellow marble is $P(\text{yellow}) = \frac{3}{15} = \frac{1}{5}$.

(b) The probability of drawing a black marble is $P(\text{black}) = \frac{0}{15} = 0$.

(c) The probability of drawing a yellow or white marble is $P(\text{yellow} \cup \text{white}) = \frac{3+4}{15} = \frac{7}{15}$.

(d) $P(\text{yellow}) = \frac{1}{5}$; $P(\text{not yellow}) = \frac{4}{5}$. The odds in favor of drawing a yellow marble are $\frac{\frac{1}{5}}{\frac{4}{5}} = \frac{1}{4}$ or 1 to 4.

(e) $P(\text{blue}) = \frac{8}{15}$; $P(\text{not blue}) = \frac{7}{15}$. The odds against drawing a blue marble are $\frac{\frac{7}{15}}{\frac{8}{15}} = \frac{7}{8}$ or 7 to 8.

15. $P(\text{sum is } 5) = \frac{2}{10} = \frac{1}{5}$; $P(\text{sum is not } 5) = \frac{8}{10} = \frac{4}{5}$. The odds in favor of the sum being 5 are $\frac{\frac{1}{5}}{\frac{4}{5}} = \frac{1}{4}$ or 1 to 4.

17. Odds in favor of candidate are 3 to 2 $\Rightarrow P(\text{lose}) = \frac{2}{3+2} = \frac{2}{5}$. The probability the candidate will lose is $\frac{2}{5}$.

19. (a) The probability of an uncle or brother arriving first is $P(\text{uncle} \cup \text{brother}) = \frac{2+3}{10} = \frac{5}{10} = \frac{1}{2}$.

(b) The probability of a brother or cousin arriving first is $P(\text{brother} \cup \text{cousin}) = \frac{3+4}{10} = \frac{7}{10}$.

(c) The probability of a brother or her mother arriving first is $P(\text{brother} \cup \text{mother}) = \frac{3+1}{10} = \frac{4}{10} = \frac{2}{5}$.

21. (a) $P(E) = -0.1$ matches with statement F, the event is impossible, because a probability value cannot be negative.

(b) $P(E) = 0.01$ matches with statement D, the event is very unlikely to occur, because the probability value is relatively low.

(c) $P(E) = 1$ matches with statement A, the event is certain to occur.

(d) $P(E) = 2$ matches with statement F, the probability cannot occur, because a probability value cannot be greater than 1.

(e) $P(E) = 0.99$ matches with statement C, the event is very likely to occur, because the probability value is relatively high.

(f) $P(E)=0$ matches with statement B, the event is impossible.

(g) $P(E)=0.5$ matches with statement E, the event is just as likely to occur as not occur.

23. (a) The probability that a randomly selected patient will need a kidney or a heart transplant is

$$P(\text{kidney} \bigcup \text{heart}) = \frac{35,025+3774}{51,277} \approx 0.76.$$

(b) The probability that a randomly selected patient will need neither a kidney nor a heart transplant is $P(\text{not kidney} \bigcap \text{not heart}) = P'(\text{kidney} \bigcup \text{heart}) \approx 1-0.76 \approx 0.24.$

25. (a) $P(\text{less than } \$20) = 0.25+0.37 = 0.62$ .The probability of a purchase that is less then \$20 is 0.62.

(b) $P(\$40 \text{ or more}) = 0.09+0.07+0.08+0.03 = 0.27.$ The probability of a purchase that is \$40 or more is 0.27.

(c) $P(\text{more than } \$99.99) = 0.08+0.03 = 0.11.$ The probability of a purchase that is more than \$99.99 is 0.11.

(d) $P(\text{less than } \$100) = 1-0.11 = 0.89.$ The probability of a purchase that is less than \$100 is 0.89.

27. Number of picks with all cards correct except the heart $=12$.
Number of picks with all cards correct except the club $=12$.
Number of picks with all cards correct except the diamond $=12$.
Number of picks with all cards correct except the spade $=12$. Total number of picks with all cards correct but one = 48. $P(\text{all cards correct but one}) = \frac{48}{28,561} \approx 0.001681.$ The probability of getting three of the four selections correct is $\frac{48}{28,561}$ or approximately 0.001681.

29. (a) P (male selected) $=1-0.28=0.72.$ The probability that a male worker is selected is 0.72.

(b) P (5 years or less) $=1-0.3=0.7.$ The probability that a worker is selected who has worked for the company 5 years or less is 0.7.

(c) $P(\text{contribute} \bigcup \text{female}) = P(\text{contribute}) + P(\text{female}) - P(\text{contribute} \bigcap \text{female})$

$=0.65+0.28-\frac{0.28}{2}=0.79.$ The probability that a worker is selected who contributes to the retirement plan or is female is 0.79.

31. There are $5^2=32$ possible outcomes, each with probability $\frac{1}{32}$. $\binom{5}{2}=10$ outcomes with 2 girls (and 3 boys)

$\Rightarrow P(\text{2 girls and 3 boys}) = 10\left(\frac{1}{32}\right) = \frac{5}{16} = 0.3125.$ The probability of having exactly 2 girls and 3 boys is $\frac{5}{16}=0.3125.$

33. $\binom{5}{0} = 1$ outcome has no girls $\Rightarrow P(\text{no girls}) = 1\left(\frac{1}{32}\right) = \frac{1}{32} = 0.03125.$

The probability of having no girls is $\frac{1}{32} = 0.03125.$

35. $\binom{5}{5} + \binom{5}{4} + \binom{5}{3} = 1 + 5 + 10 = 16$ outcomes have at least 3 boys $\Rightarrow P(\text{at least 3 boys}) = 16\left(\frac{1}{32}\right) = \frac{1}{2} = 0.5.$

The probability of having at least 3 boys is $\frac{1}{2} = 0.5.$

37. $P(\text{1 student smokes less than 10 per day}) = 0.45 + 0.24 = 0.69.$ $P(\text{4 of 10 smoke less than 10 per day}) =$

$\binom{10}{4}(0.69)^4(1-0.69)^6 \approx 0.042246.$ The probability that 4 of 10 students selected at random smoked less than 10 cigarettes per day is approximately 0.042246.

39. $P(\text{1 student smokes between 1 and 19 per day}) = 0.023831.$

$P(\text{fewer than 2 smoke between 1 and 19 per day}) =$

$P(\text{0 smoke between 1 and 19 per day}) + P(\text{ 1 smokes between 1 and 19 per day}) =$

$= \binom{10}{0}(0.44)^0(1-0.44)^{10} + \binom{10}{1}(0.44)^1(1-0.44)^9 \approx 0.026864.$ The probability that fewer than 2 of 10 students selected at random smoked between 1 and 19 cigarettes per day is approximately 0.026864.

41. $P(\text{exactly 12 ones}) = \binom{12}{12}\left(\frac{1}{6}\right)^{12}\left(\frac{5}{6}\right)^0 = \frac{1}{6^{12}} \approx 4.6 \times 10^{-10}.$ The probability of rolling exactly 12 ones is

$\frac{1}{6^{12}} \approx 4.6 \times 10^{-10}.$

43. $P(\text{no more than 3 ones}) = P(\text{0 ones}) + P(\text{1 ones}) + P(\text{2 ones}) + P(\text{3 ones}) =$

$\binom{12}{0}\left(\frac{1}{6}\right)^0\left(\frac{5}{6}\right)^{12} + \binom{12}{1}\left(\frac{1}{6}\right)^1\left(\frac{5}{6}\right)^{11} + \binom{12}{2}\left(\frac{1}{6}\right)^2\left(\frac{5}{6}\right)^{10} + \binom{12}{3}\left(\frac{1}{6}\right)^3\left(\frac{5}{6}\right)^9 \approx 0.875.$ The probability of rolling no more than 3 ones is approximately 0.875.

45. $P(\text{fewer than 4}) = P(1) + P(\text{2 or 3}) = 0.2 + 0.29 = 0.49.$ The probability that a randomly selected student applied to fewer than 4 colleges is 0.49.

47. $P(\text{more than 3}) = P(\text{4-6}) + P(\text{7 or more}) = 0.37 + 0.14 = 0.51.$ The probability that a randomly selected student applied to more than 3 colleges is 0.51.

49. (a) $P(\text{5 of 53 are color blind}) = \binom{53}{5}(0.042)^5(1-0.042)^{48} \approx 0.047822.$ The probability that exactly 5 of 53 men are color blind is approximately 0.047822.

(b) $P(\text{no more than } 5) = P(0) + P(1) + P(2) + P(3) + P(4) + P(5) =$

$$\binom{53}{0}(0.042)^0(1-0.42)^{53} + \binom{53}{1}(0.042)^1(1-0.042)^{52} + \binom{53}{2}(0.042)^2(1-0.042)^{51} +$$

$$\binom{53}{3}(0.042)^3(1-0.042)^{50} + \binom{53}{4}(0.042)^4(1-0.042)^{49} + \binom{53}{5}(0.042)^5(1-0.042)^{48} \approx 0.976710.$$

The probability that no more than 5 of 53 men are color blind is approximately 0.976710.

(c) $P(\text{none are color blind}) = \binom{53}{0}(0.042)^0(1-0.042)^{53} \approx 0.102890.$

$P(\text{at least 1 color blind}) = 1 - P(\text{none are color blind}) \approx 0.897110.$

The probability that at least 1 man of 53 men is color blind is approximately 0.897110.

51. (a) $I = 2, S = 4, p = 0.1 \Rightarrow q(1-p)^I = (1-0.1)^2 = 0.81..$

$P(\text{3 not becoming infected}) = \binom{4}{3}(0.81)^3(1-0.81)^{4-3} \approx 0.404 = 40.4\%$. The probability of 3 family members not becoming infected is approximately 40.4%.

(b) $I = 2, S = 4, p = 0.5 \Rightarrow q = (1-p)^I = (1-0.5)^2 = 0.25.$

$P(\text{3 not becoming infected}) = \binom{4}{3}(0.25)^3(1-0.25)^{4-3} \approx 4.7\%$. The probability of 3 family members not becoming infected is approximately 4.7%.

(c) $I = 1, S = 9, p = 0.5 \Rightarrow q = (1-p)^I = (1-0.5)^1 = 0.5.$

$P(\text{0 not becoming infected}) = \binom{9}{0}(0.5)^0(1-0.5)^{9-0} \approx 0.002$

The probability that all of the other family members becoming sick is approximately 0.2%. Thus, it is unlikely that everyone in a large family would become sick even though the disease is highly infectious.

Reviewing Basic Concepts (Sections 11.6 and 11.7)

1. $4 + 8 + 12 + 16 + \cdots + 4n = 2n(n+1).$

(i) Show that the statement is true for $n = 1$: $4(1) = 2(1)(1+1) \Rightarrow 4 = 4.$

(ii) Assume that S_k is true: $4 + 8 + 12 + 16 + \cdots + 4k = 2k(k+1)$. Show that S_{k+1} is true:

$4 + 8 + 12 + 16 + \cdots + 4(k+1) = 2(k+1)(k+2).$ Add $4(k+1)$ to each side of S_k:

$4 + 8 + 12 + 16 + \cdots + 4k + 4(k+1) = 2k(k+1) + 4(k+1).$

$= 2k^2 + 2k + 4k + 4 = 2k^2 + 6k + 4 = 2(k^2 + 3k + 2) = 2(k+1)(k+2).$

Since S_k implies S_{k+1}, the statement is true for every positive integer n.

2. $n^2 \le 2^n$, if $n \ge 4$

(i) Show that the statement is true for $n=4$: $4^2 \le 2^4 \Rightarrow 16 \le 16$.

(ii) Assume that S_k is true: $k^2 \le 2^k$. Show that S_{k+1} is true: $(k+1)^2 \le 2^{(k+1)}$.

Multiply each side of S_k by 2: $k^2 \cdot 2 \le 2^k \cdot 2 \Rightarrow 2k^2 \le 2^{k+1}$.

Because $(k+1)^2 < 2k^2$ for all $k \ge 4$, we may substitute $(k+1)^2$ for $2k^2$ in the expression.

That is, $(k+1)^2 \le 2^{k+1}$.

Since S_k implies S_{k+1}, the statement is true for every positive integer $n \ge 4$.

3. The sample space S for tossing a coin twice is $S = \{(H,H),(H,T),(T,H),(T,T)\}$.

4. 36 possible outcomes, each with probability $\frac{1}{36}$. 2 outcomes yield 11, $\{(5,6),(6,5)\}$.

   Therefore, $1^3 = 1; 1^2(2(1)^2 - 1) = 1(2-1) = 1$; The probability of rolling a sum of 11 is $\frac{2}{36} = \frac{1}{18}$.

5. $\binom{4}{4} = 1$ way of drawing 4 aces; $\binom{4}{1} = 4$ ways of drawing 1 queen; $\binom{52}{5} = 2{,}598{,}960$ ways of drawing 5 cards. P (4 aces and 1 queen) $= \frac{1 \times 4}{2{,}598{,}960} \approx 0.0000015$. The probability of drawing 4 aces and 1 queen is approximately 0.0000015.

6. From number 5, the probability of drawing 4 aces and 1 queen is given by $\frac{4}{2{,}598{,}960}$. Therefore, the probability of not drawing 4 aces and 1 queen is given by $1 - \frac{4}{2{,}598{,}960} = \frac{2{,}598{,}956}{2{,}598{,}960} \approx 0.9999985$.

7. P (rain) $= \frac{3}{3+7} = \frac{3}{10}$. The probability of rain is $\frac{3}{10}$.

8. P (female) $= \frac{2.81 - 1.45}{2.81} \approx 0.484$. The probability that a randomly selected graduate is female is approximately 0.484.

Chapter 11 Review Exercises

1. $a_1 = \frac{1}{1+1} = \frac{1}{2}; a_2 = \frac{2}{2+1} = \frac{2}{3}; a_3 = \frac{3}{3+1} = \frac{3}{4}; a_4 = \frac{4}{4+1} = \frac{4}{5}; a_5 = \frac{5}{5+1} = \frac{5}{6}$. The first five terms of the sequence are $\frac{1}{2}, \frac{2}{3}, \frac{3}{4}, \frac{4}{5}$, and $\frac{5}{6}$. The sequence is neither arithmetic nor geometric.

2. $a_1 = (-2)^1 = -2; a_2 = (-2)^2 = 4; a_3 = (-2)^3 = -8; a_4 = (-2)^4 = 16; a_5 = (-2)^5 = -32$. The first five terms of the sequence are –2, 4, –8, 16, and –32. The sequence is geometric with common ratio –2.

3. $a_1 = 2(1+3) = 8$; $a_2 = 2(2+3) = 10$; $a_3 = 2(3+3) = 12$; $a_4 = 2(4+3) = 14$; $a_5 = 2(5+3) = 16$. The first five terms of the sequence are 8, 10, 12, 14, and 16. The sequence is arithmetic with common difference 2.

4. $a_1 = 1(1+1) = 2$; $a_2 = 2(2+1) = 6$; $a_3 = 3(3+1) = 12$; $a_4 = 4(4+1) = 20$; $a_5 = 5(5+1) = 30$. The first five terms of the sequence are 2, 6, 12, 20, and 30. The sequence is neither arithmetic nor geometric.

5. $a_1 = 5$; $a_2 = 5-3$; $a_3 = 2-3$; $a_4 = -1-3 = -4$; $a_5 = -4-3 = -7$. The first five terms of the sequence are 5, 2, –1, –4, and –7. The sequence is arithmetic with common difference –3.

6. $a_1 = \left(\frac{1}{5}\right)^{1-1}$; $a_2 = \left(\frac{1}{5}\right)^{2-1}$; $a_3 = \left(\frac{1}{5}\right)^{3-1}$; $a_4 = \left(\frac{1}{5}\right)^{4-1}$; $a_5 = \left(\frac{1}{5}\right)^{5-1}$. The first five terms of the sequence are ,

$1, \frac{1}{5}, \frac{1}{25}, \frac{1}{125}, \frac{1}{625}$. The sequence is geometric with common ratio $\frac{1}{5}$.

7. $a_2 = 10, d = -2 \Rightarrow a_1 = 10+2 = 12$; $a_3 = 10-2 = 8$; $a_4 = 8-2 = 6$; $a_5 = 6-2 = 4$. The first five terms of the sequence are 12, 10, 8, 6, and 4.

8. $a_3 = \pi, a_4 = 1 \Rightarrow d = 1-\pi \Rightarrow a_2 = \pi-(1-\pi) = 2\pi-1$; $a_1 = 2\pi-1-(1-\pi) = 3\pi-2$;

$a_5 = 1+(1-\pi) = -\pi+2$. The first five terms of the sequence are $3\pi-2, 2\pi-1, \pi, 1,$ and $-\pi+2$.

9. $a_1 = 6, r = 2 \Rightarrow a_2 = 6\cdot 2 = 12$; $a_3 = 12\cdot 2 = 24$; $a_4 = 24\cdot 2 = 48$; $a_5 = 48\cdot 2 = 96$. The first five terms of the sequence are 6, 12, 24, 48, and 96.

10. $a_1 = -5, a_2 = -1 \Rightarrow r = \frac{1}{5} \Rightarrow a_3 = (-1)\left(\frac{1}{5}\right) = -\frac{1}{5}$; $a_4 = \left(\frac{-1}{5}\right)\left(\frac{1}{5}\right) = -\frac{1}{25}$; $a_5 = \left(-\frac{1}{25}\right)\left(\frac{1}{5}\right) = -\frac{1}{125}$. The first five terms of the sequence are $-5, -1, -\frac{1}{5}, -\frac{1}{25}, -\frac{1}{125}$.

11. $a_5 = -3, a_{15} = 17 \Rightarrow 17 = -3+10d \Rightarrow d = 2 \Rightarrow -3 = a_1 + 4\cdot 2 \Rightarrow a_1 = -11 \Rightarrow$

$a_n = -11+2(n-1) = 2n-13$. The first term of the sequence is –11 and the nth term of the sequence is given by $a_n = 2n-13$.

12. $a_1 = -8, a_7 = -\frac{1}{8} \Rightarrow -\frac{1}{8} = (-8)r^6 \Rightarrow r^6 = \frac{1}{64} \Rightarrow r = \pm\frac{1}{2}$. Therefore, $a_4 = (-8)\left(\frac{1}{2}\right)^3 = -1$ and

$a_n = (-8)\left(\frac{1}{2}\right)^{n-1} = -\left(\frac{1}{2}\right)^{n-4}$ or $a_4 = (-8)\left(-\frac{1}{2}\right)^3 = 1$ and $a_n = (-8)\left(-\frac{1}{2}\right)^{n-1} = \left(-\frac{1}{2}\right)^{n-4}$. Either the common ratio is $\frac{1}{2}$ and the fourth term of the sequence is –1 and the nth term of the sequence is given by

$a_n = (-8)\left(\frac{1}{2}\right)^{n-1} = -\left(\frac{1}{2}\right)^{n-4}$ or the common ratio is $-\frac{1}{2}$ and the fourth term of the sequence is 1 and the nth term of the sequence is given by $a_n = (-8)\left(-\frac{1}{2}\right)^{n-1} = \left(-\frac{1}{2}\right)^{n-4}$.

13. $a_1 = 6, d = 2 \Rightarrow a_8 = 6+(8-1)(2) = 20.$ The eighth term of the sequence is 20.

14. $a_1 = 6x-9, a_2 = 5x+1 \Rightarrow d = (5x+1)-(6x-9) = -x+10 \Rightarrow a_8 = (6x-9)+(8-1)(-x+10) = -x+61.$

The eighth term of the sequence is $-x+61$.

15. $a_1 = 2, d = 3 \Rightarrow a_{12} = 2+(12-1)(3) = 35 \Rightarrow S_{12} = \frac{12}{2}(2+35) = 222.$ The sum of the first twelve terms of the sequence is 222.

16. $a_2 = 6, d = 10 \Rightarrow a_1 = 6-10 = -4 \Rightarrow a_{12} = -4+(12-1)(10) = 106 \Rightarrow S_{12} = \frac{12}{2}(-4+106) = 612.$ The sum of the first twelve terms of the sequence is 612.

17. $a_1 = -2, r = 3 \Rightarrow a_5 = (-2)(3)^{5-1} = -162.$ The fifth term of the sequence is -162.

18. $a_3 = 4, r = \frac{1}{5} \Rightarrow a_1 = \frac{4}{\left(\frac{1}{5}\right)^2} = 100 \Rightarrow a_5 = (100)\left(\frac{1}{5}\right)^{5-1} = \frac{100}{625} = \frac{4}{25}.$ The fifth term of the sequence is $\frac{4}{25}$.

19. $a_1 = 3, r = 2 \Rightarrow S_4 = \frac{3(1-2^4)}{1-2} = 45.$ The sum of the first four terms of the sequence is 45.

20. $a_1 = \frac{3}{4}, a_2 = -\frac{1}{2}, a_3 = \frac{1}{3} \Rightarrow r = -\frac{2}{3} \Rightarrow S_4 = \frac{\frac{3}{4}(1-(-\frac{2}{3})^4)}{1-(-\frac{2}{3})} = \frac{13}{36}.$ The sum of the first four terms is $\frac{13}{36}$.

21. $S = 2000\left[\frac{(1+0.03)^5-1}{0.03}\right] = 10{,}618.27.$ The future value of the annuity is \$10,618.27.

22. $a_1 = 6, r = 0.9;$ therefore $a_n = 6(0.9)^{n-1}$

23. $a_1 = (-1)^{1-1} = 1; r = -1; \sum_{i=1}^{7}(-1)^{i-1} = S_7 = \frac{1\left[1-(-1)^7\right]}{1-(-1)} = \frac{1(2)}{2} = 1.$ The value of the sum is 1.

24. $\sum_{i=1}^{5}(i^2+i) = \sum_{i=1}^{5} i^2 + \sum_{i=1}^{5} i = \frac{5(5+1)(2\cdot 5+1)}{6} + \frac{5(5+1)}{2} = 55+15 = 70.$ The value of the sum is 70.

25. $\sum_{i=1}^{4}\frac{i+1}{i} = \frac{2}{1}+\frac{3}{2}+\frac{4}{3}+\frac{5}{4} = \frac{73}{12}.$ The value of the sum is $\frac{73}{12}$.

26. $a_1 = 3(1)-4 = -1; a_{10} = 3(10)-4 = 26; \sum_{j=1}^{10}(3j-4) = S_{10} = \frac{10}{2}(-1+26) = 125.$

The value of the sum is 125.

27. $\sum_{j=1}^{2500} j = \frac{2500(2500+1)}{2} = 3{,}126{,}250.$ The value of the sum is 3,126,250.

28. $a_1 = 4\cdot 2^1 = 8;\ r = 2;\ \sum_{i=1}^{5} 4\cdot 2^i = S_5 = \frac{8(1-2^5)}{1-2} = 248.$ The value of the sum is 248.

29. $a_1 = \left(\frac{4}{7}\right)^1 = \frac{4}{7};\ r = \frac{4}{7};\ \sum_{i=1}^{\infty}\left(\frac{4}{7}\right)^i = S_\infty = \frac{\frac{4}{7}}{1-\frac{4}{7}} = \frac{4}{3}$. The value of the sum is $\frac{4}{3}$.

30. $r = \frac{6}{5} \Rightarrow |r| \geq 1 \Rightarrow \sum_{i=1}^{\infty} -2\left(\frac{6}{5}\right)^i$ does not exist.

31. $a_1 = 24,\ a_2 = 8,\ a_3 = \frac{8}{3},\ a_4 = \frac{8}{9} \Rightarrow r = \frac{1}{3} \Rightarrow S_\infty = \frac{24}{1-\frac{1}{3}} = 36$. The value of the sum is 36.

32. $a_1 = -\frac{3}{4},\ a_2 = \frac{1}{2},\ a_3 = -\frac{1}{3},\ a_4 = \frac{2}{9} \Rightarrow r = -\frac{2}{3} \Rightarrow S_\infty = \frac{-\frac{3}{4}}{1-\left(-\frac{2}{3}\right)} = -\frac{9}{20}$. The value of the sum is $-\frac{9}{20}$.

33. $a_1 = \frac{1}{12},\ a_2 = \frac{1}{6},\ a_3 = \frac{1}{3},\ a_4 = \frac{2}{3} \Rightarrow r = 2 \Rightarrow |r| \geq 1 \Rightarrow S_\infty$ diverges. The value of the sum diverges and does not exist.

34. $a_1 = 0.9,\ a_2 = 0.09,\ a_3 = 0.009,\ a_4 = 0.0009 \Rightarrow r = \frac{1}{10} \Rightarrow S_\infty = \frac{0.9}{1-\frac{1}{10}} = 1$. The value of the sum is 1.

35. $\sum_{i=1}^{4}(x_i^2 - 6) = \sum_{i=1}^{4} x_i^2 - \sum_{i=1}^{4} 6 = \sum_{i=1}^{4}(x_i + 1)^2 - 24 = \frac{3(3+1)(2\cdot 3+1)}{6} - 24 = -10$. The value of the sum is –10.

36. $\sum_{i=1}^{6} f(x_i)\Delta x = (0-2)^3(0.1) + (1-2)^3(0.1) + (2-2)^3(0.1) + (3-2)^3(0.1) +$

$(4-2)^3(0.1) + (5-2)^3(0.1) = 2.7$. The value of the sum is 2.7.

37. $a_1 = 4,\ a^2 = -1,\ a_3 = -6 \Rightarrow d = -5 \Rightarrow a_n = 4 - 5(n-1) = -5n + 9$.

$a_n = -66 = -5n + 9 \Rightarrow n = 15 \Rightarrow 4 - 1 - 6 - \ldots - 66 = \sum_{i=1}^{15}(-5i + 9)$.

The sum may be written as $\sum_{i=1}^{15}(-5i + 9)$.

38. $a_1 = 10,\ a_2 = 14,\ a_3 = 18 \Rightarrow d = 4 \Rightarrow a_n = 10 + 4(n-1) = 4n + 6$..

$a_n = 86 = 4n + 6 \Rightarrow n = 20 \Rightarrow 10 + 14 + 18 + \ldots + 86 = \sum_{i=1}^{20}(4i + 6)$. The sum may be written as $\sum_{i=1}^{20}(4i + 6)$.

39. $a_1 = 4,\ a_2 = 12,\ a_3 = 36 \Rightarrow r = 3 \Rightarrow a_n 4(3)^{n-1}$. $a_n = 972 = 4(3)^{n-1} \Rightarrow$

$n - 1 = 5 \Rightarrow n = 6 \Rightarrow 4 + 12 + 36 + \ldots + 972 = \sum_{i=1}^{6} 4(3)^{n-1}$. The sum may be written as $\sum_{i=1}^{6} 4(3)^{n-1}$.

40. $a_n = \frac{n}{n+1}$ for $5 \leq n \leq 12$; $\frac{5}{6} + \frac{6}{7} + \frac{7}{8} + \ldots + \frac{12}{13} = \sum_{i=5}^{12} \frac{i}{i+1}$. The sum may be written as $\sum_{i=5}^{12} \frac{i}{i+1}$.

41. $9! = 362{,}880$. The value of $9!$ is 362,880.

42. $P(9,2) = 9_n P_r 2 = 72$. The value of $P(9,2)$ is 72.

43. $P(6,0) = 6_n P_r 0 = 1$. The value of $P(6,0)$ is 1.

44. $C(10,5) = 10_n C_r 5 = 252.$ The value of $C(10,5)$ is 252.

45. $\binom{8}{3} = 8_n C_r 3 = 56.$ The value of $\binom{8}{3}$ is 56.

46. Order is significant in the digits of a student identification number. Thus, such a number is an example of a permutation.

47. $(x+2y)^4 = \binom{4}{0}x^4(2y)^0 + \binom{4}{1}x^3(2y)^1 + \binom{4}{2}x^2(2y)^2 + \binom{4}{3}x^1(2y)^3 + \binom{4}{4}x^0(2y)^4 =$

$x^4 + 8x^3y + 24x^2y^2 + 32xy^3 + 16y^4$. The binomial expansion of $(x+2y)^4$ is

$x^4 + 8x^3y + 24x^2y^2 + 32xy^3 + 16y^4$.

48. $(3z-5w)^3 = \binom{3}{0}(3z)^3(-5w)^0 + \binom{3}{1}(3z)^2(-5w)^1 + \binom{3}{2}(3z)^1(-5w)^2 + \binom{3}{3}(3z)^0(-5w)^3 = \backslash$

$27z^3 - 135z^2w + 225zw^2 - 125w^3$. The binomial expansion of $(3z-5w)^3$ is

$27z^3 - 135z^2w + 225zw^2 - 125w^3$.

49. $\left(3\sqrt{x} - \frac{1}{\sqrt{x}}\right)^5 = \binom{5}{0}(3x^{1/2})^5(-x^{-1/2})^0 + \binom{5}{1}(3x^{1/2})^4(-x^{-1/2})^1 + \binom{5}{2}(3x^{1/2})^3(-x^{-1/2})^2 +$

$\binom{5}{3}(3x^{1/2})^2(-x^{-1/2})^3 + \binom{5}{4}(3x^{1/2})^1(-x^{-1/2})^4 + \binom{5}{5}(3x^{1/2})^0(-x^{-1/2})^5 =$

$243x^{5/2} - 405x^{3/2} + 270x^{1/2} - 90x^{-1/2} + 15x^{-3/2} - x^{-5/2}$. The binomial expansion of $\left(3\sqrt{x} - \frac{1}{\sqrt{x}}\right)^5$ is

$243x^{5/2} - 405x^{3/2} + 270x^{1/2} - 90x^{-1/2} + 15x^{-3/2} - x^{-5/2}$.

50. $(m^3 - m^{-2})^4 = \binom{4}{0}(m^3)^4(-m^{-2})^0 + \binom{4}{1}(m^3)^3(-m^{-2})^1 + \binom{4}{2}(m^3)^2(-m^{-2})^2 +$

$\binom{4}{3}(m^3)^1(-m^{-2})^3 + \binom{4}{4}(m^3)^0(-m^{-2})^4 = m^{12} - 4m^7 + 6m^2 - 4m^{-3} + m^{-8}$. The binomial expansion of

$(m^3 - m^{-2})^4$ is $m^{12} - 4m^7 + 6m^2 - 4m^{-3} + m^{-8}$.

51. $\binom{8}{5}(4x)^3(-y)^5 = -3584x^3y^5$. The sixth term of $(4x-y)^8$ is $-3584x^3y^5$.

52. $\binom{14}{6}(m)^8(-3n)^6 = 2{,}189{,}187m^8n^6$. The seventh term of $(m-3n)^{14}$ is $2{,}189{,}187\,m^8n^6$.

53. $(x+2)^{12} = \binom{12}{0}(x)^{12}(2)^0 + \binom{12}{1}(x)^{11}(2)^1 + \binom{12}{2}(x)^{10}(2)^2 +$

$\binom{12}{3}(x)^9(2)^3 + \ldots = x^{12} + 24x^{11} + 264x^{10} + 1760x^9 + \ldots$. The first four terms of $(x+12)^{12}$ are

$x^{12} + 24x^{11} + 264x^{10} + 1760x^9$.

54. $(2a+5b)^{16} = \ldots \binom{16}{14}(2a)^2(5b)^{14} + \binom{16}{15}(2a)^1(5b)^{15} + \binom{16}{16}(2a)^0(5b)^{16} =$

$\ldots + 480 \bullet 5^{14} a^2 b^{14} + 32 \bullet 5^{15} ab^{15} + 5^{16} b^{16}$. The last three terms of $(2a+5b)^{16}$ are

$480 \bullet 5^{14} a^2 b^{14} + 32 \bullet 5^{15} ab^{15} + 5^{16} b^{16}$.

55. Statements which are defined on the natural numbers; that is, statements which have the natural numbers as their domain, are proved by mathematical induction. An example is $\sum_{i=1}^{n} i = \frac{n(n+1)}{2}$.

56. A proof by mathematical induction consists of two steps. First, show that the statement is true for $n = 1$. Then show that if the statement is true for $n = k$, it must follow that the statement is also true for $n = k+1$.

57. Prove $1+3+5+7+\ldots+(2n-1) = n^2$.

Step 1: $1 = 1^2 \Rightarrow 1+3+5+7+\ldots+(2n-1) = n^2$ for $n = 1$.

Step 2: Assume $1+3+5+7+\ldots+(2k-1) = k^2$, then

$1+3+5+7+\ldots+(2k-1)+(2k+1) = k^2+2k+1 \Rightarrow 1+3+5+7+\ldots+(2k-1)+(2(k+1)-1) = (k+1)^2$.

Thus, if $1+3+5+7+\ldots+(2n-1) = n^2$ for $n = k$, then it must follow that

$1+3+5+7+\ldots+(2n-1) = n^2$ for $n = k+1$ also.

58. Prove $2+6+10+14+\ldots+(4n-2) = 2n^2$.

Step 1: $2 = 2(1)^2 \Rightarrow 2+6+10+14+\ldots+(4n-2) = 2n^2$ for n = 1.

Step 2: Assume $2+6+10+14+\ldots+(4k-2) = 2k^2$, then

$2+6+10+14+\ldots+(4k-2)+(4k+2) = 2k^2+4k+2 \Rightarrow$

$2+6+10+14+\ldots+(4k-2)+(4(k+1)-1) = 2(k+1)^2$. Thus if $2+6+10+14+\ldots+(4n-2) = 2n^2$ for $n = k$, then it must follow that $2+6+10+14+\ldots+(4n-2) = 2n^2$ for $n = k+1$ also.

59. Prove $2+2^2+2^3+\ldots+2^n = 2(2^n-1)$.

Step 1: $2 = 2(2^1-1) \Rightarrow 2+2^2+2^3+\ldots+2^n = 2(2^n-1)$ for $n = 1$.

Step 2: Assume $2+2^2+2^3+\ldots+2^k = 2(2^k-1)$, then

$2+2^2+2^3+\ldots+2^k+2^{k+1} = 2(2^k-1)+2^{k+1} = 2^{k+1}-2+2^{k+1} = 2(2^{k+1}-1)$ Thus, if

$2+2^2+2^3+...+2^n=2(2^n-1)$ for $n=k$,then it must follow that $2+2^2+2^3+...+2^n=2(2^n-1)$ for $n=k+1$ also.

60. Prove $1^3+3^3+5^3+...+(2n-1)^2=n^2(2n^2-1)$.

Step 1: $1^3=1; 1^2(2(1)^2-1)=1(2-1)=1$, so $1^3+3^3+5^3+...+(2n-1)^3=n^2(2n^2-1)$ for n = 1.

Step 2: Assume $1^3+3^3+5^3+...+(2k-1)^3=k^2(2k^2-1)$. Then,

$1^3+3^3+5^3+...+(2k-1)^3+(2k+1)^3=k^2(2k^2-1)+(2k=1)^3$. Then

$1^3+3^3+5^3+...+(2k-1)^3+(2(k+1)-1)^3=2k^4+8k^3+11k^2+6k+1=$

$(k^2+2k+1)(2k^2+4k+1)=(k+1)^2(2(k+1)^2-1)$. Thus, if $1^3+3^3+5^3+...+(2n-1)^3=n^2(2n^2-1)$ for $n=k$, then it must follow that $1^3+3^3+5^3+...+(2n-1)^3=n^2(2n^2-1)$ for $n=k+1$ also.

61. $2\times4\times3\times2=48$. The number of wedding arrangements possible is 48.

62. $5\times3\times6=90$. The number of available couches is 90.

63. $4_n P_r 4=24$. The jobs can be assigned in 24 different ways.

64. (a) $\binom{6}{3}=20$. The number of different delegations possible is 20.

(b) $\binom{5}{2}=10$. If the president must attend, 10 different delegations are possible.

65. $9_n P_r 3=504$. There are 504 different ways the winners could be determined.

66. $26\times10\times10\times10\times26\times26\times26=456{,}976{,}000$. The number of possible different license plates is 456,976,000.

67. (a) P(green) = $\frac{4}{15}$. The probability of drawing a green marble is $\frac{4}{15}$.

(b) P(not black) = $\frac{4+6}{15}=\frac{10}{15}=\frac{2}{3}$. The probability of drawing a marble that is not black is $\frac{2}{3}$.

(c) P(blue) = $\frac{0}{15}=0$. The probability of drawing a green blue is 0.

68. (a) P(green) = $\frac{4}{15}\Rightarrow\frac{4}{15-4}=\frac{4}{11}$. The odds in favor of drawing a green marble are 4 to 11.

(b) P(not white) = $\frac{4+5}{15}=\frac{9}{15}=\frac{3}{5}\Rightarrow\frac{3}{5-3}=\frac{3}{2}$. The odds against drawing a white marble are 3 to 2.

(c) From (b), the odds in favor of drawing a marble that is not white are 3 to 2.

69. P(black king) = $\frac{2}{52}=\frac{1}{26}$. The probability of drawing a black king is $\frac{1}{26}$.

70. P(face or ace) = $\frac{16}{52}=\frac{4}{13}$. The probability of drawing a face card or an ace is $\frac{4}{13}$.

71. $P(\text{ace} \cup \text{diamond}) = P(\text{ace}) + P(\text{diamond}) - P(\text{ace} \cap \text{diamond}) = \frac{4}{52} + \frac{13}{52} - \frac{1}{52} = \frac{16}{52} = \frac{4}{13}$. The probability of drawing an ace or a diamond is $\frac{4}{13}$.

72. $P(\text{not a diamond}) = 1 - P(\text{diamond}) = 1 - \frac{1}{4} = \frac{3}{4}$.

The probability of drawing a card that is not a diamond is $\frac{3}{4}$.

73. $P(\text{no more than 3}) = 1 - [P(4) + P(5)] = 1 - (0.08 + 0.06) = 0.86$. The probability that no more than 3 filters are defective is 0.86.

74. $P(\text{at least 2}) = 1 - [P(0) + P(1)] = 1 - (0.31 + 0.25) = 0.44$. The probability that at least 2 filters are defective is 0.44.

75. $P(\text{more than 5}) = 0$. Only 5 filters were selected, so it is not possible for more than 5 to be defective. Note that, from the probabilities in the table, $P(0) + P(1) + P(2) + P(3) + P(4) + P(5) = 1$.

76. $\binom{12}{2}\left(\frac{1}{6}\right)^2\left(\frac{5}{6}\right)^{10} \approx 0.296$. The probability that exactly 2 of 12 rolls results in a 5 is approximately 0.296.

77. $\binom{10}{4}\left(\frac{1}{2}\right)^4\left(\frac{1}{2}\right)^6 \approx 0.205$. The probability that exactly 4 of 10 coins tossed result in a tail is approximately 0.205.

78. (a) $P(\text{conservation}) = \frac{56.51}{282.2} \approx 0.2002$. The probability that a randomly selected student is in the conservative group is approximately 0.2002.

(b) $P(\text{far left or far right}) = \frac{7.06 + 3.673}{282.2} \approx 0.0380$. The probability that a randomly selected student is on the far left or the far right is approximately 0.0380.

(c) $P(\text{not middle of road}) = 1 - P(\text{middle of road}) = 1 - \frac{143.5}{282.2} \approx 0.4915$. The probability that a randomly selected student is not middle of the road is approximately 0.4915.

Chapter 11 Test

1. (a) $a_1 = (-1)^1(1+2) = -3$; $a_2 = (-1)^2(2+2) = 4$; $a_3 = (-1)^3(3+2) = -5$; $a_4 = (-1)^4(4+2) = 6$; $a_5 = (-1)^5(5+2) = -7$. The first five terms of the sequence are -3, 4, -5, 6, and -7. The sequence is neither arithmetic nor geometric.

(b) $a_1 = (-3)\left(\frac{1}{2}\right)^1 = -\frac{3}{2}$; $a_2 = (-3)\left(\frac{1}{2}\right)^2 = -\frac{3}{4}$; $a_3 = (-3)\left(\frac{1}{2}\right)^3 = -\frac{3}{8}$;

$a_4 = (-3)\left(\frac{1}{2}\right)^4 = -\frac{3}{16}$; $a_5 = (-3)\left(\frac{1}{2}\right)^5 = -\frac{3}{32}$. The first five terms of the sequence are

$-\frac{3}{2}, -\frac{3}{4}, -\frac{3}{8}, -\frac{3}{16}$, and $-\frac{3}{32}$. The sequence is geometric with common ratio $\frac{1}{2}$.

(c) $a_1 = 2$; $a_2 = 3$; $a_3 = 3 + 2(2) = 7$; $a_4 = 7 + 2(3) = 13$; $a_5 = 13 + 2(7) = 27$. The first five terms of the sequence are 2, 3, 7, 13, and 27. The sequence is neither arithmetic nor geometric.

2. (a) $a_1 = 1$, $a_3 = 25 \Rightarrow 25 = 1 + (3-1)d \Rightarrow 2d = 24 \Rightarrow d = 12 \Rightarrow a_5 = 1 + (5-1)(12) = 49$. The fifth term of the sequence is 49.

(b) $a_1 = 81$, $r = -\frac{2}{3} \Rightarrow a_5 = 81\left(-\frac{2}{3}\right)^{5-1} = 16$. The fifth term of the sequence is 16

3. (a) $a_1 = -43$, $d = 12 \Rightarrow a_{10} = -43 + 12(10-1) = 65 \Rightarrow S_{10} = \frac{10}{2}(-43 + 65) = 110$. The sum of the first ten terms of the sequence is 110.

(b) $a_1 = 5$, $r = -2 \Rightarrow S_{10} = \frac{5(1-(-2)^{10})}{1-(-2)} = -1705$.

The sum of the first ten terms of the sequence is -1705.

4. (a) $a_1 = 5(1) + 2 = 7$; $a_{30} = 5(30) + 2 = 152$; $d = 5 \Rightarrow \sum_{i=1}^{30}(5i+2) = S_{30} = \frac{30}{2}(7 + 152) = 2385$. The value of the sum is 2385.

(b) $a_1 = (-3)(2)^1 = -6$; $r = 2 \Rightarrow \sum_{i=1}^{5}(-3 \cdot 2^i) = \frac{-6(1-(2)^5)}{1-2} = -186$. The value of the sum is -186.

(c) $r = 2 \Rightarrow |r| \geq 1 \Rightarrow \sum_{i=1}^{\infty}(2^i) \cdot 4$ does not exist. The value of the sum diverges and does not exist.

(d) $a_1 = 54\left(\frac{2}{9}\right)^1 = 12$; $r = \frac{2}{9} \Rightarrow \sum_{i=1}^{\infty} 54\left(\frac{2}{9}\right)^i = S_\infty = \frac{12}{1-\frac{2}{9}} = \frac{108}{7}$. The value of the sum is $\frac{108}{7}$.

5. (a) $10_n C_r 2 = 45$. The value of $10_n C_r 2 = 45$.

(b) $\binom{7}{3} = 7_n C_r 3 = 35$. The value of $\binom{7}{3}$ is 35.

(c) $7! = 5040$. The value of $7!$ is 5040.

(d) $P(11, 3) = 11_n P_r 3 = 990$. The value of $P(11, 3)$ is 990.

6. (a) $(2x-3y)^4=\binom{4}{0}(2x)^4(-3y)^0+\binom{4}{1}(2x)^3(-3y)^1+\binom{4}{2}(2x)^2(-3y)^2+$

$\binom{4}{3}(2x)^1(-3y)^3+\binom{4}{4}(2x)^0(-3y)^4 = 16x^4-96x^3y+216x^2y^2-216xy^3+81y^4.$

The binomial expansion of $(2x-3y)^4$ is $16x^4-96x^3y+216x^2y^2-216xy^3+81y^4$.

(b) $\binom{6}{2}(w)^4(-2y)^2=60w^4y^2$. The third term of the expansion of $(w-2y)^6=60w^4y^2$.

7. Prove $8+14+20+26+...+(6n+2)=3n^2+5n$.

Step 1: $3(1)^2+5(1)=8\Rightarrow 8+14+20+26+...+(6n+2)=3n^2+5n$ for $n=1$.

Step 2: Assume $8+14+20+26+...+(6k+2)=3k^2+5k$, then

$8+14+20+26+...+(6k+2)+(6k+8)=3k^2+5k+6k+8\Rightarrow$

$8+14+20+26+...+(6k+2)+[6(k+1)+2]=3k^2+11k+8=$

$3k^2+6k+3+5k+5=3(k+1)^2+5(k+1)$. Thus, if $8+14+20+26+...+(6n+2)=3n^2+5n$ for $n=k$, then it must follow that $8+14+20+26+...+(6n+2)=3n^2+5n$ for $n=k+1$ also.

8. $4\times3\times2=24$. 24 different types of shoes can be made.

9. $20_n P_r 3=6840$. The three offices can be filled in 6840 different ways.

10. $\binom{8}{2}\binom{12}{3}=28\cdot220=6160$. Two men and three women can be chosen in 6160 different ways.

11. (a) $P(\text{red } 3)=\frac{2}{52}=\frac{1}{26}$. The probability of drawing a red three is $\frac{1}{26}$.

(b) $4\times3=12$ cards are face cards $\Rightarrow 52-12=40$ cards are not face cards $\Rightarrow P(\text{not a face card})=$

$\frac{40}{52}=\frac{10}{13}$. The probability of drawing a card that is not a face card is $\frac{10}{13}$.

(c) $P(\text{king}\cup\text{spade})=P(\text{king})+P(\text{spade})-P(\text{king}\cap\text{spade})=\frac{4}{52}+\frac{13}{52}-\frac{1}{52}=\frac{16}{52}=\frac{4}{13}$.

(d) $P(\text{facecard})=\frac{12}{52}=\frac{3}{13}\Rightarrow\frac{3}{13-3}=\frac{3}{10}$. The odds in favor of drawing a face card are 3 to 10.

12. (a) $\binom{8}{3}\left(\frac{1}{6}\right)^3\left(\frac{5}{6}\right)^5\approx0.104$. The probability that exactly 3 of 8 rolls will result in a 4 is approximately 0.104.

(b) $\binom{8}{8}\left(\frac{1}{6}\right)^8\left(\frac{5}{6}\right)^0\approx0.000000595$. The probability that all 8 of 8 rolls will result in a 6 is approximately 0.000000595.

Chapter R: Reference: Basic Algebraic Concepts

R.1: Review of Exponents and Polynomials

1. $(-4)^3 \cdot (-4)^2 = (-4)^{3+2} = (-4)^5$

3. $2^0 = 1$

5. $(5m)^0 = 1$, if $m \neq 0$

7. $(2^2)^5 = 2^{2\cdot 5} = 2^{10}$

9. $(2x^5y^4)^3 = 2^3(x^5)^3(y^4)^3 = 2^3x^{15}y^{12}$ or $8x^{15}y^{12}$

11. $-\left(\dfrac{p^4}{q}\right)^2 = -\left(\dfrac{p^{4\cdot 2}}{q^2}\right) = -\dfrac{p^8}{q^2}$

13. $-5x^{11}$ is a polynomial. It is a monomial since it has one term. It has degree 11 since 11 is the highest exponent.

15. $18p^5q + 6pq$ is a polynomial. It is a binomial since it has two terms. It has degree 6 since 6 is the sum of the exponents in the term $18p^5q$. (The term $6pq$ has degree 2.)

17. $\sqrt{2}x^2 + \sqrt{3}x^6$ is a polynomial. It is a binomial since it has two terms. It has degree 6 since 6 is the highest exponent.

19. $\dfrac{1}{3}r^2s^2 - \dfrac{3}{5}r^4s^2 + rs^3$ is a polynomial. It is a trinomial since it has three terms. It has degree 6 since 6 is the sum of the exponents in the term $-\dfrac{3}{5}r^4s^2$. (The other terms have degree 4.)

21. $-5\sqrt{z} + 2\sqrt{z^3} - 5\sqrt{z^5} = -5z^{1/2} + 2z^{3/2} - 5z^{5/2}$ is not a polynomial since the exponents are not integers.

23. $(4m^3 - 3m^2 + 5) + (-3m^3 - m^2 + 5) = (4m^3 - 3m^3) + (-3m^2 - m^2) + (5+5) = m^3 - 4m^2 + 10$

25. $(8p^2 - 5p) - (3p^2 - 2p + 4) = 8p^2 - 5p - 3p^2 + 2p - 4 = 5p^2 - 3p - 4$

27. $-(8x^3 + x - 3) + (2x^3 + x^2) - (4x^2 + 3x - 1) = -8x^3 - x + 3 + 2x^3 + x^2 - 4x^2 - 3x + 1 = -6x^3 - 3x^2 - 4x + 4$

29. $(5m-6)(3m+4) = 5m\cdot 3m + 5m\cdot 4 - 6\cdot 3m - 6\cdot 4 = 15m^2 + 20m - 18m - 24 = 15m^2 + 2m - 24$

31. $\left(2m - \dfrac{1}{4}\right)\left(3m + \dfrac{1}{2}\right) = 2m\cdot 3m + 2m\cdot\dfrac{1}{2} - \dfrac{1}{4}\cdot 3m - \dfrac{1}{4}\cdot\dfrac{1}{2} = 6m^2 + m - \dfrac{3}{4}m - \dfrac{1}{8} = 6m^2 + \dfrac{1}{4}m - \dfrac{1}{8}$

33. $2b^3(b^2 - 4b + 3) = 2b^3\cdot b^2 + 2b^3\cdot(-4b) + 2b^3\cdot 3 = 2b^5 - 8b^4 + 6b^3$

35. $(m - n + k)(m + 2n - 3k) = m^2 + 2mn - 3km - mn - 2n^2 + 3kn + km + 2kn - 3k^2 =$
$m^2 + mn - 2km - 2n^2 + 5kn - 3k^2$

37. To find the square of a binomial, find the sum of the square of the first term, twice the product of the two terms, and the square of the last term.

39. $(2m+3)(2m-3)=(2m)^2-(3)^2=4m^2-9$

41. $(4m+2n)^2=(4m)^2+2(4m)(2n)+(2n)^2=16m^2+16mn+4n^2$

43. $(5r+3t^2)^2=(5r)^2+2(5r)(3t^2)+(3t^2)^2=25r^2+30rt^2+9t^4$

45. $[(2p-3)+q]^2=(2p-3)^2+2(2p-3)(q)+(q)^2=\left[(2p)^2-2(2p)(3)+3^2\right]+4pq-6q+q^2=$

$4p^2-12p+9+4pq-6q+q^2$

47. $[(3q+5)-p][(3q+5)+p]=(3q+5)^2-(p)^2=\left[(3q)^2+2(3q)(5)+(5)^2\right]-p^2=9q^2+30q+25-p^2$

49. $[(3a+b)-1]^2=(3a+b)^2-2(3a+b)(1)+(-1)^2=\left[(3a)^2+2(3a)(b)+(b)^2\right]-6a-2b+1=$

$9a^2+6ab+b^2-6a-2b+1$

51. $(6p+5q)(3p-7q)=(6p)(3p)+(6p)(-7q)+(5q)(3p)+(5q)(-7q)=18p^2-42pq+15pq-35q^2=$

$18p^2-27pq-35q^2$

53. $(p^3-4p^2+p)-(3p^2+2p+7)=p^3-4p^2+p-3p^2-2p-7=p^3-7p^2-p-7$

55. $y(4x+3y)(4x-3y)=y\left[(4x)^2-(3y)^2\right]=y\left[16x^2-9y^2\right]=16x^2y-9y^3$

57. $(2z+y)(3z-4y)=(2z)(3z)+(2z)(-4y)+(y)(3z)+(y)(-4y)=6z^2-8yz+3yz-4y^2=6x^2-5yz-4y^2$

59. $(3p+5)^2=(3p)^2+2(3p)(5)+5^2=9p^2+30p+25$

61. $p(4p-6)+2(3p-8)=4p^2-6p+6p-16=4p^2-16$

63. $-y(y^2-4)+6y^2(2y-3)=-y^3+4y+12y^3-18y^2=11y^3-18y^2+4y$

R.2: Review of Factoring

1. (a) $(x+5y)^2=x^2+10xy+25y^2$; B

   (b) $(x-5y)^2=x^2-10xy+25y^2$; C

   (c) $(x+5y)(x-5y)=x^2-25y^2$; A

   (d) $(5y+x)(5y-x)=25y^2-x^2$; D

3. $4k^2m^3+8k^4m^3-12k^2m^4=4k^2m^3(1+2k^2-3m)$

5. $2(a+b)+4m(a+b)=(a+b)(2+4m)=2(a+b)(1+2m)$

7. $(2y-3)(y+2)+(y+5)(y+2)=(y+2)(2y-3+y+5)=(y+2)(3y+2)$

9. $(5r-6)(r+3)-(2r-1)(r+3)=(r+3)(5r-6-(2r-1))=(r+3)(3r-5)$

11. $2(m-1)-3(m-1)^2+2(m-1)^3=(m-1)\left(2-3(m-1)+2(m-1)^2\right)=$

$(m-1)\left(2-3m+3+2\left(m^2-2m+1\right)\right)=(m-1)\left(5-3m+2m^2-4m+2\right)=(m-1)\left(2m^2-7m+7\right)$

13. $6st+9t-10s-15=3t(2s+3)-5(2s+3)=(2s+3)(3t-5)$

15. $10x^2-12y+15x-8xy=10x^2+15x-8xy-12y=5x(2x+3)-4y(2x+3)=(2x+3)(5x-4y)$

17. $t^3+2t^2-3t-6=t^2(t+2)-3(t+2)=(t+2)\left(t^2-3\right)$

19. $(8a-3)(2a-5)=\left[-1(-8a+3)\right]\left[-1(-2a+5)\right]=(3-8a)(5-2a)$; Both are correct.

21. $8h^2-24h-320=8\left(h^2-3h-40\right)=8(h-8)(h+5)$

23. $9y^4-54y^3+45y^2=9y^2\left(y^2-6y+5\right)=9y^2(y-1)(y-5)$

25. $14m^2+11mr-15r^2=(7m-5r)(2m+3r)$

27. $12s^2+11st-5t^2=(3s-t)(4s+5t)$

29. $30a^2+am-m^2=(5a+m)(6a-m)$

31. $18x^5+15x^4z-75^3z^2=3x^3\left(6x^2+5xz-25z^2\right)=3x^3(2x+5z)(3x-5z)$

33. $16p^2-40p+25=(4p-5)^2$

35. $20p^2-100pq+125q^2=5\left(4p^2-20pq+25q^2\right)=5(2p-5q)^2$

37. $9m^2n^2-12mn+4=(3mn-2)^2$

39. $(2p+q)^2-10(2p+q)+25=((2p+q)-5)((2p+q)-5)=(2p+q-5)^2$

41. $9a^2-16=(3a+4)(3a-4)$

43. $25s^4-9t^2=\left(5s^2+3t\right)\left(5s^2-3t\right)$

45. $(a+b)^2-16=(a+b+4)(a+b-4)$

47. $p^4-625=\left(p^2+25\right)\left(p^2-25\right)=\left(p^2+25\right)(p+5)(p-5)$

49. $x^4-1=\left(x^2+1\right)\left(x^2-1\right)=\left(x^2+1\right)(x+1)(x-1)$; B

51. $8-a^3=2^3-a^3=(2-a)\left(4+2a+a^2\right)$

53. $125x^3-27=(5x)^3-3^3=(5x-3)\left(25x^2+15x+9\right)$

55. $27y^9+125z^6=\left(3y^3\right)^3+\left(5z^2\right)^3=\left(3y^3+5z^2\right)\left(9y^6-15y^3z^2+25z^4\right)$

57. $(r+6)^3-216=(r+6)^3-6^3=(r+6-6)\left((r+6)^2+6(r+6)+36\right)=$

$r\left(r^2+12r+36+6r+36+36\right)=r\left(r^2+18r+108\right)$

59. $27-(m+2n)^3=3^3-(m+2n)^3=(3-(m+2n))(9+3(m+2n)+(m+2n)^2)$

$(3-m-2n)(9+3m+6n+m^2+4mn+4n^2)$

61. $3a^4+14a^2-5$; Let $u=a^2 \Rightarrow 3u^2+14u-5=(3u-1)(u+5)$. Then $a^2=u \Rightarrow (3a^2-1)(a^2+5)$ was not substituted back in for u.

63. a^4-2a^2-48; Let $u=a^2 \Rightarrow u^2-2u-48=(u-8)(u+6)$. Then $a^2=u \Rightarrow (a^2-8)(a^2+6)$.

65. $6(4z-3)^2+7(4z-3)-3$; Let $u=4z-3 \Rightarrow 6u^2+7u-3=(3u-1)(2u+3)$. Then

$4z-3=u \Rightarrow (3(4z-3)-1)(2(4z-3)+3)=(12z-10)(8z-3)=2(6z-5)(8z-3)$.

67. $20(4-p)^2-3(4-p)-2$; Let $u=4-p \Rightarrow 20u^2-3u-2=(5u-2)(4u+1)$. Then

$4-p=u \Rightarrow (5(4-p)-2)(4(4-p)+1)=(18-5p)(17-4p)$.

69. $4b^2+4bc+c^2-16=(4b^2+4b+c^2)-16=(2b+c)^2-16=(2b+c+4)(2b+c-4)$.

71. $x^2+xy-5y-5y=x(x+y)-5(x+y)=(x+y)(x-5)$

73. $p^4(m-2n)+q(m-2n)=(m-2n)(p^4+q)$

75. $4z^2+28z+49=(2z+7)^2$

77. $1000x^3+343y^3=(10x)^3+7(y)^3=(10x+7y)(100x^2-70xy+49y^2)$

79. $125m^6-216=(5m^2)^3-6^3=(5m^2-6)(25m^4+30m^2+36)$

81. $12m^2+16mn-35n^2=(6m-7n)(2m+5n)$

83. $4p^2+3p-1=(4p-1)(p+1)$

85. $144z^2+121$ does not factor, it is prime.

87. $(4t+5)^2+16(4t+5)+64$; Let $u=4t+5 \Rightarrow u^2+16u+64=(u+8)^2$ Then

$4t+5=u \Rightarrow (4t+5+8)^2=(4t+13)^2$

R.3: Review of Rational Expressions

1. $\dfrac{x-2}{x+6}$; $x+6=0, \{x \mid x \neq -6\}$

3. $\dfrac{2x}{5x-3}$; $5x-3=0, \left\{x \mid x \neq \dfrac{3}{5}\right\}$

5. $\dfrac{-8}{x^2+1}$; No restrictions since $x^2+1>0$. Domain: $(-\infty, \infty)$

7. $\dfrac{3x+7}{(4x+2)(x-1)}$; $4x+2=0 \Rightarrow 4x=-2 \Rightarrow x=-\dfrac{2}{4}=-\dfrac{1}{2}$ and $x-1=0 \Rightarrow x=1$; $\left\{x \mid x \neq -\dfrac{1}{2}, 1\right\}$

9. $\dfrac{25p^3}{10p^2}=\dfrac{5p^2(5p)}{5p^2(2)}=\dfrac{5p}{2}$

11. $\dfrac{8k+16}{9k+18}=\dfrac{8(k+2)}{9(k+2)}=\dfrac{8}{9}$

13. $\dfrac{3(t+5)}{(t+5)(t-3)}=\dfrac{3}{t-3}$

15. $\dfrac{8x^2+16x}{4x^2}=\dfrac{8x(x+2)}{4x^2}=\dfrac{2(x+2)}{x}=\dfrac{2x+4}{x}$

17. $\dfrac{m^2-4m+4}{m^2+m-6}=\dfrac{(m-2)^2}{(m-2)(m+3)}=\dfrac{m-2}{m+3}$

19. $\dfrac{8m^2+6m-9}{16m^2-9}=\dfrac{(4m-3)(2m+3)}{(4m-3)(4m+3)}=\dfrac{2m+3}{4m+3}$

21. $\dfrac{15p^3}{9p^2}\div\dfrac{6p}{10p^2}=\dfrac{15p^3}{9p^2}\cdot\dfrac{10p^2}{6p}=\dfrac{25p^2}{9}$

23. $\dfrac{2k+8}{6}\div\dfrac{3k+12}{2}=\dfrac{2(k+4)}{6}\times\dfrac{2}{3(k+4)}=\dfrac{2}{9}$

25. $\dfrac{x^2+x}{5}\cdot\dfrac{25}{xy+y}=\dfrac{x(x+1)}{5}\cdot\dfrac{25}{y(x+1)}=\dfrac{5x}{y}$

27. $\dfrac{4a+12}{2a-10}\div\dfrac{a^2-9}{a^2-a-20}=\dfrac{4(a+3)}{2(a-5)}\cdot\dfrac{(a+4)(a-5)}{(a-3)(a+3)}=\dfrac{2(a+4)}{a-3}$

29. $\dfrac{p^2-p-12}{p^2-2p-15}\cdot\dfrac{p^2-9p+20}{p^2-8p+16}=\dfrac{(p-4)(p+3)}{(p+3)(p-5)}\cdot\dfrac{(p-5)(p-4)}{(p-4)(p-4)}=1$

31. $\dfrac{m^2+3m+2}{m^2+5m+4}\div\dfrac{m^2+5m+6}{m^2+10m+24}=\dfrac{(m+2)(m+1)}{(m+4)(m+1)}\cdot\dfrac{(m+4)(m+6)}{(m+3)(m+2)}=\dfrac{m+6}{m+3}$

33. $\dfrac{2m^2-5m-12}{m^2-10m+24}\div\dfrac{4m^2-9}{m^2-9m+18}=\dfrac{(2m+3)(m-4)}{(m-4)(m-6)}\cdot\dfrac{(m-3)(m-6)}{(2m+3)(2m-3)}=\dfrac{m-3}{2m-3}$

35. $\dfrac{x^3+y^3}{x^2-y^2}\cdot\dfrac{x+y}{x^2-xy+y^2}=\dfrac{(x+y)(x^2-xy+y^2)}{(x+y)(x-y)}\cdot\dfrac{x+y}{x^2-xy+y^2}=\dfrac{x+y}{x-y}$

37. $\dfrac{x^3+y^3}{x^3-y^3}\cdot\dfrac{x^2-y^2}{x^2+2xy+y^2}=\dfrac{(x+y)(x^2-xy+y^2)}{(x-y)(x^2+xy+y^2)}\cdot\dfrac{(x-y)(x+y)}{(x+y)(x+y)}=\dfrac{x^2-xy+y^2}{x^2+xy+y^2}$

39. A: $\dfrac{x-4}{x+4}\neq -1$, B: $\dfrac{-x-4}{x+4}=\dfrac{-1(x+4)}{x+4}=-1$, C: $\dfrac{x-4}{4-x}=\dfrac{x-4}{-1(x-4)}=-1$, D: $\dfrac{x-4}{-x-4}=\dfrac{x-4}{-1(x+4)}\neq -1$

41. $\frac{3}{2k}+\frac{5}{3k}=\frac{9}{6k}+\frac{10}{6k}=\frac{19}{6k}$

43. $\frac{a+1}{2}-\frac{a-1}{2}=\frac{a+1-(a-1)}{2}=\frac{2}{2}=1$

45. $\frac{3}{p}+\frac{1}{2}=\frac{6}{2p}+\frac{p}{2p}=\frac{6+p}{2p}$

47. $\frac{1}{6m}+\frac{2}{5m}+\frac{4}{m}=\frac{5(1)+6(2)+30(4)}{30m}=\frac{5+12+120}{30m}=\frac{137}{30m}$

49. $\frac{1}{a+1}-\frac{1}{a-1}=\frac{a-1-(a+1)}{(a+1)(a-1)}=\frac{-2}{(a+1)(a-1)}$

51. $\frac{m+1}{m-1}+\frac{m-1}{m+1}=\frac{(m+1)^2+(m-1)^2}{(m-1)(m+1)}=\frac{m^2+2m+1+m^2-2m+1}{(m-1)(m+1)}=\frac{2m^2+2}{(m-1)(m+1)}$

53. $\frac{3}{a-2}-\frac{1}{2-a}=\frac{3}{a-2}-\frac{-1}{a-2}=\frac{3+1}{a-2}=\frac{4}{a-2}$ or $\frac{-4}{2-a}$

55. $\frac{x+y}{2x-y}-\frac{2x}{y-2x}=\frac{x+y}{2x-y}-\frac{-2x}{2x-y}=\frac{x+y+2x}{2x-y}=\frac{3x+y}{2x-y}$ or $\frac{-3x-y}{y-2x}$

57. $\frac{1}{a^2-5a+6}-\frac{1}{a^2-4}=\frac{1}{(a-2)(a-3)}-\frac{1}{(a-2)(a+2)}=\frac{1(a+2)-1(a-3)}{(a-2)(a-3)(a+2)}=$

$\frac{a+2-a+3}{(a-2)(a-3)(a+2)}=\frac{5}{(a-2)(a-3)(a+2)}$

59. $\frac{1}{x^2+x-12}-\frac{1}{x^2-7x+12}+\frac{1}{x^2-16}=\frac{1}{(x+4)(x-3)}-\frac{1}{(x-3)(x-4)}+\frac{1}{(x-4)(x+4)}=$

$\frac{1(x-4)-1(x+4)+1(x-3)}{(x+4)(x-3)(x-4)}=\frac{x-4-x-4+x-3}{(x+4)(x-3)(x-4)}=\frac{x-11}{(x+4)(x-3)(x-4)}$

61. $\frac{3a}{a^2+5a-6}-\frac{2a}{a^2+7a+6}=\frac{3a}{(a+6)(a-1)}-\frac{2a}{(a+6)(a+1)}=\frac{3a(a+1)-2a(a-1)}{(a+6)(a-1)(a+1)}=$

$\frac{3a^2+3a-2a^2+2a}{(a+6)(a-1)(a+1)}=\frac{a^2+5a}{(a+6)(a-1)(a+1)}$

63. $\frac{1+\frac{1}{x}}{1-\frac{1}{x}}=\frac{x\left(1+\frac{1}{x}\right)}{x\left(1-\frac{1}{x}\right)}=\frac{x+1}{x-1}$

65. $\frac{\frac{1}{x+1}-\frac{1}{x}}{\frac{1}{x}}=\frac{x(x+1)\left(\frac{1}{x+1}-\frac{1}{x}\right)}{x(x+1)\left(\frac{1}{x}\right)}=\frac{x-(x+1)}{x+1}=\frac{-1}{x+1}$

67. $\frac{1+\frac{1}{1-b}}{1-\frac{1}{1+b}}=\frac{(1-b)(1+b)\left(1+\frac{1}{1-b}\right)}{(1-b)(1+b)\left(1-\frac{1}{1+b}\right)}=\frac{(1-b)(1+b)+(1+b)}{(1-b)(1-b)-(1-b)}=\frac{1-b^2+1+b}{1-b^2-1+b}=\frac{-b^2+b+2}{-b^2+b}=\frac{(2-b)(1+b)}{b(1-b)}$

69. $\dfrac{m-\frac{1}{m^2-4}}{\frac{1}{m+2}}=\dfrac{(m^2-4)\left(m-\frac{1}{m^2-4}\right)}{(m^2-4)\left(\frac{1}{m+2}\right)}=\dfrac{m(m^2-4)-1}{m-2}=\dfrac{m^3-4m-1}{m-2}$

R.4: Review of Negative and Rational Exponents

1. $\left(\dfrac{4}{9}\right)^{3/2}=\left[\left(\dfrac{4}{9}\right)^{1/2}\right]^3=\left(\dfrac{2}{3}\right)^3=\dfrac{2^3}{3^3}=\dfrac{8}{27}$; E

3. $-\left(\dfrac{9}{4}\right)^{3/2}=\left[\left(\dfrac{9}{4}\right)^{1/2}\right]^3=-\left(\dfrac{3}{2}\right)^3=-\dfrac{3^3}{2^3}=-\dfrac{27}{8}$; F

5. $\left(\dfrac{8}{27}\right)^{2/3}=\left[\left(\dfrac{8}{27}\right)^{1/3}\right]^2=\left(\dfrac{2}{3}\right)^2=\dfrac{2^2}{3^2}=\dfrac{4}{9}$; D

7. $-\left(\dfrac{27}{8}\right)^{2/3}=\left[\left(\dfrac{27}{8}\right)^{1/3}\right]^2=-\left(\dfrac{3}{2}\right)^2=-\dfrac{3^2}{2^2}=-\dfrac{9}{4}$; B

9. $(-4)^{-3}=\left(-\dfrac{1}{4}\right)^3=-\dfrac{1}{64}$

11. $\left(\dfrac{1}{2}\right)^{-3}=2^3=8$

13. $-4^{1/2}=-(4^{1/2})=-2$

15. $8^{2/3}=(8^{1/3})^2=2^2=4$

17. $27^{-2/3}=\left[\left(\dfrac{1}{27}\right)^{1/3}\right]^2=\left(\dfrac{1}{3}\right)^2=\dfrac{1}{9}$

19. $\left(\dfrac{27}{64}\right)^{-4/3}=\left[\left(\dfrac{64}{27}\right)^{1/3}\right]^4=\left(\dfrac{4}{3}\right)^4=\dfrac{256}{81}$

21. $(16p^4)^{1/2}=4p^2$

23. $(27x^6)^{2/3}=[(27x^6)^{1/3}]^2=(3x^2)^2=9x^4$

25. $2^{-3}\cdot 2^{-4}=2^{-7}=\dfrac{1}{2^7}$

27. $27^{-2}\cdot 27^{-1}=27^{-3}=\dfrac{1}{27^3}$

29. $\dfrac{4^{-2}\cdot 4^{-1}}{4^{-3}}=4^{(-2+(-1)-(-3))}=4^0=1$

31. $(m^{2/3})(m^{5/3})=m^{(2/3+5/3)}=m^{7/3}$

33. $(1+n)^{1/2}(1+n)^{3/4}=(1+n)^{(2/4+3/4)}=(1+n)^{5/4}$

35. $(2y^{3/4}z)(3y^{-2}z^{-1/3}) = 6y^{(3/4-8/4)}z^{(1-1/3)} = 6y^{-5/4}z^{2/3} = \dfrac{6z^{2/3}}{y^{5/4}}$

37. $(4a^{-2}b^{7})^{1/2} \cdot (2a^{1/4}b^{3})^{5} = (2a^{-1}b^{7/2})(2^{5}a^{5/4}b^{15}) = 2^{6}a(^{-1+5/4})b^{(7/2+30/2)} = 2^{6}a^{1/4}b^{37/2}$

39. $\left(\dfrac{r^{-2}}{s^{-5}}\right)^{-3} = \dfrac{r^{6}}{s^{15}}$

41. $\left(\dfrac{-a}{b^{-3}}\right)^{-1} = \left(\dfrac{b^{-3}}{-a}\right) = \dfrac{-1}{ab^{3}}$

43. $\dfrac{12^{5/4}\cdot y^{-2}}{12^{-1}y^{-3}} = 12^{(5/3-(-1))}y^{(-2-(-3))} = 12^{9/4}y$

45. $\dfrac{8p^{-3}(4p^{2})^{-2}}{p^{-5}} = \dfrac{8p^{5}}{p^{3}(4p^{2})^{2}} = \dfrac{8p^{5}}{16p^{7}} = \dfrac{1}{2p^{2}}$

47. $\dfrac{m^{7/3}n^{-2/5}p^{3/8}}{m^{-2/3}n^{3/5}p^{-5/8}} = m^{(7/3+2/3)}n^{(-2/5-3/5)}p^{(3/8+5/8)} = m^{3}n^{-1}p^{1} = \dfrac{m^{3}p}{n}$

49. $\dfrac{-4a^{-1}a^{2/3}}{a^{-2}} = -4a^{(-3/3+2/3+6/3)} = -4a^{5/3}$

51. $\dfrac{(k+5)^{1/2}(k+5)^{-1/4}}{(k+5)^{3/4}} = (k+5)^{(1/2-1/4-3/4)} = (k+5)^{-1/2} = \dfrac{1}{(k+5)^{1/2}}$

53. $y^{5/8}(y^{3/8} - 10y^{11/8}) = y^{(5/8+3/8)} - 10y^{(5/8+11/8)} = y - 10y^{2}$

55. $-4k(k^{7/3} - 6k^{1/3}) = -4k^{(1+7/3)} + 24k^{(1+1/3)} = -4k^{10/3} + 24k^{4/3}$

57. $(x + x^{1/2})(x - x^{1/2}) = x^{2} - x^{(1+1/2)} + x^{(1+1/2)} - x^{(1/2+1/2)} = x^{2} - x$

59. $(r^{1/2} - r^{-1/2})^{2} = (r^{1/2})^{2} - 2(r^{1/2})(r^{-1/2}) + (r^{-1/2})^{2} = r - 2r^{0} + r^{-1} = r - 2 + \dfrac{1}{r}$

61. $4k^{-1} + k^{-2} = 4k^{1}\cdot k^{-2} + k^{-2} = k^{-2}(4k+1)$

63. $9z^{-1/2} + 2z^{1/2} = 9z^{-1/2} + 2z^{-1/2}\cdot z^{2/2} = z^{-1/2}(9+2z)$

65. $p^{-3/4} - 2p^{-7/4} = p^{-7/4}\cdot p^{4/4} - 2p^{-7/4} = p^{-7/4}(p-2)$

67. $(p+4)^{-3/2} + (p+4)^{-1/2} + (p+4)^{1/2} = (p+4)^{-3/2}(1 + (p+4) + (p+4)^{2}) =$

$(p+4)^{-3/2}(1 + p + 4 + p^{2} + 8p + 16) = (p+4)^{-3/2}(p^{2} + 9p + 21)$

R.5: Review of Radicals

1. $(-3x)^{1/3} = \sqrt[3]{-3x}$; F

3. $(-3x)^{-1/3} = \dfrac{1}{(-3x)^{1/3}} = \dfrac{1}{\sqrt[3]{-3x}}$; H

5. $(3x)^{1/3} = \sqrt[3]{3x}$; G

7. $(3x)^{-1/3} = \dfrac{1}{(3x)^{1/3}} = \dfrac{1}{\sqrt[3]{3x}}$; C

9. $(-m)^{2/3} = \sqrt[3]{(-m)^2}$ or $(\sqrt[3]{-m})^2$

11. $(2m+p)^{2/3} = \sqrt[3]{(2m+p)^2}$ or $(\sqrt[3]{(2m+p)})^2$

13. $\sqrt[5]{k^2} = k^{2/5}$

15. $-3\sqrt{5p^3} = -3(5p^3)^{1/2} = -3 \cdot 5^{1/2} p^{3/2}$

17. A: $\sqrt{ab} = \sqrt{a} \cdot \sqrt{b}$ is true for $a > 0, b > 0$.

19. $\sqrt{9ax^2} = 3x\sqrt{a}$ is true for all $x \geq 0$.

21. $\sqrt[3]{125} = \sqrt[3]{5^3} = 5$

23. $\sqrt[5]{-3125} = \sqrt[5]{(-5)^5} = -5$

25. $\sqrt{50} = \sqrt{25 \cdot 2} = 5\sqrt{2}$

27. $\sqrt[3]{81} = \sqrt[3]{3^4} = 3\sqrt[3]{3}$

29. $-\sqrt[4]{32} = -\sqrt[4]{2^5} = -2\sqrt[4]{2}$

31. $-\sqrt{\dfrac{9}{5}} = -\dfrac{\sqrt{9}}{\sqrt{5}} \cdot \dfrac{\sqrt{5}}{\sqrt{5}} = -\dfrac{3\sqrt{5}}{5}$

33. $-\sqrt[3]{\dfrac{4}{5}} = -\dfrac{\sqrt[3]{4}}{\sqrt[3]{5}} \cdot \dfrac{\sqrt[3]{5^2}}{\sqrt[3]{5^2}} = -\dfrac{\sqrt[3]{100}}{5}$

35. $\sqrt[3]{16(-2)^4(2)^8} = \sqrt[3]{(2)^4(-2)^4(2)^8} = \sqrt[3]{(2)^{12}(-2)^4} = 2^4(-2)\sqrt[3]{-2} = -32\sqrt[3]{-2}$ or $32\sqrt[3]{2}$

37. $\sqrt{8x^5z^8} = \sqrt{2^3x^5z^8} = 2x^2z^4\sqrt{2x}$

39. $\sqrt[3]{16z^5x^8y^4} = \sqrt[3]{2^4z^5x^8y^4} = 2zx^2y\sqrt[3]{2z^2x^2y}$

41. $\sqrt[4]{m^2n^7p^8} = np^2\sqrt[4]{m^2n^3}$

43. $\sqrt[4]{x^4 + y^4}$ cannot be simplified.

45. $\sqrt{\dfrac{2}{3x}} = \dfrac{\sqrt{2}}{\sqrt{3x}} \cdot \dfrac{\sqrt{3x}}{\sqrt{3x}} = \dfrac{\sqrt{6x}}{3x}$

47. $\sqrt{\dfrac{x^5y^3}{z^2}} = \dfrac{\sqrt{x^5y^3}}{\sqrt{z^2}} = \dfrac{x^2y\sqrt{xy}}{z}$

49. $\sqrt[3]{\dfrac{8}{x^2}} = \dfrac{\sqrt[3]{2^3}}{\sqrt[3]{x^2}} \cdot \dfrac{\sqrt[3]{x}}{\sqrt[3]{x}} = \dfrac{2\sqrt[3]{x}}{x}$

51. $\sqrt[4]{\dfrac{g^3h^5}{9r^6}} = \dfrac{\sqrt[4]{g^3h^5}}{\sqrt[4]{3^2r^6}} = \dfrac{h\sqrt[4]{g^3h}}{r\sqrt[4]{3^2r^2}} \cdot \dfrac{\sqrt[4]{3^2r^2}}{\sqrt[4]{3^2r^2}} = \dfrac{h\sqrt[4]{9g^3hr^2}}{3r^2}$

53. $\dfrac{\sqrt[3]{mn} \cdot \sqrt[3]{m^2}}{\sqrt[3]{n^2}} \cdot \dfrac{\sqrt[3]{n}}{\sqrt[3]{n}} = \dfrac{\sqrt[3]{m^3n^2}}{n} = \dfrac{m\sqrt[3]{n^2}}{n}$

55. $\dfrac{\sqrt[4]{32x^5y} \cdot \sqrt[4]{2xy^4}}{\sqrt[4]{4x^3y^2}} = \dfrac{\sqrt[4]{2^6x^6y^5}}{\sqrt[4]{2^2x^3y^2}} \cdot \dfrac{\sqrt[4]{2^2xy^2}}{\sqrt[4]{2^2xy^2}} = \dfrac{\sqrt[4]{2^8x^7y^7}}{2xy} = \dfrac{4x\sqrt[4]{x^3y^3}}{2xy} = 2\sqrt[4]{x^3y^3}$

57. $\sqrt[3]{\sqrt{4}} = \sqrt[3]{2}$

59. $\sqrt[6]{\sqrt[3]{x}} = (x^{1/3})^{1/6} = x^{1/18} = \sqrt[18]{x}$

61. $4\sqrt{3} - 5\sqrt{12} + 3\sqrt{75} = 4\sqrt{3} - 5(2\sqrt{3}) + 3(5\sqrt{3}) = 4\sqrt{3} - 10\sqrt{3} + 15\sqrt{3} = 9\sqrt{3}$

63. $3\sqrt{28p} - 4\sqrt{63p} + \sqrt{112p} = 3(2\sqrt{7p}) - 4(3\sqrt{7p}) + 4\sqrt{7p} = 6\sqrt{7p} - 12\sqrt{7p} + 4\sqrt{7p} = -2\sqrt{7p}$

65. $2\sqrt[3]{3} + 4\sqrt[3]{24} - \sqrt[3]{81} = 2\sqrt[3]{3} + 4(2\sqrt[3]{3}) - 3\sqrt[3]{3} = 2\sqrt[3]{3} + 8\sqrt[3]{3} - 3\sqrt[3]{3} = 7\sqrt[3]{3}$

67. $\dfrac{1}{\sqrt{3}} - \dfrac{2}{\sqrt{12}} + 2\sqrt{3} = \dfrac{1}{\sqrt{3}} \cdot \dfrac{\sqrt{3}}{\sqrt{3}} - \dfrac{2}{2\sqrt{3}} \cdot \dfrac{\sqrt{3}}{\sqrt{3}} + 2\sqrt{3} = \dfrac{\sqrt{3}}{3} - \dfrac{2\sqrt{3}}{6} + 2\sqrt{3} = \dfrac{2\sqrt{3} - 2\sqrt{3} + 12\sqrt{3}}{6} = \dfrac{12\sqrt{3}}{6} = 2\sqrt{3}$

69. $\dfrac{5}{\sqrt[3]{2}} - \dfrac{2}{\sqrt[3]{16}} + \dfrac{1}{\sqrt[3]{54}} = \dfrac{5}{\sqrt[3]{2}} \cdot \dfrac{\sqrt[3]{4}}{\sqrt[3]{4}} - \dfrac{2}{\sqrt[3]{16}} \cdot \dfrac{\sqrt[3]{4}}{\sqrt[3]{4}} + \dfrac{1}{\sqrt[3]{54}} \cdot \dfrac{\sqrt[3]{4}}{\sqrt[3]{4}} = \dfrac{5\sqrt[3]{4}}{2} - \dfrac{\sqrt[3]{4}}{2} + \dfrac{\sqrt[3]{4}}{6} = \dfrac{15\sqrt[3]{4} - 3\sqrt[3]{4} + \sqrt[3]{4}}{6} = \dfrac{13\sqrt[3]{4}}{6}$

71. $(\sqrt{2} + 3)(\sqrt{2} - 3) = (\sqrt{2})^2 - 3^2 = 2 - 9 = -7$

73. $(\sqrt[3]{11} - 1)(\sqrt[3]{11^2} + \sqrt[3]{11} + 1) = \sqrt[3]{11} \cdot \sqrt[3]{11^2} + \sqrt[3]{11} \cdot \sqrt[3]{11} + \sqrt[3]{11} \cdot 1 - \sqrt[3]{11^2} - \sqrt[3]{11} - 1 =$

$11 + \sqrt[3]{11^2} + \sqrt[3]{11} - \sqrt[3]{11^2} - \sqrt[3]{11} - 1 = 11 - 1 = 10$

75. $(\sqrt{3} + \sqrt{8})^2 = (\sqrt{3})^2 + 2\sqrt{3} \cdot \sqrt{8} + (\sqrt{8})^2 = 3 + 2\sqrt{24} + 8 = 11 + 2(2\sqrt{6}) = 11 + 4\sqrt{6}$

77. $(3\sqrt{2} + \sqrt{3})(2\sqrt{3} - \sqrt{2}) = 3\sqrt{2} \cdot 2\sqrt{3} + 3\sqrt{2}(-\sqrt{2}) + \sqrt{3} \cdot 2\sqrt{3} + \sqrt{3}(-\sqrt{2}) = 6\sqrt{6} - 6 + 6 - \sqrt{6} = 5\sqrt{6}$

79. $\dfrac{8}{\sqrt{5}} = \dfrac{8}{\sqrt{5}} \cdot \dfrac{\sqrt{5}}{\sqrt{5}} = \dfrac{8\sqrt{5}}{5}$

81. $\dfrac{6}{\sqrt[3]{x^2}} = \dfrac{6}{\sqrt[3]{x^2}} \cdot \dfrac{\sqrt[3]{x}}{\sqrt[3]{x}} = \dfrac{6\sqrt[3]{x}}{x}$

83. $\dfrac{\sqrt{3}}{\sqrt{5} + \sqrt{3}} \cdot \dfrac{\sqrt{5} - \sqrt{3}}{\sqrt{5} - \sqrt{3}} = \dfrac{\sqrt{3} \cdot \sqrt{5} - \sqrt{3} \cdot \sqrt{3}}{(\sqrt{5})^2 - (\sqrt{3})^2} = \dfrac{\sqrt{15} - 3}{5 - 3} = \dfrac{\sqrt{15} - 3}{2}$

85. $\dfrac{1 + \sqrt{3}}{3\sqrt{5} + 2\sqrt{3}} \cdot \dfrac{3\sqrt{5} - 2\sqrt{3}}{3\sqrt{5} - 2\sqrt{3}} = \dfrac{3\sqrt{5} - 2\sqrt{3} + \sqrt{3} \cdot 3\sqrt{5} - 2(\sqrt{3})^2}{(3\sqrt{5})^2 - (2\sqrt{3})^2} = \dfrac{3\sqrt{5} - 2\sqrt{3} + 3\sqrt{15} - 6}{45 - 12} = \dfrac{3\sqrt{5} - 2\sqrt{3} + 3\sqrt{15} - 6}{33}$

87. $\dfrac{p}{\sqrt{p} + 2} \cdot \dfrac{\sqrt{p} - 2}{\sqrt{p} - 2} = \dfrac{p\sqrt{p} - 2p}{(\sqrt{p})^2 - (2)^2} = \dfrac{p(\sqrt{p} - 2)}{p - 4}$

89. $\frac{a}{\sqrt{a+b}-1}\cdot\frac{\sqrt{a+b}+1}{\sqrt{a+b}+1}=\frac{a(\sqrt{a+b}+1)}{(\sqrt{a+b})^2-(1)^2}=\frac{a(\sqrt{a+b}+1)}{a+b-1}$

Chapter R-Test

1. (a) $(-3)^4\cdot(-3)^5=(-3)^{4+5}=(-3)^9$

   (b) $\left(2x^3y\right)^2=2^2\left(x^3\right)^2y^2=4x^6y^2$

   (c) $(-5z)^0=1$

   (d) $\left(-\frac{4}{5}\right)^2=\frac{4^2}{5^2}$

   (e) $-\left(\frac{m^2}{p}\right)^3=-\frac{m^6}{p^3}$

2. (a) $\left(x^2-3x+2\right)-\left(x-4x^2\right)+3x(2x+1)=\left(x^2-3x+2\right)-\left(x-4x^2\right)+6x^2+3x=$

   $x^2-3x+2-x+4x^2+6x^2+3x=11x^2-x+2$

   (b) $(6r-5)^2=(6r-5)(6r-5)=36r^2-60r+25$

   (c) $(u+2)\left(3u^2-u+4\right)=3u^3-u^2+4u+6u^2-2u+8=3u^3+5u^2+2u+8$

   (d) $(4x-5)(4x+5)=16x^2-25$

   (e) $[(5p-1)+4]^2=[5p+3]^2=(5p+3)(5p+3)=25p^2+30p+9$

3. (a) $6x^2-17x+7=(3x-7)(2x-1)$

   (b) $x^4-16=\left(x^2\right)^2-4^2=\left(x^2-4\right)\left(x^2+4\right)=(x-2)(x+2)\left(x^2+4\right)$

   (c) $z^2-6zk-16k^2=(z+2k)(z-8k)$

   (d) $x^3y^2-9x^3-8y^2+72=x^3\left(y^2-9\right)-8\left(y^2-9\right)=\left(y^2-9\right)\left(x^3-8\right)=$

   $(x-2)\left(x^2+2x+4\right)(y+3)(y-3)$

4. (a) $\frac{16x^3}{4x^5}=\frac{4}{x^2};\{x\,|\,x\neq 0\}$

   (b) $\frac{1+k}{k^2-1}=\frac{1+k}{(k+1)(k-1)}=\frac{1}{k-1};\{k\,|\,k\neq -1,1\}$

(c) $\dfrac{x^2+x-2}{x^2+5x+6}=\dfrac{(x+2)(x-1)}{(x+2)(x+3)}=\dfrac{x-1}{x+3};\{x\,|\,x\neq -3,-2\}$

5. (a) $\dfrac{5x^2-9x-2}{30x^3+6x^2}\cdot\dfrac{2x^8+6x^7+4x^6}{x^4-3x^2-4}=\dfrac{(5x+1)(x-2)}{6x^2(5x+1)}\cdot\dfrac{2x^6(x+2)(x+1)}{(x+2)(x-2)\left(x^2+1\right)}=\dfrac{x^4(x+1)}{3\left(x^2+1\right)}$

(b) $\dfrac{x}{x^2+3x+2}+\dfrac{2x}{2x^2-x-3}=\dfrac{x}{(x+2)(x+1)}+\dfrac{2x}{(2x-3)(x+1)}=\dfrac{x(2x-3)+2x(x+2)}{(2x-3)(x+2)(x+1)}=$

$\dfrac{4x^2+x}{(2x-3)(x+2)(x+1)}=\dfrac{x(4x+1)}{(2x-3)(x+2)(x+1)}$

(c) $\dfrac{a+b}{2a-3}-\dfrac{a-b}{3-2a}=\dfrac{a+b}{2a-3}+\dfrac{a-b}{2a-3}=\dfrac{2a}{2a-3}=\dfrac{-2a}{3-2a}$

(d) $\dfrac{g-\dfrac{2}{g}}{g-\dfrac{4}{g}}=\dfrac{g^2-2}{g^2-4}=\dfrac{g^2-2}{(g-2)(g+2)}$

6. (a) $(-7)^{-2}=\dfrac{1}{(-7)^2}=\dfrac{1}{49}$

(b) $\left(16x^8\right)^{3/4}=16^{3/4}\left(x^8\right)^{3/4}=8x^6$

(c) $\left(\dfrac{64}{27}\right)^{-2/3}=\left(\dfrac{27}{64}\right)^{2/3}=\dfrac{9}{16}$

7. (a) $\dfrac{5^{-3}\cdot 5^{-1}}{5^{-2}}=\dfrac{5^{-4}}{5^{-2}}=5^{-2}=\dfrac{1}{5^2}$

(b) $(x+2)^{-1/5}(x+2)^{-7/10}=(x+2)^{-9/10}=\dfrac{1}{(x+2)^{9/10}}$

(c) $\left(m^{-1}n^{1/2}\right)^4\left(4m^3n^{-2}\right)^{-1/2}=\left(m^{-4}n^2\dfrac{1}{2}m^{-3/2}n\right)=\dfrac{n^3}{2m^{11/2}}$

8. (a) $\sqrt[4]{16}=2$

(b) $\sqrt[3]{\dfrac{2}{3}}=\dfrac{\sqrt[3]{2}}{\sqrt[3]{3}}=\dfrac{\sqrt[3]{2\cdot 9}}{\sqrt[3]{3\cdot 9}}=\dfrac{\sqrt[3]{18}}{3}$

(c) $\sqrt{18x^5y^8}=\sqrt{9x^4y^8}\sqrt{2x}=3x^2y^4\sqrt{2x}$

(d) $\dfrac{\sqrt[4]{pq}\cdot\sqrt[4]{q^2}}{\sqrt[4]{p^3}}=\sqrt[4]{\dfrac{pq^3}{p^3}}=\sqrt[4]{\dfrac{q^3}{p^2}}=\dfrac{\sqrt[4]{p^2q^3}}{p}$

9. (a) $\sqrt{32x}+\sqrt{2x}-\sqrt{18x}=4\sqrt{2x}+\sqrt{2x}-3\sqrt{2x}=2\sqrt{2x}$

(b) $\left(\sqrt[3]{2}+4\right)\left(\sqrt[3]{4}+1\right)=\sqrt[3]{8}+\sqrt[3]{2}+4\sqrt[3]{4}+4=6+\sqrt[3]{2}+4\sqrt[3]{4}$

(c) $\left(\sqrt{x}-\sqrt{y}\right)\left(\sqrt{x}+\sqrt{y}\right)=x-\sqrt{xy}+\sqrt{xy}-y=x-y$

10. $\dfrac{14}{\sqrt{11}-\sqrt{7}}=\dfrac{14}{\sqrt{11}-\sqrt{7}}\cdot\dfrac{\sqrt{11}+\sqrt{7}}{\sqrt{11}+\sqrt{7}}=\dfrac{14\left(\sqrt{11}+\sqrt{7}\right)}{11-7}=\dfrac{7\left(\sqrt{11}+\sqrt{7}\right)}{2}$

Appendix B: Vectors in Space

1. The plane determined by the x-axis and the z-axis is called the xz-plane.

3. The component form of the position vector with terminal point $(5,3,-2)$ is $(5,3,-2)$.

5. $d=\sqrt{(2-0)^2+(-2-0)^2+(5-0)^2}=\sqrt{4+4+25}=\sqrt{33}$

7. $d=\sqrt{(8-10)^2+(3-15)^2+(-4-9)^2}=\sqrt{4+144+169}=\sqrt{317}$

9. $d=\sqrt{(5-20)^2+(5-25)^2+(6-16)^2}=\sqrt{225+400+100}=\sqrt{725}=5\sqrt{29}$

11. $PQ=\langle(2-0),(-2-0),(5-0)\rangle=\langle 2,-2,5\rangle;2i-2j+5k$

13. $PQ=\langle(8-10),(3-15),(-4-0)\rangle=\langle -2,-12,-4\rangle;-2i-12j-4k$

15. $PQ=\langle(5-20),(5-25),(16-6)\rangle=\langle -15,-20,10\rangle;\ -15i-20j+10k$

17. $QP=\langle(0-2),(0-(-2)),(0-5)\rangle=\langle -2,2,-5\rangle$; **QP** is the opposite of **PQ**. That is **QP = -PQ.**

19. $u-w=(2-4)i+(4-(-3))j+(7-(-6))k=-2i+7j+13k$

21. $4u+5v=[4(2)+5(-3)]i+[4(4)+[5(5)]j+[4(7)+5(2)]k=-7i+41j+38k$

23. $|\mathbf{u}|=\sqrt{2^2+4^2+7^2}=\sqrt{69}$

25. $w+u=(4+2)i(-(-3)+4)j+((-6)+7)k=6i+j+k\Rightarrow|w+u|=\sqrt{6^2+1^2+1^2}=\sqrt{38}$

27. $v\cdot w=(-3)(4)+(5)(-3)+(2)(-6)=-39$

29. $v\cdot v=(-3)(-3)+(5)(5)+(2)(2)=38$

31. Let $\mathrm{v}=\langle 2,-2,0\rangle$ and $w=\langle 5,-2,-1\rangle$. The dot product is $v\cdot w=(2)(5)+(-2)(-2)+(0)(-1)=14$. The required magnitudes are $|v|=\sqrt{2^2+(-2)^2+0^2}=\sqrt{8}$ and $|w|=\sqrt{5^2+(-2)^2+(-1)^2}=\sqrt{30}$.

$$\cos\theta=\frac{v\cdot w}{|v||w|}\Rightarrow\cos\theta=\frac{14}{\sqrt{8}/\sqrt{30}}\Rightarrow\theta=\cos^{-1}\frac{14}{\left(\sqrt{8}/\sqrt{30}\right)}\approx 25.4^\circ$$

33. Let $\mathbf{v}=\langle 6,0,0\rangle$ and $\mathbf{w}=\langle 8,3,-4\rangle$. The dot product is $v\cdot w=(6)(8)+(0)(3)+(0)(-4)=48$.

The required magnitudes are $|\mathbf{v}|=\sqrt{6^2+0^2+0^2}=6$ and $|w|=\sqrt{8^2+3^2+(-4)^2}=\sqrt{89}$.

$$\cos\theta=\frac{v\cdot w}{|v||w|}\Rightarrow\cos\theta=\frac{48}{6\sqrt{89}}\Rightarrow\theta=\cos^{-1}\left(\frac{48}{6\sqrt{89}}\right)\approx 32.0^\circ$$

35. Let $\mathbf{v}=\langle 1,0,0\rangle$ and $\mathbf{w}=\langle 0,1,0\rangle$. The dot product is $v\cdot w=(-1)(0)+(0)(1)+(0)(0)=0$. The required magnitudes are $|v|=\sqrt{1^2+0^2+0^2}=1$ and $|w|=\sqrt{0^2+1^2+0^2}=1$.

$$\cos\theta = \frac{v \cdot w}{|v||w|} \Rightarrow \cos\theta = \frac{0}{1(1)} \Rightarrow \theta = \cos^{-1} 0 = 90^\circ$$

37. $|\mathbf{u}| = \sqrt{2^2 + 4^2 + 7^2} = \sqrt{69}; \cos\alpha = \frac{a}{|\mathbf{u}|} = \frac{2}{\sqrt{69}} \Rightarrow \alpha = \cos^{-1}\left(\frac{2}{\sqrt{69}}\right) \approx 76.1^\circ$

$$\cos\beta = \frac{b}{|\mathbf{u}|} = \frac{4}{\sqrt{69}} \Rightarrow \beta = \cos^{-1}\left(\frac{4}{\sqrt{69}}\right) \approx 61.2^\circ; \cos\gamma = \frac{c}{|\mathbf{u}|} = \frac{7}{\sqrt{69}} \Rightarrow \gamma = \cos^{-1}\left(\frac{7}{\sqrt{69}}\right) \approx 32.6^\circ$$

39. $|\mathbf{w}| = \sqrt{4^2 + (-3)^2 + (-6)^2} = \sqrt{61}; \cos\alpha = \frac{a}{|\mathbf{w}|} = \frac{4}{\sqrt{61}} \Rightarrow \alpha = \cos^{-1}\left(\frac{4}{\sqrt{61}}\right) \approx 59.2^\circ$

$$\cos\beta = \frac{b}{|\mathbf{w}|} = \frac{-3}{\sqrt{61}} \Rightarrow \beta = \cos^{-1}\left(\frac{-3}{\sqrt{61}}\right) \approx 112.6^\circ; \cos\gamma = \frac{c}{|\mathbf{w}|} = \frac{-6}{\sqrt{61}} \Rightarrow \gamma = \cos^{-1}\left(\frac{-6}{\sqrt{61}}\right) \approx 140.2^\circ$$

41. Here $\cos\alpha = \cos 45^\circ = \frac{\sqrt{2}}{2}$ and $\cos\beta = \cos 120^\circ = -\frac{1}{2}$. Since $\cos^2\alpha + \cos^2\beta + \cos^2\gamma = 1$,

$$\cos^2\gamma = 1 - \cos^2\alpha - \cos^2\beta = 1 - \frac{1}{2} - \frac{1}{4} = \frac{1}{4} \Rightarrow \cos\gamma = \frac{1}{2} \Rightarrow \gamma = \cos^{-1}\frac{1}{2} = 60^\circ$$

43. Two vectors are parallel if the position vector for one is a scalar multiple of the position vector of the other.

45. $\mathbf{PQ} = \langle (1-0), (3-0), (2-0) \rangle = \langle 1,3,2 \rangle; \mathbf{F} \cdot \mathbf{PQ} = (2)(1) + (0)(3) + (5)(2) = 12$ work units

47. $\mathbf{PQ} = \langle (5-2), (7-(-1)), (8-2) \rangle = \langle 3,8,6 \rangle; \mathbf{F} \cdot \mathbf{PQ} = (1)(3) + (2)(8) + (-1)(6) = 13$ work units

49. $\cos^2\alpha + \cos^2\beta + \cos^2\gamma = \left(\frac{a}{|\mathbf{v}|}\right)^2 + \left(\frac{b}{|\mathbf{v}|}\right)^2 + \left(\frac{c}{|\mathbf{v}|}\right)^2 = \frac{a^2}{|\mathbf{v}|^2} + \frac{b^2}{|\mathbf{v}|^2} + \frac{c^2}{|\mathbf{v}|^2} =$

$$\frac{a^2}{(\sqrt{a^2+b^2+c^2})^2} + \frac{b^2}{(\sqrt{a^2+b^2+c^2})^2} + \frac{c^2}{(\sqrt{a^2+b^2+c^2})^2} = \frac{a^2+b^2+c^2}{a^2+b^2+c^2} = 1$$

Appendix C: Polar Form of Conic Sections

1. Only the calculator graph is shown here. See Figure 1.
3. Only the calculator graph is shown here. See Figure 3.

[-10,10] by [-10,10]
Xscl = 1 Yscl = 1

r1=6/(3+3sin(θ))

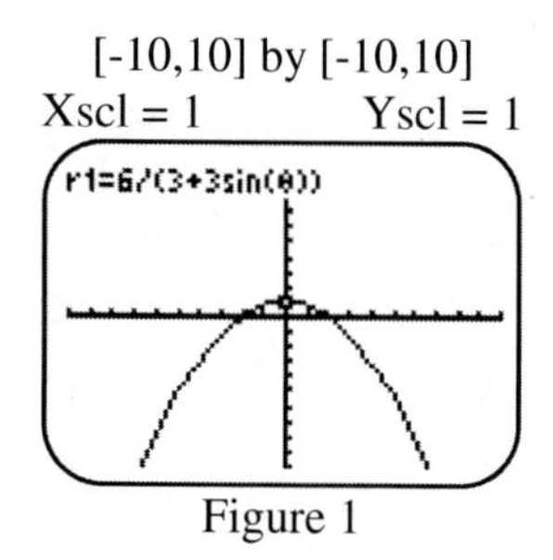

Figure 1

[-2.75,3.25] by [-2.8,2.8]
Xscl = 1 Yscl = 1

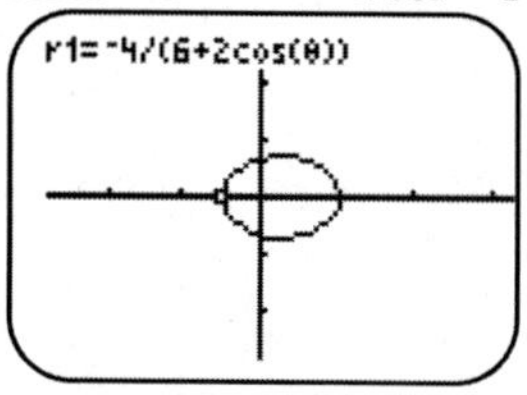

r1=-4/(6+2cos(θ))

Figure 3

[-4.7,4.7] by [-3.1,3.1]
Xscl= 1 Yscl = 1

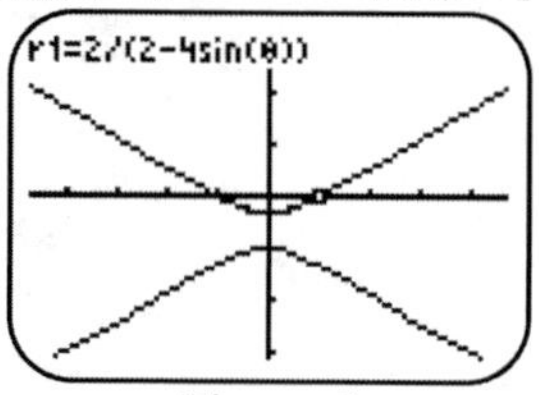

r1=2/(2-4sin(θ))

Figure 5

[-9.4,9.4] by [-6.2,6.2]
Xscl = 1 Yscl= 1

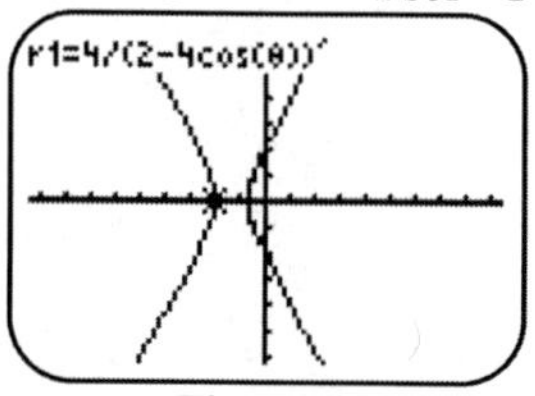

r1=4/(2-4cos(θ))

Figure 7

5. Only the calculator graph is shown here. See Figure 5.
7. Only the calculator graph is shown here. See Figure 7.

9. Only the calculator graph is shown here. See Figure 9.

11. Only the calculator graph is shown here. See Figure 11.

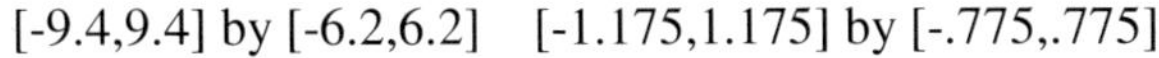

[-9.4,9.4] by [-6.2,6.2] [-1.175,1.175] by [-.775,.775]
Xscl = 1 Yscl = 1 Xscl = 1 Yscl = 1

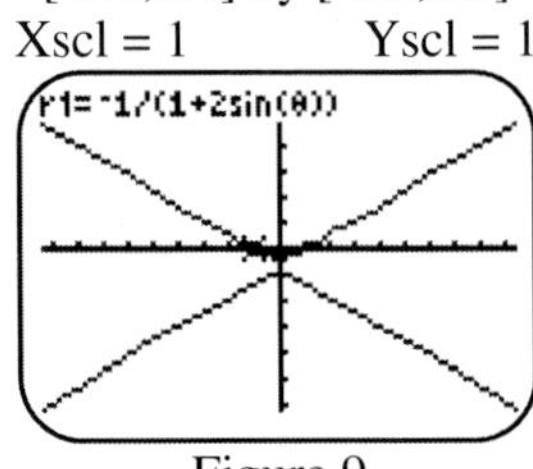

Figure 9

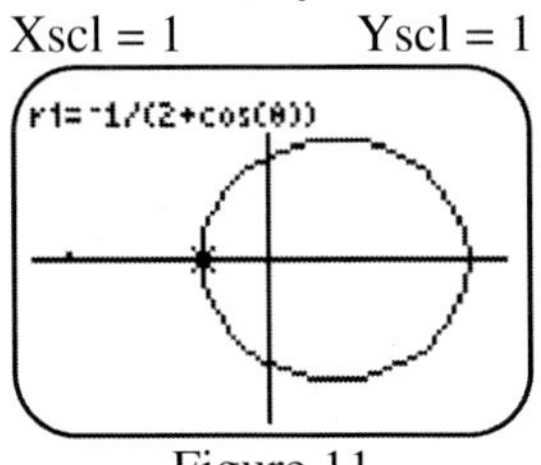

Figure 11

13. The conic is a parabola, so e = 1. Since the vertical directrix is 3 units to the right of the pole, the equation is of the form $r=\dfrac{ep}{1+e\cos\theta}$ with p = 3. The equation for this parabola is $r=\dfrac{3}{1+\cos\theta}$.

15. The conic is a parabola, so e = 1. Since the horizontal directrix is 5 units below the pole, the equation is of the form $r=\dfrac{ep}{1-e\sin\theta}$ with p = 5. The equation for this parabola is $r=\dfrac{5}{1-\sin\theta}$.

17. Since the vertical directrix is 5 units to the right of the pole, the equation is of the form $r=\dfrac{ep}{1+e\cos\theta}$ with p = 5. Since $e=\dfrac{4}{5}<1$, the conic is an ellipse with equation

$$r=\frac{\left(\frac{4}{5}\right)5}{1+\frac{4}{5}\cos\theta}\Rightarrow r=\frac{4}{1+\frac{4}{5}\cos\theta}\Rightarrow r=\frac{4}{1+\frac{4}{5}\cos\theta}\cdot\frac{5}{5}\Rightarrow r=\frac{20}{5+4\cos\theta}.$$

19. Since the horizontal directrix is 8 units below the pole, the equation is of the form $r=\dfrac{ep}{1-e\sin\theta}$ with p = 8.

Since $e=\dfrac{5}{4}>1$, the conic is an hyperbola with equation

$$r=\frac{\left(\frac{5}{4}\right)8}{1-\frac{5}{4}\sin\theta}\Rightarrow r=\frac{10}{1-\frac{5}{4}\sin\theta}\Rightarrow r=\frac{10}{1-\frac{5}{4}\sin\theta}\cdot\frac{4}{4}\Rightarrow r=\frac{40}{4-5\sin\theta}.$$

21. $r=\dfrac{6}{3-\cos\theta}\cdot\dfrac{\frac{1}{3}}{\frac{1}{3}}=\dfrac{2}{1-\frac{1}{3}\cos\theta}$. Since $e=\dfrac{1}{3}<1$, this equation represents an ellipse.

$$r=\frac{6}{3-\cos\theta}\Rightarrow r(3-\cos\theta)=6\Rightarrow 3r-r\cos\theta=6\Rightarrow 3r=r\cos\theta+6\Rightarrow(3r)^2=(r\cos\theta+6)^2\Rightarrow$$

$$(3r)^2=(x+6)^2\Rightarrow 9r^2=x^2+12x+36\Rightarrow 9(x^2+y^2)=x^2+12x+36\Rightarrow 8x^2+9y^2-12x-36=0$$

The rectangle form of the equation is $8x^2+9y^2-12x-36=0$.

23. $r = \frac{-2}{1+2\cos\theta}$; Since $e = 2 > 1$, this equation represents a hyperbola

$r = \frac{-2}{1+2\cos\theta} \Rightarrow r(1+2\cos\theta) = -2 \Rightarrow r + 2r\cos\theta = -2 \Rightarrow r = -2r\cos\theta - 2 \Rightarrow$

$r^2 = (-2r\cos\theta - 2)^2 \Rightarrow r^2 = (-2x-2)^2 \Rightarrow r^2 = 4x^2 + 8x + 4 \Rightarrow x^2 + y^2 = 4x^2 + 8x + 4 \Rightarrow$

$3x^2 - y^2 + 8x + 4 = 0;$

The rectangular form of equation is $3x^2 - y^2 + 8x + 4 = 0$.

25. $r = \frac{-6}{4+2\sin\theta} \cdot \frac{\frac{1}{4}}{\frac{1}{4}} = \frac{-\frac{3}{2}}{1+\frac{1}{2}\sin\theta}$; Since $e = \frac{1}{2} < 1$, this equation represents an ellipse.

$r = \frac{-6}{4+2\sin\theta} \Rightarrow r(4+2\sin\theta) = -6 \Rightarrow 4r + 2r\sin\theta = -6 \Rightarrow 4r = -2r\sin\theta - 6 \Rightarrow$

$(4r)^2 = (-2r\sin\theta - 6)^2 \Rightarrow (4r)^2 = (-2y-6)^2 \Rightarrow 16r^2 = 4y^2 + 24y + 36 \Rightarrow$

$16(x^2 + y^2) = 4y^2 + 24y + 36 = 0 \Rightarrow 16x^2 + 12y^2 - 24y - 36 = 0 \Rightarrow 4x^2 + 3y^2 - 6y - 9 = 0$

The rectangular form of equation is $4x^2 + 3y^2 - 6y - 9 = 0$.

27. $r = \frac{10}{2-2\sin\theta} \cdot \frac{\frac{1}{2}}{\frac{1}{2}} = \frac{5}{1-\sin\theta}$; Since $e = 1$, this equation represents a parabola.

$r = \frac{10}{2-2\sin\theta} \Rightarrow r(2-2\sin\theta) = 10 \Rightarrow 2r - 2r\sin\theta = 10 \Rightarrow 2r = 2r\sin\theta + 10 \Rightarrow$

$(2r)^2 = (2r\sin\theta + 10)^2 \Rightarrow (2r)^2 = (2y+10)^2 \Rightarrow 4r^2 = 4y^2 + 40y + 100 \Rightarrow$

$4(x^2 + y^2) = 4y^2 + 40y + 100 = 0 \Rightarrow 4x^2 - 40y - 100 = 0 \Rightarrow x^2 - 10y - 25 = 0$

The rectangular form of equation is $x^2 - 10y - 25 = 0$.

Appendix D: Rotation of Axes

1. $4x^2 + 3y^2 + 2xy - 5x = 8 \Rightarrow B^2 - 4AC = 2^2 - 4(4)(3) = 4 - 48 < 0$. Circle, ellipse, or a point.

3. $2x^2 + 3xy - 4y^2 = 0 \Rightarrow B^2 - 4AC = 3^2 - 4(2)(-4) = 9 + 32 > 0$. Hyperbola or 2 intersecting lines.

5. $4x^2 + 4xy + y^2 + 15 = 0 \Rightarrow B^2 - 4AC = 4^2 - 4(4)(1) = 16 - 16 = 0$. Parabola, one line, or 2 parallel lines.

7. $2x^2 + \sqrt{3}xy + y^2 + x = 5 \Rightarrow \cot 2\theta = \frac{A-C}{B} = \frac{2-1}{\sqrt{3}} \Rightarrow \cot 2\theta = \frac{1}{\sqrt{3}} \Rightarrow 2\theta = 60° \Rightarrow \theta = 30°$

9. $3x^2 + \sqrt{3}xy + 4y^2 + 2x - 3y = 12 \Rightarrow \cot 2\theta = \frac{A-C}{B} = \frac{3-4}{\sqrt{3}} \Rightarrow \cot 2\theta = -\frac{1}{\sqrt{3}} \Rightarrow 2\theta = 120° \Rightarrow \theta = 60°$

11. $x^2 - 4xy + 5y^2 = 18 \Rightarrow \cot 2\theta = \frac{A-C}{B} = \frac{1-5}{-4} = 1 \Rightarrow \cot 2\theta = 1 \Rightarrow 2\theta = 45° \Rightarrow \theta = 22.5°$

13. $x^2 - xy + y^2 = 6[1]; \theta = 45°$; $x = x'\cos\theta - y'\sin\theta = \frac{\sqrt{2}}{2}x' - \frac{\sqrt{2}}{2}y'[2]$; $y = x'\sin\theta + y'\cos\theta = \frac{\sqrt{2}}{2}x' + \frac{\sqrt{2}}{2}y'[3]$.

Substitute [2] and [3] in [1].

$$\left(\frac{\sqrt{2}}{2}x'-\frac{\sqrt{2}}{2}y'\right)^2-\left(\frac{\sqrt{2}}{2}x'-\frac{\sqrt{2}}{2}y'\right)\left(\frac{\sqrt{2}}{2}x'+\frac{\sqrt{2}}{2}y'\right)+\left(\frac{\sqrt{2}}{2}x'+\frac{\sqrt{2}}{2}y'\right)^2=6\Rightarrow$$

See Figure 13.

$$\frac{1}{2}x'^2-x'y'+\frac{1}{2}y'^2-\frac{1}{2}x'^2+\frac{1}{2}y'^2+\frac{1}{2}x'^2+x'y'+\frac{1}{2}y'^2=6\Rightarrow\frac{1}{2}x'^2+\frac{3}{2}y'^2=6\Rightarrow\frac{x'^2}{12}+\frac{y'^2}{4}=1.$$

15. $8x^2-4xy+5y^2=36[1];\sin\theta=\frac{2}{\sqrt{5}},\ y=2,\ r=\sqrt{5},\ x=\sqrt{5-4}=1\Rightarrow\cos\theta=\frac{1}{\sqrt{5}};$

$$x=x'\cos\theta-y'\sin\theta=\frac{1}{\sqrt{5}}x'-\frac{2}{\sqrt{5}}y'[2];\ y=x'\sin\theta+y'\cos\theta=\frac{2}{\sqrt{5}}x'+\frac{1}{\sqrt{5}}y'[3].$$

Substitute [2] and [3] in [1].

$$8\left(\frac{1}{\sqrt{5}}x'-\frac{2}{\sqrt{5}}y'\right)^2-4\left(\frac{1}{\sqrt{5}}x'-\frac{2}{\sqrt{5}}y'\right)\left(\frac{2}{\sqrt{5}}x'+\frac{1}{\sqrt{5}}y'^2\right)+5\left(\frac{2}{\sqrt{5}}x'+\frac{1}{\sqrt{5}}y'\right)^2=36\Rightarrow$$

$$8\left(\frac{1}{5}x'^2-\frac{4}{5}x'y'+\frac{4}{5}y'^2\right)-4\left(\frac{2}{5}x'^2-\frac{3}{5}x'y'-\frac{2}{5}y'^2\right)5\left(\frac{4}{5}x'^2+\frac{4}{5}x'y'+\frac{1}{5}y'^2\right)=36\Rightarrow$$

See Figure 15.

$$\frac{8}{5}x'^2-\frac{32}{5}x'y'+\frac{32}{5}y'^2-\frac{8}{5}x'^2+\frac{12}{5}x'y'+\frac{8}{5}y'^2+4x'^2+4x'y'+y'^2=36\Rightarrow4x'^2+9y'^2=36\Rightarrow$$

$$\frac{x'^2}{9}+\frac{y'^2}{4}=1.$$

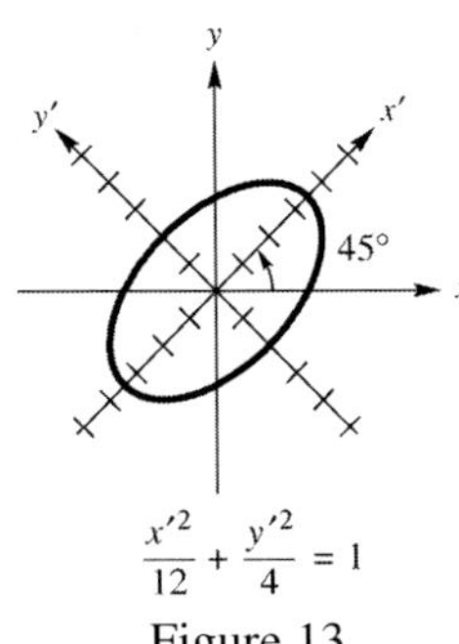

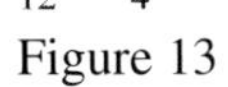

Figure 13

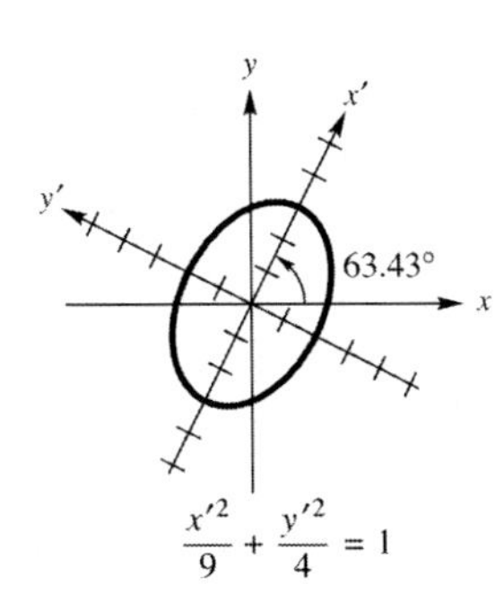

Figure 15

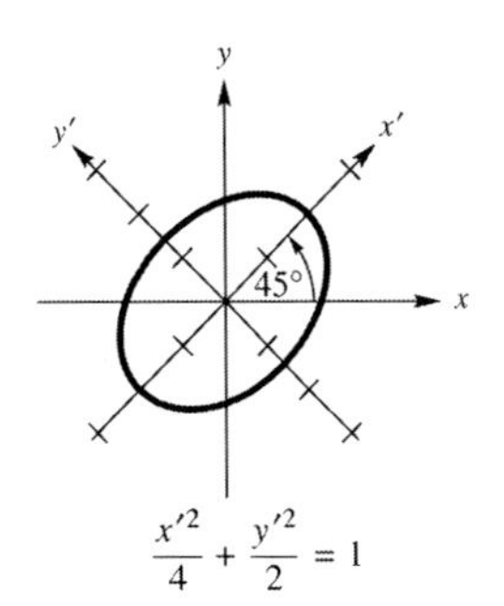

Figure 17

17. $3x^2-2xy+3y^2=8[1];\cot2\theta=\frac{A-C}{B}=\frac{3-3}{-2}=0\Rightarrow2\theta=90°\Rightarrow\theta=45°;$

$$x=x'\cos\theta-y'\sin\theta=\frac{\sqrt{2}}{2}x'-\frac{\sqrt{2}}{2}y'[2];\ y=x'\sin\theta+y'\cos\theta=\frac{\sqrt{2}}{2}x'+\frac{\sqrt{2}}{2}y'[3].$$

Substitute [2] and [3] in [1].

$$3\left(\frac{\sqrt{2}}{2}x'-\frac{\sqrt{2}}{2}y'\right)^2-2\left(\frac{\sqrt{2}}{2}x'-\frac{\sqrt{2}}{2}y'\right)\left(\frac{\sqrt{2}}{2}x'+\frac{\sqrt{2}}{2}y'\right)+3\left(\frac{\sqrt{2}}{2}x'+\frac{\sqrt{2}}{2}y'\right)^2=8\Rightarrow$$

$$3\left(\frac{1}{2}x'^2-x'y'+\frac{1}{2}y'^2\right)-2\left(\frac{1}{2}x'^2-\frac{1}{2}y'^2\right)+3\left(\frac{1}{2}x'^2+x'y'+\frac{1}{2}y'^2\right)=8\Rightarrow$$

$$\frac{3}{2}x'^2-3x'y'+\frac{3}{2}y'^2-x'^2+y'^2+\frac{3}{2}x'^2+3x'y'+\frac{3}{2}y'^2=8\Rightarrow2x'^2+4y'^2=8\Rightarrow\frac{x'^2}{4}+\frac{y'^2}{2}=1$$

See Figure 17.

19. $x^2-4xy+y^2=-5[1];\cot 2\theta=\dfrac{A-C}{B}=\dfrac{1-1}{-4}=0\Rightarrow 2\theta=90°\Rightarrow\theta=45°;$

$x=x'\cos\theta-y'\sin\theta=\dfrac{\sqrt{2}}{2}x'-\dfrac{\sqrt{2}}{2}y'[2];\ y=x'\sin\theta+y'\cos\theta=\dfrac{\sqrt{2}}{x}x'+\dfrac{\sqrt{2}}{2}y'[3].$

Substitute [2] and 3 in [1].

$$\left(\frac{\sqrt{2}}{2}x'-\frac{\sqrt{2}}{2}y'\right)-4\left(\frac{\sqrt{2}}{2}x'-\frac{\sqrt{2}}{2}y'\right)\left(\frac{\sqrt{2}}{2}x'+\frac{\sqrt{2}}{2}y'\right)+\left(\frac{\sqrt{2}}{2}x'+\frac{\sqrt{2}}{2}y'\right)^2=-5\Rightarrow$$

$$\frac{1}{2}x'^2-x'y'+\frac{1}{2}y'^2-4\left(\frac{1}{2}x'^2-\frac{1}{2}y'^2\right)+\frac{1}{2}x'^2+x'y'+\frac{1}{2}y'^2=-5\Rightarrow$$

$$\frac{1}{2}x'^2-x'y'+\frac{1}{2}y'^2-2x'^2+2y'^2+\frac{1}{2}x'^2+x'y'+\frac{1}{2}y'^2=-5\Rightarrow-x'^2+3y'^2=-5\Rightarrow\frac{x'^2}{5}-\frac{3y'^2}{5}=1.$$

See Figure 19.

21. $7x^2+6\sqrt{3}xy+13y^2=64[1];\cot 2\theta=\dfrac{A-C}{B}=\dfrac{7-13}{6\sqrt{3}}=\dfrac{-6}{6\sqrt{3}}=-\dfrac{1}{\sqrt{3}}\Rightarrow 2\theta=120°\Rightarrow\theta=60°;$

$x=x'\cos\theta-y'\sin\theta=\dfrac{1}{2}x'-\dfrac{\sqrt{3}}{2}y'[2];\ y=x'\sin\theta+y'\cos\theta=\dfrac{\sqrt{3}}{2}x'+\dfrac{1}{2}y'[3].$

Substitute [2] and 3 in [1].

$$7\left(\frac{1}{2}x'-\frac{\sqrt{3}}{2}y'\right)^2+6\sqrt{3}\left(\frac{1}{2}x'-\frac{\sqrt{3}}{2}y'\right)\left(\frac{\sqrt{3}}{2}x'+\frac{1}{2}y'\right)+13\left(\frac{\sqrt{3}}{2}x'+\frac{1}{2}y'\right)^2=64\Rightarrow$$

$$7\left(\frac{1}{4}x'^2-\frac{\sqrt{3}}{2}x'y'+\frac{3}{4}y'^2\right)^2+6\sqrt{3}\left(\frac{\sqrt{3}}{4}x'^2-\frac{1}{2}x'y'-\frac{\sqrt{3}}{4}y'^2\right)+13\left(\frac{3}{4}x'^2+\frac{\sqrt{3}}{2}x'y'+\frac{1}{4}y'^2\right)=64\Rightarrow$$

$$\frac{7}{4}x'^2-\frac{7\sqrt{3}}{2}x'y'+\frac{21}{4}y'^2+\frac{18}{4}x'^2-\frac{6\sqrt{3}}{2}x'y'-\frac{18}{4}y'^2+\frac{39}{4}x'^2+\frac{13\sqrt{3}}{2}x'y'+\frac{13}{4}y'=64\Rightarrow$$

$$16x'^2+4y'^2=64\Rightarrow\frac{x'^2}{4}+\frac{y'^2}{16}=1.$$

See Figure 21.

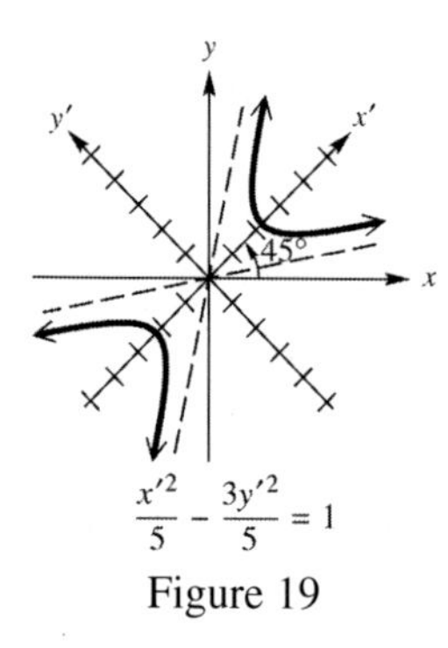

Figure 19

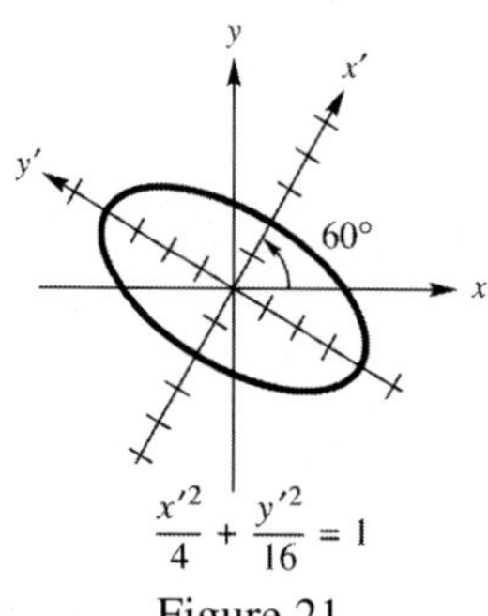

Figure 21

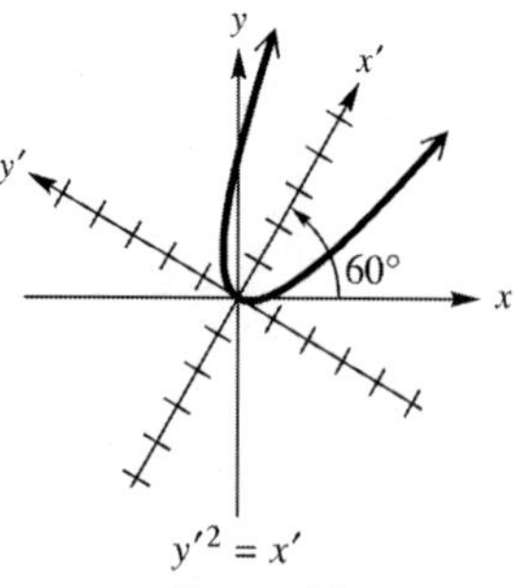

Figure 23

23. $3x^2-2\sqrt{3}xy+y^2-2x-2\sqrt{3}y=0[1];\cot 2\theta=\dfrac{A-C}{B}=\dfrac{3-1}{-2\sqrt{3}}=-\dfrac{1}{\sqrt{3}}\Rightarrow 2\theta=120°\Rightarrow$

$\theta=60°;x=x'\cos\theta-y'\sin\theta=\dfrac{1}{2}x'-\dfrac{\sqrt{3}}{2}y'[2];\ y=x'\sin\theta+y'\cos\theta=\dfrac{\sqrt{3}}{2}x'+\dfrac{1}{2}y'[3].$

Substitute [2] and 3 in [1].

$$3\left(\frac{1}{2}x'-\frac{\sqrt{3}}{2}y'\right)^2-2\sqrt{3}\left(\frac{1}{2}x'-\frac{\sqrt{3}}{2}y'\right)\left(\frac{\sqrt{3}}{2}x'+\frac{1}{2}y'\right)+$$

$$\left(\frac{\sqrt{3}}{2}x'+\frac{1}{2}y'\right)^2-2\left(\frac{1}{2}x'-\frac{\sqrt{3}}{2}y'\right)-2\sqrt{3}\left(\frac{\sqrt{3}}{2}x'+\frac{1}{2}y'\right)=0\Rightarrow 3\left(\frac{1}{4}x'^2-\frac{\sqrt{3}}{2}x'y'+\frac{3}{4}y'^2\right)-$$

$$2\sqrt{3}\left(\frac{\sqrt{3}}{4}x'^2-\frac{1}{2}x'y'-\frac{\sqrt{3}}{4}y'^2\right)+\left(\frac{3}{4}x'^2+\frac{\sqrt{3}}{2}x'y'+\frac{1}{4}y'^2\right)-x'+\sqrt{3}y'-3x-\sqrt{3}y'=0\Rightarrow$$

$$\frac{4}{3}x'^2-\frac{3\sqrt{3}}{2}x'y'+\frac{9}{4}y'^2-\frac{3}{2}x'^2+\sqrt{3}x'y'+\frac{3}{2}y'^2+\frac{3}{4}x'^2+\frac{\sqrt{3}}{2}x'y'+\frac{1}{4}y'^2-4x=0\Rightarrow$$

$4y'^2-4x'=0\Rightarrow 4y'^2=4x'\Rightarrow y'^2=x'.$

See Figure 23.

25. $x^2+3xy+y^2-5\sqrt{2}y=15[1];\cot 2\theta=\dfrac{A-C}{B}=\dfrac{1-1}{3}=0\Rightarrow 2\theta=90°\Rightarrow\theta=45°;$

$x=x'\cos\theta-y'\sin\theta=\dfrac{\sqrt{2}}{2}x'-\dfrac{\sqrt{2}}{2}y'[2];\ y=x'\sin\theta+y'\cos\theta=\dfrac{\sqrt{2}}{2}x'+\dfrac{\sqrt{2}}{2}y'[3].$

Substitute [2] and [3] in [1].

$$\left(\frac{\sqrt{2}}{2}x'-\frac{\sqrt{2}}{2}y'\right)^2+3\left(\frac{\sqrt{2}}{2}x'-\frac{\sqrt{2}}{2}y'\right)\left(\frac{\sqrt{2}}{2}x'+\frac{\sqrt{2}}{2}y'\right)+$$

$$\left(\frac{\sqrt{2}}{2}x'+\frac{\sqrt{2}}{2}y'\right)^2-5\sqrt{2}\left(\frac{\sqrt{2}}{2}x'+\frac{\sqrt{2}}{2}y'\right)=15\Rightarrow$$

$$\frac{1}{2}x'^2-x'y'+\frac{1}{2}y'^2+3\left(\frac{1}{2}x'^2-\frac{1}{2}y'^2\right)+\frac{1}{2}x'^2+x'y'+\frac{1}{2}y'^2-5x'-5y'=15\Rightarrow$$

$$\frac{1}{2}x'^2-x'y'+\frac{1}{2}y'^2+\frac{3}{2}x'^2-\frac{3}{2}y'^2+\frac{1}{2}x'^2+x'y'+\frac{1}{2}y'^2-5x'-5y'=15\Rightarrow$$

$$\frac{5}{2}x'-\frac{1}{2}y'^2-5x'-5y'=15\Rightarrow 5x'^2-10x'-y'^2-10y'=30\Rightarrow$$

$$5\left(x'^2-2x'+1\right)-\left(y'^2+10y'+25\right)=30+5-25\Rightarrow 5\left(x'-1\right)^2-\left(y'+5\right)^2=10\Rightarrow$$

$$\frac{\left(x'-1\right)^2}{2}-\frac{\left(y'+5\right)^2}{10}=1.$$

See Figure 25.

The graph of the equation is a hyperbola with its center at (1, –5). By translating the axes of the $x'y'$ –system down 5 units and right 1 unit, we get an $x''y''$ –coordinate system, in which the hyperbola is centered at the origin. Thus $\dfrac{x''^2}{2}-\dfrac{y''^2}{10}=1.$

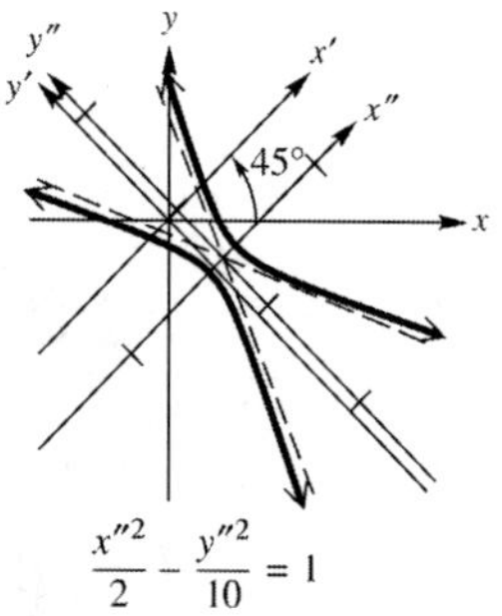

$\frac{x''^2}{2} - \frac{y''^2}{10} = 1$

Figure 25

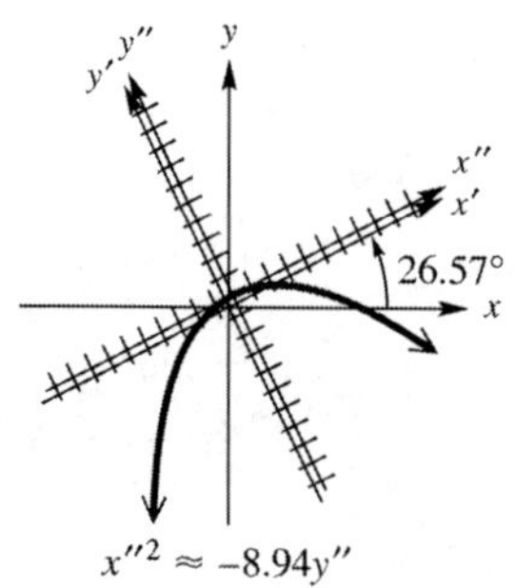

Figure 27

27. $4x^2 + 4xy + y^2 - 24x + 38y - 19 = 0[1]; \cot 2\theta = \frac{A-C}{B} = \frac{4-1}{4} = \frac{3}{4} \Rightarrow 2\theta \approx 53.13^\circ \Rightarrow \theta \approx 26.57^\circ$. For

$2\theta; x = 3, y = 4, r = \sqrt{3^2 + 4^2} = 5 \Rightarrow \cos 2\theta = \frac{3}{5}$.

$\sin\theta = \sqrt{\frac{1-\cos 2\theta}{2}} = \sqrt{\frac{1-\frac{3}{5}}{2}} = \sqrt{\frac{2}{10}} = \frac{\sqrt{5}}{5}; \cos\theta = \sqrt{\frac{1+\cos 2\theta}{2}} = \sqrt{\frac{1+\frac{3}{5}}{2}} = \sqrt{\frac{8}{10}} = \frac{2\sqrt{5}}{5};$

$x = x'\cos\theta - y'\sin\theta = \frac{2\sqrt{5}}{5}x' - \frac{\sqrt{5}}{5}y'[2]; y = x'\sin\theta + y'\cos\theta = \frac{\sqrt{5}}{5}x' + \frac{2\sqrt{5}}{5}y'[3].$

Substitute [2] and [3] in [1]. $4\left(\frac{2\sqrt{5}}{5}x' - \frac{\sqrt{5}}{5}y'\right)^2 + 4\left(\frac{2\sqrt{5}}{5}x' - \frac{\sqrt{5}}{5}y'\right)\left(\frac{\sqrt{5}}{5}x' + \frac{2\sqrt{5}}{5}y'\right) +$

$\left(\frac{\sqrt{5}}{5}x' + \frac{2\sqrt{5}}{5}y'\right)^2 - 24\left(\frac{2\sqrt{5}}{5}x' - \frac{\sqrt{5}}{5}y'\right)^2 + 38\left(\frac{\sqrt{5}}{5}x' + \frac{2\sqrt{5}}{5}y'\right)^2 = 19 \Rightarrow$

$4\left(\frac{4}{5}x'^2 - \frac{4}{5}x'y' + \frac{1}{5}y'^2\right) + 4\left(\frac{2}{5}x'^2 + \frac{3}{5}x'y' - \frac{1}{5}y'^2\right) + \left(\frac{1}{5}x'^2 + \frac{4}{5}x'y' + \frac{4}{5}y'^2\right) - \frac{48\sqrt{5}}{5}x' + \frac{25\sqrt{5}}{5}y' +$

$\frac{38\sqrt{5}}{5}x' + \frac{76\sqrt{5}}{5}y' = 19 \Rightarrow 5x'^2 - 2\sqrt{5}x' + 20\sqrt{5}y' = 19 \Rightarrow$

$5\left(x'^2 - \frac{2\sqrt{5}}{5}x' + \frac{1}{5}\right) + 20\sqrt{5}y' = 19 + 1 \Rightarrow 5\left(x' - \frac{\sqrt{5}}{5}\right)^2 = 20 - 20\sqrt{5}y \Rightarrow$

$\left(x' - \frac{\sqrt{5}}{5}\right)^2 = 4 - 4\sqrt{5}y' \Rightarrow \left(x' - \frac{\sqrt{5}}{5}\right)^2 = -4\sqrt{5}\left(y' - \frac{\sqrt{5}}{5}\right).$

See Figure 27.

29. $16x^2 + 24xy + 9y^2 - 130x + 90y = 0[1]; \cot 2\theta = \frac{A-C}{B} = \frac{16-9}{24} = -\frac{7}{24} \Rightarrow$

$2\theta \approx 73.74^\circ \Rightarrow \theta \approx 36.87^\circ$. For $2\theta; x = 7, y = 24, r = \sqrt{7^2 + 24^2} = 25 \Rightarrow \cos 2\theta = \frac{7}{25}$.

$\sin\theta = \sqrt{\frac{1-\cos 2\theta}{2}} = \sqrt{\frac{1-\frac{7}{25}}{2}} = \sqrt{\frac{18}{50}} = \frac{3}{5}; \cos\theta = \sqrt{\frac{1+\cos 2\theta}{2}} = \sqrt{\frac{1+\frac{7}{25}}{2}} = \sqrt{\frac{32}{50}} = \frac{4}{5};$

$x = x'\cos\theta - y'\sin\theta = \frac{4}{5}x' - \frac{3}{5}y'[2]; y = x'\sin\theta + y'\cos\theta = \frac{3}{5}x' + \frac{4}{5}y'[3].$

Substitute [2] and [3] in [1].

$$16\left(\frac{4}{5}x'-\frac{3}{5}y'\right)+24\left(\frac{4}{5}x'-\frac{3}{5}y'\right)\left(\frac{3}{5}x'+\frac{4}{5}y'\right)+$$

$$9\left(\frac{3}{5}x'+\frac{4}{5}y'\right)^2-130\left(\frac{4}{5}x'-\frac{3}{5}y'\right)+90\left(\frac{3}{5}x'+\frac{4}{5}y'\right)=0\Rightarrow$$

$$16\left(\frac{16}{25}x'^2-\frac{24}{25}x'y'+\frac{9}{25}y'^2\right)+24\left(\frac{12}{25}x'^2+\frac{7}{25}x'y'-\frac{12}{25}y'^2\right)+9\left(\frac{9}{25}x'^2+\frac{24}{25}x'y'+\frac{16}{25}y'^2\right)-$$

$$104x'+78y'+54x'+72y'=0\Rightarrow\frac{256}{25}x'^2-\frac{384}{25}x'y'+\frac{144}{25}y'^2+\frac{288}{25}x'^2+\frac{168}{25}x'y'-\frac{288}{25}y'^2+$$

$$\frac{81}{25}x'^2+\frac{216}{25}x'y'+\frac{144}{25}y'^2-50x'+150y'=0\Rightarrow 25x'^2-50x'+150y'=0\Rightarrow$$

$$25\left(x'^2-2x'+1\right)=-150y'+25\Rightarrow 25(x'-1)^2=-150\left(y'-\frac{1}{6}\right)\Rightarrow(x'-1)^2=-6\left(y'-\frac{1}{6}\right).$$

See Figure 29. The graph of the equation is a parabola with its vertex at $\left(1,\frac{1}{6}\right)$. By translating the axes of the $x'y'$–system up $\frac{1}{6}$ units and right 1 unit, we get an $x''y''$–coordinate system, in which the parabola is centered at the origin. Thus $x''^2=-6y''$.

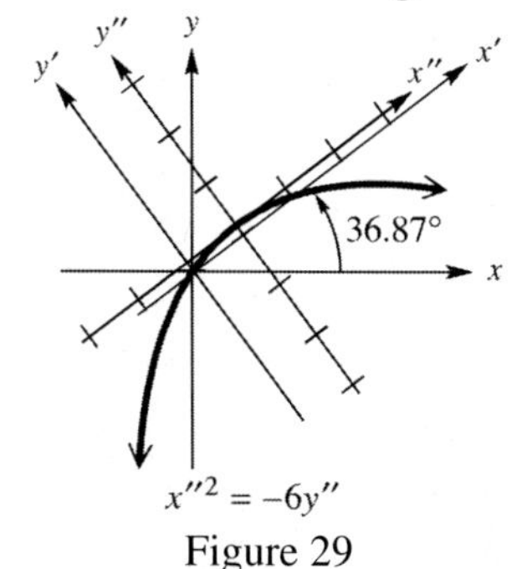

Figure 29

31. If $B=0$, $\cot 2\theta$ is undefined, and the graph may be translated but not rotated.

32. If $A=C$, then $A-C=0, \cot 2\theta=0, 2\theta=90^\circ$ and $\theta=45^\circ$.